Statistical Record OF Black America

ISSN 1051-0000

Statistical Record OF Black America

Jessie Carney Smith and
Robert L. Johns, Editors

An International Thomson Publishing Company

NEW YORK • LONDON • BONN • BOSTON • DETROIT • MADRID
MELBOURNE • MEXICO CITY • PARIS • SINGAPORE • TOKYO
TORONTO • WASHINGTON • ALBANY NY • BELMONT CA • CINCINNATI OH

Jessie Carney Smith and Robert L. Johns, *Editors*

Gale Research Inc. Staff

Mary Beth Trimper, *Production Director*
Shanna Heilveil, *Production Assistant*

Cynthia Baldwin, *Product Design Manager*
Bernadette M. Gornie, *Cover Design*

Editorial Code and Data Inc. Staff

Nancy Ratliff, Sherae R. Carroll, *Data Entry*
Gary Alampi, *Data Processing*

Library of Congress Catalog Card Number 84-643570
ISBN 0-8103-8419-1
ISSN 0740-2880

Printed in the United States of America
Published simultaneously in the United Kingdom
by Gale Research International Limited
(An affiliated company of Gale Research Inc.)
10 9 8 7 6 5 4 3 2 1

I(T)P™

The trademark **ITP** is used under license.

TABLE OF CONTENTS

CHAPTER 2 - ATTITUDES, VALUES, AND BEHAVIOR continued:

CHAPTER 4 - CRIME, LAW ENFORCEMENT, AND LEGAL JUSTICE continued:

CHAPTER 4 - CRIME, LAW ENFORCEMENT, AND LEGAL JUSTICE continued:

x

CHAPTER 5 - EDUCATION continued:

CHAPTER 5 - EDUCATION continued:

CHAPTER 5 - EDUCATION continued:

CHAPTER 7 - HEALTH AND MEDICAL CARE continued:

CHAPTER 9 - INCOME, SPENDING, AND WEALTH continued:

CHAPTER 10 - LABOR AND EMPLOYMENT continued:

CHAPTER 10 - LABOR AND EMPLOYMENT continued:

CHAPTER 10 - LABOR AND EMPLOYMENT continued:

CHAPTER 10 - LABOR AND EMPLOYMENT continued:

CHAPTER 12 - POLITICS AND ELECTIONS continued:

CHAPTER 17 - VITAL STATISTICS continued:

Abbreviations

ACT	American College Testing
AP	Advanced Placement
CMSA	Consolidated Metropolitan Statistical Area
EOEI	Equal Opportunity Educational Institutions
HBCU	Historically Black Colleges and Universities
HIED	Higher Education
MSA	Metropolitan Statistical Area
NAACP	National Association for the Advancement of Colored People
NAEP	National Assessment of Educational Progress
NAFEO	National Association for Equality of Opportunity in Higher Education
NRC	National Research Council
PBI	Predominantly Black Institutions
PMSA	Primary Metropolitan Statistical Area
PWI	Predominantly White Institutions
TBI	Traditionally Black Institutions
TWI	Traditionally White Institutions
SAT	Scholastic Aptitude Test
SEF	Southern Education Foundation
SREB	Southern Regional Education Board

Preface

In 1990 the editors of the first edition wrote:

> *Statistical Record of Black America* is designed to fill a need that has long existed and is ever increasing. Its value to librarians and researchers should be apparent, but it is also a volume that contains data of interest to civic and community groups as well as to individuals who are curious about the general and/or specific aspects of conditions and attitudes in the United States.

As the United States continues to become a more highly industrialized and technological society and as its population continues to grow in diversity, the need for information that assists in determining policy and facilitates understanding of the nature and quality of life of United States residents grows exponentially. Interest in black Americans and other minorities is particularly acute. The United States government is a prolific publisher of statistics that range in concern from broad issues such as population characteristics to minute details such as the number of children left alone after school, with data presented separately for population subgroups. Government sources are widely used by private organizations—foundation, corporation, and others—to combine data in different ways and rework them to emphasize particular groups and points of interest. Private organizations also collect their own data, and though these data are not always national in coverage, the information provided is equally useful. The problem has been that the majority of separate data sources provide information on a narrow range of subject areas. The broader the interests of researchers and policy-makers whose interests overlap subject areas, the greater the number of sources they must consult. This volume does not eliminate that problem, but it does make the task of researching and understanding black America easier by bringing together in a single source the published information that is most readily accessible and available. It also combines the most recent data available with historical information, providing a perspective on changes that have occurred though the years. Wherever available, comparative information on the total, on the majority, and on other minority groups is included to provide a frame of reference.

The reception of the first two editions of *Statistical Record of Black America* has been gratifying and indicates that the work does fill a genuine need. The aims and the overall format of the book remain unchanged for this edition. It contains seventeen chapters. The chapters themselves show variation in length due to several factors. The access to information varies since there is not a uniform rate of government publication in all fields, and various agencies of the federal government are major collectors of data. The analysis and the publication of government data tends to lag behind collection while preliminary data are corrected in later publications. In addition, data are often re-analyzed to answer new questions and so find place in different sections. There are also areas where the placement of data has led to editorial consultation and reassignment since some lie on the boundaries of chapters. Finally, there are shifts in concerns of researchers who both focus on areas deemed important by society at the moment and also sometimes try to set an agenda for addressing problems. Thus the entries in this edition continue to reflect broad trends in public policy discussion in the United States.

How to Use This Book

Chapters and subheading within the chapters are a first guide to locating material. The index provides additional entry into the book.

There is a deliberate attempt not to duplicate material from one edition to the next unless it is a question of updating tables. Therefore, it is sometimes worthwhile to consult the earlier editions which may present data in a different light or even in some cases may contain the latest material on some matters. Greater historical depth, going back to earliest times, is also available in the companion volume *Historical Statistics of Black America* (1994).

Space limitations have meant that tables are not always reprinted in their entirety. More complete information in terms of such items as span of years and related variables may be available in the original source. Users are therefore urged to consult the original sources if greater detail is needed.

All tables present numerical data exactly as presented in the original source. Data on the same major variable may lead to different interpretations when two or more tables are compared. The nature of the samples used and the time of data collection are important considerations when seeking to reconcile differences.

Acknowledgments

Our editors at Gale Research, Sandra C. Davis and James E. Person Jr., have been very helpful. Davis, Person, and former co-editor Carrell Peterson Horton have been especially helpful to the new co-editor.

Many of our colleagues at Fisk have provided assistance over an extended period covering all three editions, among them Henry Ponder, President, and George Neely Jr., Executive Vice-President. Jacqueline R. London, Fisk Library staff, has been stalwart on the last two editions. Among those who have contributed to this edition must be mentioned Ronald Stover, Fisk student, and Dixie Jernigan and John Henderson, Fisk Library staff.

- Jessie Carney Smith, University Librarian and Cosby Professor, Fisk University

- Robert L. Johns, Modern Foreign Languages, Fisk University

Statistical
Record
OF Black
America

Chapter 1
THE ARTS

Cultural Centers

★1★

Annual Income of Ethnic Centers, to 1989

	Total annual income	Average annual income	Median annual income
African American (n=211)	$37,311,693	$176,833	$49,000
Asian American (n=74)	7,834,873	105,877	25,250
Latino American (n=119)	16,139,088	135,623	40,000
Native American (n=57)	23,636,092	414,668	74,850
Multi-ethnic (n=13)	3,263,203	251,016	85,700

Source: "Ethnic Organizations by Regions." Elinor Bowles, *Cultural Centers of Color*, p. 46. *Notes:* There was a "no response" rate of 13 percent for this survey item. In many cases, Native-American organizations include funds for programs other than arts and culture.

★ 2 ★
Cultural Centers

Ethnic Communities by Aesthetic Orientation, 1992-93

Aesthetic orientation	Ethnic communities				
	African American (n=239)	Asian American (n=82)	Latino American (n=136)	Native American (n=72)	Multi-ethnic (n=14)
Contemporary	74%	40%	58%	38%	93%
Traditional	51	63	68	88	57
Classical	31	29	21	3	29
Experimental	31	16	27	13	57
Other	22	13	18	13	7

Source: "Ethnic Organizations by Regions." Elinor Bowles, *Cultural Centers of Color*, p. 44. *Notes:* This item had a 13 percent "no response" rate. Due to rounding, percentages may not total 100.

★ 3 ★
Cultural Centers

Ethnic Communities by Artistic Disciplines, 1992-93

Artistic Discipline	Ethnic communities				
	African American (n=239)	Asian American (n=82)	Latino American (n=136)	Native American (n=72)	Multi-ethnic (n=14)
Dance	16%	13%	17%	3%	21%
Music	14	16	10	0	0
Theater	17	10	14	4	7
Visual arts	11	5	10	21	21
Multidisciplinary	29	23	31	35	21
Other	14	33	19	38	29

Source: "Ethnic Organizations by Regions." Elinor Bowles, *Cultural Centers of Color*, p. 42. *Note:* Due to rounding, percentages do not total 100.

★ 4 ★

Cultural Centers

Ethnic Communities by Date of Organizations' Founding, Pre-1940-1989

Decade founded	Ethnic communities				
	African American (n=239)	Asian American (n=82)	Latino American (n=136)	Native American (n=72)	Multi-ethnic (n=14)
1980s	37%	45%	40%	36%	21%
1985-89	9	17	13	15	0
1980-84	28	28	27	21	21
1970s	41	32	42	31	43
1960s	15	16	13	19	21
1940s-50s	3	5	2	4	0
Pre-1940s	2	2	2	4	14
Unknown	2	0	3	6	0

Source: "Ethnic Organizations by Regions." Elinor Bowles, *Cultural Centers of Color*, p. 41. *Note:* Due to rounding, percentages do not total 100.

★ 5 ★

Cultural Centers

Ethnic Communities by Organizational Activities, 1992-93

Organizational activities	Ethnic communities				
	African American (n=239)	Asian American (n=82)	Latino American (n=136)	Native American (n=72)	Multi-ethnic (n=14)
Artistic production	51%	62%	51%	19%	29%
Visual arts exhibitions	12	5	8	26	21
Cultural events	10	7	19	24	21
Workshops	11	6	5	4	21
Technical assistance	3	2	2	3	0
Publications	3	2	3	7	0
Readings	1	0	1	0	0
Support services	1	2	1	3	0
Other	5	9	7	17	0

Source: "Ethnic Organizations by Regions." Elinor Bowles, *Cultural Centers of Color*, p. 43. *Note:* Due to rounding, percentages do not total 100.

★ 6 ★

Cultural Centers

Geographic Location of Ethnic Cultural Centers, 1992-93

Regions	African American (n=239)	Asian American (n=82)	Latino American (n=136)	Native American (n=72)	Multi-ethnic (n=14)
Mid-America	10%	1%	23%	6%	0%
Mid-Atlantic	35	26	21	10	29
Mid-West	18	4	10	15	14
New England	3	1	2	0	0
South	16	0	8	11	7
West	19	68	40	58	50

Source: "Ethnic Organizations by Regions." Elinor Bowles, *Cultural Centers of Color*, p. 41. *Note:* Due to rounding, percentages do not total 100.

★ 7 ★

Cultural Centers

Income of Ethnic Centers, 1992-93

Income	Ethnic communities				
	African American (n=211)	Asian American (n=74)	Latino American (n=119)	Native American (n=57)	Multi-ethnic (n=14)
Less than $50,000	50%	68%	54%	35%	46%
$50-$100,000	12	15	20	23	8
$101-$250,000	21	8	9	11	15
$251-$500,000	10	5	8	9	15
$500,000+	7	4	8	21	15

Source: "Ethnic Organizations by Regions." Elinor Bowles, *Cultural Centers of Color*, p. 46. *Notes:* This item had a 13 percent "no response" rate. Due to rounding, percentages may not total 100.

★ 8 ★

Cultural Centers

Multiethnic Focus of Cultural Centers, 1992-93

Other ethnic groups	Ethnically specific organizations				
	African American (n = 239)	Asian American (n = 82)	Latino American (n = 136)	Native American (n = 72)	Multi-ethnic (n = 14)
African American	-	27%	43%	6%	100%
Asian American	26	-	24	15	86
Latino American	56	20	-	24	100
Native American	20	13	34	-	71
European American	75	54	64	58	100

Source: "Ethnic Organizations by Regions." Elinor Bowles, *Cultural Centers of Color*, p. 47.

★ 9 ★

Cultural Centers

Regional Comparisons of Ethnic Organizations and Populations, 1992-93

Regions	Ethnic organizations	Ethnic populations
Mid-American region	(n = 59)	(Total = 8,852,606)
African American	39%	38%
Asian American	2	5
Latino American	53	52
Native American	7	4
Mid-Atlantic region	(n = 144)	(Total = 13,086,025)
African American	58%	61%
Asian American	15	11
Latino American	20	27
Native American	5	1
Midwestern region	(n = 73)	(Total = 7,467,729)
African American	59%	66%
Asian American	4	9
Latino American	19	21
Native American	15	4
New England region	(n = 11)	(Total = 1,460,237)
African American	64%	43%
Asian American	9	16
Latino American	27	39
Native American	0	2

[Continued]

★ 9 ★

Regional Comparisons of Ethnic Organizations and Populations, 1992-93
[Continued]

Regions	Ethnic organizations	Ethnic populations
Southern region	(n = 51)	(Total = 12,884,620)
African American	73%	80%
Asian American	0	3
Latino American	9	15
Native American	16	2
Western region	(n = 205)	(Total = 17,915,576)
African American	22%	16%
Asian American	27	23
Latino American	26	56
Native American	21	5

Source: "Ethnic Organizations by Regions." Elinor Bowles, *Cultural Centers of Color*, p. 54. *Notes:* Because of rounding and the exclusion of multi-ethnic organizations, percentage of organizations in each region does not total 100.

★ 10 ★

Cultural Centers

Size of Cultural Centers, 1992-93

| Number of employees | Ethnic communities | | | | |
	African American (n = 239)	Asian American (n = 82)	Latino American (n = 136)	Native American (n = 72)	Multi-ethnic (n = 14)
0-5	20%	17%	22%	21%	14%
6-10	18	7	17	17	43
11-20	23	27	24	24	7
21-30	13	17	13	13	7
31-50	12	23	10	10	21
51-100	9	6	8	8	0
101+	6	2	7	7	7

Source: "Ethnic Organizations by Regions." Elinor Bowles, *Cultural Centers of Color*, p. 42. *Note:* Due to rounding, percentages do not total 100.

★ 11 ★

Cultural Centers

Sources of Funding

Sources of income	Ethnic communities				
	African American (n=211)	Asian American (n=74)	Latino American (n=119)	Native American (n=57)	Multi-ethnic (n=14)
Earned income	28%	34%	21%	8%	26%
Private funding	27	23	26	6	20
Individuals	6	10	3	1	5
Corporations	7	7	12	3	9
Foundations	14	6	11	1	6
Public funding	33	23	39	61	45
Federal	8	8	10	46	8
State	12	10	12	13	16
Local	13	5	17	2	21
Other sources	11	21	14	26	10

Source: "Ethnic Organizations by Regions." Elinor Bowles, *Cultural Centers of Color*, p. 47. *Notes:* There was a 13 percent "no response" rate for this survey item. Due to rounding, percentages may not total 100.

Chapter 2
ATTITUDES, VALUES, AND BEHAVIOR

Attitudes

★ 12 ★

Attitudes Toward Banning the Possession of Handguns Except by the Police and Other Authorized Persons, 1993

	Should	Should not	Don't know or refused
"Do you think there should or should not be a law that would ban the possession of handguns except by the police and other authorized persons?"			
National	42%	54%	4%
Race			
White	39	57	4
Black	69	28	3
Nonwhite[a]	64	31	5

Source: "Attitudes Toward Banning the Possession of Handguns Except by the Police and Other Authorized Persons," *Sourcebook of Criminal Justice Statistics-1992*, p. 211. Primary source: Table constructed by *Sourcebook* staff from data provided by The Gallup Organization. Published by permission. *Note:* A. Includes black respondents.

★ 13 ★
Attitudes

Attitudes Toward Child Abuse Prevention, 1991

	A lot/some	A little	Nothing	Not sure
"How much do you think, as an individual, can do to prevent child abuse?"				
National	57%	28%	11%	4%

[Continued]

★ 13 ★

Attitudes Toward Child Abuse Prevention, 1991

[Continued]

	A lot/some	A little	Nothing	Not sure
Race, ethnicity				
White	57	29	11	4
Black	63	23	11	4
Hispanic	80	15	1	4

Source: "Attitudes Toward Child Abuse Prevention," *Sourcebook of Criminal Justice Statistics-1992*, p. 240. Primary source: Table constructed by *Sourcebook* Staff from data provided by the National Committee for Prevention of Child Abuse. *Notes:* These data are from a telephone survey of 1,250 randomly selected adults across the country. This research was conducted by Schulman, Ronca, and Bucuvalas for the Nationa Committee for Prevention of Child Abuse (NCPCA). A. Percents may not add to 100 because of rounding.

★ 14 ★

Attitudes

Attitudes Toward Crime in Own Neighborhood, 1992

	More	Less	Same[a]	No opinion
"Is there more crime in your area than there was a year ago, or less?"				
National	54%	19%	23%	4%
Race				
White	52	19	25	4
Nonwhite	66	23	6	5

Source: "Attitudes Toward Crime in Own Neighborhood" Department of Justice, *Sourcebook of Criminal Justice Statistics-1992*, p. 188. Primary source: George Gallup, Jr., *The Gallup Poll Monthly* 318 (March 1992). Table adapted by *Sourcebook* staff. Published by permission. *Note:* A. Response volunteered.

★ 15 ★

Attitudes

Attitudes Toward Death Penalty For Persons Convicted of Murder, 1980, 1986, 1991

	1980			1986			1991		
	Favor	Oppose	Don't know	Favor	Oppose	Don't know	Favor	Oppose	Don't know
"Do you favor or oppose the death penalty for persons convicted of murder?"									
National	67%	27%	6%	71%	23%	5%	72%	22%	6%

[Continued]

★ 15 ★

Attitudes Toward Death Penalty For Persons Convicted of Murder, 1980, 1986, 1991
[Continued]

	1980			1986			1991		
	Favor	Oppose	Don't know	Favor	Oppose	Don't know	Favor	Oppose	Don't know
Race									
White	70	24	6	75	20	5	75	19	6
Black/other	40	51	9	49	43	8	53	37	10

Source: "Attitudes Toward Death Penalty For Persons Convicted of Murder," Department of Justice, *Sourcebook of Criminal Justice Statistics-1992*, pp. 206-07. Primary source: Table constructed by *Sourcebook* staff from data provided by the National Opinion Research Center. Data were made available through The Roper Center for Public Opinion Research. *Note:* A. Percents may not add to 100 because of rounding.

★ 16 ★

Attitudes

Attitudes Toward Discriminatory Application of the Death Penalty, 1991

	Blacks more likely to receive death penalty			Poor people more likely to receive death penalty		
	Agree	Disagree	No opinion	Agree	Disagree	No opinion
"As I read off each of these statements would you tell me if you agree or disagree with it: (a) A black person is more likely than white person to receive the death penalty for the same crime; (b) A poor persons is more likely than a person of average or above average income to receive the death penalty for the same crime."						
National	45%	50%	5%	60%	36%	4%
Race						
White	41	54	5	59	37	4
Black	73	20	7	72	22	6

Source: "Attitudes Toward Discriminatory Application of the Death Penalty," Department of Justice, *Sourcebook of Criminal Justice Statistics-1992*, p. 208. Primary source: George Gallup Jr., *The Gallup Poll Monthly* 309 (June 1991). Published by permission.

★ 17 ★

Attitudes

Attitudes Toward Drug Use In Respondent's Neighborhood, 1990

	Very serious	Some-what serious	Not too serious	Not at all serious	Don't know/ no answer
"In your opinion, how much of a problem is illegal drug use in your neighborhood--very serious, somewhat serious, not too serious, or not at all serious?"					
National	18%	30%	31%	19%	2%
Race, ethnicity					
White	17	30	32	19	2
Black	35	22	24	20	0
Hispanic	16	48	30	0	7
Other	40	27	30	4	0

Source: "Attitudes Toward Drug Use In Respondent's Neighborhood," *Sourcebook of Criminal Justice Statistics-1992*, p. 232. Primary source: Table adapted by *Sourcebook* Staff from table provided by the Media General/Associated Press Poll. Published by permission. *Note:* A. Percents may not add to 100 because of rounding.

★ 18 ★

Attitudes

Attitudes Toward Laws Covering the Sale of Firearms, 1993

	More strict	Less strict	Kept as they are now	No opinion
"In general, do you feel the laws covering the sale of firearms should be made more strict, less strict, or kept as they are now?"				
National	70%	4%	24%	2%
Race				
White	68	4	26	2
Black	84	3	13	0
Nonwhite[b]	82	4	13	1

Source: "Attitudes Toward Laws Covering the Sale of Firearms," *Sourcebook of Criminal Justice Statistics-1992*, p. 210. Primary source: George Gallup Jr., *The Gallup Poll Monthly* 330 (March 1993), and from data provided by the Gallup Organization. Published by permission. *Note:* A. Includes black respondents.

★ 19 ★

Attitudes

Attitudes Toward Laws Regulating the Distribution of Pornography, 1980, 1986, and 1991

	1986				1991			
	Laws forbidding distribution		No laws forbidding distribution	Don't know	Laws forbidding distribution		No laws forbidding distribution	Don't know
	Whatever the age	To persons under 18			Whatever the age	To persons under 18		
"Which of these statements comes closest to your feelings about pornography laws: There should be laws against the distribution of pornography whatever the age; there should be laws against the distribution of pornography to persons under 18; or there should be no laws forbidding the distribution of pornography?"								
National	43%	53%	4%	1%	40%	55%	4%	2%
Race								
White	43	53	3	1	41	54	4	1
Black/other	38	53	6	3	31	60	6	2

Source: "Attitudes Toward Laws Regulating the Distribution of Pornography," *Sourcebook of Criminal Justice Statistics-1992*, pp. 236-37. Primary source: Table constructed by *Sourcebook* Staff from data provided by the National Opinion Research Center. Data were made available through The Roper Center for Public Opinion Research. *Notes:* A. Percents may not add to 100 because of rounding. B. One-half of 1 percent or less.

★ 20 ★

Attitudes

Attitudes Toward Personal Danger From Gun Violence, 1993

	Yes	No	No opinion
"Do you, personally, feel any sense of danger from gun violence where you live and work, or not?"			
National	36%	64%	(a)
Race			
White	34	66	(a)
Black	46	54	0
Nonwhite[b]	45	55	(a)

Source: "Attitudes Toward Personal Danger From Gun Violence," Department of Justice, *Sourcebook of Criminal Justice Statistics-1992*, pp. 186. Primary source: Table constructed by *Sourcebook* staff from data provided by The Gallup Organization. Published by permission. *Notes:* A. Less than 1 percent. B. Includes black respondents.

★ 21 ★

Attitudes

Attitudes Toward Police Brutality in Own Area, 1991

	Yes	No	No opinion
"In some places in the nation, there have been charges of police brutality. Do you think there is any police brutality in your area, or not?"			
National	35%	60%	5%
Race ethnicity			
White	33	62	5
Black	45	46	9
Other	43	56	1

Source: "Attitudes Toward Police Brutality in Own Area," Department of Justice, *Sourcebook of Criminal Justice Statistics-1992*, p. 173. Primary source: *The Gallup Poll Monthly* 306 (March 1991). Published by permission.

★ 22 ★

Attitudes

Attitudes Toward Proposals to Reduce Illegal Drug Use 1990

	Punishing	Putting into treatment programs	Don't know/ no answer
"Which of these do you think will do more to reduce the use of illegal drugs--punishing the drug user or putting them into drug treatment programs?"			
National	33%	57%	10%
Race, ethnicity			
White	32	58	10
Black	38	57	5
Hispanic	37	57	7
Other	44	46	10

Source: "Attitudes Toward Proposals to Reduce Illegal Drug Use," *Sourcebook of Criminal Justice Statistics-1992*, p. 232. Primary source: Table adapted by *Sourcebook* Staff from table provided by the Media General/ Associated Press Poll. Published by permission. *Note:* A. Percents may not add to 100 because of rounding.

★ 23 ★

Attitudes

Attitudes Toward the Death Penalty For Drug Traffickers, 1990

	Strongly favor	Favor	Oppose	Strongly oppose	Don't know
"The following is a list of some programs and proposals that are being discussed in this country today. For each one, please tell me whether you strongly favor, favor, oppose, or strongly oppose it. The first one is...A mandatory death penalty for major drug traffickers."					
Total	42.3%	30.4%	18.3%	5.5%	3.2%
Race					
White	42.9	31.0	17.5	5.1	3.2
Nonwhite	9.2	27.2	22.4	7.7	3.5

Source: "Attitudes Toward The Death Penalty For Drug Traffickers," Department of Justice, *Sourcebook of Criminal Justice Statistics-1992,* p. 208. Primary source: Table adapted from tables provided by Princeton Research Associates, Inc. Data are from Times Mirror Center for the People and the Press. *Notes:* These data are derived from interviews conducted among a sample of adult Americans on two separate occasions. The surveys were designed and analyzed by Princeton Survey Research Associates Inc. (PSRA) for the Times Mirror Center for The People and The Press.

★ 24 ★

Attitudes

Attitudes Toward the Legality of Homosexual Relations, 1992

	Legal	Not legal	No opinion
"Do you think homosexual relations between consenting adults should or should not be legal?"			
National	48%	44%	8%
Race			
White	49	44	7
Nonwhite	47	43	10

Source: "Attitudes Toward The Legality of Homosexual Relations," *Sourcebook of Criminal Justice Statistics-1992,* p. 241. Primary source: George Gallup Jr., The Gallup Poll Montly 321 (June 1992). Published by permission.

★ 25 ★

Attitudes

Attitudes Toward the Level of Spending to Deal With Drug Addition, 1985 and 1991

	1985				1991			
	Too little	About right	Too much	Don't known	Too little	About right	Too much	Don't known
"We are faced with many problems in this country, none of which can be solved easily or inexpensively. "I'm going to name some of these problems, and for each one I'd like you to tell me whether you think we're spending too much money on it, too little money, or about the right amount. First (dealing with drug addiction) are we spending too much, too little, or about the right amount on (dealing with drug addiction)?"								
National	62%	28%	5%	4%	58%	32%	7%	4%
Race								
White	62	28	5	5	54	34	8	3
Black/other	59	30	9	2	72	19	4	5

Source: "Attitudes Toward The Level of Spending to Deal With Drug Addition," Department of Justice, *Sourcebook of Criminal Justice Statistics-1992*, pp. 180-81. Primary source: Table constructed by *Sourcebook* staff from data provided by the National Opinion Research Center: data were made available through The Roper Center for Public Opinion Research.

★ 26 ★

Attitudes

Attitudes Toward the Legalization of the Use of Marijuana, 1980, 1986, and 1991

	1980			1986			1991		
	Should	Should not	Don't know	Should	Should not	Don't know	Should	Should not	Don't know
"Do you think the use of marijuana should be made legal or not?"									
National	25%	72%	3%	18%	80%	2%	18%	78%	4%
Race									
White	25	72	3	18	81	1	18	78	3
Black/other	27	71	2	19	77	4	16	76	8

Source: "Attitudes Toward the Legalization of the Use of Marijuana," *Sourcebook of Criminal Justice Statistics-1992*, pp. 230-31. Primary source: Table constructed by Sourcebook Staff from data provided by the National Opinion Research Center. Data were made available through The Roper Center for Public Opinion Research. *Note:* A. Percents may not add to 100 because of rounding.

★ 27 ★

Attitudes

Awareness of Health Problems Related to Dietary Components, 1989

Nutrient	Race	
	White	Black
Percent of female main meal planners and preparers		
Salt or sodium	89	90
Cholesterol	91	76
Fat	81	71
Saturated fat	64	39
Fiber	57	36

Source: "Awareness of Health Problems Related to Dietary Components, by Race, 1989," Interagency Board for Nutrition Monitoring and Related research, *Nutrition Monitoring in the United States: Chartbook I: Selected Findings From the National Nutrition Monitoring and Related Research Programs,* p. 115. Primary source: U.S. Department of Agriculture, Human Nutrition Information Service, Diet and Health Knowledge Survey, 1989.

★ 28 ★

Attitudes

Confidence in the U.S. Supreme Court, 1991

	Great deal/quite a lot	Some	Very little/none[a]
"I am going to read you a list of institutions in American society. Please tell me how much confidence you, yourself have in each one--a great deal, quite a lot, some, or very little: The U.S. Supreme Court?"			
National	39%	39%	17%
Race			
White	42	39	15
Nonwhite	28	42	26

Source: "Reported Confidence in the U.S. Supreme Court," Department of Justice, *Sourcebook of Criminal Justice Statistics-1992,* p. 165. Primary source: George Gallup Jr., *The Gallup Poll Monthly* 313 (October 1991). Published by permission. *Notes:* Responses of "no opinion" have been omitted by the source. A. Response volunteered.

★ 29 ★

Attitudes

Fear of Walking Alone at Night in Own Neighborhood 1980, 1985, and 1991

	1980			1985			1991		
	Yes	No	Don't know	Yes	No	Don't know	Yes	No	Don't know
"Is there any area right here--that is, within a mile--where you would be afraid to walk alone at night?"									
National	43%	56%	1%	40%	59%	1%	43%	56%	(b)
Race									
White	42	58	1	38	61	1	41	59	1
Black/other	52	47	1	60	39	1	56	44	0

Source: "Respondents Reporting Whether They Feel Afraid to Walk Alone At Night in Their Own Neighborhood" Department of Justice *Sourcebook of Criminal Justice Statistics-1992*, pp. 190-91. Primary source: Table constructed by *Sourcebook* staff from data provided by the National Opinion Survey Research Center: data were made available through the Roper Center for Public Opinion Research. *Notes:* A. Percents may not add to 100 because of rounding. B. One-half of 1 percent or less.

★ 30 ★

Attitudes

High School Seniors Reporting Positive Attitudes toward the Performance of the Courts and Justice System in General, 1991 to 1992

	Class of 1991 (N-2,582)	Class of 1992 (N-2,684)
"Now we'd like to make some ratings of how good or bad a job you feel each of the following organizations is doing for the country as a whole...How good or bad a job is being done for the country as a whole by...the courts and the justice system in general?" (Percent responding "good" or "very good")		
Total	31.2%	23.4%
Race		
White	32.5	24.6
Black	23.5	18.6

Source: "High School Seniors Reporting Positive Attitudes toward the Performance of the Courts and Justice System In General," *Sourcebook of Criminal Justice Statistics-1992*, p. 220. Primary source: Lloyd D. Johnston, Jerald G. Bachman, and Patrick M. O'Malley, *Monitoring the Future 1981, 1983, 1985, 1987*, Jerald G. Bachman, Lloyd D. Johnson, and Patrick M. O'Malley, *Monitoring the Future, 1980, 1982, 1984, 1986, 1988;* and data provided by the Monitoring the Future Project (University of Michigan). Table adapted by *Sourcebook* staff. Published by permission. *Notes:* Response categories were "very poor," "poor," "fair," "good," and "no opinion."

★ 31 ★

Attitudes

High School Seniors Reporting Positive Attitudes toward the Performance of the Police and Other Law Enforcement Agencies, 1991 to 1992

	Class of 1991 (N-2,582)	Class of 1992 (N-2,684)
"Now we'd like you to make some ratings of how good or bad a job feel each of the following organizations is doing for the country as a whole...How good or bad a job is being done for the country as a whole by...the police and other law enforcement agencies?" (Percent responding "good" or "very good")		
Total	28.0%	26.9%
Race		
White	31.5	30.0
Black	11.0	12.4

Source: "High School Seniors Reporting Positive Attitudes toward the Performance of the Police and Other Law Enforcement Agencies," *Sourcebook of Criminal Justice Statistics-1992*, p. 218. Primary source: Lloyd D. Johnston, Jerald G. Bachman, and Patrick M. O'Malley, *Monitoring the Future 1981, 1983, 1985, 1987*, Jerald G. Bachman, Lloyd D. Johnson, and Patrick M. O'Malley, *Monitoring the Future, 1980, 1982, 1984, 1986, 1988;* and data provided by the Monitoring the Future Project (University of Michigan). Table adapted by *Sourcebook* staff. Published by permission. *Notes:* See Notes, tables 2.68 and 2.69. Response categories were "very poor," "poor," "fair," "good," "very good," and "no opinion."

★ 32 ★

Attitudes

High School Seniors Reporting Positive Attitudes toward the Performance of the U.S. Supreme Court, 1991 to 1992

	Class of 1991 (N-2,582)	Class of 1992 (N-2,684)
"Now we'd like to make some ratings of how good or bad a job you feel each of the following organizations is doing for the country as a whole...How good or bad a job is being done for the country as a whole by...the U.S. Supreme Court?" (Percent responding "good" or "very good")		
Total	44.1%	35.7%

[Continued]

★ 32 ★

High School Seniors Reporting Positive Attitudes toward the Performance of the U.S. Supreme Court, 1991 to 1992

[Continued]

	Class of 1991 (N-2,582)	Class of 1992 (N-2,684)
Race		
White	47.1	38.4
Black	29.9	27.8

Source: "High School Seniors Reporting Positive Attitudes toward the Performance of the U.S. Supreme Court," *Sourcebook of Criminal Justice Statistics-1992*, p. 219. Primary source: Lloyd D. Johnston, Jerald G. Bachman, and Patrick M. O'Malley, *Monitoring the Future 1981, 1983, 1985, 1987*, Jerald G. Bachman, Lloyd D. Johnson, and Patrick M. O'Malley, *Monitoring the Future, 1980, 1982, 1984, 1986, 1988;* and data provided by the Monitoring the Future Project (University of Michigan). Table adapted by *Sourcebook* staff. Published by permission. *Notes:* Response categories were "very poor," "poor," "fair," "good," and "no opinion."

★ 33 ★

Attitudes

High School Seniors Reporting That They Worry About Crime and Violence, 1991 to 1992

	Class of 1991 (N-2,595)	Class of 1992 (N-2,736)
"Of all the problems facing the nation today, how often do you worry about...crime and violence?" (Percent responding "often" or "sometimes")		
Total	88.1%	91.6%
Race		
White	86.6	90.5
Black	94.5	96.9

Source: "High School Seniors Reporting That They Worry About Crime and Violence," *Sourcebook of Criminal Justice Statistics-1992*, p. 216. Primary source: Lloyd D. Johnston, Jerald G. Bachman, and Patrick M. O'Malley, *Monitoring the Future 1981, 1983, 1985, 1987*; Jerald G. Bachman, Lloyd D. Johnson, and Patrick M. O'Malley, *Monitoring the Future, 1980, 1982, 1984, 1986, 1988*; and data provided by the Monitoring the Future Project (University of Michigan). Table adapted by *Sourcebook* staff. Published by permission. *Notes:* Data are given for those who identify themselves as White or Caucasian and those who identify themselves as Black or African-American because these are the two largest racial/ethnic subgroups in the population. Data are not given for the other ethnic categories because these groups comprise less than 3 percent of the sample in any given year.

★ 34 ★

Attitudes

Percentage of High School Sophomores Aspiring to Various Levels Of Education, 1980 and 1990

Student characteristics	High school diploma or less		Two years or less of college or vocational school		College graduate		Postgraduate degree	
	1980	1990	1980	1990	1980	1990	1980	1990
All sophomores	26.5	10.2	32.9	30.3	22.7	32.1	17.9	27.4
Asian	11.7	8.2	21.5	21.7	32.4	31.4	34.3	38.7
Hispanic	33.7	14.3	33.7	38.5	17.0	25.5	15.6	21.7
Black	26.3	11.1	32.7	30.2	21.8	28.2	19.2	30.5
White	25.9	9.4	33.1	29.5	23.4	33.9	17.7	27.3
American Indian	35.7	18.8	32.9	43.0	17.2	21.8	14.2	16.5

Source: "Percentages of 1980 and 1990 Sophomores Aspiring to Various Levels of Post-Secondary Education, by Student Characteristics," U.S. Department of Education, National Center for Educational Statistics, *America's High School Sophomores: A Ten Year Comparison*, p. 44. Primary source: National Center for Education Statistics, High School and Beyond base year sophomore cohort and NELS:88 first follow-up. *Note:* Owing to rounding, percentages may not sum to 100.

★ 35 ★

Attitudes

Problems in Neighborhood, 1990

Neighborhood	Total	White	Black
"Now here is a list of things that are problems in some neighborhoods. (Card shown respondent) Would you go down the list and call off each you feel is a real problem in this neighborhood? Any others?"			
Crime	29%	27%	49%
Unemployment	23	20	39
Juvenile delinquency	22	20	32
Availability of places for working mothers to leave their children during the day	21	21	24
Street cleaning and street repairs	20	19	28
Public transportation	19	19	26
Upkeep of houses and yards by people who live here	17	16	24
Schools and education	17	17	20
The supply of good housing	15	13	28
Street lighting	15	14	22
Inadequate parks and recreation facilities	13	12	19
Concern of public officials about the neighborhood	11	10	18
Inadequate supermarkets and shopping facilities	9	8	16
Treatment by police	9	7	18
Garbage collections	7	6	12

[Continued]

★ 35 ★

Problems in Neighborhood, 1990
[Continued]

Neighborhood	Total	White	Black
None[a]	25	26	15
Don't know	4	4	4

Source: "Respondents Reporting Problems in Own Neighborhood," Department of Justice, *Sourcebook of Criminal Justice Statistics-1992*, p. 163. Primary source: Table provided to *Sourcebook* staff by the Roper Organization. *Notes:* Rankings based on results for total sample. A. Response volunteered.

★ 36 ★

Attitudes

Ratings of Local Police Response Time, 1992

	Excellent	Pretty good	Only fair	Poor	Not sure/ refused
"How would you rate the police in your community on the following...responding quickly to calls for help and assistance?"					
National	28%	41%	20%	9%	2%
Race ethnicity					
White	30	43	18	7	2
Black	14	36	33	16	1
Hispanic	24	36	29	8	3

Source: "Respondents' Ratings of Local Police Response Time," Department of Justice, *Sourcebook of Criminal Justice Statistics-1992*, p. 170. Primary source: Table adapted by *Sourcebook* staff from data provided by Lou Harris and Associates. Published by permission. *Notes:* A. Percents may not add to 100 because of rounding. B. Less than 0.5 percent.

★ 37 ★

Attitudes

Ratings of Local Police Solving Crimes, 1992

	Excellent	Pretty good	Only fair	Poor	Not sure/ refused
"How would you rate the police in your community on the following...solving crime?"					
National	13%	45%	30%	9%	3%
Race ethnicity					
White	15	46	28	8	3

[Continued]

★ 37 ★

Ratings of Local Police Solving Crimes, 1992

[Continued]

	Excellent	Pretty good	Only fair	Poor	Not sure/ refused
Black	2	41	38	17	2
Hispanic	11	46	31	9	2

Source: "Respondents' Ratings of Local Police Solving Crimes," Department of Justice, *Sourcebook of Criminal Justice Statistics-1992*, p. 169. Primary source: Table adapted by *Sourcebook* staff from data provided by Lou Harris and Associates. Published by permission. *Notes:* A. Percents may not add to 100 because of rounding. B. Less than 0.5 percent.

★ 38 ★

Attitudes

Ratings of Local Police Treating People Fairly, 1992

	Excellent	Pretty good	Only fair	Poor	Not sure/ refused
"How would you rate the police in your community on the following...treating people fairly?"					
National	20%	43%	24%	11%	2%
Race ethnicity					
White	22	46	21	9	2
Black	9	29	36	26	0
Hispanic	17	37	20	13	3

Source: "Respondents' Ratings of Local Police Treating People Fairly," Department of Justice, *Sourcebook of Criminal Justice Statistics-1992*, p. 172. Primary source: Table adapted by *Sourcebook* staff from data provided by Lou Harris and Associates. Published by permission.

★ 39 ★

Attitudes

Ratings of Local Policing Preventing Crimes, 1992

	Excellent	Pretty good	Only fair	Poor	Not sure/ refused
"How would you rate the police in your community on the following...preventing crime?"					
National	16%	42%	28%	13%	1%
Race ethnicity					
White	17	42	28	11	1

[Continued]

★ 39 ★

Ratings of Local Policing Preventing Crimes, 1992
[Continued]

	Excellent	Pretty good	Only fair	Poor	Not sure/ refused
Black	10	46	27	18	0
Hispanic	15	40	24	19	1

Source: "Respondents' Ratings of Local Police Preventing Crimes," Department of Justice, *Sourcebook of Criminal Justice Statistics-1992*, p. 170. Primary source: Table adapted by *Sourcebook* staff from data provided by Lou Harris and Associates. Published by permission. *Note:* A. Percents may not add to 100 because of rounding.

★ 40 ★

Attitudes

Ratings of the Honesty and Ethical Standards of Lawyers, 1992

	Very high	High	Average	Low	Very low	Don't know
"How would you rate the honesty and ethical standards of people in these different fields--very high, high, average, low, or very low: Lawyers?"						
National	3%	15%	43%	25%	11%	3%
Race						
White	3	14	43	26	12	3
Nonwhite	6	21	45	17	8	3
Black	7	18	50	14	6	5

Source: "Respondents' Ratings of the Honesty and Ethical Standards of Lawyers," Department of Justice, *Sourcebook of Criminal Justice Statistics-1992*, p. 168. Primary source: Table constructed by *Sourcebook* staff from data provided by the Gallup Organization. Published by permission. *Note:* A. Percents may not add to 100 because of rounding.

★ 41 ★

Attitudes

Ratings of the Honesty and Ethical Standards of Policemen, 1992

	Very high	High	Average	Low	Very low	Don't know
"How would you rate the honesty and ethnical standards of people in these different fields--very high, high, average, low, or very low: Policemen?"						
National	8%	34%	42%	10%	4%	2%

[Continued]

★ 41 ★

Ratings of the Honesty and Ethical Standards of Policemen, 1992

[Continued]

	Very high	High	Average	Low	Very low	Don't know
Race						
White	8	35	44	8	2	2
Nonwhite	4	28	28	20	15	5
Black	3	26	29	21	16	6

Source: "Respondents' Ratings of the Honesty and Ethical Standards of Policemen," Department of Justice, *Sourcebook of Criminal Justice Statistics-1992*, p. 169. Primary source: Table constructed by *Sourcebook* staff from data provided by the Gallup Organization. Published by permission. *Notes:* A. Percents may not add to 100 because of rounding. B. Less than 1 percent.

★ 42 ★

Attitudes

Recommendations to High School Sophomores to Attend College, 1980 and 1990

Student characteristics	Father		Mother		Guidance counselor		Teachers	
	1980	1990	1980	1990	1980	1990	1980	1990
All sophomores	59.1	77.0	64.8	82.9	32.3	65.2	32.3	65.5
Asian	78.7	87.9	81.1	88.8	32.9	68.6	34.6	72.0
Hispanic	56.3	75.3	63.2	81.1	32.2	64.8	34.5	65.2
Black	56.6	69.4	67.2	76.6	37.1	66.1	42.0	70.0
White	59.7	78.2	64.5	84.3	31.4	65.1	30.4	64.6
American Indian	46.8	62.4	51.9	70.3	31.7	52.4	29.6	59.9

Source: "1980 and 1990 Sophomores' Reports of Percentages of Fathers, Mothers, Guidance Counselors, and Teachers Who Recommend Attending College After High School, by Student Characteristics," U.S. Department of Education, National Center for Educational Statistics, *America's High School Sophomores: A Ten Year Comparison*, p. 45. Primary source: National Center for Education Statistics, High School and Beyond base year sophomore cohort and NELS:88 first follow-up.

★ 43 ★

Attitudes

Regional of Local Police Not Using Excessive Force, 1992

	Excellent	Pretty good	Only fair	Poor	Not sure/ refused
"How would you rate the police in your community on the following...not using excessive force?"					
National	26%	41%	21%	9%	3%

[Continued]

★ 43 ★

Regional of Local Police Not Using Excessive Force, 1992

[Continued]

	Excellent	Pretty good	Only fair	Poor	Not sure/ refused
Race ethnicity					
White	29	42	19	7	3
Black	11	43	30	16	0
Hispanic	22	39	24	13	2

Source: "Respondents' Ratings of Local Police Not Using Excessive Force," Department of Justice, *Sourcebook of Criminal Justice Statistics-1992*, p. 172. Primary source: Table adapted by *Sourcebook* staff from data provided by Lou Harris and Associates. Published by permission. *Note:* A. Percents may not add to 100 because of rounding.

★ 44 ★

Attitudes

Respect For Police in Own Area, 1991

	Great deal	Some	Hardly any	No opinion
"How much respect do you have for the police in your area--a great deal, some, or hardly any?"				
National	60%	32%	7%	1%
Race ethnicity				
White	62	31	6	1
Black	51	32	17	0
Other	51	37	9	3

Source: "Reported Respect for Police in Own Area," Department of Justice, *Sourcebook of Criminal Justice Statistics-1992*, p. 171. Primary source: George Gallup Jr., The Gallup Pole Monthly 306 (March 1991). Published by permission. *Note:* A. Less than 1 percent.

★ 45 ★

Attitudes

School Safety: High School Sophomores, 1980 and 1990

Student characteristics	1980	1990
All sophomores	12.2	8.1
Asian	13.9	9.9
Hispanic	16.2	10.8
Black	17.7	12.9

[Continued]

★ 45 ★

School Safety: High School Sophomores, 1980 and 1990
[Continued]

Student characteristics	1980	1990
White	10.7	6.7
American Indian	13.3	10.1

Source: "Percent of 1980 and 1990 Sophomores Who Report That They Do Not Feel Safe at Their Schools by Student Characteristics," U.S. Department of Education, National Center for Educational Statistics, *America's High School Sophomores: A Ten Year Comparison,* p. 21. Primary source: National Center for Education Statistics, High School and Beyond base year sophomore cohort and NELS:88 first follow-up.

Behavior

★ 46 ★

Alcohol Consumption As A Source of Family Trouble, 1992

	Yes	No	No opinion/refused
"Has drinking ever been a cause of trouble in your family?"			
National	24%	76%	(a)
Race			
White	25	75	(a)
Nonwhite	21	79	0

Source: "Respondents Reporting Whether Drinking Has Ever Been a Source of Family Trouble," Department of Justice, *Sourcebook of Criminal Justice Statistics-1992,* p. 354. Primary source: George Gallup Jr., *The Gallup Poll Monthly* 317 (February 1992). Reprinted by permission. *Note:* A. Less than 1 percent.

★ 47 ★

Behavior

Alcohol Consumption among Females, 1988

Drinking level	Percent		
	Total	White	Black
Abstainer	60	57	72
Light	24	25	18
Moderate	13	14	8
Heavy	3	4	2

Source: "Alcohol Consumption among Females, by Race, 1988," Interagency Board for Nutrition Monitoring and Related Research, *Nutrition Monitoring in the United States: Chartbook I: Selected Findings From the National Nutrition Monitoring and Related Research Program,* p. 106. Primary source: Data from the National Institutes of Health, National Institute on Alcohol Abuse and Alcoholism. *Notes:* Drinking levels of individuals are defined as: abstainer: less than 12 drinks per year; light drinker: 3 drinks per week to 12 or more drinks per year; moderate drinker: 4-13 drinks per week; and heavy drinker: two or more drinks per day or 14 or more drinks per week. A drink is defined as a 12-ounce can or bottle of beer, a 4-ounce glass of wine, or a 1-ounce shot of distilled spirits. The total category includes individuals of all races. Data from the other race category are not shown because individuals in this category are racially diverse with divergent drinking patterns, making aggregation of the data meaningless.

★ 48 ★

Behavior

Alcohol Consumption among Males, 1988

Drinking level	Percent		
	Total	White	Black
Abstainer	37	35	48
Light	26	26	21
Moderate	25	26	21
Heavy	12	13	10

Source: "Alcohol Consumption among Males, by Race, 1988," Interagency Board for Nutrition Monitoring and Related Research, *Nutrition Monitoring in the United States: Chartbook I: Selected Findings From the National Nutrition Monitoring and Related Research Program,* p. 106. Primary source: Data from the National Institutes of Health, National Institute on Alcohol Abuse and Alcoholism. *Notes:* Drinking levels of individuals are defined as: abstainer: less than 12 drinks per year; light drinker: 3 drinks per week to 12 or more drinks per year; moderate drinker: 4-13 drinks per week; and heavy drinker: two or more drinks per day or 14 or more drinks per week. A drink is defined as a 12-ounce can or bottle of beer, a 4-ounce glass of wine, or a 1-ounce shot of distilled spirits. The total category includes individuals of all races. Data from the other race category are not shown because individuals in this category are racially diverse with divergent drinking patterns, making aggregation of the data meaningless.

★ 49 ★

Behavior

Alcohol Use, 1991

By demographic characteristics, United States, 1991.

	Alcohol			
			Most recent use	
	Never used	Ever used	Within last year	Within last 30 days
Total (N=32,594)	15.4%	84.6%	68.0%	50.9%
Race, ethnicity				
White	13.3	86.7	69.9	52.7
Black	21.0	79.0	59.7	43.7
Hispanic	22.6	77.4	64.9	47.5

Source: "Estimated Prevalence and Most Recent Use of Alcohol, Marijuana, and Cocaine," Department of Justice, Bureau of Criminal Statistics, *Sourcebook of Criminal Justice Statistics, 1992*, p. 339. Primary source: U.S. Department of Health and Human Services, Substance Abuse and Mental Health Services Administration, *National Household Survey on Drug Abuse: Main Findings 1991*. Table constructed by *Sourcebook* staff.

★ 50 ★

Behavior

Apparent Motivation of High School Sophomores as Shown in Behavior, 1980 and 1990

Student characteristics	come to school without books		Come to school without paper, pen, or pencil		Come to school without homework	
	1980	1990	1980	1990	1980	1990
All sophomores	8.5	6.3	15.1	10.5	22.1	18.1
Asian	13.0	9.5	14.6	11.0	17.1	17.6
Hispanic	13.8	10.9	20.1	13.5	27.7	20.6
Black	13.7	8.1	17.6	9.6	22.9	16.0
White	6.7	5.1	13.9	10.2	21.2	18.1
American Indian	17.5	11.1	25.9	11.8	30.9	21.9

Source: "Percentages of 1980 and 1990 Sophomores Saying That They Usually or Often Come to School Without Paper and Pencil, Books, and/or Homework, by Student Characteristics," U.S. Department of Education, National Center for Educational Statistics, *America's High School Sophomores: A Ten Year Comparison*, p. 20. Primary source: National Center for Education Statistics, High School and Beyond base year sophomore cohort and NELS:88 first follow-up.

★ 51 ★

Behavior

Characteristics of Drug Users, 1991

Drug type	Number of persons who used during the--		Age group with highest percent of users	Characteristics of current users				
				Percent of users who are:			Percent of users who are	
	Past month	Past year		White	Black	Hispanic	Male	Female
Heroin[1]	[1]	701,000	35+	74.5%	14.8%	10.7%	42.7%	57.3%
Cocaine	1,892,000	6,383,000	18-25	62.0	23.4	14.6	70.1	29.9
Crack cocaine	479,000	1,021,000	18-25 & 26-34	49.9	35.9	14.2	82.0	18.0
Marijuana	9,721,000	19,549,000	18-25	75.1	17.5	7.3	63.6	36.4

Source: "What Are The Characteristics of Users of Different Drugs," *Drugs, Crime, and the Justice System*, p. 28. Primary source: NIDA, *National Household Survey on Drug Abuse: Population Estimates 1991*, 1991. *Notes:* 1. For heroin, characteristics are for persons who used in the past year. Past month data are not available.

★ 52 ★

Behavior

Inhalant and Hallucinogen Use, 1991

	Inhalants					Hallucinogens				
	Total all ages	Age group				Total all ages	Age group			
		12 to 17 years	18 to 25 years	26 to 34 years	35 years and older		12 to 17 years	18 to 25 years	26 to 34 years	35 years and older
Total (N = 32,594)	5.4%	7.0%	10.9%	9.2%	2.5%	8.1%	3.3%	13.1%	15.5%	5.2%
Race, ethnicity										
White	5.6	7.6	12.7	10.3	2.3	8.9	3.8	15.8	18.0	5.4
Black	3.8	5.1	4.5	4.6	2.9	4.1	1.2	5.4	5.9	3.7
Hispanic	4.8	6.6	6.5	6.3	2.6	6.4	3.5	7.5	9.2	5.5

Source: "Estimate Prevalence Inhalant and Hallucinogen Use," Department of Justice Bureau of Criminal Statistics, *Sourcebook of Criminal Justice Statistics, 1992*, p. 340. Primary source: U.S. Department of Health and Human Services, Substance Abuse and Mental Health Services Administration, *National Household Survey on Drug Abuse: Main Findings 1991*. Table constructed by *Sourcebook* staff.

★ 53 ★

Behavior

Marijuana and Cocaine Use, 1991

By demographic characteristics, United States, 1991.

	Marijuana				Cocaine			
			Most recent use				Most recent use	
	Never used	Ever used	Within last year	Within last 30 days	Never used	Ever used	Within last year	Within last 30 days
Total (N=32,594)	66.8%	33.2%	9.5%	4.8%	88.5%	11.5%	3.0%	0.9%
Race, ethnicity								
White	66.2	33.6	9.2	4.5	88.2	11.8	2.8	0.7
Black	64.3	35.7	12.2	7.2	88.8	11.2	3.9	1.8
Hispanic	72.8	27.2	8.7	4.3	88.9	11.1	3.8	1.6

Source: "Estimated Prevalence and Most Recent Use of Alcohol, Marijuana, and Cocaine," Department of Justice, Bureau of Criminal Statistics, *Sourcebook of Criminal Justice Statistics, 1992*, p. 339. Primary source: U.S. Department of Health and Human Services, Substance Abuse and Mental Health Services Administration, *National Household Survey on Drug Abuse: Main Findings 1991*. Table constructed by *Sourcebook* staff.

★ 54 ★

Behavior

PCP and Heroin Use, 1991

	PCP					Heroin				
	Total all ages	Age group				Total all ages	Age group			
		12 to 17 years	18 to 25 years	26 to 34 years	35 years and older		12 to 17 years	18 to 25 years	26 to 34 years	35 years and older
Total (N=32,594)	3.6%	1.1%	4.2%	8.0%	2.4%	1.3%	0.3%	0.8%	1.8%	1.5%
Race, ethnicity										
White	3.8	1.2	4.8	9.0	2.4	1.2	0.2	0.8	1.7	1.3
Black	2.5	0.4	3.2	4.3	2.2	1.9	0.4	0.9	2.2	2.5
Hispanic	3.1	1.4	2.5	5.8	2.4	1.5	0.5	0.8	1.9	2.0

Source: "Estimate Prevalence of PCP and Heroin Use," Department of Justice Bureau of Criminal Statistics, *Sourcebook of Criminal Justice Statistics, 1992*, p. 341. Primary source: U.S. Department of Health and Human Services, Substance Abuse and Mental Health Services Administration, *National Household Survey on Drug Abuse: Main Findings 1991*. Table constructed by *Sourcebook* staff.

★ 55 ★

Behavior

Reported Alcohol Use, 1992

	Yes	No, total abstainer
"Do you have occasion to use alcoholic beverages such as liquor, wine, or beer or are you a total abstainer?"		
National	64%	35%
Race		
White	66	33
Nonwhite	50	50

Source: "Reported Alcohol Use," Department of Justice, *Sourcebook of Criminal Justice Statistics-1992,* p. 352. Primary source: George Gallup Jr., *The Gallup Poll Month* 317 (February 1992). Reprinted by permission. *Note:* Responses of "no opinion" were omitted by the source.

★ 56 ★

Behavior

Reported Behavior Changes Because of Fear of Crime, 1991

	Yes	No	Not sure
"Has fear of crime caused you to...?"			
Limit the places or times			
that you go shopping	32%	68%	0%
White	30	70	0
Black	44	56	0
Hispanic	37	61	1
Limit the places or times			
that you work	22	76	2
White	19	78	2
Black	33	65	3
Hispanic	37	63	0
Limit the places you will			
go by yourself	60	40	1
White	60	40	1
Black	63	37	0
Hispanic	64	36	0
Purchase a weapon			
for self-protection	18	82	0
White	16	84	0
Black	27	72	1

[Continued]

★ 56 ★

Reported Behavior Changes Because of Fear of Crime, 1991
[Continued]

	Yes	No	Not sure
Hispanic	25	75	0
Install a home security			
system	25	75	1
White	22	77	1
Black	34	66	0
Hispanic	41	58	2

Source: "Reported Behavior Changes Because of Fear of Crime," Department of Justice, *Sourcebook of Criminal Justice Statistics-1992*, p. 192. Primary source: National Victim Center, "America Speaks Out: Citizens' Attitudes About Victims' Rights and Violence," April 1991 (mimeographed). Table constructed by *Sourcebook* staff. *Note:* A. Percents may not add to 100 because of rounding.

★ 57 ★

Behavior

Reported Overindulgence in Alcohol, 1992

	Yes	No
"Do you sometimes drink more than you think you should?"		
National	29%	71%
Race		
White	29	71
Nonwhite	24	76

Source: "Respondents Reporting Whether They Drink More Than They Should," Department of Justice. *Sourcebook of Criminal Justice Statistics-1992*, p. 353. Primary source: Table constructed by *Sourcebook* staff from data provided by the Gallup Organization. Reprinted by permission. *Notes:* This question was presented to the 64 percent of respondents answering "yes" to the question: "Do you have occasion to use alcohol beverages such as liquor, wine or beer, or are you a total abstainer? asked in February 1992. Responses of "don't known/refused" were omitted.

★ 58 ★

Behavior

Reported Plans To Change Alcohol Consumptions, 1992

	Yes, cut down	Yes quit	No
"Do you plan to cut down or quit drinking within the next year?"			
National	17%	9%	74%
Race			
White	16	7	76
Nonwhite	20	22	56

Source: "Respondents Reporting Whether They Plan to Cut Down or Quit Drinking," Department of Justice, *Sourcebook of Criminal Justice Statistics-1992*, p. 353. Primary source: Table constructed by *Sourcebook* staff from data provided by the Gallup Organization. Reprinted by permission. *Notes:* This question was presented to the 64 percent of respondents answering "yes" to the question: "Do you have occasion to use alcohol beverages such as liquor, wine, or beer, or are you a total abstainer?" asked in February 1992. Responses of "don't know/or refused" were omitted.

★ 59 ★

Behavior

Respondents Reporting a Firearm in Their Home, 1980 to 1991

	1980	1982	1984	1985	1987	1988	1989	1990	1991
"Do you happen to have in your home (or garage) any guns or revolvers?" (Percent reporting have guns)									
National	48%	45%	45%	44%	46%	40%	46%	43%	40%
Race									
White	50	48	48	46	49	43	50	45	42
Black/other	29	30	30	29	33	28	23	29	29

Source: "Respondents Reporting a Firearm in Their Home," Department of Justice, *Sourcebook of Criminal Justice Statistics-1992*, p. 209. Primary source: Table constructed by *Sourcebook* staff from data provided by the National Opinion Research Center. Data were made avilable through The Roper Center for Public Opinion Research.

★ 60 ★

Behavior

Type of Firearm in Their Home, 1991

	Guns in the home			
	All types	Type of firearm		
		Pistol	Shotgun	Rifle
"Do you happen to have in your home (or garage) any guns or revolvers?" If yes, "Is it a pistol, shotgun, rifle, or what?" (Percent of respondents reporting having guns)				
National	40%	20%	26%	25%
Race				
White	42	22	28	28
Black/other	29	15	15	11

Source: "Respondents Reporting a Firearm in Their Home," Department of Justice, *Sourcebook of Criminal Justice Statistics-1992*, p. 210. Primary source: Table constructed by *Sourcebook* staff from data provided by the National Opinion Research Center. Data were made available through The Roper Center for Public Opinion Research.

★ 61 ★

Behavior

Use of Alcohol During the Last Month, 1991 and Preliminary 1992

By age group and other demographic characteristics, United States, 1991 and preliminary 1992.

	Total all ages		Age group							
			12 to 17 years		18 to 25 years		26 to 34 years		35 years and older	
	1991	1992	1991	1992	1991	1992	1991	1992	1991	1992
Total	50.9%	47.8%	20.3%	15.7%	63.6%	59.2%	61.7%	61.2%	49.5%	46.5%
Race, ethnicity										
White	52.7	49.7	20.4	16.7	67.2	62.9	63.8	63.7	50.9	47.8
Black	43.7	39.8	20.1	13.2	56.0	50.9	57.1	55.6	40.3	37.2
Hispanic	47.5	45.0	22.5	16.2	52.8	52.8	57.2	56.1	47.8	44.9
Other	41.4	38.4	12.8	7.8	52.7	38.2	42.8	44.1	43.6	43.0

Source: "Estimate Use of Alcohol During the Past Month," Department of Justice, Bureau of Criminal Statistics, *Sourcebook of Criminal Justice Statistics, 1992*, p. 345. Primary source: U.S. Department of Health and Human Services, Substance Abuse and Mental Health Services Administration, *National Household Survey on Drug Abuse: Main Findings 1991*.

★ 62 ★

Behavior

Use of Any Illicit Drug During the Last Month, 1991 and Preliminary 1992

By age group and other demographic characteristics, United States, 1991 and preliminary 1992.

| | Total all ages | | Age group | | | | | | | |
| | | | 12 to 17 years | | 18 to 25 years | | 26 to 34 years | | 35 years and older | |
	1991	1992	1991	1992	1991	1992	1991	1992	1991	1992
Total	6.3%	5.5%	6.8%	6.1%	15.4%	13.0%	9.0%	10.1%	3.1%	2.2%
Race, ethnicity										
White	5.9	5.5	6.6	6.1	16.0	13.7	8.7	10.6	2.7	2.2
Black	9.4	6.6	7.0	6.1	16.9	12.1	13.7	10.3	5.8	3.5
Hispanic	6.3	5.3	7.9	7.1	11.6	10.2	5.9	7.8	3.8	1.3
Other	5.4	3.6	4.3	4.2	9.2	11.2	7.5	6.2	[1]	[1]

Source: "Estimate Use of Any Illicit Drug During the Past Month," Department of Justice, Bureau of Criminal Statistics, *Sourcebook of Criminal Justice Statistics, 1992*, p. 342. Primary source: U.S. Department of Health and Human Services, Substance Abuse and Mental Health Services Administration, *National Household Survey on Drug Abuse: Main Findings 1991. Notes:* These data are from the National Household Survey on Drug Abuse (NHSDA), an ongoing series of surveys measuring the prevalence of drug abuse among the American household population aged 12 and older. The 1991 NHSDA was sponsored by the U.S. Department of Health and Human Services, National Institute on Drug Abuse. The 1992 NHSDA was sponsored by the newly created Substance Abuse and Mental Health Service Administration of the U.S. Department of Health and Human Services. Households were randomly sampled from all households in the United States. A sample of 32,594 persons were interviewed in 1991 and a sample of 28,832 persons were interviewed in 1992. The data for 1992 are preliminary and subject to revision. 1. Estimates based on only a few respondents are omitted because one cannot place a high degree of confidence in their statistical accuracy.

★ 63 ★

Behavior

Use of Cocaine During the Last Month, 1991 and Preliminary 1992

By age group and other demographic characteristics, United States, 1991 and preliminary 1992.

| | Total all ages | | Age group | | | | | | | |
| | | | 12 to 17 years | | 18 to 25 years | | 26 to 34 years | | 35 years and older | |
	1991	1992	1991	1992	1991	1992	1991	1992	1991	1992
Total	0.9%	0.6%	0.4%	0.3%	2.0%	1.8%	1.8%	1.4%	0.5%	0.2%
Race, ethnicity										
White	0.7	0.5	0.3	0.1	1.7	2.0	1.6	1.2	0.2	0.1
Black	1.8	1.0	0.5	0.2	3.1	1.4	2.7	1.7	1.3	0.8
Hispanic	1.6	1.2	1.3	1.2	2.7	1.8	2.0	2.4	1.0	0.4
Other	2.0	0.1	[1]	[1]	[1]	[1]	1.9	[1]	[1]	[1]

Source: "Estimate Use of Cocaine During the Past Month," Department of Justice, Bureau of Criminal Statistics, *Sourcebook of Criminal Justice Statistics, 1992*, p. 344. Primary source: U.S. Department of Health and Human Services, Substance Abuse and Mental Health Services Administration, *National Household Survey on Drug Abuse: Main Findings 1991. Notes:* 1. Estimates based on only a few respondents are omitted because one cannot place a high degree of confidence in their statistical accuracy.

★ 64 ★

Behavior

Use of Marijuana During the Last Month, 1991 and Preliminary 1992

By age group and other demographic characteristics, United States, 1991 and preliminary 1992.

	Total all ages		Age group							
			12 to 17 years		18 to 25 years		26 to 34 years		35 years and older	
	1991	1992	1991	1992	1991	1992	1991	1992	1991	1992
Total	4.8%	4.4%	4.3%	4.0%	13.0%	11.0%	7.0%	8.2%	2.1%	1.6%
Race, ethnicity										
White	4.5	4.4	4.4	4.1	13.7	11.6	6.6	8.8	1.9	1.6
Black	7.2	5.2	4.5	3.4	14.6	11.2	11.9	8.2	3.5	2.5
Hispanic	4.3	3.7	4.6	4.8	9.1	8.0	4.2	5.6	2.3	0.7
Other	3.4	2.4	1.2	2.9	4.8	6.4	6.5	4.8	1	1

Source: "Estimate Use of Marijuana During the Past Month," Department of Justice, Bureau of Criminal Statistics, *Sourcebook of Criminal Justice Statistics, 1992,* p. 343. Primary source: U.S. Department of Health and Human Services, Substance Abuse and Mental Health Services Administration, *National Household Survey on Drug Abuse: Main Findings 1991. Notes:* 1. Estimates based on only a few respondents are omitted because one cannot place a high degree of confidence in their statistical accuracy.

High School Students

★ 65 ★

High School Sophomores Who Participate in Various Activities At Least Once a Week and Amount of TV Watching, Percentages: 1980 and 1990

Student characteristic	Just driving or riding around		Visiting with friends at local hangout		Talking with friends on the telephone		Reading for pleasure		Hours of television on school nights	
									5 or more hours,	More than 5 hours,
	1980	1990	1980	1990	1980	1990	1980	1990	1980	1990
All sophomores	47.1	56.1	67.2	66.3	76.6	80.1	41.1	41.0	27.3	9.1
Male	51.0	57.9	69.4	69.5	66.5	72.5	34.3	33.8	29.1	10.2
Female	43.3	54.3	65.2	63.1	86.2	87.7	47.9	48.2	24.9	8.0
Race/ethnicity										
White	49.0	58.9	68.7	68.7	78.4	81.7	40.4	41.5	25.1	6.7
Black	38.0	50.1	64.8	59.1	73.3	79.6	46.6	41.2	39.8	23.0
Hispanic	46.6	47.6	60.2	59.3	68.6	72.4	36.3	38.2	27.3	10.2
Asian	31.5	44.0	55.3	57.1	67.7	78.3	50.4	40.2	23.5	6.9
American Indian	51.6	53.3	62.2	70.4	59.4	65.1	41.8	39.5	26.5	15.8

Source: "Percent of High School Sophomores Who Say They Engage in Various Activities at Least Once or Twice a Week and Amount of TV Watching, by Student Characteristics: 1980 and 1990." *Digest of Educational Statistics,* 1993, p. 138. Primary source: U.S. Department of Education, National Center for Education Statistics, "High School and Beyond," Base Year Survey, 1980 Sophomore Cohort; and "National Education Longitudinal Study of 1988," First Follow- up Student Survey. (This table was prepared April 1993.).

★ 66 ★

High School Students

Tenth Graders' Attitude toward School Attendance, 1990

Reason for going to school	Percent of 10th graders					
	All 10th graders	Race/ethnicity				
		White	Black	Hispanic	Asian	American Indian
Think subjects are interesting	71.0	68.8	79.1	74.5	77.3	81.2
Get a feeling of satisfaction	76.9	74.8	85.8	81.3	79.6	81.6
Nothing else to do	30.3	30.1	29.0	31.1	32.4	31.3
Need education to get a job	96.6	96.5	96.7	96.8	97.1	93.4
To meet friends	82.7	85.5	66.1	80.1	84.9	80.8
Play on a team or belong to a club	53.6	55.3	49.3	45.3	56.3	46.2
Teachers care and expect student to succeed	74.0	72.4	81.6	76.0	74.6	79.4

Source: "Tenth Graders' Who Agree or Strongly Agree with Statements on Why They Go to School: 1990." *Digest of Educational Statistics*, 1993, p. 143. Primary source: U.S. Department of Education, National Center for Education Statistics, "National Education Longitudinal Study of 1988," First Followup survey. (This table was prepared February 1993.).

★ 67 ★

High School Students

Tenth Graders' Attitudes About Academic Classes, 1990

Class subject and opinion	Percent of tenth graders who answered, "a few times a week" or more often					
	All 10th graders	Race/ethnicity				
		White	Black	Hispanic	Asian	American Indian
Mathematics class						
Understood the material	60.8	59.5	68.5	60.7	59.6	61.8
Try very hard	80.2	79.0	84.9	81.8	83.1	84.4
Feel challenged	74.4	73.7	77.7	75.1	75.9	70.9
English class						
Understood the material	50.9	49.0	59.7	52.0	53.4	48.0
Try very hard	79.3	77.7	84.8	82.5	81.7	82.0
Feel challenged	59.8	56.7	72.6	66.0	60.8	56.4
History class						
Understood the material	32.5	31.7	32.6	36.7	32.0	43.0
Try very hard	53.5	52.9	54.4	55.4	52.1	64.2
Feel challenged	42.1	40.5	48.0	45.1	40.7	54.7
Science class						
Understood the material	48.3	47.4	53.0	47.6	49.5	53.4

[Continued]

★ 67 ★

Tenth Graders' Attitudes About Academic Classes, 1990
[Continued]

Class subject and opinion	Percent of tenth graders who answered, "a few times a week" or more often					
	All 10th graders	Race/ethnicity				
		White	Black	Hispanic	Asian	American Indian
Try very hard	72.8	72.8	73.5	70.4	74.2	77.8
Feel challenged	66.6	66.1	69.9	63.9	69.7	67.9

Source: "Tenth Graders' Attitudes About Academic Classes, by Selected Student and School Characteristics: 1990." *Digest of Educational Statistics*, 1993, p. 135. Primary source: U.S. Department of Education, National Center for Education Statistics, "National Education Longitudinal Study of 1988," First Followup survey. (This table was prepared February 1993.).

★ 68 ★

High School Students

Tenth Graders' Attitudes About School Climate, 1990

Statements about school climate	Percent who strongly agree or agree with statement					
	Total	Race/ethnicity				
		White	Black	Hispanic	Asian	American Indian
Students get along well with teachers	74.9	77.0	62.8	73.3	82.9	68.3
There is real school spirit	70.4	71.7	67.5	67.1	67.9	64.9
Rules for behavior are strict	63.8	64.4	62.3	63.7	61.4	53.6
Discipline is fair	70.2	70.4	65.1	72.5	77.2	64.2
Other students often disrupt class	70.7	69.9	76.3	69.1	69.6	75.3
Teaching is good	81.9	81.1	83.1	84.9	85.1	81.2
Teachers are interested in students	76.0	75.6	76.0	77.3	79.8	82.2
Teachers praise my effort when I work hard	57.2	54.4	64.6	64.6	66.5	55.1
I often feel "put down" by my teachers	16.0	16.2	16.3	15.7	12.5	11.4
Teachers listen to what I have to say	70.1	68.8	74.8	73.4	71.0	72.5
I don't feel safe at this school	8.0	6.7	12.8	10.8	9.7	8.4
Disruptions by other students interfere with my learning	39.9	36.7	51.1	44.6	45.7	53.1
Misbehaving students often get away with it	52.7	53.4	45.5	53.8	57.1	60.1

Source: "Eighth and Tenth Graders' Attitudes About School Climate, by Student and School Characteristics: 1988 and 1990." *Digest of Educational Statistics*, 1993, p. 136. Primary source: U.S. Department of Education, National Center for Education Statistics, "National Education Longitudinal Study of 1988," Base Year and First Followup surveys. (This table was prepared February 1993.).

★ 69 ★

High School Students

Tenth Graders' Expected Occupations, 1990

Expected occupation at age 30	Total	Race/ethnicity				
		White	Black	Hispanic	Asian	American Indian
Total	100.0	100.0	100.0	100.0	100.0	100.0
Craftsperson or operator	5.6	5.7	5.7	5.0	2.3	8.6
Farmer or farm manager	1.1	1.4	0.4	0.6	0.2	1.0
Housewife/homemaker	2.0	1.9	1.3	3.8	1.3	1.5
Laborer or farm worker	0.8	0.8	0.8	0.8	0.9	2.0
Military police, or security officer	5.7	5.6	6.4	6.6	3.2	8.9
Professional, business, or managerial	45.7	45.3	50.3	40.2	56.3	42.9
Teacher	4.1	4.7	2.6	2.6	1.7	5.0
Business owner	5.3	5.2	4.4	6.8	7.1	4.6
Technical	4.7	4.2	5.1	7.0	5.6	5.5
Salesperson, clerical, or office worker	4.9	4.6	6.0	6.2	4.1	4.3
Service worker	1.8	1.7	3.2	0.9	0.4	1.2
Other employment	7.7	8.7	4.4	6.8	4.6	7.8
Don't know	10.5	10.3	9.6	12.7	12.5	6.8

Source: "Expected Occupations of 8th and 9th Graders at Age 30, by Selected Student and School Characteristics: 1988 and 1990." *Digest of Educational Statistics*, 1993, p. 135. Primary source: U.S. Department of Education, National Center for Education Statistics, "National Education Longitudinal Study of 1988," Base Year and First Followup surveys. (This table was prepared February 1993.).

★ 70 ★

High School Students

Tenth Graders' Participation in Extracurricular Activities, 1990

Extracurricular activities	Percent who participated in extracurricular activities					
	Total	Race/ethnicity				
		White	Black	Hispanic	Asian	American Indian
Athletics						
Baseball/softball	15.6	16.0	13.7	15.7	13.9	19.8
Basketball	19.9	18.2	30.9	16.6	22.8	21.8
Football	15.9	14.7	22.6	16.0	16.2	14.5
Soccer	7.6	7.9	4.0	8.5	10.2	5.9
Swim team	3.9	4.1	2.8	3.3	5.2	3.7
Other team sport	14.2	14.5	11.3	13.5	19.7	18.0
Individual sports	23.2	23.9	21.9	17.9	28.4	20.4
Performing arts						
Cheerleading	5.9	5.3	9.9	5.2	3.8	14.6

[Continued]

★ 70 ★

Tenth Graders' Participation in Extracurricular Activities, 1990
[Continued]

Extracurricular activities	Percent who participated in extracurricular activities					
	Total	Race/ethnicity				
		White	Black	Hispanic	Asian	American Indian
Drill team	4.5	3.6	9.4	5.2	3.6	5.7
School band or orchestra	20.9	21.7	22.3	14.1	20.1	19.5
School play or musical	11.0	11.0	12.1	9.3	13.0	9.8
School government/clubs						
Student government	7.3	7.3	7.4	5.9	9.8	8.6
Academic honor society	7.7	7.3	8.1	7.5	13.9	7.7
School yearbook/newspaper	8.8	8.5	10.5	7.3	12.7	13.1
School service clubs	11.5	11.7	10.4	9.9	18.1	9.2
School academic clubs	30.7	31.1	25.1	26.7	35.9	31.2
School hobby clubs	7.3	7.4	5.2	6.4	11.5	7.2
School FTA, FHA, and FFA	11.7	12.3	13.8	8.0	5.0	19.3

Source: "Participation of 10th Graders in Extracurricular Activities, by Selected Student Characteristics: 1990." *Digest of Educational Statistics*, 1993, p. 137. Primary source: U.S. Department of Education, National Center for Education Statistics, "National Education Longitudinal Study of 1988," First Follow- up survey. (This table was prepared May 1992.).

Chapter 3
BUSINESS AND ECONOMICS

Banks, Financial Institutions, Finances

★ 71 ★

Banks: Percent Change, 1992

Black owned banks	1991	1992	Percent change
Number of banks	38	36	-5.3
Number of employees	1,666	1,637	-2.9
Assets[1]	2,010.853	2,107.865	4.8
Capital[1]	147.568	158.952	7.7
Deposits[1]	1,778.716	1,879.617	5.7
Loans[1]	827.976	902.854	9.0

Source: "1993 Bank Summary," *Black Enterprise* 24 (June 1993), p. 144. Published by permission. Primary source: B.E. Research. *Note:* 1. In millions of dollars, to the nearest thousand.

★ 72 ★

Banks, Financial Institutions, Finances

Financial Companies: Characteristics, 1992

Company	Location	Chief executive	Year started	Staff	Assets[1]	Capital[1]	Deposits[1]	Loans[1]
Carver Federal Savings Bank	New York, New York	Richard T. Greene	1948	100	320.862	13.741	252.684	246.097
Independence Federal Savings Bank	Washington, D.C.	William B. Fitzgerald	1968	73	239.223	12.852	181.708	216.293
Seaway National Bank of Chicago	Chicago, Illinois	Walter E. Grady	1965	165	202.093	14.136	168.076	56.570
Industrial Bank of Washington	Washington, D.C.	B. Doyle Mitchell Jr.	1934	107	186.808	12.299	173.807	71.116
Family Savings Bank, FSB	Los Angeles, California	Wayne-Kent Bradshaw	1948	51	140.113	7.279	114.806	115.270
Independence Bank of Chicago	Chicago, Illinois	Alvin J. Boutte	1964	119	137.278	10.593	121.726	41.375
Citizens Trust Bank	Atlanta, Georgia	William L. Gibbs	1921	150	128.152	8.543	117.401	53.770
Drexel National Bank	Chicago, Illinois	Alvin J. Boutte	1989	90	127.754	8.463	118.482	35.210
First texas Bank	Dallas, Texas	William E. Stahnke	1975	55	110.314	9.672	99.709	44.967
Mechanics and Farmers Bank	Durham, North Carolina	Julia W. Taylor	1908	70	107.154	11.156	94.584	69.083
Illinois Service/Federal S&L Assn. of Chicago	Chicago, Illinois	Thelma J. Smith	1934	34	104.347	5.940	97.027	64.556
Broadway Federal Savings and Loan Assn.	Los Angeles, California	Paul C. Hudson	1946	46	97.784	4.345	91.976	70.501
Consolidated Bank and Trust Co.	Richmond, Virginia	Vernard W. Henley	1903	68	94.495	5.550	88.000	52.550
Liberty Bank and Trust Co.	New Orleans, Louisiana	Alden J. McDonald Jr.	1972	91	89.977	5.746	82.070	47.942

[Continued]

★ 72 ★

Financial Companies: Characteristics, 1992
[Continued]

Company	Location	Chief executive	Year started	Staff	Assets[1]	Capital[1]	Deposits[1]	Loans[1]
Boston Bank of Commerce	Boston, Massachusetts	Ronald A. Homer	1982	50	85.636	4.447	75.402	58.579
First Independence National Bank of Detroit	Detroit, Michigan	Richard W. Shealey	1970	71	81.480	3.306	73.900	31.348
Citizens Federal Savings Bank	Birmingham, Alabama	Bunny Stokes Jr.	1957	29	76.745	6.156	69.199	59.455
Founders National Bank of Los Angeles	Los Angeles, California	John P. Kelly Jr.	1991	45	74.201	5.201	61.725	32.910
Tri-State Bank of Memphis	Memphis, Tennessee	Jesse H. Turner Jr.	1946	67	73.839	7.829	65.307	27.942
City National Bank of New Jersey	Newark, New Jersey	Louis Prezeau	1973	35	61.911	3.356	57.853	19.431
The Harbor Bank of Maryland	Baltimore, Maryland	Joseph Haskins, Jr.	1982	31	56.575	4.046	52.037	31.951
Mutual Community Savings Bank, SSB	Durham, North Carolina	Ferdinand V. Allison Jr.	1921	24	47.174	4.125	42.053	37.493
North Milwaukee State Bank	Milwaukee, Wisconsin	James Jackson	1971	22	36.908	2.279	34.497	13.542
First Tuskegee Bank	Tuskegee, Alabama	James W. Wright	1894	25	36.424	3.244	25.000	19.475
Mutual Federal Savings & Loan Assn. of Atlanta	Atlanta, Georgia	Hamilton Glover	1925	15	35.912	2.244	31.618	21.943

Source: "B.E. Financial Companies," *Black Enterprise* 24 (June 1993), p. 149. Published by permission. Primary source: B.E. Research. *Notes:* 1. In millions of dollars, to the nearest thousand. Ranked by total assets as of Dec.31, 1992. Prepared by B.E. Research. Reviewed by Mitchell/Titus & Co.

★ 73 ★

Banks, Financial Institutions, Finances

Investment Banks: Issues, 1992

Company	Rank	Location	Chief executive	Year started	Senior-managed issues[1] (millions of dollars)	Co-managed issues[1] (billions of dollars)
Grigsby Brandford & Co. Inc.	1	San Francisco, CA	Calvin B. Grigsby	1981	577.0	32.9
Pryor, McClendon, Counts & Co. Inc.	2	Philadelphia, PA	Malcomn D. Pryor	1981	554.8	48.1
W.R. Lazard & Co.	3	New York, NY	Wardell R. Lazard	1985	349.7	28.4
M.R. Beal & Co.	4	New York, NY	Bernard B. Beal	1988	117.8	32.8
Apex Securities Inc.	5	Houston, TX	Richard M. Ramirez	1987	51.2	6.8
Charles A. Bell Securities Corp.	6	San Francisco, CA	Charles A. Bell	1986	9.9	1.9
Howard Gary & Co.	7	Miami, FL	Howard V. Gary	1980	4.7	2.1
Doley Securities Inc.	8	New Orleans, LA	Harold E. Doley	1976	_[2]	6.2
The Chapman Co.	9	Baltimore, MD	Nathan A. Chapman	1986	_[2]	5.1
Weldon, Sullivan, Carmichael & Co.	10	Denver, CO	Ennis Hudson	1988	_[2]	2.5
United Daniels Securities Inc.	11	New York, NY	Willie L. Daniels	1984	_[2]	2.4
Sturdivant & Co. Inc.	12	Camden, NJ	Ralph A. Sturdivant	1988	_[2]	1.6
Ward & Associates Inc.	13	Atlanta, GA	Felker W. Ward Jr.	1988	_[2]	1.3

Source: "B.E. Investment Banks," *Black Enterprise* 24 (June 1993), p. 178. Published by permission. Primary source: B.E. Research. *Notes:* 1. This is for all issues, including municipal, agency, corporate and mortgage-backed securities for year ending Dec. 31, 1992. 2. These investment banks did not participate as senior managers for municipal, agency, corporate and mortgage-backed securities for year ending Dec. 31, 1992. Source: Securities Data Co. Inc., Newark, N.J., 1993.

★ 74 ★

Banks, Financial Institutions, Finances

Mortgage Load-denial Rate in Largest Metropolitan Areas, 1990-1992

Ranked by black rejection ration, 1992.

	1990 population (millions)	Blacks as percentage of population	Black applications as % of total applications		Ratio of black rejection rate to white rejection rate	
			1991	1992	1991	1992
Chicago	6.07	21.9%	8.3%	7.6%	3.08	3.04
Philadelphia	4.46	20.9	8.9	8.5	2.76	2.82
Houston	3.30	18.5	6.6	6.7	2.50	2.53
Atlanta	2.83	25.9	16.2	13.3	2.63	2.50
St. Louis	2.44	17.3	6.7	7.0	2.56	2.43
Dallas	2.55	16.1	7.3	6.3	2.55	2.32
Detroit	4.38	21.5	9.0	7.5	2.61	2.20
Boston	2.87	7.3	4.0	3.9	2.27	2.12
New York	8.55	26.0	12.9	13.3	1.89	1.69
Los Angeles	8.86	11.2	4.5	5.7	1.40	1.50

Source: "State by State: How Blacks and White Compare." *Wall Street Journal*, 21 December 1993, A-8. Compiled by Edward P. Foldessy, *Wall Street Journal*, from analysis if Federal Reserve data. Published by permission.

★ 75 ★

Banks, Financial Institutions, Finances

Mortgage Load-denial Rate, 1991, 1992

Black mortgage-rejection rates as a ratio of white rejection rates, by income level, where loans sought were fairly typical 1.75 to 2.25 times annual income.

Income	Rejection ratio	
	1991	1992
Under $25,000	1.63	1.65
$25,000 to $49,999	2.39	2.33
$50,000 to $74,999	2.58	2.40
$75,000 to $99,999	2.53	2.48
$100,000 or more	1.95	1.85
Average	2.36	2.41

Source: "The Impact of Race on Mortgage Lending." *Wall Street Journal*, 21 December 1993, A-2. Primary source: *Wall Street Journal* analysis of Federal Reserve data. Published by permission.

★ 76 ★

Banks, Financial Institutions, Finances

Mortgage Loan-denial Rate in Selected States, 1990-1992. Part 1 - Alabama

Lender	Black to white rejection ratio		
	1992	1991	1990
U.S. average	2.06	1.91	2.27
Alabama	2.08	2.02	2.05
Amsouth Bank	2.75	2.64	2.27
Amsouth Mtge. Co.	4.12	3.44	2.94
BancBoston Mtge. Corp.	4.49	3.48	3.20
Central Bank of the South	3.73	3.38	5.33
Chemical Residential Mtge.	1.68	-	-
Collateral Mtge. Ltd.	4.20	3.71	4.17
Colonial Mtge. Co.	4.14	-	3.63
First Union Mtge. Corp.	2.52	3.12	3.10
Fleet Bank R.I.	2.66	-	-
Fleet Mtge. Corp.	2.50	2.74	2.65
Goldome Credit Corp.	1.50	-	-
Green Tree Fin. Corp.	1.50	-	-
Magnolia Fed. Bank for Savings	2.07	-	-
Molton, Allen & Williams	2.33	-	3.49
Mortgage America	2.98	3.96	3.82
Real estate Financing	2.32	2.03	2.06
Sec. Pacific Housing Srv.	1.36	1.15	1.31
Secor Bank	4.75	3.00	3.96
Southtrust Mobile Srv.	1.45	1.21	1.24
Southtrust Mtge. Corp.	4.23	3.49	4.07
Troy & Nichols	4.06	3.30	4.12
United Cos. Lending Corp.	0.78	-	-

Source: "State by State: How Blacks and White Compare." *Wall Street Journal*, 21 December 1993, A-8. Compiled by Edward P. Foldessy, *Wall Street Journal*, from analysis of Federal Reserve data. Published by permission.

★ 77 ★

Banks, Financial Institutions, Finances

Mortgage Loan-denial Rate in Selected States, 1990-1992. Part 2 - Arkansas

Lender	Black to white rejection ratio		
	1992	1991	1990
U.S. average	2.06	1.91	2.27
Arkansas	2.31	2.78	2.41
First Commercial Mtge. Co.	2.42	-	2.83

[Continued]

★ 77 ★

Mortgage Loan-denial Rate in Selected States, 1990-1992. Part 2 - Arkansas

[Continued]

Lender	Black to white rejection ratio		
	1992	1991	1990
Green Tree Fin. Corp.	1.34	-	-
Sears Mtge. Corp.	4.13	-	4.79
Simmons First Nat. Bank	1.91	3.39	2.41
Union Modern Mtge. Corp.	3.52	2.44	3.08
Worthen Mtge. Co.	2.64	3.21	1.00
Worthen Nat. Bank	1.70	2.63	3.72

Source: "State by State: How Blacks and White Compare." *Wall Street Journal,* 21 December 1993, A-8. Compiled by Edward P. Foldessy, *Wall Street Journal,* from analysis of Federal Reserve data. Published by permission.

★ 78 ★

Banks, Financial Institutions, Finances

Mortgage Loan-denial Rate in Selected States, 1990-1992. Part 3 - California

Lender	Black to white rejection ratio		
	1992	1991	1990
U.S. average	2.06	1.91	2.27
California	1.70	1.53	1.72
Accubanc Mtge. Corp.	3.27	-	-
All PAcific Mtge. Co.	2.36	1.87	1.90
Amer. City Mtge. Corp.	1.08	-	-
Amer. Residential Mtge.	0.99	-	1.04
Amer. Savings Bank	1.53	1.30	1.55
ARCS Mtge. Inc.	2.17	3.87	4.18
Bank of Amer.	1.67	1.69	1.54
Bank of United Texas	2.35	-	-
Cal-Bay Mtge. Group	3.01	-	-
Calif. Fed. Bank	1.91	2.26	1.96
Citibank	2.12	2.20	2.31
Coast Fed. Bank	2.08	1.89	1.94
Commerce Security Bank	2.26	-	-
Community West Mtge.	1.83	1.08	1.06
Countrywide Funding Corp.	1.86	-	-
CTX Mtge.	4.36	-	1.96
Cypress Fin. Corp.	1.64	-	1.52
Directors Mtge. Loan Corp.	2.46	2.62	2.53
First Calif. Mtge. Co.	0.98	1.00	1.85
First Fed. Bank of Calif.	1.40	1.36	0.96
First Interstate Bank of Calif	1.19	0.68	1.76
First Mtge. Corp.	2.21	-	-

[Continued]

★ 78 ★

Mortgage Loan-denial Rate in Selected States, 1990-1992. Part 3 - California

[Continued]

Lender	Black to white rejection ratio		
	1992	1991	1990
First Nationwide Bank	1.91	2.80	1.76
Glendale Fed. Bank	2.54	1.39	1.54
GMAC Mtge. Corp.	2.62	-	3.21
Great Western Bank	1.64	2.32	1.55
Greater Suburban Mtge.	2.25	-	1.70
Guild Mtge. Co.	2.21	-	1.99
Hammond Co.	2.17	3.72	1.06
Home Savings of Amer.	1.86	2.18	1.83
Kaufman & Broad Mtge. Co.	2.43	1.40	1.95
Long Beach Bank	1.14	1.29	1.25
Medallion Mtge. Co.	1.35	-	1.55
Mission Hills Mtge.	1.86	-	-
Nat. Pacific Mtge. Corp.	1.71	1.73	1.67
NewAmer. S.B.	0.95	-	1.27
North Amer. Mtge. Co.	1.68	-	-
Norwest Mtge.	2.48	1.31	4.21
Pacific Nat. Bank	1.57	5.31	1.23
Quality Mtge.	0.95	-	6.00
Rancho Mtge. & Inv. Corp.	1.19	2.65	0.86
RNG Mtge. Srv.	1.20	-	-
Ryland Mtge. Co.	2.74	-	3.39
San Francisco Fed. Savings	1.73	2.64	2.86
Sears Mtge. Corp.	3.91	-	3.28
Sec. Pacific Housing Srv.	1.53	1.11	1.22
United S.B.	1.92	3.65	2.25
Wells Fargo Bank	1.18	1.58	1.36
Western Bank	1.56	2.76	1.52
Western Cities Mtge. Corp.	2.72	2.14	2.06
Western Fed. S&L	1.39	2.01	1.62
Western Sunrise Mtge.	1.9	1.84	1.57
Weyerhaeuser Mtge. Co.	2.44	-	-
World S&L Assn.	2.11	1.90	2.20

Source: "State by State: How Blacks and White Compare." *Wall Street Journal,* 21 December 1993, A-8. Compiled by Edward P. Foldessy, *Wall Street Journal,* from analysis of Federal Reserve data. Published by permission.

★ 79 ★

Banks, Financial Institutions, Finances

Mortgage Loan-denial Rate in Selected States, 1990-1992. Part 4 - Connecticut

Lender	Black to white rejection ratio		
	1992	1991	1990
U.S. average	2.06	1.91	2.27
Connecticut	2.35	2.60	2.90
Bank of Boston Conn.	4.55	-	-
Bristol Mtge. Corp.	3.33	-	-
Centerbank Mtge. Co.	2.09	-	-
GMAC Mtge. Corp.	3.27	-	4.23
Great Western Mtge. Corp.	1.70	1.90	2.00
McCue Mtge. Co.	2.07	2.26	2.13
New Haven S.B.	2.48	-	-
People's Bank	3.08	-	-
Shawmut Mtge. Co.	2.47	2.72	2.92

Source: "State by State: How Blacks and White Compare." *Wall Street Journal*, 21 December 1993, A-8. Compiled by Edward P. Foldessy, *Wall Street Journal*, from analysis of Federal Reserve data. Published by permission.

★ 80 ★

Banks, Financial Institutions, Finances

Mortgage Loan-denial Rate in Selected States, 1990-1992. Part 5 - District of Columbia

Lender	Black to white rejection ratio		
	1992	1991	1990
U.S. average	2.06	1.91	2.27
District of Columbia	2.82	2.34	2.78
Crestar Mtge. Corp.	W	2.30	7.39
Dominion Bankshares Mtge.	1.74	2.88	2.64
PaineWebber Mtge. Finance	4.00	-	-

Source: "State by State: How Blacks and White Compare." *Wall Street Journal*, 21 December 1993, A-8. Compiled by Edward P. Foldessy, *Wall Street Journal*, from analysis of Federal Reserve data. Published by permission.

★ 81 ★
Banks, Financial Institutions, Finances

Mortgage Loan-denial Rate in Selected States, 1990-1992.
Part 6 - Florida

Lender	Black to white rejection ratio		
	1992	1991	1990
U.S. average	2.06	1.91	2.27
Florida	2.09	1.95	1.92
Amer. Home Funding	2.58	2.90	2.15
Amsouth Bank of Fla.	4.42	2.72	3.84
Andrew Jackson S.B.	1.90	5.36	6.03
BancBoston Mtge. Corp.	2.79	2.50	1.99
Bank United of Texas	2.47	-	-
Barnett Bank of Cent. Fla.	3.10	3.04	2.07
Barnett Bank of Jacksonville	2.11	4.14	2.92
Barnett Bank of Pinellas City	4.55	-	-
Barnett Bank of South Fla.	2.82	2.52	2.74
Barnett Bank of Volusia County	1.34	-	-
Chase Fed. Bank	0.99	3.01	1.28
Chemical Residential Mtge.	1.56	-	-
Citibank Fed. S.B.	1.06	1.09	1.12
Community First Bank	3.08	1.10	2.25
Coral Gables Fed S&L	2.45	2.82	2.50
Countrywide Funding Corp.	2.08	-	-
Crossland Mtge. Corp.	1.35	-	0.69
CTX Mtge.	3.35	-	2.30
Empire of Amer. Realty Credit	0.87	-	0.58
First Union Fla.	2.01	2.95	2.13
First Wachovia Mtge. Co.	2.56	3.34	4.33
Medallion Mtge. Co.	1.35	-	1.55
Mission Hills Mtge.	1.86	-	-
Nat. Pacific Mtge. Corp.	1.71	1.73	1.67
NewAmer S.B.	0.95	-	1.27
North Amer. Mtge. Co.	1.68	-	-
Norwest Mtge.	2.48	1.31	4.21
Pacific Nat. Bank	1.57	5.31	1.23
Quality Mtge.	0.95	-	6.00
Rancho Mtge. & Inv. Corp.	1.19	2.65	0.86
RNG Mtge. Srv.	1.20	-	-
Ryland Mtge. Co.	2.74	-	3.39
San Francisco Fed. Savings	1.73	2.64	2.86
Sears Mtge. Corp.	3.91	-	3.28
Sec. Pacific Housing Srv.	1.53	1.11	1.22
United S.B.	1.92	3.65	2.25
Wells Fargo Bank	1.18	1.58	1.36
Western Bank	1.56	2.76	1.52
Western Cities Mtge. Corp.	2.72	2.14	2.06
Western Fed. S&L	1.39	2.01	1.62
Western Sunrise Mtge.	1.09	1.84	1.57

[Continued]

★ 81 ★

Mortgage Loan-denial Rate in Selected States, 1990-1992.
Part 6 - Florida
[Continued]

Lender	Black to white rejection ratio		
	1992	1991	1990
Weyerhaeuser Mtge. Co.	2.44	-	-
World S&L Assn.	2.11	1.90	2.20

Source: "State by State: How Blacks and White Compare." *Wall Street Journal*, 21 December 1993, A-8. Compiled by Edward P. Foldessy, *Wall Street Journal*, from analysis of Federal Reserve data. Published by permission.

★ 82 ★

Banks, Financial Institutions, Finances

Mortgage Loan-denial Rate in Selected States, 1990-1992. Part 7 - Georgia

Lender	Black to white rejection ratio		
	1992	1991	1990
U.S. average	2.06	1.91	2.27
Georgia	2.22	2.46	2.52
Allatoona Fed. S.B.	2.37	2.75	2.54
Bank South Mtge.	1.96	-	2.86
Bank United of Texas	1.61	-	-
Bankers First Mtge. Corp.	4.67	3.70	4.65
Barnett Bank of SW Ga.	9.93	-	-
Chemical Residential Mtge.	2.42	-	-
Collateral Mtge. Ltd.	2.20	2.44	4.82
Decatur Fed.	2.47	3.29	2.59
Eagle Service Corp.	7.61	11.92	-
Entrust Funding Co.	3.52	2.59	2.96
First Town Mtge. Corp.	1.55	-	-
First Union Ga.	3.49	5.46	3.21
Georgia Fed. Bank	2.40	3.26	4.13
Goldome Credit Corp.	1.19	-	-
Great Western Mtge. Corp.	2.10	1.42	1.94
Green Tree Fin. Corp.	1.39	-	-
Griffin Fed. S.B.	3.08	4.88	3.17
Gulf States Mtge. Co.	4.17	-	2.78
Homebanc Fed. S.B.	3.88	3.36	8.16
ICM Mtge. Corp.	1.74	-	2.23
Liberty Mtge. Corp.	1.84	2.37	3.92
Liberty S.B.	4.05	4.18	4.53
Nat. Mtge.	4.24	-	-
NationsBanc Mtge. Corp.	4.67	-	-
Norwest Mtge.	1.79	1.82	3.12
Oakwood Acceptance Corp.	1.29	-	1.07

[Continued]

★ 82 ★

Mortgage Loan-denial Rate in Selected States, 1990-1992. Part 7 - Georgia

[Continued]

Lender	Black to white rejection ratio		
	1992	1991	1990
Prime Bank	4.29	2.50	-
Prime Lending	3.45	-	-
Prudential Home Mtge.	1.62	2.14	4.12
Ryland Mtge. Co.	B	-	-
Sears Mtge. Corp.	3.07	-	39.32
Sec. Pacific Housing Srv.	1.66	1.51	1.15
Southeastern Mtge. Corp.	2.78	-	-
Sunshine Mtge. Corp.	2.55	2.83	1.75
Suntrust Mtge.	2.11	3.08	1.77
Unity Mtge. Corp.	2.05	-	2.09

Source: "State by State: How Blacks and White Compare." *Wall Street Journal*, 21 December 1993, A-8. Compiled by Edward P. Foldessy, *Wall Street Journal*, from analysis of Federal Reserve data. Published by permission.

★ 83 ★

Banks, Financial Institutions, Finances

Mortgage Loan-denial Rate in Selected States, 1990-1992. Part 8 - Illinois

Lender	Black to white rejection ratio		
	1992	1991	1990
U.S. average	2.06	1.91	2.27
Illinois	2.51	2.74	2.81
Amcore Mtge.	2.55	-	-
Bancplus Mtge. Corp.	5.63	-	1.39
Bank United of Texas	4.76	-	-
Beverly Bank	9.58	12.46	14.64
Champion Fed. S&L Assn.	3.22	2.80	2.54
Chemical Residential Mtge.	3.51	-	-
Citibank	2.10	-	2.28
Countrywide Funding Corp.	2.66	-	-
Crown Mtge. Corp.	2.00	-	1.46
CTX Mtge.	2.31	-	2.62
Draper & Kramer Inc.	2.67	0.70	1.84
First Nat. Bank of Chicago	3.44	6.04	3.21
First Nationwide Bank	3.21	3.51	2.82
Fleet Bank R.I.	4.44	-	-
Fleet Mtge. Corp.	2.35	2.17	4.23
GMAC Mtge. Corp.	4.19	-	5.47
Great Western Mtge. Corp.	2.45	2.47	1.45
Green Tree Fin. Corp.	1.52	-	-

[Continued]

★ 83 ★

Mortgage Loan-denial Rate in Selected States, 1990-1992.
Part 8 - Illinois
[Continued]

Lender	Black to white rejection ratio		
	1992	1991	1990
Harris Trust & S.B.	8.79	8.22	14.63
Home Savings of Amer.	2.68	2.34	1.96
Household Bank	2.63	1.47	2.62
Huntington Mtge. Co.	2.57	-	3.49
Independence One Mtge. Corp.	4.74	6.15	6.74
LaSalle Talman Bank	3.31	4.53	3.11
Magna Bank of St. Clair	4.10	1.32	5.80
Margaretten & Co.	3.35	5.65	4.75
Midwest Funding Corp.	3.09	3.26	-
Northern Trust Co.	10.67	6.22	20.12
Norwest Mtge.	2.77	2.64	1.88
Sears Mtge. Corp.	5.79	-	3.88
Source One Mtge.	1.85	6.72	4.26
St. Paul Fed Bank for Savings	18.43	5.80	3.94

Source: "State by State: How Blacks and White Compare." *Wall Street Journal*, 21 December 1993, A-8. Compiled by Edward P. Foldessy, *Wall Street Journal*, from analysis of Federal Reserve data. Published by permission.

★ 84 ★

Banks, Financial Institutions, Finances

Mortgage Loan-denial Rate in Selected States, 1990-1992. Part 9 - Louisiana

Lender	Black to white rejection ratio		
	1992	1991	1990
U.S. average	2.06	1.91	2.27
Louisiana	1.95	2.33	2.34
Bank of LaPlace of St. John	1.32	3.70	1.43
Deposit Guaranty Mtge. Co.	3.19	3.58	3.13
Equitable Trust S&L	3.68	4.67	3.14
First Amer. Bank & Trust	5.62	2.43	1.40
First Nat. Bank of Commerce	4.01	2.14	4.63
First Nat. Bank	3.53	2.93	1.75
Fleet Bank R.I.	1.36	-	-
Fleet Mtge. Corp.	1.72	1.40	1.84
Green Tree Fin. Corp.	1.53	-	-
Gulf Coast Bank & Trust Co.	1.81	-	-
Hibernia Nat. Bank	2.21	2.14	1.98
Iberia Savings Bank	2.11	-	-
Norwest Mtge.	2.59	-	3.60
Oak Tree Mtge. Corp.	2.25	2.41	2.01

[Continued]

★ 84 ★

Mortgage Loan-denial Rate in Selected States, 1990-1992. Part 9 - Louisiana

[Continued]

Lender	Black to white rejection ratio		
	1992	1991	1990
Parish Nat. Bank	2.69	-	2.91
Pioneer Mtge. Corp.	3.10	1.60	1.33
Premier Bank	2.42	1.83	2.80
Premier Mtge. Co.	3.33	2.43	3.07
Secor Bank	5.76	2.63	4.24
Standard Mtge. Corp.	2.91	2.19	2.81
State Nat. Mtge. Corp.	2.79	-	2.91
Sunburst Mtge. Corp.	3.04	-	3.33
Troy & Nichols	2.06	2.48	2.23
United Cos. Lending Corp.	0.76	-	-

Source: "State by State: How Blacks and White Compare." *Wall Street Journal*, 21 December 1993, A-8. Compiled by Edward P. Foldessy, *Wall Street Journal*, from analysis of Federal Reserve data. Published by permission.

★ 85 ★

Banks, Financial Institutions, Finances

Mortgage Loan-denial Rate in Selected States, 1990-1992. Part 10 - Maryland

Lender	Black to white rejection ratio		
	1992	1991	1990
U.S. average	2.06	1.91	2.27
Maryland	2.27	2.16	1.99
Ahmanson Mtge. Co.	1.95	2.79	1.74
Amer. Home Funding	4.17	4.02	4.86
Amer. Residential Mtge.	1.40	-	1.11
Ameribanc S.B.	0.86	2.94	2.50
Atlantic Home Mtge. Corp.	1.58	7.23	3.28
Atlantic Residential Mtge.	2.75	-	2.32
B.F. Saul Mtge. Co.	3.78	-	2.55
Banc One Mtge. Corp.	3.22	6.42	4.43
Carl I. Brown & Co.	2.17	-	-
Colonial Savings	4.06	2.49	2.15
Columbia First	1.27	2.83	1.16
Crestar Mtge. Corp.	2.94	2.41	9.68
Developers Mtge. Corp.	3.56	4.04	1.42
Dominion Bankshares Mtge.	3.20	2.59	2.76
Fairfax Mtge. Corp.	1.97	2.36	1.79
First Advantage Mtge. Corp.	3.74	2.65	2.30
First Nat. Mtge. Corp.	3.70	6.66	3.61
First Security	1.05	-	-

[Continued]

★ 85 ★

Mortgage Loan-denial Rate in Selected States, 1990-1992.
Part 10 - Maryland
[Continued]

Lender	Black to white rejection ratio		
	1992	1991	1990
First Town Mtge. Corp.	3.91	-	-
First Virginia Mtge. Co.	3.14	3.49	0.88
1st Washington Mtge. Corp.	2.97	0.46	5.54
Fleet Real Estate Funding	0.86	-	-
Great Western Mtge. Corp.	2.11	2.43	1.42
Green Tree Fin. Corp.	1.31	-	-
HomeAmer. Mtge. Corp.	1.14	-	8.54
ICM Mtge. Corp.	5.38	-	4.11
Independence One Mtge. Corp.	1.87	1.33	4.34
Loyola Fed. S.B.	5.70	2.45	7.26
Margaretten & Co.	3.66	2.21	3.36
Market Street Mtge. Corp.	3.70	-	4.45
Md. Nat. Mtge.	2.93	1.64	3.11
Mortgage Group	2.88	-	2.54
Nat. city Mtge. Co.	1.57	1.27	2.63
NationsBanc Mtge. Corp.	3.43	4.50	5.66
North Amer. Mtge. Co.	5.55	-	-
Norwest Mtge.	4.15	7.00	2.38
NVR Mortage	4.95	4.21	3.30
PaineWebber Mtge. Finance	2.53	-	-
Ryland Mtge. Co.	3.74	-	4.33
Sears Mtge. Corp.	2.75	-	4.92
Shelter Mtge. Corp.	0.74	-	-
Signet Mtge. Corp.	3.00	2.01	5.47
Southern Atlantic Mtge.	B	-	1.82
Temple-Inland Mtge. Co.	3.45	-	1.82
UFSB of Indianapolis	2.42	2.62	4.59
Weyerhaeuser Mtge. Co.	4.99	-	-

Source: "State by State: How Blacks and White Compare." *Wall Street Journal*, 21 December 1993, A-8. Compiled by Edward P. Foldessy, *Wall Street Journal*, from analysis of Federal Reserve data. Published by permission.

★ 86 ★

Banks, Financial Institutions, Finances

Mortgage Loan-denial Rate in Selected States, 1990-1992. Part 11 - Michigan

Lender	Black to white rejection ratio		
	1992	1991	1990
U.S. average	2.06	1.91	2.27
Michigan	1.73	2.27	2.41
Citizens Commercial & Savings	1.25	2.15	1.57
Comerica Bank	2.31	-	1.51
Comerica Mtge. Corp.	2.53	2.84	2.70
DMR Fin. Srv.	3.42	2.18	2.22
First Fed. of Mich.	2.70	2.37	3.60
First of Amer. Bank	1.32	3.66	2.55
First Security S.B.	2.65	3.11	2.70
Fleet Mtge. Corp.	3.88	2.28	3.27
GMAC Mtge. Corp.	5.67	-	2.97
Green Tree Fin. Corp.	1.28	-	-
independence One Mtge. Corp.	1.41	3.88	1.18
Marathon Mtge. Corp.	1.63	2.09	2.20
Michigan Nat. Bank	0.53	2.12	0.46
Mortgage Corp. of Amer.	1.53	-	2.21
NBD Mtge. Co.	2.36	3.22	2.12
Norwest Mtge.	4.77	5.63	3.19
Sears Mtge. Corp.	3.25	-	4.36
Standard Fed. Bank	4.25	4.04	2.43
Waterfield Fin. Corp.	3.62	3.72	2.77

Source: "State by State: How Blacks and White Compare." *Wall Street Journal*, 21 December 1993, A-8. Compiled by Edward P. Foldessy, *Wall Street Journal*, from analysis of Federal Reserve data. Published by permission.

★ 87 ★

Banks, Financial Institutions, Finances

Mortgage Loan-denial Rate in Selected States, 1990-1992. Part 12 - Mississippi

Lender	Black to white rejection ratio		
	1992	1991	1990
U.S. average	2.06	1.91	2.27
Mississippi	2.27	2.61	2.68
Bank of Miss.	3.54	3.69	3.50
Deposit Guaranty Mtge. Co.	6.11	3.84	3.92
Green Tree Fin. Corp.	1.43	-	-
Magnolia Fed. Bank	1.90	2.42	2.25

[Continued]

★ 87 ★

Mortgage Loan-denial Rate in Selected States, 1990-1992. Part 12 - Mississippi

[Continued]

Lender	Black to white rejection ratio		
	1992	1991	1990
Merchants & Farmers Bank	3.71	6.06	-
Sunburst Bank	5.43	3.89	3.64
Trustmark Nat. Bank	3.18	2.51	2.42
United Southern Bank	2.28	-	1.67

Source: "State by State: How Blacks and White Compare." *Wall Street Journal,* 21 December 1993, A-8. Compiled by Edward P. Foldessy, *Wall Street Journal,* from analysis of Federal Reserve data. Published by permission.

★ 88 ★

Banks, Financial Institutions, Finances

Mortgage Loan-denial Rate in Selected States, 1990-1992. Part 13 - Missouri

Lender	Black to white rejection ratio		
	1992	1991	1990
U.S. average	2.06	1.91	2.27
Missouri	1.98	2.35	2.51
Delmar Fin. Co.	2.27	3.13	3.53
Equality Mtge. Corp.	2.28	1.63	3.02
First Bank S.B.	1.21	2.53	2.94
First Nationwide Bank	2.12	2.83	2.22
Gershman Invest. Corp.	3.26	1.88	-
Green Tree Fin. Corp.	1.50	-	-
Home Savings of Amer.	1.71	2.36	2.26
Leader Fin. Corp.	2.67	4.37	-
Mark Twain Mtge. Co.	2.42	-	-
Mercantile Bank of St. Louis	3.01	4.87	3.11
Norwest Mtge.	4.40	2.39	3.38
Pulaski Bank	1.86	2.67	3.83
Sears Mtge. Corp.	3.41	-	1.52
Security Fin. & Mtg. Corp.	4.21	-	-
United Postal Savings Assn.	3.33	5.15	4.17

Source: "State by State: How Blacks and White Compare." *Wall Street Journal,* 21 December 1993, A-8. Compiled by Edward P. Foldessy, *Wall Street Journal,* from analysis of Federal Reserve data. Published by permission.

★ 89 ★

Banks, Financial Institutions, Finances

Mortgage Loan-denial Rate in Selected States, 1990-1992.
Part 14 - Ohio

Lender	Black to white rejection ratio		
	1992	1991	1990
U.S. average	2.06	1.91	2.27
Ohio	2.00	2.21	2.30
Bank One Cleveland	2.29	3.12	2.49
Central Trust Co.	2.76	5.84	3.61
Century Bank	1.43	2.15	1.35
Chemical Residential Mtge.	3.09	-	-
CitFed Mtge. Corp. of Amer.	3.38	3.35	3.91
Countrywide Funding Corp.	3.02	-	-
Developers Mtge. Co.	3.12	-	-
Fifth Third Bank	2.15	2.86	2.42
First Nat. Bank Dayton	0.95	3.49	2.06
First Nationwide Bank	2.13	5.27	2.28
Gateway Fed. S.B.	2.07	1.83	2.15
Huntington Mtge. Co.	1.91	-	4.48
Mahoning Nat. Bank	2.01	-	2.84
Marathon Mtge. Corp.	1.81	-	1.95
Mid. Amer. Nat. Bank	3.33	2.67	6.15
Nat. City Bank	1.47	0.78	1.73
Nat. City Mtge. Co.	3.27	2.30	2.74
Norwest Mtge.	4.10	2.30	2.53
Ohio S.B.	3.03	3.67	3.32
Principal Mutual Life Ins. Co.	2.97	3.18	2.01
Provident Bank	0.48	3.41	3.07
Sears Mtge. Corp.	2.26	-	2.96
Society Mtge. Co.	2.14	-	-
Society Nat. Bank	2.40	2.06	2.70
Star Bank	2.76	1.67	1.49
State S.B.	2.73	2.90	3.62
Third Fed. S&L	1.73	3.87	2.72
Trustcorp Mtge. Co.	2.77	-	-
UFSB of Indianapolis	2.57	2.30	2.18
United Cos. Lending Corp.	1.13	-	-

Source: "State by State: How Blacks and White Compare." *Wall Street Journal,* 21 December 1993, A-8. Compiled by Edward P. Foldessy, *Wall Street Journal,* from analysis of Federal Reserve data. Published by permission.

★ 90 ★

Banks, Financial Institutions, Finances

Mortgage Loan-denial Rate in Selected States, 1990-1992.
Part 15 - New Jersey

Lender	Black to white rejection ratio		
	1992	1991	1990
U.S. average	2.06	1.91	2.27
New Jersey	2.67	2.59	2.59
Amer. Home Funding	4.18	2.75	1.96
Amer. Residential Mtge.	1.19	-	1.39
Arbor Nat. Mtge. Inc.	2.65	-	-
Bank United of Texas	2.25	-	-
Carteret S.B.	1.20	2.25	1.69
Collective Fed. Savings	2.91	5.06	2.95
Eastern Mtge. Srv.	3.08	-	-
Fleet Bank R.I.	1.83	-	-
Fleet Real Estate Funding	2.33	1.11	1.40
GMAC Mtge. Corp.	6.06	-	3.09
K. Hovnanian Mtge.	1.41	2.39	2.19
Lumbermens Mtge. Corp.	2.54	-	2.20
Margaretten & Co.	2.69	2.25	4.98
Maryland Nat. Mtge.	3.66	5.03	3.59
Meridian Mtge. Corp.	4.01	1.64	3.63
Midcoast Mtge. Corp.	1.58	-	-
Mtge. Access Corp.	3.52	-	1.88
Murray Fin. Assn.	2.79	2.52	2.51
Nat. State Bank	4.29	-	-
PHH U.S. Mtge. Corp.	3.20	4.67	4.11
Prudential Home Mtge.	2.45	2.39	3.07
Sears Mtge. Corp.	3.29	-	2.95
United Jersey Bank	1.65	2.62	2.23

Source: "State by State: How Blacks and White Compare." *Wall Street Journal*, 21 December 1993, A-8.
Compiled by Edward P. Foldessy, *Wall Street Journal*, from analysis of Federal Reserve data. Published
by permission.

★ 91 ★

Banks, Financial Institutions, Finances

Mortgage Loan-denial Rate in Selected States, 1990-1992.
Part 16 - New York

Lender	Black to white rejection ratio		
	1992	1991	1990
U.S. average	2.06	1.91	2.27
New York	1.78	2.16	2.45
Amer. Residential Mtge.	0.82	-	1.62
Apple Bank for Savings	2.63	2.66	1.67
Arbor Nat. Mtge. Inc.	2.03	3.32	-
Arcs Mtge. Inc.	3.07	-	1.46
Bank United of Texas	0.59	-	-
Centerbank Mtge. Co.	2.38	-	-
Chase Home Mtge. Corp.	2.74	-	3.88
Chase Manhattan Bank	1.21	3.02	2.00
Chemical Bank	2.04	1.88	2.26
Citibank	1.93	1.47	1.57
Columbia Equities Ltd.	1.98	0.91	2.53
Dale Mtge. Bankers Corp.	2.33	-	2.08
Dime S.B. of NY	2.29	2.10	1.20
Empire of Amer. Realty Credit	1.00	-	1.90
Exchange Mtge. Corp.	1.95	-	-
First FS&LA of Rochester	3.26	2.54	2.71
Fleet Bank R.I.	1.56	-	-
Fleet Real Estate Funding	1.55	1.28	1.45
GMAC Mtge. Corp.	2.30	-	2.84
Great Western Mtge. Corp.	1.48	2.91	1.74
Home Savings of Amer.	2.82	-	-
Kadilac Mtge.	1.33	-	-
Keycorp Mtge. Inc.	2.65	3.36	-
Manhattan S.B.	1.00	2.87	1.81
Manufacturers & Traders Trust	3.01	3.37	3.86
Marine Midland Mtge. Corp.	1.89	-	2.04
Midcoast Mtge. Corp.	2.84	-	-
Prudential Home Mtge.	2.04	2.70	2.32
Residential Mtge. Banking	1.05	-	-
Roosevelt S.B.	5.93	5.08	7.60
Sears Mtge. Corp.	4.43	-	1.58
Sibley Mtge. Corp.	2.00	4.90	2.51
Statewide Funding Corp.	2.58	-	2.90

Source: "State by State: How Blacks and White Compare." *Wall Street Journal*, 21 December 1993, A-8. Compiled by Edward P. Foldessy, *Wall Street Journal*, from analysis of Federal Reserve data. Published by permission.

★ 92 ★

Banks, Financial Institutions, Finances

Mortgage Loan-denial Rate in Selected States, 1990-1992.
Part 17 - North Carolina

Lender	Black to white rejection ratio		
	1992	1991	1990
U.S. average	2.06	1.91	2.27
North Carolina	2.16	2.25	2.69
Amsouth Mtge. Co.	5.06	20.55	4.81
Atlantic Residential Mtge.	7.59	-	-
Bancplus Mtge. Corp.	2.62	-	4.18
BarclaysAmer./Mtge. Corp.	1.93	-	2.49
Branch Banking & Trust Co.	2.63	3.24	4.54
Central Carolina Bank	2.89	2.47	3.69
Centura Bank	3.90	3.68	3.69
Chemical Residential Mtge.	3.78	-	-
Countrywide Funding Corp.	5.95	-	-
CTX Mtge.	1.93	-	4.13
First Union Mtge. Corp.	3.35	2.32	3.35
First Union N.C.	2.59	4.47	1.97
First Wachovia Mtge. Co.	3.77	4.77	4.38
First-Citizens Bank & Trust Co.	2.54	9.39	4.19
Green Tree Fin. corp.	1.34	-	-
ICM Mtge. Corp.	4.09	-	3.90
Nat. City Mtge. Co.	1.96	4.96	3.24
NationsBanc Mtge. Corp.	5.91	3.91	3.76
NationsBanc N.C.	1.82	1.58	1.46
Norwest Mtge.	2.71	3.09	5.03
NVR Mtge.	3.71	0.40	4.06
Oakwood Acceptance Corp.	1.29	-	1.38
Old Stone Bank	3.03	3.75	2.89
One Mtge. Corp.	2.54	3.23	3.11
Pioneer S.B.	2.25	-	2.31
Principal Mutual Life Ins. Co.	1.15	3.20	3.70
Raleigh Fed. S.B.	2.27	4.29	2.76
Sears Mtge. Corp.	2.84	-	6.63
Sec. Pacific Housing Srv.	3.44	1.20	1.13
Sunbelt Nat. Mtge. Corp.	2.87	-	-
United Carolina Bank	1.60	1.50	1.58
Vanderbilt Mtge.	1.37	-	-

Source: "State by State: How Blacks and White Compare." *Wall Street Journal*, 21 December 1993, A-8. Compiled by Edward P. Foldessy, *Wall Street Journal*, from analysis of Federal Reserve data. Published by permission.

★ 93 ★

Banks, Financial Institutions, Finances

Mortgage Loan-denial Rate in Selected States, 1990-1992. Part 19 - Pennsylvania

Lender	Black to white rejection ratio		
	1992	1991	1990
U.S. average	2.06	1.91	2.27
Pennsylvania	1.92	2.04	2.36
Amer. Residential Mtge.	0.77	-	0.85
Arbor Nat. Mtge. Inc.	0.99	-	-
Avstar Mtge. Corp.	1.50	2.59	1.57
Bank United of Texas	3.10	-	-
Comnet Mtge. Srv.	4.02	2.08	3.37
Continental Bank	2.70	1.75	2.07
Corestates Bank	1.29	1.66	1.76
Eastern Mtge. Srv.	2.21	-	-
Equibank	1.20	2.76	2.98
Fleet Real Estate Funding	2.36	14.50	-
GMAC Mtge. Corp.	3.88	-	3.39
Huntington Mtge. Co.	2.29	-	3.63
Integra Nat. Bank/Pittsburgh	0.89	1.59	1.32
Lumbermens Mtge. Corp.	1.81	-	1.32
Margaretten & Co.	1.74	2.69	-
Maryland Nat. Mtge.	2.19	3.88	2.25
Mellon Bank	2.75	2.59	-
Meridian Bank	2.22	3.68	2.45
Meridian Mtge. Corp.	3.37	2.16	2.57
Pittsburgh Nat. Bank	1.08	1.37	1.24
Police & Fire Fed. Credit	2.12	3.55	3.61
Provident Nat. Bank	3.41	2.28	2.41
Sears Mtge. Corp.	4.04	-	3.01

Source: "State by State: How Blacks and White Compare." *Wall Street Journal*, 21 December 1993, A-8. Compiled by Edward P. Foldessy, *Wall Street Journal*, from analysis of Federal Reserve data. Published by permission.

★ 94 ★

Banks, Financial Institutions, Finances

Mortgage Loan-denial Rate in Selected States, 1990-1992. Part 20 - South Carolina

Lender	Black to white rejection ratio		
	1992	1991	1990
U.S. average	2.06	1.91	2.27
South Carolina	2.06	2.43	2.50
Amer. Fed. Bank	3.80	1.99	4.69
CTX Mtge.	4.25	-	3.44
First Fed. S&L Assn.	2.00	2.14	2.44
First S.B.	3.51	3.67	4.20
First Union Mtge. Corp.	3.41	2.25	3.91
First Union S.C.	2.01	-	1.43
First Wachovia Mtge. Co.	3.93	3.54	2.66
Fleet Bank R.I.	2.87	-	-
Fleet Mtge. Corp.	2.34	2.10	3.40
Green Tree Fin. Corp.	1.17	-	-
Nat. City Mtge. Co.	1.51	3.08	1.52
NationsBanc Mtge. Corp.	4.23	8.46	1.94
NationsBanc S.C.	1.87	14.13	7.66
Norwest Mtge.	1.24	1.76	2.37
Oakwood Acceptance Corp.	1.44	-	1.36
Palmetto Fed. S.B.	2.26	3.75	2.93
Pee Dee State Bank	1.90	3.92	2.36
Sears Mtge. Corp.	1.70	-	3.49
Sec. Pacific Housing Srv.	1.25	1.34	1.19
S.C. Nat. Bank	2.77	3.28	2.69
United Carolina Bank	1.50	1.52	1.48
Vanderbilt Mtge.	1.23	-	-

Source: "State by State: How Blacks and White Compare." *Wall Street Journal*, 21 December 1993, A-8. Compiled by Edward P. Foldessy, *Wall Street Journal*, from analysis of Federal Reserve data. Published by permission.

★ 95 ★

Banks, Financial Institutions, Finances

Mortgage Loan-denial Rate in Selected States, 1990-1992. Part 21 - Tennessee

Lender	Black to white rejection ratio		
	1992	1991	1990
U.S. average	2.06	1.91	2.27
Tennessee	1.98	2.14	2.14
Barclays Amer./Mtge. Corp.	3.71	-	2.27

[Continued]

★ 95 ★

Mortgage Loan-denial Rate in Selected States, 1990-1992. Part 21 - Tennessee

[Continued]

Lender	Black to white rejection ratio		
	1992	1991	1990
Bartlett Mtge.	4.58	3.22	5.30
Community Mtge. Corp.	1.18	-	2.61
Countrywide Funding Corp.	2.68	-	-
First Am. Nat. Bank	2.25	3.59	2.42
First Commercial Mtge. Co.	2.44	-	2.58
First Fed. Bank	4.59	4.51	8.42
First Tennessee Bank	4.50	3.21	2.63
Green Tree Fin. Corp.	1.28	-	-
Guaranty Trust Co.	1.91	1.41	1.79
Leader Fed. Bank for Saving	0.68	0.59	0.42
Leader Fed. Mtge.	1.10	1.26	1.02
Nat. Bank of Commerce	16.59	6.00	2.56
Nat. Mtge. Co.	4.53	-	-
NationsBanc Mtge. Corp.	5.96	-	-
Norwest Mtge.	2.96	6.89	1.85
Pulaski Mtge. Co.	2.62	-	2.71
Real Estate Financing	2.01	2.92	2.98
Third Nat. Bank Nashville	2.31	4.64	4.00
Union Planters Nat. Bank	3.26	2.38	3.48

Source: "State by State: How Blacks and White Compare." *Wall Street Journal,* 21 December 1993, A-8. Compiled by Edward P. Foldessy, *Wall Street Journal,* from analysis of Federal Reserve data. Published by permission.

★ 96 ★

Banks, Financial Institutions, Finances

Mortgage Loan-denial Rate in Selected States, 1990-1992. Part 22 - Texas

Lender	Black to white rejection ratio		
	1992	1991	1990
U.S. average	2.06	1.91	2.27
Texas	2.24	2.35	2.35
Accubanc Mtge. Corp.	2.91	-	-
Amer. Residential Mtge.	1.61	-	2.15
Banc One Mtge. Corp.	2.70	3.14	2.88
Bancplus Mtge. Corp.	3.26	-	2.89
Bank One Texas	3.17	3.91	2.35
Bank United of Texas	2.49	1.36	2.34
Benchmark Mtge. Co.	2.44	-	2.88
Colonial Savings	2.86	2.75	2.09
Countrywide Funding Corp.	2.71	-	-

[Continued]

★ 96 ★

Mortgage Loan-denial Rate in Selected States, 1990-1992.
Part 22 - Texas
[Continued]

Lender	Black to white rejection ratio		
	1992	1991	1990
CTX Mtge.	3.61	-	3.14
Empire of Amer. Realty	1.01	-	1.14
Farm & Home Savings Assn.	3.04	3.11	4.53
First Community Mtge. Co.	1.64	1.18	1.60
First Gibraltar Mtge. Corp.	1.96	2.76	2.63
First Interstate Bank of texas	1.36	2.22	1.89
Goldome Credit Corp.	1.30	-	-
Great Western Mtge. Corp.	1.45	1.44	0.94
Green Tree Fin. Corp.	1.40	-	-
Harbor Fin. Mtge. Corp.	3.17	-	-
Home Savings of Amer.	2.80	2.34	2.42
ICM Mtge. Corp.	3.14	-	2.35
Independence One Mtge. Corp.	1.83	3.44	2.07
Knutson Mtge. Corp.	3.10	12.85	2.77
Mellon Mtge. Co.	3.82	-	3.37
NationsBanc Mtge. Corp.	3.07	-	2.22
North Amer. Mtge. Co.	3.48	-	-
Norwest Mtge.	3.39	4.12	3.68
Principal Mutual Life Ins. Co.	2.66	2.54	1.88
Prudential Home Mtge.	3.13	2.65	2.15
Sears Mtge. Corp.	1.95	-	1.79
Sec. Pacific Housing Srv.	1.82	1.86	1.33
STM Mtge. Co.	2.02	-	-
Sunbelt Nat. Mtge. Corp.	3.00	-	-
Temple-Inland Mtge. Co.	2.33	-	2.94
Texas Commerce Mtge. Co.	2.23	3.74	2.10
Troy & Nichols	2.29	1.61	2.23
Union Modern Mtge. Corp.	2.08	2.65	1.60
Waterfield Fin. corp.	3.19	2.66	2.58
Weyerhaeuser Mtge. Co.	2.05	-	-

Source: "State by State: How Blacks and White Compare." *Wall Street Journal*, 21 December 1993, A-8. Compiled by Edward P. Foldessy, *Wall Street Journal*, from analysis of Federal Reserve data. Published by permission.

★ 97 ★

Banks, Financial Institutions, Finances

Mortgage Loan-denial Rate in Selected States, 1990-1992.
Part 23 - Virginia

Lender	Black to white rejection ratio		
	1992	1991	1990
U.S. average	2.06	1.91	2.27
Virginia	2.57	2.46	2.46
Amer. Home Funding	4.93	3.43	3.68
Ameribanc S.B.	5.11	2.20	2.81
Atlantic Coast Mtge. Co.	2.22	1.78	1.73
B.F. Saul Mtge. Co.	2.21	-	4.00
Banc One Mtge. Corp.	5.42	4.67	3.64
Beach Fed. Mtge. Corp.	3.26	3.81	3.61
Central Fidelity Bank	2.53	2.86	2.47
Columbia First	2.18	-	2.51
Countrywide Funding Corp.	5.06	-	-
Crestar Mtge. Corp.	5.51	2.61	3.58
Dominion Bankshares Mtge.	3.82	3.14	3.96
Essex First Mtge. Corp.	8.86	1.51	4.63
First Nat. Mtge. Corp.	3.09	2.01	7.07
First Security	4.07	-	-
Green Tree Fin. Corp.	1.18	-	-
Guild Mtge. Co.	5.93	-	3.05
ICM Mtge. Corp.	2.63	-	1.65
Margaretten & Co.	5.78	3.54	5.85
Market Street Mtge. Corp.	3.59	4.42	-
Nat. City Mtge. Co.	1.87	1.69	2.00
NationsBanc Mtge. Corp.	4.57	7.03	3.11
NationsBanc Mtge. Corp. of Va.	2.68	-	-
North Amer. Mtge. Co.	4.75	-	-
Norwest Mtge.	4.94	7.87	5.50
NVR Mortage	5.97	4.02	2.79
Oakwood Acceptance Corp.	1.28	-	1.37
Ryland Mtge. Co.	6.62	-	3.12
Signet Mtge. Corp.	3.57	3.17	3.21
Southern Atlantic Mtge.	B	14.43	-
Temple-Inland Mtge. Co.	1.51	-	-
Tidemark Bank	3.05	1.12	2.64
UFSB of Indianapolis	5.36	1.44	3.29
Va. First S.B.	6.25	-	-
Weyerhaeuser Mtge. Co.	4.77	-	-

Source: "State by State: How Blacks and White Compare." *Wall Street Journal*, 21 December 1993, A-8. Compiled by Edward P. Foldessy, *Wall Street Journal*, from analysis of Federal Reserve data. Published by permission.

★ 98 ★

Banks, Financial Institutions, Finances

Overview of Black Commercial Banks, Selected Years, 1973-1992

	1973	1980	1983	1989	1991	1992
Black banks						
Number	37	48	47	37	38	36
Assets ($billion)	0.66	1.26	1.55	1.88	2.01	2.11
All commercial banks						
Number	14,161	14,435	14,460	12,705	12,163	11,461
Assets ($billion)	806.4	1,702.8	2,113.1	3,245.8	3,545.4	3,652.7
Black banks						
% Number	0.26	0.32	0.31	0.29	0.31	0.31
% Assets	0.08	0.07	0.07	0.05	0.06	0.06
Annual compound rate of change, 1973-92						
Black banks						
Number -0.1%						
Assets +6.3%						
All commercial banks						
Number -1.1%						
Assets +8.3%						
Projected for January 1, 2000, based upon 1973-92 experience						
Black banks						
Number 36						
(-0.1% annual change)						
Assets $3.23 billion						
(+6.3% annual change)						

Source: "Overview of Black Commercial Banks." *The State of Black America 1994*, p. 38. Published by permission. Primary source: *Black Enterprise*, Federal Reserve *Bulletin*, various issues, 1973- 1993.

★ 99 ★

Banks, Financial Institutions, Finances

Overview of Black S&L's

	Overview of Black S&L's					
	1973	1980	1983	1989	1991	1992
Black's						
Number	43	42	36	24	18	18
Assets ($ billions)	0.45	0.95	1.16	1.33	1.16	1.23
All S&L's						
Number	4,177	4,042	3,391	2,878	2,096	1,894
Assets ($ billions)	272.4	62938	771.7	1,249.0	875.8	808.2

[Continued]

★ 99 ★

Overview of Black S&L's
[Continued]

	Overview of Black S&L's					
	1973	1980	1983	1989	1991	1992
Black S&L's						
% number	1.02	1.04	1.06	0.83	0.86	0.95
% assets	0.17	0.15	0.15	0.11	0.13	0.15

Annual compound rate of change, 1973-92
Black S&L's
  Number   -4.5%
  Assets   +5.4%

All S&L's
  Number   -4.0%
  Assets   +5.8%

Projected for January 1, 2000
Black S&L's
  Number   13
    (-4.5% annual change)
  Assets $1.78 billion
    (+5.4% annual change)

Source: "Overview of Black S&L's," *The State of Black America 1994*, p. 41. Primary source: *Black Enterprise*, Federal Reserve *Bulletin*, various issues, 1973- 1993.

★ 100 ★

Banks, Financial Institutions, Finances

Racial Disparities in Mortgage Lending, 1992

Ranked by the ratio of black rejection rates to white rejection rates. Based on nationwide data for 1992. Lenders that didn't receive 50 or more applications from blacks and from whites in a given state aren't included in the state-by-state table.

Lender	Lenders' comments	State	Applications		White rejection rate	Black-to-white rej. ratio
			White	Black		
St. Paul Fed. Bank for Savings	An "aggressive" two-year campaign, including home-buying seminars, spurred a sharp rise in minority applicants and loans. But "a lot more marginal" ones were rejected.	Ill.	910	69	1.21%	21.58
National Bank of Commerce	Began vigorous black-community advertising and better training in minority lending. Home loan applications rose 72% for blacks vs. 12% for whites; 50% more black loans were granted than in 1991.	Tenn.	324	109	1.85	17.34
Northern Trust Co.	The bank says its loan approvals for blacks soared more than fivefold from 1990 to 1992, despite the high rejection rate.	Ill.	782	98	3.58	11.40
Harris Trust & Savings Bank	Ran 27 home-seeker seminars in 1992; added many minority loan originators and a second review for denied loans. Aggressively seeking borrowers can mean a higher black-white rejection ratio.	Ill.	245	55	3.27	10.58

[Continued]

★ 100 ★

Racial Disparities in Mortgage Lending, 1992
[Continued]

Lender	Lenders' comments	State	Applications		White rejection rate	Black-to-white rej. ratio
			White	Black		
Beverly Bank	Program to attract low and moderate-income families makes firm a leader in loans to blacks in Chicago area, but also means an "unavoidable" higher denial ratio.	Ill.	297	118	5.39	10.38
Guaranty Mortgage	Says it was the largest minority lender in the Milwaukee area for the third straight year, with a black denial rate below 10%. Advertises in minority areas, and holds home-buyer seminars there.	Wis.	868	96	1.27	7.40
Essex First Mortgage Corp.	The firm's "profile" is better than the figures would indicate. A program to educate low and moderate-income borrowers regarding special loan programs should improve 1993 numbers.	Va.	672	51	5.65	7.28
Roosevelt Savings Bank	Joins in partnership to promote affordable housing for first-time buyers, helps New York set up housing-service offices in areas like Bedford-Stuyvesant and offers seminar on state lending programs.	N.Y.	165	51	14.55	5.93
Eagle Service Corp.	Is making "exceptional outreach efforts" such as marketing to minority realtors. finds that some people "can't qualify for a loan, though a "second look" panel helps rejectees try again.	Ga.	495	59	4.44	5.72
Developers Mortgage Corp.	Some 20% of the lender's loan portfolio comes from Virginia program for low and moderate-income housing. It also joins in first-time homebuyer programs.	Va.	720	116	1.53	5.64
Capitol Federal	Tries to teach blacks about mortgages, including more radio and TV ads. Disparity is partly a statistical anomaly, due low numbers of black applicants.	Kan.	2,859	61	7.80	5.47
Sunburst Bank	The black-to-white denial ratio is for the core bank. A separate rural-urban unit issues more mortgages, and has a lower ratio. Sunburst was praised by Fannie Mae for its minority lending.	Miss.	727	127	8.25	5.44
Virginia First Savings Bank	The bank manages to approve "about 98%" of its mortgage loans "one way or another." Applicants initially turned down are steered toward alternate products such as adjustable-rate mortgages.	Va.	1,163	113	1.72	5.15
Investors Savings Bank	The small number of black applications makes meaningful comparison difficult. Minority loan-rejection files are reviewed, for consistent compliance with the bank's underwriting standards.	Minn.	4,813	74	5.26	5.14
East Coast Savings Bank	The company's ratio "baffles" an official. The lender sought minority borrowers, including offers of 100% financing; it used mortgage seminars, with savings bond drawings to encourage participation.	N.C.	438	70	4.79	5.07

Source: "Institutions with Largest Racial Disparities in Mortgage Lending." *Wall Street Journal,* 21 December 1993, A-8. Published by permission.

★ 101 ★

Banks, Financial Institutions, Finances

Savings and Loan Companies: Percent Change, 1992

Black savings and loan associations	1991	1992	Percent change
Number of Savings and Loan Associations	13	13	0
Number of employees	501	487	-2.8
Assets[1]	1,157.291	1,226.104	5.9
Equity capital[1]	62.815	68.896	9.7
Savings capital/deposits[1]	1,012.038	1,025.930	1.4
Loans[1]	921.038	945.269	2.6

Source: "1993 Savings and Loans Summary," *Black Enterprise* 24 (June 1993), p. 146. Published by permission. Primary source: B.E. Research. *Note:* 1. In millions of dollars, to the nearest thousand.

★ 102 ★

Banks, Financial Institutions, Finances

Survival Analysis of Black Commercial Banks: Selected Years, 1978-1993

	1978		1980		1983		1989		1993	
	No.	%	No.	%	No.	%	No.	%	No.	%
Those existing in										
1973 37	34	92	30	81	27	73	20	54	13	35
1978 49			44	90	39	80	27	55	16	33
1981 46					40	87	27	59	16	35
1982 44					43	98	29	66	16	36

Source: "Survival Analysis of Black Commercial Banks," *The State of Black America 1994*, p. 39. Published by permission. Primary source: *Black Enterprise*, various issues, 1973-1993.

★ 103 ★

Banks, Financial Institutions, Finances

Survival Analysis of Black S&L's, 1978-1993

		Which still exist in									
		1978		1980		1983		1989		1993	
		No.	%	No.	%	No.	%	No.	%	No.	%
Those existing in											
1973	43	37	86	36	84	33	77	23	53	16	37
1978	40			39	98	34	85	24	60	16	40
1981	40					35	88	24	60	16	40
1982	37					35	95	24	65	16	43

Source: "Survival Analysis of Black S&L's," *The State of Black America 1994,* p. 42. Primary source: *Black Enterprise,* various issues, 1973-1993.

Business Growth and Sales

★ 104 ★

Sales by Industry

		Percent
Media	$912.724[1]	10.1
Technology	668.828[1]	7.4
Manufacturing	492.027[1]	5.4
Construction	349.498	4.0
Health and Beauty Aids	290.074[1]	3.2
Engineering	130,580[1]	1.4
Other	385.988[1]	4.2
Commodities		
Entertainment		
Health care		
Security & maintenance		
Transportation		
Miscellaneous		
Total sales	$9,028.449[1]	

Source: "The B.E. 100s Sales by Industry," *Black Enterprise* 24 (June 1993), p. 80. Published by permission. Primary source: B.E. Research As of December 31, 1992. *Notes:* 1. Sales in millions of dollars to the nearest thousand. As of December 31, 1992. Prepared by B.E. Research.

Insurance Companies

★ 105 ★

Black-Owned Insurance Companies: Summary: 1993

Black-owned insurance companies	1991	1992	Percent change
Number of companies	27	23	-14.8
Number of employees	3,599	3,417	-5.1
Assets[1]	736.583	720.122	-2.2
Capital reserves[1]	546.094	499.470	-8.5
Insurance in force[1]	24,984.724	23,225.999	-7.0
Premium income[1]	187.167	161.227	-13.9
Net investment income[1]	51.645	46.996	-9.0

Source: "1993 Insurance Summary," *Black Enterprise* 24 (June 1993), p. 147. Published by permission. Primary source: B.E. Research. *Notes:* Prepared by B.E. Research. Reviewed by Mitchell/Titus & Co. 1. In millions of dollars to the nearest thousand.

★ 106 ★

Insurance Companies

Insurance Companies: Characteristics, 1992

Company/location	Chief executive	Year started	Staff	Assets[1]	Statutory reserves[1]	Insurance in force[1]	Premium income[1]	Net investment income[1]
North Carolina Mutual Life Insurance Co. Durham, North Carolina	Bert Collins	1899	571	220.133	134.820	9,406.671	55.091	12.779
Atlanta Life Insurance Co. Atlanta, Georgia	Jesse Hill Jr.	1905	835	163.093	123.294	3,223.000	25.672	11.517
Golden State Mutual Life Insurance Co. Los Angeles, California	Larkin Teasley	1925	189	105.171	73.706	5,855.082	15.177	9.583
Universal Life Insurance Co. Memphis, Tennessee	Gerald P. Howell	1923	615	63.970	56.061	760.873	21.644	4.320
Chicago Metropolitan Assurance Co. Chicago, Illinois	Anderson M. Schweich	1927	157	58.511	35.778	2,664.238	17.856	2.296
Booker T. Washington Insurance Co. Birmingham, Alabama	Louis J. Willie	1932	176	40.986	33.543	945.332	10.965	2.275
Protective Industrial Ins. Co. of Alabama Inc. Birmingham, Alabama	Paul E. Harris	1923	108	14.637	10.286	72.622	3.102	0.902
Golden Circle Life Insurance Co. Brownsville, Tennessee	William D. Rawls Sr.	1958	75	9.072	4.125	23.632	1.554	0.675
Winnfield Life Insurance Co. Natchitoches, Louisiana	Ben D. Johnson	1936	12	8.173	6.840	50.643	1.734	0.453
Williams-Progressive Life & Accident Insurance Co. Opelousas, Louisiana	Borel C. Dauphin	1947	51	5.892	4.843	35.791	1.283	0.439
Wright Mutual Insurance Co. Detroit, Michigan	Wardell C. Croft	1942	36	4.686	1.133	39.477	1.088	0.270
Reliable Life Insurance Co. Monroe, Louisiana	Joseph H. Miller Jr.	1940	73	4.500	3.631	60.883	1.456	0.105
Gertrude Geddes Willis Life Insurance Co. New Orleans, Louisiana	Joseph O. Misshore Jr.	1941	45	4.421	3.885	40.500	1.156	0.320
Benevolent Life Insurance Co. Inc. Shreveport, Louisiana	Granville L. Smith	1934	85	3.782	1.977	16.547	0.643	0.236

[Continued]

★ 106 ★

Insurance Companies: Characteristics, 1992
[Continued]

Company/location	Chief executive	Year started	Staff	Assets[1]	Statutory reserves[1]	Insurance in force[1]	Premium income[1]	Net investment income[1]
Majestic Life Insurance Co. New Orleans, Louisiana	Cecilia Roberts	1947	16	3.089	0.885	9.989	0.414	0.198

Source: "B.E. Insurance Companies," *Black Enterprise* 24 (June 1993), p. 151. Published by permission. Primary source: B.E. Research. *Notes:* 1. In millions of dollars, to the nearest thousand. Ranked by assets as of Dec.31, 1992. Prepared by B.E. research. Reviewed by Mitchell/Titus & Co.

★ 107 ★

Insurance Companies

Overview of Black Insurance Companies, 1973-1992

	Overview of Black insurance companies					
	1973	1980	1983	1989	1991	1992
Black insurance companies						
Number	41	38	38	30	27	23
Assets ($ billions)	0.52	0.70	0.77	0.80	0.74	0.72
All life insurance companies						
Number	1,766	1,958	2,117	2,770	2,065	2,005
Assets ($ billions)	252.1	479.2	663.0	1,300.0	1,551.2	1,664.5
Black insurance companies						
% number	2.32	1.94	1.79	1.08	1.31	1.15
% assets	0.21	0.15	0.12	0.06	0.05	0.04
Annual compound rate of change, 1973-92						
Black insurance companies						
Number -3.0%						
Assets +1.7%						
All life insurance companies						
Number +0.7%						
Assets +10.4%						
Projected for January 1, 2000						
Black insurance companies						
Number 19						
(+3.0% annual change)						
Assets $1.16 billion						
(-1.7% annual change)						

Source: "Overview of Black Insurance Companies," *The State of Black America 1994*, p. 46. Primary source: *Black Enterprise*, Federal Reserve *Bulletin*, various issues, 1973- 1993.

★ 108 ★

Insurance Companies

Survival Analysis of Black Insurance Companies, 1978-1993

	Which still exist in									
	1978		1980		1983		1989		1993	
	No.	%	No.	%	No.	%	No.	%	No.	%
Those existing in										
1973 41	38	93	36	88	34	83	28	68	16	39
1978 39			37	95	35	90	28	72	17	44
1981 38					35	92	29	76	19	50
1982 38					36	95	29	76	19	50

Source: "Survival Analysis of Black Insurance Companies," *The State of Black America 1994*, p. 47. Primary source: *Black Enterprise*, various issues, 1973-1993.

Leading Businesses

★ 109 ★

Automobile Dealer of the 1992

Year founded: 1979.

Year	Staff	Sales in millions of dollars
1988	65	15.7
1989	73	23.9
1990	74	29.0
1991	69	33.8
1992	64	45.7

Source: "Alan Young Buick-GM Truck," *Black Enterprise* 24 (June 1993), p. 126. Published by permission. Primary source: B.E. Research.

★ 110 ★

Leading Businesses

Automobile Dealers and Industrial Service Companies: Total Sales and Staff, 1992

	1991	1992	Difference	Percent change
Total sales[1]	$7,911.192	$9,028.449	$1,117.257	14.1
Total staff	38,723	38,020	-703	-1.8

1993 B.E. Industrial/Service 100

	1991	1992	Difference	Percent change
Total sales[1]	4,996.809	5,691.557	694.748	13.9
Total staff	32,590	31,668	-992	-2.8

1992 B.E. Auto Dealer 100

	1991	1992	Difference	Percent change
Total sales[1]	2,914.383	3,336.892	422.509	14.5
Total staff	6,133	6,352	219	3.6

Source: "The 1993 Black Enterprise 100s," *Black Enterprise* 24 (June 1993), p. 82. Published by permission. Primary source: B.E. Research As of December 31, 1992. *Notes:* 1. In millions of dollars to the nearest thousand. Prepared by B.E. Research.

★ 111 ★

Leading Businesses

Businesses Located by State, 1992

State	Number
Michigan	18
New York	18
California	16
Illinois	15
Georgia	14
Ohio	12
Texas	11
Virginia	10
Maryland	8
Alabama	7
Florida	6
Indiana	5
New Jersey	5
North Carolina	5
Washington	4
Wisconsin	4
Tennessee	4
South Carolina	3
Kentucky	3

[Continued]

★ 111 ★

Businesses Located by State, 1992
[Continued]

State	Number
Distric of Columbia	3
Massachusetts	3
Pennsylvania	3
Iowa	3
Connecticut	2
Oregon	2
Oklahoma	2
Nebraska	2
Louisiana	2
Missouri	2
West Virginia	2
Delaware	1
Idaho	1
Arkansas	1
Arizona	1
New Mexico	1
Mississippi	1

Source: "1993 B.E. 100s by State," *Black Enterprise* 24 (June 1993), p. 88. Published by permission.
Primary source: B.E. Research As of December 31, 1992.

★ 112 ★

Leading Businesses

Chicago Businesses, 1992

Company	Chief executive	Year started	Staff	Type of business	Sales
Johnson Publishing Co.	John H. Johnson	1942	2,785	Publishing; broadcasting; cosmetics; hair care	274.1
Soft Sheen Products	Edward G. Gardner	1964	547	Hair care products	91.7
Burrell Communications	Thomas J. Burrell	1971	115	Advertising; public relations; promotions	77.0
Johnson Products Co.	Joan B. Johnson	1954	215	Hair & personal care products	46.0
Luster Products Co.	Jory Luster	1957	315	Hair care products	46.0
Southside Ford Truck Sales	Carl Statham	1984	75	Ford automobile dealer	41.0
UBM Inc.	Sandra Dixon Jiles	1975	53	General contracting, construction	16.7

Source: "Chicago's Leading Black Business & Financial Companies," *Black Enterprise* 24 (June 1993), p. 196. Published by permission. Primary source: B.E. Research.

★ 113 ★

Leading Businesses

Financial Companies: Leading Institutions in Chicago, 1992

Financial companies	Chief executive	Year started	Staff	Assets[1]
Seaway National Bank of Chicago	Walter E. Grady	1965	165	202.0
Independence Bank of Chicago	Alvin J. Boutte	1964	119	137.2
Drexel National Bank	Alvin J. Boutte	1989	90	127.7
Illinois Service/Federal S&L Assn. of Chicago	Thelma J. Smith	1934	34	104.3
Chicago Metropolitan Assurance Co.	Anderson M. Schweich	1927	157	58.5

Source: "Chicago's Leading Black Business & Financial Companies," *Black Enterprise* 24 (June 1993), p. 196. Published by permission. Primary source: B.E. Research. *Black Enterprise* Research Department. *Notes:* 1. In millions of dollars, to the nearest thousand. As of December 31, 1992.

★ 114 ★

Leading Businesses

Financial Company of the Year, 1992

Chief executive officer: John P. Kelly Jr. Year founded: 1991.

	Year ending	
	1991	1992
Staff	45	45
Assets[1]	61.437	74.201
Capital[1]	4.224	5.201
Deposits[1]	48.102	61.725
Loans[1]	18.210	32.910

Source: "Founders National Bank of Los Angeles," *Black Enterprise* 24 (June 1993), p. 158. Published by permission. Primary source: B.E. research. *Note:* 1. In millions of dollars to the nearest thousand.

★ 115 ★

Leading Businesses

Franchise Companies: 50 Companies by Industry, 1992

Industry	Percent
Fast Food	26.0
Automobile Services	18.0
Business/Postal Services	12.0
Cleaning Services	10.0
Personal Services	8.0
Employment Services	4.0

[Continued]

★ 115 ★

Franchise Companies: 50 Companies by Industry, 1992
[Continued]

Industry	Percent
Specialty Foods	4.0
Travel Services	4.0
Automobile Rental	2.0
Other	12.0

Source: "The 1993 B.E. Franchise 50." *Black Enterprise* 24 (September 1993), p. 46. Published by permission. Primary source: B.E. Research As of December 31, 1992.

Minority-Owned Businesses

★ 116 ★

Federal Contract Actions: Small and Minority-Owned Businesses, 1990 and 1991

In millions of dollars, except percent. For fiscal year. Excludes Guam, Puerto Rico, and Virgin Islands. Represents contract awards of $25,000 or more awarded to establishments. A contract may consist of more than one action. Minus sign (-) indicates decrease.

Region, division and state	Total contract actions			Small business share				Small minority-owned share			
	1990	1991	Percent change, 1990-91	1990	1991		Percent change, 1990-91	1990	Amount	Percent of total	Percent change, 1990-91
					Amount	Percent of total					
U.S.	163,775	186,387	14	25,399	29,205	16	15	5,757	6,684	4	16
Northeast	31,730	34,514	9	3,859	4,755	14	23	655	621	2	-5
New England	15,399	15,801	3	1,263	1,460	9	16	212	255	2	20
Maine	921	1,149	25	59	81	7	37	4	13	1	225
New Hampshire	419	481	15	61	62	13	2	5	6	1	20
Vermont	96	100	4	19	27	27	42	1	4	4	300
Massachusetts	9,008	8,207	-9	715	867	11	21	141	190	2	35
Rhode Island	580	443	-24	137	109	25	-20	31	13	3	-58
Connecticut	4,375	5,421	24	272	314	6	15	30	29	1	-3
Middle Atlantic	16,331	18,713	15	2,596	3,295	18	27	443	366	2	-17
New York	8,238	9,666	17	1,098	1,318	14	20	203	80	1	-61
New Jersey	4,127	4,507	9	784	1,022	23	30	121	167	4	38
Pennsylvania	3,966	4,540	14	714	955	21	34	119	119	3	-
Midwest	24,250	29,449	21	3,526	3,954	13	12	653	681	2	4
East North Central	12,929	16,210	25	2,559	2,661	16	4	438	467	3	7
Ohio	5,879	6,676	14	925	998	15	8	263	250	4	-5
Indiana	1,888	2,430	29	222	198	8	-11	31	33	1	6
Illinois	2,357	3,791	61	574	839	22	46	112	123	3	10
Michigan	1,597	1,919	20	421	381	20	-10	26	52	3	100
Wisconsin	1,208	1,394	15	417	245	18	-41	6	9	1	50
West North Central	11,321	13,239	17	967	1,293	10	34	215	214	2	(Z)
Minnesota	2,015	2,142	6	140	146	7	4	13	18	1	38

[Continued]

★ 116 ★

Federal Contract Actions: Small and Minority-Owned Businesses, 1990 and 1991
[Continued]

Region, division and state	Total contract actions			Small business share				Small minority-owned share			
	1990	1991	Percent change, 1990-91	1990	1991		Percent change, 1990-91	1990	Amount	Percent of total	Percent change, 1990-91
					Amount	Percent of total					
Iowa	652	670	3	58	109	16	88	4	2	(Z)	-50
Missouri	6,884	8,392	22	349	464	6	33	51	72	1	41
North Dakota	159	230	45	94	137	60	46	44	61	27	39
South Dakota	120	219	83	53	98	45	85	56	10	5	-82
Nebraska	402	434	8	134	143	33	7	10	15	3	50
Kansas	1,089	1,152	6	139	196	17	41	37	36	3	-3
South	58,205	67,653	16	12,069	13,817	20	14	3,060	3,740	6	22
South Atlantic	33,528	38,084	14	6,972	8,856	23	27	2,184	2,624	7	20
Delaware	116	155	34	36	50	32	39	10	9	6	-10
Maryland	6,798	7,728	14	1,269	1,767	23	39	559	749	10	34
District of Columbia	3,096	4,157	34	874	1,179	28	35	424	562	14	33
Virginia	10,434	10,222	-2	2,403	3,019	30	26	768	900	9	17
West Virginia	305	668	119	151	180	27	19	88	28	4	-68
North Carolina	1,533	2,142	40	361	545	25	51	41	56	3	37
South Carolina	2,664	3,188	20	377	383	12	2	39	57	2	46
Georgia	2,147	2,540	18	456	545	21	20	80	113	4	41
Florida	6,435	7,284	13	1,045	1,188	16	14	175	150	2	-14
East South Central	10,156	12,546	24	2,576	2,225	18	-14	355	511	4	44
Kentucky	1,972	2,010	2	1,075	727	36	-32	49	159	8	224
Tennessee	3,241	4,880	51	603	493	10	-18	134	85	2	-37
Alabama	3,352	3,562	6	713	695	20	-3	133	189	5	42
Mississippi	1,591	2,094	32	185	310	15	68	39	78	4	100
West South Central	14,521	17,023	17	2,521	2,736	16	9	521	605	4	16
Arkansas	361	439	22	152	173	39	14	15	12	3	-20
Louisiana	2,332	2,130	-9	586	479	22	-18	51	56	3	10
Oklahoma	845	1,029	22	292	406	39	39	159	199	19	25
Texas	10,983	13,425	22	1,491	1,678	12	13	296	338	3	14
West	49,590	54,771	10	5,945	6,679	12	12	1,389	1,642	3	18
Mountain	15,004	15,522	3	1,687	1,794	12	6	459	543	3	18
Montana	156	173	11	122	135	78	11	29	29	17	(Z)
Idaho	854	1,042	22	83	107	10	29	15	16	2	7
Wyoming	131	125	-5	89	78	62	-12	6	4	3	-33
Colorado	4,651	4,598	-1	437	472	10	8	125	131	3	5
New Mexico	3,245	3,460	7	303	346	10	14	128	161	5	26
Arizona	3,625	3,052	-16	290	318	10	10	68	84	3	24
Utah	1,529	1,927	26	264	224	12	-15	56	73	4	30
Nevada	813	1,145	41	99	114	10	15	32	45	4	41
Pacific	34,586	39,249	13	4,258	4,885	12	15	930	1,099	3	18
Washington	3,878	3,953	2	419	567	14	35	80	105	3	31
Oregon	530	566	7	201	302	53	50	20	28	5	40
California	29,110	33,160	14	3,211	3,398	10	6	687	792	2	15

[Continued]

★ 116 ★

Federal Contract Actions: Small and Minority-Owned Businesses, 1990 and 1991

[Continued]

Region, division and state	Total contract actions			Small business share				Small minority-owned share			
	1990	1991	Percent change, 1990-91	1990	1991		Percent change, 1990-91	1990	Amount	Percent of total	Percent change, 1990-91
					Amount	Percent of total					
Alaska	529	822	55	190	340	41	79	48	58	7	21
Hawaii	539	748	39	237	278	37	17	95	116	16	22

Source: "Federal Contract Actions—Small and Minority-Owned Small Business Share, by State of Principal Place of Performance: 1990 and 1991," U.S. Bureau of the Census, *Statistical Abstract of the United States*, 1993, p. 542. Primary source: U.S. Small Business Administration. *The State of Small Business: A Report of the President*, annual. Data from Federal Procurement Data System, Special Report S92132B, June 18, 1992. *Note:* - Represents or rounds to zero. Z Less than 0.5 percent.

Types of Business

★ 117 ★

Automobile Dealers: 50 Companies Below the 50 Leaders, 1992

Company	Location	Chief executive	Year started	Staff	Type of business	Sales[1]
Ferndale Honda Inc.	Ferndale, Michigan	Barbara J. Wilson	1983	34	Honda	21.500
Heritage Cadillac Inc.	Forrest Park, Georgia	Ernest M. Hodge	1991	35	GM	21.403
Noble Ford-Mercury Inc.	Indianola, Iowa	Dimaggio Nichols	1985	58	Ford	21.138
Barron Chevrolet-Geo Inc.	Danvers, Masssachusetts	Reginald Barron	1984	59	GM	20.468
Utica Chrysler-Plymouth Inc.	Yorkville, New York	William E. Norris	1986	40	Chrysler	20.373
Winter Haven Ford Inc.	Winter Haven, Florida	Johnny Mac Brown	1989	51	Ford	20.247
Hayes-Franklin Ford Inc.	Crosby, Texas	Elvin E. Hayes	1991	50	Ford	20.000
Al Meyer Ford Inc.	Lufkin, Texas	Alton J. Meyer	1987	65	Ford	20.000
Beddingfield Buick-GMC Truck-BMW Inc.	Decatur, Illinois	Edward C. Beddingfield	1989	46	GM-BMW	19.865
Highland Lincoln-Mercury Inc. d/b/a/ Tyson Lincoln Mercury	Highland, Indiana	Nathan Z. Cain	1991	40	Ford	19.830
Pasadena Linciln-Mercury Inc.	Pasadena, California	Lester C. Jones	1991	50	Ford	19.487
Thomas A. Moorehead Buick Inc. d/b/a/ Sentry Buick Isuzu	Omaha, Nebraska	Thomas A. Moorehead	1988	50	GM-Isuzu	19.288
Cardinal Dodge Inc.	Louisville, Kentucky	Winston R. Pittman Sr.	1988	45	Chrysler	19.232
Brandon Dodge Inc.	Tampa, Florida	Sanford L. Woods	1989	39	Chrysler	18.894
Smokey Point Sales & Service Inc.	Arlington, Washington	Henry F. Taylor	1981	48	GM	18.773
Red Bluff Ford-Mercury Inc.	Red Bluff, California	Phillip G. Price	1989	37	Ford	18.315
Southland Chrysler-Plymouth Inc.	Memphis, Tennessee	John Willie Roy	1986	48	Chrysler	17.920
Mike Pruitt's Lima Ford Inc.	Lima, Ohio	Michael Pruitt	1990	58	Ford	17.600
Wilson Buick-Pontiac-GMC-Truck Inc.	Jackson, Tennessee	Sidney Wilson Jr.	1990	48	GM-Hyundai	17.577
Gresham Dodge Inc.	Gresham, Oregon	Dorian S. Boyland	1987	40	Chrysler	17.520
Heritage Lincoln-Mercury Inc.	Hackensack, New Jersey	T. Errol Harper	1983	35	Ford	17.518
Conway Ford Inc.	Conway, South Carolina	Samuel H. Frink	1986	52	Ford	17.493
Midfield Dodge Inc.	Midfield, Alabama	Jordan A. Frazier	1989	44	Chrysler	17.477
West Covina Lincoln-Mercury Inc.	West Covina, California	Boyd Harrison Jr.	1986	49	Ford	17.112
Puget Sound Chrysler-Plymouth Inc.	Renton, Washington	B. Edward Fitzpatrick	1986	46	Chrysler	17.000
Mountain Home Ford-Lincoln-Mercury Inc.	Mountain-Home, Idaho	Robert E. Montgomery	1988	31	Ford	16.998
Macon Chrysler-Plymouth Inc.	Macon, Georgia	James B. Jones	1988	35	Chrysler	16.967
All American Ford Inc.	Saginaw, Michigan	Laval Perry	1988	60	Ford	16.500
Vision Ford-Lincoln-Mercury Inc.	Alamagordo, New Mexico	Wayne Martin	1989	57	Ford	16.235
Bay City Chrysler-Plymouth Inc.	Green Bay, Wisconsin	Larry L. Hovell	1987	44	Chrysler	16.046
Shamrock Lincoln-Mercury-Nissan-Saab Inc.	Mishawaka, Indiana	Theodore Williams Jr.	1988	36	Ford-Saab-	16.000

[Continued]

★ 117 ★

Automobile Dealers: 50 Companies Below the 50 Leaders, 1992
[Continued]

Company	Location	Chief executive	Year started	Staff	Type of business	Sales[1]
Freedom Ford-Lincoln-Mercury Inc.	Wise, Virginia	Bobby H. Dawson	1990	35	Nissan Ford	15.889
Gresham Chrysler-Plymouth Inc.	Gresham, Oregon	Clarence E. Parker	1991	38	Chrysler	15.860
Mike Branker Buick Inc.	Lincoln, Nebraska	Julian Michael Branker	1985	45	GM-Hyundai	15.750
Mission Blvd. Lincoln-Mercury Inc.	Hayward, California	Austin O. Chucks-Orji	1986	38	Ford	15.349
George Hughes Chevrolet Inc.	Freehold, New Jersey	George Hughes	1978	30	GM	15.025
Pittsburgh Ford Inc.	Pittsburgh, California	Laroy S. Doss	1974	45	Ford	14.919
Ross Park Dodge Inc.	Pittsburgh, Pennsylvania	David A. Eaton	1987	42	Chrysler	14.916
Shoals Ford Inc.	Muscle Shoals, Alabama	Fred D. Lee Jr.	1986	47	Ford	14.786
Ray Wilkinson Buick-Cadillac-Isuzu Inc.	Racine, Wisconsin	Raymond M. Wilkinsin Jr.	1984	31	GM-Isuzu	14.775
Classic Cadillac-GMC Truck Inc.	Winston-Salem, North Carolina	Chandler B. Lee	1991	38	GM	14.749
Gene Moon Buick-Pontiac Inc. - Signature Toyota Inc.	Paw Paw, Michigan	Gene E. Moon	1976	40	GM-Toyota	14.593
Courtesy Ford-Lincoln-Mercury Inc.	Danville, Illinois	G. Michael McDonald	1987	45	Ford	14.550
Huntsville Dodge Inc.	Huntsville, Alabama	Ellenae L. Henry-Fairhurst	1990	36	Chrysler	14.486
Southland Chrysler Products Inc.	Marion, Ohio	Eugene Turner	1985	36	Chrysler	13.798
Vicksburg Chrysler-Plymouth-Dodge Inc.	Vicksburg, Michigan	Monti M. Long	1990	42	Chrysler	13.792
Prestige Ford	Eustis, Florida	Irving J. Matthews	1991	42	Ford	13.661
Auburn Ford-Lincoln-Mercury Inc.	Auburn, Alabama	Andrew L. Ferguson	1985	34	Ford	13.580
Dyersburg Ford Inc.	Dyersburg, Tennessee	George L. Mitchell	1985	30	Ford	13.354
Plaza Ford-Lincoln-Mercury Inc.	Lexington, North Carolina	Archie Kindle	1987	40	Ford	13.226

Source: "B.E. 100s Auto Dealers," *Black Enterprise* 24 (June 1993), pp. 104, 106, 108. Published by permission. Primary source: B.E. Research. *Notes:* 1. In millions of dollars, to the nearest thousand. As of Dec. 31, 1992. Prepared by B.E. Research. Reviewed by Mitchell/Titus & Co.

★ 118 ★

Types of Business

Automobile Dealers: Sales by Manufacturer

Manufacturer	Number of dealerships	Sales[1]	Percent of total
Ford	46	$1,284.468	38.49%
General Motors	16	765.442	22.94
Chrysler	21	438.711	13.15
Ford/Volkswagen	1	228.300	6.84
Ford/Toyota	2	145.431	4.36
Ford/Subaru	1	112.199	3.36
General Motors/Acura/Ford	1	69.292	2.08
General Motors/Chrysler/Isuzu	1	44.994	1.35
General Motors/Mercedes-Benz	1	41.407	1.24
General Motors/Ford	1	37.500	1.12
General Motors/Isuzu	2	34.063	1.02
General Motors/Hyundai	2	33.327	1.00
General Motors/Saab/Avanti	1	29.800	.89
Honda	1	21.500	.64
General Motors/BMW	1	19.865	.60
Ford/Saab/Nissan	1	16.000	.48

[Continued]

★ 118 ★

Automobile Dealers: Sales by Manufacturer

[Continued]

Manufacturer	Number of dealerships	Sales[1]	Percent of total
General Motors/Toyota	1	14.593	.44
Total	100	3,336.892	100.00

Source: "B.E. Auto Dealer 100, Sales by Manufacturer," *Black Enterprise* 24 (June 1993), p. 84. Published by permission. Primary source: B.E. Research. *Notes:* 1. In millions of dollars, to the nearest thousand. As of Dec. 31, 1992. Prepared by B.E. Research.

★ 119 ★

Types of Business

Automobile Dealers: Top 50 Companies, 1992

Company	Location	Chief executive	Year started	Staff	Type of business	Sales[1]
Trainer Oldsmobile-Cadillac-Pontiac-GMC Truck Inc.	Warner Robins, Georgia	James E. Trainer	1991	49	GM	254.574
Shack-Woods & Associates	Long Beach, California	William E. Shack, Jr.	1977	280	Ford-Volkswagen	228.300
Pavilion Lincoln-Mercury Inc.	Austin, Texas	J. Michael Chargois	1988	95	Ford	223.788
Peninsula Pontiac-Oldsmobile Inc.	Torrance, California	Cecil B. Willis	1979	51	GM	128.912
Mel Farr Automotive Group	Oak Park, Michigan	Mel Farr	1975	243	Ford-Toyota	118.000
S&J Enterprises	Charlotte, North Carolina	Sam Johnson	1973	279	Ford-Subaru	112.199
The Baranco Automotive Group	Decatur, Georgia	Gregory T. Baranco	1978	171	GM-Acura-Ford	69.292
Avis Ford Inc.	Southfield, Michigan	Walter E. Douglas Sr.	1986	120	Ford	66.384
Rountree Cadillac-Oldsmobile Co. Inc.	Shreveport, Louisiana	Lonnie M. Bennett	1991	112	GM	47.000
Tropical Ford Inc.	Orlando, Florida	Hamilton W. Massey	1985	90	Ford	46.665
Alan Young Buick-GMC Truck Inc.	Fort Worth, Texas	Alan Young	1979	64	GM	45.661
Martin Automotive Group	Bowling Green, Kentucky	Cornelius A. Martin	1985	91	GM-Chrysler-Isuzu	44.994
Metrolina Dodge Inc.	Charlotte, North Carolina	Reginald T. Hubbard	1986	76	Chrysler	43.466
Sidney Moncrief Pontiac-Buick-GMC Truck Inc.	Sherwood, Arkansas	Sidney A. Moncrief	1987	55	GM	42.018
Bob Ross Buick-Mercedes-GMC Inc.	Centerville, Ohio	Robert P. Ross	1974	104	GM-Mercedes-Benz	41.407
Southside Ford Truck Sales Inc.	Chicago, Illinois	Carl Statham	1984	75	Ford	41.172
Leader Motors Inc.	St. Louis, Missouri	Jesse Morrow	1983	75	Ford	37.823
Dick Gidron Cadillac & Ford Inc.	Bronx, New York	Richard D. Gidron	1972	100	GM-Ford	37.500
Ford Mercury Inc.	Batavia, Ohio	Clarence Warren	1990	50	Ford	35.118
Gulf Freeway Pontiac-GMC Truck	Houston, Texas	Carl L. Barnett Sr.	1991	68	GM	32.298
Shelby Dodge Inc.	Memphis, Tennessee	H. Steve Harrell	1987	71	Chrysler	31.563
Varsity Ford-Lincoln-Mercury Inc.	Bryan, Texas	Tony Majors	1988	54	Ford	30.134
Conyers Riverside Ford Inc.	Detroit, Michigan	Nathan G. Conyers	1970	80	Ford	30.100
Al Johnson Cadillac-Avanti-Saab Inc.	Tinley Park, Illinois	Albert W. Johnson Sr.	1967	69	GM-Saab-Avanti	29.800
Campus Ford Inc.	Okemos, Michigan	Wendell Barron	1986	70	Ford	29.792
Olympia Fields Ford Sales Inc.	Olympia Fields, Illinois	Nathaniel K. Sutton	1989	82	Ford	29.244
Jim Bradley Pontiac-Cadillac-GMC Inc.	Ann Arbor, Michigan	James H. Bradley Jr.	1973	70	M	29.192
Team Ford Inc.	Sioux City, Iowa	Arthur P. Silva	1986	92	Ford	29.056
Bob Johnson Chevrolet Inc.	Rochester, New York	Robert Johnson	1981	65	GM	28.855
Northwestern Dodge Inc.	Ferndale, Michigan	Theresa Jones	1980	70	Chrysler	28.791
Quality Ford Inc.	W. Des Moines, Iowa	Franklin D. Greene	1989	75	Ford	28.714
North Seattle Chrysler-Plymouth Inc.	Seattle, Washington	William E. McIntosh Jr.	1985	50	Chrysler	28.000
Duryea Ford Inc.	Brockport, New York	Jesse Thompson	1985	88	Ford-Toyota	27.431
University Ford of Peoria Inc.	Peoria, Illinois	James L. Oliver	1985	77	Ford	26.612
Deerbrook Forest Chrysler-Plymouth Inc.	Kingwood, Texas	Ezzard Dale Early	1987	55	Chrysler	26.443
River View Ford-Mercury Inc.	Columbia, Illinois	John Carthen Sr.	1988	42	Ford	26.260
Spalding Ford-Lincoln-Mercury Inc.	Griffin, Georgia	Alan M. Reeves	1981	60	Ford	26.057
Republic Ford Inc.	Republic, Missouri	Franklin D. Greene	1983	55	Ford	25.939
Empire Ford Inc.	Spokane, Washington	Nathaniel D. Greene	1986	69	Ford	25.600
Coastal Ford Inc.	Mobile, Alabama	Delmont O. Dapremont Jr.	1984	85	Ford	25.000
Fairway Ford of Augusta Inc.	Augusta, Georgia	James H. Brown	1989	65	Ford	24.743
Kemper Dodge Inc.	Cincinnati, Ohio	Paul C. Keels	1986	42	Chrysler	24.167
Chino Hills Ford Sales Inc.	Chino, California	Timothy L. Woods	1982	66	Ford	23.042
R.H. Peters Chevrolet Inc.	Hurricane, West Virginia	R. H. Peters Jr.	1982	45	GM	22.663
University Motors	Athens, Georgia	Ronald Hill	1991	66	Ford-Mazda	22.082
Fred Jones Pontiac-GMC Truck Inc	Brookfield, Wisconsin	Frederick E. Jones	1984	54	GM	22.079
Hill Top Chrysler Plymouth Inc.	Lanchaster, Texas	Eric V. Wilkins	1991	40	Chrysler	22.000

[Continued]

★ 119 ★

Automobile Dealers: Top 50 Companies, 1992
[Continued]

Company	Location	Chief executive	Year started	Staff	Type of business	Sales[1]
Southwest Ford Sales Inc.	Oklahoma City, Oklahoma	Roger L. Williams	1990	79	Ford	21.856
Prestige Pontiac-Oldsmobile Inc.	Mt. Morris, Michigan	Freddie J. Poe	1989	52	GM	21.772
Broadway Ford Inc.	Edmond, Oklahoma	LeMon Henderson	1981	52	Ford	21.500

Source: "B.E. 100s Auto Dealers," *Black Enterprise* 24 (June 1993), pp. 101-102, 104. Published by permission. Primary source: B.E. Research. *Notes:* 1. In millions of dollars, to the nearest thousands. As of Dec. 31, 1992. Prepared by B.E. Research. Reviewed by Mitchell/Titus & Co.

★ 120 ★

Types of Business

Franchise Companies: Black and Percent Black of Total Units: 1992

Company	Location	Type	Black units	Total units	Black % of units	Start-up costs
Subway Sandwiches & Salads	Milford, Conn.	Fast food	1,343	7,900	17.00	$48,200 to $80,400
Coverall North America Inc.	San Diego, Calif.	Commercial cleaning	829	2,755	30.09	$350 to $3,000
McDonald's Corp.	Oak Brook, Ill	Fast food	613	9,000	6.81	$440,000 to $550,00
Burger King Corp.	Miami, Fla	Fast food	168	5,797	2.90	$73,000 to $511,000
D&K Enterprises Inc.	Carrollton, Texas	Personalized books	143	1,600	8.94	$6,000
O.P.E.N. Cleaning Systems	Phoenix, Ariz	Janitorial/commercial cleaning	139	598	23.24	$3,800 and up
KFC Corp.	Louisville, KY	Fast food	139	5,012	2.77	$150,000
Mister Softee Inc.	Runnemede, N.J.	Mobile soft ice cream units	87	729	11.93	$20,000 to $30,000
The Southland Corp. (7-Eleven)	Dallas, Texas	Convenience stores	80	6,302	1.27	$54,000
Wendy's International Inc.	Dublin, Ohio	Fast food	62	4,026	1.54	$700,000 to $1,400,0
Mail Boxes Etc.	San Diego, Calif	Postal, business & communication services	44	1,904	2.31	$50,730 to $88,330
RACS International	Indianapolis, Ind.	Commercial cleaning	35	100	35.00	$2,500 to $31,000
Merry Maids	Omaha, Neb.	Residential cleaning	28	657	4.26	$30,000 to $35,000
Goodyear Tire & Rubber Co.	Akron, Ohio	Retail tires, automotive services	26	1,448	1.80	$75,000 and up
Blimpie International Inc.	New York, N.Y.	Fast food	25	570	4.39	$75,000 to $100,000
Travel Network Ltd.	Englewood Cliffs, N.J.	Travel agencies	23	285	8.07	$85,000 to $100,000
Blockbuster Video	Ft. lauderdale, Fla	Home video sales and rentals	23	3,258	0.71	$285,000 to $775,00
Wash on Wheels	Sanford, Fla.	Mobile cleaning services	22	261	8.43	$3,500 to $7,500
Hardee's Food Systems Inc.	Rocky Mount, N.C.	Fast food	22	4,016	0.55	$50,000
Popeye's Famous Fried Chicken & Biscuits	Atlanta, Ga.	Fast food	21	811	2.59	$300,000 to $500,00
Jackson Hewitt Inc.	Virginia Beach, Va.	Tax services	20	682	2.93	$13,000
Church's Fried Chicken	Atlanta, Ga.	Fast food	20	1,083	1.85	$300,000 to $700,00
Checkers Drive-In Restaurants Inc.	Clearwater, Fla.	Fast food	18	290	6.21	$425,000 to $525,00
Meineke Discount Muffler Shops Inc.	Charlotte, N.C.	Automotive aftermarket services	17	983	1.73	$127,500
Shoney's Inc.	Nashville, Tenn.	Restaurants	16	1,078	1.48	$23,000 to $32,500

Source: "B.E. Franchise." *Black Enterprise* 24 (September 1993), p. 53. Published by permission. Primary source: B.E. Research.

★ 121 ★

Types of Business

Industrial Service Companies: Top 50 in Sales: 1992

Company	Location	Chief executive	Year started	Staff	Type of business	Sales[1]
TLC Beatrice International Holdings Inc.	New York, New York	Jean S. Fugett Jr.	1987	5,000	International food processor and distributor	1,665.000
Johnson Publishing Co. Inc.	Chicago, Illinois	John H. Johnson	1942	2,785	Publishing; broadcasting; TV production; cosmetics; hair care	274.197
Philadelphia Coca-Cola Bottling Co. Inc.	Philadelphia, Pennsylvania	J. Bruce Llewellyn	1985	1,000	Soft-drink bottler	266.000
H.J. Russell & Co.	Atlanta, Georgia	Herman J. Russell	1952	825	Construction & development; food services	145.610
The Anderson-Bubose Co.	Solon, Ohio	Warren Anderson	1991	80	Food distributor	110.000
RMS Technologies Inc.	Marlton, New Jersey	David W. Huggins	1977	1,176	Computer & technical services	103.300
Gold Line Refining Ltd.	Houston, Texas	Earl Thomas	1990	51	Oil refinery	91.880
Soft Sheen Products Inc.	Chicago, Illinois	Edward G. Gardner	1964	547	Hair care products manufacturer	91.700
Garden State Cable TV	Cherry Hill, New Jersey	J. Bruce Llewellyn	1989	300	Cable TV operator	91.000
Threads 4 Life Corp. d/b/a/ Cross Colours	Los Angeles, California	Carl Jones	1990	250	Apparel manufacturer	89.000
Barden Communications Inc.	Detroit, Michigan	Don H. Barden	1981	328	Communications & real estate development	78.600
The Bing Group	Detroit, Michigan	David Bing	1980	210	Steel processing & metal stamping distribution	77.634
Burrell Communications Group	Chicago, Illinois	Thomas J. Burrell	1971	115	Advertising; public relations; consumer promotions	77.007
Uniworld Group Inc.	New York, New York	Byron E. Lewis	1969	85	Advertising	72.419
Pulsar Systems Inc.	New Castle, Delaware	William W. Davis Sr.	1982	65	Systems integration; office automation; computer resaler	67.000
Stop Shop and Save	Baltimore, Maryland	Henry T. Baines Sr.	1978	600	Supermarkets	66.000
Black Entertainment Television Holdings	Washington, D.C.	Robert Johnson	1980	328	Cable TV network & magazine publishing	61.655
Mays Chemical Co. Inc.	Indianapolis, Indiana	William G. Mays	1980	75	Industrial chemical distributors	60.800
Essence Communications Inc.	New York, New York	Edward lewis	1969	87	Magazine publishing; TV production; direct-mail catalog	56.345
Community Foods Inc.	Baltimore, Maryland	Oscar A. Smith Jr.	1970	400	Supermarkets	47.500
Technology Applications Inc.	Alexandria, Virginia	James I. Chatman	1977	525	Systems integration & software engineering	46.500
Surface Protection Industries Inc.	Los Angeles, California	Robert C. Davidson Jr.	1978	200	Paint & specialty coatings manufacturer	46.200
Johnson Products Co. Inc.	Chicago, Illinois	Joan B. Johnson	1954	215	Hair & personal care products manufacturer	46.000
Luster Products Co.	Chicago, Illinois	Jory Luster	1957	315	Hair care products manufacturer & distributor	46.000
The Maxima Corp.	Lanham, Maryland	Joshua I. Smith	1978	752	Systems engineering & computer facilities management	45.098
Wesley Industries Inc.	Flint, Michigan	Delbert W. Mullens	1983	395	Makers of industrial coatings & foundry products	45.000
Pepsi-Cola of Washington, DC, L.P.	Washington, D.C.	Earl G. Graves	1990	138	Soft-drink distributor	43.869
Integrated Systems Analysts Inc.	Arlington, Virginia	C. Michael Gooden	1980	595	Systems engineering; computer systems services	43.600
Granite Broadcasting Corp.	New York, New York	W. Don Cornwell	1988	364	Network TV affiliates	43.108
The Mingo Group	New York, New York	Samuel J. Chisholm	1977	40	Advetising & public relations	42.733
Crest Computer Supply	Skokie, Illinois	Gale Sayers	1984	60	Computer hardware & software supplier	42.000
Beauchamp Distributing Co.	Compton, California	Patrick L. Beauchamp	1971	100	Beverage distributor	40.200
Rush Communications	New York, New York	Russell Simmons	1990	65	Music publishing; TV, film, radio production	40.000
Grimes Oil Co. Inc.	Boston, Massachusetts	Calvin M. Grimes	1940	18	Petroluem products distributor	38.700
Westside Distributors	South Gate, California	Edison R. Lara Sr.	1974	115	Beer & snack foods distributor	37.131
Pro-Line Corp.	Dallas, Texas	Comer J. Cottrell Jr.	1970	236	Hair care products manufacturer & distributor	36.874
Thacker Engineering Inc.	Atlanta, Georgia	Floyd Thacker	1970	140	Construction; construction management; engineering	36.500
Calhoun Enterprises	Montgomery, Alabama	Greg Calhoun	1984	578	Supermarket	36.479
The Gourmet Cos.	Atlanta, Georgia	Nathaniel R. Goldston III	1975	813	Food services; golf facilities management	36.200
Drew Pearson Cos.	Addison, Texas	Drew Pearson	1985	85	Sports licensing & sportswear manufacturing	36.000
Capsonic Group Division of Gabriel Inc.	Elgin, Illinois	Jim Liautaud	1968	232	Maker of electrical components	36.000
Trumark Inc.	Lansing, Michigan	Carlton L. Guthrie	1985	300	Metal stampings; manufacturing; welding	35.300
Network Solutions Inc.	Herndon, Virginia	Emmit J. McHenry	1979	380	Systems integration	35.000
Am-Pro Protective Agency Inc.	Columbia, South Carolina	John E. Brown	1982	1,082	Security guard services	32.127
Metters Industries Inc.	McLean, Virginia	Samuel Metters	1981	494	Systems engineering; telecommunications	31.597
Input Output Computer Services Inc.	Waltham, Massachusetts	Thomas A. Farrington	1969	200	Computer software & systems integrations	31.000
Advantage Enterprises Inc.	Toledo, Ohio	Levi Cook Jr.	1980	250	Project integrator for health care & construction	30.134
Automated Sciences Group Inc.	Silver Spring, Maryland	Arthur Holmes Jr.	1974	300	Maker of information & sensor technologies	30.000
Dudley Products Inc.	Greensboro, North Carolina	Joe Louis Dudley Sr.	1968	501	Beauty products manufacturer	30.000
Brooks Sausage Co. Inc.	Kenosha, Wisconsin	Frank B. Brooks	1985	148	Sausage manufacturer	29.000

Source: "B.E. Industrial/Service Companies," *Black Enterprise* 24 (June 1993), pp. 91-92, 94. Published by permission. Primary source: B.E. Research. *Notes:* 1. In millions of dollars, to the nearest thousand. As of Dec. 31, 1992. Prepared by B.E. Research. Reviewed by Mitchell/Titus & Co.

★ 122 ★

Types of Business

Industrial Service Companies: 50 Companies Ranked in Sales Below the 50 Leaders: 1992

Company	Location	Chief executive	Year started	Staff	Type of business	Sales[1]
Inner City Broadcasting Corp.	New York, New York	Pierre Sutton	1972	200	Radio, TV, cable TV franchise	28.000
Yancy Minerals	Woodbridge, Connecticut	Earl Yancy	1977	8	Industrial metals, minerals & coal distributors	28.000
Cimarron Express, Inc.	Genoa, Ohio	Glenn G. Grady	1984	85	Interstate trucking	27.773
Queen City Broadcasting Inc.	Buffalo, New York	J. Bruce Llewellyn	1985	130	Network TV affiliates	26.350
Premium Distributors Inc. of Washington, D.C.	Washington, D.C.	Henry Neloms	1984	75	Beer distributor	26.000
Integrated Steel Inc.	Detroit, Michigan	Geralda L. Dodd	1990	305	Automotive stamping & steel services	26.000
African Development Public Investment Corp.	Hollywood, California	Dick Griffey	1985	12	African commodities & air charter service	25.500
Restoration Supermarket Corp.	Brooklyn, New York	Roderick B. Mitchell	1977	178	Supermarket & drugstore	25.457
NAVACOM Systems Inc.	Manassas, Virginia	Elijah "Zeke" Jackson	1986	139	Electronic engineering system design & integration	25.000
Lockhart & Pettus	New York, New York	Keith E. Lockhart	1977	32	Advertising agency	24.893
Parks Sausage Co.	Baltimore, Maryland	Raymond V. Haysbert Sr.	1951	230	Sausage manufacturer	24.800
Dick Griffey Productions	Hollywood, California	Dick Griffey	1975	78	Entertainment	24.200
R.O.W. Sciences Inc.	Rockville, Maryland	Ralph Williams	1983	365	Biomedical & health services; research	24.000
American Development Corp.	N. Charleston, South Carolina	W. Melvin Brown Jr.	1972	175	Manufacturing & sheet metal fabrication	23.000
Sylvest Management Systems Corp.	Lanham, Maryland	Gary S. Murray	1987	42	Computer systems & engineering	22.600
Regal Plastics Co. Inc.	Roseville, Michigan	William F. Pickard	1985	222	Custom plastic injection molding	21.711
Simmons Enterprises Inc.	Cincinnati, Ohio	Carvel Simmons Inez Simmons	1970	52	Trucking; farm operations; day care centers	21.475
Earl G. Graves Ltd.	New York, New York	Earl G. Graves	1970	65	Magazine publishing	21.418
H.F. Henderson Industries Inc.	West Caldwell, New Jersey	Henry F. Henderson Jr.	1954	150	Industrial process controls & defense electronics	20.662
Stephens Engineering Co. Inc.	Lanham, Maryland	Wallace O. Stephens	1979	140	System irrigation, facility & computer maintenance	20.500
D-Orum Hair Products	Gary, Indiana	Ernest Daurham Jr.	1979	150	Minority hair products manufacturer	20.000
Bronner Brothers	Atlanta, Georgia	Nathaniel Bronner Sr.	1947	250	Hair care products manufacturer	19.500
Dual Inc.	Arlington, Virginia	J. Fred Dual Jr.	1983	241	Engineering & technical services	19.306
C.H. James & Co.	Charleston, West Virginia	Charles H. James III	1883	22	Wholesale food distribution	18.702
Consolidated Beverage Corp.	New York, New York	Albert N. Thompson	1978	24	Beverage exporter & importer	18.500
Watiker & Son Inc.	Zanesville, Ohio	Al Watiker Jr.	1973	200	Heavy highway, bridges, mine reclamation	18.000
Terry Manufacturing Co. Inc.	Roanoke, Alabama	Roy Terry	1963	300	Apparel manufacturer	17.500
J.E. Ethridge Construction Inc.	Fresno, California	John E. Ethridge	1971	25	Commercial construction	17.300
Burns Enterprises	Louisville, Kentucky	Tommie Burns Jr.	1969	460	Janitorial services & supermarkets	17.000
Ozanne Construction Co. Inc.	Cleveland, Ohio	Leroy Ozanne	1956	130	General construction & construction management	17.000
UBM Inc.	Chicago, Illinois	Sandra Dixon Jiles	1975	53	General contracting & construction management	16.674
AMSCO Wholesalers Inc.	Norcross, Georgia	Thurmond B. Woodard	1990	86	Wholesale distributor to apartment industry	16.200
Systems Engineering & Management Associates Inc.	Alexandria, Virginia	James C. Smith	1985	260	ADP technical support services	16.000
Mid-Delta Home Health Inc.	Belzoni, Mississippi	Clara Taylor Reed	1978	345	Home health care medical equipment & supplies	15.000
Urban Constructors Inc.	Miami, Florida	Jacque E. Thermilus	1988	60	General contracting & construction management	15.000
American Urban Radio Networks	New York, New York	Sydney Small	1973	65	Radio network; radio station; telemarketing	15.000
Specialized Packaging International Inc.	Hamden, Connecticut	Carlton L. Highsmith	1983	7	Packaging design; engineering; brokerage	14.860
A Minority Entity Inc.	Norco, Louisiana	Burnell K. Moliere	1978	1,200	Janitorial & food services	14.753
Tresp Associates Inc.	Alexandria, Virginia	Lillian B. Handy	1981	220	Military logistics; systems engineering; computers	14.000
Solo Construction Corp.	N. Miami Beach, Florida	Randy Pierson	1978	46	General engineering construction	13.959
RPM Supply Co. Inc.	Philadelphia, Pennsylvania	Robert P. Mapp	1977	20	Electrical & electric components distributor	13.891
Powers & Sons Construction Co. Inc.	Gary, Indiana	Mamon Powers Sr.	1967	60	Construction	13.721
Williams-Russell and Johnson Inc.	Atlanta, Georgia	Pelham Williams	1976	125	Engineering, architecture & construction management	13.600
Black River Mfg. Inc.	Port Huron, Michigan	Isaac Lang. Jr.	1977	77	Auto parts manufacturer	13.400
Eltrex Industries	Rochester, New York	Matthew Augustine	1968	155	Office furniture manufacturer fulfillment services	12.976
Advanced Systems Technology Inc.	Atlanta, Georgia	Wayne H. Knox	1981	200	Nuclear, environmental & corrosion technology	12.700
Advanced Consumer Marketing Corp.	Burlingame, California	Harry W. Brooks Jr.	1984	35	Information systems & telecommunications	12.380
Spiral Distribution Inc.	Phoenix, Arizona	Reggie Fowler	1987	24	Packaging suppliers to the grocery industry	12.300
Wise Construction Co. Inc.	Dayton, Ohio	Warren C. Wise	1983	75	General construction	12.000
Systems Management American Corp.	Norfolk, Virginia	Herman Valentine	1970	130	Computer systems integration	12.000

Source: "B.E. Industrial/Service Companies," *Black Enterprise* 24 (June 1993), pp. 94, 96, 98. Published by permission. Primary source: B.E. Research. *Notes:* 1. In millions of dollars, to the nearest thousand. As of Dec. 31, 1992. Prepared by B.E. Research. Reviewed by Mitchell/Titus & Co.

Chapter 4
CRIME, LAW ENFORCEMENT, AND LEGAL JUSTICE

Crime

★ 123 ★

Arrests of Persons 18 and Older, Black and White, 1991

By offense charged, age group, and race, United States, 1991.

Offense charged	Total arrests			Percent[1]		
	Total	White	Black	Total	White	Black
Total	8,807,080	6,031,024	2,604,958	100.0%	68.5%	29.6%
Murder and nonnegligent manslaughter	15,560	6,824	8,471	100.0	43.9	54.4
Forcible rape	25,066	13,707	10,920	100.0	54.7	43.6
Robbery	101,373	38,090	62,123	100.0	7.6	61.3
Aggravated assault	312,337	189,025	117,997	100.0	60.5	37.8
Burglary	215,222	139,926	72,118	100.0	65.0	33.5
Larceny-theft	827,186	527,153	281,775	100.0	63.7	34.1
Motor vehicle theft	90,020	52,545	35,989	100.0	58.4	40.0
Arson	7,843	5,598	2,111	100.0	71.4	26.9
Violent crime[2]	454,336	247,646	199,511	100.0	54.5	43.9
Property crime[3]	1,140,271	725,222	391,993	100.0	63.6	34.4
Total Crime Index[4]	1,594,607	972,868	591,504	100.0	61.0	37.1
Other assaults	652,408	423,598	215,443	100.0	64.9	33.0
Forgery and counterfeiting	68,686	43,486	24,257	100.0	63.3	35.3
Fraud	280,611	191,341	86,888	100.0	68.2	31.0
Embezzlement	9,781	6,638	2,998	100.0	67.9	30.7
Stolen property: buying, receiving, possessing	94,622	53,483	40,097	100.0	56.5	42.4
Vandalism	142,654	102,434	37,517	100.0	71.8	26.3
Weapons: carrying, possessing, etc.	137,010	75,997	59,015	100.0	55.5	43.1
Prostitution and commercialized vice	77,761	46,860	29,608	100.0	60.3	38.1
Sex offenses (except forcible rape and prostitution)	66,652	52,907	12,378	100.0	79.4	18.6
Drug abuse violations	705,083	415,168	283,825	100.0	58.9	40.3
Gambling	11,648	5,417	5,209	100.0	46.5	44.7

[Continued]

★ 123 ★

Arrests of Persons 18 and Older, Black and White, 1991
[Continued]

Offense charged	Total arrests			Percent[1]		
	Total	White	Black	Total	White	Black
Offenses against family and children	68,079	45,202	20,293	100.0	66.4	29.8
Driving under the influence	1,257,412	1,117,463	115,129	100.0	88.9	9.2
Liquor laws	345,615	296,659	38,608	100.0	85.8	11.2
Drunkenness	609,605	493,706	100,958	100.0	81.0	16.6
Disorderly conduct	462,260	300,758	152,475	100.0	65.1	33.0
Vagrancy	28,515	14,005	13,854	100.0	49.1	48.6
All other offenses (except traffic)	2,185,778	1,370,344	769,330	100.0	62.7	35.2
Suspicion	8,293	2,690	5,572	100.0	32.4	67.2
Curfew and loitering law violations	X	X	X	X	X	X
Runaways	X	X	X	X	X	X

Source: "Arrests," Department of Justice, *Sourcebook of Criminal Justice Statistics-1992*, pp. 434-36. Primary source: U.S. Department of Justice, Federal Bureau of Investigation, *Crime in the United States, 1991. Notes:* Estimates by the U.S. Bureau of the Census indicate that on July 1, 1991, whites comprised 83.6 percent, blacks 12.4 percent, and other racial categories 4.0 percent of the total U.S. resident population. For data on Native Americans and Asian or Pacific Islanders, see source table. 1. Because of rounding, percents may not add to total. 2. Violent crimes are offenses of murder, forcible rape, robbery, and aggravated assault. 3. Property crimes are offenses of burglary, larceny-theft, motor vehicle theft, and arson. 4. Includes arson.

★ 124 ★

Crime

Arrests of Persons Under 18, Black and White, 1991

By offense charged, age group, and race, United States, 1991.

Offense charged	Total arrests			Percent[1]		
	Total	White	Black	Total	White	Black
Total	1,709,319	1,220,838	444,341	100.0%	71.4%	26.0%
Murder and nonnegligent manslaughter	2,536	1,037	1,453	100.0	40.9	57.3
Forcible rape	4,701	2,599	2,040	100.0	55.3	43.4
Robbery	34,803	13,127	21,023	100.0	37.7	60.4
Aggravated assault	51,913	29,603	21,410	100.0	57.0	41.2
Burglary	108,448	82,891	22,570	100.0	76.4	20.8
Larceny-theft	362,851	265,742	86,278	100.0	73.2	23.8
Motor vehicle theft	70,083	41,183	26,929	100.0	58.8	38.4
Arson	6,895	5,711	1,053	100.0	82.8	15.3
Violent crime[2]	93,953	46,366	45,926	100.0	49.4	48.9
Property crime[3]	548,277	395,527	136,830	100.0	72.1	25.0
Total Crime Index[4]	642,230	441,893	182,756	100.0	68.8	28.5
Other assaults	119,608	74,899	41,678	100.0	62.6	34.8
Forgery and counterfeiting	6,183	5,049	1,007	100.0	81.7	16.3
Fraud	10,917	6,302	4,342	100.0	57.7	39.8
Embezzlement	784	564	197	100.0	71.9	25.1

[Continued]

★ 124 ★

Arrests of Persons Under 18, Black and White, 1991

[Continued]

Offense charged	Total arrests			Percent[1]		
	Total	White	Black	Total	White	Black
Stolen property: buying, receiving, possessing	34,987	20,425	13,914	100.0	58.4	39.8
Vandalism	106,598	87,040	17,497	100.0	81.7	16.4
Weapons: carrying, possessing, etc.	36,480	22,612	13,122	100.0	62.0	36.0
Prostitution and commercialized vice	1,018	657	335	100.0	64.5	32.9
Sex offenses (except forcible rape and prostitution)	14,186	10,278	3,607	100.0	72.5	25.4
Drug abuse violations	58,257	28,428	29,172	100.0	48.8	50.1
Gambling	816	164	634	100.0	20.1	77.7
Offenses against family and children	2,866	2,102	649	100.0	73.3	22.6
Driving under the influence	13,301	12,412	595	100.0	93.3	4.5
Liquor laws	103,265	95,332	4,968	100.0	92.3	4.8
Drunkenness	15,522	13,865	1,349	100.0	89.3	8.7
Disorderly conduct	96,244	65,007	29,939	100.0	67.5	31.1
Vagrancy	2,240	1,730	487	100.0	77.2	21.7
All other offenses (except traffic)	242,262	172,546	62,527	100.0	71.2	25.8
Suspicion	1,891	1,460	392	100.0	77.2	20.7
Curfew and loitering law violations	72,037	55,389	14,819	100.0	76.9	20.6
Runaways	127,627	102,683	20,355	100.0	80.5	15.9

Source: "Arrests," Department of Justice, *Sourcebook of Criminal Justice Statistics-1992*, pp. 434-36. Primary source: U.S. Department of Justice, Federal Bureau of Investigation, *Crime in the United States, 1991. Notes:* Estimates by the U.S. Bureau of the Census indicate that on July 1, 1991, whites comprised 83.6 percent, blacks 12.4 percent, and other racial categories 4.0 percent of the total U.S. resident population. For data on Native Americans and Asian or Pacific Islanders, see source table. 1. Because of rounding, percents may not add to total. 2. Violent crimes are offenses of murder, forcible rape, robbery, and aggravated assault. 3. Property crimes are offenses of burglary, larceny-theft, motor vehicle theft, and arson. 4. Includes arson.

★ 125 ★

Crime

Arrests, Black and White, 1991

By offense charged, age group, and race, United States, 1991. (10,075 agencies; 1991 estimated population 186,621,000).

Offense charged	Total arrests			Percent[1]		
	Total	White	Black	Total	White	Black
Total	10,516,399	7,251,862	3,049,299	100.0%	69.0%	29.0%
Murder and nonnegligent manslaughter	18,096	7,861	9,924	100.0	43.4	54.8
Forcible rape	29,767	16,306	12,960	100.0	54.8	43.5
Robbery	136,176	51,217	83,146	100.0	37.6	61.1
Aggravated assault	364,250	218,628	139,407	100.0	60.0	38.3
Burglary	323,670	222,817	94,688	100.0	68.8	29.3
Larceny-theft	1,190,037	792,895	368,053	100.0	66.6	30.9
Motor vehicle theft	160,103	93,728	62,918	100.0	58.5	39.3
Arson	14,738	11,309	3,164	100.0	76.7	21.5

[Continued]

★ 125 ★

Arrests, Black and White, 1991
[Continued]

Offense charged	Total arrests			Percent[1]		
	Total	White	Black	Total	White	Black
Violent crime[2]	548,289	294,012	245,437	100.0	53.6	44.8
Property crime[3]	1,688,548	1,120,749	528,823	100.0	66.4	31.3
Total Crime Index[4]	2,236,837	1,414,761	774,260	100.0	63.2	34.6
Other assaults	772,016	498,497	257,121	100.0	64.6	33.3
Forgery and counterfeiting	74,869	48,535	25,264	100.0	64.8	33.7
Fraud	291,528	197,643	91,230	100.0	67.8	31.3
Embezzlement	10,565	7,202	3,195	100.0	68.2	30.2
Stolen property: buying, receiving, possessing	129,609	73,908	54,011	100.0	57.0	41.7
Vandalism	249,252	189,474	55,014	100.0	76.0	22.1
Weapons: carrying, possessing, etc.	173,490	98,609	72,137	100.0	56.8	41.6
Prostitution and commercialized vice	78,799	47,517	29,943	100.0	60.3	38.0
Sex offenses (except forcible rape and prostitution)	80,838	63,185	15,985	100.0	78.2	19.8
Drug abuse violations	763,340	443,596	312,997	100.0	58.1	41.0
Gambling	12,464	5,581	5,843	100.0	44.8	46.9
Offenses against family and children	70,945	47,304	20,942	100.0	66.7	29.5
Driving under the influence	1,270,713	1,129,876	115,724	100.0	88.9	9.1
Liquor laws	448,880	391,991	43,576	100.0	87.3	9.7
Drunkenness	625,127	507,571	102,307	100.0	81.2	16.4
Disorderly conduct	558,504	365,765	182,414	100.0	65.5	32.7
Vagrancy	30,755	15,735	14,341	100.0	51.2	46.6
All other offenses (except traffic)	2,428,040	1,542,890	831,857	100.0	63.5	34.3
Suspicion	10,184	4,150	5,964	100.0	40.8	58.6
Curfew and loitering law violations	72,037	55,389	14,819	100.0	76.9	20.6
Runaways	127,627	102,683	20,355	100.0	80.5	15.9

Source: "Arrests," Department of Justice, *Sourcebook of Criminal Justice Statistics-1992*, pp. 434-36. Primary source: U.S. Department of Justice, Federal Bureau of Investigation, *Crime in the United States, 1991. Notes:* Estimates by the U.S. Bureau of the Census indicate that on July 1, 1991, whites comprised 83.6 percent, blacks 12.4 percent, and other racial categories 4.0 percent of the total U.S. resident population. For data on Native Americans and Asian or Pacific Islanders, see source table. 1. Because of rounding, percents may not add to total. 2. Violent crimes are offenses of murder, forcible rape, robbery, and aggravated assault. 3. Property crimes are offenses of burglary, larceny-theft, motor vehicle theft, and arson. 4. Includes arson.

★ 126 ★
Crime

Characteristics of Juvenile Murder and Nonnegligent Manslaughter Victims and Offenders, Aggregate of 1976 to 1991

By selected characteristics of the victim and offense, United States, 1976-91 (aggregate)[1].

Characteristics of victim and offense	Total	Characteristics of offender							
		Sex		Race		Male		Female	
		Male	Female	White	Black	White	Black	White	Black
Age									
17 years and younger	28.7%	28.1%	35.6%	31.3%	26.3%	31.0%	25.4%	34.6%	36.1%
18 to 29 years	36.3	37.1	28.1	33.3	38.9	34.2	39.5	24.0	31.9
30 to 49 years	22.1	22.0	23.2	22.3	22.0	21.9	22.2	26.3	20.5
50 years and older	12.9	12.8	13.2	13.1	12.8	12.9	12.9	15.1	11.6
Race									
White	56.1	56.8	49.1	92.3	22.6	92.1	24.0	94.4	8.7
Black	41.7	41.1	48.9	6.5	76.3	6.7	74.9	4.6	90.7
Other	2.1	2.1	2.0	1.2	1.1	1.2	1.1	1.1	0.6
Sex									
Male	83.6	84.9	70.0	81.8	85.3	83.0	86.8	69.8	70.5
Female	16.4	15.1	30.0	18.2	14.7	17.0	13.2	30.2	29.5
Type of weapon									
Firearm	64.7	66.8	42.4	60.7	68.8	61.6	72.0	50.4	35.7
Knife	19.3	18.1	32.1	22.2	16.4	22.4	13.9	20.7	42.8
Blunt object	6.5	6.6	5.2	7.2	5.9	7.3	6.1	6.0	4.2
Personal weapon	6.2	5.8	10.8	5.9	6.3	5.3	6.1	12.3	8.9
Other	3.3	2.7	9.5	4.0	2.5	3.4	2.0	10.6	8.5
Relationship to offender									
Family	15.5	12.8	41.2	20.7	10.5	18.1	7.7	46.8	36.2
Other known	52.6	53.3	45.7	51.9	53.3	53.4	53.4	37.6	53.0
Stranger	31.9	33.9	13.2	27.4	36.2	28.6	38.9	15.6	10.8

Source: "Characteristics of Juvenile Murder and Nonnegligent Manslaughter Offenders," Department of Justice, Bureau of Criminal Statistics, *Sourcebook of Criminal Justice Statistics, 1992*, p. 393. Primary source: James Alan Fox, "Teenage Males are Committing Murder at an Increasing Rate." National Crime Analysis Program, Northwestern University (Mimeographed). *Note:* 1. Percents may not add to 100 because of rounding.

★ 127 ★

Crime

Circumstances Surrounding Murders in the 75 Largest Counties, 1988

Circumstances	Percent of victims		
	Total	Race	
		White	Black
Criminal activity	22%	20%	24%
Drugs	11	8	13
Other than drugs	12	12	12
Felony-murder	16	20	12
Robbery	12	15	9
Sexual assault	2	2	1
Burglary	1	1	1
Arson	1	1	1
Personal conflict	44	42	45
Property dispute	18	14	22
Love/sex dispute	19	19	20
Domestic issues	17	14	19
Redress of insult	10	10	10
On-going feud	3	3	4
Dispute at the scene	6	7	5
Other activity	16	16	15
Act of retaliation	5	4	5
Child abuse	3	3	3
Premeditated violence	4	4	3
Circumstances not known	5	5	3

Source: "Circumstances Surrounding Murders in the 75 Largest Counties," *Sourcebook of Criminal Justice Statistics-1992*, p. 387. Primary source: U.S. Department of Justice, Bureau of Justice Statistics, *Murder in Large Urban Counties, 1988*, Special Report NCJ-140614. *Notes:* The number in a cell equals the percent of victims of that race or sex who were murdered under the specified circumstances. In some cases more than one type of circumstance was identified; hence, an individual may be counted in more than one cell of the table. Percents may sum to more than 100 in some columns of the table. Victims counted in detail categories are also counted in the summary category.

★ 128 ★

Crime

High School Seniors Reporting Traffic Ticket or Warning for Moving Violations, 1990 to 1992

Number of tickets/warnings	Class of 1990		Class of 1991		Class of 1992	
	White (N=11,410)	Black (N=1,614)	White (N=10,754)	Black (N=1,757)	White (N=11,029)	Black (N=2,244)
Question: "Within the last 12 months, how many times, if any, have you received a ticket (or been stopped and warned) for moving violations such as speeding, running a stop light, or improper passing?"						
None	64.3%	82.9%	65.4%	81.8%	65.4%	82.9%
One	21.5	11.2	21.0	11.1	21.2	10.3
Two	8.5	3.8	7.4	4.2	7.5	4.8
Three	3.3	0.9	3.5	1.8	3.3	1.4
Four or more	2.5	1.2	2.6	1.1	2.6	0.6

Source: "High School Seniors Reporting Receiving Traffic Ticket or Warning in Last 12 Months," *Sourcebook of Criminal Justice Statistics-1992*, pp. 320-321. Primary source: Data provided by Monitoring the Future Project, Survey Research Center, University of Michigan, Lloyd D. Johnson and Jerald G. Bachman, principal investigators. Table adapted by *Sourcebook* staff. Published by permission.

★ 129 ★

Crime

Homicides Per 100,000 Persons by Firearms and Nonfirearms, 1987-90

Characteristics	Firearm homicides				Nonfirearm homicides			
	1987	1988	1989	1990	1987	1988	1989	1990
10 to 14 years of age								
Total	1.1	1.1	1.4	1.5	0.6	0.6	0.6	0.6
White male	0.8	0.9	1.2	1.3	0.2	0.4	0.3	0.3
White female	0.4	0.4	0.4	0.4	0.5	0.4	0.6	0.5
Black male	5.3	4.7	6.8	6.9	1.7	1.3	0.8	1.2
Black female	1.1	2.6	1.8	3.1	1.4	1.9	2.0	1.6
15 to 19 years of age								
Total	7.0	9.0	11.1	14.0	2.8	2.6	2.4	3.1
White male	5.2	6.0	7.5	9.7	2.1	1.9	1.9	2.9
White female	1.2	1.3	1.7	2.0	1.8	1.7	1.5	1.6
Black male	50.1	69.2	85.5	105.3	10.6	9.1	8.4	10.5
Black female	7.3	7.2	8.7	10.4	4.8	4.5	3.1	5.1
20 to 24 years of age								
Total	12.4	13.2	14.5	17.1	5.0	5.3	5.0	5.4
White male	10.2	10.1	11.1	12.9	4.2	4.2	4.0	5.3
White female	2.3	2.3	2.2	2.3	2.3	2.4	2.2	2.2
Black male	90.4	102.5	113.7	140.7	20.3	23.1	21.7	21.6
Black female	12.1	11.8	13.1	12.4	11.2	11.4	9.7	9.7

[Continued]

★ 129 ★

Homicides Per 100,000 Persons by Firearms and Nonfirearms, 1987-90
[Continued]

Characteristics	Firearm homicides				Nonfirearm homicides			
	1987	1988	1989	1990	1987	1988	1989	1990
25 to 34 years of age								
Total	10.0	11.0	11.2	12.2	5.2	5.3	5.3	5.5
White male	9.3	9.5	9.8	10.8	4.1	3.9	4.2	4.3
White female	2.4	2.4	2.3	2.4	2.2	2.1	2.0	2.0
Black male	71.2	82.4	85.3	94.4	29.4	28.5	29.5	30.8
Black female	11.8	12.7	11.7	12.7	10.9	13.1	11.9	12.6

Source: "Homicide Rate (Per 100,000 Persons in Each Age Group) Due to Firearms and Nonfirearms," *Sourcebook of Criminal Justice Statistics-1992*, p. 385. Primary source: Lois A Fingerhut, "Firearm Mortality Among Children, Youth, and Young Adults 1-34 Year of Age," *Advance Data*, No. 231, U.S. Department of Health and Human Services, Centers for Disease Control and Prevention. *Notes:* These data are based on information from all death certificates filed in the 50 states and the District of Columbia. Mortality statistics are based on information coded by the National Center for Health Statistics (NCHS) of the U.S. Department of Health and Human Services, Centers for Disease Control and Prevention from copies of the original death certificates received from state registration offices and on state-coded data provided to NCHS through the Vital Statistics Cooperative Program. The mortality statistics were compiled in accordance with World Health Organization regulations. Homicides are deaths purposely inflicted by other persons and include deaths resulting from legal intervention. (Source, p. 17). Firearm death rates for 1985-89 are based on intercensal population estimates and death rates for 1990 are based on postcensal estimates of July 1, 1990. Population estimates were provided to the Source by the U.S. Bureau of the Census. totals include races not shown separately.

★ 130 ★

Crime

Homicides and Suicides Resulting From Firearms, 1990

By age, race, and sex of victim, United States, 1990.

Age	Total	White		Black	
		Male	Female	Male	Female
Percent of all homicides due to firearms					
10 to 14 years	72.5%	80.3%	45.2%	85.2%	66.1%
15 to 19 years	81.7	76.7	54.8	90.9	67.0
20 to 24 years	75.9	70.8	50.6	86.7	56.0
25 to 34 years	69.1	71.8	54.5	75.4	50.1
Percent of all suicides due to firearms					
10 to 14 years	55.0	53.7	56.1	71.4	62.5
15 to 19 years	67.3	69.4	57.3	76.4	65.4
20 to 24 years	63.4	65.2	54.2	69.2	51.4
25 to 34 years	57.6	61.1	48.9	55.9	38.0

Source: "Homicides and Suicides Resulting From Firearms," *Sourcebook of Criminal Justice Statistics-1992*, p. 385. Primary source: Lois A Fingerhut, "Firearm Mortality Among Children, Youth, and Young Adults 1-34 Years of Age," *Advance Data*, No. 231, U.S. Department of Health and Human Services, Centers for Disease Control and Prevention. Table adapted by *Sourcebook* staff.

★ 131 ★

Crime

Murders and Nonnegligent Manslaughters Known to Police by Race and Sex of Offender, 1991

Year	Total victims/ offenders	Characteristics of offender						
		Race				Sex		
		White	Black	Other	Unknown	Male	Female	Unknown
Total	10,924	4,838	5,778	204	104	9,490	1,330	104
Race								
White	5,194	4,399	691	60	44	4,627	523	44
Black	5,433	347	5,035	18	33	4,619	781	33
Other	239	72	36	125	6	211	22	6
Unknown	58	20	16	1	21	33	4	21
Sex								
Male	8,149	3,447	4,507	141	54	7,052	1,043	54
Female	2,717	1,371	1,255	62	29	2,405	283	29
Unknown	58	20	16	1	21	33	4	21

Source: "Murders and Nonnegligent Manslaughters Known to Police," Department of Justice, Bureau of Criminal Statistics, *Sourcebook of Criminal Justice Statistics, 1992,* p. 395. Primary source: U.S. Department of Justice, Federal Bureau of Investigation, *Crime in the United States, 1991. Notes:* These data pertain to the 10,924 murders and nonnegligent manslaughters that involved a single offender and a single victim.

★ 132 ★

Crime

Murders and Nonnegligent Manslaughters Known to Police, 1991

By sex, race, and age of victim, United States, 1991.

Age of victim	Total	Race of victim			
		White	Black	Other	Unknown
Total	21,505	10,135	10,660	531	179
Infant (under 1)	304	178	116	5	5
1 to 4	371	178	177	15	1
5 to 9	110	66	39	5	0
10 to 14	290	152	129	7	2
15 to 19	2,702	1,035	1,600	49	18
20 to 24	3,948	1,571	2,278	76	23
25 to 29	3,362	1,450	1,806	90	16
30 to 34	2,898	1,355	1,455	76	12
35 to 39	2,145	1,074	994	66	11
40 to 44	1,496	775	666	48	7

[Continued]

★ 132 ★

Murders and Nonnegligent Manslaughters Known to Police, 1991
[Continued]

Age of victim	Total	Race of victim			
		White	Black	Other	Unknown
45 to 49	981	574	374	26	7
50 to 54	658	399	226	27	6
55 to 59	459	289	155	11	4
60 to 64	421	255	154	12	0
65 to 69	321	206	110	4	1
70 to 74	241	148	87	5	1
75 and older	424	176	141	3	4
Unknown	374	154	153	6	61

Source: "Murders and Nonnegligent Manslaughters Known to Police," *Sourcebook of Criminal Justice Statistics-1992*, p. 391. Primary source: U.S. Department of Justice, Bureau of Justice Statistics, *Crime in the United States, 1991.* Table adapted by *Sourcebook* staff.

★ 133 ★

Crime

Percent Distribution of Murders and Nonnegligent Manslaughters Known to Police, 1972-1991

By race of victim, United States, 1972-91.

Year	Total number of murders and nonnegligent manslaughters	Total[1]	White	Black	All other (including race unknown)
1972	15,832	100%	45%	53%	2%
1973	17,123	100	47	52	1
1974	18,632	100	48	50	2
1975	18,642	100	51	47	2
1976	16,605	100	51	47	2
1977	18,033	100	52	45	2
1978	18,714	100	54	44	2
1979	20,591	100	54	43	2
1980	21,860	100	53	42	4
1981	20,053	100	54	44	2
1982	19,485	100	55	42	2
1983	18,673	100	55	42	3
1984	16,689	100	56	41	3
1985	17,545	100	56	42	3
1986	19,257	100	53	44	3
1987	17,859	100	52	45	3
1988	18,269	100	49	48	3
1989	18,954	100	48	49	3

[Continued]

★ 133 ★

Percent Distribution of Murders and Nonnegligent Manslaughters Known to Police, 1972-1991

[Continued]

Year	Total number of murders and nonnegligent manslaughters	Total[1]	White	Black	All other (including race unknown)
1990	20,045	100	48	49	3
1991	21,505	100	47	50	2

Source: "Percent Distribution of Murders and Nonnegligent Manslaughters Known to Police," *Sourcebook of Criminal Justice Statistics-1992,* p. 390. Primary source: U.S. Department of Justice, Bureau of Justice Statistics, *Crime in the United States, 1964 to 1991,* Table constructed by *Sourcebook* staff. *Note:* 1. Because of rounding, percents may not add to total.

★ 134 ★

Crime

Race of Offenders Committing Murder and Nonnegligent Manslaughter, 1976 to 1991

Year	Race		
	White	Black	Other[1]
1976	46.2%	52.0%	1.8%
1977	47.5	50.6	2.0
1978	48.0	50.3	1.6
1979	48.9	49.0	2.1
1980	49.7	49.0	1.3
1981	50.0	48.5	1.5
1982	51.2	47.0	1.8
1983	51.6	46.4	2.1
1984	53.5	44.5	2.0
1985	52.6	45.4	2.0
1986	50.2	47.6	2.2
1987	50.9	47.2	1.9
1988	48.6	49.6	1.9
1989	47.9	50.2	1.9
1990	48.1	50.2	1.6
1991	46.1	51.8	2.1

Source: "Characteristics of Murder and Nonnegligent Manslaughter Offenders Known to Police," Department of Justice, Bureau of Criminal Statistics, *Sourcebook of Criminal Justice Statistics, 1992,* p. 395. Primary source: James Alan Fox, National Crime Analysis Program, Northwestern University. *Notes:* These data include only those incidents for which age, sex, and race of offender were available. 1. Includes American Indians, Asians, Pacific Islanders, and all other races.

★ 135 ★

Crime

Rate Per 1,000 Persons of Whites and Blacks Committing Murder and Nonnegligent Manslaughter, 1976 to 1991

Rate (per 100,000 persons in each group) of offenders committing murder and nonnegligent manslaughter.

	Race	
	White	Black
1976	4.1	35.3
1977	3.9	34.1
1978	4.3	34.5
1979	4.6	36.9
1980	5.1	33.4
1981	4.4	34.2
1982	4.5	31.3
1983	4.4	28.6
1984	4.2	22.0
1985	4.1	25.5
1986	4.0	25.9
1987	3.6	23.1
1988	3.4	24.5
1989	3.5	25.8
1990	4.1	27.5
1991	3.6	31.2

Source: "Rate (Per 100,000 Persons in Each Group) of Offenders Committing Murder and Nonnegligent Manslaughter," Department of Justice, Bureau of Criminal Statistics, *Sourcebook of Criminal Justice Statistics, 1992*, p. 395. Primary source: James Alan Fox, National Crime Analysis Program, Northwestern University.

★ 136 ★

Crime

Rate Per 100,000 Persons in Each Group of Murder and Nonnegligent Homicide Committed by Juveniles, 1976 to 1991

	Age							
	10 to 13 years				14 to 17 years			
	Male		Female		Male		Female	
	White	Black	White	Black	White	Black	White	Black
1976	0.6	2.7	0.1	0.9	7.6	47.3	0.9	7.2
1977	0.8	2.5	0.1	0.5	7.8	44.1	0.9	4.3
1978	0.8	2.8	0.1	0.4	7.9	44.3	0.9	5.8
1979	0.7	2.8	0.1	0.9	9.5	47.7	0.9	5.9
1980	0.7	3.2	0.1	0.6	9.4	49.4	0.7	5.1
1981	0.8	1.9	0.1	0.4	8.2	51.2	0.9	5.8
1982	0.6	2.2	0.1	0.6	8.2	44.6	0.9	4.5
1983	0.6	1.8	0.1	0.4	7.9	37.0	1.1	5.3

[Continued]

★ 136 ★

Rate Per 100,000 Persons in Each Group of Murder and Nonnegligent Homicide Committed by Juveniles, 1976 to 1991

[Continued]

	Age							
	10 to 13 years				14 to 17 years			
	Male		Female		Male		Female	
	White	Black	White	Black	White	Black	White	Black
1984	0.7	1.5	0.1	0.5	7.0	32.0	0.9	4.4
1985	0.8	2.5	0.1	0.8	7.2	43.6	0.7	4.7
1986	0.8	2.0	0.1	0.2	9.3	49.8	0.9	4.3
1987	0.7	2.3	0.1	0.6	7.6	50.4	1.0	4.7
1988	0.7	2.8	0.1	0.7	9.3	65.8	0.7	4.7
1989	1.0	3.5	0.0	0.7	10.9	78.1	0.7	4.9
1990	0.8	2.5	0.1	0.3	13.2	102.5	1.0	5.1
1991	0.5	4.3	0.1	0.8	13.6	111.8	0.8	7.0

Source: "Rate (Per 100,000 Persons in Each Group) of Murder and Nonnegligent Manslaughter Committed by Juveniles," *Sourcebook of Criminal Justice Statistics-1992*, p. 392. Primary source: James Alan Fox, "Teenage Males Are Committing Murder at an Increasing Rate," National Crime Analysis Program, Northeastern University (Mimeographed). Published by permission.

★ 137 ★

Crime

Weapon or Method of Homicide in the 75 Largest Counties, 1988

Weapon or method	Percent of victims		
	Total	Race	
		White	Black
Total	100%	100%	100%
Guns			
Handgun	50	44	55
Shotgun	5	4	5
Rifle	4	4	4
Knife	21	21	22
Blunt instrument	5	8	4
Personal weapon	5	6	4
Strangulation	3	3	2
Vehicle	2	3	1

[Continued]

★ 137 ★

Weapon or Method of Homicide in the 75 Largest Counties, 1988
[Continued]

Weapon or method	Percent of victims		
	Total	Race	
		White	Black
Fire	2	1	2
Other	3	7	1

Source: "Type of Weapon or Method Used in Murders in the 75 Largest Counties," *Sourcebook of Criminal Justice Statistics-1992,* p. 383. Primary source: U.S. Department of Justice, Bureau of Justice Statistics, *Murder in Large Urban Counties, 1988,* Special Report NCJ-140614. *Notes:* These data were collected for the U.S. Department of Justice, Bureau of Justice Statistics by Abt Associates, Inc. through the Prosecution of Felony Arrests project. The data are based on a sample of 33 of the 75 most populous counties in the United States. A total of 2,539 murder cases were studied, which yielded data on 3,119 defendants and 2,655 victims. These cases were a sample of about half of all those with a murder charge brought to the prosecutors in 1988, or earlier, and that were disposed during 1988. During 1988 prosecutors and courts in the 75 largest counties disposed of murder cases involving an estimated (after statistical weighing) 9,576 defendants and 8,063 victims. These data are derived from a sample and therefore subject to sampling variation. "Other" includes asphyxiation, drowning, throwing from height, neglect, scalding, and use of a machine gun. 1. Less than 1 percent.

Juvenile Delinquency

★ 138 ★

High School Students Reporting Delinquent Activities, 1990 to 1992 – Part I

Delinquent activity	Class of 1990		Class of 1991		Class of 1992	
	White (N=1,907)	Black (N=277)	White (N=1,818)	Black (N=289)	White (N=1,806)	Black (N=368)
Question: "During the last 12 months, how often..."						
Argued or had a fight with either of your parents?						
Not at all	6.3%	21.7%	6.8%	22.4%	5.5%	23.9%
Once	6.9	14.4	7.7	8.4	7.5	11.1
Twice	12.0	13.7	11.9	15.0	11.1	12.3
3 or 4 times	24.8	21.8	26.1	24.4	24.3	24.0
5 or more times	50.0	28.4	47.6	29.9	51.5	28.7
Hit an instructor or supervisor?						
Not at all	97.7	95.9	97.3	95.9	97.2	96.4
Once	1.1	2.7	1.5	1.9	1.8	2.2
Twice	0.8	0.4	0.5	0.8	0.5	0.8
3 or 4 times	0.2	0.2	0.1	0.5	0.2	0.5
5 or more times	0.2	0.8	0.5	0.8	0.4	0.1
Gotten into a serious fight in school or at work?						
Not at all	80.4	82.2	83.1	76.8	82.1	80.6

[Continued]

★ 138 ★

High School Students Reporting Delinquent Activities, 1990 to 1992 – Part I

[Continued]

Delinquent activity	Class of 1990		Class of 1991		Class of 1992	
	White (N = 1,907)	Black (N = 277)	White (N = 1,818)	Black (N = 289)	White (N = 1,806)	Black (N = 368)
Once	11.8	12.4	9.7	13.6	10.8	12.7
Twice	5.0	2.2	4.0	5.3	4.3	2.7
3 or 4 times	2.0	1.3	1.7	2.1	1.7	1.9
5 or more times	0.9	1.9	1.6	2.2	1.2	2.1
Taken part in a fight where a group of your friends were against another group?						
Not at all	78.4	80.1	80.8	76.5	79.3	76.3
Once	12.0	8.8	11.3	9.7	11.6	12.8
Twice	4.8	3.8	4.6	6.6	4.1	4.4
3 or 4 times	3.2	3.7	2.2	3.9	2.8	3.7
5 or more times	1.7	3.6	1.2	3.3	2.2	2.8

Source: "High School Seniors Reporting Involvement in Selected Delinquent Activities in Last 12 Months," *Sourcebook of Criminal Justice Statistics-1992*, pp. 314-316. Primary source: Data provided by Monitoring the Future Project, Survey Research Center, University of Michigan, Lloyd D. Johnson and Jerald G. Bachman, principal investigators. Table adapted by *Sourcebook* staff. Published by permission.

★ 139 ★

Juvenile Delinquency

High School Students Reporting Delinquent Activities, 1990 to 1992 – Part II

Delinquent activity	Class of 1990		Class of 1991		Class of 1992	
	White (N = 1,907)	Black (N = 277)	White (N = 1,818)	Black (N = 289)	White (N = 1,806)	Black (N = 368)
Question: "During the last 12 months, how often..."						
Hurt someone badly enough to need bandages or a doctor?						
Not at all	87.7%	85.3%	88.2%	84.4%	87.9%	84.7%
Once	7.6	9.3	7.7	10.0	7.3	7.8
Twice	2.6	3.6	2.0	1.7	2.9	2.9
3 or 4 times	1.2	1.0	1.1	0.7	1.3	1.9
5 or more times	0.9	0.9	0.9	3.2	0.6	2.8
Used a knife or gun or some other thing (like a club) to get something from a person?						
Not at all	97.2	84.0	97.4	94.1	97.1	93.2
Once	1.6	3.0	1.4	1.7	1.5	2.9
Twice	0.6	1.7	0.3	2.1	0.9	1.5
3 or 4 times	0.1	0.4	0.1	0.5	0.3	1.3
5 or more times	0.4	0.9	0.8	1.6	0.2	1.0
Taken something not belonging to you worth under $50?						
Not at all	64.8	78.7	67.2	74.9	65.3	79.0
Once	14.4	8.6	13.9	11.2	14.9	7.8
Twice	7.2	5.7	7.9	6.5	9.2	3.9
3 or 4 times	6.8	2.5	3.8	3.0	5.7	5.3

[Continued]

★ 139 ★

High School Students Reporting Delinquent Activities, 1990 to 1992 – Part II

[Continued]

Delinquent activity	Class of 1990		Class of 1991		Class of 1992	
	White (N = 1,907)	Black (N = 277)	White (N = 1,818)	Black (N = 289)	White (N = 1,806)	Black (N = 368)
5 or more times	6.8	4.5	7.2	4.5	5.0	4.0
Taken something not belonging to you worth over $50?						
Not at all	89.6	91.4	90.5	93.2	89.9	92.2
Once	4.8	4.7	4.4	3.0	5.2	3.4
Twice	2.2	0.8	2.1	1.4	1.7	1.6
3 or 4 times	1.4	1.3	1.3	1.6	1.4	1.7
5 or more times	2.0	1.9	1.7	0.8	1.8	1.2

Source: "High School Seniors Reporting Involvement in Selected Delinquent Activities in Last 12 Months," *Sourcebook of Criminal Justice Statistics- 1992*, pp. 314-316. Primary source: Data provided by Monitoring the Future Project, Survey Research Center, University of Michigan, Lloyd D. Johnson and Jerald G. Bachman, principal investigators. Table adapted by *Sourcebook* staff. Published by permission.

★ 140 ★

Juvenile Delinquency

High School Students Reporting Delinquent Activities, 1990 to 1992 – Part III

Delinquent activity	Class of 1990		Class of 1991		Class of 1992	
	White (N = 1,907)	Black (N = 277)	White (N = 1,818)	Black (N = 289)	White (N = 1,806)	Black (N = 368)
Question: "During the last 12 months, how often..."						
Taken something from a store without paying for it?						
Not at all	66.9%	74.3%	68.3%	74.5%	70.0%	74.0%
Once	14.1	10.0	12.1	9.4	12.3	10.6
Twice	6.2	6.9	7.1	6.8	6.6	5.8
3 or 4 times	5.5	3.6	5.3	4.1	5.5	5.0
5 or more times	7.3	5.2	7.1	5.2	5.6	4.6
Taken a car that didn't belong to someone in your family without permission of the owner?						
Not at all	93.5	93.7	94.4	92.2	95.1	91.9
Once	3.3	2.4	3.2	4.2	2.3	4.6
Twice	1.7	2.4	1.1	1.2	1.4	1.0
3 or 4 times	0.7	0.1	0.9	1.4	0.5	1.4
5 or more times	0.7	1.4	0.5	1.1	0.8	1.1
Gone into some house or building when you weren't supposed to be there?						
Not at all	72.3	80.7	75.0	78.6	71.7	81.1
Once	11.5	6.1	11.6	8.0	13.1	8.0
Twice	8.5	7.3	6.8	5.2	8.1	3.6
3 or 4 times	4.6	3.1	3.2	3.6	4.1	4.1
5 or more times	3.0	2.8	3.4	4.6	2.9	3.2
Set fire to someone's property on purpose?						
Not at all	98.1	97.2	98.1	98.3	97.3	98.2
Once	1.0	1.3	1.0	0.8	1.8	0.7

[Continued]

★ 140 ★

High School Students Reporting Delinquent Activities, 1990 to 1992 – Part III

[Continued]

Delinquent activity	Class of 1990		Class of 1991		Class of 1992	
	White (N = 1,907)	Black (N = 277)	White (N = 1,818)	Black (N = 289)	White (N = 1,806)	Black (N = 368)
Twice	0.5	0.4	0.3	0.6	0.4	0.3
3 or 4 times	0.2	0.3	[1]	0.3	0.3	0.3
5 or more times	0.2	0.8	0.5	0.0	0.2	0.6
Damaged school property on purpose?						
Not at all	86.0	87.9	87.4	88.0	85.8	88.2
Once	6.6	5.7	6.9	4.0	8.1	5.6
Twice	4.3	2.7	2.4	4.3	3.0	3.9
3 or 4 times	1.9	1.2	1.1	2.0	1.0	1.4
5 or more times	1.2	2.6	2.1	1.6	2.0	0.9
Damaged property at work on purpose?						
Not at all	93.1	95.1	93.4	95.7	93.8	96.3
Once	3.1	2.2	3.1	2.1	2.8	1.7
Twice	2.2	1.3	1.2	1.7	1.4	0.5
3 or 4 times	0.8	0.0	0.9	0.0	1.2	0.9
5 or more times	0.9	1.4	1.4	0.5	0.8	0.6
Gotten into trouble with police because of something you did?						
Not at all	73.2	85.3	76.3	80.3	75.8	84.2
Once	15.0	8.3	12.9	12.1	13.4	7.9
Twice	6.5	4.4	6.0	5.0	5.5	4.5
3 or 4 times	3.9	1.2	3.0	2.2	3.2	2.1
5 or more times	1.5	0.8	1.7	0.4	2.2	1.2

Source: "High School Seniors Reporting Involvement in Selected Delinquent Activities in Last 12 Months," *Sourcebook of Criminal Justice Statistics- 1992,* pp. 314-316. Primary source: Data provided by Monitoring the Future Project, Survey Research Center, University of Michigan, Lloyd D. Johnson and Jerald G. Bachman, principal investigators. Table adapted by *Sourcebook* staff. Published by permission. *Note:* 1. Less than 0.05 percent.

Law Enforcement

★ 141 ★

Average Sentence Length of Offenders Sentenced to Incarceration in U.S. District Courts, 1989

By offense and demographic characteristics, United States, 1989. In months.

Offender characteristics	Average sentence length for offenders convicted of:						
	All offenses	Violent offenses	Property offenses		Drug offenses	Public-order offenses	
			Fraudulent	Other		Regulatory	Other
All offenders[1]	54.5	90.6	26.1	25.7	74.9	24.0	28.1
Race							
White	51.9	88.0	29.3	27.4	70.0	26.2	27.5
Black	65.4	98.4	18.7	24.1	89.4	33.1	41.2
Other	49.9	68.3	23.5	13.3	69.6	[2]	19.4
Ethnicity							
Hispanic	51.9	75.2	23.0	23.4	69.8	25.0	15.7
Non-Hispanic	56.4	89.9	26.6	26.0	77.7	27.3	36.0

Source: "Average Sentence Length Imposed on Offenders Sentenced to Incarceration in U.S. District Courts," Department of Justice, *Sourcebook of Criminal Statistics-1992*, p. 492. Primary source: U.S. Department of Justice, Bureau of Justice Statistics, *Compendium of Federal Justice Statistics, 1989*, NCJ-134730. *Notes:* Data exclude corporations, offenders sentenced to life sentences, and indeterminate sentences for youthful or drug offenders. Includes prison portion of split or mixed sentences. 1. Includes offenders for whom these characteristics were unknown. 2. Too few cases to obtain statistically reliable data.

★ 142 ★

Law Enforcement

Defendants Convicted in U.S. District Courts, 1989

By offense and demographic characteristics, United States, 1989.

Defendant characteristics	Total number of defendants	Percent of offenders convicted of:						
		All offenses	Violent offenses	Property offenses		Drug offenses	Public-order offenses	
				Fraudulent	Other		Regulatory	Other
Race								
White	23,900	71.3	59.4	68.5	61.1	71.8	84.3	78.1
Black	8,660	25.8	29.6	28.7	35.3	26.7	11.6	19.4
Other	941	2.8	11.1	2.8	3.7	1.5	4.1	2.5

[Continued]

★ 142 ★

Defendants Convicted in U.S. District Courts, 1989
[Continued]

Defendant characteristics	Total number of defendants	Percent of offenders convicted of:						
		All offenses	Violent offenses	Property offenses		Drug offenses	Public-order offenses	
				Fraudulent	Other		Regulatory	Other
Ethnicity								
Hispanic	6,261	18.7	5.5	8.4	5.9	27.9	10.5	24.1
Non-Hispanic	27,237	81.3	94.5	91.6	94.1	72.1	89.5	75.9

Source: "Defendants Convicted in U.S. District Courts," Department of Justice, *Sourcebook of Criminal Statistics-1992*, p. 487. Primary source: U.S. Department of Justice, Bureau of Justice Statistics, *Compendium of Federal Justice Statistics, 1989*, NCJ-134730. *Notes:* This table was created by matching the Administrative Office master data files with the Pretrial Services Agency data files and probation and parole data files. Records were included in the table if the relevant information was available from any source. Some items are available only from one of the files. The number of records for these items is lower than those for items which might have come from two or three files. Moreover, many records omit data on certain items, such as ethnicity. Table indicates the number of records on which relevant data were available.

★ 143 ★

Law Enforcement

Defendants Sentenced for Drug Trafficking Under the U.S. Sentencing Commission Guidelines, Fiscal Year 1992

Drug type	Total cases	Race, ethnicity[1]							
		White		Black		Hispanic		Other[2]	
		Number	Percent	Number	Percent	Number	Percent	Number	Percent
Total	13,370	4,889	36.5%	4,131	30.9%	4,167	31.1%	183	1.4%
Cocaine	5,794	1,859	32.0	1,572	27.1	2,312	39.9	51	0.9
Cocaine base ("crack")	2,070	61	3.0	1,895	91.5	110	5.3	4	0.2
Heroin	1,038	170	16.4	454	43.7	372	35.8	42	4.1
Marijuana	3,347	1,851	55.2	153	4.6	1,302	38.9	41	1.2
Methamphetamine	638	547	85.6	6	0.9	53	8.3	32	5.0
LSD	159	150	94.3	3	1.9	3	1.9	3	1.9
Other	324	251	77.5	48	14.8	15	4.6	10	3.1

Source: "Defendants Sentenced for Drug Trafficking Under the U.S. Sentencing Commission Guidelines," Department of Justice, *Sourcebook of Criminal Justice Statistics-1992*, p. 525. Primary source: U.S. Sentencing Commission, *Annual Report 1992. Notes:* This table includes only the 13,511 cases where drug trafficking was the primary offense. Of these 13,511 cases, 141 were excluded due to one or both of the following conditions: missing information on race of the defendant, 15; and missing information on drug type, 127. 1. The Hispanic category includes both black and white Hispanics. As such, the numbers reported underrepresent black defendants. 2. Includes Native Americans, Alaska Natives, Asian, or Pacific Islanders.

★ 144 ★

Law Enforcement

Drug Use by Female Arrestees in 24 U.S. Cities, Black and White, 1991

By type of drug, race, ethnicity, and sex, 1991. Percent testing positive.

City	Any drug[1]		Marijuana		cocaine		Heroin	
	Black	White	Black	White	Black	White	Black	White
Female								
Atlanta, GA	71%	66%	8%	8%	68%	59%	3%	8%
Birmingham, AL	63	60	10	10	53	30	5	21
Cleveland, OH	81	73	7	7	78	67	6	11
Dallas, TX	58	55	12	10	51	37	8	10
Denver, CO	68	45	15	16	60	32	3	2
Detroit, MI	66	76	4	4	61	67	8	20
Fort Lauderdale, FL	66	61	12	15	63	48	3	4
Houston, TX	65	60	8	10	559	51	4	5
Indianapolis, IN	57	51	18	25	40	12	9	12
Kansas City, MO	67	53	11	17	62	42	3	6
Los Angeles, CA	80	76	12	11	73	55	10	22
Manhattan, NY	75	82	10	14	64	69	15	31
New Orleans, LA	49	62	6	16	43	38	6	11
Philadelphia, PA	76	67	14	11	68	40	5	26
Phoenix, AZ	81	59	12	15	72	39	9	18
Portland, OR	73	69	18	35	54	35	12	20
St. Louis, MO	54	54	5	18	50	38	6	9
San Antonio, TX	51	49	3	21	43	19	19	18
San Diego, CA	84	72	20	25	72	28	14	20
San Jose, CA	69	50	19	13	63	20	6	9
Washington, DC	75	72	5	15	70	59	15	23

Source: "Drug Use by Arrestees in 24 U.S. Cities," Department of Justice, *Sourcebook of Criminal Justice Statistics-1992,* p. 462. Primary source: U.S. Department of Justice, National Institute of Justice, *Drug Use Forecasting 1991 Annual Report,* NCJ-137776. Table was adapted by *Sourcebook* staff. *Notes:* 1. Includes cocaine, opiates, marijuana, phencyclidine (PCP), methadone, benzodiazepine, methaqualone, propoxyphene, barbiturates, and amphetamines.

★ 145 ★

Law Enforcement

Drug Use by Male Arrestees in 24 U.S. Cities, Black and White, 1991

By type of drug, race, ethnicity, and sex, 1991. Percent testing positive.

City	Any drug[1]		Marijuana		cocaine		Heroin	
	Black	White	Black	White	Black	White	Black	White
Male								
Atlanta, GA	65%	48%	12%	20%	59%	33%	3%	4%
Birmingham, AL	66	55	13	27	59	27	2	13
Chicago, IL	75	72	23	20	63	60	23	16

[Continued]

★ 145 ★

Drug Use by Male Arrestees in 24 U.S. Cities, Black and White, 1991

[Continued]

City	Any drug[1] Black	Any drug[1] White	Marijuana Black	Marijuana White	cocaine Black	cocaine White	Heroin Black	Heroin White
Cleveland, OH	61	41	10	21	56	21	3	2
Dallas, TX	59	55	18	24	50	35	4	6
Denver, CO	61	41	22	25	47	16	1	[2]
Detroit, MI	56	49	19	9	41	40	8	12
Fort Lauderdale, FL	71	53	29	28	58	32	1	2
Houston, TX	77	59	16	17	70	49	3	4
Indianapolis, IN	45	44	18	29	30	10	4	2
Kansas City, MO	56	41	17	20	43	16	1	2
Los Angeles, CA	77	65	17	28	63	27	10	11
Manhattan, NY	77	74	17	17	68	59	12	20
Miami, FL	76	57	26	17	70	44	2	7
New Orleans, LA	60	51	16	18	54	28	4	4
Omaha, NE	44	32	27	25	23	6	1	2
Philadelphia, PA	75	65	17	24	66	41	8	15
Phoenix, AZ	53	43	16	24	41	17	4	6
Portland, OR	66	58	27	37	46	18	9	8
St. Louis, MO	60	54	12	36	53	26	6	5
San Antonio, TX	55	48	22	26	45	18	2	9
San Diego, CA	79	74	29	35	59	22	11	12
San Jose, CA	72	59	32	32	54	25	5	8
Washington, DC	60	49	11	18	51	26	10	22

Source: "Drug Use by Arrestees in 24 U.S. Cities," Department of Justice, *Sourcebook of Criminal Justice Statistics-1992*, p. 462. Primary source: U.S. Department of Justice, National Institute of Justice, *Drug Use Forecasting 1991 Annual Report*, NCJ-137776. Table was adapted by *Sourcebook* staff. *Notes:* 1. Includes cocaine, opiates, marijuana, phencyclidine (PCP), methadone, benzodiazepine, methaqualone, propoxyphene, barbiturates, and amphetamines. 2. Less than 20 cases.

★ 146 ★

Law Enforcement

Jail Inmates by Race and Hispanic Origin, 1990 and 1991

By race and Hispanic origin, United states, 1990 and 1991[1].

Characteristic	Percent of jail inmates 1990	Percent of jail inmates 1991
Total	100%	100%
Race, Hispanic origin		
White, non-Hispanic	41.8	41.1
Black, non-Hispanic	42.5	43.4

[Continued]

★ 146 ★

Jail Inmates by Race and Hispanic Origin, 1990 and 1991

[Continued]

Characteristic	Percent of jail inmates	
	1990	1991
Hispanic	14.3	14.2
Other[2]	1.3	1.2

Source: "Jail Inmates," Department of Justice, *Sourcebook of Criminal Justice Statistics-1992*, p. 594. Primary source: U.S. Department of Justice, Bureau of Justice Statistics, *Jail Inmates, 1991*, Bulletin NCJ-134726. *Notes:* Data are for June 29, 1990 and June 28, 1991. Race was reported for 99 percent of the inmates in both years. 1. Percents may not add to total because of rounding. 2. Native Americans, Aleuts, Asians, and Pacific Islanders.

★ 147 ★

Law Enforcement

Jail Inmates by Sex, Race and Hispanic Origin, 1983 and 1989

By sex, United States, 1983 and 1989[1].

Characteristic	Percent of female inmates		Percent of male inmates	
	1983	1989	1983	1989
Race and ethnicity				
White, non-Hispanic	41.8%	37.8%	46.9%	38.7%
Black, non-Hispanic	42.2	43.4	37.1	41.5
Hispanic	12.7	16.3	14.3	17.5
Other[2]	3.2	2.5	1.7	2.3

Source: "Characteristics of Jail Inmates," Department of Justice, *Sourcebook of Criminal Justice Statistics-1992*, p. 595. Primary source: U.S. Department of Justice, Bureau of Justice Statistics, *Women in Jail 1989*, Special Report NCJ-134732. *Notes:* A jail is defined as a confinement facility administered by a local government agency that holds persons pending adjudication and persons committed after adjudication, usually for sentences of a year or less. 1. Detail may not add to total because of rounding. 2. Includes Asians, Pacific Islanders, American Indians, Alaska Natives, and other racial groups.

★ 148 ★

Law Enforcement

Killers of Law Enforcement Officers, Aggregate 1982 to 1991 and 1991

Characteristics of persons identified	1982 to 1991		1991	
	Number	Percent	Number	Percent
Race, ethnicity				
White	563	56	52	54
Black	406	41	43	45
Other[1]	30	3	1	1

Source: "Persons Identified in the Killing of Law Enforcement Officers," Department of Justice, *Sourcebook of Criminal Justice Statistics-1992,* p. 409. Primary source: U.S. Department of Justice, Federal Bureau of Investigation, *Law Enforcement Officers Killed and Assaulted, 1991.* Table constructed by *Sourcebook* staff. *Notes:* 1. Other includes Asian, Pacific Islander, American Indian, and Alaska Native.

★ 149 ★

Law Enforcement

Law Enforcement Officers Killed, 1982 to 1991

By selected characteristics of officers, United States, 1982-91[1].

Characteristics of officers killed	1982 (N=92)	1983 (N=80)	1984 (N=72)	1985 (N=78)	1986 (N=66)	1987 (N=73)	1988 (N=78)	1989 (N=66)	1990 (N=65)	1991
Race										
White	84%	84%	85%	88%	89%	90%	91%	89%	80%	87%
Black	15	13	14	10	11	10	9	11	18	13
Other	1	4	1	1	0	0	0	0	2[2]	0

Source: "Percent Distribution of Law Enforcement Officers Killed," Department of Justice, *Sourcebook of Criminal Justice Statistics-1992,* p. 408. Primary source: U.S. Department of Justice, Federal Bureau of Investigation, *Law Enforcement Officers Killed,* annuals 1982 to 1991. Table constructed by *Sourcebook* staff. *Notes:* 1. Percents may not add to 100 because of rounding. 2. For 1990, other was specified as Asian.

★ 150 ★

Law Enforcement

Most Serious Offense of Felony Offenders Convicted in State Courts, 1990

Most serious conviction offense	Estimated total number of convictions	Total	Race		
			White	Black	Other
All offenses	829,344	100%	52%	47%	1%
Violent offenses	147,766	100	50	48	2
Murder[1]	10,895	100	42	56	2
Rape	18,024	100	65	33	2
Robbery	47,446	100	36	63	1
Aggravated assault	53,861	100	53	44	3
Other violent[2]	17,540	100	72	24	4
Property offenses	280,748	100	57	42	1
Burglary	109,750	100	57	42	1
Larceny[3]	113,094	100	57	42	1
Fraud[4]	57,904	100	58	41	1
Drug offenses	274,613	100	43	56	1
Possession	106,253	100	45	54	1
Trafficking	168,360	100	42	57	1
Weapons offenses	20,733	100	42	57	1
Other offenses[5]	105,484	100	65	33	2

Source: "Most Serious Offense of Felony Offenders Convicted in State Courts," Department of Justice, *Sourcebook of Criminal Justice Statistics-1992*, p. 528. Primary source: U.S. Department of Justice, Bureau of Justice Statistics, *Felony Sentences in State Courts, 1992*. Bulletin NCJ-140186. Table adapted by *Sourcebook* staff. *Notes:* Figures on sex are based on 88 percent of the estimated total of 829,344 convicted felons; figures on race, 65 percent of the total; figures on age, 80 percent of the total. 1. Includes nonnegligent manslaughter. 2. Includes offenses such as nonnegligent manslaughter, sexual assault, and kidnaping. 3. Includes motor vehicle theft. 4. Includes forgery and embezzlement. 5. Composed of nonviolent offenses such as receiving stolen property and driving while intoxicated.

★ 151 ★

Law Enforcement

Offenders Sentenced Under the U.S. Sentencing Commission Guidelines, Fiscal Year 1992

Primary offense	Total cases	Race, ethnicity[1]							
		White		Black		Hispanic		Other[2]	
		Number	Percent	Number	Percent	Number	Percent	Number	Percent
Total	37,506	17,022	45.4%	10,621	28.3%	8,598	22.9%	1,265	3.4%
Murder	59	21	35.6	7	11.9	5	8.5	26	44.1
Manslaughter	50	3	6.0	4	8.0	2	4.0	41	82.0
Kidnaping, hostage-taking	53	25	47.2	13	24.5	8	15.1	7	13.2
Sexual abuse	168	29	17.3	14	8.3	5	3.0	120	71.4
Assault	351	136	38.8	94	26.8	36	10.3	85	24.2
Robbery	1,615	894	55.4	578	35.8	120	7.4	23	1.4
Arson	79	70	88.6	6	7.6	1	1.3	2	2.5
Drug offenses									
Trafficking	15,559	5,612	36.1	4,800	30.8	4,900	31.5	247	1.6
Communication facility	287	138	48.1	46	16.0	94	32.8	9	3.1
Simple possession	878	369	42.0	194	22.1	301	34.3	14	1.6
Firearms	3,102	1,489	48.0	1,202	38.8	337	10.9	74	2.4
Burglary, breaking and entering	71	33	46.5	18	25.4	4	5.6	16	22.5
Auto theft	223	171	76.7	37	16.6	14	6.3	1	0.4
Larceny	2,574	1,308	53.6	92852	36.0	166	6.4	100	3.9
Fraud	4,443	2,674	60.2	1,252	28.2	365	8.2	152	3.4
Embezzlement	1,176	786	66.8	283	24.1	56	4.8	51	4.3
Forgery, counterfeiting	935	492	52.6	298	31.9	125	13.4	20	2.1
Bribery	272	132	48.5	58	21.3	34	12.5	48	17.6
Tax	649	509	78.4	72	11.1	46	7.1	22	3.4
Money laundering	664	326	49.1	148	22.3	164	24.7	26	3.9
Racketeering, extortion	318	204	64.2	75	23.6	27	8.5	12	3.8
Gambling, lottery	197	161	81.7	24	12.2	1	0.5	11	5.6
Civil rights	126	104	82.5	17	13.5	4	3.2	1	0.8
Immigration	1,873	158	8.4	130	6.9	1,521	81.2	64	3.4
Pornography, prostitution	138	112	81.2	21	15.2	3	2.2	2	1.4
Prison offenses	216	111	51.4	63	29.2	40	18.5	2	0.9
Administration of justice offenses	547	284	51.9	114	20.8	130	23.8	19	3.5
Environmental, wildlife	110	79	71.8	1	0.9	16	14.6	14	12.7
National defense	46	16	34.8	2	4.4	26	56.5	2	4.4
Anti-trust	20	20	100.0	0	X	0	X	0	X
Food and drug	45	34	75.6	5	11.1	3	6.7	3	6.7
Other	662	450	68.0	117	17.7	44	6.6	51	7.7

Source: "Offenders Sentenced Under the U.S. Sentencing Commission Guidelines," Department of Justice, *Sourcebook of Criminal Justice Statistics-1992*, p. 520. Primary source: U.S. Sentencing Commission, *Annual Report 1992*. *Notes:* Given the nature of the data file and reporting requirements, the following types of cases are not included in the data presented here: cases initiated but for which no convictions were obtained; defendants convicted for whom no sentences were yet issued; defendants sentenced but for whom no data were submitted to the Commission; and cases that solely involved petty offenses. A case or defendant is defined as a single sentencing event for a single defendant. Multiple defendants in a single sentencing event are treated as separate cases. If an individual defendant is sentenced more than once during the time period of interest, each sentencing event is identified as a separate case. Of the 38,258 guideline cases some were excluded due to missing information. 1. The Hispanic category includes both black and white Hispanics. As such, the numbers reported underrepresent black defendants. 2. Includes Native Americans, Alaska Natives, Asian, or Pacific Islanders.

★ 152 ★

Law Enforcement

Offenders Sentenced to Incarceration in U.S. District Courts, 1989

By offense and demographic characteristics, United States, 1989.

Offender characteristics	Total number of defendants	Of all offenders convicted in cases terminated in 1989, the percent who were incarcerated						
		All offenses	Violent offenses	Property offenses		Drug offenses	Public-order offenses	
				Fraudulent	Other		Regulatory	Other
All offenders[1]	45,014	58.5%	87.0%	44.4%	43.3%	85.0%	37.0%	41.1%
Race								
White	23,900	67.8	88.6	48.1	48.5	87.9	41.9	62.2
Black	8,660	69.6	91.6	45.3	45.6	93.8	37.3	64.2
Other	941	54.3	83.2	34.8	40.6	80.4	28.1	36.4
Ethnicity								
Hispanic	6,261	81.3	88.7	51.1	49.4	93.0	59.6	73.8
Non-Hispanic	27,237	64.8	88.9	46.5	47.1	88.0	38.5	58.2

Source: "Offenders Sentenced to Incarceration in U.S. District Courts," Department of Justice, *Sourcebook of Criminal Statistics-1992*, p. 489. Primary source: U.S. Department of Justice, Bureau of Justice Statistics, *Compendium of Federal Justice Statistics, 1989*, NCJ-134730. *Notes:* Data exclude corporations. Offenders are classified by the most serious offenses of conviction. 1. Includes offenders for whom these characteristics were unknown.

★ 153 ★

Law Enforcement

Persons Arrested by Charge and Race, Part 1, Numbers

Represents arrests (not charges) reported by 10,075 agencies with a total 1991 population 186,621,000 as estimated by FBI.

Offense charged	Total arrests (1,000)				
	Total	White	Black	American Indian or Alaskan Native	Asian or Pacific Islander
Total	10,517	7,252	3,049	115	100
Serious crimes[1]	2,237	1,415	774	21	26
Murder and nonnegligent manslaughter	18	8	10	(Z)	(Z)
Forcible rape	30	16	13	(Z)	(Z)
Robbery	136	51	83	1	1
Aggravated assault	364	219	139	3	3
Burglary	324	223	95	3	3
Larceny/theft	1,190	793	368	13	16
Motor vehicle theft	160	94	63	1	2
Arson	15	11	3	(Z)	(Z)

[Continued]

★ 153 ★

Persons Arrested by Charge and Race, Part 1, Numbers

[Continued]

Offense charged	Total arrests (1,000)				
	Total	White	Black	American Indian or Alaskan Native	Asian or Pacific Islander
All other nonserious crimes					
Other assaults	772	498	257	10	7
Forgery and counterfeiting	75	49	25	(Z)	1
Fraud	292	198	91	1	1
Embezzlement	11	7	3	(Z)	(Z)
Stolen property--buying, receiving, possessing	130	74	54	1	1
Vandalism	249	189	55	2	2
Weapons: carrying, possessing, etc.	173	99	72	1	2
Prostitution and commercialized vice	79	48	30	(Z)	1
Sex offenses (except forcible rape and prostitution)	81	63	16	1	1
Drug abuse violations	763	444	313	3	4
Gambling	12	6	6	(Z)	1
Offenses against family and children	71	47	21	1	2
Driving under the influence	1,271	1,130	116	15	10
Liquor laws	449	392	44	11	3
Drunkenness	625	508	102	14	2
Disorderly conduct	559	366	182	7	3
Vagrancy	31	16	14	1	(Z)
All other offenses (except traffic)	2,428	1,543	832	25	28
Suspicion	10	4	6	(Z)	(Z)
Curfew and loitering law violations	72	55	15	1	1
Runaways	128	103	20	1	3

Source: "Persons Arrested, by Charge and Race: 1991," *Statistical Abstract of the United States, 1993*, p. 198. Primary source: U.S. Federal Bureau of Investigation, *Crime in the United States*, annual. *Notes:* Z less than 500. 1. Includes arson.

★ 154 ★

Law Enforcement

Persons Arrested by Charge and Race, Part 2, Percent Distribution

Represents arrests (not charges) reported by 10,075 agencies with a total 1991 population 186,621,000 as estimated by FBI.

Offense charged	Percent distribution				
	Total	White	Black	American Indian or Alaskan Native	Asian or Pacific Islander
Total	100.0	69.0	29.0	1.1	0.9
Serious crimes[1]	100.0	63.2	34.6	1.0	1.2
Murder and nonnegligent manslaughter	100.0	43.4	54.8	0.8	0.9
Forcible rape	100.0	54.8	43.5	0.9	0.8
Robbery	100.0	37.6	61.1	0.4	0.9
Aggravated assault	100.0	60.0	38.3	0.9	0.8
Burglary	100.0	68.8	29.3	0.9	1.0
Larceny/theft	100.0	66.6	30.9	1.1	1.4
Motor vehicle theft	100.0	58.5	39.3	0.8	1.4
Arson	100.0	76.7	21.5	0.9	0.9
All other nonserious crimes					
Other assaults	100.0	64.6	33.3	1.3	0.9
Forgery and counterfeiting	100.0	64.8	33.7	0.6	0.9
Fraud	100.0	67.8	31.3	0.4	0.5
Embezzlement	100.0	68.2	30.2	0.7	0.9
Stolen property--buying, receiving, possessing	100.0	57.0	41.7	0.5	0.8
Vandalism	100.0	76.0	22.1	1.0	0.9
Weapons: carrying, possessing, etc.	100.0	56.8	41.6	0.5	1.1
Prostitution and commercialized vice	100.0	60.3	38.0	0.6	1.1
Sex offenses (except forcible rape and prostitution)	100.0	78.2	19.8	1.0	1.0
Drug abuse violations	100.0	58.1	41.0	0.3	0.5
Gambling	100.0	44.8	46.9	0.1	8.2
Offenses against family and children	100.0	66.7	29.5	1.1	2.7
Driving under the influence	100.0	88.9	9.1	1.2	0.8
Liquor laws	100.0	87.3	9.7	2.4	0.6
Drunkenness	100.0	81.2	16.4	2.2	0.3
Disorderly conduct	100.0	65.5	32.7	1.3	0.6
Vagrancy	100.0	51.2	46.6	2.0	0.2
All other offenses (except traffic)	100.0	63.5	34.3	1.0	1.2
Suspicion	100.0	40.8	58.6	0.5	0.2
Curfew and loitering law violations	100.0	76.9	20.6	0.8	1.8
Runaways	100.0	80.5	15.9	1.1	2.5

Source: "Persons Arrested, by Charge and Race: 1991," *Statistical Abstract of the United States, 1993,* p. 198. Primary source: U.S. Federal Bureau of Investigation, *Crime in the United States,* annual. *Note:* 1. Includes arson.

★ 155 ★

Law Enforcement

Police Departments, Full-Time Sworn Employees by Size of Population Served and by Race/Ethnicity, 1990

By size of population served, United States, 1990[1].

Population served	Total	White		Black		Hispanic		Other[2]	
		Male	Female	Male	Female	Male	Female	Male	Female
All sizes	100%	77.5%	5.5%	8.5%	2.0%	4.7%	0.5%	1.2%	0.1%
1,000,000 or more	100	65.4	7.0	12.4	4.2	8.4	1.5	0.9	0.1
500,000 to 999,999	100	62.9	5.7	15.6	4.4	5.6	0.5	5.0	0.4
250,000 to 499,999	100	68.0	6.8	13.0	3.2	7.2	0.7	1.0	0.1
100,000 to 249,999	100	76.7	6.0	9.1	1.8	4.1	0.4	1.9	0.1
50,000 to 99,999	100	84.2	4.9	5.7	0.8	3.4	0.2	0.7	[3]
25,000 to 49,999	100	85.8	4.4	5.6	0.5	2.9	0.1	0.6	[3]
10,000 to 24,999	100	89.4	4.3	3.5	0.3	2.1	0.2	0.3	0.1
2,500 to 9,999	100	88.2	4.6	3.6	0.4	2.6	[3]	0.5	0.1
Under 2,500	100	87.6	3.8	4.5	0.2	2.4	0.3	1.2	0.0

Source: "Characteristics of Full-Time Sworn Personnel in Local Police Departments," *Sourcebook of Criminal Justice Statistics-1992,* p. 46. Primary source: U.S. Department of Justice, Bureau of Justice Statistics, *State and Local Police Departments, 1990,* NCJ-133284. *Notes:* Black and white racial categories do not include Hispanics. 1. Detail may not add to total because of rounding. 2. Includes American Indians, Alaskan Natives, Asians, and Pacific Islanders. 3. Less than 0.05 percent.

★ 156 ★

Law Enforcement

Police Departments: Black Officers in the 50 Largest Cities, 1990

1983 and 1992.

City	Total number of officers		Black officers				Index of Black representation		
			1983		1992		1983	1992	Percent change
	1983	1992	Number	Percent	Number	Percent			
New York, NY	23,408	27,154	2,395	10.2%	3,121	11.4%	0.40	0.40	0.0%
Los Angeles, CA	6,928	8,020	657	9.4	1,127	14.1	0.55	1.00	81.8
Chicago, IL	12,472	12,291	2,508	20.1	3,063	24.9	0.51	0.64	25.4
Houston, TX	3,629	4,056	355	9.7	595	14.7	0.35	0.52	48.5
Philadelphia, PA	7,265	6,280	1,201	16.5	1,615	25.7	0.44	0.64	45.4
San Diego, CA	1,363	1,937	76	5.5	146	7.5	0.62	0.80	29.0
Detroit, MI	4,032	4,787	1,238	30.7	2,556	53.3	0.49	0.70	42.8
Dallas, TX	2,053	2,878	169	8.2	546	19.0	0.28	0.64	128.5
Phoenix, AZ	1,660	1,644	48	2.8	66	4.0	0.58	0.77	32.7
San Antonio, TX[1]	1,164	1,606	54	4.6	90	5.6	NA	0.80	NA
San Jose, CA	915	1,223	20	2.1	50	4.1	0.46	0.85	84.7
Baltimore, MD	3,056	2,822	537	17.5	851	30.2	0.32	0.51	59.3
Indianapolis, IN	936	979	123	13.1	174	17.8	0.60	0.78	30.0

[Continued]

★ 156 ★

Police Departments: Black Officers in the 50 Largest Cities, 1990

[Continued]

City	Total number of officers		Black officers				Index of Black representation		
			1983		1992		1983	1992	Percent change
	1983	1992	Number	Percent	Number	Percent			
San Francisco, CA	1,957	1,818	159	8.1	170	9.4	0.64	0.85	32.8
Jacksonville, FL[1]	1,263	1,205	78	6.1	232	19.2	0.24	0.76	216.6
Columbus, OH	1,197	1,444	133	11.1	256	17.7	0.50	0.78	56.0
Milwaukee, WI	1,438	1,971	168	11.6	283	14.4	0.50	0.47	-6.0
Memphis, TN	1,216	1,403	268	22.0	481	34.3	0.46	0.62	34.7
Washington, DC	3,851	4,396	1,931	50.1	2,980	67.8	0.71	1.03	45.0
Boston, MA	1,871	1,972	248	13.2	404	20.5	0.59	0.80	35.5
Seattle, WA	1,011	1,231	42	4.1	105	8.5	0.43	0.84	95.3
El Paso, TX	650	787	13	2.0	17	2.2	0.63	0.62	-1.5
Cleveland, OH	2,091	1,668	238	11.3	439	26.3	0.26	0.56	115.3
New Orleans, LA	1,317	1,551	276	20.9	608	39.2	0.38	0.63	65.7
Nashville, TN	969	1,058	114	11.7	139	13.1	0.50	0.54	8.0
Denver, CO	1,379	1,348	82	5.9	130	9.2	0.49	0.72	46.9
Austin, TX	607	830	43	7.0	81	9.8	0.57	0.78	36.8
Fort Worth, TX	766	967	43	5.6	112	11.6	0.25	0.52	108.0
Oklahoma City, OK	662	932	27	4.0	69	7.4	0.27	0.47	74.0
Portland, OR	688	877	19	2.7	32	3.6	0.36	0.46	27.7
Kansas City, MO	1,140	1,166	123	10.7	156	13.4	0.39	0.45	15.3
Long Beach, CA	637	696	20	3.1	39	5.6	0.27	0.41	51.8
Tucson, AZ	549	771	17	3.0	25	3.2	0.81	0.74	-8.6
St. Louis, MO	1,763	1,552	346	19.6	437	28.2	0.43	0.59	37.2
Charlotte, NC	644	872	144	22.3	167	19.2	0.72	0.60	-16.6
Atlanta, GA	1,313	1,223	602	45.8	668	54.6	0.69	0.81	17.3
Virginia Beach, VA	NA	599	NA	NA	50	8.3	NA	0.60	NA
Albuquerque, NM	561	765	14	2.4	16	2.0	0.96	0.67	-30.2
Oakland, CA	636	549	147	23.1	144	26.2	0.49	0.60	22.4
Pittsburgh, PA	1,222	1,128	175	14.3	289	25.6	0.60	0.99	65.0
Sacramento, CA	NA	607	NA	NA	38	6.3	NA	0.41	NA
Minneapolis, MN	672	840	20	2.9	46	5.5	0.38	0.42	10.5
Tulsa, OK	695	718	30	4.3	68	9.5	0.36	0.69	91.6
Honolulu, HI	1,557	1,870	11	0.7	28	1.4	0.58	1.07	84.4
Cincinnati, OH	971	927	89	9.1	176	19.0	0.27	0.50	85.1
Miami, FL	1,051	1,032	181	17.2	231	22.4	0.69	0.81	17.3
Fresno, CA	NA	412	NA	NA	33	8.0	NA	0.96	NA
Omaha, NE	551	610	46	8.3	70	11.5	0.69	0.87	26.0

[Continued]

★ 156 ★

Police Departments: Black Officers in the 50 Largest Cities, 1990

[Continued]

City	Total number of officers		Black officers				Index of Black representation		
			1983		1992		1983	1992	Percent change
	1983	1992	Number	Percent	Number	Percent			
Toledo, OH	757	639	139	18.3	119	18.8	1.05	0.94	-10.4
Buffalo, NY	1,018	963	86	8.4	195	20.2	0.37	0.66	78.3

Source: "Number of Police Officers and Number of Black Police Officers in the 50 Largest Cities," *Sourcebook of Criminal Justice Statistics-1992,* p. 47. Primary source: Samuel Walker, "Employment of Black and Hispanic Police Officers," *Review of Applied Urban Research XI* (October 1983), p. 3; and Samuel Walker and K. B. Turner, "A Decade of Modest Progress: Employment of Black and Hispanic Officers, 1983-1992," Department of Criminal Justice, University of Nebraska at Omaha, 1992. (Mimeographed). Table adapted by *Sourcebook* staff. Published by permission. *Notes:* Data for 1983 were obtained through a questionnaire mailed to the office of the chief of police and the office of the municipal director of personnel (or equivalent position) in the 50 largest cities in the United States. Forty-seven cities returned completed questionnaires in 1983; all 50 cities returned completed questionnaires in 1992. Cities are listed in rank order of size based on the 1990 census of the population. The index of Black representation is calculated by dividing the percent of Black police officers in a department by the percent of Blacks in the local population. An index approaching 1.0 indicates that a city is closer to achieving a representation of Black police officers equal to their proportion in the local population. The Black population of a city is derived from the 1990 census of the population. 1. Data for 1983 are based on 1980-81 information from the Police Executive Research Forum, *Survey of Police Operational and Administrative Practices 1981* (Washington, DC: Police Executive Research Forum, 1981).

★ 157 ★

Law Enforcement

Race of Felony Probationers in 32 Counties, Aggregate 1986 to 1989

By conviction offense, selected sentence and demographic characteristics.

Most serious conviction offense	Number	White	Black
All offenses	79,043	59%	38%
Violent offenses	9,965	52	45
Murder[1]	247	51	46
Rape	1,406	72	26
Robbery	4,035	37	61
Assault[2]	4,277	61	36
Property offenses	26,670	59	38
Burglary	10,380	59	37
Larceny[3]	12,458	58	38
Fraud[4]	3,832	58	37
Drug offenses	27,052	60	39
Trafficking	15,480	59	40
Possession	11,572	62	37

[Continued]

★ 157 ★

Race of Felony Probationers in 32 Counties, Aggregate 1986 to 1989

[Continued]

Most serious conviction offense	Number	White	Black
Weapons offenses	2,117	45	54
Other offenses[5]	13,239	66	31

Source: "Characteristics of Felony Probationers in 32 Counties," Department of Justice, *Sourcebook of Criminal Justice Statistics-1992*, p. 570. Primary source: U.S. Department of Justice, Bureau of Justice Statistics, *Recidivism of Felons on Probation, 1986-89*, Special Report NCJ-134177. *Notes:* These data were compiled by the U.S. Department of Justice, Bureau of Justice Statistics. The data were drawn mainly from two surveys: a survey of felons sentenced to probation in 100 counties nationwide in 1986, and a followup survey of felons sentenced to probation in 32 of the original 100 counties. The followup survey comprised 12,370 sample cases representing 79,043 felons placed on probation in 32 counties from 17 states. These data are derived from a sample and therefore subject to sampling variation. Any person convicted of multiple offenses received the designation of the most serious felony conviction offense. The hierarchy from most to least serious is generally the order in which offense categories are displayed in the table. Conviction offense was ascertained in 100 percent of cases; jail confinement in original sentence, 99 percent; prior felony conviction, 76 percent; sentence recommendation, 50 percent; drug abuser, 69 percent; intensive supervision, 61 percent; sex, 99 percent; race and age, 97 percent. 1. Includes murder and nonnegligent manslaughter. 2. Aggravated assault only. 3. Includes larceny and motor vehicle theft. 4. Includes forgery, fraud, and embezzlement. 5. Includes receiving stolen property, sexual assault (not including rape), kidnaping, negligent manslaughter, and other felonies.

★ 158 ★

Law Enforcement

Sex and Race of Felony Defendants in the 75 Largest Counties, 1990

By most serious arrest charge, 1990[1].

Most serious conviction offense	Number of defendants	Percent of felony defendants							
		Male				Female			
		Total	Black	White	Other	Total	Black	White	Other
All offenses	50,444	86%	47%	37%	2%	14%	8%	7%	2
Violent offenses	12,978	90	54	34	2	10	7	3	2
Murder	547	92	57	30	6	8	4	4	0%
Rape	705	98	49	47	3	2	1	1	0
Robbery	4,374	93	68	24	1	7	5	2	0
Assault	5,953	86	48	36	2	14	10	4	2
Other violent	1,399	91	38	49	4	9	4	5	2
Property offenses	17,183	85	42	41	2	15	7	8	2
Burglary	5,126	94	48	44	2	6	2	4	0
Theft	7,294	82	41	40	2	18	9	9	0
Other property	4,762	80	38	38	3	20	10	10	2
Drug offenses	16,467	82	48	34	1	18	9	8	0
Sales/trafficking	9,458	84	49	35	2	16	10	7	0
Other drug	7,009	81	47	33	1	19	9	10	0
Public-order offenses	3,815	88	36	50	2	12	5	7	2

[Continued]

★ 158 ★

Sex and Race of Felony Defendants in the 75 Largest Counties, 1990

[Continued]

Most serious conviction offense	Number of defendants	Percent of felony defendants							
		Male				Female			
		Total	Black	White	Other	Total	Black	White	Other
Driving-related	1,131	89	12	76	1	11	3	8	0
Other public-order	2,684	88	47	39	3	12	6	6	2

Source: "Sex and Race of Felony Defendants in the 75 Largest Counties," Department of Justice, *Sourcebook of Criminal Justice Statistics-1992*, p. 531. Primary source: U.S. Department of Justice, Bureau of Justice Statistics, *Felony Sentences in State Courts, 1992*. Bulletin NCJ-140186. *Notes:* Data on both sex and race of defendants were available for 89 percent of all cases. 1. Detail may not add because of rounding. 2. Less than 0.5 percent.

★ 159 ★

Law Enforcement

Sheriffs' Departments: Full-Time Sworn Employees by Size of Population Served and by Race/Ethnicity, 1990

By size of population served, United States, 1990[1].

Population served	Total	Percent of full-time sworn personnel							
		White		Black		Hispanic		Other[2]	
		Male	Female	Male	Female	Male	Female	Male	Female
All sizes	100%	72.6%	11.9%	7.2%	2.6%	3.9%	0.8%	0.9%	0.1%
1,000,000 or more	100	64.0	14.2	6.8	2.9	8.2	1.9	1.9	0.2
500,000 to 999,999	100	68.9	9.7	11.8	4.0	4.1	0.5	0.9	0.1
250,000 to 499,999	100	74.7	10.6	6.1	2.0	5.0	1.0	0.6	0.1
100,000 to 249,999	100	72.7	12.4	8.1	3.1	2.3	0.5	0.8	3
50,000 to 99,999	100	78.8	12.1	5.6	2.2	0.8	0.2	0.3	0.1
25,000 to 49,999	100	77.7	11.4	5.4	1.5	2.0	1.0	0.8	0.2
10,000 to 24,999	100	79.5	11.5	4.6	1.0	2.3	0.2	0.8	0.1
Under 10,000	100	76.4	14.0	4.7	0.9	2.9	0.4	0.5	0.2

Source: "Characteristics of Full-Time Sworn Personnel in Sheriffs' Departments," *Sourcebook of Criminal Justice Statistics-1992*, p. 46. Primary source: U.S. Department of Justice, Bureau of Justice Statistics, *Sheriffs' Departments 1990*, Bulletin NCJ-133283. *Notes:* Black and white racial categories do not include Hispanics. 1. Detail may not add to total because of rounding. 2. Includes American Indians, Alaska Natives, Asians, and Pacific Islanders. 3. Less than 0.05 percent.

★ 160 ★
Law Enforcement

State Law Enforcement Agencies; Full-Time Sworn Employees by Race/Ethnicity, 1990

By agency and sex, race, and ethnicity of employee, 1990[1].

Name of agency	Race and ethnicity of full-time sworn employees				
	White, non-Hispanic	Black, non-Hispanic	Hispanic, any race	American Indian/ Alaskan Native	Asian/ Pacific Islander
Alabama Department of Public Safety	68.5%	31.5%	0.0%	0.0%	0.0%
Alaska State troopers	93.7	1.6	2.0	2.4	0.4
Arizona Department of Public Safety	82.9	2.0	12.7	2.1	0.3
Arkansas State police	88.4	10.5	0.4	0.2	0.4
California Highway Patrol	82.3	3.9	11.1	0.1	2.6
Colorado State Patrol	87.5	2.0	9.3	0.6	0.6
Connecticut State Police	89.2	5.7	4.6	0.3	0.2
Delaware State Police	88.1	9.8	2.1	0.0	0.0
Florida Highway Patrol	81.6	12.2	5.8	0.2	0.2
Georgia State Police	89.2	10.6	0.1	0.1	0.0
Idaho State Police	96.8	0.0	2.6	0.5	0.0
Illinois State Police	82.6	11.6	4.8	0.1	0.8
Indiana State police	91.3	8.2	0.5	0.0	0.0
Iowa State Patrol	98.1	1.0	0.7	0.0	0.2
Kansas Highway Patrol	97.3	1.0	1.2	0.5	0.0
Kentucky State Police	96.3	3.2	0.3	0.0	0.1
Louisiana State Police	86.9	12.9	0.3	0.0	0.0
Maine State police	100.0	0.0	0.0	0.0	0.0
Maryland State Police	82.5	16.7	0.2	0.2	0.2
Massachusetts State Police	92.3	5.4	1.8	0.4	0.0
Michigan State Police	87.5	8.7	2.7	0.9	0.2
Minnesota State Patrol	97.5	1.0	0.4	1.2	0.0
Mississippi Highway Safety Patrol	77.5	22.5	0.0	0.0	0.0
Missouri State Highway Patrol	90.7	7.3	0.8	1.0	0.1
Montana Highway Patrol	98.0	0.0	0.0	2.0	0.0
Nebraska State Patrol	97.8	0.6	1.3	0.2	0.0
Nevada Highway Patrol	90.6	2.3	4.5	1.1	1.5
New Hampshire State police	99.2	0.4	0.0	0.4	0.0
New Jersey State Police	86.5	7.9	4.2	0.5	0.8
New Mexico State Police	70.7	1.1	26.7	1.6	0.0
New York State Police	83.1	10.0	6.6	0.3	0.1
North Carolina State Highway Patrol	86.2	12.5	0.0	1.2	0.1
North Dakota Highway Patrol	96.5	0.0	0.0	3.5	0.0
Ohio State Highway Patrol	90.5	7.9	1.0	0.1	0.5
Oklahoma Highway Patrol	88.8	4.6	0.6	5.9	0.1
Oregon State Police	96.7	0.6	1.1	0.6	1.0
Pennsylvania State Police	89.6	8.3	1.5	0.1	0.4
Rhode Island State Police	95.6	2.2	2.2	0.0	0.0
South Carolina Highway Patrol	86.5	13.3	0.2	0.0	0.0
South Dakota Highway Patrol	98.6	0.0	0.0	1.4	0.0
Tennessee Department of Safety	93.0	6.4	0.4	0.0	0.2
Texas Department of Public Safety	76.8	5.9	16.7	0.4	0.2

[Continued]

★ 160 ★

State Law Enforcement Agencies; Full-Time Sworn Employees by Race/Ethnicity, 1990

[Continued]

Name of agency	Race and ethnicity of full-time sworn employees				
	White, non-Hispanic	Black, non-Hispanic	Hispanic, any race	American Indian/ Alaskan Native	Asian/ Pacific Islander
Utah Highway Patrol	96.3	0.0	2.2	0.9	0.6
Vermont Department of Public Safety	98.0	2.0	0.0	0.0	0.0
Virginia State Police	93.0	6.8	0.1	0.1	0.0
Washington State Patrol	92.6	3.0	1.5	1.7	1.1
West Virginia State Police	98.1	1.7	0.2	0.0	0.0
Wisconsin State Patrol	93.8	2.1	1.7	2.1	0.4
Wyoming Highway Patrol	99.3	0.7	0.0	0.0	0.0

Source: "Full-Time Sworn Employees in State Law Enforcement Agencies," *Sourcebook of Criminal Justice Statistics-1992*, p. 45. Primary source: U.S. Department of Justice, Bureau of Justice Statistics, *Law Enforcement Management and Administrative Statistics, 1990: Data for Individual State and Local Agencies with 100 or More Officers*, NCJ-134436. *Notes:* Percents are based on employee counts for the pay period that included June 15, 1990. 1. Percents may not total 100 due to rounding.

Prisons and Prisoners

★ 161 ★

Admissions to State and Federal Prisons, 1950 to 1986

Year	Number of admissions to State prisons	Percent of State prison admissions				Number of admissions to State prisons	Percent of Federal prison admissions			
		Total	White	Black	Other		Total	White	Black	Other
1950	46,496	100	69	30	1	11,492	100	70	28	2
1960	69,235	100	65	34	1	14,833	100	71	25	4
1964	67,879	100	63	35	2	13,220	100	73	25	2
1970	37,437	100	57	43[1]	X	11,060	100	73	27[1]	X
1974	37,064	100	54	41	5	15,181	100	71	29[1]	X
1975	25,796	100	60	38	2	16,555	100	70	30[1]	X
1976	51,035	100	58	37	5	18,711	100	71	29[1]	X
1977	54,023	100	59	40	1	18,160	100	70	30[1]	X
1978	77,017	100	55	44	1	18,485	100	69	31[1]	X
1979	79,535	100	58	41	1	15,293	100	73	27[1]	X
1980	117,251	100	57	42	1	17,383	100	73	27[1]	X
1981	121,211	100	55	44	1	14,400	100	74	26[1]	X
1982	114,391	100	53	46	1	17,226	100	75	25[1]	X

[Continued]

★ 161 ★

Admissions to State and Federal Prisons, 1950 to 1986

[Continued]

Year	Number of admissions to State prisons	Percent of State prison admissions				Number of admissions to State prisons	Percent of Federal prison admissions			
		Total	White	Black	Other		Total	White	Black	Other
1983	103,588	100	55	44	1	18,987	100	76	24[1]	X
1984	119,042	100	55	44	1	18,541	100	77	23[1]	X
1985	146,862	199	54	45	1	19,881	100	76	24[1]	X
1986	167,474	100	53	46	1	16,295	100	77	21	2

Source: "Admissions To State and Federal Prisons," Department of Justice, *Sourcebook of Criminal Justice Statistics-1992*, p. 618. Primary source: U.S. Department of Justice, Bureau of Justice Statistics, *Race of Prisoners Admitted to State and Federal Institutions, 1926-86*, NCJ-125618. *Notes:* These data were collected by the U.S. Department of Justice, Bureau of Justice Statistics through the National Prisoners Statistics (NPS) program. The NPS is now administered by the U.S. Bureau of Justice Statistics and data are collected and processed by the U.S. Bureau of the Census. Where admission data are available on "sentenced felons admitted to prison as new court commitments," these data are used. Where there are not data on new court commitments, data on a more broadly defined category of admitted prisoners are the source of the numbers for race. The more broadly defined category or prisoners admitted applies to the years 1980-82. For 1980, the data cover all types of sentenced felons admitted to prison, not just new court commitments. For 1981 and 1982, the figures for race cover sentenced felons who were new court commitments or returned conditional release violators. From 1937-60, felons were defined as prisoners with maximum sentences of 6 months or longer. No data were available for 1961-63. From 1964-70, felons were defined as prisoners with maximum sentences of 1 year or longer. After 1970, felons were defined as prisoners with maximum sentences longer than 1 year. No statistics were available for 1951-59, 1961-63, and 1971-73. The number of admissions to State and Federal prisons included prisoners whose race was unknown. Percentages are based on figures that exclude cases whether races was unknown. The "other races" category consists of Asians, American Indians, Alaska Natives, and Pacific Islanders. 1. Includes blacks plus "other" races.

★ 162 ★

Prisons and Prisoners

Characteristics of State Prison Inmates, 1986 and 1991

Characteristics	Percent of prison inmates					
	1986			1991		
	Total	Male	Female	Total	Male	Female
Number of inmates	450,416	430,604	19,812	711,643	672,847	38,796
Race, Hispanic origin						
White non-Hispanic	39.6%	39.6%	39.7%	35.4%	35.4%	36.2%
Black non-Hispanic	45.3	45.3	46.0	45.6	45.5	46.0
Hispanic	12.6	12.7	11.7	16.7	16.8	14.2
Other[b]	2.5	2.5	2.5	2.4	2.3	3.6

Source: "Characteristics of State Prison Inmates," Department of Justice, *Sourcebook of Criminal Justice Statistics-1992*, p. 622. Primary source: U.S. Department of Justice, Bureau of the Justice Statistics, *Correctional Populations in the United States, 1991*, NCJ-142729. *Notes:* These data were collected by the U.S. Bureau of the Census for the U.S. Department of Justice, Bureau of Justice Statistics through the 1991 Survey of Inmates in State Correctional Facilities. Similar surveys were conducted in 1974, 1979, and 1986. A. Percents may not add to 100 because of rounding. B. Includes Asians, Pacific Islanders, American Indians, Alaska Natives, and other racial groups.

★ 163 ★

Prisons and Prisoners

Commitment Offenses of Federal Prisoners, 1992, Part 1, Men

Offense	Total		Male					
			White		Black		Other[a]	
	Number	Percent	Number	Percent	Number	Percent	Number	Percent
Total	70,465	100.0%	42,379	100.0%	20,322	100.0%	1,797	100.0%
Federal offenses	67,720	96.1	41,533	98.0	19,434	95.6	922	51.3
Drug	41,208	58.5	24,932	58.8	11,925	58.7	491	27.3
Robbery	7,128	10.1	4,035	9.5	2,808	13.8	70	3.9
Property	3,672	5.2	2,432	5.7	815	4.0	92	5.1
Extortion, fraud, bribery	5,327	7.6	3,418	8.1	1,022	5.0	83	4.6
Violent[b]	871	1.2	545	1.3	219	1.1	55	3.1
Firearms, explosives, arson	5,366	7.6	3,057	7.2	2,071	10.2	83	4.6
White collar	988	1.4	613	1.4	161	0.8	11	0.6
Immigration	1,473	2.1	1,350	3.2	60	0.3	11	0.6
Court, corrections[d]	573	0.8	368	0.9	95	0.5	9	0.5
Sex offenses	49	0.1	23	0.1	20	0.1	0	X
National security	77	0.1	50	0.1	15	0.1	3	0.2
Continuing criminal enterprise	619	0.9	449	1.1	148	0.7	6	0.3
Other	369	0.5	261	0.6	75	0.4	8	0.4

Source: "Type of Commitment Offense Among Federal Prisoners," Department of Justice, *Sourcebook of Criminal Justice Statistics-1992*, p. 636. Primary source: U.S. Department of Justice, Federal Bureau of Prisoners, *Federal Bureau of Prisons Annual Statistical Report Calendar Year 1992. Notes:* A. Includes Asians and Native Americans. B. Includes crimes such as homicide and kidnaping. C. Value to small to display. D. Includes crimes such as harboring a fugitive, possessing or bringing contraband into a prison, and perjury.

★ 164 ★

Prisons and Prisoners

Commitment Offenses of Federal Prisoners, 1992, Part 2, Women

Offense	Total		Female					
			White		Black		Other[a]	
	Number	Percent	Number	Percent	Number	Percent	Number	Percent
Total	70,465	100.0%	3,479	100.0%	2,328	100.0%	160	100.0%
Federal offenses	67,720	96.1	3,431	98.6	2,290	98.4	110	68.8
Drug	41,208	58.5	2,218	63.8	1,569	67.4	73	45.6
Robbery	7,128	10.1	140	4.0	73	3.1	2	1.3
Property	3,675	5.2	184	5.3	149	6.4	0	X
Extortion, fraud, bribery	5,327	7.6	455	13.1	327	14.0	22	13.8
Violent[b]	871	1.2	36	1.0	11	0.5	5	3.1
Firearms, explosives, arson	5,366	7.6	87	2.5	66	2.8	2	1.3
White collar	988	1.4	153	4.4	48	2.1	2	1.3
Immigration	1,473	2.1	51	1.5	1	(c)	0	X
Court, corrections[d]	573	0.8	69	2.0	30	1.3	2	1.3

[Continued]

★ 164 ★

Commitment Offenses of Federal Prisoners, 1992, Part 2, Women
[Continued]

Offense	Total		Female					
			White		Black		Other[a]	
	Number	Percent	Number	Percent	Number	Percent	Number	Percent
Sex offenses	49	0.1	5	0.1	1	(c)	0	X
National security	77	0.1	8	0.2	0	X	1	0.6
Continuing criminal enterprise	619	0.9	12	0.3	4	0.2	0	X
Other	369	0.5	13	0.4	11	0.5	1	0.6

Source: "Type of Commitment Offense Among Federal Prisoners," Department of Justice, *Sourcebook of Criminal Justice Statistics-1992*, p. 636. Primary source: U.S. Department of Justice, Federal Bureau of Prisons, *Federal Bureau of Prisons Annual Statistical Report Calendar Year 1992*. *Notes:* A. Includes Asians and Native Americans. B. Includes crimes such as homicide and kidnaping. C. Value to small to display. D. Includes crimes such as harboring a fugitive, possessing or bringing contraband into a prison, and perjury.

★ 165 ★

Prisons and Prisoners

Correctional Officers in the Federal Bureau of Prisons, 1992

By race and ethnicity, 1992.

	Total		Race and ethnicity							
			White		Black		Hispanic		Other[1]	
	Number	Percent	Number	Percent	Number	Percent	Number	Percent	Number	Percent
Total	9,217	100.0%	6,058	100.0%	2,052	100.0%	974	100.0%	133	100.0%
Sex										
Male	8,160	88.5	5,479	90.4	1,662	81.0	901	92.5	118	88.7
Female	1,057	11.5	579	9.6	390	19.0	73	7.5	15	11.3
Age										
18 to 24 years	431	4.7	320	5.3	62	3.0	40	4.1	9	6.8
25 to 29 years	2,566	27.8	1,698	28.0	568	27.7	277	28.4	23	17.3
30 to 34 years	2,886	31.3	1,788	29.5	710	34.6	348	35.7	40	30.1
35 to 39 years	2,005	21.8	1,344	22.2	422	20.6	206	21.1	33	24.8
40 to 44 years	846	9.2	581	9.6	175	8.5	71	7.3	19	14.3
45 to 49 years	341	3.7	247	4.1	68	3.3	19	2.0	7	5.3
50 to 55 years	71	0.8	47	0.8	23	1.1	1	0.1	0	X
56 years and older	71	0.8	33	0.5	24	1.2	12	1.2	2	1.5
Education										
High school	3,834	41.6	2,603	43.0	764	37.2	414	42.5	53	39.8
Technical school	452	4.9	318	5.2	88	4.3	41	4.2	5	3.8
Some college	3,234	35.1	2,066	34.1	703	34.3	409	42.0	56	42.1
College degree	1,477	16.0	938	15.5	428	20.9	93	9.5	18	13.5
Some graduate school	124	1.4	85	1.4	31	1.5	10	1.0	0	X

[Continued]

★ 165 ★

Correctional Officers in the Federal Bureau of Prisons, 1992
[Continued]

	Total		Race and ethnicity							
			White		Black		Hispanic		Other[1]	
	Number	Percent	Number	Percent	Number	Percent	Number	Percent	Number	Percent
Professional degree	15	0.2	8	0.1	6	0.3	1	0.1	0	X
Master's degree	79	0.9	40	0.7	32	1.6	6	0.6	1	0.8

Source: "Characteristics of Federal Bureau of Prisons Correctional Officers," *Sourcebook of Criminal Justice Statistics-1992,* p. 127. Primary source: U.S. Department of Justice, Federal Bureau of Prisons, Federal Bureau of Prisons *Annual Statistical Report Calendar Year 1992. Notes:* These data refer to staff who are in current pay status and excludes staff who are on leave without pay. 1. Includes Asians and Native Americans.

★ 166 ★

Prisons and Prisoners

Correctional Personnel in Adult Systems, 1992

Jurisdiction	Total number of employees	Adult systems							
		White		Black		Hispanic		All others	
		Male	Female	Male	Female	Male	Female	Male	Female
Total	309,361	163,189	60,715	38,386	21,907	12,340	4,489	4,726	1,946
Alabama	3,196	1,272	382	996	531	0	0	12	3
Alaska	790	480	99	51	10	13	4	98	35
Arizona	5,890	3,022	1,324	210	102	817	295	85	35
Arkansas	2,303	1,125	297	691	186	1	0	3	0
California	30,217	12,172	5,514	2,901	2,004	3,811	1,813	1,322	680
Colorado	2,814	1,591	662	97	27	264	96	61	16
Connecticut	4,776	2,441	879	794	288	292	53	20	9
Delaware	1,404	686	249	310	126	10	4	13	6
District of Columbia	4,235	343	85	2,066	1,175	36	9	322	199
Florida	20,305	9,853	5,001	2,215	2,163	544	277	173	79
Georgia	10,680	4,993	2,278	2,232	1,053	63	11	40	10
Hawaii	993[1]	167	23	71	13	95	17	540	67
Idaho	901[2]	NA	NA	NA	NA	NA	NA	NA	NA
Illinois	11,407	7,227	2,322	1,127	484	134	37	55	21
Indiana	5,838	3,333	1,502	491	425	48	10	19	10
Iowa	1,709	1,218	380	37	14	28	6	20	6
Kansas	2,927	1,913	707	154	57	45	16	25	10
Kentucky	2,984	1,890	850	141	89	4	2	7	1
Louisiana	4,376[1]	2,088	740	1,033	509	4	1	1	0
Maine	1,336	1,160	175	1	0	0	0	0	0
Maryland	5,385	2,508	535	1,449	851	19	3	11	9
Massachusetts	5,224	3,698	957	275	127	116	21	25	5
Michigan	13,966	8,018	3,052	1,217	1,150	151	58	236	84
Minnesota	2,660	1,633	848	57	22	21	8	46	25
Mississippi	2,544	587	354	836	755	2	0	5	5
Missouri	5,911	3,778	1,770	177	142	17	2	16	9
Montana	624	471	125	0	0	7	4	14	3

[Continued]

★ 166 ★

Correctional Personnel in Adult Systems, 1992
[Continued]

Jurisdiction	Total number of employees	Adult systems							
		White		Black		Hispanic		All others	
		Male	Female	Male	Female	Male	Female	Male	Female
Nebraska	1,446	850	450	74	30	27	7	6	2
Nevada	1,571	981	354	79	32	59	20	37	9
New Hampshire	700	524	163	4	1	3	1	3	1
New Jersey	10,125	4,592	1,502	2,413	1,037	390	114	50	27
New Mexico	1,945	476	151	44	5	974	249	30	16
New York	29,450	20,362	4,834	1,949	931	753	208	303	110
North Carolina	10,089	5,455	1,242	2,463	761	31	4	110	23
North Dakota	135	110	20	0	0	0	0	5	0
Ohio	8,935	5,348	2,046	894	545	37	12	41	12
Oklahoma	1,732	1,292	147	109	27	14	2	120	21
Oregon	2,218	1,357	616	47	22	56	27	63	30
Pennsylvania	7,464	5,445	1,153	565	224	54	5	13	5
Rhode Island	1,557	1,147	231	72	19	29	7	40	12
South Carolina	5,731	1,814	969	1,854	1,020	0	0	56	18
South Dakota	588[1]	391	184	1	1	0	0	10	1
Tennessee	5,073	2,894	1,183	580	403	12	1	0	0
Texas	21,833	9,806	5,393	2,504	1,833	1,556	590	101	50
Utah	1,785[3]	1,196	589	0	0	0	0	0	0
Vermont	762[1,2]	NA	NA	NA	NA	NA	NA	NA	NA
Virginia	8,381	3,572	1,583	2,047	1,108	37	10	14	10
Washington	5,884	3,125	1,925	275	94	141	68	138	118
West Virginia	776	546	217	8	4	1	0	0	0
Wisconsin	2,210	1,734	320	42	25	38	3	40	8
Wyoming	243	154	38	0	0	41	7	3	0
Federal Bureau of Prisons	23,333	12,351	4,295	2,733	1,482	1,545	407	374	146

Source: "Correctional Personnel in Adult and Juvenile Systems," *Sourcebook of Criminal Justice Statistics-1992,* pp. 96-97. Primary source: American Correctional Association, *1993 Directory of Juvenile and Adult Correctional Departments, Institutions, Agencies, and Paroling Authorities* (Laurel, MD: American Correctional Association, 1993). Published by permission. *Notes:* This information was collected through a mail survey sent to the director of each State department of corrections. Questionnaires may have been forwarded to another office. 1. Data as of June 30, 1991. 2. Race, ethnicity, and sex breakdowns not available. 3. Race breakdown: 1,681 white, 68 non-white, 36 unknown.

★ 167 ★

Prisons and Prisoners

Correctional Personnel in Juvenile Systems, 1992

Jurisdiction	Total number of employees	Juvenile systems							
		White		Black		Hispanic		All others	
		Male	Female	Male	Female	Male	Female	Male	Female
Total	36,127	10,940	8,204	6,171	3,982	1,578	760	379	265
Alabama	502	109	76	227	90	0	0	0	0
Alaska	294[1]	151	100	12	6	3	2	7	13
Arizona	667	287	189	53	29	59	34	8	8
Arkansas	333[1]	63	74	135	61	0	0	0	0
California	5,529	1,714	1,070	766	513	737	380	200	149
Colorado	544	216	150	55	24	64	23	9	3
Connecticut	377	139	105	64	26	25	14	1	3
Delaware	184	39	41	64	37	2	1	0	0
District of Columbia	580[1]	28	16	339	181	4	2	5	5
Florida	2,697[2]	NA	NA	NA	NA	NA	NA	NA	NA
Georgia	2,380	516	512	754	582	6	3	6	1
Hawaii	62	1	1	3	0	1	0	35	21
Idaho	190[3]	117	67	1	1	2	0	1	1
Illinois	[4]	X	X	X	X	X	X	X	X
Indiana	[4]	X	X	X	X	X	X	X	X
Iowa	211	152	56	1	0	1	1	0	0
Kansas	502	184	194	76	23	13	2	4	6
Kentucky	791	550	156	58	27	0	0	0	0
Louisiana	[4]	X	X	X	X	X	X	X	X
Maine	[4]	X	X	X	X	X	X	X	X
Maryland	1,083	332	246	208	289	2	2	2	2
Massachusetts	554	272	131	75	20	35	15	3	3
Michigan	957[2]	NA	NA	NA	NA	NA	NA	NA	NA
Minnesota	[4]	X	X	X	X	X	X	X	X
Mississippi	368	68	93	106	100	0	0	1	0
Missouri	311	135	98	54	24	0	0	0	0
Montana	194[2]	NA	NA	NA	NA	NA	NA	Na	NA
Nebraska	[4]	X	X	X	X	X	X	X	X
Nevada	162	108	43	1	0	4	2	4	0
New Hampshire	193	116	73	4	0	0	0	0	0
New Jersey	[4]	X	X	X	X	X	X	X	X
New Mexico	373	75	57	3	6	169	62	1	0
New York	3,274	1,096	820	774	373	136	52	14	9
North Carolina	900	238	210	275	166	4	1	3	3
North Dakota	84	49	33	1	0	0	0	1	0
Ohio	1,797	496	446	473	360	8	3	7	4
Oklahoma	504	214	134	122	12	1	3	10	8
Oregon	533	273	196	21	6	17	7	7	6
Pennsylvania	847	400	147	204	89	2	3	1	1
Rhode Island	122	80	9	21	5	4	0	3	0
South Carolina	971	167	269	265	267	1	1	0	1
South Dakota	[4]	X	X	X	X	X	X	X	X
Tennessee	984	343	321	180	140	0	0	0	0

[Continued]

★ 167 ★

Correctional Personnel in Juvenile Systems, 1992
[Continued]

Jurisdiction	Total number of employees	Juvenile systems							
		White		Black		Hispanic		All others	
		Male	Female	Male	Female	Male	Female	Male	Female
Texas	1,980	575	465	342	219	238	125	10	6
Utah	655	332	226	27	7	26	9	22	6
Vermont	377[3]	112	264	1	0	0	0	0	0
Virginia	1,587	413	515	367	280	2	7	2	1
Washington	819[1]	405	321	39	19	12	6	12	5
West Virginia	[4]	X	X	X	X	X	X	X	X
Wisconsin	537[5]	322	215	0	0	0	0	0	0
Wyoming	118	53	65	0	0	0	0	0	
Federal Bureau of Prisons	[6]	X	X	X	X	X	X	X	X

Source: "Correctional Personnel in Adult and Juvenile Systems," *Sourcebook of Criminal Justice Statistics-1992*, pp. 96-97. Primary source: American Correctional Association, *1993 Directory of Juvenile and Adult Correctional Departments, Institutions, Agencies, and Paroling Authorities* (Laurel, MD: American Correctional Association, 1993). Published by permission. *Notes:* This information was collected through a mail survey sent to the director of each State department of corrections. Questionnaires may have been forwarded to another office. 1. Data as of June 30, 1991. 2. Race, ethnicity, and sex breakdowns not available. 3. Data as of June 30, 1990. 4. Combined adult and juvenile departments. 5. Race breakdown: 478 white and 59 non-white. 6. The Federal Bureau of Prisons does not operate facilities for juveniles.

★ 168 ★

Prisons and Prisoners

Drug Use History of State Prison Inmates, 1991

Characteristics	Number of inmates	Drug use history			
		Ever used drugs	Used drugs in the month before the offense		Under the influence of drugs at time of offense
			At all	Daily	
All inmates	710,798	79.4%	49.9%	36.0%	30.9%
Race, Hispanic origin					
White non-Hispanic	251,916	79.9	49.4	38.2	32.4
Black non-Hispanic	323,677	78.9	49.0	33.7	28.8
Hispanic	118,457	80.1	53.6	37.8	33.6
Other[a]	16,748	76.6	48.4	36.1	32.3

Source: "Drug Use History of State Prison Inmates," Department of Justice, *Sourcebook of Criminal Justice Statistics-1992*, p. 627. Primary source: U.S. Department of Justice, Bureau of the Statistics, *Correctional Populations in the United States, 1991*, NCJ-142729 *Notes:* A. Includes Asians, Pacific Islanders, American Indians, Alaska Natives, and other racial groups.

★ 169 ★

Prisons and Prisoners

Federal Prisoners, 1990

	Federal prison admissions		
	Total	Total maximum sentence length	
		12 months or less	More than 12 months
Number of admissions	32,825	13,186	18,476
Total	100%	100%	100%
Race[a]			
White	72.9	81.2	66.0
Black	24.9	16.8	31.7
Other[b]	2.2	2.0	2.3
Hispanic origin[c]			
Hispanic	35.0	48.9	23.3
Non-Hispanic	65.0	51.1	76.7

Source: "Federal Prison Admissions," Department of Justice, *Sourcebook of Criminal Justice Statistics-1992*, p. 636. Primary source: U.S. Department of Justice, Federal Bureau of Prisons, Federal Bureau of Prisons, *National Correctional Reporting Program, 1990*, NCH- 141879. *Notes:* Sentence length refers to the total maximum sentence that an offender may be required to serve all offenses. A. Includes persons of Hispanic origin. B. Includes American Indians, Alaska Natives, Asians, and Pacific Islanders. C. Includes persons of all races.

★ 170 ★

Prisons and Prisoners

Federal Prisoners, 1992, Part 1, Men

Total	Total		Male					
			White		Black		Other[a]	
	Number	Percent	Number	Percent	Number	Percent	Number	Percent
Total	79,859	100.0%	48,165	100.0%	23,304	100.0%	1,991	100.0%
Security Level								
High	8,466	10.6	4,399	9.1	3,739	16.0	244	12.3
Medium	25,437	31.9	15,221	31.6	8,619	37.0	748	37.6
Low	10,923	13.7	7,310	15.2	3,276	14.1	337	16.9
Minimum	16,413	20.6	10,268	21.3	3,651	15.7	188	9.4

[Continued]

★ 170 ★

Federal Prisoners, 1992, Part 1, Men
[Continued]

Total	Total		Male					
			White		Black		Other[a]	
	Number	Percent	Number	Percent	Number	Percent	Number	Percent
Administrative[b]	10,432	13.1	5,549	11.5	2,426	10.4	234	11.8
Contract[c]	8,188	10.3	5,418	11.2	1,593	6.8	240	12.1

Source: "Federal Prisoners" Department of Justice, *Sourcebook of Criminal Justice Statistics-1992*, p. 636. Primary source: U.S. Department of Justice, Federal Bureau of Prisons, *Federal Bureau of Prisons Annual Statistical Report Calendar Year 1992. Notes:* A. Includes Asians and Native Americans. B. Includes special populations such as individuals requiring medical treatment or those in pretrial status regardless of security level. C. Facilities operated by an entity other than the Federal Bureau of Prisons that house Bureau prisoners under contract, e.g., community corrections centers.

★ 171 ★

Prisons and Prisoners

Federal Prisoners, 1992, Part 2, Women

Total	Total		Female					
			White		Black		Other[a]	
	Number	Percent	Number	Percent	Number	Percent	Number	Percent
Total	79,859	100.0%	3,767	100.0%	2,459	100.0%	173	100.0%
Security level								
High	8,466	10.6	51	1.4	31	1.3	2	1.2
Medium	25,437	31.9	556	14.8	241	9.8	52	30.1
Low	10,923	13.7	0	X	0	X	0	X
Minimum	16,413	20.6	1,310	34.8	963	39.2	33	19.1
Administrative[b]	10,432	13.1	1,275	33.8	915	37.3	33	19.1
Contract[c]	8,188	10.3	575	15.3	309	12.6	53	30.6

Source: "Federal Prisoners" Department of Justice, *Sourcebook of Criminal Justice Statistics-1992*, p. 636. Primary source: U.S. Department of Justice, Federal Bureau of Prisons, *Federal Bureau of Prisons Annual Statistical Report Calendar Year 1992. Notes:* A. Includes Asians and Native Americans. B. Includes special populations such as individuals requiring medical treatment or those in pretrial status regardless of security level. C. Facilities operated by an entity other than the Federal Bureau of Prisons that house Bureau prisoners under contract, e.g., community corrections centers.

★ 172 ★

Prisons and Prisoners

Jail Inmates 1985 to 1991

Excludes Federal and State prisons or other correctional institutions; institutions exclusively for juveniles; State-operated jails in Alaska, Connecticut, Delaware, Hawaii, Rhode Island, and Vermont; and other facilities which retain persons for less than 48 hours. As of June 30. For 1978, 1983, and 1988, data based on National Jail Census; for other years, based on sample survey and subject to sampling variability.

Characteristic	1985	1986	1987	1988	1989	1990	1991
Total inmates[1]	256,615	274,444	295,873	343,569	395,553	405,320	426,479
White	151,403	159,178	168,647	166,302	201,732	186,989	190,333
Black	102,646	112,522	124,267	141,979	185,910	174,335	187,617
Other races	2,566	2,744	2,959	3,932	7,911	5,321	5,391
Hispanic[2]	35,926	38,422	41,422	51,455	55,377	57,449	60,129
Non-Hispanic	220,689	236,022	254,451	292,114	340,176	347,871	366,350

Source: "Jail Inmates, by Race and Detention Status: 1978 to 1991," *Statistical Abstract of the United States, 1993,* p. 210. Primary source: U.S. Bureau of Justice Statistics, *Profile of Jail Inmates, 1978 and 1989, Jail Inmates,* annual, and *1988 Census of Local Jails. Notes:* 1. For 1985 to 1987, 1989, and 1991, includes juveniles not shown separately by sex, and for 1988, 1990, and 1991 includes 31,356, 38,671, and 43,138 persons, respectively, of unknown race not shown separately. 2. Hispanic persons may be of any race.

★ 173 ★

Prisons and Prisoners

Most Serious Offense of State Prison Inmates, 1991

Most serious offense	Percent of prison inmates 1991				
	Total	White non-Hispanic	Black non-Hispanic	Hispanic	Other[b]
Number of inmates	704,181	248,705	321,217	117,632	16,627
Violent offenses	46.6%	49.0%	47.2%	38.6%	53.6%
Murder	10.6	11.8	10.3	8.8	12.0
Negligent manslaughter	1.8	2.0	1.7	1.9	0.7
Kidnaping	1.2	1.5	1.0	0.8	1.8
Rape	3.5	4.8	3.1	1.7	6.3
Other sexual assault	5.9	10.5	2.9	4.3	5.9
Robbery	14.8	10.3	19.2	12.6	12.8
Assault	8.2	7.4	8.5	8.1	13.8
Other violent	0.6	0.7	0.5	0.5	0.4
Property offenses	24.8	30.2	22.0	20.6	28.4
Burglary	12.4	15.3	10.5	11.6	11.9
Larceny/theft	4.9	5.4	5.0	3.5	4.3
Motor vehicle theft	2.2	2.4	1.9	2.3	2.3
Arson	0.7	1.1	0.4	0.5	0.9

[Continued]

★ 173 ★

Most Serious Offense of State Prison Inmates, 1991
[Continued]

Most serious offense	Percent of prison inmates 1991				
	Total	White non-Hispanic	Black non-Hispanic	Hispanic	Other[b]
Fraud	2.8	3.8	2.5	1.2	5.0
Stolen property	1.4	1.5	1.3	1.0	3.3
Other property	0.4	0.5	0.4	0.4	0.6
Drug offenses	21.3	12.0	24.9	33.0	9.7
Possession	7.6	4.0	8.6	13.1	2.0
Trafficking	13.3	7.7	15.7	19.5	7.4
Other and unspecified	0.5	0.3	0.6	0.4	0.3
Public-order offenses	6.9	8.4	5.4	7.5	8.2
Weapons	1.8	1.2	2.2	2.0	1.1
Other public-order	5.1	7.1	3.2	5.5	7.1
Other offenses	0.4	0.5	0.4	0.3	0.1

Source: "Most Serious Offence of State Prison Inmates," Department of Justice, *Sourcebook of Criminal Justice Statistics-1992*, p. 623. Primary source: U.S. Department of Justice, Bureau of the Justice Statistics, *Correctional Populations in the United States, 1991*, NCJ-142729. *Notes:* "Murder" includes non-negligent manslaughter. "Other violent" includes blackmail, extortion, hit-and-run driving with bodily injury, child abuse, and criminal endangerment against a person. "Other property" includes destruction of property, vandalism, criminal tampering, trespassing, entering without breaking, and possession of burglary tools. "Other public-order" includes escape from custody, court offenses, obstruction, driving while intoxicated, other traffic offenses, drunkenness, disorderly conduct, morals and decency violations, commercialized vice, and liquor violations. "Other offenses" includes juvenile offenses and unspecified offenses. A. Subcategories may not add to total because of rounding. B. Includes Asians, Pacific Islanders, American Indians, Alaska Natives, and other racial groups.

★ 174 ★

Prisons and Prisoners

New Court Commitments to Prisons in 35 States, 1990

Most serious offense	All new court commitments	Total	Race[2]			Hispanic origin[1]	
			White	Black	Other[3]	Hispanic	Non-Hispanic
All offenses	267,394	100%	45.2%	54.0%	0.8%	18.3%	81.7%
Violent offenses	71,778	100	44.6	54.3	1.1	17.2	82.8
Homicide	10,359	100	47.4	51.1	1.5	20.5	79.5
Murder and nonnegligent manslaughter	7,072	100	43.5	54.9	1.6	22.9	77.1
Murder	5,865	100	44.2	54.1	1.7	21.6	78.4
Nonnegligent manslaughter	1,207	100	39.9	58.8	1.3	27.7	72.3
Negligent manslaughter	3,079	100	56.3	42.3	1.3	16.6	83.4
Unspecified homicide	208	100	39.6	59.9	0.5	4.9	95.1
Kidnaping	1,572	100	53.7	44.7	1.6	17.1	82.9
Rape	6,400	100	60.3	38.6	1.1	17.0	83.0
Other sexual assault	8,224	100	73.3	25.0	1.0	14.9	85.1

[Continued]

★ 174 ★

New Court Commitments to Prisons in 35 States, 1990

[Continued]

Most serious offense	All new court commitments	Total	Race[2]			Hispanic origin[1]	
			White	Black	Other[3]	Hispanic	Non-Hispanic
Robbery	24,721	100	29.4	69.9	0.7	16.3	83.7
Assault	18,725	100	42.3	56.4	1.3	17.8	82.2
Other violent	1,777	100	59.4	38.5	2.0	14.5	85.5
Property offenses	86,405	100	52.4	46.8	0.8	14.1	85.9
Burglary	38,667	100	54.4	44.7	0.9	17.0	83.0
Larceny-theft	21,588	100	48.1	51.1	0.8	12.6	87.4
Motor vehicle theft	6,954	100	51.3	47.4	1.2	18.0	82.0
Arson	1,678	100	63.0	36.1	0.9	11.1	88.9
Fraud	10,501	100	55.4	43.9	0.6	6.1	93.9
Stolen property	5,063	100	46.4	53.2	0.4	10.6	89.4
Other property	1,954	100	53.6	45.6	0.8	9.1	90.9
Drug offenses	84,679	100	33.7	66.0	0.4	22.5	77.5
Possession	24,358	100	28.3	71.5	0.2	14.6	85.4
Trafficking	47,637	100	34.2	65.4	0.5	26.6	73.4
Other drug	12,684	100	41.4	58.2	0.4	23.3	76.7
Public-order offenses	21,682	100	59.4	39.6	1.0	19.6	80.4
Weapons	5,136	100	34.4	64.8	0.8	15.4	84.6
Driving while intoxicated	7,377	100	82.5	16.1	1.4	35.3	64.7
Other public-order	9,169	100	54.8	44.5	0.8	6.5	93.5
Other offenses	2,850	100	61.8	37.2	1.0	21.1	78.9

Source: "New Court Commitments to Prisons in 35 States," Department of Justice, *Sourcebook of Criminal Justice Statistics-1992*, p. 629. Primary source: U.S. Department of Justice, Bureau of the Statistics, *National Corrections Reporting Program, 1990*, NCJ-141879. *Notes:* Numbers by offense are based on new court commitments with sentences of more than 1 year and valid sex data. Base numbers of race and Hispanic origin by offense may differ from those for sex. 1. Includes persons of all races. 2. Includes persons of Hispanic origin. 3. Includes American Indians, Alaska Natives, Asians, and Pacific Islanders.

★ 175 ★

Prisons and Prisoners

Parole Discharge in 27 States, 1990

Method of parole discharge	All discharges	Race[1]			
		White	Black	Other[2]	Hispanic[3]
Number of discharges	159,279	73,295	64,643	794	30,414
All methods	100%	100%	100%	100%	100%
Successful completion	36.4	38.6	34.1	34.9	31.3
Absconder	0.5	0.4	0.8	0.5	0.2
Return to jail or prison[4]	61.6	59.5	63.5	63.1	67.4
Transfer	0.2	0.3	0.2	0.3	0.1

[Continued]

★ 175 ★

Parole Discharge in 27 States, 1990
[Continued]

Method of parole discharge	All discharges	Race[1]			
		White	Black	Other[2]	Hispanic[3]
Death	0.8	0.7	1.0	0.9	0.8
Other	0.4	0.5	0.4	0.4	0.2

Source: "Parole Discharges in 27 States," Department of Justice, *Sourcebook of Criminal Justice Statistics-1992*, p. 662. Primary source: U.S. Department of Justice, Federal Bureau of Prisons, Federal Bureau of Prisons, *National Correctional Reporting Program, 1990*, NCH- 141879. *Notes:* Data were reported for 97.7 percent of the 162,959 State parole discharges who entered prison with a sentence of more than a year and include those on supervised release even if not technically termed "parole." 1. Includes persons of Hispanic origin. 2. Includes American Indians, Alaska Natives, Asians, and Pacific Islanders. 3. Includes persons of all races. 4. Includes those returned to prison with a new sentence, technical parole violators, and those returned pending parole revocation.

★ 176 ★

Prisons and Prisoners

Prisoners Executed, 1930 to 1991

Excludes executions by military authorities. The Army (including the Air Force) carried out 160 (148 between 1942 and 1950, 3 each in 1954, 1955, and 1957, and 1 each in 1958, 1959, and 1961). Of the total, 106 were executed for murder (including 21 involving rape), 53 for rape and 1 for desertion. The Navy carried out no executions during the period.

Year or period	Total[1]	White	Black	Executed for murder			Executed for rape			Executed other offenses[2]		
				Total[1]	White	Black	Total[1]	White	Black	Total[1]	White	Black
All years	4,016	1,845	2,129	3,491	1,758	1,693	455	48	405	70	39	31
1930 to 1939	1,667	827	816	1,514	803	687	125	10	115	28	14	14
1940 to 1949	1,284	490	781	1,064	458	595	200	19	179	20	13	7
1950 to 1959	717	336	376	601	316	280	102	13	89	14	7	7
1960 to 1967	191	98	93	155	87	68	28	6	22	8	5	3
1968 to 1976	-	-	-	-	-	-	-	-	-	-	-	-
1977 to 1980	3	3	-	3	3	-	-	-	-	-	-	-
1981	1	1	-	1	1	-	-	-	-	-	-	-
1982	2	1	1	2	1	1	-	-	-	-	-	-
1983	5	4	1	5	4	1	-	-	-	-	-	-
1984	21	13	8	21	13	8	-	-	-	-	-	-
1985	18	11	7	18	11	7	-	-	-	-	-	-
1986	18	11	7	18	11	7	-	-	-	-	-	-
1987	25	13	12	25	13	12	-	-	-	-	-	-
1988	11	6	5	11	6	5	-	-	-	-	-	-
1989	16	8	8	16	8	8	-	-	-	-	-	-
1990	23	16	7	23	16	7	-	-	-	-	-	-
1991	14	7	7	14	7	7	-	-	-	-	-	-

Source: "Prisoners Under Sentence of Death: 1980 to 1991," *Statistical Abstract of the United States, 1993*, p. 213. Primary source: Through 1978, U.S. Law Enforcement Assistance Administration; thereafter, U.S. Bureau of Justice Statistics, *Correctional Population in the United States*, annual. *Notes:* - represents zero. 1. Includes other races other than White or Black. 2. Includes 25 armed robbery, 20 kidnaping, 11 burglary, 8 espionage (6 in 1942 and 2 in 1953), and 6 aggravated assault.

★ 177 ★

Prisons and Prisoners

Prisoners Under Jurisdiction of State and Federal Correctional Authorities, 1991 Part 1, Northeast and Midwest

By race, region, and jurisdiction, on dec. 31, 1991.

Region and jurisdiction	Total	White	Black	American Indian or Alaska Native	Asian or Pacific Islander	Not known
United States, total	824,133	385,347	395,245	7,407	3,423	32,711
Federal institutions, total	71,608	46,868	22,727	1,222	791	0
State institutions, total	752,525	338,479	372,518	6,185	2,632	32,711
Northeast	131,866	56,815	66,442	214	338	8,057
Connecticut[1,2]	10,977	3,053	5,144	7	26	2,747
Maine	1,579	1,522	37	16	4	0
Massachusetts[2,3]	9,155	4,410	3,036	14	51	1,644
New Hampshire	1,533	1,443	80	5	5	0
New Jersey[2]	23,483	6,762	15,005	4	41	1,671
New York[4]	57,862	28,181	29,151	135	155	240
Pennsylvania[2]	23,388	8,470	13,090	28	45	1,755
Rhode Island[1]	2,771	1,856	899	5	11	0
Vermont[1,5]	1,118	1,118	NA	NA	NA	0
Midwest	155,917	71,227	79,217	1,394	130	3,949
Illinois[2,3]	29,115	8,055	18,306	49	28	2,677
Indiana[3]	13,008	8,000	4,971	30	7	0
Iowa[3]	4,145	3,089	940	69	15	32
Kansas[2]	5,903	3,329	2,145	81	33	315
Michigan[2,3]	36,423	14,586	20,985	137	25	690
Minnesota[2]	3,472	1,960	1,051	287	1	173
Missouri	15,897	8,547	7,317	30	3	0
Nebraska	2,495	1,564	830	95	0	0
North Dakota	492	397	4	88	3	0
Ohio[5]	35,744	16,433	19,311	0	0	0
South Dakota	1,374	992	32	350	NA	0
Wisconsin	7,849	4,275	3,325	178	15	56

Source: "Prisoners Under Jurisdiction of State and Federal Correctional Authorities," Department of Justice, *Sourcebook of Criminal Justice Statistics-1992,* p. 613. Primary source: U.S. Department of Justice, Bureau of Justice Statistics, *Correctional Populations In The United States, 1991,* NCJ-142729. *Notes:* 1. Figures include both jail and prison inmates; jails and prisons are combined in one system. 2. Hispanic prisoners were classified as persons of unknown race. 3. All data for Arizona, California, Florida, Georgia, Illinois, Indiana, Iowa, Massachusetts, Michigan, Texas, and Wyoming are custody rather than jurisdiction counts. 4. Hispanic persons were classified as white. 5. Racial group membership of the population was estimated.

★ 178 ★

Prisons and Prisoners

Prisoners Under Jurisdiction of State and Federal Correctional Authorities, 1991, Part 2, South and West

Region and jurisdiction	Total	White	Black	American Indian or Alaska Native	Asian or Pacific Islander	Not known
United States, total	824,133	385,347	395,245	7,407	3,423	32,711
Federal institutions, total	71,608	46,868	22,727	1,222	791	0
State institutions, total	752,525	338,479	372,518	6,185	2,632	32,711
South	301,866	104,969	181,341	1,249	374	13,933
Alabama[b]	16,760	5,958	10,793	6	2	1
Arkansas[b]	7,766	3,302	4,437	3	1	23
Delaware[a,b]	3,717	1,175	2,449	2	3	88
District of Columbia[a,e]	10,455	218	10,237	0	0	0
Florida[b,c]	46,533	18,383	27,185	0	105	860
Georgia[c]	23,644	7,613	15,931	20	6	74
Kentucky	9,799	6,672	3,123	2	0	2
Louisiana[f]	20,003	5,168	14,834	NA	NA	1
Maryland	19,291	4,581	14,638	6	0	66
Mississippi[b]	8,904	2,437	6,410	7	9	41
North Carolina	18,903	6,747	11,522	421	11	202
Oklahoma[b]	13,340	7,522	4,652	760	0	406
South Carolina	18,269	6,099	12,120	13	2	35
Tennessee	11,474	5,857	5,503	NA	NA	114
Texas[b,c]	51,677	15,013	24,520	6	193	11,945
Virginia[b]	19,829	6,942	12,769	2	41	75
West Virginia	1,502	1,282	218	1	1	0
West	162,876	105,468	45,518	3,328	1,790	6,772
Alaska[a,e]	2,706	1,488	339	847	32	0
Arizona[c]	15,415	12,261	2,633	498	12	1
California[c]	101,808	61,594	35,205	662	NA	4,347
Colorado[e]	8,392	5,990	1,937	108	27	330
Hawaii[a,b,c]	2,700	642	155	34	1,470	399
Idaho[e]	2,143	1,997	32	94	15	5
Montana	1,478	1,189	20	269	0	0
Nevada[b]	5,503	3,141	1,719	77	50	516
New Mexico	3,119	2,680	316	97	4	22
Oregon	6,732	4,994	923	147	51	617
Utah	2,625	2,264	222	67	36	36

[Continued]

★ 178 ★

Prisoners Under Jurisdiction of State and Federal Correctional Authorities, 1991, Part 2, South and West

[Continued]

Region and jurisdiction	Total	White	Black	American Indian or Alaska Native	Asian or Pacific Islander	Not known
Washington	9,156	6,345	1,966	372	91	382
Wyoming[b,c]	1,099	873	51	56	2	117

Source: "Prisoners Under Jurisdiction of State and Federal Correctional Authorities," Department of Justice, *Sourcebook of Criminal Justice Statistics-1992*, p. 613. Primary source: U.S. Department of Justice, Bureau of Justice Statistics, *Correctional Populations in the United States, 191*, NCJ-142729. *Notes:* A. Figures include both jail and prison inmates; jails and prisons are combined in one system. B. Hispanic prisoners were classified as persons of unknown race. C. All data for Arizona, California, Florida, Georgia, Illinois, Indiana, Iowa, Massachusetts, Michigan, Texas, and Wyoming are custody, rather than jurisdiction counts. D. Hispanic prisoners were classified as white. E. Racial group membership of the population was estimated. F. Louisiana and Tennessee reported persons whose race is neither black nor white under "other race," here reported under "unknown race."

★ 179 ★

Prisons and Prisoners

Prisoners Under Sentence of Death, 1980 to 1991

As of December 31. Excludes prisoners under sentence of death who remained within local correctional systems pending exhaustion of appellate process or who had not been committed to prison.

Race	1980	1984	1985	1986	1987	1989	1990	1991
Total[1]	688	1,405	1,575	1,781	1,984	2,243	2,346	2,482
White	418	804	896	1,006	1,138	1,308	1,368	1,464
Black and other	270	601	679	775	846	935	978	1,018

Source: "Prisoners Under Sentence of Death: 1980 to 1991," *Statistical Abstract of the United States, 1993*, p. 212. Primary source: U.S. Bureau of Justice Statistics, *Capital Punishment*, annual. *Notes:* 1. For 1980 to 1990, revisions to the total number of prisoners were not carried to the characteristics except for race.

★ 180 ★

Prisons and Prisoners

Prisoners Under Sentence of Death, 1980 to 1991

Prisoners reported under sentence of death by civil authorities. The term "under sentence of death" begins when the court pronounces the first sentence of death for a capital offense.

Status	1980	1981	1982	1983	1984	1985	1986	1987	1988	1989	1990	1991
Under sentence of death, Jan. 1	595	697	856	1,063	1,209	1,420	1,575	1,800	1,967	2,117	2,243	2,346
Received death sentence[1]	203	250	287	263	296	281	297	299	296	251	244	266
White	125	131	166	156	173	165	164	190	196	133	147	163
Black	77	115	117	105	119	114	123	106	91	114	94	101

[Continued]

★ 180 ★

Prisoners Under Sentence of Death, 1980 to 1991
[Continued]

Status	1980	1981	1982	1983	1984	1985	1986	1987	1988	1989	1990	1991
Under sentence of death, Dec. 31[1]	688	864	1,063	1,209	1,420	1,575	1,800	1,967	2,117	2,243	2,346	2,482
White	425	499	615	694	809	896	1,006	1,128	1,238	1,308	1,368	1,464
Black	268	357	446	508	595	664	750	813	853	898	940	982

Source: "Movement of Prisoners Under Sentence of Death: 1980 to 1991," *Statistical Abstract of the United States, 1993*, p. 213. Primary source: U.S. bureau of Justice Statistics, *Capital Punishment*, annual. *Notes:* - represents zero. 1. Includes races other than White or Black.

★ 181 ★

Prisons and Prisoners

Prisoners Under Sentence of Death, Changes During 1991

Region and jurisdiction	Changes during 1991								
	Received under sentence of death			Removed from death row (excluding executions)[a]			Executed		
	Total[b]	White	Black	Total[b]	White	Black	Total[b]	White	Black
United States, total	266	163	101	116	60	52	14	7	7
Federal[c]	1	1	0	0	0	0	0	0	0
State	265	162	101	116	60	52	14	7	7
Northeast	21	6	14	10	2	8	0	0	0
Connecticut	2	0	2	0	0	0	0	0	0
New Hampshire	0	0	0	0	0	0	0	0	0
New Jersey	0	0	0	6	2	4	0	0	0
Pennsylvania	19	6	12	4	0	4	0	0	0
Midwest	37	21	16	17	5	12	1	0	1
Illinois	7	1	6	3	0	3	0	0	0
Indiana	3	2	1	2	1	1	0	0	0
Missouri	13	9	4	6	3	3	1	0	1
Nebraska	1	1	0	0	0	0	0	0	0
Ohio	13	8	5	6	1	5	0	0	0
South Dakota	0	0	0	0	0	0	0	0	0
South	158	104	53	73	40	31	13	7	6
Alabama	6	4	2	4	2	2	0	0	0
Arkansas	2	1	1	1	0	1	0	0	0
Delaware	1	1	0	0	0	0	0	0	0
Florida	45	29	16	23	12	11	2	1	1
Georgia	7	2	5	4	1	3	1	0	1
Kentucky	3	3	0	0	0	0	0	0	0
Louisiana	7	3	4	1	0	1	1	0	1
Maryland	1	1	0	2	2	0	0	0	0
Mississippi	5	3	2	0	0	0	0	0	0

[Continued]

★ 181 ★

Prisoners Under Sentence of Death, Changes During 1991

[Continued]

Region and jurisdiction	Changes during 1991								
	Received under sentence of death			Removed from death row (excluding executions)[a]			Executed		
	Total[b]	White	Black	Total[b]	White	Black	Total[b]	White	Black
North Carolina	17	10	7	26	13	11	1	1	0
Oklahoma	12	6	5	4	3	1	0	0	0
South Carolina	8	7	1	2	2	0	1	1	0
Tennessee	12	10	2	0	0	0	0	0	0
Texas	26	19	7	4	3	1	5	3	2
Virginia	6	5	1	2	2	0	2	1	1
West	49	31	18	16	13	1	0	0	0
Arizona	13	11	2	3	3	0	0	0	0
California	24	11	13	3	2	0	0	0	0
Colorado	1	1	0	1	1	0	0	0	0
Idaho	2	2	0	0	0	0	0	0	0
Montana	0	0	0	0	0	0	0	0	0
Nevada	4	2	2	3	3	0	0	0	0
New Mexico	0	0	0	0	0	0	0	0	0
Oregon	3	3	0	4	3	1	0	0	0
Utah	1	1	0	0	0	0	0	0	0
Washington	1	0	1	1	0	0	0	0	0
Wyoming	0	0	0	1	1	0	0	0	0

Source: "Prisoners Under Sentence of Death," Department of Justice, *Sourcebook of Criminal Justice Statistics-1992*, p. 671. Primary source: U.S. Department of Justice, Federal Bureau of Prisons, Federal Bureau of Prisons, *Capital Punishment, 1991*, NCJ-136946. *Notes:* Thirty-six states and the Federal Government had death penalty statutes in effect on Dec. 31, 1990 and on Dec. 31, 1991. Some figures shown for year end 1990 have been revised from previous presentations by the source. A. Includes 6 deaths due to natural causes (2 each in Pennsylvania and FLorida, and 1 each in Missouri and California) and 1 suicide in Nevada. B. Totals include persons of other races. C. Excludes 5 males held under Armed Forces jurisdiction with a military death sentence for murder.

★ 182 ★

Prisons and Prisoners

Prisoners Under Sentence of Death, December 31, 1990 and 1991

Region and jurisdiction	Prisoners under sentence of death on Dec. 31, 1990			Prisoners under sentence of death on Dec. 31, 1991		
	Total[a]	White	Black	Total[a]	White	Black
United States, total	2,346	1,368	940	2,482	1,464	982
Federal[b]	0	0	0	1	1	0
State	2,346	1,368	940	2,481	1,463	982
Northeast	134	53	80	145	57	86
Connecticut	2	2	0	4	2	2
New Hampshire	0	0	0	0	0	0
New Jersey	10	4	6	4	2	2

[Continued]

★ 182 ★

Prisoners Under Sentence of Death, December 31, 1990 and 1991
[Continued]

Region and jurisdiction	Prisoners under sentence of death on Dec. 31, 1990			Prisoners under sentence of death on Dec. 31, 1991		
	Total[a]	White	Black	Total[a]	White	Black
Pennsylvania	122	47	74	137	53	82
Midwest	362	169	191	381	185	194
Illinois	128	47	81	132	48	84
Indiana	48	32	16	49	33	16
Missouri	71	39	32	77	45	32
Nebraska	11	7	3	12	8	3
Ohio	104	44	59	111	51	59
South Dakota	0	0	0	0	0	0
South	1,362	801	540	1,434	858	566
Alabama	117	58	58	119	60	58
Arkansas	33	21	12	34	22	12
Delaware	6	2	4	7	3	4
Florida	291	188	103	311	204	107
Georgia	99	53	46	101	54	47
Kentucky	27	21	6	30	24	6
Louisiana	32	14	18	37	17	20
Maryland	17	2	15	16	1	15
Mississippi	46	18	28	51	21	30
North Carolina	84	45	35	74	41	31
Oklahoma	117	80	28	125	83	32
South Carolina	40	17	23	45	21	24
Tennessee	85	57	23	97	67	28
Texas	323	201	117	340	214	121
Virginia	45	24	21	47	26	21
West	488	330	122	521	383	146
Arizona	87	77	7	97	85	9
California	280	173	99	301	182	112
Colorado	3	3	0	3	3	0
Idaho	19	19	0	21	21	0
Montana	6	4	0	6	4	0
Nevada	1	1	0	1	1	0
Oregon	10	8	2	9	8	1
Utah	11	8	3	12	9	3
Washington	10	8	1	10	8	2
Wyoming	2	2	0	1	1	0

Source: "Prisoners Under Sentence of Death," Department of Justice, *Sourcebook of Criminal Justice Statistics-1992*, p. 671. Primary source: U.S. Department of Justice, Federal Bureau of Prisons, Federal Bureau of Prisons, *Capital Punishment, 1991*, NCJ-136946. *Notes:* Thirty-six States and the Federal Government had death penalty statutes in effect on Dec. 31, 1990 and on Dec. 31, 1991. Some figures shown for year end 1990 have been revised from previous presentations by the source. A. Totals include persons of other races. B. Excludes 5 males held under Armed Forces jurisdiction with a military death sentence for murder.

★ 183 ★

Prisons and Prisoners

State Prison Inmates by Criminal History, 1991

Violent/nonviolent refers to the current or past criminal offense for which the inmate is or was incarcerated. Data is based on a sample survey of 13,986 inmates; subject to sampling variability.

| Characteristic | Total | Criminal History of prison inmates | | | | | | | |
| | | First-timers | | | Recidivists[1] | | | | |
		Total	Nonviolent	Violent	Total	Nonviolent	Prior violent	Current violent only	Current and prior violent
Prison inmates, total	697,653	134,131	45,559	88,572	563,722	223,117	88,689	131,289	120,626
Percent of total	100.0	19.2	6.5	12.7	80.8	32.0	12.7	18.8	17.3
Percent distribution	100.0	100.0	100.0	100.0	100.0	100.0	100.0	100.0	100.0
Male	94.5	92.0	89.7	93.2	95.1	92.1	96.5	96.9	97.8
Female	5.5	8.0	10.3	6.8	4.9	7.9	3.5	3.1	2.2
White	49.0	52.6	50.9	53.4	48.2	52.9	40.1	50.4	42.9
Black	47.5	43.6	45.7	42.6	48.4	43.7	56.5	45.6	54.0
Other races	3.5	3.8	3.4	4.0	3.5	3.4	3.4	4.0	3.1
Median age (years)	30	31	30	31	30	29	30	30	32
Median age at first arrest (years)	18	24	25	23	17	18	16	17	16

Source: "State Prison Inmates, by Criminal History and Selected Characteristics of the Inmate: 1991," *Statistical Abstract of the United States,* 1993, p. 211. Primary source: U.S. Bureau of Justice Statistics, *Survey of State Prison Inmates, 1991. Notes:* 1. An individual who has previously sentenced to probation or incarceration as a juvenile or adult.

★ 184 ★

Prisons and Prisoners

State Prison Inmates. 1986 and 1991

Based on a sample survey of about 13,986 inmates in 1991 and 13,711 inmates in 1986; subject to sampling variability.

| Characteristic | Number | | Percent of prison inmates | |
	1986	1991	1986	1991
Total[1]	450,416	711,643	100.0	100.0
White	223,648	349,628	49.7	49.1
Black	211,021	336,920	46.9	47.3
Other races	15,412	25,094	3.4	3.5

Source: "State Prison Inmates—Selected Characteristics: 1986 and 1991," *Statistical Abstract of the United States, 1993,* p. 210. Primary source: U.S. Bureau of Justice Statistics, *Profile of State Prison Inmates, 1986,* and *Survey of State Prison Inmates, 1991. Notes:* 1. For 1986, includes data not reported for all characteristics except sex. For 1991, includes data not reported for marital status, re-arrest, employment status and years of school.

Victims of Crime

★ 185 ★

Average Annual Victimization Rates Per 1,000 of Persons Age 65 and Older

By type of crime and demographic characteristics, United States, 1987-90 (aggregate).

	Crimes of violence[1]		Crimes of theft[2]		Household crimes[3]	
	65 to 74 years	75 years and older	65 to 74 years	75 years and older	65 to 74 years	75 years and older
Race						
White	4.2	2.6	23.1	14.2	77.6	61.4
Black	13.9	6.5	36.7	16.1	156.8	149.6

Source: "Average Annual Victimization Rates (Per 1,000) of Persons Age 65 and Older," *Sourcebook of Criminal Justice Statistics-1992*, p. 249. Primary source: U.S. Department of Justice, Federal Bureau of Prisons, Federal Bureau of Prisons *Elderly Victims*, Special Report NCJ-138330. *Notes:* 1. Crimes of violence include rape, robbery, and assault. 2. Crimes of theft include personal larceny with contact and personal larceny without contact. 3. Household crimes include burglary, household larceny, and motor vehicle theft.

★ 186 ★

Victims of Crime

Characteristics of Juvenile Murders and Nonnegligent Manslaughter Victims, Aggregate of 1976 to 1991

By selected characteristics of the offender and offense, United States, 1976-91 (aggregate)[1].

Characteristics of offender and offense	Total	Characteristics of victim					
		Race		Male		Female	
		White	Black	White	Black	White	Black
Age							
17 years and younger	24.4%	23.9%	25.2%	28.6%	29.0%	16.1%	16.6%
18 to 29 years	52.4	21.1	54.3	47.8	51.2	56.7	61.2
30 to 49 years	20.5	22.2	17.8	20.6	17.0	24.8	19.6
50 years and older	2.8	2.8	2.7	3.0	2.8	2.4	2.7
Race							
White	54.2	93.2	7.3	92.8	9.0	93.8	3.6
Black	43.5	5.9	92.4	6.3	90.8	5.1	96.1
Other	2.2	1.0	0.3	0.9	0.3	1.1	0.3
Sex							
Male	80.5	80.9	80.1	83.6	85.1	76.3	68.6
Female	19.5	19.1	19.9	16.4	14.9	23.7	31.4

[Continued]

★ 186 ★

Characteristics of Juvenile Murders and Nonnegligent Manslaughter Victims, Aggregate of 1976 to 1991

[Continued]

Characteristics of offender and offense	Total	Characteristics of victim					
		Race		Male		Female	
		White	Black	White	Black	White	Black
Type of weapon							
Firearm	50.0	46.1	54.8	54.5	63.9	31.8	32.6
Knife	12.8	13.5	11.8	13.5	11.6	13.4	12.4
Blunt object	5.1	5.4	4.7	4.2	3.7	7.4	7.2
Personal weapon	19.6	20.7	18.5	17.1	13.8	26.8	29.7
Other	12.5	14.3	10.2	10.6	7.0	20.5	18.1
Relationship to victim							
Family	40.1	43.5	35.1	38.3	28.6	52.4	49.5
Other known	45.3	42.3	49.6	45.5	52.9	36.9	42.2
Stranger	14.6	14.1	15.3	16.1	18.5	10.7	8.2

Source: "Characteristics of Juvenile Murders and Nonnegligent Manslaughter Victims," *Sourcebook of Criminal Justice Statistics-1992*, p. 392. Primary source: James Alan Fox, "Children are Slain by their Parents and Teenagers by their Peers," National Crime Analysis Program, Northeastern University (Mimeographed). Published by permission. *Note:* 1. Percents may not add to 100 because of rounding.

★ 187 ★

Victims of Crime

Child Abuse and Neglect Cases Substantiated: 1990 and 1991

Based on reports alleging child abuse and neglect that were referred for investigation by the respective child protective services agency in each state. The reporting period may be either calendar or fiscal year. The majority of states were unable to provide unduplicated counts. Also, varying number of states reported the various characteristics presented below. Excludes the Armed Forces. A substantiated case represents a type of investigation disposition that determines that there is sufficient evidence under state law to conclude that maltreatment occurred or that the child is at risk of maltreatment.

Item	1990		1991	
	Number	Percent	Number	Percent
Race/ethnic group of victim[1]				
Victims, total	775,409	100.0	817,718	100.0
White	424,470	54.7	453,955	55.5
Black	197,400	25.5	218,026	26.7
Asian and Pacific Islander	6,408	0.8	6,564	0.8
American Indian, Eskimo, and Aleut	10,283	1.3	10,875	1.3
Other races	11,749	1.5	13,005	1.6

[Continued]

★ 187 ★

Child Abuse and Neglect Cases Substantiated: 1990 and 1991
[Continued]

Item	1990		1991	
	Number	Percent	Number	Percent
Hispanic origin	73,132	9.4	78,025	9.5
Unknown	51,967	6.7	37,268	4.6

Source: "Child Abuse and Neglect Cases Substantiated and Indicated-Victim Characteristics: 1990 and 1991," *Statistical Abstract of the United States, 1993,* p. 209. Primary source: U.S. Department of Health and Human Services, National Center on Child Abuse and Neglect, National Child Abuse and Neglect Data System, Working Paper 2, 1991 Summary Data Component, May 1993. *Notes:* 1. Some states were unable to report on the number of Hispanic victims, thus it is probable that nationwide the percentage of Hispanic victims is higher than 9 percent.

★ 188 ★

Victims of Crime

Estimated Number and Rate of Motor Vehicle Theft Per 1,000 Households and Per 1,000 Vehicles, 1991

Race of head of household	Household			Vehicles owned		
	Number	Number of thefts	Rate per 1,000	Number	Number of thefts	Rate per 1,000
All races	96,839,300	2,112,330	21.8	172,258,570	2,165,980	12.6
White	82,952,520	1,608,450	19.4	154,584,020	1,655,770	10.7
Black	11,283,680	416,050	36.9	13,509,060	418,620	31.0
Other	2,603,100	87,820	33.7	4,165,490	91,580	22.0

Source: "Estimated Number and Rate (Per 1,000 Households and Per 1,000 Vehicles Owned) of Motor Vehicle Theft," *Sourcebook of Criminal Justice Statistics- 1992,* p. 286. Primary source: U.S. Department of Justice, Bureau of Justice Statistics, *Criminal Victimization in the United States: 1991,* NCJ-139564. *Notes:* The number of thefts based on vehicles owned is equal to or higher than the corresponding figure based on households because the former includes all completed or attempted vehicle thefts, regardless of the final classification of the event; personal crimes of contact and burglary occurring in conjunction with motor vehicle thefts take precedence in determining the final classification based on the number of households.

★ 189 ★

Victims of Crime

Estimated Percent Distribution of Personal Victimization by Lone Offenders, 1991

By type of victimization, race of victim, and perceived race of offender, United States, 1991[1].

Type of victimization and race of victim	Number of victimizations	Perceived race of lone offender				
		Total	White	Black	Other	Not known and not available
Crimes of violence						
White	3,634,310	100%	74.5%	17.1%	7.1%	1.3%
Black	672,120	100	8.3	85.2	5.9	0.6[2]
Completed						
White	1,200,190	100	73.8	18.7	5.5	2.0[2]
Black	308,490	100	2.6[2]	92.3	4.4[2]	0.7[2]
Attempted						
White	2,434,110	100	74.9	16.3	7.8	1.0[2]
Black	363,620	100	13.1	79.2	7.1[2]	0.6[2]
Rape						
White	141,050	100	84.2	14.3[2]	0.0[2]	1.5[2]
Black	13,510[2]	100	0.0[2]	100.0[2]	0.0[2]	0.0[2]
Robbery						
White	378,430	100	39.5	44.4	11.3	4.8[2]
Black	173,180	100	4.0[2]	89.1	5.7[2]	1.2[2]
Completed						
White	229,560	100	40.5	47.6	5.7[2]	6.2[2]
Black	122,080	100	0.0[2]	92.3	6.0[2]	1.6[2]
With injury						
White	66,160	100	50.3	39.0[2]	2.7[2]	8.0[2]
Black	34,160	100	0.0[2]	78.5	21.5[2]	0.0[2]
Without injury						
White	163,400	100	36.5	51.1	6.9[2]	5.5[2]
Black	87,910	100	0.0[2]	97.7	0.0[2]	2.3[2]
Attempted						
White	148,860	100	37.8	39.5	20.0	2.7[2]
Black	51,090	100	13.6[2]	81.4	5.0[2]	0.0[2]
With injury						
White	35,290	100	37.2[2]	30.1[2]	32.7[2]	0.0[2]
Black	7,280[2]	100	0.0[2]	100.0[2]	0.0[2]	0.0[2]
Without injury						
White	113,570	100	38.0	42.4	16.1[2]	3.5[2]
Black	43,810	100	15.9[2]	78.3	5.8[2]	0.0[2]
Assault						
White	3,114,820	100	78.4	13.9	6.9	0.9
Black	485,420	100	10.1	83.4	6.1	0.5[2]
Aggravated						
White	848,270	100	76.2	13.6	9.5	0.7[2]
Black	168,030	100	10.8[2]	82.2	5.7[2]	1.4[2]

[Continued]

★ 189 ★

Estimated Percent Distribution of Personal Victimization by Lone Offenders, 1991

[Continued]

Type of victimization and race of victim	Number of victimizations	Perceived race of lone offender				
		Total	White	Black	Other	Not known and not available
Simple						
White	2,266,540	100	79.2	13.9	5.9	1.0[2]
Black	317,380	100	9.7	84.0	6.3[2]	0.0[2]

Source: "Estimated Percent Distribution of Personal Victimization by Lone Offenders," *Sourcebook of Criminal Justice Statistics-1992,* p. 288. Primary source: U.S. Department of Justice, Bureau of Justice Statistics, *Criminal Victimization in the United States, 1991,* NCJ-139563. *Notes:* 1. Subcategories may not sum to total because of rounding. 2. Estimate is based on about 10 or fewer sample cases.

★ 190 ★

Victims of Crime

Estimated Percent Distribution of Personal Victimization by Multiple Offenders, 1991

By type of victimization, and perceived races of offenders, United States, 1991[1].

Type of victimization and race of victim	Number of victimizations	Perceived races of multiple offenders					
		Total	All white	All black	All other	Mixed races	Not known and not available
Crimes of violence	1,879,010	100%	34.8%	37.5%	7.9%	16.0%	3.9%
Completed	859,540	100	30.7	41.2	7.2	17.3	3.5
Attempted	1,019,460	100	38.2	34.4	8.5	14.9	4.1
Rape	14,400[2]	100	47.8[2]	39.8[2]	0.0[2]	0.0[2]	12.4[0]
Robbery	555,910	100	16.1	56.3	5.9	16.2	5.4
Completed	370,800	100	13.1	59.4	6.9[2]	15.3	5.4[2]
With injury	152,260	100	11.4[2]	55.4	2.5[2]	20.2	10.5[2]
Without injury	218,530	100	14.3	62.2	9.9[2]	11.8[2]	1.8[2]
Attempted	185,100	100	22.1	50.2	4.1[2]	18.1	5.5[2]
With injury	82,910	100	22.5[2]	61.3	0.0[2]	9.3[2]	7.0[2]
Without injury	102,190	100	21.8[2]	41.3	7.4[2]	25.2[2]	4.3[2]
Assault	1,308,700	100	42.5	29.5	8.8	16.1	3.1
Aggravated	508,360	100	40.8	27.2	9.5	18.0	4.5[2]
Simple	800,330	100	43.7	30.9	8.3	14.9	2.2[2]

Source: "Estimated Percent Distribution of Personal Victimization by Multiple Offenders," *Sourcebook of Criminal Justice Statistics-1992,* p. 290. Primary source: U.S. Department of Justice, Bureau of Justice Statistics, *Criminal Victimization in the United States, 1991,* NCJ-139563. Table adapted by *Sourcebook* staff. *Notes:* 1. Subcategories may not sum to total because of rounding. 2. Estimate is based on about 10 or fewer sample cases.

★ 191 ★

Victims of Crime

Estimated Percent Distribution of Personal Victimization by Multiple Offenders, by Race of Victim, 1991

By type of victimization, race of victim, and perceived races of offenders, United States, 1991[1].

Type of victimization and race of victim	Number of victimizations	Perceived races of multiple offenders					
		Total	All white	All black	All other	Mixed races	Not known and not available
Crimes of violence[2]							
White	1,435,100	100%	42.3%	26.9%	8.2%	18.8%	3.7%
Black	374,130	100	7.7	80.2	0.0[3]	7.0	5.1[3]
Robbery							
White	382,430	100	21.1	45.1	8.0	20.6	5.3[3]
Black	151,590	100	5.7[3]	83.5	0.0[3]	4.2[3]	6.6[3]
Assault							
White	1,040,150	100	50.2	20.0	8.5	18.4	3.0
Black	222,540	100	9.1[3]	77.9	0.0[3]	8.9[3]	4.1[3]

Source: "Estimate Percent Distribution of Personal Victimization by Multiple Offenders," *Sourcebook of Criminal Justice Statistics-1992*, p. 291. Primary source: U.S. Department of Justice, Bureau of Justice Statistics, *Criminal Victimization in the United States, 1991*, NCJ-139563. *Notes:* 1. Subcategories may not sum to total because of rounding. 2. Includes data on rape, not shown separately. 3. Estimate is based on about 10 or fewer sample cases.

★ 192 ★

Victims of Crime

Estimated Percent Distribution of Self-protective Measures Against Violent Crime, 1991

By sex and race of victim and by type of measure, United States, 1991[1].

Type of self-protective measure	Race	
	White	Black
Total	100%	100%
Attacked offender with weapon	1.4	2.1[2]
Attacked offender without weapon	11.0	14.5
Threatened offender with weapon	1.5	4.3
Threatened offender without weapon	1.8	1.6[2]
Resisted or captured offender	23.0	25.3
Scared or warned offender	8.0	5.2
Persuaded or appeased offender	14.0	13.6
Ran away or hid	16.0	16.3
Got help or gave alarm	11.0	7.8
Screamed from pain or fear	3.4	2.1[2]

[Continued]

★ 192 ★

Estimated Percent Distribution of Self-protective Measures Against Violent Crime, 1991

[Continued]

Type of self-protective measure	Race	
	White	Black
Employed another method	8.9	7.2
Total number of self-protective measures[3]	5,750,090	980,330

Source: "Estimated Percent Distribution of Self-protective Measures Employed by Victims of Violent Crime," *Sourcebook of Criminal Justice Statistics-1992*, p. 277. Primary source: U.S. Department of Justice, Bureau of Justice Statistics, *Criminal Victimization in the United States, 1991*, NCJ-139563. Notes: 1. Subcategories may not sum to total because of rounding. 2. Estimate is based on about 10 or fewer sample cases. 3. Some respondents may have reported more than one self-protective measure employed.

★ 193 ★

Victims of Crime

Estimated Rate of Household Victimization Per 1,000 Households, 1991

By type of victimization, race, and ethnicity of head of household, United States, 1991[1].

Type of victimization	Total[2] (N=96,839,300)	Race			Ethnicity	
		White (N=82,952,520)	Black (N=11,283,680)	Other (N=2,603,100)	Hispanic (N=6,772,470)	Non-hispanic (N=89,878,800)
Household crimes	162.9	156.6	207.6	170.7	239.9	157.0
Completed	138.1	133.3	172.8	140.0	198.7	133.4
Attempted	24.8	23.3	34.8	30.7	41.2	23.6
Burglary	53.1	50.2	74.5	51.9	74.8	51.3
Completed	41.4	39.5	55.1	42.1	58.3	40.1
Forcible entry	17.2	15.1	32.8	17.1	34.3	16.0
Unlawful entry without force	24.1	24.4	22.3	25.0	24.0	24.1
Attempted forcible entry	11.7	10.7	19.4	9.8[3]	16.5	11.3
Household larceny	88.0	87.0	96.2	85.1	123.1	85.3
Completed	82.7	81.6	92.0	79.8	114.5	80.3
Less than $50	34.7	35.2	32.6	26.5	35.5	34.6
$50 or more	43.6	42.3	51.3	49.2	71.9	41.4
Amount not available	4.5	4.0	8.1	4.1[3]	7.1	4.3
Attempted	5.3	5.4	4.3	5.3[3]	8.6	5.0
Motor vehicle theft	21.8	19.4	36.9	33.7	41.9	20.3
Completed	13.9	12.2	25.8	18.1	25.9	13.0
Attempted	7.9	7.2	11.1	15.7	16.0	7.3

Source: "Estimated Rate (Per 1,000 Households) of Household Victimization," *Sourcebook of Criminal Justice Statistics-1992*, p. 281. Primary source: U.S. Department of Justice, Bureau of Justice Statistics, *Criminal Victimization in the United States, 1991*, NCJ-139563. Notes: 1. Subcategories may not sum to total because of rounding. 2. Total includes households where the ethnicity of the household head could not be determined. 3. Estimate is based on about 10 or fewer sample cases.

★ 194 ★

Victims of Crime

Estimated Rate of Personal Victimization Per 1,000 Persons Age 12 and Older, by Sex, 1991

By type of victimization, sex, and race of victim, United States, 1991[1].

Type of victimization	Male		Female	
	White (N = 84,632,280)	Black (N = 11,059,660)	White (N = 89,844,340)	Black (N = 13,077,650)
Personal crimes	102.0	130.8	80.5	84.2
Crimes of violence	37.7	60.9	22.0	30.5
Completed	13.2	26.1	8.3	16.8
Attempted	24.4	34.8	13.7	13.8
Rape	0.2[2]	0.0[2]	1.5	1.0[2]
Robbery	6.1	20.0	2.8	7.9
Completed	3.9	11.9	1.8	6.8
With injury	1.5	3.4	0.6	2.3
Without injury	2.3	8.5	1.2	4.4
Attempted	2.2	8.1	1.0	1.1[2]
With injury	0.6	3.0	0.4	0.2[2]
Without injury	1.6	5.1	0.6	1.0[2]
Assault	31.3	40.9	17.6	21.6
Aggravated	10.8	17.0	4.2	6.2
Completed with injury	3.8	8.7	1.4	2.8
Attempted with weapon	6.9	8.2	2.8	3.4
Simple	20.6	23.9	13.4	15.4
Completed with injury	5.5	5.4	4.4	6.8
Attempted with weapon	15.1	18.5	9.0	8.6
Crimes of theft	64.3	69.9	58.6	53.7
Completed	60.4	64.0	54.4	50.7
Attempted	3.9	5.9	4.2	3.0
Personal with larceny with contact	1.9	3.5	2.1	5.6
Personal larceny without contact	62.5	66.4	56.5	48.1
Completed	58.6	60.5	52.6	45.1
Attempted	3.9	5.9	3.8	3.0

Source: "Estimated Rate (Per 1,000 Persons Age 12 and Older) of Personal Victimization," *Sourcebook of Criminal Justice Statistics-1992*, p. 263. Primary source: U.S. Department of Justice, Bureau of Justice Statistics, *Criminal Victimization in the United States*, NCJ-139563. *Notes:* 1. Subcategories may not sum to total because of rounding. 2. Estimate is based on about 10 or fewer sample cases.

★ 195 ★

Victims of Crime

Estimated Rate of Personal Victimization Per 1,000 Persons Aged 12 and Older

By type of victimization, race, and ethnicity of victim, United States, 1991[1].

Type of victimization	Race			Ethnicity	
	White (N = 174,476,630)	Black (N = 24,137,310)	Other (N = 6,730,960)	Hispanic (N = 16,989,400)	Non-Hispanic (N = 187,961,530)
Personal crimes	90.9	105.6	80.2	95.6	91.9
Crimes of violence	29.6	44.4	28.1	36.2	30.8
Completed	10.7	21.0	10.9	16.2	11.5
Attempted	18.9	23.4	17.3	20.0	19.3
Rape	0.9	0.6[2]	0.3[2]	1.0[2]	0.8
Robbery	4.4	13.5	7.4	10.0	5.2
Completed	2.8	9.1	6.4	6.7	3.4
With injury	1.0	2.8	0.9[2]	2.7	1.1
From serious assault	0.5	1.7	0.3[2]	1.3[2]	0.6
From minor assault	0.5	1.1	0.6[2]	1.4[2]	0.5
Without injury	1.8	6.3	5.5	3.9	2.3
Attempted	1.6	4.3	1.0[2]	3.3	1.8
With injury	0.5	1.4	0.0[2]	1.8	0.5
From serious assault	0.2	1.1	0.0[2]	0.5[2]	0.3
From minor assault	0.3	0.4[2]	0.0[2]	1.2[2]	0.2
Without injury	1.1	2.9	1.0[2]	1.5[2]	1.3
Assault	24.3	30.4	20.5	25.2	24.8
Aggravated	7.4	11.1	8.2	11.8	7.5
Completed with injury	2.6	5.5	1.4[2]	3.8	2.8
Attempted with weapon	4.8	5.6	6.8	7.9	4.7
Simple	16.9	19.3	12.3	13.4	17.3
Completed with injury	4.9	6.2	3.1[2]	5.5	5.0
Attempted with weapon	12.0	13.1	9.2	8.0	12.3
Crimes of theft	61.4	61.1	52.0	59.4	61.2
Completed	57.3	56.8	47.0	54.2	57.2
Attempted	4.0	4.3	5.0	5.2	4.0
Personal with larceny with contact	2.0	4.6	3.5[2]	4.6	2.1
Purse snatching	0.6	1.2	1.0[2]	1.7	0.6
Pocket picking	1.4	3.4	2.5[2]	2.9	1.6
Personal larceny without contact	59.4	56.5	48.5	54.9	59.0
Completed	55.5	52.2	43.5	50.3	55.1
Less than $50	21.7	19.3	17.7	15.4	21.8
$50 or more	31.1	30.0	23.5	31.2	30.7
Amount not available	2.8	2.8	2.3[2]	3.7	2.7
Attempted	3.9	4.3	5.0	4.6	3.9

Source: "Estimated Rate (Per 1,000 Persons Age 12 and Older) of Personal Victimization," *Sourcebook of Criminal Justice Statistics-1992,* p. 262. Primary source: U.S. Department of Justice, Bureau of Justice Statistics, *Criminal Victimization in the United States,* NCJ-139563. Table adapted by *Sourcebook* staff. *Notes:* 1. Subcategories may not sum to total because of rounding. 2. Estimate is based on about 10 or fewer sample cases.

★ 196 ★

Victims of Crime

Estimated Rate of Personal Victimization Per 1,000 Persons by Victim-Offender Relationship: Part 1, Crimes of Violence, 1991

Characteristics of victim	Total population	Crimes of violence[1]			
		Relatives	Well known	Casual acquaintances	Strangers
Race					
White	174,476,630	2.1	6.4	3.9	16.0
Black	24,137,310	1.8	9.6	6.4	24.3
Other	6,730,960	1.4[2]	2.7[2]	1.6[2]	20.6

Source: "Estimated Rate (Per 1,000 Persons Age 12 and Over) of Personal Victimization," *Sourcebook of Criminal Statistics-1992,* p. 268. Primary source: U.S. Department of Justice, Bureau of Justice Statistics, *Criminal Victimization in the United States,* NCJ-139563. *Notes:* This table combines victimizations committed by single and multiple offenders. 1. Includes data on rape and robbery not shown separately. 2. Estimate is based on about 10 or fewer sample cases.

★ 197 ★

Victims of Crime

Estimated Rate of Personal Victimization Per 1,000 Persons by Victim-Offender Relationship: Part 2, Assault and Aggravated Assault, 1991

Characteristics of victim	Total population	Assault				Aggravated assault			
		Relatives	Well known	Casual acquaintances	Strangers	Relatives	Well known	Casual acquaintances	Strangers
Race									
White	174,476,630	1.9	5.7	3.6	12.1	0.5	1.5	0.8	4.3
Black	24,137,310	1.7	7.8	5.2	13.8	0.3[1]	2.3	1.4	5.9
Other	6,730,960	1.4[1]	2.7[1]	1.6[1]	13.7	0.3	1.0[1]	1.0[1]	5.4

Source: "Estimated Rate (Per 1,000 Persons Age 12 and Over) of Personal Victimization," *Sourcebook of Criminal Statistics-1992,* p. 268. Primary source: U.S. Department of Justice, Bureau of Justice Statistics, *Criminal Victimization in the United States,* NCJ-139563. *Notes:* This table combines victimizations committed by single and multiple offenders. 1. Estimate is based on about 10 or fewer sample cases.

★ 198 ★

Victims of Crime

Estimated Rate of Personal Victimization Per 1,000 Persons in Each Age Group, 1991

Sex, age, and race of victim	Total population	Crimes of violence	Crimes of theft
White			
Male			
12 to 15 years	5,610,640	84.3	107.6
16 to 19 years	5,394,180	118.7	97.7
20 to 24 years	7,396,080	93.5	122.2
25 to 34 years	17,977,320	37.1	77.1
35 to 49 years	22,877,720	23.7	57.0
50 to 64 years	13,919,410	10.3	36.5
65 years and older	11,456,900	2.7	18.6
Female			
12 to 15 years	5,392,640	35.2	95.1
16 to 19 years	5,292,160	60.1	105.6
20 to 24 years	7,479,530	53.5	120.4
25 to 34 years	17,780,590	29.6	67.0
35 to 49 years	23,017,390	15.8	55.8
50 to 64 years	14,947,140	8.1	35.2
65 years and older	15,880,860	3.4	17.9
Black			
Male			
12 to 15 years	1,137,690	133.9	133.3
16 to 19 years	1,060,590	147.5	69.5
20 to 24 years	1,060,630	118.5	89.3
25 to 34 years	2,571,740	41.4	82.0
35 to 49 years	2,638,450	31.9	64.5
50 to 64 years	1,555,720	19.6	22.8
65 years and older	1,034,810	17.7[1]	35.5
Female			
12 to 15 years	1,081,950	28.4	63.5
16 to 19 years	1,054,050	65.1	54.6
20 to 24 years	1,348,880	56.2	79.1
25 to 34 years	2,947,870	53.2	66.8
35 to 49 years	3,211,190	11.9	47.9
50 to 64 years	1,895,370	9.3[1]	38.1
65 years and older	1,538,320	7.2[1]	29.7

Source: "Estimated Rate (Per 1,000 Persons in Each Age Group) of Personal Victimization," *Sourcebook of Criminal Justice Statistics-1992,* p. 264. Primary source: U.S. Department of Justice, Bureau of Justice Statistics, *Criminal Victimization in the United States,* NCJ-139563. *Note:* 1. Estimate is based on about 10 or fewer sample cases.

★ 199 ★

Victims of Crime

Estimated Rate of Personal Victimization by Crimes of Theft Per 1,000 Persons in Age Group, 1991

By type of victimization, race, and age of victim, United States, 1991[1].

Race and age of victim	Total population	Crimes of theft			Personal larceny	
		Total	Completed	Attempted	With contact	Without contact
White						
12 to 15 years	11,003,290	101.5	99.1	2.4[2]	1.9[2]	99.5
16 to 19 years	10,686,350	101.6	95.6	6.0	2.9	98.7
20 to 24 years	14,875,610	121.3	11.1	10.2	3.2	118.1
25 to 34 years	35,757,910	72.1	66.3	5.8	2.7	69.4
35 to 49 years	45,949,110	56.4	52.9	3.5	1.4	55.0
50 to 64 years	28,866,550	35.8	33.3	2.5	1.2	34.7
65 years and older	27,337,760	18.2	17.5	0.7[2]	1.8	16.4
Black						
12 to 15 years	2,219,640	99.3	97.1	2.2[2]	4.3[2]	95.0
16 to 19 years	2,114,640	62.1	59.9	2.1[2]	2.6[2]	59.5
20 to 24 years	2,409,510	83.6	76.5	7.1[2]	5.3[2]	78.3
25 to 34 years	5,519,610	73.9	66.9	7.0	3.5[2]	70.4
35 to 49 years	5,849,650	55.4	50.7	4.7	2.1[2]	53.4
50 to 64 years	3,451,090	31.2	27.7	3.6[2]	7.4[2]	23.8
65 years and older	2,573,130	32.1	32.1	0.0[2]	10.6	21.5

Source: "Estimated Rate (Per 1,000 Persons in Each Age Group) of Personal Victimization," *Sourcebook of Criminal Justice Statistics-1992,* p. 263. Primary source: U.S. Department of Justice, Bureau of Justice Statistics, *Criminal Victimization in the United States,* NCJ-139563. *Notes:* 1. Subcategories may not sum to total because of rounding. 2. Estimate is based on about 10 or fewer sample cases.

★ 200 ★

Victims of Crime

Estimated Rate of Personal Victimization by Violent Crime Per 1,000 Persons in Age Group, 1991

By type of victimization, race, and age of victim, United States, 1991[1].

Race and age of victim	Total population	Crimes of violence				Robbery			Assault		
		Total	Completed	Attempted	Rape	Total	With injury	Without injury	Total	Aggravated	Simple
White											
12 to 15 years	11,003,290	60.3	21.6	38.7	1.4[2]	7.8	2.9	4.8	51.1	12.8	38.3
16 to 19 years	10,686,350	89.7	31.2	58.5	3.7	9.2	3.5	5.7	76.7	24.6	52.1
20 to 24 years	14,875,610	73.4	27.0	46.4	1.9	11.5	4.4	7.0	60.0	23.6	36.4
25 to 34 years	35,757,910	33.4	13.6	19.7	1.1	5.8	2.0	3.8	26.5	7.7	18.7
35 to 49 years	45,949,110	19.7	6.2	13.6	0.6	2.9	1.0	1.8	16.2	3.7	12.5
50 to 64 years	28,866,550	9.1	3.1	6.0	0.2[2]	1.3	0.3[2]	1.0	7.6	2.1	5.4
65 years and older	27,337,760	3.1	1.2	1.9	0.0[2]	1.4	0.4[2]	1.0	1.7	0.7[2]	1.0

[Continued]

★ 200 ★

Estimated Rate of Personal Victimization by Violent Crime Per 1,000 Persons in Age Group, 1991
[Continued]

| Race and age of victim | Total population | Crimes of violence | | | | | | | | | |
| | | Total | Completed | Attempted | Rape | Robbery | | | Assault | | |
						Total	With injury	Without injury	Total	Aggravated	Simple
Black											
12 to 15 years	2,219,640	82.4	34.8	47.6	0.0[2]	19.5	2.1[2]	17.4	62.9	16.1	46.8
16 to 19 years	2,114,640	106.5	40.2	66.3	3.5[2]	5.2[2]	2.6[2]	1.6[2]	97.8	29.5	68.3
20 to 24 years	2,409,510	83.6	38.8	44.8	0.9[2]	29.8	7.2[2]	22.6	53.0	19.6	33.4
25 to 34 years	5,519,610	47.7	27.5	20.2	0.7[2]	16.6	4.5[2]	12.1	30.4	12.1	18.3
35 to 49 years	5,849,650	20.9	11.3	9.6	0.0[2]	11.4	7.5	4.0[2]	9.5	5.0	4.5
50 to 64 years	3,451,090	14.0	5.4[2]	8.5	0.0[2]	5.8[2]	0.5[2]	5.3[2]	8.2	5.3[2]	2.9[2]
65 years and older	2,573,130	11.4	6.0[2]	5.4[2]	0.0[2]	7.9[2]	1.3[2]	6.6[2]	3.5[2]	3.5[2]	0.0[2]

Source: "Estimated Rate (Per 1,000 Persons in Each Age Group) of Personal Victimization," *Sourcebook of Criminal Justice Statistics-1992*, p. 263. Primary source: U.S. Department of Justice, Bureau of Justice Statistics, *Criminal Victimization in the United States*, NCJ-139563. *Notes:* 1. Subcategories may not sum to total because of rounding. 2. Estimate is based on about 10 or fewer sample cases.

★ 201 ★

Victims of Crime

High School Students as Victims of Crime, 1990 to 1992

| Type of victimization | Class of 1990 | | Class of 1991 | | Class of 1992 | |
	White (N=1,907)	Black (N=277)	White (N=1,818)	Black (N=289)	White (N=1,806)	Black (N=368)
Question: "During the last 12 months, how often..."						
Has something of yours (worth under $50) been stolen?						
Not at all	54.1%	54.0%	57.9%	47.3%	58.2%	52.0%
Once	25.4	24.6	25.4	25.3	26.2	25.0
Twice	12.5	11.7	10.2	15.6	9.7	11.5
3 or 4 times	5.7	8.4	4.4	7.8	4.6	7.6
5 or more times	2.3	1.3	2.1	3.9	1.4	3.8
Has something of yours (worth over $50) been stolen?						
Not at all	79.9	71.4	80.4	68.8	80.6	71.3
Once	14.3	19.9	14.3	20.5	14.1	18.3
Twice	3.9	5.6	4.0	5.7	3.4	6.4
3 or 4 times	1.4	2.3	1.0	3.4	1.6	2.4
5 or more times	0.5	0.8	0.3	1.6	0.2	1.7
Has someone deliberately damaged your property (your car, clothing, etc.)?						
Not at all	67.3	69.4	66.3	67.3	67.3	73.4
Once	19.7	15.3	21.3	22.8	20.7	14.8
Twice	8.7	9.6	7.8	4.7	8.5	8.0
3 or 4 times	3.2	4.6	3.5	3.6	3.1	2.5
5 or more times	1.1	1.0	1.1	1.6	0.4	1.3
Has someone injured you with a weapon (like a knife, gun, or club)?						
Not at all	95.3	94.4	95.1	92.1	96.0	93.3

[Continued]

★ 201 ★

High School Students as Victims of Crime, 1990 to 1992

[Continued]

Type of victimization	Class of 1990		Class of 1991		Class of 1992	
	White (N = 1,907)	Black (N = 277)	White (N = 1,818)	Black (N = 289)	White (N = 1,806)	Black (N = 368)
Once	3.1	4.3	3.7	5.7	3.0	4.9
Twice	1.0	0.8	0.4	1.8	0.8	1.6
3 or 4 times	0.3	0.3	0.3	0.0	0.1	0.2
5 or more times	0.4	0.2	0.4	0.4	0.1	0.1
Has someone threatened you with a weapon, but not actually injured you?						
Not at all	82.6	79.7	83.5	71.2	83.1	74.1
Once	10.1	11.4	10.3	15.7	9.9	14.0
Twice	3.7	4.1	3.3	6.9	3.5	4.5
3 or 4 times	2.1	2.4	1.3	3.8	2.0	3.7
5 or more times	1.6	2.4	1.6	2.4	1.5	3.7
Has someone injured you on purpose without using a weapon?						
Not at all	83.0	83.9	83.7	83.1	83.9	87.3
Once	10.2	11.4	9.7	9.3	9.8	6.6
Twice	3.5	1.8	3.2	2.3	3.2	2.5
3 or 4 times	2.1	0.8	1.9	2.6	1.9	1.2
5 or more times	1.2	2.2	1.5	2.7	1.2	2.3
Has an unarmed person threatened you with injury but not actually injured you?						
Not at all	65.1	69.4	68.6	65.7	68.0	73.8
Once	15.6	17.2	12.7	16.1	13.5	12.6
Twice	8.6	7.6	7.0	6.7	7.2	3.0
3 or 4 times	5.1	3.4	5.2	5.7	5.6	4.2
5 or more times	5.6	2.4	6.4	5.7	5.7	6.4

Source: "High School Seniors Reporting Victimization Experiences in Last 12 Months," *Sourcebook of Criminal Justice Statistics-1992,* pp. 300-301. Primary source: Data provided by Monitoring the Future Project, Survey Research Center, University of Michigan, Lloyd D. Johnson and Jerald G. Bachman, principal investigators. Table adapted by *Sourcebook* staff. Published by permission. *Notes:* Data are given for those who identify themselves as White or Caucasian and those who identify themselves as Black or African-American because these are the two largest racial/ethnic subgroups in the population. Data are not given for the other ethnic categories because these groups comprise less than 3 percent of the sample in any given year (Source *1982,* p.9).

★ 202 ★

Victims of Crime

Number and Rate of Burglary Victimization Per 1,000 Households, 1973 to 1990

	Race of head of household					
	White		Black		Other[1]	
	Number	Rate	Number	Rate	Number	Rate
1973	5,429,200	86.8	950,800	132.5	78,700	109.2
1974	5,637,200	88.3	1,015,400	135.4	68,000	86.2
1975	5,651,500	87.1	1,014,200	129.4	78,000	95.5
1976	5,552,300	84.0	1,047,500	130.8	63,700	71.9

[Continued]

★ 202 ★

Number and Rate of Burglary Victimization Per 1,000 Households, 1973 to 1990
[Continued]

| | Race of head of household | | | | | |
| | White | | Black | | Other[1] | |
	Number	Rate	Number	Rate	Number	Rate
1977	5,644,200	83.9	1,009,900	122.4	110,800	122.4
1978	5,661,700	82.6	970,300	114.7	72,000	73.2
1979	5,587,400	80.1	982,500	114.0	115,500	102.5
1980	5,838,700	80.5	1,028,600	115.4	105,300	80.2
1981	6,074,900	82.7	1,218,900	133.6	100,200	68.1
1982	5,461,200	73.4	1,085,100	117.2	116,600	75.9
1983	5,042,880	66.7	925,830	97.9	94,440	59.2
1984	4,641,880	60.6	887,140	91.7	113,840	63.5
1985	4,688,500	60.5	820,380	83.4	85,540	45.2
1986	4,513,730	57.5	921,330	91.6	121,540	64.0
1987	4,553,890	57.2	1,009,230	98.2	141,420	70.8
1988	4,635,570	57.4	997,150	95.6	144,060	66.0
1989	4,261,060	52.1	957,640	88.4	133,600	58.3
1990	4,047,010	49.1	932,050	85.4	168,680	67.7

Source: "Number and Rate (Per 1,000 Households) of Burglary Victimization," *Sourcebook of Criminal Justice Statistics-1992*, p. 283. Primary source: U.S. Department of Justice, Bureau of Justice Statistics, *Criminal Victimization in the United States: 1973-70 Trends*, NCJ-139564. *Note:* 1. Includes mainly Asians and American Indians.

★ 203 ★

Victims of Crime

Number and Rate of Household Larceny Victimization Per 1,000 Households, 1973 to 1990

| | Race of head of household | | | | | |
| | White | | Black | | Other[1] | |
	Number	Rate	Number	Rate	Number	Rate
1973	6,733,700	107.7	744,400	103.7	59,200	82.2
1974	7,975,400	124.9	841,400	112.2	116,400	147.6
1975	8,213,900	126.6	898,100	114.6	111,000	135.9
1976	8,311,400	125.8	897,700	112.1	91,800	103.7
1977	8,341,900	124.0	959,600	116.3	116,900	129.0
1978	8,190,600	119.5	1,019,700	120.6	141,700	143.9
1979	9,309,100	133.5	1,148,700	133.2	172,400	153.0
1980	9,072,700	125.1	1,196,900	134.3	198,400	151.2
1981	8,710,900	118.5	1,291,900	141.6	173,200	117.7
1982	8,288,600	111.4	1,222,600	132.0	193,400	125.9
1983	7,809,900	103.3	1,122,220	118.7	181,580	113.9
1984	7,457,410	97.4	1,109,040	114.7	183,750	102.6
1985	7,355,340	94.9	1,181,380	120.1	166,190	87.9
1986	7,253,590	92.5	1,022,940	101.7	178,700	94.1
1987	7,425,340	93.3	1,188,340	115.7	174,550	87.4

[Continued]

★ 203 ★

Number and Rate of Household Larceny Victimization Per 1,000 Households, 1973 to 1990

[Continued]

| | Race of head of household | | | | | |
| | White | | Black | | Other[1] | |
	Number	Rate	Number	Rate	Number	Rate
1988	7,062,560	87.5	1,175,920	112.7	180,530	82.8
1989	7,581,310	92.7	1,181,250	109.0	192,900	84.2
1990	6,975,460	84.7	1,103,570	101.1	225,150	90.4

Source: "Number and Rate (Per 1,000 Households) of Household Larceny Victimization," *Sourcebook of Criminal Justice Statistics-1992*, p. 283. Primary source: U.S. Department of Justice, Bureau of Justice Statistics, *Criminal Victimization in the United States: 1973-70 Trends*, NCJ-139564. Table adapted by *Sourcebook* staff. *Note:* 1. Includes mainly Asians and American Indians.

★ 204 ★

Victims of Crime

Number and Rate of Motor Vehicle Theft Per 1,000 Households, 1973-90

| | Race of head of household | | | | | |
| | White | | Black | | Other[1] | |
	Number	Rate	Number	Rate	Number	Rate
1973	1,145,000	18.3	175,500	24.5	23,400	32.4
1974	1,155,300	18.1	195,100	26.0	8,000	10.2[2]
1975	1,204,000	18.6	210,600	26.9	18,400	22.6
1976	1,050,400	15.9	171,700	21.5	12,600	14.2
1977	1,105,000	16.4	174,300	21.1	17,400	19.3
1978	1,156,000	16.9	181,500	21.5	27,600	28.0
1979	1,183,100	17.0	188,800	21.9	21,000	18.6
1980	1,130,900	15.6	223,500	25.1	26,400	20.1
1981	1,200,600	16.3	219,100	24.0	19,300	13.1
1982	1,109,100	14.9	232,700	25.1	35,000	22.8
1983	1,002,400	13.3	238,550	25.2	22,670	14.2
1984	1,064,550	13.9	248,840	25.7	26,920	15.0
1985	1,018,380	13.1	219,730	22.3	32,060	17.0
1986	1,089,770	13.9	237,850	23.7	28,240	14.9
1987	1,205,670	15.2	220,690	21.5	46,470	23.3
1988	1,193,850	14.8	375,810	36.0	64,410	29.5
1989	1,419,940	17.4	334,050	30.8	66,120	28.9
1990	1,509,360	18.3	397,640	36.4	60,540	24.3

Source: "Number and Rate (Per 1,000 Households) of Motor Vehicle Theft," *Sourcebook of Criminal Justice Statistics-1992*, p. 286. Primary source: U.S. Department of Justice, Bureau of Justice Statistics, *Criminal Victimization in the United States: 1973-90 Trends*, NCJ-139564. Table adapted by *Sourcebook* staff. *Notes:* 1. Includes mainly Asians and American Indians. 2. Estimate based on about 10 or fewer sample cases.

★ 205 ★

Victims of Crime

Number and Rate of Rape Per 1,000 Females Age 12 and Older, 1973 to 1990

| | Total rape victimization | | Female rape victimization | | | | | |
| | | | Total | | White | | Black | |
	Number	Rate	Number	Rate	Number	Rate	Number	Rate
1973	155,730	0.9	151,700	1.8	125,700	1.7	24,900	2.6
1974	163,010	1.0	159,400	1.8	118,400	1.6	36,600	3.8
1975	153,740	0.9	146,400	1.7	126,100	1.6	18,200	1.8
1976	145,190	0.8	129,300	1.4	96,500	1.2	32,900	3.2
1977	154,240	0.9	141,900	1.6	123,900	1.6	16,400	1.6
1978	171,050	1.0	153,000	1.7	112,800	1.4	40,100	3.8
1979	191,740	1.1	171,200	1.8	141,700	1.8	28,100	2.6
1980	173,770	0.9	151,400	1.6	126,800	1.5	20,300	1.8
1981	177,540	1.0	169,700	1.8	135,600	1.6	31,100	2.8
1982	152,570	0.8	140,500	1.4	123,700	1.5	16,800	1.5
1983	154,170	0.8	137,900	1.4	114,990	1.3	19,790	1.7
1984	179,890	0.9	164,480	1.6	117,050	1.4	39,290	3.3
1985	138,490	0.7	130,850	1.3	90,260	1.0	36,910	3.1
1986	129,940	0.7	122,200	1.2	95,900	1.1	24,370	2.0
1987	148,450	0.8	134,300	1.3	84,380	1.0	40,410	3.3
1988	127,370	0.6	119,780	1.2	83,420	0.9	32,360	2.6
1989	135,410	0.7	122,740	1.2	101,470	1.1	21,260	1.7
1990	130,260	0.6	106,660	1.0	88,690	1.0	12,380	1.0[1]

Source: "Number and Rate (Per 1,000 Persons and per 1,000 Females Age 12 and Older) of Rape Victimizations," *Sourcebook of Criminal Justice Statistics- 1992*, p. 269. Primary source: U.S. Department of Justice, Bureau of Justice Statistics, *Criminal Victimization in the United States: 1973-90 Trends*, NCJ-139564. Table adapted by *Sourcebook* staff. *Note:* 1. Estimate is based on about 10 or fewer sample cases.

★ 206 ★

Victims of Crime

Number and Rate of Robberies Per 1,000 Females Age 12 and Older, 1973 to 1990

| | Race of victim | | | |
| | White | | Black | |
	Number	Rate	Number	Rate
1973	869,500	6.0	225,500	12.9
1974	914,300	6.2	270,500	15.1
1975	870,600	5.8	260,900	14.1
1976	832,500	5.5	256,100	13.6
1977	822,100	5.4	251,300	13.0
1978	801,300	5.2	224,800	11.4
1979	852,300	5.5	246,900	12.5
1980	917,800	5.7	283,000	14.0

[Continued]

★ 206 ★

Number and Rate of Robberies Per 1,000 Females Age 12 and Older, 1973 to 1990

[Continued]

| | Race of victim | | | |
| | White | | Black | |
	Number	Rate	Number	Rate
1981	995,400	6.2	347,300	16.9
1982	986,800	6.0	302,600	14.4
1983	834,020	5.1	286,320	13.4
1984	832,400	5.0	256,800	11.8
1985	709,420	4.2	240,990	10.9
1986	781,680	4.6	193,040	8.6
1987	733,740	4.3	293,180	12.9
1988	801,750	4.7	215,690	9.4
1989	763,650	4.4	302,280	12.9
1990	786,560	4.5	309,350	13.0

Source: "Number and Rate (Per 1,000 Persons Age 12 and Older) of Robbery Victimizations," *Sourcebook of Criminal Justice Statistics-1992,* p. 269. Primary source: U.S. Department of Justice, Bureau of Justice Statistics, *Criminal Victimization in the United States: 1973-90 Trends,* NCJ-139564. Table adapted by *Sourcebook* staff.

★ 207 ★

Victims of Crime

Percent of Household Experiencing Crime During the Last 12 Months, 1991

By type of victimization, race and ethnicity of head of household, family income, and place of residence, United States, 1991[1].

| | Race of head of household | | | Ethnicity of head of household | |
	White	Black	Other	Non-Hispanic	Hispanic
Any crime	23.2%	26.7%	24.3%	23.2%	30.4%
Violent crime	4.7	5.7	5.3	4.8	6.1
Rape	0.2	0.2	0.2	0.2	0.2
Robbery	0.8	2.1	1.5	0.9	2.0
Assault	4.0	3.8	3.7	4.0	4.3
Aggravated	1.4	1.7	1.2	1.4	2.0
Simple	2.9	2.3	2.8	2.9	2.5
Total theft	16.6	16.7	17.0	16.4	19.8
Personal	10.5	9.5	11.3	10.3	11.4
Household	7.6	8.6	7.6	7.5	10.4
Burglary	4.4	6.8	4.8	4.6	6.8
Motor vehicle theft	1.6	3.3	2.4	1.7	3.5

[Continued]

★ 207 ★

Percent of Household Experiencing Crime During the Last 12 Months, 1991

[Continued]

	Race of head of household			Ethnicity of head of household	
	White	Black	Other	Non-Hispanic	Hispanic
Serious violent crime[2]	2.3	3.8	2.9	2.3	4.1
Crimes of high concern[3]	6.9	9.5	8.2	6.9	10.9

Source: "Percent of Households Experiencing Crime During the Last 12 Months," *Sourcebook of Criminal Justice Statistics-1992*, p. 280. Primary source: U.S. Department of Justice, Bureau of Justice Statistics, *Crime and the Nation's Households, 1991*, NCJ-136950. *Notes:* The figures for each racial and ethnic subgroup are computed as a percent of the total number of households comprising that subgroup. "Other" refers to those households headed by an individual whose racial identification is other than white or black. 1. Detail may not add to total because of overlap in households experiencing different crimes. 2. Rape, robbery, or aggravated assault. 3. Rape, robbery, and assault by stranger, or burglary.

★ 208 ★

Victims of Crime

Rate Per 100,000 in Each Group of Juvenile Murders and Nonnegligent Manslaughters, 1976 to 1991

	Age							
	10 to 13 years				14 to 17 years			
	Male		Female		Male		Female	
	White	Black	White	Black	White	Black	White	Black
1976	1.0	3.5	0.8	1.3	3.9	24.3	1.9	5.4
1977	1.0	3.1	0.7	2.4	4.5	23.3	2.3	8.1
1978	0.9	2.8	0.9	2.7	4.8	20.8	2.4	7.0
1979	1.0	2.9	0.7	1.8	5.4	25.8	2.3	7.4
1980	1.0	2.5	0.8	2.5	5.6	29.1	2.6	6.4
1981	0.9	3.4	0.9	2.1	4.5	24.7	2.2	5.8
1982	0.9	2.2	1.1	1.4	4.4	24.2	2.0	7.3
1983	1.0	3.2	0.8	1.2	4.3	23.6	2.1	5.1
1984	0.8	2.3	0.8	1.7	3.9	19.2	2.1	6.2
1985	1.2	2.9	0.7	1.1	4.3	23.9	1.8	6.8
1986	0.8	3.1	1.0	1.5	4.4	27.0	2.3	6.3
1987	0.7	3.3	0.8	1.8	3.7	33.7	2.1	6.8
1988	0.9	3.4	0.9	3.6	4.2	39.5	2.1	6.4
1989	1.1	3.7	0.9	2.6	5.4	52.1	2.1	8.2
1990	1.1	4.4	0.6	3.8	7.5	54.1	2.3	9.1
1991	1.2	3.7	0.6	1.7	8.5	65.9	2.3	8.8

Source: "Rate (Per 100,000 Persons in Each Group) of Juvenile Murders and Nonnegligent Manslaughters Victimization," *Sourcebook of Criminal Justice Statistics-1992*, p. 391. Primary source: James Alan Fox, "Children are Slain by their Parents and Teenagers by their Peers," National Crime Analysis Program, Northeastern University (Mimeographed). Published by permission.

★ 209 ★

Victims of Crime

Victims of Violent State Prison Inmates, 1991

Victim characteristics	All violent offenses[b]	Most serious offense						
		Murder	Negligent man-slaughter	Rape	Other sexual assault	Robbery	Assault	Other violent
Number of inmates	327,958	74,693	12,786	24,833	41,649	104,136	57,558	12,303
Race of victim(s)								
White	54.6	51.8	41.9	63.3	71.5	53.8	45.1	59.8
Black	27.3	33.8	41.6	25.4	17.2	20.9	36.2	20.0
Hispanic	11.2	11.5	13.2	7.4	7.8	11.9	13.0	13.3
Other	2.7	1.4	2.1	2.5	1.7	4.2	2.5	4.2
Mixed	4.1	1.5	1.2	1.4	1.7	9.2	3.2	2.7

Source: "Characteristics of Victims of Violent State Prison Inmates," Department of Justice, *Sourcebook of Criminal Justice Statistics-1992*, p. 629. Primary source: U.S. Department of Justice, Bureau of the Statistics, *Correctional Populations in the United States, 1991*, NCJ-142729. *Notes:* This table excludes the following number of inmates because of missing data; race of victims, 33,717. A. Percents may not add to 100 because of rounding. B. Total includes all violent offenders.

Chapter 5
EDUCATION

★ 210 ★

Participants in Adult Education 17 Years old and older, 1991

Numbers in thousands.

Characteristics of participants	Number of adults in population[1]	Ever a participant in adult education		Participated in adult education[2] in past 3 years		Participated in adult education[2] in past year	
		Number	Percent of population	Number	Percent of population	Number	Percent of population
Racial/ethnic group							
White, non-Hispanic	143,144	80,099	56	56,715	40	47,401	33
Black, non-Hispanic	20,141	8,213	41	5,552	28	4,586	23
Hispanic	13,804	6,905	50	5,396	39	4,032	29
Other races, non-Hispanic	4,711	2,180	46	1,698	36	1,371	29

Source: Participants in adult education 17 years old and older, by selected characteristics of participants: 1991. *Digest of Educational Statistics,* 1993, p. 347. Primary source: U.S. Department of Education, National Center for Education Statistics, "Participation in Adult Education," unpublished data. (This table was prepared July 1991.) *Notes:* Data are based upon a sample survey of the civilian noninstitutional population. Because of rounding and survey item nonresponse, details may not add to totals. 1. Persons 17 years of age and over on the date of the survey. 2. Adult education is defined as all non-full-time education activities such as part-time college attendance, classes or seminars given by employers, and classes taken for adult literacy purposes, or for recreation and enjoyment.

Advanced Placement

★ 211 ★

Number of 11th and 12th Grade Advanced Placement Examinations Taken by 1,000 11th and 12th Graders and by Race/Ethnicity, 1992

	Number of examinations taken					
	Social studies	English	Foreign language	Calculus	Computer science	Science
Total	24	23	7	14	1	13
Race/ethnicity						
White	25	24	5	14	1	13
Black	6	6	1	3	0	3
Hispanic	11	10	18	5	1	4
Other[1]	62	49	17	58	6	56

Source: "Number of 11th and 12th Grade Advanced Placement Examinations Taken and 11th and 12th Graders Scoring 3 or Above Per 1,000 11th and 12th Graders, by Subject Area, and by Sex and Race/Ethnicity: 1992," U.S. Department of Education, National Center for Education Statistics, *The Condition of Education, 1993*, p. 72. Primary source: The College Board, Advanced Placement Program, National Summary Reports, 1984-1992, (Copyright 1992 by College Entrance Board.); Educational Testing Service, unpublished tabulations; U.S. Department of Commerce, Bureau of the Census, October Current Population Survey. Published by permission. *Notes:* Since, on average, AP candidates take more than one examination, there is not a 1:1 correspondence between candidates and examinations. 1. Includes individuals who are not Hispanic, black, or white; most are Asian and a few are American Indian.

★ 212 ★

Advanced Placement

Number of 11th and 12th Grade Students Scoring 3 or Above on Advanced Placement Examinations by 1,000 11th and 12th Graders and by Race/Ethnicity, 1992

| | Number of examinations taken | | | | | |
	Social studies	English	Foreign language	Calculus	Computer science	Science
Race/ethnicity						
White	16	17	3	9	1	8
Black	2	2	1	1	0	1
Hispanic	5	5	16	3	0	2
Other[1]	39	33	11	43	4	39

Source: "Number of 11th and 12th Grade Advanced Placement Examinations Taken and 11th and 12th Graders Scoring 3 or Above Per 1,000 11th and 12th Graders, by Subject Area, and by Sex and Race/Ethnicity: 1992," U.S. Department of Education, National Center for Education Statistics, *The Condition of Education, 1993*, p. 72. Primary source: The College Board, Advanced Placement Program, National Summary Reports, 1984-1992, (Copyright 1992 by College Entrance Board.); Educational Testing Service, unpublished tabulations; U.S. Department of Commerce, Bureau of the Census, October Current Population Survey. Published by permission. *Notes:* Since, on average, AP candidates take more than one examination, there is not a 1:1 correspondence between candidates and examinations. 1. Includes individuals who are not Hispanic, black, or white; most are Asian and a few are American Indian.

★ 213 ★

Advanced Placement

Number of 11th and 12th Grade Students Taking Advanced Placement Examinations by 1,000 11th and 12th Graders and by Race/Ethnicity, 1984 to 1992

| | Year | | | | | | | | |
	1984	1985	1986	1987	1988	1989	1990	1991	1992
Total	24	29	33	36	39	44	48	53	57
Race/ethnicity									
White	23	29	32	34	40	45	48	54	58
Black	4	5	6	8	9	11	13	15	14
Hispanic	10	14	14	17	22	31	32	32	37
Other[1]	56	64	80	79	104	108	133	142	149

Source: "Number of 11th and 12th Grade Students Taking Advanced Placement Examinations per 1,000 11th and 12th Graders by Sex and Race/Ethnicity: 1984-1992," U.S. Department of Education, National Center for Education Statistics, *The Condition of Education, 1993*, p. 72. Primary source: The College Board, Advanced Placement Program, National Summary Reports, 1984-1992, (Copyright 1992 by College Entrance Board.); Educational Testing Service, unpublished tabulations; U.S. Department of Commerce, Bureau of the Census, October Current Population Survey. Published by permission. *Notes:* 1. Includes individuals who are not Hispanic, black, or white; most are Asian and a few are American Indian.

Associate Degrees

★ 214 ★

Associate Degrees By Racial/Ethnic Group and Sex, 1976-91

Year and sex of student	Total	White, non-Hispanic	Black, non-Hispanic	Hispanic	Asian or Pacific Islander	American Indian/ Alaskan Native	Nonresident alien
1976-77							
Total[1]	404,956	342,290	33,159	16,636	7,044	2,498	3,329
Men	209,672	178,236	15,330	9,105	3,630	1,216	2,155
Women	195,284	164,054	17,829	7,531	3,414	1,282	1,174
1978-79							
Total[2]	396,745	331,092	34,979	16,269	7,518	2,336	4,551
Men	187,284	156,671	14,425	8,135	4,058	1,069	2,926
Women	209,461	174,421	20,554	8,134	3,460	1,267	1,625
1980-81							
Total[3]	410,174	339,167	35,330	17,800	8,650	2,584	6,643
Men	183,819	151,242	14,290	8,327	4,557	1,108	4,295
Women	226,355	187,925	21,040	9,473	4,093	1,476	2,348
1984-85							
Total[4]	429,815	355,343	35,791	19,407	9,914	2,953	6,407
Men	190,409	157,278	14,184	8,561	5,492	1,198	3,696
Women	239,406	198,065	21,607	10,846	4,422	1,755	2,711
1986-87							
Total[5]	436,299	361,819	35,457	19,345	11,794	3,196	4,688
Men	190,832	158,126	13,947	8,764	6,172	1,263	2,560
Women	245,467	203,693	21,510	10,581	5,622	1,933	2,128
1988-89							
Total[6]	432,144	354,813	34,722	20,381	12,531	3,335	6,362
Men	183,963	150,950	12,913	9,212	6,375	1,325	3,188
Women	248,181	203,863	21,809	11,169	6,156	2,010	3,174
1989-90							
Total[7]	450,263	369,546	35,341	22,216	13,478	3,530	6,152
Men	188,631	154,719	13,161	9,869	6,478	1,434	2,970
Women	261,632	214,827	22,180	12,347	7,000	2,096	3,182
1990-91							
Total[8]	462,030	376,069	37,659	24,255	13,729	3,675	6,643

[Continued]

★ 214 ★

Associate Degrees By Racial/Ethnic Group and Sex, 1976-91
[Continued]

Year and sex of student	Total	White, non-Hispanic	Black, non-Hispanic	Hispanic	Asian or Pacific Islander	American Indian/ Alaskan Native	Nonresident alien
Men	190,221	155,320	13,720	10,213	6,444	1,374	3,150
Women	271,809	220,749	23,939	14,042	7,285	2,301	3,493

Source: "Associate Degrees Conferred by Institutions of Higher Education, by Racial/Ethnic Group and Sex of Student: 1976-77 to 1990-91." *Digest of Educational Statistics*, 1993, p. 272. Primary source: U.S. Department of Education, National center for Education Statistics, "Degrees and Other Formal Awards Conferred" survey, and Integrated Postsecondary Education Data System (IPEDS), "Completions" survey. (This table was prepared April 1993.)
Notes: 1. Excludes 1,170 men and 251 women whose racial/ethnic group was not available. 2. Excludes 4,807 men and 1,150 women whose racial/ethnic group was not available. 3. Excludes 4,819 men and 1,384 women whose racial/ethnic group was not available. 4. Racial/ethnic data were imputed for approximately 45,400 men and 55,400 women. This tabulation excludes 11,490 men and 10,862 women whose racial/ethnic group could not be imputed. In addition, data for 1,033 men and 1,512 women were not available by field of study and were not imputed by race. 5. Excludes 693 men and 146 women whose racial/ethnic group was not available. 6. Reported racial/ethnic distributions of students by level of degree, field of degree and sex were used to estimate race/ethnicity for students whose race/ethnicity was not reported. Excludes 2,353 men and 2,267 women whose racial/ethnic group and field of study were not available. 7. Reported racial/ethnic distributions of students by level of degree, field of degree and sex were used to estimate race/ethnicity for students whose race/ethnicity was not reported. Excludes 2,564 men and 2,275 women whose racial/ethnic group and field of study were not available. Revised from previously published data. 8. Reported racial/ethnic distributions of students by level of degree, field of degree and sex were used to estimate race/ethnicity for students whose race/ethnicity was not reported. Excludes 8,413 men and 11,277 women whose racial/ethnic group and field of study were not available. Preliminary data.

★ 215 ★
Associate Degrees

Associate Degrees by Racial/Ethnic Group, Major Field of Study, and Sex: 1990-91

Associate degrees conferred by institutions of higher education, by racial/ethnic group, major field of study, and sex of student: 1990-91[1].

Major field of study	Total						Men Black, non-Hispanic	Women Black, non-Hispanic
	White, non-Hispanic	Black, non-Hispanic	Hispanic	Asian/Pacific Islander	American Indian/ Alaskan Native	Nonresident alien		
All fields, total[2]	376,069	37,659	24,255	13,729	3,675	6,643	13,720	23,939
Agriculture and natural resources	4,694	29	63	22	60	42	17	12
Architecture and environmental design	1,665	49	160	88	7	62	1	48
Area and ethnic studies	7	7	0	1	4	0	2	5
Business and management	80,973	10,657	5,582	3,359	767	1,640	2,858	7,799
Communications	3,200	329	175	53	31	68	171	158
Computer and information sciences	5,782	927	398	335	81	154	338	589
Education	6,271	633	506	101	153	131	235	398
Engineering	1,973	133	109	154	22	69	112	21
Engineering technologies	40,783	3,442	2,502	1,967	341	602	2,969	473
Foreign languages	244	6	53	12	6	6	2	4
Health sciences	60,540	5,443	2,383	1,467	468	518	668	4,775
Home economics	8,488	1,124	609	427	63	168	206	918
Law	4,701	439	235	47	42	20	66	373
Letters	375	39	30	11	10	8	17	22
Liberal/general studies	110,908	10,485	8,661	4,144	1,111	2,595	4,096	6,389
Library and archival sciences	96	7	3	3	7	0	0	7
Life sciences	778	59	109	129	22	22	22	37
Mathematics	490	28	56	67	13	16	17	11
Military sciences	66	11	4	4	0	0	8	3
Multi/interdisciplinary studies	10,937	614	374	225	53	64	293	321
Parks and recreation	371	33	6	1	2	2	22	11

[Continued]

★ 215 ★

Associate Degrees by Racial/Ethnic Group, Major Field of Study, and Sex: 1990-91

[Continued]

Major field of study	Total						Men Black, non-Hispanic	Women Black, non-Hispanic
	White, non-Hispanic	Black, non-Hispanic	Hispanic	Asian/Pacific Islander	American Indian/ Alaskan Native	Nonresident alien		
Philosophy and religion	81	2	0	3	0	3	1	1
Physical sciences	1,725	152	128	122	17	41	70	82
Protective services	11,360	1,111	774	167	101	51	601	510
Psychology	754	105	90	24	16	8	32	73
Public affairs	4,220	601	296	139	96	60	204	397
Social sciences	1,703	347	276	67	63	49	153	194
Theology	495	35	12	18	1	17	26	9
Visual and performing arts	12,389	812	661	572	118	227	513	299

Source: "Associate Degrees Conferred by Institutions of Higher Education, by Racial/Ethnic Group, Major Field of Study, and Sex of Student: 1990-91." *Digest of Educational Statistics,* 1993, p. 274. Primary source: U.S. Department of Education, National Center for Education Statistics, Integrated Postsecondary Education Data System (IPEDS), "Completions" survey. (This table was prepared April 1993.) *Notes:* To facilitate trend comparisons, certain aggregations have been made of the degree fields as reported in the IPEDS "Completions" survey: "Agriculture and natural resources" includes Agribusiness and agriculture production, Agricultural sciences, and Renewable natural resources; "Business and management" includes Business and management, Business and office, Marketing and distribution, and Consumer and personal services; "Engineering and related technologies" includes Engineering and related technologies, Mechanics and repairers, and Construction trades; "Physical sciences" includes Physical sciences and Science technologies; "Public affairs" includes Public affairs and Transportation and material moving; and "Visual and performing arts" includes Visual and performing arts and Precision production. 1. Preliminary data. 2. Reported racial/ethnic distributions of students by level of degree, field of degree, and sex were used to estimate race/ethnicity for students whose race/ethnicity was not reported. Excludes 423 men and 324 women whose racial/ethnic group and field of study were not available.

Attainment

★ 216 ★

Average Mathematics Proficiency by Age and Race/Ethnicity, 1973 to 1990

	Age 9			Age 13			Age 17		
	All races	White	Black	All races	White	Black	All races	White	Black
1973	219[1]	225[1]	190[1]	266[1]	274	228[1]	304	310	270[1]
1978	219[1]	224[1]	192[1]	264[1]	272[1]	230[1]	300[1]	306[2]	268[1]
1982	219[1]	224[1]	195[1]	269	274	240[1,2]	299[1,2]	304[1,2]	272[1]
1986	222[1]	227[1]	202[2]	269	274	249[2]	302	308	279[1,2]
1990	230[2]	235[2]	208[2]	270[2]	276	249[2]	305	310	289[2]

Source: "Average Mathematics Proficiency (Scale Score), by Age and Race/Ethnicity: 1970-1990," U.S. Department of Education, National Center for Education Statistics, *The Condition of Education,* 1993, p. 44. Primary source: National Assessment of Educational Progress, *Trends in Academic Progress: Achievement of American Students in Science, 1969-70 to 1990, Mathematics, 1973 to 1990, Reading, 1971 to 1990, Writing, 1984 to 1990,* 1991. *Notes:* Mathematics Proficiency Scale has a range from 0 to 500. Level 150: Knows simple arithmetic facts; Level 200: Beginning skills and understandings; Level 250: Numerical operations and beginning problem solving; Level 300: Moderately complex procedures and reasoning; Level 350: Multi-step problem solving and algebra. 1. Statistically significant difference from 1990. 2. Statistically significant difference from 1973.

★ 217 ★

Attainment

Average Reading Proficiency by Age and Race/Ethnicity, 1971 to 1990

	Age 9			Age 13			Age 17		
	All races	White	Black	All races	White	Black	All races	White	Black
1971	208	214	170[1]	255	261	222[1]	285[1]	291[1]	239[1]
1975	210	217	181[2]	256	262	226[1]	286[1]	293	241[1]
1980	215[1,2]	221[2]	189[2]	259	264[2]	233[1,2]	286	293	243[1]
1984	211	218[2]	186[2]	257	263	236[2]	289	295[2]	264[2]
1988	212	218	189[2]	258	261	243[2]	290[2]	295	274[2]
1990	209	217	182[2]	257	262	242[2]	290[2]	297[2]	267[2]

Source: "Average Reading Proficiency (Scale Score), by Age and Race/Ethnicity: 1971-1990," U.S. Department of Education, National Center for Education Statistics, *The Condition of Education, 1993*, p. 40. Primary source: National Assessment of Educational Progress, *Trends in Academic Progress: Achievement of American Students in Science, 1969-70 to 1990, Mathematics, 1973 to 1990, Reading, 1971 to 1990, Writing, 1984 to 1990*, 1991. *Notes:* Reading Proficiency Scale has a range from 0 to 500. Level 150: Simple discrete reading tasks; Level 200: Partial skills and understanding; Level 250: Interrelated ideas, and make generalizations; Level 300: Understands relatively complicated information; Level 350: Learns from specialized reading materials. 1. Statistically significant difference from 1990. 2. Statistically significant difference from 1971 for all except Hispanics. Statistically significant difference from 1975 for Hispanics.

★ 218 ★

Attainment

Average Science Proficiency by Age and Race/Ethnicity, 1970 to 1990

	Age 9			Age 13			Age 17		
	All races	White	Black	All races	White	Black	All races	White	Black
1970	225	236	179[1]	255	263	215	305[1]	312[1]	258
1973	220[1,2]	231[1,2]	177[1]	250[1,2]	259[1,2]	205[1]	296[1,2]	304[2]	250[2]
1977	220[1,2]	230[1,2]	175[1]	247[1,2]	256[1,2]	208[1]	290[2]	298[2]	240[1,2]
1982	221[1]	229[1,2]	187	250[1]	257[1,2]	217	283[1,2]	293[1,2]	235[1,2]
1986	224[1]	232[1]	296[2]	251	259[1]	222	289[2]	298[2]	253
1990	229	238	196[2]	255	264	226	290[2]	301[2]	253

Source: "Average Science Proficiency (Scale Score), by Age and Race/Ethnicity: 1970-1990," U.S. Department of Education, National Center for Education Statistics, *The Condition of Education, 1993*, p. 46. Primary source: National Assessment of Educational Progress, *Trends in Academic Progress: Achievement of American Students in Science, 1969-70 to 1990, Mathematics, 1973 to 1990, Reading, 1971 to 1990, Writing, 1984 to 1990*, 1991. *Notes:* Science Proficiency Scale has a range from 0 to 500. Level 150: Knows everyday science facts; Level 200: Understands simple scientific principles; Level 250: Applies general scientific information; Level 300: Analyzes scientific procedures and data; Level 350: Integrates specialized scientific information. 1. Statistically significant difference from 1990. 2. Statistically significant difference from 1970 for all except Hispanics. Statistically significant difference from 1977 for Hispanics.

★ 219 ★

Attainment

Average Writing Proficiency by Age and Race/Ethnicity, 1984 to 1990

	Grade 4			Grade 8			Grade 11		
	All races	White	Black	All races	White	Black	All races	White	Black
1984	179	186	154	206[1]	210[1]	190	212	218	195
1988	186	193	154	203[1]	207[1]	190	214	219	200
1990	183	191	155	198	202	182	212	217	194

Source: "Average Writing Proficiency (Scale Score), by Age and Race/Ethnicity: 1970-1990," U.S. Department of Education, National Center for Education Statistics, *The Condition of Education, 1993,* p. 42. Primary source: National Assessment of Educational Progress, *Trends in Academic Progress: Achievement of American Students in Science, 1969-70 to 1990, Mathematics, 1973 to 1990, Reading, 1971 to 1990, Writing, 1984 to 1990,* 1991. *Notes:* Writing Proficiency Scale has a range from 0 to 400. Level 100: Unsatisfactory-Failed to reflect a basic understanding of the task.; Level 200: Minimal-Recognized the elements needed to complete the task, but these were not managed well enough to insure the intended purpose. Level 300: Adequate-included features critical to accomplishing the purpose of the task and were likely to have the intended effect;Level 400: Elaborated- Reflected a higher level of coherence and elaboration; beyond adequate. Average NAEP writing assessment scores were produced using the Average Response Method (ARM). The ARM provides an estimate of average writing achievement for each respondent as if he or she took 11 of the 12 writing tasks given, and as if NAEP had computed average achievement across that set of tasks. 1. Statistically significant difference from 1990.

★ 220 ★

Attainment

Educational Level of Persons Age 18 and Over, Spring 1990

Numbers in thousands.

Sex, race, and age	Total	Not high school graduate[1]	High school graduate only	Some college, no degree or certificate	Vocational certificate	Associate degree	Bachelor's degree	Master's degree	Professional degree	Doctor's degree
Total population, 18 and over	182,591	38,012	65,291	33,191	4,973	7,570	22,845	7,599	2,054	1,056
White, total[2]	156,385	30,270	56,240	28,608	4,541	6,677	20,381	6,813	1,898	956
Men	75,262	14,425	25,556	14,076	1,588	3,242	10,629	3,552	1,449	744
Women	81,123	15,845	30,684	14,532	2,953	3,435	9,752	3,261	449	212
Black, total[2]	20,401	6,510	7,495	3,534	284	670	1,367	462	46	34
Men	9,158	3,045	3,483	1,441	87	257	581	199	38	28
Women	11,242	3,465	4,012	2,094	197	413	786	262	8	6
Hispanic, total[3]	13,548	5,934	4,091	1,933	208	316	734	245	55	32
Men	6,708	2,950	1,961	976	89	153	388	121	44	27
Women	6,841	2,984	2,130	958	119	163	346	124	11	5
Percentage distribution, by highest degree earned										
Total population, 18 and over	100.0	20.8	35.8	18.2	2.7	4.1	12.5	4.2	1.1	0.6
White, total[2]	100.0	19.4	36.0	18.3	2.9	4.3	13.0	4.4	1.2	0.6
Men	100.0	19.2	34.0	18.7	2.1	4.3	14.1	4.7	1.9	1.0
Women	100.0	19.5	37.8	17.9	3.6	4.2	12.0	4.0	0.6	0.3
Black, total[2]	100.0	31.9	36.7	17.3	1.4	3.3	6.7	2.3	0.2	0.2
Men	100.0	33.2	36.0	15.7	0.9	2.8	6.3	2.2	0.4	0.3
Women	100.0	30.8	35.7	18.6	1.7	3.7	7.0	2.3	0.1	0.1

[Continued]

★ 220 ★

Educational Level of Persons Age 18 and Over, Spring 1990
[Continued]

Sex, race, and age	Total	Not high school graduate[1]	High school graduate only	Some college, no degree or certificate	Vocational certificate	Associate degree	Bachelor's degree	Master's degree	Professional degree	Doctor's degree
Hispanic, total[3]	100.0	43.8	30.2	14.3	1.5	2.3	5.4	1.8	0.4	0.2
Men	100.0	44.0	29.2	14.5	1.3	2.3	5.8	1.8	0.7	0.4
Women	100.0	43.6	31.1	14.0	1.7	2.4	5.1	1.8	0.2	0.1

Source: "Highest Education Level and Degree Earned by Persons Age 18 and Over, by Sex, Race, and Age: Spring 1990." *Digest of Educational Statistics,* 1993, p. 20. Primary source: U.S. Department of Commerce, Bureau of the Census, *Current Population Reports,* Series P-70, No. 32, "What's It Worth? Educational Background and Economic Status: Spring 1990." (This table was prepared February 1993). *Notes:* Data are based on sample surveys of the civilian noninstitutional population. Because of rounding, details may not add to totals. 1. Some people are still enrolled in high school. 2. Includes persons of Hispanic origin. 3. Persons of Hispanic origin may be of any race.

★ 221 ★

Attainment

Fields of Study of persons Age 18 and Over with Bachelor's Degree or Higher, Spring 1990

Numbers in thousands.

Field of study	Total	Race		Percentage distribution of degree holders, by field		
		White[1]	Black[1]	Total	White[1]	Black[1]
Total population, 18 and over	182,591	156,385	20,401	100.0	100.0	100.0
Number of persons with bachelor's or higher degree	33,554	30,049	1,908			
Percent of population	18.4	19.2	9.4			
Agriculture and forestry	371	351	6	1.1	1.2	0.3
Biology	857	767	34	2.6	2.6	1.8
Business and management	6,189	5,531	368	18.4	18.4	19.3
Economics	691	581	40	2.1	1.9	2.1
Education	5,879	5,296	478	17.5	17.6	25.1
Engineering	3,090	2,635	154	9.2	8.8	8.1
English and journalism	1,369	1,306	40	4.1	4.3	2.1
Home economics	385	350	14	1.1	1.2	0.7
Law	1,004	948	15	3.0	3.2	0.8
Liberal arts and humanities	3,002	2,703	160	8.9	9.0	8.4
Mathematics and statistics	699	648	13	2.1	2.2	0.7
Medicine and dentistry	1,046	893	36	3.1	3.0	1.9
Nursing, pharmacy, and health technologies	1,913	1,717	83	5.7	5.7	4.4
Physical and earth sciences	856	781	35	2.6	2.6	1.8
Police science and law enforcement	238	201	25	0.7	0.7	1.3
Psychology	1,103	1,001	80	3.3	3.3	4.2
Religion and theology	488	452	24	1.5	1.5	1.3
Social sciences	1,960	1,769	125	5.8	5.9	6.6
Vocational and technical studies	179	155	19	0.5	0.5	1.0
Other fields	2,233	1,963	162	6.7	6.5	8.5

Source: "Number of Persons Age 18 and Over Who Hold a Bachelor's or Higher Degree, by Field of Study, Sex, Race, and Age: Spring 1990." *Digest of Educational Statistics,* 1993, p. 19. Primary source: U.S. Department of Commerce, Bureau of the Census, *Current Population Reports,* Series P-70, No. 32, "What's It Worth? Educational Background and Economic Status: Spring 1990." (This table was prepared February 1993). *Notes:* Data are based on sample surveys of the civilian noninstitutional population. Because of rounding, details may not add to totals. 1. Includes Hispanic origin.

★ 222 ★

Attainment

Population Aged 25 Years or Older with Bachelor's Degree or Over, April 1990

State	Percent with bachelor's degree or higher					
	Total	White[1]	Black[1]	Hispanic[2]	Asian/Pacific Islander[1]	American Indian or Alaskan Native[1]
United States	20.3	21.5	11.4	9.2	36.6	9.3
Alabama	15.7	17.3	9.3	20.1	43.7	11.6
Alaska	23.0	26.8	14.1	14.6	20.5	4.1
Arizona	20.3	22.2	14.3	6.9	37.5	4.6
Arkansas	13.3	14.1	8.4	11.1	24.6	9.8
California	23.4	25.4	14.8	7.1	34.1	11.1
Colorado	27.0	28.3	17.1	8.6	32.1	12.1
Connecticut	27.2	28.5	12.3	12.1	50.8	12.5
Delaware	21.4	23.0	10.6	16.5	55.9	10.2
District of Columbia	33.3	69.0	15.3	24.0	50.9	17.7
Florida	18.3	19.3	9.8	14.2	33.6	11.5
Georgia	19.3	21.8	11.0	20.5	38.6	12.5
Hawaii	22.9	30.2	15.2	10.3	19.4	17.7
Idaho	17.7	18.0	15.8	6.6	27.6	7.2
Illinois	21.0	22.4	11.4	8.0	49.8	13.4
Indiana	15.6	17.6	9.3	10.8	53.1	8.4
Iowa	16.9	16.7	12.8	13.7	47.3	9.7
Kansas	21.1	21.7	11.6	10.1	39.9	10.8
Kentucky	13.6	13.9	7.7	18.9	44.2	8.0
Louisiana	16.1	18.7	9.1	16.6	31.4	5.5
Maine	18.8	18.8	22.3	23.6	44.9	7.7
Maryland	26.5	28.9	16.1	25.2	50.3	19.7
Massachusetts	27.2	27.7	17.0	13.6	44.9	14.9
Michigan	17.4	18.1	10.1	11.6	54.1	7.6
Minnesota	21.8	21.9	17.5	17.2	33.5	7.7
Mississippi	14.7	17.2	8.8	17.1	35.1	8.1
Missouri	17.8	18.3	11.2	18.0	47.3	11.0
Montana	19.8	20.3	18.4	10.9	32.1	7.9
Nebraska	18.9	19.2	12.4	9.4	39.5	8.8
Nevada	15.3	15.9	9.0	7.0	21.9	8.0
New Hampshire	24.4	24.2	25.7	25.5	26.1	16.0
New Jersey	24.9	25.8	13.6	10.8	57.1	14.8
New Mexico	20.4	23.4	14.2	8.7	38.7	5.8
New York	23.1	25.3	12.6	9.3	38.7	13.4
North Carolina	17.4	19.3	9.5	17.9	39.3	7.9
North Dakota	18.1	18.3	17.1	15.9	37.8	8.3
Ohio	17.0	17.6	9.1	14.2	53.2	8.3
Oklahoma	17.8	18.7	12.0	10.5	34.7	10.8
Oregon	20.6	20.8	9.1	10.1	32.3	8.3
Pennsylvania	17.9	18.5	10.0	11.8	45.2	12.0
Rhode Island	21.3	21.8	12.7	8.9	30.6	8.3
South Carolina	16.6	19.8	7.6	19.8	34.4	10.9

[Continued]

★ 222 ★

Population Aged 25 Years or Older with Bachelor's Degree or Over, April 1990

[Continued]

State	Percent with bachelor's degree or higher					
	Total	White[1]	Black[1]	Hispanic[2]	Asian/Pacific Islander[1]	American Indian or Alaskan Native[1]
South Dakota	17.2	17.6	24.1	13.4	33.1	6.8
Tennessee	16.0	16.7	10.2	21.9	42.6	10.5
Texas	20.3	22.6	12.0	7.3	41.3	13.9
Utah	22.3	22.7	15.9	9.1	29.4	6.4
Vermont	24.3	24.2	30.5	28.2	52.1	11.1
Virginia	24.5	27.0	11.1	22.4	40.2	14.7
Washington	22.9	23.3	15.4	11.0	30.2	9.1
West Virginia	12.3	12.2	10.9	17.6	63.3	6.5
Wisconsin	17.7	18.1	8.3	10.0	40.4	5.5
Wyoming	18.8	19.3	9.5	4.8	28.6	6.2

Source: "Educational Attainment of Persons 25 Years Old and Over, by State and Race: April 1990." *Digest of Educational Statistics*, 1993, p. 22. Primary source: U.S. Department of Commerce, Bureau of the Census, Decennial Census, Minority Economic Profiles, unpublished data. (This table was prepared June 1993). *Notes:* 1. Includes persons of Hispanic origin. 2. Persons of Hispanic origin may be of any race.

★ 223 ★
Attainment

Population Aged 25 Years or Older with High School Diploma or Over, April 1990

State	Percent with high school diploma or higher					
	Total	White[1]	Black[1]	Hispanic[2]	Asian/Pacific Islander[1]	American Indian or Alaskan Native[1]
United States	75.2	77.9	63.1	49.8	77.5	65.5
Alabama	66.9	70.3	54.6	73.8	78.9	64.9
Alaska	86.6	91.1	88.2	80.4	75.4	63.1
Arizona	78.7	82.4	75.1	51.7	80.2	52.1
Arkansas	66.3	68.6	51.5	59.1	66.4	65.4
California	76.2	81.1	75.6	45.0	77.2	71.4
Colorado	84.4	86.1	80.8	58.3	78.3	73.9
Connecticut	79.2	80.9	67.0	53.5	81.9	68.9
Delaware	77.5	80.3	63.2	60.1	86.1	62.0
District of Columbia	73.1	93.1	63.8	52.6	80.2	66.3
Florida	74.4	77.0	9.8	57.2	77.8	68.2
Georgia	70.9	74.9	58.6	66.2	77.5	71.6
Hawaii	80.1	89.3	94.2	73.9	74.7	84.4
Idaho	79.7	80.9	82.8	43.4	80.3	68.1
Illinois	76.2	79.1	65.2	45.0	83.9	71.4

[Continued]

★ 223 ★

Population Aged 25 Years or Older with High School Diploma or Over, April 1990

[Continued]

State	Percent with high school diploma or higher					
	Total	White[1]	Black[1]	Hispanic[2]	Asian/Pacific Islander[1]	American Indian or Alaskan Native[1]
Indiana	75.6	76.5	65.4	62.6	85.8	65.0
Iowa	80.1	80.3	70.1	64.2	76.4	67.6
Kansas	81.3	82.4	71.0	58.1	73.6	75.4
Kentucky	64.6	64.7	61.7	74.0	77.9	59.8
Louisiana	68.3	74.2	53.1	67.6	68.1	49.1
Maine	78.8	78.9	87.6	83.8	74.3	69.9
Maryland	78.4	80.8	70.6	70.3	84.8	73.4
Massachusetts	80.0	81.2	70.0	52.0	74.1	71.1
Michigan	76.8	78.6	64.9	60.9	83.3	67.8
Minnesota	82.4	82.8	76.2	71.1	69.7	68.2
Mississippi	64.3	71.7	47.3	67.7	68.2	57.4
Missouri	73.9	74.9	65.1	71.0	81.5	65.1
Montana	81.0	81.7	80.9	66.4	78.5	68.1
Nebraska	81.8	82.4	73.2	60.0	80.0	69.0
Nevada	78.8	80.9	70.8	53.7	74.1	69.8
New Hampshire	82.2	82.2	86.1	78.2	82.7	65.9
New Jersey	76.7	78.6	67.0	53.9	86.8	66.9
New Mexico	75.1	78.6	74.7	59.6	80.8	58.2
New York	76.7	78.5	64.7	50.4	72.4	65.2
North Carolina	70.0	73.1	58.1	71.0	77.9	51.5
North Dakota	76.7	76.9	95.9	75.2	83.7	64.3
Ohio	75.7	76.9	64.6	63.3	83.5	65.3
Oklahoma	74.6	75.7	70.1	55.9	76.1	68.1
Oregon	81.5	82.3	75.0	53.0	79.4	71.0
Pennsylvania	74.7	75.9	63.5	52.2	77.1	67.8
Rhode Island	72.0	73.0	65.9	46.8	59.6	64.5
South Carolina	68.3	73.6	53.3	71.8	77.4	62.5
South Dakota	77.1	77.8	82.2	71.3	74.3	62.5
Tennessee	67.1	68.2	59.4	71.5	79.3	63.1
Texas	72.1	76.2	66.1	44.6	79.1	70.9
Utah	85.1	86.2	77.0	61.0	80.7	59.3
Vermont	80.8	80.8	82.9	84.7	87.1	66.8
Virginia	75.2	78.3	60.3	70.5	82.1	70.7
Washington	83.8	85.0	81.2	56.7	77.3	72.3
West Virginia	66.0	66.0	64.7	70.3	88.8	57.9
Wisconsin	78.6	79.6	61.3	54.1	71.5	66.8
Wyoming	83.0	83.9	81.2	59.3	77.5	68.2

Source: "Educational Attainment of Persons 25 Years Old and Over, by State and Race: April 1990." *Digest of Educational Statistics*, 1993, p. 22. Primary source: U.S. Department of Commerce, Bureau of the Census, Decennial Census, Minority Economic Profiles, unpublished data. (This table was prepared June 1993). *Notes:* 1. Includes persons of Hispanic origin. 2. Persons of Hispanic origin may be of any race.

★ 224 ★

Attainment

Years of School Completed by Persons 25 and Over, 1940 to 1991

Race, age, and date	Percent, by years of school completed			Median school years completed
	Less than 5 years of elementary school	4 years of high school or more	4 or more years of college	
White[1]				
25 and over				
April 1940	10.9	26.1	4.9	8.7
April 1950	8.9	36.4	6.6	9.7
April 1960	6.7	43.2	8.1	10.8
March 1970	4.2	57.4	11.6	12.2
March 1980	2.6	70.5	17.8	12.5
March 1985	2.2	75.5	20.0	12.7
March 1988	2.0	77.7	20.9	12.7
March 1989	2.0	78.4	21.8	12.7
March 1990	2.0	79.1	22.0	12.7
March 1991	2.0	79.9	22.2	12.8
Black and other races[2]				
25 and over				
April 1940	41.8	7.7	1.3	5.7
April 1950	32.6	13.7	2.2	6.9
April 1960	23.5	21.7	3.5	8.2
March 1970	14.7	36.1	6.1	10.1
MArch 1980	8.8	54.6	11.1	12.2
March 1985	6.0	63.2	15.4	12.4
March 1988	5.1	66.7	16.4	12.5
March 1989	5.6	67.3	16.9	12.5
March 1990	5.4	68.7	16.5	12.5
March 1991	5.0	69.6	16.7	12.5

Source: "Years of School Completed by Persons Age 25 and Over and 25 to 29, by Race: 1910 to 1991." *Digest of Educational Statistics,* 1993, p. 17. Primary source: U.S. Department of Commerce, Bureau of the Census, *U.S. Census of Population, 1960,* Vol. 1, part 1: *Current Population Reports,* Series P-20; Series P-19, No. 4; *1960 Census Monograph,* "Education of the American Population," by John K. Folger and Charles B. Nam; and unpublished data from the Current Population Survey; and U.S. Department of Labor, Bureau of Labor Statistics, Office of Employment and Unemployment Statistics, "Educational Attainment of Workers, March 1991." (This table was prepared April 1993). *Notes:* Data for 1975 and subsequent years are for the noninstitutional population. Some data have been revised from previously published figures. 1. Persons of Hispanic origin are included, as appropriate, in the "white" or in the "black and other races" category.

★ 225 ★

Attainment

Years of School Completed by Persons Age 18 and Over, 1991

In thousands.

Age, sex, and race	Total population[1]	Elementary level		High school		College		
		Less than 8 years	8 years	1 to 3 years	4 years	1 to 3 years	4 years	5 years or more
White[2]								
18 and over	156,682	7,884	6,469	17,275	61,815	31,325	19,133	12,782
18 and 19 years old	5,555	101	97	1,834	2,748	770	5	-
20 to 24 years old	14,828	406	222	1,558	5,817	5,197	1,395	231
25 years old and over	136,299	7,376	6,150	13,883	53,250	25,357	17,733	12,551
25 to 29 years old	17,252	475	283	1,687	6,825	3,742	3,078	1,162
30 to 34 years old	18,650	536	265	1,523	7,693	3,938	3,054	1,640
35 to 39 years old	17,227	482	255	1,167	6,687	4,031	2,650	1,955
40 to 49 years old	27,808	1,055	577	1,999	10,505	5,837	4,125	3,732
50 to 59 years old	19,128	1,108	797	2,107	7,958	3,212	2,029	1,915
60 to 64 years old	9,316	665	620	1,239	3,902	1,364	840	688
65 years old and over	26,898	3,054	3,354	4,162	9,679	3,234	1,958	1,458
Black[2]								
18 and over	20,645	1,942	770	4,013	8,093	3,751	1,384	692
18 and 19 years old	1,068	11	32	471	491	63	-	-
20 to 24 years old	2,481	24	38	457	1,154	698	79	31
25 years old and over	17,096	1,907	700	3,085	6,448	2,990	1,305	661
25 to 29 years old	2,730	31	34	437	1,266	662	231	70
30 to 34 years old	2,705	56	32	396	1,243	616	267	96
35 to 39 years old	2,365	45	37	350	1,009	516	293	116
40 to 49 years old	3,340	147	88	565	1,363	724	272	180
50 to 59 years old	2,376	386	114	552	832	248	130	112
60 to 64 years old	1,033	252	65	276	287	86	39	29
65 years old and over	2,547	992	330	510	447	137	74	58
Hispanic origin[3]								
18 and over	13,948	3,466	837	2,421	4,253	1,823	685	464
18 and 19 years old	710	71	38	315	231	56	-	-
20 to 24 years old	2,030	338	96	411	736	389	46	14
25 years old and over	11,208	3,058	703	1,696	3,285	1,380	639	449
25 to 29 years old	2,124	367	97	453	697	312	139	56
30 to 34 years old	2,096	421	101	339	702	306	143	84
35 to 39 years old	1,598	352	88	220	509	248	100	81
40 to 49 years old	2,270	619	132	300	682	304	127	105
50 to 59 years old	1,446	491	104	188	396	134	71	61

[Continued]

★ 225 ★

Years of School Completed by Persons Age 18 and Over, 1991
[Continued]

| Age, sex, and race | Total population[1] | Elementary level | | High school | | College | | |
		Less than 8 years	8 years	1 to 3 years	4 years	1 to 3 years	4 years	5 years or more
60 to 64 years old	582	241	54	80	124	30	25	28
65 years old and over	1,091	567	126	114	175	45	34	32

Source: "Years of School Completed by Persons Age 18 and Over, by Age, Sex, and Race/Ethnicity: 1991." *Digest of Educational Statistics,* 1993, p. 20. Primary source: U.S. Department of Commerce, Bureau of the Census, *Current Population Reports,* Series P-20, No. 426. (This table was prepared April 1993). *Notes:* Data are based on sample surveys of the noninstitutional population. Although cells with fewer than 75,000 people are subject to relatively wide sampling variation, they are included in the table to permit various types of aggregations. Because of rounding, details may not add to totals. - Data not applicable or not available. 1. Civilian noninstitutional population. 2. Includes persons of Hispanic origin. 3. Persons of Hispanic origin may be of any race.

★ 226 ★

Attainment

Years of School Completed, by Age and Race, 1950 to 1992

| Item | Persons 25 years old and over | | | | Persons 25 to 29 years old | | | |
	1950	1970	1990	1992	1950	1970	1990	1992
All persons								
Not high school graduates	65.7	47.7	22.4	20.6	47.2	26.2	14.3	13.7
With less than 5 years of school	11.1	5.5	2.4	2.1	4.7	1.7	1.2	0.9
High school graduates or more	34.3	52.3	77.6	79.4	52.8	73.8	85.7	86.3
College or more	6.2	10.7	21.3	21.4	7.7	16.3	23.2	23.6
Median school years completed[1]	9.3	12.1	12.7	(NA)	12.0	12.6	12.9	(NA)
Black persons								
Not high school graduates	87.1	68.6	33.8	32.3	77.8	44.6	18.4	19.1
With less than 5 years of school	32.9	14.6	5.1	3.9	16.8	3.2	1.0	0.9
High school graduates or more	12.9	31.4	66.2	67.7	22.2	55.4	81.7	80.9
College or more	2.1	4.4	11.3	11.9	2.7	6.0	13.4	11.3
Median school years completed[1]	6.8	9.8	12.4	(NA)	8.6	12.1	12.7	(NA)

Source: "Years of School Completed, by Age and Race: 1950 to 1992," U.S. Department of Commerce, *Statistical Abstract of the United States, 1993,* p. 152. Primary source: U.S. Bureau of the Census, *U. S. Census of Population, 1950, 1960, 1970, and 1980,* vol. I; and *Current Population Reports,* series P20-462; and unpublished data. *Notes:* In percent, except median. Through 1990, as of April 1; beginning 1990 as of March. Excludes Armed Forces, except members living off post or with their families on post. Beginning 1980, excludes inmates of institutions. 1950 based on 20-percent sample; 1960 on 25-percent sample; 1970 on 20- percent sample; and 1980 on 17-percent sample. Beginning 1990, based on the Current Population Survey. NA Not available. 1. Due to revised methodology, median school years completed not available beginning 1992.

Bachelor's Degrees

★ 227 ★

Bachelor's Degrees By Racial/Ethnic Group and Sex, 1976-91

Year and sex of student	Total	White, non-Hispanic	Black, non-Hispanic	Hispanic	Asian or Pacific Islander	American Indian/ Alaskan Native	Nonresident alien
1976-77							
Total[1]	917,900	807,688	58,636	18,743	13,793	3,326	15,714
Men	494,424	438,161	25,147	10,318	7,638	1,804	11,356
Women	423,476	369,527	33,489	8,425	6,155	1,522	4,358
1978-79							
Total[2]	919,540	802,542	60,246	20,096	15,407	3,410	17,839
Men	476,065	418,215	24,659	10,418	8,261	1,736	12,776
Women	443,475	384,327	35,587	9,678	7,146	1,674	5,063
1980-81							
Total[3]	934,800	807,319	60,673	21,832	18,794	3,593	22,589
Men	469,625	406,173	24,511	10,810	10,107	1,700	16,324
Women	465,175	401,146	36,162	11,022	8,687	1,893	6,265
1984-85							
Total[4]	968,311	826,106	57,573	25,874	25,395	4,246	29,217
Men	476,148	405,085	23,018	12,402	13,554	1,998	20,091
Women	492,163	421,021	34,455	13,472	11,841	2,248	9,126
1986-87							
Total[5]	991,260	841,820	56,555	26,990	32,618	3,971	29,306
Men	480,780	406,751	22,499	12,864	17,249	1,819	19,598
Women	510,480	435,069	34,056	14,126	15,369	2,152	9,708
1988-89							
Total[6]	1,016,350	859,699	58,065	29,910	37,686	3,954	27,036
Men	481,946	407,142	22,363	13,947	19,271	1,731	17,492
Women	534,404	452,557	35,702	15,963	18,415	2,223	9,544
1989-90							
Total[7]	1,048,631	884,372	61,065	32,846	39,247	4,393	26,708
Men	490,317	413,571	23,264	14,941	19,719	1,861	16,961
Women	558,314	470,801	37,801	17,905	19,528	2,532	9,747
1990-91							
Total[8]	1,081,280	904,061	65,338	36,612	41,622	4,513	29,134

[Continued]

★ 227 ★

Bachelor's Degrees By Racial/Ethnic Group and Sex, 1976-91

[Continued]

Year and sex of student	Total	White, non-Hispanic	Black, non-Hispanic	Hispanic	Asian or Pacific Islander	American Indian/ Alaskan Native	Nonresident alien
Men	496,424	415,506	24,326	16,157	20,681	1,901	17,853
Women	584,856	488,555	41,012	20,455	20,941	2,612	11,281

Source: "Bachelor's Degrees Conferred by Institutions of Higher Education, by Racial/Ethnic Group and Sex of Student: 1976-77 to 1990-91." *Digest of Educational Statistics*, 1993, p. 276. Primary source: U.S. Department of Education, National center for Education Statistics, "Degrees and Other Formal Awards Conferred" survey, and Integrated Postsecondary Education Data System (IPEDS), "Completions" survey. (This table was prepared April 1993.) *Notes:* 1. Excludes 1,121 men and 528 women whose racial/ethnic group was not available. 2. Excludes 1,279 men and 571 women whose racial/ethnic group was not available. 3. Excludes 258 men and 82 women whose racial/ethnic group was not available. 4. Excludes 6,380 men and 4,786 women whose racial/ethnic group was not available. 5. Reported racial/ethnic distributions of students by level of degree, field of degree, and sex were used to estimate race/ethnicity for students whose race/ethnicity was not reported. Excludes 74 men and 5 women whose racial/ethnic group and field of study were not available. 6. Reported racial/ethnic distributions of students by level of degree, field of degree and sex were used to estimate race/ethnicity for students whose race/ethnicity was not reported. Excludes 1,400 men and 1,005 women whose racial/ethnic group and field of study were not available. 7. Reported racial/ethnic distributions of students by level of degree, field of degree and sex were used to estimate race/ethnicity for students whose race/ethnicity was not reported. Excludes 1,379 men and 1,334 women whose racial/ethnic group and field of study were not available. Revised from previously published data. 8. Reported racial/ethnic distributions of students by level of degree, field of degree and sex were used to estimate race/ethnicity for students whose race/ethnicity was not reported. Excludes 7,621 men and 5,637 women whose racial/ethnic group and field of study were not available. Preliminary data.

★ 228 ★

Bachelor's Degrees

Bachelor's Degrees by Racial/Ethnic Group, Major Field of Study, and Sex: 1990-91

Bachelor's degrees conferred by institutions of higher education, by racial/ethnic group, major field of study, and sex of student: 1990-91[1].

Major field of study	Total						Men Black, non-Hispanic	Women Black, non-Hispanic
	White, non-Hispanic	Black, non-Hispanic	Hispanic	Asian/Pacific Islander	American Indian/ Alaskan Native	Nonresident alien		
All fields, total[2]	904,061	65,338	36,612	41,622	4,513	29,134	24,326	41,012
Agriculture and natural resources	11,863	341	233	238	65	384	207	134
Architecture and environmental design	7,934	329	458	527	33	500	214	115
Area and ethnic studies	3,488	362	262	353	24	134	135	227
Business and management	206,856	16,689	7,852	9,115	871	8,577	6,469	10,220
Communications	45,674	3,637	1,568	958	155	807	1,236	2,401
Computer and information sciences	17,903	2,063	917	2,075	82	2,043	1,001	1,062
Education	100,325	4,825	3,510	891	619	840	1,199	3,626
Engineering	46,192	2,279	2,057	6,361	161	4,582	1,519	760
Engineering technologies	13,913	1,203	583	777	75	681	987	216
Foreign languages	9,744	344	1,254	388	41	324	85	259
Health sciences	50,041	4,220	1,715	2,028	286	978	540	3,680
Home economics	13,581	932	284	413	61	203	116	816
Law	1,515	119	63	44	11	6	33	86
Letters	47,228	2,325	1,322	1,262	167	576	576	1,749
Liberal/general studies	22,115	1,941	1,365	602	142	527	779	1,162
Library and archival sciences	84	6	1	0	0	2	0	6
Life sciences	30,994	2,154	1,503	3,634	180	1,065	712	1,442
Mathematics	11,908	825	380	926	45	577	387	438
Military sciences	367	14	7	11	1	18	12	2
Multi/interdisciplinary studies	18,081	1,465	683	911	130	383	555	910
Parks and recreation	3,670	209	77	45	15	46	128	81
Philosophy and religion	6,438	240	202	280	21	134	144	96

[Continued]

★ 228 ★

Bachelor's Degrees by Racial/Ethnic Group, Major Field of Study, and Sex: 1990-91
[Continued]

| Major field of study | Total | | | | | | Men Black, non-Hispanic | Women Black, non-Hispanic |
	White, non-Hispanic	Black, non-Hispanic	Hispanic	Asian/Pacific Islander	American Indian/ Alaskan Native	Nonresident alien		
Physical sciences	13,500	772	390	1,004	70	608	402	370
Protective services	13,036	2,470	837	230	102	131	1,225	1,245
Psychology	49,428	3,786	2,379	1,926	241	691	903	2,883
Public affairs	13,404	2,086	757	341	186	202	486	1,600
Social sciences	104,198	8,099	4,681	4,632	520	2,763	3,511	4,588
Theology	4,316	134	103	132	12	116	105	29
Visual and performing arts	36,265	1,469	1,169	1,518	197	1,236	660	809

Source: "Bachelor's Degrees Conferred by Institutions of Higher Education, by Racial/Ethnic Group, Major Field of Study, and Sex of Student: 1976-77 to 1990-91." *Digest of Educational Statistics*, 1993, p. 276. Primary source: U.S. Department of Education, National Center for Education Statistics, Integrated Postsecondary Education Data System (IPEDS), "Completions" survey. (This table was prepared April 1993.) *Notes:* To facilitate trend comparisons, certain aggregations have been made of the degree fields as reported in the IPEDS "Completions" survey: "Agriculture and natural resources" includes Agribusiness and agriculture production, Agricultural sciences, and Renewable natural resources; "Business and management" includes Business and management, Business and office, Marketing and distribution, and Consumer and personal services; "Engineering and related technologies" includes Engineering and related technologies, Mechanics and repairers, and Construction trades; "Physical sciences" includes Physical sciences and Science technologies; "Public affairs" includes Public affairs and Transportation and material moving; and "Visual and performing arts" includes Visual and performing arts and Precision production. 1. Preliminary data. 2. Reported racial/ethnic distributions of students by level of degree, field of degree, and sex were used to estimate race/ethnicity for students whose race/ethnicity was not reported. Excludes 423 men and 324 women whose racial/ethnic group and field of study were not available.

Computers

★ 229 ★

The Black-White Computer Gap at Selected Elite Private Colleges and Universities

College	Undergraduate students	PC terminals	Students per terminal
Black			
Howard	8,100	140	57.9
Morehouse	3,200	175	18.0
Spelman	2,000	140	14.0
Tuskegee	3,687	92	40.1
White			
Dartmouth	4,275	1,000	4.3
Michigan	23,126	4,200	5.5
MIT	4,389	1,200	3.7
Vassar	2,218	295	7.5

Source: "The Black-White Computer Gap at Private Colleges and Universities," *The Journal of Blacks in Higher Education*, Vol. 1, No. 1, Autumn 1993, p. 91. Primary source: Survey by *The Journal of Blacks in Higher Education*. Published by permission.

★ 230 ★

Computers

The Black-White Computer Gap at Selected Southern State Universities

Institution	Undergraduate students	Percent black	PC terminals	Students per terminal
Alabama				
Alabama State University	4,500	99.0	84	53.6
University of Alabama	15,943	10.0	1,000	15.9
Louisiana				
Southern University	8,000	96.0	100	80.0
Louisiana State University	21,243	8.0	1,694	12.5
Mississippi				
Alcorn State University	3,041	96.0	50	60.8
University of Mississippi	8,791	8.0	950	9.3

Source: "Why Unequal Treatment Cries Out for Action," *The Journal of Blacks in Higher Education*, Vol. 1, No. 1, Autumn 1993, p. 93. Primary source: Survey by *The Journal of Blacks in Higher Education*. Published by permission.

Doctorates

★ 231 ★

Baccalaureate-Origin Institutions of Black Science and Engineering Doctorate Recipients, 1985-90

Baccalaureate origin institution	Field of doctoral study								
	S&E	Phys Sci	Math	Comp	Agri Sci	Bio Sci	Psych	Soc Sci	Eng
Howard University	53	4	0	0	1	6	20	18	4
Spelman Coll	25	2	2	0	0	4	10	7	0
CUNY-City Coll	22	3	0	1	0	1	8	7	2
Tuskegee Univ	22	0	0	1	3	11	2	2	3
Hampton Univ	21	0	2	0	0	3	10	5	1
Fisk Univ	20	3	0	0	1	3	10	4	0
Harvard Univ	20	1	0	0	0	1	10	5	3
North Carolina A&T Univ	18	1	0	0	2	3	6	6	2
Jackson State Univ	17	3	0	0	2	8	2	3	0
Southern Univ	16	2	1	0	0	4	3	4	3
Univ of Maryland	16	0	2	0	0	0	6	7	1
Morehouse Coll	14	3	2	0	0	2	4	4	1
Massachusetts Inst. of Technology	14	5	0	0	0	1	1	0	7
Princeton Univ	14	2	0	0	0	2	9	0	1
Tennessee State Univ	14	1	0	0	1	4	4	1	3

[Continued]

★ 231 ★

Baccalaureate-Origin Institutions of Black Science and Engineering Doctorate Recipients, 1985-90

[Continued]

Baccalaureate origin institution	Field of doctoral study								
	S&E	Phys Sci	Math	Comp	Agri Sci	Bio Sci	Psych	Soc Sci	Eng
Univ of California-Berkeley	14	1	1	0	0	1	5	6	0
Virginia State Univ	13	1	0	0	0	1	4	6	1
North Carolina Central Univ	13	1	0	0	0	2	2	7	1
Brown Univ	12	1	0	0	0	3	6	0	2
Yale Univ	12	1	1	0	0	0	8	1	1
Cornell Univ	12	1	0	0	1	3	5	2	0
Lincoln Univ	12	1	0	0	0	1	7	2	1
Univ of Michigan	12	1	0	1	0	0	5	4	1
Total (Top 23)	413	38	11	3	11	64	147	101	38
Total (All institutions)	1534	136	29	7	38	232	586	392	114
Top 23 as a percent of all institutions	27%	28%	38%	43%	29%	28%	25%	26%	33%

Source: "Top 23 Baccalaureate-Origin Institutions of 1985-90 African American Science and Engineering Doctorate Recipients," *Black Issues in Higher Education*, Vol. 10, No. 10, 15 July 1993, p. 50. Primary source: National Science Foundation, Undergraduate Origins of Recent Science and Engineering Doctorate Recipients, 1992.

★ 232 ★

Doctorates

Doctorates Conferred by Field, 1990-91

Doctor's degrees conferred by institutions of higher education, by racial/ethnic group, major field of study, and sex of student: 1990-91[1].

Major field of study	Total						Men Black, non-Hispanic	Women Black, non-Hispanic
	White, non-Hispanic	Black, non-Hispanic	Hispanic	Asian/Pacific Islander	American Indian/ Alaskan Native	Nonresident alien		
All fields, total[2]	25,328	1,212	732	1,458	102	9,715	582	630
Agriculture and natural resources	632	17	5	33	1	497	14	3
Architecture and environmental design	65	2	2	5	0	61	1	1
Area and ethnic studies	112	13	6	12	1	23	8	5
Business and management	749	25	6	56	2	405	18	7
Communications	186	24	1	8	0	55	11	13
Computer and information sciences	332	4	6	39	1	294	0	4
Education	5,313	479	160	131	37	577	174	305
Engineering	2,053	47	53	372	7	2,730	39	8
Engineering technologies	5	0	0	0	0	5	0	0
Foreign languages	319	5	57	14	0	131	1	4
Health sciences	1,159	59	29	63	3	301	26	33
Home economics	187	11	4	8	1	44	3	8
Law	24	2	2	10	0	52	2	0
Letters	1,104	39	22	34	3	214	14	25
Liberal/general studies	27	2	0	0	0	7	2	0
Library and archival sciences	38	5	1	1	0	11	1	4
Life sciences	2,764	46	66	206	5	1,006	23	23
Mathematics	401	10	13	39	1	514	6	4
Military sciences	0	0	0	0	0	0	0	0
Multi/interdisciplinary studies	177	10	4	7	0	60	5	5

[Continued]

★ 232 ★

Doctorates Conferred by Field, 1990-91
[Continued]

Major field of study	Total						Men Black, non-Hispanic	Women Black, non-Hispanic
	White, non-Hispanic	Black, non-Hispanic	Hispanic	Asian/Pacific Islander	American Indian/ Alaskan Native	Nonresident alien		
Parks and recreation	15	2	0	0	0	10	2	0
Philosophy and religion	373	12	5	13	0	53	11	1
Physical sciences	2,566	38	67	177	9	1,433	28	10
Protective services	20	5	1	0	0	2	3	2
Psychology	2,973	132	114	62	16	125	50	82
Public affairs	307	36	13	10	2	62	14	22
Social sciences	1,919	106	66	86	10	825	62	44
Theology	824	67	21	46	1	116	59	8
Visual and performing arts	684	14	8	26	2	102	5	9

Source: "Doctor's Degrees Conferred by Institutions of Higher Education, by Racial/Ethnic Group, Major Field of Study, and Sex of Student: 1990-91." *Digest of Educational Statistics*, 1993, p. 282. Primary source: U.S. Department of Education, National Center for Education Statistics, Integrated Postsecondary Education Data System (IPEDS), "Completions" survey. (This table was prepared April 1993.) *Notes:* To facilitate trend comparisons, certain aggregations have been made of the degree fields as reported in the IPEDS "Completions" survey: "Agriculture and natural resources" includes Agribusiness and agriculture production, Agricultural sciences, and Renewable natural resources; "Business and management" includes Business and management, Business and office, Marketing and distribution, and Consumer and personal services; "Engineering and related technologies" includes Engineering and related technologies, Mechanics and repairers, and Construction trades; "Physical sciences" includes Physical sciences and Science technologies; "Public affairs" includes Public affairs and Transportation and material moving; and "Visual and performing arts" includes Visual and performing arts and Precision production. 1. Preliminary data. 2. Reported racial/ethnic distributions of students by level of degree, field of degree, and sex were used to estimate race/ethnicity for students whose race/ethnicity was not reported. Excludes 423 men and 324 women whose racial/ethnic group and field of study were not available.

★ 233 ★

Doctorates

Doctorates Conferred by Racial/Ethnic Group Numbers: 1976-91

Doctor's degrees[1] conferred by institutions of higher education, by racial/ethnic group and sex of student: 1976-77 to 1990-91.

Year and sex of student	Total	White, non-Hispanic	Black, non-Hispanic	Hispanic	Asian or Pacific Islander	American Indian/ Alaskan Native	Nonresident alien
1976-77							
Total[2]	33,126	26,851	1,253	522	658	95	3,747
Men	25,036	20,032	766	383	540	67	3,248
Women	8,090	6,819	487	139	118	28	499
1978-79							
Total[3]	32,675	26,138	1,268	439	811	104	3,915
Men	23,488	18,433	734	294	646	69	3,312
Women	9,187	7,705	534	145	165	35	639
1980-81							
Total[4]	32,839	25,908	1,265	456	877	130	4,203
Men	22,595	17,310	694	277	655	95	3,564
Women	10,244	8,598	571	179	222	35	639
1984-85							
Total[5]	32,307	23,934	1,154	677	1,106	119	5,317
Men	21,296	15,017	561	431	802	64	4,421
Women	11,011	8,917	593	246	304	55	896

[Continued]

★ 233 ★

Doctorates Conferred by Racial/Ethnic Group Numbers: 1976-91

[Continued]

Year and sex of student	Total	White, non-Hispanic	Black, non-Hispanic	Hispanic	Asian or Pacific Islander	American Indian/ Alaskan Native	Nonresident alien
1986-87							
Total[6]	34,033	24,435	1,060	750	1,097	104	6,587
Men	22,059	14,813	488	439	795	58	5,466
Women	11,974	9,622	572	311	302	46	1,121
1988-89							
Total[7]	35,659	24,882	1,065	628	1,324	85	7,675
Men	22,597	14,540	490	350	946	50	6,221
Women	13,062	10,342	575	278	378	35	1,454
1989-90							
Total[8]	38,113	25,880	1,152	788	1,235	99	8,959
Men	24,248	15,104	533	423	871	49	7,268
Women	13,865	10,776	619	365	364	50	1,691
1990-91							
Total[9]	38,547	25,328	1,212	732	1,458	102	9,715
Men	24,333	14,564	582	387	987	58	7,755
Women	14,214	10,764	630	345	471	44	1,960

Source: "Doctor's Degrees Conferred by Institutions of Higher Education, by Racial/Ethnic Group and Sex of Student: 1976-77 to 1990-91." *Digest of Educational Statistics*, 1993, p. 281. Primary source: U.S. Department of Education, National Center for Education Statistics, "Degrees and Other formal Awards Conferred" survey, and Integrated Postsecondary Education Data System (IPEDS), "Completions" survey. (This table was prepared April 1993.) *Notes:* 1. Includes Ph.D., Ed.D, and comparable degrees at the doctoral level. Excludes first-professional degrees. 2. Excludes 106 men whose racial/ ethnic group was not available. 3. Excludes 53 men and 2 women whose racial/ethnic group was not available. 4. Excludes 116 men and 3 women whose racial/ethnic group was not available. 5. Excludes 404 men and 232 women whose racial/ ethnic group was not available. 6. Reported racial/ethnic distributions of students by level of degree, field of degree, and sex were used to estimate race/ethnicity for students whose race/ethnicity was not reported. Excludes 40 men and 47 women whose racial/ethnic group and field of study were not available. 7. Reported racial/ethnic distributions of students by level of degree, field of degree and sex were used to estimate race/ethnicity for students whose race/ethnicity was not reported. Excludes 51 men and 10 women whose racial/ethnic group and field of study were not available. 8. Reported racial/ethnic distributions of students by level of degree, field of degree and sex were used to estimate race/ethnicity for students whose race/ethnicity was not reported. Excludes 153 men and 105 women whose racial/ethnic group and field of study were not available. Revised from previously published data. 9. Reported racial/ethnic distributions of students by level of degree, field of degree and sex were used to estimate race/ethnicity for students whose race/ethnicity was not reported. Excludes 423 men and 324 women whose racial/ethnic group and field of study were not available. Preliminary data.

★ 234 ★

Doctorates

Doctorates Conferred by Racial/Ethnic Group Percentages: 1976-91

Doctor's degrees[1] conferred by institutions of higher education, by racial/ethnic group and sex of student: 1976-77 to 1990-91.

Year and sex of student	Total	White, non-Hispanic	Black, non-Hispanic	Hispanic	Asian or Pacific Islander	American Indian/ Alaskan Native	Nonresident alien
1976-77							
Total[2]	100.0	81.1	3.8	1.6	2.0	0.3	11.3
Men	100.0	80.0	3.1	1.5	2.2	0.3	13.0
Women	100.0	84.3	6.0	1.7	1.5	0.3	6.2
1978-79							
Total[3]	100.0	80.0	3.9	1.3	2.5	0.3	12.0
Men	100.0	78.5	3.1	1.3	2.8	0.3	14.1
Women	100.0	83.9	5.8	1.6	1.8	0.4	6.6
1980-81							
Total[4]	100.0	78.9	3.9	1.4	2.7	0.4	12.8
Men	100.0	76.6	3.1	1.2	2.9	0.4	15.8
Women	100.0	83.9	5.6	1.7	2.2	0.3	6.2
1984-85							
Total[5]	100.0	74.1	3.6	2.1	3.4	0.4	16.5
Men	100.0	70.5	2.6	2.0	3.8	0.3	20.8
Women	100.0	81.0	5.4	2.2	2.8	0.5	8.1
1986-87							
Total[6]	100.0	71.8	3.1	2.2	3.2	0.3	19.4
Men	100.0	67.2	2.2	2.0	3.6	0.3	24.8
Women	100.0	80.4	4.8	2.6	2.5	0.4	9.4
1988-89							
Total[7]	100.0	69.8	3.0	1.8	3.7	0.2	21.5
Men	100.0	64.3	2.2	1.5	4.2	0.2	27.5
Women	100.0	79.2	4.4	2.1	2.9	0.3	11.1
1989-90							
Total[8]	100.0	67.9	3.0	2.1	3.2	0.3	23.5
Men	100.0	62.3	2.2	1.7	3.6	0.2	30.0
Women	100.0	77.7	4.5	2.6	2.6	0.4	12.2
1990-91							
Total[9]	100.0	65.7	3.1	1.9	3.8	0.3	25.2

[Continued]

★ 234 ★

Doctorates Conferred by Racial/Ethnic Group Percentages: 1976-91

[Continued]

Year and sex of student	Total	White, non-Hispanic	Black, non-Hispanic	Hispanic	Asian or Pacific Islander	American Indian/ Alaskan Native	Nonresident alien
Men	100.0	59.9	2.4	1.6	4.1	0.2	31.9
Women	100.0	75.7	4.4	2.4	3.3	0.3	13.8

Source: "Doctor's Degrees Conferred by Institutions of Higher Education, by Racial/Ethnic Group and Sex of Student: 1976-77 to 1990-91." *Digest of Educational Statistics*, 1993, p. 281. Primary source: U.S. Department of Education, National Center for Education Statistics, "Degrees and Other formal Awards Conferred" survey, and Integrated Postsecondary Education Data System (IPEDS), "Completions" survey. (This table was prepared April 1993.) *Notes:* 1. Includes Ph.D., Ed.D, and comparable degrees at the doctoral level. Excludes first-professional degrees. 2. Excludes 106 men whose racial/ethnic group was not available. 3. Excludes 53 men and 2 women whose racial/ethnic group was not available. 4. Excludes 116 men and 3 women whose racial/ethnic group was not available. 5. Excludes 404 men and 232 women whose racial/ethnic group was not available. 6. Reported racial/ethnic distributions of students by level of degree, field of degree, and sex were used to estimate race/ethnicity for students whose race/ethnicity was not reported. Excludes 40 men and 47 women whose racial/ethnic group and field of study were not available. 7. Reported racial/ethnic distributions of students by level of degree, field of degree and sex were used to estimate race/ethnicity for students whose race/ethnicity was not reported. Excludes 51 men and 10 women whose racial/ethnic group and field of study were not available. 8. Reported racial/ethnic distributions of students by level of degree, field of degree and sex were used to estimate race/ethnicity for students whose race/ethnicity was not reported. Excludes 153 men and 105 women whose racial/ethnic group and field of study were not available. Revised from previously published data. 9. Reported racial/ethnic distributions of students by level of degree, field of degree and sex were used to estimate race/ethnicity for students whose race/ethnicity was not reported. Excludes 423 men and 324 women whose racial/ethnic group and field of study were not available. Preliminary data.

Enrollment

★ 235 ★

Enrollment Status by Race, Hispanic Origin, and Sex, 1975 and 1991

As of October. For persons 18 to 21 years old. For the civilian noninstitutional population. Based on the Current Population Survey.

Characteristic	Total persons 18 to 21 years old (1,000)		Percent distribution									
			Enrolled in high school		High school graduates						Not high school graduate	
					Total		In college		Not in college			
	1975	1991	1975	1991	1975	1991	1975	1991	1975	1991	1975	1991
Total[1]	15,693	13,906	5.7	8.2	78.0	77.7	33.5	42.2	44.5	35.5	16.3	14.1
White	13,448	11,187	4.7	6.9	80.6	79.1	34.6	44.1	46.0	35.0	14.7	13.9
Black	1,997	2,138	12.5	14.2	60.4	69.3	24.9	28.3	35.6	41.0	27.0	16.6
Hispanic[2]	899	1,637	12.0	13.4	57.2	51.6	24.4	23.9	32.8	27.7	30.8	35.1
Male[1]	7,584	6,772	7.4	10.5	76.6	75.0	35.4	40.1	41.3	34.9	15.9	14.6
White	6,545	5,459	6.2	8.9	79.7	76.1	36.9	41.4	42.8	34.7	14.1	15.0
Black	911	1,026	15.9	18.9	55.0	66.3	23.9	26.8	31.1	39.5	29.0	14.7
Hispanic[2]	416	826	17.3	13.9	54.6	45.3	25.2	17.8	29.3	27.5	27.9	41.0
Female[1]	8,109	7,134	4.2	6.0	79.2	80.3	31.8	44.2	47.4	36.1	16.6	13.6
White	6,903	5,728	3.2	5.0	81.4	82.1	32.4	46.8	49.0	35.3	15.3	12.8
Black	1,085	1,112	9.7	9.6	65.0	71.9	25.8	29.7	39.2	42.3	25.4	18.4
Hispanic[2]	484	812	7.6	12.8	59.3	58.1	23.6	30.0	35.7	28.1	33.1	29.1

Source: "Enrollment Status by Race, Hispanic Origin, and Sex: 1970 and 1991," U.S. Department of Commerce, *Statistical Abstract of the United States, 1993*, p. 169. Primary source: U.S. Bureau of the Census, *Current Population Reports*, series P20- 469. *Notes:* 1. Includes other races not shown separately. 2. Persons of Hispanic origin may be of any race.

Enrollment in Public Elementary and Secondary Schools, Fall 1991

State	Total	White[1]	Black[1]	Hispanic	Asian or Pacific Islander	American Indian/ Alaskan Native
United States	100.0	67.4[2]	16.4[2]	11.2[2]	3.4[2]	1.0[2]
Alabama	100.0	62.8	35.5	0.3	0.5	0.9
Alaska	100.0	66.9	4.4	2.2	3.9	22.6
Arizona	100.0	62.4	4.2	25.0	1.5	6.9
Arkansas	100.0	74.5	24.0	0.6	0.6	0.3
California	100.0	44.5	8.6	35.3	10.8	0.8
Colorado	100.0	74.9	5.2	16.6	2.3	1.0
Connecticut	100.0	74.3	12.8	10.4	2.2	0.2
Delaware	100.0	67.3	27.8	3.1	1.6	0.2
District of Columbia	100.0	4.0	89.5	5.3	1.1	[3]
Florida	100.0	61.2	24.2	12.9	1.6	0.2
Georgia	-	-	-	-	-	-
Hawaii	100.0	23.9	2.6	5.2	67.9	0.3
Idaho	-	-	-	-	-	-
Illinois	100.0	65.4	21.4	10.3	2.8	0.1
Indiana	100.0	86.4	10.9	1.9	0.7	0.1
Iowa	100.0	94.0	2.9	1.4	1.4	0.4
Kansas	100.0	84.6	8.1	4.7	1.7	0.9
Kentucky	100.0	89.8	9.4	0.2	0.5	[3]
Louisiana	100.0	52.7	44.7	1.0	1.2	0.4
Maine	-	-	-	-	-	-
Maryland	100.0	60.4	33.2	2.5	3.6	0.3
Massachusetts	100.0	80.5	7.8	8.1	3.5	0.2
Michigan	100.0	78.2	17.2	2.4	1.3	1.0
Minnesota	100.0	89.9	3.6	1.4	3.2	1.8
Mississippi	100.0	48.3	50.7	0.1	0.5	0.4
Missouri	100.0	82.5	15.7	0.8	0.9	0.2
Montana	100.0	88.4	0.4	1.3	0.7	9.2
Nebraska	100.0	89.4	5.5	2.9	1.1	1.1
Nevada	100.0	73.2	9.0	12.1	3.7	2.0
New Hampshire	100.0	97.0	0.8	1.0	1.0	0.2
New Jersey	100.0	64.4	18.6	12.2	4.7	0.1
New Mexico	100.0	41.2	2.3	45.3	0.9	10.4
New York	100.0	59.4	20.1	15.8	4.4	0.3
North Carolina	100.0	66.4	30.2	0.9	1.0	1.6
North Dakota	100.0	91.2	0.7	0.6	0.7	6.8
Ohio	100.0	83.6	14.1	1.3	0.9	0.1
Oklahoma	100.0	73.5	10.0	3.0	1.1	12.4
Oregon	100.0	88.1	2.4	4.9	2.9	1.8
Pennsylvania	100.0	82.2	13.2	2.9	1.7	0.1
Rhode Island	100.0	82.7	6.5	7.2	3.1	0.4
South Carolina	100.0	57.7	41.1	0.5	0.6	0.1
South Dakota	-	-	-	-	-	-
Tennessee	100.0	76.6	22.2	0.3	0.7	0.1
Texas	100.0	49.0	14.3	34.4	2.1	0.2
Utah	100.0	91.9	0.7	4.0	1.9	1.4

[Continued]

★ 236 ★

Enrollment in Public Elementary and Secondary Schools, Fall 1991

[Continued]

State	Total	White[1]	Black[1]	Hispanic	Asian or Pacific Islander	American Indian/ Alaskan Native
Vermont	100.0	97.9	0.6	0.3	0.7	0.6
Virginia	-	-	-	-	-	-
Washington	100.0	81.4	4.2	6.1	5.8	2.5
West Virginia	100.0	95.5	3.9	0.2	0.4	0.1
Wisconsin	100.0	85.2	8.8	2.7	2.1	1.3
Wyoming	100.0	89.6	0.9	6.0	0.7	2.8
Other areas						
American Samoa	100.0	[3]	[3]	[3]	100.0	[3]
Guam	100.0	10.3	1.6	0.3	87.8	[3]
Northern Marianas	100.0	0.1	[3]	[3]	99.9	[3]
Puerto Rico	-	-	-	-	-	-
Virgin Islands	100.0	0.9	86.8	11.8	0.5	[3]

Source: "Enrollment in Public Elementary and Secondary School, by Race and State: Fall 1986 and Fall 1991." *Digest of Educational Statistics,* 1993, p. 61. Primary source: U.S. Department of Education, Office for Civil Rights, *1986 State Summaries of Elementary and Secondary School Civil Rights Survey;* and National Center for Education Statistics, Common Core of Data Survey. (This table was prepared April 1993). *Notes:* State estimates may differ from other data sources because of variations in survey methodology. Because of rounding, details may not add to totals. - Data not available. 1. Excludes persons of Hispanic origin. 2. Includes estimate for nonresponding states. 3. Less than 0.05 percent.

★ 237 ★

Enrollment

School Enrollment, Age 3 to 34 Years, Percentage, 1992

Percent of the population 3 to 34 years old enrolled in school[1], by race/ethnicity, sex, and age: October 1975 to October 1992.

Year and age	Total			Male			Female		
	White, non-Hispanic	Black, non-Hispanic	Hispanic origin	White, non-Hispanic	Black, non-Hispanic	Hispanic origin	White, non-Hispanic	Black, non-Hispanic	Hispanic origin
1992									
Total, 3 to 34 years	50.9	53.2	48.9	51.3	55.1	47.7	50.5	51.5	50.2
3 and 4 years	42.3	37.8	27.9	42.7	40.0	24.4	41.9	35.3	31.2
5 and 6 years	95.4	95.3	96.0	94.9	97.3	96.6	95.9	93.4	95.3
7 to 9 years	99.3	99.3	99.6	99.4	99.7	100.0	99.2	98.9	99.2
10 to 13 years	99.4	99.7	99.1	99.5	99.7	99.2	99.2	99.6	99.1
14 and 15 years	99.2	99.4	98.8	99.1	99.9	98.1	99.3	98.8	99.6
16 and 17 years	95.3	93.0	87.2	96.5	94.7	89.2	94.1	91.4	85.0
18 and 19 years	63.1	56.3	53.7	62.2	60.9	52.6	63.9	51.8	54.9
20 and 21 years	47.6	33.2	30.1	46.8	27.2	24.3	48.4	38.4	35.6
22 to 24 years	24.6	20.3	14.5	25.2	18.5	13.8	24.0	21.9	15.4
25 to 29 years	10.0	8.0	6.7	9.1	7.7	5.3	10.9	8.2	8.2
30 to 34 years	6.1	5.3	4.7	5.4	3.3	3.5	6.8	7.0	6.0

Source: "Percent of the Population 3 to 34 Years Old enrolled in School, by Race/Ethnicity, Sex, and Age: 1992." *Digest of Educational Statistics,* 1993, p. 16. Primary source: U.S. Department of Commerce, Bureau of the Census, Current Population Survey, and unpublished data. (This table was prepared June 1993). *Notes:* Data are based upon sample surveys of the civilian noninstitutional population. 1. Includes enrollment in any type of graded public, parochial, or other private schools. Includes nursery schools, kindergartens, elementary schools, high schools, colleges, universities, and professional schools. Attendance may be on either a full-time or part-time basis and during the day or night. Enrollments in "special" schools, such as trade schools, business colleges, or correspondence schools, are not included.

First-professional degrees

★ 238 ★

First-professional Degrees Conferred, Numbers: 1976-91

Year and sex of student	Total	White, non-Hispanic	Black, non-Hispanic	Hispanic	Asian or Pacific Islander	American Indian/ Alaskan Native	Nonresident alien
1976-77, total[1]	63,953	58,422	2,537	1,076	1,021	196	701
Men	51,980	47,777	1,761	893	776	159	614
Women	11,973	10,645	776	183	245	37	87
1978-79, total[2]	68,611	62,430	2,836	1,283	1,205	216	641
Men	52,425	48,123	1,783	989	860	150	520
Women	16,186	14,307	1,053	294	345	66	121
1980-81, total[3]	71,340	64,551	2,931	1,541	1,456	192	669
Men	52,194	47,629	1,772	1,131	991	134	537
Women	19,146	16,922	1,159	410	465	58	132
1984-85, total[4]	71,057	63,219	3,029	1,884	1,816	248	861
Men	47,501	42,630	1,623	1,239	1,152	176	681
Women	23,556	20,589	1,406	645	664	72	180
1986-87, total[5]	71,617	62,688	3,420	2,051	2,270	304	884
Men	46,522	41,149	1,835	1,303	1,420	183	632
Women	25,095	21,539	1,585	748	850	121	252
1988-89, total	70,856	61,214	3,148	2,269	2,976	264	985
Men	45,046	39,399	1,618	1,374	1,819	148	688
Women	25,810	21,815	1,530	895	1,157	116	297
1989-90, total[6]	70,744	60,240	3,410	2,427	3,362	257	1,048
Men	43,778	37,850	1,672	1,450	1,963	135	708
Women	26,966	22,390	1,738	977	1,399	122	340
1990-91, total[7]	71,515	60,327	3,575	2,527	3,755	261	1,070
Men	43,601	37,348	1,672	1,506	2,178	144	753
Women	27,914	22,979	1,903	1,021	1,577	117	317

Source: "First-professional Degrees Conferred by Institutions of Higher Education by Racial/Ethnic Group and Sex of Student: 1976-77 to 190-91." *Digest of Education Statistics*, 1993, p. 284. Primary source: U.S. Department of Education, National Center for Education Statistics, "Degrees and Other Formal Awards Conferred" survey, and Integrated Postsecondary Education Data System (IPEDS). "Completions" survey. (This table was prepared April 1993.) *Notes:* For years 1984-85 to 1990-91, reported racial/ethnic distributions of students by level of degree, field of degree, and sex were used to estimate race/ethnicity for students whose race/ethnicity was not reported. 1. Excludes 394 men and 12 women whose racial/ethnic group was not available. 2. Excludes 227 men and 10 women whose racial/ethnic group was not available. 3. Excludes 598 men and 18 women whose racial/ethnic group was not available. 4. Excludes 2,954 men and 1,052 women whose racial/ethnic group was not available. 5. Excludes 938 men and 195 women whose racial/ethnic group was not available. 6. Excludes 183 men and 61 women whose racial/ethnic group was not available. Revised from previously published data. 7. Excludes 245 men and 188 women whose racial/ethnic group was not available. Preliminary data.

★ 239 ★

First-professional degrees

First-professional Degrees Conferred, Percentages: 1976-91

Year and sex of student	Total	White, non-Hispanic	Black, non-Hispanic	Hispanic	Asian or Pacific Islander	American Indian/ Alaskan Native	Nonresident alien
1976-77, total[1]	100.0	91.4	4.0	1.7	1.6	0.3	1.1
Men	100.0	91.9	3.4	1.7	1.5	0.3	1.2
Women	100.0	88.9	6.5	1.5	2.0	0.3	0.7
1978-79, total[2]	100.0	91.0	4.1	1.9	1.8	0.3	0.9
Men	100.0	91.8	3.4	1.9	1.6	0.3	1.0
Women	100.0	88.4	6.5	1.8	2.1	0.4	0.7
1980-81, total[3]	100.0	90.5	4.1	2.2	2.0	0.3	0.9
Men	100.0	91.3	3.4	2.2	1.9	0.3	1.0
Women	100.0	88.4	6.1	2.1	2.4	0.3	0.7
1984-85, total[4]	100.0	89.0	4.3	2.7	2.6	0.3	1.2
Men	100.0	89.7	3.4	2.6	2.4	0.4	1.4
Women	100.0	87.4	6.0	2.7	2.8	0.3	0.8
1986-87, total[5]	100.0	87.5	4.8	2.9	3.2	0.4	1.2
Men	100.0	88.5	3.9	2.8	3.1	0.4	1.4
Women	100.0	85.8	6.3	3.0	3.4	0.5	1.0
1988-89, total	100.0	86.4	4.4	3.2	4.2	0.4	1.4
Men	100.0	87.5	3.6	3.1	4.0	0.3	1.5
Women	100.0	84.5	5.9	3.5	4.5	0.4	1.2
1989-90, total[6]	100.0	85.2	4.8	3.4	4.8	0.4	1.5
Men	100.0	86.5	3.8	3.3	4.5	0.3	1.6
Women	100.0	83.0	6.4	3.6	5.2	0.5	1.3
1990-91, total[7]	100.0	84.4	5.0	3.5	5.3	0.4	1.5
Men	100.0	85.7	3.8	3.5	5.0	0.3	1.7
Women	100.0	82.3	6.8	3.7	5.6	0.4	1.1

Source: "First-professional Degrees Conferred by Institutions of Higher Education by Racial/Ethnic Group and Sex of Student: 1976-77 to 190-91." *Digest of Education Statistics,* 1993, p. 284. Primary source: U.S. Department of Education, National Center for Education Statistics, "Degrees and Other Formal Awards Conferred" survey, and Integrated Postsecondary Education Data System (IPEDS). "Completions" survey. (This table was prepared April 1993.) *Notes:* For years 1984-85 to 1990-91, reported racial/ethnic distributions of students by level of degree, field of degree, and sex were used to estimate race/ ethnicity for students whose race/ethnicity was not reported. 1. Excludes 394 men and 12 women whose racial/ethnic group was not available. 2. Excludes 227 men and 10 women whose racial/ethnic group was not available. 3. Excludes 598 men and 18 women whose racial/ethnic group was not available. 4. Excludes 2,954 men and 1,052 women whose racial/ethnic group was not available. 5. Excludes 938 men and 195 women whose racial/ethnic group was not available. 6. Excludes 183 men and 61 women whose racial/ethnic group was not available. Revised from previously published data. 7. Excludes 245 men and 188 women whose racial/ethnic group was not available. Preliminary data.

High School Dropouts

★ 240 ★

High School Dropout, Completion, and Enrollment Rates for 19- to 20-year-olds, 1972 to 1991

Year	Status dropout rate			High school completion rate			High school enrollment rate		
	Total[1]	White	Black	Total[1]	White	Black	Total[1]	White	Black
1972	16.1	13.1	26.8	80.7	84.7	66.3	3.2	2.2	6.9
1973	15.3	12.2	25.8	82.2	85.9	68.2	2.5	1.9	5.9
1974	16.4	13.8	24.8	80.6	84.6	65.6	2.9	1.5	9.5
1975	16.2	13.5	26.4	81.0	84.7	66.0	2.8	1.8	7.5
1976	15.9	13.2	24.1	81.1	85.2	67.6	3.0	1.7	8.3
1977	15.7	13.3	22.0	81.4	84.9	69.1	2.9	1.7	8.8
1978	16.0	12.8	24.9	80.9	85.2	67.1	3.1	1.9	8.1
1979	16.7	13.8	26.7	80.4	83.8	68.5	2.9	2.3	4.8
1980	16.4	12.7	23.5	81.1	85.6	71.0	2.5	1.8	5.4
1981	15.8	12.9	21.1	80.8	84.8	71.8	3.4	2.3	7.1
1982	16.3	13.4	23.0	80.6	84.7	69.4	3.1	1.9	7.6
1983	15.2	12.2	21.3	81.2	85.2	73.2	3.6	2.6	5.5
1984	15.0	12.8	18.1	82.0	85.4	75.3	3.1	1.9	6.5
1985	13.6	11.1	18.7	83.1	87.0	73.8	3.3	2.0	7.5
1986	12.9	10.2	17.6	83.8	87.8	75.0	3.3	2.0	7.5
1987	13.9	11.4	15.5	82.9	86.4	79.3	3.2	2.2	5.1
1988	14.9	10.8	20.2	82.1	87.1	73.5	3.0	2.1	6.2
1989	15.1	11.6	18.6	81.8	86.8	74.8	3.2	1.6	6.6
1990	13.6	10.4	15.6	82.8	87.3	77.6	3.5	2.3	6.8
1991	14.3	10.7	16.9	81.4	87.0	72.5	4.3	2.4	10.5

Source: "High School Dropout, Completion, and Enrollment Rates for 19- to 20- year-olds, 1972-1991," U.S. Department of Education, National Center for Education Statistics, *The Condition of Education, 1993*, p. 58. Primary source: U.S. Department of Commerce, Bureau of the Census, October Current Population Surveys. *Notes:* The status dropout rate is the percentage of 19- to 20-year-olds who had not completed high school and were not currently enrolled in school. The high school completion rates is the percentage of individuals 19 to 20 years old who had completed 12 or more years of school. The high school enrollment rate is the percentage of 19- to 20-year-olds who were enrolled in school below the college level. The 3 rates sum to 100 percent. Data for 1987 through 1991 reflect new editing procedures instituted by the Bureau of the Census in 1986 for cases with missing data on school enrollment items. 1. Included in the total are individuals who are not Hispanic, white, or black; most of these individuals are Asian and some are American Indian.

★ 241 ★

High School Dropouts

Reasons for Dropping Out, Grades 10 to 12

	Male	Female	Hispanic	Black	White
School-related					
Dislike school	43.6	42.2	48.0	28.8	5.5
Not getting along with teachers	24.6	21.1	24.6	27.8	21.5
Not getting along with students	17.7	11.6	15.6	18.4	13.6
Did not feel safe at school	7.0	5.1	8.3	8.5	4.8
Felt didn't belong	25.8	22.7	16.0	25.9	26.6
Could not keep up with schoolwork	32.7	29.9	35.0	25.6	30.3
Was failing school	43.4	34.5	40.6	39.5	36.6
Did not like new school	10.5	10.7	12.3	9.1	10.2
Suspended/expelled	21.6	10.0	10.1	24.4	15.4
Job-related					
Not able to work and go to school	26.9	19.1	20.4	15.4	24.6
Found a job	35.9	21.8	34.1	19.1	27.5
Family-related					
Had to support family	10.4	11.9	15.8	11.8	9.9
Wanted a family	6.4	8.4	9.1	4.6	8.2
Was pregnant	-	26.8	30.6	34.5	25.6
Became a parent	7.7	21.0	19.6	21.0	12.4
Got married	3.7	19.7	13.4	2.0	15.1
Had to care for family member	9.5	14.0	8.5	14.7	10.7
Other					
Wanted to travel	8.2	8.0	6.6	7.3	7.1
Friends dropped out	8.5	7.5	7.6	6.7	8.6
Had a drug/alcohol problem	6.1	2.8	1.8	2.1	5.9

Source: "Reasons for Dropping Out, Grades 10 to 12," *Black Issues in Higher Education*, Vol. 10, No. 24, 27 January 1994, p. 40. Primary source: U.S. Department of Education, National Center for Education Statistics, 1992.

★ 242 ★

High School Dropouts

Workforce Participation of High School Dropouts 16 to 24 Years Old, 1980-1992

Numbers in thousands.

Year, sex, and race	Dropouts		Dropouts in civilian labor force[1]					
	Number	Percent of total	Number	Labor force participation rate	Employed		Unemployed	
					Number	Percent of dropouts	Number	Unemployment rate
White[10]								
1979-80 dropouts in October[2]	580	78.5	392	67.6	286	49.3	106	27.0
1984-85 dropouts in October[3]	458	74.8	330	72.1	214	46.7	116	35.2
1988-89 dropouts in October[6]	324	72.6	228	70.6	176	54.3	52	22.9
1989-90 dropouts in October[7]	303	74.8	211	69.8	156	51.4	56	26.3
1990-91 dropouts in October[8]	273	71.8	177	65.1	109	40.0	68	38.5
1991-92 dropouts in October[9]	319	78.6	190	59.7	128	40.3	62	32.5
Black[10]								
1979-80 dropouts in October[2]	146	19.8	73	50.0	33	22.6	40	[11]
1984-85 dropouts in October[3]	132	21.6	69	52.3	39	29.5	30	[11]
1988-89 dropouts in October[6]	112	25.1	59	52.2	31	27.7	27	[11]
1989-90 dropouts in October[7]	86	21.2	56	65.3	26	29.9	30	[11]
1990-91 dropouts in October[8]	98	25.8	54	55.0	28	28.4	26	[11]
1991-92 dropouts in October[9]	66	16.3	35	[11]	7	[11]	28	[11]
Hispanic[12]								
1979-80 dropouts in October[2]	91	12.3	60	65.9	43	47.3	17	[11]
1984-85 dropouts in October[3]	106	17.3	73	68.9	40	37.7	33	[11]
1988-89 dropouts in October[6]	65	14.6	36	[11]	26	[11]	11	[11]
1989-90 dropouts in October[7]	67	16.5	32	[11]	22	[11]	10	[11]
1990-91 dropouts in October[8]	61	16.1	48	[11]	30	[11]	18	[11]
1991-92 dropouts in October[9]	80	19.7	40	49.9	23	28.4	17	[11]

Source: Labor status of 1979-80 to 1991-92 high school dropouts, by sex and race/ethnicity: October 1980 to October 1992. *Digest of Educational Statistics*, 1993, p. 395. Primary source: U.S. Department of Labor, Bureau of Labor Statistics, Students, Graduates, and Dropouts, October 1980-82; and *Employment Status of School Age Youth, High School Graduates and Dropouts*, various years. (This table was prepared July 1993.) *Notes:* Data are based upon sample surveys of the civilian noninstitutional population. Includes dropouts from any grade, including a small number from elementary and middle schools. Percents are only shown when the base is 75,000 or greater. Even though the standard errors are large, smaller estimates are shown to permit users to combine categories in various ways. Because of rounding, details may not add to totals. 1. The labor force includes all employed persons plus those seeking employment. The labor force participation rate is the percentage of persons either employed or seeking employment. 2. Persons who dropped out of school between October 1979 and October 1980. 3. Persons who dropped out of school between October 1984 and October 1985. 4. Persons who dropped out of school between October 1985 and October 1986. 5. Persons who dropped out of school between October 1986 and October 1987. 6. Persons who dropped out of school between October 1988 and October 1989. 7. Persons who dropped out of school between October 1989 and October 1990. 8. Persons who dropped out of school between October 1990 and October 1991. 9. Persons who dropped out of school between October 1991 and October 1992. 10. Includes persons of Hispanic origin. 11. Data not shown where base is less than 75,000. 12. Persons of Hispanic origin may be of any race.

Education   Statistical Record of Black America, 3rd Edition

High School Graduates

★ 243 ★

College Enrollment and Labor Force Status of 1991 High School Graduates 18 to 24 Years Old

Numbers in thousands.

Item	Civilian noninstitutional population			Civilian labor force[1]				
	Number	Percent	Percent of high school graduates	Number	Labor force participation rate	Employed	Unemployed	
							Number	Unemployment rate
1991 high school graduates[2]								
Total	2,276	100.0	100.0	1,359	59.7	1,107	252	18.5
White[3]	1,867	82.0	82.0	1,145	61.3	973	173	15.2
Black[3]	320	14.1	14.1	172	53.7	96	75	43.9
Hispanic origin[4]	154	6.8	6.8	102	66.0	92	10	9.8
Enrolled in college, October 1991	1,420	100.0	62.4	675	47.5	596	78	11.6
Men	656	46.2	28.8	306	46.6	274	31	10.2
Women	763	53.7	33.5	369	48.3	322	47	12.8
Full-time students	1,288	90.7	56.6	556	43.2	481	75	13.5
Part-time students	132	9.3	5.8	119	90.2	116	3	2.6
White[3]	1,207	85.0	53.0	594	49.3	532	62	10.5
Black[3]	146	10.3	6.4	53	36.0	38	15	[5]
Hispanic origin[4]	88	6.2	3.9	54	61.3	50	4	[5]
Not enrolled in college, October 1991	857	100.0	37.7	685	79.9	511	173	25.3
Men	483	56.4	21.2	408	84.4	301	107	26.2
Women	374	43.6	16.4	277	74.2	210	67	24.1
White[3]	660	77.0	29.0	551	83.4	440	110	20.0
Black[3]	173	20.2	7.6	119	68.6	58	61	51.2
Hispanic origin[4]	66	7.7	2.9	48	[5]	41	6	[5]

Source: College enrollment and labor force status of 1991 and 1992 high school graduates 16 to 24 years old by sex and race/ethnicity: October 1991 and October 1992. *Digest of Educational Statistics*, 1993, p. 394. Primary source: U.S. Department of Labor, Bureau of Labor Statistics, *Employment Status of School Age Youth, High School Graduates and Dropouts*, various years. (This table was prepared July 1993.) *Notes:* Data are based upon sample surveys of the civilian noninstitutional population. Percents are only shown when the base is 75,000 or greater. Even though the standard errors are large, smaller estimates are shown to permit users to combine categories in various ways. Because of rounding, details may not add to totals. 1. The labor force includes all employed persons plus those seeking employment. The labor force participation rate is the percentage of persons either employed or seeking employment. 2. Includes persons who graduated from high school between January and October 1991. 3. Includes persons of Hispanic origin. 4. Persons of Hispanic origin may be of any race. 5. Data not shown where base is less than 75,000.

★ 244 ★

High School Graduates

College Enrollment and Labor Force Status of 1992 High School Graduates 18 to 24 Years Old

Numbers in thousands.

Item	Civilian noninstitutional population			Civilian labor force[1]				
	Number	Percent	Percent of high school graduates	Number	Labor force participation rate	Employed	Unemployed	
							Number	Unemployment rate
1992 high school graduates[2]								
Total	2,398	100.0	100.0	1,449	60.4	1,204	245	16.9
White[3]	1,900	79.2	79.2	1,193	62.8	1,039	153	12.9
Black[3]	353	14.7	14.7	172	48.6	104	68	39.6
Hispanic origin[4]	199	8.3	8.3	127	63.8	90	37	28.8
Enrolled in college, October 1992	1,479	100.0	61.7	735	49.7	628	106	14.5
Men	725	49.0	30.2	357	49.3	311	47	13.1
Women	754	51.0	31.4	377	50.0	318	60	15.8
2-year	552	37.3	23.0	381	69.1	330	51	13.3
4-year	928	62.7	38.7	353	38.1	298	54	15.9
Full-time students	1,372	92.8	57.2	645	47.0	553	92	14.3
Part-time students	108	7.3	4.5	90	83.2	76	14	15.4
White[3]	1,204	81.4	50.2	619	51.4	552	67	10.8
Black[3]	169	11.4	7.0	58	34.1	35	22	5
Hispanic origin[4]	109	7.4	4.5	62	57.2	42	21	5
Not enrolled in college, October 1992	919	100.0	38.3	714	77.8	576	139	19.4
Men	491	53.4	20.5	418	85.2	339	80	19.1
Women	428	46.6	17.8	296	69.2	237	59	19.8
White[3]	696	75.7	29.0	574	82.5	487	86	15.1
Black[3]	184	20.0	7.7	114	62.0	69	46	40.0
Hispanic origin[4]	90	9.8	3.8	64	71.8	48	16	5

Source: College enrollment and labor force status of 1991 and 1992 high school graduates 16 to 24 years old by sex and race/ethnicity: October 1991 and October 1992. *Digest of Educational Statistics*, 1993, p. 394. Primary source: U.S. Department of Labor, Bureau of Labor Statistics, *Employment Status of School Age Youth, High School Graduates and Dropouts*, various years. (This table was prepared July 1993.) *Notes:* Data are based upon sample surveys of the civilian noninstitutional population. Percents are only shown when the base is 75,000 or greater. Even though the standard errors are large, smaller estimates are shown to permit users to combine categories in various ways. Because of rounding, details may not add to totals. 1. The labor force includes all employed persons plus those seeking employment. The labor force participation rate is the percentage of persons either employed or seeking employment. 2. Includes persons who graduated from high school between January and October 1992. 3. Includes persons of Hispanic origin. 4. Persons of Hispanic origin may be of any race. 5. Data not shown where base is less than 75,000.

High School Students

★ 245 ★

Tenth Graders' Plans to Attend College, 1980 and 1990

Percent of high school sophomores who plan to go to college after graduation, by student characteristics: 1980 and 1990.

Student characteristics	Right after high school		After a year		After more than a year		No or don't know	
	1980	1990	1980	1990	1980	1990	1980	1990
Race/ethnicity								
White	48.4	60.3	15.2	17.0	20.8	9.1	15.6	13.7
Black	51.5	62.2	17.9	15.5	21.0	10.0	9.6	12.3
Hispanic	43.8	52.7	18.3	22.9	25.1	12.9	12.8	11.5
Asian	73.2	78.2	13.3	10.1	11.5	4.6	2.0	7.1
American Indian	33.0	45.4	22.5	17.5	30.1	15.3	14.5	21.7

Source: "Percent of High School Students Who Plan to Go to College After Graduation, by Student Characteristics: 1980 and 1990." *Digest of Educational Statistics*, 1993, p. 137. Primary source: U.S. Department of Education, National Center for Education Statistics, "High School and Beyond," Base Year Survey, 1980 Sophomore Cohort; and "National Education Longitudinal Study of 1988," First Follow-up Student Survey. (This table was prepared April 1993.).

High Schools

★ 246 ★

High School Dropout Ages 16-24, Percentages: 1969-1991

Year	Total[1]				Men				Women			
	All races	White[2]	Black[2]	Hispanic origin	All races	White[2]	Black[2]	Hispanic origin	All races	White[2]	Black[2]	Hispanic origin
1969	15.2	13.6	26.7	-	14.3	12.6	26.9	-	16.0	14.6	26.7	-
1972	14.6	13.7	21.5	34.3	14.1	13.1	22.3	3.6	15.1	14.2	20.8	35.0
1975	13.9	12.6	22.8	29.2	13.3	12.0	22.8	26.6	14.5	13.2	22.8	31.5
1978	14.2	13.4	20.2	33.1	14.6	13.6	22.5	33.2	13.9	13.2	18.2	33.0
1981	13.9	13.8	18.5	33.1	15.1	14.5	20.0	35.9	12.8	13.2	17.2	30.4
1984	13.1	12.7	15.6	29.8	14.0	13.5	16.7	30.6	12.3	11.8	14.5	29.1
1987	12.7	12.5	14.5	28.6	13.3	13.0	15.7	29.0	12.2	12.0	13.5	28.1
1990	12.1	12.0	13.2	32.4	12.2	12.7	11.8	34.3	11.6	11.4	14.4	30.3
1991	12.5	8.9	13.6	35.3	13.0	8.9	13.5	39.2	11.9	8.9	13.7	31.1

Source: "Percent of High School Dropouts among Persons 16 to 24 Years Old, by Sex and Race/Ethnicity: 1969 to 1991." *Digest of Educational Statistics*, 1993, p. 110. Primary source: U.S. Department of Commerce, Bureau of the Census, *Current Population Survey*, unpublished tabulations; and U.S. Department of Education, National Center for Education Statistics, *Dropout Rates in the United States*. (This table was prepared April 1992.) Notes: "Status" dropouts are persons who are not enrolled in school and who are not high school graduates. People who have received GED credentials are counted as graduates. Data are based upon sample surveys of the civilian noninstitutional population. - Data not available. 1. "Status" dropouts. 2. White and black includes persons of Hispanic origin.

★ 247 ★

High Schools

High School Graduates' Credits in Subject Fields: 1982, 1987, and 1990

Characteristic	Total	English	History/ social studies	Math	Computer science	Science	Foreign language	Vocational education[1]	Arts	Physical education	Other[2]
1982 graduates	21.2	3.80	3.10	2.54	0.11	2.19	1.05	3.98	1.39	1.93	1.14
Race/ethnicity											
White	21.4	3.78	3.15	2.59	0.12	2.27	1.13	3.89	1.45	1.89	1.12
Black	20.5	3.90	2.97	2.44	0.10	1.99	0.73	4.15	1.18	1.98	1.07
Hispanic	20.8	3.79	2.94	2.22	0.07	1.79	0.78	4.55	1.27	2.13	1.25
Asian	22.0	3.94	3.04	3.11	0.19	2.56	1.81	2.56	1.22	2.21	1.34
1987 graduates	23.0	4.03	3.33	2.97	0.43	2.59	1.46	3.65	1.43	1.97	1.14
Race/ethnicity											
White	23.1	3.99	3.30	2.98	0.45	2.64	1.50	3.69	1.48	1.94	1.11
Black	22.5	4.14	3.31	2.90	0.35	2.39	1.12	4.01	1.20	2.01	1.11
Hispanic	22.9	4.23	3.23	2.77	0.36	2.33	1.27	3.57	1.35	2.40	1.37
Asian	24.5	4.31	3.64	3.72	0.57	3.17	2.17	2.08	1.12	2.57	1.14
1990 graduates	23.58	4.09	3.47	3.11	0.46	2.82	1.62	3.23	1.53	2.01	1.24
Race/ethnicity											
White	23.61	4.03	3.45	3.10	0.45	2.87	1.68	3.27	1.59	1.97	1.20
Black	23.39	4.17	3.43	3.09	0.52	2.73	1.26	3.49	1.34	2.06	1.30
Hispanic	23.91	4.43	3.45	3.07	0.48	2.55	1.55	3.19	1.48	2.27	1.44
Asian	24.19	4.42	3.70	3.50	0.49	3.04	2.15	2.10	1.31	2.15	1.33
Other	23.36	4.25	3.40	3.03	0.58	2.72	1.45	3.27	1.01	2.04	1.61

Source: "Average Number of Carnegie Units Earned by High School Graduates in Various Subject Fields, by Student Characteristics: 1982, 1987, and 1990." *Digest of Educational Statistics*, 1993, p. 132. Primary source: U.S. Department of Education, National Center for Education Statistics, "1990 High School Transcript Study." (This table was prepared April 1993.) *Notes:* The Carnegie unit is a standard of measurement that represents one credit for the completion of a 1-year course. 1. Includes non-occupational vocational education, vocational general introduction, agriculture, business, marketing, health, occupational home economics, trade and industry, and technical courses. 2. Includes personal and social courses, religion and theology, and courses not included in the other subject field.

★ 248 ★

High Schools

High School Graduates' Credits in Vocational Education Courses: 1982, 1987, and 1990

Student characteristic	Total	Non-occupa-tional vocational education	Vocational general introduction	Agriculture	Business	Marketing	Health	Occupational home economics	Trade and industry	Technical
1982 graduates	3.98	1.84	0.37	0.17	0.78	0.08	0.04	0.09	0.60	0.01
Race/ethnicity										
White	3.89	1.78	0.36	0.18	0.80	0.08	0.03	0.09	0.55	0.02
Black	4.15	1.96	0.41	0.06	0.74	0.10	0.10	0.10	0.67	0.01

[Continued]

★ 248 ★

High School Graduates' Credits in Vocational Education Courses: 1982, 1987, and 1990

[Continued]

Student characteristic	Total	Non-occupa-tional vocational education	Vocational general introduction	Agriculture	Business	Marketing	Health	Occupational home economics	Trade and industry	Technical
Hispanic	4.55	2.17	0.43	0.18	0.73	0.07	0.05	0.10	0.81	0.01
Asian	2.56	1.37	0.18	0.05	0.45	0.03	0.03	0.03	0.41	0.01
Other	4.33	1.96	0.33	0.16	0.66	0.09	0.05	0.07	1.00	0.01
1987 graduates	3.65	1.64	0.34	0.17	0.68	0.10	0.05	0.10	0.56	0.01
Race/ethnicity										
White	3.69	1.66	0.33	0.20	0.69	0.10	0.04	0.09	0.57	0.01
Black	4.01	1.83	0.44	0.09	0.74	0.11	0.09	0.19	0.50	0.02
Hispanic	3.57	1.64	0.30	0.06	0.70	0.11	0.05	0.09	0.62	0.00
Asian	2.08	1.01	0.20	0.01	0.44	0.08	0.03	0.05	0.25	0.01
Other	4.11	1.90	0.42	0.21	0.64	0.06	0.05	0.10	0.72	0.01
1990 graduates	3.23	1.50	0.35	0.14	0.59	0.08	0.02	0.09	0.44	0.02
Race/ethnicity										
White	3.27	1.47	0.37	0.17	0.57	0.08	0.02	0.09	0.48	0.02
Black	3.49	1.80	0.30	0.05	0.71	0.09	0.02	0.15	0.35	0.02
Hispanic	3.19	1.52	0.29	0.10	0.65	0.10	0.02	0.11	0.39	0.01
Asian	2.10	1.11	0.16	0.04	0.44	0.02	0.01	0.02	0.30	0.00
Other	3.27	1.54	0.43	0.19	0.58	0.07	0.01	0.08	0.37	0.00

Source: "Average Number of Carnegie Units Earned by High School Graduates in Vocational Education, by Student Characteristics: 1982, 1987, and 1990." *Digest of Educational Statistics*, 1993, p. 133. Primary source: U.S. Department of Education, National Center for Education Statistics, "1990 High School Transcript Study." (This table was prepared April 1993.) *Notes:* The Carnegie unit is a standard of measurement that represents one credit for the completion of a 1-year course.

★ 249 ★

High Schools

High School Sophomores by Race/Ethnicity, 1980 and 1990

Race/ethnicity	1980	1990
All sophomores	100.0	100.0
Asian	1.3	3.9
Hispanic	8.3	10.1
Black	14.2	12.5
White	75.3	72.3
American Indian	1.0	1.2

Source: "Percentages of 1980 and 1990 Sophomores in Each Racial/Ethnic Category," U.S. Department of Education, National Center for Educational Statistics, *America's High School Sophomores: A Ten Year Comparison*, p. 6. Primary source: National Center for Education Statistics High School and Beyond base year sophomore cohort and NELS:88 first follow-up. *Note:* Percentages may not sum to 100 because of rounding.

★ 250 ★

High Schools

Mathematics Proficiency of 17-year-olds, 1977-78 and 1989-1990

| Sex and race/ethnicity | Percent of students | Proficiency by highest mathematics course taken | | | | | | Percentage of students at or above anchor points | | Level 300[3] | Level 350[4] |
		Average proficiency, all areas	Prealgebra or general mathematics	Algebra I	Geometry	Algebra II	Precalculus or calculus	Level 200[1]	Level 250[2]		
1977-78											
Total	-	300.4	267.0	286.0	307.0	321.0	334.0	99.8	92.0	51.5	7.3
White	-	305.9	272.0	291.0	310.0	325.0	338.0	100.0	95.6	57.6	8.5
Black	-	268.4	247.0	264.0	281.0	292.0	297.0	98.8	70.7	16.8	0.5
Hispanic	-	276.3	256.0	273.0	294.0	303.0	306.0	99.3	78.3	23.4	1.4
Other[5]	-	312.9	-	-	-	-	-	100.0	94.5	64.7	15.4
1989-90											
Total	100.0	304.6	273.0	288.0	299.0	319.0	344.0	100.0	96.0	56.1	7.2
White	73.9	309.5	277.0	292.0	304.0	323.0	347.0	100.0	97.6	63.2	8.3
Black	14.0	288.5	264.0	278.0	285.0	302.0	329.0	99.9	92.4	32.8	2.0
Hispanic	7.9	283.5	259.0	278.0	286.0	306.0	323.0	99.6	85.8	30.1	1.9
Other[5]	4.2	312.5	-	-	-	-	-	100.0	97.9	61.6	15.9

Source: "Mathematics Proficiency of 17-years-old, by Highest Mathematics Course Taken, Sex, and Race/Ethnicity: 1977-7 and 1989-90." *Digest of Educational Statistics*, 1993, p. 123. Primary source: U.S. Department of Education, National Center for Education Statistics, National Assessment of Educational Progress, *Trends in Academic Progress*, prepared by Educational Testing Service. (This table was prepared February 1992.) *Notes:* Scale ranges from 0 to 500. - Data not available. 1. Indicates ability to perform simple additive reasoning and problem solving. 2. Indicates ability to perform simple multiplicative reasoning and 2- step problem solving. 3. Indicates ability to perform reasoning and problem solving involving fractions, decimals, percents, elementary geometry, and simple algebra. 4. Indicates ability to perform reasoning and problem solving involving geometry, algebra, and beginning statistics and probability. 5. Includes Asian/Pacific Islanders and American Indians/Alaskan Natives.

★ 251 ★

High Schools

Percent High School Sophomores in Each Sector by Race/Ethnicity, 1980 and 1990

| Race/ethnicity | Public | | Catholic | | Other private | |
	1980	1990	1980	1990	1980	1990
All sophomores	90.6	90.3	6.1	6.1	3.3	3.6
Asian	91.1	84.6	5.9	8.1	2.9	7.3
Hispanic	92.3	92.8	5.8	5.5	1.9	1.7
Black	97.0	93.8	2.5	5.3	0.5	0.1
White	89.2	89.5	6.9	6.2	4.0	4.2
American Indian	97.1	98.3	1.1	1.7	1.8	0.0

Source: "Percentages of 1980 and 1990 Sophomores in Each Sector by Race/Ethnicity," U.S. Department of Education, National Center for Educational Statistics, *America's High School Sophomores: A Ten Year Comparison*, p. 8. Primary source: National Center for Education Statistics, High School and Beyond base year sophomore cohort and NELS:88 first follow-up. *Note:* Owing to rounding, percentages may not sum to 100.

★ 252 ★

High Schools

Percent High School Sophomores in Each Socioeconomic Category by Race/Ethnicity, 1980 and 1990

Race/ethnicity	Low SES		Middle SES		High SES	
	1980	1990	1980	1990	1980	1990
All sophomores	25.0	25.1	50.0	50.4	25.0	24.6
Asian	23.2	18.3	45.4	49.8	31.5	32.0
Hispanic	48.2	51.6	40.8	37.7	11.1	10.7
Black	45.7	42.2	43.5	48.5	10.9	9.4
White	18.8	18.7	52.2	52.4	29.0	28.9
American Indian	38.0	41.4	50.9	52.2	11.1	6.3

Source: "Percentages of 1980 and 1990 Sophomores in Each Socioeconomic Category by Race/Ethnicity," U.S. Department of Education, National Center for Educational Statistics, *America's High School Sophomores: A Ten Year Comparison*, p. 7. Primary source: National Center for Education Statistics, High School and Beyond base year sophomore cohort and NELS:88 first follow-up. *Note:* Owing to rounding, percentages may not sum to 100.

★ 253 ★

High Schools

Percent of 17-Year-Old Students Taking Science Courses for One Year or More: 1981-82, 1985-86, and 1989-90

Percent of 17-year-old students[1] taking science courses for one year or more, by selected student characteristics: 1981-82, 1985-86, and 1989- 90.

Selected characteristics of students	Biology	General science	Chemistry	Physical science	Earth and space science	Life science	Physics
1981-82							
All students	76	61	31	33	27	27	11
White, non-Hispanic	78	61	33	32	28	27	11
Black, non-Hispanic	66	66	19	34	28	27	12
Hispanic	62	58	13	35	20	31	9
1985-86							
All students	80	69	33	41	38	40	11
White, non-Hispanic	81	71	35	41	38	40	11
Black, non-Hispanic	77	62	23	45	44	40	9
Hispanic	70	64	16	37	23	41	7
1989-90							
All students	85	56	42	41	35	30	10
White, non-Hispanic	86	56	44	39	34	28	9

[Continued]

★ 253 ★

Percent of 17-Year-Old Students Taking Science Courses for One Year or More: 1981-82, 1985-86, and 1989-90

[Continued]

Selected characteristics of students	Biology	General science	Chemistry	Physical science	Earth and space science	Life science	Physics
Black, non-Hispanic	79	58	36	47	35	35	13
Hispanic	78	69	26	55	38	44	11

Source: "Percent of 17-Year-Old Students Taking Science Courses for One Year or More, by Selected Student Characteristics: 1981-82, 1985-86, and 1989-90." *Digest of Educational Statistics*, 1993, p. 134. Primary source: U.S. Department of Education, National Center for Education Statistics, National Assessment of Educational Progress, *Trends in Academic Progress*, prepared by Educational Testing Service. (This table was prepared January 1992.) *Note:* 1. Excludes persons not enrolled in school.

★ 254 ★

High Schools

Percentage of High School Graduates Earning Minimum Credits in Selected Combination of Academic Courses: 1982, 1987, and 1990

Year of graduation and course combinations taken[1]	students	Race/ethnicity			
		White	Black	Hispanic	Asian
1990 graduates					
4 Eng., 3 S.S., 3 Sci., 3 Math, .5 Comp., & 2 F.L.[2]	17.3	18.1	14.4	15.7	23.8
4 Eng., 3 S.S., 3 Sci., 3 Math, .5 Comp.[3]	22.7	22.7	25.1	20.3	27.8
4 Eng., 3 S.S., 3 Sci., 3 Math, 2 F.L.	30.6	32.6	23.4	24.8	44.1
4 Eng., 3 S.S., 3 Sci., 3 Math	39.9	40.6	41.3	32.7	51.2
4 Eng., 3 S.S., 2 Sci., 2 Math	66.8	65.4	71.8	70.4	75.5
	Increase from 1982 to 1987, in percentage points				
Difference from 1982 to 1987					
4 Eng., 3 S.S, 3 Sci., 3 Math, .5 Comp., 2 F.L.[2]	10.1	10.5	7.6	5.0	18.3
	Increase from 1982 to 1990, in percentage points				
Difference from 1982 to 1990					
4 Eng., 3 S.S, 3 Sci., 3 Math, .5 Comp., 2 F.L.[2]	5.3	5.4	6.1	10.2	-0.5

Source: "Percent of 17-Year-Old Students Taking Science Courses for One Year or More, by Selected Student Characteristics: 1981-82, 1985-86, and 1989-90." *Digest of Educational Statistics*, 1993, p. 134. Primary source: U.S. Department of Education, National Center for Education Statistics, "1990 High School Transcript Study." (This table was prepared January 1993.) *Notes:* Calculations based on unrounded figures. 1. Eng.=English; S.S. = social studies; Sci = science; Comp. = computer science; and F.L.=foreign language. 2. The National Commission on Excellence in Education recommended that all college-bound high school students take these courses as a minimum. 3. The National Commission on Excellence in Education recommended that all high school students take these courses as a minimum.

★ 255 ★

High Schools

Percentage of High School Graduates Earning Recommended Units in Core Courses and Percentage Point Change, 1982, 1987, 1990

	1982	1987	1990	Change 1982-1987	Change 1987-1990
Total	13.4	28.6	39.8	15.2	11.2
Race/ethnicity					
White	14.9	29.7	40.5	14.8	10.8
Black	10.1	24.4	41.1	14.2	16.8
Hispanic	6.3	17.9	32.7	11.6	14.8
Asian	21.0	48.3	51.0	27.3	2.8
Other	5.9	28.9	26.0	23.0	-2.9

Source: "Percentage of High School Graduates Earning Recommended Units in Core Courses in 1982, 1987, and 1990, and Percentage Point Change in Core Courses Taken: 1982-1987 and 1987-1990," U.S. Department of Education, National Center for Education Statistics, *The Condition of Education, 1993*, p. 70. Primary source: U.S. Department of Education, National Center for educational Statistics, The 1990 High School Transcript Study Tabulations, 1993. *Notes:* The core curriculum is four units of English, three units of science, three units of social studies, three units of mathematics, and one half- year of computer science. The computer science component is not included in the table.

★ 256 ★

High Schools

Percentage of High School Sophomores in General, College Preparatory, and Vocational Programs, 1980 and 1990

Student characteristics	General		College preparatory or academic		Vocational	
	1980	1990	1980	1990	1980	1990
All sophomores	46.0	50.8	33.1	41.3	21.0	7.9
Race/ethnicity						
White	47.4	51.7	35.0	42.0	17.6	6.3
Black	39.0	42.9	26.9	40.9	34.1	6.2
Hispanic	46.1	55.0	24.6	35.1	29.2	9.9
Asian	37.1	42.3	48.8	49.2	14.1	8.5
American Indian	51.6	58.5	19.8	22.9	28.7	8.6

Source: "Percent of High School Sophomores in General, College Preparatory, and Vocational Programs, by Student Characteristics: 1980 and 1990." *Digest of Educational Statistics*, 1993, p. 131. Primary source: U.S. Department of Education, National Center for Education Statistics, High School and Beyond, Base Year Survey, 1980 Sophomore Cohort; and National Education Longitudinal Study of 1988, First Follow-up Student Survey. (This table was prepared April 1993).

★ 257 ★

High Schools

Percentage of High School Students in Various High School Programs, 1982, 1986, 1990

Year	White			Black			Hispanic		
	Academic/ college prep	Vocational/ technical	General	Academic/ college prep	Vocational/ technical	General	Academic/ college prep	Vocational/ technical	General
1982	45.6	11.3	43.1	36.8	16.9	46.3	28.0	16.5	55.5
1986	55.1	8.6	36.2	38.0	17.9	44.1	35.9	14.2	50.0
1990	56.0	7.3	36.7	51.3	14.9	33.9	43.0	11.4	45.6

Source: "Percentage of High School Graduates Earning Recommended Units in Core Courses in 1982, 1987, and 1990, and Percentages Point Change in Core Courses Taken: 1982-1987 and 1987-1990," U.S. Department of Education, National Center for Education Statistics, *The Condition of Education, 1993*, p. 70. Primary source: U.S. Department of Education, National Center for Educational Statistics, The 1990 High School Transcript Study Tabulations, 1993. *Notes:* As part of the National Assessment of educational Progress mathematics background questionnaire, respondents were asked: "Which best describes high school program: 1) General, 2) Academic/College prep, 3) Vocational/Technical?" The question was identical in each survey year.

★ 258 ★

High Schools

Tenth Graders' Attendance Patterns, 1990

Reason for going to school	Percent of 10th graders					
	All 10th graders	Race/ethnicity				
		White	Black	Hispanic	Asian	American Indian
Number of days missed first half of current school year						
None	14.3	13.0	21.2	12.5	23.1	12.0
1 or 2 days	23.2	22.8	27.2	20.6	28.6	12.5
3 or 4 days	27.7	28.8	24.5	25.0	23.9	33.7
5 or more days	34.8	35.4	27.1	41.9	24.4	41.9
Number of times late first half of current school year						
None	25.2	27.8	17.8	17.8	22.0	18.6
1 or 2 days	38.2	38.0	41.1	36.7	39.7	31.3
3 or more days	36.7	34.2	41.1	45.5	38.3	50.1
Cut classes						
Never or almost never	84.8	85.8	86.5	75.8	87.1	81.4
At least sometimes	15.2	14.2	13.5	24.2	12.9	18.6

Source: "Tenth Graders' Attendance Patterns, by Selected Student and School Characteristics: 1990." *Digest of Educational Statistics*, 1993, p. 143. Primary source: U.S. Department of Education, National Center for Education Statistics, "National Education Longitudinal Study of 1988," First Followup survey. (This table was prepared February 1993.).

Higher Education

★ 259 ★

Baccalaureate Degrees Conferred on Blacks by Leading Universities, 1990

University of Calif., Los Angeles	276
University of Calif. Berkeley	243
University of Michigan	223
University of Virginia	201
University of Pennsylvania	122
Stanford University	113
Northwestern University	97
Harvard University	93
Yale University	86
Emory University	86
Brown University	83
Georgetown University	80
Duke University	69
Columbia University	68
Princeton University	66
Cornell University	60
Dartmouth College	40
Massachusetts Inst. of Technology	38
Vanderbilt University	36
Johns Hopkins University	34
George Washington University	34
Carnegie Mellon University	32
Rice University	24
University of Chicago	22
California Institute of Technology	3

Source: "College and Graduate Degrees Awarded African Americans by the Nation's Leading Universities," *Journal of Blacks in Higher Education*, Vol. 1, No. 1, Autumn 1993, p. 28. Primary source: U.S. Department of Education, Office of Educational Research and Improvement, Integrated Postsecondary Education Data System. Published by permission. *Notes:* This survey is limited to the 25 most prestigious national universities as identified in a poll of college presidents and other senior administrators conducted by *U.S. News & World Report* in 1993.

Baccalaureate Degrees Conferred on Blacks by the Twenty Most Productive Historically Black Colleges and Universities, 1989-90

Institution	State	Women	Men	Total	%
Hampton Univ	VA	454	1024	1478	94.7
Southern Univ-Baton Rouge	LA	542	788	1330	89.3
North Carolina A&T Univ	NC	594	562	1156	87.3
Jackson State Univ	MS	396	726	1122	97.6
Grambling State Univ	LA	362	682	1044	93.0
Florida A&M Univ	FL	354	586	940	75.3
Chicago State Univ	IL	270	610	880	86.8
Howard Univ	DC	331	545	876	70.6
Univ of The District of Columbia	DC	388	480	868	87.9
Prairie View A&M Univ	TX	402	460	862	85.3
South Carolina State Coll	SC	348	504	852	97.0
Norfolk State Univ	VA	294	554	848	78.5
Morgan State Univ	MD	302	494	796	92.6
North Carolina Central Univ	NC	238	554	792	89.2
Tennessee State Univ	TN	280	482	762	66.0
Morehouse Coll	GA	678	0	678	100.0
Wayne State Univ	MI	182	462	644	12.9
Coll of New Rochelle	NY	66	578	644	58.4
Texas Southern Univ	TX	222	396	618	64.5
Virginia State Univ	VA	226	388	614	87.2

Source: "Baccalaureate Degrees Conferred 1989-90, All Disciplines, HBCUs, African Americans," *Black Issues in Higher Education,* Vol. 10, No. 6, 20 May 1993, p. 65. Primary source: Department of Education.

Baccalaureate Degrees Conferred on Blacks by the Twenty Most Productive Institutions, 1989-90

Institution	State	Women	Men	Total	%
Hampton Univ	VA	454	1024	1478	94.7
Southern Univ-Baton Rouge	LA	542	788	1330	89.3
North Carolina A&T Univ	NC	594	562	1156	87.3
Jackson State Univ	MS	396	726	1122	97.6
Grambling State Univ	LA	362	682	1044	93.0
Florida A&M Univ	FL	354	586	940	75.3
Chicago State Univ	IL	270	610	880	86.8
Howard Univ	DC	331	545	876	70.6
Univ of The District of Columbia	DC	388	480	868	87.9

[Continued]

★ 261 ★

Baccalaureate Degrees Conferred on Blacks by the
Twenty Most Productive Institutions, 1989-90
[Continued]

Institution	State	Women	Men	Total	%
Prairie View A&M Univ	TX	402	460	862	85.3
South Carolina State Coll	SC	348	504	852	97.0
Norfolk State Univ	VA	294	554	848	78.5
Morgan State Univ	MD	302	494	796	92.6
North Carolina Central Univ	NC	238	554	792	89.2
Southern Illinois Univ-Carbondale	IL	446	336	782	8.2
Tennessee State Univ	TN	280	482	762	66.0
Rutgers Univ New Brunswick	NJ	268	488	756	7.0
Univ of South Carolina-Columbia	SC	232	480	712	12.2
Univ of Maryland-Coll Park	MD	276	434	710	6.3
Temple Univ	PA	214	470	684	10.0

Source: "Baccalaureate Degrees Conferred 1989-90, All Disciplines, African Americans," *Black Issues in Higher Education*, Vol. 10, No. 6, 20 May 1993, p. 64. Primary source: Department of Education.

★ 262 ★

Higher Education

Baccalaureate Degrees Conferred on Blacks by the Twenty
Most Productive Predominately White Colleges and
Universities, 1989-90

Institution	State	Women	Men	Total	%
Southern Illinois Univ-Carbondale	IL	446	336	782	8.2
Rutgers Univ New Brunswick	NJ	268	488	756	7.0
Univ of South Carolina-Columbia	SC	232	480	712	12.2
Univ of Maryland Coll-Park	MD	276	434	710	6.3
Temple Univ	PA	214	470	684	10.0
Michigan State Univ	MI	194	466	660	4.7
PA State Univ Main Campus	PA	234	330	564	3.6
Univ of California-Los Angeles	CA	184	368	552	5.4
Memphis State Univ	TN	160	366	526	14.1
Univ of Maryland Univ Coll	MD	226	298	524	15.7
Univ of Pittsburgh Main Campus	PA	246	270	516	8.0
Georgia State Univ	GA	152	360	512	12.1
Virginia Commonwealth Univ	VA	146	346	492	11.8
Long Island Univ Brooklyn Campus	NY	124	364	488	46.5
Univ of California-Berkeley	CA	232	254	486	4.3
North Carolina State Univ-Raleigh	NC	180	282	462	7.2
Univ of Florida	FL	202	256	458	4.2
Univ of Michigan-Ann Arbor	MI	170	276	446	4.0

[Continued]

★ 262 ★

Baccalaureate Degrees Conferred on Blacks by the Twenty Most Productive Predominately White Colleges and Universities, 1989-90

[Continued]

Institution	State	Women	Men	Total	%
Univ of North Carolina-Chapel Hill	NC	124	314	438	6.1
Univ. of Illinois-Urbana Campus	IL	178	246	424	3.3

Source: "Baccalaureate Degrees Conferred 1989-90, All Disciplines, (Predominately White Schools), African Americans," *Black Issues in Higher Education*, Vol. 10, No. 6, 20 May 1993, p. 65. Primary source: Department of Education.

★ 263 ★

Higher Education

Baccalaureate Degrees in Business Conferred on Blacks by the Twenty Most Productive Institutions, 1989-90

Institution	State	Women	Men	Total	%
Hampton Univ	VA	110	208	318	97.2
CUNY-Baruch	NY	106	194	300	22.8
Howard Univ	DC	90	140	230	81.3
Jackson State Univ	MS	65	128	193	98.5
North Carolina A&T Univ	NC	79	86	165	93.8
Morehouse Coll	GA	159	0	159	100.0
Southern Univ-Baton Rouge	LA	62	91	153	93.9
Morgan State Univ	MA	57	81	138	94.5
Grambling State Univ	LA	51	70	121	100.0
Virginia State Univ	VA	43	68	111	97.4
Univ of The District of Columbia	DC	52	56	108	94.7
Pace Univ-New York	NY	35	69	104	16.2
Temple Univ	PA	41	61	102	10.7
City Univ of New York-York	NY	36	60	96	56.1
Georgia State Univ	GA	36	59	95	10.2
Tennessee State Univ	TN	37	56	93	78.2
Winston-Salem State Univ	NC	30	58	88	95.7
North Carolina Central Univ	NC	21	62	83	93.3
Florida A&M Univ	FL	28	54	82	98.8
National Univ	CA	51	28	79	8.8

Source: "Baccalaureate Degrees Conferred 1989-90, Business Degrees, African Americans," *Black Issues in Higher Education*, Vol. 10, No. 6, 20 May 1993, p. 76. Primary source: Department of Education.

★ 264 ★

Higher Education

Baccalaureate Degrees in Education Conferred on Blacks by the Twenty Most Productive Institutions, 1989-90

Institution	State	Women	Men	Total	%
North Carolina A&T	NC	83	39	122	84.7
Southern Illinois Univ-Carbondale	IL	53	37	90	9.9
Norfolk State Univ	VA	9	60	69	59.5
Southern Univ-Baton Rouge	LA	22	46	68	93.2
Jackson State Univ	MS	10	54	64	92.8
Chicago State Univ	IL	9	50	59	85.5
Florida A&M Univ	FL	9	47	56	81.2
Southern Univ-New Orleans	LA	10	43	53	93.0
CUNY-York Coll	NY	3	49	52	69.3
South Carolina State Coll	SC	13	37	50	96.2
CUNY-Brooklyn	NY	5	41	46	23.0
Prairie View A&M Univ	TX	16	29	45	69.2
North Carolina Central Univ	NC	7	37	44	83.0
CUNY-City Coll	NY	3	30	43	41.3
Mississippi State Univ	MS	9	33	42	11.6
East Carolina Univ	NC	13	29	42	8.3
Grambling State Univ	LA	7	30	37	88.1
Univ of Arkansas Pine Bluff	AR	10	24	34	56.7
Univ of The District of Columbia	DC	5	28	33	97.1
Mississippi Valley State Univ	MS	7	26	33	97.1

Source: "Baccalaureate Degrees Conferred 1989-90, Education, African Americans," *Black Issues in Higher Education*, Vol. 10, No. 6, 20 May 1993, p. 82. Primary source: Department of Education.

★ 265 ★

Higher Education

Baccalaureate Degrees in Engineering Conferred on Blacks by the Nineteen Most Productive Institutions, 1991-92

Institution	Total
North Carolina A&T Univ	124
Howard Univ	105
Prairie View A&M Univ	100
Georgia Inst of technology	87
Tuskegee Univ	83
North Carolina St Univ-Raleigh	61
CUNY-City Coll	59
Southern Univ	58
Massachusetts Inst of Technology	46
Pratt Institute	38

[Continued]

★ 265 ★

Baccalaureate Degrees in Engineering Conferred on Blacks by the Nineteen Most Productive Institutions, 1991-92

[Continued]

Institution	Total
Purdue Univ	38
Tennessee State Univ	34
US Air Force Academy	33
Univ of Michigan-Ann Arbor	32
Michigan State Univ	30
Florida A&M Univ/Florida State Coll	28
Clemson Univ	27
Stanford Univ	27
US Military Academy	27
Morgan State Univ	25

Source: "Baccalaureate Degrees Conferred 1991-92, Engineering, African Americans," *Black Issues in Higher Education*, Vol. 10, No. 6, 20 May 1993, p. 86. Primary source: Engineering Manpower Commission. Published by permission.

★ 266 ★

Higher Education

Baccalaureate Degrees in Engineering, Computer Science, and Mathematics Conferred on Blacks by the Twenty Most Productive Institutions, 1989-90

Institution	State	Women	Men	Total	%
Southern Univ-Baton Rouge	LA	95	53	148	73.3
North Carolina A&T Univ	NC	81	55	136	81.9
Prairie View A&M Univ	TX	83	41	124	81.0
DeVry Institute of Tech	GA	76	36	112	40.3
CUNY-City	NY	82	18	100	26.3
Tuskegee Univ	AL	60	39	99	92.5
Tennessee State Univ	TN	57	41	98	60.1
Univ of The District of Columbia	DC	57	38	95	77.2
Howard Univ	DC	58	36	94	62.7
Grambling State Univ	LA	42	51	93	86.9
Jackson State Univ	MS	49	31	80	95.2
Georgia Institute of Tech	GA	48	23	71	5.8
South Carolina State Coll	SC	35	33	68	98.6
Florida A&M Univ	FL	35	32	67	70.5
Alabama A&M Univ	AL	31	32	63	80.8
North Carolina State-Raleigh	NC	32	31	63	5.5
Hampton Univ	VA	30	33	63	100.0
Norfolk State Univ	VA	39	24	63	82.9

[Continued]

★ 266 ★

Baccalaureate Degrees in Engineering, Computer Science, and Mathematics Conferred on Blacks by the Twenty Most Productive Institutions, 1989-90

[Continued]

Institution	State	Women	Men	Total	%
DeVry Institute of Tech	IL	34	27	61	17.4
CUNY-Baruch	NY	22	28	50	30.1

Source: "Baccalaureate Degrees Conferred 1989-90, Engineering, Computer Science & Math, African Americans," *Black Issues in Higher Education*, Vol. 10, No. 6, 20 May 1993, p. 84. Primary source: Department of Education.

★ 267 ★

Higher Education

Baccalaureate Degrees in English Conferred on Blacks by the Twenty Most Productive Institutions, 1989-90

Institution	State	Men	Women	Total	%
Spelman Coll	GA	0	58	58	98.3
Univ of Virginia Main Campus	VA	12	26	38	8.9
Univ of Pittsburgh Main Campus	PA	17	19	36	9.1
Univ of Illinois-Urbana	IL	10	22	32	6.6
Rutgers Univ New Brunswick	NJ	9		31	7.2
North Carolina State-Raleigh	NC	8	22	30	12.0
Ohio Univ Main Campus	OH	12	16	28	9.4
Univ of California-Los Angeles	CA	6	18	24	4.6
SUNY-Albany	NY	7	15	22	4.9
Univ of North Carolina-Greensboro	NC	5	17	22	11.6
Alcorn State Univ	MS	3	18	21	100.0
Univ of North Carolina-Chapel Hill	NC	6	15	21	6.1
CUNY-Lehman	NY	5	16	21	37.5
Northern Illinois Univ	IL	5	14	19	5.4
CUNY-Hunter	NY	3	16	19	13.7
Hampton Univ	VA	5	14	19	100.0
North Carolina Central Univ	NC	5	13	18	90.0
CUNY-Queens	NY	7	11	18	11.0
Univ of Maryland Coll Park	MD	4	13	17	5.2
SUNY Coll-New Paltz	NY	7	10	17	9.7

Source: "Baccalaureate Degrees Conferred 1989-90, English Degrees, African Americans," *Black Issues in Higher Education*, Vol. 10, No. 6, 20 May 1993, p. 92. Primary source: Department of Education.

★ 268 ★

Higher Education

Baccalaureate Degrees in Health Sciences Conferred on Blacks by the Twenty Most Productive Institutions, 1989-90

Institution	State	Men	Women	Total	%
Long Island Univ Brooklyn Campus	NY	19	88	107	43.9
CUNY-City Coll	NY	10	57	67	67.7
SUNY Health Science Center-Brooklyn	NY	4	61	65	68.4
CUNY-Lehman	NY	3	60	63	50.4
Florida A&M Univ	FL	10	51	61	74.4
Xavier Univ	LA	9	51	60	75.0
Southern Illinois Univ-Carbondale	IL	26	32	58	25.3
Howard Univ	DC	12	40	52	70.3
Chicago State Univ	IL	2	43	45	80.4
Hampton Univ	VA	3	39	42	76.4
CUNY-Hunter	NY	1	40	41	24.3
Texas Southern Univ	TX	17	24	41	49.4
Univ of Alabama-Birmingham	AL	1	36	37	22.2
Tennessee State Univ	TN	2	35	37	68.5
CUNY-Medgar Evers	NY	1	35	36	100.0
Saint Josephs Coll Main Campus	NY	2	34	36	42.9
Florida International Univ	FL	2	33	35	28.2
Coppin State Coll	MD	2	32	34	85.0
Virginia Commonwealth Univ	VA	6	27	33	12.5
Univ of The District of Columbia	DC	4	27	31	91.2
Saint Xavier Coll	IL	0	31	31	30.4
Texas Woman's Univ	TX	0	31	31	14.6

Source: "Baccalaureate Degrees Conferred 1989-90, Health Sciences Degrees, African American," *Black Issues in Higher Education*, Vol. 10, No. 6, 20 May 1993, p. 96. Primary source: Department of Education.

★ 269 ★

Higher Education

Baccalaureate Degrees in Life Sciences Conferred on Blacks by the Twenty Most Productive Institutions, 1989-90

Institution	State	Men	Women	Total	%
Howard Univ	DC	22	33	55	70.0
Prairie View A&M Univ	TX	15	30	45	90.0
Hampton Univ	VA	11	31	42	93.3
Jackson State Univ	MS	8	29	37	100.0
North Carolina Central Univ	NC	8	20	28	96.6
Spelman Coll	GA	0	26	26	96.3
Xavier Univ	LA	9	15	24	100.0
Univ of Maryland-Coll Park	MD	10	14	24	7.9
Rutgers Univ-New Brunswick	NJ	9	14	23	5.4

[Continued]

★ 269 ★

Baccalaureate Degrees in Life Sciences Conferred on Blacks by the Twenty Most Productive Institutions, 1989-90
[Continued]

Institution	State	Men	Women	Total	%
Univ of California-Berkeley	CA	11	11	22	4.3
South Carolina State Coll	SC	9	11	20	100.0
Tennessee State Univ	TN	7	13	20	100.0
North Carolina State-Raleigh	NC	6	11	17	9.0
Univ of Texas-Austin	TX	6	11	17	5.4
Texas Southern Univ	TX	7	9	16	84.2
Savannah State Coll	GA	4	11	15	93.8
Alcorn State Univ	MS	5	10	15	100.0
CUNY-City Coll	NY	8	7	15	16.5
Univ of California-Davis	CA	7	7	14	2.7
Grambling State Univ	LA	7	7	14	93.3

Source: "Baccalaureate Degrees Conferred 1989-90, Life Science Degrees, African American," *Black Issues in Higher Education*, Vol. 10, No. 6, 20 May 1993, p. 96. Primary source: Department of Education.

★ 270 ★

Higher Education

Baccalaureate Degrees in Physical Sciences Conferred on Blacks by the Twenty Most Productive Institutions, 1989-90

Institution	State	Men	Women	Total	%
Xavier Univ	LA	12	18	30	100.0
Spelman Coll	GA	0	26	26	100.0
Howard Univ	DC	13	12	25	71.4
United States Naval Academy	MD	14	1	15	8.7
Morehouse Coll	GA	12	0	12	100.0
Tougaloo Coll	MS	4	8	12	100.0
Jackson State Univ	MS	5	6	11	100.0
Morgan State Univ	MD	3	7	10	100.0
Lincoln Univ	PA	3	7	10	90.9
Dillard Univ	LA	4	5	9	100.0
Hampton Univ	VA	4	5	9	100.0
CUNY-Brookland	NY	3	5	8	44.4
CUNY-City Coll	NY	5	3	8	33.3
Benedict Coll	SC	2	6	8	88.9
Rust Coll	MS	1	6	7	87.5
SUNY-Stony Brook	NY	3	4	7	12.5
Talladega Coll	AL	4	2	6	100.0
Tuskegee Univ	AL	3	3	6	85.7
Grambling State Univ	LA	3	3	6	100.0
Temple Univ	PA	2	4	6	18.8

Source: "Baccalaureate Degrees Conferred 1989-90, Physical Science Degrees, African American," *Black Issues in Higher Education*, Vol. 10, No. 6, 20 May 1993, p. 98. Primary source: Department of Education.

★ 271 ★

Higher Education

Black Enrollment in Elite Liberal Arts Colleges, 1991-92

College	State	Percent black students
Amherst College	MA	7.5
Bowdoin College	ME	3.5
Bryn Mawr College	PA	4.7
Carleton College	MN	3.0
Claremont McKenna College	CA	4.8
Colby College	ME	1.6
Colgate University	NY	4.2
Davidson College	NC	4.4
Grinnell College	IA	4.9
Haverford College	PA	5.4
Holy Cross	MA	3.9
Middlebury College	VT	2.8
Mount Holyoke	MA	3.9
Oberlin College	OH	7.6
Occidental College	CA	6.0
Pomona College	CA	4.9
Smith College	MA	3.7
Swarthmore College	PA	4.8
Vassar College	NY	6.8
Washington & Lee University	VA	3.8
Wellesley College	MA	7.1
Wesleyan University	CT	7.4
Williams College	MA	7.9

Source: "Blacks at Elite Liberal Arts Colleges," *Journal of Blacks in Higher Education*, Vol. 1, No. 1, 7 Autumn 1993, p. 19. Primary source: U.S. Department of Education. Published by permission. *Notes:* The top twenty-five liberal arts colleges in the United States as ranked by *U.S. News and World Report.*

★ 272 ★

Higher Education

Black Enrollment in Highly Selective Institutions, 1991-92

Institution	% Black students
Brown	5.8
CalTech	1.1
Carnegie Mellon University	2.5
Columbia	5.2
Cornell	3.9

[Continued]

★ 272 ★

Black Enrollment in Highly Selective Institutions, 1991-92

[Continued]

Institution	% Black students
Dartmouth	5.4
Duke	6.4
Georgetown University	7.0
Harvard	5.4
Johns Hopkins	5.3
MIT	3.6
Northwestern	6.0
Princeton	5.4
Rice	4.2
Stanford	5.4
University of Calif.-Berkeley	7.0
University of Calif.-Los Angeles	6.0
University of Chicago	3.7
University of Michigan	7.2
University of North Carolina	8.5
University of Pennsylvania	5.2
University of Virginia	9.1
Vanderbilt University	4.9
Washington University	5.5*
Yale University	5.8

Source: "Black Enrollment at America's Highly Selective Institutions," *Journal of Blacks in Higher Education,* Vol. 1, No. 1, 7 Autumn 1993, p. 18. Primary source: U.S. Department of Education. Published by permission. *Notes:* The top twenty-five universities in the United States as ranked by *U.S. News and World Report* in 1992.

★ 273 ★

Higher Education

Black Faculty at Ivy League Institutions

Institution	Number of blacks	% black	Tenured blacks	Black % of all tenured
Brown University	18	3.3%	11	2.7%
Columbia University	163	7.1	33	2.8
Cornell University	40	2.6	27	2.2
Dartmouth College	13	1.5	6	2.1
Harvard University	54	1.4	22	2.0
Princeton University	16	1.8	11	2.3
Univ. of Pennsylvania	50	2.6	21	1.9
Yale University	51	2.0	27	7.8

Source: "Minor Inroads at the Prestigious Institutions," *Journal of Blacks in Higher Education,* Vol. 1, No. 1, Autumn 1993, p. 24. Primary source: U.S. Equal Employment Opportunity Commission, American Council on Education, Public Affairs Office of Ivy League Institutions, and the American Association of University Professors. Published by permission.

★ 274 ★

Higher Education

College Enrollment and Percent High School Graduates Enrolled in or Completed One or More Years of College, by Sex and Race/Ethnicity, 1960 to 1991

As of October, except as noted. Covers civilian noninstitutional population 14 to 24 years old, except as noted.

Item and year	All persons total[1]	Male			Female		
		White	Black	Hispanic origin[2]	White	Black	Hispanic origin[2]
College enrollment (1,000)							
1960[3]	2,279	1,297	68[4]	(NA)	841	73[4]	(NA)
1970	6,065	3,213	202	(NA)	2,322	236	(NA)
1975	7,228	3,437	308	148	2,931	392	160
1980	7,475	3,303	292	156	3,243	426	168
1985	7,799	3,374	355	178	3,357	400	211
1990	8,142	3,355	442	226	3,413	486	222
1991	8,304	3,311	382	216	3,607	461	310
Percent of high school graduates enrolled							
1960[3]	23.8	31.1	21.1[4]	(NA)	18.1	16.9[4]	(NA)
1970	33.3	42.9	29.5	(NA)	26.3	24.7	(NA)
1975	33.1	36.9	33.4	37.9	29.4	32.0	34.8
1980	32.3	34.3	27.0	31.2	30.9	29.2	29.4
1985	34.3	36.6	28.2	26.4	33.6	25.1	28.4
1990	39.6	40.7	35.1	29.4	38.9	32.4	29.5
1991	41.4	41.9	32.2	29.7	42.1	31.4	39.2
Percent of high school graduates enrolled in college or completed 1 or more years of college							
1960[3]	40.4	47.1	33.5[4]	(NA)	35.6	31.8[4]	(NA)
1970	52.3	60.9	41.4	(NA)	47.2	39.3	(NA)
1975	52.5	56.6	50.5	55.4	49.1	46.4	46.7
1980	51.1	51.8	44.1	49.5	47.4	45.4	
1985	54.3	55.5	43.6	44.9	55.2	44.0	48.0
1990[5]	58.9	58.7	48.9	46.5	61.4	47.3	43.0
1991[5]	60.7	59.9	46.1	42.2	64.5	47.0	52.4

Source: "College Enrollment and Percent of High School Graduates Enrolled in or Completed One or More Years of College, by Sex, Race, and Hispanic Origin: 1960 to 1991," *Statistical Abstract of the United States, 1993*, p. 173. Primary source: U.S. Bureau of the Census, *U.S. Census of the Population: 1960*, Vol. 1, *Characteristics of the Population*, Part 1; *Current Population Reports*, series P20-469, and earlier reports, and unpublished data. *Notes:* NA Not available. 1. Includes other races, not shown separately. 2. Persons of Hispanic origin may be of any race. 3. As of April. 4. Black and other races. 5. Population 16 to 24 years old.

★ 275 ★

Higher Education

Degrees Conferred, 1985 to 1990

For school year ending in year shown. Data exclude some institutions not reporting field of study and are slight undercounts of degrees awarded.

Level of degree and race/ethnicity	Total				Percent distribution			
	1985	1987	1989	1990	1985	1987	1989	1990
All degrees								
Total[1]	1,781,911	1,822,550	1,864,779	1,926,635	100.0	100.0	100.0	100.0
White, non-Hispanic	1,492,230	1,519,632	1,543,364	1,589,127	83.7	83.4	82.8	82.5
Black, non-Hispanic	111,386	110,359	111,096	116,217	6.3	6.1	6.0	6.0
Hispanic	54,706	56,180	60,470	65,863	3.1	3.1	3.2	3.4
Bachelor's degree								
Total	968,311	991,260	1,016,350	1,046,930	100.0	100.0	100.0	100.0
White, non-Hispanic	826,106	841,820	859,699	882,996	85.3	84.9	84.6	84.3
Black, non-Hispanic	57,473	56,555	58,065	61,074	5.9	5.7	5.7	5.8
Hispanic	25,874	26,990	29,910	32,686	2.7	2.7	2.9	3.1
Master's degree								
Total	280,421	289,341	309,770	321,992	100.0	100.0	100.0	100.0
White, non-Hispanic	223,628	228,870	242,756	251,518	79.7	79.1	78.4	78.1
Black, non-Hispanic	13,939	13,867	14,096	15,331	5.0	4.8	4.6	4.8
Hispanic	6,864	7,044	7,282	7,095	2.4	2.4	2.4	2.5
Doctor's degree								
Total	32,307	34,033	35,659	37,980	100.0	100.0	100.0	100.0
White, non-Hispanic	23,934	24,435	24,882	25,793	74.1	71.8	69.8	67.9
Black, non-Hispanic	1,154	1,060	1,065	1,145	3.6	3.1	3.0	3.0
Hispanic	677	750	628	783	2.1	2.2	1.8	2.1
First-professional degrees								
Total	71,057	71,617	70,856	70,736	100.0	100.0	100.0	100.0
White, non-Hispanic	63,219	62,688	61,214	60,291	89.0	87.5	86.4	85.2
Black, non-Hispanic	3,029	3,420	3,148	3,389	4.3	4.8	4.4	4.8
Hispanic	1,884	2,051	2,269	2,427	2.7	2.9	3.2	3.4

Source: "Degrees Conferred, by Level and Race/Ethnicity: 1981 to 1990," *Statistical Abstract of the United States, 1993*, p. 185. Primary source: U.S. National Center for Education Statistics, *Race/Ethnicity Trends in Degrees Conferred by Institutions of Higher Education: 1980-81 through 1989-90*, NECS 92-039. *Note:* 1. Includes associates degrees, not shown separately.

★ 276 ★

Higher Education

Doctor's Degrees Conferred on Blacks by Leading Universities, 1990

Columbia University	32
Vanderbilt University	22
University of Michigan	17
University of Calif., Berkeley	12
Harvard University	13
Emory University	8
Stanford University	8
Northwestern University	6
Yale University	6
University of Calif., Los Angeles	4
Cornell University	4
University of Virginia	4
University of Pennsylvania	4
Massachusetts Inst. of Technology	4
University of Chicago	3
Johns Hopkins University	3
Princeton University	2
George Washington University	2
Carnegie Mellon University	2
Rice University	1
Brown University	1
Georgetown University	1
California Institute of Technology	0
Duke University	0
Dartmouth College	0

Source: "College and Graduate Degrees Awarded African Americans by the Nation's Leading Universities," *Journal of Blacks in Higher Education*, Vol. 1, No. 1, Autumn 1993, p. 28. Primary source: U.S. Department of Education, Office of Educational Research and Improvement, Integrated Postsecondary Education Data System. Published by permission. *Notes:* This survey is limited to the 25 most prestigious national universities as identified in a poll of college presidents and other senior administrators conducted by *U.S. News & World Report* in 1993.

★ 277 ★

Higher Education

Doctoral Degrees Conferred on Blacks by the Twenty Most Productive Institutions, 1989-90

Institution	State	Women	Men	Total	%
Clark Atlanta Univ	GA	38	72	110	84.6
Univ of Maryland-Coll Park	MD	14	40	54	5.8
Howard Univ	DC	30	18	48	64.0
Vanderbilt Univ	TN	8	36	44	9.6

[Continued]

★ 277 ★

Doctoral Degrees Conferred on Blacks by the Twenty Most Productive Institutions, 1989-90

[Continued]

Institution	State	Women	Men	Total	%
Univ of Michigan Ann Arbor	MI	18	16	34	2.9
Temple Univ	PA	10	22	32	6.4
Univ of Illinois Urbana Campus	IL	16	14	30	2.1
Michigan State Univ	MI	26	4	30	3.5
United Theological Seminary	OH	28	2	30	44.1
Texas Southern Univ	TX	6	24	30	57.7
Southern Illinois Univ-Carbondale	IL	8	20	28	8.2
Univ of Massachusetts-Amherst	MA	14	14	28	3.9
Univ of Texas-Austin	TX	10	18	28	2.2
Harvard Univ	MA	12	14	26	2.6
New York Univ	NY	12	14	26	3.3
Univ of North Carolina Chapel Hill	NC	2	24	26	3.9
Virginia Polytechnic Institute	VA	10	16	26	3.8
Rutgers Univ New Brunswick	NJ	10	16	26	3.8
Univ of California-Berkeley	CA	12	12	24	1.5
Florida State Univ	FL	6	18	24	4.8
Georgia State Univ	GA	6	18	24	9.0
Pennsylvania State Univ	PA	14	10	24	2.9
South Carolina State Coll	SC	14	10	24	80.0

Source: "Doctoral Degrees Conferred 1989-90, All Disciplines, African Americans," *Black Issues in Higher Education*, Vol. 10, No. 6, 20 May 1993, p. 73. Primary source: Department of Education.

★ 278 ★

Higher Education

Enrollment in Institutions of Higher Education by Racial/Ethnic Group, 1972 to 1991

In thousands.

Sex, age, and race	1972	1975	1980	1985	1987	1988	1989	1990	1991
Total	9,095	10,880	11,387	12,524	12,719	13,116	13,180	13,621	14,057
White	8,147	9,547	9,926	10,782	10,731	11,140	11,243	11,488	11,686
Male	4,723	5,263	4,804	5,101	5,104	5,078	5,136	5,235	5,304
Female	3,427	4,285	5,123	5,681	5,627	6,063	6,107	6,253	6,382
Black	727	1,099	1,163	1,208	1,351	1,321	1,287	1,393	1,477
Male	384	523	476	518	587	494	480	587	629
Female	343	577	686	689	764	827	807	807	848
Hispanic origin[1]	242	411	443	665	739	747	754	748	830
Male	126	218	222	301	390	355	353	364	347
Female	117	193	221	363	349	391	401	384	483

Source: "College Enrollment, by Sex, Age, Race, and Hispanic Origin: 1972 to 1991," *Statistical Abstract of the United States, 1993*, p. 174. Primary source: U.S. Bureau of the Census, *Current Population Reports*, Series P20- 469, and earlier reports. *Note:* 1. Persons of Hispanic origin may be of any race.

★ 279 ★

Higher Education

Enrollment in Institutions of Higher Education by Region and Racial/Ethnic Group, 1991

Number of institutions beginning in academic year. Opening fall enrollment of resident and extension students attending full-time or part-time. Excludes students taking courses for credit by mail, radio, or TV, and students in branches of U.S. institutions operated in foreign countries.

State or other area	Total enrollment, 1980 (1,000)	Number of institutions[1]	1991					
			Enrollment prel. (1,000)					
			Total	White[2]	Minority enrollment			Nonresident alien
					Total[3]	Black[2]	Hispanic	
United States	12,097	3,601	14,359	10,990	2,953	1,335	867	416
Northeast	2,587	857	2,834	2,232	503	240	142	99
New England	766	255	823	703	87	34	23	32
Middle Atlantic	1,822	602	2,011	1,529	416	206	119	67
Midwest	3,044	932	3,570	3,011	466	270	91	93
East North Central	2,169	571	2,490	2,050	379	226	78	62
West North Central	875	361	1,080	961	87	45	14	32
South	3,437	1,162	4,428	3,287	1,031	643	260	110
South Atlantic	1,724	592	2,259	1,688	515	357	93	57
East South Central	614	277	776	622	140	127	4	14
West South Central	1,099	293	1,393	978	376	158	163	39
West	2,978	640	3,474	2,418	943	175	371	113
Mountain	658	191	920	745	152	23	86	23
Pacific	2,321	449	2,554	1,673	791	152	285	90

Source: "Institutions of Higher Education-Number, 1991, and Enrollment 1980 and 1991, and by Selected Characteristics, 1991-States and Other Areas," *Statistical Abstract of the United States, 1993*, p. 176. Primary source: U.S. National Center for Education Statistics, *Digest of Education Statistics* annual. *Notes:* 1. Branch campuses counted as separate institutions. 2. Non-Hispanic. 3. Includes other races not shown separately.

★ 280 ★

Higher Education

Master's Degrees Conferred on Blacks by Leading Universities, 1990

University of Michigan	146
Columbia University	132
Harvard University	114
Northwestern University	81
University of Chicago	80
University of Calif., Berkeley	75
Johns Hopkins University	71
University of Pennsylvania	70
University of Virginia	65
University of Calif., Los Angeles	61
Stanford University	58

[Continued]

★ 280 ★

Master's Degrees Conferred on Blacks by Leading Universities, 1990

[Continued]

Yale University	31
George Washington University	29
Massachusetts Inst. of Technology	25
Duke University	22
Emory University	21
Georgetown University	15
Vanderbilt University	10
Brown University	8
Princeton University	7
Dartmouth College	6
Cornell University	5
Carnegie Mellon University	4
Rice University	2
California Institute of Technology	2

Source: "College and Graduate Degrees Awarded African Americans by the Nation's Leading Universities," *Journal of Blacks in Higher Education,* Vol. 1, No. 1, Autumn 1993, p. 28. Primary source: U.S. Department of Education, Office of Educational Research and Improvement, Integrated Postsecondary Education Data System. Published by permission. *Notes:* This survey is limited to the 25 most prestigious national universities as identified in a poll of college presidents and other senior administrators conducted by *U.S. News & World Report* in 1993.

★ 281 ★

Higher Education

Master's Degrees Conferred on Blacks by the Twenty Most Productive Institutions, 1989-90

Institution	State	Women	Men	Total	%
Webster Univ	MO	288	306	594	12.2
Central Michigan Univ	MI	178	292	470	11.6
Clark Atlanta Univ	GA	130	282	412	74.9
Trevecca Nazarene Coll	TN	68	342	410	43.2
Chicago State Univ	IL	90	260	350	65.1
Wayne State Univ	MI	64	262	326	10.0
Long Island Univ-Brooklyn Campus	NY	100	208	308	26.0
Univ of Michigan-Ann Arbor	MI	120	172	292	5.4
Texas Southern Univ	TX	40	240	280	59.6
Univ of South Carolina-Columbia	SC	54	202	256	7.2
Golden Gate Univ	CA	146	98	244	9.4
Jackson State Univ	MS	78	166	244	71.8
National-Louis Univ	IL	40	204	244	11.8
Univ of The District of Columbia	DC	106	136	242	83.4
Adelphi Univ	NY	56	186	242	8.9
National Univ	CA	148	94	242	7.0
Prairie View A&M Univ	TX	68	170	238	72.1

[Continued]

★ 281 ★

Master's Degrees Conferred on Blacks by the Twenty Most Productive Institutions, 1989-90

[Continued]

Institution	State	Women	Men	Total	%
New York Univ	NY	74	158	232	3.2
Southern Univ-Baton Rouge	LA	108	124	232	76.3
Harvard Univ	MA	108	120	228	4.2

Source: "Master's Degrees Conferred 1989-90, All Disciplines, African Americans," *Black Issues in Higher Education*, Vol. 10, No. 6, 20 May 1993, p. 69. Primary source: Department of Education.

★ 282 ★

Higher Education

Master's Degrees in Engineering, Computer Science, and Mathematics Conferred on Blacks by the Eighteen Most Productive Institutions, 1989-90

Institution	State	Men	Women	Total	%
Howard Univ	DC	18	9	27	52.9
Georgia Institute of Technology	GA	12	11	23	4.3
Polytechnic Univ	NY	16	4	20	4.9
Clark Atlanta Univ	GA	7	11	18	54.5
Bowie State Univ	MD	13	4	17	35.4
Stanford Univ	CA	13	3	16	2.0
Univ of Michigan-Ann Arbor	MI	11	3	14	2.6
Univ of California-Berkeley	CA	7	6	13	3.1
DePaul Univ	IL	8	5	13	5.7
Johns Hopkins Univ	MD	9	4	13	1.9
Tuskegee Univ	AL	10	2	12	75.0
George Washington Univ	DC	9	3	12	2.5
Univ of Virginia-Main Campus	VA	8	4	12	6.1
Southern Univ-Baton Rouge	LA	10	2	12	50.0
Massachusetts Institute of Technology	MA	10	1	11	1.8
North Carolina State-Raleigh	NC	7	4	11	3.8
Univ of Florida	FL	6	4	10	3.3
Cornell Univ-Endowed Colls	NY	4	6	10	2.9

Source: "Master's Degrees Conferred 1989-90, Engineering, Computer Science & Math Degrees, African Americans," *Black Issues in Higher Education*, Vol. 10, No. 6, 20 May 1993, p. 88. Primary source: Department of Education.

★ 283 ★

Higher Education

Master's Degrees in English Conferred on Blacks by the Eight Most Productive Institutions, 1989-90

Institution	State	Men	Women	Total	%
Southern Illinois Univ-Edwardsville	IL	1	4	5	12.5
Morgan State Univ	MD	0	5	5	100.0
Chicago State Univ	IL	0	4	4	44.4
Arkansas State Univ	AR	2	1	3	23.1
Northeastern Illinois Univ	IL	1	2	3	8.3
Cornell Univ-Endowed Colls	NY	2	1	3	8.3
Univ of North Carolina-Chapel Hill	NC	0	3	3	3.9
Governors State Univ	IL	1	2	3	18.8

Source: "Master's Degrees Conferred 1989-90, English Degrees, African Americans," *Black Issues in Higher Education*, Vol. 10, No. 6, 20 May 1993, p. 90. Primary source: Department of Education.

★ 284 ★

Higher Education

Percentage of African American State Population and Percentage Enrolled in College, 1990, Part I

State	Absolute #	% of total pop.	% of total enroll
California	114,388	7.4	6.6
New York	111,000	15.9	10.9
Illinois	83,090	14.8	11.7
Texas	75,478	11.9	8.6
North Carolina	58,267	22.0	16.7
Michigan	51,494	13.9	9.2
Florida	48,396	13.6	8.4
Virginia	44,164	18.8	12.8
Georgia	43,029	27.0	18.0
Louisiana	41,213	30.8	23.0
Maryland	39,530	24.9	15.5
Alabama	38,978	25.3	18.7
Pennsylvania	38,415	9.2	6.3
Ohio	38,130	10.6	6.9
Mississippi	30,367	35.6	26.0
South Carolina	29,247	29.8	20.0
New Jersey	28,831	13.4	9.2
Tennessee	28,494	16.0	13.0
District of Columbia	23,926	65.8	30.0
Missouri	21,110	10.7	7.2
Massachusetts	17,777	5.0	4.2
Indiana	14,723	7.8	5.3

[Continued]

★ 284 ★

Percentage of African American State Population and Percentage Enrolled in College, 1990, Part I
[Continued]

State	Absolute #	% of total pop.	% of total enroll
Oklahoma	11,777	7.4	6.6
Arkansas	11,361	15.9	12.8
Kentucky	9,296	7.1	5.6

Source: "States Ranked by Number of Minorities Enrolled in College, African American," *Black Issues in Higher Education*, Vol. 9, No. 5, 7 May 1992, p. 64. Primary source: U.S. Census Bureau 1990.

★ 285 ★

Higher Education

Percentage of African American State Population and Percentage Enrolled in College, 1990, Part II

State	Absolute #	% of total pop.	% of total enroll
Wisconsin	9,060	5.0	3.1
Connecticut	8,930	8.3	5.3
Arizona	7,263	3.0	2.9
Washington	6,504	3.1	2.5
Kansas	6,300	5.8	4.0
Colorado	5,078	4.0	2.5
Delaware	4,313	16.9	10.6
Iowa	3,511	1.7	2.1
Minnesota	3,274	2.2	1.3
West Virginia	2,872	3.1	3.5
Nebraska	2,520	3.6	2.3
Nevada	2,242	6.6	4.0
Rhode Island	2,185	3.9	2.9
Oregon	2,013	1.6	1.2
New Mexico	1,667	2.0	2.0
Alaska	1,048	4.1	3.7
Hawaii	957	2.5	1.8
Utah	619	0.7	0.5
New Hampshire	611	0.6	0.1
Idaho	280	0.3	0.6
Vermont	277	0.3	0.8
Wyoming	267	0.8	0.9
Maine	263	0.4	0.5
South Dakota	226	0.5	0.7

[Continued]

★ 285 ★

Percentage of African American State Population and Percentage Enrolled in College, 1990, Part II

[Continued]

State	Absolute #	% of total pop.	% of total enroll
North Dakota	215	0.6	0.5
Montana	141	0.3	0.4

Source: "States Ranked by Number of Minorities Enrolled in College, African American," *Black Issues in Higher Education*, Vol. 9, No. 5, 7 May 1992, p. 64. Primary source: U.S. Census Bureau 1990.

★ 286 ★

Higher Education

Percentage of College Graduates Completing the Baccalaureate Degree Within Various Years of Starting College by Race/Ethnicity, Year of College Graduation 1990

Characteristic	4 or fewer years	5 or fewer years	6 or fewer years	More than 6 years
Total	43.3	70.8	81.0	19.0
Race/ethnicity				
White	44.4	71.6	81.5	18.5
Black	37.0	65.1	77.6	22.4
Hispanic	31.1	60.3	72.9	27.1
Asian	44.4	76.1	85.7	14.3
American Indian	26.6	47.7	59.0	41.0

Source: "Percentage of College Graduates Completing the Baccalaureate Degree Within Various Years of Starting College by Sex, Control of Institution, and Race/Ethnicity: Year of College Graduation 1990," U.S. Department of Education, National Center for Education Statistics, *The Condition of Education, 1993*, p. 26. Primary source: U.S. Department of Education, National Center for Education Statistics, Recent College Graduate surveys. *Notes:* Revised from previously published data. For the calculation of elapsed times, the actual month of the award of the baccalaureate degree was used. The month of high school graduation was assumed to be June. The month of starting college was assumed to be September.

★ 287 ★

Higher Education

Percentage of High School Graduates Enrolled in College the October Following Graduation, 1974 to 1990

Year	Race/ethnicity[1]			
	White	Black	Hispanic	Other[2]
1974	48.7	40.5	53.1	69.3
1975	49.1	44.5	52.7	67.7
1976	50.3	45.3	53.6	57.3
1977	50.1	46.8	48.8	61.1
1978	50.4	47.5	46.1	56.4
1979	50.1	45.2	46.3	60.5
1980	51.5	44.0	49.6	64.3
1981	52.4	40.3	48.7	72.7
1982	54.2	38.8	49.4	69.0
1983	55.5	38.0	46.7	60.9
1984	57.9	39.9	49.3	60.1
1985	58.6	39.5	46.1	66.2
1986	58.5	43.5	42.3	72.5
1987	58.8	44.2	45.0	73.4
1988	60.1	49.7	48.5	73.9
1989	61.6	48.0	52.7	72.6
1990	63.0	48.9	52.5	72.6

Source: "Percentage of High School Graduates Who Were Enrolled in College the October Following Graduation, by Type of College, Sex, and Race/Ethnicity: Year of College Graduation 1990," U.S. Department of Education, National Center for Education Statistics, *The Condition of Education, 1993*, p. 24. Primary source: U.S. Department of Commerce, Bureau of the Census, October Current Population Surveys. *Notes:* 1. Due to small sample sizes for the Black, Hispanic, and Other categories, 3-year averages are calculated. The 3-year average for 1990 is the average percentage enrolling in college in 1989, 1990, and 1991. 2. Includes individuals who are not Hispanic, white, or black; most are Asian and some are American Indian.

★ 288 ★

Higher Education

Percentage of High School and College Students Enrolled in the Previous October Who Are Enrolled Again the Following October, 1972 to 1991

October	High school students, grades 10-12, ages 15-24				College students, 1st-3rd years, ages 16-24			
	Total	White	Black	Hispanic	Total	White	Black	Hispanic
1972	93.9	94.7	90.5	88.8	77.7	78.1	71.3	78.1
1973	93.7	94.5	90.1	90.0	76.7	76.8	77.2	73.8
1974	93.3	94.2	88.4	90.1	77.5	77.4	74.3	76.0
1975	94.2	95.0	91.3	89.1	79.3	79.9	77.0	72.8
1976	94.1	94.4	92.6	92.7	79.2	79.3	81.3	74.9
1977	93.5	93.9	91.4	92.2	79.2	79.3	79.1	75.9
1978	93.3	94.2	89.8	87.7	77.7	77.8	75.3	76.7

[Continued]

★ 288 ★

Percentage of High School and College Students Enrolled in the Previous October Who Are Enrolled Again the Following October, 1972 to 1991

[Continued]

October	High school students, grades 10-12, ages 15-24				College students, 1st-3rd years, ages 16-24			
	Total	White	Black	Hispanic	Total	White	Black	Hispanic
1979	93.3	94.0	90.1	90.2	77.8	78.4	73.6	72.4
1980	93.9	94.8	91.8	88.3	79.0	80.2	71.0	69.2
1981	94.1	95.2	90.3	89.3	78.0	79.4	72.3	72.5
1982	94.5	95.3	92.2	90.8	80.4	81.2	74.6	77.4
1983	94.8	95.6	93.0	89.9	80.3	81.1	74.8	74.4
1984	94.9	95.6	94.3	88.9	79.1	79.8	74.2	72.8
1985	94.8	95.7	92.2	90.2	79.7	81.0	71.4	67.7
1986	95.3	96.3	94.6	88.1	80.2	80.5	74.4	81.7
1987	95.9	96.5	93.6	94.6	81.3	82.9	69.6	74.9
1988	95.2	95.8	94.1	89.6	83.0	83.7	78.0	77.0
1989	95.5	96.5	92.2	92.2	83.8	84.3	79.0	81.1
1990	96.0	96.7	95.0	92.1	81.8	81.7	79.4	79.7
1991	96.0	96.8	94.0	92.7	84.1	84.4	77.8	80.8

Source: "Percentage of High School and College Students Enrolled the Previous October Who Are Enrolled Again the Following October: 1972 to 1991," U.S. Department of Education, National Center for Education Statistics, *The Condition of Education, 1993*, p. 22. Primary source: U.S. Department of Commerce, Bureau of the Census, October Current Population Surveys. *Notes:* High school students were either enrolled again the following October or had graduated. Not shown separately but included in the total are non-Hispanics who are neither black nor white. Data for 1987 through 1991 reflect new editing procedures instituted by the Bureau of the Census for cases involving missing school enrollment items.

★ 289 ★

Higher Education

Percentage of Total Enrollment in Higher Education Institutions, by Race/Ethnicity, Fall 1976 to 1991

Fall of year and type of institution	White	Minority					Nonresident alien
		Total minority	Black	Hispanic	Asian	American Indian	
All institutions, by fall of year							
1976	82.6	15.4	9.4	3.5	1.8	0.7	2.0
1978	81.9	15.9	9.4	3.7	2.1	0.7	2.2
1980	81.4	16.1	9.2	3.9	2.4	0.7	2.5
1982	80.7	16.6	8.9	4.2	2.8	0.7	2.7
1984	80.2	17.0	8.8	4.4	3.2	0.7	2.7
1986	79.3	17.9	8.7	4.9	3.6	0.7	2.8
1988	78.8	18.4	8.7	5.2	3.8	0.7	2.8
1990	77.9	19.2	8.9	5.5	4.0	0.7	2.9
1991	76.5	20.6	9.3	6.0	4.4	0.8	2.9
By type and control of institution: Fall 1991							
Public	76.2	21.3	9.3	6.6	4.6	0.9	2.4

[Continued]

★ 289 ★

Percentage of Total Enrollment in Higher Education Institutions, by Race/Ethnicity, Fall 1976 to 1991

[Continued]

Fall of year and type of institution	White	Minority					Nonresident alien
		Total minority	Black	Hispanic	Asian	American Indian	
Private	77.6	17.7	9.2	4.1	4.0	0.4	4.6
4-year	78.0	18.1	8.7	4.4	4.4	0.6	3.9
2-year	74.3	24.4	10.2	8.6	4.5	1.1	1.3

Source: "Percentage of Total Enrollment in Higher Education Institutions, by Race/Ethnicity: Fall 1976-1991," U.S. Department of Education, National Center for Education Statistics, *The Condition of Education, 1993,* p. 118. Primary source: U.S. Department of Education, National Center for Education Statistics, IPEDS/HEGIS surveys of fall enrollment, various years. *Note:* Detail may not sum to totals due to rounding.

★ 290 ★

Higher Education

Proportional Representation of Blacks in Higher Education, 1980-1990

	As a percent of college-age (18-24) population	As a percent of U.S. college enrollment
1980	12.6	9.2
1982	13.4	8.9
1984	13.8	8.8
1986	14.0	8.7
1988	14.4	8.7
1990	14.9	8.9

Source: "African-American Proportional Representation in U.S. Colleges 1980- 1990," United Negro College Fund, *Statistical Report,* 1992, p. 3. Primary source: U.S. Census Bureau and U.S. Department of Education. Published by permission.

★ 291 ★

Higher Education

Racial Make-up of Higher Education Faculty, 1991

Job title	Race				
	White	Black	Hispanic	Asian	Other
Professor	91.5%	2.5%	1.4%	4.5%	0.1%
Associate professor	89.1	4.2	1.8	4.6	0.3
Assistant professor	84.3	6.0	2.6	6.9	0.2
Instructor	86.5	6.7	3.2	3.0	0.6

Source: "Currently There Are Few Blacks in Senior Academic Posts," *Journal of Blacks in Higher Education*, Vol. 1, No. 1, Autumn 1993, p. 24. Primary source: U.S. Equal Employment Opportunity Commission, American Council on Education, Public Affairs Office of Ivy League Institutions, and the American Association of University Professors.

★ 292 ★

Higher Education

Student Graduation Rates at Selective Colleges and Universities

College or university	Black student graduation rate
Harvard University	92%
Wesleyan University	90
Yale University	90
Dartmouth College	90
Princeton University	89
Brown University	88
Williams College	88[1]
Amherst College	87[2]
Vassar College	85
Stanford University	84
Vanderbilt University	84
Davidson College	82
Duke University	82
Colby College	80
Georgetown University	80
Lafayette College	80
Columbia University	79
Rice University	78
Carleton College	77
Northwestern University	76
Cornell University	74
University of Pennsylvania	73

[Continued]

★ 292 ★

Student Graduation Rates at Selective Colleges and Universities

[Continued]

College or university	Black student graduation rate
University of Virginia	72
University of Michigan	68

Source: "At the Nation's Most Selective Liberal Arts Colleges and Universities, the Black Student Graduation Rates Are Very Strong and Much Higher Than the National Average," *Journal of Blacks in Higher Education*, Vol. 1, No. 1, Autumn 1993, p. 73. Primary source: *Journal of Blacks in Higher Education* survey of college and university admission officers; *1992-93 NCAA Division 1 Graduation Rates Report*. The NCAA report covers all students, not just athletes. The graduation rate is the percentage of entering freshmen who graduate within six years. The black student graduation rate national average is 56% according to the American Council on Education. Published by permission. *Notes:* All institutions named above are classified among the 25 leading universities or liberal arts colleges in the annual survey of the opinions of academic administrators conducted by *U.S. News & World Report*. 1. Varying from 79 percent to 88 percent over the past five years. 2. 87 percent or better over the past six years.

★ 293 ★

Higher Education

Student Graduation Rates at Somewhat Less Selective State Universities

College or University	Black student graduation rate
University of South Carolina	59%
University of Delaware	55
University of Illinois	51
University of Wisconsin	50
University of Georgia	48
Ohio University	46
Michigan State University	43
University of Massachusetts	40
University of Iowa	40
University of Missouri	39
West Virginia University	39
University of Colorado	38
University of Kentucky	38
University of Maryland-College Park	37
University of Connecticut	34
University of Tennessee	34
University of Oklahoma	30
Iowa State University	28
Ohio State University	27
University of Washington	27
Indiana University	21
California State University-Fullerton	20

[Continued]

★ 293 ★

Student Graduation Rates at Somewhat Less Selective
State Universities
[Continued]

College or University	Black student graduation rate
Louisiana State University	17
Indiana State University	11

Source: "The Black Graduation Rate at the Somewhat Less Selective State Universities Is Generally Lower than the National Average for Blacks and far Lower Than That Prevailing at the Most Prestigious Colleges and Universities," *Journal of Blacks in Higher Education*, Vol. 1, No. 1, Autumn 1993, p. 73. Primary source: *Journal of Blacks in Higher Education* survey of college and university admission officers; *1992-93 NCAA Division 1 Graduation Rates Report*. The NCAA report covers all students, not just athletes. The graduation rate is the percentage of entering freshmen who graduate within six years. The black student graduation rate national average is 56% according to the American Council on Education. Published by permission. *Notes:* All institutions named above are classified as less selective than the 25 leading universities or liberal arts colleges in the annual survey of the opinions of academic administrators conducted by *U.S. News & World Report*.

★ 294 ★

Higher Education

Time to Completion of Baccalaureate Degree, 1990

Characteristic	Total with bachelor's degrees (1,000)	Years to BA Degree from end of high school						Mean duration
		Number (1,000)			Percent			
		4 years or less	5 years or less	6 years or less	4 years or less	5 years or less	6 years or less	
All persons	33,553	14,509	21,905	24,822	43.2	65.3	74.0	6.21
Male	18,145	6,893	10,914	12,710	38.0	60.1	70.0	6.23
Female	15,409	7,616	10,991	12,111	49.4	71.3	78.6	6.19
White	30,048	13,267	19,845	22,347	44.2	66.0	74.4	6.18
Male	16,374	6,297	9,920	11,509	38.5	60.6	70.3	6.20
Female	13,675	6,970	9,926	10,838	51.0	72.6	79.3	6.16
Black	1,908	623	1,052	1,266	32.7	55.1	66.4	7.01
Male	846	242	428	536	28.6	50.6	63.4	7.12
Female	1,062	382	624	731	36.0	58.8	68.8	6.92
Hispanic origin[1]	1,065	395	610	733	37.1	57.3	68.8	6.58
Male	579	210	343	410	36.3	59.2	70.8	6.05
Female	487	184	267	323	37.8	54.8	66.3	7.21

Source: "Time Spent Earning Bachelor's Degree, by Selected Characteristic: 1990," *Statistical Abstract of the United States, 1993*, p. 182. Primary source: U.S. Bureau of the Census, *Current Population Reports*, Series P70- 32. *Note:* 1. Persons of Hispanic origin may be of any race.

★ 295 ★

Higher Education

Women College and University Heads, 1993

Sector	All women Number	African-American women Number
2-year private	30	-
2-year public	107	14
4-year private	154	3
4-year public	56	9

Source: "Women University CEOs, by Race and Sector," *State of Black America, 1993*, p. 143. Primary source: Featherman, S. "Gender and Caste in Higher Education Leadership: Women's Voices in Setting the Agenda for Higher Education." Published by permission.

Higher Education Faculty

★ 296 ★

Full-Time Instructional Faculty at Institutions of Higher Education, by Race and Sex, Fall 1991

	Number		Percent	
	Male	Female	Male	Female
White	313,205	143,017	60.1	27.5
Black	13,056	11,460	2.5	2.2
Hispanic	7,353	4,069	1.4	0.8
Asian	20,481	6,029	3.9	1.2

Source: "Full-Time Instructional Faculty at Institutions of Higher Education, by Race and Sex, Fall 1991," *Black Issues in Higher Education*, Vol. 10, No. 25, 10 February 1994, p. 16. Compiled by editors from data in *Black Issues of Higher Education*, Vol. 10, No. 25, 10 February 1994, p. 16. Primary source: Department of Education, *Digest of Educational Statistics, 1993*.

★ 297 ★

Higher Education Faculty

Full-time Instructional Faculty in Institutions of Higher Education by Race/Ethnicity, Academic Rank, and Sex: Fall 1991

Academic rank and sex	Total	Race/ethnicity				
		White, non-Hispanic	Black, non-Hispanic	Hispanic	Asian or Pacific Islander	American Indian/ Alaskan Native
Men and women, all ranks	520,327	456,222	24,516	11,422	26,510	1,654
Professors	144,336	132,065	3,572	2,038	6,371	295
Associate professors	116,639	103,918	4,942	2,107	5,391	273
Assistant professors	126,344	106,557	7,524	3,246	8,649	368
Instructors	78,082	67,539	5,223	2,532	2,326	462
Lecturers	11,275	9,603	739	397	483	53
Other faculty	43,651	36,540	2,516	1,102	3,290	203
Men, all ranks	355,111	313,205	13,056	7,353	20,481	1,016
Professors	123,173	113,097	2,466	1,654	5,721	235
Associate professors	84,311	75,341	2,924	1,490	4,363	193
Assistant professors	76,129	63,573	3,884	1,964	6,511	197
Instructors	41,124	35,776	2,328	1,421	1,339	260
Lecturers	5,362	4,599	326	183	225	29
Other faculty	25,012	20,.819	1,128	641	2,322	102
Women, all ranks	165,216	143,017	11,460	4,069	6,029	638
Professors	21,163	18,968	1,106	384	650	60
Associate professors	32,328	28,577	2,018	617	1,028	80
Assistant professors	50,215	42,984	3,640	1,282	2,138	171
Instructors	36,958	31,763	2,895	1,111	987	202
Lecturers	5,913	5,004	413	214	258	24
Other faculty	18,639	15,721	1,388	461	968	101

Source: "Full-time Instructional Faculty in Institutions of Higher Education, by Race/Ethnicity, Academic Rank, and Sex: Fall 1991." *Digest of Educational Statistics*, 1993, p. 227. Primary source: U.S. Equal Employment Opportunity Commission, EEO-6 Higher Education Staff Information, 1991 (This table was prepared June 1993.) *Notes:* Data exclude faculty employed by system offices. Totals may differ from figures reported on other tables because of varying survey methodologies.

★ 298 ★

Higher Education Faculty

Full-time Regular Instructional Faculty in Institutions of Higher Education, Fall 1987

Faculty characteristics	Number in thousands	All fields	Business	Education	Fine arts	Health	Humanities	Natural sciences	Social sciences	Other
Race										
White, non-Hispanie	438	90	88	88	92	88	90	91	90	89
Black, non-Hispanic	16	3	4	6	3	2	3	2	5	5
Hispanic	11	2	1	4	3	1	5	1	3	2
Asian	21	4	6	1	1	7	2	6	2	3
American Indian	4	1	1	1	1[1]	1	1	1[1]	1	1

Source: "Full-time Regular Instructional Faculty in Institutions of Higher Education, by Faculty Characteristics and Field: Fall 1987." *Digest of Educational Statistics,* 1993, p. 229. Primary source: U.S. Department of Education, *National Survey of Postsecondary Faculty* (NSOPF), 1987-88. (This table was prepared April 1991.) *Notes:* Because of rounding and survey item nonresponse, details may not add to totals. Data for Agriculture and Home Economics (0% blacks) and Engineering (less than .5% black) omitted. 1. Less than 0.5 percent.

★ 299 ★

Higher Education Faculty

Ratio of Students to Full-Time Faculty Members of Higher Education by Race and Sex, 1991

	Enrollment	Faculty	Ratio
White			
Male	4,323,000	313,205	13.8:1
Female	4,594,000	143,017	32.1:1
Black			
Male	523,000	13,056	40.1:1
Female	667,000	11,460	58.2:1
Hispanic			
Male	310,000	7,353	42.2:1
Female	411,000	4,069	101.1:1

Source: "Ratio of Students to Full-Time Faculty Members of Higher Education by Race and Sex, 1991," *Black Issues in Higher Education,* Vol. 10, No. 25, 10 February 1994, p. 19. Primary source: Department of Education, *Digest of Educational Statistics,* 1993.

★ 300 ★

Higher Education Faculty

Regular and Temporary Instructional Faculty in Higher Educational Institutions Granting Doctorates, Fall 1987

Selected characteristics	Public research	Private research	Public doctoral	Private doctoral	Medical
Race					
White, non-Hispanic	91	85	93	91	82
Black, non-Hispanic	1	7	2	[1]	2
Hispanic	2	5	1	4	[1]
Asian	4	4	4	4	15
American Indian	1	[1]	1	1	1

Source: "Total Regular and Temporary Instructional Faculty in Institutions of Higher Education, by Selected Characteristics and Type and Control of Institution: Fall 1987." *Digest of Educational Statistics*, 1993, p. 230. Primary source: U.S. Department of Education, National Center for Education Statistics, National Survey of postsecondary Faculty (NSOPF), 1988. (This table was prepared June 1990.) *Notes:* Data may not add to totals because of rounding or missing data. 1. Less than 0.5 percent.

★ 301 ★

Higher Education Faculty

Regular and Temporary Instructional Faculty in Higher Educational Institutions Not Granting Doctorates, Fall 1987

Selected characteristics	Public comprehensive	Private comprehensive	Liberal arts	Public 2-year	Private 2-year	Other
Race						
White, non-Hispanic	88	88	88	91	90	92
Black, non-Hispanic	3	3	3	3	4	4
Hispanic	2	2	2	4	2	1
Asian	6	6	6	2	2	4
American Indian	1	1	1	1	2	[1]

Source: "Total Regular and Temporary Instructional Faculty in Institutions of Higher Education, by Selected Characteristics and Type and Control of Institution: Fall 1987." *Digest of Educational Statistics*, 1993, p. 230. Primary source: U.S. Department of Education, National Center for Education Statistics, National Survey of postsecondary Faculty (NSOPF), 1988. (This table was prepared June 1990.) *Notes:* Data may not add to totals because of rounding or missing data. 1. Less than 0.5 percent.

‹‹

Higher Education, Financial Aid

‹‹

★ 302 ★

Percentage of Undergraduates Receiving Financial Aid Fall 1989 and Average Amount in 1989-90 - I

Selected student characteristics	Enrollment of under-graduates[1], in thousands	Any aid			Grants		
		Total[2]	Federal	Nonfederal	Total	Federal	Nonfederal
Percent of all undergraduates receiving aid							
All undergraduates	12,600	44.0	30.0	32.3	37.2	21.4	28.4
Race/ethnicity							
White, non-Hispanic	9,410	41.2	26.3	31.2	34.2	17.5	27.2
Black, non-Hispanic	1,142	61.2	50.0	40.5	55.3	42.2	37.0
Hispanic	840	44.2	34.4	31.6	38.7	27.6	28.1
Asian American	575	35.5	25.5	28.2	31.2	20.2	25.9
American Indian	83	51.6	31.8	44.1	46.8	27.5	38.4
Average 1989-90 award for full-time, full-year undergraduates enrolled in fall 1989 (Award averages are computed for students participating in the designated program)							
All full-time, full-year undergraduates	3,947	$4,732	$3,511	$2,836	$3,095	$1,770	$2,544
Race/ethnicity							
White, non-Hispanic	3,208	4,597	3,488	2,785	2,976	1,702	2,494
Black, non-Hispanic	301	5,116	3,586	2,902	3,433	1,997	2,668
Hispanic	189	5,139	3,502	3,002	3,388	1,867	2,698
Asian American	174	5,614	3,650	3,304	3,836	1,886	2,874
American Indian	19	6,299	4,004	3,510	3,921	2,099	2,908
Average 1989-90 award for other undergraduate enrolled in fall 1989 (Award averages are computed for students participating in the designated program)							
All other undergraduates	7,285	2,798	2,728	1,577	1,715	1,370	1,324
Race/ethnicity							
White, non-Hispanic	5,465	2,699	2,758	1,544	1,619	1,341	1,282
Black, non-Hispanic	665	3,021	2,580	1,599	1,948	1,396	1,434
Hispanic	556	2,946	2,680	1,490	1,798	1,345	1,287
Asian American	339	3,624	2,904	2,274	2,388	1,654	1,758
American Indian	52	2,945	3,265	1,762	2,131	1,787	1,601

Source: Percentage of undergraduates enrolled in fall 1989 receiving aid and average amount awarded in 1989-90 per student, by type and source of aid and selected student characteristics. *Digest of Educational Statistics,* 1993, pp. 311-312. Primary source: U.S. Department of Education, National Center for Education Statistics, National Postsecondary Student Aid Study, 1989-90. (This table was prepared June 1992.) *Notes:* Because of rounding and/or the fact that some students receive aid from multiple sources, row details may not add to totals. Because of rounding and survey item nonresponse, enrollment data do not add to totals. Data include undergraduates in noncollegiate and collegiate institutions. 1. Numbers of undergraduates may not equal figures reported in other tables, since these data are based on a sample survey. 2. Includes students who reported they were awarded aid, but did not specify the source or type of aid.

★ 303 ★

Higher Education, Financial Aid

Percentage of Undergraduates Receiving Financial Aid Fall 1989 and Average Amount in 1989-90 - II

Selected student characteristics	Loans			Work study	Other		
	Total	Federal	Nonfederal	Total[1]	Total	Federal	Nonfederal
Percent of all undergraduates receiving aid							
All undergraduates	20.4	19.3	2.3	5.4	8.2	1.8	6.5
Race/ethnicity							
White, non-Hispanic	19.1	18.0	2.4	5.2	8.3	1.8	6.7
Black, non-Hispanic	28.2	27.2	2.4	8.4	7.8	2.3	5.7
Hispanic	19.9	19.2	2.4	5.3	7.9	1.6	6.4
Asian American	14.7	13.7	2.0	5.7	6.5	1.2	5.2
American Indian	16.2	15.5	1.5	6.9	11.9	3.4	9.6
Average 1989-90 award for full-time, full-year undergraduates enrolled in fall 1989 (Award averages are computed for students participating in the designated program)							
All full-time, full-year undergraduates	$2,764	$2,660	$2,252	$1,071	$2,091	$3,133	$1,694
Race/ethnicity							
White, non-Hispanic	2,783	2,671	2,305	1,033	2,028	3,222	1,602
Black, non-Hispanic	2,543	2,501	1,565	1,143	2,442	2,443	2,348
Hispanic	2,755	2,632	2,047	1,252	1,919	3,058	1,606
Asian American	2,915	2,840	2,541	1,296	2,758	3,500	2,584
American Indian	3,361	3,387	1,610	1,182	3,362	4,404	2,893
Average 1989-90 award for other undergraduate enrolled in fall 1989 (Award averages are computed for students participating in the designated program)							
All other undergraduates	2,668	2,527	2,004	1,063	1,523	2,783	1,248
Race/ethnicity							
White, non-Hispanic	2,665	2,511	2,066	989	1,537	2,840	1,261
Black, non-Hispanic	2,558	2,521	1,510	1,048	1,609	2,381	1,334
Hispanic	2,789	2,630	1,962	1,280	1,114	3,088	710
Asian American	2,795	2,568	2,065	1,549	1,902	2,434	1,917
American Indian	3,472	3,094	3,240	1,187	1,188	2,019	1,029

Source: Percentage of undergraduates enrolled in fall 1989 receiving aid and average amount awarded in 1989-90 per student, by type and source of aid and selected student characteristics. *Digest of Educational Statistics*, 1993, pp. 311-312. Primary source: U.S. Department of Education, National Center for Education Statistics, National Postsecondary Student Aid Study, 1989-90. (This table was prepared June 1992.) *Notes:* Because of rounding and/or the fact that some students receive aid from multiple sources, row details may not add to totals. Because of rounding and survey item nonresponse, enrollment data do not add to totals. Data include undergraduates in noncollegiate and collegiate institutions. 1. Details on federal and nonfederal Work Study participants are not available.

Higher Education: Disabled Students

★ 304 ★

Percentage of Postsecondary Students by Disability Status, Fall 1989

Selected student characteristics	Disabled students[1]	Nondisabled students
Race/ethnicity	100.0	100.0
White, non-Hispanic	83.2	80.6
Black, non-Hispanic	7.7	7.9
Hispanic	4.9	6.1
Asian American	3.2	4.7
American Indian	1.0	0.6

Source: "Percentage of Students Enrolled in Postsecondary Institutions, by Disability Status and Selected Student Characteristics; Fall 1989." *Digest of Educational Statistics,* 1993, p. 209. Primary source: U.S. Department of Education, National Center for Education Statistics, "The 1989-90 National Postsecondary Student Aid Study." (This table was prepared May 1992.) *Notes:* Because of rounding, details may not add to totals. 1. Disabled students are those who reported that they had one or more of the following conditions: a specific learning disability, a visual handicap, hard-of-hearing, deafness, a speech disability, an orthopedic handicap, or a health impairment.

Higher Education: Enrollment

★ 305 ★

Enrollment of 18- to 24-year-olds in Institutions of Higher Education: 1972-1991

Year	All students		White, non-Hispanic		Black, non-Hispanic		Hispanic origin	
	Enrollment as a percent of 18- to 24-year-olds	Enrollment as a percent of high school graduates	Enrollment as a percent of 18- to 24-year-olds	Enrollment as a percent of high school graduates	Enrollment as a percent of 18- to 24-year-olds	Enrollment as a percent of high school graduates	Enrollment as a percent of 18- to 24-year-olds	Enrollment as a percent of high school graduates
1972	25.5	31.1	27.2	31.9	18.3	25.2	13.4	24.1
1973	24.0	28.9	25.5	29.5	15.9	22.5	16.1	27.6
1974	24.6	29.8	25.8	29.9	17.6	24.6	18.0	30.7
1975	26.3	31.4	27.4	31.3	20.4	30.1	20.4	33.0
1976	26.7	32.3	27.6	32.1	22.5	32.1	20.0	34.7
1977	26.1	31.4	27.2	31.3	21.1	29.1	17.2	30.5
1978	25.3	30.0	26.5	30.1	20.1	27.9	15.2	25.9
1979	25.0	29.9	26.3	30.2	19.8	27.5	16.7	27.8
1980	25.7	30.5	27.3	31.0	19.4	26.0	16.1	27.6
1981	26.2	31.3	27.7	31.6	19.9	26.6	16.6	28.5

[Continued]

★ 305 ★

Enrollment of 18- to 24-year-olds in Institutions of Higher Education: 1972-1991
[Continued]

Year	All students Enrollment as a percent of 18- to 24-year-olds	All students Enrollment as a percent of high school graduates	White, non-Hispanic Enrollment as a percent of 18- to 24-year-olds	White, non-Hispanic Enrollment as a percent of high school graduates	Black, non-Hispanic Enrollment as a percent of 18- to 24-year-olds	Black, non-Hispanic Enrollment as a percent of high school graduates	Hispanic origin Enrollment as a percent of 18- to 24-year-olds	Hispanic origin Enrollment as a percent of high school graduates
1982	26.6	31.6	28.1	32.0	19.9	26.5	16.8	27.6
1983	26.2	31.3	28.0	31.8	19.2	25.3	17.3	29.9
1984	27.1	31.8	28.9	32.6	20.3	25.6	17.9	28.8
1985	27.8	32.5	30.0	33.9	19.6	24.5	16.9	25.0
1986	27.9	32.7	29.7	33.3	21.9	26.9	17.6	28.3
1987	29.7	35.4	31.9	36.6	23.0	28.2	17.7	26.6
1988	30.2	36.0	33.1	37.4	21.1	26.8	17.1	29.1
1989	30.9	36.5	34.2	38.3	23.4	28.5	16.0	26.6
1990	32.1	37.7	35.2	39.2	25.3	30.4	16.2	26.8
1991	33.3	39.3	36.8	41.0	23.4	28.2	17.8	31.4

Source: "Enrollment Rates of 18- to 24-Year-Olds in Institutions of Higher Education, by Race/Ethnicity: 1972 to 1991." *Digest of Educational Statistics*, 1993, p. 187. Primary source: U.S. Department of Commerce, Bureau of the Census, *Current Population Reports*, unpublished data. (This table was prepared May 1992.) *Notes:* Data are based upon sample surveys of the civilian noninstitutional population. Some data have been revised from previously published figures.

★ 306 ★

Higher Education: Enrollment

Fall Enrollment in Institutions of Higher Education, by Race/Ethnicity of Student and by State, 1991

Total fall enrollment in institutions of higher education, by race/ethnicity of student and by state: 1991[1].

State or other area	Total	White, non-Hispanic	Minority enrollment, by race/ethnicity Total	Minority enrollment, by race/ethnicity Black, non-Hispanic	Minority enrollment, by race/ethnicity Hispanic	Minority enrollment, by race/ethnicity Asian/ Pacific Islander	Minority enrollment, by race/ethnicity American Indian/ Alaskan Native	Nonresident alien	Percent minority[2] 1991
United States	14,358,953	10,989,776	2,952,824	1,335,388	866,572	637,151	113,713	416,353	21.2
Alabama	224,331	170,076	49,521	45,772	1,236	1,770	743	4,734	22.6
Alaska	30,793	24,956	5,310	1,123	660	756	2,771	527	17.5
Arizona	272,971	210,204	55,753	8,410	31,931	6,606	8,806	7,014	21.0
Arkansas	94,340	77,709	15,102	13,152	590	813	547	1,529	16.3
California	2,024,274	1,256,658	694,286	138,031	271,820	259,916	24,519	73,330	35.6
Colorado	235,108	195,444	34,750	7,177	19,030	6,042	2,501	4,914	15.1
Connecticut	165,824	139,725	20,974	10,144	5,951	4,496	383	5,125	13.1
Delaware	42,988	35,548	6,689	5,035	713	839	102	751	15.8
District of Columbia	77,964	39,717	29,302	23,475	2,414	3,233	180	8,945	42.5
Florida	611,781	441,926	153,789	67,165	70,657	13,752	2,215	16,066	25.8
Georgia	277,023	203,397	67,296	57,998	3,485	5,161	652	6,330	24.9
Hawaii	57,302	16,290	36,411	1,478	1,060	33,684	189	4,601	69.1
Idaho	55,397	51,019	2,863	345	1,168	771	579	1,515	5.3
Illinois	753,297	553,422	183,107	93,662	51,886	35,250	2,309	16,768	24.9
Indiana	290,301	255,894	26,496	16,431	4,874	4,310	881	7,911	9.4
Iowa	171,024	154,169	9,283	4,259	1,891	2,691	442	7,572	5.7
Kansas	167,699	145,837	16,214	7,318	3,857	2,825	2,214	5,648	10.0

[Continued]

★ 306 ★

Fall Enrollment in Institutions of Higher Education, by Race/Ethnicity of Student and by State, 1991

[Continued]

State or other area	Total	White, non-Hispanic	Minority enrollment, by race/ethnicity					Nonresident alien	Percent minority[2] 1991
			Total	Black, non-Hispanic	Hispanic	Asian/ Pacific Islander	American Indian/ Alaskan Native		
Kentucky	187,958	170,793	14,571	11,550	948	1,500	573	2,594	7.9
Louisiana	197,438	136,011	56,303	48,067	3,995	3,250	991	5,124	29.3
Maine	57,178	55,033	1,572	393	263	494	422	573	2.8
Maryland	267,931	193,046	66,783	48,181	5,165	12,532	905	8,102	25.7
Massachusetts	419,381	344,484	53,021	19,325	13,668	18,532	1,496	21,876	13.3
Michigan	568,491	473,042	80,169	55,462	9,344	11,660	3,703	15,280	14.5
Minnesota	255,054	234,036	15,041	4,874	2,310	5,759	2,098	5,977	6.0
Mississippi	125,350	86,272	36,816	35,173	447	799	397	2,262	29.9
Missouri	297,154	254,377	34,583	24,825	3,688	4,904	1,166	8,194	12.0
Montana	37,821	33,403	3,467	141	319	171	2,836	951	9.4
Nebraska	113,648	104,688	6,682	2,926	1,678	1,310	768	2,278	6.0
Nevada	62,664	51,078	10,873	3,205	3,818	2,920	930	713	17.6
New Hampshire	63,718	59,625	2,888	953	814	873	248	1,205	4.6
New Jersey	334,641	245,879	75,897	35,425	23,742	15,916	814	12,865	23.6
New Mexico	93,507	56,307	35,536	2,577	26,359	1,336	5,264	1,664	38.7
New York	1,056,487	751,702	266,504	125,896	83,860	53,340	3,408	38,281	26.2
North Carolina	371,968	287,178	79,094	67,419	3,190	5,241	3,244	5,696	21.6
North Dakota	38,739	34,582	2,658	275	222	251	1,910	1,499	7.1
Ohio	569,326	489,982	64,579	48,313	6,294	8,317	1,655	14,765	11.6
Oklahoma	183,536	146,890	30,234	12,771	3,062	3,238	11,163	6,412	17.1
Oregon	167,107	144,527	16,126	2,626	4,073	7,478	1,949	6,454	10.0
Pennsylvania	620,036	531,004	73,253	44,252	11,758	16,042	1,201	15,779	12.1
Rhode Island	79,112	69,268	7,058	2,778	1,986	2,037	257	2,786	9.2
South Carolina	164,907	126,354	35,735	32,516	1,184	1,690	345	2,818	22.0
South Dakota	36,332	33,032	2,626	290	118	239	1,979	674	7.4
Tennessee	238,042	194,412	39,495	34,684	1,593	2,642	576	4,135	16.9
Texas	917,443	617,056	274,211	84,503	155,534	30,704	3,470	26,176	30.8
Utah	130,419	118,230	6,978	710	2,682	2,267	1,319	5,211	5.6
Vermont	37,436	35,226	1,467	400	400	554	113	743	4.0
Virginia	356,325	278,071	71,604	52,056	5,521	13,069	958	6,650	20.5
Washington	274,760	230,722	38,461	9,050	7,553	17,512	4,346	5,577	14.3
West Virginia	88,602	82,390	4,495	3,159	387	770	179	1,717	5.2
Wisconsin	308,986	277,472	24,738	11,806	5,259	5,416	2,257	6,776	8.2
Wyoming	32,118	29,579	1,883	263	974	186	460	656	6.0
U.S. Service schools	52,921	42,034	10,277	7,569	1,141	1,287	280	610	19.6
Outlying areas	168,771	925	166,026	2,194	157,197	6,603	32	1,820	99.4
American Samoa	1,267	169	806	0	0	806	0	292	82.7
Federated States of Micronesia	837	0	837	0	0	837	0	0	100.0
Guam	5,016	457	4,090	46	56	3,966	22	469	89.9
Northern Marianas	847	33	620	2	2	616	0	194	94.9
Palau	355	0	355	0	0	355	0	0	100.0

[Continued]

★ 306 ★

Fall Enrollment in Institutions of Higher Education, by Race/Ethnicity of Student and by State, 1991

[Continued]

State or other area	Total	White, non-Hispanic	Minority enrollment, by race/ethnicity					Nonresident alien	Percent minority[2] 1991
			Total	Black, non-Hispanic	Hispanic	Asian/ Pacific Islander	American Indian/ Alaskan Native		
Puerto Rico	157,733	24	157,041	3	157,036	2	0	668	100.0
Virgin Islands	2,716	242	2,277	2,143	103	21	10	197	90.4

Source: "Total Fall Enrollment in Institutions of Higher Education, by Race/Ethnicity of Student and by State: 1991." *Digest of Educational Statistics*, 1993, p. 208. Primary source: U.S. Department of Education, National Center for Education Statistics, Integrated Postsecondary Education Data System (IPEDS), "Fall Enrollment" survey. (This table was prepared April 1993.) *Notes:* Because of adjustments to underreported and nonreported racial/ethnic data, figures are slightly different from corresponding data in other tables. 1. Preliminary data. 2. Percent minority based on U.S. citizen enrollment (total enrollment less enrollment of nonresident aliens).

★ 307 ★

Higher Education: Enrollment

Fall Enrollment in Institutions of Higher Education, by Sex and Race/Ethnicity of Student and by Level, Numbers: 1976-1991

Level of study, sex and race/ethnicity of student	Number, in thousands						
	1976	1980	1984	1986	1988	1990[1]	1991[2]
All students							
Total	10,985.6	12,086.8	12,233.0	12,503.5	13,043.1	13,819.5	14,359.0
White, non-Hispanic	9,076.1	9,833.0	9,814.7	9,920.6	10,283.2	10,723.0	10,989.8
Total minority	1,690.8	1,948.8	2,083.8	2,238.2	2,398.8	2,705.0	2,952.8
Black, non-Hispanic	1,033.0	1,106.8	1,075.8	1,082.3	1,129.6	1,247.1	1,335.4
Hispanic	383.8	471.7	534.9	618.0	680.0	782.6	866.6
Asian or Pacific Islander	197.9	286.4	389.5	447.8	796.7	572.5	637.2
American Indian/Alaskan Native	76.1	83.9	83.6	90.1	92.5	102.8	113.7
Nonresident alien	218.7	305.0	334.6	344.7	361.2	416.4	
Men	5,794.4	5,868.1	5,858.3	5,884.5	5,998.2	6,284.4	6,501.8
Total minority	826.6	884.4	937.9	1,004.7	1,051.3	1,176.8	1,280.3
Black, non-Hispanic	469.9	463.7	436.8	436.1	442.7	484.7	517.0
Women	5,191.2	6,218.7	6,374.7	6,619.0	7,044.9	7,535.1	7,857.1
Total minority	864.2	1,064.4	1,145.8	1,233.5	1,347.4	1,528.2	1,672.5
Black, non-Hispanic	563.1	643.0	639.0	646.2	686.9	762.4	818.4
Undergraduate							
Total minority	1,535.3	1,778.5	1,911.0	2,035.9	2,192.4	2,467.8	2,397.9
Black, non-Hispanic	943.4	1,018.8	994.9	996.2	1,038.8	1,147.2	1,229.3
Graduate							
Total minority	134.5	144.0	141.1	166.6	167.2	190.5	204.1
Black, non-Hispanic	78.5	75.1	67.4	72.0	76.5	83.9	88.9

[Continued]

★ 307 ★

Fall Enrollment in Institutions of Higher Education, by Sex and Race/Ethnicity of Student and by Level, Numbers: 1976-1991

[Continued]

Level of study, sex and race/ethnicity of student	Number, in thousands						
	1976	1980	1984	1986	1988	1990[1]	1991[2]
First-professional							
Total minority	21.1	26.3	31.7	35.7	39.1	46.8	50.8
Black, non-Hispanic	11.2	12.8	13.4	14.1	14.3	16.0	17.2

Source: "Total Fall Enrollment in Institutions of Higher Education, by Level of Study, Sex, and Race/Ethnicity of Student: 1976 to 1991." *Digest of Educational Statistics*, 1993, p. 206. Primary source: U.S. Department of Education, National Center for Education Statistics, "Fall Enrollment in Colleges and Universities," and Integrated Postsecondary Education Data System (IPEDS), "Fall Enrollment" survey. (This table was prepared January 1993.) *Notes:* Because of underreporting and nonreporting of racial/ethnic data, some figures are slightly lower than corresponding data in other tables. Because of rounding, details may not add to totals. 1. Revised from previously published data. 2. Preliminary data.

★ 308 ★

Higher Education: Enrollment

Fall Enrollment in Institutions of Higher Education, by Sex and Race/ Ethnicity of Student and by Level, Percentages: 1976-1991

Level of study, sex, and race/ethnicity of student	Percent distribution by level of study[1]						
	1976	1980	1984	1986	1988[2]	1990[2]	1991[3]
All students							
Total	100.0	100.0	100.0	100.0	100.0	100.0	100.0
White, non-Hispanic	84.3	83.5	82.5	81.6	81.1	79.9	78.8
Total minority	15.7	16.5	17.5	18.4	18.9	20.1	21.2
Black, non-Hispanic	9.6	9.4	9.0	8.9	8.9	9.3	9.6
Hispanic	3.6	4.0	4.5	5.1	5.4	5.8	6.2
Asian or Pacific Islander	1.8	2.4	3.3	3.7	3.9	4.3	4.6
American Indian/Alaskan Native	0.7	0.7	0.7	0.7	0.7	0.8	0.8
Nonresident alien	-	-	-	-	-	-	-
Men	52.4	48.0	47.3	46.5	45.4	45.0	44.8
Total minority	7.7	7.5	7.9	8.3	8.3	8.8	9.2
Black, non-Hispanic	4.4	3.9	3.7	3.6	3.5	3.6	3.7
Women	47.6	52.0	52.7	53.5	54.6	55.2	55.2
Total minority	8.0	9.0	9.6	10.1	10.6	11.4	12.0
Black, non-Hispanic	5.2	5.5	5.4	5.3	5.4	5.7	5.9
Undergraduate							
Total minority	16.6	17.3	18.4	19.2	19.8	21.0	22.1
Black, non-Hispanic	10.2	9.9	9.6	9.4	9.4	9.8	10.1
Graduate							
Total minority	10.8	11.5	11.5	12.8	12.7	13.4	14.0
Black, non-Hispanic	6.3	6.0	5.5	5.5	5.8	5.9	6.1

[Continued]

★ 308 ★

Fall Enrollment in Institutions of Higher Education, by Sex and Race/Ethnicity of Student and by Level, Percentages: 1976-1991

[Continued]

Level of study, sex, and race/ethnicity of student	Percent distribution by level of study[1]						
	1976	1980	1984	1986	1988[2]	1990[2]	1991[3]
First-professional							
Total minority	8.7	9.6	11.5	13.4	14.9	17.4	18.5
Black, non-Hispanic	4.6	4.7	4.9	5.3	5.5	5.9	6.3

Source: "Total Fall Enrollment in Institutions of Higher Education, by Level of Study, Sex, and Race/Ethnicity of Student: 1976 to 1991." *Digest of Educational Statistics,* 1993, p. 206. Primary source: U.S. Department of Education, National Center for Education Statistics, "Fall Enrollment in Colleges and Universities;" and Integrated Postsecondary Education Data System (IPEDS), "Fall Enrollment" survey. (This table was prepared January 1993.) *Notes:* Because of underreporting and nonreporting of racial/ethnic data, some figures are slightly lower than corresponding data in other tables. Because of rounding, details may not add to totals. 1. Distribution for U.S. citizens only. 2. Revised from previously published data. 3. Preliminary data.

Historically Black Colleges and Universities

★ 309 ★

Associate Degrees Conferred by Historically Black Colleges and Universities by Field of Study, 1989-90

Major field of study and sex of student	White, non-Hispanic	Black, non-Hispanic	Hispanic	Asian or Pacific Islander	American Indian/ Alaskan Native	Nonresident alien	Associate degrees from HBCUs as a percent of total associate degrees	
							Total	Black, non-Hispanic
All fields	793	1,477	153	11	13	42	0.6	4.2
Agriculture and natural resources	-	-	-	-	-	-	-	-
Architecture and environmental design	-	2	-	-	-	-	0.1	3.1
Area and ethnic studies	-	-	-	-	-	-	-	-
Business and management	131	365	48	2	5	15	0.5	3.4
Communications	6	11	-	-	-	-	0.5	4.1
Computer and information sciences	28	84	-	-	-	2	1.5	9.0
Education	8	151	16	-	-	-	2.2	20.7
Engineering	8	22	1	1	-	3	1.5	22.9
Engineering technologies	115	98	46	6	-	4	0.5	2.9
Foreign languages	-	-	-	-	-	-	-	-
Health professionals	363	258	24	2	1	4	1.0	5.3
Home economics	7	36	6	-	1	-	0.5	3.6
Law	1	11	-	-	-	-	0.3	3.2
Letters	-	6	-	-	-	-	1.1	11.8
Liberal/general studies	32	153	3	-	-	9	0.2	1.7
Library and archival science	-	1	-	-	-	-	0.9	20.0
Life sciences	1	7	1	-	1	1	1.1	9.3
Mathematics	-	1	-	-	-	-	0.4	33.3
Military sciences	-	-	-	-	-	-	-	-
Multi/interdisciplinary studies	-	-	-	-	-	-	-	-
Parks and recreation	-	2	-	-	-	-	0.4	6.7
Philosophy and religion	-	1	-	-	-	-	1.1	16.7

[Continued]

★ 309 ★

Associate Degrees Conferred by Historically Black Colleges and Universities by Field of Study, 1989-90

[Continued]

Major field of study and sex of student	White, non-Hispanic	Black, non-Hispanic	Hispanic	Asian or Pacific Islander	American Indian/ Alaskan Native	Nonresident alien	Associate degrees from HBCUs as a percent of total associate degrees	
							Total	Black, non-Hispanic
Physical sciences	12	10	-	-	1	2	1.2	9.3
Protective services	32	105	-	-	-	-	1.1	9.7
Psychology	-	5	2	-	-	-	0.6	4.5
Public affairs	8	30	-	-	-	1	0.7	5.3
Social sciences	4	29	2	-	1	-	1.3	8.6
Theology	1	-	-	-	-	-	0.2	-
Visual and performing arts	35	17	4	-	3	1	0.4	2.9
Undistributed	-	69	-	-	-	-	-	-

Source: "Associate Degrees Conferred by Historically Black Colleges and Universities, by Racial/Ethnic Group, Major Field of Study, and Sex of Student: 1989-90," Charlene Hoffman, Thomas D. Snyder, and Bill Sonnenberg, *Historically Black Colleges and Universities*, pp. 40-41. Primary source: U.S. Department of Education, National Center for Education Statistics, Integrated Postsecondary Education Data System (IPEDS), "Completions" survey. (This table was prepared January 1992). *Notes:* - Data not reported or not applicable. To facilitate trend comparisons, certain aggregations have been made of the degree field as reported in the IPEDS "Completions" survey. "Agriculture and natural resources" includes Agribusiness and agriculture production, Agricultural sciences, and Renewable natural resources; "Business and management" includes Business and management, Business and office, Marketing and distribution, and Consumer and personal services; "Engineering and related technologies" includes Engineering and related technologies, Mechanics and repairers, and Construction trades; "Physical sciences" includes Physical science technologies; "Public affairs" includes Public affairs, and Transportation and material moving; "Visual and performing arts" includes Visual and performing arts and Precision production. Information on data adjustments appears in technical appendix.

★ 310 ★

Historically Black Colleges and Universities

Associate Degrees Conferred by Historically Black Colleges and Universities, 1976-77 to 1989-90

Year and sex of student	Number of degrees conferred			Percentage distribution of degrees conferred			Degrees from historically black colleges and universities as a percent of total associate degrees		
	White, non-Hispanic	Black, non-Hispanic	Hispanic	White, non-Hispanic	Black, non-Hispanic	Hispanic	White, non-Hispanic	Black, non-Hispanic	Hispanic
Men									
1976-77	229	905	125	17.8	70.5	9.7	0.1	5.9	1.4
1978-79	310	555	109	30.7	54.9	10.8	0.2	3.8	1.3
1980-81	336	597	124	30.5	54.2	11.3	0.2	4.2	1.5
1982-83	349	526	98	34.1	51.4	9.6	-	-	-
1984-85	374	539	109	34.5	49.7	10.0	0.2	3.8	1.3
1986-87	315	524	125	31.1	51.7	12.3	0.2	3.8	1.4
1988-89	347	476	102	36.0	49.4	10.6	0.2	3.7	1.1
1989-90[2]	318	464	97	34.7	50.7	10.6	0.2	3.5	1.0
Women									
1976-77	148	1,260	33	10.1	85.7	2.2	0.1	7.1	0.4
1978-79	280	1,117	34	19.3	76.8	2.3	0.2	5.4	0.4
1980-81	371	1,106	31	24.2	72.1	2.0	0.2	5.3	0.3
1982-83	446	916	34	30.6	62.8	2.3	-	-	-
1984-85	506	1,008	53	31.5	62.8	3.3	0.3	4.7	0.5
1986-87[1]	481	1,047	49	30.1	65.5	3.1	0.2	4.9	0.5
1988-89	478	1,011	32	30.6	64.7	2.0	0.2	4.6	0.3
1989-90[2]	475	1,013	56	30.2	64.4	3.6	0.2	4.6	0.5

Source: "Associate Degrees Conferred by Historically Black Colleges and Universities, by Racial/Ethnic Group and Sex of Student: 1976-77 to 1989-90," Charlene Hoffman, Thomas D. Snyder, and Bill Sonnenberg, *Historically Black Colleges and Universities*, p. 35. Primary source: U.S. Department of Education, National Center for Education Statistics, Higher Education General Information Survey (HEGIS), "Degrees and Other Formal Awards Conferred" surveys; and Integrated Postsecondary Education Data System (IPEDS), "Completions" surveys. (This table was prepared January 1992). *Notes:* - Data not available. 1. Excludes 2 women whose racial/ethnic group was not available. 2. Data are preliminary.

★ 311 ★

Historically Black Colleges and Universities

Average Salary of Full-Time Instructional Faculty in Historically Black Colleges and Universities, 1976-77 to 1989-90

Year	Current dollars						
	All ranks	Professor	Associate professor	Assistant professor	Instructor	Lecturer	Undesignated or no academic rank
Men							
1976-77	14,879	19,706	16,037	13,725	11,223	13,306	11,239
1977-78	15,942	20,936	17,107	14,626	11,898	13,677	12,726
1978-79	17,513	23,250	18,857	15,770	12,738	14,109	13,339
1979-80	18,451	24,151	19,960	16,639	13,493	12,183	14,810
1980-81	20,599	26,837	21,817	18,230	14,723	17,898	17,147
1981-82	22,117	28,845	22,987	19,796	15,697	19,297	16,860
1982-83	23,362	30,035	24,410	20,702	16,757	20,644	18,197
1984-85	26,441	33,861	27,614	23,181	18,969	19,503	19,903
1985-86	27,905	35,887	28,901	24,446	19,784	22,779	21,229
1987-88	30,071	38,358	31,160	26,594	20,921	24,844	24,476
1989-90	33,436	42,872	34,953	29,182	23,169	28,471	28,091
Women							
1976-77	13,178	18,505	15,138	13,140	10,940	12,009	11,323
1977-78	14,211	20,241	16,166	14,179	11,629	13,460	12,660
1978-79	15,471	22,235	18,108	15,279	12,386	13,401	13,484
1979-80	16,472	23,194	19,175	16,331	13,159	11,914	14,516
1980-81	18,243	25,656	21,030	17,779	14,291	16,759	17,674
1981-82	19,402	27,607	22,629	18,616	15,220	17,574	17,697
1982-83	20,791	28,905	23,730	19,855	16,166	18,797	19,305
1984-85	23,351	32,112	26,184	22,448	17,950	20,881	19,565
1985-86	24,538	33,917	27,770	23,282	19,114	20,815	21,754
1987-88	26,584	35,967	30,004	25,221	20,545	22,257	25,949
1989-90	29,769	40,395	33,531	28,401	22,723	25,092	27,472

Source: "Average Salary of Full-Time Instructional Faculty in Historically Black Colleges and Universities, (HBCUs): 1976-77 to 1989-90," Charlene Hoffman, Thomas D. Snyder, and Bill Sonnenberg, *Historically Black Colleges and Universities*, p. 57. Primary source: U.S. Department of Education, National Center for Education Statistics, Faculty Salaries, Tenure, and Benefits; and Integrated Postsecondary Education Data System (IPEDS), "Salaries, Tenure, and Fringe Benefits of Full-Time Instructional Faculty" surveys. (This table was prepared February 1992). *Notes:* Data for years prior to 1989-90 exclude imputations for survey nonresponse.

★ 312 ★

Historically Black Colleges and Universities

Bachelor Degrees Conferred by Historically Black Colleges and Universities, 1976-77 to 1989-90

Year and sex of student	Number of degrees conferred				Percentage distribution of degrees conferred			Degrees from historically black colleges and universities as a percent of total bachelor's degrees		
	Total	White, non-Hispanic	Black, non-Hispanic	Hispanic	White, non-Hispanic	Black, non-Hispanic	Hispanic	White, non-Hispanic	Black, non-Hispanic	Hispanic
Men										
1976-77	10,201	1,064	8,362	23	10.4	82.0	0.2	0.2	33.3	0.2
1978-79	10,067	844	8,070	62	8.4	80.2	0.6	0.2	32.7	0.6
1980-81	10,142	854	7,866	38	8.4	77.6	0.4	0.2	32.1	0.4
1982-83	9,675	749	7,052	52	7.7	72.9	0.5	-	-	-
1984-85	9,188	921	6,448	142	10.0	70.2	1.5	0.2	28.0	1.1
1986-87[1]	8,828	883	6,576	55	10.0	74.5	0.6	0.2	29.2	0.4
1988-89[2]	7,809	871	6,066	50	11.2	77.7	0.6	0.2	27.1	0.4
1989-90[3]	7,774	944	6,064	57	12.1	78.0	0.7	0.2	26.1	0.4
Women										
1976-77	13,350	704	12,392	21	5.3	92.8	0.2	0.2	37.0	0.2
1978-79	13,582	875	12,238	31	6.4	90.1	0.2	0.2	34.4	0.3
1980-81	12,780	678	11,690	46	5.3	91.5	0.4	0.2	32.3	0.4
1982-83	12,530	738	10,735	56	5.9	85.7	0.4	-	-	-
1984-85	11,699	949	9,878	76	8.1	84.4	0.6	0.2	28.7	0.6
1986-87[1]	11,442	936	10,013	66	8.2	87.5	0.6	0.2	29.4	0.5
1988-89[2]	11,709	1,145	10,096	42	9.8	86.2	0.4	0.3	28.3	0.3
1989-90[3]	11,960	1,268	10,261	54	10.6	85.8	0.5	0.3	27.1	0.3

Source: "Bachelor Degrees Conferred by Historically Black Colleges and Universities, by Racial/Ethnic Group and Sex of Student: 1976-77 to 1989-90," Charlene Hoffman, Thomas D. Snyder, and Bill Sonnenberg, *Historically Black Colleges and Universities*, p. 36. Primary source: U.S. Department of Education, National Center for Education Statistics, Higher Education General Information Survey (HEGIS), "Degrees and Other Formal Awards Conferred" surveys; and Integrated Postsecondary Education Data System (IPEDS), "Completions" surveys. (This table was prepared November 1991). *Notes:* - Data not available. 1. Excludes 10 men and 11 women whose racial/ethnic group was not available. 2. Excludes 86 men and 144 women whose racial/ethnic group was not available. 3. Excludes 77 men and 103 women whose racial/ethnic group was not available.

★ 313 ★

Historically Black Colleges and Universities

Bachelor's Degrees Conferred by Historically Black Colleges and Universities by Field of Study, 1989-90

Major field of study and sex of student	White, non-Hispanic	Black, non-Hispanic	Hispanic	Asian or Pacific Islander	American Indian/ Alaskan Native	Nonresident alien	Bachelor's degrees from HBCUs as a percent of total bachelor's degrees	
							Total	Black, non-Hispanic
All fields[1]	2,212	16,325	111	176	19	891	1.9	26.7
Agriculture and natural resources	34	146	4	4	-	19	1.6	45.3
Architecture and environmental design	30	74	6	3	-	20	1.4	23.8
Area and ethnic studies	-	3	-	-	-	-	0.1	0.9
Business and management	481	4,829	18	26	3	212	2.2	30.7
Communications	51	847	1	3	1	38	1.8	24.2
Computer and information sciences	53	897	3	18	1	71	3.8	38.1
Education	519	1,601	13	3	1	59	2.1	36.5
Engineering	62	524	6	51	-	87	1.1	24.6
Engineering technologies	80	377	3	17	-	71	3.0	33.0
Foreign languages	2	35	5	-	-	4	0.4	10.5
Health professionals	236	837	20	24	1	90	2.1	20.0
Home economics	10	257	-	-	-	17	1.9	27.5

[Continued]

★ 313 ★

Bachelor's Degrees Conferred by Historically Black Colleges and Universities by Field of Study, 1989-90

[Continued]

Major field of study and sex of student	White, non-Hispanic	Black, non-Hispanic	Hispanic	Asian or Pacific Islander	American Indian/ Alaskan Native	Nonresident alien	Bachelor's degrees from HBCUs as a percent of total bachelor's degrees	
							Total	Black, non-Hispanic
Law	-	-	-	-	-	-	-	-
Letters	34	399	1	-	1	14	0.9	20.0
Liberal/general studies	12	24	1	-	-	3	0.2	1.3
Library and archival science	-	1	-	-	-	-	1.2	33.3
Life sciences	34	764	6	3	-	42	2.3	37.5
Mathematics	41	307	1	3	-	21	2.6	41.8
Military sciences	-	-	-	-	-	-	-	-
Multi/interdisciplinary studies	156	124	-	1	-	7	1.5	10.0
Parks and recreation	7	59	-	-	-	-	1.5	26.5
Philosophy and religion	3	54	-	1	-	5	0.9	20.4
Physical sciences	21	296	2	8	-	25	2.2	44.0
Protective services	86	666	5	-	1	12	5.0	30.1
Psychology	75	662	4	2	2	13	1.4	20.3
Public affairs	46	494	-	-	-	4	3.3	25.1
Social sciences	120	1,646	12	8	7	49	1.6	23.0
Theology	-	85	-	-	-	3	1.7	35.1
Visual and performing arts	19	317	-	1	1	5	0.9	22.4

Source: "Bachelor's Degrees Conferred by Historically Black Colleges and Universities, by Racial/Ethnic Group, Major Field of Study, and Sex of Student: 1989-90," Charlene Hoffman, Thomas D. Snyder, and Bill Sonnenberg, *Historically Black Colleges and Universities*, pp. 42-43. Primary source: U.S. Department of Education, National Center for Education Statistics, Integrated Postsecondary Education System (IPEDS), "Completions" survey. (This table was prepared January 1992). *Notes:* - Data not reported or not applicable. To facilitate trend comparisons, certain aggregations have been made of the degree field as reported in the IPEDS "Completions" survey. "Agriculture and natural resources" includes Agribusiness and agriculture production, Agricultural sciences, and Renewable natural resources; "Business and management" includes Business and management, Business and office, Marketing and distribution, and Consumer and personal services; "Engineering and related technologies" includes Engineering and related technologies, Mechanics and repairers, and Construction trades; "Physical sciences" includes Physical science technologies; "Public affairs" includes Public affairs, and Transportation and material moving; "Visual and performing arts" includes Visual and performing arts and Precision production. 1. Excludes 77 men and 103 women whose racial/ethnic group was not available. 2. Less than 0.05 percent.

★ 314 ★

Historically Black Colleges and Universities

Black Presidents at Historically Black Colleges and Universities by Sex, 1993

Sex	HBCUs		Other institutions	
	Number	Percent	Number	Percent
Women	5	6.6	13	22.8
Men	71	93.4	44	77.2
Total	76	100.0	57	100.0

Source: "African-American Presidents at HBCUs and Other Institutions by Sex: 1990," *State of Black America, 1993*, p. 143. Primary source: Ross, Green, and Henderson, *The American College President: 1993* Edition. Published by permission.

★ 315 ★

Historically Black Colleges and Universities

Current Fund Revenue of Historically Black Colleges and Universities, 1976-77 to 1988-89, Part 1

In thousands of current dollars.

Source	1976-77	1978-79	1980-81	1982-83	1984-85	1986-87	1988-89[1]
Total current-fund revenue	$984,241	$1,221,301	$1,488,851	$1,657,463	$1,946,310	$2,120,199	$2,521,902
Tuition and fees	177,383	201,691	245,303	292,394	344,644	429,498	502,689
Federal government	298,416	334,043	411,887	383,184	416,457	406,779	492,231
Appropriations	99,785	125,247	151,680	164,760	172,816	162,084	187,215
Unrestricted grants and contracts	10,417	9,670	16,225	13,229	15,106	21,131	20,712
Restricted grants and contracts[2]	188,214	199,126	243,983	205,196	227,415	222,610	283,421
Independent operations (FFRDC)[3]	0	0	0	0	1,119	953	883
State governments	243,741	321,462	392,424	446,259	572,719	619,092	720,709
Appropriations	233,302	302,333	372,906	426,401	544,051	591,727	680,341
Unrestricted grants and contracts	2,455	1,455	2,311	4,679	8,037	5,761	2,402
Restricted grants and contracts	7,984	17,674	17,207	15,179	20,631	21,603	37,966
Local governments	12,729	63,074	64,183	74,442	86,385	83,845	98,255
Appropriations	9,908	54,821	61,502	69,456	81,065	77,100	89,757
Unrestricted grants and contracts	118	6,747	48	1,454	1,970	473	3,787
Restricted grants and contracts	2,703	1,506	2,633	3,532	3,350	6,272	4,711
Private gifts, grants, and contracts	55,639	58,651	68,369	91,296	96,339	109,125	128,413
Unrestricted	29,565	35,063	40,744	54,118	55,064	61,378	70,235
Restricted	26,074	23,588	27,625	37,177	41,275	47,747	58,178
Endowment income	9,568	12,845	17,304	21,179	21,755	25,910	27,505
Unrestricted	7,680	10,634	13,579	16,241	16,484	19,691	22,082
Restricted	1,888	2,211	3,725	4,938	5,271	6,219	5,423
Sales and service	172,652	205,773	253,123	312,835	353,564	402,643	496,400
Educational activities	6,588	7,192	8,103	13,505	15,221	19,257	17,563
Auxiliary enterprises	125,355	142,569	176,142	202,296	219,996	225,567	281,428
Hospitals	40,708	56,012	68,878	97,034	118,348	157,819	197,409
Other sources	14,113	23,761	36,257	35,873	54,446	43,308	55,700

Source: "Current-Fund Revenue of Historically Black Colleges and Universities, by Source: 1976-77 to 1988-89," Charlene Hoffman, Thomas D. Snyder, and Bill Sonnenberg, *Historically Black Colleges and Universities,* pp. 68-69. Primary source: U.S. Department of Education, National Center for Education Statistics, Higher Education General Information Survey (HEGIS), "Financial Statistics of Institutions of Higher Education" surveys; and Integrated Postsecondary Education Data System (IPEDS), "Finance" surveys. (This table was prepared January 1992). *Notes:* Because of rounding, details may not add to totals. 1. Preliminary data. 2. Excludes Pell Grants. Federally supported student aid that is received through students is included under tuition and auxiliary enterprises. 3. Generally includes only those revenues associated with major federally funded research and development centers (FFRDC).

★ 316 ★

Historically Black Colleges and Universities

Current Fund Revenue of Historically Black Colleges and Universities, 1976-77 to 1988-89, Part 2

Percentage distribution.

Source	1976-77	1978-79	1980-81	1982-83	1984-85	1986-87	1988-89[1]
Total current-fund revenue	100.0	100.0	100.0	100.0	100.0	100.0	100.0
Tuition and fees	18.0	16.5	16.5	17.6	17.7	20.3	19.9
Federal government	30.3	27.4	27.7	23.1	21.4	19.2	19.5
Appropriations	10.1	10.3	10.2	9.9	8.9	7.6	7.4
Unrestricted grants and contracts	1.1	0.8	1.1	0.8	0.8	1.0	0.8
Restricted grants and contracts[2]	19.1	16.3	16.4	12.4	11.7	10.5	11.2
Independent operations (FFRDC)[3]	0.0	0.0	0.0	0.0	0.1	[4]	[4]
State governments	24.8	26.3	26.4	26.9	29.4	29.2	28.6
Appropriations	23.7	24.8	25.0	25.7	28.0	27.9	27.0
Unrestricted grants and contracts	0.2	0.1	0.2	0.3	0.4	0.3	0.1
Restricted grants and contracts	0.8	1.4	1.2	0.9	1.1	1.0	1.5
Local governments	1.3	5.2	4.3	4.5	4.4	4.0	3.9
Appropriations	1.0	4.5	4.1	4.2	4.2	3.6	3.6
Unrestricted grants and contracts	[4]	0.6	[4]	0.1	0.1	[4]	0.2
Restricted grants and contracts	0.3	0.1	0.2	0.2	0.2	0.3	0.2
Private gifts, grants, and contracts	5.7	4.8	4.6	5.5	4.9	5.1	5.1
Unrestricted	3.0	2.9	2.7	3.3	2.8	2.9	2.8
Restricted	2.6	1.9	1.9	2.2	2.1	2.3	2.3
Endowment income	1.0	1.1	1.2	1.3	1.1	1.2	1.1
Unrestricted	0.8	0.9	0.9	1.0	0.8	0.9	0.9
Restricted	0.2	0.2	0.3	0.3	0.3	0.3	0.2
Sales and service	17.5	16.8	17.0	18.9	18.2	19.0	19.7
Educational activities	0.7	0.6	0.5	0.8	0.8	0.9	0.7
Auxiliary enterprises	12.7	11.7	11.8	12.2	11.3	10.6	11.2
Hospitals	4.1	4.6	4.6	5.9	6.1	7.4	7.8
Other sources	1.4	1.9	2.4	2.2	2.8	2.0	2.2

Source: "Current-Fund Revenue of Historically Black Colleges and Universities, by Source: 1976-77 to 1988-89," Charlene Hoffman, Thomas D. Snyder, and Bill Sonnenberg, *Historically Black Colleges and Universities*, pp. 68-69. Primary source: U.S. Department of Education, National Center for Education Statistics, Higher Education General Information Survey (HEGIS), "Financial Statistics of Institutions of Higher Education" surveys; and Integrated Postsecondary Education Data System (IPEDS), "Finance" surveys. (This table was prepared January 1992). *Notes:* Because of rounding, details may not add to totals. 1. Preliminary data. 2. Excludes Pell Grants. Federally supported student aid that is received through students is included under tuition and auxiliary enterprises. 3. Generally includes only those revenues associated with major federally funded research and development centers (FFRDC). 4. Less than 0.05 percent.

★ 317 ★

Historically Black Colleges and Universities

Degrees Conferred by Historically Black Colleges and Universities by Type of Degree and Sex, 1976-77 to 1989-90

Year	Associate		Bachelor's		Master's		Doctor's		First-professional	
	Men	Women	Men	Women	Men	Women	Men	Women	Men	Women
1976-77	1,283	1,470	10,201	13,350	2,421	3,729	42	24	567	164
1977-78	1,190	1,411	10,210	13,421	2,341	3,683	42	31	553	194
1978-79	1,011	1,454	10,067	13,582	2,103	3,337	56	27	586	215
1979-80	1,105	1,469	9,906	13,563	1,814	2,995	48	23	583	243
1980-81	1,101	1,534	10,142	12,780	1,865	2,757	65	37	620	263
1981-82	1,128	1,452	9,737	12,542	1,768	2,709	59	28	571	316
1982-83	1,024	1,458	9,675	12,530	1,872	2,619	89	46	552	314
1983-84	962	1,424	9,383	12,047	1,791	2,338	74	44	565	348
1984-85	1,085	1,606	9,188	11,699	1,791	2,399	106	68	592	370
1985-86	963	1,676	9,195	11,604	1,730	2,286	99	83	595	413
1986-87	1,014	1,600	8,838	11,453	1,584	2,428	105	89	544	334
1987-88	952	1,607	8,215	11,699	1,616	2,440	113	91	482	359
1988-89	963	1,563	7,895	11,853	1,477	2,439	105	85	493	350
1989-90[1]	916	1,573	7,851	12,063	1,494	2,542	105	102	489	331

Source: "Degrees Conferred by Historically Black Colleges and Universities, by Type of Degree and Sex of Student: 1976-77 to 1989-90," Charlene Hoffman, Thomas D. Snyder, and Bill Sonnenberg, *Historically Black Colleges and Universities*, p. 48. Primary source: U.S. Department of Education, National Center for Education Statistics, Higher Education General Information Survey (HEGIS), "Degrees and Other formal Awards Conferred" surveys; and Integrated Postsecondary Education Data System (IPEDS), "Completions" survey. (This table was prepared November 1991). *Note:* 1. Preliminary data.

★ 318 ★

Historically Black Colleges and Universities

Degrees Conferred by Historically Black Colleges and Universities, 1989-90

Item	Degrees	HBCU degrees as a percentage of all degrees awarded	HBCU degrees to blacks as a percentage of all degrees to blacks
Associate	2,489	0.6	4.2
Bachelor's	19,914	1.9	26.7
Master's	4,036	1.3	15.3
Doctor's	207	0.5	12.5
First-professional	820	1.2	16.3

Source: "Degrees Conferred by Historically Black Colleges by Level: 1989-90," Charlene Hoffman, Thomas D. Snyder, and Bill Sonnenberg, *Historically Black Colleges and Universities*, p. 7. Primary source: U.S. Department of Education, National Center for Education Statistics, Integrated Postsecondary Education Data System (IPEDS), "Completions, 1989-90" survey.

★ 319 ★

Historically Black Colleges and Universities

Distribution of Fiscal Year 1990 Obligations by Selected Federal Agencies

[Dollars, in millions.]

	Research and Development	Evaluation	Training	Facilities and equipment	Fellowships	Tuition assistance	Other	Total
Department of Agriculture								
HBCUs	9.0%	6.6%	4.5%	17.9%	45.8%	20.9%	0.0%	9.3%
Other IHEs	91.0%	93.4%	95.5%	82.1%	54.2%	79.1%	0.0%	90.7%
Total	$338.7	$368.1	$5.7	$55.5	$14.4	$6.9	$0	$789.2
Department of Commerce								
HBCUs	0.5%	3.2%	8.5%	0.5%	18.4%	9.9%	100.0%	4.5%
Other IHEs	99.5%	96.8%	91.5%	99.5%	81.6%	90.1%	0.0%	95.5%
Total	$5.8	$0.2	$8.2	$15.1	$1.9	$0.2	$0.2	$31.5
Department of Defense								
HBCUs	1.2%	61.1%	5.7%	93.5%	14.3%	5.1%	100.0%	3.6%
Other IHEs	98.8%	38.9%	94.3%	6.5%	85.7%	94.9%	0.0%	96.4%
Total	$1,124.4	$2.1	$55.7	$20.6	$17.4	$191.2	$1.2	$1,412.5
Department of Education								
HBCUs	2.2%	0.0%	14.3%	18.5%	5.6%	4.2%	100.0%	8.5%
Other IHEs	97.8%	0.0%	85.7%	94.4%	95.8%	95.8%	0.0%	91.5%
Total	$266.4	$0	$298.7	$32.0	$34.0	$5,961.6	$277.8	$6,871.3

Source: "Distribution of FY 1900 Obligations by Selected Federal Agencies to Historically Black Colleges and Universities, and Other Higher Education Institutions," *State of Black America, 1993*, p. 136.

★ 320 ★

Historically Black Colleges and Universities

Doctor's Degrees Conferred by Historically Black Colleges and Universities by Field of Study, 1989-90

Major field of study and sex of student	White, non-Hispanic	Black, non-Hispanic	Hispanic	Asian or Pacific Islander	American Indian/ Alaskan Native	Nonresident alien	Doctor's degrees from HBCUs as a percent of total doctor's degrees	
							Total	Black, non-Hispanic
All fields	20	143	1	-	-	43	0.5	12.5
Agriculture and natural resources	-	-	-	-	-	-	-	-
Architecture and environmental design	-	-	-	-	-	-	-	-
Area and ethnic studies	-	1	-	-	-	1	1.6	25.0
Business and management	-	-	-	-	-	-	-	-
Communications	-	5	-	-	-	2	2.6	33.3
Computer and information sciences	-	-	-	-	-	-	-	-
Education	15	80	1	-	-	15	1.6	15.4
Engineering	-	1	-	-	-	1	[1]	2.9
Engineering technologies	-	-	-	-	-	-	-	-
Foreign languages	-	-	-	-	-	-	-	-
Health professionals	-	2	-	-	-	-	0.1	5.1
Home economics	-	2	-	-	-	-	0.7	13.3
Law	-	-	-	-	-	-	-	-
Letters	-	3	-	-	-	1	0.3	11.5
Liberal/general studies	-	-	-	-	-	-	-	-
Library and archival science	-	-	-	-	-	-	-	-
Life sciences	1	7	-	-	-	4	0.3	15.9
Mathematics	-	2	-	-	-	2	0.4	28.6
Military sciences	-	-	-	-	-	-	-	-
Multi/interdisciplinary studies	-	-	-	-	-	-	-	-
Parks and recreation	-	-	-	-	-	-	-	-
Philosophy and religion	-	-	-	-	-	-	-	-
Physical sciences	-	5	-	-	-	2	0.2	17.9
Protective services	-	-	-	-	-	-	-	-
Psychology	4	6	-	-	-	3	0.4	5.1
Public affairs	-	7	-	-	-	1	1.6	23.3
Social sciences	-	13	-	-	-	9	0.7	11.5
Theology	-	9	-	-	-	2	0.8	14.1
Visual and performing arts	-	-	-	-	-	-	-	-

Source: "Doctor's Degrees Conferred by Historically Black Colleges and Universities, by Racial/Ethnic Group, Major Field of Study, and Sex of Student: 1989-90," Charlene Hoffman, Thomas D. Snyder, and Bill Sonnenberg, *Historically Black Colleges and Universities*, pp. 46-47. Primary source: U.S. Department of Education, National Center for Education Statistics, Integrated Postsecondary Education Data System (IPEDS), "Completions" survey. (This table was prepared January 1992). *Notes:* - Data not reported or not applicable. To facilitate trend comparisons, certain aggregations have been made of the degree field as reported in the IPEDS "Completions" survey. "Agriculture and natural resources" includes Agribusiness and agriculture production, Agricultural sciences, and Renewable natural resources; "Business and management" includes Business and management, Business and office, Marketing and distribution, and Consumer and personal services; "Engineering and related technologies" includes Engineering and related technologies, Mechanics and repairers, and Construction trades; "Physical sciences" includes Physical science technologies; "Public affairs" includes Public affairs, and Transportation and material moving; "Visual and performing arts" includes Visual and performing arts and Precision production. 1. Less than 0.1 percent.

★ 321 ★

Historically Black Colleges and Universities

Doctor's Degrees Conferred by Historically Black Colleges and Universities, 1976-77 to 1989-90

Year and sex of student	Number of degrees conferred			Percentage distribution of degrees conferred			Degrees from historically black colleges and universities as a percent of total doctor's degrees		
	White, non-Hispanic	Black, non-Hispanic	Hispanic	White, non-Hispanic	Black, non-Hispanic	Hispanic	White, non-Hispanic	Black, non-Hispanic	Hispanic
Men									
1976-77	1	17	0	2.4	40.5	0.0	1	2.2	0.0
1978-79	3	27	1	5.4	48.2	1.8	1	3.7	0.3
1980-81	7	45	0	10.8	69.2	0.0	1	6.5	0.0
1982-83	3	54	0	3.4	60.7	0.0	-	-	-
1984-85	13	54	0	12.3	50.9	0.0	0.1	9.6	0.0
1986-87	15	49	0	14.3	46.7	0.0	0.1	10.0	0.0
1988-89[2]	7	60	0	6.8	58.3	0.0	1	12.2	0.0
1989-90[3]	9	64	1	8.6	61.0	1.0	0.1	12.0	0.2
Women									
1976-77	2	18	0	8.3	75.0	0.0	1	3.7	0.0
1978-79	1	23	0	3.7	85.2	0.0	1	4.3	0.0
1980-81	3	24	1	8.1	64.9	2.7	1	4.2	0.6
1982-83	3	31	0	6.5	67.4	0.0	-	-	-
1984-85	9	51	0	13.2	75.0	0.0	0.1	8.6	0.0
1986-87	8	65	0	9.0	73.0	0.0	0.1	11.4	0.0
1988-89[2]	4	68	0	4.8	81.0	0.0	1	11.8	0.0
1989-90[3]	11	79	0	10.8	77.5	0.0	0.1	12.9	0.0

Source: "Doctor's Degrees Conferred by Historically Black Colleges and Universities, by Racial/Ethnic Group and Sex of Student: 1976-77 to 1989-90," Charlene Hoffman, Thomas D. Snyder, and Bill Sonnenberg, *Historically Black Colleges and Universities*, p. 38. Primary source: U.S. Department of Education, National Center for Education Statistics, Higher Education General Information Survey (HEGIS), "Degrees and Other Formal Awards Conferred" surveys; and Integrated Postsecondary Education Data System (IPEDS), "Completions" surveys. (This table was prepared November 1991). *Notes:* - Data not available. 1. Less than 0.05 percent. 2. Excludes 2 men and 1 woman whose racial/ethnic group was not available. 3. Data are preliminary.

★ 322 ★

Historically Black Colleges and Universities

Enrollment in Historically Black Colleges and Universities, 1990

Item	HBCU enrollment in thousands	Enrollment in HBCUs as a percent of all institutions	Black enrollment in HBCUs as a percent of all black enrollment
Total	258	1.9	17.2
Men	106	1.7	17.5
Women	152	2.0	16.9
4-year	241	2.8	27.9
2-year	17	0.3	2.1

[Continued]

★ 322 ★

Enrollment in Historically Black Colleges and Universities, 1990
[Continued]

Item	HBCU enrollment in thousands	Enrollment in HBCUs as a percent of all institutions	Black enrollment in HBCUs as a percent of all black enrollment
Public	187	1.7	15.1
Private	71	2.4	24.3

Source: "Enrollment in Historically Black Colleges and Universities: 1990," Charlene Hoffman, Thomas D. Snyder, and Bill Sonnenberg, *Historically Black Colleges and Universities*, p. 4. Primary source: U.S. Department of Education, National Center for Education Statistics, Integrated Postsecondary Education Data System (IPEDS), "Fall Enrollment" survey.

★ 323 ★

Historically Black Colleges and Universities

Enrollment in Historically Black Colleges and Universities: 1980, 1988, and 1991

Selected statistics on historically black colleges and universities:[1] 1980, 1988, and 1991.

Item	Total	Public		Private	
		4-year	2-year	4-year	2-year
Number of institutions, fall 1991	105	40	11	49	5
Total enrollment, fall 1980	233,557	155,085	13,132	62,924	2,416
Men	106,387	70,236	6,758	28,352	1,041
Men, black	81,818	53,654	2,781	24,412	971
Women	127,170	84,849	6,374	34,572	1,375
Women, black	109,171	70,582	4,644	32,589	1,356
Total enrollment, fall 1988	239,755	158,606	15,066	64,644	1,439
Men	100,561	66,097	6,772	27,219	473
Men, black	78,268	50,545	3,192	24,081	450
Women	139,194	92,509	8,294	37,425	966
Women, black	115,883	73,893	5,894	35,145	951
Total enrollment, fall 1991	269,280	182,204	15,643	69,834	1,599
Men	110,455	74,650	6,477	28,780	548
Men, black	87,755	57,722	3,358	26,237	438
Women	158,825	107,554	9,166	41,054	1,051
Women, black	131,413	85,689	6,095	38,603	1,026
Full-time enrollment, fall 1991	206,488	133,283	9,004	62,970	1,231
Men	86,994	57,026	3,568	25,894	506
Women	119,494	76,257	5,436	37,076	725

[Continued]

★ 323 ★

Enrollment in Historically Black Colleges and Universities:
1980, 1988, and 1991

[Continued]

Item	Total	Public		Private	
		4-year	2-year	4-year	2-year
Part-time enrollment, fall 1991	62,792	48,921	6,639	6,864	368
Men	23,461	17,624	2,909	2,886	42
Women	39,331	31,297	3,730	3,978	326

Source: "Selected Statistics on Historically Black Colleges and Universities: 1980, 1988, and 1991." *Digest of Educational Statistics*, 1993, p. 221. U.S. Department of Education, National Center for Education Statistics, "Fall Enrollment in Institutions of Higher Education"; and Integrated postsecondary Education Data System (IPEDS), "Fall Enrollment," "Completions," and "Finance" surveys. (This table was prepared March 1993.) *Notes:* Enrollment data for fall 1991 are preliminary. 1. Historically black colleges and universities are accredited institutions of higher education established prior to 1964 with the principal mission of educating black Americans. Federal regulations, 20 U.S. Code, Section 1061(2), allow for certain exceptions to the founding date. Most institutions are in the southern and border states and were established prior to 1954.

★ 324 ★

Historically Black Colleges and Universities

Fall Enrollment in All Four Year Institutions and in Four-Year Historically Black Colleges and Universities, 1976 to 1990

Type and control of institution and race/ethnicity of student	All institutions of higher education				Enrollment in historically black colleges and universities as a percent of total enrollment			
	1976	1980	1988	1990[1]	1976	1980	1988	1990[1]
Total	7,106,502	7,565,401	8,175,008	8,529,132	2.9	2.9	2.7	2.8
White, non-Hispanic	5,998,982	6,274,542	6,581,628	6,756,844	0.3	0.3	0.4	0.4
Total minority	931,014	1,049,938	1,291,833	1,450,202	19.5	17.5	14.5	14.0
Black, non-Hispanic	603,732	634,299	656,307	714,571	29.8	28.6	27.9	27.9
Hispanic	173,612	216,633	296,026	344,485	0.3	0.5	0.6	0.5
Asian or Pacific Islander	118,717	162,097	297,404	342,976	0.5	0.8	0.5	0.5
American Indian/Alaskan Native	34,953	36,909	42,096	48,170	0.6	1.1	0.7	0.6
Nonresident alien	176,506	240,921	301,547	322,086	3.8	5.2	2.9	2.5
Public	4,892,942	5,127,641	5,543,987	5,802,877	2.9	3.0	2.9	3.0
White, non-Hispanic	4,120,184	4,243,017	4,454,845	4,594,718	0.4	0.5	0.6	0.6
Total minority	666,712	740,788	907,713	1,014,451	18.4	17.1	14.0	13.6
Black, non-Hispanic	421,753	438,177	448,487	487,245	28.9	28.4	27.7	27.7
Hispanic	129,283	156,394	215,798	250,390	0.3	0.4	0.6	0.6
Asian or Pacific Islander	87,488	117,175	210,156	238,521	0.5	1.0	0.6	0.6
American Indian/Alaskan Native	28,188	29,042	33,272	38,295	0.5	1.2	0.8	0.8
Nonresident alien	106,046	143,836	181,429	193,708	3.0	5.7	2.9	2.6
Private	2,213,560	2,437,760	2,631,021	2,726,255	2.9	2.6	2.5	2.5
White, non-Hispanic	1,878,798	2,031,525	2,126,783	2,162,126	0.1	[2]	0.1	0.1
Total minority	264,302	309,150	384,120	435,751	22.1	18.7	15.6	14.9
Black, non-Hispanic	181,979	196,122	207,820	227,326	31.9	29.1	28.4	28.3
Hispanic	44,329	60,239	80,228	94,095	0.3	0.7	0.7	0.4
Asian or Pacific Islander	31,229	44,922	87,248	104,455	0.5	0.5	0.2	0.3

[Continued]

★ 324 ★

Fall Enrollment in All Four Year Institutions and in Four-Year Historically Black Colleges and Universities, 1976 to 1990
[Continued]

Type and control of institution and race/ethnicity of student	All institutions of higher education				Enrollment in historically black colleges and universities as a percent of total enrollment			
	1976	1980	1988	1990[1]	1976	1980	1988	1990[1]
American Indian/Alaskan Native	6,765	7,867	8,824	9,875	0.6	0.5	0.3	0.2
Nonresident alien	70,460	97,085	120,118	128,378	5.0	4.4	3.0	2.4

Source: "Fall Enrollment in All Institutions and Historically Black Colleges and Universities, by Type and Control of Institution and Race/Ethnicity of Student: 1976 to 1990," Charlene Hoffman, Thomas D. Snyder, and Bill Sonnenberg, *Historically Black Colleges and Universities*, p. 21. Primary source: U.S. Department of Education, National Center for Education Statistics, Higher Education General Information Survey (HEGIS), "Fall Enrollment in Colleges and Universities" surveys; and Integrated Postsecondary Education Data System (IPEDS). "Fall Enrollment" surveys. (This table was prepared January 1992). *Notes:* 1. Preliminary data. 2. Less than 0.05 percent.

★ 325 ★

Historically Black Colleges and Universities

Fall Enrollment in All Two-Year Institutions and in Two-Year Historically Black Colleges and Universities, 1976 to 1990

Type and control of institution and race/ethnicity of student	All institutions of higher education				Enrollment in historically black colleges and universities as a percent of total enrollment			
	1976	1980	1988	1990[1]	1976	1980	1988	1990[1]
Total	3,879,112	4,518,059	4,868,110	5,181,018	0.4	0.3	0.3	0.3
White, non-Hispanic	3,077,149	3,558,470	3,701,548	3,917,940	0.1	0.1	0.1	0.1
Total minority	759,789	898,878	1,106,931	1,188,576	1.8	1.4	1.2	1.1
Black, non-Hispanic	429,293	472,451	473,273	508,732	2.4	2.1	2.3	2.1
Hispanic	210,178	255,084	383,936	413,569	1.4	1.1	0.6	0.5
Asian or Pacific Islander	79,161	124,349	199,284	211,827	0.1	[2]	0.1	[2]
American Indian/Alaskan Native	41,157	46,994	50,438	54,448	0.1	0.1	0.1	0.1
Nonresident alien	42,174	60,711	59,631	74,502	0.4	0.3	0.2	0.1
Public	3,748,095	4,328,782	4,612,388	4,937,663	0.4	0.3	0.3	0.3
White, non-Hispanic	2,974,337	3,413,077	3,508,988	3,744,829	0.1	0.1	0.1	0.1
Total minority	734,534	855,366	1,046,978	1,121,775	1.5	1.2	1.1	1.0
Black, non-Hispanic	409,459	437,893	432,601	465,116	1.9	1.7	2.2	2.0
Hispanic	207,535	249,756	371,069	397,868	1.4	1.1	0.6	0.5
Asian or Pacific Islander	78,228	122,535	195,533	206,836	[2]	[2]	0.1	[2]
American Indian/Alaskan Native	39,312	45,182	47,775	51,955	0.1	0.1	0.1	0.1
Nonresident alien	39,224	60,339	56,422	71,059	0.3	0.2	0.1	0.1
Private	131,017	189,277	255,722	243,355	2.0	1.3	0.6	0.6
White, non-Hispanic	102,812	145,393	192,560	173,111	[2]	[2]	[2]	[2]
Total minority	25,255	43,512	59,953	66,801	10.1	5.4	2.3	2.3
Black, non-Hispanic	19,834	34,558	40,672	43,616	12.8	6.7	3.4	3.5
Hispanic	2,643	5,328	12,867	15,701	0.1	0.1	-	-
Asian or Pacific Islander	933	1,814	3,751	4,991	0.9	0.1	0.1	-
American Indian/Alaskan Native	1,845	1,812	2,663	2,493	0.1	-	-	-
Nonresident alien	2,950	372	3,209	3,443	1.6	1.9	0.6	0.8

Source: "Fall Enrollment in All Institutions and Historically Black Colleges and Universities, by Type and Control of Institution and Race/Ethnicity of Student: 1976 to 1990," Charlene Hoffman, Thomas D. Snyder, and Bill Sonnenberg, *Historically Black Colleges and Universities*, p. 21. Primary source: U.S. Department of Education, National Center for Education Statistics, Higher Education General Information Survey (HEGIS), "Fall Enrollment in Colleges and Universities" surveys; and Integrated Postsecondary Education Data System (IPEDS). "Fall Enrollment" surveys. (This table was prepared January 1992). *Notes:* - Data not reported or not applicable. 1. Preliminary data. 2. Less than 0.05 percent.

★ 326 ★

Historically Black Colleges and Universities

Fall Enrollment in Four-Year Historically Black Colleges and Universities, 1980 to 1990

Type and control of institution and race/ethnicity of student	Enrollment				Percent distribution by type and control			
	1980	1984	1988	1990[1]	1980	1984	1988	1990[1]
Total	218,009	212,844	223,250	241,149	100.0	100.0	100.0	100.0
White, non-Hispanic	21,528	23,761	27,439	30,131	9.9	11.2	12.3	12.5
Total minority	184,066	174,662	186,968	202,978	84.4	82.1	83.7	84.2
Black, non-Hispanic	181,237	171,401	183,402	199,189	83.1	80.5	82.2	82.6
Hispanic	1,079	1,653	1,886	1,780	0.5	0.8	0.8	0.7
Asian or Pacific Islander	1,347	1,367	1,399	1,696	0.6	0.6	0.6	0.7
American Indian/Alaskan Native	403	241	281	313	0.2	0.1	0.1	0.1
Nonresident alien	12,415	14,421	8,843	8,040	5.7	6.8	4.0	3.3
Public	155,085	151,289	158,606	171,969	100.0	100.0	100.0	100.0
White, non-Hispanic	20,586	22,767	26,190	28,893	13.3	15.0	16.5	16.8
Total minority	126,362	118,958	127,202	138,068	81.5	78.6	80.2	80.3
Black, non-Hispanic	124,236	116,845	124,438	134,924	80.1	77.2	78.5	78.5
Hispanic	639	970	1,324	1,428	0.4	0.6	0.8	0.8
Asian and Pacific Islander	1,125	927	1,185	1,421	0.7	0.6	0.7	0.8
American Indian/Alaskan Native	362	216	255	295	0.2	0.1	0.2	0.2
Nonresident alien	8,137	9,564	5,214	5,008	5.2	6.3	3.3	2.9
Private	62,924	61,555	64,644	69,180	100.0	100.0	100.0	100.0
White, non-Hispanic	942	994	1,249	1,238	1.5	1.6	1.9	1.8
Total minority	57,704	55,704	59,766	64,910	91.7	90.5	92.5	93.8
Black, non-Hispanic	57,001	54,556	58,964	64,265	90.6	88.6	91.2	92.9
Hispanic	440	683	562	352	0.7	1.1	0.9	0.5
Asian and Pacific Islander	222	440	214	275	0.4	0.7	0.3	0.4
American Indian/Alaskan Native	41	25	26	18	0.1	[2]	[2]	[2]
Nonresident alien	4,278	4,857	3,629	3,032	6.8	7.9	5.6	4.4

Source: "Fall Enrollment in Historically Black Colleges and Universities by Type and Control of Institution and Race/Ethnicity of Student: 1976 to 1990," Charlene Hoffman, Thomas D. Snyder, and Bill Sonnenberg, *Historically Black Colleges and Universities*, p. 20. Primary source: U.S. Department of Education, National Center for Education Statistics, Higher Education General Information Survey (HEGIS), "Fall Enrollment in Colleges and Universities" surveys; and Integrated Postsecondary Education Data System (IPEDS). "Fall Enrollment" surveys. (This table was prepared January 1992). *Notes:* 1. Preliminary data. 2. Less than 0.05 percent.

★ 327 ★

Historically Black Colleges and Universities

Fall Enrollment in Historically Black Colleges and Universities as a Percentage of Total enrollment, 1976 to 1990

| Year | Enrollment in historically black colleges and universities as a percent of total enrollment in all institutions[2] | | | | | | | | |
| | Total enrollment | Sex of student | | Full-time | | | Part-time | | |
		Men	Women	Total	Men	Women	Total	Men	Women
1976	2.02	1.80	2.27	2.68	2.32	3.13	0.99	0.90	1.08
1977	2.00	1.80	2.22	2.67	2.31	3.09	1.00	0.93	1.06
1978	2.02	1.85	2.20	2.64	2.31	3.01	1.12	1.07	1.17
1979	1.99	1.86	2.12	2.62	2.35	2.92	1.09	1.05	1.13
1980	1.93	1.81	2.04	2.54	2.28	2.83	1.06	1.01	1.10
1981	1.88	1.77	1.98	2.47	2.24	2.72	1.06	1.01	1.10
1982	1.84	1.74	1.93	2.36	2.14	2.61	1.11	1.08	1.13
1983	1.88	1.77	1.98	2.46	2.22	2.71	1.08	1.04	1.11
1984	1.86	1.75	1.95	2.38	2.14	2.62	1.15	1.11	1.17
1985	1.84	1.73	1.95	2.41	2.15	2.68	1.06	1.04	1.08
1986	1.79	1.66	1.90	2.36	2.10	2.62	1.03	0.96	1.08
1987	1.79	1.64	1.92	2.39	2.09	2.69	1.00	0.94	1.04
1988	1.84	1.68	1.97	2.42	2.12	2.72	1.06	0.98	1.11
1989	1.84	1.66	2.00	2.47	2.13	2.79	1.02	0.93	1.09
1990[1]	1.88	1.69	2.04	2.51	2.17	2.84	1.05	0.95	1.12

Source: "Fall Enrollment in Historically Black Colleges and Universities, by Sex and Attendance Status of Student: 1976 to 1990," Charlene Hoffman, Thomas D. Snyder, and Bill Sonnenberg, *Historically Black Colleges and Universities*, p. 17. Primary source: U.S. Department of Education, National Center for Education Statistics, Higher Education General Information Survey (HEGIS), "Fall Enrollment in Colleges and Universities" surveys, and Integrated Postsecondary Education System (IPEDS), "Fall Enrollment" surveys. (This table was prepared January 1992). *Notes:* 1. Preliminary data. 2. Percentages are based on total enrollment with imputations.

★ 328 ★

Historically Black Colleges and Universities

Fall Enrollment in Historically Black Colleges and Universities by Institution and Race/Ethnic Group, 1990

Fall enrollment in historically black colleges and universities, by institution and race/ethnicity: 1990[1].

Institution	State	Total	Black, non-Hispanic	Percent black	White, non-Hispanic	Hispanic	Asian or Pacific Islander	American Indian/ Alaskan Native	Nonresident alien
Total	-	257,804	210,014	81.05	33,722	3,828	1,794	341	8,105
Alabama A&M University	AL	4,886	3,783	77.4	695	9	28	1	370
Alabama State University	AL	4,587	4,469	97.4	71	15	7	1	24
Bishop State Community College	AL	2,057	1,221	59.4	802	1	7	3	23
C.A. Fredd State Technical College	AL	279	262	93.9	17	-	-	-	-
Carver State technical College	AL	535	449	83.9	82	-	3	1	-

[Continued]

★ 328 ★

Fall Enrollment in Historically Black Colleges and Universities by Institution and Race/Ethnic Group, 1990

[Continued]

Institution	State	Total	Black, non-Hispanic	Percent black	White, non-Hispanic	Hispanic	Asian or Pacific Islander	American Indian/ Alaskan Native	Nonresident alien
Concordia College	AL	380	380	100.0	-	-	-	-	-
Daniel Payne College, Birmingham[2]	AL	-	-	-	-	-	-	-	-
J.F. Drake Technical College	AL	852	310	36.4	536	1	4	1	-
Lawson State Community College	AL	1,711	1,650	96.4	61	-	-	-	-
Lomax-Hannon Junior College[2]	AL	-	-	-	-	-	-	-	-
Miles College	AL	584	582	99.7	1	-	1	-	-
Oakwood College	AL	1,266	1,136	89.7	-	-	-	-	130
Selma University[3]	AL	316	315	99.7	-	-	-	1	-
Stillman College	AL	770	763	99.1	2	-	5	-	-
Talladega College	AL	667	648	97.2	14	3	1	1	-
Trenholm State technical College	AL	783	617	78.8	166	-	-	-	-
Tuskegee University	AL	3,510	3,236	92.2	88	92	27	-	67
Arkansas Baptist College	AR	291	285	97.9	6	-	-	-	-
Philander Smith College[3]	AR	594	497	83.7	4	-	15	-	78
Shorter College	AR	135	110	81.5	6	-	-	-	19
University of Arkansas, Pine Bluff	AR	3,672	3,075	83.7	573	6	8	5	5
Delaware State College	DE	2,606	1,553	59.6	946	32	12	10	53
Howard University[3]	DC	11,617	9,535	82.1	160	53	92	8	1,769
University of the District of Columbia	DC	11,990	9,952	83.0	480	360	239	-	959
Bethune-Cookman College	FL	2,342	2,243	95.8	13	8	8	-	70
Edward Waters College[3]	FL	597	509	85.3	-	-	-	-	88
Florida A&M University	FL	8,344	7,227	86.6	781	142	61	-	133
Florida Memorial College	FL	2,251	2,026	90.0	45	113	-	-	67
Albany State College	GA	2,405	1,977	82.2	408	4	9	7	-
Clark Atlanta University[4]	GA	3,508	3,329	-	43	6	21	-	109
Atlanta University[4]	GA	-	-	-	-	-	-	-	-
Clark College[4]	GA	-	-	-	-	-	-	-	-
Fort Valley State College	GA	2,158	2,001	92.7	112	5	3	1	36
Interdominational Theological Center	GA	294	285	96.9	-	-	-	-	9
Morehouse College	GA	2,720	2,716	99.9	-	-	-	-	4
Morehouse School of Medicine	GA	145	120	82.8	13	4	7	-	1
Morris Brown College	GA	2,049	1,953	95.3	-	-	2	-	94
Paine College[3]	GA	606	557	91.9	31	1	1	-	16
Savannah State College	GA	2,319	1,967	84.8	272	1	4	-	75
Spelman College	GA	1,710	1,678	98.1	2	-	-	-	30
Kentucky State University	KY	2,506	1,125	44.9	1,330	8	6	4	33
Dillard University	LA	1,998	1,998	100.0	-	-	-	-	-
Grambling State University	LA	6,485	6,155	94.9	228	3	2	15	82
Southern University and A&M College, Baton Rouge	LA	8,941	8,332	93.2	369	24	17	3	196
Southern University, New Orleans	LA	4,064	3,710	91.3	228	13	33	2	78
Southern University, Shreveport-Bossier City Campus	LA	1,020	933	91.5	85	-	-	-	2
Xavier University of Louisiana	LA	2,943	2,670	90.7	161	17	16	-	79
Bowie State University	MD	4,188	2,716	64.9	1,227	34	143	9	59
Coppin State College	MD	2,578	2,337	90.7	121	11	31	6	72
Morgan State University	MD	4,693	4,306	91.8	178	13	24	11	161

[Continued]

★ 328 ★

Fall Enrollment in Historically Black Colleges and Universities by Institution and Race/Ethnic Group, 1990
[Continued]

Institution	State	Total	Black, non-Hispanic	Percent black	White, non-Hispanic	Hispanic	Asian or Pacific Islander	American Indian/ Alaskan Native	Nonresident alien
University of Maryland, Eastern Shore	MD	2,067	1,439	69.6	466	13	21	3	125
Lewis College of Business	MI	233	233	100.0	-	-	-	-	-
Alcorn State University	MS	2,863	2,702	94.4	146	13	2	-	-
Coahoma Junior College	MS	1,351	1,315	97.3	36	-	-	-	-
Hinds Community College, Utica Campus	MS	668	646	96.7	22	-	-	-	-
Jackson State University	MS	6,837	6,294	92.1	272	8	137	7	119
Mary Holmes College	MS	742	734	98.9	-	-	-	-	8
Mississippi Industrial College[2]	MS	-	-	-	-	-	-	-	-
Mississippi Valley State University	MS	1,873	1,863	99.5	9	-	-	-	1
Natchez Junior College[5]	MS	-	-	-	-	-	-	-	-
Prentiss Institute[2]	MS	-	-	-	-	-	-	-	-
Rust College	MS	1,021	987	96.7	16	-	-	-	18
Tougaloo College	MS	956	953	99.7	2	-	1	-	-
Harris-Stowe State College	MO	1,973	1,512	76.6	418	6	14	4	19
Lincoln University	MO	3,619	834	23.0	2,651	17	15	23	79
Barber-Scotia College[3]	NC	422	417	98.8	1	-	-	-	4
Bennett College	NC	586	565	96.4	2	-	-	-	19
Elizabeth City State University	NC	1,746	1,314	75.3	412	4	5	2	9
Fayetteville State University	NC	3,337	2,202	66.0	1,019	49	34	32	1
Johnson College Smith University	NC	1,182	1,182	100.0	-	-	-	-	-
Livingstone College	NC	682	661	96.9	3	-	-	-	18
North Carolina Agricultural and Technical State University	NC	6,595	5,583	84.7	838	12	41	12	109
North Carolina Central University	NC	5,482	4,566	83.3	829	8	32	15	32
St. Augustine's College	NC	1,900	1,896	99.8	4	-	-	-	-
Shaw University	NC	1,846	1,696	91.9	143	2	1	4	-
Winston-Salem State University	NC	2,517	2,066	82.1	431	3	11	6	-
Central State University	OH	2,886	2,583	89.5	142	1	1	9	150
Wilberforce University	OH	809	799	98.8	3	-	1	-	6
Langston University	OK	2,792	1,440	51.6	1,250	23	15	55	9
Cheyney University of Pennsylvania	PA	1,738	1,622	93.3	57	15	11	4	29
Lincoln University	PA	1,374	1,244	90.5	85	17	1	-	27
Allen University[3]	SC	233	233	100.0	-	-	-	-	-
Benedict College	SC	1,478	1,466	99.2	2	-	-	-	10
Claflin College	SC	913	903	98.9	1	1	1	-	7
Clinton Junior College	SC	88	88	100.0	-	-	-	-	-
Denmark Technical College	SC	617	601	97.4	12	2	1	-	1
Friendship College[2]	SC	-	-	-	-	-	-	-	-
Morris College	SC	760	759	99.9	1	-	-	-	-
South Carolina State College	SC	4,822	4,538	94.1	257	2	20	-	5
Voorhees College	SC	566	560	98.9	-	-	-	-	6
Fisk University	TN	911	891	97.8	1	-	-	-	19
Knoxville College[6]	TN	1,266	1,254	99.1	7	1	4	-	-
Lane College	TN	530	524	98.9	4	-	1	-	1
Le Moyne-Owen College	TN	1,066	1,054	98.9	2	-	-	-	10
Meharry Medical College	TN	623	490	78.7	40	10	42	-	41
Morristown College[6]	TN	-	-	-	-	-	-	-	-

[Continued]

★ 328 ★

Fall Enrollment in Historically Black Colleges and Universities by Institution and Race/Ethnic Group, 1990

[Continued]

Institution	State	Total	Black, non-Hispanic	Percent black	White, non-Hispanic	Hispanic	Asian or Pacific Islander	American Indian/ Alaskan Native	Nonresident alien
Tennessee State University	TN	7,393	4,588	62.1	2,549	36	210	10	-
Bishop College[2]	TX	-	-	-	-	-	-	-	-
Huston-Tillotson College	TX	714	594	83.2	14	27	16	-	63
Jarvis Christian College	TX	598	586	98.0	10	2	-	-	-
Paul Quinn College	TX	997	954	95.7	30	5	-	-	8
Prairie View A&M University	TX	4,990	4,183	83.8	445	50	37	1	274
St. Philip's College	TX	5,204	1,276	24.5	1,766	2,044	83	23	12
Southwestern Christian College	TX	225	196	87.1	1	-	-	-	28
Texas College	TX	478	456	95.4	2	-	-	-	20
Texas Southern University	TX	9,427	7,317	77.6	308	367	100	1	1,334
Wiley College	TX	463	448	96.8	4	1	-	-	10
Hampton University[3]	VA	5,305	4,846	91.3	320	4	10	3	122
Norfolk State University	VA	8,008	6,673	83.3	1,151	35	37	16	96
St. Paul's College	VA	574	540	94.1	28	2	1	-	3
Virginia College[2]	VA	-	-	-	-	-	-	-	-
Virginia State University	VA	3,988	3,618	90.7	312	18	12	3	25
Virginia Union University	VA	1,298	1,274	98.2	14	-	1	1	8
Bluefield State College	WV	2,702	189	7.0	2,482	5	8	5	13
West Virginia State College	WV	4,834	588	12.2	4,184	13	21	10	18
University of the Virgin Islands, St. Thomas Campus	VI	1,684	1,281	76.1	161	33	9	2	198

Source: "Fall Enrollment in Black Colleges and Universities by Institution and Race/Ethnicity: 1990," Charlene Hoffman, Thomas D. Snyder, and Bill Sonnenberg, *Historically Black Colleges and Universities*, pp. 32-33. Primary source: U.S. Department of Education, National Center for Education Statistics, Integrated Postsecondary Education Data System (IPEDS). "Fall Enrollment, 1990" survey. (This table was prepared January 1992). *Notes:* - Data not reported or not applicable. 1. Preliminary data. 2. School closed. 3. Estimate based on 1988 data. 4. Atlanta University and Clark College merged July 1, 1989, and became Clark Atlanta University. 5. School no longer eligible for listing. 6. Knoxville College has two campuses now. In 1989, Morristown College was annexed by Knoxville College.

★ 329 ★

Historically Black Colleges and Universities

Fall Enrollment in Historically Black Colleges and Universities, 1976 to 1990

Year	Total enrollment	Sex of student		Full-time			Part-time		
		Men	Women	Total	Men	Women	Total	Men	Women
1976	222,613	104,669	117,944	180,059	85,794	94,265	42,554	18,875	23,679
1977	226,062	104,178	121,884	181,244	84,272	96,972	44,818	19,906	24,912
1978	227,797	104,216	123,581	176,243	81,661	94,582	51,554	22,555	28,999
1979	130,124	105,494	124,630	177,925	83,118	94,807	52,199	22,376	29,823
1980	233,557	106,387	127,170	180,521	84,222	96,299	53,036	22,165	30,871
1981	232,460	106,033	126,427	177,448	83,096	94,352	55,012	22,937	32,075
1982	228,371	104,897	123,474	170,611	80,231	90,380	57,760	24,666	33,094
1983	234,446	106,884	127,562	178,265	83,389	94,876	56,181	23,495	32,686

[Continued]

★ 329 ★

Fall Enrollment in Historically Black Colleges and Universities, 1976 to 1990
[Continued]

Year	Total enrollment	Sex of student		Full-time			Part-time		
		Men	Women	Total	Men	Women	Total	Men	Women
1984	227,519	102,823	124,696	168,616	78,202	90,414	58,903	24,621	34,282
1985	225,801	100,698	125,103	170,798	77,746	93,052	55,003	22,952	32,051
1986	223,275	97,523	125,752	167,825	75,545	92,280	55,450	21,978	33,472
1987	227,994	97,085	130,909	172,752	75,336	97,416	55,242	21,749	33,493
1988	239,755	100,561	139,194	180,215	77,540	102,675	59,540	23,021	36,519
1989	249,096	102,484	146,612	189,030	79,747	109,283	60,066	22,737	37,329
1990[1]	257,804	105,538	152,266	195,437	82,240	113,197	62,367	23,298	39,069

Source: "Fall Enrollment in Historically Black Colleges and Universities, by Sex and Attendance Status of Student: 1976 to 1990," Charlene Hoffman, Thomas D. Snyder, and Bill Sonnenberg, *Historically Black Colleges and Universities*, p. 17. Primary source: U.S. Department of Education, National Center for Education Statistics, Higher Education General Information Survey (HEGIS), "Fall Enrollment in Colleges and Universities" surveys, and Integrated Postsecondary Education System (IPEDS), "Fall Enrollment" surveys. (This table was prepared January 1992). *Note:* 1. Preliminary data.

★ 330 ★
Historically Black Colleges and Universities

Fall Enrollment in Historically Black Colleges and Universities, 1976-1991

Year	Total enrollment	Type of institution		Public institutions			Private institutions		
		4-year	2-year	Total	4-year	2-year	Total	4-year	2-year
1976	222,613	206,676	15,937	156,836	143,528	13,308	65,777	63,148	2,629
1977	226,062	209,898	16,164	158,823	145,450	13,373	67,239	64,448	2,791
1978	227,797	211,651	16,146	163,237	150,168	13,069	64,560	61,483	3,077
1979	230,124	214,147	15,977	166,315	153,139	13,176	63,809	61,008	2,801
1980	233,557	218,009	15,548	168,217	155,085	13,132	65,340	62,924	2,416
1981	232,460	217,152	15,308	166,991	154,269	12,722	65,469	62,883	2,586
1982	228,371	212,017	16,354	165,871	151,472	14,399	62,500	60,545	1,955
1983	234,446	217,909	16,537	170,051	155,665	14,386	64,395	62,244	2,151
1984	227,519	212,844	14,675	164,116	151,289	12,827	63,403	61,555	1,848
1985	225,801	210,648	15,153	163,677	150,002	13,675	62,124	60,646	1,478
1986	223,275	207,231	16,044	162,048	147,631	14,417	61,227	59,600	1,627
1987	227,994	211,654	16,340	165,486	150,560	14,926	62,508	61,094	1,414
1988	239,755	223,250	16,505	173,672	158,606	15,066	66,083	64,644	1,439
1989	249,096	232,890	16,206	181,151	166,481	14,670	67,945	66,409	1,536
1990[1]	257,152	240,497	16,655	187,046	171,969	15,077	70,106	68,528	1,578
1991[2]	269,280	252,038	17,242	197,847	182,204	15,643	71,433	69,834	1,599

Source: "Fall Enrollment in Historically Black Colleges and Universities, by Type and Control of Institution: 1976 to 1991." *Digest of Educational Statistics*, 1993, p. 224. Primary source: U.S. Department of Education, National Center for Education Statistics, Higher Education General Information Survey (HEGIS), "Fall Enrollment in Colleges and Universities", and Integrated Postsecondary Education Data System (IPEDS), "Fall Enrollment" survey. (This table was prepared April 1993.) *Notes:* 1. Revised from previously published data. 2. Preliminary data.

★ 331 ★

Historically Black Colleges and Universities

Fall Enrollment in Private Historically Black Colleges and Universities by Sex and State, 1990

Fall enrollment in historically black colleges and universities, by type and control of institution, number of institutions, sex, and state: 1990[1].

State	Private 4-year			Private 2-year		
	Number of institutions	Men	Women	Number of institutions	Men	Women
Total	49	28,435	40,745	5	562	1,016
Alabama	6	3,169	3,944	1	96	284
Arkansas	2	399	486	1	58	77
Delaware	0	-	-	0	-	-
District of Columbia	1	4,935	6,682	0	-	-
Florida	3	1,881	3,309	0	-	-
Georgia	7	5,227	5,805	0	-	-
Kentucky	0	-	-	0	-	-
Louisiana	2	1,401	3,540	0	-	-
Maryland	0	-	-	0	-	-
Michigan	0	-	-	1	52	181
Mississippi	2	660	1,317	1	298	444
Missouri	0	-	-	0	-	-
North Carolina	6	2,696	3,922	0	-	-
Ohio	1	268	541	0	-	-
Oklahoma	0	-	-	0	-	-
Pennsylvania	0	-	-	0	-	-
South Carolina	5	1,478	2,472	1	58	30
Tennessee	5	1,862	2,534	0	-	-
Texas	6	1,694	1,781	0	-	-
Virginia	3	2,765	4,412	0	-	-
West Virginia	0	-	-	0	-	-
Virgin Islands	0	-	-	0	-	-

Source: "Fall Enrollment in Historically Black Colleges and Universities by Type and Control of Institutions, Sex, and State: 1990," Charlene Hoffman, Thomas D. Snyder, and Bill Sonnenberg, *Historically Black Colleges and Universities*, p. 34. Primary source: U.S. Department of Education, National Center for Education Statistics, Integrated Postsecondary Education Data System (IPEDS). "Fall Enrollment, 1990" survey. (This table was prepared January 1992). *Notes:* - Not applicable. 1. Preliminary data.

★ 332 ★

Historically Black Colleges and Universities

Fall Enrollment in Public Historically Black Colleges and Universities by Sex and State, 1990

Fall enrollment in historically black colleges and universities, by type and control of institution, number of institutions, sex, and state: 1990[1].

State	Public 4-year			Public 2-year		
	Number of institutions	Men	Women	Number of institutions	Men	Women
Total	40	70,220	101,749	11	6,321	8,756
Alabama	2	4,069	5,404	6	2,435	3,782
Arkansas	1	1,435	2,237	0	-	-
Delaware	1	1,137	1,469	0	-	-
District of Columbia	1	5,068	6,922	0	-	-
Florida	1	3,606	4,738	0	-	-
Georgia	3	2,679	4,203	0	-	-
Kentucky	1	1,039	1,467	0	-	-
Louisiana	3	7,639	11,851	1	303	717
Maryland	4	5,524	8,002	0	-	-
Michigan	0	-	-	0	-	-
Mississippi	3	4,793	6,780	2	687	1,332
Missouri	2	1,928	3,664	0	-	-
North Carolina	5	8,158	11,519	0	-	-
Ohio	1	1,343	1,543	0	-	-
Oklahoma	1	1,173	1,619	0	-	-
Pennsylvania	2	1,353	1,759	0	-	-
South Carolina	1	1,976	2,846	1	266	351
Tennessee	1	2,996	4,397	0	-	-
Texas	2	6,224	8,193	1	2,630	2,574
Virginia	2	4,528	7,468	0	-	-
West Virginia	2	3,110	4,426	0	-	-
Virgin Islands	1	442	1,242	0	-	-

Source: "Fall Enrollment in Historically Black Colleges and Universities by Type and Control of Institutions, Sex, and State: 1990," Charlene Hoffman, Thomas D. Snyder, and Bill Sonnenberg, *Historically Black Colleges and Universities*, p. 34. Primary source: U.S. Department of Education, National Center for Education Statistics, Integrated Postsecondary Education Data System (IPEDS). "Fall Enrollment, 1990" survey. (This table was prepared January 1992). *Notes:* - Not applicable. 1. Preliminary data.

★ 333 ★

Historically Black Colleges and Universities

Fall Enrollment in Two-Year Historically Black Colleges and Universities, 1980 to 1990

Type and control of institution and race/ethnicity of student	Enrollment				Percent distribution by type and control			
	1980	1984	1988	1990[1]	1980	1984	1988	1990[1]
Total	15,548	14,675	16,505	16,655	100.0	100.0	100.0	100.0
White, non-Hispanic	2,834	2,441	3,294	3,591	18.2	16.6	20.0	21.6
Total minority	12,535	12,107	13,118	12,999	80.6	82.5	79.5	78.0
Black, non-Hispanic	9,752	9,402	10,749	10,825	62.7	64.1	65.1	65.0
Hispanic	2,692	2,577	2,236	2,048	17.3	17.6	13.5	12.3
Asian or Pacific Islander	50	91	104	98	0.3	0.6	0.6	0.6
American Indian/Alaskan Native	41	37	29	28	0.3	0.3	0.2	0.2
Nonresident alien	179	127	93	65	1.2	0.9	0.6	0.4
Public	13,132	12,827	15,066	15,077	100.0	100.0	100.0	100.0
White, non-Hispanic	2,822	2,439	3,279	3,585	21.5	19.0	21.8	23.8
Total minority	10,203	10,303	11,712	11,454	77.7	80.3	77.7	76.0
Black, non-Hispanic	7,425	7,600	9,348	9,280	56.5	59.3	62.0	61.6
Hispanic	2,688	2,577	2,236	2,048	20.5	20.1	14.8	13.6
Asian and Pacific Islander	49	89	99	98	0.4	0.7	0.7	0.6
American Indian/Alaskan Native	41	37	29	28	0.3	0.3	0.2	0.2
Nonresident alien	107	85	75	38	0.8	0.7	0.5	0.3
Private	2,416	1,848	1,439	1,578	100.0	100.0	100.0	100.0
White, non-Hispanic	12	2	15	6	0.5	0.1	1.0	0.4
Total minority	2,332	1,804	1,406	1,545	96.5	97.6	97.7	97.9
Black, non-Hispanic	2,327	1,802	1,401	1,545	96.3	97.5	97.4	97.9
Hispanic	4	-	-	-	0.2	-	-	-
Asian and Pacific Islander	1	2	5	-	[2]	0.1	0.3	-
American Indian/Alaskan Native	-	-	-	-	-	-	-	-
Nonresident alien	72	42	18	27	3.0	2.3	1.3	1.7

Source: "Fall Enrollment in Historically Black Colleges and Universities by Type and Control of Institution and Race/Ethnicity of Student: 1976 to 1990," Charlene Hoffman, Thomas D. Snyder, and Bill Sonnenberg, *Historically Black Colleges and Universities*, p. 20. Primary source: U.S. Department of Education, National Center for Education Statistics, Higher Education General Information Survey (HEGIS), "Fall Enrollment in Colleges and Universities" surveys; and Integrated Postsecondary Education Data System (IPEDS). "Fall Enrollment" surveys. (This table was prepared January 1992). *Notes:* - Data not reported or not applicable. 1. Preliminary data. 2. Less than 0.05 percent.

★ 334 ★

Historically Black Colleges and Universities

Fall Enrollment, Degrees Conferred, and Expenditures in Historically Black Colleges and Universities, 1991

Institution	Type and control[1]	Enrollment, 1991		Degrees conferred, 1990-91					Expenditures, 1990-91 (in thousands)	
		Total	Black	Associate	Bachelor's	Master's	Doctor's	First-professional	Current-fund expenditures	Educational and general expenditures
Total	-	269,280	219,168	2,664	21,627	4,145	200	812	$2,784,795	$2,315,631
Alabama A&M University, AL[2]	1	5,215	4,125	2	344	245	2	-	49,573	44,279
Alabama State University, AL	1	4,822	4,733	-	252	129	-	-	34,575	28,870
Bishop State Community College, AL	2	2,144	1,258	186	-	-	-	-	7,734	7,348
C.A. Fredd State Technical College, AL	2	297	272	-	-	-	-	-	1,898	1,883
Carver State Technical College, AL	2	543	466	19	-	-	-	-	3,281	3,236
Concordia College, AL	4	383	371	62	-	-	-	-	2,521	2,245
J.F. Drake Technical College, AL	2	805	323	63	-	-	-	-	3,620	3,489
Lawson State Community College, AL	2	1,959	1,911	93	-	-	-	-	7,819	7,573
Miles College, AL	3	732	731	-	62	-	-	-	5,868	5,457
Oakwood College, AL	3	1,244	1,110	23	143	-	-	-	16,236	12,814
Selma University, AL	3	219	219	16	6	-	-	-	3,272	2,643
Stillman College, AL	3	821	817	-	112	-	-	-	9,665	8,147
Talladega College, AL	3	751	707	-	95	-	-	-	8,111	7,348
Trenholm State Technical College, AL	2	746	552	67	-	-	-	-	5,099	4,911
Tuskegee University, AL[2]	3	3,749	3,325	-	377	33	-	55	58,212	51,726
Arkansas Baptist College, AR	3	306	303	-	31	-	-	-	1,528	1,413
Philander Smith College, AR	3	776	732	-	62	-	-	-	4,554	3,950
Shorter College, AR	4	133	107	16	-	-	-	-	960	884
University of Arkansas, Pine Bluff, AR[2]	1	3,459	2,814	2	312	-	-	-	27,936	25,847
Delaware State College, DE[2]	1	2,882	1,760	-	229	89	-	-	30,663	27,019
Howard University, DC	3	10,724	8,996	-	1,384	281	72	179	449,856	284,208
University of the District of Columbia, DC[2]	1	11,422	9,373	213	564	138	-	-	97,556	96,411
Bethune-Cookman College, FL	3	2,273	2,202	-	326	-	-	-	24,518	20,768
Edward Waters College, FL	3	657	560	-	79	-	-	-	6,849	6,089
Florida A&M University, FL[2]	1	9,196	8,040	8	586	102	5	24	84,883	77,118
Florida Memorial College, FL	3	1,530	1,282	-	160	-	-	-	14,737	13,618
Albany State College, GA	1	2,746	2,299	-	204	38	-	-	20,168	17,526
Clark Atlanta University, GA	3	3,993	3,800	-	246	241	40	-	53,439	51,498
Fort Valley State College, GA[2]	1	2,368	2,171	4	175	57	-	-	23,345	20,679
Interdenominational Theological Center, GA	3	330	297	-	-	-	4	46	4,131	4,039
Morehouse College, GA	3	2,992	2,955	-	339	-	-	-	29,521	25,676
Morehouse School of Medicine, GA	3	144	121	-	-	-	-	31	29,255	29,255
Morris Brown College, GA	3	2,050	1,913	-	145	-	-	-	20,760	18,278
Paine College, GA	3	582	570	-	63	-	-	-	7,194	6,366
Savannah State College, GA	1	2,624	2,330	1	168	3	-	-	20,346	17,738
Spelman College, GA	3	1,905	1,862	-	383	-	-	-	26,659	21,266
Kentucky State University, KY[2]	1	2,533	1,170	43	149	32	-	-	29,563	26,585
Dillard University, LA	3	1,670	1,665	-	227	-	-	-	15,345	13,309
Grambling State University, LA	1	7,030	6,707	74	576	97	2	-	50,880	40,372
Southern University and A&M College, Baton Rouge, LA[2]	1	9,914	9,163	32	784	166	-	130	70,400	60,673
Southern University, New Orleans, LA	1	4,255	3,926	40	303	44	-	-	17,616	16,969
Southern University, Shreveport-Bossier, City Campus, LA	2	932	831	97	-	-	-	-	7,213	6,945
Xavier University of Louisiana, LA	3	3,071	2,782	-	311	88	-	4	30,830	28,302
Bowie State University, MD	1	4,434	2,930	-	266	198	-	-	27,245	23,537
Coppin State College, MD	1	2,816	2,560	-	208	61	-	-	21,749	20,554
Morgan State University, MD	1	5,034	4,674	-	497	96	4	-	53,282	47,518
University of Maryland, Eastern Shore, MD[2]	1	2,397	1,651	-	159	24	-	-	26,798	23,968
Lewis College of Business, MI	4	274	273	39	-	-	-	-	1,454	1,432
Alcorn State University, MS[2]	1	3,244	3,065	25	269	57	-	-	28,893	23,459
Coahoma Community College, MS	2	1,422	1,387	135	-	-	-	-	8,297	7,555
Hinds Community College, Utica Campus, MS	2	683	658	74	-	-	-	-	-	-
Jackson State University, MS	1	6,639	6,119	-	551	174	4	-	56,918	46,816
Mary Holmes College, MS	4	733	639	111	-	-	-	-	7,054	6,457
Mississippi Valley State University, MS	1	2,059	2,050	-	184	-	-	-	18,470	16,146
Rust College, MS	3	1,075	1,017	2	123	-	-	-	10,159	8,483
Tougaloo College, MS	3	1,003	999	7	98	-	-	-	9,391	8,680
Harris-Stowe State College, MO	1	1,980	1,493	-	62	-	-	-	7,630	7,630
Lincoln University, MO[2]	1	4,101	1,038	88	209	35	-	-	25,628	22,686
Barber-Scotia College, NC	3	610	602	-	40	-	-	-	5,318	4,699

[Continued]

★ 334 ★

Fall Enrollment, Degrees Conferred, and Expenditures in Historically Black Colleges and Universities, 1991

[Continued]

Institution	Type and control[1]	Enrollment, 1991		Degrees conferred, 1990-91					Expenditures, 1990-91 (in thousands)	
		Total	Black	Associate	Bachelor's	Master's	Doctor's	First-professional	Current-fund expenditures	Educational and general expenditures
Bennett College, NC	3	568	549	-	65	-	-	-	8,931	8,036
Elizabeth City State University, NC	1	1,773	1,327	-	349	-	-	-	22,634	19,219
Fayetteville State University, NC	1	3,736	2,357	9	255	96	-	-	28,727	23,951
Johnson C. Smith University, NC	3	1,256	1,256	-	144	-	-	-	15,789	13,660
Livingstone College, NC	3	646	640	-	77	-	-	9	8,444	6,760
North Carolina Agricultural and Technical State University, NC[2]	1	7,199	6,112	-	750	219	-	-	75,542	63,927
North Carolina Central University, NC	1	5,385	4,487	-	506	169	-	66	47,894	38,688
St. Augustine's College, NC	3	1,907	1,903	-	188	-	-	-	22,270	18,227
Shaw University, NC	3	2,149	1,980	25	230	-	-	-	17,015	15,736
Winston-Salem State University, NC	1	2,637	2,107	-	331	-	-	-	26,651	22,024
Central State University, OH	1	3,266	2,898	-	254	-	-	-	35,151	30,028
Wilberforce University, OH	3	809	788	-	107	-	-	-	10,887	8,899
Langston University, OK[2]	1	3,112	1,646	-	368	7	-	-	19,295	15,730
Cheyney University of Pennsylvania, PA	1	1,477	1,371	-	133	67	-	-	21,442	18,775
Lincoln University, PA	1	1,458	1,309	-	208	92	-	-	22,362	19,353
Allen University, SC	3	239	239	-	55	-	-	-	3,049	2,778
Benedict College, SC	3	1,422	1,377	-	152	-	-	-	16,492	14,808
Claflin College, SC	3	934	924	-	80	-	-	-	8,394	6,990
Clinton Junior College, SC	4	76	74	39	-	-	-	-	546	504
Denmark Technical College, SC	2	711	685	81	-	-	-	-	5,834	5,027
Morris College, SC	3	701	700	-	109	-	-	-	7,933	6,945
South Carolina State College, SC[2]	1	5,145	4,824	-	486	59	16	-	43,584	34,498
Voorhees College, SC	3	613	605	-	74	-	-	-	6,040	5,336
Fisk University, TN	3	838	833	-	134	8	-	-	11,005	9,365
Knoxville College, TN	3	1,177	1,162	6	30	-	-	-	15,006	13,541
Lane College, TN	3	562	561	-	48	-	-	-	5,801	4,711
Le Moyne-Owen College, TN	3	1,177	1,173	-	100	-	-	-	7,544	7,205
Meharry Medical College, TN	3	606	495	-	-	6	3	99	66,744	39,677
Tennessee State University, TN[2]	1	7,405	4,598	171	516	206	35	-	56,555	52,827
Huston-Tillotson College, TX	3	653	508	-	54	-	-	-	3,355	2,723
Jarvis Christian College, TX	3	551	544	-	78	-	-	-	7,935	7,043
Paul Quinn College, TX	3	934	897	-	58	-	-	-	7,609	6,938
Prairie View A&M University, TX[2]	1	5,590	4,771	-	528	168	-	-	54,373	45,108
St. Philip's College, TX	2	5,401	1,110	339	-	-	-	-	23,998	23,920
Southwestern Christian College, TX	3	244	206	36	-	-	-	-	2,611	2,275
Texas College, TX	3	488	465	-	40	-	-	-	4,358	3,755
Texas Southern University, TX	1	10,274	8,132	-	514	259	13	140	62,195	55,318
Wiley College, TX	3	438	423	-	24	-	-	-	4,902	4,462
Hampton University, VA	3	5,704	5,095	-	848	109	-	-	59,812	52,284
Norfolk State University, VA	1	8,298	6,974	53	536	161	-	-	55,134	47,080
St. Paul's College, VA	3	651	617	-	83	-	-	-	7,354	6,832
Virginia State University, VA[2]	1	4,589	4,161	-	356	60	-	-	47,836	35,795
Virginia Union University, VA	3	1,360	1,333	-	90	-	-	29	13,958	12,051
Bluefield State College, WV	1	2,907	177	194	237	-	-	-	9,771	8,846
West Virginia State College, WV	1	4,986	583	119	453	-	-	-	19,776	16,533
University of the Virgin Islands, St. Thomas Campus, VI[2]	1	1,797	1,386	50	116	31	-	-	35,781	33,412

Source: "Fall Enrollment, Degrees Conferred, and Expenditures in Historically Black Colleges and Universities, by Institution, 1976 to 1991." *Digest of Educational Statistics,* 1993, pp. 223-324. Primary source: U.S. Department of Education, National Center for Education Statistics, Integrated Postsecondary Education Data System (IPEDS), "Fall Enrollment, 1991," "Completions, 1990-91," and "Finance, 1990-91" surveys. (This table was prepared April 1993.)
Notes: - Data not reported or not applicable. 1. 1=public 4-year; 2=public 2-year; 3=private 4-year; and 4=private 2- year. 2. Land-grant institution.

★ 335 ★
Historically Black Colleges and Universities
Fall Graduate Enrollment in Historically Black Colleges and Universities by Race/Ethnic Group, 1976 to 1990

Type and control of institution and race/ethnicity of student	Year				Percentage distribution by level enrolled			
	1976	1980	1988	1990[1]	1976	1980	1988	1990[1]
Graduate								
Total	18,287	17,582	19,768	20,716	100.0	100.0	100.0	100.0
Men	7,759	7,358	7,452	7,235	42.4	41.8	37.7	34.9
Women	10,528	10,224	12,316	13,481	57.6	58.2	62.3	65.1
White, non-Hispanic	4,008	3,170	5,059	6,079	21.9	18.0	25.6	29.3
Men	1,827	1,323	1,685	1,923	10.0	7.5	8.5	9.3
Women	2,181	1,847	3,374	4,156	11.9	10.5	17.1	20.1
Black, non-Hispanic	12,740	12,024	12,074	12,060	69.7	68.4	61.1	58.2
Men	4,792	4,277	3,905	3,646	26.2	24.3	19.8	17.6
Women	7,948	7,747	8,169	8,414	43.5	44.1	41.3	40.6
Hispanic	46	95	177	214	0.3	0.5	0.9	1.0
Men	31	48	70	82	0.2	0.3	0.4	0.4
Women	15	47	107	132	0.1	0.3	0.5	0.6
Asian or Pacific Islander	79	314	265	459	0.4	1.8	1.3	2.2
Men	52	230	163	283	0.3	1.3	0.8	1.4
Women	27	84	102	176	0.1	0.5	0.5	0.8
American Indian/Alaskan Native	11	22	46	39	0.1	0.1	0.2	0.2
Men	4	7	12	11	[2]	[2]	0.1	0.1
Women	7	15	34	28	[2]	0.1	0.2	0.1
Nonresident alien	1,403	1,957	2,147	1,865	7.7	11.1	10.9	9.0
Men	1,053	1,473	1,617	1,290	5.8	8.4	8.2	6.2
Women	350	484	530	575	1.9	2.8	2.7	2.8
First-professional								
Total	2,958	3,699	3,352	3,576	100.0	100.0	100.0	100.0
Men	2,171	2,396	1,967	1,984	73.4	64.8	58.7	55.5
Women	787	1,303	1,385	1,592	26.6	35.2	41.3	44.5
White, non-Hispanic	467	503	597	634	15.8	13.6	17.8	17.7
Men	398	353	398	417	13.5	9.5	11.9	11.7
Women	69	150	199	217	2.3	4.1	5.9	6.1
Black, non-Hispanic	2,154	2,883	2,263	2,399	72.8	77.9	67.5	67.1
Men	1,508	1,796	1,225	1,203	51.0	48.6	36.5	33.6
Women	646	1,087	1,038	1,196	21.8	29.4	31.0	33.4
Hispanic	110	80	153	189	3.7	2.2	4.6	5.3
Men	92	64	109	122	3.1	1.7	3.3	3.4
Women	18	16	44	67	0.6	0.4	1.3	1.9
Asian or Pacific Islander	16	43	70	75	0.5	1.2	2.1	2.1
Men	12	34	51	51	0.4	0.9	1.5	1.4
Women	4	9	19	24	0.1	0.2	0.6	0.7
American Indian/Alaskan Native	12	12	13	3	0.4	0.3	0.4	0.1
Men	7	9	9	2	0.2	0.2	0.3	0.1
Women	5	3	4	1	0.2	0.1	0.1	[2]
Nonresident alien	199	178	256	276	6.7	4.8	7.9	7.7

[Continued]

★ 335 ★

Fall Graduate Enrollment in Historically Black Colleges and Universities by Race/Ethnic Group, 1976 to 1990

[Continued]

Type and control of institution and race/ethnicity of student	Year				Percentage distribution by level enrolled			
	1976	1980	1988	1990[1]	1976	1980	1988	1990[1]
Men	154	140	175	189	5.2	3.8	5.2	5.3
Women	45	38	81	87	1.5	1.0	2.4	2.4

Source: "Fall Enrollment in Black Colleges and Universities by Level, Sex, and Race/Ethnicity of Student: 1976 to 1990," Charlene Hoffman, Thomas D. Snyder, and Bill Sonnenberg, *Historically Black Colleges and Universities,* p. 22. Primary source: U.S. Department of Education, National Center for Education Statistics, Higher Education General Information Survey (HEGIS), "Fall Enrollment in Colleges and Universities" surveys; and Integrated Postsecondary Education Data System (IPEDS). "Fall Enrollment" surveys. (This table was prepared January 1992). *Note:* 1. Preliminary data.

★ 336 ★

Historically Black Colleges and Universities

Fall Undergraduate Enrollment in Historically Black Colleges and Universities by Race/Ethnic Group, 1976 to 1990

Type and control of institution and race/ethnicity of student	Year				Percentage distribution by level enrolled			
	1976	1980	1988	1990[1]	1976	1980	1988	1990[1]
Undergraduate								
Total	201,368	212,276	216,635	233,512	100.0	100.0	100.0	100.0
Men	94,739	96,633	91,142	96,319	47.0	45.5	42.1	41.2
Women	106,629	115,643	125,493	137,193	53.0	54.5	57.9	58.8
White, non-Hispanic	16,565	20,689	25,077	27,009	8.2	9.7	11.6	11.6
Men	9,429	9,962	11,077	11,608	4.7	4.7	5.1	5.0
Women	7,136	10,727	14,000	15,401	3.5	5.1	6.5	6.6
Black, non-Hispanic	175,411	176,082	179,814	195,555	87.1	82.9	83.0	83.7
Men	78,192	75,745	73,138	78,687	38.8	35.7	33.8	33.7
Women	97,219	100,337	106,676	116,868	48.3	47.3	49.2	50.0
Hispanic	3,286	3,596	3,792	3,425	1.6	1.7	1.8	1.5
Men	2,466	2,387	2,054	1,748	1.2	1.1	0.9	0.7
Women	820	1,209	1,738	1,677	0.4	0.6	0.8	0.7
Asian or Pacific Islander	554	1,040	1,168	1,260	0.3	0.5	0.5	0.5
Men	394	683	716	742	0.2	0.3	0.3	0.3
Women	160	357	452	518	0.1	0.2	0.2	0.2
American Indian/Alaskan Native	207	410	251	299	0.1	0.2	0.1	0.1
Men	115	205	109	121	0.1	0.1	0.1	0.1
Women	92	205	142	178	[2]	0.1	0.1	0.1
Nonresident alien	5,345	10,459	6,533	5,964	2.7	4.9	3.0	2.6
Men	4,143	7,651	4,048	3,413	2.1	3.6	1.9	1.5
Women	1,202	2,808	2,485	2,551	0.6	1.3	1.1	1.1

Source: "Fall Enrollment in Black Colleges and Universities by Level, Sex, and Race/Ethnicity of Student: 1976 to 1990," Charlene Hoffman, Thomas D. Snyder, and Bill Sonnenberg, *Historically Black Colleges and Universities,* p. 22. Primary source: U.S. Department of Education, National Center for Education Statistics, Higher Education General Information Survey (HEGIS), "Fall Enrollment in Colleges and Universities" surveys; and Integrated Postsecondary Education Data System (IPEDS). "Fall Enrollment" surveys. (This table was prepared January 1992). *Note:* 1. Preliminary data.

★ 337 ★

Historically Black Colleges and Universities

First-Professional Degrees Conferred by Historically Black Colleges and Universities by Field of Study, 1989-90

Major field of study and sex of student	Total	White, non-Hispanic	Black, non-Hispanic	Hispanic	Asian or Pacific Islander	American Indian/ Alaskan Native	Nonresident alien	First-professional degrees from HBCUs as a percent of total first-professional degrees	
								Total	Black, non-Hispanic
All fields	820	149	552	33	18	4	64	1.2	16.3
Men	489	100	306	24	14	2	43	1.1	18.5
Women	331	49	246	9	4	2	21	1.2	14.1
Dentistry (D.D.S. or D.M.D.)	78	10	46	-	2	-	20	1.9	26.6
Men	43	6	27	-	1	-	9	1.5	31.0
Women	35	4	19	-	1	-	11	2.8	22.1
Medicine (M.D.)	196	18	145	5	10	1	17	1.3	16.3
Men	105	15	67	3	7	1	12	1.1	15.8
Women	91	3	78	2	3	-	5	1.8	16.8
Optometry (O.D.)	-	-	-	-	-	-	-	-	-
Men	-	-	-	-	-	-	-	-	-
Women	-	-	-	-	-	-	-	-	-
Osteopathic medicine (D.O.)	-	-	-	-	-	-	-	-	-
Men	-	-	-	-	-	-	-	-	-
Women	-	-	-	-	-	-	-	-	-
Pharmacy (D. Phar.)	16	-	10	-	3	-	3	1.3	26.3
Men	9	-	3	-	3	-	3	1.9	23.1
Women	7	-	7	-	-	-	-	1.0	28.0
Podiatry (Pod. D. or D.P.) or podiatric medicine (D.P.M.)	-	-	-	-	-	-	-	-	-
Men	-	-	-	-	-	-	-	-	-
Women	-	-	-	-	-	-	-	-	-
Veterinary medicine (D.V.M.)	39	12	27	-	-	-	-	1.8	62.8
Men	21	7	14	-	-	-	-	2.3	70.0
Women	18	5	13	-	-	-	-	1.4	56.5
Chiropractic medicine (D.C. or D.C.M.)	-	-	-	-	-	-	-	-	-
Men	-	-	-	-	-	-	-	-	-
Women	-	-	-	-	-	-	-	-	-
Law, general (LL.B or J.D.)	346	105	194	28	2	3	14	0.9	11.3
Men	199	69	95	21	2	1	11	0.9	12.5
Women	147	36	99	7	-	2	3	1.0	10.4
Theological professions, general (B.D., M. Div., Rabbi)	145	4	130	-	1	-	10	2.5	31.7
Men	112	3	100	-	1	-	8	2.5	36.0
Women	33	1	30	-	-	-	2	2.3	22.7

Source: "First-Professional Degrees Conferred by Historically Black Colleges and Universities, by Racial/Ethnic Group, Major Field of Study, and Sex of Student: 1989-90," Charlene Hoffman, Thomas D. Snyder, and Bill Sonnenberg, *Historically Black Colleges and Universities*, p. 48. Primary source: U.S. Department of Education, National Center for Education Statistics, Integrated Postsecondary Education Data System (IPEDS), "Completions" survey. (This table was prepared November 1991). *Note:* - Data not reported or not applicable.

★ 338 ★

Historically Black Colleges and Universities

First-Professional Degrees Conferred by Historically Black Colleges and Universities, 1976-77 to 1989-90

Year and sex of student	Number of degrees conferred			Percentage distribution of degrees conferred			Degrees from historically black colleges and universities as a percent of total first-professional degrees		
	White, non-Hispanic	Black, non-Hispanic	Hispanic	White, non-Hispanic	Black, non-Hispanic	Hispanic	White, non-Hispanic	Black, non-Hispanic	Hispanic
Men									
1976-77	100	408	11	17.6	72.0	1.9	0.2	23.2	1.2
1978-79	93	422	23	15.9	72.0	3.9	0.2	23.7	2.3
1980-81	117	419	20	18.9	67.6	3.2	0.2	23.6	1.8
1982-83	69	423	19	12.5	76.6	3.4	-	-	-
1984-85	103	407	25	17.4	68.8	4.2	0.2	25.1	2.0
1986-87[1]	98	364	14	18.2	67.7	2.6	0.2	19.8	1.1

[Continued]

★ 338 ★

First-Professional Degrees Conferred by Historically Black Colleges and Universities, 1976-77 to 1989-90

[Continued]

Year and sex of student	Number of degrees conferred			Percentage distribution of degrees conferred			Degrees from historically black colleges and universities as a percent of total first-professional degrees		
	White, non-Hispanic	Black, non-Hispanic	Hispanic	White, non-Hispanic	Black, non-Hispanic	Hispanic	White, non-Hispanic	Black, non-Hispanic	Hispanic
1988-89[2]	90	273	6	21.5	65.3	1.4	0.2	16.9	0.4
1989-90[3]	100	306	24	20.4	62.6	4.9	0.3	18.5	1.7
Women									
1976-77	13	144	1	7.9	87.8	0.6	0.1	18.6	0.5
1978-79	18	179	3	8.4	83.3	1.4	0.1	17.0	1.0
1980-81	42	203	1	16.0	77.2	0.4	0.2	17.5	0.2
1982-83	28	270	6	8.9	86.0	1.9	-	-	-
1984-85	62	286	3	16.8	77.3	0.8	0.3	20.3	0.5
1986-87[1]	44	254	1	13.2	76.0	0.3	0.2	16.0	0.1
1988-89[2]	42	205	4	15.3	74.5	1.5	0.2	13.4	0.4
1989-90[3]	49	246	9	14.8	74.3	2.7	0.2	14.1	0.9

Source: "First-Professional Degrees Conferred by Historically Black Colleges and Universities, by Racial/Ethnic Group and Sex of Student: 1976-77 to 1989- 90," Charlene Hoffman, Thomas D. Snyder, and Bill Sonnenberg, *Historically Black Colleges and Universities*, p. 39. Primary source: U.S. Department of Education, National Center for Education Statistics, Higher Education General Information Survey (HEGIS), "Degrees and Other Formal Awards Conferred" surveys; and Integrated Postsecondary Education Data System (IPEDS), "Completions" surveys. (This table was prepared November 1991). *Notes:* - Data not available. 1. Excludes 6 men whose racial/ethnic group was not available. 2. Excludes 75 men and 75 women whose racial/ethnic group was not available. 3. Data are preliminary.

★ 339 ★

Historically Black Colleges and Universities

Full-Time Fall Enrollment in Historically Black Colleges and Universities by Race/Ethnic Group and Sex, 1976 to 1990

Year	Total	White, non-Hispanic		Black, non-Hispanic		Hispanic	
		Men	Women	Men	Women	Men	Women
1976	180,059	6,352	3,886	72,562	88,379	1,655	537
1978	176,243	5,369	3,845	68,076	88,082	1,463	678
1980	180,521	5,352	4,520	68,735	88,421	1,368	772
1982	170,611	4,997	4,847	64,862	81,873	1,010	696
1984	168,616	5,167	4,821	62,883	81,639	1,187	891
1986	167,825	5,764	6,215	62,118	82,499	1,179	1,036
1988	180,215	6,088	6,766	65,246	92,320	1,108	1,008
1990[1]	195,437	6,762	8,141	70,181	101,338	994	977

Source: "Full-Time Fall Enrollment in Black Colleges and Universities by Race/Ethnicity and Sex: 1976 to 1990," Charlene Hoffman, Thomas D. Snyder, and Bill Sonnenberg, *Historically Black Colleges and Universities*, p. 23. Primary source: U.S. Department of Education, National Center for Education Statistics, Higher Education General Information Survey (HEGIS), "Fall Enrollment in Colleges and Universities" surveys; and Integrated Postsecondary Education Data System (IPEDS). "Fall Enrollment" surveys. (This table was prepared January 1992). *Note:* 1. Preliminary data.

★ 340 ★

Historically Black Colleges and Universities

Master's Degrees Conferred by Historically Black Colleges and Universities by Field of Study, 1989-90

Major field of study and sex of student	White, non-Hispanic	Black, non-Hispanic	Hispanic	Asian or Pacific Islander	American Indian/ Alaskan Native	Nonresident alien	Master's degrees from HBCUs as a percent of total master's degrees	
							Total	Black, non-Hispanic
All fields	1,103	2,352	34	117	13	417	1.3	15.3
Agriculture and natural resources	8	22	-	4	-	21	1.6	44.0
Architecture and environmental design	5	27	-	2	-	19	1.5	24.1
Area and ethnic studies	-	-	-	-	-	2	0.2	[1]
Business and management	56	286	3	25	-	95	0.6	8.6
Communications	5	40	-	-	-	8	1.2	17.0
Computer and information sciences	23	64	1	27	-	44	1.6	24.7
Education	647	1,161	16	13	8	56	2.2	20.8
Engineering	12	42	-	15	1	27	0.4	10.2
Engineering technologies	8	3	-	-	-	4	1.7	8.6
Foreign languages	-	1	-	-	-	1	0.1	3.3
Health professionals	35	63	1	2	-	5	0.5	6.8
Home economics	10	30	1	-	-	1	2.0	28.6
Law	1	1	-	3	-	14	1.0	2.4
Letters	14	11	-	-	-	3	0.4	7.4
Liberal/general studies	-	-	-	-	-	-	-	-
Library and archival science	46	46	-	2	2	2	2.3	26.7
Life sciences	6	37	-	5	-	9	1.2	32.2
Mathematics	10	17	-	6	-	7	1.1	23.0
Military sciences	-	-	-	-	-	-	-	-
Multi/interdisciplinary studies	-	6	-	-	-	3	0.3	5.0
Parks and recreation	-	13	-	-	-	3	3.7	46.4
Philosophy and religion	-	-	-	-	-	-	-	-
Physical sciences	2	24	-	3	1	20	0.9	26.4
Protective services	3	21	-	2	-	5	2.7	13.7
Psychology	52	75	-	2	1	16	1.6	15.2
Public affairs	135	267	11	5	-	34	2.5	14.7
Social sciences	17	73	1	1	-	16	0.9	16.3
Theology	2	7	-	-	-	1	0.2	3.8
Visual and performing arts	6	15	-	-	-	1	0.3	6.1

Source: "Master's Degrees Conferred by Historically Black Colleges and Universities, by Racial/Ethnic Group, Major Field of Study, and Sex of Student: 1989-90," Charlene Hoffman, Thomas D. Snyder, and Bill Sonnenberg, *Historically Black Colleges and Universities*, pp. 44-45. Primary source: U.S. Department of Education, National Center for Education Statistics, Integrated Postsecondary Education Data System (IPEDS), "Completions" survey. (This table was prepared January 1992). *Notes:* - Data not reported or not applicable. To facilitate trend comparisons, certain aggregations have been made of the degree field as reported in the IPEDS "Completions" survey. "Agriculture and natural resources" includes Agribusiness and agriculture production, Agricultural sciences, and Renewable natural resources; "Business and management" includes Business and management, Business and office, Marketing and distribution, and Consumer and personal services; "Engineering and related technologies" includes Engineering and related technologies, Mechanics and repairers, and Construction trades; "Physical sciences" includes Physical science technologies; "Public affairs" includes Public affairs, and Transportation and material moving; "Visual and performing arts" includes Visual and performing arts and Precision production.

★ 341 ★

Historically Black Colleges and Universities

Master's Degrees Conferred by Historically Black Colleges and Universities, 1976-77 to 1989-90

Year and sex of student	Number of degrees conferred			Percentage distribution of degrees conferred			Degrees from historically black colleges and universities as a percent of total master's degrees		
	White, non-Hispanic	Black, non-Hispanic	Hispanic	White, non-Hispanic	Black, non-Hispanic	Hispanic	White, non-Hispanic	Black, non-Hispanic	Hispanic
Men									
1976-77	612	1,497	1	25.3	61.8	3	0.4	19.2	3
1978-79	428	1,313	17	20.4	62.4	0.8	0.3	18.6	0.6
1980-81	325	1,088	13	17.4	58.3	0.7	0.3	17.7	0.4
1982-83	337	1,025	15	18.0	54.8	0.8	-	-	-
1984-85	314	874	20	17.5	48.8	1.1	0.3	16.8	0.7
1986-87	271	782	6	17.1	49.4	0.4	0.3	15.2	0.2
1988-89	303	740	16	20.5	50.1	1.1	0.3	14.3	0.5
1989-90[2]	354	746	13	23.7	49.9	0.9	0.3	13.6	0.4
Women									
1976-77	550	3,071	7	14.7	82.4	0.2	0.4	23.2	0.2
1978-79	574	2,643	15	17.2	79.2	0.4	0.5	21.4	0.5
1980-81	507	2,097	10	18.4	76.1	0.4	0.4	19.1	0.3
1982-83	437	1,931	16	16.7	73.7	0.6	-	-	-
1984-85	485	1,681	14	20.2	70.1	0.6	0.4	19.2	0.4
1986-87	573	1,661	19	23.6	68.4	0.8	0.5	19.1	0.5
1988-89[1]	582	1,648	21	24.0	67.9	0.9	0.4	18.5	0.5
1989-90[2]	749	1,606	21	29.5	63.2	0.8	0.5	16.3	0.5

Source: "Master's Degrees Conferred by Historically Black Colleges and Universities, by Racial/Ethnic Group and Sex of Student: 1976-77 to 1989-90," Charlene Hoffman, Thomas D. Snyder, and Bill Sonnenberg, *Historically Black Colleges and Universities*, p. 37. Primary source: U.S. Department of Education, National Center for Education Statistics, Higher Education General Information Survey (HEGIS), "Degrees and Other Formal Awards Conferred" surveys; and Integrated Postsecondary Education Data System (IPEDS), "Completions" surveys. (This table was prepared November 1991). *Notes:* - Data not available. 1. Excludes 12 women whose racial/ethnic group was not available. 2. Data are preliminary. 3. Less than 0.05 percent.

★ 342 ★

Historically Black Colleges and Universities

Part-Time Fall Enrollment in Historically Black Colleges and Universities by Race/Ethnic Group and Sex, 1976 to 1990

Year	Total	White, non-Hispanic		Black, non-Hispanic		Hispanic	
		Men	Women	Men	Women	Men	Women
1976	42,554	5,302	5,500	11,930	17,434	934	316
1978	51,554	5,420	5,878	14,376	21,709	1,072	490
1980	53,036	6,286	8,204	13,083	20,750	1,131	500
1982	57,760	7,133	8,951	14,012	21,892	1,376	732
1984	58,903	6,661	9,173	14,128	22,703	1,311	841
1986	55,450	6,801	9,900	12,158	21,853	1,072	686
1988	59,540	7,072	10,807	13,022	23,563	1,125	881
1990[1]	62,367	7,186	11,633	13,355	25,140	958	899

Source: "Full-Time Fall Enrollment in Black Colleges and Universities by Race/Ethnicity and Sex: 1976 to 1990," Charlene Hoffman, Thomas D. Snyder, and Bill Sonnenberg, *Historically Black Colleges and Universities*, p. 23. Primary source: U.S. Department of Education, National Center for Education Statistics, Higher Education General Information Survey (HEGIS), "Fall Enrollment in Colleges and Universities" surveys; and Integrated Postsecondary Education Data System (IPEDS). "Fall Enrollment" surveys. (This table was prepared January 1992). *Note:* 1. Preliminary data.

★ 343 ★

Historically Black Colleges and Universities

Student Graduation Rates at Selected Historically Black Colleges and Universities

Although historically black colleges and universities offer campuses that are presumed to be more socially and academically hospitable to African Americans than the nation's large research institutions, they nonetheless have a highly disappointing student graduation rate.

Institution and state	Graduation rate
Fisk University (Tenn.)	61%
Grambling State University (La.)	60
Spelman College (Ga.)	54
Albany State College (Ga.)	52
Howard University (D.C.)	41
Florida A&M University	40
LeMoyne-Owen College (Tenn.)	40
Bethune-Cookman College (Fla.)	38
Morgan State University (Md.)	37
Bennett College (N.C.)	35
North Carolina A&T University	35
Xavier University (La.)	35
Lincoln University (Pa.)	33
Prairie View A&M University (Tex.)	28
Virginia State University	28
Jackson State University (Miss.)	27
North Carolina Central University	27
University of Maryland Eastern Shore	22
Alcorn State University (Miss.)	21
Voorhees College (S.C.)	15
Alabama State Unievrsity	13
Coppin State College (Md.)	13
Chicago State University (Ill.)	8

Source: "The Student Graduation Rates is Distressingly Low at the Historically Black Institutions," *Journal of Blacks in Higher Education*, Vol. 1, No. 1, Autumn 1993, p. 72. Primary source: *1992-93 NCAA Division 1 Graduation Rates Report* and *Peterson's Guide to Four-Year Colleges. Notes:* The NCAA report covers all students, not just athletes. The graduation rate is the percentage of entering freshmen who graduate within six years. Published by permission.

~~~

Labor Force Participants by Educational Level

~~~

★ 344 ★

Labor Force Participants by Educational Level of Persons 16 Years Old and Older, 1992

Age, sex, and race/ethnicity	Labor force participation rate[1]					
	Total	Less than high school graduate[2]	High school graduate	College		
				Some, college no degree	Associate degree	Bachelor's degree or higher
16 to 24 years old[3]	79.4	61.9	93.0	87.6	92.4	95.1
Men	87.6	75.8	91.6	92.2	96.7	96.2
Women	71.5	46.8	74.4	83.7	89.3	94.2
White[4]	81.8	64.9	85.1	89.1	93.3	95.5
Black[4]	68.2	49.7	74.8	79.9	87.1	93.3
Hispanic[5]	72.1	64.0	79.6	83.8	83.7	86.0
25 and older	66.3	40.7	66.2	73.5	79.5	81.2
Men	76.6	53.8	78.0	82.2	87.9	86.8
Women	57.0	29.4	56.7	65.6	73.0	74.7
White[4]	66.3	40.8	65.4	72.8	78.9	80.9
Black[4]	65.7	39.7	72.3	78.8	85.3	85.6
Hispanic[5]	67.9	55.9	75.2	81.0	85.1	83.0

Source: Labor force participation of persons 16 years old and over, by age, sex, race/ethnicity, and highest level of education: 1992. *Digest of Educational Statistics,* 1993, p. 390. Primary source: U.S. Department of Labor, Bureau of Labor Statistics, Office of Employment and Unemployment Statistics, unpublished data. (This table was prepared May 1993.) *Notes:* 1. Percent of the civilian population who are employed or seeking employment. 2. Includes persons reporting no school years completed. 3. Excludes persons enrolled in school. 4. Includes persons of Hispanic origin. 5. Hispanics may be of any race.

Level of Achievement

★ 345 ★

Education Level by Race for Selected Age Groups, 1982 and 1991

Age cohort/Education level	1982		1991	
	African American	White	African American	White
20-29 years				
4 years college or more	8.4	16.2	7.9 (-6.0%)	18.3 (+13.0%)
1-3 years college	25.9	24.9	26.1 (+0.8%)	27.9 (+12.0%)
High school	44.7	44.6	46.4 (+3.8%)	39.4 (-11.7%)
Less than high school	21.0	14.3	19.6 (-6.7%)	14.4 (+0.7%)
35-44 years				
4 years college or more	10.0	22.9	15.2 (+52.0%)	28.4 (+24.0%)
1-3 years college	13.0	17.4	22.3 (+71.5%)	22.9 (+31.6%)
High school	40.7	41.3	44.4 (+9.1%)	37.3 (-9.7%)
Less than high school	36.5	18.4	18.1 (-50.4%)	11.4 (-38.0%)
Total population				
4 years college or more	6.9	15.5	9.4 (+36.2%)	19.4 (+25.2%)
1-3 years college	14.1	15.9	16.9 (+19.9%)	19.0 (+19.5%)
High school	32.3	37.9	36.5 (+13.0%)	37.6 (-0.8%)
Less than high school	46.7	30.7	37.2 (-20.3%)	24.0 (-21.8%)

Source: "Educational Attainment by Race and Selected Age Cohorts: 1982 and 1991," *State of Black America, 1993*, p. 226.

★ 346 ★

Level of Achievement

Education Level of Men by Race for Selected Age Groups, 1982 and 1991

Age cohort/Education level	1982		1991	
	African American	White	African American	White
20-29 years				
4 years college or more	8.0	16.9	7.9 (-1.3%)	17.7 (+4.7%)
1-3 years college	25.7	24.8	23.5 (-8.6%)	27.2 (+9.7%)
High school	45.0	43.7	49.9 (+10.9%)	39.7 (-9.2%)
Less than high school	21.3	14.6	18.7 (-12.2%)	15.4 (+5.5%)
35-44 years				
4 years college or more	10.8	28.0	16.0 (+48.1%)	26.6 (-5.0%)
1-3 years college	12.1	17.9	23.5 (+94.2%)	21.1 (+17.9%)
High school	41.2	35.6	41.5 (+0.7%)	39.7 (+11.5%)
Less than high school	35.9	18.5	19.0 (-47.1%)	12.6 (-31.9%)
Total population				
4 years college or more	7.0	18.8	9.1 (+30.0%)	21.8 (+16.0%)
1-3 years college	14.2	16.2	16.1 (+13.4%)	18.7 (+15.4%)
High school	31.7	34.3	36.6 (+15.5%)	35.0 (+2.0%)
Less than high school	47.1	30.7	38.2 (-18.9%)	24.5 (-20.2%)

Source: "Male Educational Attainment by Race and Selected Age Cohorts: 1982 and 1991," *State of Black America, 1993*, p. 230.

★ 347 ★

Level of Achievement

Education Level of Women by Race for Selected Age Groups, 1982 and 1991

Age cohort/Education level	1982		1991	
	African American	White	African American	White
20-29 years				
4 years college or more	8.6	15.5	7.9 (-8.1%)	18.9 (+21.9%)
1-3 years college	26.2	25.0	28.3 (+8.0%)	28.5 (+14.0%)
High school	44.5	45.6	43.5 (-2.2%)	39.1 (-14.3%)
Less than high school	20.7	13.9	20.3 (-1.9%)	13.5 (-2.9%)
35-44 years				
4 years college or more	9.4	17.8	17.0 (+80.9%)	26.0 (+46.1%)
1-3 years college	13.8	17.0	21.9 (+59.0%)	23.3 (+37.1%)
High school	40.2	46.8	41.8 (+4.0%)	39.9 (-14.7%)
Less than high school	36.6	18.4	19.3 (-47.3%)	10.8 (-41.3%)
Total population				
4 years college or more	6.8	12.5	9.6 (+41.2%)	17.2 (+37.6%)
1-3 years college	14.1	15.7	17.5 (+24.1%)	19.3 (+22.9%)
High school	32.7	41.2	36.4 (+11.3%)	40.0 (-2.9%)
Less than high school	46.4	30.6	36.5 (-21.3%)	23.5 (-23.2%)

Source: "Female Educational Attainment by Race and Selected Age Cohorts: 1982 and 1991," *State of Black America, 1993*, p. 228.

★ 348 ★

Level of Achievement

Educational Attainment by Race, 1992

For persons 25 years old and over. As of March. Based on Current Population Survey.

Characteristic	Population (1,000)	Percent of population--					
		Not a high school graduate	High school graduate	With some college, no degree	With an associate's degree[1]	With a bachelor's degree	With an advanced degree
Total persons	160,827	20.6	36.0	16.2	5.9	14.2	7.2
Race							
White	137,646	19.1	36.4	16.4	6.1	14.6	7.5
Black	17,445	32.3	35.7	15.5	4.6	8.3	3.6
Other	5,736	19.7	27.8	13.9	5.3	22.1	11.2
Hispanic origin							
Hispanic	11,623	47.4	27.3	11.7	4.2	6.3	3.0
Non-Hispanic	149,204	18.5	36.7	16.5	6.0	14.8	7.5

Source: "Years of School Completed, by Selected Characteristics: 1992," U.S. Department of Commerce, *Statistical Abstract of the United States, 1993*, p. 153. Primary source: U.S. Bureau of the Census, unpublished data. *Note:* 1. Includes vocational degrees.

★ 349 ★

Level of Achievement

Educational Attainment by Race, Ethnicity, and Sex, 1960 to 1992

In percent. For persons 25 years old and over. 1960, 1970 and 1980 as of April 1 and based on sample data from the Censuses of Population. Other years as of March and based on the Current Population Survey.

Year	All races		White		Black		Hispanic[1]	
	Male	Female	Male	Female	Male	Female	Male	Female
Completed 4 years of high school or more								
1960	39.5	42.5	41.6	44.7	18.2	21.8	(NA)	(NA)
1965	48.0	49.9	50.2	52.2	25.8	28.4	(NA)	(NA)
1970	51.9	52.8	54.0	55.0	30.1	32.5	37.9	34.2
1975	63.1	62.1	65.0	64.1	41.6	43.3	39.5	36.7
1980	67.3	65.8	69.6	68.1	50.8	51.5	67.3	65.8
1985	74.4	73.5	76.0	75.1	58.4	60.8	48.5	47.4
1990	77.7	77.5	79.1	79.0	65.8	66.5	50.3	51.3
1991	78.5	78.3	79.8	79.9	66.7	66.7	51.4	51.2
1992[2]	79.7	79.2	81.1	80.7	67.0	68.2	53.7	51.5
Completed 4 years of college or more								
1960	9.7	5.8	10.3	6.0	2.8	3.3	(NA)	(NA)
1965	12.0	7.1	12.7	7.3	4.9	4.5	(NA)	(NA)
1970	13.5	8.1	14.4	8.4	4.2	4.6	7.8	4.3
1975	17.6	10.6	18.4	11.0	6.7	6.2	8.3	4.6

[Continued]

★ 349 ★

Educational Attainment by Race, Ethnicity, and Sex, 1960 to 1992
[Continued]

Year	All races		White		Black		Hispanic[1]	
	Male	Female	Male	Female	Male	Female	Male	Female
1980	20.1	12.8	21.3	13.3	8.4	8.3	9.4	6.0
1985	23.1	16.0	24.0	16.3	11.2	11.0	9.7	7.3
1990	24.4	18.4	25.3	19.0	11.9	10.8	9.8	8.7
1991	24.3	18.8	25.4	19.3	11.4	11.6	10.0	9.4
1992[2]	24.3	18.6	25.2	19.1	11.9	12.0	10.2	8.5

Source: "Educational Attainment by Race, Ethnicity, and Sex: 1960 to 1992," U.S. Department of Commerce, *Statistical Abstract of the United States, 1993*, p. 153. Primary source: U.S. Bureau of the Census, *U.S. Census of Population, 1960, 1970, and 1980*, vol. 1; and *Current Population Reports* series P20-459, P20-462; and unpublished data. *Notes:* NA Not available. 1. Persons of Hispanic origin may be of any race. 2. For 1992, persons high school graduates and those with a BA degree or higher.

★ 350 ★

Level of Achievement

Level of Education by Race/Ethnicity and Age, March 1992

Age	Total	White	Black	Hispanic
High school diploma or equivalency certificate				
20-24	86.2	89.8	79.0	65.9
25-29	87.1	90.6	80.1	64.2
30-34	87.5	90.7	81.9	61.9
35-39	88.5	91.8	80.2	63.9
40-44	89.0	92.1	80.3	63.6
45-49	85.6	89.3	71.1	59.2
50-54	79.3	83.3	63.3	51.4
55-59	75.0	79.2	56.5	46.3
60-64	70.1	75.5	39.0	39.2
Some college or associate's degree				
20-24	51.7	55.8	35.7	34.6
25-29	48.6	52.3	34.9	31.3
30-34	48.1	50.8	38.0	31.6
35-39	51.3	54.6	39.0	31.7
40-44	55.4	58.4	43.2	32.8
45-49	48.5	51.5	35.2	26.4
50-54	39.8	42.5	26.8	23.8
55-59	36.1	38.4	24.3	19.2
60-64	30.9	34.0	12.9	13.3
Bachelor's degree				
20-24	-	-	-	-
25-29	23.7	26.7	10.6	11.4
30-34	22.6	24.6	12.7	11.1
35-39	25.3	27.2	15.9	13.4
40-44	28.2	30.5	17.4	12.3
45-49	25.4	27.3	14.9	10.6
50-54	21.1	22.9	11.1	10.8
55-59	18.0	19.2	10.4	9.0

[Continued]

★ 350 ★

Level of Education by Race/Ethnicity and Age, March 1992
[Continued]

Age	Total	White	Black	Hispanic
60-64	15.8	17.7	4.2	6.3
Advanced degree				
20-24	-	-	-	-
25-29	4.4	4.8	2.2	2.0
30-34	6.2	6.7	2.1	3.9
35-39	8.3	9.0	5.1	4.7
40-44	10.9	11.8	6.2	5.2
45-49	11.0	12.0	5.1	3.9
50-54	9.2	10.1	3.9	3.0
55-59	8.0	8.5	3.4	4.0
60-64	6.0	6.8	1.7	3.1

Source: "Percentage of the Population Who Have Attained Various Levels of Education, by Race/Ethnicity, Sex, and Age," U.S. Department of Education, National Center for Education Statistics, *The Condition of Education, 1993*, p. 62. Primary source: U.S. Department of Commerce, Bureau of the Census, March Current Population Survey, 1992. *Notes:* - Age group is too young for a meaningful estimate of attainment at this level.

★ 351 ★

Level of Achievement

Mathematics Improvement of High School Sophomores, 1980 and 1990

Student characteristics	Number correct (of 55)	
	1980	1990
Asian	38.82	40.26
Black	24.51	28.74
Hispanic	25.96	30.75
White	35.41	37.96

Source: "Gains in Mathematics by Racial/Ethnic Group," U.S. Department of Education, National Center for Educational Statistics, *America's High School Sophomores: A Ten Year Comparison*, p. 24. Primary source: Table compiled by editors.

✵✵✵

Master's Degrees

★ 352 ★

Master's Degrees by Racial/Ethnic Group and Sex: 1976-91

Master's degrees conferred by institutions of higher education, by racial/ethnic group and sex of student: 1976-77 to 1990-91.

Year and sex of student	Total	White, non-Hispanic	Black, non-Hispanic	Hispanic	Asian or Pacific Islander	American Indian/ Alaskan Native	Nonresident alien
Number of degrees conferred							
1976-77							
Total[1]	316,602	266,061	21,037	6,071	5,122	967	17,344
Men	167,396	139,210	7,781	3,268	3,123	521	13,493
Women	149,206	126,851	13,256	2,803	1,999	446	3,851
1978-79							
Total[2]	300,255	249,360	19,418	5,555	5,496	999	19,427
Men	152,637	124,058	7,070	2,786	3,325	495	14,903
Women	147,618	125,302	12,348	2,769	2,171	504	4,524
1980-81							
Total[3]	294,183	241,216	17,133	6,461	6,282	1,034	22,057
Men	145,666	115,562	6,158	3,085	3,773	501	16,587
Women	148,517	125,654	10,975	3,376	2,509	533	5,470
1984-85							
Total[4]	280,421	223,628	13,939	6,864	7,782	1,256	26,952
Men	139,417	106,059	5,200	3,059	4,842	583	19,674
Women	141,004	117,569	8,739	3,805	2,940	673	7,278
1986-87							
Total[5]	289,341	228,870	13,867	7,044	8,558	1,104	29,898
Men	141,264	105,573	5,151	3,330	5,238	517	21,455
Women	148,077	123,297	8,716	3,714	3,320	587	8,443
1988-89							
Total[6]	309,770	242,756	14,096	7,282	10,336	1,086	34,214
Men	148,872	109,709	5,175	3,328	6,050	476	24,134
Women	160,898	133,047	8,921	3,954	4,286	610	10,080
1989-90							
Total[7]	322,465	251,689	15,446	7,954	10,578	1,099	35,699
Men	152,926	112,879	5,539	3,588	6,002	461	24,457
Women	169,539	138,810	9,907	4,366	4,576	638	11,242
1990-91							
Total[8]	328,645	255,286	16,136	8,382	11,180	1,136	36,525

[Continued]

★ 352 ★

Master's Degrees by Racial/Ethnic Group and Sex: 1976-91
[Continued]

Year and sex of student	Total	White, non-Hispanic	Black, non-Hispanic	Hispanic	Asian or Pacific Islander	American Indian/ Alaskan Native	Nonresident alien
Men	151,796	111,228	5,707	3,667	6,319	459	24,416
Women	176,849	144,058	10,429	4,715	4,861	677	12,109

Source: "Master's Degrees Conferred by Institutions of Higher Education, by Racial/Ethnic Group and Sex of Student: 1976-77 to 1990-91." *Digest of Educational Statistics,* 1993, p. 281. Primary source: U.S. Department of Education, National Center for Education Statistics, "Degrees and Other formal Awards Conferred" survey, and Integrated Postsecondary Education Data System (IPEDS), "Completions" survey. (This table was prepared April 1993.) *Notes:* 1. Excludes 387 men and 175 women whose racial/ethnic group was not available. 2. Excludes 733 men and 91 women whose racial/ethnic group was not available. 3. Excludes 1,377 men and 179 women whose racial/ethnic group was not available. 4. Excludes 3,973 men and 1,857 women whose racial/ethnic group was not available. 5. Reported racial/ethnic distributions of students by level of degree, field of degree, and sex were used to estimate race/ethnicity for students whose race/ethnicity was not reported. Excludes 99 men and 117 women whose racial/ethnic group and field of study were not available. 6. Reported racial/ethnic distributions of students by level of degree, field of degree and sex were used to estimate race/ethnicity for students whose race/ethnicity was not reported. Excludes 482 men and 369 women whose racial/ethnic group and field of study were not available. 7. Reported racial/ethnic distributions of students by level of degree, field of degree and sex were used to estimate race/ethnicity for students whose race/ethnicity was not reported. Excludes 727 men and 1,109 women whose racial/ethnic group and field of study were not available. Revised from previously published data. 8. Reported racial/ethnic distributions of students by level of degree, field of degree and sex were used to estimate race/ethnicity for students whose race/ethnicity was not reported. Excludes 4,686 men and 3,837 women whose racial/ethnic group and field of study were not available. Preliminary data.

★ 353 ★

Master's Degrees

Master's Degrees by Racial/Ethnic Group, Major Field of Study, and Sex, Percentages: 1990-91

Master's degrees conferred by institutions of higher education, by racial/ethnic group, major field of study, and sex of student: 1990-91[1].

Major field of study	Total						Men Black, non-Hispanic	Women Black, non-Hispanic
	White, non-Hispanic	Black, non-Hispanic	Hispanic	Asian/Pacific Islander	American Indian/ Alaskan Native	Nonresident alien		
All fields, total[2]	255,286	16,136	8,382	11,180	1,136	36,525	5,707	10,429
Agriculture and natural resources	2,241	71	47	64	8	864	47	24
Architecture and environmental design	2,447	110	112	134	10	677	68	42
Area and ethnic studies	858	57	74	61	12	188	20	37
Business and management	61,431	3,536	1,688	3,156	203	8,667	1,829	1,707
Communications	3,226	251	70	133	6	650	93	158
Computer and information sciences	4,958	303	137	1,085	15	2,826	160	143
Education	76,102	5,836	2,741	1,121	413	2,691	1,277	4,559
Engineering	13,400	421	472	2,129	42	7,520	318	103
Engineering technologies	744	46	19	46	3	117	30	16
Foreign languages	1,433	25	189	39	6	381	8	17
Health sciences	17,772	1,049	445	628	95	1,239	204	845
Home economics	1,627	102	36	50	8	198	9	93
Law	1,189	46	48	46	1	727	30	16
Letters	6,581	171	146	152	13	747	41	130
Liberal/general studies	1,539	51	32	21	6	87	12	39
Library and archival sciences	4,143	174	89	101	14	284	34	140
Life sciences	3,514	144	101	242	13	751	61	83

[Continued]

★ 353 ★

Master's Degrees by Racial/Ethnic Group, Major Field of Study, and Sex, Percentages: 1990-91

[Continued]

| Major field of study | Total | | | | | | Men Black, non-Hispanic | Women Black, non-Hispanic |
	White, non-Hispanic	Black, non-Hispanic	Hispanic	Asian/Pacific Islander	American Indian/ Alaskan Native	Nonresident alien		
Mathematics	2,171	105	71	199	9	1,060	56	49
Military sciences	0	0	0	0	0	0	0	0
Multi/interdisciplinary studies	2,145	92	58	59	8	186	48	44
Parks and recreation	340	12	6	4	1	30	4	8
Philosophy and religion	1,214	44	27	45	5	106	31	13
Physical sciences	3,351	80	86	268	14	1,510	48	32
Protective services	860	144	32	20	5	47	68	76
Psychology	8,369	477	338	179	51	317	136	341
Public affairs	14,739	1,805	712	384	99	795	556	1,249
Social sciences	8,510	568	298	357	45	2,291	256	312
Theology	3,564	185	99	167	5	488	128	57
Visual and performing arts	6,818	231	209	290	26	1,081	135	96

Source: "Master's Degrees Conferred by Institutions of Higher Education, by Racial/Ethnic Group, Major Field of Study, and Sex of Student: 1976-77 to 1990-91." *Digest of Educational Statistics*, 1993, p. 281. Primary source: U.S. Department of Education, National Center for Education Statistics, Integrated Postsecondary Education Data System (IPEDS), "Completions" survey. (This table was prepared April 1993.) *Notes:* To facilitate trend comparisons, certain aggregations have been made of the degree fields as reported in the IPEDS "Completions" survey: "Agriculture and natural resources" includes Agribusiness and agriculture production, Agricultural sciences, and Renewable natural resources; "Business and management" includes Business and management, Business and office, Marketing and distribution, and Consumer and personal services; "Engineering and related technologies" includes Engineering and related technologies, Mechanics and repairers, and Construction trades; "Physical sciences" includes Physical sciences and Science technologies; "Public affairs" includes Public affairs and Transportation and material moving; and "Visual and performing arts" includes Visual and performing arts and Precision production. 1. Preliminary data. 2. Reported racial/ethnic distributions of students by level of degree, field of degree, and sex were used to estimate race/ethnicity for students whose race/ethnicity was not reported. Excludes 423 men and 324 women whose racial/ethnic group and field of study were not available.

Military Schools

★ 354 ★

Enrollment at Selected Military Schools by Race and Sex, Fall 1993

| | White | | Black | | Hispanic | | Asian | | Nat. Am | |
	M	F	M	F	M	F	M	F	M	F
The Citadel	1,704	0	116	0	20	0	33	0	3	0
Virginia Military Institute	988	0	89	0	17	0	60	0	3	0
U.S. Air Force Academy	2,964	426	219	38	216	37	122	22	27	9
U.S. Coast Guard Academy	601	115	30	8	33	11	49	23	6	2
U.S. Naval Academy	2,920	393	263	41	233	26	148	29	38	1
U.S. Military Academy at West Point	3,116	368	216	52	153	23	187	40	21	3

Source: "Enrollment at Selected Institutions by Race and Sex, Fall 1993," *Black Issues in Higher Education*, Vol. 10, No. 25, 10 February 1994, p. 26. Primary source: *Black Issues'* Survey. Published by permission.

Outcomes

★ 355 ★

Percentage Difference Between Median Annual Earnings of Workers by Level of Education, Sex, and Racial/Ethnic Identity, 1991

Type of workers and educational attainment	Male				Female			
	Total	White	Black	Hispanic	Total	White	Black	Hispanic
Ages 25 to 34								
All workers								
Grades 9 to 11	(35)	(30)	(32)	(16)	(40)	(37)	(47)	(42)
Some college	12	13	14	26	30	32	31	24
Bachelor's degree	54	47	62	62	88	88	93	71
Full-time, full-year workers								
Grades 9 to 11	(20)	(15)	(22)	(14)	(29)	(29)	(34)	-
Some college	18	13	25	35	18	19	16	18
Bachelor's degree	57	49	65	81	56	55	46	49
Ages 45 to 54								
All workers								
Grades 9 to 11	(27)	(27)	(29)	-	(20)	(21)	(11)	-
Some college	22	17	23	43	28	26	54	-
Bachelor's degree	62	56	47	-	93	93	99	-
Full-time, full-year workers								
Grades 9 to 11	(29)	(24)	(32)	-	(22)	(25)	(15)	-
Some college	16	14	25	-	24	22	42	18
Bachelor's degree	52	53	40	-	68	68	68	-

Source: "Percentage Difference Between Median Annual Earnings of Wage and Salary Workers Who Are High School Graduates and Workers With Other Levels of Educational Attainment, by Sex, Race/Ethnicity, Type of Worker, and Age: 1991," *The Condition of Education, 1993*, p. 86. Primary source: U.S. Department of Commerce, Bureau of the Census, March Current Population Survey, 1992. *Notes:* - Too few sample observations for a reliable estimate. Parentheses are used to indicate negative numbers. Grades 9 to 11 includes those who attended grade 12 but did not receive a diploma; high school includes those who received an equivalency certificate; some college includes those who have received an associate's degree; and bachelor's degree includes those who received advance degrees. Included in the total but not shown separately are workers of other races, primarily Asians and American Indians.

Post-Baccalaureate Education

★ 356 ★

Doctorates By Field of Study, 1990-91

Statistical profile of persons receiving doctor's degrees[1], by field of study: 1990-91.

Item	All fields	Field of study								
		Education	Engineering	Humanities	Life sciences	Physical sciences[2]		Business and management	Social sciences and psychology	Other professional fields
						Total	Mathematics			
Doctor's degrees conferred (number)	37,451	6,397	5,212	4,094	6,928	6,276	1,040	1,164	6,127	1,172
Racial/ethnic group (percent)[3]										
American Indian	0.5	0.9	0.3	0.3	0.4	0.4	0.0	0.3	0.4	0.6
Asian	5.6	2.2	17.0	2.5	6.5	8.1	11.5	8.0	3.2	4.1
Black	4.1	7.8	2.3	3.0	2.3	1.4	2.2	2.9	4.9	6.8
Mexican-American	0.9	1.2	0.6	0.9	0.7	0.6	0.6	0.3	1.1	0.9
Puerto Rican	0.7	0.9	0.4	0.9	0.5	0.5	0.0	0.1	0.9	0.7
Other Hispanic	1.6	1.2	1.5	2.4	1.4	1.5	1.2	0.9	2.1	1.3
White	85.2	84.9	75.8	88.3	86.6	85.6	83.0	86.2	86.0	84.4
Other and unknown	1.5	0.8	2.1	1.7	1.6	1.9	1.4	1.2	1.3	1.3

Source: Statistical profile of persons receiving doctor's degrees, by field of study: 1990-91. *Digest of Educational Statistics*, 1993, p. 297. Primary source: National Academy of Sciences, National Research Council, Office of Scientific and Engineering Personnel, *Summary Report 1991: Doctorate Recipients From United States Universities*. (This table was prepared March 1993.) *Notes:* The above classification of degrees by field differs somewhat from that in most publications of the National Center for Education Statistics (NCES). The major differences are that history is included under humanities rather than social sciences and that psychology is included under social sciences. The number of degrees also differs slightly from that reported in the NCES "Degrees and Other Formal Awards Conferred" survey. The above tabulation excludes some non-research doctorate degrees such as doctor's degrees in theology. Because of rounding, percents may not add to 100.0 1. Includes Ph.D., Ed. D., and comparable degrees at the doctorate level. Excludes first-professional degrees, such as M.D., D.D.S., and D.V.M. 2. Includes mathematics, computer science, physics, astronomy, chemistry, and earth, atmospheric, and marine sciences. 3. Includes 2,064 individuals who did not report their citizenship at time of doctorate. Distribution by race/ethnicity based on U.S. citizens only.

★ 357 ★

Post-Baccalaureate Education

Education Doctorates: 1981-91

Item	1980-81	1981-82	1982-83	1983-84	1984-85	1985-86	1986-87	1987-88	1988-89	1989-90	1990-91
Number of doctorates	7,489	7,226	7,147	6,780	6,717	6,602	6,447	6,349	6,265	6,484	6,397
Racial/ethnic group (percent)[1]											
American Indian	0.6	0.5	0.7	0.5	0.7	0.5	0.7	0.6	0.4	0.6	0.9
Asian	1.8	1.7	1.9	1.5	1.7	1.6	1.7	2.4	1.9	1.7	2.2
Black	8.8	9.5	8.1	8.5	8.6	8.0	7.3	7.5	8.0	8.2	7.8
Mexican-American	1.1	1.2	1.3	1.2	1.2	1.4	1.3	1.2	0.9	0.9	1.2
Puerto Rican	0.6	0.7	0.7	0.6	1.0	0.9	0.9	0.8	1.0	1.0	0.9
Other Hispanic	0.7	1.0	0.9	0.8	1.0	1.3	1.3	0.9	1.2	1.4	1.2
White	83.1	83.6	84.8	85.1	84.5	84.8	85.1	85.3	85.7	85.4	84.9
Other and unknown	3.3	1.8	1.7	1.8	1.4	1.6	1.6	1.2	0.9	0.9	0.8

Source: Statistical profile of persons receiving doctor's degrees in Education: 1979-80 to 1990-91. *Digest of Educational Statistics*, 1993, p. 298. Primary source: National Academy of Sciences, National Research Council, Office of Scientific and Engineering Personnel, Doctorate Records File. (This table was prepared March 1993.) *Notes:* The National Research Council's classification of degrees by field differs somewhat from that in most publications of the National Center for Education Statistics (NCES). The number of degrees also differs slightly from that reported in the NCES "Completions" survey. Because of rounding, percents may not add to 100.0 1. Longitudinal comparisons by race/ethnicity should be done with extreme care, due to periodic changes in the survey. Distribution by race/ethnicity based on U.S. citizens only.

★ 358 ★

Post-Baccalaureate Education

Engineering Doctorates: 1981-91

Item	1980-81	1981-82	1982-83	1983-84	1984-85	1985-86	1986-87	1987-88	1988-89	1989-90	1990-91
Number of doctorates	2,528	2,644	2,780	2,915	3,165	3,376	3,716	4,190	4,536	4,892	5,212
Racial/ethnic group (percent)[1]											
American Indian	0.3	0.2	0.1	0.2	0.1	0.3	0.4	0.2	0.3	0.2	0.3
Asian	19.2	16.8	16.7	16.5	17.6	15.2	17.1	15.5	16.2	15.0	17.0
Black	1.3	1.4	1.9	1.0	2.1	1.4	1.3	1.4	1.4	1.7	2.3
Mexican-American	0.1	0.3	0.3	0.4	0.4	0.3	0.4	0.5	0.6	0.6	0.6
Puerto Rican	0.3	0.7	0.4	0.5	0.3	0.6	0.2	0.6	0.3	0.3	0.4
Other Hispanic	0.6	1.4	1.3	1.4	0.7	1.1	1.1	1.8	1.2	1.5	1.5
White	74.4	75.2	76.1	76.4	74.5	78.3	76.2	77.0	77.4	78.9	75.8
Other and unknown	3.7	4.0	3.2	3.6	4.3	2.7	3.3	2.9	2.5	1.9	2.1

Source: Statistical profile of persons receiving doctor's degrees in Engineering: 1979-80 to 1990-91. *Digest of Educational Statistics*, 1993, p. 298. Primary source: National Academy of Sciences, National Research Council, Office of Scientific and Engineering Personnel, Doctorate Records File. (This table was prepared March 1993.) *Notes:* The National Research Council's classification of degrees by field differs somewhat from that in most publications of the National Center for Education Statistics (NCES). The number of degrees also differs slightly from that reported in the NCES "Completions" survey. Because of rounding, percents may not add to 100.0 1. Longitudinal comparisons by race/ethnicity should be done with extreme care, due to periodic changes in the survey. Distribution by race/ethnicity based on U.S. citizens only.

★ 359 ★

Post-Baccalaureate Education

Humanities Doctorates: 1981-91

Item	1980-81	1981-82	1982-83	1983-84	1984-85	1985-86	1986-87	1987-88	1988-89	1989-90	1990-91
Number of doctorates	3,745	3,560	3,494	3,528	3,428	3,461	3,504	3,553	3,558	3,820	4,094
Racial/ethnic group (percent)[1]											
American Indian	0.4	0.2	0.2	0.2	0.3	0.2	0.4	0.2	0.2	0.3	0.3
Asian	1.7	1.7	1.5	1.8	2.2	1.8	2.1	2.3	2.9	2.4	2.5
Black	2.8	3.3	2.5	3.3	2.5	2.8	2.8	3.0	2.8	2.3	3.0
Mexican-American	0.5	0.8	0.7	0.8	0.8	0.8	0.6	0.8	0.8	0.6	0.9
Puerto Rican	0.7	0.9	0.7	0.7	0.6	0.5	1.1	0.9	0.8	0.9	0.9
Other Hispanic	1.9	2.5	2.2	2.1	2.3	2.1	2.5	2.1	2.1	2.6	2.4
White	88.0	87.8	89.4	88.4	88.9	89.6	88.4	89.1	88.2	89.7	88.3
Other and unknown	4.1	2.8	2.7	2.7	2.4	2.2	2.2	1.6	2.2	1.3	1.7

Source: Statistical profile of persons receiving doctor's degrees in the humanities: 1979-80 to 1990-91. *Digest of Educational Statistics*, 1993, p. 299. Primary source: National Academy of Sciences, National Research Council, Office of Scientific and Engineering Personnel, Doctorate Records File. (This table was prepared March 1993.) *Notes:* The National Research Council's classification of degrees by field differs somewhat from that in most publications of the National Center for Education Statistics (NCES). The major differences are that history is included under humanities rather than social sciences and that psychology is included under social sciences. The number of degrees also differs slightly from that reported in the NCES "Completions" survey. Because of rounding, percents may not add to 100.0 Includes American studies, archeology, art history, classics, history, letters, literature, music, philosophy, religion, and theatre. 1. Longitudinal comparisons by race/ethnicity should be done with extreme care, due to periodic changes in the survey. Distribution by race/ethnicity based on U.S. citizens only.

★ 360 ★

Post-Baccalaureate Education

Life Science Doctorates: 1981-91

Item	1981-82	1982-83	1983-84	1984-85	1985-86	1986-87	1987-88	1988-89	1989-90	1990-91
Number of doctorates	5,565	5,540	5,745	5,748	5,720	5,742	6,143	6,343	6,613	6,928
Racial/ethnic group (percent)[1]										
American Indian	0.3	0.2	0.3	0.4	0.5	0.4	0.4	0.3	0.2	0.4
Asian	4.5	5.2	4.6	4.6	4.8	5.6	4.9	5.2	5.5	6.5
Black	1.5	1.6	2.0	2.1	1.9	2.4	2.2	2.1	1.9	2.3
Mexican-American	0.4	0.2	0.3	0.4	0.4	0.4	0.4	0.5	0.8	0.7
Puerto Rican	0.3	0.3	0.3	0.4	0.4	0.6	0.6	0.6	0.7	0.5
Other Hispanic	0.8	0.7	0.8	1.1	1.3	1.0	1.3	1.0	1.2	1.4
White	89.1	89.5	89.1	89.0	88.9	87.3	88.5	88.3	88.3	86.6
Other and unknown	3.1	2.3	2.6	2.0	1.8	2.3	1.7	2.0	1.4	1.6

Source: Statistical profile of persons receiving doctor's degrees in the life sciences: 1979-80 to 1990-91. *Digest of Educational Statistics*, 1993, p. 298. Primary source: National Academy of Sciences, National Research Council, Office of Scientific and Engineering Personnel, Doctorate Records File. (This table was prepared March 1993.) *Notes:* The National Research Council's classification of degrees by field differs somewhat from that in most publications of the National Center for Education Statistics (NCES). The number of degrees also differs slightly from that reported in the NCES "Completions" survey. Because of rounding, percents may not add to 100.0 Includes agricultural, biological, and health sciences. 1. Longitudinal comparisons by race/ethnicity should be done with extreme care, due to periodic changes in the survey. Distribution by race/ethnicity based on U.S. citizens only.

★ 361 ★

Post-Baccalaureate Education

Physical Science Doctorates: 1981-91

Item	1980-81	1981-82	1982-83	1983-84	1984-85	1985-86	1986-87	1987-88	1988-89	1989-90	1990-91
Number of doctorates	3,208	3,348	3,438	3,459	3,531	3,679	3,837	4,046	3,987	4,263	4,439
Racial/ethnic group (percent)[1]											
American Indian	0.1	0.2	0.3	0.2	0.1	0.2	0.3	0.3	0.6	0.1	0.5
Asian	6.5	6.0	6.4	6.4	6.7	6.9	6.8	5.5	6.6	6.4	6.5
Black	1.2	1.1	1.0	1.3	1.2	1.0	1.0	1.3	1.3	1.0	1.2
Mexican-American	0.1	0.2	0.2	0.3	0.5	0.5	0.4	0.6	0.5	0.5	0.6
Puerto Rican	0.4	0.3	0.3	0.4	0.2	0.5	1.0	0.8	0.7	1.0	0.6
Other Hispanic	0.7	0.7	0.8	1.2	0.9	1.0	0.9	1.1	1.3	1.5	1.6
White	85.3	88.5	87.4	87.0	87.0	86.5	86.6	87.3	86.8	87.2	87.1
Other and unknown	5.7	3.0	3.6	3.2	3.2	3.4	3.0	3.1	2.1	2.3	1.9

Source: Statistical profile of persons receiving doctor's degrees in the physical sciences: 1979-80 to 1990-91. *Digest of Educational Statistics*, 1993, p. 300. Primary source: National Academy of Sciences, National Research Council, Office of Scientific and Engineering Personnel, Doctorate Records File. (This table was prepared March 1993.) *Notes:* The National Research Council's classification of degrees by field differs somewhat from that in most publications of the National Center for Education Statistics (NCES). The major differences are that history is included under humanities rather than social sciences and that psychology is included under social sciences. The number of degrees also differs slightly from that reported in the NCES "Completions" survey. Because of rounding, percents may not add to 100.0 Includes physics, chemistry, and earth, atmospheric, and marine sciences. Excludes mathematics and computer science. 1. Longitudinal comparisons by race/ethnicity should be done with extreme care, due to periodic changes in the survey. Distribution by race/ethnicity based on U.S. citizens only.

★ 362 ★

Post-Baccalaureate Education

Social Science Doctorates: 1981-91

Item	1980-81	1981-82	1982-83	1983-84	1984-85	1985-86	1986-87	1987-88	1988-89	1989-90	1990-91
Number of doctorates	6,505	6,250	6,055	5,895	5,720	5,841	5,718	5,769	5,955	6,076	6,127
Racial/ethnic group (percent)[1]											
American Indian	0.2	0.4	0.2	0.2	0.4	0.4	0.5	0.3	0.4	0.5	0.4
Asian	2.4	2.4	2.1	2.4	2.5	2.5	3.1	3.2	3.1	2.9	3.2
Black	3.9	4.6	3.8	4.5	4.3	4.0	3.7	4.3	4.3	4.3	4.9
Mexican-American	0.8	0.9	0.9	0.8	0.9	1.0	1.0	1.0	1.0	0.9	1.1
Puerto Rican	0.3	0.6	0.4	0.6	0.5	0.6	0.5	0.7	0.8	0.9	0.9
Other Hispanic	1.1	1.1	1.6	1.3	1.4	1.6	2.0	1.5	1.5	2.1	2.1
White	87.6	87.8	88.2	88.0	87.6	87.9	87.3	87.3	87.5	87.2	86.0
Other and unknown	3.7	2.2	2.8	2.1	2.4	2.0	2.0	1.7	1.6	1.2	1.3

Source: Statistical profile of persons receiving doctor's degrees in the social sciences: 1979-80 to 1990-91. *Digest of Educational Statistics*, 1993, p. 300. Primary source: National Academy of Sciences, National Research Council, Office of Scientific and Engineering Personnel, Doctorate Records File. (This table was prepared March 1993.) *Notes:* The National Research Council's classification of degrees by field differs somewhat from that in most publications of the National Center for Education Statistics (NCES). The major differences are that history is included under humanities rather than social sciences and that psychology is included under social sciences. The number of degrees also differs slightly from that reported in the NCES "Completions" survey. Because of rounding, percents may not add to 100.0 Includes anthropology, area studies, criminology, economics, geography, political science, public policy, psychology, and sociology. 1. Longitudinal comparisons by race/ethnicity should be done with extreme care, due to periodic changes in the survey. Distribution by race/ethnicity based on U.S. citizens only.

Prekindergarten

★ 363 ★

Percentage of 3- to 4-year-olds Enrolled in Prekindergarten, October 1973 to 1990

October	Race/ethnicity[1]			
	Total	White	Black	Hispanic
1973	19.1	19.5	19.0	13.8
1974	21.3	21.6	21.1	15.6
1975	23.0	23.6	22.2	15.8
1976	24.1	24.7	23.9	15.4
1977	25.4	26.1	25.8	15.4
1978	27.3	27.9	-	16.2
1979	29.2	29.8	-	20.9
1980	29.7	30.7	-	20.6
1981	30.4	32.3	28.4	18.7
1982	30.6	32.8	28.7	15.7
1983	30.7	32.9	28.9	15.3
1984	31.2	33.6	28.7	17.4
1985	31.9	34.6	28.6	19.2
1986	32.4	35.5	27.4	20.3

[Continued]

★ 363 ★

Percentage of 3- to 4-year-olds Enrolled in Prekindergarten,
October 1973 to 1990
[Continued]

October	Race/ethnicity[1]			
	Total	White	Black	Hispanic
1987	32.5	36.1	25.9	18.7
1988	33.0	36.8	26.7	18.0
1989	36.0	39.9	30.4	19.6
1990	36.5	40.3	31.4	21.0

Source: "Percentage of 3- to 4-year-olds Enrolled in Prekindergarten, by Race/Ethnicity and Family Income: October 1973 to 1991," U.S. Department of Education, National Center for Education Statistics, *The Condition of Education, 1993*, p. 20. Primary source: U.S. Department of Commerce, Bureau of the Census, October Current Population Surveys. *Notes:* - Not available. 1. Due to small sample sizes for the Black and Hispanic categories, 3-year averages are calculated. The 3-year average for 1990 is the average percentage enrolled in pre-K in 1989, 1990, and 1991.

Preprimary School

★ 364 ★

Preprimary School Enrollment, 1975 to 1991

As of October. Civilian noninstitutional population. Includes public and nonpublic nursery school and kindergarten programs. Excludes 5 year olds enrolled in elementary school. Based on Current Population Survey.

Item	1975	1980	1985	1986	1987	1988	1989	1990	1991
Number of children (1,000)									
Population of 3 to 5 years old	10,183	9,284	10,733	10,866	10,872	10,994	11,038	11,207	11,370
Total enrolled	4,954	4,878	5,865	5,971	5,932	5,977	6,026	6,659	6,334
White	4,105	3,994	4,757	4,851	4,748	4,891	4,911	5,389	5,104
Black	731	725	919	892	893	814	872	964	928
Hispanic[2]	(NA)	370	496	593	587	544	520	642	675
Enrollment Rate									
Total enrolled[1]	48.6	52.5	54.6	55.0	54.6	54.4	54.6	59.4	55.7
White	48.6	52.7	54.7	55.2	54.1	55.4	55.0	59.7	56.2
Black	48.1	51.8	55.8	54.1	54.2	48.2	54.2	57.8	53.1
Hispanic[2]	(NA)	43.3	43.3	47.8	45.5	44.2	41.6	49.0	46.4

Source: "Preprimary School Enrollment-Summary: 1970 to 1991," U.S. Department of Commerce, *Statistical Abstract of the United States, 1993*, p. 156. Primary source: U.S. Bureau of the Census, *Current Population Reports*, series P20- 469. *Notes:* NA Not available. 1. Includes races not shown separately. 2. Persons of Hispanic origin may be of any race. The method of identifying Hispanic children was changed in 1980 from allocation based on status of mother to status reported for each child. The number of Hispanic children using the new method is larger.

Presecondary Education

★ 365 ★

School Principals, 1990-1991

Selected characteristics	Total[1]	Percent of principals by highest degree earned[2]				Average years of experience			Average annual salary of principals, by length of work year		
		Bachelor's	Master's	Education specialist	Doctor's and first-professional	As a principal	Other (nonteaching) school position	Outside school position	10 months or less	11 months	12 months
Public schools											
Total	78,889	1.8	60.5	28.2	9.5	9.3	3.8	0.8	$45,126	$48,377	$52,761
Race/ethnicity											
White, non-Hispanic	67,794	1.7	60.5	28.6	9.1	9.6	3.7	0.8	44,645	48,184	52,674
Black, non-Hispanic	6,770	0.9	57.8	27.4	13.9	8.3	4.7	0.9	48,589	49,501	53,338
Hispanic	3,097	4.1	67.5	21.6	6.4	7.4	4.6	0.9	49,176	49,220	54.981[3]
Asian or Pacific Islander	529	7.1	64.8	20.6	7.5	6.7	4.5	1.0	50,857	58,652[3]	
American Indian or Alaskan Native	700	6.0	52.8	28.0	13.2	7.7	5.6	0.8	38,374		46,176
Private schools											
Total	23,881	26.9	47.4	11.5	6.8	8.7	2.8	2.4	$20,591	$29,738	$30,410
Race/ethnicity											
White, non-Hispanic	22,366	26.6	47.9	11.7	6.6	8.7	2.8	2.5	20,481[3]	29,496[3]	30,429
Black, non-Hispanic	643	24.0	44.1	4.7	13.2	6.9	3.6	2.2	[3]	[3]	29,559
Hispanic	607	44.9	36.0	12.8	3.5	7.0	3.2	1.4			29,479

Source: "Principals in Public and Private Elementary and Secondary Schools, by Selected Characteristics 1990-1991." *Digest of Educational Statistics*, 1993, p. 94. Primary source: U.S. Department of Education, National Center for Education Statistics, "Schools and Staffing Survey, 1990-91." (This table was prepared July 1993.) *Notes:* Details may not add to 100 percent because of rounding and survey item nonresponse. 1. Total differs from data appearing in other tables because of varying survey processing procedures and time period coverages. 2. Percentages for those with less than a bachelor's degree are not shown. 3. Too few cases for reliable estimates.

★ 366 ★

Presecondary Education

Science Proficiency of 9-year-olds, 13-year-olds, and 17-year-olds: 1976-77, 1981-82, 1985-86, and 1989-1990

Sex, race/ethnicity and year	9-year-olds[1]				13-year-olds[2]				17-year-olds[2]			
	Know everyday science facts	Understand simple scientific principles	Apply basic scientific information	Analyze scientific procedures and data	Understand simple scientific principles	Apply basic scientific information	Analyze scientific procedures and data	Integrate specialized scientific information	Understand simple scientific principles	Apply basic scientific information	Analyze scientific procedures and data	Integrate specialized scientific information
Total												
1976-77	93.5	68.0	25.7	3.2	86.0	48.8	11.1	0.7	97.1	81.6	41.7	8.5
1981-82	95.2	70.7	24.3	2.3	89.8	50.9	9.6	0.4	95.7	76.6	37.3	7.1
1985-86	96.2	72.0	27.5	3.0	91.6	52.5	9.1	0.2	97.1	80.7	41.3	7.9
1989-90	97.0	76.4	31.1	3.1	92.3	56.5	11.2	0.4	96.7	81.2	43.3	9.2
White[3]												
1976-77	97.7	76.8	30.8	3.9	92.2	56.5	13.4	0.8	99.2	88.2	47.5	10.0
1981-82	98.3	78.4	29.4	2.9	94.4	58.3	11.5	0.4	98.6	84.9	43.9	8.6
1985-86	98.2	78.9	32.7	3.8	96.1	61.0	11.3	0.3	98.8	87.8	48.7	9.6
1989-90	99.2	84.4	37.5	3.9	96.9	66.5	14.2	0.5	99.0	89.6	51.2	11.4
Black[3]												
1976-77	72.4	27.2	3.5	0.2	57.3	14.9	1.2	[4]	83.6	40.5	7.7	0.4
1981-82	82.1	38.9	3.9	0.1	68.6	17.1	0.8	[4]	79.7	35.0	6.5	0.2
1985-86	88.6	46.2	8.3	0.3	73.6	19.6	1.1	[4]	90.9	52.2	12.5	0.9
1989-90	88.0	46.4	8.5	0.1	77.6	24.3	1.5	0.1	88.3	51.4	15.7	1.5
Hispanic												
1976-77	84.6	42.0	8.8	0.3	62.2	18.1	1.8	[4]	93.1	61.5	18.5	1.8
1981-82	85.1	40.2	4.2	0.0	75.5	24.1	2.4	[4]	86.9	48.0	11.1	1.4

[Continued]

★ 366 ★

Science Proficiency of 9-year-olds, 13-year-olds, and 17-year-olds: 1976-77, 1981-82, 1985-86, and 1989-1990

[Continued]

Sex, race/ethnicity and year	9-year-olds[1]				13-year-olds[2]				17-year-olds[2]			
	Know everyday science facts	Understand simple scientific principles	Apply basic scientific information	Analyze scientific procedures and data	Understand simple scientific principles	Apply basic scientific information	Analyze scientific procedures and data	Integrate specialized scientific information	Understand simple scientific principles	Apply basic scientific information	Analyze scientific procedures and data	Integrate specialized scientific information
1985-86	89.6	50.1	10.7	0.2	76.7	24.9	1.5	[4]	93.3	60.0	14.8	1.1
1989-90	93.6	56.3	11.6	0.4	80.2	30.0	3.3	0.1	91.9	59.9	21.1	2.1

Source: "Percent of Students at or above Five Science Proficiency Levels, by Race/Ethnicity and Age: 1976-77, 1981-82, 1985-86, and 1989-1990." *Digest of Educational Statistics*, 1993, p. 123. Primary source: U.S. Department of Education, National assessment of Educational Progress, *Trends in Academic Progress*, prepared by Educational testing Service. (This table was prepared January 1992.) *Notes:* 1. Virtually no students were able to integrate specialized scientific information. 2. Virtually all students knew everyday science facts. Data exclude persons not enrolled in school. 3. Excludes persons of Hispanic origin. 4. Virtually no students were able to perform at this level.

★ 367 ★

Presecondary Education

Student Proficiency in Reading, 1979-80 to 1989-1990

Selected characteristics of students	9-year-olds				13-year-olds				17-year-olds[1]			
	1979-80	1983-84	1987-88	1989-90	1979-80	1983-84	1987-88	1989-90	1979-80	1983-84	1978-88	1989-90
Total	214.8	211.0	211.8	209.2	258.5	257.1	257.5	256.8	285.8	288.8	290.1	290.2
Race/ethnicity												
White	221.3	218.3	217.7	217.0	264.4	262.6	261.3	262.3	293.1	295.6	294.7	296.6
Black	189.2	185.7	188.5	181.8	232.4	236.0	242.9	241.5	242.5	264.2	274.4	267.3
Hispanic	189.5	187.2	193.7	189.4	236.8	239.6	240.1	237.8	260.7	268.1	270.8	274.8

Source: "Student Proficiency in Reading by Age and Selected Characteristics of Students: 1970-71, to 1989-1990." *Digest of Educational Statistics*, 1993, p. 113. Primary source: U.S. Department of Education, National Center for Education Statistics, National Assessment of Educational Progress, *The Reading Report Card, 1971-88* and *Trends in Academic Progress*, by Educational Testing Service. (This table was prepared April 1992.) *Notes:* The NAEP scores have been evaluated at certain performance levels. A score of 300 implies an ability to find, understand, summarize, and explain relatively complicated literary and informational material. A score of 250 implies an ability to search for specific information, interrelate ideas, and make generalizations about literature, science, and social studies materials. A score of 200 implies an ability to understand, combine ideas, and make inferences based on short uncomplicated passages about specific or sequentially related information. A score of 150 implies an ability to follow written directions and select phrases to describe pictures. Scale range from 0 to 500. 1. All participants of this age were in school.

Professional Degrees

★ 368 ★

Professional Degrees By Field of Study, 1990-91

Major field of study	Total	White, non-Hispanic	Black, non-Hispanic	Hispanic	Asian/Pacific Islander	American Indian/ Alaskan Native	Non resident	Men Black, non-Hispanic	Women Black, non-Hispanic
All fields[2]	71,515	60,327	3,575	2,527	3,755	261	1,070	1,672	1,903
Dentistry (D.D.S. or D.M.D.)	3,699	2,657	205	235	446	14	142	102	103
Medicine (M.D.)	15,043	11,847	882	578	1,540	54	142	382	500
Optometry (O.D.)	1,115	918	17	34	118	7	21	6	11
Osteopathic medicine (D.O.)	1,459	1,291	17	51	83	12	5	8	9
Pharmacy (Pharm. D.)	1,244	870	61	58	210	6	39	20	41
Podiatry (Pod. D. or D.P.) or podiatric medicine (D.P.M.)	589	460	52	28	34	3	12	32	20
Veterinary medicine (D.V.M.)	2,032	1,877	44	56	32	10	13	17	27
Chiropractic medicine (D.C. or D.C.M.)	2,640	2,367	30	55	58	5	125	18	12
Law (LL.B. or J.D.)	37,945	33,302	1,860	1,336	1,014	138	295	802	1,058
Theology (M. Div., M.H.L., B.D., or Ord.)	5,695	4,695	407	94	219	11	269	285	122
Medicine, other	54	43	0	2	1	1	7	0	0

Source: First-professional degrees conferred by institution of higher education, by racial/ethnic group, major field of study, and sex of student: 1990-91. *Digest of Educational Statistics*, 1993, p. 285. Primary source: US Department of Education, National Center for Education Statistics, Integrated Postsecondary Education Data System (IPEDS), "Completions" survey (This table was prepared April 1993) *Notes:* 1. Preliminary data. 2. Reported racial/ethnic distributions of students by level of degree, field of degree, and sex were used to estimate race/ethnicity for students whose race/ethnicity was not reported. Excludes 245 men and 188 women whose racial/ethnic group and field of study were not available.

Public Schools

★ 369 ★

Percentage of Students in Grades 1 to 12 Who Are Black, by Control of School and Metropolitan Status, 1970 to 1991

Year	Black Public schools				Private schools
	Total	Central cities	Other metropolitan	Non-metropolitan	
1970	14.8	32.5	6.2	12.0	4.7
1971	15.2	34.4	6.5	11.6	4.6
1972	14.9	31.7	6.3	11.3	5.2
1973	14.8	32.1	5.8	11.0	5.7
1974	15.4	33.2	6.6	11.8	4.3
1975	15.6	33.0	7.0	11.8	5.0

[Continued]

★ 369 ★

Percentage of Students in Grades 1 to 12 Who Are Black, by Control of School and Metropolitan Status, 1970 to 1991

[Continued]

Year	Black				Private schools
	Public schools				
	Total	Central cities	Other metropolitan	Non-metropolitan	
1976	16.0	34.0	7.6	11.7	5.8
1977	15.9	35.5	7.1	12.6	6.2
1978	16.1	35.9	7.4	12.3	6.0
1979	16.1	35.8	8.8	10.9	7.5
1980	-	-	-	-	-
1981	16.2	35.2	8.1	11.8	6.5
1982	16.2	34.0	8.6	11.9	6.6
1983	16.3	33.9	9.1	11.5	6.5
1984	16.1	-	-	-	6.3
1985	17.0	36.0	9.5	12.7	5.6
1986	16.7	32.9	8.3	14.1	6.9
1987	16.7	32.9	8.8	12.8	7.4
1988	16.8	32.4	9.8	12.2	8.2
1989	16.7	32.8	10.0	11.5	7.7
1990	16.5	33.1	8.8	12.5	7.2
1991	16.7	33.0	9.2	12.4	7.3

Source: "Percentage of Student in Grades 1 to 12 Who Are Black or Hispanic, by Control of School and Metropolitan Status: 1970-1991," U.S. Department of Education, National Center for Education Statistics, *The Condition of Education, 1993,* p. 116. Primary source: U.S. Department of Commerce, Bureau of the Census, Current Population Reports, Series P-20, "School Enrollment..." various years; October Current Population Surveys. *Notes:* - Not available. Control not available in 1980. Residence of students not available in 1984. The definition of metropolitan areas in the U.S. was changed in 1985.

★ 370 ★

Public Schools

Public School Employment, by Race, 1982 and 1992

In thousands. Covers full-time employment, 1982 excludes Hawaii, District of Columbia, and New Jersey. Based on sample survey of school districts with 250 or more students. 1990 based on sample survey of school districts with 100 or more employees; see source for sampling variability.

Occupation	1982			1990		
	Total	White[1]	Black[1]	Total	White[1]	Black[1]
All occupations	3,082	2,496	432	3,181	2,502	463
Officials, administrators	41	36	3	43	37	4
Principals and assistant principals	90	76	11	90	70	13
Classroom teachers[2]	1,680	1,435	186	1,746	1,469	192
Elementary schools	798	667	98	875	722	103
Secondary schools	706	619	67	662	570	66
Other professional staff	235	193	35	227	187	30

[Continued]

★ 370 ★

Public School Employment, by Race, 1982 and 1992
[Continued]

Occupation	1982			1990		
	Total	White[1]	Black[1]	Total	White[1]	Black[1]
Teachers aides[3]	215	146	45	324	208	69
Clerical, secretarial staff	210	177	19	226	181	24
Service workers[4]	611	434	132	524	348	129

Source: "Public School Employment, by Occupation, Sex, and Race: 1982 and 1992," U.S. Department of Commerce, *Statistical Abstract of the United States, 1993,* p. 162. Primary source: U.S. Equal Employment Opportunity Commission, *Elementary-Secondary Staff Information* (EEO-5), biennial. *Notes:* 1. Excludes individuals of Hispanic origin. 2. Includes other classroom teachers, not shown separately. 3. Includes technicians. 4. Includes craftworkers and laborers.

★ 371 ★
Public Schools

Racial/Ethnic Enrollment in the Twenty-Five Largest Public School Districts in the United States, 1990-91

District	Total	African American	White	Hispanic American	Asian/ Pacific Islander	Alaskan/ American Native
New York	955,514	363,057	181,897	334,168	75,635	757
Los Angeles	634,397	96,410	85,836	402,671	47,887	1,593
Chicago	408,714	236,914	48,367	110,707	11,994	732
Dade County	292,411	97,298	55,167	135,982	3,497	116
Philadelphia	195,530	123,216	44,763	18,847	8,451	253
Houston	194,548	74,220	27,772	87,304	5,108	144
Detroit	168,906	150,556	13,609	4,024	1,270	497
Broward Country	154,103	51,063	91,454	14,853	3,297	434
Dallas	135,436	NA	NA	NA	NA	NA
San Diego	122,984	19,542	45,349	32,891	22,720	650
Jacksonville	115,589	41,280	67,470	1,966	2,757	139
Baltimore	111,988	NA	NA	NA	NA	NA
Memphis[1]	103,618	83,079	20,438	NA	NA	NA
Milwaukee	98,371	55,231	28,612	9,180	3,129	2,219
New Orleans	82,609	NA	NA	NA	NA	NA
Washington, DC	80,694	72,474	3,123	4,199	870	28
Long Beach	71,654	13,838	20,052	22,726	14,717	321
Fresno	71,457	7,464	23,586	25,472	14,459	476
Cleveland	71,135	49,746	14,688	4,438	979	162
Nashville[1]	67,056	25,559	39,199	NA	NA	NA
Columbus	64,280	30,474	31,893	285	1,571	57
El Paso	64,092	2,941	13,578	46,917	596	60
San Francisco	63,506	15,070	9,088	12,497	26,221	393
E. Baton Rouge	61,272	29,411	13,578	46,917	596	60
Boston	60,922	29,096	12,730	13,401	5,440	255

Source: "25 Largest Public School Districts in the U.S.," *Black Issues in Higher Education,* Vol. 9, No. 5, 7 May 1992, p. 62. Primary source: Council of the Great City Schools. Published by permission. *Note:* 1. White includes other.

Safety

★ 372 ★

Students Fearing Attacks At School, 1989

Student characteristics	Total number of students	Percent of students		
			Ever fearing an attack	
		Avoiding places at school	At school	Going to and from school
Race				
White	17,306,626	6	22	13
Black	3,449,488	7	22	21
Other	797,978	6	22	18
Hispanic origin				
Yes	2,026,968	8	26	22
No	19,452,697	6	21	14
Not ascertained	74,428	14[1]	23[1]	19[1]

Source: "Students Avoiding Places at School out of Fear, or Ever Fearing an Attack," *Sourcebook of Criminal Justice Statistics-1992*, p. 295. Primary source: U.S. Department of Justice, Bureau of Justice Statistics, *School Crime*, NCJ-131645. Table adapted by *Sourcebook* staff. *Note:* 1. Estimate is based on 10 or fewer sample cases.

School Enrollment

★ 373 ★

School Enrollment of Blacks, by Age, 1985 to 1981

As of October. Covers civilian noninstitutional population enrolled in nursery school and above. Based on Current Population Survey.

Age, race, and Hispanic origin	Enrollment (1,000)				Rate			
	1985	1989	1990	1991	1985	1989	1990	1991
Black								
Total 3 to 34 years old	8,444	8,707	8,854	9,031	50.9	51.3	51.9	52.5
3 and 4 years old	469	407	452	428	42.7	38.9	41.6	37.2
5 and 6 years old	1,030	1,084	1,129	1,108	95.7	94.9	96.3	95.8
7 to 13 years old	3,549	3,761	3,832	3,941	99.1	99.2	99.8	99.8
14 and 15 years old	1,106	1,023	1,023	1,032	97.9	99.4	99.2	99.1
16 and 17 years old	994	1,033	962	959	91.7	93.7	91.7	91.7
18 and 19 years old	472	541	596	578	44.1	50.2	55.2	55.6
20 and 21 years old	298	309	305	329	27.7	30.7	28.4	30.0

[Continued]

★ 373 ★

School Enrollment of Blacks, by Age, 1985 to 1981

[Continued]

Age, race, and Hispanic origin	Enrollment (1,000)				Rate			
	1985	1989	1990	1991	1985	1989	1990	1991
22 to 24 years old	215	253	274	249	13.7	17.2	20.0	18.2
25 to 29 years old	192	168	162	229	7.4	6.4	6.1	8.7
30 to 34 years old	119	130	119	177	5.1	4.9	4.4	6.5
35 years old and over	233	167	238	289	1.9	1.5	2.1	2.5

Source: "School Enrollment by Age, Race, and Hispanic Origin: 1970 to 1991," U.S. Department of Commerce, *Statistical Abstract of the United States, 1993*, p. 151. Primary source: U.S. Bureau of the Census, *Current Population Reports*, series P20-469, and earlier reports.

School Teachers

★ 374 ★

School Teachers, Degrees and Teaching Experience, 1990-1991

Selected characteristics	Total[1]	Percent of teachers, by highest degree earned						Percent of teachers, by years of full-time teaching experience			
		No degree	Associate	Bachelor's	Master's	Education specialist	Doctor's	Less than 3	3 to 9	10 to 20	Over 20
					Public schools						
Total	2,559,488	0.5	0.2	51.9	42.1	4.6	0.8	9.7	26.0	39.0	25.3
Race/ethnicity											
White, non-Hispanic	2,214,097	0.5	0.2	51.5	42.7	4.5	0.7	9.7	26.3	39.0	25.1
Black, non-Hispanic	211,640	0.5	0.3	50.8	42.1	5.0	1.3	6.5	20.0	40.9	32.8
Hispanic	86,917	0.7	0.2	61.0	32.9	4.3	0.9	14.0	33.4	39.6	13.1
Asian or Pacific Islander	26,766	0.7	0.1	51.2	31.2	15.3	1.6	12.4	29.8	33.0	24.7
American Indian or Alaskan Native	20,070	0.5	0.5	64.4	30.8	3.7	0.2	15.3	28.1	36.9	20.1
					Private schools						
Total	256,285	5.3	1.1	61.9	27.0	2.9	1.8	27.5	36.6	25.0	10.9
Race/ethnicity											
White, non-Hispanic	328,624	5.1	1.1	61.8	27.3	3.0	1.8	27.2	36.6	25.1	11.1
Black, non-Hispanic	9,462	3.4	0.2	72.8	21.7	1.0	0.9	28.9	43.0	22.5	5.6
Hispanic	11,651	11.1	1.8	60.6	22.1	1.7	2.7	32.4	33.0	22.8	11.9
Asian or Pacific Islander	5,190	4.0	0.9	58.6	26.4	8.9	1.2	24.8	38.7	26.5	10.0
American Indian or Alaskan Native	1,360	20.1	0.9	50.2	26.3	2.5	0.0	43.4	24.9	24.4	7.3

Source: "Teachers in Public and Private Elementary and Secondary Schools, by Selected Characteristics: 1990-1991," *Digest of Educational Statistics, 1993*, p. 77. Primary source: U.S. Department of Education, National Center for Education Statistics, "Schools and Staffing Survey, 1990-91." (This table was prepared July 1993). *Notes:* Excludes prekindergarten teachers. Details may not add to totals because of survey item nonresponse and rounding. 1. Data are based upon a sample survey and may not be strictly comparable with data reported elsewhere.

Schools

★ 375 ★

Parental Involvement in School-Related Activities, Fall 1988

Characteristics of parents	Percent of parents[1] who talk with child regularly about			Percent of parents[1] who report family rules about			Percent of parents[1] who report that they			Percent of parents[1] who have contacted school about child's	
	Current school experiences	High school plans	Plans after high school	Numbers of hours of tele-vision watched on school days	Doing homework	Maintaining certain grade average	Never or seldom help with homework	Belong to a parent-teacher organization	Attend the parent-teacher organization meeting	Academic performance	Academic program
Total	79.4	47.1	38.3	61.7	92.0	72.7	29.4	31.9	36.2	52.5	34.8
Race/ethnicity											
Asian/Pacific Islander	59.8	41.7	36.5	67.1	89.3	74.7	42.8	29.4	41.2	36.0	29.4
Hispanic	67.1	52.7	44.8	68.7	92.3	79.8	44.7	15.5	43.0	48.3	34.5
Black, non-Hispanic	75.0	57.8	51.4	75.3	95.5	82.3	31.4	30.4	47.8	52.1	34.2
White, non-Hispanic	82.3	45.0	35.4	58.5	91.4	70.1	26.8	34.3	33.3	53.7	35.1
American Indian/Alaskan Native	72.5	44.6	39.9	62.9	95.9	7537	35.5	16.6	35.0	52.5	42.5

Source: "Parental Involvement in 8th Graders' School-Related Activities, by Selected Parental Characteristics: 1988," *Digest of Educational Statistics*, 1993, p. 31. Primary source: U.S. Department of Education, National Center for Education Statistics, National Education Longitudinal Study of 1988, "Base Year Parent Survey." (This table was prepared July 1990). *Notes:* Because of rounding, details may not add to totals. 1. The respondent was the parent most knowledgeable about the child's education. The responding parents reported on their own and their spouses' activities.

★ 376 ★

Schools

Percentage of First Grade Students Who Are 7-years Old or Older, 1972 to 1991

October	Total	Family income[1]			Sex		Race/ethnicity		
		Low	Middle	High	Male	Female	White	Black	Hispanic
1972	12.7	21.7	12.3	8.8	14.3	11.0	11.0	17.3	20.0
1973	12.9	20.9	12.8	9.2	14.6	11.1	12.0	14.7	20.7
1974	11.3	-	-	-	12.8	9.7	10.8	12.7	14.8
1975	12.3	21.4	11.8	8.9	14.1	10.5	11.3	14.8	17.6
1976	11.0	18.7	10.4	8.1	13.2	8.7	10.0	12.8	17.2
1977	12.2	16.9	12.1	9.6	15.1	9.2	12.0	13.5	12.3
1978	14.5	19.4	15.4	8.3	17.5	11.2	13.6	16.1	20.4
1979	14.8	19.1	16.0	7.7	17.0	12.2	14.7	14.6	-
1980	16.2	22.0	16.3	11.3	18.5	13.9	14.8	18.8	22.3
1981	15.6	22.9	15.0	11.3	18.8	12.0	15.5	17.0	15.7
1982	18.0	23.6	16.4	17.1	21.2	14.4	16.5	20.1	25.0
1983	16.3	22.0	16.1	11.7	17.5	15.0	14.9	17.6	23.8
1984	18.7	22.2	20.4	10.4	21.1	16.1	17.2	23.0	24.2
1985	19.0	23.8	17.8	18.2	22.0	15.8	18.5	20.3	21.4
1986	20.7	26.1	20.9	14.8	24.7	16.3	21.0	22.6	19.6
1987	21.2	30.6	19.5	17.3	23.9	18.3	21.5	20.6	20.5
1988	21.3	23.9	21.4	18.3	24.3	18.0	22.1	18.1	22.9
1989	22.3	25.8	21.9	20.8	26.1	18.3	23.8	19.3	21.3

[Continued]

★ 376 ★

Percentage of First Grade Students Who Are 7-years Old or Older, 1972 to 1991

[Continued]

October	Total	Family income[1]			Sex		Race/ethnicity		
		Low	Middle	High	Male	Female	White	Black	Hispanic
1990	23.3	27.1	23.7	18.8	26.2	20.2	24.0	21.5	22.1
1991	21.2	27.1	20.3	18.4	23.9	18.3	21.6	22.3	17.8

Source: "Percentage of First Grade Students Who are 7-years Old or Older, by Family Income, Sex, Race/ Ethnicity: 1972 to 1991," U.S. Department of Education, National Center for Education Statistics, *The Condition of Education, 1993*, p. 20. Primary source: U.S. Department of Commerce, Bureau of the Census, October Current Population Surveys. *Notes:* - Not available. The percentage of first graders who are 7 or older in October can be affected by changes in the minimum age of starting school set by states and school districts. For example, between 1984 and 1991, seven states (with about 8.3 percent of elementary school enrollment) increased the minimum age for starting school by an average of 2 months which could account for about a 1.3 percentage point increase in the percentage of first grade students who are age 7 or older. 1. Low income is defined as the bottom 20 percent of all family incomes; high income is defined as the top 20 percent of all family incomes; and middle income is defined as the 60 percent of incomes between high and low income.

★ 377 ★

Schools

Salaries and other Characteristics of School Teachers, 1990-91

Selected characteristics	Total earned income	Base salary	Number of full-time teachers	School year supplemental contract		Supplemental contract during summer		Number of teachers with Nonschool employment		
				Number of teachers	Supplementary salary	Number of teachers	Supplemental salary	Teaching or tutor	Education related	Noneducation related
				Public schools						
Total	$33,578	$31,296	2,348,315	788,215	$1,942	393,215	$1,993	109,923	67,072	229,670
Race/ethnicity										
White, non-Hispanic	33,611	31,293	2,021,075	702,746	1,977	321,128	1,935	95,488	58,916	203,859
Black, non-Hispanic	33,539	31,579	201,690	48,905	1,664	45,331	2,251	7,680	5,359	15,920
Hispanic	32,907	30,743	82,119	25,190	1,709	18,183	2,375	4,874	1,576	4,947
Asian or Pacific Islander	35,889	33,908	25,208	5,064	1,454	5,859	2,137	910	818	2,175
American Indian or Alaskan Native	30,167	27,322	18,222	6,310	1,567	2,714	1,681	971	403	2,768
				Private schools						
Total	$21,673	$19,783	301,257	60,038	$1,712	54,503	$1,864	21,438	9,622	31,492
Race/ethnicity										
White, non-Hispanic	21,569	19,709	277,539	56,645 [1]	1,695 [1]	49,853	1,832	19,742 [1]	8,556 [1]	29,532 [1]
Black, non-Hispanic	23,094	20,333	8,593	[1]	[1]	2,058	1,930	[1]	[1]	[1]
Hispanic	22,912	20,740	9,487	[1]	[1]	1,553	2,320	[1]	[1]	[1]
Asian or Pacific Islander	22,795	21,145	4,645	[1]	[1]	867 [1]	2,968 [1]	[1]	[1]	[1]
American Indian or Alaskan Native	21,373	20,128	994	[1]	[1]	[1]	[1]	[1]	[1]	[1]

Source: "Average Salaries for Full-Time Teachers in Public and Private Elementary and Secondary Schools, by Selected Characteristics," *Digest of Educational Statistics*, 1993, p. 82. Primary source: U.S. Department of Education, National Center for Education Statistics, "Schools and Staffing Survey, 1990-91." (This table was prepared July 1993.) *Notes:* Details may not add to totals because of rounding, or missing values in cells with too few cases, or survey item nonresponse. 1. Too few sample cases (fewer than 30) for a reliable estimate.

★ 378 ★
Schools

School Teachers/Administrators: Racial/Ethnic Distribution of Public School Teachers, 1961 to 1991

Item	1961	1966	1971	1976	1981	1986	1991
Number of teachers, in thousands	1,408	1,710	2,055	2,196	2,184	2,207	2,398
Race (percent)							
White	-	-	88.3	90.8	91.6	89.6	86.8
Black	-	-	8.1	8.0	7.8	6.9	8.0
Other	-	-	3.6	1.2	0.7	3.4	5.2

Source: "Selected Characteristics of Public School Teacher: Spring 1961 to Spring 1991." *Digest of Educational Statistics*, 1993, p. 79. Primary source: National Education Association, "Status of the American Public School Teacher, 1990-91." (Copyright ????????? 1992 by the National Education Association. All rights reserved.) (This table was prepared January 1993.) *Notes:* Data are based upon sample surveys of public school teachers. Data differ from figures appearing in other tables because of varying processing procedures and time period coverages. Because of rounding, percents may not add to 100.0. - stands for data not available.

★ 379 ★
Schools

Vocational and Nonvocational Teachers in Public High Schools, 1990-91

Characteristics of teachers	Total	Teacher type	
		Vocational	Nonvocational
Total	100.0	100.0	100.0
Race/ethnicity			
White	89.1	88.2	89.3
Black	6.6	8.3	6.3
Hispanic	2.8	2.1	2.9
Asian	0.8	0.7	0.8
American Indian or Alaskan Native	0.7	0.8	0.7

Source: "Percent of Vocational and Non-Vocational Public School Teachers of Grades 9 to 12, by Selected Demographic and Educational Characteristics: 1990-91." *Digest of Educational Statistics*, 1993, p. 80. Primary source: U.S. Department of Education, National Center for Education Statistics, "School and Staffing Survey, 1990-91." (This table was prepared July 1993.).

Standardized Tests

★ 380 ★

Scholastic Aptitude Test Scores, 1978-1992 - I

Racial/ethnic background	1977-78	1978-79	1979-80	1980-81	1981-82	1982-83	1983-84
SAT-Verbal							
All students	429	427	424	424	426	425	426
White	446	444	442	442	444	443	445
Black	332	330	330	332	341	339	342
Mexican-American	370	370	372	373	377	375	376
Puerto Rican	349	345	350	353	360	358	358
Asian-American	401	396	396	397	398	395	398
American Indian	387	386	390	391	388	388	390
Other	399	393	394	388	392	386	388
SAT-Mathematical							
All students	468	467	466	466	467	468	471
White	485	483	482	483	483	484	487
Black	354	358	360	362	366	369	373
Mexican-American	402	410	413	415	416	417	420
Puerto Rican	388	388	394	398	403	403	405
Asian-American	510	511	509	513	513	514	519
American Indian	419	421	426	425	424	425	427
Other	450	447	449	447	449	446	450

Source: "Scholastic Aptitude Test Score Averages, by Race/Ethnicity: 1975-76 to 1991-90." *Digest of Educational Statistics,* 1993, p. 126. Primary source: College Entrance Examination Board, *National Report on College-Bound Seniors,* various years. (Copyright 1992 by the College Entrance Examination Board. All rights reserved.) (This table was prepared January 1993.) *Notes:* Possible scores on each part of the SAT range from 200 to 800. No racial/ethnic group data are available prior to 1975-76. No data are available for 1985-86 due to changes in the Student Descriptive Questionnaire completed when students registered for the test.

★ 381 ★

Standardized Tests

Scholastic Aptitude Test Scores, 1978-1992 - II

Racial/ethnic background	1984-85	1986-87	1987-88	1988-89	1989-90	1990-91	1992
SAT-Verbal							
All students	431	430	428	427	424	422	423
White	449	447	445	446	442	441	442
Black	346	351	353	351	352	351	352
Mexican-American	382	379	382	381	380	377	372

[Continued]

★ 381 ★

Scholastic Aptitude Test Scores, 1978-1992 - II

[Continued]

Racial/ethnic background	1984-85	1986-87	1987-88	1988-89	1989-90	1990-91	1992
Puerto Rican	368	360	355	360	359	361	366
Asian-American	404	405	408	409	410	411	413
American Indian	392	393	393	384	388	393	395
Other	391	405	410	414	410	411	417
SAT-Mathematical							
All students	475	476	476	476	476	474	476
White	490	489	490	491	491	489	491
Black	376	377	384	386	385	385	385
Mexican-American	426	424	428	430	429	427	425
Puerto Rican	409	400	402	406	405	406	406
Asian-American	518	521	522	525	528	530	532
American Indian	428	432	435	428	437	437	442
Other	448	455	460	467	467	466	473

Source: "Scholastic Aptitude Test Score Averages, by Race/Ethnicity: 1975-76 to 1991-90." *Digest of Educational Statistics*, 1993, p. 126. Primary source: College Entrance Examination Board, *National Report on College-Bound Seniors*, various years. (Copyright 1992 by the College Entrance Examination Board. All rights reserved.) (This table was prepared January 1993.) *Notes:* Possible scores on each part of the SAT range from 200 to 800. No racial/ethnic group data are available prior to 1975-76. No data are available for 1985-86 due to changes in the Student Descriptive Questionnaire completed when students registered for the test.

Unemployment by Educational Level

★ 382 ★

Unemployment by Educational Level of Persons 16 Years Old and Older, 1992

Sex, race/ethnicity, and highest degree attained	Percent unemployed[1]			
	Persons 16 to 24 years old[2]			25 years and over
	Total	16 to 19 years	20 to 24 years	
White[3]				
All education levels	12.0	18.4	10.0	5.5
Less than a high school graduate	21.5	24.0	19.2	10.7
High school graduate, no college	11.5	15.6	10.3	6.0
Some college, no degree	7.8	9.5	7.5	5.4
Bachelor's degree or higher	6.3	-	6.3	3.0
Black[3]				
All education levels	28.8	41.8	24.8	10.9
Less than a high school graduate	44.4	49.4	40.1	15.1

[Continued]

★ 382 ★

Unemployment by Educational Level of Persons 16 Years Old and Older, 1992

[Continued]

Sex, race/ethnicity, and highest degree attained	Percent unemployed[1]			
	Persons 16 to 24 years old[2]			25 years and over
	Total	16 to 19 years	20 to 24 years	
High school graduate, no college	26.6	37.2	24.1	12.3
Some college, no degree	21.6	26.1	21.0	10.3
Bachelor's degree or higher	7.6	-	7.6	4.4
Hispanic origin[4]				
All education levels	16.7	26.5	13.7	9.8
Less than a high school graduate	20.3	29.0	16.5	12.8
High school graduate, no college	14.7	23.2	12.4	9.0
Some college, no degree	11.4	18.8	10.5	8.4
Bachelor's degree or higher	10.3	-	10.3	5.0

Source: Unemployment rate of persons 16 years old and over, by age, sex, race/ethnicity, and highest degree attained: 1992. *Digest of Educational Statistics,* 1993, p. 391. Primary source: U.S. Department of Labor, Bureau of Labor Statistics, Office of Employment and Unemployment Statistics, unpublished data. (This table was prepared May 1993.) *Notes:* - Data not available. Table excludes racial/ethnic data on associate degrees. 1. The unemployment rate is the percent of individuals in the labor force who are not working and who made specific efforts to find employment sometime during the prior 4 weeks. The labor force includes both employed and unemployed persons. 2. Excludes persons enrolled in school. 3. Includes persons of Hispanic origin. 4. Persons of Hispanic origin may be of any race.

United Negro College Fund Schools

★ 383 ★

Bachelor's Degrees Awarded by Area at United Negro College Fund Schools, 1989-90 and 1990-91

Shown by number and percent of total.

	1989-90		1990-91	
	No.	%	No.	%
Business & Management	1,524	30.8	1,520	29.2
Communications	228	4.6	195	3.7
Computer & Info. Science	240	4.9	180	3.4
Criminal Justice	142	2.9	144	2.8
Education	418	8.5	481	9.2
Health Professions	213	4.3	247	4.7
Humanities	262	5.3	276	5.3
Life Sciences	319	6.5	366	7.0
Mathematics	132	2.7	149	2.9
Physical Sciences	174	3.5	164	3.1
Psychology	209	4.2	271	5.2

[Continued]

★ 383 ★

Bachelor's Degrees Awarded by Area at United Negro College Fund Schools, 1989-90 and 1990-91

[Continued]

	1989-90		1990-91	
	No.	%	No.	%
Social Sciences	791	1.0	845	16.2
Visual & Performing Arts	84	1.7	88	1.7
All Other Areas	200	4.0	277	5.3
Total	4,936		5,229	

Source: "Bachelor's Degrees Awarded by Area At UNCF Colleges, 1989-90 and 1990- 91," United Negro College Fund, *Statistical Report*, 1992, p. 12. Published by permission.

★ 384 ★

United Negro College Fund Schools

Distribution of Endowment Funds at United Negro College Fund Schools, 1990-91

By percent of total expenditures.

Value of endowment	Number of UNCF institutions	Percentage of all UNCF institutions
Over $50 million	1	2
$40 - 49 million	1	2
$30 - 39 million	1	2
$20 to 29 million	1	2
$15 - 19 million	4	10
$10 - 14 million	4	10
$5 - 9 million	10	25
$1 - 4 million	18	44
Under $1 million	1	2

Source: "Distribution of Endowment Funds UNCF Member Institutions, 1990-91," United Negro College Fund, *Statistical Report*, 1992, p. 34. Published by permission.

★ 385 ★

United Negro College Fund Schools

Endowment Funds Per Student at United Negro College Fund Schools, and at All Four-Year Private Institutions, 1983-84 to 1990-91

By percent of total expenditures.

	UNCF institutions		All four year private inst.	
	Endowment per student	Percent increase since 1983-84	Endowment per student	Percent increase since 1983-84
1983-84	$4,367	-	$12,997	-
1984-85	5,416	24	15,744	21
1985-86	6,192	42	19,918	53
1987-88	7,113	63	23,784	83
1989-90	7,769	78	26,795	106
1990-91	8,117	86	30,856	137

Source: "Endowment Fund Per Student, Selected Years, 1983-84 through 1990-91 UNCF and All Four Year Private Institutions," United Negro College Fund, *Statistical Report*, 1992, p. 35. Published by permission.

★ 386 ★

United Negro College Fund Schools

Enrollment by Gender at United Negro College Fund Schools, 1990-91

UNCF institutions	Male		Female		Total
	Number	Percent	Number	Percent	
Barber-Scotia College	282	47	323	53	605
Benedict College	474	33	948	67	1,422
Bennett College	-	0	568	100	568
Bethune-Cookman College	903	40	1,370	60	2,273
Claflin College	341	40	514	60	855
Clark Atlanta University	1,257	31	2,739	69	3,996
Dillard University	411	25	1,254	75	1,665
Edward Waters College	228	36	397	64	625
Fisk University	219	26	619	74	838
Florida Memorial College	523	34	1007	66	1,530
Huston-Tillotson College	323	49	330	51	653
Interdenom. Theo. Center	230	70	101	30	331
Jarvis Christian College	253	46	295	54	548
Johnson C. Smith University	498	40	758	60	1,256
Knoxville College	721	61	456	39	1,177
Lane College	272	48	290	52	562
Lemoyne-Owen College	304	31	669	69	973

[Continued]

★ 386 ★

Enrollment by Gender at United Negro College Fund Schools, 1990-91

[Continued]

UNCF institutions	Male		Female		Total
	Number	Percent	Number	Percent	
Livingstone College	346	56	298	46	644
Miles College	366	50	366	50	732
Morehouse College	2,985	>99	7	<1	2,992
Morris College	228	33	473	67	701
Morris Brown College	868	42	1,181	58	2,049
Oakwood College	519	42	725	58	1,244
Paine College	197	34	385	66	582
Paul Quinn College	435	47	498	53	933
Philander Smith College	337	43	439	57	776
Rust College	386	36	689	64	1,075
St. Augustine's College	803	42	1,104	58	1,907
St. Paul's College	242	37	409	63	651
Shaw University	924	43	1225	57	2,149
Spelman College	-	0	1,905	100	1,905
Stillman College	256	31	566	69	822
Talladega College	321	43	430	57	751
Texas College	192	54	161	46	353
Tougaloo College	298	30	705	70	1,003
Tuskegee University	1,701	46	2,001	54	3,702
Virginia Union University	676	50	684	50	1,360
Voorhees College	287	47	326	53	613
Wilberforce University	299	36	538	64	837
Wiley College	187	43	243	57	430
Xavier University	982	32	2,117	68	3,099
Total	21,074	41	30,113	59	51,187

Source: "1991 Fall Enrollment by Gender," United Negro College Fund, *Statistical Report*, 1992, p. 43. Published by permission.

★ 387 ★

United Negro College Fund Schools

Enrollment by Race/Ethnic Group at United Negro College Fund Schools, Fall 1991

UNCF institutions	U.S. Citizens					Non-res alien	Total
	African American	White	Hispanic	Asian	Other		
Barber-Scotia College	601	2	2	-	-	-	605
Benedict College	1,375	-	-	-	-	47	1,422
Bennett College	549	1	-	-	-	18	568
Bethune-Cookman Coll.	2,191	9	7	7	10	49	2,273
Claflin College	839	5	1	1	-	9	855
Clark Atlanta Univ.	3,330	30	6	21	483	126	3,996

[Continued]

★ 387 ★

Enrollment by Race/Ethnic Group at United Negro College Fund Schools, Fall 1991
[Continued]

UNCF institutions	U.S. Citizens					Non-res alien	Total
	African American	White	Hispanic	Asian	Other		
Dillard University	1,656	5	3	1	-	-	1,665
Edward Waters College	603	2	-	1	-	19	625
Fisk University	832	1	-	-	-	5	838
Florida Memorial Coll.	1,281	17	140	-	-	92	1,530
Huston-Tillotson Coll.	508	14	45	34	-	52	653
Interdenom. Theo. Cen.	298	11	-	2	-	20	331
Jarvis Christian Coll.	542	3	1	-	1	1	548
Johnson C. Smith Univ.	1,256	-	-	-	-	-	1,256
Knoxville College	1,162	8	-	5	2	-	1,177
Lane College	561	-	-	1	-	-	562
Lemoyne-Owen College	966	2	1	-	-	4	973
Livingstone College	626	4	-	-	4	10	644
Miles College	731	1	-	-	-	-	732
Morehouse College	2,955	-	-	-	-	37	2,992
Morris College	700	1	-	-	-	-	701
Morris Brown College	1,893	-	-	2	61	94	2,049
Oakwood College	1,110	-	-	-	-	134	1,244
Paine College	570	5	1	-	-	6	582
Paul Quinn College	893	4	3	5	-	28	933
Philander Smith Coll.	732	4	-	28	-	12	776
Rust College	1,017	37	-	-	-	21	1,075
St. Augustine's Coll.	1,901	6	-	-	-	-	1,907
St. Paul's College	617	23	2	-	-	9	651
Shaw University	1,978	129	4	1	18	19	2,149
Spelman College	1,861	4	-	1	-	39	1,905
Stillman College	812	10	-	2	-	-	822
Talladega College	707	28	12	1	3	-	751
Texas College	339	1	2	-	-	11	353
Tougaloo College	999	1	-	-	-	3	1,003
Tuskegee University	3,278	126	110	39	8	141	3,702
Virginia Union Univ.	1,333	16	-	1	-	10	1,360
Voorhees College	607	-	-	6	-	-	613
Wilberforce Univ.	805	19	-	1	-	12	837
Wiley College	415	5	-	-	-	10	430
Xavier University	2,766	181	17	59	46	30	3,099
Total	49,128	715	357	219	636	1,901	51,187
Percent of total	95	1	1	<1	1	2	

Source: "Enrollment by Racial/Ethnic/Citizenship Background, Fall 1991," United Negro College Fund, *Statistical Report,* 1992, p. 41. Published by permission. *Note:* Estimated.

★ 388 ★

United Negro College Fund Schools

Expenditures by Area at United Negro College Fund Schools, 1985-86 to 1990-91

By percent of total expenditures.

	Instruction	Research	Acad supt	Pub serv	Instl supt	Student aid	Plant	Student serv	Aux enter	All otr
1985-86	23	2	5	2	19	12	10	8	13	5
1986-87	24	2	6	1	20	11	12	8	12	4
1987-88	23	2	5	1	19	15	11	8	12	4
1988-89	22	2	5	2	19	16	11	8	11	4
1989-90	21	2	5	2	19	20	10	7	11	3
1990-91	22	2	6	2	18	20	9	7	11	3

Source: "Funds Expended by Area at UNCF Institutions, 1985-86 through 1990-91 (by Percent of Total Expenditures)," United Negro College Fund, *Statistical Report,* 1992, p. 31. Published by permission.

★ 389 ★

United Negro College Fund Schools

Faculty Distribution by Area at United Negro College Fund Schools, 1985 to 1991

	Fall 1985	Fall 1987	Fall 1989	Fall 1991
Business	12	12	11	12
Humanities	23	23	22	25
Science & Math	22	22	21	23
Education	14	14	13	12
Social Sciences	15	15	15	15
Fine Arts	7	8	8	8
Health professions	4	3	2	2
Computer Science	3	3	4	3

Source: "Faculty Distribution by Area, 1985 through 1991, UNCF Member Institutions," United Negro College Fund, *Statistical Report,* 1992, p. 18. Published by permission.

★ 390 ★

United Negro College Fund Schools

Freshman Application, Admissions, and Enrollment at United Negro College Fund Schools, 1986 to 1991

Year	Completed applications	No. offered admission	Percent of applicants admitted	No. enrolled	Of those admitted, percent enrolled
1986	36,199	25,865	71	12,116	47
1987	38,973	27,439	70	13,060	48
1988	44,783	30,376	68	13,617	45
1989	43,799	34,453	79	14,394	42
1990	48,195	33,129	69	13,930	42
1991	45,906	33,040	72	13,645	41

Source: "Freshman Applications, Admissions and Enrollment at UNCF Institutions, 1986-1991." United Negro College Fund, *Statistical Report*, 1992, p. 8. Published by permission.

★ 391 ★

United Negro College Fund Schools

Gender and Racial Composition of Full-Time Faculty at United Negro College Fund Schools, 1991

	Male		Female		Total	
	No.	%	No.	%	No.	%
Black	1,000	33	915	27	1,915	60
Nonblack	852	28	350	12	1,202	40
Total	1,852	61	1,165	39	3,017	100

Source: "Gender/Racial Composition of Full-Time Faculty at UNCF Institutions, Fall 1991" United Negro College Fund, *Statistical Report*, 1992, p. 16. Published by permission.

★ 392 ★

United Negro College Fund Schools

Money Needed to Supplement Family Contributions at United Negro College Fund Schools, Fall 1990

Family income category	No. of students from category	Percent from category	Average amount needed	Total amount needed
Less than $20,000	17,344	34.8	$9,300	$161,299,200
$20,000-$24,999	4,585	9.2	9,114	41,787,690
$25,000-$29,999	4,485	9.0	8,184	36,705,240
$30,000-$39,999	8,273	16.6	6,882	56,934,786
$40,000-$49,999	3,887	7.8	4,743	18,436,041
$50,000-$59,999	2,990	6.0	1,674	5,005,260
$60,000-$74,999	3,788	7.6	186	704,568
$75,000 and above	4,487	8.9	0	0
Total	49,839		6,438	320,872,725

Source: "Amount Needed to Supplement Family Contributions of UNCF Students, Fall, 1990 (by Family Income Categories)," United Negro College Fund, *Statistical Report*, 1992, p. 21. Published by permission.

★ 393 ★

United Negro College Fund Schools

Number of Students Receiving Student Aid at United Negro College Fund Schools, 1986-87 to 1990-91

	1986-87	1987-88	1988-89	1989-90	1990-91	Percent change 1986-87 to 1990-91
Stafford Loans	22,070	22,236	21,782	23,688	25,583	30.7
Pell Grants	25,819	25,450	27,505	29,944	30,359	17.6
College Wk-Study	16,544	16,734	15,706	15,794	14,373	(13.1)
SEOG	15,414	15,410	13,902	15,409	14,781	(4.1)
State Schlrshps	17,074	15,049	13,064	13,560	14,094	(17.4)
Inst. Schlrshps	9,376	11,270	10,481	12,773	13,313	41.2

Source: "Numbers of Students Receiving Aid from Various Sources UNCF Member Institutions," United Negro College Fund, *Statistical Report*, 1992, p. 24. Published by permission.

★ 394 ★

United Negro College Fund Schools

Percentage Distribution of Degrees Awarded by Area at United Negro College Fund Schools, 1979-80 to 1990-91

By percent.

	Business	Social Sciences	Education	Humanities	Life Sciences	Math/ Engineering/ Physical Sciences	Computer Sciences	Health Professions
1979-80	24.2	22.4	19.5	11.2	6.5	6.1	<1	4.3
1981-82	26.7	18.2	17.6	10.8	6.2	6.9	<1	5.0
1983-84	31.2	16.8	14.5	7.5	6.0	7.2	1.2	4.8
1985-86	29.2	12.6	13.4	8.1	6.0	9.9	4.0	5.0
1987-88	28.7	13.0	9.9	11.6	5.5	8.7	5.9	5.9
1989-90	30.1	16.0	8.5	5.3	6.5	7.8	4.9	4.3
1990-91	29.2	16.2	9.2	5.3	7.0	8.0	3.4	4.7

Source: "Degrees Awarded by Area of Study 1979-80 through 1990-91 (by percent)," United Negro College Fund, *Statistical Report*, 1992, p. 14. Published by permission.

★ 395 ★

United Negro College Fund Schools

Source of Revenue at United Negro College Fund Schools, 1985-86 to 1990-91

	Tuition & fees	Government	Private gifts	Endowment income	Aux. enter.	Other income
1985-86	38	20	17	4	16	3
1986-87	37	22	20	4	15	3
1987-88	36	22	19	4	16	3
1988-89	37	23	19	3	14	3
1989-90	35	27	18	4	14	2
1990-91	36	27	18	3	14	2

Source: "Source of Revenues at UNCF Institutions, 1985-86 through 1990-91 (by Percent of Total Revenues)," United Negro College Fund, *Statistical Report*, 1992, p. 30. Published by permission.

★ 396 ★

United Negro College Fund Schools

Staff Composition at United Negro College Fund Schools, 1989-90 to 1991-92

	1989-90		1990-91		1991-92	
	No.	% of total	No.	% of total	No.	% of total
Administrators						
Senior admin.	327	4	276	3	269	3
Middle level admin.	792	9	910	11	814	9
Other admin.	757	9	707	8	914	10
Total administrators	1,876	22	1,893	22	1,997	22
Full-time faculty	2,796	32	2,868	33	3,036	34
Support personnel	4,024	46	3,943	45	3,963	44
Total staff	8,686	100	8,710	100	8,996	100

Source: "Staff Composition at UNCF Member Institutions, 1989-90, 1990-91, and 1991-92," United Negro College Fund, *Statistical Report*, 1992, p. 19. Published by permission.

★ 397 ★

United Negro College Fund Schools

Top Fifteen States in Students At United Negro College Fund Schools, 1988 to 1991

	1988	1989	1990	1991	Percent change 1988-1991
Alabama	3,046	3,005	2,985	3,068	0.7
California	1,078	1,184	1,412	1,562	44.9
Florida	4,625	5,206	5,278	4,852	4.9
Georgia	5,233	5,281	5,142	5,560	6.2
Illinois	1,918	1,997	2,035	2,013	4.9
Louisiana	2,865	3,232	3,356	3,470	21.1
Michigan	1,437	1,555	1,561	1,505	4.7
Mississippi	1,851	1,967	2,048	2,199	18.8
New York	1,752	1,892	1,927	1,997	14.0
North Carolina	3,309	3,398	3,694	4,128	24.7
Ohio	1,080	1,044	1,161	1,178	9.1
South Carolina	3,935	4,169	4,049	3,999	1.6
Tennessee	2,278	2,300	2,384	2,306	1.2
Texas	2,332	2,534	3,196	3,019	29.4
Virginia	1,506	1,639	1,657	1,714	13.8

Source: "15 States Sending Largest Numbers of Students to UNCF Institutions," United Negro College Fund, *Statistical Report*, 1992, p. 6. Published by permission.

★ 398 ★

United Negro College Fund Schools

Total Amount of Aid and Average Award Per Student at United Negro College Fund Schools, 1986-87 to 1990-91

	1986-87	1987-88	1988-89	1989-90	1990-91	Percent change 1986-87 to 1990-91
Total amount (in 000's)						
Stafford Loans	$46,614	$53,061	$51,579	$56,808	$74,022	58.8
Pell Grants	39,176	41,591	45,264	50,340	50,998	30.2
College Wk-Study	17,003	16,932	14,459	15,268	15,813	(7.0)
SEOG	13,237	13,447	13,290	14,038	13,510	2.1
State Schlrshps	18,239	15,553	15,799	19,584	18,248	0.1
Inst. Schlrshps	17,070	17,363	18,906	24,711	30,620	79.4
Average amounts						
Stafford Loans	2,112	2,386	2,367	2,398	2,590	22.6
Pell Grants	1,517	1,634	1,645	1,681	1,680	10.7
College Wk-Study	1,027	1,012	920	967	1,100	7.0
SEOG	859	870	955	911	914	6.4
State Schlrshps	1,068	1,033	1,209	1,444	1,296	21.2
Inst. Schlrshps	1,820	1,564	1,803	1,934	2,300	26.4

Source: "Total Amount of Aid (in 000's) from Various Sources UNCF Member Institutions," United Negro College Fund, *Statistical Report*, 1992, p. 25, and "Average Amount of Financial Award from Various Sources UNCF Member Institutions," p. 26. Published by permission.

★ 399 ★

United Negro College Fund Schools

United Negro College Fund Schools Enrollment, 1986 to 1992

	Enrollment	Percent increase over previous year	Percent increase since 1986
1986	42,613	-	-
1987	43,984	3.2	3.2
1988	45,987	4.6	7.9
1989	48,396	5.2	13.6
1990	49,816	2.9	16.9
1991	51,181	2.0	20.0
1992	53,179	3.9	24.8

Source: "Enrollment Growth, UNCF Member Institutions," 1986-1992, United Negro College Fund *Statistical Report*, 1992, p. 1. Published by permission.

★ 400 ★

United Negro College Fund Schools

United Negro College Fund Schools with Enrollment of 2000, 1992

Institution	Enrollment
Clark Atlanta	4,405
Tuskegee	3,792
Xavier	3,333
Morehouse	2,990
Shaw	2,415
Bethune-Cookman	2,301
Spelman	2,030
Morris Brown	2,019

Source: Untitled, United Negro College Fund *Statistical Report*, 1992, p. 1. Published by permission.

Chapter 6
THE FAMILY

Children

★ 401 ★

Children Below Poverty Level, by Race and Hispanic Origin, 1972 to 1991

[Persons as of **March of the following year**. Covers only related children in families under 18 years old].

YEAR	NUMBER BELOW POVERTY LEVEL (1,000)				PERCENT BELOW POVERTY LEVEL			
	All races[1]	White	Black	Hispanic[2]	All races[1]	White	Black	Hispanic[2]
1972	10,082	5,784	4,025	(NA)	14.9	10.1	42.7	(NA)
1973	9,453	5,462	3,822	1,364	14.2	9.7	40.6	27.8
1974	9,967	6,079	3,713	1,414	15.1	11.0	39.6	28.6
1975	10,882	6,748	3,884	1,619	16.8	12.5	41.4	33.1
1976	10,081	6,034	3,758	1,424	15.8	11.3	40.4	30.1
1977	10,028	5,943	3,850	1,402	16.0	11.4	41.6	28.0
1978	9,722	5,674	3,781	1,354	15.7	11.0	41.2	27.2
1979	9,993	5,909	3,745	1,505	16.0	11.4	40.8	27.7
1980	11,114	6,817	3,906	1,718	17.9	13.4	42.1	33.0
1981	12,068	7,429	4,170	1,874	19.5	14.7	44.9	35.4
1982	13,139	8,282	4,388	2,117	21.3	16.5	47.3	38.9
1983[3]	13,427	8,534	4,273	2,251	21.8	17.0	46.2	37.7
1984	12,929	8,086	4,320	2,317	21.0	16.1	46.2	38.7
1985	12,483	7,838	4,057	2,512	20.1	15.6	43.1	39.6
1986	12,257	7,714	4,039	2,413	19.8	15.3	42.7	37.1
1987[4]	12,275	7,398	4,234	2,606	19.7	14.7	44.4	38.9
1988	11,935	7,095	4,148	2,576	19.0	14.0	42.8	37.3
1989	12,001	7,165	4,257	2,496	19.0	14.1	43.2	35.5
1990	12,715	7,696	4,412	2,750	19.9	15.1	44.2	39.7
1991	13,658	8,316	4,637	2,977	21.1	16.1	45.6	39.8

Source: "Children Below Poverty Level, by Race and Hispanic Origin, 1970 to 1991," *Statistical Record of the United States, 1993*, p. 469. Primary source: U.S. Bureau of the Census, *Current Population Reports*, P60-181. *Notes:* NA Not available. 1. Includes persons of other races, not shown separately. 2. Persons of Hispanic origin may be of any race. 3. Beginning 1983, data based on revised Hispanic population controls and not directly comparable with prior years. 4. Beginning 1987, data based on revised processing procedures and not directly comparable with prior years.

★ 402 ★
Children

Children Living in the Home of Their Grandparents, 1992

[Excludes children whose parent(s) maintain the home, even though grandparents are living with them. Excludes inmates of institutions].

LIVING ARRANGEMENT	1992			
	Total[1]	White	Black	Hispanic origin[2]
Grandchild of householder under 18 years (1,000)	3,253	1,887	1,208	458
Percent of all children under 18 years	4.9	3.6	11.6	6.0
With both parents present (1,000)	502	428	25	118
With mother only present (1,000)	1,740	957	719	220
With father only present (1,000)	144	96	40	23
With neither parent present (1,000)	867	407	424	97
Total, percent distribution	100.0	100.0	100.0	100.0
With both parents present	15	23	2	26
With mother only present	53	51	60	48
With father only present	4	5	3	5
With neither parent present	27	22	35	21

Source: "Children Living in the Home of Their Grandparents, by Presence of Parents, Race, and Hispanic Origin: 1970 to 1992," *Statistical Abstract of the United States, 1993*, p. 63. Primary source: U.S. Bureau of the Census, *Current Population Reports*, P20-468, and earlier reports. *Notes:* 1. Includes other races not shown separately. 2. Persons of hispanic origin may be of any race.

★ 403 ★
Children

Children Living with Biological, Step, and Adoptive Married Couple Parents, 1980 to 1990

[**As of June**. See head note, table 77].

TYPE OF PARENT	ALL RACES[1]			WHITE			BLACK		
	1980	1985	1990	1980	1985	1990	1980	1985	1990
NUMBER (1,000)									
Own children under 18 years, total	47,248	45,347	45,448	42,329	39,942	39,732	3,775	3,816	3,671
Biological mother and father	39,523	37,213	37,026	35,852	33,202	32,975	2,698	2,661	2,336
Biological mother-stepfather	5,355	6,049	6,643	4,362	4,918	5,258	877	952	1,149
Stepmother-biological father	727	740	608	664	676	549	46	50	38
Adoptive mother and father	1,350	868	974	1,209	754	815	119	76	97
Unknown mother or father	293	479	197	242	391	135	35	77	51

[Continued]

★ 403 ★

Children Living with Biological, Step, and Adoptive Married Couple Parents, 1980 to 1990

[Continued]

TYPE OF PARENT	ALL RACES[1]			WHITE			BLACK		
	1980	1985	1990	1980	1985	1990	1980	1985	1990
PERCENT DISTRIBUTION									
Own children under 18 years, total	100.0	100.0	100.0	100.0	100.0	100.0	100.0	100.0	100.0
Biological mother and father	83.7	82.1	81.5	84.7	83.1	83.0	71.5	69.7	63.6
Biological mother-stepfather	11.3	13.3	14.6	10.3	12.3	13.2	23.2	24.9	31.3
Stepmother-biological father	1.5	1.6	1.3	1.6	1.7	1.4	1.2	1.3	1.0
Adoptive mother and father	2.9	1.9	2.1	2.9	1.9	2.1	3.1	2.0	2.6
Unknown mother or father	0.6	1.1	0.4	0.6	1.0	0.3	0.9	2.0	1.4

Source: "Children Living with Biological, Step, and Adoptive Married Couple Parents, by Race and Hispanic Origin: 1980 to 1990," *Statistical Abstract of the United States, 1993,* p. 62. Primary source: U.S. Bureau of the Census, *Current Population Reports,* P23-180. *Note:* 1. Includes other races not shown separately.

★ 404 ★

Children

Children Under 18 Years Old, by Presence of Parents, 1970 to 1992

[As of **March**. Excludes persons under 18 years old who maintained households or family groups. Based on Current Population Survey.].

RACE, HISPANIC ORIGIN, AND YEAR	Number (1,000)	PERCENT LIVING WITH-							
		Both parents	Mother only					Father only	Neither parent
			Total	Divorced	Married, spouse absent	Never married (single)	Widowed		
ALL RACES[1]									
1970	69,162	85	11	3	5	1	2	1	3
1980	63,427	77	18	8	6	3	2	2	4
1985	62,475	74	21	9	5	6	2	3	3
1990	64,137	73	22	8	5	7	2	3	3
1992	65,965	71	23	8	6	8	1	3	3
WHITE									
1970	58,790	90	8	3	3	(Z)	2	1	2
1980	52,242	83	14	7	4	1	2	2	2
1985	50,836	80	16	8	4	2	1	2	2
1990	51,390	79	16	8	4	3	1	3	2
1992	52,493	77	18	8	5	4	1	3	2
BLACK									
1970	9,422	59	30	5	16	4	4	2	10
1980	9,375	42	44	11	16	13	4	2	12
1985	9,479	40	51	11	12	25	3	3	7
1990	10,018	38	51	10	12	27	2	4	8
1992	10,427	36	54	10	12	31	1	3	7

[Continued]

★ 404 ★

Children Under 18 Years Old, by Presence of Parents, 1970 to 1992

[Continued]

RACE, HISPANIC ORIGIN, AND YEAR	Number (1,000)	Both parents	PERCENT LIVING WITH-					Father only	Neither parent
				Mother only					
			Total	Divorced	Married, spouse absent	Never married (single)	Widowed		
HISPANIC[2]									
1970	4,006[3]	78	(NA)	(NA)	(NA)	(NA)	(NA)	(NA)	(NA)
1980	5,459	75	20	6	8	4	2	2	4
1985	6,057	68	27	7	11	7	2	2	3
1990	7,174	67	27	7	10	8	2	3	3
1992	7,619	65	28	8	9	10	1	4	3

Source: "Children Under 18 Years Old, by Presence of Parents: 1970 to 1992," *Statistical Abstract of the United States, 1993*, p. 64. Primary source: U.S. Bureau of the Census, *Current Population Reports*, P20-468 and earlier reports. *Notes:* NA Not available. Z Less than 0.5 percent. 1. Includes other races not shown separately. 2. Hispanic persons may be of any race. 3. All persons under 18 years old.

★ 405 ★

Children

Children of Female Jail Inmates, 1989

Characteristic	Percent of female inmates		
	All[a]	White	Black
Have children			
No	26.2%	28.6%	23.3%
Yes	73.8	71.5	76.7
Any under age of 18	67.9	64.9	71.3
Adult only	5.9	6.6	5.4
Number of inmates	37,071	19,306	16,513
Number of children under age 18[b]			
1	37.8	38.9	35.5
2	33.4	37.0	31.3
3	17.9	14.9	21.1
4	6.4	5.3	7.9
5 or more	4.4	3.9	4.2
Live with child(ren) under 18 before entering jail[b]			
No	32.8	35.9	27.7
Yes	67.2	64.1	72.3
Where child(ren) under 18 live(s) now[b,c]	23.5	30.1	15.8
Maternal grandparents	41.6	34.9	50.0

[Continued]

★ 405 ★

Children of Female Jail Inmates, 1989
[Continued]

Characteristic	Percent of female inmates		
	All[a]	White	Black
Paternal grandparents	8.7	9.3	7.1
Other relative	22.9	18.6	27.0
Friends	4.3	4.2	4.5
Foster home	6.5	7.2	6.1
Agency/institution	1.6	2.2	0.9
Other	4.0	5.0	1.9
Plan to live with child(ren) under 18 after release from jail[b]			
Yes	84.5	77.7	91.7
No	12.4	18.1	6.3
Don't know	3.1	4.3	1.9

Source: "Children of Female Jail Inmates," *Sourcebook of Criminal Justice Statistics, 1992*, p. 599. Primary source: U.S. Department of Justice, Bureau of Criminal Statistics, *Women in Jail 1989*, Special Report NCJ-134732. *Notes:* Female inmates had an estimated total of 52,267 children under age 18. A. Includes Asians, Pacific Islanders, American Indians, Alaska Natives, and other racial groups. B. Percents are based on those inmates with children under age 18. C. Percents add to more than 100 because inmates with more than one child may have provided multiple responses.

★ 406 ★

Children

Distribution of All Children and of Poor Children, 6 Years Old, by Family Type and Race, 1991

[Excludes unrelated and foster children. Based on published and unpublished tabulations from the 1992 March Supplement to the Current Population Survey. Numbers and percentages may not add due to rounding].

RACE AND FAMILY TYPE	ALL CHILDREN		CHILDREN, BELOW POVERTY LEVEL		Poverty rate
	Number (mil.)	Percent	Number (mil.)	Percent	
All races:					
All family types	22.9	100.0	5.3	100.0	23.0
Married-couple	17.2	74.7	2.1	39.5	12.1
Single-parent	5.8	25.2	3.2	60.4	55.0
Mother-only	5.1	22.4	3.0	57.5	59.0
White, non-Hispanic:					
All family types	15.7	100.0	2.2	100.0	14.2
Married-couple	13.1	83.7	1.1	51.1	8.6
Single-parent	2.5	16.3	1.1	49.0	42.7
Mother-only	2.1	13.7	1.1	45.7	47.3

[Continued]

★ 406 ★

Distribution of All Children and of Poor Children, 6 Years Old, by Family Type and Race, 1991

[Continued]

RACE AND FAMILY TYPE	ALL CHILDREN		CHILDREN, BELOW POVERTY LEVEL		Poverty rate
	Number (mil.)	Percent	Number (mil.)	Percent	
Black, non-Hispanic:					
All family types	3.5	100.0	1.8	100.0	49.5
Married-couple	1.3	35.5	0.3	14.5	20.2
Single-parent	2.3	64.5	0.3	85.5	65.7
Mother-only	2.2	60.8	1.4	80.3	67.9
Hispanic:[2]					
All family types	2.8	100.0	1.1	100.0	40.1
Married-couple	2.0	72.0	0.6	53.6	29.9
Single-parent	0.7	28.0	0.6	46.4	66.4
Mother-only	0.7	24.6	0.4	36.8	70.4

Source: "Distribution of All Children and of Poor Children, 6 Years Old, by Family Type and Race: 1991," *Statistical Record of the United States, 1993*, p. 470. Primary source: National Center for Children in Poverty, Columbia University School of Public Health, New York, N.Y., unpublished data. Notes: 1. Includes father-only, relative-only, and nonrelative-only families. 2. Persons of Hispanic origin may be of any race.

★ 407 ★

Children

Families, by Number of Own Children Under 18 Years Old, 1970 to 1992, Part 1, Numbers

[Except as noted, as of **March** and based on Current Population Survey].

RACE, HISPANIC ORIGIN, AND YEAR	NUMBER OF FAMILIES (1,000)				
	Total	No children	One child	Two children	Three or more children
ALL FAMILIES[1]					
1970	51,586	22,774	9,398	8,969	10,445
1980	59,550	28,528	12,443	11,470	7,109
1985	62,706	31,594	13,108	11,645	6,359
1990	66,090	33,801	13,530	12,263	6,496
1992	67,173	34,427	13,615	12,364	6,768
Married couple	52,457	28,037	9,520	9,728	5,173
Male householder[2]	3,025	1,742	768	391	123
Female householder[2]	11,692	4,648	3,327	2,244	1,472
WHITE FAMILIES					
1970	46,261	20,719	8,437	8,174	8,931
1980	52,243	25,769	10,727	9,977	5,769
1985	54,400	28,169	11,174	9,937	5,120

[Continued]

★ 407 ★

Families, by Number of Own Children Under 18 Years Old, 1970 to 1992, Part 1, Numbers

[Continued]

RACE, HISPANIC ORIGIN, AND YEAR	NUMBER OF FAMILIES (1,000)				
	Total	No children	One child	Two children	Three or more children
1990	56,590	29,872	11,186	10,342	5,191
1992	57,224	30,178	11,204	10,477	5,365
BLACK FAMILIES					
1970	4,887	1,903	858	726	1,401
1980	6,184	2,364	1,449	1,235	1,136
1985	6,778	2,887	1,579	1,330	982
1990	7,470	3,093	1,894	1,433	1,049
1992	7,716	3,271	1,870	1,429	1,146
HISPANIC FAMILIES[3]					
1970	2,004	597	390	388	629
1980	3,029	946	680	698	706
1985	3,939	1,337	904	865	833
1990	4,840	1,790	1,095	1,036	919
1992	5,177	1,843	1,160	1,139	1,035

Source: "Families, by Number of Own Children Under 18 Years Old: 1970 to 1992," *Statistical Abstract of the United States, 1993,* p. 61. Primary source: U.S. Bureau of the Census, *U.S. Census of Population, 1970,* and *Current Population Reports,* P-20-476 and earlier reports. *Notes:* 1. Includes other races, not shown separately. 2. No spouse present. 3. Hispanic persons may be of any race. 1970 Hispanic data as of April and based on Census of Population.

★ 408 ★
Children

Families, by Number of Own Children Under 18 Years Old, 1970 to 1992, Part 2, Percentages and Average Number of Children

[Except as noted, as of **March** and based on Current Population Survey].

RACE, HISPANIC ORIGIN, AND YEAR	PERCENT DISTRIBUTION					Average size of family
	Total	No children	One child	Two children	Three or more children	
ALL FAMILIES[1]						
1970	100	44	18	17	20	3.58
1980	100	48	21	19	12	3.29
1985	100	50	21	19	20	3.23
1990	100	51	20	19	10	3.17
1992	100	51	20	18	10	3.17
Married couple	100	53	18	19	10	3.23
Male householder[2]	100	58	25	13	4	2.77

[Continued]

★ 408 ★

Families, by Number of Own Children Under 18 Years Old, 1970 to 1992, Part 2, Percentages and Average Number of Children

[Continued]

RACE, HISPANIC ORIGIN, AND YEAR	PERCENT DISTRIBUTION					Average size of family
	Total	No children	One child	Two children	Three or more children	
Female householder[2]	100	40	28	19	13	2.98
WHITE FAMILIES						
1970	100	45	18	18	19	3.52
1980	100	49	21	19	11	3.23
1985	100	52	21	18	9	3.16
1990	100	53	20	18	9	3.11
1992	100	53	20	18	9	3.11
BLACK FAMILIES						
1970	100	39	18	15	29	4.13
1980	100	38	23	20	18	3.67
1985	100	43	23	20	15	3.60
1990	100	41	25	19	14	3.46
1992	100	42	24	19	15	3.43
HISPANIC FAMILIES[3]						
1970	100	30	20	19	31	4.28
1980	100	31	22	23	23	3.90
1985	100	34	23	22	21	3.88
1990	100	37	23	21	19	3.83
1992	100	36	22	22	20	3.81

Source: "Families, by Number of Own Children Under 18 Years Old: 1970 to 1992," *Statistical Abstract of the United States, 1993*, p. 61. Primary source: U.S. Bureau of the Census, *U.S. Census of Population, 1970*, and *Current Population Reports*, P20-476 and earlier reports. *Notes:* 1. Includes other races, not shown separately. 2. No spouse present. 3. Hispanic persons may be of any race. 1970 Hispanic data as of April and based on Census of Population.

★ 409 ★
Children

Families, by Size and Presence of Children, 1992

[**In thousands, except as indicated**. As of **March**. Excludes members of Armed Forces except those living off post or with their families on post. Based on Current Population Survey].

CHARACTERISTIC	1992					
	White		Black		Hispanic[2]	
	Total[1]	Married couple	Total[2]	Married couple	Total[1]	Married couple
Total	57,224	47,124	7,716	3,631	5,177	3,532
Size of family:						
Two persons	24,828	19,360	2,680	1,026	1,266	696
Three persons	13,097	10,250	2,007	889	1,209	725
Four persons	11,997	10,840	1,582	886	1,238	927
Five persons	4,953	4,564	836	485	752	614
Six persons	1,534	1,393	341	219	387	314
Seven or more persons	815	717	272	126	324	255
Average per family	3.11	3.18	3.43	3.62	3.81	4.02
Own children under age 18:						
None	30,178	25,607	3,271	1,705	1,843	1,210
One	11,204	8,355	1,870	766	1,160	754
Two	10,477	8,668	1,429	685	1,139	817
Three	3,926	3,297	740	332	655	481
Four or more	1,439	1,197	406	144	380	269
Own children under age 6:						
None	44,612	36,643	5,557	2,706	3,388	2,210
One	8,722	7,135	1,471	661	1,174	871
Two or more	3,891	3,346	688	263	615	450
PERCENT DISTRIBUTION						
Total	100	100	100	100	100	100
Size of family:						
Two persons	43	41	35	28	24	20
Three persons	23	22	26	24	23	21
Four persons	21	23	21	24	24	26
Five persons	9	10	11	13	15	17
Six persons	3	3	4	6	7	9
Seven or more persons	1	2	4	3	6	7
Own children under age 18:						
None	53	54	42	47	36	34
One	20	18	24	21	22	21
Two	18	18	19	19	22	23
Three	7	7	10	9	13	14
Four or more	3	3	5	4	7	8
Own children under age 6:						
None	78	78	72	75	65	63

[Continued]

★ 409 ★

Families, by Size and Presence of Children, 1992
[Continued]

CHARACTERISTIC	1992					
	White		Black		Hispanic[2]	
	Total[1]	Married couple	Total[2]	Married couple	Total[1]	Married couple
One	15	15	19	18	23	25
Two or more	7	7	9	7	12	13

Source: "Families, by Size and Presence of Children: 1980 to 1992," *Statistical Abstract of the United States, 1993*, p. 60. Primary source: U.S. Bureau of the Census, *U.S. Census of Population, 1970*, and *Current Population Reports*, P20-476 and earlier reports. *Notes:* 1. Includes other types of families, not shown separately. 2. Hispanic persons may be of any race.

★ 410 ★

Children

Female Family Householders With No Spouse Present, 1980 to 1992

[**As of March**. Covers persons 15 years old and over. Based on Current Population Survey].

CHARACTERISTIC	Unit	WHITE			BLACK		
		1980	1990	1992	1980	1990	1992
Female family householder	1,000	6,052	7,306	7,726	2,495	3,275	3,582
Marital status:							
Never married (single)	Percent	11	15	17	27	39	42
Married, spouse absent	Percent	17	16	17	29	21	19
Widowed	Percent	33	26	23	22	17	16
Divorced	Percent	40	43	43	22	23	23
Presence of children under 18:							
No own children	Percent	41	43	42	28	32	35
With no children	Percent	59	58	58	72	68	65

Source: "Female Family Householders With No Spouse Present-Characteristics by Race and Hispanic Origin: 1980 to 1992," *Statistical Abstract of the United States, 1993*, p. 62. Primary source: U.S. Bureau of the Census, *Current Population Reports*, P20-467, and earlier reports.

★ 411 ★

Children

Living Arrangements of Children Under 18 Years Old by Age of Parents, 1992

[In thousands. As of March. Covers only those persons under 18 years old who are living with one or both parents. Characteristics are shown for the householder or reference person in married-couple situations.].

CHARACTERISTICS OF PARENT	ALL RACES[1]				WHITE				BLACK			
	Total	Living with-			Total	Living with-			Total	Living with-		
		Both parents	Mother only	Father only		Both parents	Mother only	Father only		Both parents	Mother only	Father only
Children under 18 years old	64,216	46,638	15,396	2,182	51,606	40,635	9,250	1,721	9,648	3,714	5,607	327
Age:												
15 to 24 years old	3,709	1,345	2,188	176	2,417	1,144	1,139	133	1,163	149	984	31
25 to 29 years old	8,466	5,010	3,132	325	6,337	4,328	1,761	249	1,858	527	1,266	65
30 to 34 years old	14,771	10,559	3,785	427	12,061	9,385	2,331	344	2,105	701	1,353	51
35 to 39 years old	15,733	12,140	3,152	440	13,045	10,739	1,951	355	2,038	905	1,074	59
40 to 44 years old	12,213	9,905	1,915	393	10,080	8,482	1,270	329	1,387	780	558	48
45 to 54 years old	7,904	6,516	1,090	299	6,609	5,670	712	228	851	481	332	38
55 to 64 years old	1,233	1,043	104	87	922	807	64	52	203	138	33	32
65 years old and over	188	121	30	36	134	80	22	32	43	34	6	4

Source: "Living Arrangements of Children Under 18 Years Old, by Selected Characteristics of Parents, 1992," *Statistical Abstract of the United States, 1993,* p. 63. Primary source: U.S. Bureau of the Census, *Current Population Reports,* P20-468. *Note:* 1. Includes other races, not shown separately.

★ 412 ★

Children

Living Arrangements of Children Under 18 Years Old by Family Income and Home Ownership, 1992

[In thousands. As of March. Covers only those persons under 18 years who are living with one or both parents. Characteristics are shown for the householder or reference persons in married-couple situations.].

CHARACTERISTICS OF PARENT	ALL RACES[1]				WHITE				BLACK			
	Total	Living with-			Total	Living with-			Total	Living with-		
		Both parents	Mother only	Father only		Both parents	Mother only	Father only		Both parents	Mother only	Father only
Family income:												
Under $5,000	3,789	717	2,901	171	2,113	587	1,439	88	1,526	94	1,367	65
$5,000 to $9,999	5,403	1,338	3,850	215	3,223	1,035	2,022	166	1,874	165	1,671	37
$10,000 to $14,999	4,774	2,431	2,124	219	3,460	1,993	1,294	173	1,069	271	767	33
$15,000 to $24,999	9,469	6,045	2,904	519	7,289	4,993	1,831	466	1,684	684	956	44
$25,000 to $29,999	4,886	3,783	893	210	4,043	3,226	657	160	621	360	224	38
$30,000 to $39,999	9,333	7,714	1,283	336	7,957	6,813	910	234	1,027	623	333	71
$40,000 to $49,999	8,149	7,313	634	202	7,173	6,496	496	181	671	546	112	12
$50,000 and over	18,414	17,296	808	310	16,348	15,491	602	255	1,176	972	176	27
Tenure:[2]												
Owned	39,702	33,703	4,885	1,114	34,612	30,180	3,497	935	3,432	2,104	1,200	128
Rented	24,514	12,935	10,511	1,068	16,994	10,455	5,753	786	6,215	1,609	4,407	199

Source: "Living Arrangements of Children Under 18 Years Old, by Selected Characteristics of Parents, 1992," *Statistical Abstract of the United States, 1993,* p. 63. Primary source: U.S. Bureau of the Census, *Current Population Reports,* P20-468. *Notes:* X Not applicable. 1. Includes other races, not shown separately. 2. Refers to the tenure of the householder (who may not be the child's parent).

★ 413 ★

Children

Living Arrangements of Children Under 18 Years Old by Parental Education and Employment Status, 1992

[**In thousands.** As of **March.** Covers only those persons under 18 years who are living with one or both parents. Characteristics are shown for the householder or reference person in married-couple situations.].

CHARACTERISTICS OF PARENT	ALL RACES[1]				WHITE				BLACK			
		Living with-				Living with-				Living with-		
	Total	Both parents	Mother only	Father only	Total	Both parents	Mother only	Father only	Total	Both parents	Mother only	Father only
Educational attainment:												
Less than 9th grade	4,138	2,765	1,182	191	3,504	2,460	929	169	399	173	214	12
9th to 12th grade, no diploma	7,947	4,061	3,548	337	5,471	3,372	1,836	264	2,188	506	1,618	64
High school graduate	22,858	16,122	5,905	831	17,996	14,055	3,282	659	4,016	1,447	2,431	138
Some college, no degree or associate degree	15,480	11,360	3,594	527	12,712	10,008	2,301	403	2,128	919	1,141	68
Bachelor's degree	8,652	7,573	857	222	7,442	6,611	652	179	618	440	150	28
Graduate or professional degree	5,141	4,757	310	74	4,480	4,184	250	46	299	229	52	18
Employment status:[2]												
In the civilian labor force	53,634	42,385	9,345	1,904	44,882	37,376	5,965	1,542	6,442	3,100	3,083	259
Employed	49,539	39,819	8,009	1,711	41,890	35,191	5,300	1,398	54,502	2,844	2,441	217
Both parents employed	25,692	25,692	(X)	(X)	22,435	22,435	(X)	(X)	2,107	2,107	(X)	(X)
Unemployed	4,095	2,566	1,335	193	2,992	2,185	664	143	941	256	642	43
Not in the labor force	9,552	3,251	6,046	254	5,961	2,519	3,280	162	3,003	415	2,524	64

Source: "Living Arrangements of Children Under 18 Years Old, by Selected Characteristics of Parents, 1992," *Statistical Abstract of the United States, 1993,* p. 63. Primary source: U.S. Bureau of the Census, *Current Population Reports,* P20-468. *Notes:* X Not applicable. 1. Includes other races, not shown separately. 2. Excludes children whose parent is the Armed Forces.

Families

★ 414 ★

Families Below Poverty Level, 1960 to 1991

[Families as of **March** of the following year].

YEAR	NUMBER BELOW POVERTY LEVEL (1,000)				PERCENT BELOW POVERTY LEVEL			
	All races[1]	White	Black	Hispanic[2]	All races[1]	White	Black	Hispanic[2]
1990	7,098	4,622	2,193	1,244	10.7	8.1	29.3	25.0
1991	7,712	5,022	2,343	1,372	11.5	8.8	30.4	26.5

Source: "Families Below Poverty Level: 1959 to 1991," *Statistical Abstract of the United States, 1993,* p. 473. Primary source: U.S. Bureau of the Census, *Current Population Reports,* P60-181.

★ 415 ★

Families

Families, and Nonfamily Households, by Race, Hispanic Origin, and Type, 1980 to 1992

[As of **March**, except as noted. Based on Current Population Survey].

RACE, HISPANIC ORIGIN, AND TYPE	NUMBER (1,000)				PERCENT DISTRIBUTION			
	1980	1985	1990	1992	1980	1985	1990	1992
FAMILY HOUSEHOLDS								
White, total	52,243	54,400	56,590	57,224	100	100	100	100
Married couple	44,751	45,643	46,981	47,124	86	84	83	82
Male householder[1]	1,441	1,816	2,303	2,374	3	3	4	4
Female householder[1]	6,052	6,941	7,306	7,726	12	13	13	14
Black, total	6,184	6,778	7,470	7,716	100	100	100	100
Married couple	3,433	3,469	3,750	3,631	56	51	50	47
Male householder[1]	256	344	446	504	4	5	6	7
Female householder[1]	2,495	2,964	3,275	3,582	40	44	44	46
Asian or Pacific Islander, total[2]	818	(NA)	1,531	1,624	100	(NA)	100	100
Married couple	691	(NA)	1,256	1,284	84	(NA)	82	79
Male householder[1]	39	(NA)	86	103	5	(NA)	6	6
Female householder[1]	88	(NA)	188	237	11	(NA)	12	15
Hispanic, total[3]	3,029	3,939	4,840	5,177	100	100	100	100
Married couple	2,282	2,824	3,395	3,532	75	72	70	68
Male householder[1]	138	210	329	383	5	5	7	7
Female householder[1]	610	905	1,116	1,261	20	23	23	24
NONFAMILY HOUSEHOLDS								
White, total	18,522	20,928	23,573	24,451	100	100	100	100
Male householder	7,499	8,608	9,951	10,476	40	41	42	43
Female householder	11,023	12,320	13,622	13,975	60	59	58	57
Black, total	2,402	2,730	3,015	3,367	100	100	100	100
Male householder	1,146	1,244	1,313	1,594	48	46	44	47
Female householder	1,256	1,459	1,702	1,773	52	54	56	53
Hispanic, total[3]	654	944	1,093	1,202	100	100	100	100
Male householder	365	509	587	660	56	54	54	55
Female householder	289	435	506	542	44	46	46	45

Source: "Families, and Nonfamily Households, by Race, Hispanic Origin, and Type, 1970 to 1992" *Statistical Abstract of the United States, 1993*, p. 56. Primary source: U.S. Bureau of the Census, *Persons of Spanish Origins*, PC (2)- 1C, and *Current Population Reports*, P20-476 and earlier reports. *Notes:* NA Not available. 1. No spouse present. 2. 1980 data as of April and are from 1980 Census of Population. 3. Hispanic persons may be of any race.

★ 416 ★

Families

Median Family Income, by Type of Family and Race of Householder, 1979 to 1991

| | Married Couple | | Female householder, no spouse present | |
	White	Black	White	Black
1979	$40,040	30,920	21,040	12,640
1991	41,510	33,310	19,550	11,410

Source: "Median Family Income, by Type of Family and Race of Householder: 1979 to 1991," *The Black Population in the United States: March 1992*, p. 13. Primary source: U.S. Department of of Commerce, Bureau of the Census, *Current Population Reports*, P20-471. Table compiled by editors.

★ 417 ★

Families

Percent of Families, by Type and Race of Householder, 1970 to 1992

| | Married Couple | | Female householder, no spouse present | |
	White	Black	White	Black
1970	89	68	9	28
1980	86	56	12	40
1990	83	50	13	44
1992	82	47	14	46

Source: "Percent of Families, by Type and Race of Householder: 1970 to 1992," *The Black Population in the United States: March 1992*, p. 7. Primary source: U.S. Department of of Commerce, Bureau of the Census, *Current Population Reports*, P20-471. Table compiled by editors.

★ 418 ★

Families

Selected Characteristics of Families Below the Poverty Level, 1974 to 1991

[Numbers in thousands. Families as of March of the following year].

| Characteristic | 1991 | | 1989 | | 1982 | | 1974 | |
	Black	White	Black	White	Black	White	Black	White
TYPE OF FAMILY								
All families	7,716	57,224	7,470	56,590	6,530	53,407	5,491	49,440
Number below poverty level	2,343	5,022	2,077	4,409	2,158	5,118	1,479	3,352
Percent below poverty level	30.4	8.8	27.8	7.8	33.0	9.6	26.9	6.8

[Continued]

★ 418 ★

Selected Characteristics of Families Below the Poverty Level, 1974 to 1991
[Continued]

Characteristic	1991		1989		1982		1974	
	Black	White	Black	White	Black	White	Black	White
Married couple	3,631	47,124	3,750	46,981	3,486	45,252	3,357	43,049
Number below poverty level	399	2,573	443	2,329	543	3,104	435	1,977
Percent below poverty level	11.0	5.5	11.8	5.0	15.6	6.9	13.0	4.6
Female householder, no spouse present	3,582	7,726	3,275	7,306	2,734	6,507	1,934	5,208
Number below poverty level	1,834	2,192	1,524	1,858	1,535	1,813	1,010	1,289
Percent below poverty level	51.2	28.4	46.5	25.4	56.2	27.9	52.2	24.8
Male householder, no spouse present	503	2,374	446	2,303	309	1,648	200	1,182
Number below poverty level	110	257	110	223	79	201	35	86
Percent below poverty level	21.9	10.8	24.7	9.7	25.6	12.2	17.4	7.3
Families with related children								
under 18 years	5,143	28,368	5,031	27,977	4,470	27,118	3,915	26,890
Number below poverty level	2,016	3,880	1,783	3,290	1,819	3,709	1,293	2,430
Percent below poverty level	39.2	13.7	35.4	11.8	40.7	13.7	33.0	9.0
Married couple	2,129	22,213	2,179	22,271	2,093	22,390	2,187	(NA)
Number below poverty level	263	1,715	291	1,457	360	2,005	317	(NA)
Percent below poverty level	12.4	7.7	13.3	6.5	17.2	9.0	14.5	(NA)
Female householder, no spouse present	2,771	4,967	2,624	4,627	2,199	4,037	1,623	3,244
Number below poverty level	1,676	1,969	1,415	1,671	1,401	1,584	949	1,180
Percent below poverty level	60.5	39.6	53.9	36.1	63.7	39.3	58.5	36.4
Male householder, no spouse present	243	1,188	228	1,079	178	692	105	(NA)
Number below poverty level	77	196	77	162	58	120	27	(NA)
Percent below poverty level	31.7	16.5	33.8	15.0	32.7	17.4	26.2	(NA)
Householder 65 years old and over	969	9,949	880	9,643	813	8,635	641	7,319
Number below poverty level	229	469	173	510	239	632	177	567
Percent below poverty level	23.7	4.7	19.6	5.3	29.4	7.3	27.7	7.7

Source: "Selected Characteristics of Families Below the Poverty Level: 1992, 1989, 1982, 1979, and 1974," *The Black Population in the United States: March 1992,* p. 19. Primary source: U.S. Department of of Commerce, Bureau of the Census, *Current Population Reports,* P20-471.

★ 419 ★
Families

Selected Characteristics of Families By Race of Householder, March 1992, The North and West, Part 1

[Numbers in thousands.].

Characteristic	Black				White			
	Total	Married couple families	Other families		Total	Married couple families	Other families	
			Female householder, no spouse present	Male householder, no spouse present			Female householder, no spouse present	Male householder, no spouse present
NORTH AND WEST								
Total, all families	3,463	1,536	1,696	231	38,203	31,188	5,331	1,684
Size of Family								
Percent	100.0	100.0	100.0	100.0	100.0	100.0	100.0	100.0
Two persons	35.9	28.7	38.6	63.5	42.6	40.4	49.4	61.8
Three persons	25.8	24.1	27.3	26.2	22.3	20.8	30.5	24.7
Four persons	20.1	24.8	17.6	6.7	21.2	23.4	12.8	8.0
Five persons	10.5	12.4	9.7	2.7	9.3	10.5	4.7	2.7
Six persons	3.7	5.7	2.3	0.9	3.0	3.3	1.6	1.7
Seven or more persons	4.0	4.2	4.4	-	1.6	1.7	1.0	1.2
Age of Householder								
Percent	100.0	100.0	100.0	100.0	100.0	100.0	100.0	100.0
15 to 34 years	33.5	23.5	42.7	32.6	24.2	22.5	31.6	32.4
35 to 44 years	26.7	28.4	25.6	23.0	26.4	26.1	28.7	26.2
45 to 54 years	17.2	19.8	14.7	17.7	18.3	18.7	15.9	18.8
55 years and over	22.7	28.4	17.0	26.7	31.1	32.8	23.8	22.5
Number of Earners								
Percent	100.0	100.0	100.0	100.0	100.0	100.0	100.0	100.0
No earners	22.6	11.9	33.8	12.2	14.3	13.3	21.5	8.6
One earner	35.1	21.1	45.0	55.2	26.5	21.8	47.4	48.6
Two earners	32.1	49.3	17.4	25.8	45.0	49.1	24.2	33.7
Three or more earners	10.2	17.7	3.8	6.8	14.3	15.8	6.9	9.1

Source: "Selected Characteristics of Families, by Type, Region, and Race of Householder," *The Black Population in the United States: March 1992*, p. 36. Primary source: U.S. Department of of Commerce, Bureau of the Census, *Current Population Reports*, P20-471.

★ 420 ★
Families

Selected Characteristics of Families By Race of Householder, March 1992, The North and West, Part 2

[Numbers are in thousands.].

Characteristic	Black				White			
	Total	Married couple families	Other families		Total	Married couple families	Other families	
			Female householder, no spouse present	Male householder, no spouse present			Female householder, no spouse present	Male householder, no spouse present
NORTH AND WEST Total, all families	3,463	1,536	1,696	231	38,203	31,188	5,331	1,684
Related Children Under 18 Years								
Percent	100.0	100.0	100.0	100.0	100.0	100.0	100.0	100.0
No related children	33.3	42.6	21.5	58.1	49.7	52.3	34.0	50.0
With related children	66.7	57.4	78.5	41.9	50.3	47.7	66.0	50.0
One child	27.8	23.1	31.9	29.1	20.2	17.7	31.3	30.2
Two children	20.5	20.0	22.3	10.3	19.2	19.0	22.2	14.5
Three children	11.3	9.2	14.5	1.9	7.8	7.9	8.5	3.5
Four or more children	7.1	5.1	9.7	0.6	3.2	3.1	3.9	1.7
Own Children Under 18 Years								
Percent	100.0	100.0	100.0	100.0	100.0	100.0	100.0	100.0
No own children	41.3	48.2	31.7	65.2	51.8	53.6	39.9	55.8
With own children	58.7	51.8	68.3	34.8	48.2	46.4	60.1	44.2
One child	24.7	20.2	28.9	24.6	19.2	17.2	28.4	27.1
Two children	18.7	18.9	19.8	9.1	18.6	18.6	20.2	13.4
Three children	9.5	8.5	11.6	1.1	7.5	7.7	7.9	2.7
Four or more children	5.8	4.3	8.0	-	2.9	2.9	3.5	1.0
Own Children Under 6 Years								
Percent	100.0	100.0	100.0	100.0	100.0	100.0	100.0	100.0
No own children	70.2	75.3	64.0	82.9	76.9	76.9	76.0	81.5
With own children	29.8	24.7	36.0	17.1	23.1	23.1	24.0	18.5
One child	20.3	17.3	23.8	15.1	15.6	15.4	17.4	14.3
Two children	7.5	6.5	9.2	1.4	6.4	6.7	5.5	3.7
Three children	1.5	0.8	2.1	0.7	1.0	1.0	1.0	0.6
Four or more children	0.5	0.1	1.0	-	0.1	0.1	0.2	-

Source: "Selected Characteristics of Families, by Type, Region, and Race of Householder," *The Black Population in the United States: March 1992*, p. 36. Primary source: U.S. Department of of Commerce, Bureau of the Census, *Current Population Reports*, P20-471.

★ 421 ★
Families

Selected Characteristics of Families By Race of Householder, March 1992, The South, Part 1

[Numbers in thousands.].

| Characteristic | Black | | | | White | | | |
| | | | Other families | | | | Other families | |
	Total	Married couple families	Female householder, no spouse present	Male householder, no spouse present	Total	Married couple families	Female householder, no spouse present	Male householder, no spouse present
SOUTH								
Total, all families	4,253	2,094	1,886	273	19,023	15,936	2,396	690
Size of Family								
Percent	100.0	100.0	100.0	100.0	100.0	100.0	100.0	100.0
Two persons	33.8	27.9	37.6	52.7	45.0	42.5	56.4	64.6
Three persons	26.1	24.8	28.0	24.0	24.1	23.7	27.6	21.2
Four persons	20.8	24.1	18.6	11.3	20.4	22.3	10.8	12.0
Five persons	11.1	14.0	8.3	8.5	7.3	8.1	3.5	1.0
Six persons	5.0	6.2	4.1	1.6	2.1	2.3	0.8	1.1
Seven or more persons	3.1	3.0	3.5	1.8	1.1	1.1	0.9	0.1
Age of Householder								
Percent	100.0	100.0	100.0	100.0	100.0	100.0	100.0	100.0
15 to 34 years	30.9	25.5	37.6	26.2	24.9	24.2	27.2	32.4
35 to 44 years	28.1	28.5	27.8	27.0	24.3	24.0	25.6	27.2
45 to 54 years	14.5	16.2	13.2	10.7	18.5	18.7	17.8	15.9
55 years and over	26.5	29.8	21.4	36.0	32.3	33.1	29.4	24.4
Number of Earners								
Percent	100.0	100.0	100.0	100.0	100.0	100.0	100.0	100.0
No earners	17.9	10.6	26.1	17.9	15.3	14.7	20.4	9.9
One earner	33.5	18.7	48.1	46.4	26.9	22.3	49.8	53.1
Two earners	37.9	53.9	21.2	29.9	46.1	50.1	23.7	31.3
Three or more earners	10.7	16.8	4.6	5.7	11.7	12.8	6.1	5.7

Source: "Selected Characteristics of Families, by Type, Region, and Race of Householder," *The Black Population in the United States: March 1992*, p. 35. Primary source: U.S. Department of of Commerce, Bureau of the Census, *Current Population Reports*, P20-471.

★ 422 ★

Families

Selected Characteristics of Families By Race of Householder, March 1992, The South, Part 2

[Numbers in thousands.].

Characteristic	Black				White			
			Other families				Other families	
	Total	Married couple families	Female householder, no spouse present	Male householder, no spouse present	Total	Married couple families	Female householder, no spouse present	Male householder, no spouse present
SOUTH								
Total, all families	4,253	2,094	1,886	273	19,023	15,936	2,396	690
Related Children Under 18 Years								
Percent	100.0	100.0	100.0	100.0	100.0	100.0	100.0	100.0
No related children	33.4	40.4	23.6	46.2	51.9	53.9	39.4	50.0
With related children	66.6	59.6	76.4	53.8	48.1	46.1	60.6	50.0
One child	26.9	23.9	30.3	25.4	21.9	19.9	33.6	28.0
Two children	21.8	20.6	23.7	17.1	18.4	18.4	18.8	17.2
Three children	11.1	10.1	12.9	6.6	5.9	5.9	6.0	4.6
Four or more children	6.9	4.9	9.4	4.6	1.9	2.0	2.1	0.2
Own Children Under 18 Years								
Percent	100.0	100.0	100.0	100.0	100.0	100.0	100.0	100.0
No own children	43.3	46.1	37.6	62.0	54.7	55.8	46.3	57.6
With own children	56.7	53.9	62.4	38.0	45.3	44.2	53.7	42.4
One child	23.8	21.8	26.5	20.8	20.3	18.8	29.6	23.7
Two children	18.4	18.8	18.9	11.3	17.7	17.9	17.0	14.4
Three children	9.7	9.6	10.5	3.7	5.6	5.7	5.5	4.3
Four or more children	4.8	3.7	6.4	2.2	1.7	1.9	1.5	-
Own Children Under 6 Years								
Percent	100.0	100.0	100.0	100.0	100.0	100.0	100.0	100.0
No own children	73.5	74.0	71.4	83.8	80.0	79.5	81.3	86.9
With own children	26.5	26.0	28.6	16.2	20.0	20.5	18.7	13.1
One child	18.0	18.9	18.1	11.0	14.5	14.7	14.6	9.7
Two children	5.8	5.2	6.6	4.2	4.9	5.1	3.6	3.2
Three children	2.5	1.8	3.4	0.6	0.5	0.6	0.4	0.2
Four or more children	0.3	0.1	0.5	0.4	0.1	0.1	0.1	-

Source: "Selected Characteristics of Families, by Type, Region, and Race of Householder," *The Black Population in the United States: March 1992*, p. 35. Primary source: U.S. Department of of Commerce, Bureau of the Census, *Current Population Reports*, P20-471.

★ 423 ★

Families

Selected Characteristics of Families By Race of Householder, March 1992, The United States, Part 1

[Numbers in thousands.].

Characteristic	Black				White			
			Other families				Other families	
	Total	Married couple families	Female householder, no spouse present	Male householder, no spouse present	Total	Married couple families	Female householder, no spouse present	Male householder, no spouse present
UNITED STATES								
Total, all families	7,716	3,631	3,582	504	57,225	47,124	7,727	2,374
Size of Family								
Percent	100.0	100.0	100.0	100.0	100.0	100.0	100.0	100.0
Two persons	34.7	28.2	38.1	57.6	43.4	41.1	51.5	62.6
Three persons	26.0	24.5	27.7	25.0	22.9	21.8	29.6	23.7
Four persons	20.5	24.4	18.1	9.2	21.0	23.0	12.2	9.1
Five persons	10.8	13.4	9.0	5.8	8.7	9.7	4.4	2.2
Six persons	4.4	6.0	3.2	1.3	2.7	3.0	1.4	1.5
Seven or more persons	3.5	3.5	3.9	1.0	1.4	1.5	1.0	0.9
Age of Householder								
Percent	100.0	100.0	100.0	100.0	100.0	100.0	100.0	100.0
15 to 34 years	32.1	24.6	40.0	29.1	24.4	23.1	30.3	32.4
35 to 44 years	27.5	28.5	26.8	25.2	25.7	25.4	27.7	26.5
45 to 54 years	15.7	17.7	13.9	13.9	18.4	18.7	16.5	18.0
55 years and over	24.8	29.2	19.3	31.7	31.5	32.9	25.6	23.1
Number of Earners								
Percent	100.0	100.0	100.0	100.0	100.0	100.0	100.0	100.0
No earners	20.1	11.2	29.7	15.3	14.6	13.8	21.2	9.0
One earners	34.2	19.7	46.7	50.5	26.6	21.9	48.1	49.9
Two earners	35.3	52.0	19.4	28.1	45.3	49.5	24.0	33.0
Three or more earners	10.4	17.2	4.2	6.2	13.4	14.8	6.7	8.1

Source: "Selected Characteristics of Families, by Type, Region, and Race of Householder," *The Black Population in the United States: March 1992*, p. 34. Primary source: U.S. Department of of Commerce, Bureau of the Census, *Current Population Reports*, P20-471.

★ 424 ★

Families

Selected Characteristics of Families By Race of Householder, March 1992, The United States, Part 2

[Numbers in thousands.].

Characteristic	Black				White			
			Other families				Other families	
	Total	Married couple families	Female householder, no spouse present	Male householder, no spouse present	Total	Married couple families	Female householder, no spouse present	Male householder, no spouse present
Related Children Under 18 Years								
Percent	100.0	100.0	100.0	100.0	100.0	100.0	100.0	100.0
No related children	33.3	41.4	22.6	51.6	50.4	52.9	35.7	50.0
With related children	66.7	58.6	77.4	48.4	49.6	47.1	64.3	50.0
One child	27.3	23.6	31.1	27.1	20.7	18.4	32.1	29.5
Two children	21.2	20.3	23.0	14.0	19.0	18.8	21.2	15.3
Three children	11.2	9.7	13.7	4.4	7.1	7.2	7.7	3.9
Four or more children	7.0	5.0	9.6	2.8	2.8	21.7	3.3	1.3
Own Children Under 18 Years								
Percent	100.0	100.0	100.0	100.0	100.0	100.0	100.0	100.0
No own children	42.4	46.9	34.8	63.4	52.8	54.3	41.9	56.3
With own children	57.6	53.0	65.2	36.6	47.2	45.7	58.1	43.7
One child	24.2	21.1	27.6	22.6	19.6	17.7	28.8	26.1
Two children	18.5	18.9	19.3	10.3	18.3	18.4	19.2	13.7
Three children	9.6	9.1	11.0	2.5	6.9	7.0	7.2	3.2
Four or more children	5.3	3.9	7.2	1.2	2.5	2.5	2.9	0.7
Own Children Under 6 Years								
Percent	100.0	100.0	100.0	100.0	100.0	100.0	100.0	100.0
No own children	72.0	74.5	67.9	83.4	78.0	77.8	77.6	83.1
With own children	28.0	25.5	32.1	16.6	22.0	22.2	22.4	16.9
One child	19.1	18.2	20.8	12.9	15.2	15.1	16.6	12.9
Two children	6.5	5.8	7.8	2.9	5.9	6.2	4.9	3.5
Three children	2.0	1.4	2.8	0.6	0.9	0.9	0.8	0.5
Four or more children	0.4	0.1	0.7	0.2	0.1	0.1	0.1	-

Source: "Selected Characteristics of Families, by Type, Region, and Race of Householder," *The Black Population in the United States: March 1992*, p. 34. Primary source: U.S. Department of of Commerce, Bureau of the Census, *Current Population Reports*, P20-471.

★ 425 ★

Families

Selected Characteristics of Families, by Race, 1992

[Numbers in thousands].

Characteristic	1992	
	Black	White
TYPE OF FAMILY		
All families	7,716	57,225
Percent	100.0	100.0
Married couple	47.1	82.3
Female householder, no spouse present	46.4	13.5
Male householder, no spouse present	6.5	4.1
CHILDREN UNDER 18 YEARS		
BY PRESENT OF PARENTS[1]		
Children in families	10,427	52,493
Percent living with-		
Both parents	35.6	77.4
Mother only	53.8	17.6
Father only	3.1	3.3
Neither parent	7.5	1.7

Source: "Selected Characteristics of Families, by Race: March 1992, 1990, 1980, and 1970," *The Black Population in the United States: March 1992*, p. 7. Primary source: U.S. Department of of Commerce, Bureau of the Census, *Current Population Reports*, P20-471. *Notes:* 1. Excludes persons under 18 years old who were maintaining households of family groups.

Income

★ 426 ★

Median Weekly Earnings of Families, 1989 to 1993

[Numbers in thousands].

Type of family, number of earners, race, and Hispanic origin	Number of families		Median weekly earnings	
	1992	1993	1992	1993
TOTAL				
Total families with earners[1]	44,137	44,383	$688	$707
WHITE				
Total families with earners[1]	37,378	37,458	716	739
Married-couple families	30,337	30,288	791	816
One earner	10,852	10,790	483	492

[Continued]

★ 426 ★

Median Weekly Earnings of Families, 1989 to 1993
[Continued]

Type of family, number of earners, race, and Hispanic origin	Number of families		Median weekly earnings	
	1992	1993	1992	1993
Husband	7,844	7,755	561	583
Wife	2,384	2,383	296	313
Two or more earners	19,485	19,497	954	984
Husband and wife	17,432	17,572	975	1,004
Families maintained by women	5,226	5,355	409	415
Families maintained by men	1,814	1,816	545	547
BLACK				
Total families with earners'	5,188	5,268	478	490
Married-couple families	2,723	2,698	646	674
One earner	895	909	309	344
Husband	486	539	359	381
Wife	330	287	279	321
Two or more earners	1,828	1,789	806	846
Husband and wife	1,609	1,590	834	873
Families maintained by women	2,079	2,168	328	334
Families maintained by men	386	403	412	413

Source: "Median Weekly Earnings of Families By Type of Family, Number of Earners, Race, and Hispanic Origin," *Employment and Earnings,* p. 239. Primary source: U.S. Department of of Labor, Bureau of Labor Statistics, *Employment and Earnings* 41 (January 1994). *Notes:* 1. Data exclude families in which there is no wage or salary earner or in which the husband, wife or other person maintaining the family is either self-employed or in the Armed Forces.

★ 427 ★

Income

Persons Below Poverty Level, by Race of Householder and Family Status, 1979 to 1991

[Persons as of March of following year].

RACE OF HOUSEHOLDER AND FAMILY STATUS	NUMBER BELOW POVERTY LEVEL (mil.)				PERCENT BELOW POVERTY LEVEL			
	1979	1989[2]	1990	1991	1979[1]	1989[2]	1990	1991
All persons[3]	26.1	31.5	33.6	35.7	11.7	12.8	13.5	14.2
In families	20.0	24.1	25.2	27.1	10.2	11.5	12.0	12.8
Householder	5.5	6.8	7.1	7.7	9.2	10.3	10.7	11.5
Related children under 18 years	10.0	12.0	12.7	13.7	16.0	19.0	19.9	21.1
Other family members	4.5	5.3	5.4	5.8	6.1	6.6	6.7	7.2
Unrelated individuals	5.7	6.8	7.4	7.8	21.9	19.2	20.7	21.1
White	17.2	20.8	22.3	23.7	9.0	10.0	10.7	11.3
In families	12.5	15.2	15.9	17.3	7.4	8.6	9.0	9.7
Householder	3.6	4.4	4.6	5.0	6.9	7.8	8.1	8.8
Related children under 18 years	5.9	7.2	7.7	8.3	11.4	14.1	15.1	16.1
Other family members	3.0	3.6	3.6	3.9	4.7	5.3	5.2	5.7
Unrelated individuals	4.5	5.1	5.7	5.9	19.7	16.9	18.6	18.8
Black	8.1	9.3	9.8	10.2	31.0	30.7	31.9	32.7
In families	6.8	7.7	8.2	8.5	30.0	29.7	31.0	32.0
Householder	1.7	2.1	2.2	2.3	27.8	27.8	29.3	30.4
Related children under 18 years	3.7	4.3	4.4	4.6	40.8	43.2	44.2	45.6

[Continued]

★ 427 ★

Persons Below Poverty Level, by Race of Householder and Family Status, 1979 to 1991

[Continued]

RACE OF HOUSEHOLDER AND FAMILY STATUS	NUMBER BELOW POVERTY LEVEL (mil.)				PERCENT BELOW POVERTY LEVEL			
	1979	1989[2]	1990	1991	1979[1]	1989[2]	1990	1991
Other family members	1.3	1.4	1.6	1.5	18.2	15.9	17.6	17.6
Unrelated individuals	1.2	1.5	1.5	1.6	37.3	35.2	35.1	35.3

Source: "Persons Below Poverty Level, by Race of Householder and Family Status: 1979 to 1991," *Statistical Record of the United States, 1993*, p. 471. Primary source: U.S. Bureau of the Census, *Current Population Reports,* P60-181 and unpublished data. *Notes:* 1. Population controls based on 1980 census; see text section 14. 2. Beginning 1987, data based on revised processing procedures and not directly comparable with prior years. 3. Includes races, members of unrelated subfamilies, and unrelated individuals (for families with female householder) not shown separately.

Structure and Composition

★ 428 ★

Characteristics of Families with Children under 18, 1991

Numbers in thousands.

Family characteristics	Characteristics of families with own children under 18, by family status[1]											
	White[2]				Black[2]				Hispanic origin[3]			
	Total	Married-couple families	Other families		Total	Married-couple families	Other families		Total	Married-couple families	Other families	
			Male householder, no spouse present	Female householder, no spouse present			Male householder, no spouse present	Female householder, no spouse present			Male householder, no spouse present	Female householder, no spouse present
Total families	56,803	47,014	2,276	7,512	7,471	3,569	472	3,430	4,981	3,454	342	1,186
Total families with												
own children under 18	26,794	21,531	925	4,337	4,380	1,884	202	2,294	3,203	2,273	131	799
Percent of all families	47.2	45.8	40.6	57.7	58.6	52.8	42.8	66.9	64.3	65.8	38.3	67.4
Percent distribution	100.0	3.5	3.5	16.2	100.0	43.0	4.6	52.4	100.0	71.0	4.1	24.9
Families with												
1 child under 18	10,955	8,181	548	2,226	1,847	762	116	968	1,105	727	69	309
2 children under 18	10,381	8,671	285	1,425	1,431	646	63	722	1,136	816	44	277
3 children under 18	3,972	3,392	71	509	691	312	14	364	596	453	15	129
4 children under 18	1,086	939	17	130	258	105	5	149	253	192	2	59
5 children under 18	264	224	5	34	103	47	3	53	78	58	1	19
6 or more under 18	137	123	-	13	50	12	-	38	34	27	-	7
Total own children under 18	48,803	40,285	1,384	7,134	8,245	3,583	314	4,348	6,796	4,949	217	1,630
Average number of children												
per family with children	1.82	1.87	1.50	1.64	1.99	1.90	1.55	1.90	2.12	2.18	1.65	2.04
Total families with												
own children under 6	12,727	10,681	337	1,709	2,089	933	90	1,066	1,732	1,305	65	362
Percent of all families	22.4	22.7	14.8	22.8	28.0	26.1	19.1	31.1	34.8	37.8	19.0	30.5
Percent distribution	100.0	83.9	2.6	13.4	100.0	44.7	4.3	51.0	100.0	75.3	3.8	20.9
Families with												
1 child under 6	8,639	7,096	262	1,281	1,422	683	70	668	1,145	858	48	239
2 children under 6	3,511	3,083	68	359	503	214	18	271	475	366	16	94
3 children under 6	532	460	7	65	128	28	2	99	101	72	1	28
4 or more under 6	45	41	-	3	36	9	-	27	11	9	-	1
Total own children under 6	16,955	14,4230	414	2,121	2,779	1,168	106	1,505	2,466	1,864	84	518
Average number of children												
per family with children	1.33	1.35	1.23	1.24	1.33	1.25	1.18	1.41	1.42	1.43	-	1.43
Total families with												
own children under 3	7,593	6,507	193	892	1,236	535	49	651	1,104	831	50	223
Percent of families	13.4	13.8	8.5	11.9	16.5	15.0	10.4	19.0	22.2	24.1	14.6	18.8
Percent distribution	100.0	85.7	2.6	11.7	100.0	43.3	4.0	52.7	100.0	75.3	4.5	20.2
Families with												
1 child under 3	6,695	5,720	180	795	1,015	464	38	512	945	710	44	190
2 or more under 3	898	788	13	97	221	71	11	139	159	121	5	33
Total own children under 3	8,566	7,368	210	989	1,417	592	57	768	1,325	1,001	57	267

[Continued]

★ 428 ★

Characteristics of Families with Children under 18, 1991

[Continued]

Family characteristics	Characteristics of families with own children under 18, by family status[1]											
	White[2]				Black[2]				Hispanic origin[3]			
			Other families				Other families				Other families	
	Total	Married-couple families	Male householder, no spouse present	Female householder, no spouse present	Total	Married-couple families	Male householder, no spouse present	Female householder, no spouse present	Total	Married-couple families	Male householder, no spouse present	Female householder, no spouse present
Average number of children per family with children	1.13	1.13	1.09	1.11	1.15	1.11	-	1.18	1.20	1.20	-	1.19

Source: "Characteristics of Families with Own Children under 18, by Family Status and Race/Ethnicity: 1991." *Digest of Educational Statistics*, 1993, p. 27. Primary source: U.S. Department of Commerce, Bureau of the Census, *Current Population Reports*, unpublished data. (This table was prepared April 1992). *Notes:* Averages and percents are only shown when the base is 75,000 or greater. Even though the standard errors are large, smaller estimated numbers are shown to permit users to combine categories in various ways. Because of rounding, details may not add to totals. - Data not available. 1. Race of family is defined as race of head of household. 2. Includes persons of Hispanic origin. 3. Persons of Hispanic origin may be of any race.

Chapter 7
HEALTH AND MEDICAL CARE

Births

★ 429 ★

Anemia Rates During Pregnancy per 1,000 Live Births, By Maternal Age and Race, 1989

Age	Rate	
	White	Black
Less than 20 years	24.4	42.2
20-24 years	18.3	39.7
25-29 years	13.6	32.2
30-34 years	12.1	29.8
35-39 years	12.5	28.1
40-49 years	14.3	26.8

Source: "Anemia rates During Pregnancy per 1,000 Live Births, By Maternal Age and Race, 1989," Interagency Board for Nutrition Monitoring and Related Research, *Nutrition Monitoring in the United States: Chartbook I: Selected Findings From the National Nutrition Monitoring and Related Research Program*, p. 97. Primary source: Centers for Disease Control, National Center for Health Statistics. *Notes:* Anemia was diagnosed during pregnancy, and was defined as a hemoglobin less than 10.0 grams per deciliter or a hematocrit less than 30 percent. Information about anemia on U.S. certificates of live birth was collected from the medical record. Findings exclude data for Louisiana, Nebraska, and Oklahoma, which did not report medical risk factors.

★ 430 ★

Births

Low Birth Weight as a Percent of Total Live Births of 40 Weeks Gestation or Longer, by Maternal Weight Gain and Race of Mother, 1989

Maternal weight gain	Percent of births	
	White	Black
Less than 16 pounds	2.8	6.8
16-20 pounds	2.2	5.4
21-25 pounds	1.5	3.7
26-30 pounds	1.1	3.0
31-35 pounds	0.9	2.4
36-40 pounds	0.9	2.3
41 or more pounds	0.8	1.8

Source: "Low Birth Weight as a Percent of Total Live Births of 40 Weeks Gestation or Longer, by Maternal Weight Gain and Race of Mother, 1989," Interagency Board for Nutrition Monitoring and Related Research, *Nutritional Monitoring in the United States: Chartbook I: Selected Findings From the National Nutrition Monitoring and Related Research Program*, p. 99. Primary source: Centers for Disease Control and Prevention, National Center for Health Statistics. *Notes:* Refers to married and unmarried mothers. Data were obtained from 1989 U.S. certificates of live births. The Institute of Medicine recommends that women of normal pregnancy body mass index (BMI) gain 25 to 35 pounds during pregnancy and that young adolescents and black women strive for weight gains at the upper end of this range. Body mass index is an index that relates weight to stature (BMI = weight/height2.). Normal BMI was defined as 19.8 to 26.0 kilograms per meter2. . Findings exclude data for California, Louisiana, Nebraska, and Oklahoma, which did not require reporting of weight gain during pregnancy.

★ 431 ★

Births

Low Birth Weight as a Percent of Total Live Births, by Age of Mother and Race of Infant, 1986-88

Age of mother	Percent			
	All races	White	Black	American Indian and Alaska Native
Less than 15 years	13.7	10.4	16.2	7.7
15-19 years	9.3	7.7	13.1	6.0
20-24 years	7.1	5.8	12.3	5.2
25-29 years	6.1	5.1	12.5	5.6
30-34 years	6.2	5.2	13.0	6.5

[Continued]

★ 431 ★

Low Birth Weight as a Percent of Total Live Births, by Age of Mother and Race of Infant, 1986-88

[Continued]

Age of mother	Percent			
	All races	White	Black	American Indian and Alaska Native
35-39 years	6.9	6.0	13.4	7.2
40 years and over	7.9	7.1	12.9	8.7

Source: "Low Birth Weight as a Percent of Total Live Births, by Age of Mother and Race of Infant, 1986-88," Interagency Board for Nutrition Monitoring and Related Research, *Nutritional Monitoring in the United States: Chartbook I: Selected Findings From the National Nutrition Monitoring and Related Research Program*, p. 99. Primary source: Data analyzed by the Indian Health Service from data compiled by the Centers for Disease Control and Prevention, National Center for Health Statistics. *Notes:* Data were obtained from U.S. certificates of live birth. Information for American Indians and Alaska Natives are for mothers who reside in IHS service areas that include counties containing reservations of Federally recognized tribes and in some cases surrounding counties. The racial designation of the infant, used in this report, is determined from the race of the parents as entered on the birth certificate. Data for all races, whites, and blacks are from 1987; data for American Indians and Alaska Natives are from 1986-88. The number of American Indian and Alaska Native vital events that occur each year is usually small when categorized by age of mother, geographic location, and other factors. For this population, 3- year natality rates were used to reduce the low frequency effect. The 3-year percent shown above is calculated based on the total number of low- birth weight births that occurred during the 3-year period under study divided by the total number of live births during these 3 years. The data exclude 165 American Indian and Alaska Native live births, 4,885 all-races live births, and 3,571 white births where birth weight was not stated.

★ 432 ★

Births

Percent of Babies Breastfed at All, by Year of Birth and Race of Mother, 1970-87

Race	Year of birth								
	1970-71	1972-73	1974-75	1976-77	1978-79	1980-81	1982-83	1984-85	1986-87
	Percent								
Total	24.36	23.6	32.4	42.2	44.3	52.5	57.3	55.5	55.0
White	25.8	25.6	35.5	46.2	48.1	57.2	62.3	59.9	60.3
Black	12.8	12.9	16.4	19.5	24.4	24.5	27.0	22.9	23.6

Source: "Percent of Babies Breastfed at All, by Year of Birth and Race of Mother, 1970-87," Interagency Board for Nutrition Monitoring and Related Research, *Nutrition Monitoring in the United States: Chartbook I: Selected Findings From the National Nutrition Monitoring and Related Research Program*, p. 98. Primary source: Centers for Disease Control and Prevention, National Center for Health Statistics, National Survey of Family Growth, 1973, 1976, 1982, and 1988.

★ 433 ★

Births

Percent of Live Births That Were Low Birth Weight For Women With or Without Anemia, By Maternal Age and Race, 1989

Race and ethnicity	Trimester	
	Second	Third
	Percent	
Non-Hispanic white	9.3	24.6
Non-Hispanic black	21.4	45.8
Hispanic	11.4	31.9
American Indian	11.9	32.8
Asian or Pacific Islander	11.8	26.8

Source: "Percent of Live Births That Were Low Birth Weight For Women With or Without Anemia, By Maternal Age and Race, 1989," Interagency Board for Nutrition Monitoring and Related Research, *Nutrition Monitoring in the United States: Chartbook I: Selected Findings From the National Nutrition Monitoring and Related Research Program*, p. 98. Primary source: Centers for Disease Control, National Center for Health Promotion, Pregnancy Nutrition Surveillance System. *Notes:* Hemoglobin and/or hematocrit levels are routinely measured for women receiving prenatal care at public health clinics. Hemoglobin and hematocrit values change substantially during the course of a normal pregnancy, and anemia must be characterized according to the specific stage of pregnancy. The Centers for Disease Control and Prevention use month-specific and trimester-specific hemoglobin and hematocrit cut-off values for defining anemia. In addition, the Pregnancy Nutrition Surveillance System adjusts the hemoglobin and hematocrit values for altitude and for smokers.

★ 434 ★

Births

Percent of Live Births That Were Low Birth Weight For Women With or Without Anemia, By Maternal Age and Race, 1989

Age	Percent	
	With anemia	Without anemia
White		
Less than 20 years	9.8	7.5
20-24 years	8.6	5.7
25-29 years	9.4	5.0
30-34 years	9.8	5.4
35-49 years	10.1	6.3
Black		
Less than 20 years	14.2	13.4
20-24 years	15.3	12.7

[Continued]

★ 434 ★

Percent of Live Births That Were Low Birth Weight For Women With or Without Anemia, By Maternal Age and Race, 1989

[Continued]

Age	Percent	
	With anemia	Without anemia
25-29 years	16.9	13.4
30-34 years	18.9	14.4
35-49 years	20.6	14.5

Source: "Percent of Live Births That Were Low Birth Weight For Women With or Without Anemia, By Maternal Age and Race, 1989," Interagency Board for Nutrition Monitoring and Related Research, *Nutrition Monitoring in the United States: Chartbook I: Selected Findings From the National Nutrition Monitoring and Related Research Program*, p. 98. Primary source: Centers for Disease Control, National Center for Health Statistics. *Notes:* Anemia was diagnosed during pregnancy, and was defined as a hemoglobin less than 10.0 grams per deciliter or a hematocrit less than 30 percent. Information about anemia on U.S. certificates of live birth was collected from the medical record. Findings exclude data for Louisiana, Nebraska, and Oklahoma, which did not report medical risk factors.

★ 435 ★

Births

Percent of Live Births of 40 Weeks Gestation or Longer by Maternal Weight Gain and Race of Mother, 1989

Maternal weight gain	Percent of births	
	White	Black
Less than 16 pounds	7.0	13.0
16-20 pounds	9.7	13.5
21-25 pounds	15.3	14.8
26-30 pounds	21.4	18.5
31-35 pounds	16.1	12.0
36-40 pounds	13.5	11.3
41 or more pounds	17.0	17.0

Source: "Percent of Live Births of 40 Weeks Gestation or Longer, By Maternal Weight Gain and Race of Mother, 1989," Interagency Board for *Nutrition Monitoring and Related Research, Nutrition Monitoring in the United States: Chartbook I: Selected Findings From the National Nutrition Monitoring and Related Research Program*, p. 97. Primary source: Centers for Disease Control, National Center for Health Statistics. *Notes:* Refers to married and unmarried mothers. The Institute of Medicine recommends that women of normal pregnancy body mass index (BMI) gain 25 to 35 pounds during pregnancy and that young adolescents and black women strive for weight gains at the upper end of this range. Body mass index is an index that relates weight to stature (BMI=weight/height2.). normal BMI was defined as 19.8 to 26.0 kilograms per meter2. . Findings exclude data for California, Louisiana, Nebraska, and Oklahoma, which did not require reporting of weight gain during pregnancy.

★ 436 ★

Births

Weight Gain Advice for Married Mothers, by Race, 1980 and 1988

Weight gain advice	Percent of mothers			
	White		Black	
	1980	1988	1980	1988
Less than 22 pounds	27	11	34	27
22-27 pounds	49	37	40	37
23-34 pounds	22	38	24	25
35 or more pounds	3	14	2	11

Source: "Weight Gain Advice for Married Mothers, by Race, 1980 and 1988," Interagency Board for Nutrition Monitoring and Related Research, *Nutrition Monitoring in the United States: Chartbook I: Selected Findings From the National Nutrition Monitoring and Related Research Program,* p. 97. Primary source: Centers for Disease Control, National Center for Health Statistics *Notes:* Refers to married mothers starting care in the first trimester of pregnancy. The medical community's advice on weight gain prior to 1990 was not based on a woman's body mass index (BMI). BMI is an index that relates weight to stature (BMI = Weight/height2.).

Diseases

★ 437 ★

AIDS Cases Reported, 1981 to 1992

[**Provisional**. For cases reported in the year shown. Data are subject to retrospective changes].

CHARACTERISTIC	NUMBER OF CASES										PERCENT DISTRIBUTION	
	Total	1981-1984	1985	1986	1987	1988	1989	1990	1991	1992	1981-1984	1992
Total	244,939	7,354	8,210	13,147	21,088	30,719	33,595	41,653	43,701	45,472	100.0	100.0
Race/ethnic group:												
White[1]	132,573	4331	4,968	7,837	12,962	17,063	18,569	22,325	22,193	22,325	58.9	49.1
Black[1]	75,929	1951	2,081	3,388	5,379	9,118	10,293	13,220	14,609	15,890	26.5	34.9
Hispanic	33,827	1039	1,099	1,803	2,543	4,260	4,358	5,658	6,405	6,662	14.1	14.7
Other/unknown	2,610	33	62	119	204	278	375	450	494	595	0.4	1.3

Source: "AIDS Cases Reported, by Patient Characteristics: 1981 to 1992," *Statistical Abstract of the United States 1993,* p. 134. Primary source: U.S. Centers for Disease Control, unpublished data. *Note:* 1. Non-Hispanic.

★ 438 ★
Diseases

AIDS Cases by Sex, Age at Diagnosis, and Race/Ethnicity, United States, June 1993

Age at diagnosis (years)	White, not Hispanic		Black, not Hispanic		Hispanic		Total[1]	
	No.	%	No.	%	No.	%	No.	%
Males								
Under 5	313	(0)	1,117	(1)	466	(1)	1,916	(1)
5-12	242	(0)	178	(0)	143	(0)	570	(0)
13-19	434	(0)	284	(0)	174	(0)	913	(0)
20-24	4,466	(3)	3,051	(4)	1,829	(4)	9,446	(3)
25-29	22,109	(15)	11,251	(15)	7,196	(16)	40,991	(15)
30-34	35,468	(23)	17,682	(23)	10,899	(24)	64,693	(23)
35-39	33,713	(22)	17,928	(23)	9,903	(22)	62,200	(22)
40-44	24,075	(16)	11,980	(16)	6,544	(15)	43,082	(16)
45-49	14,268	(9)	6,232	(8)	3,472	(8)	24,255	(9)
50-54	7,661	(5)	3,466	(5)	1,878	(4)	13,149	(5)
55-59	4,407	(3)	1,942	(3)	1,096	(2)	7,536	(3)
60-64	2,628	(2)	1,066	(1)	546	(1)	4,272	(2)
65 or older	2,217	(1)	794	(1)	397	(1)	3,453	(1)
Male subtotal	152,001	(100)	76,971	(100)	44,543	(100)	276,476	(100)
Females								
Under 5	309	(3)	1,100	(5)	436	(5)	1,859	(5)
5-12	78	(1)	179	(1)	101	(1)	365	(1)
13-19	96	(1)	231	(1)	58	(1)	388	(1)
20-24	610	(6)	1,218	(6)	541	(7)	2,394	(6)
25-29	1,712	(18)	3,463	(17)	1,559	(20)	6,786	(17)
30-34	2,196	(23)	5,094	(24)	1,933	(24)	9,308	(24)
35-39	1,707	(18)	4,560	(22)	1,526	(19)	7,855	(20)
40-44	987	(10)	2,514	(12)	879	(11)	4,433	(11)
45-49	526	(5)	1,033	(5)	400	(5)	1,987	(5)
50-54	329	(3)	628	(3)	246	(3)	1,218	(3)
55-59	314	(3)	352	(2)	144	(2)	821	(2)
60-64	239	(2)	218	(1)	81	(1)	552	(1)
65 or older	614	(6)	233	(1)	84	(1)	948	(2)
Female subtotal	9,717	(100)	20,823	(100)	7,988	(100)	38,914	(100)
Total	161,718		97,794		52,531		315,390	

Source: "AIDS Cases by Sex, Age at Diagnosis, and Race/Ethnicity, Reported Through June 1993, United States" U.S. Department of Health and Human Service, Centers for Disease Control and Prevention, *HIV/Aids Surveillance Report*, 5 (July 1993), p. 11. *Notes:* 1. Includes Asians, Pacific Islanders, American Indians, Alaska Natives, and 575 males and 79 females whose race/ethnicity is unknown.

★ 439 ★

Diseases

Acute Conditions, by Type, 1990

[Covers civilian noninstitutional population. Estimate include only acute conditions which were medically attended or caused at least one day of restricted activity].

| | NUMBER OF CONDITIONS (mil.) | | | | | RATE PER 100 POPULATION | | | | |
| | Infective and parasitic | Respiratory | | Digestive system | Injuries | Infective and parasitic | Respiratory | | Digestive system | Injuries |
		Common cold	Influenza				Common cold	Influenza		
1990, total[1]	51.7	61.5	106.8	13.0	60.1	21.0	25.0	43.4	5.3	24.4
White	44.4	51.2	95.8	10.4	52.8	21.4	24.7	46.3	5.0	25.5
Black	6.1	8.1	7.3	2.1	5.9	20.2	26.8	24.1	6.8	19.5

Source: "Acute Conditions, by Type, 1970 to 1990, and by Selected Characteristics, 1990," *Statistical Abstract of the United States 1993*, p. 135. Primary source: U.S. National Center for Health Statistics, *Vital and Health Statistics*, series 10 (181) and unpublished data. *Note:* 1. Includes other races and unknown income not shown separately.

★ 440 ★

Diseases

Cancer-Five Year Survival Rates, 1974-76 to 1983-88

[The 5-year relative survival rate, which is derived by adjusting the observed survival rate for expected mortality, represents the likelihood that a person will not die from causes directly related to their cancer within 5 years. Survival data shown are based on those patients diagnosed while residents of an area listed below during the time periods shown].

| | 5-YEAR RELATIVE SURVIVAL RATES (percent) | | | | | | | |
| | White | | | | Black | | | |
SITE	1974 -76	1977 -79	1980 -82	1983 -88	1974 -76	1977 -79	1980 -82	1983 -88
All sites[1]	50.2	50.7	51.5	53.5	38.7	38.8	39.1	38.3
Lung	11.0	12.0	12.0	11.8	11.3	11.0	12.1	10.5
Breast[2]	74.8	75.1	76.7	79.3	62.9	62.5	65.6	62.1
Colon	50.2	52.7	55.3	59.1	45.5	47.6	48.7	47.9
Prostate	67.6	71.7	74.0	77.6	57.7	61.9	63.9	62.9
Bladder	73.6	75.6	78.6	79.1	47.8	54.8	57.5	58.5
Rectum	48.6	50.4	52.7	56.8	41.6	37.7	37.7	46.1
Corpus uteri	88.6	86.1	82.5	84.2	60.4	57.8	53.6	54.3
Non-Hodgkin's lymphoma[3]	47.5	48.1	51.4	51.9	47.8	49.6	50.0	42.6
Oral cavity and pharnyx	54.8	54.1	54.9	53.7	35.7	36.0	30.4	31.7
Leukemia[3]	34.5	37.2	37.5	37.6	31.1	30.1	31.5	29.4
Melanoma of skin	79.8	81.5	81.9	82.6	68.5	51.7	59.5	68.3
Pancreas	2.7	2.2	2.8	2.8	2.2	3.8	4.8	4.7
Kidney	51.5	50.6	50.7	55.2	48.6	52.4	55.1	53.0
Stomach	14.4	16.1	16.1	16.0	16.4	15.3	19.1	17.1

[Continued]

★ 440 ★

Cancer-Five Year Survival Rates, 1974-76 to 1983-88

[Continued]

| SITE | 5-YEAR RELATIVE SURVIVAL RATES (percent) | | | | | | | |
| | White | | | | Black | | | |
	1974 -76	1977 -79	1980 -82	1983 -88	1974 -76	1977 -79	1980 -82	1983 -88
Ovary	36.2	37.5	38.6	39.3	40.7	39.5	37.3	37.1
Cervix uteri[4]	69.2	68.8	67.3	68.0	63.2	61.8	60.1	55.3

Source: "Current-Estimated New Cases, 1992, and Survival Rates, 1974-76 to 1983- 88," *Statistical Abstract of the United States 1993*, p. 138. Primary source: U.S. National Institutes of Health, National Cancer Institute, *Cancer Statistics Review*, annual. *Notes:* 1. Includes other sites not shown separately. 2. Survival rates for female only. 3. All types combined. 4. Invasive cancer only.

★ 441 ★

Diseases

Children Immunized Against Specified Diseases, 1991

[**In percent**. Covers civilian noninstitutional population].

| VACCINE | ALL RESPONDENTS | | | | | |
| | 2 years old 1991 | | | 1 to 4 years old 1991 | | |
	Total	White	Black and other	Total	White	Black and other
Diphtheria-tetanus-pertussis[1]	66.6	70.1	53.0	65.8	68.6	54.6
Polio[1]	52.2	55.9	38.1	50.6	52.7	42.1
Measles containing[2]	80.4	81.9	74.9	77.6	77.9	76.4
Hib[3]	56.4	59.6	44.5	56.9	58.5	50.4
Up-to-date for age	37.2	41.3	21.5	42.1	44.6	31.7

Source: "Children Immunized Against Specified Diseases, by Age Group: 1985 and 1991," *Statistical Abstract of the United States 1993*, p. 133. Primary source: U.S. Centers for Disease Control, United States Immunization Survey, and the National Health Interview Survey. *Notes:* NA Not available. 1. Three or more doeses. 2. Measles, measles/ rubella, measles/mumps, and measles/mumps/rubella. 3. Haemophilus B.

Drugs and Drug Use

★ 442 ★

Characteristics of Drug Abuse-Related Emergency Room Episodes, 1991

Episode characteristics	Total[a]	Patient characteristics Race, ethnicity			
		White	Black	Hispanic	Other[b]
Total number of episodes	400,079	224,906	108,576	33,575	4,368
Number of drugs					
Single-drug episode	50.2%	47.2%	52.8%	56.8%	55.4%
Multi-drug episode	49.8%	52.8	47.2	43.2	44.6
Drug use motive					
Psychic effects	15.3	16.2	12.8	17.8	17.9
Recreational use	7.7	6.7	8.0	11.0	9.0
Other psychic effects	7.6	9.4	4.8	6.8	8.9
Dependence	29.0	16.8	53.7	35.0	11.7
Suicide	43.8	56.6	20.8	29.7	58.3
Other[d]	1.4	1.7	0.8	1.5	(c)
Unknown/no response	10.5	8.6	11.9	16.0	8.2
Reason for emergency room contact					
Unexpected reaction	10.5	7.9	15.3	12.0	7.8
Overdose	56.9	71.3	29.4	46.8	70.8
Chronic effects	11.2	5.5	20.1	19.1	6.0
Withdrawal	1.8	2.0	1.7	1.8	0.7
Seeking detoxification	9.3	5.4	19.1	7.1	3.3
Accident/Injury	2.7	1.9	4.2	3.0	1.4
Other	3.4	3.1	4.1	4.7	5.1
Unknown/no response	4.2	3.0	6.2	5.5	5.0
Patient disposition					
Treated and released	45.0	37.3	55.9	55.2	52.3
Admitted to hospital	51.5	59.3	40.6	40.5	45.5
Left against medical advice	2.1	1.8	2.5	2.5	1.1
Died	0.3	0.3	0.2	0.2	0.2
Unknown/no response	1.1	1.2	0.8	1.4	0.8

Source: "Drug Abuse-Related Emergency Room Episodes," Department of Justice, *Sourcebook of Criminal Justice Statistics-1992*, p. 349. Primary source: U.S. Department of Health and Human Services, National Institute on Drug Abuse, *Annual Emergency Room Data, 1991*. Table adapted by *Sourcebook* staff. *Notes:* A. Includes episodes for which sex, race, ethnicity, and age were unknown or not reported. B. Includes American Indians, Alaska Natives, Asians, and Pacific Islanders. C. Estimates does not meet standard of precision (estimates with a relative standard error of 50 percent or higher are suppressed). D. Includes self-medication for physical ailment, to prevent pregnancy or induce abortion, accident, used unknowingly, etc.

★ 443 ★

Drugs and Drug Use

Drug Abuse-Related Emergency Room Episodes, 1991

| | Total | | Sex | | | |
| | | | Male | | Female | |
	Number	Percent	Number	Percent	Number	Percent
Total[a]	400,079	100.0%	192,524	100.0%	203,964	100.0%
Race, ethnicity						
White	224,906	56.2	94,105	48.9	127,786	63.1
Black	108,576	27.1	64,550	33.5	43,285	21.2
Hispanic	33,575	8.4	18,884	9.8	14,313	7.0
Other[b]	4,368	1.1	1,582	0.8	2,757	1.4
Unknown/no response	28,654	7.2	13,402	7.0	14,822	7.3

Source: "Drug Abuse-Related Emergency Room Episodes," Department of Justice, *Sourcebook of Criminal Justice Statistics-1992*, p. 348. Primary source: U.S. Department of Health and Human Services, National Institute on Drug Abuse, *Annual Emergency Room Data, 1991*. Table adapted by *Sourcebook* staff. *Notes:* These data were gathered through the Drug Abuse Warning Network (DAWN) sponsored by the National Institute on Drug Abuse. The data are weighted estimates representing all drug abuse-related emergency room episodes from a stratified random sample of hospitals in the 48 contiguous states, the District of Columbia, and 21 metropolitan areas. A. Includes episodes for which sex of patient was unknown or not reported. B. Includes American Indians, Alaska Natives, Asians, and Pacific Islanders.

★ 444 ★

Drugs and Drug Use

Motives for Drug Abuse-Related Emergency Room Episodes, 1991

| Patient and episode characteristics | Total | Drug use motive | | | | | |
		Recreational use	Other psychic effects	Dependence	Suicide	Other[a]	Unknown
Total number of episodes	400,079	30,830	30,490	116,007	175,203	5,541	42,009
Race, ethnicity							
White	56.2	49.1	69.6	32.6	72.7	70.6	46.2
Black	27.1	28.2	17.1	50.3	12.9	15.2	30.8
Hispanic	8.4	12.0	7.5	10.1	5.7	8.8	12.8
Other[b]	1.1	1.3	1.3	0.4	1.5	3.1	0.9
Unknown/no response	7.2	9.4	4.5	6.5	7.3	2.2	9.3

Source: "Drug Abuse-Related Emergency Room Episodes," Department of Justice, *Sourcebook of Criminal Justice Statistics-1992*, p. 350. Primary source: U.S. Department of Health and Human Services, National Institute on Drug Abuse, *Annual Emergency Room Data, 1991*. Table adapted by *Sourcebook* staff. *Notes:* A. Includes self-medication for physical ailment, to percent pregnancy or induce abortion, accident, used unknowingly, etc. B. Includes American Indians, Alaska Natives, Asians, and Pacific Islanders.

Facilities and Care

★ 445 ★

Mental Retardation Facilities, 1986

[Mental Retardation (MR) facilities include intermediate care facilities (ICF-MR), foster homes, group residence semi-independent living programs, State institutions and other kinds of MR places].

CHARACTERISTIC	RESIDENTS						Average number of residents
	Number			Percent distribution			
	Total[1]	Black	Hispanic[2]	Total[1]	Black	Hispanic[2]	
Total	250,472	29,442	10,181	100.0	100.0	100.0	17
BED SIZE							
1 to 5 beds	15,234	1,726	623	6.1	5.9	6.1	3
6 to 9 beds	34,815	3,159	1,587	13.9	10.7	15.6	6
10 to 15 beds	23,444	1,950	937	9.4	6.6	9.2	11
16 to 99 beds	54,090	5,485	2,333	21.6	18.6	22.9	35
100 beds or more	122,889	17,122	4,701	49.1	58.2	46.2	303
100 to 199 beds	24,539	3,698	995	9.8	12.6	9.8	124
200 to 499 beds	38,322	4,904	1,465	15.3	16.7	14.4	304
500 beds or more	60,028	8,520	2,241	24.0	28.9	22.0	741
GEOGRAPHIC REGION							
Northeast	61,707	5,634	1,578	24.6	19.1	15.5	16
Midwest	77,193	6,438	1,079	30.8	21.9	10.6	16
South	66,767	14,538	2,668	26.7	49.4	26.2	26
West	44,805	2,832	4,856	17.9	9.6	47.7	12
TYPE OF OWNERSHIP							
Profit	60,560	6,503	3,495	24.2	22.1	34.3	10
Nonprofit	75,193	6,938	2,316	30.0	23.6	22.7	12
Government	114,719	16,001	4,370	45.8	54.3	42.9	6

Source: "Mental Retardation Facilities-Summary by Selected Characteristics: 1986," *Statistical Abstract of the United States 1993*, p. 130. Primary source: U.S. National Center for Health Statistics, *Advance Data*, 143. (September 1987). *Notes:* 1. Includes races not shown separately. 2. Persons of Hispanic origin may be of any race.

★ 446 ★

Facilities and Care

Patients Discharged From Hospitals-Principal Source of Expected Payment, 1991

[Covers non-Federal short-stay hospitals. Discharges exclude newborn infants].

CHARACTERISTIC	Total discharges[1] (1,000)	PRINCIPAL SOURCE OF EXPECTED PAYMENT- PERCENT DISTRIBUTION							
		Private insurance	Government				Self-pay	No charge	Other[2]
			Medicare	Medicaid	Worker's compensation	Other			
Total	31,098	36.6	35.6	12.9	1.2	2.0	5.2	0.3	3.1
RACE									
White	20,816	38.1	39.8	8.7	1.3	1.6	4.3	0.2	2.7
All other	4,753	29.9	23.4	28.8	0.9	2.9	7.6	0.8	3.5
Not stated	5,528	37.1	30.2	15.2	1.1	3.1	6.8	0.2	3.2

Source: "Patients Discharged From Hospital-Principal Source of Expected Payment, by Selected Characteristic: 1991," *Statistical Abstract of the United States 1993*, p. 128. Primary source: U.S. National Center for Health Statistics, unpublished data. *Notes:* 1. Includes discharges for whom excepted source of payment was unknown. 2. Includes all other nonprofit source of payment such as church, welfare, or united way.

General Health Status

★ 447 ★

Days of Disability, 1989 to 1990

[Covers civilian noninstitutional population].

ITEM	TOTAL DAYS OF DISABILITY (millions)		DAYS PER PERSON	
	1989	1990	1989	1990
White[2]	3,087	3,057	15.0	14.8
Black[2]	511	536	17.1	17.7
Hispanic[3]	278	(NA)	13.2	(NA)

Source: "Days of Disability, by Type and Selected Characteristics: 1970 to 1990," *Statistical Abstract of the United States 1993*, p. 132. Primary source: U.S. National Center for Health Statistics, *Advance Data from Vital and Health Statistics*, 10 (181) and unpublished data. *Notes:* NA Not available. 1. A day when a person cuts down on his activities for more than half a day because of illness or injury. Includes bed-disability, work-loss, and school-loss days. Total includes other races and unknown income, not shown separately. 2. Beginning 1980, race was determined by asking the household respondent to report his race. In earlier years the racial classification of respondents was determined by interviewer observation. 3. Persons of Hispanic origin may be of any race.

★ 448 ★

General Health Status

Mean Calcium Intakes for Females 11-74 Years of Age, 1976-80 and 1982-84

Age	Non-Hispanic white	Non-Hispanic black	Mexican American	Puerto Rican	Cuban
	Milligrams per day				
11-17 years	842	700	853	886	774
18-39 years	642	467	650	601	615
40-54 years	617	423	561	583	584
55-74 years	564	460	582	557	616

Source: "Mean Calcium Intakes for Females 11-74 Years of Age, By Age, Race, and Ethnicity, 1976-80 and 1982-84," Interagency Board for Nutrition Monitoring and Related Research, *Nutrition Monitoring in the United states: Chartbook I: Selected Findings From the National Nutrition Monitoring and Related research Program*, p. 111. Primary source: Centers for Disease Control and Prevention, National Center Health Statistics, Division of Health Examination Statistics, National Health and Nutrition Survey II, 1976-80, and Hispanic Health and Nutrition Examination Survey, 1982-84. *Notes:* Data for non-Hispanic white and non-Hispanic black subjects were from the 1976-80 National Health and Nutrition Examination Survey (NHANES II), which was a national probability sample. Data for Mexican Americans, Puerto Ricans, and Cubans were from the 1982-84 Hispanic Health and Nutrition Examination Survey (HHANES), which was not a national probability sample.

★ 449 ★

General Health Status

Mean Calcium Intakes for Males 11-74 Years of Age, 1976-80 and 1982-84

Age	Non-Hispanic white	Non-Hispanic black	Mexican American	Puerto Rican	Cuban
	Milligrams per day				
11-17 years	1,332	887	1,195	1,255	1,185
18-39 years	1,092	797	985	922	959
40-54 years	868	563	742	699	729
55-74 years	785	628	743	701	792

Source: "Mean Calcium Intakes for Males 11-74 Years of Age, By Age, Race, and Ethnicity, 1976-80 and 1982-84," Interagency Board for Nutrition Monitoring and Related Research, *Nutrition Monitoring in the United states: Chartbook I: Selected Findings From the National Nutrition Monitoring and Related research Program*, p. 111. Primary source: Centers for Disease Control and Prevention, National Center Health Statistics, Division of Health Examination Statistics, National Health and Nutrition Survey II, 1976-80, and Hispanic Health and Nutrition Examination Survey, 1982-84. *Notes:* Data for non-Hispanic white and non-Hispanic black subjects were from the 1976-80 National Health and Nutrition Examination Survey (NHANES II), which was a national probability sample. Data for Mexican Americans, Puerto Ricans, and Cubans were from the 1982-84 Hispanic Health and Nutrition Examination Survey (HHANES), which was not a national probability sample.

★ 450 ★

General Health Status

Percent of Adults 20-74 Years of Age Needing Blood Cholesterol Analysis Who Would be Candidates For Intervention, 1976-80 and 1982-84

[Percent]

Race and ethnicity	Total	High-risk low-density lipoprotein cholesterol values	Borderline high risk low density lipoprotein cholesterol values plus coronary heart disease (CHD) or at least two other CHD risk factors
White	88	62	26
Black	88	63	25
Mexican American	78	41	37

Source: "Percent of Adults 20-74 Years of Age Needing Lipoprotein Analysis Who Would Be Candidates For Intervention, by Race and Ethnicity, 1976-80 and 1982-84," Interagency Board for Nutrition Monitoring and Related Research, *Nutrition Monitoring in the United States: Chartbook I: Selected Findings From the National Nutrition Monitoring and Related Research Program,* p. 96. Primary source: Centers for Disease Control, National Center for Health Statistics. *Notes:* Data for white and black subjects were from the 1976-80 National Health and Nutrition Examination Survey (NHANES II), which was a national probability sample. The white and black population in NHANES II included Hispanics. Data for Mexican Americans, Cubans, and Puerto Ricans were from the 1982-84 Hispanic Health and Nutrition Examination Survey (HHANES), which was not a national probability sample. The numbers of Cubans and Puerto Ricans referred for Lipoprotein analysis with known low-density lipoprotein cholesterol (LDL-C), determinations were too small to present statistically reliable results. High-risk LDL-C values were defined as LDL-C levels greater than or equal to 160 milligrams per deciliter; borderline-high-risk LDL-C values were LDL-C levels between 130 and 159 milligrams per deciliter. Based on statistical analyses, the investigators concluded that although the prevalence estimates were based on a single measurement these estimates do not seem appreciably biased by not having repeated blood lipid measurements.

★ 451 ★

General Health Status

Percent of Adults 20-74 Years of Age Needing Blood Cholesterol Analysis, 1976-80 and 1982-84

Race and ethnicity	Total	High blood cholesterol values	Borderline-high cholesterol values plus coronary heart disease (CHD) or at least two other CHD risk factors
		Percent	
White	41	27	14
Black	39	24	15
Mexican American	30	17	13
Puerto Rican	30	18	12
Cuban	34	18	16

Source: "Percent of Adults 20-74 Years of Age needing Lipoprotein Analysis, by Race and Ethnicity, 1976-80 and 1982-84," Interagency Board for Nutrition Monitoring and Related Research, *Nutrition Monitoring in the United States: Chartbook I: Selected Findings From the National Nutrition Monitoring and Related Research Program*, p. 96. Primary source: Centers for Disease Control, National Center for Health Statistics. *Notes:* Data for white and black subjects were from the 1976-80 National Health and Nutrition Examination Survey (NHANES II), which was a national probability sample. The white and black populations in NHANES II included Hispanics. Data for Mexican Americans, Cubans, and Puerto Ricans were from the 1982-84 Hispanic Health and Nutrition Examination Survey (HHANES), which was not a national probability sample. High blood cholesterol values were defined as serum total cholesterol levels between 200 and 239 milligrams per deciliter.

★ 452 ★

General Health Status

Percent of Children 2-5 Years of Age With Height-for-age Less Than the 5th Percentile, by Race and Ethnicity, 1980-91

Year	Percent				
	Non-Hispanic white	Non-Hispanic black	Hispanic	American Indian	Asian
1980	10.3	10.9	12.0	10.7	22.3
1981	11.1	12.7	13.6	9.8	23.8
1982	11.0	12.9	13.7	11.3	23.8
1983	11.0	12.5	13.3	10.2	20.5
1984	10.5	11.3	12.4	9.5	18.3
1985	10.1	11.0	11.1	8.1	15.5
1986	10.2	11.2	11.0	9.3	14.9
1987	10.0	11.4	10.3	7.7	14.3
1988	9.4	10.8	9.8	7.3	12.9

[Continued]

★ 452 ★

Percent of Children 2-5 Years of Age With Height-for-age Less Than the 5th Percentile, by Race and Ethnicity, 1980-91

[Continued]

Year	Percent				
	Non-Hispanic white	Non-Hispanic black	Hispanic	American Indian	Asian
1989	8.8	10.3	9.4	8.7	11.3
1990	9.2	11.2	9.9	7.6	10.8
1991	8.9	11.0	9.4	8.8	10.7

Source: "Percent of Children 2-5 Years of Age With Height-for-age Less Than the 5th Percentile, By Race and Ethnicity, 1980-91," Interagency Board for Nutrition Monitoring and Related Research, *Nutrition Monitoring in the United States: Chartbook I: Selected Findings From the National Nutritional Monitoring and Related Research Program*, p. 100. Primary source: Centers for Disease Control and Prevention, National Center for Chronic Disease Prevention and Health Promotion, Division of Nutrition, Pediatric Nutrition Surveillance System. *Notes:* The sample consisted of low-income, high-risk children. Low income was defined as income at or below 185 percent of the Federal poverty income guidelines, the criteria used for participation in the Supplemental Food Program for Women, Infants, and Children (WIC). The criteria for high risk includes certain nutritional and medical risk factors that would make a child eligible for WIC benefits, such as risk for iron deficiency, growth problems, poor diet, or previous poor medical history. At least 95 percent of the sample met the U.S. Bureau of Census' definition of a refugee.

★ 453 ★

General Health Status

Percent of Overweight Females 20-74 Years of Age, 1976-80 and 1982-84

Age	Non-Hispanic white	Non-Hispanic black	Mexican American
20-29 years	12.2	27.5	24.4
30-39 years	20.6	36.4	39.5
40-49 years	27.4	50.1	47.9
50-59 years	32.7	62.6	53.4
60-69 years	35.6	61.4	59.9
70-74 years	34.9	54.0	43.0

Source: "Percent of Overweight Females 20-74 Years of Age, by Age, Race, and Ethnicity, 1976-80 and 1982-84," Interagency Board for Nutrition Monitoring and Related Research, *Nutrition Monitoring in the United States: Chartbook I: Selected Findings From the National Nutrition Monitoring and Related Research Program*, p. 94. Primary source: Centers for Disease Control, National Center for Health Statistics. *Notes:* Overweight was defined as a body mass index (BMI) at or above the sex-specific 85th percentile of the 1976-80 National Health and Nutrition Examination Survey (NHANES II) reference population 20-29 years of age. For men, this was a BMI greater than or equal to 27.8 kilograms per meter2. . For women, it was BMI greater than or equal to 27.3 kilograms per meter2. . Data for non-Hispanic white and non-Hispanic black subjects were from NHANES II, which was a national probability sample. Data for Mexican-American subjects were from the 1982-84 Hispanic Health and Nutrition Examination Survey (HHANES), which was not a national probability sample.

★ 454 ★

General Health Status

Percent of Overweight Males 20-74 Years of Age, 1976-80 and 1982-84

Age	Non-Hispanic white	Non-Hispanic black	Mexican American
	Percent		
20-29 years	15.1	12.4	20.4
30-39 years	24.5	24.3	34.6
40-49 years	31.7	46.0	38.2
50-59 years	28.9	30.5	36.0
60-69 years	28.1	29.2	34.0
70-74 years	24.7	24.3	26.8

Source: "Percent of Overweight Males 20-74 Years of Age, by Age, Race, and Ethnicity, 1976-80 and 1982-84," Interagency Board for Nutrition Monitoring and Related Research, *Nutrition Monitoring in the United States: Chartbook I: Selected Findings From the National Nutrition Monitoring and Related Research Program*, p. 93. Primary source: Centers for Disease Control, National Center for Health Statistics. *Notes:* Overweight was defined as a body mass index (BMI) at or above the sex- specific 85th percentile of the 1976-80 National Health and Nutrition Examination Survey (NHANES II) reference population 20-29 years of age. For men, this was a BMI greater than or equal to 27.8 kilograms per meter2. ; for women, it was a BMI greater or equal to 27.3 kilograms per meter2. . Data for non-Hispanic white and non-Hispanic black subjects were from NHANES II, which was a national probability sample. Data for Mexican- American subjects were from the 1982-84 Hispanic Health and Nutrition Examination Survey (HHANES), which was not a national probability sample.

★ 455 ★

General Health Status

Personal Health Practices, 1990

[In percent, except total persons. For persons 18 years of age and over].

CHARACTERISTIC	Total persons (1,000)	Eats breakfast[1]	Rarely snacks	Exercised regularly[2]	Had two or more drinks any day[3]	Current smoker	20 percent or more above weight[4]
All persons[5]	181,447	56.4	25.5	40.7	5.5	25.5	27.5
Race:							
White	155,301	57.8	25.8	41.5	5.8	25.6	26.7
Black	20,248	46.9	22.7	34.3	4.3	26.2	38.0
Hispanic Origin:							
Hispanic	14,314	52.5	29.3	34.9	4.6	23.0	27.6
Non-Hispanic	166,599	56.7	25.2	41.2	5.6	25.7	27.5

Source: "Personal Health Practices, by Selected Characteristics: 1990," *Statistical Abstract of the United States 1993*, p. 139. Primary source: U.S. National Center for Health Statistics, *Health Promotion and Disease Prevention, United States: 1990, Vital and Health Statistics*, series 10 (185). *Notes:* 1. Almost every day. 2. Or played sports regularly. 3. On average per day in the past two weeks. 4. Above desirable weight. Based on 1983 Metropolitan Life Insurance Company standards. Height and weight data are self-reported. 5. Includes persons whose characteristics are unknown.

★ 456 ★

General Health Status

Women – Selected Health Practices, 1990

| CHARACTERISTIC | POPULATION 18 YEARS AND OVER | | | | | POPULATION 35 YEARS AND OVER | | |
| | Total (1,000) | Percent who- | | | | Total (1,000) | Percent who had a mammogram | |
		Had a breast exam	Know how to do BSE[2]	Did a BSE monthly[2,3]	Had a PAP smear[4]		Ever	In the past years
Total[5]	95,169	53.1	88.1	43.1	50.1	59,934	57.7	50.5
Race:								
White	81,255	53.1	88.8	42.3	49.7	52,188	58.9	51.5
Black	11,212	55.3	86.0	50.9	54.3	6,282	51.3	44.9
Hispanic origin:								
Hispanic	7,709	50.4	74.7	44.1	49.1	3,704	49.4	42.3
Non-Hispanic	87,151	53.4	89.2	43.1	50.1	56,029	58.2	51.0

Source: "Women—Selected Health Practices: 1990," *Statistical Abstract of the United States 1993*, p. 139. Primary source: U.S. National Center for Health Statistics, *Health Promotion and Disease Prevention, United States 1990, Vital and Health Statistics*, series 10 (185). *Notes:* 1. In the past year by a professional. 2. BSE = Breast self-examination. 3. On average over the past year. 4. In the past year. 5. Includes persons whose characteristics are unknown.

Insurance Coverage

★ 457 ★

Children-Health Insurance and Medical Care, 1988

| CHARACTERISTIC | Total | AGE GROUP | | | | | |
		Under 1 year	1 to 4 years	5 to 7 years	8 to 11 years	12 to 14 years	15 to 17 years
All children (1,000)[1]	63,569	3,850	14,536	11,037	13,635	9,872	10,639
PERCENT COVERED BY HEALTH INSURANCE							
All children	83.1	80.1	83.7	83.3	83.8	83.0	82.3
Race: White	83.7	80.7	84.4	84.3	83.6	83.7	83.5
Black	80.9	81.2	80.5	79.7	84.1	81.7	77.6
Hispanic origin: Hispanic	70.0	62.2	75.2	76.3	65.0	68.1	68.3
Non-Hispanic	84.9	82.8	85.2	84.3	86.3	85.1	83.8
PERCENT WHO VISITED A DOCTOR FOR HEALTH CARE[2]							
All children	63.9	93.8	81.5	66.0	49.6	54.8	53.9

[Continued]

★ 457 ★

Children-Health Insurance and Medical Care, 1988
[Continued]

CHARACTERISTIC	Total	AGE GROUP					
		Under 1 year	1 to 4 years	5 to 7 years	8 to 11 years	12 to 14 years	15 to 17 years
Race: White	63.7	95.1	81.4	65.7	47.9	56.0	54.0
Black	65.0	87.6	82.6	66.6	55.7	51.7	55.7
Hispanic origin: Hispanic	63.4	93.5	82.3	70.4	49.5	51.9	45.5
Non-hispanic	63.6	93.5	81.1	65.2	49.5	55.0	54.8

Source: "Children-Health Insurance and Medical Care: 1988," *Statistical Abstract of the United States 1993,* p. 117. Primary source: U.S. National Center for Health Statistics, *Advance Data from Vital and Health Statistics,* 188 (1 October 1990). *Notes:* 1. Includes other races and unknown income. 2. For routine health care within the past year.

★ 458 ★

Insurance Coverage

Health Insurance Coverage Status, 1991, Part 1, Numbers

[Government health insurance includes Medicare, Medicaid, military plans].

CHARACTERISTIC	NUMBER (mil.)					
	Total	Covered by private or Government health insurance				Not covered by health insurance
		Total[1]	Private insurance		Covered by Medicaid	
			Total	Related to employment[2]		
1991, total[3]	249.0	215.5	186.9	150.5	21.3	33.5
Male	121.2	103.1	91.0	75.4	8.2	18.1
Female	127.7	112.4	95.9	75.1	13.1	15.4
White	208.5	182.5	164.0	131.8	12.7	26.0
Black	31.0	25.2	16.8	13.9	7.4	5.7
Hispanic origin[4]	21.4	15.2	10.9	9.2	3.8	6.2

Source: "Health Insurance Coverage Status, by Selected Characteristics: 1985 to 1992," *Statistical Abstract of the United States, 1993,* p. 115. Primary source: U.S. Bureau of the Census, *Current Population Reports,* P70-29 and unpublished data. *Notes:* 1. Includes other Government insurance not shown separately. 2. Related to current or prior employment of self or other family members. 3. Includes other races not shown separately. 4. Persons for Hispanic origin may be of any race.

★ 459 ★

Insurance Coverage

Health Insurance Coverage Status, 1991, Part 2, Percentages

[Government health insurance includes Medicare, Medicaid, and military plans].

CHARACTERISTIC	PERCENT				
		Covered by private or Government health insurance			Not covered health insurance
	Total	Total[1]	Private	Covered by Medicaid	
1991, total[2]	100.0	86.6	75.1	8.6	13.5
Male	100.0	85.1	75.1	6.8	14.9
Female	100.0	88.0	75.1	10.3	12.1
White	100 0	87.5	78.7	6.1	12.5
Black	100.0	81.5	54.2	23.9	18.4
Hispanic origin[3]	100.0	71.0	50.9	17.8	29.0

Source: "Health Insurance Coverage Status, by Selected Characteristics: 1985 to 1992," *Statistical Abstract of the United States, 1993*, p. 115. Primary source: U.S. Bureau of the Census, *Current Population Reports*, P70-29 and unpublished data. *Notes:* 1. Includes other Government insurance not shown separately. 2. Includes other races not shown separately. 3. Persons for Hispanic origin may be of any race.

★ 460 ★

Insurance Coverage

Persons Without Health Insurance Coverage By Age, 1989

[**In percent, except as indicated**. Annual average of monthly figures. Based on Current Population Survey].

CHARACTERISTIC	Total	UNDER 65 YEARS OLD					65 years and over
		Total	Under 18 years	18 to 24 years	25 to 44 years	45 to 64 years	
All persons (1,000)	243,532	214,313	64,003	25,401	78,795	46,114	29,219
Persons not covered[1]	13.9	15.7	14.9	27.4	15.5	10.5	1.2
RACE							
White	12.8	14.5	14.0	26.3	14.4	9.4	1.0
Black	20.2	21.9	18.9	34.3	22.5	17.5	2.5
Other	19.7	20.4	18.9	27.8	20.7	17.5	8.4

Source: "Persons Without Health Insurance Coverage, by Selected Characteristics: 1989," *Statistical Abstract of the United States, 1993*, p. 116. Primary source: U.S. National Center for Health Statistics, *Advance Data from Vital and Health Statistics*, 201 (18 June 1991). *Notes:* 1. Excludes 9.7 million persons for whom insurance coverage was unknown. Includes persons whose demographic coverage was unknown.

Medical and Dental Care

★ 461 ★

Blacks With Disability, by Selected Characteristics, 1991-92, Part 1

Numbers in thousands.

Sex and age	Total	With a disability			
		Number	Percent	With a severe disability	
				Number	Percent
Both sexes, all ages					
Total	31,420	6,277	20.0	3,836	12.2
Age					
Less than 6 years	3,703	126	3.4	16	0.4
6 to 14 years	5,165	302	5.9	44	0.8
15 to 17 years	1,700	184	10.9	94	5.5
18 to 24 years	3,416	384	11.2	156	4.6
25 to 34 years	5,475	910	16.6	548	10.0
35 to 44 years	4,457	879	19.7	523	11.7
45 to 54 years	2,734	867	31.7	528	19.3
55 to 64 years	2,033	897	44.1	667	32.8
65 to 74 years	1,584	822	51.9	537	33.9
75 years and over	1,152	905	78.5	723	62.8

Source: "Blacks With a Disability, by Selected Characteristics: 1991-92," U.S. Department of Commerce, Bureau of the Census, *Americans With Disabilities: 1991-92*, Current Population Reports, P70-33, p. 56.

★ 462 ★

Medical and Dental Care

Blacks With Disability, by Selected Characteristics, 1991-92, Part 2

Numbers in thousands.

Sex and age	Total	With a disability			
		Number	Percent	With a severe disability	
				Number	Percent
Family Relationship					
Family householder or spouse	11,042	2,883	26.1	1,784	16.2
With children under 18 years	6,303	1,178	18.7	624	9.9
Married, spouse present	7,165	1,648	23.0	974	13.6
With children under 18 years	3,903	584	15.0	297	7.6
No spouse present	3,877	1,234	31.8	810	20.9
With children under 18 years	2,400	594	24.8	327	13.6
Nonfamily householder	3,336	1,381	41.4	926	27.6
Lives alone	2,961	1,301	43.9	875	29.6

[Continued]

★ 462 ★

Blacks With Disability, by Selected Characteristics, 1991-92, Part 2

[Continued]

Sex and age	Total	With a disability		With a severe disability	
		Number	Percent	Number	Percent
Child of householder	12,536	1,190	9.5	587	4.7
Other relative of householder	3,191	563	17.7	382	12.0
Member of secondary family	204	16	7.7	7	3.5
Secondary individual	1,111	243	21.9	149	13.4

Source: "Blacks With a Disability, by Selected Characteristics: 1991-92," U.S. Department of Commerce, Bureau of the Census, *Americans With Disabilities: 1991-92*, Current Population Reports, P70-33, p. 56.

★ 463 ★

Medical and Dental Care

Blacks With Disability, by Selected Characteristics, 1991-92, Part 3

Numbers in thousands.

Sex and age	Total	With a disability		With a severe disability	
		Number	Percent	Number	Percent
Years of school completed					
Persons 25 years old and over	17,436	5,281	30.3	3,526	20.2
Less than 12	5,807	2,886	49.7	2,096	36.1
12	6,457	1,571	24.3	1,002	15.5
13 to 15	3,112	564	18.1	285	9.2
16 and over	2,060	259	12.6	143	6.9
Region					
Northeast	5,650	962	17.0	600	10.6
Midwest	5,564	1,249	22.4	776	14.0
South	17,390	3,556	20.5	2,151	12.4
West	2,815	510	18.1	308	10.9
Residence					
In central city	17,573	3,469	19.7	2,154	12.3
In metro area, not central city	7,819	1,357	17.4	776	9.9
Not in metro area	6,028	1,451	24.1	906	15.0
Ratio of income to low-income threshold					
Less than 1.00	11,281	2,663	23.6	1,692	15.0
1.0 to 1.49	4,629	1,132	24.5	763	16.5
1.50 to 1.99	3,679	815	22.1	530	14.4
2.00 to 2.99	4,938	899	18.2	519	10.5
3.00 to 3.99	2,846	353	12.4	157	5.5
4.00 and over	3,999	407	10.2	171	4.3

Source: "Blacks With a Disability, by Selected Characteristics: 1991-92," U.S. Department of Commerce, Bureau of the Census, *Americans With Disabilities: 1991-92*, Current Population Reports, P70-33, p. 56.

★ 464 ★

Medical and Dental Care

Blacks With Disability, by Selected Characteristics, 1991-92, Part 4

Numbers in thousands.

Sex and age	Total	With a disability		With a severe disability	
		Number	Percent	Number	Percent
Health insurance coverage status					
Covered by private insurance	17,358	2,457	14.2	1,239	7.1
Not covered by private insurance	14,061	3,820	27.2	2,597	18.5
Covered by Medicaid	6,740	2,132	31.6	1,580	23.4
Covered by Medicare	3,111	2,183	70.2	1,730	55.6
Covered by Medicaid	7,537	2,359	31.3	1,726	22.9
Not covered by health insurance	5,649	904	16.0	413	7.3
Means-tested assistance					
Received cash assistance	2,897	1,657	57.2	1,407	48.6
Received food stamps	2,736	1,265	46.2	979	35.8
Received housing assistance	4,348	1,117	25.7	729	16.8
Received both cash assistance and food stamps	1,954	932	47.7	765	39.2
Received all 3 benefits	798	364	45.6	290	36.4
Received food stamps only	620	265	42.7	174	28.0
Did not receive assistance	24,584	3,780	15.4	1,971	8.0
Tenure of housing unit					
Owner occupied	14,797	3,130	21.2	1,993	13.5
Renter occupied	16,623	3,147	18.9	1,843	11.1

Source: "Blacks With a Disability, by Selected Characteristics: 1991-92," U.S. Department of Commerce, Bureau of the Census, *Americans With Disabilities: 1991-92,* Current Population Reports, P70-33, p. 56.

★ 465 ★

Medical and Dental Care

Medical Device Implants, 1988

[In thousands, except percent].

DEVICE	Total[1]	Race	
		White	Black
Artificial joints[2]	1,625	92.3	6.2
Hip joints	816	93.5	5.5
Knee joints	521	88.1	10.0
Fixation devices[4]	4,890	91.7	4.8[3]
Head	351	93.2	4.8[3]
Torso	563	92.2	6.9[3]
Upper extremities	646	91.6	5.6[3]
Lower extremities	2,690	91.1	7.7
Other	622	93.6	3.7[3]
Other devices:			
Ear vent tubes	1,494	92.3	6.2
Silicone implants	620	97.6	0.8[3]
Breast implants	544	98.5	-
Shunt or catheter	321	85.7	12.5[3]
Dental implants	275	93.8	4.0[3]
Heart valve	279	89.2	10.0
Pacemaker	460	93.7	3.5
Eye lens	3,765	95.6	3.5

Source: "Medical Device Implants, by Age, Sex, and Race: 1988," *Statistical Abstract of the United States 1993*, p. 135. Primary source: U.S. National Center for Health Statistics, *National Health Interview Survey*, 1988; and publishing data. *Notes:* - Represents or rounds to zero. 1. Includes other races not shown separately. 2. Includes other devices not shown separately. 3. Figure does not meet standards of reliability or precision. 4. Includes sites unknown. Each device represents a single body part regardless of the number of pins, screws, nails, wire, rods or plates.

★ 466 ★

Medical and Dental Care

Persons With Disabilities, by Race and Hispanic Origin and Type of Disability, 1991-92, Part 1

[Numbers in thousands].

Characteristics	White		Black		Hispanic origin	
	Number	Percent distribution	Number	Percent distribution	Number	Percent distribution
ALL AGES						
Total	210,873	100.0	31,420	100.0	21,905	100.0
With a disability	41,521	19.7	6,277	20.0	3,343	15.3
Severe	19,736	9.4	3,836	12.2	1,838	8.4
Not severe	21,785	10.3	2,441	7.8	1,505	6.9

[Continued]

★ 466 ★

Persons With Disabilities, by Race and Hispanic Origin and Type of Disability, 1991-92, Part 1

[Continued]

Characteristics	White		Black		Hispanic origin	
	Number	Percent distribution	Number	Percent distribution	Number	Percent distribution
PERSONS 0 TO 14 YEARS Total	44,704	100.0	8,868	100.0	6,506	100.0
With a disability	2,403	5.4	428	4.8	202	3.1
Severe	451	1.0	59	0.7	26	0.4
Not severe	1,952	4.4	369	4.2	228	3.5

Source: "Persons With Disabilities, by Race and Hispanic Origin and Type of Disability," U.S. Department of Commerce, Bureau of the Census, *Americans With Disabilities: 1991-92,* Current Population Reports, P70-33, p. 22.

★ 467 ★

Medical and Dental Care

Persons With Disabilities, by Race and Hispanic Origin and Type of Disability, 1991-92, Part 2

Numbers in thousands.

Characteristics	White		Black		Hispanic origin	
	Number	Percent distribution	Number	Percent distribution	Number	Percent distribution
Persons 15 years old and over Total	166,169	100.0	22,551	100.0	15,399	100.0
With a disability	39,118	23.5	5,849	25.9	3,141	20.4
Severe	19,285	11.6	3,776	16.8	1,812	11.8
Not severe	19,833	11.9	2,073	9.2	1,329	8.6
With a functional limitation	28,931	17.4	4,480	19.9	2,359	15.3
Severe	12,595	7.6	2,338	10.4	1,124	7.3
Seeing words and letters	7,789	4.7	1,703	7.6	767	5.0
Unable	1,234	0.7	329	1.5	162	1.1
Hearing normal conversation	9,846	5.9	878	3.9	478	3.1
unable	795	0.5	103	0.5	57	0.4
Having speech understood	1,762	1.1	454	2.0	175	1.1
Unable	200	0.1	34	0.2	15	0.1
Lifting and carrying 10 lbs.	13,478	8.1	2,359	10.5	1,141	7.4
Unable	6,323	3.8	1,228	5.5	565	3.7
Climbing stairs without resting	14,373	8.7	2,754	12.2	1,266	8.2
Unable	7,442	4.5	1,487	6.6	682	4.4
Walking 3 city blocks	14,410	8.7	2,614	11.6	1,108	7.2
Unable	7,468	4.5	1,370	6.1	537	3.5

Source: "Persons With Disabilities, by Race and Hispanic Origin and Type of Disability," U.S. Department of Commerce, Bureau of The Census, *Americans With Disabilities: 1991-92,* Current Population Reports, P70-33, p. 22.

★ 468 ★

Medical and Dental Care

Persons With Disabilities, by Race and Hispanic Origin and Type of Disability, 1991-92, Part 3

Numbers in thousands.

Characteristics	White		Black		Hispanic origin	
	Number	Percent distribution	Number	Percent distribution	Number	Percent distribution
Number of functional limitations						
1	12,582	7.6	1,510	6.7	1,049	6.8
2	5,958	3.6	976	4.3	459	3.0
3 or more	10,391	6.3	1,994	8.8	851	5.5
Number of severe functional limitations						
1	5,874	3.5	984	4.4	558	3.6
2	3,199	1.9	675	3.0	297	1.9
3 or more	3,522	2.1	679	3.0	269	1.8
With ADL limitation	6,585	4.0	1,216	5.4	492	3.2
Needs personal assistance	3,171	1.9	650	2.9	246	1.6
Getting around inside the home	2,955	1.8	646	2.9	216	1.4
Needs personal assistance	1,352	0.8	318	1.4	113	0.7
Getting in and out of bed or a chair	4,384	2.6	833	3.7	328	2.1
Needs personal assistance	1,626	1.0	367	1.6	140	0.9
Taking a bath or shower	3,743	2.3	695	3.1	213	1.4
Needs personal assistance	2,233	1.3	444	2.0	151	1.0
Dressing	2,705	1.6	487	2.2	188	1.2
Needs personal assistance	1,706	1.0	324	1.4	127	0.8
Eating	887	0.5	170	0.8	57	0.4
Needs personal assistance	388	0.2	87	0.4	17	0.1
Using the toilet, including getting to the toilet	1,713	1.0	343	1.5	102	0.7
Needs personal assistance	948	0.6	190	0.8	62	0.4

Source: "Persons With Disabilities, by Race and Hispanic Origin and Type of Disability," U.S. Department of Commerce, Bureau of The Census, *Americans With Disabilities: 1991-92*, Current Population Reports, P70-33, p. 22.

★ 469 ★

Medical and Dental Care

Persons With Disabilities, by Race and Hispanic Origin and Type of Disability, 1991-92, Part 4

Numbers in thousands.

Characteristics	White		Black		Hispanic origin	
	Number	Percent distribution	Number	Percent distribution	Number	Percent distribution
Number of ADL limitations						
1	2,797	1.7	476	2.1	217	1.4
2	1,170	0.7	210	0.9	116	0.8
3 or more	2,617	1.6	530	2.4	159	1.0
Number of ADL's for which personal assistance needed						
1	1,220	0.7	242	1.1	101	0.7
2	627	0.4	136	0.6	45	0.3
3 or more	1,324	0.8	271	1.2	99	0.6
With an IADL limitation	9,626	5.8	1,832	8.1	812	5.3
Needs personal assistance	7,106	4.3	1,405	6.2	646	4.2
Going outside the home, for example to shop or visit a doctor's office	6,429	3.9	1,236	5.5	516	3.4
Needs personal assistance	4,929	3.0	954	4.2	425	2.8
Keeping track of money and bills	3,078	1.9	734	3.3	271	1.8
Needs personal assistance	2,710	1.6	637	2.8	239	1.6
Preparing meals	3,664	2.2	778	3.5	299	1.9
Needs personal assistance	2,994	1.8	609	2.7	262	1.7
Doing light housework, such as washing dishes or sweeping a floor	5,166	3.1	1,030	4.6	457	3.0
Needs personal assistance	3,900	2.4	742	3.3	343	2.2
Using the telephone	2,648	1.6	421	1.9	189	1.2
Unable to use	728	0.4	178	0.8	81	0.5

Source: "Persons With Disabilities, by Race and Hispanic Origin and Type of Disability," U.S. Department of Commerce, Bureau of The Census, *Americans With Disabilities: 1991-92*, Current Population Reports, P70-33, p. 23. *Notes:* ADLs are activities of daily living such as getting around inside the home, getting in or out of bed or a chair, bathing, dressing, eating, and using the toilet. IADL's are instrumental activities of daily living such as going outside the home, keeping track of money or bills, preparing meals, doing light housework, and using the telephone.

★ 470 ★

Medical and Dental Care

Persons With Disabilities, by Race and Hispanic Origin and Type of Disability, 1991-92, Part 5

Numbers in thousands.

Characteristics	White		Black		Hispanic origin	
	Number	Percent distribution	Number	Percent distribution	Number	Percent distribution
Number of IADL limitations						
1	4,167	2.5	752	3.3	356	2.3
2	2,056	1.2	366	1.6	185	1.2
3 or more	3,403	2.1	714	3.2	271	1.8
Number of IADL's for which personal assistance needed						
1	3,021	1.8	570	2.5	291	1.9
2	1,609	1.0	318	1.4	142	0.9
3 or more	2,476	1.5	518	2.3	213	1.4
Uses a wheelchair	1,269	0.8	185	0.8	70	0.5
Does not use a wheelchair but has used a cane, crutches, or a walker for 6 months or longer	4,015	2.4	820	3.6	268	1.7
Needs personal assistance with an ADL or IADL						
With a mental or emotional disability	5,636	3.4	1,099	4.9	521	3.4
Mental retardation	907	0.6	299	1.3	133	0.9
Persons 16 to 67 years old						
Total	141,612	100.0	19,858	100.0	14,027	100.0
With a work disability	16,276	11.5	2,722	13.7	1,559	11.1
Prevented from working	6,664	4.7	1,713	8.6	853	6.1
Persons 16 years old and over						
Total	163,542	100.0	21,982	100.0	14,975	100.0
With a housework disability	15,233	9.3	2,497	11.4	1,266	8.5
Unable to do housework	2,935	1.8	567	2.6	258	1.7

Source: "Persons With Disabilities, by Race and Hispanic Origin and Type of Disability," U.S. Department of Commerce, Bureau of The Census, *Americans With Disabilities: 1991-92*, Current Population Reports, P70-33, p. 23. *Notes:* ADLs are activities of daily living such as getting around inside the home, getting in or out of bed or a chair, bathing, dressing, eating, and using the toilet. IADL's are instrumental activities of daily living such as going outside the home, keeping track of money or bills, preparing meals, doing light housework, and using the telephone.

★ 471 ★

Medical and Dental Care

Physician Contacts, by Place of Contact, 1990

CHARACTERISTIC	NUMBER (1,000)					VISITS PER PERSON				
	All places[1]	Telephone	Office	Hospital[2]	Other	All places[1]	Telephone	Office	Hospital[2]	Other
All persons[3]	1,363	164	808	182	199	5.5	0.7	3.3	0.7	0.8
RACE										
White	1,178	147	717	142	163	5.7	0.7	3.5	0.7	0.8
Black	148	13	72	35	29	4.9	0.4	2.4	1.1	0.9

Source: "Physician Contacts, by Place of Contact and Selected Patient Characteristics: 1990," *Statistical Abstract of the United States, 1993*, p. 120. Primary source: U.S. National Center for Health Statistics, *Vital and Health Statistics*, series 10 (181). *Notes:* 1. Includes unknown place and contact. 2. Excludes contact while an overnight patient in a hospital. 3. Includes other races and unknown income.

★ 472 ★

Medical and Dental Care

Physician and Dental Contacts, 1970 to 1990

TYPE OF VISIT AND YEAR	TOTAL VISITS (mil.) Race[1]		VISITS PER PERSON YEAR Race[1]	
	White	Black	White	Black
1989	1,148	140	5.6	4.7
1990	1,178	148	5.7	4.9

Source: "Physician and Dental Contacts, by Patient Characteristics: 1970 to 1990," *Statistical Abstract of the United States, 1993*, p. 119. Primary source: U.S. National Center for Health Statistics, *Vital and Health Statistics*, Series 10, No. 181 and unpublished data.

Substance Abuse

★ 473 ★

Current Cigarette Smoking, by Sex, Age, and Race, 1965 to 1991

[**In percent**. A current smoker is a person who has smoked at least 100 cigarettes and who now smokes; includes occasional smokers. Excludes unknown smoking status].

SEX, AGE, AND RACE	1965	1974	1979	1983	1985	1987	1988	1990	1991
Total smokers, 18 years old and over	42.4	37.1	33.5	32.1	30.1	28.8	28.1	25.5	25.6
Male, total	51.9	43.1	37.5	35.1	32.6	31.2	30.8	28.4	28.1
White, total	51.1	41.9	36.8	34.5	31.7	30.5	30.1	28.0	27.4
18 to 24 years	53.0	40.8	34.3	32.5	28.4	29.2	26.7	27.4	25.1
25 to 34 years	60.1	49.5	43.6	38.6	37.3	33.8	35.4	31.6	32.1
35 to 44 years	57.3	50.1	41.3	40.8	36.6	36.2	35.8	33.5	32.1
45 to 64 years	51.3	41.2	38.3	35.0	32.1	32.4	30.0	28.7	28.0
65 years and over	27.7	24.3	20.5	20.6	18.9	16.0	16.9	13.7	13.7
Black, total	60.4	54.3	44.1	40.6	39.9	39.0	36.5	32.5	35.0
18 to 24 years	62.8	54.9	40.2	34.2	27.2	24.9	18.6	21.3	15.0
25 to 34 years	68.4	58.5	47.5	39.9	45.6	44.9	41.6	33.8	39.4
35 to 44 years	67.3	61.5	48.6	45.5	45.0	44.0	42.5	42.0	44.4
45 to 64 years	57.9	57.8	50.0	44.8	46.1	44.3	43.2	36.7	42.0
65 years and over	36.4	29.7	26.2	38.9	27.7	30.3	29.8	21.5	24.3
Female, total	33.9	32.1	29.9	29.5	27.9	26.5	25.7	22.8	23.5
White, total	34.0	31.7	30.1	29.4	27.7	26.7	25.7	23.4	23.7
18 to 24 years	38.4	34.0	34.5	36.5	31.8	27.8	27.5	25.4	25.1
25 to 34 years	43.4	38.6	34.1	32.2	32.0	31.9	31.0	28.5	28.4
35 to 44 years	43.9	39.3	37.2	34.8	31.0	29.2	28.3	25.0	27.0
45 to 64 years	32.7	33.0	30.6	30.6	29.7	29.0	27.7	25.4	25.3
65 years and over	9.8	12.3	13.8	13.2	13.3	13.9	12.6	11.5	12.1
Black, total	33.7	36.4	31.1	32.2	31.0	28.0	27.8	21.2	24.4
18 to 24 years	37.1	35.6	31.8	32.0	23.7	20.4	21.8	10.0	11.8
25 to 34 years	47.8	42.2	35.2	38.0	36.2	35.8	37.2	29.1	32.4
35 to 44 years	42.8	46.4	37.7	32.7	40.2	35.3	27.6	25.5	35.3
45 to 64 years	25.7	38.9	34.2	36.3	33.4	28.4	29.5	22.6	23.4
65 years and over	7.1	8.9	8.5	13.1	14.5	11.7	14.8	11.1	9.6

Source: "Current Cigarette Smoking, by Sex, Age, and Race: 1965 to 1991," *Statistical Abstract of the United States 1993*, p. 138. Primary source: U.S. National Health Center, *Health, United States, 1992*.

★ 474 ★

Substance Abuse

Percentage of Persons 65 Years and Over Who Smoked Cigarettes, by Sex and Race, 1965 to 1987

Year	Male			Female		
	Total	White	Black	Total	White	Black
1987	17.2	16.0	30.3	13.7	13.9	11.7
1979	20.9	20.5	26.2	13.2	13.8	8.5
1974	24.8	24.3	29.7	12.0	12.3	8.9
1965	28.5	27.7	36.4	9.6	9.8	7.1

Source: "Percentage of Persons 65 Years and Over Who Smoked Cigarettes at Time of Survey, by Sex and Race: 1965 to 1987," *Sixty-Five Plus in America*, p. 3-9. Primary source: U.S. National Center for Health Statistics, *Health United States 1990*. *Notes:* Civilian noninstitutional population who has smoked at least 100 cigarettes and who smoked at the time of the survey; includes occasional smokers. Excludes unknown smoking status.

★ 475 ★

Substance Abuse

Use of Selected Drugs, by Age of User, 1991, Part 1 Current Users

SUBSTANCE AND AGE GROUP	Total[1]	RACE/ETHNICITY		
		White[2]	Black[2]	Hispanic
CURRENT USERS				
Cigarettes: Total	27.0	27.3	27.9	24.7
12 to 17 years old	10.8	12.7	4.4	8.7
18 to 25 years old	32.2	35.8	21.8	24.8
26 to 34 years old	32.9	33.3	36.9	28.6
35 years old and over	26.6	25.8	32.6	27.7
Alcohol: Total	50.9	52.7	43.7	47.5
12 to 17 years old	20.3	20.4	20.1	22.5
18 to 25 years old	63.6	67.2	56.0	52.8
26 to 34 years old	61.7	63.8	57.1	57.2
35 years old and over	49.5	50.9	40.3	47.8
Marijuana: Total	4.8	4.5	7.2	4.3
12 to 17 years old	4.3	4.4	4.5	4.6
18 to 25 years old	13.0	13.7	14.6	9.1
26 to 34 years old	7.0	6.6	11.9	4.2
35 years old and over	2.1	1.9	3.5	2.3
Cocaine: Total	0.9	0.7	1.8	1.6
12 to 17 years old	0.4	0.3	0.5	1.3
18 to 25 years old	2.0	1.7	3.1	2.7

[Continued]

★ 475 ★

Use of Selected Drugs, by Age of User, 1991, Part 1
Current Users
[Continued]

SUBSTANCE AND AGE GROUP	Total[1]	RACE/ETHNICITY White[2]	Black[2]	Hispanic
26 to 34 years old	1.8	1.6	2.7	2.0
35 years old and over	0.5	0.2	1.3	1.0
Smokeless tobacco: Total	3.4	3.9	2.0	0.8
12 to 17 years old	3.0	3.9	0.8	1.1
18 to 25 years old	5.8	7.4	21.2	1.7
26 to 34 years old	3.6	4.3	1.5	0.8
35 years old and over	2.8	3.0	2.8	0.4

Source: "Use of Selected Drugs, by Age of User: 1991," *Statistical Abstract of the United States 1993,* p. 137. Primary source: Substance Abuse and Mental Health Services Administration, *National Household Survey on Drug Abuse,* 1991. *Notes:* 1. Includes other races, not shown separately. 2. Non-Hispanic. 3. Nonmedical use; does not included over-the-counter drugs.

★ 476 ★

Substance Abuse

Use of Selected Drugs, by Age of User, 1991, Part 2 Ever Used

SUBSTANCE AND AGE GROUP	Total[1]	RACE/ETHNICITY White[2]	Black[2]	Hispanic
EVER USED				
Crack: Total	1.9	1.5	4.3	2.1
12 to 27 years old	0.9	0.8	1.1	1.3
18 to 25 years old	3.8	3.7	4.7	3.3
26 to 34 years old	3.7	2.8	9.3	3.7
35 years old and over	1.0	0.8	3.0	1.1
Inhalants: Total	5.4	5.6	3.8	4.8
12 to 17 years old	7.0	7.6	5.1	6.6
18 to 25 years old	10.9	12.7	4.5	6.5
26 to 34 years old	9.2	10.3	4.6	6.3
35 years old and over	2.5	2.3	2.9	2.6
Hallucinogens: Total	8.1	8.9	4.1	6.4
12 to 17 years old	3.3	3.8	1.2	3.5
18 to 25 years old	13.1	15.8	5.4	7.5
26 to 34 years old	15.5	18.0	5.9	9.2
35 years old and over	5.2	5.4	3.7	5.5
Stimulants: Total[4]	7.0	7.9	3.3	4.8
12 to 17 years old	3.0	3.5	0.8	2.1
18 to 25 years old	9.4	11.3	3.2	5.1

[Continued]

★ 476 ★

Use of Selected Drugs, by Age of User, 1991, Part 2 Ever Used

[Continued]

SUBSTANCE AND AGE GROUP	Total[1]	RACE/ETHNICITY		
		White[2]	Black[2]	Hispanic
26 to 34 years old	12.2	14.3	5.0	6.3
35 years old and over	5.4	5.8	3.3	4.7
Sedatives: Total[4]	4.3	4.6	3.0	3.0
12 to 17 years old	2.4	2.7	1.2	2.2
18 to 25 years old	4.3	5.1	2.4	2.3
26 to 34 years old	7.5	8.6	3.8	3.4
35 years old and over	3.5	3.5	3.4	3.3
Tranquilizers: Total[4]	5.6	6.1	3.1	3.9
12 to 17 years old	2.1	2.6	1.1	1.0
18 to 25 years old	7.4	8.8	3.9	3.7
26 to 34 years old	10.0	11.4	5.7	5.4
35 years old and over	4.2	4.4	2.4	4.2
Analgesics: Total[4]	6.1	6.5	4.7	3.9
12 to 17 years old	4.4	4.8	3.9	3.6
18 to 25 years old	10.2	11.4	6.8	6.3
26 to 34 years old	9.8	11.1	6.8	4.1
35 years old and over	4.1	4.2	3.3	3.0

Source: "Use of Selected Drugs, by Age of User: 1991," *Statistical Abstract of the United States 1993*, p. 137. Primary source: Substance Abuse and Mental Health Services Administration, *National Household Survey on Drug Abuse*, 1991. *Notes:* 1. Includes other races, not shown separately. 2. Non-Hispanic. 3. Nonmedical use; does not included over-the-counter drugs.

Chapter 8
HOUSING

Homeless

★ 477 ★

Homeless in Cities, 1987

Characteristic	Large cities 1987	All local surveys 1981-1988
Demographic		
Male	84%	74%
Black	45	44
Hispanic	10	12
Over 65	3	na
Mental health		
Spent time in mental hospital	22	24
Attempted suicide	24	na
Diagnosed as currently mentally ill	na	33
Substance abuse		
Currently addicted to alcohol	na	27
Spent time in residential treatment program	na	29
Social ties		
Never married	53	na
Not currently with spouse	97[1]	na
No friends	na	36
No contact with relatives	na	31

[Continued]

★ 477 ★

Homeless in Cities, 1987
[Continued]

Characteristic	Large cities 1987	All local surveys 1981-1988
Spent time in jail or prison	41	41
Current health "fair" or "poor" (self-report)	44	38

Source: Christopher Jencks, "The Homeless." *The New York Review of Books* 41 (21 April 1994): 23. Published by permission. Primary source: Column 1 is from Martha Burt and Barbara Cohen, *America's Homeless: Numbers, Characteristics, and the Programs That service Them* (Washington: Urban Institute Press, 1989), pp. 69-71, and combines their data on homeless service users and nonusers. Column 2 is from Anne Shlay and Peter Rossi, "Social Science Research and Contemporary Studies of Homelessness," *Annual Review of Sociology,* Vol. 18 (1992), pp. 129-160, and somewhat overrepresents shelter residents. *Note:* 1. Adults using shelters or soup kitchens.

Homeownership

★ 478 ★

Homeownership Rates, by Characteristic of Homeowner, 1991

[Percent]

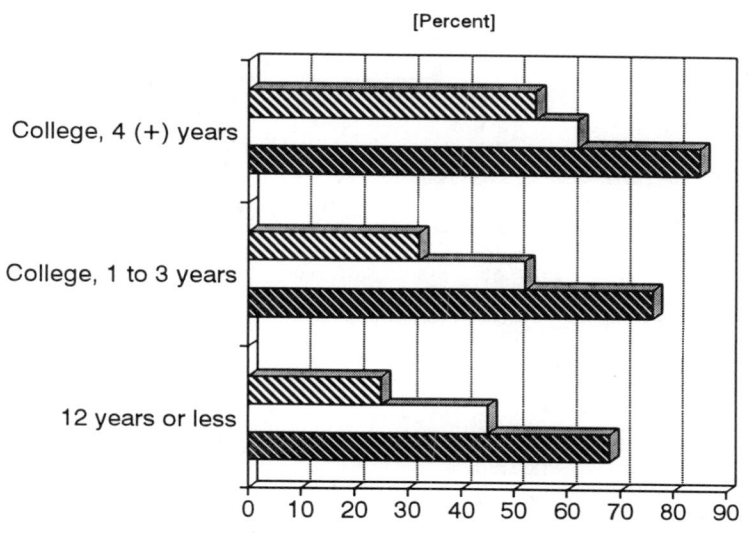

Married couples ☐ Other 2 (+) adults 1 adult

In percent.

	Married couples	Other 2 (+) adults	1 adult
12 years or less	68	45	25
College, 1 to 3 years	76	52	32
College, 4 (+) years	85	62	54

Source: "Homeownership Rates, Households with Children by Type and Educational Attainment of Householder: 1991." Jeanne Woodward, *Housing America's Children*, p. 8.

★ 479 ★
Homeownership

Homeownership Rates, by Characteristic of Homeowner, 1991

[Percent]

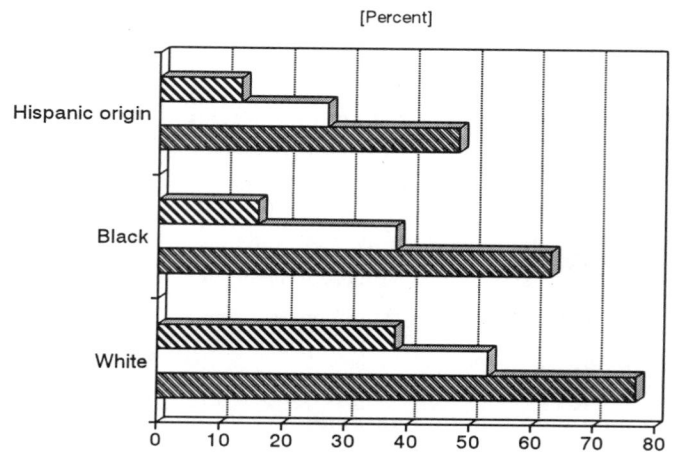

| | Married couples | Other 2 (+) adults | 1 adult |

In percent.

	Married couples	Other 2 (+) adults	1 adult
White	77	53	38
Black	63	38	16
Hispanic origin	48	27	13

Source: "Homeownership Rates, Households with Children by Type and Educational Attainment of Householder: 1991." Jeanne Woodward, *Housing America's Children*, p. 8. *Note:* Hispanic-origin householders may be of any race.

<hr>

Housing of the Elderly

<hr>

★ 480 ★

Householders 65 Years of Age or Older Living in Units Built Before 1950

	Percent
Race	
Black	55
White	36
Other races	24
Household type	
2 or more persons	51
Persons living alone or other male householders	43
Married couples	31
Income	
Less than $10,000	46
$10,000 to $24,999	39
$25,000 or more	30

Source: "Percent of Owners 65 Years of Age or Older Living in Housing Units Built Before 1950: Selected Demographic Characteristics," Mary L. Naifeh, *Housing of the Elderly: 1991*, p. 18.

★ 481 ★

Housing of the Elderly

Householders 65 Years of Age or Older Who Are Owners, 1991

	Percent
Race	
White	79
Black and other races	64
Hispanic origin	
White, non-Hispanic	79
Hispanic	59
Income	
Less than 10,000	61
$10,000 to $24,999	80
$25,000 or more	91

[Continued]

★ 481 ★

Householders 65 Years of Age or Older Who Are
Owners, 1991
[Continued]

	Percent
Household type	
Married couples	91
2 or more persons, spouse not present	78
Persons living alone	64

Source: "Percent of Householders 65 Years of Age or Older Who Are Owners, by Demographic Characteristics," Mary L. Naifeh, *Housing of the Elderly: 1991,* p. 5.

★ 482 ★

Housing of the Elderly

Householders 65 Years of Age or Older Who Occupied
Homes Before 1960

	Percent
Race	
Black	35
White	35
Other races	21
Household type	
Female householders	41
Married couples or male householders	32
Income	
Less than $25,000	38
$25,000 or more	30

Source: "Percent of Owners 65 Years of Age or Older Living Who Moved into Their Homes Before 1960: Selected Demographic Characteristics," Mary L. Naifeh, *Housing of the Elderly: 1991,* p. 14.

★ 483 ★

Housing of the Elderly

Owners and Renters 65 Years of Age or Older with Central Air Conditioning

Race and income	Owners %	Renters %
Race		
White	44	31
Black and other races	23	17
Hispanic origin		
White, non-Hispanic	45	
Hispanic	35	
Income		
Less than $10,000	30	24
$10,000 to $24,999	40	30
$25,000 or more	54	42

Source: "Percent of Households 65 Years of Age or Older With Central Air Conditioning, by Tenure and Demographic Characteristics," Mary L. Naifeh, *Housing of the Elderly: 1991*, p. 21.

★ 484 ★

Housing of the Elderly

Owners and Renters 65 Years of Age or Older with Central Heating, 1991

Race	Owners %	Renters %
White, non-Hispanic	88	91
Hispanic	73	73
White and other races	89	
White	88	
Black	76	
Black and other races	69	

Source: "Percent of Households 65 Years of Age or Older With Central Heating, by Tenure and Demographic Characteristics," Mary L. Naifeh, *Housing of the Elderly: 1991*, p. 19.

★ 485 ★

Housing of the Elderly

Renters 65 Years of Age or Older Living in Units Built Before 1950

	Percent
Race	
Black	52
White	32
Other races	29
Household type	
2 or more persons	50
Married couples	35
Female living alone	30
Income	
Less than $10,000	37
$10,000 or more	33

Source: "Percent of Renters 65 Years of Age or Older Living in Housing Units Built Before 1950: Selected Demographic Characteristics," Mary L. Naifeh, *Housing of the Elderly: 1991*, p. 18.

Physical Characteristics

★ 486 ★

Cars and Trucks Available to Householders, 1991

Characteristic	All occupied units				Owner-occupied				Renter-occupied			
		Age of householder				Age of householder				Age of householder		
	Total	Under 40 years	40 to 64 years	65 years and over	Total	Under 40 years	40 to 64 years	65 years and over	Total	Under 40 years	40 to 64 years	65 years and over
White												
Cars and trucks available												
Percent	100.0	100.0	100.0	100.0	100.0	100.0	100.0	100.0	100.0	100.0	100.0	100.0
No cars, trucks, or vans	8.1	5.8	4.6	17.9	4.0	1.1	1.5	11.0	17.0	10.5	16.1	43.5
With 1 car, truck or van	61.6	62.3	59.5	64.3	61.4	60.1	58.6	67.8	62.0	64.4	62.8	51.4
2 or more cars, trucks, or vans	30.2	31.9	35.9	17.8	34.6	38.8	39.9	21.2	21.0	25.2	21.1	5.1
Black												
Cars and trucks available												
Percent	100.0	100.0	100.0	100.0	100.0	100.0	100.0	100.0	100.0	100.0	100.0	100.0
No cars, trucks, or vans	29.2	29.7	22.7	42.9	13.5	8.9	7.8	29.1	40.9	36.2	40.4	67.3
With 1 car, truck or van	51.6	52.4	52.9	46.4	57.8	58.3	58.7	55.7	46.9	50.6	46.0	29.9
2 or more cars, trucks, or vans	19.3	17.9	24.4	10.7	28.7	32.8	33.5	15.2	12.2	13.2	13.6	2.7
Hispanic												
Cars and trucks available												
Percent	100.0	100.0	100.0	100.0	100.0	100.0	100.0	100.0	100.0	100.0	100.0	100.0

[Continued]

★ 486 ★

Cars and Trucks Available to Householders, 1991

[Continued]

Characteristic	All occupied units				Owner-occupied				Renter-occupied			
		Age of householder				Age of householder				Age of householder		
	Total	Under 40 years	40 to 64 years	65 years and over	Total	Under 40 years	40 to 64 years	65 years and over	Total	Under 40 years	40 to 64 years	65 years and over
No cars, trucks, or vans	19.8	19.4	16.9	31.7	4.9	1.6	2.8	18.3	29.2	25.4	32.2	51.0
With 1 car, truck or van	57.5	58.2	57.0	55.7	62.0	60.3	63.0	62.7	54.6	57.5	50.5	45.8
2 or more cars, trucks, or vans	22.7	22.3	26.1	12.6	33.0	38.1	34.2	19.1	16.1	17.1	17.3	3.2

Source: "Selected Social Characteristics of Householders in Occupied Units, by Tenure, Race and Hispanic Origin of Householder: 1991." Timothy S. Grail, *Our Nation's Housing in 1991*, pp. 39, 40, 43. Adapted by the editors.

★ 487 ★

Physical Characteristics

Housing Equipment and Amenities, 1991. Part I.

Characteristic	All occupied units				Owner-occupied				Renter-occupied			
		Age of householder				Age of householder				Age of householder		
	Total	Under 40 years	40 to 64 years	65 years and over	Total	Under 40 years	40 to 64 years	65 years and over	Total	Under 40 years	40 to 64 years	65 years and over
White												
Percent with												
Complete kitchen facilities	99.2	99.1	99.2	99.3	99.5	99.3	99.6	99.6	98.4	98.8	97.7	98.4
Clothes washer	79.1	70.5	86.8	79.4	94.4	94.1	96.1	91.9	46.7	47.7	52.2	32.8
Clothes dryer	74.0	67.2	82.6	69.8	89.4	91.8	92.6	81.6	41.3	43.3	45.5	26.1
Telephone	94.8	92.1	96.0	96.9	97.5	96.9	97.6	98.0	89.0	87.4	90.3	92.8
All selected equipment	71.1	63.3	80.0	68.0	86.9	88.6	90.1	79.8	37.5	38.8	42.2	24.1
Physical problems												
Percent of units with physical problems	6.6	7.4	6.1	6.3	5.4	5.9	4.9	5.8	9.3	8.9	10.8	8.2
Severe physical problems	2.8	2.8	2.9	2.7	2.5	2.6	2.4	2.5	3.5	2.9	4.8	3.5
Moderate physical problems	3.8	4.6	3.2	3.6	2.9	3.3	2.5	3.3	5.8	5.9	6.0	4.7
Equipment failures												
With hot and cold piped water	78,979	29,115	31,654	18,210	53,688	14,342	24,975	14,370	25,292	14,773	6,679	3,840
Percent with stoppage in last 3 months	4.5	5.0	4.4	3.9	4.1	4.6	4.2	3.5	5.3	5.4	5.3	5.2
With at least 1 flush toilet	78,933	29,097	31,628	18,208	53,685	14,343	24,969	14,373	25,248	14,754	6,659	3,835
Percent with breakdown in last 3 months	4.5	5.4	4.0	4.1	3.6	4.0	3.3	3.8	6.5	6.9	6.6	5.0

Source: "Selected Social Characteristics of Householders in Occupied Units, by Tenure, Race and Hispanic Origin of Householder: 1991." Timothy S. Grail, *Our Nation's Housing in 1991*, p. 38. Adapted by the editors.

★ 488 ★

Physical Characteristics

Housing Equipment and Amenities, 1991. Part II.

Characteristic	All occupied units				Owner-occupied				Renter-occupied			
		Age of householder				Age of householder				Age of householder		
	Total	Under 40 years	40 to 64 years	65 years and over	Total	Under 40 years	40 to 64 years	65 years and over	Total	Under 40 years	40 to 64 years	65 years and over
Black												
Percent with												
Complete kitchen facilities	97.8	98.1	97.6	97.4	98.6	99.3	98.6	97.8	97.1	97.7	96.3	96.8
Clothes washer	58.0	46.4	68.2	63.2	84.8	83.7	87.8	79.8	37.9	34.6	45.1	33.8

[Continued]

★ 488 ★

Housing Equipment and Amenities, 1991. Part II.

[Continued]

| Characteristic | All occupied units | | | | Owner-occupied | | | | Renter-occupied | | | |
| | Total | Age of householder | | | Total | Age of householder | | | Total | Age of householder | | |
		Under 40 years	40 to 64 years	65 years and over		Under 40 years	40 to 64 years	65 years and over		Under 40 years	40 to 64 years	65 years and over
Clothes dryer	42.0	35.8	51.2	35.9	66.4	70.4	72.5	50.5	23.7	24.9	26.2	10.3
Telephone	86.8	81.6	90.0	92.3	94.7	93.1	95.3	94.8	80.9	78.0	83.7	87.8
All selected equipment	38.3	31.5	47.6	33.7	62.4	64.9	68.9	47.0	20.3	21.0	22.5	10.3
Physical problems												
Percent of units with physical problems	17.4	17.0	15.9	22.0	15.3	15.0	11.6	23.1	18.9	17.6	20.9	20.2
Severe physical problems	4.9	4.8	4.9	4.8	3.4	3.0	2.8	4.8	6.0	5.4	7.5	4.8
Moderate physical problems	12.5	12.2	11.0	17.2	12.0	11.9	8.8	18.3	13.0	12.2	13.5	15.4
Equipment failures												
With hot and cold piped water	10,740	4,650	4,276	1,814	4,603	1,118	2,326	1,158	6,137	3,532	1,950	656
Percent with stoppage in last 3 months	3.8	4.6	3.5	2.4	1.8	1.7	1.8	1.7	5.3	5.5	5.4	3.7
With at least 1 flush toilet	10,699	4,630	4,255	1,814	4,605	1,118	2,328	1,158	6,095	3,512	1,927	656
Percent with breakdown in last 3 months	7.5	9.7	5.0	8.1	5.3	8.0	3.3	6.7	9.2	10.2	7.0	10.6

Source: "Selected Social Characteristics of Householders in Occupied Units, by Tenure, Race and Hispanic Origin of Householder: 1991." Timothy S. Grail, *Our Nation's Housing in 1991*, p. 40. Adapted by the editors.

★ 489 ★

Physical Characteristics

Housing Equipment and Amenities, 1991. Part III.

| Characteristic | All occupied units | | | | Owner-occupied | | | | Renter-occupied | | | |
| | Total | Age of householder | | | Total | Age of householder | | | Total | Age of householder | | |
		Under 40 years	40 to 64 years	65 years and over		Under 40 years	40 to 64 years	65 years and over		Under 40 years	40 to 64 years	65 years and over
Hispanic												
Percent with												
Complete kitchen facilities	98.1	98.0	98.1	99.0	99.3	98.8	99.5	100.0	97.4	97.7	96.6	97.5
Clothes washer	56.5	47.3	66.9	64.4	89.9	88.7	90.7	89.6	35.3	33.5	41.0	28.1
Clothes dryer	41.5	35.2	49.1	45.0	72.8	74.3	74.3	65.3	21.6	22.2	21.7	15.8
Telephone	86.9	83.0	90.3	93.3	93.7	91.7	94.0	97.1	82.5	80.2	86.2	87.9
All selected equipment	37.9	30.9	45.9	43.6	68.0	67.2	69.9	63.9	18.7	18.7	19.8	14.4
Physical problems												
Percent of units with physical problems	13.3	14.2	11.7	14.8	10.7	14.6	7.0	14.5	15.0	14.1	16.8	15.1
Severe physical problems	4.3	3.9	4.6	4.9	2.6	2.9	2.2	3.3	5.4	4.3	7.2	7.2
Moderate physical problems	9.0	10.3	7.1	9.9	8.2	11.7	4.8	11.2	9.6	9.8	9.5	7.9
Equipment failures												
With hot and cold piped water	6,210	3,212	2,333	665	2,418	806	1,217	395	3,793	2,406	1,116	270
Percent with stoppage in last 3 months	4.7	5.5	3.8	3.6	3.7	5.0	2.8	3.6	5.3	5.7	4.8	3.5
With at least 1 flush toilet	6,193	3,205	2,324	665	2,420	807	1,218	395	3,774	2,398	1,106	270
Percent with breakdown in last 3 months	8.7	9.5	8.0	7.0	7.4	9.4	6.1	7.3	9.5	9.5	10.1	6.6

Source: "Selected Social Characteristics of Householders in Occupied Units, by Tenure, Race and Hispanic Origin of Householder: 1991." Timothy S. Grail, *Our Nation's Housing in 1991*, p. 43. Adapted by the editors.

★ 490 ★

Physical Characteristics

Housing Size, Age, and Bathrooms, 1991. Part I.

Numbers in thousands, except percents and derived measures.

| Characteristic | All occupied units | | | | Owner-occupied | | | | Renter-occupied | | | |
| | Total | Age of householder | | | Total | Age of householder | | | Total | Age of householder | | |
		Under 40 years	40 to 64 years	65 years and over		Under 40 years	40 to 64 years	65 years and over		Under 40 years	40 to 64 years	65 years and over
White												
Total	79,140	29,175	31,712	18,253	53,748	14,361	24,993	14,395	25,391	14,814	6,720	3,858
Units in structure												
Percent	100.0	100.0	100.0	100.0	100.0	100.0	100.0	100.0	100.0	100.0	100.0	100.0
1 unit, detached	64.5	52.6	73.8	67.4	82.4	79.3	85.3	80.7	26.5	26.7	31.2	17.8
1 unit, attached	5.3	6.5	4.5	4.8	4.1	4.7	3.7	4.1	7.8	8.2	7.3	7.3
2 to 4 units	9.3	13.7	6.6	7.3	3.1	2.9	2.8	3.8	22.5	24.1	20.4	20.2
5 to 49 units	11.3	17.3	7.3	8.6	1.7	1.9	1.3	2.4	31.5	32.3	29.5	31.6
50 units or more	3.1	2.6	2.0	5.7	0.8	0.7	0.5	1.6	7.8	4.4	7.6	21.0
Mobile home or trailer	6.5	7.4	5.9	6.3	7.8	10.5	6.4	7.4	3.9	4.4	3.9	2.1
Cooperatives and condominiums												
Percent coop or condo	4.5	4.3	4.2	5.6	4.7	4.7	3.9	5.9	4.3	3.9	5.1	4.2
Year structure built												
Median age in years	26.0	22.0	24.6	33.2	25.8	18.3	23.6	34.6	26.6	25.8	28.5	25.8
Standard error	0.2	0.3	0.2	0.3	0.2	0.4	0.3	0.3	0.3	0.5	0.6	0.7
Percent new construction	5.7	8.6	5.0	2.4	6.5	12.7	5.4	2.1	4.2	4.6	3.7	3.6
Rooms in unit												
Median rooms	5.5	5.1	6.0	5.3	6.1	6.0	6.4	5.6	4.2	4.3	4.4	3.7
Standard error	0.01	0.02	0.02	0.02	0.01	0.02	0.02	0.02	0.01	0.02	0.03	0.04
Bedrooms in unit												
Median bedrooms	2.7	2.5	2.9	2.4	2.9	2.9	3.1	2.7	1.9	2.0	2.0	1.5
Standard error	0.01	0.01	0.01	0.01	0.01	0.01	0.01	0.01	0.01	0.01	0.02	0.02
Complete bathrooms												
Percent	100.0	100.0	100.0	100.0	100.0	100.0	100.0	100.0	100.0	100.0	100.0	100.0
None	0.5	0.5	0.5	0.6	0.3	0.2	0.2	0.4	1.1	0.8	1.5	1.2
1	45.8	52.8	36.0	51.4	33.5	34.1	27.7	43.1	71.6	70.9	67.0	82.3
More than 1	53.7	46.7	63.5	48.0	66.2	65.7	72.1	56.5	27.3	28.2	31.5	16.5
Persons per room												
1.01 or more persons per room	1,627	1.034	551	42	615	293	292	31	1,011	741	259	11
Percent of total	2.1	3.5	1.7	0.2	1.1	2.0	1.2	0.2	4.0	5.0	3.9	0.3
Square footage of unit												
Single detached and mobile homes	52,004	16,010	23,575	12,420	45,183	11,937	21,480	11,765	6,821	4,073	2,095	654
Median square footage	1,723	1,575	1,875	1,614	1,793	1,705	1,924	1,640	1,276	1,247	1,355	1,222
Standard error	7	13	11	15	8	15	11	15	15	19	29	49
Median square feet per person	690	503	701	970	720	529	719	975	492	437	538	842
Standard error	3	5	5	9	4	6	6	9	7	9	16	49

Source: "Selected Social Characteristics of Householders in Occupied Units, by Tenure, Race and Hispanic Origin of Householder: 1991." Timothy S. Grail, *Our Nation's Housing in 1991*, pp. 37-38. Adapted by the editors.

★ 491 ★

Physical Characteristics

Housing Size, Age, and Bathrooms, 1991. Part II.

Numbers in thousands, except percents and derived measures.

Characteristic	All occupied units				Owner-occupied				Renter-occupied			
		Age of householder				Age of householder				Age of householder		
	Total	Under 40 years	40 to 64 years	65 years and over	Total	Under 40 years	40 to 64 years	65 years and over	Total	Under 40 years	40 to 64 years	65 years and over
Black												
Total	10,832	4,670	4,323	1,839	4,635	1,120	2,341	1,174	6,197	3,550	1,982	665
Units in structure												
Percent	100.0	100.0	100.0	100.0	100.0	100.0	100.0	100.0	100.0	100.0	100.0	100.0
1 unit, detached	46.2	31.5	56.1	60.2	79.2	68.9	83.9	79.5	21.5	19.7	23.3	26.0
1 unit, attached	9.7	11.0	8.7	8.8	8.9	11.0	7.2	10.0	10.4	11.0	10.5	6.5
2 to 4 units	15.2	19.2	12.4	11.8	3.9	4.5	3.0	5.1	23.7	23.9	23.5	23.6
5 to 49 units	19.2	28.0	14.3	8.7	1.2	2.1	1.0	0.8	32.7	36.1	29.9	22.5
50 units or more	6.2	6.2	5.5	7.7	0.6	0.4	0.6	0.5	10.4	8.0	11.3	20.5
Mobile home or trailer	3.5	4.1	3.0	2.9	6.3	13.0	4.2	4.0	1.3	1.3	1.5	0.9
Cooperatives and condominiums												
Percent coop or condo	2.3	2.1	2.8	1.6	2.2	2.7	2.5	1.3	2.3	1.9	3.2	2.1
Year structure built												
Median age in years	32.5	27.7	32.6	43.5	32.4	21.1	30.4	43.9	32.7	29.4	36.1	42.5
Standard error	0.6	0.8	0.9	1.2	0.8	1.7	1.0	1.4	0.9	0.9	1.6	2.6
Percent new construction	3.1	4.3	2.6	1.4	4.1	10.1	2.6	1.3	2.4	2.5	2.6	1.4
Rooms in unit												
Median rooms	5.0	4.6	5.4	5.0	5.9	5.7	6.1	5.6	4.2	4.3	4.3	3.7
Standard error	0.03	0.04	0.06	0.07	0.04	0.08	0.05	0.08	0.03	0.04	0.06	0.11
Bedrooms in unit												
Median bedrooms	2.4	2.2	2.7	2.4	2.9	2.9	3.0	2.8	1.9	2.0	2.0	1.5
Standard error	0.02	0.03	0.03	0.06	0.02	0.04	0.03	0.05	0.02	0.03	0.05	0.08
Complete bathrooms												
Percent	100.0	100.0	100.0	100.0	100.0	100.0	100.0	100.0	100.0	100.0	100.0	100.0
None	1.6	1.1	1.8	2.2	1.0	0.2	0.7	2.4	2.0	1.4	3.2	1.8
1	63.6	68.3	56.0	69.3	44.0	42.8	38.6	55.8	78.2	76.3	76.6	93.0
More than 1	34.8	30.6	42.1	28.6	55.0	57.0	60.7	41.8	19.8	22.2	20.2	5.2
Persons per room												
1.01 or more persons per room	566	309	220	36	180	54	98	27	386	255	122	9
Percent of total	5.2	6.6	5.1	2.0	3.9	4.8	4.2	2.3	6.2	7.2	6.2	1.3
Square footage of unit												
Single detached and mobile homes	4,660	1,417	2,249	994	3,501	796	1,848	857	1,159	621	401	137
Median square footage	1,415	1,322	1,558	1,283	1,519	1,468	1,648	1,345	1,127	1,169	1,164	877
Standard error	20	34	38	42	30	52	40	43	39	46	78	62
Median square feet per person	528	409	546	730	577	468	566	745	410	359	460	624
Standard error	13	16	17	25	14	22	18	27	17	16	28	84

Source: "Selected Social Characteristics of Householders in Occupied Units, by Tenure, Race and Hispanic Origin of Householder: 1991." Timothy S. Grail, *Our Nation's Housing in 1991*, p. 39. Adapted by the editors.

★ 492 ★

Physical Characteristics

Housing Size, Age, and Bathrooms, 1991. Part III.

Numbers in thousands, except percents and derived measures.

Characteristic	All occupied units				Owner-occupied				Renter-occupied			
		Age of householder				Age of householder				Age of householder		
	Total	Under 40 years	40 to 64 years	65 years and over	Total	Under 40 years	40 to 64 years	65 years and over	Total	Under 40 years	40 to 64 years	65 years and over
Hispanic												
Total	6,239	3,224	2,346	669	2,423	807	1,221	395	3,816	2,417	1,125	274
Units in structure												
Percent	100.0	100.0	100.0	100.0	100.0	100.0	100.0	100.0	100.0	100.0	100.0	100.0
1 unit, detached	45.6	36.2	55.4	56.7	81.8	79.6	83.5	81.3	22.6	21.7	24.9	21.4
1 unit, attached	5.9	6.7	4.8	6.3	4.5	5.3	3.7	5.1	6.8	7.1	5.9	8.0
2 to 4 units	16.4	19.4	13.8	10.9	3.9	2.4	4.7	4.1	24.4	25.1	23.7	20.7
5 to 49 units	22.9	29.9	16.6	11.8	2.5	3.4	1.8	2.8	35.9	38.7	32.7	24.7
50 units or more	5.8	4.8	6.0	10.3	1.0	0.8	0.9	1.5	8.9	6.1	11.4	22.8
Mobile home or trailer	3.3	3.1	3.5	4.1	6.4	8.5	5.4	5.2	1.4	1.3	1.4	2.4
Cooperatives and condominiums												
Percent coop or condo	3.9	3.6	4.3	4.2	5.4	6.0	5.0	5.6	3.0	2.8	3.6	2.3
Year structure built												
Median age in years	30.7	28.9	31.4	35.8	27.1	21.6	26.4	36.2	33.0	30.6	37.0	35.1
Standard error	0.7	1.0	1.0	1.8	1.1	2.1	1.3	2.0	0.9	1.1	1.4	3.7
Percent new construction	3.5	4.4	2.6	2.3	5.2	9.7	3.0	2.5	2.4	2.6	2.1	2.1
Rooms in unit												
Median rooms	4.6	4.4	5.1	4.7	5.8	5.7	6.0	5.3	4.1	4.1	4.2	3.7
Standard error	0.04	0.04	0.07	0.11	0.06	0.09	0.08	0.12	0.03	0.04	0.06	0.17
Bedrooms in unit												
Median bedrooms	2.3	2.1	2.6	2.2	2.9	2.9	3.0	2.6	1.9	1.8	2.0	1.5
Standard error	0.03	0.03	0.04	0.08	0.03	0.05	0.04	0.09	0.03	0.03	0.05	0.11
Complete bathrooms												
Percent	100.0	100.0	100.0	100.0	100.0	100.0	100.0	100.0	100.0	100.0	100.0	100.0
None	1.2	1.0	1.4	1.2	0.6	0.4	0.6	1.1	1.5	1.2	2.3	1.5
1	63.8	69.8	55.1	65.7	41.3	44.4	35.7	52.5	78.1	78.2	76.1	84.7
More than 1	35.0	29.3	43.5	33.1	58.1	55.2	63.7	46.5	20.4	20.6	21.6	13.8
Persons per room												
1.01 or more persons per room	918	612	293	13	199	94	99	7	719	518	194	6
Percent of total	14.7	19.0	12.5	1.9	8.2	11.6	8.1	1.7	18.8	21.4	17.3	2.3
Square footage of unit												
Single detached and mobile homes	2,783	1,146	1,254	383	1,974	648	998	328	810	499	256	55
Median square footage	1,371	1,292	1,456	1,358	1,467	1,399	1,548	1,397	1,143	1,148	1,122	-
Standard error	22	31	34	75	26	41	51	84	40	48	81	-
Median square feet per person	423	344	453	771	478	366	493	793	321	302	318	-
Standard error	14	11	19	45	16	12	21	44	17	17	29	-

Source: "Selected Social Characteristics of Householders in Occupied Units, by Tenure, Race and Hispanic Origin of Householder: 1991." Timothy S. Grail, *Our Nation's Housing in 1991*, p. 42. Adapted by the editors.

★ 493 ★

Physical Characteristics

Telephone Subscribers: 1984-1992

For March. Based on Current Population Survey.

Characteristic	1984				1992			
	All races	White	Black	Hispanic[1]	All races	White	Black	Hispanic[1]
Total	91.8	93.3	80.1	80.7	93.9	95.3	83.8	86.6
Age of householder								
16 to 24 years old	77.8	80.3	57.9	59.0	82.0	85.3	63.1	78.0
25 to 54 years old	91.9	93.5	80.4	83.2	93.2	94.7	82.9	85.8
55 to 59 years old	94.9	95.7	87.6	88.7	96.2	97.0	91.2	91.3
60 to 64 years old	94.2	95.9	81.7	87.4	96.6	97.6	90.1	92.2
65 to 69 years old	96.1	97.0	87.8	85.8	95.9	97.1	85.8	90.3
70 years old and over	95.3	96.2	87.2	82.2	97.6	98.3	92.0	95.4
Household size								
1 person	88.6	90.7	73.9	72.2	91.9	93.8	80.1	79.6
2 to 3 persons	93.3	94.5	82.4	80.7	95.3	96.5	86.2	88.5
4 to 5 persons	92.7	94.1	82.9	85.4	93.5	95.0	83.3	87.2
6 or more persons	86.4	88.6	78.8	78.8	90.6	91.8	85.2	86.6
Labor force status of persons								
16 years old and over								
Total civilian noninstitutional								
population	93.0	94.2	83.5	83.3	94.7	95.9	86.5	88.5
Employed	94.5	95.3	87.6	87.1	96.0	96.8	89.8	90.2
Unemployed	82.0	83.8	75.5	73.3	88.5	90.0	82.8	85.9
Not in labor force	92.0	93.8	80.2	79.6	93.4	95.1	82.8	86.0
Income level								
Under $5,000	71.4	74.7	62.8	53.6	71.7	76.0	63.0	66.9
$5,000 to $7,499	83.6	85.8	74.6	70.0	83.7	85.9	76.7	72.8
$7,500 to $9,999	85.8	87.7	75.9	72.2	88.3	90.4	79.3	78.4
$10,000 to $12,499	90.0	91.3	82.5	81.8	90.0	91.0	83.5	81.2
$12,500 to $14,999	92.7	93.6	84.6	88.5	91.2	92.3	85.4	84.1
$15,000 to $17,499	93.6	94.3	87.6	89.4	93.2	94.3	86.4	87.3
$17,500 to $19,999	95.3	95.4	94.8	87.1	95.6	96.4	89.7	91.6
$20,000 to $24,999	97.1	97.3	94.6	90.0	97.2	97.8	91.5	96.0
$25,000 to $29,999	98.1	98.5	93.5	96.2	98.2	98.4	96.9	96.1
$30,000 to $34,999	98.8	98.8	97.5	99.2	98.6	99.2	95.7	96.7
$35,000 to $39,999	99.4	99.5	96.3	100.0	99.0	99.3	96.4	96.0
$40,000 to $49,999	99.4	99.5	98.0	100.0	99.5	99.5	98.9	99.1
$50,000 to $74,999	99.2	99.3	97.0	100.0	99.4	99.4	98.7	100.0
$75,000 and over	98.9	99.0	94.0	95.1	99.4	99.5	95.3	99.1

Source: "Percent of Households with Telephone Service: 1984 and 1992." U.S. Bureau of the Census, *Statistical Abstract of the United States*, 1993, p. 563. Primary source: Federal Communications Commission, *Telephone Subscribership in the U.S.*, November 1992. *Note:* 1. Persons of Hispanic origin may be of any race.

Social Characteristics

★ 494 ★

Education of Householder, 1991. Part I.

Numbers in thousands, except percents.

Characteristic	All occupied units				Owner-occupied				Renter-occupied			
		Age of householder				Age of householder				Age of householder		
	Total	Under 40 years	40 to 64 years	65 years and over	Total	Under 40 years	40 to 64 years	65 years and over	Total	Under 40 years	40 to 64 years	65 years and over
White												
Total	79,140	29,175	31,712	18,253	53,748	14,361	24,993	14,395	25,391	14,814	6,720	3,858
Less than 12 years	15,381	3,319	5,170	6,892	9,789	1,040	3,561	5,189	5,591	2,279	1,609	1,704
High school graduate	28,663	10,899	11,424	6,340	19,630	5,460	9,108	5,063	9,033	5,440	2,316	1,277
1 to 3 years of college	15,463	6,894	6,219	2,350	10,166	3,380	4,902	1,883	5,298	3,514	1,317	466
4 or more years of college	19,632	8,062	8,900	2,670	14,163	4,481	7,422	2,260	5,469	3,581	1,478	410
Percent	100.0	100.0	100.0	100.0	100.0	100.0	100.0	100.0	100.0	100.0	100.0	100.0
Less than 12 years	19.4	11.4	16.3	37.8	18.2	7.2	14.2	36.0	22.0	15.4	23.9	44.2
High school graduate	36.2	37.4	36.0	34.7	36.5	38.0	36.4	35.2	35.6	36.7	34.5	33.1
1 to 3 years of college	19.5	23.6	19.6	12.9	18.9	23.5	19.6	13.1	20.9	23.7	19.6	12.1
4 or more years of college	24.8	27.6	28.1	14.6	26.4	31.2	29.7	15.7	21.5	24.2	22.0	10.6

Source: "Selected Social Characteristics of Householders in Occupied Units, by Tenure, Race and Hispanic Origin of Householder: 1991." Timothy S. Grail, *Our Nation's Housing in 1991*, p. 29. Adapted by the editors.

★ 495 ★

Social Characteristics

Education of Householder, 1991. Part II.

Numbers in thousands, except percents.

Characteristic	All occupied units				Owner-occupied				Renter-occupied			
		Age of householder				Age of householder				Age of householder		
	Total	Under 40 years	40 to 64 years	65 years and over	Total	Under 40 years	40 to 64 years	65 years and over	Total	Under 40 years	40 to 64 years	65 years and over
Black												
Total	10,832	4,670	4,323	1,839	4,635	1,120	2,341	1,174	6,197	3,550	1,982	665
Less than 12 years	3,313	710	1,325	1,278	1,468	88	615	765	1,845	621	710	514
High school graduate	4,006	2,037	1,622	347	1,571	457	886	228	2,435	1,580	736	119
1 to 3 years of college	2,051	1,168	776	107	812	301	428	83	1,239	867	348	23
4 or more years of college	1,462	755	600	107	784	274	412	98	678	482	188	9
Percent	100.0	100.0	100.0	100.0	100.0	100.0	100.0	100.0	100.0	100.0	100.0	100.0
Less than 12 years	30.6	15.2	30.6	69.5	31.7	7.9	26.3	65.2	29.8	17.5	35.8	77.2
High school graduate	37.0	43.6	37.5	18.9	33.9	40.8	37.8	19.4	39.3	44.5	37.2	17.9
1 to 3 years of college	18.9	25.0	17.9	5.8	17.5	26.9	18.3	7.1	20.0	24.4	17.6	3.5
4 or more years of college	13.5	16.2	13.9	5.8	16.9	24.4	17.6	8.3	10.9	13.6	9.5	1.3

Source: "Selected Social Characteristics of Householders in Occupied Units, by Tenure, Race and Hispanic Origin of Householder: 1991." Timothy S. Grail, *Our Nation's Housing in 1991*, p. 30. Adapted by the editors.

★ 496 ★

Social Characteristics

Education of Householder, 1991. Part III.

Numbers in thousands, except percents.

Characteristic	All occupied units				Owner-occupied				Renter-occupied			
	Total	Age of householder			Total	Age of householder			Total	Age of householder		
		Under 40 years	40 to 64 years	65 years and over		Under 40 years	40 to 64 years	65 years and over		Under 40 years	40 to 64 years	65 years and over
Hispanic												
Total	6,239	3,224	2,346	669	2,423	807	1,221	395	3,816	2,417	1,125	274
Less than 12 years	2,655	1,183	1,051	421	909	210	472	228	1,746	973	579	194
High school graduate	1,900	1,127	626	148	726	305	330	90	1,174	821	295	58
1 to 3 years of college	908	508	343	57	405	143	217	46	503	365	126	12
4 or more years of college	775	407	327	42	382	150	201	31	393	257	125	11
Percent	100.0	100.0	100.0	100.0	100.0	100.0	100.0	100.0	100.0	100.0	100.0	100.0
Less than 12 years	42.6	36.7	44.8	63.0	37.5	26.0	38.7	57.7	45.7	40.3	51.5	70.6
High school graduate	30.5	34.9	26.7	22.2	30.0	37.8	27.1	22.9	30.8	34.0	26.2	21.1
1 to 3 years of college	14.6	15.8	14.6	8.6	16.7	17.7	17.8	11.6	13.2	15.1	11.2	4.3
4 or more years of college	12.4	12.6	13.9	6.3	15.8	18.5	16.5	7.9	10.3	10.6	11.1	4.0

Source: "Selected Social Characteristics of Householders in Occupied Units, by Tenure, Race and Hispanic Origin of Householder: 1991." Timothy S. Grail, *Our Nation's Housing in 1991*, p. 34. Adapted by the editors.

★ 497 ★

Social Characteristics

Homeownership Rates by Presence of Children

Numbers in thousands.

Characteristic	All households	Households with children	Households without children
United States	93,147	34,588	58,559
Owner occupied	59,796	21,937	37,859
Percent owner occupied	64	63	65
Renter occupied	33,351	12,651	20,700
Race/ethnicity of householder			
White	79,140	27,864	51,276
Percent owner occupied	68	69	68
Black	10,832	5,124	5,708
Percent owner occupied	43	40	46
Hispanic origin[1]	6,239	3,437	2,802
Percent owner occupied	39	38	40
Non-Hispanic origin	86,907	31,150	55,757
Percent owner occupied	66	66	66

Source: "Home Ownership Rates by Presence of Children, 1991." Jeanne Woodward, *Housing America's Children in 1991*, p. 4. *Note:* 1. Persons of Hispanic origin may be of any race.

★ 498 ★

Social Characteristics

Homeownership Rates for Selected Race and Ethnic Groups, 1991

[Percent]

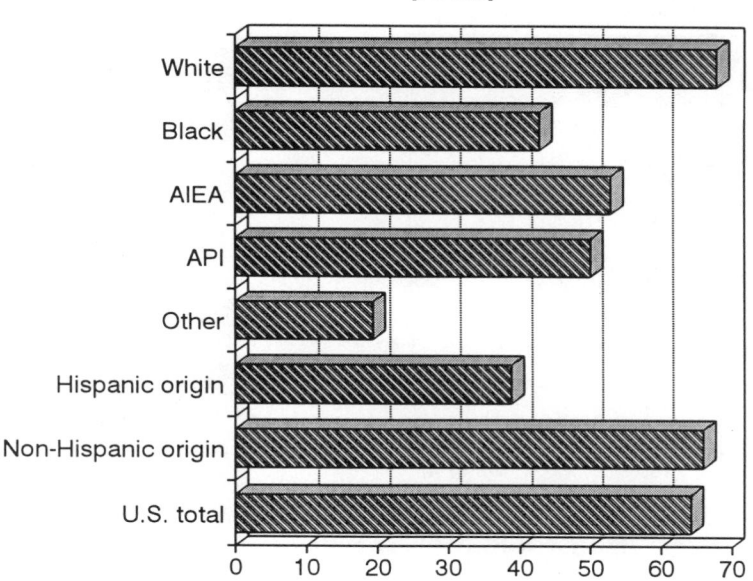

In percent.

White	67.9
Black	42.8
AIEA	52.9
API	50.0
Other	19.3
Hispanic origin	38.8
Non-Hispanic origin	66.0
U.S. total	64.2

Source: "Homeownership Rates for Selected Races and Ethnic Groups: 1991." Timothy S. Grail, "Our Nation's Housing in 1991," p. 6. *Notes:* Hispanic origin may be of any race. AIEA - American Indian, Eskimo, or Aleut. API - Asian or Pacific Islander.

★ 499 ★

Social Characteristics

Homeownership and Households with Children

Numbers in thousands, excluding percents.

| Characteristics | All households | Households with children | | | | Percent owner occupied | | | | |
| | | Total | Married couples | Other households with two or more adults | Households with one adult | All households | Households with children | | | |
							Total	Married couples	Other households with two or more adults	Households with one adult
Total	93,147	34,588	24,034	4,724	5,830	64	63	74	48	30
Race and origin										
White	79,140	27,864	20,916	3,189	3,759	68	69	77	53	38
Black	10,832	5,124	1,990	1,255	1,878	43	40	63	38	16
Hispanic origin	6,239	3,437	2,113	726	598	39	38	48	27	13
Non-Hispanic origin	86,907	31,150	21,921	3,998	5,232	66	66	77	52	32

Source: "Home Ownership Rates by Presence of Children, 1991." Jeanne Woodward, *Housing America's Children in 1991,* p. 7.

★ 500 ★

Social Characteristics

Household Moves, 1991

| Characteristic | All occupied units | | | | Owner-occupied | | | | Renter-occupied | | | |
| | Total | Age of householder | | | Total | Age of householder | | | Total | Age of householder | | |
		Under 40 years	40 to 64 years	65 years and over		Under 40 years	40 to 64 years	65 years and over		Under 40 years	40 to 64 years	65 years and over
White												
Total units where householder moved in last year	13,235	9,276	3,234	725	3,740	2,140	1,321	280	9,494	7,136	1,913	445
Percent total	16.7	31.8	10.2	4.0	7.0	14.9	5.3	1.9	37.4	48.2	28.5	11.5
Black												
Total units where householder moved in last year	2,300	1,708	506	86	283	183	91	9	2,017	1,525	416	77
Percent of total	21.2	36.6	11.7	4.7	6.1	16.3	3.9	0.8	32.5	42.9	21.0	11.6
Hispanic												
Total units where householder moved in last year	1,720	1,276	392	52	232	133	88	11	1,488	1,142	304	41
Percent of total	27.6	39.6	16.7	7.8	9.6	16.5	7.2	2.8	39.0	47.3	27.0	15.0

Source: "Selected Social Characteristics of Householders in Occupied Units, by Tenure, Race and Hispanic Origin of Householder: 1991." Timothy S. Grail, *Our Nation's Housing in 1991,* pp. 29, 30, 34. Adapted by the editors.

★ 501 ★

Social Characteristics

Householders in Occupied Units: Composition, 1991. Part I.

Numbers in thousands, except percents and derived measures.

Characteristic	All occupied units				Owner-occupied				Renter-occupied			
	Total	Age of householder			Total	Age of householder			Total	Age of householder		
		Under 40 years	40 to 64 years	65 years and over		Under 40 years	40 to 64 years	65 years and over		Under 40 years	40 to 64 years	65 years and over
White												
Household composition												
Total	79,140	29,175	31,712	18,253	53,748	14,361	24,993	14,395	25,391	14,814	6,720	3,858
Households with children	27,864	16,561	10,763	540	19,137	9,920	8,768	448	8,727	6,640	1,995	92
Family households	27,407	16,280	10,647	480	18,938	9,837	8,701	400	8,469	6,444	1,946	80
Married couples	20,916	12,303	8,414	199	16,032	8,545	7,315	172	4,884	3,758	1,098	28
Other male householder	1,601	956	596	50	923	460	422	41	678	496	174	8
Other female householder	4,890	3,021	1,638	231	1,983	832	964	187	2,907	2,189	674	44
Nonfamily households	457	280	116	60	199	84	67	48	258	197	49	12
Households with no children	51,276	12,614	20,949	17,713	34,612	4,441	16,224	13,947	16,664	8,173	4,725	3,766
Family households	28,766	4,603	14,594	9,569	23,739	2,455	12,711	8,573	5,026	2,148	1,883	996
Married couples	24,076	3,936	12,205	7,935	20,405	2,194	10,966	7,245	3,671	1,742	1,239	690
Other male householder	1,667	417	764	486	1,164	166	573	425	503	251	190	62
Other female householder	3,022	250	1,626	1,147	2,171	95	1,172	904	852	155	454	243
Nonfamily households	22,510	8,011	6,355	8,144	10,872	1,985	3,514	5,373	11,638	6,026	2,841	2,771
2-or-more persons	3,502	2,542	759	201	1,080	532	422	127	2,421	2,010	337	75
Male householder	2,115	1,577	448	90	642	359	227	56	1,474	1,218	221	34
Female householder	1,387	965	311	111	439	173	195	71	948	792	116	40
1-person	19,008	5,469	5,596	7,943	9,792	1,453	3,092	5,247	9,216	4,016	2,504	2,696
Male householder	7,279	3,216	2,427	1,636	3,141	889	1,178	1,073	4,138	2,327	1,249	562
Female householder	11,729	2,253	3,169	6,307	6,651	564	1,914	4,173	5,078	1,689	1,255	2,134

Source: "Selected Social Characteristics of Householders in Occupied Units, by Tenure, Race and Hispanic Origin of Householder: 1991." Timothy S. Grail, "Our Nation's Housing in 1991," pp. 27-28. Adapted by the editors.

★ 502 ★

Social Characteristics

Householders in Occupied Units: Composition, 1991. Part II.

Numbers in thousands.

Characteristic	All occupied units				Owner-occupied				Renter-occupied			
	Total	Age of householder			Total	Age of householder			Total	Age of householder		
		Under 40 years	40 to 64 years	65 years and over		Under 40 years	40 to 64 years	65 years and over		Under 40 years	40 to 64 years	65 years and over
Black												
Total	10,832	4,670	4,323	1,839	4,635	1,120	2,341	1,174	6,197	3,550	1,982	665
Household composition												
Total	10,832	4,670	4,323	1,839	4,635	1,120	2,341	1,174	6,197	3,550	1,982	665
Households with children	5,124	3,115	1,804	205	2,028	789	1,093	146	3,096	2,326	711	59
Family households	5,046	3,069	1,781	195	2,009	782	1,088	139	3,037	2,287	693	57
Married couples	1,990	1,055	866	70	1,252	498	691	63	738	557	175	7
Other male householder	326	200	112	15	128	69	46	13	199	131	65	2
Other female householder	2,729	1,814	804	111	628	216	351	62	2,100	1,599	453	48
Nonfamily households	79	46	23	10	20	7	5	7	59	38	18	3
Households with no children	5,708	1,555	2,519	1,634	2,607	331	1,248	1,028	3,101	1,224	1,271	606
Family households	2,489	450	1,325	714	1,635	166	879	590	855	284	446	124

[Continued]

★ 502 ★

Householders in Occupied Units: Composition, 1991. Part II.

[Continued]

Characteristic	All occupied units				Owner-occupied				Renter-occupied			
	Total	Age of householder			Total	Age of householder			Total	Age of householder		
		Under 40 years	40 to 64 years	65 years and over		Under 40 years	40 to 64 years	65 years and over		Under 40 years	40 to 64 years	65 years and over
Married couples	1,452	281	746	425	1,039	117	552	369	413	164	193	55
Other male householder	259	91	92	75	129	28	47	55	129	63	45	21
Other female householder	779	77	488	214	467	20	280	166	313	57	208	48
Nonfamily households	3,219	1,105	1,193	920	972	165	369	438	2,246	940	824	482
2-or-more persons	345	220	90	35	74	35	24	14	271	184	66	21
Male householder	219	139	59	21	41	19	17	5	178	120	42	16
Female householder	126	81	31	14	33	16	7	9	93	64	24	4
1 person	2,874	886	1,103	885	898	130	345	424	1,975	756	758	461
Male householder	1,326	506	550	269	345	72	147	126	981	434	403	143
Female householder	1,548	379	553	616	553	58	198	298	994	321	355	318

Source: "Selected Social Characteristics of Householders in Occupied Units, by Tenure, Race and Hispanic Origin of Householder: 1991." Timothy S. Grail, "Our Nation's Housing in 1991," p. 28. Adapted by the editors.

★ 503 ★

Social Characteristics

Householders in Occupied Units: Composition, 1991. Part III.

Numbers in thousands.

Characteristic	All occupied units				Owner-occupied				Renter-occupied			
	Total	Age of householder			Total	Age of householder			Total	Age of householder		
		Under 40 years	40 to 64 years	65 years and over		Under 40 years	40 to 64 years	65 years and over		Under 40 years	40 to 64 years	65 years and over
Hispanic												
Total	6,239	3,224	2,346	669	2,423	807	1,221	395	3,816	2,417	1,125	274
Household composition												
Total	6,239	3,224	2,346	669	2,423	807	1,221	395	3,816	2,417	1,125	274
Households with children	3,437	2,196	1,175	66	1,291	640	619	32	2,147	1,557	556	34
Family households	3,383	2,161	1,157	64	1,287	640	617	30	2,096	1,522	540	34
Married couples	2,113	1,357	721	35	1,019	531	470	18	1,094	826	251	17
Other male householder	262	176	81	5	74	44	27	3	188	132	54	2
Other female householder	1,008	628	356	24	194	65	121	9	814	564	235	15
Nonfamily households	55	35	18	2	4	-	1	2	51	35	16	-
Households with no children	2,802	1,028	1,171	603	1,132	168	602	362	1,670	860	569	240
Family households	1,491	414	758	319	830	106	485	239	661	308	273	80
Married couples	1,040	269	519	251	677	85	389	203	363	184	131	48
Other male householder	193	96	85	12	41	9	23	9	152	87	61	4
Other female householder	258	48	154	56	112	11	73	28	146	37	81	28
Nonfamily households	1,311	614	413	283	302	62	117	123	1,009	552	296	160
2-or-more persons	279	212	58	9	37	22	11	4	242	190	47	5
Male householder	191	154	34	4	19	12	5	2	173	141	29	2
Female householder	87	59	24	5	18	10	6	2	69	49	18	3
1 person	1,032	402	356	274	265	40	106	119	767	363	250	155
Male householder	530	269	187	74	96	25	46	25	434	244	142	48
Female householder	502	133	168	201	169	15	60	94	333	118	108	107

Source: "Selected Social Characteristics of Householders in Occupied Units, by Tenure, Race and Hispanic Origin of Householder: 1991." Timothy S. Grail, "Our Nation's Housing in 1991," pp. 32-33. Adapted by the editors. *Note:* - represents zero or rounds to zero.

★ 504 ★

Social Characteristics

Marital Status of Householder, 1991. Part I.

Numbers in thousands, except percents.

Characteristic	All occupied units				Owner-occupied				Renter-occupied			
	Total	Age of householder			Total	Age of householder			Total	Age of householder		
		Under 40 years	40 to 64 years	65 years and over		Under 40 years	40 to 64 years	65 years and over		Under 40 years	40 to 64 years	65 years and over
White												
Total	79,140	29,175	31,712	18,253	53,748	14,361	24,993	14,395	25,391	14,814	6,720	3,858
Married	45,787	16,543	20,920	8,323	36,899	10,855	18,468	7,576	8,887	5,687	2,453	747
Widowed	9,936	184	2,113	7,640	7,252	103	1,606	5,544	2,684	81	507	2,096
Divorced or separated	12,601	4,560	6,642	1,398	6,122	1,536	3,816	770	6,479	3,024	2,826	628
Never-married	10,816	7,888	2,038	891	3,474	1,866	1,104	504	7,342	6,021	934	387
Percent	100.0	100.0	100.0	100.0	100.0	100.0	100.0	100.0	100.0	100.0	100.0	100.0
Married	57.9	56.7	66.0	45.6	68.7	75.6	73.9	52.6	35.0	38.4	36.5	19.4
Widowed	12.6	0.6	6.7	41.9	13.5	0.7	6.4	38.5	10.6	0.5	7.5	54.3
Divorced or separated	15.9	15.6	20.9	7.7	11.4	10.7	15.3	5.4	25.5	20.4	42.1	16.3
Never-married	13.7	27.0	6.4	4.9	6.5	13.0	4.4	3.5	28.9	40.6	13.9	10.0

Source: "Selected Social Characteristics of Householders in Occupied Units, by Tenure, Race and Hispanic Origin of Householder: 1991." Timothy S. Grail, *Our Nation's Housing in 1991*, pp. 28-29. Adapted by the editors.

★ 505 ★

Social Characteristics

Marital Status of Householder, 1991. Part II.

Numbers in thousands, except percents.

Characteristic	All occupied units				Owner-occupied				Renter-occupied			
	Total	Age of householder			Total	Age of householder			Total	Age of householder		
		Under 40 years	40 to 64 years	65 years and over		Under 40 years	40 to 64 years	65 years and over		Under 40 years	40 to 64 years	65 years and over
Black												
Total	10,832	4,670	4,323	1,839	4,635	1,120	2,341	1,174	6,197	3,550	1,982	665
Married	3,607	1,430	1,669	508	2,343	625	1,276	442	1,264	805	393	66
Widowed	1,445	63	486	895	823	18	254	551	622	45	233	344
Divorced or separated	2,990	1,077	1,609	304	989	207	656	126	2,001	870	953	178
Never-married	2,791	2,100	560	132	481	270	156	55	2,310	1,830	404	76
Percent	100.0	100.0	100.0	100.0	100.0	100.0	100.0	100.0	100.0	100.0	100.0	100.0
Married	33.3	30.6	38.6	27.6	50.5	55.8	54.5	37.6	20.4	22.7	19.8	10.0
Widowed	13.3	1.4	11.2	48.7	17.8	1.7	10.8	46.9	10.0	1.3	11.7	51.8
Divorced or separated	27.6	23.1	37.2	16.5	21.3	18.5	28.0	10.7	32.3	24.5	48.1	26.8
Never-married	25.8	45.0	12.9	7.2	10.4	24.1	6.7	4.7	37.3	51.5	20.4	11.5

Source: "Selected Social Characteristics of Householders in Occupied Units, by Tenure, Race and Hispanic Origin of Householder: 1991." Timothy S. Grail, *Our Nation's Housing in 1991*, p. 30. Adapted by the editors.

★ 506 ★

Social Characteristics

Marital Status of Householder, 1991. Part III.

Numbers in thousands, except percents.

| Characteristic | All occupied units | | | | Owner-occupied | | | | Renter-occupied | | | |
| | | Age of householder | | | | Age of householder | | | | Age of householder | | |
	Total	Under 40 years	40 to 64 years	65 years and over	Total	Under 40 years	40 to 64 years	65 years and over	Total	Under 40 years	40 to 64 years	65 years and over
Hispanic												
Total	6,239	3,224	2,346	669	2,423	807	1,221	395	3,816	2,417	1,125	274
Married	3,297	1,715	1,291	291	1,716	620	873	224	1,581	1,096	418	67
Widowed	437	30	186	221	232	7	100	126	204	23	86	96
Divorced or separated	1,321	564	639	119	315	88	194	33	1,006	476	445	85
Never-married	1,184	915	231	38	159	92	55	12	1,025	823	176	26
Percent	100.0	100.0	100.0	100.0	100.0	100.0	100.0	100.0	100.0	100.0	100.0	100.0
Married	52.8	53.2	55.0	43.5	70.8	76.8	71.5	56.7	41.4	45.3	37.2	24.6
Widowed	7.0	0.9	7.9	33.1	9.6	0.8	8.2	31.9	5.4	0.9	7.6	34.9
Divorced or separated	21.2	17.5	27.2	17.7	13.0	10.9	15.9	8.5	26.4	19.7	39.5	31.0
Never-married	19.0	28.4	9.8	5.7	6.6	11.4	4.5	3.0	26.9	34.0	15.7	9.5

Source: "Selected Social Characteristics of Householders in Occupied Units, by Tenure, Race and Hispanic Origin of Householder: 1991." Timothy S. Grail, *Our Nation's Housing in 1991*, p. 33. Adapted by the editors.

★ 507 ★

Social Characteristics

Owner-Occupied Housing by Presence of Children, 1991

Numbers in thousands. Persons of Hispanic origin may be of any race.

| Characteristics | All households | Households with children | | | | Households without children | | | |
		Total	Married couples	Other households with two or more adults	Households with one adult	Total	Married couples	Other households with two or more adults	Households with one adult
All occupied units									
Total	93,147	34,588	24,034	4,724	5,830	58,559	26,116	10,054	22,389
Race and origin									
White	79,140	27,864	20,916	3,189	3,759	51,276	24,076	8,188	19,012
Non-Hispanic	73,624	24,799	18,946	2,583	3,269	48,826	23,123	7,578	18,125
Hispanic	5,515	3,065	1,970	605	490	2,450	954	610	886
Black	10,832	5,124	1,990	1,255	1,878	5,708	1,452	1,383	2,874
Other	3,175	1,599	1,127	279	193	1,575	588	483	504
Total Hispanic	6,239	3,437	2,113	726	598	2,802	1,040	730	1,032
Owner occupied units									
Total	59,796	21,937	17,903	2,276	1,758	37,859	21,821	5,227	10,811
Race and origin									
White	53,748	19,137	16,032	1,683	1,423	34,612	20,405	4,411	9,795
Non-Hispanic	51,465	17,913	15,051	1,509	1,354	33,552	19,765	4,240	9,547

[Continued]

★ 507 ★

Owner-Occupied Housing by Presence of Children, 1991
[Continued]

Characteristics	All households	Households with children				Households without children			
		Total	Married couples	Other households with two or more adults	Households with one adult	Total	Married couples	Other households with two or more adults	Households with one adult
Hispanic	2,284	1,224	981	174	69	1,060	640	171	249
Black	4,635	2,028	1,252	481	295	2,607	1,039	670	898
Other	1,412	771	619	112	41	641	377	146	117
Total Hispanic	2,423	1,291	1,019	194	78	1,132	677	190	265

Source: "Selected Social Characteristics of the Householder by Presence of Children, Household Type, and Tenure: 1991." Jeanne Woodward, *Housing America's Children*, p. 26.

★ 508 ★

Social Characteristics

Percent Householders With or Without Children, 1991. Part I

Characteristic	All occupied units				Owner-occupied				Renter-occupied			
	Total	Age of householder			Total	Age of householder			Total	Age of householder		
		Under 40 years	40 to 64 years	65 years and over		Under 40 years	40 to 64 years	65 years and over		Under 40 years	40 to 64 years	65 years and over
White												
Percent	100.0	100.0	100.0	100.0	100.0	100.0	100.0	100.0	100.0	100.0	100.0	100.0
Households with children	35.2	56.8	33.9	3.0	35.6	69.1	35.1	3.1	34.4	44.8	29.7	2.4
Family households	98.4	98.3	98.9	88.9	99.0	99.2	99.2	89.2	97.0	97.0	97.5	87.3
Married couples	76.3	75.6	79.0	41.5	84.7	86.9	84.1	42.9	57.7	58.3	56.4	34.6
Other male householder	5.8	5.9	5.6	10.3	4.9	4.7	4.8	10.4	8.0	7.7	8.9	10.2
Other female householder	17.8	18.6	15.4	48.1	10.5	8.5	11.1	46.7	34.3	34.0	34.6	55.2
Nonfamily households	1.6	1.7	1.1	11.1	1.0	0.8	0.8	10.8	3.0	3.0	2.5	12.7
Households with no children	64.8	43.2	66.1	97.0	64.4	30.9	64.9	96.9	65.6	55.2	70.3	97.6
Family households	56.1	36.5	69.7	54.0	68.6	55.3	78.3	61.5	30.2	26.3	39.9	26.4
Married couples	83.7	85.5	83.6	82.9	86.0	89.4	863.3	84.5	73.0	81.1	65.8	69.3
Other male householder	5.8	9.1	5.2	5.1	4.9	6.8	4.5	5.0	10.0	11.7	10.1	6.2
Other female householder	10.5	5.4	11.1	12.0	9.1	3.9	9.2	10.5	16.9	7.2	24.1	24.4
Nonfamily households	43.9	63.5	30.3	46.0	31.4	44.7	21.7	38.5	69.8	73.7	60.1	73.6
2-or-more persons	15.6	31.7	11.9	2.5	9.9	26.8	12.0	2.4	20.8	33.4	11.9	2.7
Male householder	60.4	62.0	59.0	44.9	59.4	67.4	53.8	44.2	60.9	60.6	65.6	46.1
Female householder	39.6	38.0	41.0	55.1	50.6	32.6	46.2	55.8	39.1	39.4	34.4	53.9
1-person	84.4	68.3	88.1	97.5	90.1	73.2	88.0	97.6	79.2	66.6	88.1	97.3
Male householder	38.3	58.8	43.4	20.6	32.1	61.2	38.1	20.5	44.9	57.9	49.9	20.9
Female householder	61.7	41.2	56.6	79.4	67.9	38.8	61.9	79.5	55.1	42.1	50.1	79.1
Household size												
Median persons	2.3	2.8	2.4	1.6	2.4	3.4	2.6	1.8	2.0	2.3	1.9	1.2
Standard error	0.01	0.02	0.01	0.01	0.01	0.03	0.02	0.01	0.02	0.02	0.03	0.02

Source: "Selected Social Characteristics of Householders in Occupied Units, by Tenure, Race and Hispanic Origin of Householder: 1991." Timothy S. Grail, *Our Nation's Housing in 1991*, p. 28. Adapted by the editors.

★ 509 ★

Social Characteristics

Percent Householders With or Without Children, 1991. Part II

Characteristic	All occupied units				Owner-occupied				Renter-occupied			
	Total	Age of householder			Total	Age of householder			Total	Age of householder		
		Under 40 years	40 to 64 years	65 years and over		Under 40 years	40 to 64 years	65 years and over		Under 40 years	40 to 64 years	65 years and over
Black												
Percent	100.0	100.0	100.0	100.0	100.0	100.0	100.0	100.0	100.0	100.0	100.0	100.0
Households with children	47.3	66.7	41.7	11.1	43.8	70.5	46.7	12.4	50.0	65.5	35.9	8.9
Family households	98.5	98.5	98.7	95.1	99.0	99.1	99.5	95.0	98.1	98.4	97.5	95.6
Married couples	39.4	34.4	48.6	35.7	62.3	63.6	63.5	45.6	24.3	24.3	25.2	11.6
Other male householder	6.5	6.5	6.3	7.5	6.4	8.8	4.2	9.4	6.5	5.7	9.4	2.9
Other female householder	54.1	59.1	45.1	56.7	31.3	27.6	32.2	45.0	69.2	69.9	65.3	85.5
Nonfamily households	1.5	1.5	1.3	4.9	1.0	0.9	0.5	5.0	1.9	1.6	2.5	4.4
Households with no children	52.7	33.3	58.3	88.9	56.2	29.5	53.3	87.6	50.0	34.5	64.1	91.1
Family households	43.6	28.9	52.6	43.7	62.7	50.1	70.4	57.4	27.6	23.2	35.1	20.5
Married couples	58.3	62.5	56.3	59.4	63.5	70.8	62.8	62.6	48.3	57.7	43.3	44.6
Other male householder	10.4	20.3	6.9	10.5	7.9	17.0	5.3	9.3	15.1	22.3	10.1	16.6
Other female householder	31.3	17.1	36.8	30.0	28.5	12.3	31.9	28.1	36.6	19.9	46.5	38.9
Nonfamily households	56.4	71.1	47.4	56.3	37.3	49.9	29.6	42.6	72.4	76.8	64.9	79.5
2-or-more persons	10.7	19.9	7.6	3.8	7.6	21.3	6.6	3.3	12.1	19.6	8.0	4.3
Male householder	63.5	63.2	65.3	61.2	56.1	53.6	71.5	36.1	65.6	65.0	63.0	78.8
Female householder	36.5	36.8	34.7	38.8	43.9	46.4	28.5	63.9	34.4	35.0	37.0	21.2
1-person	89.3	80.1	92.4	96.2	92.4	78.7	93.4	96.7	87.9	80.4	92.0	95.7
Male householder	46.1	57.2	49.9	30.4	38.4	55.5	42.6	29.8	49.7	57.5	53.2	31.0
Female householder	53.9	42.8	50.1	69.6	61.6	44.5	57.4	70.2	50.3	42.5	46.8	69.0

Source: "Selected Social Characteristics of Householders in Occupied Units, by Tenure, Race and Hispanic Origin of Householder: 1991." Timothy S. Grail, *Our Nation's Housing in 1991*, pp. 29-30. Adapted by the editors.

★ 510 ★

Social Characteristics

Percent Householders With or Without Children, 1991. Part III

Characteristic	All occupied units				Owner-occupied				Renter-occupied			
	Total	Age of householder			Total	Age of householder			Total	Age of householder		
		Under 40 years	40 to 64 years	65 years and over		Under 40 years	40 to 64 years	65 years and over		Under 40 years	40 to 64 years	65 years and over
Hispanic												
Percent	100.0	100.0	100.0	100.0	100.0	100.0	100.0	100.0	100.0	100.0	100.0	100.0
Households with children	55.1	68.1	50.1	9.9	53.3	79.2	50.7	8.2	56.3	64.4	49.4	12.4
Family households	98.4	98.4	98.5	96.6	99.7	100.0	99.8	92.9	97.6	97.8	97.1	100.0
Married couples	62.5	62.8	62.3	55.2	79.2	83.0	76.1	59.9	52.2	54.3	46.5	51.1
Other male householder	7.7	8.2	7.0	7.8	5.8	6.9	4.3	11.5	9.0	8.7	10.0	4.5
Other female householder	29.8	29.1	30.7	37.0	15.1	10.1	19.6	28.7	38.8	37.0	43.5	44.3
Nonfamily households	1.6	1.6	1.5	3.4	0.3	0.0	0.2	7.1	2.4	2.2	2.9	-
Households with no children	44.9	31.9	49.9	90.1	46.7	20.8	49.3	91.8	43.7	35.6	50.6	87.6
Family households	53.2	40.2	64.7	53.0	73.3	63.0	80.5	66.0	39.6	35.8	47.9	33.3
Married couples	69.8	65.2	68.5	78.7	81.6	80.8	80.1	84.9	54.9	59.8	47.9	60.1
Other male householder	12.9	23.2	11.2	3.9	4.9	8.5	4.8	3.6	23.0	28.2	22.5	4.8
Other female householder	17.3	11.7	20.3	17.4	13.5	10.7	15.0	11.5	22.1	12.0	29.7	35.0

[Continued]

★ 510 ★

Percent Householders With or Without Children, 1991. Part III

[Continued]

Characteristic	All occupied units				Owner-occupied				Renter-occupied			
	Total	Age of householder			Total	Age of householder			Total	Age of householder		
		Under 40 years	40 to 64 years	65 years and over		Under 40 years	40 to 64 years	65 years and over		Under 40 years	40 to 64 years	65 years and over
Nonfamily households	46.8	59.8	35.3	47.0	26.7	37.0	19.5	34.0	60.4	64.2	52.1	66.7
2-or-more persons	21.3	34.5	14.0	3.1	12.3	36.1	9.5	3.1	23.9	34.4	15.8	3.1
Male householder	68.7	72.3	59.2	42.8	50.7	54.4	46.5	41.1	71.4	74.4	62.2	44.0
Female householder	31.3	27.7	40.8	57.2	49.3	45.6	53.5	58.9	28.6	25.6	37.8	56.0
1-person	78.7	65.5	86.0	96.9	87.7	63.9	90.5	96.9	76.1	65.6	84.2	96.9
Male householder	51.3	66.8	52.7	26.8	36.2	62.0	43.2	21.3	56.6	67.3	56.8	31.1
Female householder	48.7	33.2	47.3	73.2	63.8	38.0	56.8	78.7	43.4	32.7	43.2	68.9

Source: "Selected Social Characteristics of Householders in Occupied Units, by Tenure, Race and Hispanic Origin of Householder: 1991." Timothy S. Grail, *Our Nation's Housing in 1991*, p. 33. Adapted by the editors.

★ 511 ★

Social Characteristics

Race and Ethnicity of Householder, by Tenure, 1991

Numbers in thousands, except percents.

	Occupied housing units		
	Total	Owner	Renter
Race and ethnicity of householder			
Total	93,147	59,796	33,351
White	79,140	53,748	25,391
Black	10,832	4,635	6,197
American Indian, Eskimo, or Aleut	486	257	229
Asian or Pacific Islander	2,066	1,034	1,032
Other	623	120	502
Hispanic origin	6,239	2,423	3,816
Not of Hispanic origin	86,907	57,373	29,534
Percent	100.0	100.0	100.0
White	85.0	89.9	76.1
Black	11.6	7.8	18.6
American Indian, Eskimo or Aleut	0.5	0.4	0.7
Asian or Pacific Islander	2.2	1.7	3.1
Other	0.7	0.2	1.5
Hispanic origin	6.7	4.1	11.4
Not of Hispanic origin	93.3	95.9	88.6

Source: "Race and Ethnicity of Householder, by Tenure: 1991." Timothy S. Grail, "Our Nation's Housing in 1991," p. 6. *Note:* Hispanic origin may be of any race.

★ 512 ★

Social Characteristics

Renter-Occupied Housing by Presence of Children, 1991

Numbers in thousands. Persons of Hispanic origin may be of any race.

Characteristics	All households	Households with children				Households without children			
		Total	Married couples	Other households with two or more adults	Households with one adult	Total	Married couples	Other households with two or more adults	Households with one adult
Renter occupied units									
Total	33,351	12,651	6,131	2,448	4,072	20,700	4,295	4,826	11,578
Race and origin									
White	25,391	8,727	4,884	1,507	2,336	16,664	3,671	3,776	9,216
Non-Hispanic	22,160	6,886	3,896	1,075	1,916	15,274	3,358	3,337	8,579
Hispanic	3,231	1,841	989	432	421	1,390	314	439	638
Black	6,197	3,096	738	774	1,583	3,101	413	713	1,975
Other	1,763	828	509	167	152	935	211	337	386
Total Hispanic	3,816	2,147	1,094	532	521	1,670	363	540	767

Source: "Selected Social Characteristics of the Householder by Presence of Children, Household Type, and Tenure: 1991." Jeanne Woodward, *Housing America's Children*, p. 27.

Tenants

★ 513 ★

Households in Public Housing

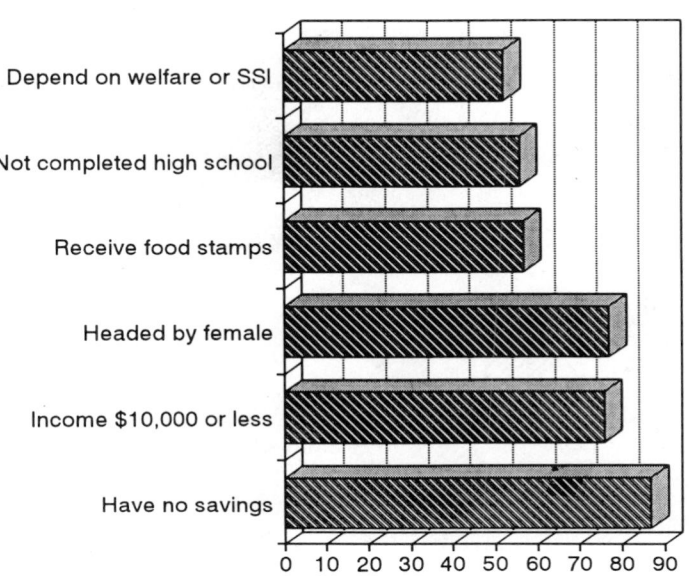

	Percent
Have no savings	87.0
Income $10,000 or less	76.0
Headed by female	77.0
Receive food stamps	57.0
Head of household has not completed high school	56.0
Depend on welfare or SSI	52.0

Source: "Minority Households in Public Housing Rental Units." *Black Issues in Higher Education*
10 (June 3, 1993), p. 18. Primary source: U.S. Department of Housing and Urban Development,
1992.

★ 514 ★
Tenants

Public Housing: Characteristics

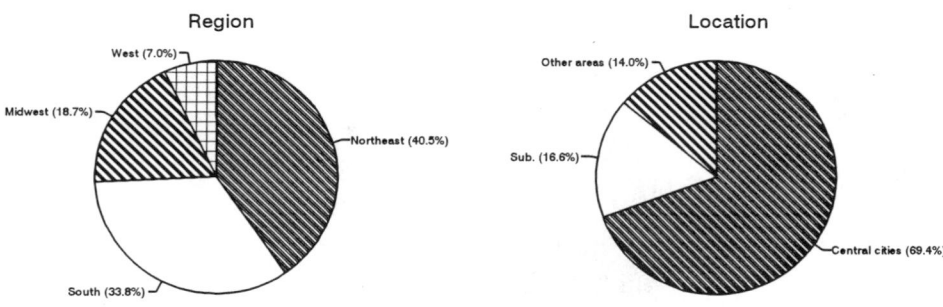

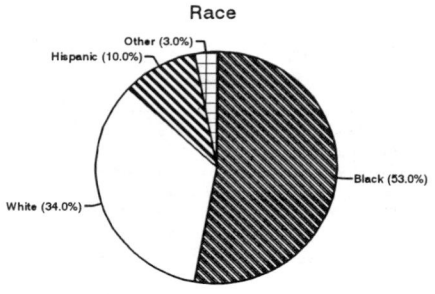

	Number	Percent
Region		
Northeast	551,000	40.51
South	460,000	33.82
Midwest	254,000	18.68
West	95,000	6.99
Location		
Central cities	944,000	69.41
Suburbs	226,000	16.62
Other areas	190,000	13.97
Race/ethnicity		
African American	720,000	53.0
White	466,000	34.0
Hispanic	132,000	10.0
Other	43,000	3.0

Source: "Public Housing." *Black Issues in Higher Education* 10 (June 3, 1993), p. 19. Primary source: U.S. Department of Housing and Urban Development, 1992.

Tenure in Housing

★ 515 ★

Occupied Housing by Race of Householder, 1920-1990

In thousands, except as indicated. As of April 1. Prior to 1960, excludes Alaska and Hawaii. Statistics on the number of occupied units are essentially comparable although identified by various terms.

Race and tenure	1920	1930	1940	1950	1960	1970	1980	1990
All races								
Occupied units, total	24,352	29,905	34,855	42,826	53,024	63,445	80,390	91,947
Owner occupied	11,114	14,280	15,196	23,560	32,797	39,886	51,795	59,025
Percent of occupied	45.6	47.8	43.6	55.0	61.9	62.9	64.4	64.2
Renter occupied	13,238	15,624	19,659	19,266	20,227	23,560	28,595	32,923
White								
Occupied units, total	21,826	26,983	31,561	39,044	47,880	56,606	68,810	76,880
Owner occupied	10,511	13,544	14,418	22,241	30,823	37,005	46,671	52,433
Percent of occupied	48.2	50.2	45.7	57.0	64.4	65.4	67.8	68.2
Renter occupied	11,315	13,439	17,143	16,803	17,057	19,601	22,139	24,447
Black and other								
Occupied units, total	2,526	2,922	3,293	3,783	5,144	6,839	11,580	15,067
Owner occupied	603	737	778	1,319	1,974	2,881	5,124	6,592
Percent of occupied	23.9	25.2	23.6	34.9	38.4	42.1	44.2	43.8
Renter occupied	1,923	2,185	2,516	2,464	3,170	3,959	6,456	8,475

Source: "Occupied Housing Units—Tenure, by Race and Hispanic Origin of Householder: 1980 and 1990." U.S. Bureau of the Census, *Statistical Abstract of the United States*, 1993, p. 724. Primary source: U.S. Bureau of the Census, *Census of Housing: 1960*, vol. 1; *1970*, vol. 1; *1980 Census of Housing*, vol. 1, chapter A (HC80-1-A); and *1990 Census of Housing, General Housing Characteristics*, series CH-90-1.

★ 516 ★

Tenure in Housing

Occupied Housing, 1980 and 1990

As of April 1. Based on the Census of Population and Housing.

Race and Hispanic origin of householder	All households			Owner occupied		Percent owner occupied		Renter occupied	
	1980	1990	Percent change 1980-1990	1980	1990	1980	1990	1980	1990
Total units	80,389,673	91,947,410	14.4	51,794,545	59,024,811	64.4	64.2	28,595,128	32,922,599
White	68,810,123	76,880,105	11.7	46,670,775	52,432,648	67.8	68.2	22,139,348	24,447,457
Black	8,381,668	9,976,161	19.0	3,724,265	4,327,265	44.4	43.4	4,657,417	5,648,896
American Indian, Eskimo, or Aleut	397,252	591,372	48.9	212,209	318,001	53.4	53.8	185,043	273,371
Asian or Pacific Islander	993,458	2,013,735	102.7	521,230	1,050,182	52.5	52.2	472,228	963,553

[Continued]

★ 516 ★

Occupied Housing, 1980 and 1990

[Continued]

Race and Hispanic origin of householder	All households			Owner occupied		Percent owner occupied		Renter occupied	
	1980	1990	Percent change 1980-1990	1980	1990	1980	1990	1980	1990
Other race	1,807,172	2,486,037	37.6	666,080	896,715	36.9	36.1	1,141,092	1,589,322
Hispanic origin[1]	4,007,896	6,001,718	49.7	1,738,920	2,545,584	43.4	42.4	2,268,976	3,456,134

Source: "Occupied Housing Units—Tenure, by Race and Hispanic Origin of Householder 1980 and 1990." U.S. Bureau of the Census, *Statistical Abstract of the United States*, 1993, p. 724. Primary source: U.S. Bureau of the Census, *1980 Census of Housing*, vol. 1, chapter A (HC80-1-A); and *1990 Census of Housing, General Housing Characteristics*, series CH-1. *Note:* 1. Persons of Hispanic origin may be of any race.

Chapter 9
INCOME, SPENDING, AND WEALTH

Expenditures

★ 517 ★

Average Annual Expenditures and Share of Aggregate Expenditures, Consumer Expenditure Survey, Part 1 Foods (Part)

[Aggregate in millions of dollars unless otherwise indicated].

Item	All consumer units	Race of reference person		Aggregate	Race of reference person	
		White and other	Black		White and other	Black
Average annual expenditures	$29,614	$30,794	$20,091	$2,897,455	92.5	7.5
Food	4,271	4,387	3,352	416,981	91.2	8.8
Food at home	2,651	2,676	2,448	258,723	89.6	10.4
Cereals and bakery products	404	412	347	39,476	90.3	9.7
Cereals and cereal products	145	146	139	14,173	89.2	10.8
Bakery products	259	266	208	25,303	91.0	9.0
Meats, poultry, fish, and eggs	709	682	922	69,162	85.4	14.6
Beef	228	223	269	22,287	86.7	13.3
Pork	144	135	215	14,076	83.2	16.8
Other meals	102	100	116	9,979	87.2	12.8
Poultry	122	116	169	11,884	84.4	15.6
Fish and seafood	81	77	114	7,914	84.1	15.9
Eggs	31	30	38	3,022	86.1	13.9
Dairy products	294	306	200	28,687	92.3	7.7
Fresh milk and cream	129	134	94	12,620	91.8	8.2
Other daily products	165	172	105	16,067	92.8	7.2
Fruits and vegetables	429	436	373	41,891	90.2	8.8
Fresh fruits	133	137	105	13,021	91.2	8.8
Fresh vegetables	127	129	109	12,370	90.3	9.7

[Continued]

★ 517 ★

Average Annual Expenditures and Share of Aggregate Expenditures, Consumer Expenditure Survey, Part 1 Foods (Part)

[Continued]

Item	All consumer units	Race of reference person		Aggregate	Race of reference person	
		White and other	Black		White and other	Black
Processed fruits	97	98	87	9,438	89.9	10.1
Processed vegetables	72	72	72	7,062	88.8	11.2

Source: "Housing Tenure, Race of Reference Person, and Type of Area: Average Annual Expenditures and Characteristics, Consumer Expenditure Survey, 1991," and "Housing Tenure, Race of Reference Persons, and Type of Area: Share of Annual Aggregate Expenditures and Sources of Income. Consumer Expenditure Survey, 1991," U.S. Department of Labor, Bureau of Labor Statistics, *Consumer Expenditure Survey, 1990-91,* Bulletin 2435 (September 1993), pp. 38 and 227. Tables combined by editors.

★ 518 ★

Expenditures

Average Annual Expenditures and Share of Aggregate Expenditures, Consumer Expenditure Survey, Part 2: Foods (Part), Alcoholic Beverages, and Housing

[Aggregate in millions of dollars unless otherwise indicated].

Item	All consumer units	Race of reference person		Aggregate	Race of reference persons	
		White and other	Black		White and other	Black
Other food at home	$815	$841	$607	$79,507	91.6	8.4
Sugar and other sweets	101	103	82	9,861	90.8	9.2
Fats and oils	72	72	66	7,003	89.6	10.4
Miscellaneous foods	375	389	268	36,600	92.0	8.0
Nonalcoholic beverages	224	230	180	21,889	91.0	9.0
Food prepared by consumer unit on out-of-town trips	42	46	10	4,154	97.3	2.7
Food away from home	1,620	1,711	904	158,259	93.7	6.3
Alcoholic beverages	297	314	159	28,970	94.0	6.0
Housing	9,252	9,570	6,692	905,668	92.0	8.0
Shelter	5,191	5,389	3,585	508,299	92.4	7.6
Owned dwellings	3,280	3,512	1,409	321,216	95.3	4.7
Mortgage interest and charges	1,932	2,074	782	189,186	95.6	4.4
Property taxes	789	840	380	77,259	94.7	5.3
Maintenance, repairs, insurance, other expenses	559	598	247	54,771	95.1	4.9

[Continued]

★ 518 ★

Average Annual Expenditures and Share of Aggregate Expenditures, Consumer Expenditure Survey, Part 2: Foods (Part), Alcoholic Beverages, and Housing

[Continued]

Item	All consumer units	Race of reference person		Aggregate	Race of reference persons	
		White and other	Black		White and other	Black
Rented dwellings	1,588	1,525	2,093	155,452	85.5	14.5
Other lodging	323	353	83	31,630	97.2	2.8

Source: "Housing Tenure, Race of Reference Person, and Type of Area: Average Annual Expenditures and Characteristics, Consumer Expenditure Survey, 1991," and "Housing Tenure, Race of Reference Persons, and Type of Area: Share of Annual Aggregate Expenditures and Sources of Income. Consumer Expenditure Survey, 1991," U.S. Department of Labor, Bureau of Labor Statistics, *Consumer Expenditure Survey, 1990-91,* Bulletin 2425 (September 1993), pp. 38 and 227. Tables combined by editors.

★ 519 ★

Expenditures

Average Annual Expenditures and Share of Aggregate Expenditures, Consumer Expenditure Survey, Part 3, Utilities, Household Operations, Housekeeping Supplies, Household Furnishings an

[Aggregate in millions of dollars unless otherwise indicated]

Item	All consumer units	Race of reference person		Aggregate	Race of reference persons	
		White and other	Black		White and other	Black
Utilities, fuels, and public services	1,990	2,005	2,032	194,820	89.7	10.3
Natural gas	250	244	96	24,440	87.0	13.0
Electricity	803	815	943	78,641	90.4	9.6
Fuel oil and other fuels	102	110	211	9,996	95.7	4.3
Telephone services	618	613	601	60,507	88.3	11.7
Water and other public services	217	223	180	21,235	91.3	8.7
Household operations	448	480	288	43,886	95.3	4.7
Personal services	211	223	130	20,686	93.8	6.2
Other household expenses	237	257	159	23,201	96.6	3.4
Housekeeping supplies	424	441	375	41,345	92.3	7.7
Laundry and cleaning supplies	116	116	113	11,346	88.4	11.6
Other household products	185	193	170	18,027	92.6	7.4
Postage and stationary	123	132	92	11,972	95.5	4.5
Household furnishings and equipment	1,200	1,255	1,049	117,317	93.0	7.0
Household textiles	100	105	80	9,773	93.4	6.6
Furniture	294	303	223	28,800	91.7	8.3
Floor coverings	115	118	106	11,265	90.6	9.4
Major appliances	132	134	150	12,953	89.9	10.1

[Continued]

★ 519 ★

Average Annual Expenditures and Share of Aggregate Expenditures, Consumer Expenditure Survey, Part 3, Utilities, Household Operations, Housekeeping Supplies, Household Furnishings an

[Continued]

Item	All consumer units	Race of reference person		Aggregate	Race of reference persons	
		White and other	Black		White and other	Black
Small appliances, miscellaneous housewares	81	87	66	7,873	95.8	4.2
Miscellaneous household equipment	477	509	424	46,652	94.6	5.4

Source: "Housing Tenure, Race of Reference Person, and Type of Area: Average Annual Expenditures and Characteristics, Consumer Expenditure Survey, 1991," and "Housing Tenure, Race of Reference Persons, and Type of Area: Share of Annual Aggregate Expenditures and Sources of Income. Consumer Expenditure Survey, 1991," U.S. Department of Labor, Bureau of Labor Statistics, *Consumer Expenditure Survey, 1990-91,* Bulletin 2415 (September 1993), pp. 38 and 227. Tables combined by editors.

★ 520 ★

Expenditures

Average Annual Expenditures and Share of Aggregate Expenditures, Consumer Expenditure Survey, Part 4 Apparel and Services and Transportation (Part)

[Aggregate in millions of dollars unless otherwise indicated].

Item	All consumer units	Race of reference person		Aggregate	Race of reference persons	
		White and other	Black		White and other	Black
Apparel and services	1,735	1,726	1,803	169,494	88.4	11.6
Men and boys	429	431	405	41,852	89.4	10.6
Men, 16 and over	345	354	273	33,651	91.1	8.9
Boys, 2 to 15	84	78	132	8,201	82.3	17.7
Women and girls	706	706	703	68,932	88.8	11.2
Women, 16 and over	607	608	607	59,300	88.8	11.2
Girls, 2 to 15	99	99	96	9,632	89.2	10.8
Children under 2	82	82	86	8,026	88.3	11.7
Footwear	242	226	364	23,592	83.1	16.9
Other apparel products and services	277	281	246	27,092	90.2	9.8
Transportation	5,151	5,413	3,029	504,293	93.5	6.5
Vehicle purchases (net outlay)	2,111	2,257	931	206,709	95.2	4.8
Cars and trucks, new	1,078	1,152	477	105,557	95.1	4.9
Cars and trucks, used	1,013	1,082	451	99,180	95.1	4.9
Other vehicles	20	22	3	1,973	98.1	1.9
Gasoline and motor oil	995	1,031	708	97,457	92.2	7.8

Source: "Housing Tenure, Race of Reference Person, and Type of Area: Average Annual Expenditures and Characteristics, Consumer Expenditure Survey, 1991," and "Housing Tenure, Race of Reference Persons, and Type of Area: Share of Annual Aggregate Expenditures and Sources of Income. Consumer Expenditure Survey, 1991," U.S. Department of Labor, Bureau of Labor Statistics, *Consumer Expenditure Survey, 1990-91,* Bulletin 2425 (September 1993), pp. 38 and 228. Tables combined by editors.

★ 521 ★

Expenditures

Average Annual Expenditures and Share of Aggregate Expenditures, Consumer Expenditure Survey, Part 5 Transportation (Part), Health Care, Entertainment

[Aggregate in millions of dollars unless otherwise indicated].

Item	All consumer units	Race of reference person		Aggregate	Race of reference person	
		White and other	Black		White and other	Black
Other vehicle expenses	$1,741	$1,813	$1,158	$170,397	92.7	7.3
Vehicle finance charges	282	291	204	27,586	92.0	8.0
Maintenance and repairs	614	640	406	60,075	92.7	7.3
Vehicle insurance	614	637	434	60,161	92.2	7.8
Vehicle rental, leases, licenses, other charges	231	245	114	22,575	94.5	5.5
Public transportation	304	313	232	29,729	91.6	8.4
Health care	1,554	1,640	857	152,143	93.9	6.1
Health insurance	656	688	400	64,229	93.3	6.7
Medical services	555	590	272	54,316	94.6	5.4
Drugs	252	264	149	24,611	93.5	6.5
Medical supplies	92	99	37	8,986	95.6	4.4
Entertainment	1,472	1,578	620	144,099	95.4	4.6
Fees and admissions	378	410	114	36,997	96.7	3.3
Television, radios, sound equipment	468	482	351	45,804	91.7	8.3
Pets, toys, and playground equipment	271	291	105	26,455	95.7	4.3
Other entertainment supplies, equipment, and services	356	394	50	34,843	98.4	1.6

Source: "Housing Tenure, Race of Reference Person, and Type of Area: Average Annual Expenditures and Characteristics, Consumer Expenditure Survey, 1991," and "Housing Tenure, Race of Reference Persons, and Type of Area: Share of Annual Aggregate Expenditures and Sources of Income. Consumer Expenditure Survey, 1991," U.S. Department of Labor, Bureau of Labor Statistics, *Consumer Expenditure Survey, 1990-91*, Bulletin 2425 (September 1993), pp. 40 and 229. Tables combined by editors.

★ 522 ★

Expenditures

Average Annual Expenditures and Share of Aggregate Expenditures, Consumer Expenditure Survey, Part 6 Personal Care Products and Services, Reading, Education, Tobacco, Misc.

[Aggregate in millions of dollars unless otherwise indicated].

Item	All consumer units	Race of reference person		Aggregate	Race of reference person	
		White and other	Black		White and other	Black
Personal care products and services	399	404	352	38,923	90.1	9.9
Reading	163	174	74	15,964	95.0	5.0
Education	447	476	206	43,727	94.9	5.1
Tobacco products and smoking supplies	276	282	228	27,067	90.9	9.1
Miscellaneous	860	903	517	84,196	93.4	6.6
Cash contributions	950	1,001	536	93,052	93.8	6.2
Personal insurance and pensions	2,787	2,925	1,665	272,880	93.4	6.6
Life and other personal insurance	356	363	302	34,847	90.7	9.3
Pensions and Social Security	2,431	2,563	1,364	238,032	93.8	6.2

Source: "Housing Tenure, Race of Reference Person, and Type of Area: Average Annual Expenditures and Characteristics, Consumer Expenditure Survey, 1991," and "Housing Tenure, Race of Reference Persons, and Type of Area: Share of Annual Aggregate Expenditures and Sources of Income. Consumer Expenditure Survey, 1991," U.S. Department of Labor, Bureau of Labor Statistics, *Consumer Expenditure Survey, 1990-91,* Bulletin 2425 (September 1993), pp. 40 and 229. Tables combined by editors.

★ 523 ★

Expenditures

Characteristics of Persons in the Consumer Expenditure Survey, 1991

Item	All consumer units	Race of reference person	
		White and other	Black
Number of consumer units (in thousands)	97,918	87,155	10,763
Consumer unit characteristics:			
Income before taxes[1]	$33,901	$35,311	$21,929
Income after taxes[1]	30,729	31,929	20,544
Age of reference persons	47.5	47.7	46.3

[Continued]

★ 523 ★

Characteristics of Persons in the Consumer Expenditure Survey, 1991
[Continued]

Item	All consumer units	Race of reference person	
		White and other	Black
Average number in consumer unit:			
Persons	2.6	2.5	2.8
Children under 18	.7	.7	1.0
Persons 65 and over	.3	.3	.2
Earners	1.4	1.4	1.3
Vehicles	2.0	2.1	1.1
Percent distribution:			
Sex of reference person:			
Male	65	68	46
Female	35	32	54
Housing tenure:			
Homeowner	63	66	41
With mortgage	39	40	25
Without mortgage	25	26	17
Renter	37	34	59
Race of reference person:			
Black	11	n.a.	100
White and other	89	100	n.a.
Education of reference person:			
Elementary (1-8)	10	9	15
High school (9-12)	43	42	53
College	46	48	32
Never attended and other	1	1	1
At least one vehicle owned	86	89	66

Source: "Housing Tenure, Race of Reference Person, and Type of Area: Average Annual Expenditures and Characteristics, Consumer Expenditure Survey, 1991," U.S. Department of Labor, Bureau of Labor Statistics, *Consumer Expenditure Survey, 1990-91,* Bulletin 2425 (September 1993), p. 38. *Note:* 1. Persons of Hispanic origin may be of any race.

★ 524 ★

Expenditures

Food as a Share of Total Expenditure, 1990-91

Expenditure components	Non-Hispanic white	Non-Hispanic black	Hispanic
Share of total expenditures			
Total food	14.4	16.4	17.8
Food at home	8.5	11.6	11.8
Food away from home	6.0	4.8	6.0
Expenditure levels in dollars			
Total food	4,402	3,186	4,366
Food at home	2,578	2,258	2,890
Food away from home	1,825	927	1,476
Total expenditures	30,474	19,398	24,488

Source: "Food as a Share of Expenditure, by Race, and Ethnicity, 1990-91," Interagency Board for Nutrition Monitoring and Related Research, *Nutrition Monitoring in the United States: Chartbook I: Selected Findings From the National Nutrition Monitoring and Related Research Program*, p. 113. Primary source: Department of Commerce, Bureau of Labor Statistics, Consumer Expenditure Survey *Notes:* A consumer is defined as a single person living alone or sharing a household with others, but who is financially independent; members of a sample household related by blood, marriage, adoption, or other legal arrangement; or two or more persons living together who share responsibility for at least two out of the three major types of expenses (food, housing, and other expenses).

★ 525 ★

Expenditures

Shares of Food at Home, 1990-91

Food groups	Non-Hispanic white	Non-Hispanic black	Hispanic
Share of food at home			
Cereal and bakery products	15.3	13.9	13.6
Meats, poultry, fish, and eggs	25.6	36.6	29.3
Dairy products	11.7	8.6	11.7
Fruits	8.7	8.3	9.7
Vegetables	7.4	7.6	8.3

[Continued]

★ 525 ★

Shares of Food at Home, 1990-91
[Continued]

Food groups	Non-Hispanic white	Non-Hispanic black	Hispanic
Other food at home	22.6	17.7	18.8
Nonalcoholic beverages	8.6	7.3	8.6

Source: "Shares of Food at Home, by Race, and Ethnicity, 1990-91," Interagency Board for Nutrition Monitoring and Related Research, *Nutrition Monitoring in the United States: Chartbook I: Selected Findings From the National Nutrition Monitoring and Related Research Program*, p. 113. Primary source: Department of Commerce, Bureau of Labor Statistics, Consumer Expenditure Survey *Notes:* A consumer is defined as a single person living alone or sharing a household with others, but who is financially independent; members of a sample household related by blood, marriage, adoption, or other legal arrangement; or two or more persons living together who share responsibility for at least two out of the three major types of expenses (food, housing, and other expenses). Other food at home includes sugar, fats and oils, and miscellaneous foods such as frozen food, soups, and potato chips.

Income

★ 526 ★

Mean and Median Money Income of Families, by Race, 1992

[Numbers in thousands. Families as of March 1993.].

Characteristic	Median income		Mean income	
	Value (dol.)	Standard error (dol.)	Value (dol.)	Standard error (dol.)
All families	36,312	186	44,483	205
RACE AND HISPANIC ORIGIN OF HOUSEHOLDER				
White	38,909	217	46,674	226
Black	21,161	449	28,050	442
Hispanic origin[1]	23,901	574	30,332	546

Source: "Selected Characteristics of Families—Total Money Income of Families in 1992," U.S. Department of Commerce, Bureau of the Census, *Money Income of Households, Families, and Persons in the United States*, Current Population Reports, P60-184, p. 44. *Note:* 1. Persons of Hispanic origin may be of any race.

★ 527 ★

Income

Median Income of Black Families, by Selected Characteristics, 1991 and 1992, Part 1

[Families as of March of the following year. An asterisk (*) preceding change indicates statistically significant change at the 90-percent confidence level].

Characteristic	1992			1991			Percent change in real median income (1991-92)
	Number (thous.)	Median income		Number (thous.)	Median income		
		Value (dollars)	Standard error (dollars)		Value (dollars)	Standard error (dollars)	
All families	7,888	21,161	449	7,716	21,548	445	*-4.7
Type of Residence							
Inside metropolitan areas	6,648	22,440	520	6,486	23,226	586	*-6.2
One million or more	4,514	24,133	725	4,507	24,741	701	*-5.3
Inside central cities	3,019	21,166	724	3,018	20,559	810	-.1
Outside central cities	1,495	32,192	1,318	1,489	32,722	1,439	-4.5
Under 1 million	2,133	19,730	745	1,979	20,659	732	*-7.3
Inside central cities	1,363	18,595	861	1,225	19,591	1,131	-7.9
Outside central cities	771	22,061	1,277	754	23,695	2,000	-9.6
Outside metropolitan areas	1,240	15,316	881	1,230	15,709	968	-5.4
Region							
Northeast	1,242	23,364	1,301	1,291	25,533	1,100	*-11.2
Midwest	1,621	20,181	987	1,554	20,860	1,411	-6.1
South	4,361	20,429	619	4,253	20,124	559	-1.5
West	663	24,827	1,964	618	28,298	1,561	*-14.8
Type of Family							
Married-couple families	3,748	34,196	946	3,631	33,307	758	-3.
Wife in paid labor force	2,425	41,799	901	2,420	41,353	843	-1.9
Wife not in paid labor force	1,323	21,035	825	1,210	20,288	747	.7
Male householder, no wife present	460	20,678	1,157	504	24,508	2,118	*-18.1
Female householder, no husband present	3,680	11,956	379	3,582	11,414	414	1.7

Source: "Median Income of Families, by Selected Characteristics, Race, and Hispanic Origin of Householder: 1992, 1991, and 1990," U.S. Department of Commerce, Bureau of the Census, *Money Income of Households, Families, and Persons in the United States,* Current Population Reports, P60-184, p. 42.

★ 528 ★

Income

Median Income of Black Families, by Selected Characteristics, 1991 and 1992, Part 2

Characteristic	1992			1991			Percent change in real median income (1991-92)
	Number (thous.)	Median income		Number (thous.)	Median income		
		Value (dollars)	Standard error (dollars)		Value (dollars)	Standard error (dollars)	
Age of Householder							
Under 65 years	6,902	22,032	485	6,747	22,888	557	-6.6
15 to 24 years	512	6,916	526	490	6,915	554	-2.9
25 to 34 years	2,039	16,210	668	1,964	16,633	807	-5.4
35 to 44 years	2,142	24,961	863	2,119	27,195	991	-10.8
45 to 54 years	1,295	32,551	1,516	1,211	31,252	1,458	1.1
55 to 64 years	914	26,346	1,702	942	26,649	1,391	-4.0
65 years and over	986	16,627	846	969	15,877	631	1.7
65 to 74 years	683	18,570	941	653	17,210	1,052	4.7
75 years and over	303	12,650	1,039	317	14,321	991	-14.2
Size of Family							
Two persons	2,724	17,583	608	2,680	18,056	701	-5.5
Three persons	2,070	21,148	883	2,007	22,270	976	-7.8
Four persons	1,658	26,555	934	1,582	26,158	1,197	-1.4
Five persons	812	24,118	1,886	836	23,337	1,821	.3
Six persons	396	23,672	1,953	341	24,871	3,533	-7.6
Seven persons or more	226	12,859	3,095	272	22,938	1,589	-45.6
Number of Earners							
No earners	1,630	6,532	207	1,547	6,480	255	-2.1
One earner	2,828	16,131	393	2,640	15,912	471	-1.6
Two earners or more	3,430	38,972	911	3,529	37,616	760	.6
Two earners	2,589	34,950	782	2,724	34,060	756	-.4
Three earners	694	50,770	2,269	647	46,221	1,602	6.6
Four earners or more	167	67,700	3,843	159	60,126	3,781	9.3

Source: "Median Income of Families, by Selected Characteristics, Race, and Hispanic Origin of Householder: 1992, 1991, and 1990," U.S. Department of Commerce, Bureau of the Census, *Money Income of Households, Families, and Persons in the United States,* Current Population Reports, P60-184, p. 42.

★ 529 ★

Income

Median Income of Black Households, 1991 and 1992, Part 2 Type of Household and Age of Householder

[Households as of March of the following year. An asterisk (*) preceding change indicates statistically significant change at the 90-percent confidence level].

Characteristic	1992			1991			Percent change in real median income (1991-92)
	Number (thous.)	Median income		Number (thous.)	Median income		
		Value (dollars)	Standard error (dollars)		Value (dollars)	Standard error (dollars)	
Type of Household							
Family households	7,888	21,761	453	7,716	22,203	470	*-4.9
Married-couple families	3,748	34,290	911	3,631	33,369	758	-.2
Male householder, no wife present	460	23,439	1,218	504	26,426	1,758	*-13.9
Female householder, no husband present	3,680	12,606	400	3,582	12,196	427	.3
Nonfamily households	3,302	12,062	387	3,367	12,202	477	-4.0
Male householder	1,484	15,267	882	1,594	15,223	746	-2.6
Living alone	1,244	13,369	746	1,319	13,665	638	-5.0
Female householder	1,818	10,005	524	1,773	9,520	491	2.0
Living alone	1,648	9,092	555	1,596	8,492	493	3.9
Age of Householder							
Under 65 years	9,282	20,997	393	9,135	21,606	393	*-5.7
15 to 24 years	718	8,705	627	703	8,603	582	-1.8
25 to 34 years	2,715	17,894	700	2,661	19,284	709	*-9.9
35 to 44 years	2,657	24,928	709	2,636	26,233	762	*-7.8
45 to 54 years	1,777	28,342	1,028	1,718	27,526	954	-
55 to 64 years	1,416	19,118	1,056	1,418	20,103	1,115	-7.7
65 years and over	1,908	10,396	442	1,948	10,466	479	-3.6
65 to 74 years	1,204	12,334	619	1,245	11,555	616	3.6
75 years and over	703	7,931	491	702	9,151	488	*-15.9

Source: "Median Income of Households, by Selected Characteristics, Race, and Hispanic Origin of Householder: 1992, 1991, and 1990," U.S. Department of Commerce, Bureau of the Census, *Money Income of Households, Families, and Persons in the United States: 1992*, Current Population Reports, P60-184, p. 3.

★ 530 ★

Income

Median Income of Black Households, 1991 and 1992, Part 3 Size of Householder, Numbers of Earners, Work Experience of Householder, Tenure

[Household as of March of the following year. An asterisk (*) preceding percent change indicates statistically significant change at the 90-percent condifence level].

Characteristic	1992			1991			Percent change in real median income (1991-92)
		Median income			Median income		
	Number (thous.)	Value (dollars)	Standard error (dollars)	Number (thous.)	Value (dollars)	Standard error (dollars)	
Size of Household							
One person	2,892	10,933	397	2,915	10,650	499	-.3
Two persons	2,895	18,638	570	2,887	19,230	674	*-5.9
Three persons	2,155	21,952	874	2,069	23,285	1,045	*-8.5
Four persons	1,721	26,432	943	1,661	26,187	1,246	-2.0
Five persons	863	24,744	1,759	878	24,371	1,803	-1.4
Six persons	418	24,443	1,866	368	25,251	3,097	-6.0
Seven persons or more	246	14,249	3,061	304	23,806	1,166	*-41.9
Number of Earners							
No earners	2,936	6,433	109	2,820	6,332	128	-1.4
One earner	4,400	17,644	374	4,301	17,230	362	-.6
Two earners or more	3,854	37,888	799	3,962	37,268	657	-1.3
Two earners	2,933	34,456	781	3,078	33,991	742	-1.6
Three earners	745	49,250	2,358	707	45,882	1,505	4.2
Four earners or more	177	68,392	4,041	176	59,256	3,930	12.0
Work Experience of Householder							
Total	11,190	18,660	386	11,083	18,807	394	-3.7
Worked	7,226	26,500	387	7,328	26,424	415	-2.6
Worked year-round, full-time	4,912	31,974	488	4,928	31,866	508	-2.6
Did not work	3,964	7,464	172	3,755	7,494	202	-3.3
Tenure							
Owner occupied	4,726	28,023	779	4,683	27,052	662	.6
Renter occupied	6,235	13,703	411	6,183	14,169	449	*-6.1
Occupier paid no cash rent	229	11,230	1,117	217	9,958	1,237	9.5

Source: "Median Income of Households, by Selected Characteristics, Race, and Hispanic Origin of Householder: 1992, 1991, and 1990," U.S. Department of Commerce, Bureau of the Census, *Money Income of Households, Families, and Persons in the United States: 1992,* Current Population Reports, P60-184, p. 3.

★ 531 ★

Income

Median Weekly Earning of Full-time Wage and Salary Workers, 1992 and 1993

Characteristic	Number of workers (in thousands)		Median weekly earnings	
	1992	1993	1992	1993
RACE, HISPANIC ORIGIN, AND SEX				
White	71,629	72,395	$462	$478
Men	41,439	41,825	518	531
Women	30,190	30,571	388	403
Black	9,537	9,729	357	370
Men	4,766	4,855	380	392
Women	4,771	4,873	336	349
Hispanic origin	6,986	7,106	324	335
Men	4,390	4,495	345	352
Women	2,596	2,611	303	314

Source: "Median Weekly Earning of Full-time Wage and Salary Workers by Selected Characteristics," U.S. Department of Labor, Bureau of Labor Statistics, *Employment and Earnings*, January 1994, p. 241. *Notes:* Detail for the above race and Hispanic-origin groups will not sum to total because data for the "other races" group are not presented and Hispanics are included in both the white and black population groups.

★ 532 ★

Income

Median Weekly Earning of Part-time Wage and Salary Workers, 1992 and 1993

Characteristic	Number of workers (in thousands)		Median weekly earnings	
	1992	1993	1992	1993
RACE, HISPANIC ORIGIN, AND SEX				
White	16,995	17,248	133	136
Men	5,293	5,361	121	122
Women	11,702	11,887	139	142
Black	1,879	1,883	123	128
Men	714	733	119	124
Women	1,165	1,150	125	130
Hispanic origin	1,355	1,469	134	134

[Continued]

★ 532 ★

Median Weekly Earning of Part-time Wage and Salary Workers, 1992 and 1993

[Continued]

Characteristic	Number of workers (in thousands)		Median weekly earnings	
	1992	1993	1992	1993
Men	564	590	136	136
Women	791	879	133	132

Source: "Median Weekly Earning of Part-time Wage and Salary Workers by Selected Characteristics," U.S. Department of Labor, Bureau of Labor Statistics, *Employment and Earnings*, January 1994, p. 242. *Notes:* Detail for the above race and Hispanic-origin groups will not sum to total because data for the "other races" group are not presented and Hispanics are included in both the white and black population groups.

★ 533 ★

Income

Median and Mean Income by Educational Attainment, 1991

[Persons as of March 1992].

Total money income, race, region, and sex	Total	Elementary Less than 9th grade	High school		College	
			9th to 12th grade (no diploma)	High school graduate	Some college or associate degree	Bachelor's degree or more
BLACK						
United States						
Both sexes...thousands	17,445	2,317	3,324	6,220	3,502	2,080
Median income...(dollars)	12,116	6,363	7,742	12,352	18,140	27,296
Standard error...(dollars)	178	142	300	264	440	679
Mean income...(dollars)	16,476	8,202	10,905	15,520	20,111	30,512
Standard error...(dollars)	189	238	304	276	404	754
Male...thousands	7,803	1,127	1,446	2,842	1,462	926
Median income...(dollars)	15,374	7,356	11,168	15,685	21,562	31,144
Standard error...(dollars)	328	287	504	463	716	1,050
Mean income...(dollars)	19,444	10,118	14,005	18,626	23,785	34,129
Standard error...(dollars)	312	395	539	437	701	1,280
Female...thousands	9,641	1,190	1,878	3,379	2,041	1,154
Median income...(dollars)	9,969	5,625	6,440	10,403	15,942	25,782
Standard error...(dollars)	222	178	193	315	659	644
Mean income...(dollars)	14,029	6,398	8,389	12,877	17,447	27,547
Standard error...(dollars)	222	244	300	333	451	858

Source: "Total Money Income in 1991 of Persons 25 Years Old and Over, by Educational Attainment, Sex, Region and Race," U.S. Department of Commerce, Bureau of the Census, *The Black Population in the United States, March 1992*, Current Population Reports, P20-471, p. 42-43.

★ 534 ★

Income

Median and Mean Income of Persons 15 Years Old and Older, 1991, Part 1: All Persons

[Persons as of March 1992].

Total money income and region	All persons					
	Black			White		
	Both sexes	Male	Female	Both sexes	Male	Female
Median income...(dollars)	10,542	12,962	8,814	15,333	21,395	10,722
Standard error...(dollars)	170	335	195	76	112	75
Mean income...(dollars)	14,870	17,307	12,838	21,208	27,693	14,900
Standard error...(dollars)	166	273	196	85	141	82

Source: "Total Money Income in 1991 of Persons 15 Years Old and Over, by Sex, Region ad Race," U.S. Department of Commerce, Bureau of the Census, *The Black Population in the United States, March 1992,* Current Population Reports, P20-471, p. 40.

★ 535 ★

Income

Median and Mean Income of Persons 15 Years Old and Older, 1991, Part 2: Year-Round Full-Time Workers

[Persons as of March 1992].

Total money income and region	Year-round, full-time workers					
	Black			White		
	Both sexes	Male	Female	Both sexes	Male	Female
Median income...(dollars)	20,823	22,628	19,134	26,501	30,953	21,554
Standard error...(dollars)	237	493	329	96	135	111
Mean income...(dollars)	24,077	26,331	21,738	32,394	37,245	24,923
Standard error...(dollars)	276	413	353	140	201	152

Source: "Total Money Income in 1991 of Persons 15 Years Old and Over, by Sex, Region and Race," U.S. Department of Commerce, Bureau of the Census, *The Black Population in the United States, March 1992,* Current Population Reports, P20-471, p. 40.

★ 536 ★

Income

Median and Mean Money Income of Persons 25 Years Old and Over, by Educational Attainment, 1991

[Persons as of March 1992].

Total money income	Total	Elementary Less than 9th grade	High school 9th to 12th grade (no diploma)	High school graduate	College Some college or associate degree	Bachelor's degree or more
Median income...(dollars)	12,116	6,363	7,742	12,352	18,140	27,296
Standard error...(dollars)	178	142	300	264	440	679
Mean income...(dollars)	16,476	8,202	10,905	15,520	20,111	30,512
Standard error...(dollars)	189	238	304	276	404	754

Source: "Total Money Income in 1991 of Persons 15 Years Old and Over, by Educational Attainment, Sex, Region and Race," U.S. Department of Commerce, Bureau of the Census, *The Black Population in the United States, March 1992*, Current Population Reports, P20-471, p. 42.

★ 537 ★

Income

Money Income of Black Households, 1991 and 1992, Part 1 Type of Residence and Region

[Households as of March of the following year. An asterisk (*) preceding percent change indicates statistically significant change at the 90-percent confidence level].

Characteristic	1992 Number (thous.)	1992 Median income Value (dollars)	1992 Median income Standard error (dollars)	1991 Number (thous.)	1991 Median income Value (dollars)	1991 Median income Standard error (dollars)	Percent change in real median income (1991-92)
BLACK							
All households	11,190	18,660	386	11,083	18,807	394	-3.7
TYPE OF RESIDENCE							
Inside metropolitan areas	9,562	19,674	407	9,402	20,211	425	*-5.5
One million or more	6,558	20,647	527	6,595	21,534	525	*-6.9
Inside central cities	4,522	17,704	585	4,514	18,243	563	*-5.8
Outside central cities	2,035	29,434	1,170	2,081	29,776	898	-4.0
Under 1 million	3,004	18,059	600	2,808	17,260	640	1.6
Inside central cities	1,959	16,720	791	1,829	15,963	655	1.7
Outside central cities	1,045	20,289	1,081	978	20,066	1,352	-1.8
Outside metropolitan areas	1,628	13,821	744	1,680	13,120	712	2.3
Region							
Northeast	1,865	19,683	847	1,907	21,284	690	*-10.2
Midwest	2,320	18,126	815	2,238	18,280	960	-3.7

[Continued]

★ 537 ★

Money Income of Black Households, 1991 and 1992, Part 1 Type of Residence and Region

[Continued]

Characteristic	1992			1991			Percent change in real median income (1991-92)
	Number (thous.)	Median income		Number (thous.)	Median income		
		Value (dollars)	Standard error (dollars)		Value (dollars)	Standard error (dollars)	
South	6,045	18,108	503	5,972	17,247	463	1.9
West	960	22,595	1,757	966	25,656	1,465	*-14.5

Source: "Median Income of Households, by Selected Characteristics, Race, and Hispanic Origin of Householder: 1992, 1991, and 1990," U.S. Department of Commerce, Bureau of the Census, *Money Income of Households, Families, and Persons in the United States: 1992*, Current Population Reports, P60-184, p. 3.

★ 538 ★

Income

Money Income of Household, Families, and Persons, 1989, 1991, and 1992

[Households, families, and persons as of March 1993].

Characteristic	1992		Median income (in 1992 dollars)		Percent change in real income	
	Number (thous.)	Median income (dols.)	1991	1989	1991 to 1992	1989 to 1992
HOUSEHOLDS						
All households	96,391	30,786	31,034	32,706	-0.8	*-5.9
Race and Hispanic origin of householder:						
White	82,083	32,368	32,520	34,403	-0.5	*-5.9
White, not Hispanic	75,859	33,388	33,296	35,143	0.3	*-5.0
Black	11,190	18,660	19,373	20,460	-3.7	*-8.8
Hispanic origin[1]	6,626	22,848	23,374	24,803	-2.3	*-7.9
FAMILIES						
All families	68,144	36,812	37,020	38,710	-0.6	*-4.9
Race and Hispanic origin of householder:						
White	57,858	38,909	38,919	40,704	-0.03	*-4.4
Black	7,888	21,161	22,197	22,866	*-4.7	*-7.5
Hispanic origin[1]	5,318	23,901	24,614	26,528	-2.9	*-9.9
Type of family:						
All races:						
Married-couple families	53,171	42,064	42,229	43,614	-0.4	*-3.6
Female householder, no husband present	11,947	17,221	17,197	18,603	0.1	*-7.4
White:						
Married-couple families	47,601	42,738	42,755	44,362	-0.04	*-3.7
Female householder, no husband present	7,848	20,130	20,141	21,436	-0.1	*-6.1

[Continued]

★ 538 ★

Money Income of Household, Families, and Persons, 1989, 1991, and 1992

[Continued]

Characteristic	1992 Number (thous.)	1992 Median income (dols.)	Median income (in 1992 dollars) 1991	Median income (in 1992 dollars) 1989	Percent change in real income 1991 to 1992	Percent change in real income 1989 to 1992
Black:						
Married-couple families	3,748	34,196	34,310	34,679	-0.3	-1.4
Female householder, no husband present	3,680	11,956	11,758	13,159	1.7	*-9.1
Hispanic origin:[1]						
Married-couple families	3,674	28,515	29,455	30,981	-3.2	*-8.0
Female householder, no husband present	1,238	12,894	12,497	13,289	3.2	-3.0
PER CAPITA INCOME						
All races	254,241	15,033	15,057	15,904	-0.2	*-5.5
White	211,955	15,981	15,977	16,854	0.03	*-5.2
Black	32,036	9,296	9,446	9,897	-1.6	*-6.1
Hispanic origin[1]	22,752	8,874	8,923	9,493	-0.5	*-6.5

Source: "Comparison of Income Summary Measures by Selected Characteristics: 1989, 1991, and 1992, " U.S. Department of Commerce, Bureau of the Census, *Money Income of Households, Families, and Persons in the United States: 1992,* Current Population Reports, P60-184, p. xii. *Notes:* *Statistically significant change at the 90-percent confidence level. 1. Persons of Hispanic origin may be of any race.

★ 539 ★

Income

Percent Distribution of Racial Ethnic Households Within Income Quintile and Top 5 Percent, 1992

[Households as of March 1993].

Characteristic	Total	Lowest fifth	Second fifth	Middle fifth	Fourth fifth	Highest fifth	Top 5 percent
Number...thous	96,391	19,278	19,278	19,278	19,278	19,278	4,820
Lower limit...dollars	(X)	(X)	12,664	24,300	38,000	58,200	99,372
RACE AND HISPANIC ORIGIN OR HOUSEHOLDER							
Total	100.0	100.0	100.0	100.0	100.0	100.0	100.0
White	85.2	74.8	84.0	88.8	88.9	91.2	92.3
Black	11.6	21.9	13.0	10.2	8.0	5.0	3.6
Hispanic origin[1]	6.9	9.5	8.7	7.2	5.4	3.6	2.6

Source: "Percent Distribution of Households Within Income Quintile and Top 5 Percent 1992," U.S. Department of Commerce, Bureau of the Census, *Money Income of Households, Families, and Persons in the United States: 1992,* Current Population Reports, P60-184, p. 7. *Note:* 1. Persons of Hispanic origin may be of any race.

★ 540 ★

Income

Sources of Income, 1991

[Aggregate in millions of dollars unless otherwise indicated].

Item	All consumer units	Race of reference person		Aggregate	Race of reference person	
		White and other	Black		White and other	Black
Sources of income and personal taxes:						
Money income before taxes	33,901	35,311	21,929	2,848,372	93.2	6.8
Wages and salaries	25,553	26,498	17,532	2,146,984	92.8	7.2
Self-employment income	2,552	2,803	422	214,399	98.3	1.7
Social Security, private and government retirement	3,584	3,764	2,058	301,169	93.9	6.1
Interest, dividends, rental income, other property income	1,164	1,284	146	97,816	98.7	1.3
Unemployment and workers' compensation, veterans' benefits	245	251	189	20,566	91.9	8.1
Public assistance, supplemental security income, food stamps	389	283	1,286	32,678	65.1	34.9
Regular contributions for support	284	300	146	23,863	94.6	5.4
Other income	130	127	151	10,896	87.8	12.2
Personal taxes	3,172	3,382	1,385	266,499	95.4	4.6
Federal income taxes	2,431	2,602	981	204,296	95.7	4.3
State and local income taxes	612	643	352	51,422	93.9	6.1
Other taxes	128	137	52	10,781	95.7	4.3
Income after taxes	30,729	31,929	20,544	2,581,873	93.0	7.0

Source: "Housing Tenure, Race of Reference Person, and Type of Area: Average Annual Expenditures and Characteristics, Consumer Expenditure Survey, 1991," and "Housing Tenure, Race of Reference Persons, and Type of Area: Share of Annual Aggregate Expenditures and Sources of Income, Consumer Expenditure Survey, 1991," U.S. Department of Labor, Bureau of Labor Statistics, *Consumer Expenditure Survey, 1990-91,* Bulletin 2425 (September 1993), pp. 40 and 229. Tables combined by editors.

★ 541 ★

Income

Total Money Earnings of Black Men 25 Years Old and Over, By Educational Attainment, 1991, Part 1, All Persons

Persons as of March 1992.

Total money earnings, race, region and sex	All persons				
	Total	Not a high school graduate	High school graduate	Some college or associate degree	Bachelor's degree or more
Black					
United States					
Male... thousands	7,803	2,574	2,842	1,462	926
Total with earnings...thousands	5,699	1,354	2,254	1,240	851
Percent	100.0	100.0	100.0	100.0	100.0
$1 to $2,499 or loss	7.1	13.2	5.3	6.1	3.8
$2,500 to $4,999	6.6	11.3	5.9	3.9	5.1
$5,000 to $7,499	6.4	8.3	7.4	4.5	3.4
$7,500 to $9,999	5.3	8.4	5.5	4.5	1.3
$10,000 to $12,499	9.2	11.3	11.4	7.1	3.0
$12,500 to $14,999	6.1	7.3	7.4	4.6	3.3
$15,000 to $17,499	8.1	9.3	9.0	7.4	4.6
$17,500 to $19,999	5.4	5.1	6.7	6.0	1.7
$20,000 to $22,499	7.3	6.6	7.5	7.8	7.2
$22,500 to $24,999	4.0	2.7	3.9	5.4	4.6
$25,000 to $29,999	8.9	4.6	9.9	11.0	10.0
$30,000 to $34,999	7.5	4.0	7.5	9.5	10.4
$35,000 to $39,999	6.7	3.5	6.4	7.8	10.7
$40,000 to $44,999	4.2	1.9	2.4	5.2	11.1
$45,000 to $49,999	2.9	1.4	1.8	4.5	5.7
$50,000 to $59,999	2.0	0.7	0.9	2.4	6.6
$60,000 to $74,999	1.2	0.1	0.8	1.4	3.8
$75,000 and over	0.9	0.2	0.3	0.8	3.9

Source: "Total Money Earnings in 1991 of Persons 25 Years Old and Older, by Educational Attainment, Sex, Region, and Race," U.S. Department of Commerce, Bureau of the Census, *The Black Population in the United States*, Current Population Reports, P20-471, p. 54.

★ 542 ★

Income

Total Money Earnings of Black Men 25 Years Old and Over, By Educational Attainment, 1991, Part 2, All Persons – Median and Mean

Persons as of March 1992.

Total money earnings, race, region and sex	All persons				
	Total	Not a high school graduate	High school graduate	Some college or associate degree	Bachelor's degree or more
Black					
United States					
Median earnings (dollars)	18,007	11,955	16,961	21,872	30,770
Standard error (dollars)	491	482	468	751	1,075
Mean earnings (dollars)	21,103	14,581	19,349	23,693	32,343
Standard error (dollars)	343	526	449	700	1,213

Source: "Total Money Earnings in 1991 of Persons 25 Years Old and Older, by Educational Attainment, Sex, Region, and Race," U.S. Department of Commerce, Bureau of the Census, *The Black Population in the United States,* Current Population Reports, P20-471, p. 54.

★ 543 ★

Income

Total Money Earnings of Black Men 25 Years Old and Over, By Educational Attainment, 1991, Part 3, Year-round, Full-time Workers

Persons as of March 1992.

Total money earnings, race, region and sex	All persons				
	Total	Not a high school graduate	High school graduate	Some college or associate degree	Bachelor's degree or more
Black					
United States					
Male... thousands	3,802	681	1,547	913	662
Total with earnings...thousands	3,802	681	1,547	913	662
$1 to $2,499 or loss	0.8	0.8	1.0	0.5	0.9
$2,500 to $4,999	0.7	0.5	0.7	0.9	0.4
$5,000 to $7,499	2.5	3.1	2.8	2.4	1.5
$7,500 to $9,999	3.6	7.2	4.3	1.7	0.7
$10,000 to $12,499	7.8	13.9	10.1	4.4	1.0
$12,500 to $14,999	6.9	9.3	8.9	4.3	3.6
$15,000 to $17,499	10.0	14.5	12.0	7.8	3.7
$17,500 to $19,999	6.7	8.1	7.4	7.6	2.1
$20,000 to $22,499	9.2	11.2	9.6	8.7	6.9
$22,500 to $24,999	5.1	3.9	4.3	7.2	5.5
$25,000 to $29,999	11.8	7.4	12.5	13.9	11.8
$30,000 to $34,999	10.3	6.6	9.7	12.1	13.2
$35,000 to $39,999	9.1	6.4	8.7	9.8	12.1
$40,000 to $44,999	5.8	2.8	3.4	6.8	13.4

[Continued]

★ 543 ★

Total Money Earnings of Black Men 25 Years Old and Over, By Educational Attainment, 1991, Part 3, Year-round, Full-time Workers
[Continued]

Total money earnings, race, region and sex	All persons				
	Total	Not a high school graduate	High school graduate	Some college or associate degree	Bachelor's degree or more
$45,000 to $49,999	3.8	2.2	2.1	5.7	6.6
$50,000 to $59,999	2.8	1.4	1.2	3.1	7.7
$60,000 to $74,999	1.6	0.3	0.9	1.9	4.2
$75,000 and over	1.4	0.4	0.5	1.1	4.8

Source: "Total Money Earnings in 1991 of Persons 25 Years Old and Older, by Educational Attainment, Sex, Region, and Race," U.S. Department of Commerce, Bureau of the Census, *The Black Population in the United States*, Current Population Reports, P20-471, p. 54.

★ 544 ★

Income

Total Money Earnings of Black Men 25 Years Old and Over, By Educational Attainment, 1991, Part 4, Year-round, Full-time Workers – Median and Mean

Persons as of March 1992.

Total money earnings, race, region and sex	All persons				
	Total	Not a high school graduate	High school graduate	Some college or associate degree	Bachelor's degree or more
Black					
United States					
Median earnings (dollars)	23,372	17,689	20,731	26,209	34,342
Standard error (dollars)	637	780	531	722	1,544
Mean earnings (dollars)	26,382	20,885	23,109	28,315	37,021
Standard error (dollars)	410	726	518	757	1,324

Source: "Total Money Earnings in 1991 of Persons 25 Years Old and Older, by Educational Attainment, Sex, Region, and Race," U.S. Department of Commerce, Bureau of the Census, *The Black Population in the United States*, Current Population Reports, P20-471, p. 54.

★ 545 ★

Income

Total Money Earnings of Black Persons 25 Years Old and Over, By Educational Attainment, 1991, Part 1, All Persons

Persons as of March 1992.

Total money earnings, race, region and sex	All persons				
	Total	Not a high school graduate	High school graduate	Some college or associate degree	Bachelor's degree or more
Black					
United States					
Both sexes... thousands	17,445	5,642	6,220	3,502	2,080
Total with earnings...thousands	11,586	2,379	4,537	2,846	1,823
$1 to $2,499 or loss	8.3	15.8	7.5	6.6	2.9
$2,500 to $4,999	7.0	13.3	6.1	5.0	4.0
$5,000 to $7,499	7.3	9.4	9.4	4.9	2.8
$7,500 to $9,999	7.1	10.7	8.2	5.4	2.3
$10,000 to $12,499	10.1	11.4	12.4	10.0	3.0
$12,500 to $14,999	6.2	7.0	7.3	5.6	3.2
$15,000 to $17,499	8.3	8.7	9.5	7.6	6.2
$17,500 to $19,999	5.9	3.8	6.7	8.2	3.3
$20,000 to $22,499	7.3	5.8	7.5	8.3	7.5
$22,500 to $24,999	4.5	2.7	3.6	6.1	6.8
$25,000 to $29,999	8.7	4.0	8.3	10.5	13.1
$30,000 to $34,999	5.9	2.5	5.0	7.6	10.1
$35,000 to $39,999	5.4	2.2	4.7	5.8	10.9
$40,000 to $44,999	3.0	1.1	1.5	3.2	8.7
$45,000 to $49,999	1.9	0.8	0.9	2.7	4.2
$50,000 to $59,999	1.5	0.5	0.5	1.2	5.6
$60,000 to $74,999	0.8	0.2	0.4	1.0	2.7
$75,000 and over	0.7	0.2	0.4	0.4	2.6

Source: "Total Money Earnings in 1991 of Persons 25 Years Old and Older, by Educational Attainment, Sex, Region, and Race," U.S. Department of Commerce, Bureau of the Census, *The Black Population in the United States*, Current Population Reports, P20-471, p. 54.

★ 546 ★

Income

Total Money Earnings of Black Persons 25 Years Old and Over, By Educational Attainment, 1991, Part 2, All Persons – Median and Mean

Persons as of March 1992.

Total money earnings, race, region and sex	All persons				
	Total	Not a high school graduate	High school graduate	Some college or associate degree	Bachelor's degree or more
Black					
United States					
Both sexes... thousands	17,445	5,642	6,220	3,502	2,080
Total with earnings...thousands	11,586	2,379	4,537	2,846	1,823
Median earnings (dollars)	16,227	10,176	14,660	19,023	27,469
Standard error (dollars)	224	367	399	461	745
Mean earnings (dollars)	18,977	12,600	16,881	20,566	30,032
Standard error (dollars)	224	367	311	406	727

Source: "Total Money Earnings in 1991 of Persons 25 Years Old and Older, by Educational Attainment, Sex, Region, and Race," U.S. Department of Commerce, Bureau of the Census, *The Black Population in the United States,* Current Population Reports, P20-471, p. 54.

★ 547 ★

Income

Total Money Earnings of Black Persons 25 Years Old and Over, By Educational Attainment, 1991, Part 3, Year-round, Full-time Workers

Persons as of March 1992.

Total money earnings, race, region and sex	All persons				
	Total	Not a high school graduate	High school graduate	Some college or associate degree	Bachelor's degree or more
Black					
United States					
Both sexes... thousands	7,516	1,206	2,947	1,991	1,372
Total with earnings...thousands	7,516	1,206	2,947	1,991	1,372
Percent	100.0	100.0	100.0	100.0	100.0
$1 to $2,499 or loss	0.7	1.0	0.9	0.5	0.4
$2,500 to $4,999	0.8	2.0	0.6	0.7	0.3
$5,000 to $7,499	2.8	5.2	3.3	1.7	1.0
$7,500 to $9,999	5.4	12.0	6.2	3.2	1.0
$10,000 to $12,499	10.2	15.9	13.1	8.2	1.6
$12,500 to $14,999	7.3	10.4	9.0	5.9	3.0
$15,000 to $17,499	10.8	13.9	13.0	9.1	5.5
$17,500 to $19,999	7.7	6.1	8.7	9.9	3.6
$20,000 to $22,499	9.7	9.6	10.5	10.0	7.8
$22,500 to $24,999	6.1	4.4	4.8	8.0	7.6
$25,000 to $29,999	11.7	6.7	11.2	13.2	15.1

[Continued]

★ 547 ★

Total Money Earnings of Black Persons 25 Years Old and Over, By Educational Attainment, 1991, Part 3, Year-round, Full-time Workers
[Continued]

Total money earnings, race, region and sex	All persons				
	Total	Not a high school graduate	High school graduate	Some college or associate degree	Bachelor's degree or more
$30,000 to $34,999	8.3	4.2	6.9	10.2	12.3
$35,000 to $39,999	7.7	4.1	6.7	7.7	12.9
$40,000 to $44,999	4.2	1.6	2.2	4.4	10.3
$45,000 to $49,999	2.5	1.2	1.2	3.7	4.9
$50,000 to $59,999	2.1	1.1	0.6	1.7	6.9
$60,000 to $74,999	1.1	0.3	0.5	1.4	2.7
$75,000 and over	1.0	0.2	0.6	0.5	3.2

Source: "Total Money Earnings in 1991 of Persons 25 Years Old and Older, by Educational Attainment, Sex, Region, and Race," U.S. Department of Commerce, Bureau of the Census, *The Black Population in the United States,* Current Population Reports, P20-471, p. 54.

★ 548 ★

Income

Total Money Earnings of Black Persons 25 Years Old and Over, By Educational Attainment, 1991, Part 4, Year-round, Full-time Workers – Median and Mean

Persons as of March 1992.

Total money earnings, race, region and sex	All persons				
	Total	Not a high school graduate	High school graduate	Some college or associate degree	Bachelor's degree or more
Black					
United States					
Both sexes... thousands	7,516	1,206	2,947	1,991	1,372
Total with earnings...thousands	7,516	1,206	2,947	1,991	1,372
Median earnings (dollars)	21,145	15,645	18,620	22,751	30,914
Standard error (dollars)	238	416	424	533	642
Mean earnings (dollars)	24,026	18,130	21,147	25,038	33,924
Standard error (dollars)	274	503	385	455	810

Source: "Total Money Earnings in 1991 of Persons 25 Years Old and Older, by Educational Attainment, Sex, Region, and Race," U.S. Department of Commerce, Bureau of the Census, *The Black Population in the United States,* Current Population Reports, P20-471, p. 54.

★ 549 ★

Income

Total Money Earnings of Black Women 25 Years Old and Over, By Educational Attainment, 1991, Part 1, All Persons

Persons as of March 1992.

Total money earnings, race, region and sex	All persons				
	Total	Not a high school graduate	High school graduate	Some college or associate degree	Bachelor's degree or more
Black					
United States					
Female...thousands	9,641	3,068	3,379	2,041	1,154
Total with earnings...thousands	5,886	1,026	2,283	1,606	972
$1 to $2,499 or loss	9.4	19.1	9.7	7.0	2.2
$2,500 to $4,999	7.3	15.9	6.4	5.7	3.1
$5,000 to $7,499	8.2	11.0	11.5	5.2	2.3
$7,500 to $9,999	8.8	13.8	10.9	6.1	3.1
$10,000 to $12,499	11.0	11.4	13.3	12.2	3.0
$12,500 to $14,999	6.3	6.5	7.3	6.4	3.2
$15,000 to $17,499	8.6	7.8	10.0	7.7	7.5
$17,500 to $19,999	6.4	2.0	6.7	9.8	4.7
$20,000 to $22,499	7.4	4.6	7.5	8.6	7.8
$22,500 to $24,999	5.0	2.6	3.4	6.6	8.8
$25,000 to $29,999	8.6	3.3	6.7	10.1	15.9
$30,000 to $34,999	4.4	0.6	2.6	6.1	9.9
$35,000 to $39,999	4.2	0.6	2.9	4.3	11.0
$40,000 to $44,999	1.8	-	0.6	1.6	6.6
$45,000 to $49,999	0.9	-	0.1	1.3	2.9
$50,000 to $59,999	0.9	0.4	-	0.3	4.8
$60,000 to $74,999	0.5	0.2	-	0.7	1.7
$75,000 and over	0.5	0.2	0.5	-	1.5

Source: "Total Money Earnings in 1991 of Persons 25 Years Old and Older, by Educational Attainment, Sex, Region, and Race," U.S. Department of Commerce, Bureau of the Census, *The Black Population in the United States,* Current Population Reports, P20-471, p. 55.

★ 550 ★

Income

Total Money Earnings of Black Women 25 Years Old and Over, By Educational Attainment, 1991, Part 2, All Persons – Median and Mean

Persons as of March 1992.

Total money earnings, race, region and sex	All persons				
	Total	Not a high school graduate	High school graduate	Some college or associate degree	Bachelor's degree or more
Black					
United States					
Median earnings (dollars)	14,660	8,224	12,182	17,384	26,124
Standard error (dollars)	407	453	316	591	683
Mean earnings (dollars)	16,918	9,985	14,444	18,151	28,007
Standard error (dollars)	284	464	415	453	842

Source: "Total Money Earnings in 1991 of Persons 25 Years Old and Older, by Educational Attainment, Sex, Region, and Race," U.S. Department of Commerce, Bureau of the Census, *The Black Population in the United States*, Current Population Reports, P20-471, p. 55.

★ 551 ★

Income

Total Money Earnings of Black Women 25 Years Old and Over, By Educational Attainment, 1991, Part 3, Year-round, Full-time Workers

Persons as of March 1992.

Total money earnings, race, region and sex	All persons				
	Total	Not a high school graduate	High school graduate	Some college or associate degree	Bachelor's degree or more
Black					
United States					
Female...thousands	3,714	525	1,400	1,078	710
Total with earnings...thousands	3,714	525	1,400	1,078	710
$1 to $2,499 or loss	0.6	1.3	0.7	0.6	-
$2,500 to $4,999	0.9	3.9	0.5	0.5	0.2
$5,000 to $7,499	3.0	7.8	3.8	1.1	0.6
$7,500 to $9,999	7.2	18.1	8.3	4.4	1.2
$10,000 to $12,499	12.6	18.4	16.5	11.4	2.3
$12,500 to $14,999	7.7	11.8	9.0	7.3	2.4
$15,000 to $17,499	11.5	13.2	14.2	10.2	7.1
$17,500 to $19,999	8.7	3.5	10.2	11.8	5.0
$20,000 to $22,499	10.3	7.6	11.5	11.1	8.6
$22,500 to $24,999	7.1	5.1	5.4	8.6	9.6
$25,000 to $29,999	11.6	5.9	9.7	12.5	18.2
$30,000 to $34,999	6.3	1.2	3.8	8.6	11.4
$35,000 to $39,999	6.2	1.1	4.6	6.0	13.6
$40,000 to $44,999	2.4	-	0.9	2.4	7.4

[Continued]

★ 551 ★

Total Money Earnings of Black Women 25 Years Old and Over, By Educational Attainment, 1991, Part 3, Year-round, Full-time Workers
[Continued]

Total money earnings, race, region and sex	All persons				
	Total	Not a high school graduate	High school graduate	Some college or associate degree	Bachelor's degree or more
$45,000 to $49,999	1.2	-	0.1	2.0	3.2
$50,000 to $59,999	1.4	0.7	-	0.5	6.2
$60,000 to $74,999	0.6	0.4	-	1.0	1.3
$75,000 and over	0.6	-	0.8	-	1.6

Source: "Total Money Earnings in 1991 of Persons 25 Years Old and Older, by Educational Attainment, Sex, Region, and Race," U.S. Department of Commerce, Bureau of the Census, *The Black Population in the United States*, Current Population Reports, P20-471, p. 55.

★ 552 ★
Income

Total Money Earnings of Black Women 25 Years Old and Over, By Educational Attainment, 1991, Part 4, Year-round, Full-time Workers – Median and Mean

Persons as of March 1992.

Total money earnings, race, region and sex	All persons				
	Total	Not a high school graduate	High school graduate	Some college or associate degree	Bachelor's degree or more
Black					
United States					
Median earnings (dollars)	19,363	12,605	16,957	20,608	28,132
Standard error (dollars)	379	628	378	552	984
Mean earnings (dollars)	21,614	14,562	18,980	22,264	31,037
Standard error (dollars)	351	584	560	504	929

Source: "Total Money Earnings in 1991 of Persons 25 Years Old and Older, by Educational Attainment, Sex, Region, and Race," U.S. Department of Commerce, Bureau of the Census, *The Black Population in the United States*, Current Population Reports, P20-471, p. 55.

★ 553 ★

Income

Total Money Earnings of Persons 15 Years Old and Over, by Sex and Race, 1991, Part 1, All Persons

Persons as of March 1992.

Total money earnings and region	All persons					
	Black			White		
	Both sexes	Male	Female	Both sexes	Male	Female
United States						
Total...thousands	22,542	10,252	12,290	165,571	80,049	85,522
Total with earnings...thousands	14,109	6,963	7,146	114,965	62,477	52,488
Percent	100.0	100.0	100.0	100.0	100.0	100.0
$1 to $2,499 or loss	13.1	12.2	13.9	11.0	8.0	14.5
$2,500 to $4,999	8.9	8.5	9.3	6.8	5.2	8.8
$5,000 to $7,499	8.1	7.2	9.0	7.1	5.5	9.0
$7,500 to $9,999	7.1	5.8	8.3	5.7	4.5	7.0
$10,000 to $12,499	9.8	8.8	10.7	7.7	6.2	9.5
$12,500 to $14,999	6.0	5.9	6.1	4.9	4.0	6.0
$15,000 to $17,499	7.8	7.8	7.8	6.4	5.6	7.4
$17,500 to $19,999	5.4	4.9	5.9	4.7	4.3	5.2
$20,000 to $22,499	6.5	6.8	6.3	6.5	6.3	6.8
$22,500 to $24,999	3.9	3.5	4.3	3.7	3.6	3.8
$25,000 to $29,999	7.5	7.6	7.3	8.4	9.3	7.3
$30,000 to $34,999	5.0	6.3	3.7	6.8	8.1	5.3
$35,000 to $39,999	4.5	5.5	3.5	5.2	6.6	3.5
$40,000 to $44,999	2.5	3.5	1.5	3.9	5.4	2.0
$45,000 to $49,999	1.5	2.3	0.7	2.5	3.6	1.3
$50,000 to $59,999	1.2	1.7	0.8	3.5	5.4	1.3
$60,000 to $74,999	0.7	1.0	0.4	2.3	3.6	0.7
$75,000 and over	0.6	0.8	0.5	2.8	4.7	0.6

Source: "Total Money Earnings in 1991 of Persons 15 Years Old and Older, by Sex, Region, and Race," U.S. Department of Commerce, Bureau of the Census, *The Black Population in the United States,* Current Population Reports, P20-471, p. 52.

★ 554 ★

Income

Total Money Earnings of Persons 15 Years Old and Over, by Sex and Race, 1991, Part 2, All Persons – Mean and Median

Persons as of March 1992.

Total money earnings and region	All persons					
	Black			White		
	Both sexes	Male	Female	Both sexes	Male	Female
United States						
Median earnings (dollars)	13,771	15,494	12,210	17,687	22,732	12,994
Standard error (dollars)	281	309	221	118	208	136
Mean earnings (dollars)	16,889	18,626	15,197	22,840	28,266	16,382
Standard error (dollars)	199	304	254	100	156	99

Source: "Total Money Earnings in 1991 of Persons 15 Years Old and Older, by Sex, Region, and Race," U.S. Department of Commerce, Bureau of the Census, *The Black Population in the United States*, Current Population Reports, P20-471, p. 52.

★ 555 ★

Income

Total Money Earnings of Persons 15 Years Old and Over, by Sex and Race, 1991, Part 3, Year-round, Full-time Workers

Persons as of March 1992.

Total money earnings and region	Year-round, full-time workers					
	Black			White		
	Both sexes	Male	Female	Both sexes	Male	Female
United States						
Total.. thousands	8,167	4,159	4,008	69,401	42,072	27,329
Total with earnings...thousands	8,167	4,159	4,008	69,343	69,343	27,280
$1 to $2,499 or loss	1.0	1.1	0.9	1.2	1.1	1.5
$2,500 to $4,999	0.8	0.6	0.9	0.8	0.6	1.0
$5,000 to $7,499	3.1	2.8	3.1	2.0	1.6	2.7
$7,500 to $9,999	5.8	4.1	7.6	3.3	2.5	4.5
$10,000 to $12,499	10.8	8.5	13.2	7.1	5.1	10.2
$12,500 to $14,999	7.7	7.5	8.0	5.7	4.1	8.1
$15,000 to $17,499	11.1	10.7	11.6	7.8	5.9	10.6
$17,500 to $19,999	7.9	6.7	9.1	6.1	4.9	8.0
$20,000 to $22,499	9.7	9.5	9.8	8.8	7.4	10.8
$22,500 to $24,999	5.9	5.0	6.8	5.1	4.5	6.2
$25,000 to $29,999	11.2	11.2	11.2	11.8	11.7	12.0
$30,000 to $34,999	7.8	9.6	5.9	9.9	10.7	8.8
$25,000 to $39,999	7.1	8.4	5.8	7.6	8.9	5.7
$40,000 to $44,999	3.8	5.3	2.3	5.8	7.4	3.3
$45,000 to $49,999	2.3	3.4	1.2	3.9	5.0	2.2
$50,000 to $59,999	2.0	2.6	1.3	5.4	7.4	2.2

[Continued]

★ 555 ★

Total Money Earnings of Persons 15 Years Old and Over, by Sex and Race, 1991, Part 3, Year-round, Full-time Workers

[Continued]

| Total money earnings and region | Year-round, full-time workers | | | | | |
| | Black | | | White | | |
	Both sexes	Male	Female	Both sexes	Male	Female
$60,000 to $74,999	1.0	1.5	0.5	3.5	5.1	1.2
$75,000 and over	0.9	1.3	0.6	4.3	6.4	1.0

Source: "Total Money Earnings in 1991 of Persons 15 Years Old and Older, by Sex, Region, and Race," U.S. Department of Commerce, Bureau of the Census, *The Black Population in the United States,* Current Population Reports, P20-471, p. 52.

★ 556 ★

Income

Total Money Earnings of Persons 15 Years Old and Over, by Sex and Race, 1991, Part 4, Year-round, Full-time Workers – Median and Mean

Persons as of March 1992.

| Total money earnings and region | Year-round, full-time workers | | | | | |
| | Black | | | White | | |
	Both sexes	Male	Female	Both sexes	Male	Female
United States						
Median earnings (dollars)	20,453	22,075	18,720	25,721	30,266	20,792
Standard error (dollars)	229	326	349	92	125	105
Mean earnings (dollars)	23,287	25,434	21,057	30,991	35,782	23,604
Standard error (dollars)	258	386	332	134	193	140

Source: "Total Money Earnings in 1991 of Persons 15 Years Old and Older, by Sex, Region, and Race," U.S. Department of Commerce, Bureau of the Census, *The Black Population in the United States,* Current Population Reports, P20-471, p. 52.

★ 557 ★

Income

Total Money Income by Marital Status, 1992, Black Men, Part 1

[Numbers in thousands. Persons 18 years old and over as of March 1993].

| Total money income | Total | Single (never married) | Married | | | Widowed | Divorced |
			Total	Spouse present	Spouse absent		
MALE-BLACK							
Total							
Total	9,622	3,930	4,431	3,865	566	426	836
Without income	789	571	153	111	42	21	44

[Continued]

★ 557 ★

Total Money Income by Marital Status, 1992, Black Men, Part 1
[Continued]

Total money income	Total	Single (never married)	Married Total	Married Spouse present	Married Spouse absent	Widowed	Divorced
With income	8,833	3,359	4,278	3,754	524	404	792
$1 to $2,499 or loss	774	482	201	170	31	30	62
$2,500 to $4,999	845	518	233	192	40	36	56
$5,000 to $7,499	1,118	531	376	289	87	121	90
$7,500 to $9,999	706	311	271	233	38	64	60
$10,000 to $12,499	786	303	373	315	58	57	53
$12,500 to $14,999	477	160	226	189	37	17	75
$15,000 to $17,499	640	232	324	303	22	22	63
$17,500 to $19,999	411	140	206	173	33	7	58
$20,000 to $22,499	566	158	342	315	27	11	56
$22,500 to $24,999	273	91	165	148	17	2	14
$25,000 to $29,999	605	157	372	328	44	16	60
$30,000 to $34,999	478	94	339	310	30	3	42
$35,000 to $39,999	371	86	255	238	17	3	28
$40,000 to $44,999	277	31	232	211	21	-	14
$45,000 to $49,999	157	16	104	101	3	8	29
$50,000 to $54,999	102	11	73	71	3	5	12

Source: "Marital Status—Total Money Income in 1992 of Persons 18 Years Old and Over, by Race, Hispanic Origin, Sex, and Work Experience in 1992," U.S. Department of Commerce, Bureau of the Census, *Money Income of Households, Families, and Persons in the United States*, Current Population Reports, P60-184, p. 113.

★ 558 ★

Income

Total Money Income by Marital Status, 1992, Black Men, Part 2

[Numbers in thousands. Persons 18 years old and over as of March 1993].

Total money income	Total	Single (never married)	Married Total	Married Spouse present	Married Spouse absent	Widowed	Divorced
MALE-BLACK							
Total							
Total	9,622	3,930	4,431	3,865	566	426	836
Without income	789	571	153	111	42	21	44
With income	8,833	3,359	4,278	3,754	524	404	792
Median income...dollars	13,485	8,687	19,151	20,114	13,018	8,021	15,046
Standard error...dollars	395	374	638	504	1,105	605	833
Mean income...dollars	17,835	12,610	22,448	23,062	18,055	10,896	18,621
Standard error...dollars	306	445	454	488	1,193	807	1,041

[Continued]

★ 558 ★

Total Money Income by Marital Status, 1992, Black Men, Part 2

[Continued]

Total money income	Total	Single (never married)	Married Total	Married Spouse present	Married Spouse absent	Widowed	Divorced
Year-Round, Full-Time Workers							
Number of income recipients	4,164	1,296	2,458	2,218	241	62	348
Median income...dollars	22,942	18,058	26,748	27,042	23,451	(B)	21,932
Standard error...dollars	563	610	548	606	2,271	(B)	1,729
Mean income...dollars	26,921	21,654	29,769	30,160	26,172	(B)	26,874
Standard error...dollars	504	944	614	652	1,770	(B)	1,919

Source: "Marital Status—Total Money Income in 1992 of Persons 18 Years Old and Over, by Race, Hispanic Origin, Sex, and Work Experience in 1991," U.S. Department of Commerce, Bureau of the Census, *Money Income of Households, Families, and Persons in the United States,* Current Population Reports, P60-184, p. 113.

★ 559 ★

Income

Total Money Income by Marital Status, 1992, Black Women, Part 1

[Numbers in thousands. Persons 18 years old and over as of March 1993. For meaning of symbols, see text].

Total money income	Total	Single (never married)	Married Total	Married Spouse present	Married Spouse absent	Widowed	Divorced
FEMALE-BLACK							
Total							
Total	11,695	4,076	4,811	3,717	1,084	1,401	1,406
Without income	978	428	467	391	76	38	45
With income	10,717	3,648	4,344	3,327	1,018	1,363	1,361
$1 to $2,499 or loss	1,259	559	550	466	84	93	58
$2,500 to $4,999	1,616	698	524	365	160	224	170
$5,000 to $7,499	1,853	573	599	355	243	476	204
$7,500 to $9,999	963	336	349	268	81	172	106
$10,000 to $12,499	939	309	402	301	101	105	122
$12,500 to $17,499	599	216	248	191	57	71	64
$15,000 to $17,499	656	198	327	253	74	40	91
$17,500 to $19,999	364	105	160	139	22	30	69
$20,000 to $22,499	460	128	212	183	29	38	83
$22,500 to $24,999	326	103	131	110	21	15	77
$25,000 to $29,999	644	175	310	258	52	49	110
$30,000 to $34,999	382	90	205	159	46	12	76
$35,000 to $39,999	211	59	102	80	21	13	37
$40,000 to $44,999	179	47	82	78	4	17	33

[Continued]

★ 559 ★

Total Money Income by Marital Status, 1992, Black Women, Part 1
[Continued]

Total money income	Total	Single (never married)	Married			Widowed	Divorced
			Total	Spouse present	Spouse absent		
$45,000 to $49,999	102	28	63	51	12	2	9
$50,000 to $54,999	63	7	25	23	2	4	27

Source: "Marital Status—Total Money Income in 1992 of Persons 18 Years Old and Over, by Race, Hispanic Origin, Sex, and Work Experience in 1991," U.S. Department of Commerce, Bureau of the Census, *Money Income of Households, Families, and Persons in the United States,* Current Population Reports, P60-184, p. 115.

★ 560 ★
Income

Total Money Income by Marital Status, 1992, Black Women, Part 2

[Numbers in thousands. Persons 18 years old and over as of March 1993. For meaning of symbols, see text].

Total money income	Total	Single (never married)	Married			Widowed	Divorced
			Total	Spouse present	Spouse absent		
FEMALE-BLACK							
$55,000 to $64,999	73	12	40	34	6	2	19
$65,000 to $74,999	8	-	8	6	1	-	-
$75,000 to $84,999	12	3	4	3	1	-	5
$85,000 to $99,999	5	1	4	4	-	-	-
$100,000 and over	2	1	-	-	-	-	1
Median income...dollars	9,135	7,479	10,931	11,736	8,175	6,909	13,287
Standard error...dollars	216	279	329	385	751	155	1,072
Mean income...dollars	12,992	11,196	14,359	14,999	12,269	9,580	16,857
Standard error...dollars	186	290	309	365	551	362	597
Year Round, Full-Time Workers							
Number of income recipients	4,066	1,229	1,971	1,596	374	185	682
Median income...dollars	20,299	18,943	20,301	20,534	18,239	20,249	22,029
Standard error...dollars	367	796	616	644	1,905	1,529	725
Mean income...dollars	22,032	20,720	22,289	22,600	20,965	22,075	23,644
Standard error...dollars	303	524	444	500	947	1,222	769

Source: "Marital Status—Total Money Income in 1992 of Persons 18 Years Old and Over, by Race, Hispanic Origin, Sex, and Work Experience in 1991," U.S. Department of Commerce, Bureau of the Census, *Money Income of Households, Families, and Persons in the United States,* Current Population Reports, P60-184, p. 115.

★ 561 ★

Income

Total Money Income by Presence of Elderly, 1992, Part 1

[Numbers in thousands. Households as of March 1993.].

Total money income	Total	No persons 65 years and over	With persons 65 years and over					
			Total	All members elderly	Some, but not all elderly			
					Total	Elderly householder or spouse only	Elderly other relative only	Elderly nonrelative only
BLACK								
Total	11,190	9,080	2,110	1,210	900	727	148	7
Less than $5,000	1,318	1,111	207	164	43	41	2	-
$5,000 to $9,999	2,090	1,347	744	597	146	131	9	6
$10,000 to $14,999	1,367	1,033	334	187	147	121	21	-
$15,000 to $19,999	1,095	887	208	102	105	75	26	-
$20,000 to $24,999	950	819	131	36	96	84	10	2
$25,000 to $29,999	827	719	109	43	66	50	12	-
$30,000 to $34,999	653	548	105	22	82	68	14	-
$35,000 to $39,999	570	494	77	24	53	38	11	-
$40,000 to $44,999	511	460	51	15	36	31	5	-
$45,000 to $49,999	354	335	19	7	11	6	6	-

Source: "Presence of the Elderly—Total Money Income of Households in 1992, by Race and Hispanic Origin of Householder," U.S. Department of Commerce, Bureau of the Census, *Money Income of Households, Families, and Persons in the United States,* Current Population Reports, P60-184, p. 39.

★ 562 ★

Income

Total Money Income by Presence of Elderly, 1992, Part 2

[Numbers in thousands. Households as of March 1993.].

Total money income	Total	No persons 65 years and over	With persons 65 years and over					
			Total	All members elderly	Some, but not all elderly			
					Total	Elderly householder or spouse only	Elderly other relative only	Elderly nonrelative only
BLACK								
Total	11,190	9,080	2,110	1,210	900	727	148	7
$50,000 to $54,999	313	286	27	8	19	8	11	-
$55,000 to $59,999	248	219	30	2	28	18	10	-
$60,000 to $64,999	197	174	22	1	21	21	-	-
$65,000 to $69,999	122	114	8	-	8	5	3	-
$70,000 to $74,999	109	102	7	-	7	5	1	-
$75,000 to $79,999	80	75	5	-	5	3	2	-
$80,000 to $84,999	67	60	7	-	7	4	3	-

[Continued]

★ 562 ★

Total Money Income by Presence of Elderly, 1992, Part 2
[Continued]

Total money income		No persons 65 years and over	With persons 65 years and over						
				All members elderly	Some, but not all elderly				
	Total		Total		Total	Elderly householder or spouse only	Elderly other relative only	Elderly nonrelative only
$85,000 to $89,999	55	52	3	-	3	1	2	-
$90,000 to $94,999	70	67	3	2	2	2	-	-
$95,000 to $99,999	30	28	2	-	2	2	-	-
$100,000 and over	163	150	12	-	12	12	-	-
Median income...dollars	18,660	20,906	11,334	7,857	20,423	19,665	27,717	(B)
Standard error...dollars	386	397	441	357	1,169	1,287	4,412	(B)
Mean income...dollars	25,409	27,132	17,996	11,522	26,701	25,944	31,425	(B)
Standard error...dollars	347	395	652	440	1,287	1,500	2,449	(B)
Income per household member...dollars	8,946	9,160	7,768	8,895	7,237	7,329	7,379	(B)
Standard error...dollars	130	147	366	626	444	530	888	(B)

Source: "Presence of the Elderly—Total Money Income of Households in 1992, by Race and Hispanic Origin of Householder," U.S. Department of Commerce, Bureau of the Census, *Money Income of Households, Families, and Persons in the United States,* Current Population Reports, P60-184, p. 39.

★ 563 ★

Income

Total Money Income of Black Families, 1991, Part 1, United States

Families as of March 1992.

Total money income	Black					
	All families	Total[1]	Husband only earner	Husband and wife earners	Female house-holder, no spouse present	Male house-holder, no spouse present
United States						
Total...thousands	7,716	3,631	438	1,714	3,582	504
Percent	100.0	100.0	100.0	100.0	100.0	100.0
Under $5,000	11.4	2.4	3.2	0.4	20.7	9.4
$5,000 to $9,999	15.0	5.8	10.8	1.2	25.0	10.1
$10,000 to $14,999	11.1	8.3	14.3	5.1	13.7	13.3
$15,000 to $19,999	9.5	8.3	15.6	6.5	10.9	8.6
$20,000 to $24,999	8.8	9.5	17.3	8.8	8.1	9.1
$25,000 to $34,999	14.4	18.5	17.9	21.5	9.8	18.4
$35,000 to $49,999	14.8	21.0	15.0	25.9	8.0	19.0
$50,000 to $59,999	5.4	9.1	3.5	11.4	1.6	5.4
$60,000 to $74,999	4.9	8.9	2.1	10.8	1.1	3.8
$75,000 and over	4.6	8.3	0.4	8.4	1.1	2.9
Median income (dollars)	21,548	33,307	21,935	38,395	11,414	24,508
Standard error (dollars)	445	758	1,081	1,091	414	2,118

[Continued]

★ 563 ★

Total Money Income of Black Families, 1991, Part 1, United States

[Continued]

Total money income	Black					
	All families	Total[1]	Husband only earner	Husband and wife earners	Female house-holder, no spouse present	Male house-holder, no spouse present
Mean income (dollars)	28,011	39,167	24,827	43,345	16,729	27,823
Standard error (dollars)	430	687	1,106	921	417	1,439

Source: "Total Money Income of Black Families in 1991, by Family Type, Earner Status, Region, and Race," U.S. Department of Commerce, Bureau of the Census, *The Black Population in the United States,* Current Population Reports, P20-471, p. 51. *Notes:* 1. Includes other combinations of earners such as wife only, wife and children, or no earners.

★ 564 ★

Income

Total Money Income of Black Families, 1991, Part 2, South

Families as of March 1992.

Total money income	Black					
	All families	Total[1]	Husband only earner	Husband and wife earners	Female house-holder, no spouse present	Male house-holder, no spouse present
South						
Total...thousands	4,253	2,094	241	1,015	1,886	273
Percent	100.0	100.0	100.0	100.0	100.0	100.0
Under $5,000	12.3	2.5	3.1	0.5	23.2	12.6
$5,000 to $9,999	14.9	6.8	14.2	1.6	23.8	15.7
$10,000 to $14,999	12.2	10.4	16.7	6.9	15.0	7.2
$15,000 to $19,999	10.2	9.3	17.1	9.4	11.3	9.3
$20,000 to $24,999	9.6	11.2	13.9	11.3	7.7	10.2
$25,000 to $34,999	14.9	19.4	20.6	22.7	9.2	20.1
$35,000 to $49,999	12.8	18.0	11.3	22.7	6.8	14.3
$50,000 to $59,999	4.6	8.2	1.3	10.4	0.9	3.6
$60,000 to $74,999	4.6	8.0	1.9	8.9	1.0	3.3
$75,000 and over	3.7	6.1	-	5.6	1.1	3.8
Median income (dollars)	20,124	29,886	19,765	33,547	11,005	21,698
Standard error (dollars)	542	852	1,391	1,297	578	2,121
Mean income (dollars)	26,106	35,505	22,302	38,690	15,683	26,002
Standard error (dollars)	534	816	1,212	1,056	548	2,016

Source: "Total Money Income of Black Families in 1991, by Family Type, Earner Status, Region, and Race," U.S. Department of Commerce, Bureau of the Census, *The Black Population in the United States,* Current Population Reports, P20-471, p. 51. *Notes:* 1. Includes other combinations of earners such as wife only, wife and children, or no earners.

★ 565 ★
Income

Total Money Income of Black Families, 1991, Part 3, North and West

Families as of March 1992.

Total money income	All families	Black				
		Total[1]	Husband only earner	Husband and wife earners	Female house-holder, no spouse present	Male house-holder, no spouse present
North and west						
Total...thousands	3,463	1,536	197	698	1,696	231
Percent	100.0	100.0	100.0	100.0	100.0	100.0
Under $5,000	10.2	2.3	3.3	0.2	17.9	5.6
$5,000 to $9,999	15.1	4.4	6.6	0.7	26.4	3.4
$10,000 to $14,999	9.8	5.3	11.3	2.5	12.3	20.6
$15,000 to $19,999	8.7	6.9	13.8	2.3	10.5	7.8
$20,000 to $24,999	7.9	7.1	21.5	5.2	8.5	7.8
$25,000 to $34,999	13.8	17.3	14.6	19.6	10.4	16.4
$35,000 to $49,999	17.3	25.0	19.5	30.6	9.3	24.6
$50,000 to $59,999	6.2	10.4	6.0	12.8	2.3	7.5
$60,000 to $74,999	5.4	10.1	2.4	13.6	1.2	4.5
$75,000 and over	5.6	11.2	1.0	12.5	1.1	1.8
Median income (dollars)	24,007	38,297	23,585	43,891	11,847	27,240
Standard error (dollars)	801	1,166	1,194	1,408	595	2,783
Mean income (dollars)	30,351	44,157	27,918	50,112	17,892	29,981
Standard error (dollars)	699	1,162	1,929	1,591	638	2,025

Source: "Total Money Income of Black Families in 1991, by Family Type, Earner Status, Region, and Race," U.S. Department of Commerce, Bureau of the Census, *The Black Population in the United States,* Current Population Reports, P20-471, p. 51. *Notes:* 1. Includes other combinations of earners such as wife only, wife and children, or no earners.

★ 566 ★
Income

Total Money Income of Black Households by Numbers of Earners, 1992, Part 1

[Numbers in thousands. Households as of March 1993].

Total money income	Total	Households having specified number of earners					
		No earners	One earner	Total	Two earner	Three earner	Four earners or more
BLACK							
Total	11,190	2,936	4,400	3,854	2,933	745	177
Less than $5,000	1,318	932	358	29	29	-	-
$5,000 to $9,999	2,090	1,283	689	113	106	7	-
$10,000 to $14,999	1,367	361	772	235	220	13	2
$15,000 to $19,999	1,095	155	683	256	234	21	2
$20,000 to $24,999	950	66	516	368	326	38	3

[Continued]

★ 566 ★

Total Money Income of Black Households by Numbers of Earners, 1992, Part 1

[Continued]

Total money income	Total	Households having specified number of earners					
		No earners	One earner	Total	Two earner	Three earner	Four earners or more
$25,000 to $29,999	827	50	427	350	289	58	3
$30,000 to $34,999	653	30	265	357	290	62	5
$35,000 to $39,999	570	25	216	329	258	65	6
$40,000 to $44,999	511	13	191	306	218	75	13
$45,000 to $49,999	354	5	88	261	212	38	11

Source: "Number of Earners—Total Money Income of Households in 1992, by Race, and Hispanic Origin of Householder," U.S. Department of Commerce, Bureau of the Census, *Money Income of Households, Families, and Persons in the United States,* Current Population Reports, P60-184, p. 31.

★ 567 ★

Income

Total Money Income of Black Households by Numbers of Earners, 1992, Part 2

[Numbers in thousands. Households as of March 1993.].

Total money income	Total	Households having specified number of earners					
		No earners	One earner	Two earners or more			
				Total	Two earner	Three earner	Four earners or more
BLACK							
$50,000 to $54,999	313	3	75	235	152	67	16
$55,000 to $59,999	248	1	48	200	135	57	7
$60,000 to $64,999	197	4	22	171	115	42	15
$65,000 to $69,999	122	-	4	118	80	29	9
$70,000 to $74,999	109	-	11	98	58	18	22
$75,000 to $79,999	80	-	7	73	38	31	4
$80,000 to $84,999	67	-	4	63	37	18	8
$85,000 to $89,999	55	-	9	46	30	4	12
$90,000 to $94,999	70	2	7	62	31	18	13
$95,000 to $99,999	30	-	1	29	9	16	4
$100,000 and over	163	-	8	154	68	66	21
Median income...dollars	18,660	6,433	17,646	37,888	34,456	49,250	68,392
Standard error...dollars	386	109	374	799	781	2,358	4,041
Mean income...dollars	25,409	8,281	20,852	43,657	38,671	56,996	70,148
Standard error...dollars	347	16	349	700	645	2,162	3,470
Income per household member...dollars	8,946	3,642	8,772	11,487	11,126	12,407	12,005
Standard error...dollars	130	132	219	251	282	669	1,086

Source: "Number of Earners—Total Money Income of Households in 1992, by Race, and Hispanic Origin of Householder," U.S. Department of Commerce, Bureau of the Census, *Money Income of Households, Families, and Persons in the United States,* Current Population Reports, P60-184, p. 31.

★ 568 ★

Income

Total Money Income of Black Households by Size of Household, 1992, Part 1

[Numbers in thousands. Households as of March 1993].

Total money income	Total	Households having specified number of persons						
		One person	Two persons	Three persons	Four persons	Five persons	Six persons	Seven persons or more
BLACK								
Total	11,190	2,892	2,895	2,155	1,721	863	418	246
Less than $5,000	1,318	460	307	265	177	55	34	20
$5,000 to $9,999	2,090	891	480	293	185	123	42	76
$10,000 to $14,999	1,367	398	402	239	157	92	49	31
$15,000 to $19,999	1,095	258	355	203	138	79	48	13
$20,000 to $24,999	950	208	244	200	157	86	39	16
$25,000 to $29,999	827	219	227	142	121	70	37	12
$30,000 to $34,999	653	143	160	125	131	56	28	9
$35,000 to $39,999	570	109	157	123	92	54	22	14
$40,000 to $44,999	511	76	144	127	93	42	21	9
$45,000 to $49,999	354	48	90	89	87	22	9	9

Source: "Size of Household—Total Money Income of Households in 1992, by Race, and Hispanic Origin of Householder," U.S. Department of Commerce, Bureau of the Census, *Money Income of Households, Families, and Persons in the United States,* Current Population Reports, P60-184, p. 28.

★ 569 ★

Income

Total Money Income of Black Households by Size of Household, 1992, Part 2

[Numbers in thousands. Households as of March 1993].

Total money income	Total	Households having specified number of persons						
		One person	Two persons	Three persons	Four persons	Five persons	Six persons	Seven persons or more
BLACK								
$50,000 to $54,999	313	24	85	68	73	40	18	6
$55,000 to $59,999	248	20	72	55	58	29	12	2
$60,000 to $64,999	197	13	37	59	51	25	9	2
$65,000 to $69,999	122	-	24	43	31	19	3	4
$70,000 to $74,999	109	2	36	22	31	14	1	3
$75,000 to $79,999	80	5	13	20	15	22	2	3
$80,000 to $84,999	67	2	12	12	24	8	5	4
$85,000 to $89,999	55	7	6	9	22	4	5	-
$90,000 to $94,999	70	4	14	11	19	3	13	6
$95,000 to $99,999	30	1	2	10	9	1	6	-
$100,000 and over	163	5	26	41	52	18	14	6
Median income...dollars	18,660	10,933	18,638	21,952	26,432	24,744	24,443	14,249
Standard error...dollars	386	397	570	875	943	1,759	1,866	3,061

[Continued]

★ 569 ★

Total Money Income of Black Households by Size of Household, 1992, Part 2

[Continued]

Total money income	Total	Households having specified number of persons						
		One person	Two persons	Three persons	Four persons	Five persons	Six persons	Seven persons or more
Mean income...dollars	25,409	16,108	24,401	28,577	33,888	31,307	32,562	26,727
Standard error...dollars	347	434	580	865	1,119	1,290	2,087	2,784
Income per household member...dollars	8,946	16,108	11,555	9,104	8,232	6,150	5,399	3,266
Standard error...dollars	130	832	412	356	340	329	433	393

Source: "Size of Household—Total Money Income of Households in 1992, by Race, and Hispanic Origin of Householder," U.S. Department of Commerce, Bureau of the Census, *Money Income of Households, Families, and Persons in the United States,* Current Population Reports, P60-184, p. 28.

★ 570 ★

Income

Total Money Income of Families, by Race, 1992

[Numbers in thousands. Families as of March 1993. For meaning of symbols, see text].

Characteristic	Total	Less than $5,000	$5,000 to $9,999	$10,000 to $14,999	$15,000 to $24,999	$25,000 to $34,999	$35,000 to $49,999	$50,000 to $74,999	$75,000 to $99,999	$100,000 and over
All families	68,144	2,540	3,972	4,954	10,595	10,189	13,115	13,335	5,234	4,210
RACE AND HISPANIC ORIGIN OR HOUSEHOLDER										
White	57,858	1,547	2,608	3,842	8,780	8,826	11,571	12,030	4,758	3,897
Black	7,888	888	1,186	933	1,480	1,028	1,104	853	271	145
Hispanic origin[1]	5,318	320	620	671	1,152	864	801	603	180	106

Source: "Selected Characteristics of Families—Total Money Income of Families in 1992," U.S. Department of Commerce, Bureau of the Census, *Money Income of Households, Families, and Persons in the United States,* Current Population Reports, P60-184, p. 44.

★ 571 ★

Income

Total Money Income of Family Households by Type, 1992, Part 1

[Numbers in thousands. Households as of March 1993.].

Total money income	Total	Family households			
			Type of family		
		Total	Married-couple families	Male householder, no wife present	Female householder, no husband present
BLACK					
Total	11,190	7,888	3,748	460	3,680
Less than $5,000	1,318	841	106	30	705
$5,000 to $9,999	2,090	1,146	251	53	842
$10,000 to $14,999	1,367	925	314	45	566
$15,000 to $19,999	1,095	774	321	65	388
$20,000 to $24,999	950	714	347	59	308
$25,000 to $29,999	827	569	293	49	227
$30,000 to $34,999	653	483	277	39	168
$35,000 to $39,999	570	442	282	29	131
$40,000 to $44,999	511	426	282	31	113
$45,000 to $49,999	354	280	197	16	67

Source: "Type of Households—Total Money Income of Households in 1992, by Race, and Hispanic Origin of Householder," U.S. Department of Commerce, Bureau of the Census, *Money Income of Households, Families, and Persons in the United States,* Current Population Reports, P60-184, p. 12.

★ 572 ★

Income

Total Money Income of Family Households by Type, 1992, Part 2

[Numbers in thousands. Households as of March 1993. For meaning of symbols, see text].

Total money income		Family households			
			Type of family		
	Total	Total	Married-couple families	Male householder, no wife present	Female householder, no husband present
BLACK					
Total	11,190	7,888	3,748	460	3,680
$50,000 to $54,999	313	269	206	13	49
$55,000 to $59,999	248	205	157	9	39
$60,000 to $64,999	197	176	147	1	27
$65,000 to $69,999	122	115	102	8	5

[Continued]

★ 572 ★

Total Money Income of Family Households by Type, 1992, Part 2

[Continued]

Total money income	Total	Family households			
			Type of family		
		Total	Married-couple families	Male householder, no wife present	Female householder, no husband present
$70,000 to $74,999	109	96	88	-	8
$75,000 to $79,999	80	74	68	-	5
$80,000 to $84,999	67	62	59	2	2
$85,000 to $89,999	55	48	44	-	4
$90,000 to $94,999	70	63	50	9	5
$95,000 to $99,999	30	29	27	-	1
$100,000 and over	163	151	129	3	19
Median income...dollars	18,660	21,761	34,290	23,439	12,606
Standard error...dollars	386	453	911	1,218	400
Mean income...dollars	25,409	28,548	36,649	26,747	17,468
Standard error...dollars	347	444	722	1,331	429
Income per household member...dollars	8,946	8,097	10,832	7,915	5,127
Standard error...dollars	130	140	259	596	151

Source: "Type of Households—Total Money Income of Households in 1992, by Race, and Hispanic Origin of Householder," U.S. Department of Commerce, Bureau of the Census, *Money Income of Households, Families, and Persons in the United States,* Current Population Reports, P60-184, p. 12.

★ 573 ★

Income

Total Money Income of Households by Age of Householder to 44 years, 1992, Part 1

[Numbers in thousands. Households as of March 1993].

Total money income	Total	Under 65 years							
		Total	15 to 24 years	25 to 34 years			35 to 44 years		
				Total	25 to 29 years	30 to 34 years	Total	35 to 39 years	40 to 44 years
BLACK									
Total	11,190	9,282	718	2,715	1,246	1,469	2,657	1,411	1,246
Less than $5,000	1,318	1,113	225	373	178	194	246	118	128
$5,000 to $9,999	2,090	1,370	175	471	228	243	282	171	111
$10,000 to $14,999	1,367	1,058	83	317	153	164	313	178	135
$15,000 to $19,999	1,095	917	82	307	176	131	211	123	88
$20,000 to $24,999	950	839	57	257	137	120	281	160	120
$25,000 to $29,999	827	734	34	245	97	147	203	100	103
$30,000 to $34,999	653	563	18	161	72	90	192	111	81
$35,000 to $39,999	570	508	9	136	47	89	175	108	67

[Continued]

★ 573 ★

Total Money Income of Households by Age of Householder to 44 years, 1992, Part 1

[Continued]

Total money income	Total	Under 65 years							
		Total	15 to 24 years	25 to 34 years			35 to 44 years		
				Total	25 to 29 years	30 to 34 years	Total	35 to 39 years	40 to 44 years
$40,000 to $44,999	511	465	9	113	36	76	169	92	76
$45,000 to $49,999	354	343	12	96	35	63	114	37	77

Source: "Age of Householder—Total Money Income of Households in 1992, by Race, and Hispanic Origin of Householder," U.S. Department of Commerce, Bureau of the Census, *Money Income of Households, Families, and Persons in the United States,* Current Population Reports, P60-184, p. 24.

★ 574 ★

Income

Total Money Income of Households by Age of Householder to 44 years, 1992, Part 2

[Numbers in thousands. Households as of March 1993].

Total money income	Total	Under 65 years							
		Total	15 to 24 years	25 to 34 years			35 to 44 years		
				Total	25 to 29 years	30 to 34 years	Total	35 to 39 years	40 to 44 years
BLACK									
$50,000 to $54,999	313	298	2	77	29	49	94	45	49
$55,000 to $59,999	248	232	4	45	22	23	94	41	52
$60,000 to $64,999	197	174	-	26	6	21	77	33	44
$65,000 to $69,999	122	120	-	17	8	10	44	20	24
$70,000 to $74,999	109	103	3	14	8	6	24	4	20
$75,000 to $79,999	80	79	-	8	2	6	30	10	20
$80,000 to $84,999	67	64	-	13	5	8	23	11	12
$85,000 to $89,999	55	55	-	5	-	5	20	12	9
$90,000 to $94,999	70	68	2	22	4	18	16	10	6
$95,000 to $99,999	30	28	-	3	9	3	4	1	3
$100,000 and over	163	152	4	7	4	3	44	23	21
Median income...dollars	18,660	20,997	8,705	17,894	16,667	20,052	24,928	23,747	26,654
Standard error...dollars	386	393	627	700	692	1,182	709	781	1,122
Mean income...dollars	25,409	27,235	13,675	22,460	20,436	24,176	30,232	28,793	31,863
Standard error...dollars	347	390	1,135	544	743	776	706	959	1,038
Income per household member...dollars	8,946	9,112	5,339	7,749	7,301	8,107	9,065	8,974	9,159
Standard error...dollars	130	144	522	256	376	368	284	412	421

Source: "Age of Householder—Total Money Income of Households in 1992, by Race, and Hispanic Origin of Householder," U.S. Department of Commerce, Bureau of the Census, *Money Income of Households, Families, and Persons in the United States,* Current Population Reports, P60-184, p. 24.

★ 575 ★

Income

Total Money Income of Households by Age of Householder 45 to 64 years, 1992, Part 1

[Numbers in thousands. Households as of March 1993.].

| | Under 65 years | | | | | |
| | 45 to 54 years | | | 55 to 64 years | | |
Total money income	Total	45 to 49 years	50 to 54 years	Total	55 to 59 years	60 to 64 years
BLACK						
Total	1,777	961	816	1,416	739	676
Less than $5,000	135	66	69	135	76	58
$5,000 to $9,999	178	96	81	264	128	136
$10,000 to $14,999	168	75	93	177	85	91
$15,000 to $19,999	161	92	70	156	68	88
$20,000 to $24,999	134	69	65	111	48	63
$25,000 to $29,999	164	95	68	88	46	42
$30,000 to $34,999	129	70	59	62	34	28
$35,000 to $39,999	96	52	43	93	50	43
$40,000 to $44,999	105	52	53	70	40	30
$45,000 to $49,999	63	32	31	56	36	19

Source: "Age of Householder—Total Money Income of Households in 1992, by Race, and Hispanic Origin of Householder," U.S. Department of Commerce, Bureau of the Census, *Money Income of Households, Families, and Persons in the United States,* Current Population Reports, P60-184, p. 25.

★ 576 ★

Income

Total Money Income of Households by Age of Householder 45 to 64 years, 1992, Part 2

[Numbers in thousands. Households as of March 1993.].

| | Under 65 years | | | | | |
| | 45 to 54 years | | | 55 to 64 years | | |
Total money income	Total	45 to 49 years	years	Total	55 to 59 years	60 to 64 years
BLACK						
$50,000 to $54,999	94	61	33	31	14	17
$55,000 to $59,999	51	26	25	18	20	
$60,000 to $64,999	51	25	26	20	13	7
$65,000 to $69,999	35	17	19	23	16	6
$70,000 to $74,999	35	25	10	27	22	5
$75,000 to $79,999	30	24	6	10	9	1
$80,000 to $84,999	17	10	7	11	6	5
$85,000 to $89,999	22	6	16	6	2	4
$90,000 to $94,999	26	18	8	2	2	-
$95,000 to $99,999	14	8	6	6	5	2

[Continued]

★ 576 ★

Total Money Income of Households by Age of Householder 45 to 64 years, 1992, Part 2
[Continued]

| Total money income | Under 65 years | | | | | |
| | 45 to 54 years | | | 55 to 64 years | | |
	Total	45 to 49 years	years	Total	55 to 59 years	60 to 64 years
$100,000 and over	68	39	29	29	20	9
Median income...dollars	28,342	29,435	26,535	19,118	21,037	17,697
Standard error...dollars	1,028	1,106	1,132	1,057	1,889	1,150
Mean income...dollars	35,756	37,060	34,220	26,944	29,455	24,201
Standard error...dollars	1,117	1,539	1,618	1,007	1,547	1,240
Income per household member...dollars	11,490	11,528	11,442	10,460	10,800	10,041
Standard error...dollars	482	657	741	540	776	755

Source: "Age of Householder—Total Money Income of Households in 1992, by Race, and Hispanic Origin of Householder," U.S. Department of Commerce, Bureau of the Census, *Money Income of Households, Families, and Persons in the United States,* Current Population Reports, P60-184, p. 25.

★ 577 ★
Income

Total Money Income of Households by Age of Householder 65 years and Older, 1992, Part 1

[Numbers in thousands. Households as of March 1993.].

| Total money income | 65 years and over | | | | |
| | Total | 65 to 74 years | | | 75 years and over |
		Total	65 to 69 years	70 to 74 years	
BLACK					
Total	1,908	1,204	654	550	703
Less than $5,000	205	109	59	50	96
$5,000 to $9,999	720	385	195	190	335
$10,000 to $14,999	309	205	116	89	104
$15,000 to $19,999	178	123	65	59	54
$20,000 to $24,999	111	81	45	36	29
$25,000 to $29,999	94	75	40	35	19
$30,000 to $34,999	90	64	38	26	26
$35,000 to $39,999	62	48	29	19	14
$40,000 to $44,999	46	41	29	12	5
$45,000 to $49,999	11	9	5	4	2

Source: "Age of Householder—Total Money Income of Households in 1992, by Race, and Hispanic Origin of Householder," U.S. Department of Commerce, Bureau of the Census, *Money Income of Households, Families, and Persons in the United States,* Current Population Reports, P60-184, p. 25.

★ 578 ★

Income

Total Money Income of Households by Age of Householder 65 years and Older, 1992, Part 2

[Numbers in thousands. Households as of March 1993.].

| Total money income | 65 years and over | | | | |
| | Total | 65 to 74 years | | | 75 years and over |
		Total	65 to 69 years	70 to 74 years	
BLACK					
$50,000 to $54,999	15	9	4	5	5
$55,000 to $59,999	16	12	7	5	5
$60,000 to $64,999	22	17	7	10	6
$65,000 to $69,999	2	2	1	1	-
$70,000 to $74,999	6	5	4	1	1
$75,000 to $79,999	2	2	2	-	-
$80,000 to $84,999	3	1	1	-	2
$85,000 to $89,999	-	-	-	-	-
$90,000 to $94,999	3	3	2	1	-
$95,000 to $99,999	2	2	2	-	-
$100,000 and over	10	10	4	7	-
Median income...dollars	10,396	12,334	12,261	12,502	7,931
Standard error...dollars	443	619	821	1,030	491
Mean income...dollars	16,528	18,929	19,057	18,776	12,415
Standard error...dollars	651	938	1,052	1,629	669
Income per household member...dollars	7,810	8,391	8,080	8,800	6,613
Standard error...dollars	406	546	640	958	545

Source: "Age of Householder—Total Money Income of Households in 1992, by Race, and Hispanic Origin of Householder," U.S. Department of Commerce, Bureau of the Census, *Money Income of Households, Families, and Persons in the United States,* Current Population Reports, P60-184, p. 25.

★ 579 ★

Income

Total Money Income of Households by Type of Residence, 1992, Part 1

[Numbers in thousands. Households as of March 1993].

Total money income	All households	Metropolitan-nonmetropolitan residence							Outside metropolitan areas
		Inside metropolitan areas							
		Total	Inside central cities			Outside central cities			
			Total	One million or more	Under 1 million	Total	One million or more	Under 1 million	
BLACK									
Total	11,190	9,562	6,481	4,522	1,959	3,080	2,035	1,045	1,628
Less than $5,000	1,318	1,117	855	597	258	262	156	105	201

[Continued]

★ 579 ★

Total Money Income of Households by Type of Residence, 1992, Part 1
[Continued]

Total money income	All households	Metropolitan-nonmetropolitan residence							Outside metropolitan areas
		Inside metropolitan areas							
		Total	Inside central cities			Outside central cities			
			Total	One million or more	Under 1 million	Total	One million or more	Under 1 million	
$5,000 to $9,999	2,090	1,680	1,265	859	405	415	247	168	410
$10,000 to $14,999	1,367	1,106	796	552	244	310	186	124	261
$15,000 to $19,999	1,095	937	654	443	211	283	164	120	158
$20,000 to $24,999	950	790	548	367	181	242	134	108	160
$25,000 to $29,999	827	704	500	351	149	204	145	59	123
$30,000 to $34,999	653	579	341	240	101	237	161	77	74
$35,000 to $39,999	570	513	316	218	99	197	134	63	57
$40,000 to $44,999	511	472	304	228	76	168	100	67	40
$45,000 to $49,999	354	311	212	154	58	99	80	19	42

Source: "Type of Residence—Total Money Income of Households in 1992, by Race, and Hispanic Origin of Householder," U.S. Department of Commerce, Bureau of the Census, *Money Income of Households, Families, and Persons in the United States,* Current Population Reports, P60-184, p. 10. *Note:* 1. Persons of Hispanic origin may be of any race.

★ 580 ★
Income

Total Money Income of Households by Type of Residence, 1992, Part 2

[Numbers in thousands. Households as of March 1993. For meaning of symbols, see text].

Total money income	All households	Metropolitan-nonmetropolitan residence							Outside metropolitan areas
		Inside metropolitan areas							
		Total	Inside central cities			Outside central cities			
			Total	One million or more	Under 1 million	Total	One million or more	Under 1 million	
BLACK									
Total	11,190	9,562	6,481	4,522	1,959	3,080	2,035	1,045	1,628
$50,000 to $54,999	313	293	154	110	44	139	114	25	20
$55,000 to $59,999	248	225	120	98	21	105	87	18	23
$60,000 to $64,999	197	175	92	61	31	83	51	31	22
$65,000 to $69,999	122	108	56	39	17	52	46	6	14
$70,000 to $74,999	109	104	40	25	15	65	52	12	5
$75,000 to $79,999	80	73	45	29	16	28	22	6	7
$80,000 to $84,999	67	63	37	28	9	26	12	14	4
$85,000 to $89,999	55	54	26	26	-	28	19	9	1
$90,000 to $94,999	70	68	35	25	10	33	28	4	2
$95,000 to $99,999	30	30	13	11	2	17	14	2	-
$100,000 and over	163	159	72	60	12	87	81	7	3
Median income...dollars	18,660	19,674	17,384	17,704	16,720	25,602	29,434	20,289	13,821
Standard error...dollars	386	408	457	585	791	943	1,170	1,081	744
Mean income...dollars	25,409	26,495	23,654	24,309	22,143	32,474	35,903	25,791	19,032
Standard error...dollars	347	389	418	521	676	806	1,092	986	763

[Continued]

★ 580 ★

Total Money Income of Households by Type of Residence, 1992, Part 2

[Continued]

Total money income	All households	Metropolitan-nonmetropolitan residence								Outside metropolitan areas
			Inside metropolitan areas							
		Total	Inside central cities			Outside central cities			
			Total	One million or more	Under 1 million	Total	One million or more	Under 1 million	
Income per household member...dollars	8,946	9,403	8,538	8,766	8,009	11,131	12,453	8,644	6,404
Standard error...dollars	130	152	181	235	333	359	506	471	350

Source: "Type of Residence—Total Money Income of Households in 1992, by Race, and Hispanic Origin of Householder," U.S. Department of Commerce, Bureau of the Census, *Money Income of Households, Families, and Persons in the United States,* Current Population Reports, P60-184, p. 10.

★ 581 ★

Income

Total Money Income of Nonfamily Households by Type, 1992, Part 1

[Numbers in thousands. Households as of March 1993.].

Total money income	Nonfamily households				
		Sex of householder			
		Male		Female	
	Total	Total	Living alone	Total	Living alone
BLACK					
Total	3,302	1,484	1,244	1,818	1,648
Less than $5,000	477	165	157	312	303
$5,000 to $9,999	945	348	321	596	570
$10,000 to $14,999	442	221	202	221	196
$15,000 to $19,999	321	165	132	156	125
$20,000 to $24,999	236	105	91	131	117
$25,000 to $29,999	258	115	95	143	123
$30,000 to $34,999	169	87	71	82	72
$35,000 to $39,999	129	73	57	55	52
$40,000 to $44,999	85	48	43	37	33
$45,000 to $49,999	74	46	26	27	22

Source: "Type of Households—Total Money Income of Households in 1992, by Race, and Hispanic Origin of Householder," U.S. Department of Commerce, Bureau of the Census, *Money Income of Households, Families, and Persons in the United States,* Current Population Reports, P60-184, p. 12.

★ 582 ★

Income

Total Money Income of Nonfamily Households by Type, 1992, Part 2

[Numbers in thousands. Households as of March 1993.].

Total money income	Nonfamily households					Living alone
		Sex of householder				
		Male		Female		
	Total	Total	Total	Living alone	Total	
BLACK						
Total	11,190	3,302	1,484	1,244	1,818	1,648
$50,000 to $54,999	313	44	28	10	16	14
$55,000 to $59,999	248	43	31	11	13	9
$60,000 to $64,999	197	21	11	6	10	7
$65,000 to $69,999	122	7	5	-	2	-
$70,000 to $74,999	109	13	10	2	3	-
$75,000 to $79,999	80	7	5	5	1	-
$80,000 to $84,999	67	5	1	-	4	2
$85,000 to $89,999	55	7	6	6	1	1
$90,000 to $94,999	70	7	4	4	4	-
$95,000 to $99,999	30	1	1	1	-	-
$100,000 and over	163	12	10	4	1	1
Median income...dollars	18,660	12,062	15,267	13,369	10,005	9,092
Standard error...dollars	386	387	883	746	524	555
Mean income...dollars	25,409	17,911	20,998	18,434	15,392	14,351
Standard error...dollars	347	452	777	767	504	485
Income per household member...dollars	8,946	14,891	16,873	18,434	13,169	14,351
Standard error...dollars	130	657	1,111	1,477	797	978

Source: "Type of Households—Total Money Income of Households in 1992, by Race, and Hispanic Origin of Householder," U.S. Department of Commerce, Bureau of the Census, *Money Income of Households, Families, and Persons in the United States,* Current Population Reports, P60-184, p. 12.

★ 583 ★

Income

Total Money Income of Persons 15 Years Old and Over, 1991, Part 1
All Persons

[Persons as of March 1992].

Total money income and region	All persons					
	Black			White		
	Both sexes	Male	Female	Both sexes	Male	Female
UNITED STATES Total...thousands	22,542	10,252	12,290	165,571	80,049	85,522
Total with income...thousands	19,671	8,943	10,728	155,311	76,578	78,733
$1 to $2,499 or loss	11.3	10.6	11.9	10.6	6.0	15.1
$2,500 to $4,999	14.0	11.0	16.5	8.0	4.8	11.1
$5,000 to $7,499	13.9	10.9	16.5	9.3	6.3	12.3
$7,500 to $9,999	9.0	7.8	9.9	7.5	6.0	9.0
$10,000 to $12,499	8.4	8.5	8.4	8.0	7.0	9.0
$12,500 to $14,999	5.7	6.3	5.1	5.7	5.3	6.2
$15,000 to $17,499	6.3	6.9	5.8	6.2	6.1	6.3
$17,500 to $19,999	4.8	4.7	4.9	4.8	5.0	4.5
$20,000 to $22,499	4.9	5.5	4.4	5.6	6.1	5.1
$22,500 to $24,999	3.2	3.3	3.0	3.7	4.0	3.5
$25,000 to $29,999	5.8	6.5	5.2	7.2	8.8	5.8
$30,000 to $34,999	3.9	5.1	3.0	5.7	7.4	4.1
$35,000 to $39,999	3.2	4.4	2.3	4.3	6.0	2.7
$40,000 to $44,999	2.0	3.0	1.2	3.3	4.9	1.7
$45,000 to $49,999	1.3	2.0	0.6	2.3	3.4	1.2
$50,000 to $59,999	1.0	1.6	0.6	3.0	4.8	1.2
$60,000 to $74,999	0.7	1.0	0.4	2.0	3.4	0.8
$75,000 and over	0.6	0.8	0.4	2.7	4.7	0.7

Source: "Total Money Income in 1991 of Persons 15 Years Old and Over, by Sex, Region and Race," U.S. Department of Commerce, Bureau of the Census, *The Black Population in the United States, March 1992*, Current Population Reports, P20-471, p. 40.

★ 584 ★

Income

Total Money Income of Persons 15 Years Old and Over, 1991, Part 2 Year-round, Full-time Workers

[Persons as of March 1992].

Total money income and region	Year-round, full-time workers					
	Black			White		
	Both sexes	Male	Female	Both sexes	Male	Female
UNITED STATES						
Total...thousands	8,167	4,159	4,008	69,401	42,072	27,329
Total with income...thousands	8,167	4,159	4,008	69,384	42,067	27,317
$1 to $2,499 or loss	0.9	1.0	0.8	1.0	0.8	1.2
$2,500 to $4,999	0.7	0.6	0.8	0.7	0.5	0.9
$5,000 to $7,499	3.0	2.7	3.2	1.7	1.4	2.3
$7,500 to $9,999	5.3	3.7	7.0	2.9	2.2	4.0
$10,000 to $12,499	10.2	8.0	12.5	6.5	4.7	9.3
$12,500 to $14,999	7.6	7.3	7.9	5.6	4.0	8.0
$15,000 to $17,499	10.9	10.4	11.5	7.4	5.7	10.2
$17,500 to $19,999	8.3	7.1	9.6	6.1	5.0	7.9
$20,000 to $22,499	9.4	9.0	9.7	8.3	7.0	10.2
$22,500 to $24,999	6.1	5.3	7.0	5.3	4.4	6.8
$25,000 to $29,999	11.2	11.5	10.9	12.0	11.8	12.4
$30,000 to $34,999	7.9	9.3	6.5	9.9	10.6	8.9
$35,000 to $39,999	6.9	8.4	5.3	7.8	8.9	6.0
$40,000 to $44,999	4.2	5.5	2.9	6.1	7.5	3.8
$45,000 to $49,999	2.7	4.0	1.4	4.2	5.3	2.6
$50,000 to $59,999	2.2	3.0	1.4	5.6	7.6	2.6
$60,000 to $74,999	1.2	1.7	0.7	3.8	5.3	1.5
$75,000 and over	1.1	1.5	0.7	5.0	7.3	1.4

Source: "Total Money Income in 1991 of Persons 15 Years Old and Over, by Sex, Region and Race," U.S. Department of Commerce, Bureau of the Census, *The Black Population in the United States, March 1992*, Current Population Reports, P20-471, p. 40.

★ 585 ★
Income

Total Money Income of Persons 25 Years Old and Over, by Educational Level, 1991, Part 1, All Blacks

[Persons as of March 1992].

Total money income and region, and sex	Total	Elementary Less than 9th grade	High school		College	
			9th to 12th grade (no diploma)	High school graduate	Some college or associate degree	Bachelor's degree or more
BLACK						
United States						
Both sexes...thousands	17,445	2,317	3,324	6,220	3,502	2,080
Total with income...thousands	16,323	2,181	2,989	5,828	3,316	2,010
$1 to $2,499 or loss	7.4	8.5	11.8	7.7	5.1	2.4
$2,500 to $4,999	12.3	25.0	18.2	10.3	6.9	4.2
$5,000 to $7,499	13.7	30.3	18.8	11.9	7.7	3.4
$7,500 to $9,999	9.1	12.6	11.7	9.8	6.9	3.4
$10,000 to $12,499	8.9	7.7	9.8	11.0	8.2	3.3
$12,500 to $14,999	5.8	4.4	5.2	7.0	6.4	3.9
$15,000 to $17,499	6.6	3.3	7.2	8.3	6.6	4.9
$17,500 to $19,999	5.2	2.3	3.0	6.3	7.9	4.3
$20,000 to $22,499	5.4	2.3	3.5	5.8	7.8	6.9
$22,500 to $24,999	3.6	0.5	2.9	3.4	5.1	6.2
$25,000 to $29,999	6.7	0.9	2.7	6.9	10.1	12.8
$30,000 to $34,999	4.7	1.2	1.6	4.2	7.1	10.2
$35,000 to $39,999	3.9	0.3	1.4	3.6	5.4	9.5
$40,000 to $44,999	2.4	0.5	0.7	1.7	3.1	8.1
$45,000 to $49,999	1.5	0.2	0.4	0.7	2.6	5.1
$50,000 to $59,999	1.3	-	0.6	0.7	1.3	5.1
$60,000 to $74,999	0.8	0.1	0.1	0.4	1.1	3.3
$75,000 and over	0.7	-	0.2	0.4	0.5	3.3

Source: "Total Money Income in 1991 of Persons 25 Years Old and Over, by Educational Attainment, Sex, Region and Race," U.S. Department of Commerce, Bureau of the Census, *The Black Population in the United States, March 1992,* Current Population Reports, P20-471, p. 42.

★ 586 ★

Income

Total Money Income of Persons 25 Years Old and Over, by Educational Level, 1991, Part 2 Black Men

[Persons as of March 1992].

Total money income and region	Total	Elementary Less than 9th grade	High school 9th to 12th grade (no diploma)	High school graduate	College Some college or associate degree	Bachelor's degree or more
BLACK						
United States						
Male...thousands	7,803	1,127	1,446	2,842	1,462	926
Total with income...thousands	7,375	1,058	1,339	2,679	1,393	905
$1 to $2,499 or loss	5.9	6.3	10.3	5.6	4.3	2.2
$2,500 to $4,999	9.3	18.6	13.3	7.3	6.1	3.5
$5,000 to $7,499	10.3	26.6	10.5	8.8	4.8	3.8
$7,500 to $9,999	7.7	11.8	10.8	7.8	5.1	2.3
$10,000 to $12,499	9.1	10.4	10.9	10.6	6.7	4.3
$12,500 to $14,999	6.6	7.0	7.0	7.5	6.2	3.2
$15,000 to $17,499	7.1	4.7	10.1	8.4	6.1	3.5
$17,500 to $19,999	5.2	4.4	4.5	6.2	6.1	2.4
$20,000 to $22,499	5.9	3.8	5.2	6.1	7.5	6.5
$22,500 to $24,999	3.8	0.8	4.1	3.6	4.8	5.8
$25,000 to $29,999	7.5	1.5	3.7	9.2	11.2	9.6
$30,000 to $34,999	6.1	2.0	2.9	6.5	8.9	9.7
$35,000 to $39,999	5.4	0.7	2.8	5.7	8.1	9.2
$40,000 to $44,999	3.6	1.0	1.6	2.6	4.6	10.8
$45,000 to $49,999	2.5	0.4	0.9	1.5	4.4	7.2
$50,000 to $59,999	1.9	-	1.0	1.1	2.4	6.8
$60,000 to $74,999	1.2	-	0.1	0.9	1.7	4.2
$75,000 and over	1.0	-	0.3	0.4	1.0	4.9

Source: "Total Money Income in 1991 of Persons 25 Years Old and Over, by Educational Attainment, Sex, Region and Race," U.S. Department of Commerce, Bureau of the Census, *The Black Population in the United States, March 1992*, Current Population Reports, P20-471, p. 42.

★ 587 ★

Income

Total Money Income of Persons 25 Years Old and Over, by Educational Level, 1991, Part 3
Black Women

[Persons as of March 1992].

Total money income and region, and sex	Total	Elementary Less than 9th grade	High school		College	
			9th to 12th grade (no diploma)	High school graduate	Some college or associate degree	Bachelor's degree or more
BLACK						
United States						
Female...thousands	9,641	1,190	1,878	3,379	2,041	1,154
Total with income...thousands	8,947	1,123	1,650	3,148	1,922	1,105
$1 to $2,499 or loss	8.6	10.6	13.1	9.4	5.7	2.6
$2,500 to $4,999	14.7	31.0	22.2	12.8	7.5	4.7
$5,000 to $7,499	16.6	33.7	25.6	14.5	9.9	3.1
$7,500 to $9,999	10.3	13.3	12.4	11.5	8.2	4.2
$10,000 to $12,499	8.6	5.2	9.0	11.4	9.4	2.6
$12,500 to $14,999	5.2	1.9	3.7	6.7	6.6	4.4
$15,000 to $17,499	6.2.	1.9	4.8	8.2	6.9	6.0
$17,500 to $19,999	5.3	0.3	1.8	6.4	9.2	5.8
$20,000 to $22,499	5.0	0.8	2.1	5.5	8.0	7.1
$22,500 to $24,999	3.5	0.1	2.0	3.2	5.4	6.5
$25,000 to $29,999	6.0	0.4	1.9	4.9	9.4	15.4
$30,000 to $34,999	3.5	0.5	0.6	2.3	5.7	10.6
$35,000 to $39,999	2.6	-	0.4	1.8	3.4	9.7
$40,000 to $44,999	1.4	-	-	0.8	2.0	5.8
$45,000 to $49,999	0.7	-	-	0.1	1.3	3.4
$50,000 to $59,999	0.7	-	0.3	0.3	0.5	3.7
$60,000 to $74,999	0.5	0.2	-	-	0.6	2.5
$75,000 and over	0.4	-	0.2	0.4	0.1	1.9

Source: "Total Money Income in 1991 of Persons 25 Years Old and Over, by Educational Attainment, Sex, Region and Race," U.S. Department of Commerce, Bureau of the Census, *The Black Population in the United States, March 1992*, Current Population Reports, P20-471, p. 42.

Income, Earnings

★ 588 ★

Coverage of Federal Minimum Hourly Wage Rates, 1991

[Employee estimates as of **September 1991**, except as indicated. The Fair Labor Standards Act of 1938 and subsequent and amendments provide for minimum wage coverage applicable to specified nonsupervisory employment categories. Exempt from coverage are executives and administration or professionals].

SEX AND RACE	NONSUPERVISORY EMPLOYEES, 1991		
	Total (1,000)	Subject to minimum wage rates	
		Total (1,000)	Percent of total
Total	91,373	80,540	88.1
Male	45,909	40,276	87.7
Female	45,464	40,264	88.6
White	80,222	70,494	87.9
Black and other	11,151	10,046	90.1
Black only	10,150	9,148	90.1

Source: "Effective Federal Minimum Hourly Wage Rates, 1950 to 1993, and Coverage in 1991," *Statistical Record of the United States, 1993*, p. 429. Primary source: U.S. Department of Labor, Employment Standards Administration, *Minimum Wage and Maximum Hours Standards Under the Fair Labor Standard Act, 1981*, annual; and unpublished data.

★ 589 ★

Income, Earnings

Federal Government Employment, by Race and National Origin and by Pay System, 1991

[**As of Sept. 30.** Covers total employment for only Executive Branch agencies participating in OPM's Central Personnel Data File (CPDF). Excludes foreign nationals abroad and U.S. Postal Service].

PAY SYSTEM	1991				
	Total employees (1,000)	Race/national origin			
		Race and origin[1] (1,000)	Percent of total	Black Non-Hispanic (1,000)	Hispanic (1,000)
All pay systems, total[2]	2,183.4	602.0	27.6	363.9	118.8
General Schedule and equivalent[3]	1,694.9	454.9	26.8	281.6	88.1
Grades 1-4 ($10,581-$18,947)	223.8	96.4	43.1	64.4	16.3
Grades 5-8 ($16,305-$29,081)	521.0	176.5	33.9	119.6	30.9
Grades 9-12 ($24,705-$46,571)	660.8	143.8	21.8	78.1	33.0
Grades 13-15 ($42,601-$76,982)	289.3	38.2	13.2	19.5	7.9

[Continued]

★ 589 ★

Federal Government Employment, by Race and National Origin and by Pay System, 1991

[Continued]

PAY SYSTEM	1991				
	Total employees (1,000)	Race/national origin			
		Race and origin[1] (1,000)	Percent of total	Black Non-Hispanic (1,000)	Hispanic (1,000)
Executive, total	14.1	1.1	7.8	0.6	0.3
Wage pay system	360.8	122.6	34.0	71.6	26.3
Other pay systems	113.7	23.4	20.6	10.1	4.2

Source: "Federal Government Employment, by Race and National Origin and by Pay System: 1982 and 1991," Statistical Record of the United States, 1993. Primary source: U.S. Office of Management, Central Personnel Data File. Notes: 1. Includes American Indians, Alaska Natives, Asians, and Pacific Islanders, not shown separately. 2. Due to the inclusion of unspecified employee records, the pay systems listed do not add to the total. 3. Pay rates as of January 1990 for general schedule. Each grade (except Executive) includes several salary steps. Range is from lowest to highest step of grades shown. 4. Includes white-collar employment in other than General Schedule and Equivalent or Executive pay plans.

★ 590 ★

Income, Earnings

Median Income of Married-Couple Families, by Work Experiences of Husbands and Wives and Race, 1991

[As of **March 1992**. Based on Current Population Survey].

WORK EXPERIENCE OF HUSBAND	NUMBER (1,000)				MEDIAN INCOME (dollars)			
		Wife worked		Wife did not work		Wife worked		Wife did not work
	Total	Total	Worked year-round, full-time		Total	Total	Worked year-round full-time	
All families[1]	52,457	33,673	18,095	18,784	40,995	47,484	53,027	28,504
Husband worked	42,015	31,159	16,863	10,857	45,995	49,178	54,391	36,240
Worked year-round, full-time	32,424	24,496	13,625	7,928	50,092	52,707	57,092	40,518
Husband did not work	10,442	2,514	1,232	7,927	23,084	28,556	34,278	21,616
White	47,124	30,005	15,806	17,119	41,506	48,101	53,645	29,370
Husband worked	37,741	27,848	14,776	9,893	46,550	49,764	54,876	37,076
Worked year-round, full-time	29,188	21,932	11,948	7,256	50,595	53,178	57,538	41,278
Husband did not work	9,383	2,157	1,030	7,226	23,663	29,174	35,386	22,296
Black	3,631	2,539	1,629	1,092	33,307	40,470	45,060	19,365
Husband worked	2,851	2,261	1,471	590	38,536	42,451	47,357	23,780
Worked year-round, full-time	2,139	1,747	1,164	392	42,618	45,312	50,918	27,101
Husband did not work	780	278	158	502	17,229	26,136	28,343	14,245
Hispanic[2]	3,532	1,998	1,009	1,533	28,594	35,580	43,769	20,543
Husband worked	3,016	1,854	932	1,162	30,564	36,358	45,028	22,350

[Continued]

★ 590 ★

Median Income of Married-Couple Families, by Work Experiences of Husbands and Wives and Race, 1991

[Continued]

WORK EXPERIENCE OF HUSBAND	NUMBER (1,000)				MEDIAN INCOME (dollars)			
		Wife worked		Wife did not work		Wife worked		Wife did not work
	Total	Total	Worked year-round, full-time		Total	Total	Worked year-round full-time	
Worked year-round, full-time	2,181	1,336	733	845	34,234	41,510	48,333	25,114
Husband did not work	516	145	77	371	17,634	27,531	32,864	14,502

Source: "Median Income of Married-Couple Families, by Work Experiences of Husbands and Wives and Race, 1991," *Statistical Record of the United States, 1993*, p. 467. Primary source: U.S. Bureau of the Census, *Current Population Reports*, P60-180. *Notes:* 1. Includes other races not shown separately. 2. Persons of Hispanic origin may be of any race.

★ 591 ★

Income, Earnings

Median Money Income of Families and Unrelated Individuals, In Current and Constant (1991) Dollars, 1970 to 1991

[Constant dollars based on CPI-U-X1 deflator. Unrelated individuals are persons not living with any relatives.].

ITEM	1970	1980	1984[1]	1985	1987[2]	1988	1989	1990	1991
CURRENT DOLLARS									
Families:[3]									
Married-couple families	10,516	23,141	29,612	31,100	34,879	36,389	38,547	39,895	40,995
Wife in paid labor force	12,276	26,879	34,668	36,431	40,751	42,709	45,266	46,777	48,169
Wife not in paid labor force	9,304	18,972	23,582	24,556	26,340	27,220	28,747	30,265	30,075
Male householder, no wife present	9,012	17,519	23,325	22,622	25,208	26,827	27,847	29,046	28,351
Female householder, no husband present	5,093	10,408	12,803	13,660	14,683	15,346	16,442	16,932	16,692
Unrelated individuals:									
Male	4,540	10,939	13,566	14,921	16,082	16,976	17,860	17,927	18,069
Female	2,483	6,668	9,501	9,865	11,029	11,881	12,390	12,450	12,731
CONSTANT (1991) DOLLARS									
Families:[3]									
Married-couple families	34,680	38,297	38,818	39,366	41,818	41,895	42,340	41,574	40,995
Wife in paid labor force	40,484	44,483	45,445	46,114	48,858	49,171	49,720	48,745	48,169
Wife not in paid labor force	30,683	31,397	30,913	31,083	31,940	31,339	31,575	31,539	30,075
Male householder, no wife present	29,720	28,993	30,576	28,635	30,223	30,886	30,587	30,268	28,351
Female householder, no husband present	16,796	17,224	16,783	17,291	17,604	17,668	18,060	17,645	16,692

[Continued]

★ 591 ★

Median Money Income of Families and Unrelated Individuals, In Current and Constant (1991) Dollars, 1970 to 1991

[Continued]

ITEM	1970	1980	1984[1]	1985	1987[2]	1988	1989	1990	1991
Unrelated individuals:									
Male	14,972	18,103	17,783	18,887	19,281	19,545	19,617	18,681	18,069
Female	8,188	11,035	12,455	12,487	13,223	13,679	13,609	12,974	12,731

Source: "Median Money Income of Families and Unrelated Individuals, In Current and Constant (1991) Dollars: 1970 to 1991," *Statistical Record of the United States, 1993*, p. 465. Primary source: U.S. Bureau of the Census, *Current Population Reports*, P60-180 and unpublished data. *Notes:* 1. Beginning 1983, data based on revised Hispanic population controls and not directly comparable with prior years. 2. Beginning 1987, data based on revised processing procedures and not directly comparable with prior years. 3. Beginning 1980, based on householder concept. Restricted to primary families, see source.

★ 592 ★

Income, Earnings

Median Weekly Earnings of African American Families, 1980 to 1992

[**In current dollars of usual weekly earnings**. Annual average of quarterly figures based on Current Population Survey].

CHARACTERISTIC	NUMBER OF FAMILIES (1,000)					MEDIAN WEEKLY EARNINGS (dollars)				
	1980	1985	1990	1991	1992	1980	1985	1990	1991	1992
BLACK										
Total families with earners[1]	4,503	4,668	5,082	5,098	5,188	299	378	459	484	478
Married-couple families	2,802	2,671	2,724	2,735	2,723	366	487	601	625	646
One earner[2]	1,103	902	893	897	895	210	257	304	313	309
Husband	769	580	527	503	486	244	292	345	366	359
Wife	279	257	290	312	330	151	206	243	272	279
Two or more earners	1,700	1,769	1,831	1,838	1,828	472	622	748	776	806
Husband and wife only	1,238	1,258	1,297	1,362	1,327	461	603	713	756	783
Families maintained by women	1,438	1,703	1,986	2,003	2,079	192	259	314	339	328
Families maintained by man	263	294	372	360	386	307	360	397	401	412

Source: "Median Weekly Earnings of Families, by Type of Family Number of Earners, Race, and Hispanic Origin: 1980 to 1992," *Statistical Record of the United States, 1993*, p. 427. Primary source: U.S. Bureau of Labor Statistics, Bulletin 2307; and *Employment and Earnings*, monthly, January issues. *Notes:* 1. Excludes families in which there is no wage or salary earner or in which the husband, wife, or other person maintaining the family is either self-employed or in the Armed Forces. 2. Includes other earners, not shown separately.

★ 593 ★
Income, Earnings

Money Income of Families – Median Family Income, by Race and Hispanic Origin, 1991

CHARACTERISTIC	NUMBER (1,000)				MEDIAN FAMILY INCOME (dollars)			
	All families[1]	White	Black	Hispanic[2]	All families	White	Black	Hispanic[2]
All families	67,173	57,224	7,716	5,177	35,939	37,783	21,548	23,895
Region:								
Northeast	13,428	11,177	1,291	856	40,265	41,815	25,533	21,437
Midwest	16,170	14,347	1,554	355	36,759	38,224	20,860	26,942
South	23,679	19,023	4,253	1,651	31,940	35,226	20,124	23,708
West	13,897	12,078	618	2,314	37,171	37,610	28,298	24,317
Type of family:								
Married-couple families	52,457	47,124	3,631	3,532	40,995	41,506	33,307	28,594
Wife in paid labor force	30,923	27,463	2,420	1,845	48,169	48,802	41,353	35,655
Wife not in paid labor force	21,534	19,661	1,210	1,687	30,075	30,792	20,288	21,923
Male householder[3]	3,025	2,374	504	383	28,351	28,924	24,508	21,759
Female householder[3]	11,692	7,726	3,582	1,261	16,692	19,547	11,414	12,132
With related children[4]	34,861	28,368	5,143	3,621	34,990	37,699	18,822	22,064
Married couple	25,357	22,213	2,129	2,445	42,514	43,179	35,358	27,296
Male householder[3]	1,513	1,187	244	204	24,171	24,507	20,920	19,182
Female householder[3]	7,991	4,967	2,771	972	13,012	15,513	9,413	10,216
Number of earners:								
No earners	10,158	8,355	1,547	719	15,631	18,183	6,480	8,264
One earners	18,500	15,245	2,640	1,714	25,960	28,095	15,912	17,058
Two earners	29,681	25,946	2,724	1,946	43,623	44,711	34,060	30,802
Three earners	6,542	5,658	647	545	55,871	56,583	46,221	38,763
Four or more earners	2,293	2,019	159	252	70,019	70,354	60,126	51,680

Source: "Money Income of Families—Median Family Income, by Race and Hispanic Origin: 1991," *Statistical Record of the United States, 1993,* 464. Primary source: U.S. Bureau of the Census, *Current Population Reports,* P60-180. *Notes:* 1. Includes other races not shown separately. 2. Persons of Hispanic origin may be of any race. 3. No spouse present. 4. Children under 18 years old.

★ 594 ★

Income, Earnings

Money Income of Households – Aggregate and Mean Income, by Race and Income, by Race and Hispanic Origin of Householder, 1991

CHARACTERISTIC	ALL RACES[1]		WHITE		BLACK		HISPANIC[2]	
	Aggregate money income (bil. dol.)	Mean income (dol.)	Aggregate money income (bil. dol.)	Mean income (dol.)	Aggregate money income (bil. dol.)	Mean income (dol.)	Aggregate money income (bil. dol.)	Mean income (dol.)
Total	3,628	37,922	3,228	39,523	276	25,043	184	28,872
Age of householder:								
15 to 24 years old	103	21,219	90	22,560	9	13,057	11	19,470
25 to 34 years old	705	35,252	615	36,897	26	21,662	50	27,223
35 to 44 years old	985	45,253	868	47,347	82	31,081	52	31,489
45 to 54 years old	788	50,700	702	52,831	56	32,763	35	35,549
55 to 64 years old	535	42,592	481	44,660	37	26,346	24	33,343
65 years old and over	511	24,424	472	25,331	30	15,654	13	19,935
Region:								
Northeast	804	41,647	729	43,058	54	28,252	30	26,851
Midwest	856	36,715	788	38,033	55	24,613	14	30,200
South	1,147	34,685	989	37,202	139	23,328	56	28,265
West	820	41,091	722	41,394	29	30,303	84	29,901
Size of household:								
One persons	503	20,984	443	21,606	47	15,964	14	15,834
Two persons	1,221	39,724	1,119	41,282	73	25,178	40	29,420
Three persons	725	44,238	645	46,719	58	28,092	37	29,621
Four persons	723	49,138	641	51,516	52	31,417	40	31,789
Five persons	301	47,057	259	49,279	27	31,295	27	32,269
Six persons	97	45,677	79	48,241	12	13,833	13	32,070
Seven persons or more	58	43,341	42	45,934	9	28,949	13	34,427

Source: "Money Income of Households—Aggregate and Mean Income, by Race and Hispanic Origin of Householder; 1991," *Statistical Record of the United States, 1993*, p. 459. Primary source: U.S. Bureau of the Census, *Current Population Reports*, P60-180. *Notes:* 1. Includes other races not shown separately. 2. Persons of Hispanic origin may be of any race.

★ 595 ★

Income, Earnings

Money Income of Households – Median Household Income in Current and Constant (1991) Dollars, by Race and Hispanic Origin of Householder, 1970 to 1991

[See headnote, table 711. Minus sign (-) indicates decrease].

YEAR	MEDIAN INCOME IN CURRENT DOLLARS				MEDIAN INCOME IN CONSTANT (1991) DOLLARS				ANNUAL PERCENT CHANGE FOR ALL HOUSEHOLDS[3] Constant (1991) dollars
	All households[1]	White	Black	Hispanic[2]	All households[1]	White	Black	Hispanic[2]	
1970	8,734	9,097	5,537	(NA)	28,803	30,000	18,260	(NA)	2.4[4]
1975	11,800	12,340	7,408	8,865	28,597	29,906	17,953	21,484	0.1[5]
1978	15,064	15,660	9,411	11,803	30,396	31,598	18,989	23,816	3.9
1979[6]	16,461	17,259	10,133	13,042	30,297	31,766	18,650	24,004	-0.3
1980	17,710	18,684	10,764	13,651	29,309	30,921	17,814	22,591	-3.3
1981	19,074	20,153	11,309	15,300	28,833	30,464	17,095	23,128	-1.6
1982	20,171	21,117	11,968	15,178	28,737	30,085	17,051	21,624	-0.3
1983	21,018	22,035	12,473	15,794	28,741	30,132	17,056	21,598	(Z)
1984	22,415	23,647	13,471	16,992	29,383	30,998	17,659	22,274	2.2
1985	23,618	24,908	14,819	17,465	29,896	31,529	18,758	22,107	1.7
1986	24,897	26,175	15,080	18,352	30,940	32,528	18,740	22,806	3.5
1987[8]	26,061	27,458	15,672	19,336	31,246	32,921	18,790	23,183	1.0
1988	27,225	28,781	16,407	20,359	31,344	33,136	18,890	23,440	0.3
1989	28,906	30,406	18,083	21,921	31,750	33,398	19,862	24,078	1.3
1990	29,943	31,231	18,676	22,330	31,203	32,545	19,462	23,270	-1.7
1991	30,126	31,569	18,807	22,691	30,126	31,569	18,807	22,691	-3.5

Source: "Money Income of Households—Median Household Income in Current and Constant (1991) Dollars, by Race and Hispanic Origin of Householder: 1970 to 1991," *Statistical Record of the United States, 1993*, p. 457. Primary source: U.S. Bureau of the Census, *Current Population Reports*, P60-180, and unpublished data. *Notes:* NA Not available. Z Less than .05 percent. 1. Includes other races not shown separately. 2. Hispanic persons may be of any race. 3. Change from preceding year unless noted otherwise. 4. Change from 1967. 5. Change from 1970. 6. Population controls based on 1980 census. 7. Beginning 1983, data based on revised Hispanic population and not directly comparable with prior years. 8. Beginning 1987, based on revised processing procedures and not directly comparable with prior years.

★ 596 ★

Income, Earnings

Money Income of Households – Percent Distribution, by Income Level, in Constant (1991) Dollars, by Race and Hispanic Origin of Householder, 1970 to 1991

[Current dollars based on CPI-U-X1 deflator. Households as of **March** of **following year**. Based on Current Population Survey].

RACE AND HISPANIC ORIGIN OF HOUSEHOLDER YEAR	Number of households (1,000)	PERCENT DISTRIBUTION, BY INCOME LEVEL							Median income (dol.)
		Under $10,000	$10,000-$14,999	$15,000-$24,999	$25,000-$34,999	$35,000-$34,999	$50,000-$74,999	$75,000 and over	
ALL HOUSEHOLDS[1]									
1970	64,778	16.2	8.7	18.0	19.0	19.8	12.8	5.4	28,803
1980	82,368	15.4	9.4	18.3	16.1	18.8	14.7	7.2	29,309
1985[2]	88,458	15.3	9.2	17.6	15.7	17.7	15.3	9.2	29,896
1990	94,312	14.2	9.2	17.0	15.6	17.5	15.7	10.8	31,203
1991	95,669	14.9	9.4	17.4	15.2	17.3	15.4	10.4	30,126

[Continued]

★ 596 ★

Money Income of Households – Percent Distribution, by Income Level, in Constant (1991) Dollars, by Race and Hispanic Origin of Householder, 1970 to 1991

[Continued]

RACE AND HISPANIC ORIGIN OF HOUSEHOLDER YEAR	Number of households (1,000)	PERCENT DISTRIBUTION, BY INCOME LEVEL							Median income (dol.)
		Under $10,000	$10,000-$14,999	$15,000-$24,999	$25,000-$34,999	$35,000-$34,999	$50,000-$74,999	$75,000 and over	
WHITE									
1970	57,575	14.9	8.2	17.5	19.4	20.7	13.5	5.8	30,000
1980	71,872	13.6	9.0	18.1	16.3	19.6	15.6	7.9	30,921
1985[2]	76,576	13.6	8.7	17.3	16.0	18.4	16.1	9.9	31,529
1990	80,968	12.2	8.9	16.9	15.9	18.0	16.5	11.5	32,546
1991	81,675	12.8	9.1	17.3	15.4	17.9	16.3	11.2	31,569
BLACK									
1970	6,180	28.8	13.5	22.4	15.6	12.0	6.3	1.3	18,260
1980	8,847	30.7	13.4	20.2	14.0	12.1	7.6	2.1	17,814
1985[2]	9,797	29.7	12.7	20.2	13.1	12.8	8.5	2.9	18,758
1990	10,671	29.7	11.6	18.3	13.8	13.3	8.8	4.4	19,462
1991	11,083	30.8	11.6	18.2	13.8	13.4	8.4	3.7	18,807
HISPANIC[4]									
1980	3,906	19.9	13.0	22.7	16.4	15.2	9.6	3.2	22,591
1985[2]	5,213	21.6	13.5	20.2	16.2	14.6	10.0	3.9	22,107
1990	6,220	20.0	13.1	20.0	16.7	15.5	9.5	5.1	23,270
1991	6,379	20.7	12.1	21.6	15.8	14.8	10.0	5.0	22,691

Source: "Money Income of Households—Percent Distribution, by Income Level, in Constant (1991) Dollars, by Race and Hispanic Origin of Householder: 1970 to 1991," *Statistical Record of the United States, 1993,* p. 457. Primary source: U.S. Bureau of the Census, *Current Population Reports,* P60-180, and unpublished data. *Notes:* 1. Includes other races not shown separately. 2. Beginning 1983, data based on revised Hispanic population controls and not directly comparable with prior years. 3. Beginning 1987, data based on revised processing procedures and not directly for Hispanic origin households are not available prior to 1972. 4. Persons of Hispanic origin may be of any race. Income data for Hispanic origin households are not available prior to 1972.

★ 597 ★

Income, Earnings

Money Income of Households – Percent Distribution, by Income Quintile and Top 5 Percent, 1991

CHARACTERISTIC	Number (1,000)	PERCENT DISTRIBUTION					
		Lowest fifth	Second fifth	Third fifth	Fourth fifth	Highest fifth	Top 5 percent
All households	95,669	20.0	20.0	20.0	20.0	20.0	20.0
White	81,675	17.7	19.8	20.3	20.8	21.4	5.4
Black	11,083	37.2	21.8	18.1	14.0	8.9	1.4
Hispanic[1]	6,379	27.9	24.4	20.7	15.9	11.1	2.3

Source: "Money Income of Households—Percent Distribution, by Income Quintile and Top 5 Percent for Selected Characteristics: 1991," *Statistical Record of the United States, 1993,* p. 460. Primary source: U.S. Bureau of the Census, *Current Population Reports,* P60-180. *Note:* 1. Persons of Hispanic origin may be of any race.

★ 598 ★

Income, Earnings

Per Capita Money Income in Current and Constant (1991) Dollars, by Race and Hispanic Origin, 1970 to 1991

[Constant dollars based on CPI-U-X1 deflator. **In dollars**].

YEAR	CURRENT DOLLARS				CONSTANT (1991) DOLLARS			
	All races	White	Black	Hispanic[1]	All races	White	Black	Hispanic[1]
1970	3,177	3,354	1,869	(NA)	10,477	11,061	6,164	(NA)
1975	4,818	5,072	2,972	2,847	11,676	12,292	7,203	6,900
1980	7,787	8,233	4,804	4,865	12,887	13,625	7,950	8,051
1981	8,476	8,979	5,129	5,349	12,813	13,573	7,753	8,086
1982	8,980	9,527	5,360	5,448	12,794	13,573	7,636	7,762
1983[2]	9,548	10,125	5,755	5,852	13,057	13,846	7,870	8,002
1984	10,328	10,939	6,277	6,401	13,539	14,340	8,228	8,391
1985	11,013	11,671	6,840	6,613	13,940	14,773	8,658	8,371
1986	11,670	12,352	7,207	7,000	14,502	15,350	8,956	8,699
1987[3]	12,391	13,143	7,645	7,653	14,856	15,758	9,166	9,176
1988	13,123	13,896	8,271	7,956	15,109	15,999	9,522	9,160
1989	14,056	14,896	8,747	8,390	15,439	16,362	9,608	9,215
1990	14,387	15,265	9,017	8,424	14,992	15,907	9,396	8,778
1991	14,617	15,510	9,170	8,662	14,617	15,510	9,170	8,662

Source: "Per Capita Money Income in Current and Constant (1991) Dollars, by Race and Hispanic Origin: 1970 to 1991," *Statistical Record of the United States, 1993,* p. 467. Primary source: U.S. Bureau of the Census, *Current Population Reports,* P60-180. *Notes:* NA Not available. 1. Hispanic persons may be of any race. 2. Beginning 1983, data based on revised Hispanic population controls and not directly comparable with prior years. 3. Beginning 1987, data based on revised processing procedures and not directly comparable with prior years.

★ 599 ★

Income, Earnings

Per Capital Money Income, by Race for Selected Cities, 1989

[**As of April. In dollars.** Based on sample data and subject to sampling variability].

CITY RANKED BY 1990 POPULATION	All races[1]	White	Black	Asian Pacific Islander	American Indian, Eskimo, Aleut	Hispanic[2]
New York, NY	16,281	21,972	10,505	12,851	10,861	8,420
Los Angeles, CA	16,188	22,191	11,257	13,875	12,901	7,111
Chicago, IL	12,899	18,258	8,569	11,581	11,251	7,438
Houston, TX	14,261	19,817	8,366	12,250	12,239	7,021
Philadelphia, PA	12,091	15,027	9,061	8,285	10,146	6,053
San Diego, CA	16,401	19,807	10,375	10,628	11,893	8,242
Detroit, MI	9,443	11,947	8,809	8,241	7,045	7,518
Dallas, TX	16,300	22,898	8,444	12,628	10,048	7,093
Phoenix, AZ	14,096	15,497	8,734	12,643	7,076	7,213
San Antonio, TX	10,884	12,322	8,730	11,075	9,651	7,032
San Jose, CA	16,905	19,496	14,239	14,066	13,613	10,007

[Continued]

★ 599 ★

Per Capital Money Income, by Race for Selected Cities, 1989

[Continued]

CITY RANKED BY 1990 POPULATION	All races[1]	White	Black	Asian Pacific Islander	American Indian, Eskimo, Aleut	Hispanic[2]
Baltimore, MD	11,994	16,563	8,991	12,820	7,988	12,075
Indianapolis, IN	14,478	16,083	9,150	15,941	10,586	10,630
San Francisco, CA	19,695	26,222	11,829	12,665	11,485	11,400
Jacksonville, FL	13,661	15,642	8,169	13,844	11,434	10,927
Columbus, OH	13,151	14,527	9,008	10,564	10,278	10,623
Milwaukee, WI	11,106	13,645	6,833	6,357	7,448	6,253
Memphis, TN	11,682	17,569	6,982	11,761	11,110	9,393
Washington, DC	18,881	34,563	12,226	16,498	14,095	12,525
Boston, MA	15,581	18,939	10,420	9,406	9,319	8,364

Source: "Per Capita Money Income, by Race for Selected Cities: 1989," *Statistical Record of the United States, 1993*, p. 468. Primary source: U.S. Bureau of the Census, *1990 Census of Population and Housing*, Summary Tape File on 3C on CD-ROM. *Notes:* 1. Includes other races not shown separately. 2. Hispanic persons may be of any race.

★ 600 ★

Income, Earnings

Workers Paid Hourly Rates, 1992

[Annual average of monthly figures; for employed wage and salary workers. Based on Current Population Survey].

CHARACTERISTIC	NUMBER OF WORKERS (1,000)				PERCENT DISTRIBUTION At or below $4.25		PERCENT OF ALL WORKERS PAID HOURLY RATES At or below $42.5		Median hourly earnings of workers paid hourly rates[2]
	Total paid hourly rates	At or below $4.25			At $4.25	Below $4.25	At $4.25	Below $4.25	
		Total	At $4.25	Below $4.25					
White	52,452	3,872	2,269	1,603	79.2	84.5	4.3	3.1	7.87
Black	8,054	719	488	231	17.0	12.2	6.1	2.9	7.07
Hispanic origin[1]	5,984	590	454	136	15.8	7.2	7.6	2.3	6.71

Source: "Workers Paid Hourly Rates, by Selected Characteristics: 1992," *Statistical Record of the United States, 1993*, p. 429. Primary source: U.S. Bureau of Labor Statistics, unpublished data. *Note:* 1. Persons of Hispanic origin may be of any race.

Poverty and its Correlates

★ 601 ★

Persons Below Poverty Level, 1960 to 1991

[Persons as of **March of the following year**].

YEAR	NUMBER BELOW POVERTY LEVEL (mil.)				PERCENT BELOW POVERTY LEVEL			
	All races[1]	White	Black	Hispanic[2]	All races[1]	White	Black	Hispanic[2]
1960	39.9	28.3	(NA)	(NA)	22.2	17.8	(NA)	(NA)
1966	28.5	20.8	8.9	(NA)	14.7	12.2	41.8	(NA)
1970	25.4	17.5	7.5	(NA)	12.6	9.9	33.5	(NA)
1975	25.9	17.8	7.5	3.0	12.3	9.7	31.3	26.9
1976	25.0	16.7	7.6	2.8	11.8	9.1	31.1	24.7
1977	24.7	16.4	7.7	2.7	11.6	8.9	31.3	22.4
1978	24.5	16.3	7.6	2.6	11.4	8.7	30.6	21.6
1979[3]	26.1	17.2	8.1	2.9	11.7	9.0	31.0	21.8
1980	29.3	19.7	8.6	3.5	13.0	10.2	32.5	25.7
1981	31.8	21.6	9.2	3.7	14.0	11.1	34.2	26.5
1982	34.4	23.5	9.7	4.3	15.0	12.0	35.6	29.9
1983[4]	35.3	24.0	9.9	4.6	15.2	12.1	35.7	28.0
1984	33.7	23.0	9.5	4.8	14.4	11.5	33.8	28.4
1985	33.1	22.9	8.9	5.2	14.0	11.4	31.3	29.0
1986	32.4	22.2	9.0	5.1	13.6	11.0	31.1	27.3
1987[5]	32.2	21.2	9.5	5.4	13.4	10.4	32.4	28.0
1988	31.7	20.7	9.4	5.4	13.0	10.1	31.3	26.7
1989	31.5	20.8	9.3	5.4	12.8	10.0	30.7	26.2
1990	33.6	22.3	9.8	6.0	13.5	10.7	31.9	28.1
1991	35.7	23.7	10.2	6.3	14.2	11.3	32.7	28.7

Source: "Persons Below Poverty Level and Below 125 Percent of Poverty Level: 1959 to 1991," *Statistical Record of the United States, 1993*, p. 469. Primary source: U.S. Bureau of the Census, *Current Population Reports*, P60-181. *Notes:* NA Not available. 1. Includes other races not shown separately. 2. Persons of Hispanic origin may be of any race. 3. Population controls based on 1980 census. 4. Beginning 1983, data based on revised Hispanic population controls and not directly comparable with prior years. 5. Beginning 1987, data based on revised processing procedures and not directly comparable with prior years.

★ 602 ★

Poverty and its Correlates

Persons Below Poverty Level, by Definition of Income, 1991

[Persons as of **March 1992**].

DEFINITION OF NUMBER	DEFINITION OF INCOME	NUMBER BELOW POVERTY LEVEL (1,000)				PERCENT BELOW POVERTY LEVEL			
				All races[1]	White	Black	Hispanic[2]	All races[1]	White
	All persons	251,179	210,121	31,312	22,068	(X)	(X)	(X)	(X)
		INCOME BEFORE TAXES							
1	Money income excluding capital gains (current) measure[3]	35,708	23,747	10,242	6,339	14.2	11.3	32.7	28.7
2	Definition 1 less government	54,803	39,808	12,759	7,809	21.8	18.9	40.7	35.4

[Continued]

★ 602 ★

Persons Below Poverty Level, by Definition of Income, 1991
[Continued]

DEFINITION OF NUMBER	DEFINITION OF INCOME	NUMBER BELOW POVERTY LEVEL (1,000)		All races[1]	White	PERCENT BELOW POVERTY LEVEL			
						Black	Hispanic[2]	All races[1]	White
3	money transfers Definition 2 plus capital gains	54,644	39,676	12,730	7,794	21.8	18.9	40.7	35.3
4	Definition 3 plus health insurance supplements to wage or salary income[4]	53,087	38,596	12,314	7,482	21.1	18.4	39.3	33.9
	INCOME AFTER TAXES								
5	Definition 4 less Social Security payroll taxes	55,198	40,223	12,709	7,873	22.0	19.1	40.6	35.7
6	Definition 5 less Federal income taxes (excluding EITC)[5]	55,939	40,803	12,835	8,037	22.3	19.4	41.0	36.4
7	Definition 6 plus EITC[5]	54,361	39,649	12,475	7,662	21.6	18.9	39.8	34.7
8	Definition 7 less State income taxes	54,697	39,918	12,517	7,683	21.8	19.0	40.0	34.8
9	Definition 8 plus nonmeans-tested government cash transfers[6]	38,037	25,432	10,744	6,731	15.1	12.1	34.3	30.5
11	Definition 9 plus nonmeans-tested government noncash transfers	36,751	24,562	10,372	6,553	14.6	11.7	33.1	29.7
12	Definition 11 plus means-tested government cash transfers	34,188	22,843	9,697	6,080	13.6	10.9	31.0	27.6
14	Definition 12 plus means-tested government noncash transfers	28,545	19,545	7,653	5,048	11.4	9.3	24.4	22.9
15	Definition 14 plus net imputed return on equity in own home	25,772	17,382	7,141	4,791	10.3	8.3	22.8	21.7

Source: "Persons Below Poverty Level, by Definition of Income: 1991," *Statistical Record of the United States, 1993*, p. 475. Primary source: U.S. Bureau of the Census, *Current Population Reports*, P60-182 and unpublished data. *Notes:* X Not applicable. 1. Includes other races not shown separately. 2. Persons of Hispanic origin may be of any race. 3. Official definition of income based on money income before taxes and includes government cash transfers. 4. Employer contributions to the health insurance plans of employees. 5. Earned Income Tax Credit. 6. Includes Social Security and Railroad Retirement, veterans payments, unemployment and workers' compensation, Black Lung payments, Pell Grants, and other government educational assistance. 7. Includes Medicare and subsidies from regular price school lunches. 8. Includes AFDC and other public assistance or welfare payments, Supplements Security Income, and veterans payments. Households must meet certain eligibility requirements in order to quality for those benefits. 9. Includes Medicaid, food stamps, subsidies from free to reduced-price school lunches, and rent subsidies. 10. Estimated amount of income a household would receive if it chose to shift amount held as home equity into an interest bearing account.

★ 603 ★

Poverty and its Correlates

Persons Below Poverty Level, by Race, Hispanic Origin, and Region, 1991

[Persons as of **March 1992**].

AGE AND REGION	NUMBER BELOW POVERTY LEVEL (1,000)				PERCENT BELOW POVERTY LEVEL			
	All races[1]	White	Black	Hispanic[2]	All races[1]	White	Black	Hispanic[2]
Total	35,708	23,747	10,242	6,339	14.2	11.3	3.27	28.7
Under 16 years old	13,178	8,135	4,375	2,840	22.2	17.2	47.0	41.1
16 to 21 years old	3,605	2,359	1,040	719	17.9	14.5	34.0	31.5
22 to 44 years old	10,838	7,399	2,878	1,987	11.7	9.5	25.3	22.9
45 to 54 years old	2,167	1,541	544	328	8.0	6.6	19.5	18.1
55 to 64 years old	1,019	716	246	113	9.6	7.9	21.3	17.0
60 to 64 years old	1,120	796	278	115	10.6	8.7	27.5	19.6
65 years old and over	3,781	2,802	880	237	12.4	10.3	33.8	20.8
Northeast	6,177	4,389	1,546	1,217	12.2	10.0	28.1	36.3
Midwest	7,989	5,242	2,380	346	13.2	9.9	37.7	23.5
South	13,783	7,837	5,716	1,765	16.0	11.7	33.6	26.1
West	7,759	6,279	600	3,012	14.3	13.5	24.0	28.8

Source: "Persons Below Poverty Level, by Race, Hispanic Origin, and Regions; 1991," *Statistical Record of the United States, 1993*, p. 470. Primary source: U.S. Bureau of the Census, *Current Population Reports*, P60-181 and unpublished data. *Notes:* 1. Includes other races not shown separately. 2. Persons of Hispanic origin may be of any race.

★ 604 ★

Poverty and its Correlates

Persons Below Poverty Level – Alternative Inflation Adjustment, 1974 to 1991

[Based on Current Population Survey. Annual adjustment for cost-of- living changes is based on the CPI-U-X1].

YEAR	NUMBER BELOW POVERTY LEVEL (1,000)				PERCENT BELOW POVERTY LEVEL			
	All races[1]	White	Black	Hispanic[2]	All races[1]	White	Black	Hispanic[2]
1974	22,076	14,870	6,773	2,448	10.5	8.2	28.6	21.9
1975	24,232	16,547	7,170	2,787	11.5	9.0	29.8	25.1
1976	23,347	15,513	7,202	2,570	11.0	8.4	29.5	22.8
1977	22,933	15,190	7,230	2,480	10.7	8.2	29.3	20.6
1978	22,472	14,829	7,085	2,416	10.4	8.0	28.4	20.0
1979	23,504	15,382	7,388	2,614	10.5	8.0	28.5	19.5
1980	25,869	17,283	7,671	3,134	11.5	9.0	29.0	23.0
1981	27,731	18,456	8,311	3,302	12.2	9.5	31.0	23.6
1982	30,288	20,385	8,824	3,842	13.2	10.4	32.4	26.7
1983	31,649	21,180	9,130	4,215	13.7	10.7	33.0	25.5
1984	29,971	20,043	8,765	4,367	12.8	10.1	31.2	25.8
1985	29,558	20,157	8,284	4,712	12.5	10.0	29.1	26.1
1986	29,101	19,629	8,391	4,570	12.2	9.7	29.1	24.4
1987	28,890	18,777	8,744	4,899	12.0	9.2	29.8	25.3
1988	28,544	18,326	8,707	4,914	11.7	8.9	29.2	24.5
1989	27,967	18,152	8,504	4,827	11.4	8.8	28.0	23.3
1990	30,097	19,677	9,145	5,401	12.1	9.4	29.7	25.2
1991	32,009	21,027	9,421	5,695	12.7	10.0	30.1	25.8

Source: "Persons Below Poverty Level—Alternative Inflation Adjustment: 1974 to 1991," *Statistical Record of the United States, 1993*, p. 474. Primary source: U.S. Bureau of the Census, *Current Population Reports*, P60-182 and unpublished data. *Notes:* 1. Includes other races not shown separately. 2. Persons of Hispanic origin may be of any race.

★ 605 ★

Poverty and its Correlates

Characteristics of the Population Below the Poverty Line, 1991, North and West, Part 1

Numbers in thousands. Persons and families as of March 1992.

Characteristic	Black			White		
	Total	Below poverty level		Total	Below poverty level	
		Number	Percent		Number	Percent
North and west						
Age by sex						
Total persons	14,308	4,526	31.6	142,899	15,911	11.1
Under 18 years	4,723	2,174	46.0	36,376	6,173	17.0
18 to 64 years	8,465	2,045	24.2	88,334	7,990	9.0
55 years and over	2,113	496	23.5	30,370	2,681	8.8
65 years and over	1,120	307	27.4	18,189	1,748	9.6
Male	6,728	1,837	27.3	70,116	6,783	9.7

[Continued]

★ 605 ★

Characteristics of the Population Below the Poverty Line, 1991, North and West, Part 1

[Continued]

Characteristic	Black			White		
	Total	Below poverty level		Total	Below poverty level	
		Number	Percent		Number	Percent
Under 18 years	2,376	1,069	45.0	18,676	3,142	16.8
18 to 64 years	3,896	671	17.2	43,806	3,187	7.3
55 years and over	910	145	16.0	13,460	794	5.9
65 years and over	457	97	21.2	7,635	454	5.9
Female	7,579	2,689	35.5	72,783	9,128	12.5
Under 18 years	2,347	1,105	47.1	17,701	3,030	17.1
18 to 64 years	4,568	1,374	30.1	44,528	4,803	10.8
55 years and over	1,203	351	29.2	16,910	1,887	11.2
65 years and over	664	210	31.6	10,554	1,295	12.3

Source: "Characteristics of the Population Below the Poverty Line in 1991, by Region and Race," U.S. Department of Commerce, Bureau of the Census, *The Black Population in the United States-March 1992,* Current Population Reports, P20-471, p. 70.

★ 606 ★

Poverty and its Correlates

Characteristics of the Population Below the Poverty Line, 1991, North and West, Part 2

Numbers in thousands. Persons and families as of March 1992.

Characteristic	Black			White		
	Total	Below poverty level		Total	Below poverty level	
		Number	Percent		Number	Percent
North and west						
Age by sex						
Total persons[1]	14,308	4,526	31.6	142,899	15,911	11.1
In families	11,911	3,727	31.3	120,060	11,587	9.7
Householder	3,463	1,028	29.7	38,203	3,332	8.7
Related children under 18 years	4,618	2,099	45.5	35,725	5,786	16.2
Other family members	3,830	601	15.7	46,132	2,469	5.4
Unrelated individuals	2,223	683	30.7	21,863	3,858	17.6
Metropolitan-nonmetropolitan residence						
Total persons	14,308	4,526	31.6	142,899	15,911	11.1
All metropolitan areas	14,009	4,454	31.8	112,890	12,131	10.7
Inside central cities	10,134	3,706	36.6	38,340	6,141	16.0
Outside central cities	3,876	748	19.3	74,551	5,990	8.0
Nonmetropolitan areas	298	72	24.2	30,009	3,779	12.6

Source: "Characteristics of the Population Below the Poverty Line in 1991, by Region and Race," U.S. Department of Commerce, Bureau of the Census, *The Black Population in the United States-March 1992,* Current Population Reports, P20-471, p. 70. *Notes:* 1. Families and unrelated individuals will not add to total persons because unrelated subfamilies are not included.

★ 607 ★

Poverty and its Correlates

Characteristics of the Population Below the Poverty Line, 1991, North and West, Part 3

Numbers in thousands. Persons and families as of March 1992.

Characteristic	Black			White		
	Total	Below poverty level		Total	Below poverty level	
		Number	Percent		Number	Percent
North and west						
Work experience in 1991						
Both sexes, 15 years and over	10,217	2,624	25.7	111,699	10,479	9.4
Worked	6,192	746	12.0	78,250	4,418	5.6
50 to 52 weeks	4,095	187	4.6	53,796	1,598	3.0
49 weeks or less	2,098	559	26.7	24,453	2,820	11.5
Duration of unemployment						
1 to 4 weeks	164	51	31.0	1,861	210	11.3
5 to 14 weeks	328	73	22.2	3,766	363	9.6
15 to 26 weeks	274	46	16.8	2,878	431	15.0
27 weeks or more	288	117	40.7	1,922	500	26.0
Did not work	4,025	1,878	46.7	33,449	6,061	18.1

Source: "Characteristics of the Population Below the Poverty Line in 1991, by Region and Race," U.S. Department of Commerce, Bureau of the Census, *The Black Population in the United States-March 1992*, Current Population Reports, P20-471, p. 71.

★ 608 ★

Poverty and its Correlates

Characteristics of the Population Below the Poverty Line, 1991, North and West, Part 4

Numbers in thousands. Persons and families as of March 1992.

Characteristic	Black			White		
	Total	Below poverty level		Total	Below poverty level	
		Number	Percent		Number	Percent
North and west						
Work experience in 1991						
Males, 15 years and over	4,628	899	19.4	53,910	3,969	7.4
Worked	3,031	288	9.5	42,343	2,219	5.2
50 to 52 weeks	1,973	53	2.7	30,453	874	2.9
49 weeks or less	1,059	235	22.2	11,890	1,344	11.3
Duration of unemployment						
1 to 4 weeks	86	19	22.2	942	92	9.8
5 to 14 weeks	196	24	12.1	2,301	167	7.3
15 to 26 weeks	183	33	18.2	1,883	264	14.0
27 weeks or more	174	69	39.8	1,253	337	26.9
Did not work	1,597	612	38.3	11,567	1,751	15.1
Females, 15 years and over	5,589	1,725	30.9	57,789	6,509	11.3

[Continued]

★ 608 ★

Characteristics of the Population Below the Poverty Line, 1991, North and West, Part 4

[Continued]

Characteristic	Black			White		
	Total	Below poverty level		Total	Below poverty level	
		Number	Percent		Number	Percent
Worked	3,161	458	14.5	35,907	2,199	6.1
50 to 52 weeks	2,122	134	6.3	23,343	723	3.1
49 weeks or less	1,039	324	31.2	12,564	1,476	11.7
Duration of unemployment						
1 to 4 weeks	78	32	40.6	919	118	12.8
5 to 14 weeks	132	49	37.3	1,465	196	13.3
15 to 26 weeks	90	13	13.8	995	167	16.8
27 weeks or more	113	48	42.2	670	162	24.3
Did not work	2,428	1,267	52.2	21,882	4,310	19.7

Source: "Characteristics of the Population Below the Poverty Line in 1991, by Region and Race," U.S. Department of Commerce, Bureau of the Census, *The Black Population in the United States-March 1992*, Current Population Reports, P20-471, p. 71.

★ 609 ★

Poverty and its Correlates

Characteristics of the Population Below the Poverty Line, 1991, South, Part 1

Numbers in thousands. Persons and families as of March 1992.

Characteristic	Black			White		
	Total	Below poverty level		Total	Below poverty level	
		Number	Percent		Number	Percent
South						
Age by sex						
Total persons	17,006	5,716	33.6	67,234	7,837	11.7
Under 18 years	5,628	2,581	45.9	16,147	2,675	16.6
18 to 64 years	9,892	2,562	25.9	41,978	4,108	9.8
55 years and over	2,660	908	34.1	15,207	1,632	10.7
65 years and over	1,486	573	38.6	9,109	1,054	11.6
Male	8,003	2,360	29.5	32,791	3,296	10.1
Under 18 years	2,899	1,313	45.3	8,262	1,314	15.9
18 to 64 years	4,503	873	19.4	20,733	1,743	8.4
55 years and over	1,126	308	27.4	6,701	462	6.9
65 years and over	601	174	29.0	3,796	239	6.3
Female	9,002	3,356	37.3	34,443	4,540	13.2
Under 18 years	2,729	1,268	46.5	7,886	1,361	17.3
18 to 64 years	5,389	1,689	31.3	21,245	2,364	11.1

[Continued]

★ 609 ★

Characteristics of the Population Below the Poverty Line, 1991, South, Part 1

[Continued]

Characteristic	Black			White		
	Total	Below poverty level		Total	Below poverty level	
		Number	Percent		Number	Percent
55 years and over	1,534	599	39.1	8,505	1,170	13.8
65 years and over	885	399	45.1	5,312	815	15.3

Source: "Characteristics of the Population Below the Poverty Line in 1991, by Region and Race," U.S. Department of Commerce, Bureau of the Census, *The Black Population in the United States-March 1992*, Current Population Reports, P20-471, p. 69.

★ 610 ★

Poverty and its Correlates

Characteristics of the Population Below the Poverty Line, 1991, South, Part 2

Numbers in thousands. Persons and families as of March 1992.

Characteristic	Black			White		
	Total	Below poverty level		Total	Below poverty level	
		Number	Percent		Number	Percent
South						
Age by sex						
Total persons	17,006	5,716	33.6	67,234	7,737	11.7
In families	14,654	4,777	32.6	57,559	5,681	9.9
Householder	4,253	1,315	30.9	19,023	1,690	8.9
Related children under 18 years	5,560	2,538	45.6	15,902	2,530	15.9
Other family members	4,841	924	19.1	22,634	1,461	6.5
Unrelated individuals	2,281	907	39.7	9,344	2,014	21.5
Metropolitan-nonmetropolitan residence						
Total persons	17,006	5,716	33.6	67,234	7,837	11.7
All metropolitan areas	12,522	3,927	31.4	48,148	4,944	10.3
Inside central cities	7,316	2,457	33.6	16,198	2,236	13.8
Outside central cities	5,206	1,469	28.2	31,950	2,708	8.5
Nonmetropolitan areas	4,484	1,789	39.9	19,086	2,892	15.2
Work experience in 1991						
Both sexes, 15 years and over	12,172	3,467	28.5	53,194	5,525	10.4
Worked	7,764	1,283	16.5	36,218	2,359	6.5
50 to 52 weeks	4,993	427	8.5	25,457	808	3.2
49 weeks or less	2,771	856	30.9	10,761	1,552	14.4
Duration of employment						
1 to 4 weeks	204	38	18.7	858	105	12.2
5 to 14 weeks	419	84	20.0	1,528	212	13.8
15 to 26 weeks	419	158	37.7	1,211	254	21.0

[Continued]

★ 610 ★

Characteristics of the Population Below the Poverty Line, 1991, South, Part 2
[Continued]

Characteristic	Black			White		
	Total	Below poverty level		Total	Below poverty level	
		Number	Percent		Number	Percent
27 weeks or more	356	165	46.3	822	235	28.6
Did not work	4,408	2,184	49.5	16,977	3,166	18.6

Source: "Characteristics of the Population Below the Poverty Line in 1991, by Region and Race," U.S. Department of Commerce, Bureau of the Census, *The Black Population in the United States-March 1992*, Current Population Reports, P20-471, p. 69.

★ 611 ★

Poverty and its Correlates

Characteristics of the Population Below the Poverty Line, 1991, South, Part 3

Numbers in thousands. Persons and families as of March 1992.

Characteristic	Black			White		
	Total	Below poverty level		Total	Below poverty level	
		Number	Percent		Number	Percent
South						
Work experience in 1991						
Males, 15 years and over	5,477	1,185	21.6	25,510	2,147	8.4
Worked	3,784	532	14.1	19,530	1,236	6.3
50 to 52 weeks	2,408	171	7.1	14,340	451	3.1
49 weeks or less	1,376	361	26.2	5,190	785	15.1
Duration of employment						
1 to 4 weeks	91	10	10.6	409	59	14.5
5 to 14 weeks	223	28	12.4	944	133	14.1
15 to 26 weeks	261	91	34.8	774	162	21.0
27 weeks or more	208	77	37.3	537	155	28.9
Did not work	1,693	652	38.5	5,980	910	15.2
Females, 15 years and over	6,695	2,282	34.1	27,684	3,378	12.2
Worked	3,980	751	18.9	16,688	1,123	6.7
50 to 52 weeks	2,584	255	9.9	11,117	356	3.2
49 weeks or less	1,395	495	35.5	5,571	767	13.8
Duration of employment						
1 to 4 weeks	113	29	25.3	449	46	10.2
5 to 14 weeks	197	56	28.7	584	79	13.5
15 to 26 weeks	158	67	42.7	437	91	20.9
27 weeks or more	148	87	58.9	285	80	28.1
Did not work	2,716	1,531	56.4	10,997	2,255	20.5

Source: "Characteristics of the Population Below the Poverty Line in 1991, by Region and Race," U.S. Department of Commerce, Bureau of the Census, *The Black Population in the United States-March 1992*, Current Population Reports, P20-471, p. 70.

★ 612 ★

Poverty and its Correlates

Characteristics of the Population Below the Poverty Line, 1991, United States, Part 1

Numbers in thousands. Persons and families as of March 1992.

Characteristic	Black			White		
	Total	Below poverty level		Total	Below poverty level	
		Number	Percent		Number	Percent
United States						
Age by sex						
Total persons	31,313	10,242	32.7	210,133	23,747	11.3
Under 18 years	10,350	4,755	45.9	52,523	8,848	16.8
16 to 64 years	18,356	4,607	25.1	130,312	12,097	9.3
55 years and over	4,772	1,404	29.4	45,577	4,314	9.5
65 years and over	2,606	880	33.8	27,297	2,802	10.3
Male	14,731	4,197	28.5	102,907	10,079	9.8
Under 18 years	5,275	2,382	45.2	26,937	4,456	16.5
18 to 64 years	8,399	1,544	18.4	64,539	4,930	7.6
55 years and over	2,035	453	22.3	20,162	1,257	6.2
65 years and over	1,058	271	25.6	11,431	693	6.1
Female	16,582	6,044	36.5	107,226	13,668	12.7
Under 18 years	5,076	2,373	46.7	25,586	4,392	17.2
18 to 64 years	9,957	3,062	30.8	65,774	7,167	10.9
55 years and over	2,737	951	34.7	25,415	3,057	12.0
65 years and over	1,549	609	39.3	15,866	2,109	13.3

Source: "Characteristics of the Population Below the Poverty Line in 1991, by Region and Race," U.S. Department of Commerce, Bureau of the Census, *The Black Population in the United States-March 1992*, Current Population Reports, P20-471, p. 68.

★ 613 ★

Poverty and its Correlates

Characteristics of the Population Below the Poverty Line, 1991, United States, Part 2

Numbers in thousands. Persons and families as of March 1992.

Characteristic	Black			White		
	Total	Below poverty level		Total	Below poverty level	
		Number	Percent		Number	Percent
United States						
Family status						
Total persons	31,313	10,242	32.7	210,133	23,747	11.3
In families	26,565	8,504	32.0	177,619	17,268	9.7
Householder	7,716	2,343	30.4	57,225	5,022	8.8

[Continued]

★ 613 ★

Characteristics of the Population Below the Poverty Line, 1991, United States, Part 2

[Continued]

Characteristic	Black			White		
	Total	Below poverty level		Total	Below poverty level	
		Number	Percent		Number	Percent
Related children under 18 years	10,178	4,637	45.6	51,627	8,316	16.1
Other family members	8,671	1,525	17.6	68,767	3,930	5.7
Unrelated individuals	4,505	1,590	35.3	31,207	5,872	18.8
Metropolitan-nonmetropolitan residence						
Total persons	31,313	10,242	32.7	210,133	23,747	11.3
All metropolitan areas	26,531	8,380	31.6	161,038	17,076	10.6
Inside central cities	17,449	6,163	35.3	54,537	8,378	15.4
Outside central cities	9,082	2,217	24.4	106,501	8,698	8.2
Nonmetropolitan areas	4,782	1,861	38.9	49,095	6,672	13.6

Source: "Characteristics of the Population Below the Poverty Line in 1991, by Region and Race," U.S. Department of Commerce, Bureau of the Census, *The Black Population in the United States-March 1992*, Current Population Reports, P20-471, p. 68.

★ 614 ★

Poverty and its Correlates

Characteristics of the Population Below the Poverty Line, 1991, United States, Part 3

Numbers in thousands. Persons and families as of March 1992.

Characteristic	Black			White		
	Total	Below poverty level		Total	Below poverty level	
		Number	Percent		Number	Percent
United States						
Work experience in 1991						
Both sexes, 15 years and over	22,389	6,091	27.2	164,894	16,003	9.7
Worked	13,956	2,029	14.5	114,467	6,777	5.9
50 to 52 weeks	9,088	613	6.7	79,253	2,405	3.0
49 weeks or less	4,869	1,415	29.1	35,214	4,371	12.4
Duration of employment						
1 to 4 weeks	368	89	24.2	2,719	315	11.6
5 to 14 weeks	747	157	21.0	5,294	575	10.9
15 to 26 weeks	693	204	29.5	4,089	685	16.7
27 weeks or more	643	282	43.8	2,745	735	26.8
Did not work	8,433	4,062	48.2	50,426	9,226	18.3
Males, 15 years and over	10,105	2,084	20.6	79,420	6,116	7.7
Worked	6,815	820	12.0	61,873	3,455	5.6
50 to 52 weeks	4,381	224	5.1	44,793	1,326	3.0
49 weeks or less	2,434	596	24.5	17,080	2,129	12.5
Duration of employment						
1 to 4 weeks	177	29	16.2	1,351	152	11.2

[Continued]

★ 614 ★

Characteristics of the Population Below the Poverty Line, 1991, United States, Part 3

[Continued]

Characteristic	Black			White		
	Total	Below poverty level		Total	Below poverty level	
		Number	Percent		Number	Percent
5 to 14 weeks	419	51	12.2	3,246	300	9.3
15 to 26 weeks	444	124	27.9	2,657	426	16.0
27 weeks or more	382	147	38.4	1,790	493	27.5
Did not work	3,290	1,264	38.4	17,547	2,661	15.2

Source: "Characteristics of the Population Below the Poverty Line in 1991, by Region and Race," U.S. Department of Commerce, Bureau of the Census, *The Black Population in the United States-March 1992*, Current Population Reports, P20-471, p. 68.

★ 615 ★

Poverty and its Correlates

Characteristics of the Population Below the Poverty Line, 1991, United States, Part 4

Numbers in thousands. Persons and families as of March 1992.

Characteristic	Black			White		
	Total	Below poverty level		Total	Below poverty level	
		Number	Percent		Number	Percent
United States						
Work experience in 1991						
Females, 15 years and over	12,284	4,007	32.6	85,473	9,887	11.6
Worked	7,141	1,209	16.9	52,594	3,322	6.3
50 to 52 weeks	4,706	389	8.3	34,460	1,080	3.1
49 weeks or less	2,434	819	33.7	18,135	2,242	12.4
Duration of employment						
1 to 4 weeks	191	60	31.6	1,368	164	12.0
5 to 14 weeks	328	106	32.1	2,049	274	13.4
15 to 26 weeks	249	80	32.2	1,432	259	18.1
27 weeks or more	261	135	51.6	955	243	25.4
Did not work	5,144	2,798	54.4	32,879	6,565	20.0

Source: "Characteristics of the Population Below the Poverty Line in 1991, by Region and Race," U.S. Department of Commerce, Bureau of the Census, *The Black Population in the United States-March 1992*, Current Population Reports, P20-471, p. 69.

★ 616 ★

Poverty and its Correlates

Persons 65 Years Old and Older Below Poverty Level, by Selected Characteristics, 1970 to 1991

[Persons as of **March of following year**].

CHARACTERISTIC	NUMBER BELOW POVERTY LEVEL				PERCENT BELOW POVERTY LEVEL			
	1970	1985[1]	1990[2]	1991	1970	1985[1]	1990[2]	1991
Persons, 65 yr. and over[3]	4,793	3,456	3,658	3,781	24.6	12.6	12.2	12.4
White	4,011	2,698	2,707	2,802	22.6	11.0	10.1	10.3
Black	735	717	860	880	47.7	31.5	33.8	33.8
Hispanic[4]	(NA)	219	245	237	(NA)	23.9	22.5	20.8
In families	2,013	1,173	1,172	1,228	14.8	6.4	5.8	6.0
Unrelated individuals	2,779	2,281	2,479	2,553	47.2	25.6	24.7	24.9
Persons, 60 yr. and over	5,977	4,677	4,756	4,901	21.3	12.3	11.7	11.9

Source: "Persons 65 Years Old and Older Below Poverty Level, by Selected Characteristics: 1970 to 1991," *Statistical Record of the United States, 1993,* p. 470. Primary source: U.S. Bureau of the Census, *Current Population Reports,* P60-181. *Notes:* NA Not available. 1. Beginning 1983, data based on revised Hispanic population controls and not directly comparable with prior years. 2. Beginning 1987, data based on revised processing procedures are not directly comparable with prior years. 3. Beginning 1979, includes members of unrelated subfamilies not shown separately. For earlier years, unrelated subfamily members are included in the "In families" category. 4. Persons of Hispanic origin may be of any race.

Wealth

★ 617 ★

Family Net Worth, 1983 and 1989

[**Mean and median value in thousands of constant 1989 dollars**. Constant dollar figures are based on consumer price index data published by U.S. Bureau of Labor Statistics].

CHARACTERISTIC	1983			1989		
	Percent of families	Net worth		Percent of families	Net worth	
		Mean	Median		Mean	Median
All families	100	149.1	42.7	100	183.7	47.2
White	82	173.0	54.3	87	203.9	58.5
Nonwhite and Hispanic	18	37.6	6.9	13	45.9	4.0

Source: "Family Net Worth—Mean and Median of Net Worth, by Selected Characteristics: 1983 and 1989," *Statistical Record of the United States, 1993,* p. 477. Primary source: Board of Governors of the Federal Reserve System, *Federal Reserve Bulletin,* January 1992.

★ 618 ★

Wealth

Family Net Worth – Percent of Families Owning Selected Nonfinancial Assets, 1983 and 1989

[Families include one-persons units. Based on Survey of Consumer Finance].

CHARACTERISTIC	Total	Vehicles	Principal residence	Investment real estate	Business	Other assets
1983						
All families	90.3	84.4	64.4	20.9	14.2	7.4
White	94.3	88.7	68.0	23.1	16.1	8.5
Non-White and Hispanic	71.3	64.4	42.2	10.9	5.4	2.5
1989						
All families	90.2	84.0	64.7	20.4	11.5	22.1
White	93.2	87.9	67.9	21.9	12.6	23.3
Non-White and Hispanic	70.0	56.8	42.8	10.5	4.4	13.2

Source: "Family Net Worth—Percent of Families Owning Selected Nonfinancial Assets: 1983 and 1989," *Statistical Record of the United States, 1993*, p. 476. Primary source: Board of Governors of the Federal Reserve System, *Federal Reserve Bulletin*, January 1992.

★ 619 ★

Wealth

Net Change in Assets and Liabilities and Rental Value of Owned Housing, 1991

Item	All consumer units	Race of reference person	
		White and other	Black
Net change in total assets and liabilities	$478	$499	$308
Net change in total assets	3,005	3,355	167
Net change in total liabilities	2,526	2,856	-141
Other financial information:			
Mortgage principal paid on owned property	-578	-618	-255
Estimated market value of owned home	67,519	72,319	28,654
Estimated monthly rental value of owned home	413	438	212

Source: "Housing Tenure, Race of Reference Person, and Type of Area: Average Annual Expenditures and Characteristics, Consumer Expenditure Survey, 1991," U.S. Department of Labor, Bureau of Labor Statistics, *Consumer Expenditure Survey, 1990-91*, Bulletin 2425 (September 1993), p. 41.

Chapter 10
LABOR AND EMPLOYMENT

Employment and Education

★ 620 ★

Educational Attainment of Civilian Noninstitutional Population, 1992

In thousands, except rate. Annual averages of monthly figures, for civilian noninstitutional population 25 years old and over. Based on Current population Survey.

Employment status, sex, and race	Population 25 years and over	Educational attainment			
		Less than a high school diploma	High school graduates, no college	Less than a bachelor's degree	College graduates
Civilian noninstitutional population, total[1]	160,646	32,397	57,420	36,050	34,780
White					
Civilian noninstitutional population	137,617	25,858	49,699	31,243	30,817
Civilian labor force	91,242	10,543	32,488	23,273	24,938
Participation rate[2]	66.3	40.8	65.4	74.5	80.9
Employed	86,260	9,417	30,552	22,104	24,187
Unemployed	4,982	1,126	1,936	1,168	751
Unemployment rate	5.5	10.7	6.0	5.0	3.0
Black					
Civilian noninstitutional population	17,391	5,317	6,234	3,733	2,106
Civilian labor force	11,422	2,110	4,505	3,004	1,803
Participation rate[2]	65.7	39.7	72.3	80.5	85.6
Employed	10,175	1,792	3,950	2,709	1,724
Unemployed	1,246	318	555	294	79
Unemployment rate	10.9	15.1	12.3	9.8	4.4
Hispanic origin[3]					
Civilian noninstitutional population	11,774	5,592	3,136	1,939	1,107
Civilian labor force	7,993	3,124	2,358	1,592	919
Participation rate[2]	67.9	55.9	75.2	82.1	83.0
Employed	7,212	2,724	2,145	1,469	873

[Continued]

★ 620 ★

Educational Attainment of Civilian Noninstitutional Population, 1992

[Continued]

Employment status, sex, and race	Population 25 years and over	Educational attainment			
		Less than a high school diploma	High school graduates, no college	Less than a bachelor's degree	College graduates
Unemployed	781	400	213	122	46
Unemployment rate	9.8	12.8	9.0	7.7	5.0

Source: "Employment Status of the Civilian Noninstitutional Population, by Educational Attainment, Sex, Race, and Hispanic Origin: 1992," U.S. Bureau of the Census, *Statistical Abstract of the United States*, 1993, p. 398. Primary source: U.S. Bureau of Labor Statistics, unpublished data. *Notes:* 1. Includes other races, not shown separately. 2. For definition, see footnote 3 table 626. 3. Persons of Hispanic origin may be of any race.

★ 621 ★

Employment and Education

Educational Attainment of Civilian labor Force: Trends, 1970-1991

As of March, except as noted. For civilian noninstitutional population 25 to 64 years of age.

Item	Total (1,000)	Civilian labor force				Participation rate[1]				
		Percent distribution				Total	Less than high school	High school graduate	College	
		Less than high school	High school graduate	College					1-3 years	4 years or more
				1-3 years	4 years or more					
Total[2]										
1970	61,765	36.1	38.1	11.8	14.1	70.3	65.5	70.2	73.8	82.3
1980	78,010	20.6	39.8	17.6	22.0	73.9	60.7	74.2	79.5	86.1
1985	88,424	15.9	40.2	19.0	24.9	76.2	59.9	75.9	81.6	87.7
1990[3]	99,981	13.3	39.4	20.8	26.5	78.9	61.4	78.4	83.5	88.6
1991[3]	101,171	12.8	39.2	21.3	26.7	78.8	61.0	78.2	83.4	88.3
White										
1970	55,044	33.7	39.3	12.2	14.8	70.1	65.2	69.7	73.3	81.9
1980	68,509	19.1	40.2	17.7	22.9	74.2	61.4	73.7	79.2	86.0
1985	76,739	14.7	40.7	19.1	25.6	76.6	60.7	75.8	81.1	87.7
1990[3]	85,882	12.5	39.4	20.8	27.3	79.5	63.0	78.6	83.4	88.6
1991[3]	86,776	12.1	39.1	21.3	27.5	79.5	62.6	78.5	83.4	88.5
Black										
1970	6,721	55.5	28.2	8.0	8.3	72.0	67.1	76.8	81.0	87.4
1980	7,731	34.7	38.1	16.3	11.0	71.5	58.1	79.2	82.0	90.1
1985	9,157	26.2	39.5	19.2	15.0	73.4	57.0	77.2	85.6	89.9
1990[3]	10,711	19.4	43.0	21.8	15.8	75.4	55.7	78.6	84.6	91.0
1991[3]	10,863	18.4	43.0	22.5	16.0	75.0	55.1	77.4	84.6	90.4

Source: "Civilian Labor Force and Participation Rates, by Educational Attainment, Sex, and Race: 1970 to 1991," U.S. Bureau of the Census, *Statistical Abstract of the United States*, 1993, p. 394. Primary source: U.S. Bureau of Labor Statistics, Bulletin 2307; and unpublished data. *Notes:* 1. Percent of the civilian population in each group in the civilian labor force. 2. Includes other races, not shown separately. For 1970, White and Black races only. 3. Not strictly comparable with previous years. Annual averages of monthly figures.

★ 622 ★

Employment and Education

Educational Attainment of Workers by Occupation of Longest Job, 1991. Part 1

Numbers in thousands. Persons as of March 1992.

Occupation of longest job	Total		Not a high school graduate		High school graduate		Some college or associate degree		Bachelor's degree or more	
	Male	Female	Male	Female	Male	Female	Male	Female	Male	Female
Black										
Number										
Total[1]	3,669	3,711	677	525	1,486	1,400	863	1,078	642	707
Executive, administrative, and managerial workers	330	333	18	17	42	63	101	110	169	144
Professional specialty workers	299	521	9	11	28	46	61	118	201	346
Technical and related support workers	120	134	8	8	33	39	43	62	36	26
Sales workers	185	187	19	16	55	81	51	65	61	25
Administrative support workers, including clerical	349	1,222	27	59	143	492	128	521	52	149
Private household workers	-	22	-	18	-	2	-	1	-	-
Protective service workers	164	64	21	7	73	25	49	27	21	5
Service workers, except private household	414	764	113	251	190	410	89	100	21	4
Farming, fishing, and forestry workers	64	7	33	4	26	3	3	-	2	-
Precision production, craft, and repair workers	619	60	115	16	295	29	173	13	36	2
Machine operators, assemblers, and inspectors	408	313	113	97	218	160	65	48	11	7
Transportation and material moving workers	438	29	126	5	246	15	54	10	12	-
Handlers, equipment cleaners, helpers, and laborers	280	54	76	17	137	36	47	2	20	-
Percent										
Total[1]	100.0	100.0	100.0	100.0	100.0	100.0	100.0	100.0	100.0	100.0
Executive, administrative, and managerial workers	9.0	9.0	2.6	3.2	2.8	4.5	11.7	10.2	26.3	20.3
Professional specialty workers	8.1	14.1	1.3	2.1	1.9	3.3	7.0	11.0	31.4	48.9
Technical and related support workers	3.3	3.6	1.2	1.5	2.2	2.7	5.0	5.7	5.6	3.7
Sales workers	5.0	5.0	2.8	3.0	3.7	5.8	5.9	6.1	9.5	3.5
Administrative support workers, including clerical	9.5	32.9	4.0	11.3	9.6	35.2	14.8	48.4	8.0	21.0
Private household workers	-	0.6	-	3.5	-	0.2	-	0.1	-	-
Protective service workers	4.5	1.7	3.1	1.2	4.9	1.8	5.7	2.5	3.2	0.8
Service workers, except private household	11.3	20.6	16.7	47.7	12.8	29.3	10.3	9.3	3.3	0.5
Farming, fishing, and forestry workers	1.7	0.2	4.9	0.7	1.8	0.2	0.3	-	0.3	-
Precision production, craft, and repair workers	16.9	1.6	17.0	3.0	19.9	2.0	20.0	1.2	5.6	0.3
Machine operators, assemblers, and inspectors	11.1	8.4	16.7	18.5	14.7	11.4	7.5	4.5	1.8	0.9
Transportation and material moving workers	11.9	0.8	18.5	0.9	16.5	1.0	6.3	0.9	1.9	-
Handlers, equipment cleaners, helpers, and laborers	7.6	1.5	11.2	3.1	9.2	2.6	5.4	0.2	3.1	-

Source: "Educational Attainment of Year-Round, Full-Time Workers 25 Years Old and Over, by Occupation of Longest Job in 1991, Sex and Race," Claudette E. Bennett, *The Black Population in the United States: March 1992*, p. 66. *Note:* 1. Armed services not included.

★ 623 ★

Employment and Education

Educational Attainment of Workers by Occupation of Longest Job, 1991. Part 2

Numbers in thousands. Persons as of March 1992.

Occupation of longest job	Total		Not a high school graduate		High school graduate		Some college or associate degree		Bachelor's degree or more	
	Male	Female	Male	Female	Male	Female	Male	Female	Male	Female
White										
Number										
Total[1]	38,326	24,581	4,046	1,895	12,969	9,233	9,483	6,808	11,829	6,645
Executive, administrative, and managerial workers	6,865	4,228	213	105	1,212	1,302	1,705	1,285	3,734	1,536
Professional specialty workers	5,342	4,285	34	38	277	286	759	819	4,273	3,142
Technical and related support workers	1,370	1,171	24	35	340	315	616	529	390	291
Sales workers	4,905	2,457	223	173	1,465	980	1,472	729	1,745	576
Administrative support workers, including clerical	2,176	7,546	136	256	808	3,884	715	2,538	517	868
Private household workers	12	97	-	52	2	37	-	5	10	3
Protective service workers	1,097	152	66	13	406	47	450	78	174	14
Service workers, except private household	1,693	2,130	429	496	799	1,068	348	460	118	106

[Continued]

★ 623 ★

Educational Attainment of Workers by Occupation of Longest Job, 1991. Part 2
[Continued]

Occupation of longest job	Total		Not a high school graduate		High school graduate		Some college or associate degree		Bachelor's degree or more	
	Male	Female	Male	Female	Male	Female	Male	Female	Male	Female
Farming, fishing, and forestry workers	1,359	157	352	38	597	63	268	35	142	22
Precision production, craft, and repair workers	7,313	574	1,133	136	3,728	272	1,966	129	487	37
Machine operators, assemblers, and inspectors	2,543	1,317	556	436	1,407	711	488	140	92	29
Transportation and material moving workers	2,387	117	537	28	1,252	65	487	19	111	5
Handlers, equipment cleaners, helpers, and laborers	1,265	349	343	90	676	203	209	41	37	15
Percent										
Total[1]	100.0	100.0	100.0	100.0	100.0	100.0	100.0	100.0	100.0	100.0
Executive, administrative, and managerial workers	17.9	17.2	5.3	5.6	9.3	14.1	18.0	18.9	31.6	23.1
Professional specialty workers	13.9	17.4	0.8	2.0	2.1	3.1	8.0	12.0	36.1	47.3
Technical and related support workers	3.6	4.8	0.6	1.8	2.6	3.4	6.5	7.8	3.3	4.4
Sales workers	12.8	10.0	5.5	9.1	11.3	10.6	15.5	10.7	14.8	8.7
Administrative support workers, including clerical	5.7	30.7	3.4	13.5	6.2	42.1	7.5	37.3	4.4	13.1
Private household workers	-	0.4	-	2.7	-	0.4	-	0.1	0.1	0.1
Protective service workers	2.9	0.6	1.6	0.7	3.1	0.5	4.7	1.1	1.5	0.2
Service workers, except private household	4.4	8.7	10.6	26.2	6.2	11.6	3.7	6.8	1.0	1.6
Farming, fishing, and forestry workers	3.5	0.6	8.7	2.0	4.6	0.7	2.8	0.5	1.2	0.3
Precision production, craft, and repair workers	19.1	2.3	28.0	7.2	28.7	2.9	20.7	1.9	4.1	0.6
Machine operators, assemblers, and inspectors	6.6	5.4	13.8	23.0	10.8	7.7	5.1	2.1	0.8	0.4
Transportation and material moving workers	6.2	0.5	13.3	1.5	9.7	0.7	5.1	0.3	0.9	0.1
Handlers, equipment cleaners, helpers, and laborers	3.3	1.4	8.5	4.7	5.2	2.2	2.2	0.6	0.3	0.2

Source: "Educational Attainment of Year-Round, Full-Time Workers 25 Years Old and Over, by Occupation of Longest Job in 1991, Sex and Race," Claudette E. Bennett, *The Black Population in the United States: March 1992,* pp. 66-67. *Note:* 1. Armed services not included.

★ 624 ★

Employment and Education

Occupations of Civilians by Educational Level

In thousands. Annual averages of monthly figures. For civilian noninstitutional population 25 years and over. Based on Current Population Survey.

Sex, Race, and years of school	Total employed	Managerial/ professional	Tech./sales/ administrative	Service[1]	Precision production[2]	Operators/ fabricators[3]	Farming, forestry, fishing
Male, total[4]	54,658	15,653	11,156	4,595	10,753	10,161	2,340
Less than a high school diploma	7,182	373	572	932	1,941	2,571	794
High school graduates, no college	18,459	1,923	3,209	1,870	5,260	5,258	940
Less than a bachelor's degree	13,213	3,032	3,790	1,288	2,825	1,881	398
College graduates	15,803	10,324	3,585	505	728	452	209
White	47,779	14,237	9,896	3,520	9,696	8,325	2,104
Less than a high school diploma	5,938	328	482	696	1,709	2,055	668
High school graduates, no college	16,085	1,779	2,884	1,411	4,797	4,356	859
Less than a bachelor's degree	11,588	2,765	3,341	1,015	2,543	1,550	374
College graduates	14,168	9,365	3,189	399	647	365	203
Black	4,945	766	825	828	798	1,551	176
Less than a high school diploma	987	30	64	177	183	435	98
High school graduates, no college	1,926	105	250	362	363	788	60
Less than a bachelor's degree	1,239	192	330	221	205	276	16
College graduates	793	440	182	68	47	52	3
Female, total[4]	45,381	13,748	19,091	7,377	1,021	3,679	464
Less than a high school diploma	4,506	232	944	1,875	218	1,115	122
High school graduates, no college	16,957	2,172	8,511	3,576	512	1,983	204

[Continued]

★ 624 ★

Occupations of Civilians by Educational Level
[Continued]

Sex, Race, and years of school	Total employed	Managerial/ professional	Tech./sales/ administrative	Service[1]	Precision production[2]	Operators/ fabricators[3]	Farming, forestry, fishing
Less than a bachelor's degree	12,357	3,358	6,714	1,509	214	473	90
College graduates	11,561	7,986	2,922	418	78	108	48
White	38,481	12,141	16,591	5,621	840	2,852	436
Less than a high school diploma	3,479	194	819	1,320	174	866	106
High school graduates, no college	14,466	1,967	7,582	2,753	426	1,543	195
Less than a bachelor's degree	10,516	2,956	5,732	1,202	178	362	87
College graduates	10,019	7,024	2,458	347	62	81	48
Black	5,231	1,125	1,871	1,461	115	644	15
Less than a high school diploma	805	26	87	480	23	179	9
High school graduates, no college	2,023	161	742	693	59	364	5
Less than a bachelor's degree	1,471	305	791	257	24	93	-
College graduates	932	633	251	31	8	9	-

Source: "Occupation of Employed Civilians, by Sex, Race, and Educational Attainment: 1992." U.S. Bureau of the Census, *Statistical Abstract of the United States,* 1993, p. 409. Primary source: U.S. Bureau of Labor Statistics, unpublished data. *Notes:* - Represents or rounds to zero. 1. Includes private household workers. 2. Includes craft and repair. 3. Includes laborers. 4. Includes other races, not shown separately.

★ 625 ★

Employment and Education

School Enrollment and Labor Force 16 to 24 Years Old, 1980 and 1991

In thousands, except percent. As of October. Civilian noninstitutional population. Based on Current Population Survey.

Characteristic	Population		Civilian labor force			Employed		Unemployed		
	1990	1991	1980, total	1991 Total	1991 Percent[1]	1980	1991	1980, total	1991 Total	1991 Rate[2]
Total, 16 to 24 years[3]	37,103	31,188	24,918	20,248	64.9	21,454	17,524	3,464	2,724	13.5
Enrolled in school[3]	15,713	15,519	7,454	7,665	49.4	6,433	6,663	1,021	1,002	13.1
16 to 19 years	11,126	10,116	4,836	4,330	42.8	4,029	3,614	807	716	16.5
20 to 24 years	4,587	5,404	2,618	3,335	61.7	2,404	3,049	214	286	8.6
White	13,242	12,523	6,687	6,656	53.2	5,889	5,888	798	768	11.5
Below college	6,566	5,610	3,095	2,449	43.7	2,579	2,019	516	430	17.6
College level	6,678	6,913	3,592	4,207	60.9	3,310	3,869	282	338	8.0
Black	2,028	2,109	595	664	31.5	406	484	189	180	27.2
Below college	1,282	1,267	294	278	22.0	174	171	120	108	38.6
College level	747	841	300	385	45.8	230	313	70	73	18.9
Not enrolled[3]	21,390	15,669	17,464	12,583	80.3	15,021	10,861	2,443	1,722	13.7
White	18,103	12,728	15,121	10,491	82.4	13,318	9,325	1,803	1,166	11.1
Black	2,864	2,439	2,055	1,708	70.0	1,451	1,222	604	486	28.5

Source: "School Enrollment and Labor Force Status of Civilians 16 to 24 Years Old, by Selected Characteristics: 1980 and 1991," U.S. Bureau of the Census, *Statistical Abstract of the United States,* 1993, p. 398. Primary source: U.S. Bureau of Labor Statistics, Bulletin 2307; *News,* USDL 92-395, June 30, 1992; and unpublished data. *Notes:* 1. Percent of civilian noninstitutional population. 2. Percent of civilian labor force in each category. 3. Includes other races.

★ 626 ★

Employment and Education

School Enrollment and Labor Force 16 to 24 Years Old, 1993. Part I: Total Enrolled

Numbers in thousands.

Employment status, educational attainment, race and Hispanic origin	Civilian non-institutional population	1993								
		Civilin labor force								
		Total	Percent of population	Employed			Unemployed			
				Total	Full time[1]	Part time[1]	Total	Looking for full-time work	Looking for part-time work	Percent of labor force
Total enrolled										
White										
Total, 16 to 24 years	10,434	5,386	51.6	4,751	951	3,799	635	128	508	11.8
16 to 19 years	6,781	3,093	45.6	2,612	453	2,358	482	66	416	15.6
20 to 24 years	3,653	2,293	62.8	2,139	698	1,441	153	61	92	6.7
Men	5,270	2,658	50.4	2,305	486	1,819	353	70	284	13.3
Women	5,164	2,728	52.8	2,446	466	1,980	282	58	224	10.3
High school	5,085	2,155	42.4	1,765	114	1,652	390	51	339	18.1
College	5,349	3,231	60.4	2,985	838	2,147	246	77	169	7.6
Full-time students	4,454	2,438	54.7	2,239	386	1,853	199	51	148	8.2
Part-time students	894	793	88.7	747	452	294	47	26	21	5.9
Black										
Total, 16 to 24 years	1,921	618	32.2	440	103	337	178	48	131	28.8
16 to 19 years	1,360	362	26.6	230	28	202	133	26	107	36.6
20 to 24 years	561	256	45.6	210	75	135	46	22	24	17.9
Men	935	297	31.8	200	42	158	97	24	73	32.7
Women	986	321	32.6	240	61	179	81	24	58	25.3
High school	1,140	279	24.5	169	16	152	110	23	88	39.6
College	781	339	43.4	271	86	185	68	25	43	20.0
Full-time students	656	243	37.0	192	42	149	51	16	35	21.0
Part-time students	125	97	77.4	80	44	36	17	9	8	17.5
Hispanic origin										
Total, 16 to 24 years	1,230	493	40.1	387	100	287	106	25	80	21.4
16 to 19 years	877	276	31.5	199	27	172	78	14	64	28.2
20 to 24 years	353	216	61.3	189	74	115	28	11	17	12.8
Men	606	251	41.4	194	50	143	57	12	45	22.8
Women	624	242	38.8	194	50	143	48	13	36	20.0
High school	752	206	27.4	138	18	120	68	12	56	32.8
College	478	287	60.1	249	82	167	38	14	24	13.2
Full-time students	358	185	51.7	161	30	131	24	6	18	13.1
Part-time students	119	102	85.2	88	52	36	14	8	6	13.5

Source: "Employment Status of the Civilian Noninstitutional Population 16 to 24 Years Old by School Enrollment, Educational Attainment, Sex, Race, and Hispanic Origin." *Employment and Earnings* 41 (January 1994), p. 189. *Notes:* In the summer months, the educational attainment levels of youth not enrolled in school are increased by the temporary movement of high school and college students into that group. Detail for the above race and Hispanic-origin groups will not sum to totals because data for the "other races" group are not presented and Hispanics are included in both the white and black population groups. 1. Employed persons with a job but not at work and persons at work part time are distributed according to whether they usually work full or part time.

★ 627 ★

Employment and Education

School Enrollment and Labor Force 16 to 24 Years Old, 1993. Part II: Total Not Enrolled

Numbers in thousands.

Employment status, educational attainment, race and Hispanic origin	Civilian non-institutional population	1993								
		Civilian labor force								
		Total	Percent of population	Employed			Unemployed			
				Total	Full time[1]	Part time[1]	Total	Looking for full-time work	Looking for part-time work	Percent of labor force
Total not enrolled										
White										
Total, 16 to 24 years	14,433	11,805	81.8	10,505	8,289	2,216	1,299	1,097	202	11.0
16 to 19 years	3,798	2,737	72.1	2,276	1,328	947	462	345	116	16.9
20 to 24 years	10,635	9,067	85.3	8,230	6,961	1,269	838	751	86	9.2
Men	7,151	6,398	89.5	5,643	4,728	915	755	669	86	11.8
Women	7,282	5,406	74.2	4,863	3,561	1,302	544	428	116	10.1
Less than a high school diploma	3,825	2,466	64.5	1,983	1,287	696	483	389	93	19.6
High school graduates, no college	5,599	4,749	84.8	4,242	3,445	797	508	442	66	10.7
Less than a bachelor's degree	3,645	3,279	90.0	3,045	2,456	589	234	194	39	7.1
College graduates	1,364	1,310	96.1	1,235	1,101	134	75	71	4	5.7
Black										
Total, 16 to 24 years	2,664	1,847	69.3	1,353	1,016	338	494	433	61	26.8
16 to 19 years	739	414	56.0	245	120	124	169	132	38	40.9
20 to 24 years	1,925	1,434	74.5	1,109	895	213	325	302	24	22.7
Men	1,265	970	76.7	705	546	159	265	243	22	27.3
Women	1,399	877	62.7	648	470	178	229	191	39	26.1
Less than a high school diploma	840	401	47.8	226	137	89	175	140	34	43.5
High school graduates, no college	1,178	888	75.4	662	506	156	226	212	14	25.5
Less than a bachelor's degree	553	471	85.1	387	303	84	83	73	11	17.7
College graduates	94	88	93.7	78	70	8	10	8	2	11.3
Hispanic origin										
Total, 16 to 24 years	2,303	1,623	70.5	1,364	1,100	264	259	229	30	15.9
16 to 19 years	638	383	60.1	288	193	95	95	79	16	24.8
20 to 24 years	1,665	1,239	74.4	1,076	907	169	164	150	13	13.2
Men	1,175	1,029	87.6	874	746	127	156	142	14	15.1
Women	1,127	593	52.6	491	354	137	103	87	16	17.3
Less than a high school diploma	1,186	734	61.8	599	478	121	135	119	15	18.3
High school graduates, no college	737	566	76.8	475	383	92	91	81	10	16.1
Less than a bachelor's degree	324	273	84.2	243	198	46	30	25	5	10.9
College graduates	55	50	91.2	47	42	5	3	3	-	6.8

Source: "Employment Status of the Civilian Noninstitutional Population 16 to 24 Years Old by School Enrollment, Educational Attainment, Sex, Race, and Hispanic Origin." *Employment and Earnings* 41 (January 1994), p. 190. *Notes:* In the summer months, the educational attainment levels of youth not enrolled in school are increased by the temporary movement of high school and college students into that group. Detail for the above race and Hispanic-origin groups will not sum to totals because data for the "other races" group are not presented and Hispanics are included in both the white and black population groups. 1. Employed persons with a job but not at work and persons at work part time are distributed according to whether they usually work full or part time.

★ 628 ★

Employment and Education

School Enrollment and Labor Force 16 to 24 Years Old, January 1994

Numbers in thousands.

Employment status, educational attainment, race and Hispanic origin	Civilian non-institutional population	January 1994								
		Civilian labor force								
		Total	Percent of population	Employed			Unemployed			
				Total	Full time[1]	Part time[1]	Total	Looking for full-time work	Looking for part-time work	Percent of labor force
Total enrolled										
Total										
16 to 24 years	16,815	7,994	47.5	6,845	1,219	5,627	1,149	210	938	14.4
16 to 19 years	11,104	4,687	42.2	3,881	296	3,585	806	78	728	17.2
20 to 24 years	5,712	3,307	57.9	2,964	922	2,042	343	132	210	10.4
High school	8,567	3,379	39.4	2,704	131	2,573	675	68	607	20.0
College	8,249	4,615	55.9	4,141	1,087	3,054	473	143	331	10.3
Full-time students	6,765	3,417	50.5	3,057	475	2,581	360	86	274	10.5
Part-time students	1,484	1,198	80.7	1,085	612	473	113	56	56	9.4
Men										
16 to 24 years	8,503	3,948	46.4	3,273	643	2,630	676	134	542	17.1
16 to 19 years	5,668	2,358	41.6	1,882	166	1,716	476	45	431	20.2
20 to 24 years	2,835	1,590	56.1	1,391	477	914	200	89	111	12.5
High school	4,533	1,797	39.6	1,382	96	1,286	415	44	372	23.1
College	3,970	2,151	54.2	1,891	547	1,344	260	90	170	12.1
Full-time students	3,319	1,607	48.4	1,417	271	1,147	190	46	143	11.8
Part-time students	651	544	83.5	473	276	197	70	44	27	13.0
Women										
16 to 24 years	8,312	4,046	48.7	3,573	576	2,997	473	77	396	11.7
16 to 19 years	5,436	2,329	42.8	1,999	130	1,869	330	33	297	14.2
20 to 24 years	2,876	1,717	59.7	1,574	445	1,128	143	44	100	8.3
High school	4,034	1,582	39.2	1,322	35	1,287	260	24	236	16.4
College	4,279	2,464	57.6	2,251	541	1,710	213	52	161	8.6
Full-time students	3,446	1,810	52.5	1,639	204	1,435	171	40	131	9.4
Part-time students	832	654	78.6	611	336	275	42	13	30	6.5

Source: "Employment Status of the Civilian Noninstitutional Population 16 to 24 Years Old by School Enrollment, Educational Attainment, Sex, Race, and Hispanic Origin." *Employment and Earnings* 41 (January 1994), p. 55. *Notes:* In the summer months, the educational attainment levels of youth not enrolled in school are increased by the temporary movement of high school and college students into that group. Detail for the above race and Hispanic-origin groups will not sum to totals because data for the "other races" group are not presented and Hispanics are included in both the white and black population groups. Data for 1994 are not directly comparable with data for 1993 and earlier years. For additional information, see "Revisions in the Current Population Survey Effective January 1994" in this issue. 1. Data not shown where base is less than 7,500.

Families in the Labor Force

★ 629 ★

Employed Civilians by Family Relationship, 1992 and 1993

Numbers in thousands.

Family relationship, race, and Hispanic origin	1992				1993			
		Percent of employed				Percent employed		
	Total	With no other employed person in family	With another employed person in family	With another person in family employed full time	Total	With no other employed person in family	With another employed person in family	With another person in family employed full time
Total								
Total employed in families[1]	93,754	23.1	76.9	66.6	94,576	23.1	76.9	66.5
Husbands	38,274	27.7	72.3	55.1	38,521	27.5	72.5	55.4
With children under 18 years of age	21,722	29.8	70.2	50.0	21,831	29.7	70.3	50.2
Wives	29,313	10.2	89.8	85.5	29,675	10.1	89.9	85.4
With children under 18 years of age	15,552	6.5	93.5	90.4	15,703	6.4	93.6	90.3
Relatives in married-couple families	11,075	7.4	92.6	88.6	10,869	8.0	92.0	88.1
Women who maintain families	6,582	66.1	33.9	22.8	6,764	66.3	33.7	22.9
With children under 18 years of age	4,246	83.7	16.3	7.8	4,403	83.3	16.7	8.2
Relatives in families maintained by women	4,598	25.5	74.5	65.3	4,804	24.8	75.2	65.5
Men who maintain families	2,288	59.0	41.0	31.9	2,339	60.0	40.0	30.3
With children under 18 years of age	1,140	85.3	14.7	8.7	1,204	85.1	14.9	7.9
Relatives in families maintained by men	1,624	18.6	81.4	75.9	1,604	18.8	81.2	74.7
White								
Total employed in families[1]	81,313	22.2	77.8	67.1	81,861	22.1	77.9	67.0
Husbands	34,502	28.1	71.9	54.1	34,705	27.6	72.4	54.6
With children under 18 years of age	19,360	30.4	69.6	48.4	19,444	30.1	69.9	48.7
Wives	26,139	9.9	90.1	85.8	26,498	9.8	90.2	85.8
With children under 18 years of age	13,659	5.9	94.1	91.1	13,816	5.9	94.1	91.0
Relatives in married-couple families	9,682	7.2	92.8	88.9	9,445	7.8	92.2	88.4
Women who maintain families	4,551	63.5	36.5	24.3	4,656	63.6	36.4	24.0
With children under 18 years of age	2,903	82.5	17.5	7.8	2,995	81.8	18.2	8.2
Relatives in families maintained by women	3,339	25.7	74.3	64.8	3,444	25.0	75.0	65.3
Men who maintain families	1,838	58.7	41.3	31.8	1,882	60.5	39.5	29.3
With children under 18 years of age	931	84.4	15.6	9.0	995	84.6	15.4	7.6
Relatives in families maintained by men	1,261	18.2	81.8	76.3	1,232	18.3	81.7	75.3
Black								
Total employed in families[1]	8,966	31.8	68.2	60.6	9,088	32.3	67.7	60.2
Husbands	2,448	22.3	77.7	66.9	2,449	23.4	76.6	66.3
With children under 18 years of age	1,501	21.5	78.5	67.5	1,506	22.6	77.4	66.5
Wives	2,120	15.1	84.9	80.5	2,088	14.1	85.9	81.1
With children under 18 years of age	1,256	11.7	88.3	84.4	1,231	10.3	89.7	85.3
Relatives in married-couple families	904	9.8	90.2	85.0	934	10.5	89.5	84.8
Women who maintain families	1,821	73.3	26.7	18.3	1,878	73.9	26.1	18.9
With children under 18 years of age	1,225	86.3	13.7	7.8	1,273	87.3	12.7	7.6
Relatives in families maintained by women	1,096	25.5	74.5	66.6	1,164	25.4	74.6	64.9
Men who maintain families	337	65.5	34.5	27.2	337	63.7	36.3	29.2
With children under 18 years of age	170	89.9	10.1	7.7	169	88.5	11.5	7.6
Relatives in families maintained by men	240	26.1	73.9	68.2	239	27.9	72.1	64.8
Hispanic origin								
Total employed in families[1]	7,184	26.5	73.5	65.0	7,401	27.1	72.9	63.9
Husbands	2,712	35.7	64.3	51.5	2,841	35.2	64.8	52.0
With children under 18 years of age	1,954	38.1	61.9	48.7	2,013	37.6	62.4	49.0
Wives	1,743	10.9	89.1	84.0	1,831	11.4	88.6	82.8
With children under 18 years of age	1,180	9.2	90.8	86.5	1,233	10.1	89.9	84.7

[Continued]

★ 629 ★

Employed Civilians by Family Relationship, 1992 and 1993

[Continued]

Family relationship, race, and Hispanic origin	1992				1993			
		Percent of employed				Percent employed		
	Total	With no other employed person in family	With another employed person in family	With another person in family employed full time	Total	With no other employed person in family	With another employed person in family	With another person in family employed full time
Relatives in married-couple families	1,065	6.7	93.3	89.2	1,039	7.7	92.3	87.6
Women who maintain families	578	64.9	35.1	26.5	601	65.2	34.8	24.6
With children under 18 years of age	405	78.9	21.1	13.7	409	79.9	20.1	12.1
Relatives in families maintained by women	472	23.8	76.2	67.3	497	24.9	75.1	64.8
Men who maintain families	327	49.4	50.6	43.5	328	53.3	46.7	38.2
With children under 18 years of age	136	82.4	17.6	12.5	155	81.6	18.4	13.0
Relatives in families maintained by men	289	8.9	91.1	85.4	264	9.2	90.8	86.0

Source: "Employed Civilians by Family Relationship, Race, Hispanic Origin, and Presence of Employed Family Members." *Employment and Earnings* 41 (January 1994), p. 238. *Notes:* Detail for the above race and Hispanic-origin groups will not sum to totals because data for the "other races" group are not presented and Hispanics are included in both the white and black population groups. 1. Excludes persons living alone or with nonrelatives, persons in married-couple families where the husband or wife is in the Armed Forces, and persons in unrelated subfamilies. Estimates for husbands, wives, and women who maintain families are somewhat different from marital status estimates shown in other tables in this publication because of differences in definitions and weighing patterns used in aggregating the data.

★ 630 ★

Families in the Labor Force

Employed Civilians by Family Relationship: Quarterly Averages, 1992 and 1993

Numbers in thousands. Not seasonally adjusted.

Family relationship, race, and Hispanic origin	IV 1992				IV 1993			
		Percent of employed				Percent employed		
	Total	With no other employed person in family	With another employed person in family	With another person in family employed full time	Total	With no other employed person in family	With another employed person in family	With another person in family employed full time
Total								
Total employed in families[1]	94,018	23.0	77.0	66.3	95,251	22.8	77.2	66.6
Husbands	38,463	27.4	72.6	55.0	38,887	26.8	73.2	55.9
With children under 18 years of age	21,648	29.5	70.5	49.8	21,944	28.9	71.1	50.5
Wives	29,731	10.3	89.7	85.2	30,280	9.9	90.1	85.6
With children under 18 years of age	15,645	6.4	93.6	90.2	15,969	5.8	94.2	91.0
Relatives in married-couple families	10,794	7.4	92.6	88.2	10,589	7.7	92.3	88.2
Women who maintain families	6,643	66.0	34.0	21.8	6,746	66.6	33.4	22.0
With children under 18 years of age	4,345	82.3	17.7	7.8	4,419	83.3	16.7	7.7
Relatives in families maintained by women	4,587	24.7	75.3	66.7	4,725	24.9	75.1	65.3
Men who maintain families	2,236	60.5	39.5	30.1	2,393	61.6	38.4	28.4
With children under 18 years of age	1,148	85.1	14.9	8.9	1,275	85.9	14.1	6.5
Relatives in families maintained by men	1,564	21.1	78.9	73.0	1,632	19.1	80.9	74.4
White								
Total employed in families[1]	81,452	22.0	78.0	66.9	82,324	21.8	78.2	67.1
Husbands	34,622	27.6	72.4	54.2	34,999	27.0	73.0	55.0
With children under 18 years of age	19,296	30.0	70.0	48.3	19,540	29.3	70.7	49.1
Wives	26,497	9.8	90.2	85.7	27,002	9.5	90.5	86.0
With children under 18 years of age	13,750	5.7	94.3	91.2	14,020	5.3	94.7	91.8
Relatives in married-couple families	9,396	7.3	92.7	88.4	9,116	7.4	92.6	88.4
Women who maintain families	4,613	63.1	36.9	23.5	4,626	64.2	35.8	23.1
With children under 18 years of age	2,982	81.0	19.0	7.8	2,976	81.9	18.1	7.9
Relatives in families maintained by women	3,356	24.3	75.7	66.8	3,385	25.5	74.5	64.5

[Continued]

★ 630 ★

Employed Civilians by Family Relationship: Quarterly Averages, 1992 and 1993

[Continued]

Family relationship, race, and Hispanic origin	IV 1992				IV 1993			
		Percent of employed				Percent employed		
	Total	With no other employed person in family	With another employed person in family	With another person in family employed full time	Total	With no other employed person in family	With another employed person in family	With another person in family employed full time
Men who maintain families	1,777	60.9	39.1	29.0	1,931	61.6	38.4	27.6
With children under 18 years of age	937	84.5	15.5	8.8	1,048	85.1	14.9	6.5
Relatives in families maintained by men	1,190	20.9	79.1	73.1	1,266	19.2	80.8	74.3
Black								
Total employed in families[1]	8,987	32.4	67.6	59.7	9,235	31.4	68.6	60.7
Husbands	2,474	22.5	77.5	66.4	2,489	21.3	78.7	67.9
With children under 18 years of age	1,474	20.6	79.4	67.7	1,498	20.1	79.9	68.0
Wives	2,149	15.5	84.5	79.7	2,167	14.0	86.0	80.5
With children under 18 years of age	1,244	12.4	87.6	82.6	1,271	10.2	89.8	84.5
Relatives in married-couple families	879	10.2	89.8	84.4	987	10.2	89.8	85.6
Women who maintain families	1,818	74.2	25.8	16.6	1,883	73.7	26.3	17.8
With children under 18 years of age	1,247	85.5	14.5	7.7	1,297	86.9	13.1	6.6
Relatives in families maintained by women	1,077	27.1	72.9	65.6	1,141	24.6	75.4	66.4
Men who maintain families	345	65.7	34.3	27.8	340	67.9	32.1	25.5
With children under 18 years of age	178	88.8	11.2	8.5	181	91.5	8.5	4.8
Relatives in families maintained by men	244	27.4	72.6	65.4	228	27.6	72.4	65.1
Hispanic origin								
Total employed in families[1]	7,188	27.1	72.9	64.2	7,564	26.9	73.1	64.0
Husbands	2,769	36.3	63.7	50.6	2,951	34.8	65.2	51.9
With children under 18 years of age	1,970	39.0	61.0	48.0	2,098	37.1	62.9	48.9
Wives	1,726	11.1	88.9	82.8	1,848	10.5	89.5	84.1
With children under 18 years of age	1,164	9.0	91.0	85.6	1,230	8.9	91.1	86.9
Relatives in married-couple families	1,064	6.9	93.1	88.9	1,028	6.7	93.3	87.5
Women who maintain families	580	66.3	33.7	25.7	612	64.0	36.0	26.3
With children under 18 years of age	405	79.0	21.0	14.0	406	79.9	20.1	12.8
Relatives in families maintained by women	425	23.4	76.6	71.1	510	25.8	74.2	65.6
Men who maintain families	326	50.0	50.0	40.3	351	58.9	41.1	34.6
With children under 18 years of age	134	79.0	21.0	13.7	165	84.1	15.9	10.4
Relatives in families maintained by men	299	10.4	89.6	84.7	263	6.0	94.0	91.8

Source: "Employed Civilians by Family Relationship, Race, Hispanic Origin, and Presence of Employed Family Members." *Employment and Earnings* 41 (January 1994), p. 76. *Notes:* Detail for the above race and Hispanic-origin groups will not sum to totals because data for the "other races" group are not presented and Hispanics are included in both the white and black population groups. 1. Excludes persons living alone or with nonrelatives, persons in in families where the husband or wife is in the Armed Forces, and persons in unrelated subfamilies. Estimates for husbands, wives, and women who maintain families are somewhat different from marital status estimates shown in other tables in this publication because of differences in definitions and weighing patterns used in aggregating the data.

Labor Force

★ 631 ★

Civilian Labor Force, 1990, Projected to 2005, and Entrants and Leavers, 1990-2005

Group	Labor force, 1990	Entrants, 1990-2005	Leavers, 1990-2005	Labor force, 2005
Number (thousands)				
Total	124,787	55,798	29,851	150,732
Men	68,234	28,197	17,090	79,338
Women	56,554	27,601	12,761	71,394
White, Non-Hispanic	98,013	36,425	24,423	110,015
Men	53,784	17,965	14,204	57,545
Women	44,229	18,460	10,219	52,470
Black	13,340	7,250	3,144	17,447
Men	6,628	3,461	1,553	8,537
Women	6,712	3,789	1,591	8,910
Hispanic	9,576	8,768	1,556	16,790
Men	5,755	5,085	939	9,902
Women	3,821	3,683	617	6,888
Asian and other	3,855	3,354	728	6,482
Men	2,064	1,686	395	3,356
Women	1,791	1,668	333	3,126
Share (percent)				
Total	100.0	100.0	100.0	100.0
Men	54.7	50.5	57.3	52.6
Women	45.3	49.5	42.7	47.4
White, Non-Hispanic	78.5	65.3	81.8	73.0
Men	43.1	32.2	47.6	38.2
Women	35.4	33.1	34.2	34.8
Black	10.7	13.0	10.5	11.6
Men	5.3	6.2	5.2	5.7
Women	5.4	6.8	5.3	5.9
Hispanic	7.7	15.7	5.2	11.1
Men	4.6	9.1	3.1	6.6
Women	3.1	6.6	2.1	4.6
Asian and other	3.1	6.0	2.4	4.3

[Continued]

★ 631 ★

Civilian Labor Force, 1990, Projected to 2005, and Entrants and Leavers, 1990-2005

[Continued]

Group	Labor force, 1990	Entrants, 1990-2005	Leavers, 1990-2005	Labor force, 2005
Men	1.7	3.0	1.3	2.2
Women	1.4	3.0	1.1	2.1

Source: "Civilian Labor Force, 1990 and Projected to 2005, and Projected Entrants and Leavers, 1990 to 2005." U.S. Department of Labor, Employment and Training Administration, *Training and Employment Report of the Secretary of Labor,* Covering the Period July 1988-September 1990, p. 101. Primary source: U.S. Department of Labor, Bureau of Labor Statistics. *Notes:* Unlike other tables in the D series, the columns in this table are additive.

★ 632 ★

Labor Force

Civilian Labor Force: Characteristics, 1975 and 1990, and Projections to 2005

Numbers in thousands.

Group	Level			Change		Percent change		Percent distribution			Annual labor force growth rate (percent)	
	1975	1990	2005	1975-1990	1990-2005	1975-1990	1990-2005	1975	1990	2005	1975-1990	1990-2005
Total, 16 years and over	93,775	124,787	150,732	31,012	25,945	33.1	20.8	100.0	100.0	100.0	1.9	1.3
White, 16 years and over	82,831	107,177	125,785	24,346	18,608	29.4	17.4	88.3	85.9	83.4	1.7	1.1
Black, 16 years and over	9,263	13,493	17,766	4,230	4,273	45.7	31.7	9.9	10.8	11.8	2.5	1.9
Asian and other, 16 years and over[1]	1,681	4,117	7,181	2,436	3,064	144.9	74.4	1.8	3.3	4.8	6.2	3.8
Hispanic, 16 years and over[2]	[3]	9,576	16,790	[3]	7,214	[3]	75.3	[3]	7.7	11.1	5.9[4]	3.8

Source: "Civilian Labor Force by Sex, Age, Race, and Hispanic Origin, 1975 and 1990, and Moderate Growth Projections to 2005." U.S. Department of Labor, Employment and Training Administration, *Training and Employment Report of the Secretary of Labor,* Covering the Period July 1988-September 1990, p. 98. Primary source: U.S. department of Labor, Bureau of Labor Statistics. *Notes:* 1. The "Asian and other" group includes (1) Asians and Pacific Islanders and (2) American Indians and Alaskan natives. The Historic data are derived by subtracting "Black" from the "Black and other" group; projections are made directly. 2. Persons of Hispanic origin may be of any race. 3. Data on Hispanics were not available before 1980. 4. 1976-90.

★ 633 ★

Labor Force

Civilian Labor Force: Employment Status, 1992 - I

For civilian noninstitutional population 16 years old and over. Annual averages of monthly figures. Based on Current Population Survey.

Age and race	Civilian labor force			Male (1,000)			Female (1,000)		
	Total (1,000)	Percent by age		Total	Employed	Unemployed	Total	Employed	Unemployed
		Male	Female						
All workers[1]	126,982	100.0	100.0	69,184	63,805	5,380	57,798	53,793	4,005
16 to 19 years	6,751	5.1	5.5	3,547	2,786	761	3,204	2,613	591
20 to 24 years	13,703	10.5	11.2	7,242	6,357	884	6,461	5,799	662
25 to 34 years	35,103	28.0	27.2	19,355	17,847	1,508	15,748	14,594	1,154
35 to 44 years	33,603	26.3	26.7	18,162	17,067	1,095	15,441	14,595	845
45 to 54 years	22,391	17.5	17.8	12,101	11,426	675	10,290	9,820	470
55 to 64 years	11,870	9.7	8.9	6,701	6,314	387	5,169	4,953	216
65 years and over	3,562	3.0	2.6	2,077	2,008	69	1,485	1,419	66
White	108,526	100.0	100.0	59,830	55,709	4,121	48,696	45,770	2,926
16 to 19 years	5,744	5.0	5.6	3,019	2,464	555	2,726	2,297	429
20 to 24 years	11,539	10.2	11.2	6,097	5,462	634	5,442	4,993	450
25 to 34 years	29,539	27.6	26.8	16,510	15,364	1,146	13,029	12,222	807
35 to 44 years	28,715	26.3	26.6	15,765	14,907	858	12,949	12,324	625
45 to 54 years	19,411	17.8	18.0	10,652	10,106	546	8,759	8,384	375
55 to 64 years	10,377	9.9	9.2	5,906	5,584	322	4,471	4,291	180
65 years and over	3,201	3.1	2.7	1,881	1,822	59	1,319	1,260	59
Black	13,891	100.0	100.0	6,892	5,846	1,046	6,999	6,087	912
16 to 19 years	787	6.1	5.3	419	243	176	368	231	137
20 to 24 years	1,683	12.7	11.6	872	658	214	811	623	187
25 to 34 years	4,251	30.8	30.4	2,121	1,820	302	2,130	1,830	300
35 to 44 years	3,623	25.1	27.1	1,729	1,534	194	1,894	1,709	186
45 to 54 years	2,143	14.8	16.1	1,018	913	105	1,124	1,052	72
55 to 64 years	1,124	8.5	7.7	586	539	47	538	515	23
65 years and over	280	2.1	1.9	146	139	7	134	128	7
Hispanic[2]	10,131	100.0	100.0	6,091	5,388	703	4,040	3,584	456
16 to 19 years	678	6.4	7.1	391	281	110	287	211	76
20 to 24 years	1,460	14.6	14.1	890	768	122	570	500	71
25 to 34 years	3,316	34.1	30.7	2,075	1,866	209	1,241	1,103	137
35 to 44 years	2,548	24.3	26.4	1,481	1,336	145	1,066	962	104
45 to 54 years	1,358	13.0	14.1	790	720	71	568	519	49
55 to 64 years	655	6.4	6.5	392	352	40	263	246	16
65 years and over	116	1.2	1.1	71	65	6	45	41	4

Source: "Civilian Labor Force-Employment Status, by Sex, Race, and Age: 1992," U.S. Bureau of the Census, *Statistical Abstract of the United States*, 1993, p. 401. Primary source: U.S. Bureau of Labor Statistics, *Employment and Earnings*, monthly, January 1992. *Notes:* 1. Includes other races not shown separately. 2. Persons of Hispanic origin may be of any race.

★ 634 ★

Labor Force

Civilian Labor Force: Employment Status, 1992 - II

For civilian noninstitutional population 16 years old and over. Annual averages of monthly figures. Based on Current Population Survey.

Age and race	Percent of labor force			
	Employed		Unemployed	
	Male	Female	Male	Female
All workers[1]	92.2	93.1	7.8	6.9
16 to 19 years	78.5	81.6	21.5	18.5
20 to 24 years	87.8	89.8	12.2	10.2
25 to 34 years	92.2	92.7	7.8	7.3
35 to 44 years	94.0	94.5	6.0	5.5
45 to 54 years	94.4	95.4	5.6	4.6
55 to 64 years	94.2	95.8	5.8	4.2
65 years and over	96.7	95.6	3.3	4.5
White	93.1	94.0	6.9	6.0
16 to 19 years	81.6	84.3	18.4	15.7
20 to 24 years	89.6	91.7	10.4	8.3
25 to 34 years	93.1	93.8	6.9	6.2
35 to 44 years	94.6	95.2	5.4	4.8
45 to 54 years	94.9	95.7	5.1	4.3
55 to 64 years	94.5	96.0	5.5	4.0
65 years and over	96.9	95.5	3.1	4.5
Black	84.8	87.0	15.2	13.0
16 to 19 years	58.0	62.8	42.0	37.2
20 to 24 years	75.5	76.8	24.5	23.1
25 to 34 years	85.8	85.9	14.2	14.1
35 to 44 years	88.7	90.2	11.2	9.8
45 to 54 years	89.7	93.6	10.3	6.4
55 to 64 years	92.0	95.7	8.0	4.2
65 years and over	95.2	95.5	4.9	4.9
Hispanic[2]	88.5	88.7	11.5	11.3
16 to 19 years	71.9	73.5	28.2	26.4
20 to 24 years	86.3	87.7	13.7	12.4
25 to 34 years	89.9	88.9	10.1	11.1
35 to 44 years	90.2	90.2	9.8	9.7
45 to 54 years	91.1	91.4	8.9	8.5
55 to 64 years	89.8	93.5	10.3	6.2
65 years and over	91.5	91.1	7.8	8.8

Source: "Civilian Labor Force-Employment Status, by Sex, Race, and Age: 1992," U.S. Bureau of the Census, *Statistical Abstract of the United States*, 1993, p. 401. Primary source: U.S. Bureau of Labor Statistics, *Employment and Earnings*, monthly, January 1992. *Notes:* 1. Includes other races not shown separately. 2. Persons of Hispanic origin may be of any race.

★ 635 ★

Labor Force

Civilian Labor Force: Participation Rate, 1970 to 1992, and Projections, 2000 and 2005

For civilian noninstitutional population 16 years old and over. Annual averages of monthly figures. Rates are based on annual average civilian noninstitutional population of each specified group and represent proportion of each specified group in the civilian labor force. Based on Current Population Survey.

Race, sex, and age	1970	1980	1985	1990	1992	2000	2005
Civilian labor force (millions)							
Total[1]	82.8	106.9	115.5	124.8	127.0	142.9	150.7
White	73.6	93.6	99.9	107.2	108.5	120.3	125.8
Male	46.0	54.5	56.5	59.3	59.8	64.5	66.8
Female	27.5	39.1	43.5	47.9	48.7	55.8	58.9
Black[2]	9.2	10.9	12.4	13.5	13.9	16.5	17.8
Male	5.2	5.6	6.2	6.7	6.9	8.1	8.7
Female	4.0	5.3	6.1	6.8	7.0	8.4	9.1
Hispanic[3]	(NA)	6.1	7.7	9.6	10.1	14.2	16.8
Male	(NA)	3.8	4.7	5.8	6.1	8.4	9.9
Female	(NA)	2.3	3.0	3.8	4.0	5.8	6.9
Participation rate (percent)							
Total[1]	60.4	63.8	64.8	66.4	66.3	68.7	69.0
White	60.2	64.1	65.0	66.8	66.7	69.3	39.7
Male	80.0	78.2	77.0	76.9	76.4	76.7	76.2
Female	42.6	51.2	54.1	57.5	57.8	62.3	63.5
Black[2]	61.8	61.0	62.9	63.3	63.3	65.7	65.6
Male	76.5	70.3	70.8	70.1	69.7	71.0	70.2
Female	49.5	53.1	56.5	57.8	58.0	61.2	61.7
Hispanic[3]	(NA)	64.0	64.6	67.0	66.5	69.3	69.9
Male	(NA)	81.4	80.3	81.2	80.5	81.8	81.6
Female	(NA)	47.4	49.3	53.0	52.6	56.6	58.0

Source: "Civilian Labor Force Participation Rates by Race, Hispanic Origin, Sex, and Age, 1970 to 1992, and Projections, 2000 and 2005," U.S. Bureau of the Census, *Statistical Abstract of the United States*, 1993, p. 393. Primary source: U.S. Bureau of Labor Statistics, Bulletin 2307; *Employment and Earnings*, monthly, January issues; *Monthly Labor Review*, November 1991; and unpublished data. *Notes:* NA Not available. 1. Beginning 1980, includes other races not shown separately. 2. For 1970, Black and other. 3. Persons of Hispanic origin may be of any race.

★ 636 ★

Labor Force

Civilian Labor Force: Participation Rate, 1980 and 1992

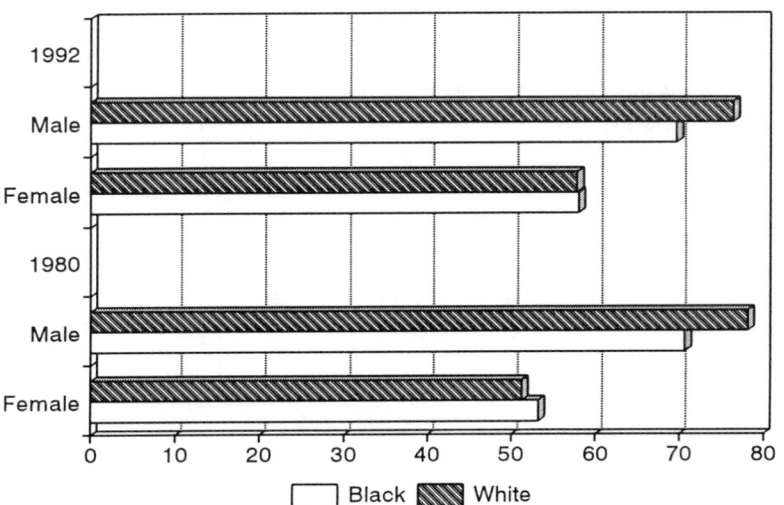

Annual averages.

	1980		1992	
	Male	Female	Male	Female
Black	70.6	53.2	69.7	58.0
White	78.2	51.2	76.4	57.8

Source: "Civilian Labor Force Participation Rates, by Sex and Race: 1980 to 1992," Claudette E. Bennett, *The Black Population in the United States: March 1992*, p. 10.

★ 637 ★

Labor Force

Civilian Labor Force: Participation Rates, 1975 and 1990, and Projections to 2005

(Percent)

Group	Participation			Annual growth rate	
	1975	1990	2005	1975-1990	1990-2005
Total					
16 years and over	61.2	66.4	69.0	0.5	0.3
16 to 24	64.6	67.3	69.5	.3	.2
25 to 54	74.1	83.5	87.3	.8	.3
55 and over	34.6	30.2	34.6	-.9	.9
Men					
16 years and over	77.9	76.1	75.4	-.2	-.1
16 to 24	72.4	71.5	73.1	-.1	.1

[Continued]

★ 637 ★

Civilian Labor Force: Participation Rates, 1975 and 1990, and Projections to 2005

[Continued]

Group	Participation			Annual growth rate	
	1975	1990	2005	1975-1990	1990-2005
25 to 54	94.4	93.5	92.4	-.1	-.1
55 and over	49.3	39.3	41.8	-1.5	.4
Women					
16 years and over	46.3	57.5	63.0	1.5	.6
16 to 24	57.2	63.1	66.0	.7	.3
25 to 54	55.1	74.1	82.3	2.0	.7
55 and over	23.1	23.0	28.7	0	1.5
White					
16 years and over	61.5	66.8	69.7	.6	.3
Black					
16 years and over	58.8	63.3	65.6	.5	.2
Asian and other					
16 years and over[1]	62.4	64.9	66.4	.3	.2
Hispanic					
16 years and over[2]	3	67.0	69.9	.7[4]	.3

Source: "Civilian Labor Force Participation Rates by Sex, Age, Race, and Hispanic Origin, 1975 and 1990, and Moderate Growth Projections to 2005." U.S. Department of Labor, Employment and Training Administration, *Training and Employment Report of the Secretary of Labor,* Covering the Period July 1988-September 1990, p. 99. Primary source: U.S. Department of Labor, Bureau of Labor Statistics. *Notes:* 1. The "Asian and other" group includes (1)Asians and Pacific Islanders and (2) American Indians and Alaskan natives. 2. Persons of Hispanic origin may be of any race. 3. Data on Hispanics were not available before 1980. 4. 1976-90.

★ 638 ★
Labor Force

Civilian Noninstitutional Employment in the U.S.: Annual Averages, 1992

Numbers in thousands.

Area and population	Civilian non-institutional population	Civilian labor force		Employment		Unemployed		
		Number	Percent of population	Number	Percent of population	Number	Rate	Error range of rate[1]
United States[2]								
Total	191,576	126,982	66.3	117,598	61.4	9,384	7.4	7.3-7.5
Men	91,541	69,184	75.6	63,805	69.7	5,380	7.8	7.6-7.9
Women	100,035	57,798	57.8	53,793	53.8	4,005	6.9	6.8-7.1
Both sexes, 16 to 19 years	13,161	6,751	51.3	5,398	41.0	1,352	20.0	19.5-20.6
White	162,658	108,526	66.7	101,479	62.4	7,047	6.5	6.4-6.6
Men	78,351	59,830	76.4	55,709	71.1	4,121	6.9	6.8-7.0
Women	84,307	48,696	57.8	45,770	54.3	2,926	6.0	5.9-6.1
Both sexes, 16 to 19 years	10,506	5,744	54.7	4,761	45.3	983	17.1	16.5-17.7

[Continued]

★ 638 ★

Civilian Noninstitutional Employment in the U.S.: Annual Averages, 1992
[Continued]

Area and population	Civilian non-institutional population	Civilian labor force		Employment		Unemployed		
		Number	Percent of population	Number	Percent of population	Number	Rate	Error range of rate[1]
Black	21,958	13,891	63.3	11,933	54.3	1,958	14.1	13.7-14.5
Men	9,888	6,892	69.7	5,846	59.1	1,046	15.2	14.6-15.8
Women	12,069	6,999	58.0	6,087	50.4	912	13.0	12.5-13.6
Both sexes, 16 to 19 years	2,074	787	37.9	474	22.9	313	39.8	37.5-42.0
Hispanic origin	15,244	10,131	66.5	8,971	58.8	1,160	11.4	11.0-11.9
Men	7,569	6,091	80.5	5,388	71.2	703	11.5	11.0-12.1
Women	7,674	4,040	52.6	3,584	46.7	456	11.3	10.7-11.9
Both sexes, 16 to 19 years	1,490	678	45.5	492	33.0	186	27.5	25.5-29.3
Single (never married)	47,284	33,498	70.8	29,454	62.3	4,044	12.1	11.9-12.3
Married, spouse present	108,255	74,148	68.5	70,440	65.1	3,708	5.0	4.9-5.1
Other marital status[3]	36,036	19,337	53.7	17,705	49.1	1,632	8.4	8.0-8.9

Source: "Census Regions and Divisions: Employment Status of the Civilian Noninstitutional Population by Sex, Age, Race, Hispanic Origin, and Marital Status, 1992 Annual Averages," U.S. Department of Labor, *Geographical Profile of Employment and Unemployment*, 1992, p.5. *Notes:* Data for demographic groups are not shown when they do not meet BLS publication standards of reliability for the particular area based on the sample in that area. Items may not add to totals or compute to displayed percentages because of rounding. Detail for race and Hispanic-origin groups will not add to totals because data for the "other races" group are not presented and Hispanics are included in both the white and black populations. 1. Error ranges are calculated at the 90-percent confidence interval, which means that if repeated samples were drawn from the same population and an error range constructed around each sample estimate, in 9 out of 10 cases the true value based on a complete census of the population would be contained within these error ranges. 2. Because of separate processing and weighing procedures, totals for the United States differ from the results obtained by aggregating the totals for regions and States. 3. "Other marital status" includes divorced, widowed, separated, and married, with spouse absent.

★ 639 ★

Labor Force

Civilian Noninstitutional Employment in the U.S.: East North Central Division, 1992

Numbers in thousands.

Area and population	Civilian non-institutional population	Civilian labor force		Employment		Unemployed		
		Number	Percent of population	Number	Percent of population	Number	Rate	Error range of rate[1]
East North Central Division								
Total	32,367	21,729	67.1	20,146	62.2	1,583	7.3	7.1-7.5
Men	15,455	11,804	76.4	10,899	70.5	905	7.7	7.4-8.0
Women	16,912	9,926	58.7	9,247	54.7	679	6.8	6.5-7.2
Both sexes, 16 to 19 years	2,295	1,300	56.6	1,048	45.7	252	19.4	18.0-20.7
White	28,244	19,235	68.1	18,038	63.9	1,198	6.2	6.0-6.4
Men	13,626	10,566	77.5	9,868	72.4	698	6.6	6.3-6.9
Women	14,619	8,669	59.3	8,170	55.9	499	5.8	5.4-6.1
Both sexes, 16 to 19 years	1,888	1,131	59.9	946	50.1	185	16.4	15.0-17.7
Black	3,520	2,097	59.6	1,745	49.6	352	16.8	15.8-17.8
Men	1,531	1,007	65.7	820	53.6	186	18.5	17.0-20.1

[Continued]

★ 639 ★

Civilian Noninstitutional Employment in the U.S.: East North Central Division, 1992
[Continued]

Area and population	Civilian non-institutional population	Civilian labor force		Employment		Unemployed		
		Number	Percent of population	Number	Percent of population	Number	Rate	Error range of rate[1]
Women	1,989	1,090	54.8	924	46.5	166	15.2	13.8-16.6
Both sexes, 16 to 19 years	348	140	40.2	80	23.1	60	42.6	36.8-48.4
Hispanic origin	904	634	70.2	563	62.3	71	11.2	9.7-12.8
Men	477	399	83.8	356	74.6	44	11.0	9.1-12.9
Women	427	235	54.9	207	46.5	27	11.6	9.1-14.2
Both sexes, 16 to 19 years	87	46	52.9	34	38.8	12	26.7	18.4-34.9
Single (never married)	8,109	5,887	72.6	5,176	63.8	711	12.1	11.6-12.6
Married, spouse present	18,242	12,663	69.4	12,059	66.1	604	4.8	4.5-5.0
Other marital status[2]	6,016	3,180	52.9	2,912	48.4	268	8.4	7.8-9.0

Source: "Census Regions and Divisions: Employment Status of the Civilian Noninstitutional Population by Sex, Age, Race, Hispanic Origin, and Marital Status, 1992 Annual Averages," U.S. Department of Labor, *Geographical Profile of Employment and Unemployment,* 1992, p.6. *Notes:* Data for demographic groups are not shown when they do not meet BLS publication standards of reliability for the particular area based on the sample in that area. Items may not add to totals or compute to displayed percentages because of rounding. Detail for race and Hispanic-origin groups will not add to totals because data for the "other races" group are not presented and Hispanics are included in both the white and black populations. 1. Error ranges are calculated at the 90-percent confidence interval, which means that if repeated samples were drawn from the same population and an error range constructed around each sample estimate, in 9 out of 10 cases the true value based on a complete census of the population would be contained within these error ranges. 2. "Other marital status" includes divorced, widowed, separated, and married, with spouse absent.

★ 640 ★
Labor Force

Civilian Noninstitutional Employment in the U.S.: East South Central Division, 1992

Numbers in thousands.

Area and population	Civilian non-institutional population	Civilian labor force		Employment		Unemployed		
		Number	Percent of population	Number	Percent of population	Number	Rate	Error range of rate[1]
East South Central Division								
Total	11,710	7,304	62.4	6,791	58.0	513	7.0	6.6-7.4
Men	5,499	3,955	71.9	3,689	67.1	265	6.7	6.2-7.2
Women	6,210	3,349	53.9	3,101	49.9	248	7.4	6.8-8.0
Both sexes, 16 to 19 years	906	418	46.2	334	36.8	85	20.2	17.7-22.8
White	9,459	5,965	63.1	5,621	59.4	344	5.8	5.4-6.2
Men	4,499	3,303	73.4	3,116	69.3	186	5.6	5.1-6.2
Women	4,960	2,662	53.7	2,505	50.5	157	5.9	5.3-6.5
Both sexes, 16 to 19 years	657	336	51.1	282	42.9	54	16.1	13.5-18.8
Black	2,172	1,294	59.6	1,127	51.9	167	12.9	11.7-14.2
Men	962	623	64.8	545	56.7	78	12.5	10.7-14.3
Women	1,210	671	55.4	582	48.0	89	13.3	11.6-15.1
Both sexes, 16 to 19 years	244	81	33.0	50	20.6	30	37.5	30.6-44.4

[Continued]

★ 640 ★

Civilian Noninstitutional Employment in the U.S.: East South Central Division, 1992

[Continued]

Area and population	Civilian non-institutional population	Civilian labor force		Employment		Unemployed		
		Number	Percent of population	Number	Percent of population	Number	Rate	Error range of rate[1]
Single (never married)	2,562	1,633	63.7	1,419	55.4	214	13.1	12.0-14.2
Married, spouse present	6,793	4,533	66.7	4,326	63.7	206	4.6	4.1-5.0
Other marital status[2]	2,355	1,138	48.3	1,046	44.4	93	8.1	7.0-9.2

Source: "Census Regions and Divisions: Employment Status of the Civilian Noninstitutional Population by Sex, Age, Race, Hispanic Origin, and Marital Status, 1992 Annual Averages," U.S. Department of Labor, *Geographical Profile of Employment and Unemployment,* 1992, p.8. *Notes:* Data for demographic groups are not shown when they do not meet BLS publication standards of reliability for the particular area based on the sample in that area. Items may not add to totals or compute to displayed percentages because of rounding. Detail for race and Hispanic-origin groups will not add to totals because data for the "other races" group are not presented and Hispanics are included in both the white and black populations. 1. Error ranges are calculated at the 90-percent confidence interval, which means that if repeated samples were drawn from the same population and an error range constructed around each sample estimate, in 9 out of 10 cases the true value based on a complete census of the population would be contained within these error ranges. 2. "Other marital status" includes divorced, widowed, separated, and married, with spouse absent.

★ 641 ★

Labor Force

Civilian Noninstitutional Employment in the U.S.: Middle Atlantic Division, 1992

Numbers in thousands.

Area and population	Civilian non-institutional population	Civilian labor force		Employment		Unemployed		
		Number	Percent of population	Number	Percent of population	Number	Rate	Error range of rate[1]
Middle Atlantic Division								
Total	29,279	18,527	63.3	17,018	58.1	1,509	8.1	7.9-8.4
Men	13,792	10,122	73.4	9,214	66.8	908	9.0	8.7-9.3
Women	15,487	8,406	54.3	7,805	50.4	601	7.2	6.9-7.4
Both sexes, 16 to 19 years	1,891	802	42.4	641	33.9	161	20.1	18.7-21.6
White	24,801	15,804	63.7	14,658	59.1	1,146	7.3	7.0-7.5
Men	11,762	8,722	74.2	8,030	68.3	691	7.9	7.6-8.2
Women	13,039	7,083	54.3	6,628	50.8	455	6.4	6.1-6.7
Both sexes, 16 to 19 years	1,501	691	46.0	570	38.0	121	17.5	16.0-19.0
Black	3,568	2,125	59.5	1,806	50.6	319	15.0	14.1-15.9
Men	1,596	1,066	66.8	875	54.8	191	17.9	16.5-19.2
Women	1,973	1,059	53.7	931	47.2	128	12.1	11.0-13.2
Both sexes, 16 to 19 years	313	90	28.7	54	17.1	36	40.4	34.0-46.8
Hispanic origin	2,091	1,233	59.0	1,067	51.0	167	13.5	12.4-14.6
Men	953	711	74.6	608	63.8	103	14.5	13.0-15.9
Women	1,138	523	45.9	459	40.3	64	12.2	10.6-10.8
Both sexes, 16 to 19 years	202	55	27.4	37	18.2	18	33.4	26.1-40.7
Single (never married)	8,059	5,373	66.7	4,699	58.3	674	12.5	12.1-13.0

[Continued]

★ 641 ★

Civilian Noninstitutional Employment in the U.S.: Middle Atlantic Division, 1992

[Continued]

Area and population	Civilian non-institutional population	Civilian labor force		Employment		Unemployed		
		Number	Percent of population	Number	Percent of population	Number	Rate	Error range of rate[1]
Married, spouse present	15,914	10,651	66.9	10,049	63.1	602	5.7	5.4-5.9
Other marital status[2]	5,306	2,504	47.2	2,271	42.8	233	9.3	8.7-9.9

Source: "Census Regions and Divisions: Employment Status of the Civilian Noninstitutional Population by Sex, Age, Race, Hispanic Origin, and Marital Status, 1992 Annual Averages," U.S. Department of Labor, *Geographical Profile of Employment and Unemployment,* 1992, p.6. *Notes:* Data for demographic groups are not shown when they do not meet BLS publication standards of reliability for the particular area based on the sample in that area. Items may not add to totals or compute to displayed percentages because of rounding. Detail for race and Hispanic-origin groups will not add to totals because data for the "other races" group are not presented and Hispanics are included in both the white and black populations. 1. Error ranges are calculated at the 90-percent confidence interval, which means that if repeated samples were drawn from the same population and an error range constructed around each sample estimate, in 9 out of 10 cases the true value based on a complete census of the population would be contained within these error ranges. 2. "Other marital status" includes divorced, widowed, separated, and married, with spouse absent.

★ 642 ★

Labor Force

Civilian Noninstitutional Employment in the U.S.: Midwest Region, 1992

Numbers in thousands.

Area and population	Civilian non-institutional population	Civilian labor force		Employment		Unemployed		
		Number	Percent of population	Number	Percent of population	Number	Rate	Error range of rate[1]
Midwest region								
Total	45,895	31,269	68.1	29,228	63.7	2,041	6.5	6.4-6.7
Men	21,919	16,905	77.1	15,741	71.8	1,164	6.9	6.6-7.1
Women	23,976	14,364	59.9	13,487	56.3	877	6.1	5.9-6.4
Both sexes, 16 to 19 years	3,289	1,923	58.4	1,595	48.5	327	17.0	16.0-18.1
White	40,906	28,211	69.0	26,634	65.1	1,577	5.6	5.4-5.8
Men	19,695	15,383	78.1	14,466	73.5	917	6.0	5.7-6.2
Women	21,211	12,827	60.5	12,168	57.4	659	5.1	4.9-5.4
Both sexes, 16 to 19 years	2,805	1,720	61.3	1,470	52.4	250	14.5	13.5-15.6
Black	4,111	2,494	60.7	2,076	50.5	418	16.8	15.8-17.7
Men	1,793	1,196	66.7	977	54.5	219	18.3	16.9-19.8
Women	2,318	1,298	56.0	1,100	47.4	199	15.3	14.0-16.6
Both sexes, 16 to 19 years	393	162	41.2	95	24.2	67	41.3	35.9-46.6
Hispanic origin	1,055	744	70.5	665	63.1	79	10.6	9.2-12.0
Men	549	460	83.8	410	74.7	50	10.8	9.0-12.6
Women	505	284	56.1	255	50.4	29	10.2	8.0-12.4
Both sexes, 16 to 19 years	99	51	51.5	37	37.9	13	26.3	18.6-34.1
Single (never married)	11,361	8,366	73.6	7,442	65.5	924	11.0	10.6-11.5

[Continued]

★ 642 ★

Civilian Noninstitutional Employment in the U.S.: Midwest Region, 1992

[Continued]

Area and population	Civilian non-institutional population	Civilian labor force		Employment		Unemployed		
		Number	Percent of population	Number	Percent of population	Number	Rate	Error range of rate[1]
Married, spouse present	26,226	18,494	70.5	17,713	67.5	781	4.2	4.0-4.4
Other marital status[2]	8,309	4,409	53.1	4,073	49.0	336	7.6	7.1-8.1

Source: "Census Regions and Divisions: Employment Status of the Civilian Noninstitutional Population by Sex, Age, Race, Hispanic Origin, and Marital Status, 1992 Annual Averages," U.S. Department of Labor, *Geographical Profile of Employment and Unemployment*, 1992, p.6. *Notes:* Data for demographic groups are not shown when they do not meet BLS publication standards of reliability for the particular area based on the sample in that area. Items may not add to totals or compute to displayed percentages because of rounding. Detail for race and Hispanic-origin groups will not add to totals because data for the "other races" group are not presented and Hispanics are included in both the white and black populations. 1. Error ranges are calculated at the 90-percent confidence interval, which means that if repeated samples were drawn from the same population and an error range constructed around each sample estimate, in 9 out of 10 cases the true value based on a complete census of the population would be contained within these error ranges. 2. "Other marital status" includes divorced, widowed, separated, and married, with spouse absent.

★ 643 ★

Labor Force

Civilian Noninstitutional Employment in the U.S.: Mountain Division, 1992

Numbers in thousands.

Area and population	Civilian non-institutional population	Civilian labor force		Employment		Unemployed		
		Number	Percent of population	Number	Percent of population	Number	Rate	Error range of rate[1]
Mountain Division								
Total	10,203	6,868	67.3	6,427	63.0	442	6.4	6.1-6.7
Men	4,947	3,762	76.1	3,511	71.0	251	6.7	6.3-7.1
Women	5,257	3,106	59.1	2,915	55.5	191	6.1	5.7-6.6
Both sexes, 16 to 19 years	723	401	55.5	320	44.2	81	20.3	18.5-22.1
White	9,542	6,431	67.4	6,036	63.3	395	6.1	5.8-6.5
Men	4,641	3,539	76.3	3,314	71.4	225	6.4	5.9-6.8
Women	4,901	2,891	59.0	2,722	55.5	170	5.9	5.4-6.3
Both sexes, 16 to 19 years	659	375	56.9	302	45.9	72	19.3	17.5-21.1
Black	254	176	69.2	1569	61.3	20	11.3	8.8-13.9
Men	126	94	74.7	84	66.4	10	11.0	7.6-14.5
Women	129	82	63.8	72	56.3	10	11.7	7.9-15.5
Hispanic origin	1,236	811	65.6	736	59.5	75	9.3	8.2-10.4
Men	607	473	78.0	430	70.9	43	9.1	7.6-10.5
Women	629	338	53.7	306	48.5	33	9.6	7.9-11.4
Both sexes, 16 to 19 years	113	50	44.0	34	30.4	15	30.9	25.1-36.6
Single (never married)	2,216	1,633	73.7	1,447	65.3	186	11.4	10.6-12.2

[Continued]

★ 643 ★

Civilian Noninstitutional Employment in the U.S.: Mountain Division, 1992

[Continued]

Area and population	Civilian non-institutional population	Civilian labor force		Employment		Unemployed		
		Number	Percent of population	Number	Percent of population	Number	Rate	Error range of rate[1]
Married, spouse present	6,059	4,097	67.6	3,922	64.7	175	4.3	3.9-4.6
Other marital status[2]	1,929	1,138	59.0	1,058	54.8	81	7.1	6.3-7.8

Source: "Census Regions and Divisions: Employment Status of the Civilian Noninstitutional Population by Sex, Age, Race, Hispanic Origin, and Marital Status, 1992 Annual Averages," U.S. Department of Labor, *Geographical Profile of Employment and Unemployment*, 1992, p.9. *Notes:* Data for demographic groups are not shown when they do not meet BLS publication standards of reliability for the particular area based on the sample in that area. Items may not add to totals or compute to displayed percentages because of rounding. Detail for race and Hispanic-origin groups will not add to totals because data for the "other races" group are not presented and Hispanics are included in both the white and black populations. 1. Error ranges are calculated at the 90-percent confidence interval, which means that if repeated samples were drawn from the same population and an error range constructed around each sample estimate, in 9 out of 10 cases the true value based on a complete census of the population would be contained within these error ranges. 2. "Other marital status" includes divorced, widowed, separated, and married, with spouse absent.

★ 644 ★

Labor Force

Civilian Noninstitutional Employment in the U.S.: New England Division, 1992

Numbers in thousands.

Area and population	Civilian non-institutional population	Civilian labor force		Employment		Unemployed		
		Number	Percent of population	Number	Percent of population	Number	Rate	Error range of rate[1]
New England Division								
Total	10,211	7,062	69.2	6,500	63.7	562	8.0	7.7-8.3
Men	4,831	3,737	77.4	3,397	70.3	340	9.1	8.7-9.5
Women	5,380	3,325	61.8	3,103	57.7	222	6.7	6.3-7.1
Both sexes, 16 to 19 years	630	350	55.5	289	45.8	61	17.4	15.6-19.3
White	9,601	6,643	69.2	6,138	63.9	505	7.6	7.3-7.9
Men	4,549	3,516	77.3	3,209	70.6	306	8.7	8.3-9.1
Women	5,052	3,128	61.9	2,929	58.0	199	6.4	6.0-6.8
Both sexes, 16 to 19 years	576	328	57.0	274	47.5	54	16.5	14.7-18.4
Black	422	300	70.9	255	60.4	45	14.9	12.8-17.0
Men	194	157	81.1	129	66.7	28	17.8	14.7-20.9
Women	229	142	62.2	126	55.0	17	11.7	9.0-14.4
Hispanic origin	289	184	63.5	156	53.9	28	15.2	12.6-17.9
Men	124	95	76.1	77	62.2	17	18.3	14.3-22.3
Women	165	89	54.1	78	47.6	11	12.0	8.6-15.4
Single (never married)	2,793	2,045	73.2	1,811	64.8	234	11.5	10.8-12.1

[Continued]

★ 644 ★

Civilian Noninstitutional Employment in the U.S.: New England Division, 1992

[Continued]

Area and population	Civilian non-institutional population	Civilian labor force		Employment		Unemployed		
		Number	Percent of population	Number	Percent of population	Number	Rate	Error range of rate[1]
Married, spouse present	5,667	4,082	72.0	3,837	67.7	244	6.0	5.6-6.3
Other marital status[23]	1,751	935	53.4	851	48.6	84	9.0	8.1-9.8

Source: "Census Regions and Divisions: Employment Status of the Civilian Noninstitutional Population by Sex, Age, Race, Hispanic Origin, and Marital Status, 1992 Annual Averages," U.S. Department of Labor, *Geographical Profile of Employment and Unemployment*, 1992, p.5. *Notes:* Data for demographic groups are not shown when they do not meet BLS publication standards of reliability for the particular area based on the sample in that area. Items may not add to totals or compute to displayed percentages because of rounding. Detail for race and Hispanic-origin groups will not add to totals because data for the "other races" group are not presented and Hispanics are included in both the white and black populations. 1. Error ranges are calculated at the 90-percent confidence interval, which means that if repeated samples were drawn from the same population and an error range constructed around each sample estimate, in 9 out of 10 cases the true value based on a complete census of the population would be contained within these error ranges. 2. "Other marital status" includes divorced, widowed, separated, and married, with spouse absent.

★ 645 ★

Labor Force

Civilian Noninstitutional Employment in the U.S.: Northeast Region, 1992

Numbers in thousands.

Area and population	Civilian non-institutional population	Civilian labor force		Employment		Unemployed		
		Number	Percent of population	Number	Percent of population	Number	Rate	Error range of rate[1]
Northeast Region								
Total	39,490	25,589	64.8	23,518	59.6	2,071	8.1	7.9-8.3
Men	18,623	13,859	74.4	12,611	67.7	1,248	9.0	8.8-9.3
Women	20,867	11,730	56.2	10,908	52.3	823	7.0	6.8-7.3
Both sexes, 16 to 19 years	2,521	1,152	45.7	929	36.9	222	19.3	18.1-20.5
White	34,402	22,448	65.3	20,796	60.5	1,615	7.4	7.2-7.5
Men	16,311	12,237	75.0	11,240	68.9	997	8.2	7.9-8.4
Women	18,092	10,210	56.4	9,557	52.8	654	6.4	6.2-6.7
Both sexes, 16 to 19 years	2,077	1,019	49.1	844	40.6	175	17.2	16.0-18.4
Black	3,991	2,424	60.7	2,061	51.6	363	15.0	14.2-15.8
Men	1,789	1,223	68.3	1,004	56.1	218	17.9	16.7-19.1
Women	2,202	1,202	54.6	1,057	48.0	145	12.1	11.0-13.1
Both sexes, 16 to 19 years	352	107	30.5	66	18.8	41	38.4	32.8-44.1
Hispanic origin	2,380	1,417	59.5	1,222	51.4	195	13.7	12.8-14.7
Men	1,077	805	74.8	685	63.6	120	14.9	13.6-16.3
Women	1,303	612	47.0	537	41.2	75	12.2	10.8-13.6
Both sexes, 16 to 19 years	227	67	29.5	46	20.1	21	31.9	25.5-38.2
Single (never married)	10,852	7,418	68.4	6,510	60.0	908	12.2	11.9-12.6

[Continued]

★ 645 ★

Civilian Noninstitutional Employment in the U.S.: Northeast Region, 1992

[Continued]

Area and population	Civilian non-institutional population	Civilian labor force		Employment		Unemployed		
		Number	Percent of population	Number	Percent of population	Number	Rate	Error range of rate[1]
Married, spouse present	21,581	14,732	68.3	13,886	64.3	846	5.7	5.5-5.9
Other marital status[23]	7,057	3,439	48.7	3,122	44.2	317	9.2	8.7-9.7

Source: "Census Regions and Divisions: Employment Status of the Civilian Noninstitutional Population by Sex, Age, Race, Hispanic Origin, and Marital Status, 1992 Annual Averages," U.S. Department of Labor, *Geographical Profile of Employment and Unemployment,* 1992, p.5. *Notes:* Data for demographic groups are not shown when they do not meet BLS publication standards of reliability for the particular area based on the sample in that area. Items may not add to totals or compute to displayed percentages because of rounding. Detail for race and Hispanic-origin groups will not add to totals because data for the "other races" group are not presented and Hispanics are included in both the white and black populations. 1. Error ranges are calculated at the 90-percent confidence interval, which means that if repeated samples were drawn from the same population and an error range constructed around each sample estimate, in 9 out of 10 cases the true value based on a complete census of the population would be contained within these error ranges. 2. "Other marital status" includes divorced, widowed, separated, and married, with spouse absent.

★ 646 ★

Labor Force

Civilian Noninstitutional Employment in the U.S.: Pacific Division, 1992

Numbers in thousands.

Area and population	Civilian non-institutional population	Civilian labor force		Employment		Unemployed		
		Number	Percent of population	Number	Percent of population	Number	Rate	Error range of rate[1]
Pacific Division								
Total	30,180	20,157	66.8	18,415	61.0	1,742	8.6	8.4-8.9
Men	14,862	11,405	76.7	10,372	69.8	1,033	9.1	8.7-9.4
Women	15,318	8,752	57.1	8,043	52.5	709	8.1	7.7-8.5
Both sexes, 16 to 19 years	2,024	1,009	49.8	777	38.4	232	23.0	21.3-24.6
White	25,360	17,026	67.1	15,601	61.5	1,425	8.4	8.1-8.7
Men	12,584	9,750	77.5	8,888	70.6	862	8.8	8.5-9.2
Women	12,776	7,276	57.0	6,713	52.5	563	7.7	7.3-8.2
Both sexes, 16 to 19 years	1,648	858	52.1	674	40.9	184	21.5	19.7-23.2
Black	1,520	952	62.6	812	53.4	140	14.7	13.1-16.4
Men	715	503	70.3	430	60.1	73	14.5	12.2-16.8
Women	805	449	55.8	382	47.4	67	15.0	12.6-17.4
Both sexes, 16 to 19 years	125	55	43.6	32	25.7	22	41.1	30.4-51.8
Hispanic origin	5,706	3,850	67.5	3,351	58.7	500	13.0	12.2-13.8
Men	2,925	2,403	82.2	2,090	71.4	313	13.0	12.1-14.0
Women	2,781	1,447	52.0	1,261	45.3	186	12.9	11.6-14.1
Both sexes, 16 to 19 years	586	272	46.4	192	32.8	80	29.3	25.2-33.3
Single (never married)	7,873	5,725	72.7	5,012	63.7	713	12.5	11.9-13.0

[Continued]

★ 646 ★

Civilian Noninstitutional Employment in the U.S.: Pacific Division, 1992
[Continued]

Area and population	Civilian non-institutional population	Civilian labor force		Employment		Unemployed		
		Number	Percent of population	Number	Percent of population	Number	Rate	Error range of rate[1]
Married, spouse present	16,622	11,187	67.3	10,474	63.0	712	6.4	6.1-6.7
Other marital status[2]	5,685	3,245	57.1	2,928	51.5	317	9.8	9.1-10.5

Source: "Census Regions and Divisions: Employment Status of the Civilian Noninstitutional Population by Sex, Age, Race, Hispanic Origin, and Marital Status, 1992 Annual Averages," U.S. Department of Labor, *Geographical Profile of Employment and Unemployment*, 1992, p.9. Published by permission. *Notes:* Data for demographic groups are not shown when they do not meet BLS publication standards of reliability for the particular area based on the sample in that area. Items may not add to totals or compute to displayed percentages because of rounding. Detail for race and Hispanic-origin groups will not add to totals because data for the "other races" group are not presented and Hispanics are included in both the white and black populations. 1. Error ranges are calculated at the 90-percent confidence interval, which means that if repeated samples were drawn from the same population and an error range constructed around each sample estimate, in 9 out of 10 cases the true value based on a complete census of the population would be contained within these error ranges. 2. "Other marital status" includes divorced, widowed, separated, and married, with spouse absent.

★ 647 ★

Labor Force

Civilian Noninstitutional Employment in the U.S.: South Atlantic Division, 1992

Numbers in thousands.

Area and population	Civilian non-institutional population	Civilian labor force		Employment		Unemployed		
		Number	Percent of population	Number	Percent of population	Number	Rate	Error range of rate[1]
South Atlantic Division								
Total	34,066	22,440	65.9	20,847	61.2	1,592	7.1	6.9-7.3
Men	16,135	11,962	74.1	11,084	68.7	878	7.3	7.0-7.6
Women	17,931	10,477	58.4	9,763	54.4	714	6.8	6.5-7.1
Both sexes, 16 to 19 years	2,194	1,081	49.3	843	38.4	238	22.0	20.5-23.5
White	26,183	17,142	65.5	16,163	61.7	979	5.7	5.5-5.9
Men	12,547	9,339	74.4	8,785	70.0	553	5.9	5.6-6.2
Women	13,636	7,803	57.2	7,377	54.1	426	5.5	5.1-5.8
Both sexes, 16 to 19 years	1,455	788	54.2	658	45.3	129	16.4	14.8-18.0
Black	7,245	4,869	67.2	4,291	59.2	577	11.9	1.2-12.5
Men	3,294	2,399	72.8	2,095	63.6	304	12.7	11.8-13.6
Women	3,951	2,469	62.5	2,196	55.6	273	11.1	10.2-11.9
Both sexes, 16 to 19 years	683	270	39.6	166	24.3	104	38.6	34.8-42.4
Hispanic origin	1,741	1,206	69.3	1,092	62.7	114	9.4	8.3-10.5
Men	860	706	82.1	646	75.1	61	8.6	7.2-9.9
Women	881	500	56.7	447	50.7	53	10.7	8.9-12.5
Both sexes, 16 to 19 years	125	66	52.7	50	39.9	16	24.2	17.3-31.2
Single (never married)	8,060	5,719	71.0	5,020	62.3	699	12.2	11.7-12.8

[Continued]

★ 647 ★

Civilian Noninstitutional Employment in the U.S.: South Atlantic Division, 1992

[Continued]

Area and population	Civilian non-institutional population	Civilian labor force		Employment		Unemployed		
		Number	Percent of population	Number	Percent of population	Number	Rate	Error range of rate[1]
Married, spouse present	19,177	12,963	67.6	12,371	64.5	592	4.6	4.3-4.8
Other marital status[2]	6,829	3,758	55.0	3,456	50.6	302	8.0	7.5-8.6

Source: "Census Regions and Divisions: Employment Status of the Civilian Noninstitutional Population by Sex, Age, Race, Hispanic Origin, and Marital Status, 1992 Annual Averages," U.S. Department of Labor, *Geographical Profile of Employment and Unemployment*, 1992, p.7. *Notes:* Data for demographic groups are not shown when they do not meet BLS publication standards of reliability for the particular area based on the sample in that area. Items may not add to totals or compute to displayed percentages because of rounding. Detail for race and Hispanic-origin groups will not add to totals because data for the "other races" group are not presented and Hispanics are included in both the white and black populations. 1. Error ranges are calculated at the 90-percent confidence interval, which means that if repeated samples were drawn from the same population and an error range constructed around each sample estimate, in 9 out of 10 cases the true value based on a complete census of the population would be contained within these error ranges. 2. "Other marital status" includes divorced, widowed, separated, and married, with spouse absent.

★ 648 ★

Labor Force

Civilian Noninstitutional Employment in the U.S.: South Region, 1992

Numbers in thousands.

Area and population	Civilian non-institutional population	Civilian labor force		Employment		Unemployed		
		Number	Percent of population	Number	Percent of population	Number	Rate	Error range of rate[1]
South Region								
Total	65,808	43,098	65.5	40,011	60.8	3,088	7.2	7.0-7.3
Men	31,190	23,253	74.6	21,570	69.2	1,683	7.2	7.0-7.5
Women	34,618	19,845	57.3	18,441	53.3	1,405	7.1	6.8-7.3
Both sexes, 16 to 19 years	4,604	2,266	49.2	1,776	38.6	490	21.6	20.6-22.6
White	52,448	34,411	65.6	32,412	61.8	1,999	5.8	5.6-6.0
Men	25,120	18,920	75.3	17,801	70.9	1,119	5.9	5.7-6.1
Women	27,327	15,491	56.7	14,611	53.5	880	5.7	5.4-5.9
Both sexes, 16 to 19 years	3,317	1,772	53.4	1,471	44.3	301	17.0	15.9-18.1
Black	12,081	7,844	64.9	6,827	56.5	1,016	13.0	12.4-13.5
Men	5,465	3,876	70.9	3,352	61.3	524	13.5	12.8-14.3
Women	6,616	3,968	60.0	3,476	52.5	492	12.4	11.7-13.1
Both sexes, 16 to 19 years	1,176	450	38.3	272	23.1	178	39.6	36.6-42.7
Hispanic origin	4,867	3,309	68.0	2,998	61.6	311	9.4	8.7-10.1
Men	2,411	1,950	80.9	1,773	73.5	177	9.1	8.2-9.9
Women	2,456	1,359	55.3	1,225	49.9	134	9.8	8.8-10.9
Both sexes, 16 to 19 years	464	238	51.3	182	39.2	57	23.7	20.4-27.0
Single (never married)	14,982	10,355	69.1	9,042	60.3	1,313	12.7	12.3-13.1

[Continued]

★ 648 ★

Civilian Noninstitutional Employment in the U.S.: South Region, 1992

[Continued]

Area and population	Civilian non-institutional population	Civilian labor force		Employment		Unemployed		
		Number	Percent of population	Number	Percent of population	Number	Rate	Error range of rate[1]
Married, spouse present	37,769	25,638	67.9	24,445	64.7	1,193	4.7	4.5-4.8
Other marital status[2]	13,507	7,106	54.4	6,524	50.0	582	8.2	7.8-8.6

Source: "Census Regions and Divisions: Employment Status of the Civilian Noninstitutional Population by Sex, Age, Race, Hispanic Origin, and Marital Status, 1992 Annual Averages," U.S. Department of Labor, *Geographical Profile of Employment and Unemployment*, 1992, p.7. *Notes:* Data for demographic groups are not shown when they do not meet BLS publication standards of reliability for the particular area based on the sample in that area. Items may not add to totals or compute to displayed percentages because of rounding. Detail for race and Hispanic-origin groups will not add to totals because data for the "other races" group are not presented and Hispanics are included in both the white and black populations. 1. Error ranges are calculated at the 90-percent confidence interval, which means that if repeated samples were drawn from the same population and an error range constructed around each sample estimate, in 9 out of 10 cases the true value based on a complete census of the population would be contained within these error ranges. 2. "Other marital status" includes divorced, widowed, separated, and married, with spouse absent.

★ 649 ★

Labor Force

Civilian Noninstitutional Employment in the U.S.: West North Central Division, 1992

Numbers in thousands.

Area and population	Civilian non-institutional population	Civilian labor force		Employment		Unemployed		
		Number	Percent of population	Number	Percent of population	Number	Rate	Error range of rate[1]
West North Central Division								
Total	13,528	9,540	70.5	9,082	67.1	458	4.8	4.5-5.1
Men	6,464	5,101	78.9	4,842	74.9	260	5.1	4.7-5.5
Women	7,064	4,439	62.8	4,240	60.0	198	4.5	4.1-4.9
Both sexes, 16 to 19 years	994	623	62.6	547	55.0	75	12.1	10.5-13.7
White	12,661	8,976	70.9	8,597	67.9	379	4.2	4.0-4.5
Men	6,069	4,817	79.4	4,598	75.8	219	4.5	4.2-4.9
Women	6,592	4,159	63.1	3,998	60.7	160	3.9	3.5-4.2
Both sexes, 16 to 19 years	916	589	64.3	524	57.2	65	11.1	9.5-12.7
Black	591	397	67.3	332	56.2	66	16.5	14.1-18.9
Men	262	190	72.5	156	59.8	33	17.5	13.9-21.0
Women	329	208	63.2	175	53.3	33	15.7	12.4-18.9
Hispanic origin	151	110	72.7	102	67.7	8	6.9	3.9-9.9
Men	73	61	83.5	55	75.4	6	9.7	5.0-14.5
Women	78	49	62.6	47	60.5	2	3.3	.1-6.5
Single (never married)	3,252	2,479	76.2	2,267	69.7	213	8.6	7.9-9.3

[Continued]

★ 649 ★

Civilian Noninstitutional Employment in the U.S.: West North Central Division, 1992

[Continued]

Area and population	Civilian non-institutional population	Civilian labor force		Employment		Unemployed		
		Number	Percent of population	Number	Percent of population	Number	Rate	Error range of rate[1]
Married, spouse present	7,984	5,831	73.0	5,654	70.8	177	3.0	2.8-3.3
Other marital status[2]	2,293	1,229	53.6	1,161	50.7	68	5.5	4.7-6.3

Source: "Census Regions and Divisions: Employment Status of the Civilian Noninstitutional Population by Sex, Age, Race, Hispanic Origin, and Marital Status, 1992 Annual Averages," U.S. Department of Labor, *Geographical Profile of Employment and Unemployment*, 1992, p.7. *Notes:* Data for demographic groups are not shown when they do not meet BLS publication standards of reliability for the particular area based on the sample in that area. Items may not add to totals or compute to displayed percentages because of rounding. Detail for race and Hispanic-origin groups will not add to totals because data for the "other races" group are not presented and Hispanics are included in both the white and black populations. 1. Error ranges are calculated at the 90-percent confidence interval, which means that if repeated samples were drawn from the same population and an error range constructed around each sample estimate, in 9 out of 10 cases the true value based on a complete census of the population would be contained within these error ranges. 2. "Other marital status" includes divorced, widowed, separated, and married, with spouse absent.

★ 650 ★

Labor Force

Civilian Noninstitutional Employment in the U.S.: West Region, 1992

Numbers in thousands.

Area and population	Civilian non-institutional population	Civilian labor force		Employment		Unemployed		
		Number	Percent of population	Number	Percent of population	Number	Rate	Error range of rate[1]
West Region								
Total	40,383	27,025	66.9	24,841	61.5	2,184	8.1	7.9-8.3
Men	19,809	15,167	76.6	13,883	70.1	1,284	8.5	8.2-8.8
Women	20,575	11,858	57.6	10,958	53.3	900	7.6	7.3-7.9
Both sexes, 16 to 19 years	2,747	1,410	51.3	1,097	39.9	313	22.2	20.9-23.5
White	34,902	23,457	67.2	21,637	62.0	1,820	7.8	7.5-8.0
Men	17,225	13,189	77.1	12,202	70.8	1,087	8.2	7.9-8.5
Women	17,677	10,167	57.5	9,435	53.4	733	7.2	6.9-7.5
Both sexes, 16 to 19 years	2,307	1,233	53.4	976	42.3	257	20.8	19.4-22.2
Black	1,775	1,128	63.6	968	54.5	160	14.2	12.8-15.6
Men	841	597	71.0	513	61.1	83	13.9	12.0-15.9
Women	934	531	56.9	454	48.7	77	14.5	12.4-16.6
Both sexes, 16 to 19 years	153	67	44.0	41	26.6	27	39.5	30.6-48.5
Hispanic origin	6,942	4,661	67.1	4,086	58.9	575	12.3	11.7-13.0
Men	3,532	2,876	81.4	2,520	71.4	356	12.4	11.6-13.2
Women	3,410	1,785	52.3	1,566	45.9	219	12.3	11.2-13.3
Both sexes, 16 to 19 years	699	322	46.0	227	32.4	95	29.5	26.1-32.9
Single (never married)	10,089	7,358	72.9	6,459	64.0	899	12.2	11.7-12.7

[Continued]

★ 650 ★

Civilian Noninstitutional Employment in the U.S.: West Region, 1992
[Continued]

Area and population	Civilian non-institutional population	Civilian labor force		Employment		Unemployed		
		Number	Percent of population	Number	Percent of population	Number	Rate	Error range of rate[1]
Married, spouse present	22,680	15,283	67.4	14,396	63.5	887	5.8	5.6-6.0
Other marital status[2]	7,614	4,383	57.6	3,986	52.3	397	9.1	8.5-9.6

Source: "Census Regions and Divisions: Employment Status of the Civilian Noninstitutional Population by Sex, Age, Race, Hispanic Origin, and Marital Status, 1992 Annual Averages," U.S. Department of Labor, *Geographical Profile of Employment and Unemployment*, 1992, p.8. *Notes:* Data for demographic groups are not shown when they do not meet BLS publication standards of reliability for the particular area based on the sample in that area. Items may not add to totals or compute to displayed percentages because of rounding. Detail for race and Hispanic-origin groups will not add to totals because data for the "other races" group are not presented and Hispanics are included in both the white and black populations. 1. Error ranges are calculated at the 90-percent confidence interval, which means that if repeated samples were drawn from the same population and an error range constructed around each sample estimate, in 9 out of 10 cases the true value based on a complete census of the population would be contained within these error ranges. 2. "Other marital status" includes divorced, widowed, separated, and married, with spouse absent.

★ 651 ★

Labor Force

Civilian Noninstitutional Employment in the U.S.: West South Central Division, 1992

Numbers in thousands.

Area and population	Civilian non-institutional population	Civilian labor force		Employment		Unemployed		
		Number	Percent of population	Number	Percent of population	Number	Rate	Error range of rate[1]
West South Central Division								
Total	20,033	13,355	66.7	12,372	61.8	983	7.4	7.1-7.7
Men	9,556	7,336	76.8	6,796	71.1	540	7.4	7.0-7.8
Women	10,477	6,018	57.4	5,576	53.2	442	7.4	6.9-7.8
Both sexes, 16 to 19 years	1,503	767	51.0	599	39.8	167	21.8	20.2-23.5
White	16,806	11,304	67.3	10,628	63.2	677	6.0	5.7-6.3
Men	8,075	6,278	77.7	5,899	73.1	379	6.0	5.7-6.4
Women	8,731	5,026	57.6	4,729	54.2	297	5.9	5.5-6.3
Both sexes, 16 to 19 years	1,205	649	53.8	531	44.1	118	18.1	16.4-19.9
Black	2,664	1,681	63.1	1,409	52.9	272	16.2	14.9-17.4
Men	1,210	854	70.6	711	58.8	143	16.7	14.9-18.5
Women	1,454	827	56.9	698	48.0	129	15.6	13.8-17.4
Both sexes, 16 to 19 years	249	99	39.8	55	22.2	44	44.2	36.7-51.7
Hispanic origin	3,085	2,073	67.2	1,879	60.9	194	9.4	8.5-10.2
Men	1,532	1,229	80.2	1,114	72.7	115	9.4	8.3-10.5
Women	1,553	844	54.4	765	49.3	79	9.3	8.0-10.7
Both sexes, 16 to 19 years	337	171	50.6	131	38.7	40	23.5	19.5-27.4
Single (never married)	4,360	3,003	68.9	2,603	59.7	400	13.3	12.6-14.1

[Continued]

★ 651 ★

Civilian Noninstitutional Employment in the U.S.: West South Central Division, 1992

[Continued]

Area and population	Civilian non-institutional population	Civilian labor force		Employment		Unemployed		
		Number	Percent of population	Number	Percent of population	Number	Rate	Error range of rate[1]
Married, spouse present	11,799	8,142	69.0	7,747	65.7	395	4.9	4.5-5.2
Other marital status[2]	3,874	2,210	57.0	2,022	52.2	188	8.5	7.7-9.2

Source: "Census Regions and Divisions: Employment Status of the Civilian Noninstitutional Population by Sex, Age, Race, Hispanic Origin, and Marital Status, 1992 Annual Averages," U.S. Department of Labor, *Geographical Profile of Employment and Unemployment,* 1992, p.8. *Notes:* Data for demographic groups are not shown when they do not meet BLS publication standards of reliability for the particular area based on the sample in that area. Items may not add to totals or compute to displayed percentages because of rounding. Detail for race and Hispanic-origin groups will not add to totals because data for the "other races" group are not presented and Hispanics are included in both the white and black populations. 1. Error ranges are calculated at the 90-percent confidence interval, which means that if repeated samples were drawn from the same population and an error range constructed around each sample estimate, in 9 out of 10 cases the true value based on a complete census of the population would be contained within these error ranges. 2. "Other marital status" includes divorced, widowed, separated, and married, with spouse absent.

★ 652 ★

Labor Force

Civilian Noninstitutional Employment: Trends, 1980 to 1992

Numbers in thousands. Annual averages.

Year	Men						Women					
	Civilian non-institutional population	Civilian labor force					Civilian non-institutional population	Civilian labor force				
		Total	Percent of population	Employed	Unemployed			Total	Percent of population	Employed	Unemployed	
					Number	Percent of labor force					Number	Percent of labor force
Black												
1992	9,888	6,892	69.7	5,846	1,046	15.2	12,069	6,999	58.0	6,087	912	13.0
1991	9,717	6,754	69.5	5,880	874	12.9	11,898	6,788	57.0	5,983	805	11.9
1990	9,567	6,708	70.1	5,915	793	11.8	11,733	6,785	57.8	6,051	734	10.8
1989	9,439	6,701	71.0	5,928	773	11.5	11,582	6,796	58.7	6,025	772	11.4
1988	9,289	6,596	71.0	5,824	771	11.7	11,402	6,609	58.0	5,834	776	11.7
1987	9,128	6,487	71.1	5,661	826	12.7	11,223	6,507	58.0	5,648	859	13.2
1986	8,956	6,374	71.2	5,428	946	14.8	11,033	6,281	56.9	5,386	895	14.2
1985	8,791	6,220	70.8	5,269	951	15.3	10,873	6,145	56.5	5,231	914	14.9
1984	8,654	6,126	70.8	5,123	1,003	16.4	10,694	5,906	55.2	4,995	911	15.4
1983	8,448	5,966	70.6	4,753	1,213	20.3	10,476	5,681	54.2	4,623	1,058	18.6
1982	8,284	5,804	70.1	4,637	1,167	20.1	10,300	5,527	53.7	4,552	975	17.6
1981	8,117	5,684	70.0	4,793	891	15.7	10,101	5,401	53.5	4,561	840	15.6
1980	7,945	5,612	70.6	4,798	815	14.5	9,881	5,253	53.2	4,515	737	14.0
White												
1992	78,351	59,830	76.4	55,709	4,121	6.9	84,307	84,696	57.8	45,770	2,926	6.0
1991	77,689	59,332	76.4	55,557	3,775	6.4	83,822	48,154	57.4	45,482	2,672	5.5
1990	77,082	59,298	76.9	56,432	2,866	4.8	83,332	47,879	57.5	45,654	2,225	4.6
1989	76,468	58,988	77.1	56,352	2,636	4.5	82,871	47,367	57.2	45,323	2,135	4.5
1988	75,855	58,317	76.9	55,550	2,766	4.7	82,340	46,439	56.4	44,262	2,177	4.7
1987	75,190	57,779	76.8	54,646	3,133	5.4	81,769	45,510	55.7	43,142	2,369	5.2
1986	74,390	57,217	76.9	53,785	3,433	6.0	81,041	44,584	55.0	41,876	2,708	6.1
1985	73,373	56,472	77.0	53,045	3,426	6.1	80,306	43,455	54.1	40,689	2,765	6.4
1984	72,723	56,061	77.1	52,462	3,600	6.4	79,624	42,430	53.3	39,658	2,772	6.5
1983	71,922	55,480	77.1	50,621	4,859	8.8	78,884	41,541	52.7	38,272	3,270	7.9
1982	71,211	55,132	77.4	50,287	4,845	8.8	78,230	41,009	52.4	37,616	3,396	8.3

[Continued]

★ 652 ★

Civilian Noninstitutional Employment: Trends, 1980 to 1992
[Continued]

Year	Men						Women					
	Civilian non-institutional population	Civilian labor force					Civilian non-institutional population	Civilian labor force				
		Total	Percent of population	Employed	Unemployed			Total	Percent of population	Employed	Unemployed	
					Number	Percent of labor force					Number	Percent of labor force
1981	70,480	54,895	77.9	51,315	3,580	6.5	77,428	40,156	51.9	37,394	2,762	6.9
1980	69,634	54,473	78.2	51,127	3,344	6.1	76,489	39,127	51.2	36,589	2,540	6.5

Source: "Employment Status of the Civilian Noninstitutional Population, by Sex and Race: 1980 to 1992," Claudette E. Bennett, *The Black Population in the United States: March 1992*, p. 9.

★ 653 ★

Labor Force

Civilian Noninstitutional Labor Force: Status and Characteristics, 1990-1993. Part I - A.

Numbers in thousands.

Employment status, race, sex, age, and Hispanic origin	1990 IV	1991				1992			
		I	II	III	IV	I	II	III	IV
White									
Civilian noninstitutional population[1]	160,830	161,095	161,357	161,646	161,947	162,223	162,486	162,788	163,135
Civilian labor force	107,258	107,298	107,609	107,399	107,693	108,149	108,565	108,706	108,689
Percent of population	66.7	66.6	66.7	66.4	66.	66.7	66.8	66.8	66.6
Employed	101,737	101,075	101,157	100,919	101,000	101,185	101,450	101,515	101,761
Employed population ratio[2]	63.3	62.7	62.7	62.4	62.4	62.4	62.4	62.4	62.4
Unemployed	5,521	6,223	6,452	6,480	6,692	6,964	7,116	7,191	6,927
Unemployment rate	5.1	5.8	6.0	6.0	6.2	6.4	6.6	6.6	6.4
Men, 20 years and over									
Civilian labor force	56,163	56,076	56,268	56,308	56,335	56,540	56,900	56,912	56,895
Percent of population	78.2	77.9	77.9	77.8	77.6	77.7	78.0	77.8	77.6
Employed	53,483	52,976	53,054	52,995	52,986	53,003	53,247	53,320	53,400
Employed population ratio[2]	74.5	73.6	73.5	73.2	73.0	72.8	73.0	72.9	72.8
Unemployed	2,680	3,100	3,214	3,313	3,349	3,537	3,653	3,592	3,495
Unemployment rate	4.8	5.5	5.7	5.9	5.9	6.3	6.4	6.3	6.1
Women, 20 years and over									
Civilian labor force	44,930	45,072	45,321	45,282	45,477	45,832	45,950	46,041	46,063
Percent of population	57.5	57.6	57.8	57.6	57.7	58.1	58.1	58.1	58.1
Employed	42,969	42,911	43,074	43,074	43,133	43,399	43,492	43,445	43,565
Employed population ratio[2]	55.0	54.8	54.9	54.8	54.8	55.0	55.0	54.9	54.9
Unemployed	1,962	2,161	2,247	2,208	2,344	2,433	2,458	2,597	2,498
Unemployment rate	4.4	4.8	5.0	4.9	5.2	5.3	5.3	5.6	5.4
Both sexes, 16 to 19 years									
Civilian labor force	6,165	6,150	6,020	5,809	5,881	5,776	5,715	5,753	5,731
Percent of population	56.4	56.8	56.2	54.6	55.5	54.8	54.5	54.9	54.5
Employed	5,285	5,189	5,028	4,850	4,881	4,782	4,711	4,750	4,796
Employed population ratio[2]	48.3	47.9	46.9	45.6	46.1	45.3	44.9	45.3	45.6

[Continued]

★ 653 ★

Civilian Noninstitutional Labor Force: Status and Characteristics, 1990-1993. Part I - A.
[Continued]

Employment status; race, sex, age, and Hispanic origin	1990 IV	1991 I	1991 II	1991 III	1991 IV	1992 I	1992 II	1992 III	1992 IV
Unemployed	880	962	992	959	1,000	994	1,005	1,002	935
Unemployment rate	14.3	15.6	16.5	16.5	17.0	17.2	17.6	17.4	16.3
Men	15.5	16.8	17.9	17.6	17.8	18.9	18.9	18.9	16.9
Women	13.0	14.3	15.0	15.3	16.1	15.4	16.1	15.8	15.6

Source: "Employment Status of the Civilian Noninstitutional Population by Race, Sex, Age, and Hispanic Origin, Seasonally Adjusted," *Employment and Earnings* 41 (January 1, 1994), p. 54. *Notes:* Detail for the above race and Hispanic-origin groups will not sum to totals because data for the "other races" group are not presented and Hispanics are included in both the white and black population groups. Seasonally adjusted data have been revised based on the experience through December 1993. 1. The population figures are not adjusted for seasonal variation. 2. Civilian employment as a percent of the civilian noninstitutional population.

★ 654 ★

Labor Force

Civilian Noninstitutional Labor Force: Status and Characteristics, 1990-1993. Part I - B.

Numbers in thousands.

Employment status, race, sex, age, and Hispanic origin	1993 I	1993 II	1993 III	1993 IV
White				
Civilian noninstitutional population[1]	163,438	163,751	164,078	164,415
Civilian labor force	108,816	109,133	109,510	109,943
Percent of population	66.6	66.6	66.7	66.9
Employed	102,119	102,508	103,036	103,581
Employed population ratio[2]	62.5	62.6	62.8	63.0
Unemployed	6,697	6,625	6,474	6,362
Unemployment rate	6.2	6.1	5.9	5.8
Men, 20 years and over				
Civilian labor force	56,960	57,059	57,143	57,264
Percent of population	77.5	77.5	77.4	77.4
Employed	53,625	53,798	53,925	54,235
Employed population ratio[2]	73.0	73.1	73.1	73.3
Unemployed	3,335	3,261	3,218	3,029
Unemployment rate	5.9	5.7	5.6	5.3
Women, 20 years and over				
Civilian labor force	46,083	46,264	46,525	46,783
Percent of population	58.0	58.1	58.4	58.6
Employed	43,673	43,863	44,188	44,390
Employed population ratio[2]	55.0	55.1	55.4	55.6
Unemployed	2,409	2,400	2,338	2,394
Unemployment rate	5.2	5.2	5.0	5.1
Both sexes, 16 to 19 years				
Civilian labor force	5,773	5,810	5,842	5,895

[Continued]

★ 654 ★

Civilian Noninstitutional Labor Force: Status and Characteristics, 1990-1993. Part I - B.
[Continued]

Employment status, race, sex, age, and Hispanic origin	1993			
	I	II	III	IV
Percent of population	54.8	55.0	55.1	55.4
Employed	4,820	4,846	4,924	4,956
Employed population ratio[2]	45.8	45.9	46.5	46.6
Unemployed	953	963	918	940
Unemployment rate	16.5	16.6	15.7	15.9
Men	17.6	18.1	17.4	17.5
Women	15.3	15.0	13.9	14.3

Source: "Employment Status of the Civilian Noninstitutional Population by Race, Sex, Age, and Hispanic Origin, Seasonally Adjusted," *Employment and Earnings* 41 (January 1, 1994), p. 54. *Notes:* Detail for the above race and Hispanic-origin groups will not sum to totals because data for the "other races" group are not presented and Hispanics are included in both the white and black population groups. Seasonally adjusted data have been revised based on the experience through December 1993. 1. The population figures are not adjusted for seasonal variation. 2. Civilian employment as a percent of the civilian noninstitutional population.

★ 655 ★

Labor Force

Civilian Noninstitutional Labor Force: Status and Characteristics, 1990-1993. Part II - A.

Numbers in thousands.

Employment status, race, sex, age, and Hispanic origin	1990 IV	1991				1992			
		I	II	III	IV	I	II	III	IV
Black									
Civilian noninstitutional population[1]	21,416	21,493	21,568	21,656	21,744	21,828	21,909	21,997	22,096
Civilian labor force	13,537	13,541	13,546	13,536	13,548	13,737	13,873	14,017	13,931
Percent of population	63.2	63.0	62.8	62.5	62.3	62.9	63.3	63.7	63.0
Employed	11,877	11,901	11,852	11,886	11,813	11,834	11,894	12,034	11,963
Employed population ratio[2]	55.5	55.4	55.4	54.9	54.3	54.2	54.36	54.7	54.1
Unemployed	1,659	1,641	1,693	1,650	1,735	1,903	1,979	1,983	1,968
Unemployment rate	12.3	12.1	12.5	12.2	12.8	13.9	14.3	14.1	14.1
Men, 20 years and over									
Civilian labor force	6,351	6,373	6,342	6,359	6,382	6,434	6,475	6,494	6,485
Percent of population	74.2	74.0	73.4	73.1	72.9	73.2	73.3	73.2	72.7
Employed	5,622	5,658	5,580	5,633	5,655	5,565	5,609	5,614	5,619
Employed population ratio[2]	65.7	65.7	64.6	64.8	64.6	63.3	63.5	63.2	63.0
Unemployed	729	714	763	726	727	869	866	880	866
Unemployment rate	11.5	11.2	12.0	11.4	11.4	13.5	13.4	13.6	13.3
Women, 20 years and over									
Civilian labor force	6,359	6,380	6,456	6,459	6,438	6,516	6,627	6,720	6,657
Percent of population	59.3	59.2	59.7	59.5	59.0	59.5	60.2	60.8	60.0
Employed	5,715	5,741	5,781	5,806	5,699	5,773	5,826	5,945	5,877

[Continued]

★ 655 ★

Civilian Noninstitutional Labor Force: Status and Characteristics, 1990-1993. Part II - A.

[Continued]

Employment status, race, sex, age, and Hispanic origin	1990 IV	1991 I	II	III	IV	1992 I	II	III	IV
Employed population ratio[2]	53.3	53.3	53.4	53.4	52.2	52.7	53.0	53.8	53.0
Unemployed	644	639	676	653	739	743	801	775	781
Unemployment rate	10.1	10.0	10.5	1.01	11.5	11.4	12.1	11.5	11.7
Both sexes, 16 to 19 years									
Civilian labor force	826	789	747	718	728	787	771	803	789
Percent of population	38.6	37.3	35.4	34.3	35.0	37.9	37.2	38.8	38.0
Employed	540	502	492	447	459	496	459	475	467
Employed population ratio[2]	25.2	23.7	23.3	21.3	22.1	23.9	22.1	22.9	22.5
Unemployed	287	287	255	271	269	291	313	328	322
Unemployment rate	34.7	36.4	34.1	37.8	36.9	36.9	40.5	40.8	40.8
Men	34.2	36.7	36.2	37.1	36.0	38.1	44.1	43.5	42.7
Women	35.3	36.1	31.8	38.5	38.0	35.6	36.8	37.9	38.6
Hispanic origin									
Civilian noninstitutional population[1]	14,474	14,593	14,711	14,829	14,948	15,066	15,184	15,303	15,421
Civilian labor force	9,554	9,633	9,721	9,834	9,875	10,021	10,105	10,184	10,219
Percent of population	66.0	66.0	66.1	66.3	66.1	66.5	66.6	66.5	66.3
Employed	8,712	8,725	8,785	8,827	8,864	8,902	8,959	8,988	9,035
Employment population ratio[2]	60.2	59.8	59.7	59.5	59.3	59.1	59.0	58.7	58.6
Unemployed	842	908	937	1,007	1,011	1,119	1,146	1,196	1,184
Unemployment rate	8.8	9.4	9.6	10.2	10.2	11.2	11.3	11.7	11.6

Source: "Employment Status of the Civilian Noninstitutional Population by Race, Sex, Age, and Hispanic Origin, Seasonally Adjusted," *Employment and Earnings* 41 (January 1, 1994), pp. 54-55. *Notes:* Detail for the above race and Hispanic-origin groups will not sum to totals because data for the "other races" group are not presented and Hispanics are included in both the white and black population groups. Seasonally adjusted data have been revised based on the experience through December 1993. 1. The population figures are not adjusted for seasonal variation. 2. Civilian employment as a percent of the civilian noninstitutional population.

★ 656 ★
Labor Force

Civilian Noninstitutional Labor Force: Status and Characteristics, 1990-1993. Part II - B.

Numbers in thousands.

Employment status, race, sex, age, and Hispanic origin	1993 I	II	III	IV
Black				
Civilian noninstitutional population[1]	22,186	22,281	22,376	22,474
Civilian labor force	13,898	13,911	13,947	14,004
Percent of population	62.6	62.4	62.3	62.3
Employed	12,004	12,060	12,187	12,329
Employed population ratio[2]	54.1	54.1	54.5	54.9
Unemployed	1,894	1,851	1,760	1,676

[Continued]

★ 656 ★

Civilian Noninstitutional Labor Force: Status and Characteristics, 1990-1993. Part II - B.

[Continued]

Employment status, race, sex, age, and Hispanic origin	1993			
	I	II	III	IV
Unemployment rate	13.6	13.3	12.6	12.0
Men, 20 years and over				
Civilian labor force	6,503	6,465	6,523	6,493
Percent of population	72.5	71.7	72.1	71.4
Employed	5,676	5,657	5,741	5,761
Employed population ratio[2]	63.3	62.8	63.4	63.4
Unemployed	826	808	782	733
Unemployment rate	12.7	12.5	12.0	11.3
Women, 20 years and over				
Civilian labor force	6,607	6,651	6,645	6,766
Percent of population	59.3	59.5	59.2	60.1
Employed	5,856	5,943	5,942	6,104
Employed population ratio[2]	52.6	53.2	53.0	54.2
Unemployed	751	708	703	662
Unemployment rate	11.4	10.6	10.6	9.8
Both sexes, 16 to 19 years				
Civilian labor force	788	795	779	745
Percent of population	37.8	38.1	37.0	35.2
Employed	472	460	504	464
Employed population ratio[2]	22.6	22.0	23.9	21.9
Unemployed	316	335	276	281
Unemployment rate	40.1	42.1	35.4	37.8
Men	41.1	4201	37.5	39.5
Women	39.0	42.2	33.0	36.0
Hispanic origin				
Civilian noninstitutional population[1]	15,542	15,682	15,824	15,966
Civilian labor force	10,270	10,255	10,380	10,595
Percent of population	66.1	65.4	65.6	66.4
Employed	9,110	9,200	9,318	9,458
Employment population ratio[2]	58.6	58.7	58.9	59.2
Unemployed	1,160	1,054	1,061	1,138
Unemployment rate	11.3	10.3	10.2	10.7

Source: "Employment Status of the Civilian Noninstitutional Population by Race, Sex, Age, and Hispanic Origin, Seasonally Adjusted," *Employment and Earnings* 41 (January 1, 1994), pp. 54-55. *Notes:* Detail for the above race and Hispanic-origin groups will not sum to totals because data for the "other races" group are not presented and Hispanics are included in both the white and black population groups. Seasonally adjusted data have been revised based on the experience through December 1993. 1. The population figures are not adjusted for seasonal variation. 2. Civilian employment as a percent of the civilian noninstitutional population.

★ 657 ★

Labor Force

Civilian Noninstitutional Population Employed: 16 Years and Over, 1960 to 1992

In thousands, except as indicated. Annual averages of monthly figures. Based on Current Population Survey.

Hispanic origin	Civilian non-institutional population	Civilian labor force							
		Total	Percent of population	Employed	Employment/ population ratio[1]	Unemployed		Not in labor force	
						Number	Percent of labor force	Number	Percent of population
Total[2]									
1960	117,245	69,628	59.4	65,778	56.1	3,852	5.5	47,617	40.6
1970	137,085	82,771	60.4	78,678	57.4	4,093	4.9	54,315	39.6
1980	167,745	106,940	63.8	99,303	59.2	7,637	7.1	60,806	36.2
1985	178,206	115,461	64.8	107,150	60.1	8,312	7.2	62,744	35.2
1988	184,613	121,669	65.9	114,968	62.3	6,701	5.5	62,944	34.1
1989	186,393	123,869	66.5	117,342	63.0	6,528	5.3	62,523	33.5
1990	188,049	124,787	66.4	117,914	62.7	6,874	5.5	63,262	33.6
1991	189,765	125,303	66.0	116,877	61.6	8,426	6.7	64,462	34.0
1992	191,576	126,982	66.3	117,598	61.4	9,384	7.4	64,593	33.7
White									
1960	105,282	61,915	58.8	58,850	55.9	3,065	5.0	43,367	41.2
1970	122,174	73,556	60.2	70,217	57.5	3,339	4.5	48,618	39.8
1980	146,122	93,600	64.1	87,715	60.0	5,884	6.3	52,523	35.9
1985	153,679	99,926	65.0	93,736	61.0	6,191	6.2	53,753	35.0
1988	158,194	104,756	66.2	99,812	63.1	4,944	4.7	53,439	33.8
1989	159,338	106,355	66.7	101,584	63.8	4,770	4.5	52,983	33.3
1990	160,415	107,177	66.8	102,087	63.6	5,091	4.7	53,237	33.2
1991	161,511	107,486	66.6	101,039	62.6	6,447	6.0	54,025	33.4
1992	162,658	108,526	66.7	101,479	62.4	7,047	6.5	54,132	33.3
Black									
1973	14,917	8,976	60.2	8,128	54.5	846	9.4	5,941	39.8
1980	17,824	10,865	61.0	9,313	52.2	1,553	14.3	6,959	39.0
1985	19,664	12,364	62.9	10,501	53.4	1,864	15.1	7,299	37.1
1988	20,692	13,205	63.8	11,658	56.3	1,547	11.7	7,487	36.2
1989	21,021	13,497	64.2	11,953	56.9	1,544	11.4	7,524	35.8
1990	21,300	13,493	63.3	11,966	56.2	1,527	11.3	7,808	36.7
1991	21,615	13,542	62.6	11,863	54.9	1,679	12.4	8,074	37.4
1992	21,958	13,891	63.3	11,933	54.3	1,958	14.1	8,067	36.7
Hispanic[3]									
1980	9,598	6,146	64.0	5,527	57.6	620	10.1	3,451	36.0
1985	11,915	7,698	64.6	6,888	57.8	811	10.5	4,217	35.4
1986	12,344	8,076	65.4	7,219	58.5	857	10.6	4,268	34.6
1989	13,791	9,323	67.6	8,573	62.2	750	8.0	4,468	32.4
1990	14,297	9,576	67.0	8,808	61.6	769	8.0	4,721	33.0
1991	14,770	9,762	66.1	8,799	59.6	963	9.9	5,008	33.9
1992	15,244	10,131	66.5	8,971	58.9	1,160	11.4	5,113	33.5
Mexican									
1986	7,377	4,941	67.0	4,387	59.5	555	11.2	2,436	33.0
1990	8,742	5,970	68.3	5,478	62.7	492	8.2	2,773	31.7
1991	8,947	5,984	66.9	5,363	59.9	621	10.4	2,963	33.1
1992	9,368	6,319	67.5	5,581	59.6	739	11.7	3,049	32.5
Puerto Rican									
1986	1,494	804	53.8	691	46.3	113	14.0	690	46.2

[Continued]

★ 657 ★

Civilian Noninstitutional Population Employed: 16 Years and Over, 1960 to 1992

[Continued]

Hispanic origin	Civilian non-institutional population	Civilian labor force								
		Total	Percent of population	Employed	Employment/ population ratio[1]	Unemployed		Not in labor force		
						Number	Percent of labor force	Number	Percent of population
1990	1,546	859	55.6	780	50.5	79	9.1	687	44.4
1991	1,629	930	57.1	822	50.5	108	11.6	699	42.9
1992	1,628	934	57.4	802	49.2	132	14.4	694	42.6
Cuban									
1986	842	570	67.7	533	63.3	36	6.4	272	32.3
1990	847	552	65.1	512	60.4	40	7.2	295	34.8
1991	849	543	63.9	499	58.8	44	8.1	306	36.0
1992	867	529	61.1	488	56.3	42	7.9	337	38.9

Source: "Employment Status of the Civilian Noninstitutional Population 16 Years Old and Over, by Sex, Race, and Hispanic Origin: 1960 to 1992," U.S. Bureau of the Census, *Statistical Abstract of the United States*, 1993, p. 629. Primary source: U.S. Bureau of Labor Statistics, Bulletin 2307; and *Employment and Earnings*, monthly, January issues. *Notes:* 1. Civilian employed as a percent of the civilian noninstitutional population. 2. Includes other races, not shown separately. 3. Persons of Hispanic origin may be of any race. Includes persons of other Hispanic origin, not shown separately.

★ 658 ★

Labor Force

Civilian Noninstitutional Population in the Midwest: Employment Status: Annual Averages, 1992

Numbers in thousands.

State and population group	Civilian non-institutional population	Civilian labor force		Employment		Unemployment		
		Number	Percent of population	Number	Percent of population	Number	Rate	Error range of rate[1]
Michigan								
Total	7,038	4,610	65.5	4,205	59.8	405	8.8	8.4 - 9.2
Men	3,374	2,529	75.0	2,306	68.4	223	8.8	8.2 - 9.4
Women	3,664	2,081	56.8	1,899	51.8	182	8.7	8.1 - 9.4
Both sexes, 16 to 19 years	500	294	58.8	234	46.7	60	20.5	18.1 - 22.8
White	6,003	4,011	66.8	3,705	61.7	305	7.6	7.2 - 8.0
Men	2,912	2,217	76.1	2,048	70.3	169	7.6	7.0 - 8.2
Women	3,091	1,793	58.0	1,657	53.6	136	7.6	6.9 - 8.2
Both sexes, 16 to 19 years	399	249	62.3	205	51.3	44	17.6	15.2 - 20.1
Black	907	515	56.8	424	46.7	91	17.8	15.9 - 19.6
Men	398	261	65.6	213	53.4	49	18.6	16.0 - 21.2
Women	509	254	49.9	211	41.5	43	16.8	14.3 - 19.4
Both sexes, 16 to 19 years	90	39	43.4	24	27.1	15	37.6	28.6 - 46.6
Hispanic origin	112	76	68.5	65	58.6	11	14.4	10.2 - 18.6
Men	59	45	76.3	38	64.3	7	15.7	10.0 - 21.4
Women	52	31	59.7	27	52.2	4	12.5	6.3 - 18.7
Single (never married)	1,837	1,323	72.0	1,144	62.3	179	13.5	12.6 - 14.5

[Continued]

★ 658 ★

Civilian Noninstitutional Population in the Midwest: Employment Status: Annual Averages, 1992

[Continued]

State and population group	Civilian non-institutional population	Civilian labor force		Employment		Unemployment		
		Number	Percent of population	Number	Percent of population	Number	Rate	Error range of rate[1]
Married spouse present	3,873	2,595	67.0	2,438	63.0	157	6.0	5.6 - 6.5
Other marital status	1,327	692	52.1	623	46.9	69	10.0	8.8 - 11.1
Minnesota								
Total	3,345	2,431	72.6	2,306	68.9	125	5.1	4.5 - 5.8
Men	1,605	1,287	80.2	1,211	75.5	76	5.9	5.0 - 6.8
Women	1,742	1,144	65.7	1,095	62.9	49	4.3	3.4 - 5.1
Both sexes, 16 to 19 years	234	158	67.2	139	59.1	19	12.1	8.5 - 15.7
White	3,190	2,335	73.2	2,228	69.9	107	4.6	4.0 - 5.2
Men	1,535	1,236	80.5	1,168	76.1	68	5.5	4.6 - 6.4
Women	1,655	1,100	66.4	1,060	64.1	39	3.6	2.8 - 4.4
Both sexes, 16 to 19 years	215	149	69.4	135	62.8	14	9.5	6.1 - 12.9
Black	85	53	62.3	38	45.1	15	27.5	18.3 - 36.7
Single (never married)	895	716	80.0	657	73.4	59	8.2	6.8 - 9.7
Married spouse present	1,895	1,420	74.9	1,371	72.3	49	3.5	2.8 - 4.2
Other marital status	557	295	53.1	279	50.0	17	5.7	3.8 - 7.6
Missouri								
Total	3,947	2,696	68.3	2,543	64.4	153	5.7	5.0 - 6.4
Men	1,850	1,437	77.7	1,353	73.1	84	5.8	4.9 - 6.8
Women	2,098	1,260	60.0	1,190	56.7	69	5.5	4.5 - 6.5
Both sexes, 16 to 19 years	291	167	57.4	147	50.5	20	12.1	8.4 - 15.9
White	3,546	2,429	68.5	2,312	65.2	117	4.8	4.2 - 5.5
Men	1,672	1,308	78.3	1,245	74.4	64	4.9	4.0 - 5.8
Women	1,874	1,121	59.8	1,068	57.0	53	4.8	3.8 - 5.7
Both sexes, 16 to 19 years	264	154	58.3	135	51.2	19	12.2	8.3 - 16.1
Black	336	226	67.2	191	56.9	35	15.4	11.6 - 19.1
Men	144	103	71.2	84	58.2	19	18.3	12.2 - 24.4
Women	192	123	64.3	108	56.0	16	12.9	8.2 - 17.7
Single (never married)	948	697	73.6	626	66.0	72	10.3	8.5 - 12.0
Married spouse present	2,299	1,622	70.5	1,563	68.0	59	3.6	2.9 - 4.4
Other marital status	700	377	53.8	355	50.7	22	5.9	4.0 - 7.7
Nebraska								
Total	1,206	856	71.0	831	68.9	25	3.0	2.5 - 3.4
Men	581	453	78.0	439	75.6	14	3.1	2.4 - 3.7

[Continued]

★ 658 ★

Civilian Noninstitutional Population in the Midwest: Employment Status: Annual Averages, 1992
[Continued]

State and population group	Civilian non-institutional population	Civilian labor force		Employment		Unemployment		
		Number	Percent of population	Number	Percent of population	Number	Rate	Error range of rate[1]
Women	625	403	64.4	391	62.6	11	2.8	2.2 - 3.5
Both sexes, 16 to 19 years	102	65	64.1	58	57.0	7	11.1	8.1 - 14.1
White	1,156	827	71.5	804	69.5	23	2.8	2.3 - 3.3
Men	561	441	78.5	428	76.3	13	2.9	2.3 - 3.5
Women	595	386	64.9	376	63.1	10	2.7	2.0 - 3.3
Both sexes, 16 to 19 years	96	63	65.8	56	58.7	7	10.7	7.7 - 13.8
Black	36	20	55.5	18	50.1	2	9.7	4.4 - 14.9
Single (never married)	278	212	76.2	199	71.7	13	6.0	4.7 - 7.3
Married spouse present	745	548	73.5	539	72.3	9	1.7	1.3 - 2.2
Other marital status	183	96	52.5	93	50.7	3	3.4	2.0 - 4.9
Illinois								
Total	8,965	6,120	68.3	5,659	63.1	461	7.5	7.1 - 7.9
Men	4,291	3,324	77.5	3,057	71.2	267	8.0	7.5 - 8.6
Women	4,674	2,796	59.8	2,602	55.7	194	6.9	6.4 - 7.5
Both sexes, 16 to 19 years	634	345	54.3	271	42.7	74	21.4	18.8 - 23.9
White	7,391	5,141	69.6	4,834	85.4	307	6.0	5.6 - 6.4
Men	3,587	2,838	79.1	2,653	74.0	185	6.5	6.0 - 7.1
Women	3,804	2,303	60.5	2,181	57.3	122	5.3	4.7 - 5.8
Both sexes, 16 to 19 years	476	279	58.7	232	48.8	47	16.7	14.2 - 19.3
Black	1,292	787	60.9	647	50.1	140	17.7	16.1 - 19.4
Men	563	376	66.7	302	53.6	74	19.7	17.1 - 22.2
Women	728	411	56.4	346	47.4	66	15.9	13.7 - 18.2
Both sexes, 16 to 19 years	130	51	39.2	27	20.9	24	46.7	37.1 - 56.3
Hispanic origin	639	461	72.1	412	64.5	49	10.6	8.9 - 12.3
Men	336	295	87.8	264	78.6	31	10.5	8.4 - 12.6
Women	303	186	54.6	148	48.8	18	10.7	7.8 - 13.5
Both sexes, 16 to 19 years	61	31	51.6	24	39.2	8	24.1	14.8 - 33.3
Single (never married)	2,433	1,775	73.0	1,554	63.9	221	12.5	11.6 - 13.4
Married spouse present	4,883	3,463	70.9	3,298	67.5	165	4.8	4.3 - 5.2
Other marital status	1,649	882	53.5	808	49.0	74	8.4	7.3 - 9.5
Indiana								
Total	4,294	2,849	66.3	2,663	62.0	186	6.5	5.8 - 7.3
Men	1,981	1,520	76.7	1,411	71.2	108	7.1	6.0 - 8.2
Women	2,313	1,329	57.5	1,251	54.1	78	5.9	4.8 - 6.9

[Continued]

★ 658 ★

Civilian Noninstitutional Population in the Midwest: Employment Status: Annual Averages, 1992
[Continued]

State and population group	Civilian non-institutional population	Civilian labor force		Employment		Unemployment		
		Number	Percent of population	Number	Percent of population	Number	Rate	Error range of rate[1]
Both sexes, 16 to 19 years	317	170	53.7	138	43.6	32	18.7	13.9 - 23.4
White	3,899	2,623	67.3	2,472	63.4	151	5.8	5.0 - 6.5
Men	1,813	1,416	78.1	1,326	73.1	90	6.3	5.3 - 7.4
Women	2,086	1,208	57.9	1,146	54.9	62	5.1	4.1 - 6.2
Both sexes, 16 to 19 years	283	159	56.1	131	46.5	27	17.1	12.3 - 21.8
Black	367	207	56.5	175	47.6	33	15.8	11.4 - 20.1
Men	155	93	60.0	75	48.8	17	18.6	11.7 - 25.6
Women	213	115	54.0	99	46.7	15	13.4	8.0 - 18.8
Single (never married)	995	693	69.7	607	61.0	87	12.5	10.5 - 14.5
Married spouse present	2,425	1,681	69.3	1,615	66.6	66	3.9	3.1 - 4.7
Other marital status	873	475	54.3	441	50.5	34	7.1	5.2 - 9.0
Iowa								
Total	2,170	1,551	71.5	1,479	68.2	72	4.6	4.0 - 5.3
Men	1,050	840	80.1	797	75.9	43	5.2	4.3 - 6.0
Women	1,120	711	63.5	682	60.9	29	4.0	3.2 - 4.9
Both sexes, 16 to 19 years	164	106	64.7	92	56.3	14	13.0	9.4 - 16.7
White	2,108	1,514	71.8	1,446	68.6	68	4.5	3.9 - 5.1
Men	1,017	818	80.4	777	76.4	41	5.0	4.1 - 5.9
Women	1,091	696	63.8	669	61.3	27	3.9	3.0 - 4.7
Both sexes, 16 to 19 years	158	104	65.8	91	57.9	13	12.1	8.5 - 15.6
Single (never married)	517	393	75.9	358	69.1	35	8.9	7.3 - 10.5
Married spouse present	1,295	973	75.1	947	73.1	26	2.6	2.0 - 3.2
Other marital status	357	186	52.1	174	48.8	12	6.3	4.3 - 8.3
Kansas								
Total	1,874	1,330	70.9	1,274	68.0	56	4.2	3.7 - 4.8
Men	905	722	79.7	693	76.5	29	4.0	3.2 - 4.7
Women	969	608	62.8	581	59.9	27	4.5	3.7 - 5.4
Both sexes, 16 to 19 years	129	80	82.5	71	55.2	9	11.7	8.2 - 15.1
White	1,730	1,226	70.8	1,183	68.4	43	3.5	3.0 - 4.0
Men	836	668	79.9	645	77.2	23	3.4	2.7 - 4.1
Women	895	558	62.4	538	60.1	20	3.6	2.8 - 4.4
Both sexes, 16 to 19 years	116	75	65.0	67	58.0	8	10.8	7.3 - 14.2
Black	108	80	73.9	69	63.4	11	14.2	10.1 - 18.3
Men	54	41	76.6	36	66.9	5	12.7	7.3 - 18.1

[Continued]

★ 658 ★

Civilian Noninstitutional Population in the Midwest: Employment Status: Annual Averages, 1992
[Continued]

State and population group	Civilian non-institutional population	Civilian labor force		Employment		Unemployment		
		Number	Percent of population	Number	Percent of population	Number	Rate	Error range of rate[1]
Women	54	39	71.1	32	59.9	6	15.8	9.7 - 21.9
Single (never married)	388	296	76.3	273	70.4	23	7.7	6.2 - 9.3
Married, spouse present	1,148	837	72.9	813	70.9	24	2.9	2.3 - 3.4
Other marital status	339	196	58.0	187	55.3	9	4.7	3.2 - 6.2
North Dakota								
Total	464	314	67.7	299	64.4	15	4.9	4.3 - 5.6
Men	220	168	76.2	159	72.2	9	5.3	4.4 - 6.3
Women	244	146	60.1	140	57.4	6	4.4	3.5 - 5.3
Both sexes, 16 to 19 years	37	23	61.5	20	52.7	3	14.4	10.7 - 18.0
White	442	302	68.3	289	65.5	12	4.0	3.4 - 4.6
Men	211	162	76.7	155	73.3	7	4.4	3.5 - 5.2
Women	231	140	60.5	135	58.3	5	3.6	2.8 - 4.5
Both sexes, 16 to 19 years	35	22	62.9	19	55.2	3	12.3	8.7 - 15.8
Single (never married)	115	83	72.6	77	66.9	7	7.8	6.3 - 9.4
Married, spouse present	278	197	70.6	190	68.2	7	3.3	2.6 - 4.0
Other marital status	70	34	48.6	32	45.3	2	6.9	4.6 - 9.1
Ohio								
Total	8,341	5,489	65.8	5,093	61.1	396	7.2	6.9 - 7.6
Men	3,970	2,996	75.5	2,765	69.6	231	7.7	7.2 - 8.2
Women	4,371	2,494	57.1	2,328	53.3	165	6.6	6.1 - 7.2
Both sexes, 16 to 19 years	589	322	54.7	258	43.8	84	19.9	17.7 - 22.2
White	7,443	4,939	66.4	4,618	62.0	321	6.5	6.1 - 6.9
Men	3,567	2,726	76.4	2,538	71.1	189	6.9	6.4 - 7.4
Women	3,875	2,212	57.1	2,080	53.7	132	6.0	5.4 - 6.5
Both sexes, 16 to 19 years	501	288	57.4	237	47.4	50	17.5	15.2 - 19.8
Black	806	493	61.2	423	52.4	71	14.4	12.6 - 16.2
Men	355	235	66.2	195	55.0	40	16.9	14.1 - 19.8
Women	451	258	57.3	227	50.4	31	12.0	9.7 - 14.3
Both sexes, 16 to 19 years	77	30	38.5	17	22.2	13	42.4	30.9 - 53.9
Hispanic origin	65	40	61.8	35	54.0	5	12.6	6.7 - 18.5
Single (never married)	1,931	1,377	71.3	1,214	62.9	164	11.9	11.0 - 12.8
Married, spouse present	4,882	3,334	68.3	3,167	64.9	167	5.0	4.6 - 5.4
Other marital status	1,528	779	51.0	713	46.6	66	8.5	7.4 - 9.5

[Continued]

★ 658 ★

Civilian Noninstitutional Population in the Midwest: Employment Status: Annual Averages, 1992
[Continued]

State and population group	Civilian non-institutional population	Civilian labor force		Employment		Unemployment		
		Number	Percent of population	Number	Percent of population	Number	Rate	Error range of rate[1]
South Dakota								
Total	520	361	69.5	350	67.3	11	3.1	2.6 - 3.6
Men	253	195	76.9	189	74.8	5	2.7	2.1 - 3.3
Women	267	167	62.4	161	60.1	6	3.6	2.9 - 4.4
Both sexes, 16 to 19 years	37	23	62.5	21	56.2	2	10.1	6.9 - 13.3
White	489	343	70.2	334	68.4	9	2.6	2.2 - 3.1
Men	238	185	77.8	181	76.1	4	2.1	1.6 - 2.7
Women	251	158	63.0	153	61.0	5	3.2	2.4 - 3.9
Both sexes, 16 to 19 years	33	22	66.1	20	59.7	2	9.7	6.4 - 12.9
Single (never married)	111	82	74.0	77	69.3	5	6.4	5.0 - 7.8
Married, spouse present	322	235	72.8	231	71.7	4	1.6	1.2 - 2.0
Other marital status	86	44	51.1	42	48.4	2	5.2	3.4 - 7.0
Wisconsin								
Total	3,729	2,661	71.4	2,526	67.7	135	5.1	4.4 - 5.7
Men	1,838	1,435	78.1	1,359	74.0	76	5.3	4.4 - 6.2
Women	1,891	1,226	64.8	1,166	81.7	59	4.8	3.9 - 5.8
Both sexes, 16 to 19 years	256	170	66.3	148	57.7	22	13.0	9.1 - 16.8
White	3,509	2,522	71.9	2,409	68.6	113	4.5	3.9 - 5.1
Men	1,746	1,369	78.4	1,303	74.6	66	4.8	3.9 - 5.7
Women	1,763	1,153	65.4	1,105	62.7	47	4.1	3.2 - 5.0
Both sexes, 16 to 19 years	230	156	68.1	140	60.7	17	10.8	7.1 - 14.5
Black	149	94	63.3	76	51.4	18	18.8	13.3 - 24.3
Women	88	52	59.2	41	46.6	11	21.3	13.6 - 29.1
Single (never married)	912	718	78.8	658	72.1	61	8.4	6.9 - 10.0
Married, spouse present	2,178	1,590	73.0	1,541	70.7	49	3.1	2.4 - 3.8
Other marital status	639	353	55.2	327	51.3	25	7.2	5.1 - 9.3

Source: "States: Employment Status of the Civilian Noninstitutional Population by Sex, Age, Race, Hispanic Origin, and Marital Status, 1992 Annual Averages," U.S. Department of Labor, *Geographical Profile of Employment and Unemployment, 1992*, pp. 38-50. Rearranged by the Editors. *Notes:* Data for demographic groups are not shown when they do not meet BLS publication standards of reliability for the particular area based on the sample in that area. See appendix B. Items may not add to totals or compute to displayed percentages because of rounding. Detail for Hispanic-origin groups will not add to totals because data for the "other races" group are not presented and Hispanics are included in both the white and black population groups. 1. Error ranges are calculated at the 90-percent confidence interval, which means that if repeated samples were drawn from the same population and an error range constructed around each sample estimate, in 9 out of 10 cases the true value based on a complete census of the population would be contained within these error ranges.

★ 659 ★

Labor Force

Civilian Noninstitutional Population in the Northeast: Employment Status: Annual Averages, 1992

Numbers in thousands.

State and population group	Civilian non-institutional population	Civilian labor force		Employment		Unemployment		
		Number	Percent of population	Number	Percent of population	Number	Rate	Error range of rate[1]
New Jersey								
Total	6,026	4,001	66.4	3,666	60.8	335	8.4	8.0 - 8.8
Men	2,876	2,206	76.7	2,006	69.7	200	9.1	8.5 - 9.6
Women	3,150	1,795	57.0	1,661	52.7	135	7.5	6.9 - 8.1
Both sexes, 16 to 19 years	384	166	43.3	134	34.8	33	19.6	17.0 - 22.3
White	5,034	3,344	66.4	3,098	61.5	246	7.3	6.9 - 7.8
Men	2,424	1,877	77.4	1,727	71.2	150	8.0	7.4 - 8.6
Women	2,610	1,467	56.2	1,371	52.5	96	6.5	5.9 - 7.1
Both sexes, 16 to 19 years	289	135	46.5	113	39.1	22	16.0	13.2 - 18.8
Black	761	499	65.6	424	55.7	75	15.0	13.5 - 16.6
Men	342	241	70.3	198	57.7	43	17.9	15.5 - 20.4
Women	418	258	61.7	226	54.1	32	12.4	10.4 - 14.4
Both sexes, 16 to 19 years	72	26	35.6	16	21.7	10	39.0	28.6 - 49.4
Hispanic origin	525	364	69.3	322	61.3	42	11.5	9.9 - 13.1
Men	250	213	85.1	188	75.1	25	11.7	9.6 - 13.8
Women	275	151	54.8	134	48.6	17	11.3	8.8 - 13.8
Single (never married)	1,693	1,190	70.3	1,045	61.7	145	12.2	11.3 - 13.0
Married spouse present	3,304	2,287	69.2	2,147	65.0	140	6.1	5.7 - 6.6
Other marital status	1,030	524	50.9	474	46.0	50	9.5	8.4 - 10.7
New York								
Total	13,809	8,522	61.7	7,798	56.5	724	8.5	8.2 - 8.8
Men	6,441	4,634	71.9	4,196	65.1	438	9.4	9.0 - 9.9
Women	7,367	3,888	52.8	3,602	48.9	286	7.4	6.9 - 7.8
Both sexes, 16 to 19 years	917	324	35.3	252	27.5	71	22.0	19.6 - 24.4
White	11,202	6,977	62.3	6,446	57.5	531	7.6	7.3 - 7.9
Men	5,270	3,834	72.7	3,513	66.7	321	8.4	7.9 - 8.8
Women	5,932	3,144	53.0	2,933	49.4	210	6.7	6.2 - 7.2
Both sexes, 16 to 19 years	693	274	39.5	219	31.6	54	19.9	17.4 - 22.4
Black	2,063	1,190	57.7	1,021	49.5	169	14.2	13.0 - 15.3
Men	914	602	65.9	500	54.6	103	17.0	15.3 - 18.8
Women	1,149	588	51.2	522	45.4	66	11.3	9.8 - 12.7
Both sexes, 16 to 19 years	182	39	21.3	24	13.3	15	37.4	27.8 - 47.0
Hispanic origin	1,459	804	55.1	689	47.2	115	14.3	13.0 - 15.7
Men	653	461	70.6	388	59.3	73	15.9	14.0 - 17.8
Women	806	343	42.6	301	37.3	42	12.2	10.3 - 14.2

[Continued]

★ 659 ★

Civilian Noninstitutional Population in the Northeast: Employment Status: Annual Averages, 1992

[Continued]

State and population group	Civilian non-institutional population	Civilian labor force		Employment		Unemployment		
		Number	Percent of population	Number	Percent of population	Number	Rate	Error range of rate[1]
Both sexes, 16 to 19 years	155	35	22.7	21	13.8	14	39.2	29.5 - 49.0
Single (never married)	4,028	2,537	63.0	2,216	55.0	321	12.6	12.0 - 13.3
Married spouse present	7,218	4,789	66.3	4,499	62.3	290	6.1	5.7 - 6.4
Other marital status	2,563	1,197	46.7	1,083	42.3	113	9.5	8.6 - 10.4
Pennsylvania								
Total	9,443	6,004	83.6	5,554	58.8	450	7.5	7.1 - 7.9
Men	4,474	3,282	73.4	3,012	67.3	270	8.2	7.7 - 8.8
Women	4,969	2,722	54.8	2,542	51.1	180	6.6	6.1 - 7.2
Both sexes, 16 to 19 years	590	312	52.9	254	43.1	58	18.5	16.1 - 20.9
White	8,565	5,483	64.0	5,114	59.7	369	6.7	6.4 - 7.1
Men	4,068	3,011	74.0	2,791	68.6	220	7.3	6.8 - 7.8
Women	4,497	2,472	55.0	2,323	51.7	149	6.0	5.5 - 6.6
Both sexes, 16 to 19 years	519	283	54.5	238	45.9	45	15.9	13.5 - 18.3
Black	745	436	58.6	361	48.5	75	17.2	15.0 - 19.4
Men	340	223	65.7	178	52.5	45	20.1	16.9 - 23.4
Women	405	213	52.5	183	45.1	30	14.1	11.3 - 16.9
Hispanic origin	107	56	61.5	56	52.7	9	14.4	9.3 - 19.4
Men	49	37	74.5	32	65.1	5	12.5	6.2 - 18.9
Women	58	29	50.5	24	42.1	5	16.7	8.6 - 24.8
Single (never married)	2,338	1,646	70.4	1,438	61.5	208	12.7	11.7 - 13.6
Married spouse present	5,392	3,575	66.3	3,402	63.1	173	4.8	4.4 - 5.2
Other marital status	1,714	783	45.7	714	41.6	70	8.9	7.8 - 10.0
Connecticut								
Total	2,508	1,791	71.4	1,656	66.0	135	7.5	6.7 - 8.3
Men	1,187	948	79.8	864	72.7	84	8.9	7.7 - 10.1
Women	1,321	843	63.8	792	60.0	51	6.0	5.0 - 7.1
Both sexes, 16 to 19 years	165	93	56.2	83	50.1	10	10.9	6.8 - 15.0
White	2,242	1,594	71.1	1,480	66.0	115	7.2	6.4 - 8.0
Men	1,061	844	79.6	773	72.9	71	8.5	7.2 - 9.7
Women	1,181	750	63.5	707	59.8	43	5.8	4.7 - 6.8
Both sexes, 16 to 19 years	140	83	59.7	75	53.6	8	10.2	6.0 - 14.3
Black	212	160	75.3	141	66.5	19	11.7	8.1 - 15.2
Men	101	82	81.9	71	70.1	12	14.4	9.0 - 19.8
Women	112	78	69.4	71	63.3	7	8.8	4.3 - 13.2

[Continued]

★ 659 ★

Civilian Noninstitutional Population in the Northeast: Employment Status: Annual Averages, 1992
[Continued]

State and population group	Civilian non-institutional population	Civilian labor force		Employment		Unemployment		
		Number	Percent of population	Number	Percent of population	Number	Rate	Error range of rate[1]
Hispanic origin	120	83	69.6	73	60.7	11	12.8	7.8 - 17.8
Men	49	39	79.8	33	66.9	6	16.2	8.1 - 24.3
Women	70	44	62.4	40	56.3	4	9.8	3.6 - 15.9
Single (never married)	664	480	72.3	427	64.2	54	11.2	9.4 - 13.0
Married spouse present	1,445	1,083	74.9	1,019	70.5	63	5.9	4.9 - 6.8
Other marital status	399	228	57.2	210	52.7	18	7.8	5.6 - 10.1
Maine								
Total	966	662	68.6	615	63.7	47	7.1	6.4 - 7.9
Men	471	357	75.8	328	69.6	29	8.1	7.0 - 9.2
Women	495	305	61.7	287	58.0	18	6.0	5.0 - 7.0
Both sexes, 16 to 19 years	71	38	54.2	31	43.9	7	19.0	14.4 - 23.6
White	953	653	68.5	607	63.7	46	7.0	6.2 - 7.7
Men	464	351	75.7	323	69.7	28	8.0	6.9 - 9.1
Women	489	302	61.7	284	58.1	18	5.8	4.8 - 6.9
Both sexes, 16 to 19 years	70	38	54.6	31	44.4	7	18.7	14.1 - 23.3
Single (never married)	227	163	71.8	144	63.7	18	11.3	9.5 - 13.1
Married spouse present	571	406	71.1	386	67.6	20	5.0	4.2 - 5.8
Other marital status	168	93	55.5	85	50.4	8	9.1	6.9 - 11.3
Massachusetts								
Total	4,630	3,126	67.5	2,862	61.8	265	8.5	8.1 - 8.9
Men	2,164	1,648	76.2	1,487	68.7	162	9.8	9.2 - 10.4
Women	2,465	1,478	60.0	1,375	55.8	103	7.0	6.5 - 7.5
Both sexes, 16 to 19 years	270	145	53.6	115	42.7	29	20.2	17.6 - 22.8
White	4,362	2,953	67.7	2,717	62.3	237	8.0	7.6 - 8.4
Men	2,045	1,558	76.2	1,412	69.1	146	9.4	8.8 - 9.9
Women	2,317	1,395	60.2	1,305	56.3	91	6.5	6.0 - 7.0
Both sexes, 16 to 19 years	247	135	54.7	110	44.5	25	18.5	15.9 - 21.2
Black	174	117	67.2	96	55.1	21	17.9	14.8 - 21.1
Men	76	62	80.6	49	64.6	12	19.8	15.3 - 24.4
Women	98	55	56.7	47	47.7	9	15.9	11.5 - 20.2
Hispanic origin	130	73	55.9	61	47.1	11	15.7	12.0 - 19.4
Men	55	38	70.6	32	58.9	6	16.6	11.4 - 21.8
Women	75	34	45.4	29	38.7	5	14.7	9.5 - 20.0
Single (never married)	1,386	1,016	73.3	899	64.9	117	11.5	10.7 - 12.3

[Continued]

★ 659 ★

Civilian Noninstitutional Population in the Northeast: Employment Status: Annual Averages, 1992
[Continued]

State and population group	Civilian non-institutional population	Civilian labor force		Employment		Unemployment		
		Number	Percent of population	Number	Percent of population	Number	Rate	Error range of rate[1]
Married spouse present	2,447	1,712	70.0	1,602	65.5	110	6.4	5.9 - 6.9
Other marital status	797	398	49.9	360	45.1	38	9.6	8.4 - 10.8
New Hampshire								
Total	880	633	72.0	586	66.6	47	7.5	6.7 - 8.2
Men	426	339	79.5	314	73.5	25	7.5	6.4 - 8.5
Women	454	295	64.9	273	60.1	22	7.5	6.3 - 8.6
Both sexes, 16 to 19 years	49	29	58.2	22	44.1	7	24.2	18.3 - 30.1
White	864	623	72.1	578	66.9	46	7.3	6.5 - 8.1
Men	417	332	79.6	308	73.9	24	7.1	6.1 - 8.2
Women	447	291	65.2	269	60.3	22	7.5	6.4 - 8.7
Both sexes, 16 to 19 years	48	28	58.8	22	44.6	7	24.1	18.3 - 30.0
Single (never married)	197	147	74.7	129	65.7	18	12.1	10.1 - 14.1
Married spouse present	522	389	74.6	369	70.7	20	5.2	4.4 - 6.1
Other marital status	161	97	60.2	88	54.4	9	9.5	7.3 - 11.7
Rhode Island								
Total	775	527	68.1	481	62.0	47	8.9	8.0 - 9.7
Men	360	276	76.6	249	69.2	27	9.7	8.5 - 11.0
Women	414	251	60.6	231	55.8	20	7.9	6.7 - 9.0
Both sexes, 16 to 19 years	45	28	62.1	23	52.0	4	16.3	11.5 - 21.0
White	732	500	68.4	459	62.7	42	8.4	7.5 - 9.2
Men	343	262	76.5	238	69.5	24	9.2	8.0 - 10.4
Women	389	239	61.3	221	56.7	18	7.5	6.3 - 8.6
Both sexes, 16 to 19 years	41	26	62.9	22	53.4	4	15.2	10.4 - 19.9
Black	25	15	82.5	12	50.0	3	19.9	12.1 - 27.7
Hispanic origin	30	21	70.6	16	52.8	5	25.2	18.0 - 32.4
Men	14	12	83.8	8	54.7	4	34.7	24.1 - 45.3
Single (never married)	207	155	75.2	136	65.9	19	12.3	10.5 - 14.1
Married spouse present	421	298	70.8	277	65.7	21	7.1	6.1 - 8.1
Other marital status	147	74	50.4	68	46.0	6	8.6	6.4 - 10.8
Vermont								
Total	452	322	71.1	300	66.4	21	6.6	5.9 - 7.4
Men	221	169	76.5	156	70.5	13	7.9	6.8 - 9.1
Women	231	152	66.0	145	62.6	8	5.2	4.2 - 6.2

[Continued]

★ 659 ★

Civilian Noninstitutional Population in the Northeast: Employment Status: Annual Averages, 1992

[Continued]

State and population group	Civilian non-institutional population	Civilian labor force		Employment		Unemployment		
		Number	Percent of population	Number	Percent of population	Number	Rate	Error range of rate[1]
Both sexes, 16 to 19 years	31	18	57.9	15	48.2	3	16.8	12.0 - 21.5
White	449	319	71.1	298	66.4	21	6.6	5.8 - 7.3
Men	220	168	76.5	155	70.6	13	7.8	6.7 - 9.0
Women	229	151	65.9	143	62.5	8	5.2	4.2 - 6.2
Both sexes, 16 to 19 years	31	18	57.7	15	48.2	3	16.6	11.8 - 21.4
Single (never married)	113	84	74.3	75	66.7	8	10.1	8.3 - 11.9
Married, spouse present	261	193	74.0	184	70.5	9	4.7	3.9 - 5.6
Other marital status	79	45	57.1	41	52.4	4	8.1	5.9 - 10.4

Source: "States: Employment Status of the Civilian Noninstitutional Population by Sex, Age, Race, Hispanic Origin, and Marital Status, 1992 Annual Averages," U.S. Department of Labor, *Geographical Profile of Employment and Unemployment, 1992*, pp. 38-50. Rearranged by the Editors. Notes: Data for demographic groups are not shown when they do not meet BLS publication standards of reliability for the particular area based on the sample in that area. See appendix B. Items may not add to totals or compute to displayed percentages because of rounding. Detail for Hispanic-origin groups will not add to totals because data for the "other races" group are not presented and Hispanics are included in both the white and black population groups. 1. Error ranges are calculated at the 90-percent confidence interval, which means that if repeated samples were drawn from the same population and an error range constructed around each sample estimate, in 9 out of 10 cases the true value based on a complete census of the population would be contained within these error ranges.

★ 660 ★

Labor Force

Civilian Noninstitutional Population in the Pacific States: Employment Status: Annual Averages, 1992

Numbers in thousands.

State and population group	Civilian non-institutional population	Civilian labor force		Employment		Unemployment		
		Number	Percent of population	Number	Percent of population	Number	Rate	Error range of rate[1]
Alaska								
Total	360	262	72.9	238	66.3	24	9.1	8.2 - 10.0
Men	174	140	80.6	126	72.3	14	10.3	9.0 - 11.5
Women	185	122	65.7	112	60.7	9	7.7	6.5 - 8.9
Both sexes, 16 to 19 years	27	13	48.7	11	39.5	2	18.8	14.3 - 23.2
White	282	214	75.9	197	69.9	17	7.8	6.9 - 8.8
Men	138	115	83.2	105	75.7	10	9.0	7.8 - 10.3
Women	144	99	68.7	92	64.3	6	6.5	5.3 - 7.7
Both sexes, 16 to 19 years	20	11	53.7	9	44.7	2	16.6	11.8 - 21.5
Black	13	10	78.7	9	69.2	1	12.0	7.3 - 16.7
Single (never married)	92	56	72.4	57	62.6	9	13.5	11.6 - 15.4
Married spouse present	203	151	74.3	141	69.3	10	6.7	5.7 - 7.7
Other marital status	64	45	69.5	40	62.1	5	10.6	8.5 - 12.8

[Continued]

★ 660 ★

Civilian Noninstitutional Population in the Pacific States: Employment Status: Annual Averages, 1992

[Continued]

State and population group	Civilian non-institutional population	Civilian labor force		Employment		Unemployment		
		Number	Percent of population	Number	Percent of population	Number	Rate	Error range of rate[1]
California								
Total	22,925	15,187	66.2	13,805	60.2	1,382	9.1	8.8 - 9.4
Men	11,317	8,686	76.7	7,865	69.5	821	9.4	9.0 - 9.9
Women	11,608	6,501	56.0	5,939	51.2	562	8.6	8.2 - 9.1
Both sexes, 16 to 19 years	1,570	748	47.6	561	35.7	187	25.1	23.0 - 27.1
White	19,183	12,778	66.6	11,656	60.8	1,123	8.8	8.4 - 9.1
Men	9,553	7,416	77.6	6,732	70.6	684	9.2	8.8 - 9.7
Women	9,631	5,363	55.7	4,924	51.1	438	8.2	7.7 - 8.7
Both sexes, 16 to 19 years	1,266	630	49.7	482	38.0	148	23.5	21.3 - 25.7
Black	1,354	831	61.4	708	52.3	123	14.8	13.0 - 16.6
Men	629	432	68.7	370	58.8	62	14.3	11.9 - 15.8
Women	725	399	55.0	337	46.6	61	15.4	12.7 - 18.0
Hispanic origin	5,504	3,694	67.1	3,216	58.4	478	13.0	12.1 - 13.8
Men	2,817	2,308	82.0	2,009	71.3	299	12.9	11.9 - 14.0
Women	2,668	1,386	51.6	1,206	44.9	180	13.0	11.6 - 14.3
Both sexes, 16 to 19 years	566	259	45.8	184	32.4	76	29.2	24.9 - 33.4
Single (never married)	6,171	4,434	71.8	3,554	62.6	570	12.9	12.2 - 13.5
Married spouse present	12,406	8,320	67.1	7,751	62.5	568	6.8	6.5 - 7.2
Other marital status	4,349	2,433	56.0	2,190	50.4	244	10.0	9.2 - 10.8
Hawaii								
Total	832	572	68.8	547	65.7	26	4.5	3.9 - 5.1
Men	393	301	76.5	287	72.9	14	4.8	3.9 - 5.6
Women	438	271	61.9	260	59.3	11	4.2	3.4 - 5.1
Both sexes, 16 to 19 years	51	26	50.9	22	42.5	4	16.4	11.6 - 21.2
White	276	196	70.9	185	67.0	11	5.4	4.2 - 6.5
Men	130	106	81.0	100	76.5	6	5.5	4.0 - 7.1
Women	146	90	61.8	85	58.6	5	5.2	3.6 - 6.9
Hispanic origin	20	14	72.5	13	64.8	2	10.6	4.4 - 16.9
Single (never married)	216	159	73.7	146	67.6	13	8.2	6.7 - 9.8
Married spouse present	470	331	70.3	320	68.1	10	3.2	2.5 - 3.8
Other marital status	145	82	56.9	80	55.4	2	2.7	1.4 - 4.0

[Continued]

★ 660 ★

Civilian Noninstitutional Population in the Pacific States: Employment Status: Annual Averages, 1992

[Continued]

State and population group	Civilian non-institutional population	Civilian labor force		Employment		Unemployment		
		Number	Percent of population	Number	Percent of population	Number	Rate	Error range of rate[1]
Oregon								
Total	2,274	1,537	67.6	1,422	62.5	115	7.5	6.7 - 8.3
Men	1,113	849	76.2	788	70.8	60	7.1	6.1 - 8.1
Women	1,161	689	59.3	834	54.6	55	8.0	6.8 - 9.2
Both sexes, 16 to 19 years	151	87	57.4	70	46.2	17	19.5	15.3 - 23.7
White	2,168	1,461	67.4	1,354	62.5	107	7.3	6.5 - 8.1
Men	1,061	807	76.1	751	70.8	56	6.9	5.9 - 7.9
Women	1,107	654	59.1	603	54.5	51	7.9	6.7 - 9.1
Both sexes, 16 to 19 years	145	84	58.3	68	47.0	16	19.3	15.1 - 23.6
Hispanic origin	87	68	77.5	56	64.6	11	16.8	11.4 - 22.2
Men	46	42	90.7	35	75.9	7	16.3	9.5 - 23.1
Single (never married)	504	376	74.6	330	65.6	45	12.0	10.2 - 13.9
Married spouse present	1,333	905	67.9	861	64.5	44	4.9	4.1 - 5.8
Other marital status	437	257	58.8	231	52.9	26	10.0	7.9 - 12.0
Washington								
Total	3,789	2,598	68.6	2,403	63.4	195	7.5	6.7 - 8.3
Men	1,864	1,429	76.7	1,306	70.0	123	8.6	7.5 - 9.8
Women	1,925	1,169	60.7	1,097	57.0	72	8.1	5.1 - 7.2
Both sexes, 16 to 19 years	224	135	60.2	114	51.0	20	15.2	10.7 - 19.7
White	3,450	2,377	68.9	2,209	64.0	168	7.1	6.2 - 7.9
Men	1,702	1,307	76.8	1,201	70.6	106	8.1	6.9 - 9.2
Women	1,749	1,070	61.2	1,008	57.7	62	5.8	4.7 - 6.9
Both sexes, 16 to 19 years	202	125	61.7	108	53.2	17	13.7	9.2 - 18.2
Black	110	79	72.2	67	60.9	12	15.6	9.1 - 22.1
Hispanic origin	87	68	78.4	60	69.3	8	11.6	5.5 - 17.8
Single (never married)	891	690	77.5	615	69.0	75	10.9	9.1 - 12.7
Married spouse present	2,208	1,480	67.0	1,401	63.4	79	5.3	4.4 - 6.2
Other marital status	690	428	61.9	387	56.1	40	9.5	7.3 - 11.6

Source: "States: Employment Status of the Civilian Noninstitutional Population by Sex, Age, Race, Hispanic Origin, and Marital Status, 1992 Annual Averages," U.S. Department of Labor, *Geographical Profile of Employment and Unemployment, 1992*, pp. 38-50. Rearranged by the Editors. *Notes:* Data for demographic groups are not shown when they do not meet BLS publication standards of reliability for the particular area based on the sample in that area. See appendix B. Items may not add to totals or compute to displayed percentages because of rounding. Detail for Hispanic-origin groups will not add to totals because data for the "other races" group are not presented and Hispanics are included in both the white and black population groups. 1. Error ranges are calculated at the 90-percent confidence interval, which means that if repeated samples were drawn from the same population and an error range constructed around each sample estimate, in 9 out of 10 cases the true value based on a complete census of the population would be contained within these error ranges.

★ 661 ★
Labor Force

Civilian Noninstitutional Population in the South: Employment Status: Annual Averages, 1992

Numbers in thousands.

State and population group	Civilian non-institutional population	Civilian labor force		Employment		Unemployment		
		Number	Percent of population	Number	Percent of population	Number	Rate	Error range of rate[1]
Alabama								
Total	3,122	1,937	62.1	1,796	57.5	142	7.3	6.5 - 8.1
Men	1,464	1,052	71.9	984	67.2	68	6.5	5.4 - 7.5
Women	1,658	885	53.4	812	49.0	74	8.3	7.0 - 9.6
Both sexes, 16 to 19 years	239	118	49.4	90	37.7	28	23.7	18.5 - 28.8
White	2,356	1,470	62.4	1,392	59.1	78	5.3	4.5 - 6.1
Men	1,127	832	73.8	790	70.0	42	5.1	4.1 - 6.1
Women	1,229	638	51.9	602	49.0	36	5.6	4.4 - 6.9
Both sexes, 16 to 19 years	152	88	58.1	74	48.6	14	16.4	11.1 - 21.6
Black	750	460	61.3	397	52.9	63	13.6	11.5 - 15.7
Men	329	215	65.3	190	57.6	25	11.8	8.9 - 14.7
Women	421	245	58.2	207	49.3	37	15.2	12.2 - 18.2
Single (never married)	681	443	65.0	379	55.7	64	14.4	12.1 - 16.6
Married spouse present	1,826	1,203	65.9	1,152	63.1	51	4.3	3.5 - 5.1
Other marital status	614	291	47.4	265	43.1	26	9.0	6.8 - 11.3
Arkansas								
Total	1,811	1,149	63.4	1,066	58.9	83	7.2	6.4 - 8.0
Men	848	616	72.6	571	67.4	45	7.2	6.2 - 8.3
Women	963	533	55.4	495	51.4	38	7.2	6.1 - 8.3
Both sexes, 16 to 19 years	148	82	55.3	64	43.5	17	21.4	17.1 - 25.6
White	1,528	976	63.9	923	60.4	53	5.4	4.7 - 6.2
Men	722	531	73.7	503	69.7	29	5.4	4.4 - 6.4
Women	807	445	55.1	420	52.1	24	5.4	4.4 - 6.5
Both sexes, 16 to 19 years	117	66	56.9	57	48.6	10	14.5	10.3 - 18.7
Black	257	157	61.2	128	50.1	29	18.2	15.1 - 21.3
Men	113	75	66.1	50	52.6	15	20.3	15.6 - 25.0
Women	143	82	57.4	69	48.0	13	16.3	12.2 - 20.4
Single (never married)	343	235	68.3	198	57.6	37	15.7	13.4 - 17.9
Married spouse present	1,114	750	67.4	717	64.4	33	4.5	3.7 - 5.2
Other marital status	354	164	46.4	151	42.8	13	7.8	5.7 - 9.9
Delaware								
Total	537	372	69.3	352	65.7	20	5.3	4.6 - 6.0
Men	256	197	77.1	185	72.5	12	5.9	5.0 - 6.9
Women	281	175	62.3	167	59.4	8	4.6	3.7 - 5.5

[Continued]

★ 661 ★

Civilian Noninstitutional Population in the South: Employment Status: Annual Averages, 1992
[Continued]

State and population group	Civilian non-institutional population	Civilian labor force		Employment		Unemployment		
		Number	Percent of population	Number	Percent of population	Number	Rate	Error range of rate[1]
Both sexes, 16 to 19 years	34	19	56.1	17	50.0	2	10.9	6.8 - 15.0
White	430	298	69.2	285	66.3	12	4.1	3.5 - 4.8
Men	209	163	78.0	155	74.4	8	4.6	3.6 - 5.6
Women	221	135	60.9	130	58.8	5	3.5	2.6 - 4.5
Both sexes, 16 to 19 years	23	15	64.2	14	60.1	1	6.4	2.7 - 10.2
Black	94	65	69.9	58	62.5	7	10.6	8.2 - 13.1
Men	40	29	72.2	25	62.2	4	13.9	9.7 - 18.0
Women	54	37	68.2	34	62.7	3	8.1	5.2 - 10.9
Hispanic origin	10	8	75.8	7	69.8	1	7.8	1.7 - 14.0
Single (never married)	136	103	75.9	95	69.6	9	8.3	6.7 - 9.8
Married spouse present	302	212	70.3	204	67.5	8	4.0	3.2 - 4.8
Other marital status	99	57	57.5	54	54.7	3	4.9	3.2 - 6.6
District of Columbia								
Total	416	276	68.3	253	60.7	23	8.4	7.5 - 9.3
Men	188	135	71.8	122	64.9	13	9.6	8.2 - 11.0
Women	228	141	61.7	131	57.2	10	7.3	6.1 - 8.5
White	143	115	80.6	111	77.8	4	3.4	2.5 - 4.4
Men	70	61	87.1	59	84.0	2	3.5	2.2 - 4.8
Women	72	54	74.3	52	71.8	2	3.4	2.0 - 4.7
Black	264	153	58.0	134	50.9	19	12.2	10.6 - 13.8
Men	113	70	61.6	59	52.5	11	15.1	12.5 - 17.7
Women	151	83	55.2	75	49.8	8	9.7	7.8 - 11.7
Hispanic origin	19	16	82.3	15	77.1	1	6.3	2.7 - 9.9
Men	10	9	91.6	8	86.0	1	6.1	1.3 - 10.8
Single (never married)	191	139	72.8	124	64.9	15	10.9	9.4 - 12.3
Married spouse present	122	83	67.7	79	64.6	4	4.7	3.4 - 5.9
Other marital status	103	54	52.5	50	48.4	4	7.9	5.9 - 9.9
Florida								
Total	10,594	6,553	61.9	6,017	56.8	536	8.2	7.8 - 8.6
Men	5,016	3,495	69.7	3,210	64.0	286	8.2	7.6 - 8.7
Women	5,578	3,058	54.8	2,808	50.3	250	8.2	7.6 - 8.7
Both sexes, 16 to 19	621	316	50.8	237	38.2	79	24.9	22.4 - 27.3

[Continued]

★ 661 ★

Civilian Noninstitutional Population in the South: Employment Status: Annual Averages, 1992
[Continued]

State and population group	Civilian non-institutional population	Civilian labor force		Employment		Unemployment		
		Number	Percent of population	Number	Percent of population	Number	Rate	Error range of rate[1]
White	8,950	5,458	61.0	5,083	56.8	395	7.2	6.8 - 7.6
Men	4,269	2,960	69.3	2,744	64.3	216	7.3	6.8 - 7.9
Women	4,681	2,498	53.4	2,319	49.5	179	7.2	6.6 - 7.7
Both sexes, 16 to 19	462	254	55.1	200	43.2	55	21.5	18.8 - 24.1
Black	1,461	978	66.9	847	58.0	130	13.3	12.0 - 14.7
Men	671	479	71.4	415	61.9	64	13.3	11.4 - 15.3
Women	791	499	63.1	432	54.7	67	13.4	11.5 - 15.3
Both sexes, 16 to 19 years	145	57	39.0	34	23.2	23	40.6	31.7 - 49.5
Hispanic origin	1,361	908	66.7	816	59.9	92	10.1	8.9 - 11.4
Men	659	524	79.4	475	72.1	48	9.2	7.7 - 10.8
Women	702	384	54.7	340	48.5	44	11.4	9.4 - 13.4
Both sexes, 16 to 19 years	98	52	53.1	39	40.1	13	24.4	17.0 - 31.8
Single (never married)	2,331	1,669	71.6	1,456	62.5	213	12.8	11.9 - 13.7
Married, spouse present	5,973	3,636	60.9	3,433	57.5	203	5.6	5.1 - 6.0
Other marital status	2,291	1,248	54.5	1,128	49.2	120	9.6	8.7 - 10.6
Georgia								
Total	4,894	3,232	66.0	3,008	61.5	224	6.9	6.2 - 7.7
Men	2,268	1,712	75.5	1,587	70.0	125	7.3	6.2 - 8.3
Women	2,625	1,520	57.9	1,421	54.1	99	6.5	5.5 - 7.8
Both sexes, 16 to 19 years	354	155	43.9	125	35.4	30	19.3	14.3 - 24.3
White	3,270	2,182	66.7	2,079	63.6	104	4.7	4.0 - 5.5
Men	1,572	1,207	76.8	1,151	73.2	56	4.7	3.7 - 5.7
Women	1,698	975	57.4	928	54.7	47	4.8	3.7 - 6.0
Both sexes, 16 to 19 years	208	103	49.4	92	44.1	11	10.7	5.7 - 15.6
Black	1,540	996	64.7	681	57.2	115	11.6	9.9 - 13.2
Men	650	475	72.0	410	62.2	65	13.6	11.0 - 16.2
Women	880	521	59.2	471	53.5	50	9.7	7.5 - 11.8
Single (never married)	1,208	800	66.2	691	57.2	109	13.6	11.7 - 15.6
Married spouse present	2,656	1,872	70.5	1,799	67.7	73	3.9	3.2 - 4.6
Other marital status	1,029	560	54.4	518	50.3	42	7.5	5.7 - 9.3
Kentucky								
Total	2,820	1,744	61.8	1,624	57.6	120	6.9	6.1 - 7.7
Men	1,344	965	71.8	890	66.2	74	7.7	6.6 - 8.9
Women	1,476	780	52.8	734	49.7	46	5.9	4.8 - 7.0
Both sexes, 16 to 19 years	208	100	47.9	84	40.4	16	15.7	10.9 - 20.4

[Continued]

★ 661 ★

Civilian Noninstitutional Population in the South: Employment Status: Annual Averages, 1992
[Continued]

State and population group	Civilian non-institutional population	Civilian labor force		Employment		Unemployment		
		Number	Percent of population	Number	Percent of population	Number	Rate	Error range of rate[1]
White	2,614	1,618	61.9	1,514	57.9	103	6.4	5.6 - 7.2
Men	1,243	893	71.8	829	66.7	64	7.2	6.0 - 8.3
Women	1,371	725	52.9	686	50.0	39	5.4	4.3 - 6.5
Both sexes, 16 to 19 years	189	91	48.1	78	41.2	13	14.3	9.5 - 19.1
Black	193	120	82.5	104	53.8	17	13.9	9.8 - 18.0
Men	95	68	70.9	57	60.1	10	15.2	9.5 - 20.8
Women	97	53	54.1	46	47.5	6	12.2	6.3 - 18.1
Single (never married)	562	363	64.6	322	57.3	41	11.2	9.0 - 13.3
Married spouse present	1,714	1,131	66.0	1,071	62.5	60	5.3	4.4 - 6.2
Other marital status	543	250	46.1	231	42.4	20	7.9	5.7 - 10.2
Louisiana								
Total	3,125	1,934	61.9	1,778	56.9	156	8.1	7.2 - 8.9
Men	1,464	1,052	71.9	967	68.0	86	8.2	7.1 - 9.3
Women	1,661	881	53.1	811	48.9	70	7.9	6.7 - 9.1
Both sexes, 16 to 19 years	238	98	41.2	78	32.8	20	20.4	15.4 - 25.3
White	2,277	1,476	64.8	1,392	61.1	85	5.7	4.9 - 6.5
Men	1,095	820	74.9	770	70.4	50	6.1	5.0 - 7.2
Women	1,182	656	55.5	621	52.6	35	5.3	4.1 - 6.4
Both sexes, 16 to 19 years	156	71	45.7	59	38.1	12	16.5	11.0 - 21.9
Black	794	431	54.3	363	45.8	68	15.7	13.0 - 18.4
Men	341	216	63.4	183	53.6	33	15.4	11.6 - 19.1
Women	452	214	47.4	180	39.9	34	16.0	12.1 - 19.8
Single (never married)	750	453	60.4	385	51.3	68	15.1	13.0 - 17.2
Married spouse present	1,781	1,185	66.5	1,131	63.5	54	4.5	3.7 - 5.3
Other marital status	594	296	49.8	262	44.1	34	11.4	9.1 - 13.8
Maryland								
Total	3,690	2,623	71.1	2,450	66.4	173	6.6	5.9 - 7.3
Men	1,758	1,370	77.9	1,271	72.3	99	7.2	6.2 - 8.3
Women	1,931	1,252	64.8	1,179	61.0	73	5.9	4.9 - 6.8
Both sexes, 16 to 19 years	204	106	52.0	78	38.3	28	26.4	20.9 - 31.9
White	2,573	1,820	70.7	1,737	67.5	83	4.6	3.8 - 5.3
Men	1,228	968	78.8	917	74.6	51	5.3	4.2 - 6.4
Women	1,345	852	63.3	820	61.0	32	3.7	2.8 - 4.7
Both sexes, 16 to 19 years	105	63	59.7	54	50.9	9	14.7	8.5 - 21.0

[Continued]

★ 661 ★

Civilian Noninstitutional Population in the South: Employment Status: Annual Averages, 1992

[Continued]

State and population group	Civilian non-institutional population	Civilian labor force		Employment		Unemployment		
		Number	Percent of population	Number	Percent of population	Number	Rate	Error range of rate[1]
Black	1,021	733	71.8	652	63.6	82	11.2	9.3 - 13.1
Men	482	364	75.6	321	66.6	43	11.9	9.1 - 14.7
Women	539	369	68.4	331	61.3	39	10.5	7.8 - 13.1
Hispanic origin	101	75	73.9	69	68.3	6	7.5	2.8 - 12.2
Single (never married)	982	744	75.8	661	67.4	83	11.2	9.5 - 12.8
Married, spouse present	2,009	1,462	72.8	1,397	69.6	64	4.4	3.6 - 5.2
Other marital status	899	417	59.6	391	55.9	25	6.1	4.4 - 7.8
Mississippi								
Total	1,919	1,182	61.6	1,086	56.6	96	8.1	7.3 - 8.9
Men	897	640	71.3	593	66.1	46	7.2	6.2 - 8.3
Women	1,022	542	53.1	492	48.2	50	9.2	7.9 - 10.5
Both sexes, 16 to 19 years	169	61	35.9	44	25.8	17	28.2	23.1 - 33.3
White	1,266	799	63.1	753	59.5	46	5.8	4.9 - 6.7
Men	608	453	74.5	431	70.9	22	4.9	3.9 - 6.0
Women	657	346	52.6	322	48.9	24	7.0	5.6 - 8.4
Both sexes, 16 to 19 years	89	37	42.3	30	33.7	8	20.2	14.0 - 26.4
Black	644	378	58.7	329	51.1	49	13.0	11.1 - 14.9
Men	285	184	64.5	160	56.3	23	12.8	10.0 - 15.5
Women	359	194	54.1	169	47.0	26	13.2	10.5 - 15.9
Single (never married)	465	275	59.3	231	49.7	44	16.1	13.9 - 18.3
Married, spouse present	1,065	715	67.1	682	64.0	33	4.6	3.8 - 5.4
Other marital status	390	192	49.2	173	44.3	19	9.8	7.7 - 12.0
North Carolina								
Total	5,127	3,487	68.0	3,281	64.0	207	5.9	5.6 - 6.3
Men	2,422	1,842	76.1	1,735	71.6	107	5.8	5.4 - 6.3
Women	2,705	1,645	60.8	1,546	57.1	100	6.0	5.6 - 6.5
Both sexes, 16 to 19 years	353	189	53.5	153	43.4	36	18.9	16.6 - 21.2
White	3,886	2,662	68.5	2,545	85.5	116	4.4	4.0 - 4.7
Men	1,855	1,432	77.2	1,370	73.9	62	4.3	3.9 - 4.8
Women	2,030	1,229	60.5	1,175	57.9	54	4.4	3.9 - 4.9
Both sexes, 16 to 19 years	232	135	58.4	118	51.1	17	12.6	10.2 - 14.9
Black	1,163	772	66.4	687	59.1	85	11.0	10.0 - 12.0
Men	524	377	71.9	335	63.8	42	11.1	9.7 - 12.6
Women	639	396	62.0	353	55.2	43	10.9	9.5 - 12.3

[Continued]

★ 661 ★

Civilian Noninstitutional Population in the South: Employment Status: Annual Averages, 1992

[Continued]

State and population group	Civilian non-institutional population	Civilian labor force		Employment		Unemployment		
		Number	Percent of population	Number	Percent of population	Number	Rate	Error range of rate[1]
Both sexes, 16 to 19 years	113	50	44.7	33	29.2	18	34.7	28.7 - 40.7
Hispanic origin	58	47	80.7	43	74.2	4	8.0	4.5 - 11.5
Men	36	33	91.5	31	86.9	2	5.0	1.7 - 8.4
Single (never married)	1,156	830	71.8	739	63.9	91	10.9	10.0 - 11.8
Married, spouse present	2,924	2,080	71.1	2,004	68.6	75	3.6	3.3 - 4.0
Other marital status	1,047	578	55.2	538	51.3	41	7.0	6.1 - 7.9
Oklahoma								
Total	2,400	1,527	63.6	1,440	60.0	86	5.7	5.0 - 6.3
Men	1,127	821	72.8	766	67.9	55	6.7	5.7 - 7.7
Women	1,272	706	55.5	674	53.0	31	4.5	3.6 - 5.3
Both sexes, 16 to 19 years	177	86	48.5	73	71.3	13	15.0	11.0 - 19.0
White	2,062	1,340	65.0	1,272	61.7	68	5.0	4.4 - 5.7
Men	972	727	74.7	684	70.3	43	5.9	4.9 - 6.9
Women	1,090	613	56.3	589	54.0	25	4.0	3.2 - 4.9
Both sexes, 16 to 19 years	144	75	52.4	65	45.5	10	13.2	9.1 - 17.3
Black	177	89	50.5	79	44.7	10	11.5	7.5 - 15.6
Men	80	43	53.4	36	45.2	7	15.4	8.8 - 12.8
Women	96	46	48.1	43	44.2	4	8.0	3.3 - 12.8
Hispanic origin	60	45	75.0	43	72.1	2	4.0	.3 - 7.6
Single (never married)	433	276	63.7	245	56.5	31	11.3	9.2 - 13.3
Married, spouse present	1,507	1,009	87.0	968	64.3	41	4.0	3.3 - 4.7
Other marital status	460	242	52.7	227	49.4	15	6.1	4.4 - 7.8
South Carolina								
Total	2,667	1,772	66.5	1,662	62.3	111	6.2	5.5 - 6.9
Men	1,265	947	74.9	891	70.4	56	6.0	5.0 - 6.9
Women	1,402	825	58.8	771	55.0	54	8.6	5.5 - 7.6
Both sexes, 16 to 19 years	203	98	48.1	81	39.7	17	17.4	12.7 - 22.0
White	1,875	1,266	67.5	1,211	64.6	54	4.3	3.6 - 5.0
Men	905	693	76.6	663	73.3	30	4.3	3.4 - 5.3
Women	970	572	59.0	548	56.5	24	4.2	3.2 - 5.3
Both sexes, 16 to 19 years	121	64	53.2	56	46.4	8	12.9	7.8 - 17.9
Black	762	436	63.8	431	56.6	55	11.3	9.6 - 13.0
Men	347	245	70.5	218	62.9	26	10.8	8.5 - 13.1

[Continued]

★ 661 ★

Civilian Noninstitutional Population in the South: Employment Status: Annual Averages, 1992
[Continued]

State and population group	Civilian non-institutional population	Civilian labor force		Employment		Unemployment		
		Number	Percent of population	Number	Percent of population	Number	Rate	Error range of rate[1]
Women	415	241	58.1	213	51.3	28	11.8	9.4 - 14.2
Single (never married)	832	413	65.4	367	58.1	46	11.2	9.3 - 13.0
Married, spouse present	1,566	1,117	71.3	1,071	68.4	46	4.1	3.4 - 4.8
Other marital status	469	242	51.7	223	47.6	19	7.8	5.7 - 9.9
Tennessee								
Total	3,849	2,440	63.4	2,285	59.4	155	6.4	5.6 - 7.1
Men	1,794	1,299	72.4	1,222	68.1	76	5.9	4.9 - 6.8
Women	2,055	1,142	55.6	1,063	51.7	79	6.9	5.8 - 8.0
Both sexes, 16 to 19 years	290	140	48.3	116	40.0	24	17.2	12.7 - 21.7
White	3,223	2,078	64.5	1,962	60.9	116	5.6	4.8 - 6.3
Men	1,520	1,124	74.0	1,067	70.2	57	5.1	4.1 - 6.1
Women	1,703	953	56.0	895	52.6	58	6.1	5.0 - 7.2
Both sexes, 16 to 19 years	227	119	52.4	100	44.0	19	16.1	11.4 - 20.9
Black	585	336	57.4	297	50.7	39	11.6	9.1 - 14.1
Men	252	156	62.1	138	54.6	19	12.0	8.3 - 15.7
Women	333	179	53.8	159	47.8	20	11.2	7.8 - 14.5
Single (never married)	854	551	64.5	486	56.9	65	11.8	9.8 - 13.8
Married, spouse present	2,188	1,484	67.8	1,422	65.0	82	4.2	3.4 - 5.0
Other marital status	807	405	50.2	378	46.8	27	6.8	4.9 - 8.6
Texas								
Total	12,698	8,745	68.9	8,088	63.7	657	7.5	7.1 - 7.9
Men	6,117	4,847	79.2	4,492	73.4	355	7.3	6.8 - 7.8
Women	6,580	3,898	59.2	3,596	54.6	303	7.8	7.2 - 8.3
Both sexes, 16 to 19 years	940	501	53.3	383	40.8	117	23.4	21.3 - 25.4
White	10,939	7,512	68.7	7,041	64.4	471	6.3	5.9 - 6.7
Men	5,287	4,200	79.4	3,842	74.6	258	6.1	5.6 - 6.6
Women	5,652	3,312	58.6	3,098	54.8	213	6.4	5.9 - 7.0
Both sexes, 16 to 19 years	789	436	55.2	350	44.3	86	19.8	17.6 - 22.0
Black	1,437	1,004	69.9	839	58.4	165	16.5	14.7 - 18.2
Men	674	520	77.1	432	64.1	88	16.9	14.5 - 19.3
Women	763	484	63.5	407	53.3	78	16.0	13.6 - 18.5
Both sexes, 16 to 19 years	123	52	42.4	26	21.1	26	50.3	39.0 - 61.6
Hispanic origin	2,955	1,983	67.1	1,792	60.6	191	9.6	8.7 - 10.6
Men	1,465	1,177	80.3	1,064	72.6	113	9.6	8.4 - 10.8

[Continued]

★ 661 ★

Civilian Noninstitutional Population in the South: Employment Status: Annual Averages, 1992

[Continued]

State and population group	Civilian non-institutional population	Civilian labor force		Employment		Unemployment		
		Number	Percent of population	Number	Percent of population	Number	Rate	Error range of rate[1]
Women	1,490	806	54.1	728	48.9	78	9.7	8.2 - 11.1
Both sexes, 16 to 19 years	322	165	51.1	125	38.9	39	23.9	19.7 - 28.2
Single (never married)	2,834	2,039	72.0	1,775	62.7	264	12.9	12.0 - 13.9
Married, spouse present	7,398	5,198	70.3	4,931	66.7	267	5.1	4.7 - 5.6
Other marital status	2,466	1,508	61.1	1,382	56.0	126	8.4	7.4 - 9.3
Virginia								
Total	4,734	3,359	70.9	3,145	66.4	213	6.4	5.7 - 7.0
Men	2,291	1,819	79.4	1,698	74.1	121	6.7	5.7 - 7.6
Women	2,444	1,540	63.0	1,447	59.2	92	6.0	5.0 - 7.0
Both sexes, 16 to 19 years	308	156	50.8	122	39.7	34	21.8	17.2 - 26.5
White	3,707	2,607	70.3	2,477	66.8	130	5.0	4.3 - 5.7
Men	1,791	1,424	79.5	1,352	75.5	72	5.0	4.1 - 6.0
Women	1,916	1,183	61.8	1,125	58.7	59	5.0	4.0 - 6.0
Both sexes, 16 to 19 years	208	116	55.8	98	47.2	18	15.5	10.6 - 20.5
Black	889	656	73.8	578	65.0	78	11.9	10.0 - 13.8
Men	437	349	79.8	302	69.3	46	13.2	10.5 - 15.9
Women	452	308	68.0	276	60.9	32	10.5	7.8 - 13.1
Hispanic origin	105	86	82.4	78	74.4	8	9.7	4.9 - 14.4
Single (never married)	1,152	861	74.7	757	65.7	104	12.1	10.3 - 13.8
Married, spouse present	2,761	2,002	72.5	1,925	69.7	77	3.8	3.2 - 4.5
Other marital status	821	496	60.4	463	56.4	33	6.6	4.8 - 8.3
West Virginia								
Total	1,406	766	54.4	679	48.3	86	11.3	10.3 - 12.3
Men	671	444	86.3	385	57.4	59	13.3	11.9 - 14.7
Women	736	321	43.7	294	40.0	27	8.5	7.1 - 9.8
Both sexes, 16 to 19 years	96	36	38.1	26	26.9	11	29.5	23.2 - 35.8
White	1,349	734	54.4	654	48.5	81	11.0	9.9 - 12.0
Men	646	430	66.5	374	57.8	56	13.1	11.6 - 14.5
Women	702	305	43.4	280	39.9	24	8.0	6.6 - 9.4
Both sexes, 16 to 19 years	91	35	38.7	25	27.6	10	28.7	22.3 - 35.1
Black	51	28	55.1	23	43.9	6	20.2	13.5 - 26.9
Single (never married)	271	160	58.9	130	47.9	30	18.7	16.0 - 21.4

[Continued]

★ 661 ★

Civilian Noninstitutional Population in the South: Employment Status: Annual Averages, 1992

[Continued]

State and population group	Civilian non-institutional population	Civilian labor force		Employment		Unemployment		
		Number	Percent of population	Number	Percent of population	Number	Rate	Error range of rate[1]
Married, spouse present	864	500	57.8	458	53.0	42	8.3	7.2 - 9.4
Other marital status	271	106	39.1	91	33.6	15	14.1	11.2 - 17.1

Source: "States: Employment Status of the Civilian Noninstitutional Population by Sex, Age, Race, Hispanic Origin, and Marital Status, 1992 Annual Averages," U.S. Department of Labor, *Geographical Profile of Employment and Unemployment, 1992*, pp. 38-50. Rearranged by the Editors. *Notes:* Data for demographic groups are not shown when they do not meet BLS publication standards of reliability for the particular area based on the sample in that area. See appendix B. Items may not add to totals or compute to displayed percentages because of rounding. Detail for Hispanic-origin groups will not add to totals because data for the "other races" group are not presented and Hispanics are included in both the white and black population groups. 1. Error ranges are calculated at the 90-percent confidence interval, which means that if repeated samples were drawn from the same population and an error range constructed around each sample estimate, in 9 out of 10 cases the true value based on a complete census of the population would be contained within these error ranges.

★ 662 ★

Labor Force

Civilian Noninstitutional Population in the West: Employment Status: Annual Averages, 1992

Numbers in thousands.

State and population group	Civilian non-institutional population	Civilian labor force		Employment		Unemployment		
		Number	Percent of population	Number	Percent of population	Number	Rate	Error range of rate[1]
Arizona								
Total	2,752	1,734	63.0	1,605	58.3	129	7.4	6.6 - 8.2
Men	1,334	952	71.4	874	65.5	79	8.3	7.2 - 9.4
Women	1,418	782	55.1	732	51.6	50	6.4	5.4 - 7.5
Both sexes, 16 to 19 years	176	89	50.5	62	35.4	27	29.9	25.3 - 34.6
White	2,506	1,638	62.8	1,521	58.4	117	7.1	6.3 - 7.9
Men	1,272	906	71.2	835	65.6	71	7.8	6.7 - 8.9
Women	1,334	732	54.9	686	51.4	46	6.3	5.2 - 7.4
Both sexes, 16 to 19 years	162	83	51.3	60	36.8	23	28.1	23.3 - 33.0
Hispanic origin	445	285	64.0	254	57.1	31	10.8	8.5 - 13.2
Men	218	170	78.3	151	69.4	19	11.3	8.2 - 14.4
Women	228	115	50.5	103	45.4	12	10.1	6.5 - 13.7
Single (never married)	593	431	72.7	372	62.7	59	13.7	11.7 - 15.7
Married spouse present	1,632	1,001	61.3	955	58.5	46	4.6	3.7 - 5.4
Other marital status	527	303	57.4	278	52.8	24	8.0	6.1 - 9.9
Colorado								
Total	2,506	1,756	70.1	1,652	65.9	104	5.9	5.2 - 6.6
Men	1,199	939	78.3	884	73.7	55	5.9	4.9 - 6.8
Women	1,307	818	62.6	768	58.8	49	6.0	5.0 - 7.0
Both sexes, 16 to 19 years	155	86	55.7	67	43.4	19	22.0	17.0 - 27.0

[Continued]

★ 662 ★

Civilian Noninstitutional Population in the West: Employment Status: Annual Averages, 1992
[Continued]

State and population group	Civilian non-institutional population	Civilian labor force		Employment		Unemployment		
		Number	Percent of population	Number	Percent of population	Number	Rate	Error range of rate[1]
White	2,291	1,612	70.4	1,521	66.4	91	5.6	4.9 - 6.3
Men	1,097	864	78.7	816	74.4	48	5.5	4.6 - 6.5
Women	1,194	748	62.7	705	59.1	43	5.7	4.7 - 6.7
Both sexes, 16 to 19 years	136	79	58.1	62	45.7	17	21.4	16.2 - 26.6
Black	120	81	67.8	71	59.5	10	12.2	7.7 - 16.8
Men	59	44	73.9	39	65.8	5	11.0	5.0 - 16.9
Women	61	38	61.8	33	53.4	5	13.7	6.6 - 20.8
Hispanic origin	247	173	69.8	160	64.8	12	7.2	4.6 - 9.8
Men	117	94	80.1	89	75.6	5	5.7	2.5 - 8.9
Women	130	78	60.4	71	55.1	7	8.9	4.6 - 13.2
Single (never married)	575	439	76.4	392	68.2	47	10.8	9.0 - 12.5
Married spouse present	1,450	1,023	70.6	982	67.7	41	4.0	3.3 - 4.8
Other marital status	481	293	60.9	278	57.7	15	5.3	3.7 - 6.8
Idaho								
Total	756	518	68.6	485	64.2	33	6.5	5.7 - 7.2
Men	367	288	78.5	269	73.3	19	6.6	5.6 - 7.6
Women	389	230	59.2	216	55.5	14	6.3	5.2 - 7.4
Both sexes, 16 to 19 years	57	35	61.8	29	51.5	6	16.6	12.5 - 20.8
White	739	507	68.7	475	64.3	32	6.4	5.6 - 7.2
Men	360	283	78.6	264	73.5	19	6.6	5.5 - 7.6
Women	379	224	59.2	210	55.5	14	6.2	5.1 - 7.3
Both sexes, 16 to 19 years	55	34	62.2	28	51.8	6	16.6	12.4 - 20.9
Hispanic origin	34	27	80.3	25	74.0	2	7.8	4.3 - 11.4
Men	18	16	87.2	15	80.1	1	8.1	3.5 - 12.7
Women	15	11	71.9	10	66.6	1	7.4	2.1 - 12.8
Single (never married)	132	98	73.9	86	65.3	11	11.7	9.5 - 13.9
Married spouse present	478	336	70.4	322	67.3	15	4.4	3.6 - 5.2
Other marital status	146	84	57.9	77	53.0	7	8.4	6.4 - 10.5
Montana								
Total	600	411	68.6	384	64.0	28	6.7	5.9 - 7.5
Men	293	218	74.5	203	69.2	16	7.1	6.0 - 8.2
Women	307	193	63.0	181	59.1	12	6.2	5.1 - 7.3
Both sexes, 16 to 19 years	48	27	57.0	23	48.6	4	14.7	10.7 - 18.8
White	563	387	68.7	365	64.7	23	5.8	5.1 - 6.6

[Continued]

★ 662 ★

Civilian Noninstitutional Population in the West: Employment Status: Annual Averages, 1992

[Continued]

State and population group	Civilian non-institutional population	Civilian labor force		Employment		Unemployment		
		Number	Percent of population	Number	Percent of population	Number	Rate	Error range of rate[1]
Men	275	205	74.7	193	70.3	12	5.9	4.8 - 6.9
Women	289	182	63.0	171	59.4	10	5.7	4.6 - 6.8
Both sexes, 16 to 19 years	43	25	59.0	22	51.3	3	13.1	9.0 - 17.1
Single (never married)	127	90	70.9	79	62.7	10	11.6	9.5 - 13.7
Married spouse present	364	258	70.7	245	67.3	12	4.8	4.0 - 5.7
Other marital status	109	64	58.9	59	54.5	5	7.5	5.4 - 9.5
New Mexico								
Total	1,133	723	63.6	674	59.5	49	6.8	6.1 - 7.6
Men	548	409	74.7	382	69.8	27	6.6	5.7 - 7.6
Women	585	314	53.6	291	49.8	22	7.1	6.0 - 8.2
Both sexes, 16 to 19 years	84	37	44.7	30	35.8	7	19.9	16.8 - 23.0
White	1,019	656	64.3	613	60.2	43	6.5	5.7 - 7.2
Men	497	374	75.3	351	70.6	24	6.3	5.3 - 7.2
Women	522	282	53.9	262	50.3	19	6.8	5.6 - 7.9
Both sexes, 16 to 19 years	70	32	45.9	26	37.4	6	18.4	14.9 - 21.9
Hispanic origin	359	216	60.1	195	54.2	21	9.9	8.2 - 11.6
Men	173	124	71.6	112	64.8	12	9.6	7.4 - 11.8
Women	187	92	49.5	83	44.3	10	10.4	7.7 - 13.0
Single (never married)	263	177	67.2	157	59.6	20	11.3	9.7 - 12.9
Married spouse present	662	433	65.4	415	62.7	18	4.1	3.4 - 4.9
Other marital status	207	113	54.7	102	49.2	11	10.1	8.1 - 12.0
Nevada								
Total	964	673	69.8	629	65.2	45	6.6	5.9 - 7.3
Men	482	376	78.0	350	72.5	26	7.0	6.1 - 7.9
Women	482	297	61.7	279	57.9	18	6.1	5.2 - 7.1
Both sexes, 16 to 19 years	62	34	55.5	29	46.6	5	16.0	12.7 - 19.3
White	867	606	69.9	567	65.4	38	6.3	5.6 - 7.1
Men	435	342	78.7	319	73.3	24	6.9	5.9 - 7.8
Women	432	263	60.9	248	57.5	15	5.6	4.6 - 6.6
Both sexes, 16 to 19 years	53	30	56.2	26	48.2	4	14.2	10.6 - 17.8
Black	53	37	69.4	33	62.3	4	10.3	6.4 - 14.3
Hispanic origin	90	66	72.9	61	68.3	4	6.4	4.1 - 8.7
Men	49	43	87.3	40	81.1	3	7.1	4.1 - 10.2

[Continued]

★ 662 ★

Civilian Noninstitutional Population in the West: Employment Status: Annual Averages, 1992

[Continued]

State and population group	Civilian non-institutional population	Civilian labor force		Employment		Unemployment		
		Number	Percent of population	Number	Percent of population	Number	Rate	Error range of rate[1]
Single (never married)	213	162	76.1	147	68.9	15	9.4	7.9 - 11.0
Married spouse present	527	366	69.5	348	66.1	18	5.0	4.2 - 5.8
Other marital status	224	145	64.5	134	59.6	11	7.6	6.1 - 9.1
Utah								
Total	1,151	811	70.4	771	67.0	40	4.9	4.3 - 5.6
Men	554	446	80.6	425	76.7	22	4.9	4.0 - 5.7
Women	597	365	61.0	346	58.0	18	5.0	4.1 - 6.0
Both sexes, 16 to 19 years	115	76	66.6	66	57.7	10	13.5	10.6 - 16.3
White	1,123	790	70.3	751	66.9	39	4.9	4.3 - 5.6
Men	539	434	80.5	413	76.6	21	4.8	4.0 - 5.7
Women	584	356	61.0	338	57.9	18	5.0	4.0 - 5.9
Both sexes, 16 to 19 years	112	75	67.2	65	58.3	10	13.3	10.5 - 16.2
Hispanic origin	41	30	75.1	27	67.1	3	10.7	5.9 - 15.4
Single (never married)	247	189	76.5	172	69.6	17	9.1	7.4 - 10.7
Married spouse present	730	524	71.8	506	69.3	18	3.5	2.8 - 4.2
Other marital status	174	98	56.3	93	53.6	5	4.8	3.0 - 6.5
Wyoming								
Total	341	240	70.4	227	66.5	14	5.6	4.9 - 6.4
Men	170	133	78.5	126	73.8	8	5.9	4.9 - 6.9
Women	171	107	62.5	101	59.2	6	5.3	4.2 - 6.3
Both sexes, 16 to 19 years	27	16	57.9	13	47.3	3	18.4	14.0 - 22.8
White	334	235	70.5	222	66.6	13	5.5	4.8 - 6.3
Men	166	130	78.5	123	73.9	8	5.8	4.8 - 6.8
Women	168	105	62.6	100	59.4	5	5.1	4.1 - 6.2
Both sexes, 16 to 19 years	27	16	58.0	13	47.5	3	18.2	13.7 - 22.6
Hispanic origin	13	9	71.6	8	65.5	1	8.6	3.8 - 13.4
Single (never married)	65	47	72.3	41	63.5	6	12.2	10.0 - 14.5
Married spouse present	215	155	72.0	149	69.3	6	3.7	3.0 - 4.5
Other marital status	61	39	63.0	37	59.8	2	5.1	3.4 - 6.8

Source: "States: Employment Status of the Civilian Noninstitutional Population by Sex, Age, Race, Hispanic Origin, and Marital Status, 1992 Annual Averages," U.S. Department of Labor, *Geographical Profile of Employment and Unemployment, 1992*, pp. 38-50. Rearranged by the Editors. *Notes:* Data for demographic groups are not shown when they do not meet BLS publication standards of reliability for the particular area based on the sample in that area. See appendix B. Items may not add to totals or compute to displayed percentages because of rounding. Detail for Hispanic-origin groups will not add to totals because data for the "other races" group are not presented and Hispanics are included in both the white and black population groups. 1. Error ranges are calculated at the 90-percent confidence interval, which means that if repeated samples were drawn from the same population and an error range constructed around each sample estimate, in 9 out of 10 cases the true value based on a complete census of the population would be contained within these error ranges.

★ 663 ★

Labor Force

Economic Characteristics of Persons and Families: Gender and Race, 1991. Part I

Numbers in thousands.

Characteristic	All races	Black	White
Labor force status in 1992[1]			
Both sexes, 16 years and over	191,576	21,958	162,658
In civilian labor force	26,982	13,891	108,526
Percent in civilian labor force	66.3	63.3	66.7
Employed	117,598	11,933	101,479
Unemployed	9,384	1,958	7,047
Percent unemployed	7.4	14.1	6.5
Not in labor force	64,593	8,067	54,132
Males, 16 years and over	91,541	9,888	78,351
In civilian labor force	69,184	6,892	59,830
Percent in civilian labor force	75.6	69.7	76.4
Employed	63,805	5,846	55,709
Unemployed	5,380	1,046	4,121
Percent unemployed	7.8	15.2	6.9
Not in labor force	22,356	2,997	18,521
Females, 16 years and over	100,035	12,069	84,307
In civilian labor force	57,798	6,999	48,696
Percent in civilian labor force	57.8	58.0	57.8
Employed	53,793	6,087	45,770
Unemployed	4,005	912	2,923
Percent unemployed	6.9	13.0	6.0
Not in labor force	42,237	5,070	35,610
Occupation in 1992[1]			
Employed males, 16 years and over	63,805	5,846	55,709
Percent	100.0	100.0	100.0
Managerial and professional specialty	25.7	14.1	26.8
Technical, sales, and administrative support	20.8	17.3	21.1
Service	10.2	19.0	9.1
Farming, forestry, and fishing	4.6	3.6	4.7
Precision production, craft, and repair	18.8	14.8	19.5
Operators, fabricators, and laborers	19.9	31.3	18.9
Employed females, 16 years and over	53,793	6,087	45,770
Percent	100.0	100.0	100.0
Managerial and professional specialty	27.4	19.5	28.5
Technical, sales, and administrative support	43.8	38.2	44.6
Service	17.9	27.9	16.5
Farming, forestry, and fishing	1.0	0.3	1.1
Precision production, craft, and repair	2.1	2.1	2.0
Operators, fabricators, and laborers	7.9	12.0	7.2
Class of worker in 1992[2]			
Employed persons, 16 years and over	115,724	11,544	99,992

[Continued]

★ 663 ★

Economic Characteristics of Persons and Families: Gender and Race, 1991. Part I

[Continued]

Characteristic	All races	Black	White
Percent	100.0	100.0	100.0
Private wage and salary workers	75.8	74.1	76.0
Federal government workers	2.8	5.2	2.6
State government workers	4.2	5.9	3.9
Local government workers	8.5	11.3	8.3
Self-employed workers	8.3	3.5	8.9
Unpaid family workers	0.3	-	0.3

Source: "Selected Economic Characteristics of Persons and Families, by Sex and Race: March 1992." Claudette E. Bennett, *The Black Population in the United States: March 1992*, p. 25. *Notes:* 1. Annual averages for labor force status and occupation of civilian noninstitutional persons. Data are from the "Employment and Earnings," Vol. 40, No. 1, January 1993. 2. Data for class of worker shown in this report reflect characteristics of the population for March 1991 and are not adjusted for seasonal changes. Data released by the Department of Labor, Bureau of Labor Statistics, may not agree entirely with data shown in this report due to differences in methodological procedures and seasonal adjustment.

★ 664 ★

Labor Force

Economic Characteristics of Persons and Families: Gender and Race, 1991. Part II

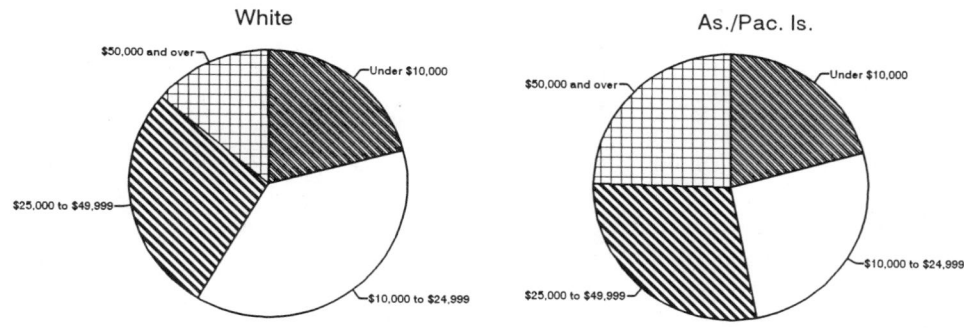

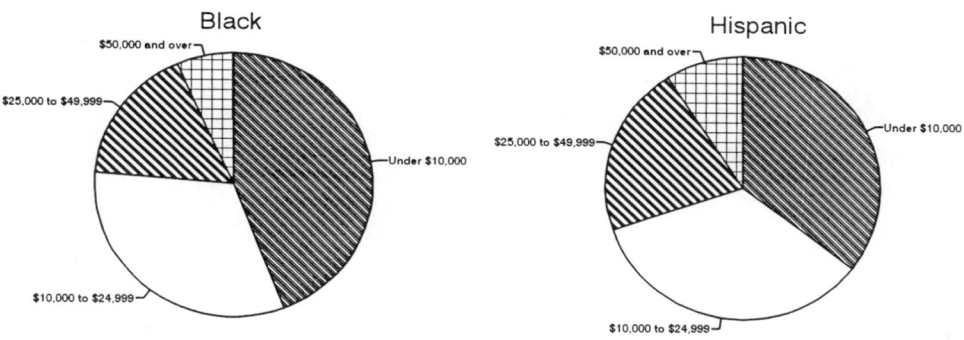

Numbers in thousands.

Characteristic	All races	Black	White
Income of persons in 1991			
Males with income[1]	88,653	8,943	76,578
Percent	100.0	100.0	100.0
$1 to $4,999 or loss	12.1	21.6	10.8
$5,000 to $9,999	13.0	18.7	12.3
$10,000 to $19,999	23.8	26.4	23.5
$20,000 to $29,999	18.3	15.3	18.8
$30,000 and over	32.8	17.9	34.6
Median income (dollars)	20,469	12,962	21,395
Standard error (dollars)	105	335	112
Females with income[1]	92,582	10,728	78,733
Percent	100.0	100.0	100.0
$1 to $4,999 or loss	26.6	28.3	26.2
$5,000 to $9,999	21.7	26.4	21.2
$10,000 to $19,999	25.8	24.3	26.0
$20,000 to $29,999	14.0	12.6	14.3

[Continued]

★ 664 ★

Economic Characteristics of Persons and Families: Gender and Race, 1991. Part II
[Continued]

Characteristic	All races	Black	White
$30,000 and over	11.9	8.5	12.3
Median income (dollars)	10,477	8,814	10,722
Standard error (dollars)	70	195	75
Per capita income in 1991			
Per capita income (dollars)	14,617	9,169	15,510
Income of families in 1991			
Total families	67,175	7,716	57,225
Percent	100.0	100.0	100.0
Under $10,000	9.7	26.4	7.4
$10,000 to $24,999	23.2	29.5	22.4
$25,000 to $34,999	15.6	14.4	15.9
$35,000 to $49,999	19.5	14.8	20.3
$50,000 and over	31.9	14.9	34.1
Median income (dollars)	35,938	21,548	37,782
Standard error (dollars)	179	445	210

Source: "Selected Economic Characteristics of Persons and Families, by Sex and Race: March 1992." Claudette E. Bennett, *The Black Population in the United States: March 1992*, pp. 25-26. *Note:* 1. Persons 15 years old and over.

★ 665 ★
Labor Force

Employed Persons in Metropolitan Areas, 1992 and 1993

Numbers in thousands.

Employment status, sex, age, race, and Hispanic origin	Total United States		Metropolitan areas					
			Total		Central cities		Suburbs	
	1992	1993	1992	1993	1992	1993	1992	1993
Total								
Civilian noninstitutional population	191,576	193,550	149,120	150,551	57,741	57,763	91,378	92,788
Civilian labor force	126,982	128,040	100,058	100,777	37,523	37,260	62,536	63,517
Percent of population	66.3	66.2	67.1	66.9	65.0	64.5	68.4	68.5
Employed	117,598	119,306	92,595	93,826	34,195	34,216	58,400	59,609
Unemployed	9,384	8,734	7,463	6,951	3,327	3,044	4,135	3,907
Unemployment rate	7.4	6.8	7.5	6.9	8.9	8.2	6.6	6.2
Not in labor force	64,593	65,509	49,061	49,774	20,219	20,502	28,842	29,271
White								
Civilian noninstitutional population	162,658	163,921	124,506	125,213	42,515	42,202	81,991	83,011
Civilian labor force	108,526	109,359	84,171	84,635	28,130	27,866	56,041	56,769

[Continued]

★ 665 ★

Employed Persons in Metropolitan Areas, 1992 and 1993
[Continued]

Employment status, sex, age, race, and Hispanic origin	Total United States		Metropolitan areas					
			Total		Central cities		Suburbs	
	1992	1993	1992	1993	1992	1993	1992	1993
Percent of population	66.7	66.7	67.6	67.6	66.2	66.0	68.3	68.4
Employed	101,479	102,812	78,721	79,574	26,111	26,039	52,610	53,534
Unemployed	7,047	6,547	5,450	5,061	2,019	1,826	3,431	3,235
Unemployment rate	6.5	6.0	6.5	6.0	7.2	6.6	6.1	5.7
Not in labor force	54,132	54,562	40,335	40,578	14,385	14,336	25,950	26,241
Black								
Civilian noninstitutional population	21,958	22,329	18,479	18,932	12,249	12,462	6,230	6,470
Civilian labor force	13,891	13,943	11,826	11,942	7,459	7,436	4,367	4,506
Percent of population	63.3	62.4	64.0	63.1	60.9	59.7	70.1	69.7
Employed	11,933	12,146	10,138	10,391	6,309	6,394	3,829	3,997
Unemployed	1,958	1,796	1,689	1,551	1,150	1,042	538	509
Unemployment rate	14.1	12.9	14.3	13.0	15.4	14.0	12.3	11.3
Not in labor force	8,067	8,386	6,652	6,990	4,790	5,026	1,863	1,964
Hispanic origin								
Civilian noninstitutional population	15,244	15,753	14,177	14,585	7,914	8,105	6,263	6,481
Civilian labor force	10,131	10,377	9,438	9,632	5,101	5,133	4,337	4,500
Percent of population	66.5	65.9	66.6	66.0	64.5	63.3	69.3	69.4
Employed	8,971	9,272	8,363	8,598	4,500	4,564	3,863	4,034
Unemployed	1,160	1,104	1,076	1,034	601	569	475	465
Unemployment rate	11.4	10.6	11.4	10.7	11.8	11.1	10.9	10.3
Not in labor force	5,113	5,377	4,739	4,953	2,813	2,972	1,925	1,981

Source: "Employment Status of the Civilian Noninstitutional Population in Metropolitan, Nonmetropolitan, Urban, and Rural Areas by Sex, Race, and Hispanic Origin." *Employment and Earnings* 41 (January 1994), p. 254. *Notes:* Detail for the above race and Hispanic-origin groups will not sum to totals because data for the "other races" group are not presented and Hispanics are included in both the white and black population groups. These data are based on 1980 census designations and are not comparable to data published through the first quarter of 1984. Beginning with the first quarter of 1993, separate data are no longer available for farm and nonfarm areas.

★ 666 ★

Labor Force

Employed Persons in Nonmetropolitan, Urban, and Rural Areas, 1992 and 1993

Numbers in thousands.

Employment status, sex, age, race, and Hispanic origin	Nonmetropolitan areas		Urban areas		Rural areas	
	1992	1993	1992	1993	1992	1993
Total						
Civilian noninstitutional population	42,456	42,999	139,350	140,380	52,225	53,170
Civilian labor force	26,924	27,263	92,564	92,943	34,418	35,097
Percent of population	63.4	63.4	66.4	66.2	65.9	66.0

[Continued]

★ 666 ★

Employed Persons in Nonmetropolitan, Urban, and Rural Areas, 1992 and 1993

[Continued]

Employment status, sex, age, race, and Hispanic origin	Nonmetropolitan areas		Urban areas		Rural areas	
	1992	1993	1992	1993	1992	1993
Employed	25,003	25,480	85,410	86,278	32,188	33,028
Unemployed	1,922	1,782	7,154	6,665	2,230	2,069
Unemployment rate	7.1	6.5	7.7	7.2	6.5	5.9
Not in labor force	15,532	15,736	46,786	47,437	17,807	18,073
White						
Civilian noninstitutional population	38,152	38,708	114,330	114,611	48,327	49,310
Civilian labor force	24,356	24,724	76,527	76,618	31,999	32,741
Percent of population	63.8	63.9	66.9	66.9	66.2	66.4
Employed	22,758	23,239	71,449	71,902	30,030	30,910
Unemployed	1,597	1,485	5,078	4,715	1,969	1,831
Unemployment rate	6.6	6.0	6.6	6.2	6.2	5.6
Not in labor force	13,796	13,984	37,803	37,993	16,329	16,569
Black						
Civilian noninstitutional population	3,479	3,398	18,972	19,464	2,985	2,866
Civilian labor force	2,064	2,001	12,041	12,211	1,850	1,731
Percent of population	59.3	58.9	63.5	62.7	62.0	60.4
Employed	1,795	1,755	10,287	10,600	1,646	1,547
Unemployed	269	246	1,754	1,612	204	185
Unemployment rate	13.0	123	14.6	13.2	11.0	10.7
Not in labor force	1,415	1,397	6,932	7,252	1,135	1,134
Hispanic origin						
Civilian noninstitutional population	1,066	1,168	13,897	14,366	1,346	1,387
Civilian labor force	693	744	9,218	9,423	913	954
Percent of population	64.9	63.7	66.3	65.6	67.8	68.8
Employed	609	674	8,163	8,422	809	851
Unemployed	84	70	1,056	1,001	104	103
Unemployment rate	12.1	9.4	11.5	10.6	11.4	10.8
Not in labor force	374	424	4,679	4,943	433	433

Source: "Employment Status of the Civilian Noninstitutional Population in Metropolitan, Nonmetropolitan, Urban, and Rural Areas by Sex, Race, and Hispanic Origin." *Employment and Earnings* 41 (January 1994), p. 254. *Notes:* Detail for the above race and Hispanic-origin groups will not sum to totals because data for the "other races" group are not presented and Hispanics are included in both the white and black population groups. These data are based on 1980 census designations and are not comparable to data published through the first quarter of 1984. Beginning with the first quarter of 1993, separate data are no longer available for farm and nonfarm areas.

★ 667 ★

Labor Force

Employed Persons in Nonmetropolitan, Urban, and Rural Areas: Quarterly Averages, 1992 and 1993

Numbers in thousands.

Employment status, sex, age, race, and Hispanic origin	Nonmetropolitan areas		Urban areas		Rural areas	
	IV 1992	IV 1993	IV 1992	IV 1993	IV 1992	IV 1993
Total						
Civilian noninstitutional population	42,737	43,083	139,458	140,653	52,861	53,662
Civilian labor force	27,181	27,327	92,222	93,228	34,723	35,261
Percent of population	63.6	63.4	66.1	66.3	65.7	65.7
Employed	25,460	25,732	85,470	87,222	32,716	33,349
Unemployed	1,722	1,595	6,751	6,006	2,007	1,912
Unemployment rate	6.3	5.8	7.3	6.4	5.8	5.4
Not in labor force	15,556	15,756	47,236	47,424	18,138	18,400
White						
Civilian noninstitutional population	38,375	38,770	114,246	114,618	48,889	49,797
Civilian labor force	24,566	24,783	76,156	76,797	32,265	32,910
Percent of population	64.0	63.9	66.7	67.0	66.0	66.1
Employed	23,156	23,458	71,423	72,545	30,505	31,208
Unemployed	1,411	1,326	4,733	4,252	1,760	1,703
Unemployment rate	5.7	5.3	6.2	5.5	5.5	5.2
Not in labor force	13,809	13,987	38,090	37,821	16,624	16,886
Black						
Civilian noninstitutional population	3,575	3,457	19,038	19,549	3,058	2,925
Civilian labor force	2,129	2,022	12,010	12,214	1,886	1,757
Percent of population	59.6	58.5	63.1	62.5	61.7	60.1
Employed	1,863	1,803	10,322	10,787	1,690	1,593
Unemployed	266	219	1,688	1,427	196	164
Unemployment rate	12.5	10.8	14.1	11.7	10.4	9.3
Not in labor force	1,446	1,435	7,028	7,335	1,172	1,168
Hispanic origin						
Civilian noninstitutional population	1,150	1,227	14,031	14,591	1,390	1,375
Civilian labor force	725	792	9,223	9,569	933	964
Percent of population	63.0	64.5	65.7	65.6	67.1	70.1
Employed	628	711	8,174	8,582	837	850
Unemployed	97	81	1,049	987	95	114
Unemployment rate	13.4	10.2	11.4	10.3	10.2	11.8
Not in labor force	425	435	4,808	5,022	457	411

Source: "Employment Status of the Civilian Noninstitutional Population in Metropolitan, Nonmetropolitan, Urban, and Rural Areas by Sex, Race, and Hispanic Origin." *Employment and Earnings* 41 (January 1994), p. 83. *Notes:* Detail for the above race and Hispanic-origin groups will not sum to totals because data for the "other races" group are not presented and Hispanics are included in both the white and black population groups. These data are based on 1980 census designations and are not comparable to data published through the first quarter of 1984. Beginning with the first quarter of 1993, separate data are no longer available for farm and nonfarm areas.

★ 668 ★

Labor Force

Employed Persons in Poverty and Nonpoverty and Areas, 1992 and 1993, Part I

Numbers in thousands.

Employment status, race, and Hispanic origin	Total United States				Metropolitan areas				Nonmetropolitan areas			
	Poverty areas		Nonpoverty areas		Poverty areas		Nonpoverty areas		Poverty areas		Nonpoverty areas	
	1992	1993	1992	1993	1992	1993	1992	1993	1992	1993	1992	1993
Total												
Civilian noninstitutional population	26,797	26,467	164,778	167,082	17,122	16,963	131,998	133,587	9,676	9,504	32,780	33,495
Civilian labor force	15,364	14,925	111,618	113,115	9,666	9,354	90,393	91,423	5,698	5,571	21,226	21,692
Percent of population	57.3	56.4	57.7	57.7	56.5	55.1	68.5	68.4	58.9	58.6	64.8	64.8
Employed	13,520	13,227	104,078	106,079	8,313	8,113	84,283	85,713	5,207	5,114	19,795	20,366
Unemployed	1,884	1,698	7,540	7,036	1,353	1,241	6,110	5,710	491	457	1,431	1,325
Unemployment rate	12.0	11.4	6.8	6.2	14.0	13.3	6.8	6.2	8.6	8.2	6.7	6.1
Men, 20 years and over	11.2	10.6	6.5	5.8	13.2	12.6	6.5	5.9	7.8	7.2	6.3	5.4
Women, 20 years and over	10.5	9.9	5.7	5.3	12.2	11.3	5.7	5.3	7.6	7.5	5.6	5.4
Both sexes, 16 to 19 years	30.5	28.8	18.4	17.5	34.7	33.1	18.7	18.0	23.4	21.8	17.5	16.0
Men	30.6	30.0	20.0	19.0	35.4	34.9	20.5	19.8	23.2	21.6	18.3	16.2
Women	30.3	27.5	16.7	16.0	33.9	30.8	16.7	16.1	23.7	22.2	16.7	15.7
Not in labor force	11,434	11,542	53,160	53,967	7,456	7,610	41,605	42,164	3,977	3,933	11,555	11,803

Source: "Employment Status of the Civilian Noninstitutional Population in Poverty and Nonpoverty Areas by Race and Hispanic Origin," *Employment and Earnings* 41 (January 1994), p. 256. *Notes:* Detail for the above race and Hispanic-origin groups will not sum to totals because data for the "other races" group are not presented and Hispanics are included in both the white and black population groups. These data are based on 1980 census designations and are not comparable to data published through the first quarter of 1984.

★ 669 ★

Labor Force

Employed Persons in Poverty and Nonpoverty and Areas: Quarterly Averages, 1992 and 1993, Part II

Numbers in thousands.

Employment status, race, and Hispanic origin	Total United States				Metropolitan areas				Nonmetropolitan areas			
	Poverty areas		Nonpoverty areas		Poverty areas		Nonpoverty areas		Poverty areas		Nonpoverty areas	
	IV 1992	IV 1993	IV 1992	IV 1993	IV 1992	IV 1993	IV 1992	IV 1993	IV 1992	IV 1993	IV 1992	IV 1993
White												
Civilian noninstitutional population	16,253	16,207	146,882	148,208	8,983	8,986	115,776	116,659	7,269	7,222	31,106	31,549
Civilian labor force	9,604	9,532	98,817	100,176	5,217	5,207	78,638	79,718	4,387	4,325	20,179	20,458
Percent of population	59.1	58.8	67.3	67.6	58.1	57.9	67.9	68.3	60.3	59.9	64.9	64.8
Employed	8,788	8,744	93,139	95,009	4,687	4,676	74,086	75,619	4,102	4,068	19,054	19,390
Unemployed	815	788	5,678	5,167	530	531	4,552	4,098	285	257	1,126	1,068
Unemployment rate	8.5	8.3	5.7	5.2	10.2	10.2	5.8	5.1	6.5	5.9	5.6	5.2
Men, 20 years and over	8.2	7.5	5.5	4.7	9.8	9.6	5.6	4.7	6.3	5.0	5.2	4.7
Women, 20 years and over	7.4	7.7	4.9	4.6	9.0	9.0	5.0	4.6	5.6	6.3	4.7	4.6
Both sexes, 16 to 19 years	18.8	19.5	15.2	14.7	21.9	24.3	15.2	14.6	15.1	13.4	15.4	14.8
Men	19.7	21.5	16.1	16.4	21.4	25.1	16.3	16.9	17.7	16.3	15.5	14.8
Women	17.8	17.0	14.2	12.9	22.5	23.0	14.0	12.3	11.9	9.9	15.2	14.7
Not in labor force	6,649	6,676	48,065	48,032	3,767	3,779	37,138	36,942	2,882	2,897	10,927	11,090
Black												
Civilian noninstitutional population	9,339	9,004	12,756	13,469	7,207	6,990	11,313	12,027	2,132	2,015	1,443	1,443
Civilian labor force	5,010	4,720	8,886	9,251	3,816	3,605	7,951	8,344	1,194	1,115	936	907
Percent of population	53.6	52.4	69.7	68.7	53.0	51.6	70.3	69.4	56.0	55.3	64.8	62.9
Employed	4,150	4,079	7,862	8,301	3,109	3,083	7,039	7,494	1,041	996	822	807
Unemployed	860	641	1,025	950	707	522	911	850	153	120	113	100
Unemployment rate	17.2	13.6	11.5	10.3	18.5	14.5	11.5	10.2	12.8	10.7	12.1	11.0
Men, 20 years and over	15.8	12.8	10.7	9.4	17.6	13.9	10.4	9.5	10.4	9.3	13.5	8.6
Women, 20 years and over	15.2	11.5	9.6	8.7	16.0	12.3	9.7	8.4	12.6	8.7	8.4	11.3
Both sexes, 16 to 19 years	43.0	39.4	38.9	35.7	45.7	40.5	39.2	36.2	33.8	36.5	1	1
Men	42.7	38.5	42.6	39.5	44.0	41.5	42.1	40.5	1	1	1	1

[Continued]

★ 669 ★

Employed Persons in Poverty and Nonpoverty and Areas: Quarterly Averages, 1992 and 1993, Part II

[Continued]

Employment status, race, and Hispanic origin	Total United States				Metropolitan areas				Nonmetropolitan areas			
	Poverty areas		Nonpoverty areas		Poverty areas		Nonpoverty areas		Poverty areas		Nonpoverty areas	
	IV 1992	IV 1993	IV 1992	IV 1993	IV 1992	IV 1993	IV 1992	IV 1993	IV 1992	IV 1993	IV 1992	IV 1993
Women	43.4	40.1	34.9	31.5	47.8	39.5	35.7	31.4	1	1	1	1
Not in labor force	4,330	4,285	3,870	4,218	3,391	3,385	3,363	3,683	939	900	507	535
Hispanic origin												
Civilian noninstitutional population	4,438	4,593	10,983	11,373	4,050	4,170	10,220	10,569	388	424	762	803
Civilian labor force	2,499	2,587	7,657	7,946	2,273	2,353	7,157	7,388	226	234	499	558
Percent of population	56.3	56.3	69.7	69.9	56.1	56.4	70.0	69.9	58.2	55.2	65.5	69.5
Employed	2,159	2,242	6,852	7,190	1,967	2,032	6,416	6,688	192	210	436	502
Unemployed	340	345	804	756	306	321	741	700	34	24	63	56
Unemployment rate	13.6	13.3	10.5	9.5	13.5	13.6	10.4	9.5	15.0	10.3	12.7	10.1
Men, 20 years and over	12.2	12.3	9.0	8.0	12.1	12.5	8.9	8.0	13.6	10.7	9.9	7.3
Women, 20 years and over	13.6	12.5	9.9	9.3	13.5	13.0	9.7	9.2	14.2	8.9	11.7	10.7
Both sexes, 16 to 19 years	24.2	25.9	27.9	25.0	23.3	27.1	27.0	24.3	1	1	1	1
Men	24.5	24.9	25.0	22.9	23.7	25.8	23.3	22.9	1	1	1	1
Women	23.7	27.5	31.8	27.2	1109	29.5	31.6	25.8	1	1	1	1
Not in labor force	1,939	2,007	3,326	3,427	1,777	1,817	3,063	3,181	162	190	263	245

Source: "Employment Status of the Civilian Noninstitutional Population in Poverty and Nonpoverty Areas by Race and Hispanic Origin," *Employment and Earnings* 41 (January 1994), p. 84. *Notes:* Detail for the above race and Hispanic-origin groups will not sum to totals because data for the "other races" group are not presented and Hispanics are included in both the white and black population groups. These data are based on 1980 census designations and are not comparable to data published through the first quarter of 1984. 1. Data not shown where base is less than 60,000.

★ 670 ★

Labor Force

Employed Persons in Selected Metropolitan Areas and Cities: Nonagricultural Industries and Annual Averages, 1992 - I

Population group and area	Total employed[1]	Private nonagricultural wage and salary workers				
		Total[2]	Construction	Manufacturing		
				Total	Durable goods	Nondurable goods
White						
Metropolitan areas[3]						
Anaheim-Santa Ana PMSA	100.0	81.6	4.0	24.0	16.2	7.8
Atlanta	100.0	81.3	4.1	12.6	6.5	6.1
Baltimore	100.0	74.5	6.8	10.0	5.7	4.3
Bergen-Passaic PMSA	100.0	82.0	4.4	19.9	8.4	11.5
Boston PMSA	100.0	79.7	2.9	13.2	8.7	4.5
Buffalo-Niagara Falls CMSA	100.0	78.3	3.7	17.2	10.3	6.9
Charlotte-Gastonia-Rock Hill	100.0	84.1	6.2	23.6	9.6	14.0
Chicago PMSA	100.0	83.8	4.2	19.2	10.7	8.6
Cincinnati PMSA	100.0	81.4	4.4	19.7	10.1	9.6
Cleveland PMSA	100.0	84.0	4.0	23.1	15.3	7.8
Columbus, Ohio	100.0	76.7	3.2	13.9	8.6	5.4
Dallas-Fort Worth CMSA	100.0	82.0	4.2	16.0	11.2	4.8
Dayton-Springfield	100.0	79.5	4.8	22.3	17.9	4.4
Denver-Boulder CMSA	100.0	77.4	4.7	10.8	7.4	3.5
Detroit PMSA	100.0	83.7	4.0	24.5	20.3	4.2

[Continued]

★ 670 ★

Employed Persons in Selected Metropolitan Areas and Cities: Nonagricultural Industries and Annual Averages, 1992 - I

[Continued]

Population group and area	Total employed[1]	Private nonagricultural wage and salary workers				
		Total[2]	Construction	Manufacturing		
				Total	Durable goods	Nondurable goods
Fort Lauderdale-Hollywood-Pompano Beach PMSA	100.0	79.7	5.1	7.4	5.4	2.0
Hartford-New Britain-Middletown CMSA	100.0	83.7	2.6	20.1	16.8	3.3
Houston PMSA	100.0	82.3	8.0	13.2	7.3	6.0
Indianapolis	100.0	80.3	3.7	17.3	12.2	5.2
Kansas City	100.0	79.0	5.0	13.1	7.5	5.7
Los Angeles-Long Beach PMSA	100.0	79.3	4.2	20.8	12.2	8.6
Louisville	100.0	82.9	6.0	18.9	10.1	8.8
Memphis	100.0	79.5	4.3	15.0	6.1	8.8
Miami-Hialeah PMSA	100.0	82.0	5.0	10.8	4.6	6.2
Milwaukee PMSA	100.0	82.6	4.0	26.1	18.4	7.7
Minneapolis-St. Paul	100.0	81.4	3.5	18.5	11.0	7.5
Nassau-Suffolk PMSA	100.0	76.9	5.4	11.2	6.0	5.2
New Orleans	100.0	77.9	6.2	7.6	4.3	3.3
New York PMSA	100.0	76.8	3.8	11.2	3.4	7.8
Newark PMSA	100.0	81.3	3.2	19.3	8.3	11.0
Norfolk-Virginia Beach-Newport News	100.0	69.8	9.2	9.4	5.8	3.6
Oakland PMSA	100.0	71.8	4.5	11.5	6.3	5.2
Oklahoma City	100.0	72.9	4.6	12.4	7.0	5.4
Philadelphia PMSA	100.0	81.0	4.6	15.6	7.4	8.2
Phoenix	100.0	79.5	6.1	14.5	10.7	3.8
Pittsburgh-Beaver Valley CMSA	100.0	83.3	4.7	12.5	9.4	3.2
Portland, Ore. PMSA	100.0	77.3	3.9	15.5	10.9	4.6
Providence-Pawtucket-Fall River CMSA	100.0	79.8	3.6	22.7	14.8	7.9
Riverside-San Bernardino PMSA	100.0	72.9	5.5	14.3	10.0	4.3
Rochester	100.0	79.5	2.8	25.7	20.8	4.9
Sacramento	100.0	64.5	3.8	6.3	2.9	3.5
St. Louis	100.0	81.9	4.3	15.6	9.3	6.3
Salt Lake City-Ogden	100.0	72.2	4.6	12.5	8.0	4.5
San Antonio	100.0	72.7	6.3	8.5	5.0	3.5
San Diego	100.0	70.6	5.5	12.2	7.9	4.3
San Francisco PMSA	100.0	75.1	3.5	9.5	4.7	4.8
San Jose PMSA	100.0	77.4	3.8	22.9	19.4	3.5
Seattle PMSA	100.0	77.1	4.6	15.9	12.4	3.5
Tampa-St. Petersburg-Clearwater	100.0	80.8	5.8	10.7	6.9	3.8
Washington D.C.	100.0	68.8	5.8	4.4	2.7	1.6
Cities						
Baltimore	100.0	74.0	6.0	14.8	4.3	10.6
Chicago	100.0	83.1	3.0	19.6	10.5	9.1
Cleveland	100.0	85.4	4.6	23.1	13.7	9.4
Dallas	100.0	83.9	3.9	13.8	8.0	5.7
Detroit	100.0	77.1	5.6	14.4	10.4	4.0

[Continued]

★ 670 ★

Employed Persons in Selected Metropolitan Areas and Cities: Nonagricultural Industries and Annual Averages, 1992 - I

[Continued]

Population group and area	Total employed[1]	Private nonagricultural wage and salary workers				
		Total[2]	Construction	Manufacturing		
				Total	Durable goods	Nondurable goods
District of Columbia	100.0	66.4	1.2	5.3	.5	4.7
Houston	100.0	82.6	8.3	12.1	6.8	5.3
Indianapolis	100.0	83.6	1.9	15.2	11.3	3.9
Los Angeles	100.0	79.4	3.4	17.0	8.1	9.0
Milwaukee	100.0	82.8	4.3	21.6	15.7	5.9
New York	100.0	76.8	3.4	10.8	3.0	7.9
Philadelphia	100.0	77.5	2.7	13.1	5.5	7.6
Phoenix	100.0	79.9	6.2	14.1	9.5	4.6
St. Louis	100.0	75.1	1.1	8.1	4.6	3.4
San Antonio	100.0	75.0	6.9	8.9	4.8	4.1
San Diego	100.0	68.6	2.9	13.5	8.7	4.8
San Francisco	100.0	75.0	2.6	6.9	1.0	5.9
Black						
Metropolitan areas[3]						
Atlanta	100.0	83.3	1.5	13.2	4.3	9.0
Baltimore	100.0	68.6	3.1	11.1	6.1	4.9
Bergen-Passaic PMSA	100.0	77.8	.7	11.1	4.4	6.8
Boston PMSA	100.0	79.8	.9	12.4	8.5	3.9
Buffalo-Niagara Falls CMSA	100.0	66.3	.9	10.4	5.0	5.4
Charlotte-Gastonia-Rock Hill	100.0	77.4	2.0	25.8	7.1	18.7
Chicago PMSA	100.0	71.5	.8	13.6	5.9	7.7
Cincinnati PMSA	100.0	68.8	3.1	16.1	6.3	9.8
Cleveland PMSA	100.0	69.3	1.7	11.4	7.9	3.5
Columbus, Ohio	100.0	68.6	.4	13.6	7.4	6.3
Dallas-Fort Worth CMSA	100.0	78.8	2.2	15.0	9.3	5.7
Dayton-Springfield	100.0	74.4	.9	17.1	12.6	4.6
Denver-Boulder CMSA	100.0	79.0	3.1	10.5	3.7	6.9
Detroit PMSA	100.0	78.9	1.5	24.2	21.0	3.2
Fort Lauderdale-Hollywood-Pompano Beach PMSA	100.0	72.9	3.2	7.6	6.3	1.3
Hartford-New Britain-Middletown CMSA	100.0	80.9	4.0	16.7	16.1	.6
Houston PMSA	100.0	68.7	2.3	8.4	3.9	4.5
Indianapolis	100.0	66.2	3.0	9.7	4.9	4.9
Kansas City	100.0	80.0	1.1	10.0	3.3	6.7
Los Angeles-Long Beach PMSA	100.0	63.1	1.7	11.9	7.6	4.3
Louisville	100.0	84.9	.6	22.4	13.8	8.6
Memphis	100.0	74.6	5.2	11.1	6.6	4.6
Miami-Hialeah PMSA	100.0	65.6	4.7	10.0	3.5	6.6
Milwaukee PMSA	100.0	84.1	3.0	16.0	7.5	8.5
Minneapolis-St. Paul	100.0	79.5	1.3	21.8	13.5	8.3
Nassau-Suffolk PMSA	100.0	69.8	2.8	8.6	5.9	2.8
New Orleans	100.0	65.3	3.0	4.8	2.8	2.0

[Continued]

★ 670 ★

Employed Persons in Selected Metropolitan Areas and Cities: Nonagricultural Industries and Annual Averages, 1992 - I

[Continued]

Population group and area	Total employed[1]	Private nonagricultural wage and salary workers				
		Total[2]	Construction	Manufacturing		
				Total	Durable goods	Nondurable goods
New York PMSA	100.0	67.4	2.3	7.6	2.3	5.3
Newark PMSA	100.0	75.5	1.3	17.4	6.3	11.1
Norfolk-Virginia Beach-Newport News	100.0	63.1	4.3	10.9	8.3	2.7
Oakland PMSA	100.0	80.8	3.1	9.2	4.7	4.5
Oklahoma City	100.0	74.7	3.5	18.6	14.3	4.3
Philadelphia PMSA	100.0	73..3	2.8	10.9	6.0	4.9
Pittsburgh-Beaver Valley CMSA	100.0	76.2	1.2	12.6	6.5	6.0
Providence-Pawtucket-Fall River CMSA	100.0	67.2	1.4	18.6	8.9	9.9
Riverside-San Bernardino PMSA	100.0	58.7	6.6	15.3	9.7	5.6
Rochester	100.0	84.6	5.5	32.7	29.4	3.3
St. Louis	100.0	81.9	.3	16.3	8.0	8.2
San Diego	100.0	60.3	3.0	8.8	5.8	3.0
Seattle PMSA	100.0	69.5	1.4	15.4	6.1	9.3
Tampa-St. Petersburg-Clearwater	100.0	68.9	3.5	10.4	6.9	3.5
Washington D.C.	100.0	59.6	2.8	2.8	.7	2.0
Cities						
Baltimore	100.0	69.8	2.8	9.7	4.8	4.9
Chicago	100.0	69.0	.6	12.6	5.3	7.4
Cleveland	100.0	67.3	1.1	11.8	6.3	5.4
Dallas	100.0	84.3	3.0	10.5	5.8	4.7
Detroit	100.0	77.8	1.7	22.6	19.3	3.3
District of Columbia	100.0	57.8	2.1	2.4	.7	1.7
Houston	100.0	73.2	2.4	9.0	2.6	6.4
Indianapolis	100.0	66.8	3.1	10.0	5.0	5.0
Los Angeles	100.0	63.0	1.4	9.3	3.7	5.6
Milwaukee	100.0	83.8	2.6	13.8	5.5	8.4
New York	100.0	67.1	2.4	7.6	2.0	5.6
Philadelphia	100.0	76.5	2.8	8.9	5.3	3.6
St. Louis	100.0	81.5	[4]	20.1	12.9	7.2
Hispanic origin						
Metropolitan areas[3]						
Anaheim-Santa Ana PMSA	100.0	90.5	7.0	32.0	20.8	11.2
Bergen-Passaic PMSA	100.0	87.6	2.4	33.5	14.8	18.6
Boston PMSA	100.0	8.0	1.7	14.5	9.7	4.8
Chicago PMSA	100.0	89.1	2.6	35.3	20.5	14.7
Dallas-Fort Worth CMSA	100.0	86.9	9.8	21.4	11.6	9.8
Denver-Boulder CMSA	100.0	71.2	7.3	8.8	4.6	4.2
Detroit PMSA	100.0	87.7	3.0	38.0	35.0	3.0
Fort Lauderdale-Hollywood-Pompano Beach PMSA	100.0	91.4	4.9	21.4	20.4	1.0
Houston PMSA	100.0	88.0	11.9	17.0	9.3	7.6

[Continued]

★ 670 ★

Employed Persons in Selected Metropolitan Areas and Cities: Nonagricultural Industries and Annual Averages, 1992 - I

[Continued]

Population group and area	Total employed[1]	Private nonagricultural wage and salary workers				
		Total[2]	Construction	Manufacturing		
				Total	Durable goods	Nondurable goods
Los Angeles-Long Beach PMSA	100.0	86.7	5.9	29.8	14.8	15.0
Miami-Hialeah PMSA	100.0	84.4	5.9	13.3	5.3	8.0
Nassau-Suffolk PMSA	100.0	82.6	4.6	19.8	9.0	10.8
New York PMSA	100.0	80.7	3.0	17.3	5.2	12.1
Newark PMSA	100.0	89.3	2.0	29.8	11.7	18.1
Oakland PMSA	100.0	85.4	7.2	14.6	6.2	8.4
Philadelphia PMSA	100.0	82.0	2.0	26.7	10.4	16.4
Phoenix	100.0	84.6	8.8	19.4	14.1	5.3
Providence-Pawtucket-Fall River CMSA	100.0	93.9	.4	64.9	47.0	17.9
Riverside-San Bernardino PMSA	100.0	79.8	6.8	21.0	15.0	5.9
Sacramento	100.0	61.5	2.9	2.7	2.7	[4]
Salt Lake City-Ogden	100.0	77.3	.2	16.5	8.1	8.5
San Antonio	100.0	75.0	6.9	10.8	6.4	4.3
San Diego	100.0	77.8	10.8	15.4	8.1	7.3
San Francisco PMSA	100.0	85.6	4.9	16.0	2.8	13.2
San Jose PMSA	100.0	82.3	3.4	18.7	14.0	4.8
Tampa-St. Petersburg-Clearwater	100.0	84.3	.3	19.8	8.3	11.4
Washington D.C.	100.0	86.5	8.3	1.6	[4]	1.6
Cities						
Chicago	100.0	87.8	2.3	33.6	17.7	16.0
Dallas	100.0	86.2	11.6	19.4	11.2	8.2
District of Columbia	100.0	84.8	2.0	2.8	[4]	2.8
Houston	100.0	89.9	13.6	12.8	7.1	5.7
Los Angeles	100.0	88.8	5.0	28.2	11.1	17.1
New York	100.0	80.5	3.1	17.4	5.1	12.3
Phoenix	100.0	85.6	10.6	13.6	10.2	3.4
San Antonio	100.0	74.7	7.3	10.8	6.3	4.6
San Diego	100.0	77.1	8.8	14.4	5.1	9.3
San Francisco	100.0	89.4	5.7	14.8	1.1	13.7

Source: "Selected Metropolitan Areas and Cities: Percent Distribution of Employed Persons in Nonagricultural Industries by Sex, Race, Hispanic Origin, 1992 Annual Averages," U.S. Department of Labor, *Geographical Profile of Employment and Unemployment*, 1992, pp. 122-124. *Notes:* Data for demographic groups are not shown when they do not meet BLS publication standards of reliability for the particular area based on the sample in that area. Detail may not add to totals because of rounding. 1. Includes self-employed and unpaid family workers and mining. 2. Includes mining. 3. All are Metropolitan Statistical Areas (MSA's) except St. Louis and those labeled Consolidated Metropolitan Statistical Areas (CMSA's) or Primary Metropolitan Statistical Areas (PMSA's). The differences are discussed in appendix C, "Geographical Boundary Definitions." 4. Less than 500 persons employed or less than 0.05 percent of total employed.

★ 671 ★

Labor Force

Employed Persons in Selected Metropolitan Areas and Cities: Nonagricultural Industries and Annual Averages, 1992 - II

Population group and area	Private nonagricultural wage and salary workers				
	Transportation communication, and public utilities	Trade	Finance, insurance, real estate	Services[1]	Government
White					
Metropolitan areas[2]					
Anaheim-Santa Ana PMSA	4.8	19.7	9.4	19.4	9.0
Atlanta	10.3	21.8	8.7	23.6	11.5
Baltimore	6.2	18.5	6.5	26.4	20.0
Bergen-Passaic PMSA	6.2	19.5	8.5	23.5	11.5
Boston PMSA	4.7	18.5	8.9	31.4	12.5
Buffalo-Niagara Falls CMSA	5.7	21.8	6.5	23.3	17.1
Charlotte-Gastonia-Rock Hill	7.9	22.0	5.7	18.6	9.3
Chicago PMSA	6.4	19.8	8.5	25.6	9.5
Cincinnati PMSA	6.8	18.6	6.8	25.0	12.1
Cleveland PMSA	4.6	19.2	5.5	27.4	10.8
Columbus, Ohio	5.3	21.7	9.1	23.4	15.7
Dallas-Fort Worth CMSA	8.0	22.1	7.7	23.4	10.1
Dayton-Springfield	3.7	22.5	3.7	22.4	15.3
Denver-Boulder CMSA	7.6	19.6	7.6	25.8	14.2
Detroit PMSA	4.3	21.3	6.8	22.8	10.7
Fort Lauderdale-Hollywood-Pompano Beach PMSA	8.5	24.0	8.4	26.2	13.9
Hartford-New Britain-Middletown CMSA	6.0	18.4	13.5	23.0	11.0
Houston PMSA	6.5	21.3	5.8	23.0	11.2
Indianapolis	4.9	21.3	7.4	25.6	14.6
Kansas City	8.6	20.7	6.8	24.8	15.4
Los Angeles-Long Beach PMSA	4.6	18.9	6.3	24.4	10.4
Louisville	7.5	21.1	6.8	22.6	11.0
Memphis	10.5	19.7	9.0	21.0	13.5
Miami-Hialeah PMSA	8.3	24.0	6.9	27.0	10.0
Milwaukee PMSA	4.7	17.6	6.8	23.3	11.9
Minneapolis-St. Paul	5.8	21.0	7.9	24.6	13.3
Nassau-Suffolk PMSA	6.9	19.0	10.1	24.2	16.7
New Orleans	5.7	25.4	6.3	24.5	14.2
New York PMSA	5.9	16.0	11.2	28.7	14.6
Newark PMSA	7.9	18.2	9.9	22.7	12.4
Norfolk-Virginia Beach-Newport News	5.2	20.4	5.5	20.1	22.6
Oakland PMSA	5.5	22.1	7.5	20.6	14.3
Oklahoma City	5.3	20.8	8.1	19.9	17.4
Philadelphia PMSA	5.7	19.1	9.0	27.1	12.8
Phoenix	5.7	20.9	8.0	24.0	12.9
Pittsburgh-Beaver Valley CMSA	6.5	22.0	7.4	29.5	10.8
Portland, Ore. PMSA	4.4	21.9	7.2	24.5	11.9
Providence-Pawtucket-Fall River CMSA	4.4	18.2	6.5	24.4	14.3
Riverside-San Bernardino PMSA	7.7	22.7	5.1	17.3	17.6

[Continued]

★ 671 ★

Employed Persons in Selected Metropolitan Areas and Cities: Nonagricultural Industries and Annual Averages, 1992 - II
[Continued]

Population group and area	Private nonagricultural wage and salary workers				
	Transportation communication, and public utilities	Trade	Finance, insurance, real estate	Services[1]	Government
Rochester	4.7	18.5	4.9	22.9	13.1
Sacramento	6.6	19.8	5.2	22.4	22.9
St. Louis	7.6	23.2	5.7	25.2	11.1
Salt Lake City-Ogden	6.2	21.4	6.9	19.6	20.3
San Antonio	5.5	24.7	5.1	22.2	20.3
San Diego	3.0	19.4	7.8	22.6	17.2
San Francisco PMSA	7.7	19.9	8.8	25.7	9.7
San Jose PMSA	6.1	18.0	3.2	23.3	12.6
Seattle PMSA	6.1	20.6	6.3	23.6	14.2
Tampa-St. Petersburg-Clearwater	5.2	22.5	8.9	27.5	12.1
Washington D.C.	5.1	17.1	5.6	30.8	24.4
Cities					
Baltimore	8.0	16.0	3.8	25.4	22.1
Chicago	4.7	18.4	7.7	29.6	10.5
Cleveland	5.8	22.4	5.4	24.2	12.5
Dallas	7.4	22.2	9.7	26.1	7.9
Detroit	3.6	26.5	5.1	22.0	14.1
District of Columbia	3.4	8.5	5.5	42.5	24.7
Houston	5.6	22.5	6.2	25.5	10.6
Indianapolis	4.3	23.1	7.5	31.5	13.7
Los Angeles	4.4	20.5	6.7	27.2	8.6
Milwaukee	5.7	18.5	6.6	26.1	13.4
New York	5.9	15.6	11.9	29.1	14.7
Philadelphia	5.6	16.6	9.7	29.7	17.1
Phoenix	5.1	22.5	7.6	24.0	13.7
St. Louis	6.9	19.6	4.1	35.3	17.2
San Antonio	4.8	24.0	6.1	23.8	19.7
San Diego	3.3	17.2	8.9	22.9	19.3
San Francisco	4.6	21.8	9.2	29.9	10.8
Black					
Metropolitan areas[2]					
Atlanta	9.8	20.2	8.2	30.4	13.4
Baltimore	4.7	16.6	4.4	28.6	29.4
Bergen-Passaic PMSA	8.9	13.9	12.5	30.7	19.7
Boston PMSA	5.7	12.5	8.2	40.1	16.4
Buffalo-Niagara Falls CMSA	3.2	13.0	9.1	29.6	33.2
Charlotte-Gastonia-Rock Hill	8.2	19.6	4.7	17.0	20.2
Chicago PMSA	6.6	17.1	7.9	25.4	24.1
Cincinnati PMSA	7.4	13.5	1.7	27.0	28.4
Cleveland PMSA	3.8	13.4	4.5	34.3	27.7

[Continued]

★ 671 ★

Employed Persons in Selected Metropolitan Areas and Cities: Nonagricultural Industries and Annual Averages, 1992 - II

[Continued]

Population group and area	Private nonagricultural wage and salary workers				
	Transportation communication, and public utilities	Trade	Finance, insurance, real estate	Services[1]	Government
Columbus, Ohio	4.9	13.5	12.7	23.5	29.7
Dallas-Fort Worth CMSA	7.4	20.8	5.9	27.1	19.7
Dayton-Springfield	4.8	23.6	4.7	23.3	22.0
Denver-Boulder CMSA	11.2	16.6	8.6	27.7	16.9
Detroit PMSA	4.5	15.7	5.9	27.2	17.6
Fort Lauderdale-Hollywood-Pompano Beach PMSA	8.9	22.2	5.7	25.4	22.7
Hartford-New Britain-Middletown CMSA	6.0	16.2	12.8	25.2	18.2
Houston PMSA	8.4	15.3	6.7	26.7	27.2
Indianapolis	10.0	22.5	1.0	19.9	33.7
Kansas City	7.5	17.4	8.7	35.4	18.5
Los Angeles-Long Beach PMSA	7.6	9.7	6.0	26.2	28.8
Louisville	10.2	16.4	4.3	31.0	10.5
Memphis	8.3	24.9	3.4	21.7	23.2
Miami-Hialeah PMSA	8.5	17.0	2.9	22.5	29.6
Milwaukee PMSA	7.4	17.1	6.1	34.6	15.0
Minneapolis-St. Paul	4.6	14.1	10.8	26.8	19.1
Nassau-Suffolk PMSA	7.6	10.8	9.7	30.2	26.5
New Orleans	4.5	21.1	3.7	27.7	27.8
New York PMSA	6.6	10.2	8.9	31.8	28.4
Newark PMSA	8.3	14.4	9.7	24.4	22.2
Norfolk-Virginia Beach-Newport News	7.0	19.3	3.6	18.1	35.3
Oakland PMSA	11.2	20.9	10.9	25.4	14.2
Oklahoma City	3.1	17.3	4.1	26.8	21.4
Philadelphia PMSA	6.4	14.0	7.8	31.4	23.3
Pittsburgh-Beaver Valley CMSA	8.4	9.8	7.3	36.9	22.3
Providence-Pawtucket-Fall River CMSA	3.4	9.6	6.1	27.9	27.7
Riverside-San Bernardino PMSA	5.2	17.4	2.6	11.6	29.8
Rochester	10.1	6.1	5.0	25.2	15.0
St. Louis	10.8	18.0	3.1	33.4	16.7
San Diego	[3]	19.8	6.7	21.9	28.3
Seattle PMSA	2.4	22.2	4.5	23.5	26.9
Tampa-St. Petersburg-Clearwater	8.4	19.7	5.0	21.8	24.9
Washington D.C.	6.8	15.0	5.4	26.7	35.6
Cities					
Baltimore	4.4	15.6	4.8	32.6	28.6
Chicago	5.1	15.9	7.9	26.9	25.9
Cleveland	3.6	10.5	3.1	37.2	28.8
Dallas	9.2	21.5	7.9	31.7	14.9
Detroit	4.4	15.6	5.8	27.7	18.5
District of Columbia	5.2	13.4	4.0	30.7	37.3

[Continued]

★ 671 ★

Employed Persons in Selected Metropolitan Areas and Cities: Nonagricultural Industries and Annual Averages, 1992 - II

[Continued]

Population group and area	Private nonagricultural wage and salary workers				
	Transportation communication, and public utilities	Trade	Finance, insurance, real estate	Services[1]	Government
Houston	7.2	19.4	7.1	26.8	24.1
Indianapolis	10.3	23.1	1.1	19.3	33.2
Los Angeles	5.5	9.6	6.8	30.4	28.6
Milwaukee	7.9	17.8	6.5	35.1	15.3
New York	5.8	10.1	8.9	32.4	28.7
Philadelphia	6.2	14.9	8.1	35.6	20.0
St. Louis	6.2	16.6	4.9	33.7	15.2
Hispanic origin					
Metropolitan areas[2]					
Anaheim-Santa Ana PMSA	3.1	21.2	5.2	22.1	6.4
Bergen-Passaic PMSA	2.9	24.6	3.3	20.9	8.5
Boston PMSA	1.4	32.0	4.4	33.5	8.7
Chicago PMSA	3.8	19.3	4.8	23.2	7.5
Dallas-Fort Worth CMSA	5.9	24.7	6.3	18.7	8.4
Denver-Boulder CMSA	8.9	20.4	4.8	21.0	25.2
Detroit PMSA	2.4	20.8	3.7	19.8	10.6
Fort Lauderdale-Hollywood-Pompano Beach PMSA	9.2	25.7	10.5	19.7	5.0
Houston PMSA	2.7	28.3	4.3	20.4	7.8
Los Angeles-Long Beach PMSA	4.7	22.2	4.5	19.6	7.5
Miami-Hialeah PMSA	7.9	25.0	6.9	25.4	7.9
Nassau-Suffolk PMSA	8.2	13.3	14.3	22.4	15.3
New York PMSA	5.5	17.8	9.4	27.8	13.9
Newark PMSA	10.2	22.1	4.3	20.9	5.9
Oakland PMSA	11.0	24.4	7.6	20.6	9.8
Philadelphia PMSA	3.6	17.6	5.1	26.9	10.9
Phoenix	3.3	23.9	7.7	21.5	11.6
Providence-Pawtucket-Fall River CMSA	1.3	9.3	6.1	11.8	1.6
Riverside-San Bernardino PMSA	8.2	22.1	2.3	19.1	13.1
Sacramento	7.0	26.4	3.0	17.9	30.4
Salt Lake City-Ogden	7.7	23.2	10.7	14.6	18.8
San Antonio	4.6	25.5	5.0	21.9	19.2
San Diego	4.4	24.6	3.3	19.3	13.2
San Francisco PMSA	5.2	30.0	7.0	22.5	6.2
San Jose PMSA	7.8	26.4	4.6	21.4	14.6
Tampa-St. Petersburg-Clearwater	7.7	26.1	4.1	26.2	8.2
Washington D.C.	4.2	28.0	7.3	37.1	8.4
Cities					
Chicago	4.0	17.8	4.1	25.8	8.7
Dallas	5.7	25.1	6.0	18.4	10.7

[Continued]

★ 671 ★

Employed Persons in Selected Metropolitan Areas and Cities: Nonagricultural Industries and Annual Averages, 1992 - II

[Continued]

Population group and area	Private nonagricultural wage and salary workers				
	Transportation communication, and public utilities	Trade	Finance, insurance, real estate	Services[1]	Government
District of Columbia	2.5	31.0	10.8	35.7	10.2
Houston	3.3	31.0	5.3	22.8	6.0
Los Angeles	5.1	24.6	4.2	21.5	5.6
New York	5.6	17.2	9.7	27.6	14.3
Phoenix	3.7	26.4	10.5	20.9	10.5
San Antonio	4.2	22.9	5.5	23.5	19.9
San Diego	8.7	19.3	3.1	22.8	13.2
San Francisco	5.4	33.1	6.7	23.8	3.6

Source: "Selected Metropolitan Areas and Cities: Percent Distribution of Employed Persons in Nonagricultural Industries by Sex, Race, Hispanic Origin, 1992 Annual Averages," U.S. Department of Labor, *Geographical Profile of Employment and Unemployment,* 1992, pp. 122-124. *Notes:* Data for demographic groups are not shown when they do not meet BLS publication standards of reliability for the particular area based on the sample in that area. Detail may not add to totals because of rounding. 1. Excludes private household workers. 2. All are Metropolitan Statistical Areas (MSA's) except St. Louis and those labeled Consolidated Metropolitan Statistical Areas (CMSA's) or Primary Metropolitan Statistical Areas (PMSA's). 3. Less than 500 persons employed or less than 0.05 percent of total employed.

★ 672 ★

Labor Force

Employed Persons in Selected Metropolitan Areas and Cities: Occupations and Annual Averages, 1992 - I

Population group and area	Total employed	Managerial and professional specialty		Technical, sales, and administrative support		
		Executive administra- tive, and managerial	Professional specialty	Technicians and related support	Sales	Administrative support, including clerical
White						
Metropolitan areas[1]						
Anaheim-Santa Ana PMSA	100.0	17.9	12.0	2.7	13.0	17.4
Atlanta	100.0	17.3	17.5	3.8	15.0	17.2
Baltimore	100.0	17.1	17.8	4.4	14.0	16.4
Bergen-Passaic PMSA	100.0	18.2	15.8	3.7	13.4	17.5
Boston PMSA	100.0	18.3	21.0	4.0	12.6	17.2
Buffalo-Niagara Falls CMSA	100.0	10.2	14.8	3.4	13.1	16.8
Charlotte-Gastonia-Rock Hill	100.0	15.2	11.7	3.1	14.8	16.5
Chicago PMSA	100.0	15.8	14.9	2.9	13.3	18.2
Cincinnati PMSA	100.0	13.7	16.6	4.0	13.1	15.7
Cleveland PMSA	100.0	15.2	17.1	4.0	11.4	14.7
Columbus, Ohio	100.0	16.0	14.6	3.9	13.6	18.6
Dallas-Fort Worth CMSA	100.0	16.4	15.0	3.9	13.6	17.8

[Continued]

★ 672 ★

Employed Persons in Selected Metropolitan Areas and Cities: Occupations and Annual Averages, 1992 - I

[Continued]

Population group and area	Total employed	Managerial and professional specialty		Technical, sales, and administrative support		
		Executive administrative, and managerial	Professional specialty	Technicians and related support	Sales	Administrative support, including clerical
Dayton-Springfield	100.0	11.7	15.1	3.6	12.4	13.9
Denver-Boulder CMSA	100.0	15.4	17.9	4.0	11.8	17.5
Detroit PMSA	100.0	13.9	14.7	4.0	13.0	15.7
Fort Lauderdale-Hollywood-Pompano Beach PMSA	100.0	16.7	13.2	3.3	16.7	18.9
Hartford-New Britain-Middletown CMSA	100.0	15.9	14.6	3.3	13.0	16.8
Houston PMSA	100.0	15.0	17.2	4.1	12.4	17.5
Indianapolis	100.0	14.1	19.2	3.4	12.3	16.2
Kansas City	100.0	15.2	14.0	3.7	13.3	18.9
Los Angeles-Long Beach PMSA	100.0	13.4	14.7	2.7	11.7	15.6
Louisville	100.0	14.3	11.7	4.3	12.8	18.4
Memphis	100.0	18.2	13.6	3.1	16.1	16.5
Miami-Hialeah PMSA	100.0	13.0	14.1	2.8	14.6	16.4
Milwaukee PMSA	100.0	10.3	15.6	3.6	11.6	16.8
Minneapolis-St. Paul	100.0	14.4	14.2	4.2	13.5	18.8
Nassau-Suffolk PMSA	100.0	15.5	16.3	3.3	14.1	18.7
New Orleans	100.0	15.7	16.7	4.6	16.8	15.8
New York PMSA	100.0	17.9	19.5	2.6	11.3	16.6
Newark PMSA	100.0	16.6	15.6	3.5	13.5	16.9
Norfolk-Virginia Beach-Newport News	100.0	15.2	16.2	3.0	12.7	14.6
Oakland Pmsa	100.0	17.0	14.6	4.6	14.1	16.4
Oklahoma City	100.0	15.2	13.0	5.6	13.9	16.2
Philadelphia PMSA	100.0	15.8	16.3	4.2	13.5	18.1
Phoenix	100.0	16.3	12.2	4.5	13.4	17.9
Pittsburgh-Beaver Valley CMSA	100.0	11.7	16.0	3.8	13.1	17.6
Portland, Ore. PMSA	100.0	16.0	16.9	3.7	14.3	14.3
Providence-Pawtucket-Fall River CMSA	100.0	12.5	14.8	3.7	11.2	15.9
Riverside-San Bernardino PMSA	100.0	12.1	11.9	2.6	12.4	16.0
Rochester	100.0	13.6	18.0	4.7	13.0	14.5
Sacramento	100.0	16.2	14.7	3.0	15.5	19.6
St. Louis	100.0	13.0	15.4	3.5	14.1	16.7
Salt Lake City-Ogden	100.0	13.5	12.7	3.8	13.8	20.1
San Antonio	100.0	12.0	9.8	2.4	11.9	17.1
San Diego	100.0	15.6	15.1	4.0	14.0	16.4
San Francisco PMSA	100.0	21.5	18.4	3.0	15.8	12.7
San Jose PMSA	100.0	15.9	22.8	4.3	12.3	16.4
Seattle PMSA	100.0	14.9	19.9	3.9	14.3	16.7
Tampa-St. Petersburg-Clearwater	100.0	16.7	13.5	4.3	14.5	16.6
Washington D.C.	100.0	21.9	23.9	3.9	10.4	15.2
Baltimore	100.0	11.9	18.0	1.5	7.8	21.5
Chicago	100.0	14.5	14.6	2.4	10.7	15.9

[Continued]

★ 672 ★

Employed Persons in Selected Metropolitan Areas and Cities: Occupations and Annual Averages, 1992 - I

[Continued]

Population group and area	Total employed	Managerial and professional specialty		Technical, sales, and administrative support		
		Executive administrative, and managerial	Professional specialty	Technicians and related support	Sales	Administrative support, including clerical
Cleveland	100.0	9.2	7.1	4.8	10.1	16.5
Dallas	100.0	14.7	15.6	4.1	13.7	16.6
Detroit	100.0	3.4	8.9	3.4	15.6	12.5
District of Columbia	100.0	24.7	43.2	4.2	6.2	10.1
Houston	100.0	13.7	18.0	4.0	12.2	15.6
Indianapolis	100.0	13.5	20.7	4.0	12.1	17.2
Los Angeles	100.0	13.0	16.4	2.6	12.4	12.7
Milwaukee	100.0	8.9	12.1	3.9	11.3	17.7
New York	100.0	16.7	19.5	2.7	11.0	17.1
Philadelphia	100.0	13.0	17.0	3.6	10.4	20.6
Phoenix	100.0	16.6	11.5	5.0	12.5	18.4
St. Louis	100.0	11.9	27.9	4.1	8.5	11.6
San Antonio	100.0	10.5	8.3	2.1	10.8	17.2
San Diego	100.0	16.5	19.8	4.7	13.9	13.6
San Francisco	100.0	18.7	25.3	1.9	12.2	11.0
Black						
Metropolitan areas[1]						
Atlanta	100.0	9.3	9.3	5.2	8.6	20.4
Baltimore	100.0	7.9	13.1	3.7	4.8	22.0
Bergen-Passaic PMSA	100.0	12.0	11.8	6.3	3.9	21.1
Boston PMSA	100.0	6.4	8.6	6.4	6.7	20.8
Buffalo-Niagara Falls CMSA	100.0	13.2	7.4	4.4	11.3	20.2
Charlotte-Gastonia-Rock Hill	100.0	6.2	8.5	1.5	6.9	15.3
Chicago PMSA	100.0	10.5	13.7	3.5	10.1	21.3
Cincinnati PMSA	100.0	6.3	14.0	5.1	8.1	18.7
Cleveland PMSA	100.0	8.0	10.3	5.4	7.9	22.7
Columbus, Ohio	100.0	12.9	11.7	3.2	6.9	27.1
Dallas-Fort Worth CMSA	100.0	7.4	8.9	3.6	10.3	18.1
Dayton-Springfield	100.0	10.4	7.2	9.2	7.3	10.4
Denver-Boulder CMSA	100.0	8.2	10.8	2.0	7.0	25.1
Detroit PMSA	100.0	7.0	9.8	2.9	8.1	19.2
Fort Lauderdale-Hollywood-Pompano Beach PMSA	100.0	6.8	6.0	2.6	7.4	18.7
Hartford-New Britain-Middletown CMSA	100.0	7.1	4.7	3.2	4.7	24.9
Houston PMSA	100.0	7.8	11.2	4.0	9.4	21.5
Indianapolis	100.0	[2]	3.1	2.1	3.8	19.1
Kansas City	100.0	5.3	4.9	3.4	10.6	22.8
Los Angeles-Long Beach PMSA	100.0	10.0	15.0	4.7	9.5	26.9
Louisville	100.0	6.0	5.7	6.2	7.5	13.9
Memphis	100.0	3.0	6.3	1.7	9.0	15.9

[Continued]

★ 672 ★

Employed Persons in Selected Metropolitan Areas and Cities: Occupations and Annual Averages, 1992 - I

[Continued]

Population group and area	Total employed	Managerial and professional specialty		Technical, sales, and administrative support		
		Executive administrative, and managerial	Professional specialty	Technicians and related support	Sales	Administrative support, including clerical
Miami-Hialeah PMSA	100.0	6.2	9.4	1.4	6.0	14.7
Milwaukee PMSA	100.0	6.2	8.4	2.7	6.5	17.5
Minneapolis-St. Paul	100.0	4.5	12.5	1.4	3.9	26.6
Nassau-Suffolk PMSA	100.0	11.9	14.5	5.3	7.9	20.4
New Orleans	100.0	8.9	13.4	3.3	6.9	15.3
New York PMSA	100.0	9.6	12.1	3.1	6.1	21.8
Newark PMSA	100.0	6.6	10.9	3.0	5.6	20.2
Norfolk-Virginia Beach-Newport News	100.0	5.8	9.0	2.1	9.1	18.6
Oakland PMSA	100.0	12.7	10.2	3.2	13.8	18.4
Oklahoma City	100.0	2.2	3.8	2.3	10.4	11.0
Philadelphia PMSA	100.0	8.7	11.3	3.6	6.9	23.1
Pittsburgh-Beaver Valley CMSA	100.0	10.4	12.3	2.8	3.5	25.4
Providence-Pawtucket-Fall River CMSA	100.0	5.8	6.6	2.6	5.8	19.8
Riverside-San Bernardino PMSA	100.0	15.6	15.6	7.8	13.4	17.5
Rochester	100.0	4.0	9.7	3.7	3.1	21.2
St. Louis	100.0	5.2	10.4	3.2	5.4	17.3
San Diego	100.0	9.1	4.9	3.5	7.9	21.5
San Francisco PMSA	100.0	7.7	16.3	6.0	13.1	16.6
Seattle PMSA	100.0	6.1	16.6	[2]	15.8	15.4
Tampa-St. Petersburg-Clearwater	100.0	7.3	11.6	4.4	6.6	18.2
Washington D.C.	100.0	13.7	13.2	4.5	7.3	23.6
Cities						
Baltimore	100.0	6.6	10.0	5.1	4.0	23.4
Chicago	100.0	9.6	13.8	3.8	9.7	21.1
Cleveland	100.0	8.8	6.6	5.9	5.9	22.8
Dallas	100.0	7.3	6.9	4.7	9.3	16.0
Detroit	100.0	5.5	9.5	3.1	8.2	19.1
District of Columbia	100.0	10.5	9.9	4.6	7.0	28.3
Houston	100.0	7.2	10.1	3.4	11.1	21.7
Indianapolis	100.0	[2]	2.2	2.2	3.8	18.2
Los Angeles	100.0	8.9	14.1	4.3	9.6	26.8
Milwaukee	100.0	5.1	6.5	1.9	6.9	18.5
New York	100.0	9.2	11.9	3.2	5.9	22.2
Philadelphia	100.0	8.1	9.3	3.6	6.5	23.3
St. Louis	100.0	6.9	9.7	1.3	1.6	18.4
Hispanic origin						
Metropolitan areas[1]						
Anaheim-Santa Ana PMSA	100.0	5.2	3.3	1.6	6.1	8.6
Bergen-Passaic PMSA	100.0	6.0	3.6	3.9	10.4	15.4
Boston PMSA	100.0	7.2	15.6	3.1	5.9	11.3

[Continued]

★ 672 ★

Employed Persons in Selected Metropolitan Areas and Cities: Occupations and Annual Averages, 1992 - I

[Continued]

Population group and area	Total employed	Managerial and professional specialty		Technical, sales, and administrative support		
		Executive administra- tive, and managerial	Professional specialty	Technicians and related support	Sales	Administrative support, including clerical
Chicago PMSA	100.0	5.2	6.1	1.2	7.2	11.9
Dallas-Fort Worth CMSA	100.0	5.0	3.4	2.2	6.9	16.1
Denver-Boulder CMSA	100.0	8.9	8.7	2.7	7.0	19.7
Detroit PMSA	100.0	4.5	10.2	.7	12.9	10.8
Fort Lauderdale-Hollywood- Pompano Beach PMSA	100.0	12.1	.4	2.2	18.1	22.1
Houston PMSA	100.0	7.8	6.4	2.4	10.2	13.7
Los Angeles-Long Beach PMSA	100.0	6.3	4.7	1.9	7.8	13.6
Miami-Hialeah PMSA	100.0	13.0	9.4	2.3	13.1	15.6
Nassau-Suffolk PMSA	100.0	11.1	9.0	1.8	9.5	17.6
New York PMSA	100.0	6.7	7.7	2.2	7.2	16.1
Newark PMSA	100.0	5.3	5.0	1.8	8.1	10.7
Oakland PMSA	100.0	6.3	8.3	1.0	9.9	16.0
Philadelphia PMSA	100.0	5.5	13.1	.4	11.6	16.9
Phoenix	100.0	6.8	5.0	1.2	9.2	17.9
Portland, Ore. PMSA	100.0	5.1	.3	[2]	14.2	6.2
Providence-Pawtucket-Fall River CMSA	100.0	1.9	6.8	[2]	7.7	10.2
Riverside-San Bernardino PMSA	100.0	8.1	5.0	1.9	8.7	13.2
Sacramento	100.0	14.1	5.6	[2]	11.1	26.3
Salt Lake City-Ogden	100.0	4.6	3.2	2.3	12.3	20.4
San Antonio	100.0	6.7	6.6	2.4	9.9	16.1
San Diego	100.0	7.0	7.7	1.5	7.6	12.7
San Francisco PMSA	100.0	7.5	2.4	1.4	12.6	17.0
San Jose PMSA	100.0	6.7	8.6	3.1	8.5	18.6
Tampa-St. Petersburg-Clearwater	100.0	8.7	9.3	2.0	9.6	13.1
Washington D.C.	100.0	10.2	8.8	1.4	7.0	11.0
Cities						
Chicago	100.0	4.7	6.7	1.3	6.6	12.4
Dallas	100.0	4.0	3.6	2.6	7.0	14.7
District of Columbia	100.0	8.3	14.6	1.3	4.5	4.6
Houston	100.0	5.0	5.6	1.6	9.6	11.9
Los Angeles	100.0	5.1	4.6	1.5	7.0	9.5
New York	100.0	6.2	7.9	2.3	7.7	17.0
Phoenix	100.0	7.8	4.8	2.0	9.1	18.7
San Antonio	100.0	7.6	6.9	1.8	8.9	15.9

[Continued]

★ 672 ★

Employed Persons in Selected Metropolitan Areas and Cities: Occupations and Annual Averages, 1992 - I
[Continued]

Population group and area	Total employed	Managerial and professional specialty		Technical, sales, and administrative support		
		Executive administra-tive, and managerial	Professional specialty	Technicians and related support	Sales	Administrative support, including clerical
San Diego	100.0	4.9	10.4	2.9	8.8	11.7
San Francisco	100.0	3.0	3.8	2	9.0	14.7

Source: "Selected Metropolitan Areas and Cities: Percent Distribution of Employed Persons by Sex, Race, Hispanic Origin, and Occupation, 1992 Annual Averages," U.S. Department of Labor, *Geographical Profile of Employment and Unemployment*, 1992, pp. 115-117. *Notes:* Data for demographic groups are not shown when they do not meet BLS publication standards of reliability for the particular area based on the sample in that area. Detail may not add to totals because of rounding. 1. All are Metropolitan Statistical Areas (MSA's) except St. Louis and those labeled Consolidated Metropolitan Statistical Areas (CMSA's) or Primary Metropolitan Statistical Areas (PMSA's). 2. Less than 500 persons employed or less than 0.05 percent of total employed.

★ 673 ★
Labor Force

Employed Persons in Selected Metropolitan Areas and Cities: Occupations and Annual Averages, 1992 - II

Population group and area	Service occupations	Precision production, craft and repair	Operators, fabricators, and laborers		
			Machine operators, assemblers and inspectors	Transportation and material moving	Handlers, equipment cleaners helpers, and laborers
White					
Metropolitan areas[1]					
Anaheim-Santa Ana PMSA	9.5	11.6	6.4	3.4	2.8
Atlanta	8.8	10.1	3.1	3.3	2.4
Baltimore	9.5	11.6	2.1	3.1	3.0
Bergen-Passaic PMSA	10.4	9.4	5.4	2.8	2.8
Boston PMSA	11.1	7.9	2.2	2.8	2.1
Buffalo-Niagara Falls CMSA	14.8	11.6	6.1	4.6	4.0
Charlotte-Gastonia-Rock Hill	9.2	13.2	8.6	3.2	3.1
Chicago PMSA	11.2	10.9	5.3	3.3	3.3
Cincinnati PMSA	12.3	11.0	5.9	3.3	2.9
Cleveland PMSA	11.4	11.4	7.0	2.8	3.7
Columbus, Ohio	11.4	7.9	5.3	3.4	4.0
Dallas-Fort Worth CMSA	11.0	9.2	4.8	3.1	3.6
Dayton-Springfield	11.4	13.3	8.6	4.2	4.6
Denver-Boulder CMSA	12.9	9.4	3.1	3.3	3.4
Detroit PMSA	11.9	11.7	6.8	3.5	3.9
Fort Lauderdale-Hollywood-Pompano Beach PMSA	11.9	11.0	2.4	3.0	2.1
Hartford-New Britain-Middletown CMSA	13.1	8.7	6.3	3.4	2.8

[Continued]

★ 673 ★

Employed Persons in Selected Metropolitan Areas and Cities: Occupations and Annual Averages, 1992 - II

[Continued]

Population group and area	Service occupations	Precision production, craft and repair	Operators, fabricators, and laborers		
			Machine operators, assemblers and inspectors	Transportation and material moving	Handlers, equipment cleaners helpers, and laborers
Houston PMSA	10.1	11.6	3.8	3.7	3.6
Indianapolis	10.7	12.1	4.3	2.4	4.0
Kansas City	11.1	11.5	4.4	3.5	3.7
Los Angeles-Long Beach PMSA	12.9	11.4	8.9	3.7	3.4
Louisville	10.6	12.3	4.9	5.1	4.1
Memphis	8.6	13.1	3.2	3.5	2.6
Miami-Hialeah PMSA	12.4	11.6	5.6	3.1	4.5
Milwaukee PMSA	11.9	11.8	9.1	4.2	4.3
Minneapolis-St. Paul	12.2	9.0	6.4	3.0	3.4
Nassau-Suffolk PMSA	10.9	11.0	2.2	3.9	2.8
New Orleans	11.4	10.3	1.6	2.3	2.8
New York PMSA	13.8	8.1	3.6	3.2	2.7
Newark PMSA	10.8	9.9	5.3	3.3	3.5
Norfolk-Virginia Beach-Newport News	10.0	14.7	3.5	5.4	2.5
Oakland Pmsa	10.9	10.9	4.1	3.7	2.8
Oklahoma City	10.6	12.6	4.1	3.9	3.2
Philadelphia PMSA	9.9	10.8	3.8	3.7	2.6
Phoenix	12.4	11.4	3.3	3.1	2.8
Pittsburgh-Beaver Valley CMSA	14.5	9.8	3.7	3.7	4.7
Portland, Ore. PMSA	11.5	9.5	4.3	2.9	3.9
Providence-Pawtucket-Fall River CMSA	13.5	11.3	9.0	3.5	3.6
Riverside-San Bernardino PMSA	13.3	14.2	5.6	5.0	4.1
Rochester	11.6	10.7	5.8	4.6	2.3
Sacramento	12.3	7.9	2.4	4.1	2.9
St. Louis	13.0	11.4	4.7	3.1	3.4
Salt Lake City-Ogden	11.2	11.0	5.0	3.9	4.0
San Antonio	19.1	12.0	3.8	3.9	6.0
San Diego	14.3	10.1	3.0	2.4	2.7
San Francisco PMSA	12.8	8.5	2.1	1.8	3.0
San Jose PMSA	9.3	9.1	3.5	3.1	2.6
Seattle PMSA	9.9	9.4	3.8	2.7	3.7
Tampa-St. Petersburg-Clearwater	12.2	10.0	3.3	2.6	3.3
Washington D.C.	10.3	8.5	1.0	2.2	2.0
Baltimore	11.9	7.2	7.2	2.5	9.0
Chicago	15.3	10.8	7.4	3.2	4.1
Cleveland	17.4	12.0	11.6	4.2	6.9
Dallas	10.7	8.2	5.7	3.9	5.2
Detroit	21.5	12.7	9.4	6.5	5.9
District of Columbia	8.5	2.3	.2	.4	.1
Houston	13.6	9.7	4.7	3.6	4.4
Indianapolis	10.1	11.7	3.5	2.6	3.9

[Continued]

★ 673 ★

Employed Persons in Selected Metropolitan Areas and Cities: Occupations and Annual Averages, 1992 - II

[Continued]

Population group and area	Service occupations	Precision production, craft and repair	Operators, fabricators, and laborers		
			Machine operators, assemblers and inspectors	Transportation and material moving	Handlers, equipment cleaners helpers, and laborers
Los Angeles	15.3	10.2	9.4	3.9	2.9
Milwaukee	14.8	13.5	7.5	5.0	4.8
New York	14.6	7.8	4.1	3.4	2.9
Philadelphia	12.2	12.0	4.5	4.1	2.1
Phoenix	12.1	12.3	3.1	3.3	3.2
St. Louis	16.0	11.9	3.2	2.3	1.2
San Antonio	21.5	12.6	4.7	3.9	6.8
San Diego	15.0	7.3	2.6	2.5	2.0
San Francisco	16.3	7.2	2.1	2.2	3.2
Black					
Metropolitan areas[1]					
Atlanta	23.4	5.6	7.3	5.9	4.5
Baltimore	21.4	7.4	3.4	8.2	6.3
Bergen-Passaic PMSA	22.7	7.3	6.6	2.7	5.7
Boston PMSA	29.2	5.0	5.3	5.1	5.8
Buffalo-Niagara Falls CMSA	31.6	1.5	7.4	1.0	.4
Charlotte-Gastonia-Rock Hill	23.6	6.7	13.5	8.3	8.5
Chicago PMSA	18.7	7.6	5.8	4.5	4.2
Cincinnati PMSA	26.9	4.7	8.4	3.6	4.3
Cleveland PMSA	21.0	6.7	7.0	6.3	4.2
Columbus, Ohio	18.4	3.3	5.7	3.4	6.9
Dallas-Fort Worth CMSA	21.1	9.2	8.0	8.7	4.5
Dayton-Springfield	17.4	6.5	14.7	10.9	5.0
Denver-Boulder CMSA	20.3	6.0	6.0	6.4	6.6
Detroit PMSA	20.2	9.3	14.0	3.2	5.4
Fort Lauderdale-Hollywood- Pompano Beach PMSA	29.6	9.3	1.7	6.2	8.2
Hartford-New Britain-Middletown CMSA	29.3	11.6	5.2	2.6	6.7
Houston PMSA	20.1	8.1	2.8	7.5	5.8
Indianapolis	34.3	10.6	4.9	7.7	10.8
Kansas City	28.4	7.0	5.5	5.4	4.8
Los Angeles-Long Beach PMSA	14.7	6.7	2.9	4.7	3.8
Louisville	27.3	9.1	10.8	4.6	8.2
Memphis	27.7	7.9	10.1	6.4	9.6
Miami-Hialeah PMSA	29.7	9.3	6.9	7.8	7.3
Milwaukee PMSA	33.2	4.2	11.3	5.1	4.6
Minneapolis-St. Paul	20.6	9.5	10.3	2.1	8.4
Nassau-Suffolk PMSA	20.6	5.8	2.9	5.2	4.8
New Orleans	29.7	4.2	2.5	7.3	6.0
New York PMSA	26.0	7.2	3.8	6.1	3.9

[Continued]

★ 673 ★

Employed Persons in Selected Metropolitan Areas and Cities: Occupations and Annual Averages, 1992 - II

[Continued]

Population group and area	Service occupations	Precision production, craft and repair	Operators, fabricators, and laborers		
			Machine operators, assemblers and inspectors	Transportation and material moving	Handlers, equipment cleaners helpers, and laborers
Newark PMSA	24.4	7.8	10.1	5.7	5.0
Norfolk-Virginia Beach-Newport News	22.1	12.7	5.4	9.6	4.5
Oakland PMSA	14.4	9.3	2.7	6.6	8.8
Oklahoma City	40.0	11.6	9.6	2.7	4.7
Philadelphia PMSA	22.3	9.7	5.1	4.6	4.5
Pittsburgh-Beaver Valley CMSA	21.2	5.1	6.0	2.5	9.5
Providence-Pawtucket-Fall River CMSA	30.0	10.7	8.1	1.7	5.0
Riverside-San Bernardino PMSA	13.8	9.0	2.4	3.2	1.7
Rochester	14.0	21.7	15.6	3.8	3.2
St. Louis	30.4	4.9	12.2	4.0	3.8
San Diego	30.7	9.0	1.4	3.7	7.1
San Francisco PMSA	22.0	2.6	1.5	12.9	1.3
Seattle PMSA	25.7	10.6	2.8	3.1	3.9
Tampa-St. Petersburg-Clearwater	18.4	10.3	7.2	10.2	5.0
Washington D.C.	18.6	7.6	.8	5.5	3.9
Cities					
Baltimore	26.5	6.1	3.8	6.4	7.0
Chicago	20.4	6.6	5.8	4.4	4.5
Cleveland	23.9	6.9	9.2	5.2	3.9
Dallas	26.4	7.7	7.9	9.2	4.2
Detroit	21.3	9.6	13.4	3.7	5.8
District of Columbia	24.0	5.1	1.5	5.1	3.3
Houston	19.9	5.8	3.0	8.6	7.1
Indianapolis	35.1	10.9	5.0	7.8	11.1
Los Angeles	17.7	5.7	3.1	4.3	3.9
Milwaukee	34.4	4.2	12.0	5.4	4.6
New York	26.4	7.3	3.8	6.0	4.0
Philadelphia	23.9	11.2	5.0	4.6	4.3
St. Louis	28.7	2.3	20.3	3.9	3.3
Hispanic origin					
Metropolitan areas[1]					
Anaheim-Santa Ana PMSA	19.0	14.8	18.0	6.5	5.2
Bergen-Passaic PMSA	21.1	8.2	24.5	1.6	4.6
Boston PMSA	32.0	6.6	10.4	3.4	3.8
Chicago PMSA	22.6	13.8	18.7	3.5	7.7
Dallas-Fort Worth CMSA	17.8	15.7	16.8	4.1	9.7
Denver-Boulder CMSA	20.1	11.8	6.3	6.9	7.0
Detroit PMSA	16.4	18.4	17.3	1.0	7.6
Fort Lauderdale-Hollywood-Pompano Beach PMSA	16.5	13.5	10.1	1.4	2.9

[Continued]

★ 673 ★

Employed Persons in Selected Metropolitan Areas and Cities: Occupations and Annual Averages, 1992 - II

[Continued]

Population group and area	Service occupations	Precision production, craft and repair	Operators, fabricators, and laborers		
			Machine operators, assemblers and inspectors	Transportation and material moving	Handlers, equipment cleaners helpers, and laborers
Houston PMSA	21.2	14.8	9.5	4.0	9.2
Los Angeles-Long Beach PMSA	19.4	14.7	17.8	5.4	5.4
Miami-Hialeah PMSA	14.3	13.4	7.6	3.3	5.8
Nassau-Suffolk PMSA	18.1	12.6	8.8	2.9	5.2
New York PMSA	27.9	9.5	11.1	4.6	5.4
Newark PMSA	20.3	12.9	16.9	7.0	10.5
Oakland PMSA	22.6	15.2	6.3	4.1	7.5
Philadelphia PMSA	18.5	12.0	15.0	3.5	2.7
Phoenix	16.1	17.3	6.5	4.0	7.8
Portland, Ore. PMSA	15.4	11.6	10.7	[2]	13.3
Providence-Pawtucket-Fall River CMSA	6.8	10.5	46.4	2.0	7.7
Riverside-San Bernardino PMSA	17.5	15.9	11.5	5.6	6.9
Sacramento	21.6	5.0	1.8	4.5	5.2
Salt Lake City-Ogden	23.0	12.9	5.7	6.3	5.9
San Antonio	24.4	13.4	5.7	4.4	7.9
San Diego	23.9	15.1	7.0	4.5	5.9
San Francisco PMSA	32.1	12.0	5.8	3.2	6.0
San Jose PMSA	21.0	11.8	7.3	6.8	4.3
Tampa-St. Petersburg-Clearwater	12.5	9.4	8.8	5.0	6.5
Washington D.C.	42.3	10.6	.9	4.1	2.2
Cities					
Chicago	24.3	13.4	17.7	3.4	7.1
Dallas	16.6	15.2	14.3	3.7	14.1
District of Columbia	54.3	9.5	.7	1.4	.8
Houston	26.2	14.2	10.1	4.5	10.4
Los Angeles	24.6	14.0	19.7	6.7	4.6
New York	26.6	9.9	11.5	4.8	5.8
Phoenix	16.8	18.2	4.7	4.4	8.9
San Antonio	25.6	12.7	6.4	4.7	7.4
San Diego	21.9	15.3	6.6	7.3	4.8
San Francisco	41.6	11.6	5.6	4.5	6.1

Source: "Selected Metropolitan Areas and Cities: Percent Distribution of Employed Persons by Sex, Race, Hispanic Origin, and Occupation, 1992 Annual Averages," U.S. Department of Labor, *Geographical Profile of Employment and Unemployment,* 1992, pp. 115-117. *Notes:* Data for demographic groups are not shown when they do not meet BLS publication standards of reliability for the particular area based on the sample in that area. Detail may not add to totals because of rounding. 1. All are Metropolitan Statistical Areas (MSA's) except St. Louis and those labeled Consolidated Metropolitan Statistical Areas (CMSA's) or Primary Metropolitan Statistical Areas (PMSA's). 2. Less than 500 persons employed or less than 0.05 percent of total employed.

★ 674 ★
Labor Force

Employed Persons in States: Percent Distribution and Industry: Annual Averages, 1992 - I

In thousands.

Population group and State	Total employed[1]		Nonagricultural industries					
				Private nonagricultural wage and salary workers				
			Total[2]	Total[3]	Construction	Manufacturing		
	Number (in thousands)	Percent				Total	Durable goods	Nondurable goods
White								
Alabama	1,392	100.0	96.7	74.0	5.5	22.5	11.3	11.2
Alaska	197	100.0	98.3	62.6	4.0	3.4	1.5	1.9
Arizona	1,521	100.0	96.1	72.8	5.6	11.1	8.1	3.1
Arkansas	923	100.0	94.2	69.3	4.0	20.1	10.2	9.9
California	11,656	100.0	95.1	70.8	4.5	15.5	9.4	6.0
Colorado	1,521	100.0	97.1	72.5	4.5	12.0	8.1	3.9
Connecticut	1,480	100.0	98.2	80.7	3.5	21.2	15.3	5.9
Delaware	285	100.0	97.5	78.5	6.6	18.1	4.9	13.2
District of Columbia	111	100.0	98.4	65.4	1.2	5.2	.5	4.7
Florida	5,063	100.0	96.3	75.1	5.2	9.1	5.6	3.5
Georgia	2,079	100.0	96.2	73.7	4.4	16.9	6.5	10.4
Hawaii	185	100.0	96.8	71.0	7.2	3.3	.7	2.6
Idaho	475	100.0	92.5	65.6	4.9	14.4	8.1	6.3
Illinois	4,834	100.0	97.4	79.2	4.2	19.0	10.8	8.2
Indiana	2,472	100.0	96.3	78.4	4.1	26.3	18.8	7.6
Iowa	1,446	100.0	90.7	68.8	4.2	15.7	8.7	7.0
Kansas	1,183	100.0	93.1	68.8	3.6	15.8	8.9	6.9
Kentucky	1,514	100.0	94.4	70.3	4.3	16.6	9.9	6.6
Louisiana	1,392	100.0	96.2	69.8	7.0	10.4	4.0	6.4
Maine	607	100.0	97.3	70.9	3.8	17.1	7.5	9.6
Maryland	1,737	100.0	97.8	71.0	7.0	7.8	4.3	3.5
Massachusetts	2,717	100.0	98.5	78.9	3.1	16.7	10.7	6.0
Michigan	3,705	100.0	96.8	77.2	4.1	23.0	17.3	5.7
Minnesota	2,228	100.0	92.9	72.0	3.5	16.3	9.5	6.8
Mississippi	753	100.0	95.9	71.3	5.4	17.8	10.3	7.5
Missouri	2,312	100.0	95.9	75.7	4.3	16.9	9.9	7.0
Montana	365	100.0	89.4	57.3	3.8	5.4	3.6	1.8
Nebraska	804	100.0	90.6	67.4	3.8	11.8	6.0	5.8
Nevada	567	100.0	97.8	76.4	6.8	4.6	2.8	1.8
New Hampshire	578	100.0	97.7	77.0	3.5	22.0	16.5	5.5
New Jersey	3,098	100.0	98.5	78.7	4.3	16.4	6.9	9.6
New Mexico	613	100.0	96.3	61.9	5.1	7.1	3.3	3.8
New York	6,446	100.0	97.8	72.6	3.8	14.0	7.8	6.1
North Carolina	2,545	100.0	96.7	75.9	5.3	24.9	10.5	14.4
North Dakota	289	100.0	86.1	60.9	3.3	5.7	3.0	2.7
Ohio	4,618	100.0	97.3	78.3	4.0	22.9	15.6	7.3
Oklahoma	1,272	100.0	94.3	66.5	3.7	12.9	8.1	4.8
Oregon	1,354	100.0	95.1	69.1	4.0	14.1	9.7	4.5
Pennsylvania	5,114	100.0	97.6	78.5	4.2	18.6	10.7	7.8
Rhode Island	459	100.0	98.7	78.2	3.0	20.3	13.5	6.9
South Carolina	1,211	100.0	97.6	75.3	5.9	22.4	7.5	14.9
South Dakota	334	100.0	86.8	62.7	3.1	10.6	6.1	4.5
Tennessee	1,962	100.0	96.1	75.2	4.3	24.0	12.6	11.4

[Continued]

★ 674 ★

Employed Persons in States: Percent Distribution and Industry: Annual Averages, 1992 - I

[Continued]

Population group and State	Total employed[1]		Nonagricultural industries					
			Total[2]	Private nonagricultural wage and salary workers				
				Total[3]	Construction	Manufacturing		
	Number (in thousands)	Percent				Total	Durable goods	Nondurable goods
Texas	7,041	100.0	96.0	72.7	5.3	12.6	7.1	5.6
Utah	751	100.0	96.9	69.1	4.7	13.4	8.1	5.3
Vermont	298	100.0	95.4	69.6	4.1	16.0	11.0	5.0
Virginia	2,477	100.0	97.4	71.5	6.2	11.9	6.2	5.8
Washington	2,209	100.0	96.9	69.6	4.9	13.8	10.1	3.7
West Virginia	654	100.0	96.7	72.7	4.5	14.8	8.2	6.6
Wisconsin	2,409	100.0	94.4	74.1	4.8	23.8	13.9	9.9
Wyoming	222	100.0	90.7	61.1	4.7	4.8	2.0	2.8
Black								
Alabama	397	100.0	96.7	71.8	2.8	25.8	12.8	12.9
Alaska	9	100.0	100.0	50.3	1.1	1.6	[4]	1.6
Arkansas	128	100.0	97.0	74.6	2.7	29.6	12.4	17.2
California	708	100.0	98.6	66.0	2.6	12.5	7.7	4.8
Colorado	71	100.0	98.5	77.5	4.7	9.5	3.3	6.2
Connecticut	141	100.0	98.7	83.1	2.7	23.5	17.4	6.2
Delaware	58	100.0	98.6	79.7	2.7	23.6	5.9	17.7
District of Columbia	134	100.0	97.3	56.2	2.0	2.3	.6	1.7
Florida	847	100.0	94.4	64.7	3.7	10.1	4.7	5.4
Georgia	881	100.0	96.4	75.6	2.5	22.7	6.6	16.0
Illinois	647	100.0	99.2	72.7	.8	15.7	7.0	8.7
Indiana	175	100.0	96.6	71.4	1.7	17.6	11.7	5.9
Kansas	69	100.0	99.6	76.5	2.7	24.3	14.6	9.7
Kentucky	104	100.0	94.5	76.8	.7	27.2	18.8	8.4
Louisiana	363	100.0	91.8	60.9	3.6	9.4	5.5	3.9
Maryland	652	100.0	96.8	62.3	3.2	8.2	3.2	5.0
Massachusetts	96	100.0	99.0	78.8	1.4	15.7	10.7	5.0
Michigan	424	100.0	98.3	76.2	1.4	26.0	21.2	4.8
Mississippi	329	100.0	92.7	67.4	3.1	27.8	13.8	14.0
Missouri	191	100.0	98.6	82.2	.1	11.7	4.8	6.9
Nebraska	18	100.0	98.8	81.6	.2	9.6	5.8	3.8
Nevada	33	100.0	99.3	74.7	1.7	4.1	2.3	1.9
New Jersey	424	100.0	98.6	72.1	1.4	15.3	6.4	8.9
New York	1,021	100.0	97.8	66.9	2.3	9.0	4.1	5.0
North Carolina	687	100.0	94.8	72.4	2.6	31.1	11.1	20.1
Ohio	423	100.0	98.7	70.2	1.5	18.1	11.9	6.2
Oklahoma	79	100.0	97.7	71.2	2.0	13.6	9.4	4.2
Pennsylvania	361	100.0	99.1	75.1	2.5	12.7	6.9	5.9
Rhode Island	12	100.0	94.7	60.6	1.3	17.0	8.0	8.9
South Carolina	431	100.0	94.0	74.8	3.3	32.5	8.4	24.2
Tennessee	297	100.0	96.9	73.1	4.3	15.4	7.7	7.7
Texas	839	100.0	96.0	69.8	2.1	14.3	8.0	6.3
Virginia	578	100.0	96.7	67.7	3.1	21.5	9.1	12.4
Washington	67	100.0	100.0	67.6	.9	13.9	6.5	7.4
West Virginia	23	100.0	98.1	80.6	[4]	17.8	9.0	8.8

[Continued]

569

★ 674 ★

Employed Persons in States: Percent Distribution and Industry: Annual Averages, 1992 - I

[Continued]

Population group and State	Total employed[1]		Nonagricultural industries					
	Number (in thousands)	Percent	Total[2]	Private nonagricultural wage and salary workers				
				Total[3]	Construction	Manufacturing		
						Total	Durable goods	Nondurable goods
Wisconsin	76	100.0	99.0	83.6	3.1	19.4	11.5	7.9
Hispanic origin								
Arizona	254	100.0	91.4	71.1	8.0	12.7	8.4	4.3
California	3,216	100.0	89.1	74.3	5.3	21.8	11.2	10.6
Colorado	160	100.0	97.0	70.2	5.8	15.3	8.2	7.1
Connecticut	73	100.0	99.1	88.2	1.3	31.2	25.3	6.0
District of Columbia	15	100.0	90.3	76.6	1.8	2.5	[4]	2.5
Florida	816	100.0	91.6	76.9	4.6	12.1	6.0	6.2
Hawaii	13	100.0	86.5	68.3	3.9	2.8	1.7	1.1
Idaho	25	100.0	77.4	64.0	4.0	30.4	3.6	26.8
Illinois	412	100.0	97.0	86.5	2.4	34.8	19.9	14.9
Maryland	69	100.0	90.7	77.9	11.8	[4]	.1	[4]
Massachusetts	61	100.0	98.9	83.3	4.3	21.4	8.2	13.2
Michigan	65	100.0	96.9	77.4	3.0	32.8	27.5	5.3
Nevada	61	100.0	96.6	85.7	6.5	3.8	2.1	1.8
New Jersey	322	100.0	97.4	85.5	2.8	27.1	10.7	16.4
New Mexico	195	100.0	95.1	62.8	7.9	7.8	3.5	4.4
New York	689	100.0	95.9	77.0	3.0	16.8	5.5	11.4
North Carolina	43	100.0	87.8	76.6	4.5	42.3	10.6	31.8
Ohio	35	100.0	95.6	75.5	2.2	23.7	11.9	11.8
Oklahoma	43	100.0	91.6	78.9	3.7	28.1	9.3	18.8
Oregon	56	100.0	80.1	69.5	1.0	18.7	9.4	9.3
Pennsylvania	56	100.0	96.0	81.5	1.4	23.9	11.6	12.3
Rhode Island	16	100.0	98.7	92.5	.4	65.1	47.1	18.0
Texas	1,792	100.0	94.7	74.9	7.5	14.2	6.9	7.3
Utah	27	100.0	91.4	68.9	1.6	14.0	7.1	6.8
Virginia	78	100.0	97.9	85.5	10.6	3.7	[4]	3.7
Wyoming	8	100.0	83.8	66.6	4.3	6.3	2.6	3.7

Source: "States: Percent Distribution of Employed Persons by Sex, Race, Hispanic Origin, and Industry, 1992 Annual Averages," U.S. Department of Labor, *Geographical Profile of Employment and Unemployment, 1992*, pp. 71-73. *Notes:* Data for demographic groups are not shown when they do not meet BLS publication standards of reliability for the particular area based on the sample in that area. Items may not add to totals or compute to displayed percentages because of rounding. Detail for Hispanic-origin groups will not add to totals because data for the "other races" group are not presented and Hispanics are included in both the white and black population groups. 1. Includes private household workers, self-employed and unpaid family workers, and mining. 2. Includes self-employed and unpaid family workers and mining. 3. Includes mining. 4. Less than 500 persons employed or less than 0.05 percent of total employed.

★ 675 ★

Labor Force

Employed Persons in States: Percent Distribution and Industry: Annual Averages, 1992 - II

In thousands.

Population group and State	Private nonagricultural wage and salary workers					
	Transportation communications, and public utilities	Trade	Finance, insurance, real estate	Service industries[1]	Government	Agriculture
White						
Alabama	5.8	17.5	5.0	17.3	13.8	2.6
Alaska	8.5	17.3	4.8	20.8	24.2	.8
Arizona	5.2	21.2	6.4	22.6	14.9	2.9
Arkansas	5.2	19.0	4.7	15.9	15.6	4.9
California	5.2	18.9	6.0	20.5	13.7	3.7
Colorado	6.5	18.4	6.7	23.1	15.4	2.1
Connecticut	5.7	17.4	9.0	23.9	10.2	1.3
Delaware	4.7	19.0	9.3	20.7	13.4	2.3
District of Columbia	3.3	8.4	5.4	41.9	24.3	.1
Florida	5.8	22.3	7.5	25.0	13.4	2.9
Georgia	7.4	18.8	6.3	19.4	14.5	3.3
Hawaii	8.2	18.5	6.4	27.4	17.3	2.9
Idaho	5.4	19.0	3.4	17.9	16.2	6.3
Illinois	5.6	20.4	7.2	22.5	11.6	2.0
Indiana	4.6	18.5	4.6	19.8	11.8	2.8
Iowa	4.5	20.2	5.5	18.7	14.8	8.5
Kansas	5.8	18.8	5.8	18.5	15.9	6.1
Kentucky	5.8	18.9	5.0	17.8	14.7	4.5
Louisiana	5.4	19.0	5.4	19.3	17.3	2.6
Maine	4.1	20.1	5.1	20.5	14.1	1.8
Maryland	5.3	19.3	6.0	25.5	19.7	1.5
Massachusetts	4.6	19.5	7.2	27.8	12.4	.9
Michigan	3.9	21.0	5.3	19.8	13.6	2.3
Minnesota	4.8	19.9	5.4	21.8	14.3	6.2
Mississippi	5.7	18.0	4.4	18.5	16.1	3.5
Missouri	6.8	21.3	4.6	21.8	12.9	3.6
Montana	6.0	20.1	3.4	17.9	19.7	9.6
Nebraska	5.3	21.2	6.3	18.8	15.4	8.6
Nevada	5.6	18.3	5.7	33.6	13.9	1.6
New Hampshire	4.9	17.6	5.5	23.4	11.6	1.4
New Jersey	6.9	19.3	8.4	23.4	14.1	1.0
New Mexico	5.2	18.4	3.9	19.1	22.9	2.8
New York	5.4	17.1	8.0	24.3	17.0	1.3
North Carolina	5.2	18.8	4.7	16.8	12.7	2.7
North Dakota	4.8	21.7	4.1	20.1	18.1	12.9
Ohio	4.4	19.8	5.0	21.9	13.0	2.0
Oklahoma	5.4	17.6	5.8	18.4	17.3	5.0
Oregon	4.8	20.2	5.6	20.3	15.2	3.9

[Continued]

571

★ 675 ★

Employed Persons in States: Percent Distribution and Industry: Annual Averages, 1992 - II

[Continued]

| Population group and State | Private nonagricultural wage and salary workers | | | | | |
	Transportation communications, and public utilities	Trade	Finance, insurance, real estate	Service industries[1]	Government	Agriculture
Pennsylvania	5.6	19.5	6.6	23.6	12.2	1.7
Rhode Island	4.8	17.9	7.2	25.1	14.4	.9
South Carolina	5.3	18.7	5.3	17.6	15.3	1.9
South Dakota	3.9	20.6	5.1	18.4	15.4	12.4
Tennessee	4.9	19.2	4.5	18.0	12.9	3.1
Texas	6.1	20.4	5.3	20.7	15.3	2.8
Utah	5.1	19.5	5.6	20.1	19.8	2.5
Vermont	3.7	17.9	4.5	22.9	14.8	3.7
Virginia	5.6	18.1	5.3	23.9	20.1	2.1
Washington	4.8	20.5	5.2	20.3	18.0	2.1
West Virginia	5.9	20.9	3.9	18.4	17.2	2.0
Wisconsin	4.0	17.8	4.9	18.8	13.2	4.9
Wyoming	7.0	18.1	3.6	15.7	20.9	8.1
Black						
Alabama	3.0	19.4	2.4	18.1	21.4	.7
Alaska	6.7	13.3	10.3	13.3	43.6	[2]
Arkansas	5.5	20.6	1.4	14.4	20.5	2.1
California	6.8	13.3	6.7	24.0	24.8	.6
Colorado	10.6	16.2	8.7	26.8	17.9	.5
Connecticut	7.9	15.5	7.3	26.2	13.5	.2
Delaware	5.1	14.0	10.4	23.9	17.2	.4
District of Columbia	5.1	13.0	3.9	29.9	36.3	.4
Florida	5.9	17.5	4.0	23.5	25.8	2.6
Georgia	7.1	15.8	5.5	21.6	17.1	1.6
Illinois	6.6	17.3	7.0	25.2	22.8	.1
Indiana	9.3	20.5	1.1	21.3	24.9	2.1
Kansas	5.7	17.4	5.0	21.5	19.6	[2]
Kentucky	5.0	16.5	1.9	25.3	15.7	1.4
Louisiana	5.0	21.4	2.3	18.7	24.3	3.9
Maryland	4.8	16.0	4.6	25.5	30.9	.9
Massachusetts	5.4	11.3	8.6	36.5	16.5	.2
Michigan	3.8	14.7	5.3	24.9	18.8	.6
Mississippi	5.1	15.6	2.4	13.0	22.7	4.4
Missouri	9.1	19.8	5.2	36.2	15.0	1.3
Nebraska	12.0	19.1	12.3	28.4	15.0	[2]
Nevada	7.8	15.7	4.7	40.7	22.9	[2]
New Jersey	8.0	14.1	8.1	25.2	24.2	.3

[Continued]

★ 675 ★

Employed Persons in States: Percent Distribution and Industry: Annual Averages, 1992 - II

[Continued]

| Population group and State | Private nonagricultural wage and salary workers | | | | | |
	Transportation communications, and public utilities	Trade	Finance, insurance, real estate	Service industries[1]	Government	Agriculture
New York	6.4	9.8	8.4	31.0	27.2	.1
North Carolina	4.9	13.8	2.4	17.6	19.2	2.6
Ohio	5.2	13.0	5.8	26.5	26.4	.3
Oklahoma	3.9	21.3	3.9	24.8	23.1	1.2
Pennsylvania	6.8	13.0	7.1	33.0	21.1	.1
Rhode Island	3.0	7.8	3.0	28.5	29.6	3.7
South Carolina	4.1	19.5	1.8	13.2	16.7	3.7
Tennessee	5.3	21.0	4.6	22.6	20.9	1.2
Texas	6.6	17.2	4.5	24.4	23.2	1.4
Virginia	7.4	15.3	3.7	16.4	26.4	.9
Washington	5.5	17.3	3.5	26.4	29.5	[2]
West Virginia	6.2	25.7	9.0	20.7	15.3	.2
Wisconsin	6.3	17.2	5.5	32.1	14.8	[2]
Hispanic origin						
Arizona	2.5	24.1	4.6	19.0	16.1	6.4
California	4.6	21.1	3.9	17.3	9.5	7.9
Colorado	7.0	20.2	4.8	17.2	22.6	2.0
Connecticut	4.8	18.9	5.5	26.5	9.5	.9
District of Columbia	2.3	28.0	9.7	32.2	9.2	[2]
Florida	6.5	22.9	5.8	24.9	8.4	6.8
Hawaii	7.3	8.9	2.9	42.4	17.9	13.5
Idaho	8.9	9.6	.4	10.7	10.0	22.2
Illinois	4.1	18.5	4.4	22.2	7.4	1.9
Maryland	8.5	24.9	4.5	28.3	9.7	.3
Massachusetts	1.1	24.7	3.1	28.5	11.4	.6
Michigan	2.8	21.3	2.2	14.7	13.3	2.3
Nevada	3.5	12.9	2.9	54.9	5.0	3.0
New Jersey	6.0	23.3	4.8	21.3	8.3	.7
New Mexico	5.3	18.0	2.5	18.6	25.2	3.1
New York	5.4	16.3	9.1	26.3	14.1	1.5
North Carolina	3.6	9.5	2.7	13.5	2.2	11.8
Ohio	.6	20.8	5.5	22.7	17.8	4.4
Oklahoma	4.2	21.1	.9	21.0	8.1	7.8
Oregon	4.7	29.7	5.0	10.3	8.8	17.9
Pennsylvania	2.5	20.3	6.0	27.5	11.1	2.3
Rhode Island	1.3	7.4	6.1	12.1	1.6	.7
Texas	4.5	24.5	4.3	18.4	14.4	3.4
Utah	6.9	22.1	6.3	15.5	17.8	7.4

[Continued]

★ 675 ★

Employed Persons in States: Percent Distribution and Industry: Annual Averages, 1992 - II

[Continued]

| Population group and State | Private nonagricultural wage and salary workers | | | | | |
	Transportation communications, and public utilities	Trade	Finance, insurance, real estate	Service industries[1]	Government	Agriculture
Virginia	1.1	28.2	7.0	34.9	7.0	1.0
Wyoming	15.6	21.4	3.7	7.9	12.7	16.0

Source: "States: Percent Distribution of Employed Persons by Sex, Race, Hispanic Origin, and Industry, 1992 Annual Averages," U.S. Department of Labor, *Geographical Profile of Employment and Unemployment, 1992*, pp. 71-73. *Notes:* Data for demographic groups are not shown when they do not meet BLS publication standards of reliability for the particular area based on the sample in that area. Items may not add to totals or compute to displayed percentages because of rounding. Detail for Hispanic-origin groups will not add to totals because data for the "other races" group are not presented and Hispanics are included in both the white and black population groups. 1. Excludes private household workers. 2. Less than 500 persons employed or less than 0.05 percent of total employed.

★ 676 ★

Labor Force

Employed Persons in States: Percent Distribution and Occupation: Annual Averages, 1992 - I

In thousands.

| Population group and State | Total employed | | Managerial and professional specialty | | Technical, sales and administrative support | | |
	Number (in thousands)	Percent	Executive, administrative, and managerial	Professional specialty	Technicians and related support	Sales	Administrative support, including clerical
White							
Alabama	1,392	100.0	10.7	11.7	3.8	12.6	14.0
Alaska	197	100.0	16.2	16.1	4.3	11.1	17.0
Arizona	1,521	100.0	15.0	13.3	4.5	13.5	17.0
Arkansas	923	100.0	10.5	11.0	3.1	12.4	14.5
California	11,656	100.0	14.6	14.2	3.1	12.5	15.9
Colorado	1,521	100.0	14.0	17.2	4.6	11.9	16.2
Connecticut	1,480	100.0	16.0	16.9	3.6	12.7	15.7
Delaware	285	100.0	16.2	15.9	4.6	12.1	16.9
District of Columbia	111	100.0	24.7	43.2	4.2	6.2	10.1
Florida	5,063	100.0	14.1	13.1	3.8	14.6	16.2
Georgia	2,079	100.0	13.6	14.7	4.1	13.6	16.5
Hawaii	185	100.0	16.5	18.7	4.2	11.3	13.5
Idaho	475	100.0	10.0	127	3.7	10.9	13.7
Illinois	4,834	100.0	13.7	14.0	3.2	13.0	16.7
Indiana	2,472	100.0	10.5	10.7	2.9	10.7	15.3
Iowa	1,446	100.0	10.4	12.9	3.0	10.9	14.2
Kansas	1,183	100.0	12.4	13.7	3.1	11.9	16.2
Kentucky	1,514	100.0	11.6	11.7	3.2	12.0	13.4
Louisiana	1,392	100.0	13.4	15.8	5.1	13.1	15.1

[Continued]

★ 676 ★

Employed Persons in States: Percent Distribution and Occupation: Annual Averages, 1992 - I

[Continued]

Population group and State	Total employed		Managerial and professional specialty		Technical, sales and administrative support		
	Number (in thousands)	Percent	Executive, administrative, and managerial	Professional specialty	Technicians and related support	Sales	Administrative support, including clerical
Maine	507	100.0	12.4	13.0	3.6	10.7	14.2
Maryland	1,737	100.0	17.2	17.3	4.1	13.6	15.6
Massachusetts	2,717	100.0	15.6	18.2	3.8	12.1	16.4
Michigan	3,705	100.0	12.0	13.7	3.4	11.9	15.1
Minnesota	2,228	100.0	12.0	12.4	3.7	11.9	16.5
Mississippi	753	100.0	11.1	12.9	4.5	12.5	15.2
Missouri	2,312	100.0	10.8	12.9	3.1	12.9	15.8
Montana	365	100.0	12.0	13.6	3.0	11.3	13.9
Nebraska	804	100.0	11.1	12.2	2.8	12.0	15.5
Nevada	567	100.0	12.8	10.4	3.2	14.1	15.6
New Hampshire	578	100.0	14.5	15.4	3.7	13.1	15.3
New Jersey	3,098	100.0	16.5	15.3	3.5	12.9	17.3
New Mexico	613	100.0	12.1	16.3	4.5	12.4	14.5
New York	6,446	100.0	14.7	17.6	3.6	11.8	16.3
North Carolina	2,545	100.0	12.1	12.5	4.0	12.5	14.7
North Dakota	289	100.0	11.0	12.5	3.4	12.7	12.4
Ohio	4,618	100.0	11.9	14.3	3.5	11.2	15.0
Oklahoma	1,272	100.0	13.2	13.2	3.6	12.8	15.1
Oregon	1,354	100.0	14.1	14.1	3.0	12.8	14.9
Pennsylvania	5,114	100.0	11.6	13.7	3.7	12.5	16.4
Rhode Island	459	100.0	12.9	15.7	4.0	11.0	17.0
South Carolina	1,211	100.0	12.8	14.0	4.5	13.6	15.1
South Dakota	334	100.0	9.7	12.6	2.9	11.7	13.0
Tennessee	1,962	100.0	11.4	11.5	3.9	12.7	14.2
Texas	7,041	100.0	13.0	13.9	3.8	12.4	16.6
Utah	751	100.0	12.4	13.9	4.0	12.7	17.4
Vermont	298	100.0	13.7	16.1	3.8	9.4	15.0
Virginia	2,477	100.0	16.8	17.7	3.5	10.8	15.0
Washington	2,209	100.0	13.3	17.1	3.6	13.2	16.1
West Virginia	654	100.0	8.7	11.7	4.2	12.6	14.0
Wisconsin	2,409	100.0	9.7	12.8	3.1	10.4	14.8
Wyoming	222	100.0	11.0	13.2	2.8	11.4	14.3
Black							
Alabama	397	100.0	4.6	7.2	2.1	8.9	9.3
Alaska	9	100.0	12.7	12.9	2.0	10.0	26.0
Arkansas	128	100.0	5.2	7.0	1.6	9.9	10.0
California	708	100.0	10.0	13.3	4.8	10.6	22.9
Colorado	71	100.0	6.5	11.6	1.5	7.7	26.4

[Continued]

★ 676 ★

Employed Persons in States: Percent Distribution and Occupation: Annual Averages, 1992 - I

[Continued]

Population group and State	Total employed		Managerial and professional specialty		Technical, sales and administrative support		
	Number (in thousands)	Percent	Executive, administrative, and managerial	Professional specialty	Technicians and related support	Sales	Administrative support, including clerical
Connecticut	141	100.0	6.2	5.2	3.1	6.1	18.7
Delaware	58	100.0	6.9	9.2	.9	6.0	20.7
District of Columbia	134	100.0	10.5	9.9	4.6	7.0	28.3
Florida	847	100.0	6.2	8.3	3.2	6.1	16.5
Georgia	881	100.0	6.7	8.4	4.2	6.6	14.9
Illinois	647	100.0	9.6	12.8	3.6	9.7	20.4
Indiana	175	100.0	1.0	4.7	2.9	5.2	16.5
Kansas	69	100.0	5.9	9.0	6.9	6.2	17.7
Kentucky	104	100.0	6.4	8.5	4.8	6.2	10.2
Louisiana	363	100.0	6.4	9.6	3.2	7.8	12.3
Maryland	652	100.0	10.6	12.4	4.3	6.3	20.3
Massachusetts	96	100.0	7.5	8.9	6.4	8.1	18.7
Michigan	424	100.0	7.0	9.7	2.7	7.7	18.3
Mississippi	329	100.0	4.7	8.4	2.2	6.9	9.6
Missouri	191	100.0	5.9	8.7	2.6	8.3	19.6
Nebraska	18	100.0	3.1	5.5	4.6	10.4	29.7
Nevada	33	100.0	5.0	8.9	1.9	10.4	16.9
New Jersey	424	100.0	7.8	12.0	3.8	5.9	20.3
New York	1,021	100.0	9.4	11.9	3.3	6.1	21.5
North Carolina	687	100.0	4.5	7.8	2.6	5.6	11.9
Ohio	423	100.0	9.1	11.8	4.9	6.6	20.5
Oklahoma	79	100.0	5.5	4.9	1.5	10.6	14.8
Pennsylvania	361	100.0	9.1	10.0	3.7	6.0	23.2
Rhode Island	12	100.0	6.1	8.9	.8	5.5	19.5
South Carolina	431	100.0	4.0	5.8	2.8	5.7	9.4
Tennessee	297	100.0	4.9	8.2	1.9	8.0	16.4
Texas	839	100.0	7.0	8.8	3.9	8.4	15.9
Virginia	578	100.0	6.0	8.8	2.3	6.5	16.9
Washington	67	100.0	6.4	12.4	7.6	11.0	13.1
West Virginia	23	100.0	5.9	7.1	1.6	13.0	18.1
Wisconsin	76	100.0	8.5	8.7	3.9	5.7	17.8
Hispanic origin							
Arizona	254	100.0	6.4	5.7	3.0	10.9	16.6
California	3,216	100.0	6.4	4.9	1.6	7.6	13.4
Colorado	160	100.0	7.9	8.9	3.4	7.6	16.6
Connecticut	73	100.0	8.4	6.6	1.9	7.3	19.7
District of Columbia	15	100.0	8.3	14.6	1.3	4.5	4.6

[Continued]

★ 676 ★

Employed Persons in States: Percent Distribution and Occupation: Annual Averages, 1992 - I
[Continued]

Population group and State	Total employed		Managerial and professional specialty		Technical, sales and administrative support		
	Number (in thousands)	Percent	Executive, administrative, and managerial	Professional specialty	Technicians and related support	Sales	Administrative support, including clerical
Florida	816	100.0	10.9	8.8	2.2	12.1	14.8
Hawaii	13	100.0	8.7	8.8	5.1	2.6	16.9
Idaho	25	100.0	3.5	6.6	5.3	4.0	7.2
Illinois	412	100.0	5.7	5.6	1.3	6.5	11.7
Maryland	69	100.0	13.6	11.5	4.3	9.5	11.8
Massachusetts	61	100.0	6.2	14.1	2.6	6.8	12.0
Michigan	65	100.0	5.2	12.6	1.1	8.4	10.2
Nevada	61	100.0	4.1	3.4	1.9	9.6	9.7
New Jersey	322	100.0	6.1	5.1	2.2	9.0	14.0
New Mexico	195	100.0	8.9	7.8	3.2	8.9	16.6
New York	689	100.0	7.0	8.6	2.1	7.3	15.7
North Carolina	43	100.0	6.4	5.8	3.6	5.4	4.7
Ohio	35	100.0	12.9	21.0	2.1	9.3	10.7
Oklahoma	43	100.0	2.4	4.8	.3	4.6	7.0
Oregon	56	100.0	4.8	3.5	1.4	12.8	7.0
Pennsylvania	56	100.0	6.5	9.3	.4	8.7	18.4
Rhode Island	16	100.0	2.0	6.9	[1]	5.9	10.6
Texas	1,792	100.0	6.2	6.2	2.8	10.4	15.0
Utah	27	100.0	4.0	9.0	3.4	9.0	12.1
Virginia	78	100.0	9.4	7.3	1.0	4.4	9.7
Wyoming	8	100.0	8.3	1.3	1.2	8.0	13.5

Source: "States: Percent Distribution of Employed Persons by Sex, Race, Hispanic Origin, and Occupation, 1992 Annual Averages," U.S. Department of Labor, *Geographical Profile of Employment and Unemployment, 1992*, pp. 62-64. *Notes:* Data for demographic groups are not shown when they do not meet BLS publication standards of reliability for the particular area based on the sample in that area. Items may not add to totals or compute to displayed percentages because of rounding. Detail for Hispanic-origin groups will not add to totals because data for the "other races" group are not presented and Hispanics are included in both the white and black population groups. 1. Less than 500 persons employed or less than 0.05 percent of total employed.

★ 677 ★

Labor Force

Employed Persons in States: Percent Distribution and Occupation: Annual Averages, 1992 - II

In thousands.

| Population group and State | Service occupations | Precision production, craft, and repair | Operators, fabricators, and laborers | | | Farming forestry, and fishing |
			Machine operators, assemblers, and inspectors	Transportation and material moving	Handlers, equipment cleaners, helpers, and laborers	
White						
Alabama	10.0	14.4	9.8	5.5	4.3	3.2
Alaska	13.0	10.5	1.7	4.0	3.7	2.3
Arizona	12.8	11.7	3.0	3.1	3.0	3.1
Arkansas	11.2	13.6	8.5	5.7	4.4	5.1
California	12.5	11.4	5.3	3.8	3.1	3.6
Colorado	12.7	10.6	3.9	3.4	3.4	2.1
Connecticut	11.7	10.9	5.4	3.0	2.7	1.3
Delaware	9.6	10.5	4.7	4.3	2.9	2.3
District of Columbia	8.5	2.3	.2	.4	.1	1
Florida	13.7	11.3	3.5	3.3	3.3	3.2
Georgia	9.3	12.3	5.5	3.8	3.5	3.3
Hawaii	15.2	11.8	1.6	1.9	2.1	3.1
Idaho	13.1	12.9	6.4	5.5	4.6	6.4
Illinois	12.1	11.2	6.2	3.9	3.7	2.2
Indiana	12.5	15.0	11.4	4.1	3.9	2.7
Iowa	14.4	11.3	6.8	3.9	3.6	8.5
Kansas	13.0	10.1	6.1	3.9	3.3	6.3
Kentucky	12.9	13.4	7.1	6.0	4.3	4.6
Louisiana	11.4	12.1	3.8	4.3	2.8	3.1
Maine	14.0	12.6	7.3	4.0	4.6	3.7
Maryland	10.7	11.6	2.4	3.0	3.0	1.5
Massachusetts	12.5	9.4	4.8	3.0	3.0	1.1
Michigan	13.1	12.2	8.1	3.9	4.2	2.5
Minnesota	13.5	10.3	6.2	3.4	4.1	6.2
Mississippi	9.2	15.6	7.4	4.8	3.4	3.4
Missouri	12.9	12.0	7.2	4.6	3.9	3.9
Montana	15.4	10.3	2.5	5.0	3.5	9.5
Nebraska	13.3	10.8	5.0	4.4	4.3	8.5
Nevada	21.2	10.8	2.4	3.7	4.2	1.5
New Hampshire	11.8	11.0	7.1	4.0	2.6	1.5
New Jersey	11.3	10.0	4.7	3.9	3.3	1.2
New Mexico	13.1	12.6	3.1	4.9	3.5	2.9
New York	13.4	10.1	4.2	3.8	3.1	1.4
North Carolina	9.8	13.6	10.1	3.8	3.9	3.0
North Dakota	15.4	9.0	2.6	4.4	3.6	12.8
Ohio	12.5	11.5	9.0	4.4	4.4	2.2
Oklahoma	12.0	12.3	4.8	4.6	3.5	5.2

[Continued]

★ 677 ★

Employed Persons in States: Percent Distribution and Occupation: Annual Averages, 1992 - II

[Continued]

Population group and State	Service occupations	Precision production, craft, and repair	Operators, fabricators, and laborers			Farming forestry, and fishing
			Machine operators, assemblers, and inspectors	Transportation and material moving	Handlers, equipment cleaners, helpers, and laborers	
Oregon	13.5	10.2	4.8	4.2	4.1	4.2
Pennsylvania	12.8	11.6	6.8	4.5	4.3	2.0
Rhode Island	12.9	10.5	7.7	3.8	3.2	1.4
South Carolina	8.4	14.0	8.4	3.7	3.4	2.2
South Dakota	14.2	10.5	5.0	4.5	3.6	12.2
Tennessee	10.2	12.7	11.3	4.1	4.6	3.3
Texas	13.0	11.5	4.6	4.3	4.0	2.9
Utah	12.5	11.6	5.4	3.9	3.7	2.5
Vermont	14.1	11.6	5.6	3.3	3.1	4.2
Virginia	10.2	12.3	4.6	3.8	3.0	2.3
Washington	11.9	10.8	3.9	3.5	3.6	3.0
West Virginia	14.3	12.9	7.0	6.8	5.4	2.3
Wisconsin	13.9	12.6	9.1	4.0	4.8	4.7
Wyoming	14.6	12.9	2.4	6.3	3.1	8.1
Black						
Alabama	26.5	8.8	17.6	5.9	7.3	1.8
Alaska	19.5	10.1	1.8	.7	4.2	.1
Arkansas	21.6	11.5	16.8	6.4	7.3	2.7
California	17.3	7.5	2.8	5.4	4.8	.6
Colorado	18.5	8.3				
Connecticut	28.2	9.0	10.9	6.9	5.7	.1
Delaware	26.3	6.0	11.1	6.0	6.5	.3
District of Columbia	24.0	5.1	1.5	5.1	3.3	.8
Florida	27.6	8.5	6.7	7.2	6.2	3.5
Georgia	23.0	9.1	12.0	6.6	6.1	2.3
Illinois	20.0	7.3	7.3	4.3	4.4	.5
Indiana	34.9	9.3	7.5	7.0	8.4	2.4
Kansas	23.0	11.8	8.9	5.3	5.2	.1
Kentucky	29.6	7.5	13.9	4.2	7.8	.9
Louisiana	26.8	8.7	4.6	9.6	6.7	4.3
Maryland	20.5	8.9	2.5	6.5	5.6	2.1
Massachusetts	26.6	6.2	7.3	4.6	4.9	.9
Michigan	21.4	8.8	14.9	3.1	5.2	1.1
Mississippi	19.5	10.9	18.5	7.4	6.9	5.0
Missouri	29.6	4.4	8.7	4.3	4.5	3.4
Nebraska	24.2	4.3	5.8	8.3	4.1	[1]
Nevada	34.6	6.6	3.6	7.1	4.5	.3

[Continued]

★ 677 ★

Employed Persons in States: Percent Distribution and Occupation: Annual Averages, 1992 - II

[Continued]

Population group and State	Service occupations	Precision production, craft, and repair	Operators, fabricators, and laborers			Farming forestry, and fishing
			Machine operators, assemblers, and inspectors	Transportation and material moving	Handlers, equipment cleaners, helpers, and laborers	
New Jersey	23.8	6.5	8.9	5.8	4.7	.6
New York	26.0	7.3	4.5	5.6	3.9	.4
North Carolina	22.4	9.8	18.5	5.4	8.1	3.4
Ohio	19.9	5.8	10.0	5.3	5.5	.7
Oklahoma	34.4	8.1	8.3	5.5	5.2	1.2
Pennsylvania	22.6	9.0	6.4	3.9	5.6	.5
Rhode Island	31.0	10.2	7.7	1.6	4.8	3.7
South Carolina	20.9	11.2	19.9	6.5	9.3	4.5
Tennessee	26.3	6.2	11.8	5.6	8.6	2.0
Texas	25.3	8.4	6.5	8.2	5.6	2.0
Virginia	20.8	10.1	13.7	7.7	5.0	2.3
Washington	26.0	9.7	3.1	3.5	7.1	[1]
West Virginia	27.0	8.6	4.5	4.6	9.3	.2
Wisconsin	27.8	4.4	10.9	5.3	6.5	.3
Hispanic origin						
Arizona	18.6	14.3	5.3	3.9	6.3	9.1
California	19.6	13.9	13.1	5.5	5.3	8.6
Colorado	18.8	11.6	8.6	6.6	7.3	2.7
Connecticut	20.0	13.3	16.8	2.2	2.3	1.5
District of Columbia	84.3	9.5	.7	1.4	.8	.1
Florida	15.3	12.4	7.1	3.7	5.1	7.5
Hawaii	24.1	15.1	3.7	3.7	.5	10.8
Idaho	8.6	12.9	16.7	3.4	10.5	21.2
Illinois	22.1	13.9	19.9	3.1	8.0	2.3
Maryland	24.6	12.4	.9	3.3	6.4	1.6
Massachusetts	23.3	10.0	16.4	3.1	4.5	.9
Michigan	18.0	15.1	16.4	2.9	8.2	1.9
Nevada	43.3	11.2	6.1	2.1	5.2	3.5
New Jersey	20.3	11.2	18.3	4.7	7.6	1.6
New Mexico	19.6	15.4	4.4	5.8	6.0	3.5
New York	27.2	9.9	10.7	4.4	5.3	1.8
North Carolina	14.4	24.3	15.2	1.6	6.7	11.9
Ohio	15.0	8.9	9.5	2.6	4.4	3.6
Oklahoma	21.9	19.7	12.1	6.7	13.9	6.6
Oregon	18.1	6.5	8.0	4.5	11.0	22.3
Pennsylvania	23.1	9.1	13.0	2.8	6.1	2.5
Rhode Island	6.9	10.7	47.2	2.1	7.8	[1]

[Continued]

★ 677 ★

Employed Persons in States: Percent Distribution and Occupation: Annual Averages, 1992 - II
[Continued]

| Population group and State | Service occupations | Precision production, craft, and repair | Operators, fabricators, and laborers | | | Farming forestry, and fishing |
			Machine operators, assemblers, and inspectors	Transportation and material moving	Handlers, equipment cleaners, helpers, and laborers	
Texas	19.7	14.6	8.7	4.5	8.0	3.8
Utah	24.6	11.0	7.7	5.1	4.2	9.8
Virginia	44.9	11.7	1.5	4.3	3.7	2.2
Wyoming	21.0	12.8	2.3	11.1	5.9	14.7

Source: "States: Percent Distribution of Employed Persons by Sex, Race, Hispanic Origin, and Occupation, 1992 Annual Averages," U.S. Department of Labor, *Geographical Profile of Employment and Unemployment, 1992*, pp. 62-64. *Notes:* Data for demographic groups are not shown when they do not meet BLS publication standards of reliability for the particular area based on the sample in that area. Items may not add to totals or compute to displayed percentages because of rounding. Detail for Hispanic-origin groups will not add to totals because data for the "other races" group are not presented and Hispanics are included in both the white and black population groups. 1. Less than 500 persons employed or less than 0.05 percent of total employed.

★ 678 ★

Labor Force

Employed Persons: Percent Distribution in Census Regions: Annual Averages, 1992 - I

| Population group and occupation | Northeast | | | Midwest | | |
	Total	New England	Middle Atlantic	Total	East North Central	West North Central
White						
Total (in thousands)	20,796	6,138	14,658	26,634	18,038	8,597
Percent	100.0	100.0	100.0	100.0	100.0	100.0
Managerial and professional speciality	30.4	31.8	29.8	24.9	25.3	24.0
Executive, administrative, and managerial	14.3	15.0	14.0	11.7	11.9	11.2
Professional speciality	16.1	16.8	15.8	13.2	13.4	12.8
Engineers	1.7	2.2	1.5	1.3	1.5	1.0
Mathematical and computer scientists	.9	1.2	.8	.6	.8	.4
Health diagnosing occupations	.9	.8	1.0	.7	.7	.7
Health assessment and treating occupations	2.5	2.7	2.4	2.2	2.2	2.1
Teachers, except college and university	4.0	3.9	4.1	3.5	3.5	3.7
Technical, sales, and administrative support	32.3	31.7	32.5	30.6	30.5	30.8
Technicians and related support	3.7	3.8	3.6	3.3	3.3	3.2
Health technologists and technicians	1.3	1.2	1.3	1.2	1.2	1.3
Engineering and science technicians	1.0	1.0	1.0	.9	1.0	.8
Sales occupations	12.2	12.0	12.3	11.8	11.7	12.0
Supervisors and proprietors	3.5	3.5	3.5	3.3	3.2	3.5
Sales representatives, finance and business services	2.3	2.2	2.3	1.7	1.7	1.6
Sales representatives, commodities, except retail	1.5	1.4	1.6	1.4	1.5	1.4
Sales workers, retail and personal service	4.8	4.8	4.8	5.3	5.3	5.4

[Continued]

★ 678 ★

Employed Persons: Percent Distribution in Census Regions: Annual Averages, 1992 - I
[Continued]

Population group and occupation	Northeast			Midwest		
	Total	New England	Middle Atlantic	Total	East North Central	West North Central
Administrative support, including clerical	16.4	15.9	16.6	15.5	15.5	15.5
Computer equipment operators	.5	.5	.5	.6	.6	.6
Secretaries, stenographers, and typists	4.2	3.7	4.4	3.7	3.8	3.5
Financial records processing	2.1	2.3	2.1	2.1	2.0	2.2
Mail and message distributing	.8	.9	.8	.7	.6	.7
Service occupations	12.7	12.5	12.7	13.0	12.7	13.5
Private household	.5	.5	.6	.6	.6	.6
Protective service	1.9	1.4	2.1	1.3	1.4	1.2
Service, except private household and protective	10.2	10.6	10.0	11.0	10.7	11.7
Food service	4.4	4.4	4.3	4.7	4.7	4.7
Health service	1.7	2.1	1.6	1.7	1.6	2.0
Cleaning and building service	2.2	2.1	2.3	2.4	2.4	2.5
Personal service	1.9	1.9	1.9	2.2	2.0	2.5
Precision production, craft, and repair	10.6	10.4	10.6	11.8	12.2	10.9
Mechanics and repairers	3.5	3.1	3.7	3.9	4.0	3.7
Construction trades	3.9	3.9	3.9	4.0	4.0	4.0
Operators, fabricators, and laborers	12.6	12.0	12.8	15.9	16.7	14.2
Machine operators, assemblers, and inspectors	5.4	5.7	5.2	7.7	8.4	6.3
Transportation and material moving occupations	3.8	3.3	4.1	4.1	4.1	4.0
Motor vehicle operators	3.0	2.6	3.2	3.0	3.0	2.9
Handlers, equipment cleaners, helpers, and laborers	3.4	3.1	3.5	4.1	4.2	3.9
Construction laborers	.5	.4	.5	.6	.6	.5
Farming, forestry, and fishing	1.6	1.6	1.6	3.9	2.7	6.6
Farm operators and managers	.4	.3	.5	2.2	1.2	4.2
Black						
Total (in thousands)	2,061	255	1,806	2,076	1,745	332
Percent	100.0	100.0	100.0	100.0	100.0	100.0
Managerial and professional speciality	19.7	13.9	20.5	18.1	18.7	15.0
Executive, administrative, and managerial	8.7	6.7	9.0	7.5	7.9	5.4
Professional speciality	11.0	7.2	11.5	10.6	10.8	9.5
Engineers	.6	.6	.6	.7	.6	.8
Mathematical and computer scientists	.6	.7	.6	.5	.6	.1
Health diagnosing occupations	.3	.3	.3	.3	.3	.2
Health assessment and treating occupations	2.8	1.3	3.0	1.9	1.8	2.5
Teachers, except college and university	2.8	1.8	2.9	3.2	3.3	2.9
Technical, sales, and administrative support	31.0	29.7	31.1	31.0	30.9	31.5
Technicians and related support	3.6	4.2	3.5	3.7	3.7	3.7
Health technologists and technicians	1.5	1.2	1.6	1.7	1.7	2.1

[Continued]

★ 678 ★

Employed Persons: Percent Distribution in Census Regions: Annual Averages, 1992 - I

[Continued]

Population group and occupation	Northeast			Midwest		
	Total	New England	Middle Atlantic	Total	East North Central	West North Central
Engineering and science technicians	.8	1.4	.7	.6	.7	.5
Sales occupations	6.2	6.9	6.1	7.8	7.9	7.8
Supervisors and proprietors	1.0	1.3	1.0	1.3	1.2	1.8
Sales representatives, finance and business services	1.0	1.1	1.0	.8	.8	.9
Sales representatives, commodities, except retail	.5	.4	.5	.5	.5	.4
Sales workers, retail and personal service	3.6	4.1	3.6	5.1	5.2	4.7
Administrative support, including clerical	21.2	18.6	21.6	19.5	19.4	20.0
Computer equipment operators	.9	1.1	.9	.9	.9	.8
Secretaries, stenographers, and typists	4.9	2.4	5.2	3.8	3.9	3.2
Financial records processing	1.3	1.6	1.3	1.1	1.1	.9
Mail and message distributing	2.0	.9	2.2	1.8	1.7	2.2
Service occupations	25.1	27.3	24.8	22.9	22.1	26.6
Private household	1.0	.7	1.0	.7	.7	.4
Protective service	4.5	3.2	4.7	3.2	3.3	2.8
Service, except private household and protective	19.6	23.4	19.1	19.0	18.1	23.5
Food service	4.2	4.8	4.1	5.8	5.4	7.6
Health service	8.3	10.4	8.0	4.9	4.7	5.6
Cleaning and building service	4.8	5.4	4.7	6.0	5.7	7.5
Personal service	2.3	2.9	2.2	2.4	2.3	2.8
Precision production, craft, and repair	7.5	7.8	7.5	7.3	7.4	6.6
Mechanics and repairers	2.7	2.2	2.8	3.2	3.2	3.6
Construction trades	2.7	1.8	2.8	1.7	1.9	1.0
Operators, fabricators, and laborers	16.3	20.7	15.6	19.7	19.9	18.3
Machine operators, assemblers, and inspectors	6.4	9.7	5.9	9.8	10.0	8.8
Transportation and material moving occupations	5.3	5.6	5.3	4.6	4.6	4.6
Motor vehicle operators	4.4	5.0	4.4	3.7	3.6	4.2
Handlers, equipment cleaners, helpers, and laborers	4.5	5.3	4.4	5.3	5.4	4.9
Construction laborers	.6	.8	.6	.4	.4	.2
Farming, forestry, and fishing	.5	.6	.5	1.0	.9	2.0
Farm operators and managers	1	.1	1	1	1	1
Hispanic origin						
Total (in thousands)	1,222	156	1,067	665	563	102
Percent	100.0	100.0	100.0	100.0	100.0	100.0
Managerial and professional speciality	14.6	16.8	14.2	14.1	13.3	18.5
Executive, administrative, and managerial	6.7	6.8	6.7	6.3	6.0	8.0
Professional speciality	7.9	10.0	7.6	7.8	7.3	10.5
Engineers	.6	.6	.6	.5	.6	.1
Mathematical and computer scientists	.5	.1	.5	.7	.5	1.9
Health diagnosing occupations	.4	1.3	.3	1.0	.8	2.1

[Continued]

★ 678 ★

Employed Persons: Percent Distribution in Census Regions: Annual Averages, 1992 - I

[Continued]

Population group and occupation	Northeast			Midwest		
	Total	New England	Middle Atlantic	Total	East North Central	West North Central
Health assessment and treating occupations	1.3	1.5	1.3	1.1	.9	2.1
Teachers, except college and university	1.6	2.6	1.4	1.2	1.3	.6
Technical, sales, and administrative support	25.2	24.7	25.3	21.3	19.7	30.0
Technicians and related support	2.1	2.1	2.1	1.8	1.3	4.3
Health technologists and technicians	1.0	.3	1.1	.4	.4	.8
Engineering and science technicians	.4	.6	.4	.5	.4	.9
Sales occupations	7.8	7.5	7.9	7.2	6.9	9.1
Supervisors and proprietors	2.0	2.5	1.9	2.4	2.3	2.9
Sales representatives, finance and business services	1.0	.5	1.0	.6	.6	.6
Sales representatives, commodities, except retail	.3	1	.4	.4	.4	.6
Sales workers, retail and personal service	4.5	4.5	4.6	3.8	3.6	5.0
Administrative support, including clerical	15.3	15.1	15.3	12.3	11.5	16.6
Computer equipment operators	.5	.6	.5	.7	.7	1.0
Secretaries, stenographers, and typists	3.3	4.3	3.2	2.4	2.4	2.3
Financial records processing	1.6	2.3	1.5	.9	.6	2.5
Mail and message distributing	1.2	.3	1.3	.7	.5	1.8
Service occupations	24.3	20.0	24.9	20.1	20.6	17.2
Private household	1.7	.3	1.9	.9	.9	.6
Protective service	3.0	.1	3.4	1.3	1.4	.5
Service, except private household and protective	19.6	19.6	19.6	17.9	18.2	16.1
Food service	6.6	8.2	6.4	8.8	9.2	6.8
Health service	3.5	3.7	3.5	1.2	1.2	1.6
Cleaning and building service	7.0	6.6	7.1	5.9	6.1	4.9
Personal service	2.4	1.1	2.6	2.0	1.8	2.9
Precision production, craft, and repair	10.4	11.5	10.2	13.2	13.6	11.0
Mechanics and repairers	3.6	3.6	3.6	4.0	4.3	2.1
Construction trades	2.8	2.3	2.9	2.7	2.7	2.5
Operators, fabricators, and laborers	23.9	25.7	23.6	28.9	30.4	20.6
Machine operators, assemblers, and inspectors	13.9	19.4	13.1	18.2	19.8	9.8
Transportation and material moving occupations	4.2	2.5	4.4	2.7	3.0	1.4
Motor vehicle operators	3.4	2.2	3.6	2.1	2.3	1.1
Handlers, equipment cleaners, helpers, and laborers	5.8	3.9	6.0	7.9	7.7	9.4
Construction laborers	.6	.4	.7	1.0	1.0	1.2
Farming, forestry, and fishing	1.7	1.3	1.8	2.5	2.4	2.7
Farm operators and managers	1	2	1	1	1	.1

Source: "Census Regions and Divisions, Percent Distribution of Employed Persons by Occupation, Sex, Race, and Hispanic Origin, 1992 Annual Averages," U.S. Department of Labor, *Geographical Profile of Employment and Unemployment, 1992,* pp. 16-18. *Notes:* Items may not add to totals or compute to displayed percentages because of rounding. Detail for race and Hispanic-origin groups will not add to totals because data for the "other races" group are not presented and Hispanics are included in both the white and black population groups. 1. Less than 500 persons employed or less than 0.05 percent of total employed. 2. Data are not shown when the labor force base does not meet BLS publication standards of reliability for the particular area, based on the sample in that area.

★ 679 ★
Labor Force

Employed Persons: Percent Distribution in Census Regions: Annual Averages, 1992 - II

Population group and occupation	South				West		
	Total	South Atlantic	East South Central	West South Central	Total	Mountain	Pacific
White							
Total (in thousands)	32,412	16,163	5,621	10,628	21,637	6,036	15,601
Percent	100.0	100.0	100.0	100.0	100.0	100.0	100.0
Managerial and professional speciality	27.1	28.9	23.0	26.7	28.7	27.6	29.1
Executive, administrative, and managerial	13.3	14.2	11.2	12.9	14.1	13.2	14.4
Professional speciality	13.9	14.6	11.8	13.8	14.6	14.4	14.7
Engineers	1.4	1.5	1.2	1.3	1.8	1.7	1.8
Mathematical and computer scientists	.8	1.0	.4	.6	.9	.8	.9
Health diagnosing occupations	.8	.8	.6	.7	.8	.8	.8
Health assessment and treating occupations	2.1	2.2	2.5	1.9	1.7	1.9	1.7
Teachers, except college and university	3.9	3.9	2.9	4.4	3.3	3.5	3.2
Technical, sales, and administrative support	32.2	32.7	30.3	32.4	31.8	32.6	31.5
Technicians and related support	3.9	3.9	3.8	3.9	3.4	4.1	3.2
Health technologists and technicians	1.4	1.4	1.7	1.3	1.0	1.2	.9
Engineering and science technicians	1.1	1.1	1.0	1.1	1.0	1.3	.9
Sales occupations	12.9	13.2	12.5	12.5	12.6	12.5	12.6
Supervisors and proprietors	3.8	3.9	3.8	3.7	3.3	3.2	3.4
Sales representatives, finance and business services	2.0	2.2	1.7	1.9	2.4	2.2	2.4
Sales representatives, commodities, except retail	1.5	1.6	1.2	1.5	1.5	1.4	1.5
Sales workers, retail and personal service	5.4	5.4	5.6	5.3	5.3	5.6	5.1
Administrative support, including clerical	15.4	15.5	14.1	16.0	15.8	15.9	15.8
Computer equipment operators	.5	.5	.5	.5	.5	.5	.5
Secretaries, stenographers, and typists	3.9	3.9	3.5	4.2	3.1	3.3	3.1
Financial records processing	2.0	2.0	1.8	2.1	2.2	2.2	2.2
Mail and message distributing	.7	.7	.6	.6	.6	.7	.6
Service occupations	11.5	11.2	10.8	12.5	12.9	13.8	12.5
Private household	.7	.5	.7	.9	.9	.7	1.0
Protective service	1.7	1.6	1.6	1.9	1.7	1.6	1.7
Service, except private household and protective	9.2	9.0	8.5	9.8	10.3	11.5	9.9
Food service	4.0	4.1	3.5	4.1	4.7	5.1	4.5
Health service	1.2	1.1	1.2	1.3	1.1	1.1	1.1
Cleaning and building service	1.8	1.7	1.8	2.0	2.3	2.5	2.2
Personal service	2.2	2.1	2.0	2.3	2.3	2.7	2.1
Precision production, craft, and repair	12.3	12.2	13.7	11.8	11.3	11.5	11.2
Mechanics and repairers	4.4	4.4	4.9	4.2	3.7	4.0	3.6
Construction trades	4.7	5.0	4.7	4.3	4.5	4.4	4.5
Operators, fabricators, and laborers	13.7	12.4	18.6	13.1	11.8	11.2	12.0
Machine operators, assemblers, and inspectors	5.9	5.3	9.3	4.9	4.6	3.7	5.0
Transportation and material moving occupations	4.2	3.7	5.1	4.4	3.8	3.9	3.8
Motor vehicle operators	3.0	2.8	3.6	3.1	2.9	2.9	3.0
Handlers, equipment cleaners, helpers, and laborers	3.7	3.4	4.3	3.8	3.3	3.5	3.3
Construction laborers	.6	.5	.7	.6	.6	.6	.6
Farming, forestry, and fishing	3.1	2.7	3.6	3.4	3.5	3.4	3.6
Farm operators and managers	1.1	.8	1.6	1.4	.8	1.2	.7

[Continued]

★ 679 ★

Employed Persons: Percent Distribution in Census Regions: Annual Averages, 1992 - II
[Continued]

Population group and occupation	South				West		
	Total	South Atlantic	East South Central	West South Central	Total	Mountain	Pacific
Black							
Total (in thousands)	6,827	4,291	1,127	1,409	968	156	812
Percent	100.0	100.0	100.0	100.0	100.0	100.0	100.0
Managerial and professional speciality	14.9	15.3	12.8	15.2	21.8	16.7	22.7
Executive, administrative, and managerial	6.3	6.6	4.9	6.6	9.2	6.6	9.7
Professional speciality	8.6	8.7	8.0	8.7	12.5	10.1	13.0
Engineers	.4	.4	.5	.4	1.3	.7	1.4
Mathematical and computer scientists	.5	.7	.1	.2	.6	.2	.7
Health diagnosing occupations	.1	.2	1	.2	.2	1	.2
Health assessment and treating occupations	1.5	1.5	1.2	1.6	1.5	1.6	1.4
Teachers, except college and university	3.4	3.3	3.2	3.6	3.5	3.7	3.4
Technical, sales, and administrative support	24.9	25.3	21.5	26.2	36.8	33.5	37.5
Technicians and related support	3.2	3.3	2.3	3.3	4.5	2.7	4.9
Health technologists and technicians	1.5	1.6	1.4	1.2	1.9	.9	2.1
Engineering and science technicians	.7	.7	.5	.8	1.0	.5	1.1
Sales occupations	7.0	6.2	7.8	8.5	10.4	8.6	10.7
Supervisors and proprietors	1.3	1.3	1.2	1.3	1.9	1.1	2.1
Sales representatives, finance and business services	.7	.7	.6	.9	2.0	1.8	2.0
Sales representatives, commodities, except retail	.2	.2	.1	.3	.5	.6	.4
Sales workers, retail and personal service	4.7	4.0	5.9	5.9	6.0	5.1	6.2
Administrative support, including clerical	14.7	15.8	11.3	14.3	22.0	22.3	21.9
Computer equipment operators	.6	.6	.5	.5	.9	.1	1.1
Secretaries, stenographers, and typists	2.7	3.0	1.7	2.4	3.9	3.1	4.1
Financial records processing	.9	.8	.7	1.1	1.7	1.1	1.8
Mail and message distributing	1.1	1.1	.7	1.4	1.7	2.2	1.7
Service occupations	23.9	23.0	24.7	25.9	19.2	22.9	18.5
Private household	1.8	1.8	1.9	1.7	.6	1.5	.4
Protective service	2.5	2.3	2.0	3.6	4.6	2.5	5.0
Service, except private household and protective	19.5	18.9	20.8	20.6	14.0	18.9	13.1
Food service	6.8	6.5	7.8	6.7	3.9	8.7	3.0
Health service	4.1	3.7	4.6	4.9	3.2	2.6	3.3
Cleaning and building service	6.0	5.9	6.4	6.0	4.4	4.4	4.5
Personal service	2.7	2.7	2.1	3.0	2.5	3.2	2.4
Precision production, craft, and repair	9.0	9.2	8.6	8.8	7.7	7.7	7.7
Mechanics and repairers	2.7	3.0	2.0	2.6	2.8	2.7	2.9
Construction trades	3.1	3.2	3.2	2.9	2.8	2.8	2.8
Operators, fabricators, and laborers	24.6	24.3	29.7	21.4	13.9	18.4	13.1
Machine operators, assemblers, and inspectors	11.1	11.2	16.0	7.1	3.3	5.4	2.9
Transportation and material moving occupations	6.9	6.6	6.1	8.3	5.2	5.9	5.0
Motor vehicle operators	5.4	5.1	4.3	7.1	4.7	5.3	4.6
Handlers, equipment cleaners, helpers, and laborers	6.6	6.5	7.6	6.0	5.4	7.1	5.1
Construction laborers	.8	.7	1.2	.7	.6	.7	.6
Farming, forestry, and fishing	2.8	2.8	2.7	2.6	.6	.8	.5
Farm operators and managers	.2	.1	.2	.1	2	2	2

[Continued]

★ 679 ★

Employed Persons: Percent Distribution in Census Regions: Annual Averages, 1992 - II
[Continued]

Population group and occupation	South				West		
	Total	South Atlantic	East South Central	West South Central	Total	Mountain	Pacific
Hispanic origin							
Total (in thousands)	2,998	1,092	26	1,879	4,086	736	3,351
Percent	100.0	100.0	100.0	100.0	100.0	100.0	100.0
Managerial and professional speciality	14.9	19.0	24.7	12.4	11.9	14.0	11.5
Executive, administrative, and managerial	7.8	10.6	11.7	6.1	6.5	7.0	6.4
Professional speciality	7.1	8.4	13.0	6.2	5.4	6.9	5.1
Engineers	.6	.7	2.4	.5	.5	.6	.5
Mathematical and computer scientists	.2	.2	.1	.2	.3	.3	.3
Health diagnosing occupations	.5	1.1	2.4	.2	.1	.2	.1
Health assessment and treating occupations	1.3	1.8	2.9	.9	.6	.9	.5
Teachers, except college and university	2.0	1.5	3.0	2.3	1.4	1.8	1.3
Technical, sales, and administrative support	27.4	26.3	30.2	27.9	23.5	27.7	22.6
Technicians and related support	2.7	2.2	6.6	2.8	1.9	3.1	1.6
Health technologists and technicians	1.1	1.1	2.1	1.0	.8	1.2	.7
Engineering and science technicians	.9	.4	3.7	1.1	.5	.8	.4
Sales occupations	10.4	10.7	10.6	10.2	8.0	9.2	7.7
Supervisors and proprietors	2.3	3.1	2.7	1.8	1.9	2.1	1.8
Sales representatives, finance and business services	1.0	1.3	2	.8	.9	.9	.9
Sales representatives, commodities, except retail	1.0	1.1	3.8	.9	.4	.4	.4
Sales workers, retail and personal service	6.1	5.0	4.1	6.7	4.8	5.8	4.6
Administrative support, including clerical	14.3	13.3	13.0	14.8	13.6	15.5	13.2
Computer equipment operators	.4	.3	.8	.5	.5	.6	.5
Secretaries, stenographers, and typists	3.1	2.6	3.7	3.4	2.4	2.3	2.4
Financial records processing	1.4	1.9	.9	1.2	1.4	1.2	1.5
Mail and message distributing	.6	.7	2	.6	.7	1.3	.6
Service occupations	19.2	18.6	10.7	19.7	19.9	20.9	19.6
Private household	1.6	1.7	3.1	1.5	2.4	1.1	2.6
Protective service	1.5	1.4	2	1.5	1.5	2.0	1.4
Service, except private household and protective	16.1	15.5	7.6	16.6	16.0	17.7	15.6
Food service	7.3	7.2	2.4	7.4	7.5	7.6	7.4
Health service	1.5	1.2	1.8	1.6	1.3	1.8	1.2
Cleaning and building service	5.2	5.0	3.4	5.4	5.1	6.0	4.9
Personal service	2.1	2.1	2	2.2	2.1	2.3	2.1
Precision production, craft, and repair	14.3	13.7	12.5	14.7	13.7	13.5	13.8
Mechanics and repairers	4.2	3.7	8.1	4.4	3.7	4.2	3.6
Construction trades	5.8	5.4	2.4	6.1	5.2	5.4	5.1
Operators, fabricators, and laborers	19.4	16.0	22.1	21.4	22.7	17.6	23.8
Machine operators, assemblers, and inspectors	8.1	6.8	14.4	8.7	11.8	6.3	13.0
Transportation and material moving occupations	4.1	3.5	2.1	4.5	5.4	4.9	5.5
Motor vehicle operators	3.1	3.2	2	3.2	4.1	3.5	4.3
Handlers, equipment cleaners, helpers, and laborers	7.2	5.7	5.6	8.2	5.5	6.4	5.3
Construction laborers	1.6	1.4	1.5	1.7	1.2	1.3	1.2

[Continued]

★ 679 ★

Employed Persons: Percent Distribution in Census Regions: Annual Averages, 1992 - II
[Continued]

Population group and occupation	South				West		
	Total	South Atlantic	East South Central	West South Central	Total	Mountain	Pacific
Farming, forestry, and fishing	4.9	6.4	2	4.0	8.3	6.3	8.8
Farm operators and managers	.2	.1	2	.3	.3	.3	.3

Source: "Census Regions and Divisions, Percent Distribution of Employed Persons by Occupation, Sex, Race, and Hispanic Origin, 1992 Annual Averages," U.S. Department of Labor, *Geographical Profile of Employment and Unemployment, 1992,* pp. 16-18. *Notes:* Items may not add to totals or compute to displayed percentages because of rounding. Detail for race and Hispanic-origin groups will not add to totals because data for the "other races" group are not presented and Hispanics are included in both the white and black population groups. 1. Less than 500 persons employed or less than 0.05 percent of total employed. 2. Data are not shown when the labor force base does not meet BLS publication standards of reliability for the particular area, based on the sample in that area.

★ 680 ★

Labor Force

Employed and Unemployed Full- and Part-time Workers, January 1994

Not seasonally adjusted. In thousands.

Age, sex, and race	January 1994									
	Employed[1]								Unemployed	
	Full-time workers				Part-time workers					
		At work								
	Total	35 hours or more	1 to 34 hours for economic or noneconomic reasons	Not at work	Total	Part time for economic reasons	Part time for noneconomic reasons	Not at work	Looking for full-time work	Looking for part-time work
Total										
Total, 16 years and over	96,087	82,527	10,199	3,361	23,814	3,640	18,536	1,637	7,766	1,725
16 to 19 years	1,363	1,116	238	29	4,124	276	3,657	190	530	767
16 to 17 years	104	77	23	5	2,030	36	1,917	77	106	489
18 to 19 years	1,279	1,040	216	24	2,093	240	1,740	113	424	277
20 years and over	94,704	81,411	9,960	3,332	19,690	3,364	14,879	1,447	7,236	959
20 to 24 years	8,424	7,180	1,005	240	3,685	711	2,704	270	1,388	297
25 years and over	86,279	74,231	8,956	3,093	16,005	2,653	12,175	1,177	5,849	662
25 to 54 years	75,732	65,447	7,765	2,520	11,771	2,319	8,674	778	5,248	490
55 years and over	10,547	8,784	1,191	573	4,234	334	3,501	399	600	172
Men, 16 years and over	56,318	48,946	5,412	1,960	8,116	1,642	5,905	569	4,755	771
16 to 19 years	794	649	120	26	1,961	155	1,740	66	348	444
20 years and over	55,524	48,297	5,292	1,935	6,155	1,487	4,164	503	4,407	326
20 to 24 years	4,769	4,056	594	119	1,516	296	1,117	103	874	136
25 years and over	50,755	44,241	4,699	1,816	4,638	1,191	3,047	400	3,533	191
25 to 54 years	44,396	38,897	4,055	1,444	2,830	1,028	1,595	207	3,151	126
55 years and over	6,359	5,344	644	372	1,808	163	1,452	192	383	65
Women, 16 years and over	39,769	33,582	4,787	1,400	15,698	1,998	12,631	1,069	3,012	955
16 to 19 years	589	467	119	3	2,163	121	1,917	125	182	322
20 years and over	39,180	33,114	4,668	1,398	13,535	1,877	10,714	944	2,829	632
20 to 24 years	3,656	3,124	411	121	2,169	415	1,587	167	514	161
25 years and over	35,524	29,990	4,257	1,277	11,367	1,462	9,128	777	2,315	471
25 to 54 years	31,336	26,550	3,710	1,076	8,940	1,291	7,079	571	2,098	364
55 years and over	4,188	3,440	547	201	2,426	171	2,049	207	218	107
White										
Men, 16 years and over	48,955	42,621	4,665	1,669	6,922	1,290	5,117	516	3,601	615
16 to 19 years	701	576	109	16	1,719	136	1,530	52	246	368

[Continued]

★ 680 ★

Employed and Unemployed Full- and Part-time Workers, January 1994

[Continued]

Age, sex, and race	January 1994									
	Employed[1]								Unemployed	
	Full-time workers				Part-time workers					
		At work								
	Total	35 hours or more	1 to 34 hours for economic or noneconomic reasons	Not at work	Total	Part time for economic reasons	Part time for noneconomic reasons	Not at work	Looking for full-time work	Looking for part-time work
20 years and over	48,254	42,045	4,556	1,653	5,204	1,153	3,587	463	3,355	246
20 to 24 years	4,126	3,540	490	96	1,262	238	925	100	634	97
25 years and over	44,128	38,505	4,066	1,557	3,941	915	2,662	364	2,721	150
25 to 54 years	38,449	33,717	3,499	1,233	2,260	772	1,313	176	2,394	90
55 years and over	5,679	4,788	567	325	1,681	143	1,350	188	327	60
Women, 16 years and over	33,072	27,924	3,990	1,157	13,678	1,561	11,161	956	2,146	761
16 to 19 years	514	418	92	3	1,828	103	1,624	102	126	261
20 years and over	32,558	27,506	3,898	1,154	11,850	1,458	9,537	854	2,020	500
20 to 24 years	3,089	2,655	343	91	1,808	334	1,327	146	360	111
25 years and over	29,469	24,850	3,555	1,064	10,042	1,124	8,210	708	1,660	390
25 to 54 years	25,868	21,893	3,087	889	7,853	982	6,361	509	1,487	288
55 years and over	3,601	2,958	468	175	2,189	142	1,849	199	173	101
Black										
Men, 16 years and over	5,062	4,304	554	203	791	256	487	47	914	112
16 to 19 years	67	54	7	6	164	18	136	11	89	58
20 years and over	4,995	4,250	548	197	626	238	352	36	826	55
20 to 24 years	427	339	73	15	154	45	107	1	220	27
25 years and over	4,568	3,911	474	182	472	193	245	35	606	28
25 to 54 years	4,103	3,532	419	151	370	177	162	31	578	25
55 years and over	465	379	55	31	102	16	82	4	28	3
Women, 16 years and over	4,938	4,116	631	190	1,483	360	1,041	83	739	158
16 to 19 years	58	36	22	-	267	14	231	22	47	50
20 years and over	4,880	4,080	610	190	1,217	346	810	60	692	108
20 to 24 years	410	331	57	22	264	64	183	17	143	46
25 years and over	4,470	3,749	552	168	953	282	627	43	549	61
25 to 54 years	4,049	3,416	487	147	759	258	461	40	517	56
55 years and over	421	334	65	22	194	24	166	4	31	6

Source: "Employed and Unemployed Full- and Part-time Workers by Sex, Age, and Race." *Employment and Earnings* 41 (January 1994), p. 57. *Notes:* Data for 1994 are not directly comparable with data for 1993 and earlier years. 1. Employed persons are classified as full- or part-time workers based on their usual weekly hours at all jobs regardless of the number of hours they are at work during the reference week. Persons absent from work are also classified according to their usual status. 2. Includes some workers at work 35 hours or more, classified by their reason for working part time.

★ 681 ★

Labor Force

Employed and Unemployed Full- and Part-time Workers: Annual Averages, 1993

In thousands.

Sex, age, and race	1993							
	Employed						Unemployed	
	Full time			Part time				
	Total	Full-time schedules	Part time for economic reasons, usually work full time	Total	Voluntary	Part time for economic reasons, usually work part time	Looking for full-time work	Looking for part-time work
Total								
Total, 16 years and over	98,439	96,443	1,996	20,868	16,515	4,352	7,146	1,588
16 to 19 years	1,786	1,683	103	3,744	3,220	524	589	707
16 to 17 years	284	262	22	1,858	1,695	163	139	441
18 to 19 years	1,502	1,421	81	1,886	1,526	360	450	266
20 years and over	96,653	94,760	1,893	17,124	13,295	3,829	6,557	881
20 to 24 years	8,923	8,663	260	3,214	2,441	773	1,180	241
25 years and over	87,730	86,097	1,633	13,910	10,854	3,056	5,377	640
25 to 54 years	76,835	75,412	1,423	10,086	7,501	2,585	4,852	498
55 years and over	10,896	10,685	210	3,824	3,353	471	525	142
Men, 16 years and over	57,643	56,435	1,208	7,057	5,226	1,832	4,277	655
16 to 19 years	1,042	982	60	1,794	1,540	254	343	385
20 years and over	56,602	55,453	1,148	5,263	3,685	1,578	3,934	270
20 to 24 years	4,968	4,812	156	1,388	1,023	365	704	103
25 years and over	51,634	50,641	992	3,875	2,662	1,212	3,230	166
25 to 54 years	44,946	44,084	861	2,293	1,264	1,029	2,884	102
55 years and over	6,688	6,557	131	1,582	1,398	183	346	64
Women, 16 years and over	40,796	40,008	788	13,810	11,290	2,521	2,869	933
16 to 19 years	744	701	43	1,950	1,680	270	246	322
20 years and over	40,051	39,307	744	11,861	9,610	2,251	2,623	611
20 to 24 years	3,955	3,851	104	1,825	1,418	407	476	137
25 years and over	36,096	35,456	641	10,035	8,192	1,844	2,147	474
25 to 54 years	31,889	31,328	561	7,793	6,236	1,557	1,968	396
55 years and over	4,208	4,128	80	2,243	1,955	287	179	78
White								
Men, 16 years and over	50,381	49,370	1,011	6,016	4,575	1,441	3,254	499
16 to 19 years	925	874	51	1,575	1,365	210	244	291
20 years and over	49,456	48,496	960	4,441	3,210	1,231	3,009	208
20 to 24 years	4,289	4,159	130	1,159	870	289	494	79
25 years and over	45,167	44,337	830	3,282	2,340	942	2,515	129
25 to 54 years	39,199	38,484	715	1,844	1,047	798	2,238	78
55 years and over	5,968	5,853	115	1,438	1,293	145	277	51
Women, 16 years and over	34,149	33,518	631	12,266	10,192	2,074	2,060	733
16 to 19 years	656	619	37	1,731	1,501	230	167	241
20 years and over	33,493	32,899	594	10,535	8,691	1,844	1,893	492
20 to 24 years	3,370	3,288	82	1,551	1,214	337	318	99

[Continued]

★ 681 ★

Employed and Unemployed Full- and Part-time Workers: Annual Averages, 1993

[Continued]

Sex, age, and race	1993							
	Employed						Unemployed	
	Full time			Part time				
	Total	Full-time schedules	Part time for economic reasons, usually work full time	Total	Voluntary	Part time for economic reasons, usually work part time	Looking for full-time work	Looking for part-time work
25 years and over	30,123	29,611	512	8,984	7,477	1,507	1,575	393
25 to 54 years	26,507	26,064	444	6,974	5,702	1,273	1,428	325
55 years and over	3,616	3,548	68	2,010	1,775	235	146	68
Black								
Men, 16 years and over	5,211	5,056	155	746	437	309	831	123
16 to 19 years	81	74	6	167	129	38	87	79
20 years and over	5,130	4,982	148	579	308	271	745	44
20 to 24 years	507	487	20	151	87	64	180	16
25 years and over	4,623	4,494	129	429	221	207	565	28
25 to 54 years	4,103	3,987	116	316	140	176	516	18
55 years and over	521	508	13	113	82	31	48	10
Women, 16 years and over	5,079	4,950	130	1,110	761	349	683	159
16 to 19 years	68	62	6	159	127	32	71	65
20 years and over	5,012	4,888	124	951	634	317	612	94
20 to 24 years	463	444	19	198	138	60	143	31
25 years and over	4,549	4,444	105	753	495	257	469	63
25 to 54 years	4,099	4,003	96	569	350	219	444	55
55 years and over	450	441	9	184	146	38	25	7

Source: "Employed and Unemployed Full- and Part-time Workers by Sex, Age, and Race." *Employment and Earnings* 41 (January 1994), p. 191. *Notes:* Employed persons with a job but not at work are distributed according to whether they usually work full or part time.

★ 682 ★

Labor Force

Employed and Unemployed Persons in Census Regions and Divisions: Annual Averages, 1992

Population group and area	Employed						Unemployed	
	Full time			Part time				
	Total	Full-time schedules[1]	Part time for economic reasons, usually work full time	Total	Voluntary[1]	Part time for economic reasons usually work part time	Looking for full time work	Looking for part time work
White								
Northeast	16,896	16,603	293	3,901	3,165	736	1,385	266
New England	4,885	4,794	91	1,253	1,004	250	422	83
Middle Atlantic	12,011	11,809	203	2,647	2,161	486	963	183

[Continued]

★ 682 ★

Employed and Unemployed Persons in Census Regions and Divisions: Annual Averages, 1992

[Continued]

Population group and area	Employed						Unemployed	
	Full time			Part time				
	Total	Full-time schedules[1]	Part time for economic reasons, usually work full time	Total	Voluntary[1]	Part time for economic reasons usually work part time	Looking for full time work	Looking for part time work
Midwest	21,576	21,221	355	5,058	4,154	905	1,262	314
East North Central	14,648	14,409	239	3,390	2,746	644	968	229
West North Central	6,928	6,812	115	1,669	1,408	261	294	85
South	27,309	26,763	546	5,103	4,123	980	1,625	374
South Atlantic	13,605	13,318	286	2,558	2,091	467	808	171
East South Central	4,739	4,647	92	882	724	159	288	55
West South Central	8,965	8,798	167	1,663	1,308	354	529	148
West	17,711	17,206	505	3,925	3,039	687	1,505	315
Mountain	4,939	4,829	110	1,097	856	240	306	89
Pacific	12,773	12,377	395	2,829	2,182	647	1,199	226
Black								
Northeast	1,791	1,761	30	270	178	92	325	38
New England	212	208	4	43	26	17	38	6
Middle Atlantic	1,579	1,553	26	227	151	76	287	32
Midwest	1,710	1,681	30	366	234	132	353	65
East North Central	1,442	1,418	24	303	192	110	296	57
West North Central	268	262	6	63	42	21	57	9
South	5,737	5,560	177	1,091	679	412	852	164
South Atlantic	3,651	3,532	119	640	418	223	477	100
East South Central	938	907	31	189	110	78	143	24
West South Central	1,148	1,121	27	261	151	111	232	40
West	818	801	17	149	101	48	135	25
Mountain	134	131	3	22	14	8	[2]	[2]
Pacific	684	670	14	128	88	40	119	21
Hispanic origin								
Northeast	1,063	1,036	27	160	99	61	177	18
New England	129	124	4	27	15	12	25	3
Middle Atlantic	934	912	22	133	84	49	152	15
Midwest	571	553	19	94	59	34	68	11
East North Central	489	474	15	74	46	27	62	9
West North Central	82	78	4	20	13	7	[2]	[2]
South	2,508	2,423	85	490	310	180	251	60
South Atlantic	928	888	39	165	100	65	91	23
West South Central	1,556	1,511	45	323	209	114	157	38
West	3,442	3,260	182	644	400	244	495	80

[Continued]

★ 682 ★

Employed and Unemployed Persons in Census Regions and Divisions: Annual Averages, 1992

[Continued]

Population group and area	Employed						Unemployed	
	Full time			Part time				
	Total	Full-time schedules[1]	Part time for economic reasons, usually work full time	Total	Voluntary[1]	Part time for economic reasons usually work part time	Looking for full time work	Looking for part time work
Mountain	611	588	23	125	84	41	59	16
Pacific	2,831	2,872	159	519	316	203	436	64

Source: "Census Regions and Divisions: Employed and Unemployed Persons by Full- and Part-time Status, Sex, Age, Race, ad Hispanic Origin, 1992 Annual Averages," U.S. Department of Labor, *Geographical Profile of Employment and Unemployment*, 1992, p. 11. *Notes:* Items may not add to totals because of rounding. Detail for race and Hispanic-origin groups will not add to totals because data for the "other races" group are not presented and Hispanics are included in both the white and black populations. 1. Employed persons with a job but not at work are distributed according to whether they usually work full or part time. 2. Data are not shown when the labor force base does not meet BLS publication standards of reliability for the particular area, based on the sample in that area.

★ 683 ★

Labor Force

Employed and Unemployed Persons in States: Work Status and Characteristics: Annual Averages, 1992

In thousands.

Population group and State	Employed						Unemployed	
	Full time			Part time				
	Total	Full-time schedules[1]	Part time for economic reasons, usually work full time	Total	Voluntary[1]	Part time for economic reasons, usually work part time	Looking for full-time work	Looking for part-time work
White								
Alabama	1,202	1,179	24	190	163	26	66	12
Alaska	167	163	4	31	22	8	14	2
Arizona	1,266	1,232	34	255	189	67	92	25
Arkansas	776	763	13	147	122	25	43	10
California	9,598	9,279	320	2,057	1,570	487	951	171
Colorado	1,270	1,250	20	251	196	55	67	24
Connecticut	1,189	1,169	20	290	236	54	98	17
Delaware	240	236	4	45	39	6	11	2
District of Columbia	98	97	1	13	9	4	2	2
Florida	4,206	4,090	116	857	677	179	316	79
Georgia	1,782	1,760	22	296	259	38	88	15
Hawaii	152	147	5	33	30	3	9	2
Idaho	383	372	11	92	70	22	26	6
Illinois	3,964	3,886	77	870	708	162	251	55
Indiana	2,044	2,011	33	428	343	86	121	30
Iowa	1,147	1,130	17	299	249	50	50	18
Kansas	970	958	12	213	183	29	32	10
Kentucky	1,234	1,213	21	280	226	55	87	17
Louisiana	1,151	1,133	18	240	178	62	68	17
Maine	480	469	11	127	97	31	39	7

[Continued]

★ 683 ★

Employed and Unemployed Persons in States: Work Status and Characteristics: Annual Averages, 1992
[Continued]

Population group and State	Employed						Unemployed	
	Full time			Part time				
	Total	Full-time schedules[1]	Part time for economic reasons, usually work full time	Total	Voluntary[1]	Part time for economic reasons, usually work part time	Looking for full-time work	Looking for part-time work
Maryland	1,455	1,433	22	282	238	44	69	14
Massachusetts	2,150	2,113	37	567	452	115	196	40
Michigan	2,952	2,897	55	753	580	173	244	61
Minnesota	1,764	1,732	32	464	387	77	86	21
Mississippi	646	633	13	106	90	17	39	8
Missouri	1,908	1,870	39	404	342	62	96	21
Montana	286	278	7	79	62	17	16	7
Nebraska	645	637	8	159	137	22	15	8
Nevada	488	474	14	80	64	16	33	5
New Hampshire	468	457	10	110	85	25	36	9
New Jersey	2,591	2,545	46	507	429	79	215	31
New Mexico	493	482	11	120	90	31	36	7
New York	5,278	5,194	83	1,168	959	210	456	75
North Carolina	2,179	2,128	50	367	317	50	96	20
North Dakota	222	217	4	68	55	12	8	4
Ohio	3,749	3,702	47	869	715	154	262	59
Oklahoma	1,079	1,055	24	193	158	35	54	13
Oregon	1,083	1,057	26	271	208	64	88	19
Pennsylvania	4,142	4,069	73	971	774	198	292	77
Rhode Island	359	352	7	100	85	14	35	7
South Carolina	1,020	1,001	18	192	162	30	47	7
South Dakota	271	268	3	63	55	8	6	3
Tennessee	1,656	1,623	34	306	245	61	96	19
Texas	5,959	5,846	113	1,082	850	232	363	108
Utah	576	586	10	175	152	22	26	12
Vermont	239	233	5	60	49	10	17	4
Virginia	2,092	2,053	39	385	308	77	107	23
Washington	1,773	1,732	40	437	352	84	137	31
West Virginia	533	520	13	121	82	39	71	10
Wisconsin	1,940	1,913	27	469	400	69	89	25
Wyoming	178	174	4	45	34	11	10	3
Black								
Alabama	337	323	14	60	32	28	54	9
Alaska	8	8	[3]	1	[3]	[3]	[2]	[2]
Arkansas	104	99	5	24	15	9	23	6
California	597	585	11	111	76	35	104	19
Colorado	62	61	2	9	6	3	[2]	[2]
Connecticut	118	115	2	24	13	10	[2]	[2]
Delaware	49	48	2	9	6	3	[2]	[2]
District of Columbia	118	115	3	16	12	5	17	1
Florida	711	680	31	136	80	56	108	22
Georgia	757	740	16	125	82	42	102	13

[Continued]

★ 683 ★

Employed and Unemployed Persons in States: Work Status and Characteristics: Annual Averages, 1992

[Continued]

Population group and State	Employed						Unemployed	
	Full time			Part time				
	Total	Full-time schedules[1]	Part time for economic reasons, usually work full time	Total	Voluntary[1]	Part time for economic reasons, usually work part time	Looking for full-time work	Looking for part-time work
Illinois	534	527	7	113	73	40	115	25
Indiana	143	141	2	32	17	15	2	2
Kansas	58	57	1	10	7	4	2	2
Kentucky	83	81	2	21	14	7	2	2
Louisiana	279	271	7	84	46	38	60	8
Maryland	566	556	10	86	57	29	64	18
Massachusetts	80	79	1	16	10	5	17	4
Michigan	349	343	6	74	49	26	80	12
Mississippi	276	266	10	53	31	23	43	6
Missouri	150	146	4	42	27	14	2	2
Nebraska	15	15	2	3	2	1	2	2
Nevada	30	29	1	3	2	1	2	2
New Jersey	372	366	6	52	35	17	70	5
New York	893	879	14	128	86	42	154	15
North Carolina	576	555	21	112	77	34	70	15
Ohio	353	347	7	69	46	23	58	13
Oklahoma	58	55	2	21	10	11	2	2
Pennsylvania	314	308	7	47	30	17	64	11
Rhode Island	10	9	3	3	2	1	2	2
South Carolina	360	345	15	71	47	24	45	10
Tennessee	242	237	5	55	34	21	2	2
Texas	707	695	12	131	79	52	140	25
Virginia	498	477	21	80	54	27	59	19
Washington	56	54	2	11	7	4	2	2
West Virginia	17	17	1	5	3	3	2	2
Wisconsin	63	61	1	14	8	6	2	2
Hispanic origin								
Arizona	212	200	11	43	25	18	2	2
California	2,719	2,565	154	496	302	194	417	62
Colorado	133	131	2	27	21	6	2	2
Connecticut	61	60	1	12	7	4	2	2
District of Columbia	12	11	1	3	1	2	2	2
Florida	687	657	30	129	81	48	74	18
Hawaii	11	11	3	2	1	3	2	2
Idaho	22	21	1	3	2	1	2	2
Illinois	368	356	12	44	28	16	43	5
Maryland	60	59	1	9	5	4	2	2
Massachusetts	48	46	3	13	6	7	2	2
Michigan	51	49	2	15	8	7	2	2
Nevada	55	53	2	7	4	2	2	2

[Continued]

★ 683 ★
Employed and Unemployed Persons in States: Work Status and Characteristics: Annual Averages, 1992
[Continued]

| Population group and State | Employed | | | | | | Unemployed | |
| | Full time | | | Part time | | | | |
	Total	Full-time schedules[1]	Part time for economic reasons, usually work full time	Total	Voluntary[1]	Part time for economic reasons, usually work part time	Looking for full-time work	Looking for part-time work
New Jersey	288	280	8	34	22	13	40	2
New Mexico	156	151	5	39	26	13	2	2
New York	600	588	12	89	54	34	104	12
North Carolina	37	36	1	6	4	2	2	2
Ohio	30	30	3	5	4	1	2	2
Oklahoma	35	34	1	7	6	1	2	2
Oregon	47	46	1	10	5	4	2	2
Pennsylvania	46	44	2	10	8	2	2	2
Rhode Island	14	13	3	2	1	1	2	2
Texas	1,483	1,439	43	309	199	110	154	38
Utah	21	21	3	6	5	1	2	2
Virginia	66	63	3	12	4	8	2	2
Wyoming	7	7	3	1	1	3	2	2

Source: "States: Employed and Unemployed Persons by Full- and Part-time Status, Sex, Age, Race and Hispanic Origin, 1992 Annual Averages," U.S. Department of Labor, *Geographical Profile of Employment and Unemployment, 1992*, pp. 54-55. *Notes:* Items may not add to totals because of rounding. Detail for race and Hispanic-origin groups will not add to totals because data for the "other races" group are not presented and Hispanics are included in both the white and black population groups. 1. Employed persons with a job but not at work are distributed according to whether they usually work full or part time. 2. Data are not shown when the labor force base does not meet BLS publication standards of reliability for the particular area, based on the sample in that area. 3. Less than 500 persons.

★ 684 ★
Labor Force
Employment Status by Age: Civilian Noninstitutional Population 16 to 75 Years and Over, 1993. Part I

Numbers in thousands.

Age, sex, and race	1993										
	Civilian non-institutional population	Civilian labor force					Not in labor force				
		Total	Percent of population	Employed	Unemployed		Total	Keeping house	Going to school	Unable to work	Other reasons
					Number	Percent of labor force					
White											
16 years and over	163,921	109,359	66.7	102,812	6,547	6.0	54,562	20,381	5,783	3,173	25,225
16 to 19 years	10,579	5,831	55.1	4,887	943	16.2	4,748	316	3,598	18	816
16 to 17 years	5,349	2,371	44.3	1,922	449	19.0	2,977	96	2,395	4	482
18 to 19 years	5,230	3,459	66.1	2,965	494	14.3	1,771	220	1,203	14	334
20 to 24 years	14,288	11,360	79.5	10,369	991	8.7	2,928	1,014	1,299	71	544
25 to 54 years	93,066	78,594	84.4	74,525	4,069	5.2	14,472	9,167	857	1,413	3,035
25 to 34 years	34,285	28,996	84.6	27,256	1,740	6.0	5,289	3,517	536	277	960
25 to 29 years	15,822	13,421	84.8	12,537	884	6.6	2,402	1,553	316	98	434
30 to 34 years	18,463	15,575	84.4	14,720	855	5.5	2,888	1,963	220	179	526
35 to 44 years	34,057	29,190	85.7	27,777	1,413	4.8	4,866	3,164	231	505	966
35 to 39 years	18,012	15,357	85.3	14,560	798	5.2	2,654	1,782	140	242	490
40 to 44 years	16,045	13,833	86.2	13,217	616	4.5	2,212	1,381	91	263	476
45 to 54 years	24,724	20,407	82.5	19,491	916	4.5	4,316	2,487	90	631	1,109

[Continued]

★ 684 ★

Employment Status by Age: Civilian Noninstitutional Population 16 to 75 Years and Over, 1993. Part I
[Continued]

Age, sex, and race	1993										
	Civilian non-institutional population	Civilian labor force					Not in labor force				
		Total	Percent of population	Employed	Unemployed		Total	Keeping house	Going to school	Unable to work	Other reasons
					Number	Percent of labor force					
45 to 49 years	13,708	11,626	84.8	11,114	512	4.4	2,081	1,220	60	293	508
50 to 54 years	11,016	8,781	79.7	8,377	404	4.6	2,235	1,267	29	337	601
55 to 64 years	18,170	10,385	57.2	9,936	449	4.3	7,785	3,109	17	652	4,008
55 to 59 years	9,247	6,309	88.2	6,035	273	4.3	2,938	1,421	11	335	1,171
60 to 64 years	8,923	4,077	45.7	3,901	176	4.3	4,847	1,688	6	317	2,836
65 years and over	27,818	3,189	11.5	3,095	94	2.9	24,629	6,775	12	1,019	16,822
65 to 69 years	8,867	1,838	20.7	1,783	56	3.0	7,029	2,033	5	230	4,761
70 to 74 years	7,586	850	11.2	829	21	2.5	6,736	1,843	4	189	4,701
75 years and over	11,364	501	4.4	483	17	3.5	10,863	2,899	4	600	7,360

Source: "Employment Status of the Civilian Noninstitutional Population by Age, Sex, and Race." *Employment and Earnings* 41 (January 1994), p. 185.

★ 685 ★

Labor Force

Employment Status by Age: Civilian Noninstitutional Population 16 to 75 Years and Over, 1993. Part II

Numbers in thousands.

Age, sex, and race	1993										
	Civilian non-institutional population	Civilian labor force					Not in labor force				
		Total	Percent of population	Employed	Unemployed		Total	Keeping house	Going to school	Unable to work	Other reasons
					Number	Percent of labor force					
White											
Men											
16 years and over	79,080	60,150	76.1	56,397	3,753	6.2	18,929	438	2,867	1,728	13,896
16 to 19 years	5,369	3,035	56.5	2,500	535	17.6	2,333	25	1,868	11	429
16 to 17 years	2,741	1,240	45.2	990	249	20.1	1,502	11	1,245	3	243
18 to 19 years	2,627	1,796	68.3	1,510	286	15.9	832	14	623	8	186
20 to 24 years	7,053	6,021	85.4	5,448	573	9.5	1,031	24	677	43	287
25 to 54 years	46,250	43,359	93.8	41,043	2,316	5.3	2,891	208	312	899	1,472
25 to 34 years	17,124	16,217	94.7	15,211	1,006	6.2	907	74	216	185	432
25 to 29 years	7,880	7,447	94.5	6,938	509	6.8	434	32	136	65	201
30 to 34 years	9,244	8,770	94.9	8,273	497	5.7	473	43	80	120	231
35 to 44 years	16,973	16,043	94.5	15,248	795	5.0	930	74	72	320	464
35 to 39 years	9,008	8,539	94.8	8,091	448	5.2	469	44	41	154	230
40 to 44 years	7,965	7,504	94.2	7,157	347	4.6	462	30	30	167	234
45 to 54 years	12,153	11,099	91.3	10,584	516	4.6	1,053	60	24	394	575
45 to 49 years	6,760	6,273	92.8	5,993	280	4.5	487	27	17	187	257
50 to 54 years	5,393	4,826	89.5	4,590	236	4.9	566	33	8	207	319
55 to 64 years	8,695	5,861	67.4	5,588	274	4.7	2,834	55	5	381	2,393
55 to 59 years	4,460	3,540	79.4	3,371	169	4.8	920	25	4	195	696
60 to 64 years	4,235	2,322	54.8	2,217	105	4.5	1,914	30	1	185	1,697
65 years and over	11,713	1,873	16.0	1,818	55	2.9	9,840	126	5	394	9,315
65 to 69 years	4,063	1,058	26.0	1,023	35	3.3	3,005	34	2	120	2,849

[Continued]

597

★ 685 ★

Employment Status by Age: Civilian Noninstitutional Population 16 to 75 Years and Over, 1993. Part II

[Continued]

Age, sex, and race	1993										
	Civilian non-institutional population	Civilian labor force					Not in labor force				
		Total	Percent of population	Employed	Unemployed		Total	Keeping house	Going to school	Unable to work	Other reasons
					Number	Percent of labor force					
70 to 74 years	3,358	514	15.3	505	10	1.9	2,844	35	2	85	2,722
75 years and over	4,292	301	7.0	291	9	3.2	3,991	56	1	189	3,744

Source: "Employment Status of the Civilian Noninstitutional Population by Age, Sex, and Race." *Employment and Earnings* 41 (January 1994), p. 185.

★ 686 ★

Labor Force

Employment Status by Age: Civilian Noninstitutional Population 16 to 75 Years and Over, 1993. Part III

Numbers in thousands.

Age, sex, and race	1993										
	Civilian non-institutional population	Civilian labor force					Not in labor force				
		Total	Percent of population	Employed	Unemployed		Total	Keeping house	Going to school	Unable to work	Other reasons
					Number	Percent of labor force					
White											
Women											
16 years and over	84,841	49,208	58.0	46,415	2,793	5.7	35,633	19,943	2,916	1,446	11,328
16 to 19 years	5,210	2,795	53.7	2,387	408	14.6	2,415	291	1,730	7	386
16 to 17 years	2,607	1,132	43.4	932	200	17.7	1,476	85	1,151	2	238
18 to 19 years	2,603	1,664	63.9	1,456	208	12.5	939	207	579	6	148
20 to 24 years	7,236	5,339	73.8	4,921	418	7.8	1,897	990	622	28	257
25 to 54 years	46,816	35,234	75.3	33,481	1,753	5.0	11,582	8,959	545	514	1,564
25 to 34 years	17,161	12,779	74.5	12,045	734	5.7	4,382	3,442	320	92	528
25 to 29 years	7,942	5,974	75.2	5,599	375	6.3	1,968	1,522	180	33	234
30 to 34 years	9,219	6,805	73.6	6,446	359	5.3	2,414	1,921	140	59	294
35 to 44 years	17,084	13,148	77.0	12,529	619	4.7	3,936	3,090	159	185	502
35 to 39 years	9,004	6,818	75.7	6,469	350	5.1	2,186	1,738	98	89	260
40 to 44 years	8,079	6,329	78.3	6,060	269	4.2	1,750	1,351	61	96	242
45 to 54 years	12,571	9,308	74.0	8,907	400	4.3	3,263	2,427	65	237	533
45 to 49 years	6,947	5,353	77.1	5,121	232	4.3	1,594	1,193	44	107	251
50 to 54 years	5,624	3,955	70.3	3,787	168	4.3	1,669	1,234	22	131	282
55 to 64 years	9,475	4,524	47.7	4,349	175	3.9	4,951	3,054	11	271	1,614
55 to 59 years	4,787	2,769	57.8	2,665	104	3.8	2,018	1,396	7	140	475
60 to 64 years	4,688	1,755	37.4	1,684	71	4.0	2,933	1,658	4	132	1,139
65 years and over	16,104	1,316	8.2	1,277	39	3.0	14,788	6,649	8	625	7,507
65 to 69 years	4,804	780	16.2	760	20	2.6	4,023	1,999	3	110	1,912
70 to 74 years	4,228	335	7.9	324	11	3.3	3,893	1,807	2	104	1,979
75 years and over	7,073	200	2.8	192	8	4.0	6,872	2,842	3	411	3,616

Source: "Employment Status of the Civilian Noninstitutional Population by Age, Sex, and Race." *Employment and Earnings* 41 (January 1994), p. 185.

★ 687 ★

Labor Force

Employment Status by Age: Civilian Noninstitutional Population 16 to 75 Years and Over, 1993. Part IV

Numbers in thousands.

Age, sex, and race	Civilian non-institutional population	1993										
		Civilian labor force						Not in labor force				
		Total	Percent of population	Employed	Unemployed		Total	Keeping house	Going to school	Unable to work	Other reasons
					Number	Percent of labor force					
Black											
16 years and over	22,329	13,943	62.4	12,146	1,796	12.9	8,386	2,690	1,468	947	3,282
16 to 19 years	2,099	776	37.0	474	302	38.9	1,323	110	969	6	237
16 to 17 years	1,077	274	25.5	165	109	39.7	802	30	636	2	134
18 to 19 years	1,022	502	49.1	309	193	38.4	520	80	333	4	103
20 to 24 years	2,486	1,689	68.0	1,319	371	22.0	797	267	302	36	192
25 to 54 years	12,889	10,119	78.5	9,086	1,033	10.2	2,770	1,380	190	444	757
25 to 34 years	5,319	4,168	78.4	3,643	525	12.6	1,151	630	132	113	276
25 to 29 years	2,567	1,967	76.6	1,712	256	13.0	600	328	85	49	137
30 to 34 years	2,752	2,201	80.0	1,931	269	12.2	551	301	47	64	139
35 to 44 years	4,616	3,738	81.0	3,383	355	9.5	878	443	44	146	245
35 to 39 years	2,533	2,058	81.2	1,844	213	10.4	476	255	25	63	132
40 to 44 years	2,083	1,680	80.7	1,538	142	8.5	403	188	19	83	113
45 to 54 years	2,953	2,213	74.9	2,060	153	6.9	740	307	13	185	236
45 to 49 years	1,593	1,245	78.1	1,160	84	6.8	348	152	10	83	103
50 to 54 years	1,361	968	71.2	900	68	7.0	392	155	3	102	133
55 to 64 years	2,183	1,102	50.5	1,023	79	7.2	1,081	361	6	210	504
55 to 59 years	1,152	679	59.0	627	52	7.6	472	168	5	122	178
60 to 64 years	1,031	422	41.0	395	27	6.4	609	194	1	88	326
65 years and over	2,673	257	9.6	245	12	4.7	2,416	572	2	250	1,592
65 to 69 years	924	160	17.3	151	9	5.5	764	191	-	63	510
70 to 74 years	767	62	8.0	60	2	3.2	706	166	-	59	480
75 years and over	981	35	3.6	34	1	3.4	946	216	1	128	602

Source: "Employment Status of the Civilian Noninstitutional Population by Age, Sex, and Race." *Employment and Earnings* 41 (January 1994), p. 186.

★ 688 ★

Labor Force

Employment Status by Age: Civilian Noninstitutional Population 16 to 75 Years and Over, 1993. Part V

Numbers in thousands.

Age, sex, and race	Civilian non-institutional population	1993										
		Civilian labor force						Not in labor force				
		Total	Percent of population	Employed	Unemployed		Total	Keeping house	Going to school	Unable to work	Other reasons
					Number	Percent of labor force					
Black											
Men											
16 years and over	10,078	6,911	68.6	5,957	954	13.8	3,167	183	694	491	1,798
16 to 19 years	1,047	413	39.5	247	166	40.1	634	18	486	4	126
16 to 17 years	546	151	27.6	86	64	42.7	395	8	319	1	68
18 to 19 years	501	263	52.4	161	101	38.6	239	10	167	3	59

[Continued]

★ 688 ★

Employment Status by Age: Civilian Noninstitutional Population 16 to 75 Years and Over, 1993. Part V

[Continued]

Age, sex, and race	1993										
	Civilian non-institutional population	Civilian labor force					Not in labor force				
		Total	Percent of population	Employed	Unemployed		Total	Keeping house	Going to school	Unable to work	Other reasons
					Number	Percent of labor force					
20 to 24 years	1,153	854	74.1	658	196	23.0	299	16	140	26	116
25 to 54 years	5,814	4,953	85.2	4,419	534	10.8	862	98	68	267	429
25 to 34 years	2,422	2,115	87.3	1,854	261	12.3	308	40	47	70	150
25 to 29 years	1,172	1,006	85.8	881	125	12.4	166	20	35 ·	34	78
30 to 34 years	1,250	1,109	88.7	973	136	12.2	141	20	13	36	72
35 to 44 years	2,077	1,788	86.1	1,600	188	10.5	289	36	16	90	147
35 to 39 years	1,144	996	87.1	885	111	11.2	148	19	10	40	78
40 to 44 years	932	792	84.9	715	77	9.7	140	17	6	50	69
45 to 54 years	1,315	1,050	79.8	964	85	8.1	566	22	5	107	131
45 to 49 years	707	586	82.9	538	48	8.2	121	11	4	49	57
50 to 54 years	608	463	76.2	426	37	8.1	145	11	1	59	75
55 to 64 years	977	566	57.9	515	51	9.1	411	16	1	104	291
55 to 59 years	519	347	66.9	315	32	9.2	172	9	1	60	102
60 to 64 years	458	219	47.7	199	19	8.8	240	7	-	44	189
65 years and over	1,087	126	11.6	119	7	5.7	961	35	-	90	836
65 to 69 years	408	78	19.1	72	6	7.4	330	11	-	31	289
70 to 74 years	327	29	9.0	28	1	1	298	11	-	24	264
75 years and over	351	18	5.3	18	-	1	333	14	-	35	284

Source: "Employment Status of the Civilian Noninstitutional Population by Age, Sex, and Race." *Employment and Earnings* 41 (January 1994), p. 186. *Note:* 1. Data not shown where base is less than 35,000.

★ 689 ★

Labor Force

Employment Status by Age: Civilian Noninstitutional Population 16 to 75 Years and Over, 1993. Part VI

Numbers in thousands.

Age, sex, and race	1993										
	Civilian non-institutional population	Civilian labor force					Not in labor force				
		Total	Percent of population	Employed	Unemployed		Total	Keeping house	Going to school	Unable to work	Other reasons
					Number	Percent of labor force					
Black											
Women											
16 years and over	12,251	7,031	57.4	6,189	842	12.0	5,220	2,507	774	455	1,484
16 to 19 years	1,051	363	34.5	227	136	37.5	688	92	484	2	110
16 to 17 years	531	124	23.3	79	45	36.1	407	23	318	1	66
18 to 19 years	521	239	46.0	148	91	38.2	281	69	166	1	44
20 to 24 years	1,334	835	62.6	661	174	20.9	498	250	162	10	76
25 to 54 years	7,075	5,166	73.0	4,667	499	9.7	1,908	1,282	122	177	328
25 to 34 years	2,897	2,053	70.9	1,789	264	12.9	844	590	85	43	126
25 to 29 years	1,395	962	68.9	831	131	13.6	434	308	50	16	59
30 to 34 years	1,502	1,092	72.7	958	134	12.2	410	281	35	27	67
35 to 44 years	2,540	1,950	76.8	1,783	167	8.6	590	407	29	56	98
35 to 39 years	1,389	1,061	76.4	960	102	9.6	327	236	15	23	53
40 to 44 years	1,151	889	77.2	823	66	7.4	262	171	14	33	44
45 to 54 years	1,638	1,163	71.0	1,096	67	5.8	475	285	8	78	104

[Continued]

★ 689 ★

Employment Status by Age: Civilian Noninstitutional Population 16 to 75 Years and Over, 1993. Part VI

[Continued]

Age, sex, and race	Civilian non-institutional population	1993									
		Civilian labor force					Not in labor force				
		Total	Percent of population	Employed	Unemployed		Total	Keeping house	Going to school	Unable to work	Other reasons
					Number	Percent of labor force					
45 to 49 years	886	658	74.3	622	36	5.5	227	141	6	35	46
50 to 54 years	752	505	67.1	474	31	6.1	247	144	2	43	58
55 to 64 years	1,206	536	44.4	508	28	5.2	670	346	5	106	213
55 to 59 years	633	332	52.5	312	20	6.0	301	159	4	62	76
60 to 64 years	573	204	35.6	196	8	3.8	369	187	1	45	137
65 years and over	1,586	131	8.2	126	5	3.6	1,455	537	2	160	756
65 to 69 years	516	82	15.9	79	3	3.7	434	180	-	32	222
70 to 74 years	440	32	7.3	31	1	1	408	156	-	36	216
75 years and over	630	16	2.6	16	1	1	614	202	1	92	318

Source: "Employment Status of the Civilian Noninstitutional Population by Age, Sex, and Race." *Employment and Earnings* 41 (January 1994), p. 186. *Note:* 1. Data not shown where base is less than 35,000.

★ 690 ★

Labor Force

Employment Status by Age: Civilian Noninstitutional Population 16 to 75 Years and Over, January 1994. Part I

Numbers in thousands.

Age, sex, and race	Civilian non-institutional population	January 1994									
		Civilian labor force									
		Total	Percent of population	Total	Employed				Unemployed		
					Percent of population	Agriculture	Nonagricultural industries	Number	Percent of labor force	Not in labor force
White											
16 years and over	165,014	109,750	66.5	102,628	62.2	2,715	99,914	7,122	6.5	55,264
16 to 19 years	11,183	5,763	51.5	4,762	42.6	141	4,621	1,000	17.4	5,420
16 to 17 years	5,651	2,344	41.5	1,854	32.8	60	1,794	490	20.9	3,307
18 to 19 years	5,532	3,419	61.8	2,908	52.6	82	2,826	510	14.9	2,113
20 to 24 years	14,888	11,487	77.2	10,286	69.1	225	10,061	1,202	10.5	3,401
25 to 54 years	93,370	78,689	84.3	74,430	79.7	1,702	72,728	4,259	5.4	14,680
25 to 34 years	34,082	28,629	84.0	26,765	78.5	611	26,154	1,864	6.5	5,453
25 to 29 years	15,826	13,308	84.1	12,308	77.8	294	12,014	1,000	7.5	2,518
30 to 34 years	18,256	15,321	83.9	14,457	79.2	317	14,140	864	5.6	2,935
35 to 44 years	34,278	29,310	85.5	27,788	81.1	633	27,158	1,522	5.2	4,968
35 to 39 years	18,012	15,272	84.8	14,441	80.2	336	14,104	831	5.4	2,740
40 to 44 years	16,266	14,038	86.3	13,348	82.1	296	13,051	691	4.9	2,228
45 to 54 years	25,009	20,751	83.0	19,877	79.5	458	19,419	874	4.2	4,259
45 to 49 years	13,943	11,864	85.1	11,375	81.6	232	11,143	489	4.1	2,079
50 to 54 years	11,066	8,887	80.3	8,502	76.8	226	8,276	385	4.3	2,180
55 to 64 years	17,910	10,291	57.5	9,784	54.6	375	9,409	507	4.9	7,619
55 to 59 years	9,224	6,404	69.4	6,103	66.2	189	5,914	301	4.7	2,820
60 to 64 years	8,686	3,887	44.8	3,681	42.4	186	3,495	206	5.3	4,798
65 years and over	27,663	3,520	12.7	3,366	12.2	271	3,095	154	4.4	24,144
65 to 69 years	8,652	1,950	22.5	1,864	21.6	135	1,730	86	4.4	6,071

[Continued]

★ 690 ★

Employment Status by Age: Civilian Noninstitutional Population 16 to 75 Years and Over, January 1994. Part I

[Continued]

Age, sex, and race	Civilian non-institutional population	January 1994								
		Civilian labor force								
		Total	Percent of population	Employed					Unemployed	
				Total	Percent of population	Agriculture	Nonagricultural industries	Number	Percent of labor force	Not in labor force
70 to 74 years	7,562	873	11.6	831	11.0	84	747	43	4.9	6,688
75 years and over	11,450	696	6.1	671	5.9	53	618	25	3.6	10,754

Source: "Employment Status of the Civilian Noninstitutional Population by Age, Sex, and Race." *Employment and Earnings* 41 (February 1994), p. 52.

★ 691 ★

Labor Force

Employment Status by Age: Civilian Noninstitutional Population 16 to 75 Years and Over, January 1994. Part II

Numbers in thousands.

Age, sex, and race	Civilian non-institutional population	January 1994								
		Civilian labor force								
		Total	Percent of population	Employed					Unemployed	
				Total	Percent of population	Agriculture	Nonagricultural industries	Number	Percent of labor force	Not in labor force
White										
Men										
16 years and over	79,764	60,093	75.3	55,878	70.1	2,056	53,822	4,216	7.0	19,671
16 to 19 years	5,702	3,034	53.2	2,420	42.4	112	2,308	614	20.2	2,668
16 to 17 years	2,907	1,224	42.1	923	31.7	44	879	301	24.6	1,683
18 to 19 years	2,795	1,810	64.8	1,497	53.6	69	1,428	313	17.3	985
20 to 24 years	7,448	6,120	82.2	5,389	72.3	188	5,200	731	11.9	1,329
25 to 54 years	46,444	43,193	93.0	40,709	87.7	1,255	39,454	2,484	5.8	3,251
25 to 34 years	17,022	15,990	93.9	14,848	87.2	470	14,378	1,142	7.1	1,032
25 to 29 years	7,896	7,364	93.3	6,733	85.3	229	6,504	631	8.6	531
30 to 34 years	9,126	8,626	94.5	8,115	88.9	241	7,874	511	5.9	501
35 to 44 years	17,088	16,000	93.6	15,151	88.7	496	14,655	849	5.3	1,088
35 to 39 years	8,993	8,442	93.9	7,963	88.8	261	7,722	459	5.4	551
40 to 44 years	8,095	7,558	93.4	7,168	88.5	235	6,933	390	5.2	537
45 to 54 years	12,334	11,203	90.8	10,710	86.6	289	10,421	493	4.4	1,130
45 to 49 years	6,903	6,380	92.4	6,121	88.7	153	5,968	259	4.1	523
50 to 54 years	5,430	4,823	88.8	4,589	84.5	135	4,453	235	4.9	607
55 to 64 years	8,607	5,766	67.0	5,457	63.4	286	5,171	309	5.4	2,841
55 to 59 years	4,474	3,544	79.2	3,361	75.1	151	3,210	183	5.2	929
60 to 64 years	4,133	2,222	53.7	2,096	50.7	135	1,962	125	5.6	1,912
65 years and over	11,562	1,981	17.1	1,903	16.5	215	1,688	78	3.9	9,582
65 to 69 years	3,937	1,092	27.7	1,042	26.5	109	933	50	4.5	2,845
70 to 74 years	3,306	498	15.1	483	14.6	63	420	15	3.0	2,808
75 years and over	4,320	391	9.0	378	8.7	43	335	13	3.3	3,929

Source: "Employment Status of the Civilian Noninstitutional Population by Age, Sex, and Race." *Employment and Earnings* 41 (February 1994), p. 52.

★ 692 ★

Labor Force

Employment Status by Age: Civilian Noninstitutional Population 16 to 75 Years and Over, January 1994. Part III

Numbers in thousands.

Age, sex, and race	Civilian non-institutional population	January 1994									
		Civilian labor force									
		Total	Percent of population	Employed					Unemployed		
				Total	Percent of population	Agriculture	Nonagricultural industries	Number	Percent of labor force	Not in labor force
White											
Women											
16 years and over	85,250	49,657	58.2	46,750	54.6	658	46,092	2,907	5.9	35,593
16 to 19 years	5,481	2,728	49.8	2,342	42.7	29	2,313	386	14.2	2,752
16 to 17 years	2,744	1,120	40.8	931	33.9	16	915	189	16.8	1,624
18 to 19 years	2,736	1,606	58.8	1,411	51.6	13	1,398	196	12.3	1,128
20 to 24 years	7,440	5,368	72.1	4,897	65.8	37	4,861	470	8.8	2,072
25 to 54 years	46,926	35,496	75.6	33,721	71.9	447	33,274	1,775	5.0	11,430
25 to 34 years	17,060	12,639	74.1	11,917	69.9	141	11,776	722	5.7	4,421
25 to 29 years	7,930	5,944	74.9	5,575	70.3	65	5,510	369	6.2	1,986
30 to 34 years	9,130	6,695	73.3	6,342	69.5	76	6,266	353	5.3	2,434
35 to 44 years	17,190	13,310	77.4	12,638	73.5	137	12,501	673	5.1	3,880
35 to 39 years	9,019	6,829	75.7	6,456	71.6	75	6,382	372	5.4	2,189
40 to 44 years	8,171	6,481	79.3	6,180	75.6	62	6,118	301	4.6	1,691
45 to 54 years	12,676	9,547	75.3	9,167	72.3	189	8,998	381	4.0	3,129
45 to 49 years	7,040	5,484	77.9	5,254	74.6	79	5,175	230	4.2	1,556
50 to 54 years	5,636	4,063	72.1	3,913	69.4	90	3,823	150	3.7	1,573
55 to 64 years	9,303	4,526	48.6	4,327	46.5	89	4,238	198	4.4	4,777
55 to 59 years	4,750	2,860	60.2	2,742	57.7	38	2,704	118	4.1	1,891
60 to 64 years	4,553	1,666	36.6	1,585	34.8	52	1,533	81	4.8	2,887
65 years and over	16,101	1,539	9.6	1,463	9.1	57	1,406	76	4.9	14,562
65 to 69 years	4,714	858	18.2	822	17.4	25	797	36	4.2	3,856
70 to 74 years	4,256	376	8.8	348	8.2	21	327	27	7.3	3,881
75 years and over	7,130	305	4.3	293	4.1	10	283	12	4.1	6,825

Source: "Employment Status of the Civilian Noninstitutional Population by Age, Sex, and Race." *Employment and Earnings* 41 (February 1994), p. 52.

★ 693 ★

Labor Force

Employment Status by Age: Civilian Noninstitutional Population 16 to 75 Years and Over, January 1994. Part IV

Numbers in thousands.

Age, sex, and race	Civilian non-institutional population	January 1994									
		Civilian labor force									
		Total	Percent of population	Employed					Unemployed		
				Total	Percent of population	Agriculture	Nonagricultural industries	Number	Percent of labor force	Not in labor force
Black											
16 years and over	22,723	14,197	62.5	12,274	54.0	103	12,171	1,923	13.5	8,526
16 to 19 years	2,194	800	36.4	556	25.3	-	556	243	30.4	1,395

[Continued]

★ 693 ★

Employment Status by Age: Civilian Noninstitutional Population 16 to 75 Years and Over, January 1994. Part IV

[Continued]

Age, sex, and race	Civilian non-institutional population	January 1994								
		Civilian labor force							Unemployed	
				Employed						
		Total	Percent of population	Total	Percent of population	Agriculture	Nonagricultural industries	Number	Percent of labor force	Not in labor force
16 to 17 years	1,128	296	26.2	217	19.2	-	217	79	26.6	832
18 to 19 years	1,067	504	47.3	339	31.8	-	339	165	32.7	563
20 to 24 years	2,604	1,691	64.9	1,255	48.2	12	1,243	436	25.8	913
25 to 54 years	13,383	10,456	78.1	9,281	69.4	65	9,216	1,175	11.2	2,927
25 to 34 years	5,389	4,213	78.2	3,629	67.3	24	3,605	584	13.9	1,176
25 to 29 years	2,627	2,027	77.2	1,719	65.4	16	1,702	309	15.2	599
30 to 34 years	2,762	2,186	79.1	1,911	69.2	8	1,903	275	12.6	576
35 to 44 years	4,976	3,981	80.0	3,546	71.3	28	3,518	435	10.9	995
35 to 39 years	2,690	2,162	80.4	1,923	71.5	16	1,907	239	11.1	528
40 to 44 years	2,286	1,819	79.6	1,623	71.0	13	1,611	196	10.8	467
45 to 54 years	3,018	2,262	74.9	2,106	69.8	13	2,093	156	6.9	756
45 to 49 years	1,725	1,333	77.3	1,241	71.9	9	1,231	93	7.0	391
50 to 54 years	1,293	928	71.8	865	66.9	4	861	63	6.8	365
55 to 64 years	2,033	1,000	49.2	942	46.3	12	930	59	5.9	1,033
55 to 59 years	1,074	623	58.0	586	54.6	6	581	36	5.8	451
60 to 64 years	959	378	39.4	355	37.0	6	349	22	5.9	582
65 years and over	2,509	250	10.0	240	9.6	14	226	10	3.9	2,259
65 to 69 years	825	137	16.6	129	15.6	2	127	8	5.8	688
70 to 74 years	693	56	8.1	54	7.8	6	48	2	1	637
75 years and over	990	57	5.7	57	5.7	5	51	-	1	933

Source: "Employment Status of the Civilian Noninstitutional Population by Age, Sex, and Race." *Employment and Earnings* 41 (February 1994), p. 53. *Notes:* Data for 1994 are not directly comparable with data for 1993 and earlier years. 1. Data not shown where base is less than 75,000.

★ 694 ★

Labor Force

Employment Status by Age: Civilian Noninstitutional Population 16 to 75 Years and Over, January 1994. Part V

Numbers in thousands. Not seasonally adjusted.

Age, sex, and race	Civilian non-institutional population	January 1994								
		Civilian labor force							Unemployed	
				Employed						
		Total	Percent of population	Total	Percent of population	Agriculture	Nonagricultural industries	Number	Percent of labor force	Not in labor force
Black										
Men										
16 years and over	10,182	6,879	67.6	5,853	57.5	89	5,763	1,027	14.9	3,302
16 to 19 years	1,077	378	35.1	232	21.5	-	232	147	38.8	698
16 to 17 years	565	114	20.2	72	12.7	-	72	42	37.1	451
18 to 19 years	512	264	51.6	160	31.2	-	160	104	39.5	248
20 to 24 years	1,191	827	69.5	581	48.8	6	574	247	29.8	364
25 to 54 years	6,052	5,075	83.9	4,473	73.9	61	4,412	602	11.9	976
25 to 34 years	2,417	2,090	66.4	1,793	74.2	22	1,771	296	14.2	328
25 to 29 years	-	999	84.7	850	72.1	15	836	149	14.9	181
30 to 34 years	1,238	1,090	88.1	943	76.2	7	936	148	13.5	147
35 to 44 years	2,272	1,912	84.1	1,691	74.4	26	1,665	220	11.5	360
35 to 39 years	1,227	1,034	84.3	926	75.5	16	911	108	10.4	193

[Continued]

★ 694 ★

Employment Status by Age: Civilian Noninstitutional Population 16 to 75 Years and Over, January 1994. Part V

[Continued]

Age, sex, and race	Civilian non-institutional population	January 1994								
		Civilian labor force								
		Total	Percent of population	Employed					Unemployed	
				Total	Percent of population	Agriculture	Nonagricultural industries	Number	Percent of labor force	Not in labor force
40 to 44 years	1,045	878	84.0	765	73.2	10	755	113	12.9	167
45 to 54 years	1,362	1,074	78.8	989	72.6	13	975	85	8.0	289
45 to 49 years	784	621	79.3	564	72.0	9	555	57	9.2	163
50 to 54 years	579	453	78.2	424	73.4	26	420	28	6.2	126
55 to 64 years	884	478	54.1	453	51.3	12	442	25	5.2	406
55 to 59 years	471	295	62.7	277	58.9	6	272	18	6.0	176
60 to 64 years	413	183	44.3	176	42.6	6	170	7	3.9	230
65 years and over	979	121	12.3	114	11.6	10	103	7	5.6	858
65 to 69 years	325	66	20.3	61	18.8	2	59	5	1	259
70 to 74 years	297	28	9.5	26	8.8	3	23	2	1	268
75 years and over	357	27	7.4	27	7.4	5	21	-	1	331

Source: "Employment Status of the Civilian Noninstitutional Population by Age, Sex, and Race." *Employment and Earnings* 41 (February 1994), p. 53. *Notes:* Data for 1994 are not directly comparable with data for 1993 and earlier years. 1. Data not shown where base is less than 75,000.

★ 695 ★

Labor Force

Employment Status by Age: Civilian Noninstitutional Population 16 to 75 Years and Over, January 1994. Part VI

Numbers in thousands. Not seasonally adjusted.

Age, sex, and race	Civilian non-institutional population	January 1994								
		Civilian labor force								
		Total	Percent of population	Employed					Unemployed	
				Total	Percent of population	Agriculture	Nonagricultural industries	Number	Percent of labor force	Not in labor force
Black										
Women										
16 years and over	12,541	7,318	58.3	6,421	51.2	14	6,408	896	12.2	5,224
16 to 19 years	1,118	421	37.7	325	29.0	-	325	97	23.0	696
16 to 17 years	563	181	32.2	145	25.8	-	145	36	20.0	381
18 to 19 years	555	240	43.3	179	32.4	-	179	60	25.2	315
20 to 24 years	1,413	864	61.1	674	47.7	6	668	190	22.0	549
25 to 54 years	7,331	5,381	73.4	4,808	65.6	5	4,804	573	10.6	1,950
25 to 34 years	2,971	2,124	71.5	1,836	61.8	2	1,834	288	13.5	848
25 to 29 years	1,447	1,028	71.1	868	60.0	2	867	160	15.6	419
30 to 34 years	1,525	1,096	71.9	968	63.5	1	967	128	11.7	429
35 to 44 years	2,705	2,070	76.5	1,855	68.6	2	1,852	215	10.4	635
35 to 39 years	1,463	1,128	77.1	997	68.1	-	997	132	11.7	335
40 to 44 years	1,241	941	75.6	858	69.1	2	856	83	8.8	300
45 to 54 years	1,655	1,188	71.7	1,117	67.5	-	1,117	70	5.9	468
45 to 49 years	941	712	75.7	676	71.9	-	676	36	5.0	229
50 to 54 years	714	476	66.6	441	61.7	-	441	35	7.3	239
55 to 64 years	1,149	522	45.4	488	42.5	-	488	34	6.5	627
55 to 59 years	603	328	54.3	309	51.2	-	309	19	5.7	276
60 to 64 years	548	194	35.6	179	32.8	-	179	15	7.5	351
65 years and over	1,530	129	8.4	126	8.3	3	123	3	2.2	1,401
65 to 69 years	500	71	14.2	68	13.5	-	68	3	1	430

[Continued]

★ 695 ★

Employment Status by Age: Civilian Noninstitutional Population 16 to 75 Years and Over, January 1994. Part VI

[Continued]

Age, sex, and race	Civilian non-institutional population	January 1994								
		Civilian labor force								
		Total	Percent of population	Employed					Unemployed	
				Total	Percent of population	Agriculture	Nonagricultural industries	Number	Percent of labor force	Not in labor force
70 to 74 years	397	26	7.1	26	7.1	3	25	-	1	369
75 years and over	633	30	4.8	30	4.8	-	30	-	1	603

Source: "Employment Status of the Civilian Noninstitutional Population by Age, Sex, and Race." *Employment and Earnings* 41 (February 1994), p. 53. *Notes:* Data for 1994 are not directly comparable with data for 1993 and earlier years. 1. Data not shown where base is less than 75,000.

★ 696 ★

Labor Force

Employment Status by Age: Civilian Noninstitutional Population Quarterly Averages, 1992 and 1993

Not seasonally adjusted. Numbers in thousands.

Sex and age	Total		White		Black		Hispanic origin	
	IV 1992	IV 1993	IV 1992	IV 1993	IV 1992	IV 1993	IV 1992	IV 1993
Total								
16 years and over	118,186	120,571	101,928	103,753	12,012	12,380	9,011	9,432
16 to 19 years	5,208	5,356	4,600	4,769	434	428	496	496
16 to 17 years	1,944	2,095	1,764	1,894	130	158	133	152
18 to 19 years	3,264	3,260	2,836	2,876	304	270	363	345
20 to 24 years	12,103	12,056	10,353	10,241	1,318	1,359	1,261	1,282
25 years and over	100,875	103,160	86,975	88,742	10,260	10,593	7,255	7,654
25 to 54 years	86,153	88,299	74,002	75,558	8,953	9,337	6,591	6,903
55 years and over	14,722	14,861	12,973	13,184	1,307	1,256	664	751
Men								
16 years and over	64,023	65,128	55,861	56,703	5,885	6,005	5,498	5,695
16 to 19 years	2,706	2,690	2,382	2,387	229	205	287	278
16 to 17 years	992	1,050	894	945	72	76	74	90
18 to 19 years	1,714	1,641	1,487	1,442	157	128	213	188
20 to 24 years	6,351	6,319	5,425	5,375	688	681	782	776
25 years and over	54,965	56,118	48,053	48,942	4,968	5,120	4,429	4,641
25 to 54 years	46,708	47,870	40,717	41,546	4,300	4,497	4,030	4,207
55 years and over	8,257	8,248	7,337	7,396	668	623	399	434
Women								
16 years and over	54,163	55,443	46,067	47,050	6,127	6,375	3,513	3,737
16 to 19 years	2,502	2,665	2,219	2,383	205	223	208	218
16 to 17 years	952	1,046	870	949	58	82	59	62
18 to 19 years	1,549	1,619	1,349	1,434	147	141	149	157
20 to 24 years	5,751	5,737	4,927	4,867	629	677	479	505
25 years and over	45,910	47,041	38,921	39,800	5,292	5,474	2,826	3,013

[Continued]

★ 696 ★

Employment Status by Age: Civilian Noninstitutional Population Quarterly Averages, 1992 and 1993

[Continued]

Sex and age	Total		White		Black		Hispanic origin	
	IV 1992	IV 1993	IV 1992	IV 1993	IV 1992	IV 1993	IV 1992	IV 1993
25 to 54 years	39,445	40,428	33,285	34,013	4,653	4,840	2,561	2,696
55 years and over	6,465	6,613	5,636	5,788	640	634	265	317

Source: "Employed Civilians by Sex, Age, Race, and Hispanic Origin," *Employment and Earnings* 41 (January 1994), p. 71. *Notes:* Detail for the above race and Hispanic-origin groups will not sum to totals because data for the "other races" group are not presented and Hispanics are included in both the white and black populations.

★ 697 ★

Labor Force

Employment Status by Age: Civilian Noninstitutional Population, 1992 and 1993

In thousands.

Sex and age	Total		White		Black		Hispanic origin	
	1992	1993	1992	1993	1992	1993	1992	1993
Total								
16 years and over	117,598	119,306	101,479	102,812	11,933	12,146	8,971	9,272
16 to 17 years	5,398	5,530	4,761	4,887	474	474	492	487
18 to 19 years	3,349	3,388	2,921	2,965	321	309	346	350
20 to 24 years	12,157	12,137	10,455	10,369	1,281	1,319	1,268	1,265
25 years and over	100,043	101,640	86,263	87,556	10,178	10,353	7,212	7,521
25 to 54 years	85,350	86,920	73,307	74,525	8,857	9,086	6,507	6,798
55 years and over	14,693	14,720	12,957	13,031	1,320	1,267	705	723
Men								
16 years and over	63,805	64,700	55,709	56,397	5,846	5,957	5,388	5,603
16 to 19 years	2,786	2,836	2,464	2,500	243	247	281	285
16 to 17 years	1,052	1,106	952	990	77	86	83	81
18 to 19 years	1,733	1,730	1,512	1,510	166	161	198	204
20 to 24 years	6,357	6,356	5,462	5,448	658	658	768	782
25 years and over	54,662	55,508	47,783	48,449	4,945	5,052	4,339	4,536
25 to 54 years	46,340	47,239	40,377	41,043	4,267	4,419	3,922	4,122
55 years and over	8,322	8,270	7,406	7,406	578	633	417	414
Women								
16 years and over	53,793	54,606	45,770	46,415	6,087	6,189	3,584	3,669
16 to 19 years	2,613	2,694	2,297	2,387	231	227	211	202
16 to 17 years	997	1,036	888	932	76	79	63	56
18 to 19 years	1,615	1,658	1,409	1,456	155	148	148	146
20 to 24 years	5,799	5,780	4,993	4,921	623	661	500	482
25 years and over	45,381	46,132	38,480	39,107	5,233	5,301	2,873	2,985

[Continued]

★ 697 ★

Employment Status by Age: Civilian Noninstitutional Population, 1992 and 1993
[Continued]

Sex and age	Total		White		Black		Hispanic origin	
	1992	1993	1992	1993	1992	1993	1992	1993
25 to 54 years	39,010	39,682	32,930	33,481	4,590	4,667	2,585	2,676
55 years and over	6,372	6,450	5,551	643	288	309		

Source: "Employed Civilians by Sex, Age, Race, and Hispanic Origin," *Employment and Earnings* 41 (January 1994), p. 232.

★ 698 ★

Labor Force

Employment Status of Civilians in Metropolitan Areas and Cities: Annual Averages, 1992 - I

Area and population group	Civilian labor force participation rate	Employment population ratio	Unemployment	
			Rate	Error range of rate[1]
Anaheim-Santa Ana PMSA				
Total	71.5	66.0	7.7	6.7 - 8.7
Men	80.9	74.6	7.8	6.5 - 9.2
Women	61.2	56.7	7.5	5.9 - 9.0
Both sexes, 16 to 19 years	49.1	38.9	20.9	14.1 - 27.7
White	72.6	67.6	6.8	5.8 - 7.9
Men	82.3	76.8	6.8	5.5 - 8.1
Women	61.6	57.3	6.9	5.3 - 8.6
Both sexes, 16 to 19 years	51.6	41.5	19.6	12.5 - 26.7
Hispanic origin	71.2	64.6	9.3	6.9 - 11.6
Men	82.7	75.3	8.9	6.1 - 11.8
Women	56.4	50.8	9.8	5.7 - 14.0
Atlanta MSA				
Total	71.9	66.9	6.9	6.0 - 7.9
Men	80.7	74.5	7.7	6.3 - 9.0
Women	64.5	60.5	6.2	4.9 - 7.5
Both sexes, 16 to 19 years	47.6	38.0	20.2	13.2 - 27.2
White	71.5	68.8	3.7	2.9 - 4.6
Men	80.6	77.6	3.8	2.6 - 5.0
Women	62.8	60.5	3.7	2.4 - 5.0
Black	73.4	63.8	13.2	10.8 - 15.6
Men	80.9	67.1	17.0	12.9 - 21.2

[Continued]

★ 698 ★

Employment Status of Civilians in Metropolitan Areas and Cities: Annual Averages, 1992 - I

[Continued]

Area and population group	Civilian labor force participation rate	Employment population ratio	Unemployment	
			Rate	Error range of rate[1]
Women	68.6	61.6	10.2	7.3 - 13.1
Baltimore MSA				
Total	70.7	65.6	7.3	6.2 - 8.3
Men	77.6	71.3	8.1	6.6 - 9.7
Women	64.6	60.5	6.3	4.9 - 7.7
Both sexes, 16 to 19 years	52.2	41.6	20.2	12.0 - 28.3
White	70.9	67.4	5.0	3.9 - 6.0
Men	78.4	73.7	6.0	4.5 - 7.5
Women	64.2	61.7	3.8	2.5 - 5.1
Black	70.2	60.7	13.5	10.6 - 16.5
Men	75.8	65.1	14.2	9.9 - 18.5
Women	65.5	57.0	12.8	8.8 - 16.9
Bergen-Passaic PMSA				
Total	66.3	60.6	8.6	7.6 - 9.6
Men	77.6	70.6	9.1	7.7 - 10.4
Women	55.9	51.5	8.0	6.6 - 9.4
Both sexes, 16 to 19 years	38.3	29.9	21.7	14.1 - 29.3
White	66.7	61.0	8.4	7.4 - 9.4
Men	77.8	70.7	9.1	7.7 - 10.5
Women	56.3	52.0	7.5	6.1 - 9.0
Both sexes, 16 to 19 years	41.1	33.5	18.4	10.8 - 26.0
Black	63.8	56.7	11.1	6.6 - 15.6
Men	74.2	67.5	9.0	3.1 - 14.9
Women	56.3	49.0	13.1	6.3 - 19.8
Hispanic origin	77.3	67.2	13.0	9.4 - 16.5
Men	91.4	81.7	10.6	6.2 - 15.0
Women	65.6	55.3	15.7	10.1 - 21.3
Boston PMSA				
Total	68.1	62.8	7.8	7.2 - 8.3
Men	77.1	70.1	9.1	8.3 - 9.9
Women	60.4	56.5	6.4	5.6 - 7.1
Both sexes, 16 to 19 years	51.2	42.2	17.4	13.5 - 21.3

[Continued]

★ 698 ★

Employment Status of Civilians in Metropolitan Areas and Cities: Annual Averages, 1992 - I

[Continued]

Area and population group	Civilian labor force participation rate	Employment population ratio	Unemployment Rate	Error range of rate[1]
White	68.6	63.7	7.1	6.5 - 7.6
Men	77.3	70.8	8.4	7.6 - 9.2
Women	60.9	57.5	5.6	4.9 - 6.3
Both sexes, 16 to 19 years	53.4	45.1	15.4	11.5 - 19.3
Black	66.0	54.6	17.3	13.5 - 21.0
Men	79.9	64.6	19.1	13.7 - 24.6
Women	55.7	47.2	15.3	10.2 - 20.3
Hispanic origin	57.5	51.5	10.4	5.9 - 14.9
Men	72.4	63.4	12.5	5.8 - 19.2
Women	46.5	42.8	8.0	2.2 - 13.9
Buffalo-Niagra Falls CMSA				
Total	62.3	57.5	7.7	6.6 - 8.9
Men	70.2	63.4	9.6	7.9 - 11.3
Women	55.3	52.2	5.6	4.2 - 7.0
Both sexes, 16 to 19 years	60.3	47.0	22.1	15.1 - 29.1
White	64.1	59.8	6.8	5.7 - 7.9
Men	72.0	65.8	8.6	6.9 - 10.2
Women	57.0	54.3	4.8	3.4 - 6.1
Both sexes, 16 to 19 years	65.2	52.0	20.2	13.2 - 27.3
Black	43.8	34.6	21.0	13.4 - 28.6
Charlotte-Gastonia-				
Rock Hill MSA				
Total	72.5	68.5	5.5	4.8 - 6.3
Men	79.5	75.0	5.6	4.6 - 6.7
Women	66.0	62.4	5.5	4.4 - 6.5
Both sexes , 16 to 19 years	49.4	39.3	20.4	14.5 - 26.2
White	71.9	68.7	4.5	3.7 - 5.3
Men	80.2	76.4	4.8	3.7 - 5.8
Women	64.0	61.3	4.2	3.1 - 5.3
Both sexes, 16 to 19 years	53.3	44.9	15.7	9.5 - 22.0
Black	75.0	67.9	9.5	7.2 - 11.8
Men	76.7	68.9	10.1	6.6 - 13.6

[Continued]

★ 698 ★

Employment Status of Civilians in Metropolitan Areas and Cities:
Annual Averages, 1992 - I
[Continued]

Area and population group	Civilian labor force participation rate	Employment population ratio	Unemployment Rate	Unemployment Error range of rate[1]
Women	73.6	67.0	8.9	5.9 - 12.0
Chicago PMSA				
Total	68.2	62.7	8.1	7.5 - 8.6
Men	77.2	70.6	8.6	7.8 - 9.3
Women	60.0	55.5	7.5	6.7 - 8.2
Both sexes, 16 to 19 years	48.0	36.5	24.0	20.3 - 27.6
White	71.0	66.7	6.1	5.5 - 6.6
Men	80.5	75.3	6.5	5.8 - 7.2
Women	61.9	58.5	5.5	4.7 - 6.3
Both sexes, 16 to 19 years	54.9	45.5	17.2	13.3 - 21.0
Black	58.7	48.9	16.6	14.7 - 18.4
Men	64.2	52.1	18.9	16.0 - 21.8
Women	54.4	46.6	14.4	12.0 - 16.9
Both sexes, 16 to 19 years	33.4	17.5	47.6	35.5 - 59.7
Hispanic origin	70.9	63.5	10.5	8.6 - 12.4
Men	87.0	78.1	10.3	8.0 - 12.6
Women	52.8	47.1	10.9	7.7 - 14.1
Cincinnati PMSA				
Total	68.4	64.2	6.1	5.2 - 7.0
Men	78.2	73.6	5.9	4.6 - 7.1
Women	60.2	56.3	6.4	5.0 - 7.7
Both sexes, 16 to 19 years	51.9	44.2	14.7	9.4 - 20.1
White	69.8	66.5	4.7	3.8 - 5.6
Men	80.0	76.3	4.6	3.4 - 5.8
Women	60.8	57.8	4.9	3.5 - 6.2
Both sexes, 16 to 19 years	54.4	49.2	9.7	4.8 - 14.5
Black	59.5	50.1	15.8	11.4 - 20.2
Men	65.2	54.0	17.2	10.1 - 24.3
Women	56.0	47.7	14.8	9.2 - 20.4
Cleveland PMSA				
Total	64.0	59.0	7.9	7.0 - 8.8
Men	75.0	68.8	8.3	7.0 - 9.6

[Continued]

611

★ 698 ★

Employment Status of Civilians in Metropolitan Areas and Cities: Annual Averages, 1992 - I

[Continued]

Area and population group	Civilian labor force participation rate	Employment population ratio	Unemployment	
			Rate	Error range of rate[1]
Women	54.5	50.4	7.4	6.1 - 8.8
Both sexes , 16 to 19 years	54.6	41.8	23.6	17.4 - 29.8
White	64.5	60.2	6.8	5.8 - 7.8
Men	76.2	70.8	7.0	5.7 - 8.4
Women	54.2	50.7	6.5	5.1 - 7.9
Both sexes, 16 to 19 years	59.4	47.4	20.3	13.7 - 26.9
Black	62.0	53.9	13.1	10.0 - 16.2
Men	69.5	58.8	15.4	10.7 - 20.2
Women	56.1	50.1	10.8	6.8 - 14.8
Columbus, Ohio MSA				
Total	69.6	66.3	4.8	4.0 - 5.6
Men	78.4	74.0	5.7	4.4 - 6.9
Women	61.8	59.4	3.8	2.8 - 4.9
Both sexes, 16 to 19 years	56.9	49.3	13.4	8.0 - 18.8
White	70.3	67.9	3.4	2.7 - 4.2
Men	80.1	76.7	4.1	3.0 - 5.2
Women	61.5	59.9	2.6	1.6 - 3.6
Both sexes, 16 to 19 years	59.4	54.6	8.1	3.3 - 12.9
Black	65.7	55.6	15.4	10.8 - 19.9
Men	63.5	50.1	21.2	13.0 - 29.3
Women	67.3	59.7	11.3	6.1 - 16.4
Dallas-Fort Worth CMSA				
Total	74.7	69.8	6.6	6.0 - 7.3
Men	85.0	79.5	6.5	5.6 - 7.4
Women	65.0	60.6	6.8	5.8 - 7.8
Both sexes, 16 to 19 years	58.8	46.5	20.9	16.4 - 25.4
White	74.9	71.0	5.3	4.6 - 5.9
Men	85.5	81.2	5.1	4.2 - 5.9
Women	64.8	61.2	5.5	4.5 - 6.5
Both sexes, 16 to 19 years	64.3	54.1	15.8	11.3 - 20.2
Black	73.1	62.9	14.0	11.3 - 16.6
Men	81.3	68.6	15.6	11.7 - 19.5
Women	66.1	58.0	12.2	8.7 - 15.8

[Continued]

★ 698 ★

Employment Status of Civilians in Metropolitan Areas and Cities:
Annual Averages, 1992 - I
[Continued]

Area and population group	Civilian labor force participation rate	Employment population ratio	Unemployment	
			Rate	Error range of rate[1]
Hispanic origin	76.1	70.7	7.1	5.0 - 9.2
Men	91.2	85.1	6.7	4.1 - 9.2
Women	59.6	55.0	7.7	4.2 - 11.3
Dayton-Springfield MSA				
Total	66.6	62.3	6.5	5.3 - 7.6
Men	76.7	71.7	6.5	4.9 - 8.1
Women	57.1	53.4	6.4	4.6 - 8.1
White	67.2	63.4	5.7	4.6 - 6.9
Men	77.5	72.9	5.9	4.3 - 7.5
Women	57.3	54.2	5.5	3.8 - 7.2
Black	61.4	51.8	15.7	8.7 - 22.6
Denver-Boulder CMSA				
Total	72.8	69.2	5.0	4.2 - 5.7
Men	80.4	76.5	4.8	3.8 - 5.8
Women	65.7	62.3	5.2	4.1 - 6.3
Both sexes, 16 to 19 years	54.0	44.2	18.1	11.9 - 24.3
White	73.3	69.9	4.6	3.8 - 5.4
Men	81.5	77.8	4.5	3.5 - 5.5
Women	65.7	62.6	4.7	3.6 - 5.8
Both sexes, 16 to 19 years	65.7	62.6	4.7	3.6 - 5.8
Black	66.8	58.6	12.3	7.1 - 17.5
Men	70.9	62.8	11.5	4.6 - 18.4
Women	62.7	54.4	13.2	5.3 - 21.1
Hispanic origin	75.6	70.9	6.2	3.4 - 9.0
Men	84.0	81.6	2.8	.1 - 5.5
Women	68.5	61.8	9.7	4.8 - 14.6
Detroit PMSA				
Total	65.6	59.8	8.9	8.3 - 9.5
Men	76.2	69.6	8.6	7.9 - 9.4
Women	56.1	50.9	9.2	8.3 - 10.1
Both sexes, 16 to 19 years	58.5	46.7	20.1	16.9 - 23.4

[Continued]

★ 698 ★

Employment Status of Civilians in Metropolitan Areas and Cities:
Annual Averages, 1992 - I

[Continued]

Area and population group	Civilian labor force participation rate	Employment population ratio	Unemployment	
			Rate	Error range of rate[1]
White	68.1	63.3	7.1	6.5 - 7.6
Men	78.4	73.1	6.8	6.1 - 7.6
Women	58.5	54.3	7.3	6.5 - 8.2
Both sexes, 16 to 19 years	64.6	55.0	14.9	11.6 - 18.1
Black	55.1	44.9	18.4	16.2 - 20.6
Men	65.2	52.8	18.9	15.8 - 22.1
Women	47.3	38.9	17.8	14.7 - 21.0
Both sexes, 16 to 19 yuears	43.1	25.5	40.9	29.4 - 52.3
Hispanic origin	73.4	66.8	9.0	3.8 - 14.2
Fort Lauderdale-Hollywood-				
Pompano Beach PMSA				
Total	60.6	55.1	8.9	7.6 - 10.3
Men	67.9	62.2	8.5	6.7 - 10.2
Women	53.7	48.6	9.5	7.5 - 11.5
White	57.1	52.6	8.0	6.6 - 9.4
Men	65.6	60.3	8.0	6.1 - 9.9
Women	49.2	45.2	8.0	5.9 - 10.2
Black	76.3	67.3	11.9	8.2 - 15.5
Men	78.6	70.3	10.6	5.5 - 15.6
Women	74.3	64.6	13.0	7.8 - 18.3
Hispanic origin	70.0	61.7	11.8	5.8 - 17.8
Hartford-New Britain-				
Middletown CMSA				
Total	73.9	67.7	8.3	6.9 - 9.7
Men	81.9	73.2	10.6	8.5 - 12.8
Women	66.7	62.9	5.8	4.1 - 7.5
Both sexes, 16 to 19 years	60.9	55.6	8.8	2.9 - 14.7
White	73.5	67.7	7.9	6.4 - 9.3
Men	80.9	72.7	10.2	7.9 - 12.5
Women	66.9	63.3	5.4	3.6 - 7.2

[Continued]

★ 698 ★

Employment Status of Civilians in Metropolitan Areas and Cities:
Annual Averages, 1992 - I
[Continued]

Area and population group	Civilian labor force participation rate	Employment population ratio	Unemployment	
			Rate	Error range of rate[1]
Black	77.9	68.1	12.6	7.5 - 17.7
Men	88.2	74.2	15.9	7.9 - 23.8
Women	69.3	63.0	9.1	2.8 - 15.4
Houston PMSA				
Total	72.2	66.4	8.0	7.2 - 8.8
Men	82.9	76.5	7.8	6.7 - 8.8
Women	61.5	56.5	8.2	7.0 - 9.5
Both sexes, 16 to 19 years	54.7	42.4	22.5	17.2 - 27.8
White	73.3	68.9	6.0	5.2 - 6.8
Men	85.1	79.9	6.1	5.1 - 7.2
Women	61.3	57.7	5.9	4.6 - 7.1
Both sexes, 16 to 19 years	59.8	49.4	17.3	12.0 - 22.6
Black	68.6	56.8	17.3	14.2 - 20.5
Men	75.4	62.1	17.6	13.2 - 22.1
Women	62.8	52.2	17.0	12.5 - 21.5
Hispanic origin	76.4	69.6	8.9	6.9 - 11.0
Men	88.2	80.6	8.6	6.1 - 11.2
Women	62.6	56.7	9.5	6.0 - 12.9

Source: "Selected Metropolitan Areas and Cities: Civilian Labor Force Participation Rates. Employment-Population Ratios, and Unemployment Rates by Sex, Age, Race, and Hispanic Origin and Marital Status, 1992 Annual Averages," U.S. Department of Labor, *Geographical Profile of Employment and Unemployment, 1992*, pp. 95-107. *Notes:* Data for demographic groups are not shown when they do not meet BLS publication standards of reliability for the particular area based on the sample in that area. 1. Error ranges are calculated at the 90-percent confidence interval, which means that if repeated samples were drawn from the same population and an error range constructed around each sample estimate, in 9 out of 10 cases the true value based on a complete census of the population would be contained within these error ranges.

★ 699 ★

Labor Force

Employment Status of Civilians in Metropolitan Areas and Cities: Annual Averages, 1992 - II

Area and population group	Civilian labor force participation rate	Employment population ratio	Unemployment	
			Rate	Error range of rate[1]
Indianapolis MSA				
Total	68.3	63.4	7.2	5.6 - 8.7
Men	77.6	71.2	8.3	6.0 - 10.5
Women	60.2	56.7	5.9	3.9 - 7.9
White	70.5	66.3	5.9	4.4 - 7.4
Men	81.1	75.3	7.2	4.9 - 9.5
Women	60.9	58.3	4.4	2.4 - 6.3
Black	59.3	51.4	13.3	8.0 - 18.5
Men	61.1	52.0	15.0	6.5 - 23.4
Women	57.9	51.0	12.0	5.3 - 18.6
Kansas City MSA				
Total	72.3	68.6	5.2	4.2 - 6.1
Men	81.4	76.5	6.0	4.6 - 7.4
Women	64.7	61.9	4.3	3.1 - 5.5
Both sexes, 16 to 19 years	55.9	52.8	5.5	1.4 - 9.6
White	72.2	69.3	4.0	3.1 - 4.8
Men	82.1	78.3	4.7	3.3 - 6.0
Women	63.8	61.7	3.2	2.0 - 4.3
Both sexes, 16 to 19 years	57.3	54.5	4.9	.6 - 9.2
Black	76.7	65.7	14.3	9.7 - 18.8
Men	79.3	64.8	18.3	10.6 - 26.0
Women	74.7	66.4	11.2	5.8 - 16.6
Los Angeles-Long Beach PMSA				
Total	65.6	59.3	9.6	9.2 - 10.1
Men	77.0	69.2	10.2	9.5 - 10.8
Women	54.3	49.5	8.9	8.2 - 9.6
Both sexes, 16 to 19 years	41.7	30.9	26.0	22.9 - 29.0
White	66.4	60.2	9.4	8.9 - 10.0
Men	78.9	71.0	10.0	9.3 - 10.6
Women	53.8	49.2	8.7	7.9 - 9.5
Both sexes, 16 to 19 years	43.9	33.3	24.1	20.8 - 27.5
Black	58.8	50.2	14.6	12.5 - 16.7

[Continued]

★ 699 ★

Employment Status of Civilians in Metropolitan Areas and Cities:
Annual Averages, 1992 - II

[Continued]

Area and population group	Civilian labor force participation rate	Employment population ratio	Unemployment Rate	Unemployment Error range of rate[1]
Men	65.4	55.4	15.3	12.3 - 18.3
Women	53.1	45.8	13.9	11.0 - 16.8
Hispanic origin	65.5	57.3	12.5	11.6 - 13.5
Men	82.2	71.8	12.6	11.4 - 13.8
Women	48.5	42.5	12.5	10.9 - 14.0
Both sexes, 16 to 19 years	41.8	30.4	27.3	22.2 - 32.4
Louisville MSA				
Total	67.9	64.3	5.3	4.1 - 6.4
Men	76.8	72.0	6.2	4.5 - 7.9
Women	59.2	56.7	4.1	2.5 - 5.7
White	67.8	64.3	5.1	3.9 - 6.3
Men	76.1	71.6	5.8	4.1 - 7.6
Women	59.9	57.3	4.3	2.6 - 6.0
Black	69.0	64.3	6.8	2.2 - 11.4
Men	82.3	74.9	9.0	2.7 - 15.4
Memphis MSA				
Total	64.9	59.7	8.0	6.5 - 9.6
Men	75.2	69.2	8.0	5.9 - 10.1
Women	56.1	51.5	8.1	5.8 - 10.3
White	66.3	63.2	4.6	3.1 - 6.1
Men	80.0	76.3	4.7	2.6 - 6.7
Women	54.6	52.1	4.6	2.3 - 6.8
Black	62.7	54.1	13.7	10.2 - 17.2
Men	67.0	57.3	14.5	9.3 - 19.7
Women	59.1	51.4	13.0	8.2 - 17.8
Miami-Hialeah PMSA				
Total	63.0	57.4	8.9	7.9 - 9.9
Men	74.0	67.4	8.8	7.4 - 10.2
Women	53.3	48.5	9.0	7.5 - 10.6
Both sexes, 16 to 19 years	42.2	33.1	21.6	15.3 - 27.8
White	64.1	59.1	7.8	6.7 - 8.9

[Continued]

★ 699 ★

Employment Status of Civilians in Metropolitan Areas and Cities:
Annual Averages, 1992 - II

[Continued]

Area and population group	Civilian labor force participation rate	Employment population ratio	Unemployment Rate	Unemployment Error range of rate[1]
Men	75.2	69.6	7.5	6.1 - 8.9
Women	53.9	49.5	8.3	6.6 - 9.9
Both sexes, 16 to 19 years	48.8	38.7	20.6	13.7 - 27.5
Black	59.5	51.3	13.7	10.7 - 16.7
Men	69.4	59.0	14.9	10.6 - 19.2
Women	51.3	44.9	12.4	8.2 - 16.6
Hispanic origin	65.1	59.3	8.9	7.4 - 10.3
Men	78.2	71.7	8.2	6.4 - 10.1
Women	53.2	48.0	9.7	7.4 - 12.1
Both sexes, 16 to 19 years	50.2	40.2	20.0	10.8 - 29.2
Milwaukee PMSA				
Total	70.2	66.1	5.9	4.8 - 6.9
Men	76.7	72.6	5.5	4.1 - 6.8
Women	64.1	60.1	6.3	4.7 - 7.9
Both sexes, 16 to 19 years	61.6	50.1	18.6	11.5 - 25.7
White	71.1	68.3	3.9	3.0 - 4.8
Men	77.5	74.3	4.1	2.8 - 5.4
Women	64.9	62.5	3.7	2.4 - 5.0
Black	61.8	47.8	22.7	16.0 - 29.5
Women	58.5	43.8	25.2	16.0 - 34.4
Minneapolis-St. Paul MSA				
Total	75.4	71.2	5.6	4.8 - 6.4
Men	82.7	77.3	6.5	5.3 - 7.7
Women	68.9	65.7	4.6	3.6 - 5.6
Both sexes, 16 to 19 years	65.5	56.6	13.6	8.7 - 18.6
White	76.4	72.8	4.7	4.0 - 5.5
Men	83.0	78.2	5.9	4.7 - 7.0
Women	70.4	68.0	3.5	2.5 - 4.4
Both sexes, 16 to 19 years	67.8	61.9	8.7	4.4 - 13.0
Black	62.8	45.5	27.5	18.3 - 36.7

[Continued]

★ 699 ★

Employment Status of Civilians in Metropolitan Areas and Cities:
Annual Averages, 1992 - II
[Continued]

Area and population group	Civilian labor force participation rate	Employment population ratio	Unemployment	
			Rate	Error range of rate[1]
Nassau-Suffolk PMSA				
Total	64.2	59.1	7.9	7.2 - 8.7
Men	74.5	68.7	7.8	6.7 - 8.8
Women	54.6	50.2	8.2	7.0 - 9.4
Both sexes, 16 to 19 years	39.8	32.4	18.7	13.1 - 24.2
White	64.0	59.0	7.8	7.0 - 8.6
Men	74.7	69.0	7.7	6.6 - 8.7
Women	54.0	49.7	8.0	6.8 - 9.3
Both sexes, 16 to 19 years	41.7	33.9	18.7	12.9 - 24.6
Black	67.9	60.6	10.7	7.3 - 14.2
Men	73.1	65.1	10.9	6.0 - 15.7
Women	63.2	56.5	10.5	5.6 - 15.5
Hispanic origin	71.0	65.2	8.2	4.4 - 12.1
Men	80.6	71.3	11.6	5.7 - 17.4
Women	60.9	58.7	3.6	-.4 - 7.6
New Orleans MSA				
Total	61.8	56.1	9.3	7.8 - 10.8
Men	67.6	60.9	9.9	7.7 - 12.1
Women	57.0	52.1	8.6	6.6 - 10.7
White	66.6	62.2	6.6	5.1 - 8.2
Men	72.4	57.2	7.1	4.9 - 9.4
Women	61.2	57.5	6.1	3.9 - 8.3
Black	54.3	46.8	13.8	10.3 - 17.3
Men	58.4	49.7	14.9	9.3 - 20.5
Women	51.8	45.0	13.0	8.6 - 17.5
New York PMSA				
Total	58.3	52.4	10.0	9.6 - 10.5
Men	70.0	62.1	11.3	10.6 - 12.0
Women	48.5	44.4	8.6	7.9 - 9.2
Both sexes, 16 to 19 years	24.2	17.5	27.7	23.5 - 31.9
White	57.9	52.7	9.0	8.4 - 9.6
Men	70.6	63.6	9.8	9.0 - 10.6

[Continued]

★ 699 ★

Employment Status of Civilians in Metropolitan Areas and Cities:
Annual Averages, 1992 - II

[Continued]

Area and population group	Civilian labor force participation rate	Employment population ratio	Unemployment	
			Rate	Error range of rate[1]
Women	47.3	43.5	8.0	7.2 - 8.8
Both sexes, 16 to 19 years	28.1	21.2	24.5	19.7 - 29.4
Black	57.2	49.2	14.0	12.7 - 15.3
Men	66.3	55.1	16.8	14.9 - 18.8
Women	50.1	44.6	11.1	9.4 - 12.7
Hispanic origin	53.6	45.5	15.1	13.6 - 16.6
Men	69.5	58.0	16.5	14.5 - 18.6
Women	41.1	35.7	13.2	11.1 - 15.4
Newark PMSA				
Total	67.3	61.7	8.3	7.5 - 9.1
Men	77.3	70.2	9.2	8.1 - 10.3
Women	58.0	53.8	7.2	6.1 - 8.3
Both sexes, 16 to 19 years	39.3	33.3	15.1	10.3 - 19.9
White	67.5	63.2	6.4	5.6 - 7.2
Men	78.4	72.8	7.0	5.9 - 8.1
Women	57.0	53.8	5.6	4.4 - 6.7
Both sexes, 16 to 19 years	41.7	38.4	8.0	3.7 - 12.4
Black	64.9	54.7	15.7	13.2 - 18.2
Men	71.4	57.8	19.0	15.0 - 22.9
Women	59.8	52.3	12.6	9.4 - 15.7
Hispanic origin	67.7	61.9	8.6	6.1 - 11.2
Men	86.1	78.0	9.4	6.0 - 12.7
Women	49.5	45.9	7.3	3.4 - 11.3
Norfolk-Virginia Beach-				
Newport News MSA				
Total	69.9	63.9	8.6	7.1 - 10.1
Men	78.1	72.2	7.7	5.7 - 9.6
Women	62.4	56.3	9.7	7.3 - 12.0
Both sexes, 16 to 19 years	48.7	31.5	35.3	25.6 - 45.0
White	67.7	63.4	6.4	4.7 - 8.1
Men	76.8	73.1	4.9	2.9 - 6.9
Women	59.4	54.5	8.2	5.4 - 11.0

[Continued]

★ 699 ★

Employment Status of Civilians in Metropolitan Areas and Cities: Annual Averages, 1992 - II

[Continued]

Area and population group	Civilian labor force participation rate	Employment population ratio	Unemployment Rate	Error range of rate[1]
Black	74.4	65.0	12.6	9.3 - 15.8
Men	81.9	72.1	12.0	7.7 - 16.2
Women	66.9	58.1	13.3	8.3 - 18.2
Oakland PMSA				
Total	66.6	60.6	9.1	7.8 - 10.4
Men	75.7	69.2	8.7	7.0 - 10.4
Women	58.3	52.7	9.7	7.7 - 11.6
Both sexes, 16 to 19 years	47.4	27.1	42.9	32.7 - 53.1
White	67.9	62.5	7.9	6.5 - 9.3
Men	76.9	70.7	8.1	6.2 - 10.0
Women	59.5	54.9	7.8	5.7 - 9.8
Black	61.0	53.1	12.9	7.9 - 18.0
Men	68.6	62.3	9.2	3.4 - 15.1
Women	53.8	44.5	17.4	8.8 - 25.9
Hispanic origin	70.0	64.5	7.9	4.1 - 11.6
Men	80.3	74.1	7.8	2.9 - 12.6
Women	59.5	54.8	8.0	2.2 - 13.7
Oklahoma City MSA				
Total	65.2	61.7	5.4	4.3 - 6.5
Men	73.3	68.9	6.1	4.4 - 7.7
Women	58.2	55.5	4.7	3.2 - 6.1
White	66.9	64.2	4.1	3.1 - 5.1
Men	75.3	71.9	4.5	3.0 - 6.0
Women	59.5	57.3	3.6	2.2 - 5.1
Black	53.9	46.5	13.7	7.9 - 19.5
Philadelphia PMSA				
Total	64.5	59.5	7.9	7.3 - 8.4
Men	74.8	68.4	8.6	7.8 - 9.4
Women	55.6	51.7	7.1	6.3 - 7.8
Both sexes, 16 to 19 years	49.2	38.6	21.5	17.6 - 25.4
White	66.0	61.8	6.3	5.7 - 6.9

[Continued]

★ 699 ★

Employment Status of Civilians in Metropolitan Areas and Cities: Annual Averages, 1992 - II
[Continued]

Area and population group	Civilian labor force participation rate	Employment population ratio	Unemployment	
			Rate	Error range of rate[1]
Men	76.7	71.5	6.8	6.0 - 7.5
Women	56.6	53.3	5.8	5.0 - 6.6
Both sexes, 16 to 19 years	53.6	44.8	16.4	12.5 - 20.4
Black	58.2	48.8	16.2	14.0 - 18.4
Men	66.5	54.1	18.6	15.3 - 21.9
Women	51.5	44.5	13.7	10.7 - 16.7
Hispanic origin	55.5	48.6	12.4	6.7 - 18.1
Phoenix MSA				
Total	67.8	62.8	7.3	6.4 - 8.3
Men	75.8	69.4	8.5	7.1 - 9.8
Women	60.1	56.5	6.0	4.7 - 7.2
Both sexes, 16 to 19 years	54.7	38.8	29.1	22.2 - 36.1
White	67.6	62.8	7.1	6.1 - 8.0
Men	75.7	69.6	8.2	6.8 - 9.5
Women	59.7	56.2	5.7	4.4 - 7.0
Both sexes, 16 to 19 years	55.7	40.3	27.6	20.5 - 34.6
Hispanic origin	66.4	59.2	10.8	7.6 - 14.1
Men	81.1	71.6	11.7	7.4 - 15.9
Women	51.3	46.5	9.5	4.6 - 14.5
Pittsburgh-Beaver Valley CMSA				
Total	60.2	56.1	6.8	6.0 - 7.6
Men	69.8	63.9	8.4	7.2 - 9.6
Women	51.7	49.2	4.9	3.9 - 5.9
Both sexes, 16 to 19 years	53.2	46.2	13.2	8.3 - 18.2
White	60.3	56.5	6.4	5.5 - 7.2
Men	70.5	65.1	7.7	6.5 - 8.9
Women	51.3	48.9	4.7	3.6 - 5.8
Both sexes, 16 to 19 years	54.0	49.2	8.9	4.5 - 13.4
Black	57.2	49.3	13.7	9.1 - 18.4
Men	60.1	48.5	19.3	11.7 - 26.9
Women	54.4	50.1	8.0	2.8 - 13.1

[Continued]

★ 699 ★

Employment Status of Civilians in Metropolitan Areas and Cities:
Annual Averages, 1992 - II
[Continued]

Area and population group	Civilian labor force participation rate	Employment population ratio	Unemployment	
			Rate	Error range of rate[1]
Portland, Ore. PMSA				
Total	69.2	64.5	6.7	5.7 - 7.7
Men	78.4	73.4	6.4	5.1 - 7.7
Women	60.1	55.8	7.1	5.6 - 8.7
Both sexes, 16 to 19 years	60.1	47.0	21.9	15.0 - 28.7
White	68.7	64.2	6.6	5.6 - 7.7
Men	78.1	73.2	6.4	5.0 - 7.7
Women	59.5	55.4	7.0	5.4 - 8.6
Both sexes, 16 to 19 years	61.7	48.4	21.6	14.6 - 28.7
Hispanic origin	82.2	63.0	23.4	14.7 - 32.0
Providence-Pawtucket-				
Fall River CMSA				
Total	67.5	61.6	8.8	8.0 - 9.6
Men	75.9	68.2	10.1	8.9 - 11.3
Women	60.0	55.6	7.3	6.2 - 8.4
Both sexes, 16 to 19 years	62.5	52.2	16.4	12.1 - 20.6
White	67.7	62.0	8.3	7.5 - 9.1
Men	75.7	68.5	9.5	8.4 - 10.7
Women	60.4	56.2	6.9	5.9 - 8.0
Both sexes, 16 to 19 years	63.0	53.6	15.0	10.8 - 19.2
Black	63.1	49.7	21.2	12.8 - 29.6
Hispanic origin	70.3	52.8	24.9	17.5 - 32.3
Men	84.1	55.6	33.9	23.1 - 44.7

Source: "Selected Metropolitan Areas and Cities: Civilian Labor Force Participation Rates. Employment-Population Ratios, and Unemployment Rates by Sex, Age, Race, and Hispanic Origin and Marital Status, 1992 Annual Averages," U.S. Department of Labor, *Geographical Profile of Employment and Unemployment, 1992*, pp. 95-107. *Notes:* Data for demographic groups are not shown when they do not meet BLS publication standards of reliability for the particular area based on the sample in that area. 1. Error ranges are calculated at the 90-percent confidence interval, which means that if repeated samples were drawn from the same population and an error range constructed around each sample estimate, in 9 out of 10 cases the true value based on a complete census of the population would be contained within these error ranges.

★ 700 ★

Labor Force

Employment Status of Civilians in Metropolitan Areas and Cities: Annual Averages, 1992 - III

Area and population group	Civilian labor force participation rate	Employment population ratio	Unemployment	
			Rate	Error range of rate[1]
Riverside-San Bernardino PMSA				
Total	62.2	55.7	10.5	9.1 - 11.8
Men	74.5	66.0	11.4	9.6 - 13.2
Women	50.8	46.1	9.2	7.3 - 11.1
Both sexes, 16 to 19 years	49.0	32.7	33.3	25.0 - 41.5
White	62.1	55.7	10.3	8.9 - 11.7
Men	74.6	66.0	11.5	9.6 - 13.4
Women	50.5	46.1	8.7	6.7 - 10.6
Both sexes, 16 to 19 years	48.6	33.0	32.2	23.5 - 40.9
Black	68.3	58.4	14.6	7.3 - 21.9
Hispanic origin	65.6	55.4	15.6	12.4 - 18.9
Men	79.8	67.2	15.8	11.7 - 20.0
Women	50.9	43.1	15.3	10.1 - 20.5
Rochester MSA				
Total	67.8	64.0	5.5	4.5 - 6.5
Men	76.4	71.7	6.2	4.7 - 7.7
Women	59.9	57.1	4.7	3.3 - 6.1
Both sexes, 16 to 19 years	57.0	50.6	11.2	4.5 - 17.9
White	67.9	64.4	5.1	4.1 - 6.1
Men	76.5	72.2	5.6	4.1 - 7.1
Women	60.0	57.3	4.5	3.1 - 6.0
Black	66.6	59.9	10.0	4.6 - 15.3
Sacramento MSA				
Total	66.3	60.4	8.9	7.5 - 10.4
Men	73.0	65.7	10.0	7.9 - 12.1
Women	60.0	55.4	7.7	5.7 - 9.7
White	67.9	62.4	8.1	6.6 - 9.6
Men	75.0	68.3	8.9	6.8 - 11.0
Women	61.1	56.7	7.2	5.1 - 9.2
Hispanic origin	68.5	58.3	14.9	8.4 - 21.3
Men	81.0	67.2	17.1	8.4 - 25.7

[Continued]

★ 700 ★

Employment Status of Civilians in Metropolitan Areas and Cities: Annual Averages, 1992 - III
[Continued]

Area and population group	Civilian labor force participation rate	Employment population ratio	Unemployment	
			Rate	Error range of rate[1]
St. Louis MSA				
Total	67.3	62.5	7.0	6.1 - 8.0
Men	76.2	70.8	7.0	5.7 - 8.3
Women	59.0	54.9	7.1	5.6 - 8.5
Both sexes, 16 to 19 years	63.0	52.3	16.9	11.4 - 22.4
White	68.4	64.9	5.1	4.2 - 6.0
Men	77.3	73.2	5.4	4.1 - 6.6
Women	59.9	57.0	4.8	3.5 - 6.1
Both sexes, 16 to 19 years	65.3	56.1	14.2	8.8 - 19.6
Black	60.7	48.8	19.7	15.0 - 24.3
Men	67.6	54.5	19.3	12.5 - 26.1
Women	55.7	44.6	20.0	13.5 - 26.5
Salt Lake City-Ogden MSA				
Total	71.7	68.3	4.7	3.9 - 5.4
Men	80.6	76.7	4.8	6.8 - 5.8
Women	63.4	60.5	4.6	3.5 - 5.6
Both sexes, 16 to 19 years	64.6	55.2	14.6	10.4 - 18.7
White	71.7	68.3	4.6	3.9 - 5.4
Men	80.6	76.8	4.8	3.7 - 5.8
Women	63.3	60.5	4.5	3.4 - 5.6
Both sexes, 16 to 19 years	65.3	55.9	14.4	10.2 - 18.5
Hispanic origin	68.0	60.2	11.4	5.0 - 17.8
San Antonio MSA				
Total	61.9	57.0	8.0	6.6 - 9.5
Men	72.0	66.0	8.3	6.3 - 10.3
Women	52.8	48.7	7.7	5.6 - 9.9
Both sexes, 16 to 19 years	47.0	34.9	25.7	16.9 - 34.4
White	61.6	56.9	7.6	6.1 - 9.1
Men	71.8	65.9	8.3	6.2 - 10.3
Women	52.2	48.7	6.8	4.7 - 8.9
Both sexes, 16 to 19 years	47.3	36.2	23.4	14.6 - 32.1
Black	68.2	58.9	13.5	5.3 - 21.7

[Continued]

★ 700 ★

Employment Status of Civilians in Metropolitan Areas and Cities: Annual Averages, 1992 - III

[Continued]

Area and population group	Civilian labor force participation rate	Employment population ratio	Unemployment	
			Rate	Error range of rate[1]
Hispanic origin	61.2	55.4	9.6	7.2 - 11.9
Men	70.8	63.3	10.6	7.3 - 13.9
Women	52.0	47.7	8.2	4.9 - 11.5
San Diego MSA				
Total	64.3	59.2	7.9	6.7 - 9.0
Men	75.2	68.8	8.5	6.9 - 10.1
Women	55.1	51.2	7.1	5.5 - 8.7
Both sexes, 16 to 19 years	49.1	38.4	21.9	13.8 - 29.9
White	64.5	59.5	7.7	6.5 - 9.0
Men	75.0	68.6	8.5	6.8 - 10.2
Women	55.4	51.6	6.9	5.2 - 8.5
Both sexes, 16 to 19 years	53.8	42.8	20.3	12.1 - 28.6
Black	61.1	55.6	9.0	2.5 - 15.4
Hispanic origin	67.6	60.2	11.0	7.9 - 14.1
Men	84.6	74.2	12.2	8.0 - 16.4
Women	52.3	47.5	9.2	4.7 - 13.7
San Francisco PMSA				
Total	70.2	64.9	7.4	6.2 - 8.6
Men	78.8	72.8	7.6	6.0 - 9.2
Women	61.5	57.1	7.3	5.5 - 9.0
White	70.2	65.4	6.9	5.5 - 8.2
Men	80.4	74.4	7.5	5.7 - 9.4
Women	59.7	56.2	6.0	4.0 - 7.9
Black	74.6	64.1	14.2	6.2 - 22.1
Hispanic origin	75.6	65.5	13.3	9.0 - 17.7
Men	86.6	75.8	12.5	6.7 - 18.3
Women	66.0	56.5	14.3	7.7 - 20.9
San Jose PMSA				
Total	72.8	67.5	7.3	6.0 - 8.5
Men	78.7	73.2	7.0	5.3 - 8.6
Women	66.5	61.4	7.7	5.7 - 9.6

[Continued]

★ 700 ★

Employment Status of Civilians in Metropolitan Areas and Cities: Annual Averages, 1992 - III

[Continued]

Area and population group	Civilian labor force participation rate	Employment population ratio	Unemployment	
			Rate	Error range of rate[1]
White	70.7	66.1	6.5	5.1 - 7.8
Men	76.6	71.7	6.5	4.6 - 8.3
Women	64.4	60.2	6.5	4.4 - 8.5
Hispanic origin	73.7	68.9	6.5	3.2 - 9.9
Men	80.3	73.4	8.6	3.4 - 13.8
Women	67.0	64.3	4.0	[4] - 8.0
Seattle PMSA				
Total	73.8	69.4	6.1	5.1 - 7.1
Men	82.0	76.7	6.5	5.1 - 7.8
Women	65.5	61.9	5.5	4.1 - 7.0
Both sexes, 16 to 19 years	61.1	50.7	17.1	9.9 - 24.3
White	75.6	71.6	5.3	4.3 - 6.3
Men	83.5	78.9	5.5	4.2 - 6.9
Women	67.6	64.2	5.1	3.6 - 6.5
Both sexes, 16 to 19 years	65.9	56.9	13.6	6.8 - 20.4
Black	67.2	54.6	18.7	10.0 - 27.4
Tampa-St. Petersburg-Clearwater MSA				
Total	63.4	58.3	8.0	7.0 - 8.9
Men	70.4	64.9	7.8	6.5 - 9.0
Women	57.4	52.7	8.2	6.8 - 9.5
Both sexes, 16 to 19 years	61.3	47.6	22.3	16.3 - 28.3
White	62.2	57.8	7.2	6.2 - 8.1
Men	69.7	64.6	7.4	6.1 - 8.7
Women	55.7	51.8	6.9	5.5 - 8.2
Both sexes, 16 to 19 years	60.2	49.4	18.0	11.8 - 24.2
Black	73.0	62.2	14.8	10.6 - 19.1
Men	76.0	67.5	11.2	5.6 - 16.8
Women	70.9	58.3	17.7	11.6 - 23.8
Hispanic origin	69.4	59.1	14.8	10.2 - 19.4
Men	78.5	67.0	14.6	8.4 - 20.7
Women	60.6	51.4	15.1	8.1 - 22.1

[Continued]

★ 700 ★

Employment Status of Civilians in Metropolitan Areas and Cities: Annual Averages, 1992 - III

[Continued]

Area and population group	Civilian labor force participation rate	Employment population ratio	Unemployment	
			Rate	Error range of rate[1]
Washington D.C. MSA				
Total	74.3	69.9	5.9	5.3 - 6.5
Men	81.3	76.3	6.2	5.3 - 7.1
Women	67.7	64.0	5.5	4.6 - 6.4
Both sexes, 16 to 19 years	51.3	38.2	25.6	19.9 - 31.4
White	76.4	73.0	4.4	3.7 - 5.1
Men	84.5	80.4	4.8	3.8 - 5.7
Women	68.7	66.0	4.0	3.0 - 4.9
Both sexes, 16 to 19 years	61.1	51.3	16.0	10.0 - 22.0
Black	69.0	62.2	9.9	8.1 - 11.7
Men	72.8	65.3	10.3	7.7 - 12.9
Women	65.7	59.5	9.4	7.0 - 11.9
Hispanic origin	80.7	74.4	7.7	4.6 - 10.8
Men	91.2	85.0	6.7	2.9 - 10.6
Women	69.8	63.5	9.0	3.9 - 14.1

Source: "Selected Metropolitan Areas and Cities: Civilian Labor Force Participation Rates. Employment-Population Ratios, and Unemployment Rates by Sex, Age, Race, and Hispanic Origin and Marital Status, 1992 Annual Averages," U.S. Department of Labor, *Geographical Profile of Employment and Unemployment, 1992,* pp. 95-107. *Notes:* Data for demographic groups are not shown when they do not meet BLS publication standards of reliability for the particular area based on the sample in that area. 1. Error ranges are calculated at the 90-percent confidence interval, which means that if repeated samples were drawn from the same population and an error range constructed around each sample estimate, in 9 out of 10 cases the true value based on a complete census of the population would be contained within these error ranges.

★ 701 ★

Labor Force

Employment Status of Civilians in Selected Central Cities: Annual Averages, 1992

Area and population group	Civilian labor force participation rate	Employment population ratio	Unemployment	
			Rate	Error range of rate[1]
Seattle PMSA				
Total	73.8	69.4	6.1	5.1 - 7.1
Men	82.0	76.7	6.5	5.1 - 7.8
Women	65.5	61.9	5.5	4.1 - 7.0
Both sexes , 16 to 19 years	61.1	50.7	17.1	9.9 - 24.3

[Continued]

★ 701 ★

Employment Status of Civilians in Selected Central Cities:
Annual Averages, 1992
[Continued]

Area and population group	Civilian labor force participation rate	Employment population ratio	Unemployment	
			Rate	Error range of rate[1]
White	75.6	71.6	5.3	4.3 - 6.3
Men	83.5	78.9	5.5	4.2 - 6.9
Women	67.6	64.2	5.1	3.6 - 6.5
Both sexes, 16 to 19 years	65.9	56.9	13.6	6.8 - 20.4
Black	67.2	54.6	18.7	10.0 - 27.4
Single (never married)	79.4	72.6	8.6	6.5 - 10.7
Married, spouse present	73.4	70.5	4.0	2.9 - 5.1
Other marital status	65.6	59.7	9.0	5.8 - 12.1
Tampa-St. Petersburg-				
Clearwater MSA				
Total	63.4	58.3	8.0	7.0 - 8.9
Men	70.4	64.9	7.8	6.5 - 9.0
Women	57.4	52.7	8.2	6.8 - 9.5
Both sexes , 16 to 19 years	61.3	47.6	22.3	16.3 - 28..3
White	62.2	57.8	7.2	6.2 - 8.1
Men	69.7	64.6	7.4	6.1 - 8.7
Women	55.7	51.8	6.9	5.5 - 8.2
Both sexes, 16 to 19 years	60.2	49.4	18.0	11.8 - 24.2
Black	73.0	62.2	14.8	10.6 - 19.1
Men	76.0	67.5	11.2	5.6 - 16.8
Women	70.9	58.3	17.7	11.6 - 23.8
Hispanic origin	69.4	59.1	14.8	10.2 - 19.4
Men	78.5	67.0	14.6	8.4 - 20.7
Women	60.6	51.4	15.1	8.1 - 22.1
Washington, D.C. MSA				
Total	74.3	69.9	5.9	5.3 - 6.5
Men	81.3	76.3	6.2	5.3 - 7.1
Women	67.7	64.0	5.5	4.6 - 6.4
Both sexes , 16 to 19 years	51.3	38.2	25.6	19.9 - 31.4
White	76.4	73.0	4.4	3.7 - 5.1
Men	84.5	80.4	4.8	3.8 - 5.7
Women	68.7	66.0	4.0	3.0 - 4.9

[Continued]

★ 701 ★

Employment Status of Civilians in Selected Central Cities:
Annual Averages, 1992
[Continued]

Area and population group	Civilian labor force participation rate	Employment population ratio	Unemployment	
			Rate	Error range of rate[1]
Both sexes, 16 to 19 years	61.1	51.3	16.0	10.0 - 22.0
Black	69.0	62.2	9.9	8.1 - 11.7
Men	72.8	65.3	10.3	7.7 - 12.9
Women	65.7	59.5	9.4	7.0 - 11.9
Hispanic origin	80.7	74.4	7.7	4.6 - 10.8
Men	91.2	85.0	6.7	2.9 - 10.6
Women	69.8	63.5	9.0	3.9 - 14.1
Baltimore central city				
Total	66.7	57.4	14.0	11.3 - 16.7
Men	72.7	60.4	16.9	12.8 - 21.0
Women	61.7	54.8	11.1	7.7 - 14.6
White	61.0	53.8	11.8	7.1 - 16.5
Men	67.7	55.1	18.6	10.8 - 26.4
Women	55.2	52.6	4.7	.3 - 9.1
Black	69.5	59.1	15.0	11.3 - 18.6
Men	75.9	63.6	16.2	10.8 - 21.6
Women	64.3	55.4	13.8	8.9 - 18.8
Chicago central city				
Total	63.8	56.7	11.0	10.1 - 12.0
Men	73.2	64.5	11.9	10.6 - 13.3
Women	55.6	50.0	10.0	8.7 - 11.3
Both sexes , 16 to 19 years	38.6	26.0	32.5	26.3 - 38.7
White	68.9	63.7	7.6	6.5 - 8.6
Men	80.2	73.7	8.1	6.7 - 9.5
Women	58.3	54.3	6.8	5.4 - 8.3
Both sexes, 16 to 19 years	48.5	38.3	21.0	13.3 - 28.7
Black	55.6	45.7	17.9	15.7 - 20.1
Men	61.1	48.0	21.4	17.9 - 25.0
Women	51.7	44.0	14.9	12.1 - 17.7
Hispanic origin	69.0	60.9	11.8	9.5 - 14.2
Men	86.3	76.9	10.9	8.1 - 13.6
Women	49.9	43.1	13.7	9.5 - 17.9

[Continued]

★ 701 ★

Employment Status of Civilians in Selected Central Cities:
Annual Averages, 1992
[Continued]

Area and population group	Civilian labor force participation rate	Employment population ratio	Unemployment Rate	Unemployment Error range of rate[1]
Cleveland central city				
Total	60.4	54.1	10.4	8.2 - 12.6
Men	69.5	61.9	10.9	7.9 - 13.9
Women	52.5	47.4	9.8	6.8 - 12.9
White	64.1	57.5	10.3	7.6 - 13.1
Men	72.7	65.5	10.0	6.3 - 13.7
Women	56.5	50.4	10.7	6.6 - 14.8
Black	55.8	49.9	10.5	6.7 - 14.3
Men	65.6	57.5	12.3	6.7 - 18.0
Women	47.9	43.8	8.5	3.5 - 13.5
Dallas central city				
Total	71.7	65.3	8.9	7.4 - 10.5
Men	82.7	75.7	8.5	6.4 - 10.6
Women	62.1	56.3	9.4	7.1 - 11.8
White	73.1	68.1	6.8	5.1 - 8.4
Men	84.0	78.4	6.7	4.5 - 8.9
Women	63.3	58.9	6.8	4.4 - 9.3
Black	69.0	59.5	13.8	9.9 - 17.7
Men	79.6	69.4	12.8	7.5 - 18.1
Women	60.8	51.8	14.8	9.0 - 20.5
Hispanic origin	69.4	61.8	11.0	6.8 - 15.2
Men	85.1	76.6	10.0	4.9 - 15.0
Women	52.7	45.9	12.9	5.4 - 20.3
Detroit central city				
Total	52.6	43.2	17.8	16.0 - 19.7
Men	63.4	51.9	18.1	15.6 - 20.7
Women	43.9	36.2	17.5	14.8 - 20.2
Both sexes , 16 to 19 years	44.1	26.4	40.1	31.5 - 48.6
White	48.5	42.4	12.6	8.8 - 16.4
Men	60.4	53.9	10.7	6.1 - 15.3
Women	37.8	32.0	15.4	9.1 - 21.8

[Continued]

★ 701 ★

Employment Status of Civilians in Selected Central Cities:
Annual Averages, 1992
[Continued]

Area and population group	Civilian labor force participation rate	Employment population ratio	Unemployment	
			Rate	Error range of rate[1]
Black	53.5	43.2	19.3	16.9 - 21.6
Men	64.2	51.0	20.4	17.0 - 23.9
Women	45.3	37.2	18.0	14.6 - 21.4
Both sexes, 16 to 19 years	40.3	21.5	46.7	33.5 - 60.0
District of Columbia				
Total	66.3	60.7	8.4	7.5 - 9.3
Men	71.8	64.9	9.6	8.2 - 11.0
Women	61.7	57.2	7.3	6.1 - 8.5
White	80.6	77.8	3.4	2.5 - 4.4
Men	87.1	84.0	3.5	2.2 - 4.8
Women	74.3	71.8	3.4	2.0 - 4.7
Black	58.0	50.9	12.2	10.6 - 13.8
Men	61.8	52.5	15.1	12.5 - 17.7
Women	55.2	49.8	9.7	7.8 - 11.7
Hispanic origin	82.3	77.1	6.3	2.7 - 9.9
Men	91.6	86.0	6.1	1.3 - 10.8
Houston central city				
Total	69.4	62.5	10.0	8.7 - 11.3
Men	80.5	73.3	8.9	7.3 - 10.5
Women	58.1	51.4	11.5	9.4 - 13.7
Both sexes , 16 to 19 years	52.6	37.7	28.5	20.2 - 36.7
White	70.9	66.3	6.5	5.2 - 7.7
Men	83.0	78.0	6.1	4.5 - 7.6
Women	57.6	53.5	7.1	5.0 - 9.2
Black	65.3	51.8	20.6	16.6 - 24.7
Men	72.5	57.8	20.3	14.5 - 26.0
Women	59.4	46.9	21.0	15.2 - 26.9
Hispanic origin	75.9	68.9	9.2	6.6 - 11.8
Men	89.5	82.0	8.4	5.4 - 11.4
Women	58.5	52.2	10.7	6.0 - 15.5

[Continued]

★ 701 ★

Employment Status of Civilians in Selected Central Cities:
Annual Averages, 1992

[Continued]

Area and population group	Civilian labor force participation rate	Employment population ratio	Unemployment Rate	Unemployment Error range of rate[1]
Indianapolis central city				
Total	67.4	62.7	7.0	5.1 - 8.9
Men	74.8	68.2	8.8	5.9 - 11.7
Women	61.4	58.2	5.2	2.9 - 7.5
White	71.1	67.7	4.7	2.9 - 6.6
Men	80.2	74.6	7.0	4.0 - 10.0
Women	63.4	61.9	2.3	.5 - 4.2
Black	58.8	50.8	13.5	8.2 - 18.9
Men	60.6	51.4	15.3	6.7 - 23.9
Women	57.5	50.4	12.2	5.5 - 19.0
Los Angeles central city				
Total	66.4	59.3	10.7	9.9 - 11.5
Men	77.3	68.6	11.3	10.2 - 12.3
Women	55.4	49.8	9.9	8.8 - 11.1
Both sexes , 16 to 19 years	42.5	31.3	26.2	21.2 - 31.3
White	67.5	60.5	10.3	9.4 - 11.2
Men	79.5	70.9	10.8	9.6 - 11.9
Women	55.1	49.8	9.7	8.3 - 11.0
Both sexes, 16 to 19 years	42.4	32.0	24.5	18.9 - 30.2
Black	57.8	48.4	16.3	13.0 - 19.5
Men	64.0	52.9	17.3	12.6 - 21.9
Women	52.4	44.5	15.2	10.6 - 19.7
Hispanic origin	68.5	59.1	13.7	12.2 - 15.3
Men	84.5	73.0	13.5	11.5 - 15.5
Women	52.0	44.7	14.1	11.5 - 16.7
Milwaukee central city				
Total	65.3	58.8	9.9	7.7 - 12.0
Men	72.1	67.3	6.7	4.2 - 9.2
Women	59.5	51.7	13.2	9.7 - 16.6
White	66.3	63.0	5.0	3.1 - 6.8
Men	72.8	71.0	2.6	.7 - 4.4
Women	60.2	55.5	7.7	4.4 - 11.1

[Continued]

★ 701 ★

Employment Status of Civilians in Selected Central Cities:
Annual Averages, 1992
[Continued]

Area and population group	Civilian labor force participation rate	Employment population ratio	Unemployment	
			Rate	Error range of rate[1]
Black	61.1	46.5	23.9	16.9 - 31.0
Women	57.6	42.4	26.3	16.8 - 35.9
New York central city				
Total	56.3	50.2	10.8	10.3 - 11.4
Men	68.3	59.9	12.3	11.5 - 13.1
Women	48.5	42.3	9.1	8.3 - 9.9
Both sexes , 16 to 19 years	22.0	15.2	30.6	25.7 - 35.4
White	55.2	49.7	10.0	9.3 - 10.7
Men	68.3	60.9	11.0	10.0 - 11.9
Women	44.5	40.6	8.8	7.8 - 9.7
Both sexes, 16 to 19 years	25.8	18.5	28.2	22.4 - 33.9
Black	56.3	48.4	14.4	12.7 - 15.4
Men	65.6	54.5	17.0	15.0 - 19.0
Women	49.1	43.7	11.0	9.3 - 12.7
Hispanic origin	51.7	43.5	15.9	14.3 - 17.4
Men	67.4	55.6	17.5	15.3 - 19.7
Women	39.7	34.2	13.7	11.4 - 16.0
Philadelphia central city				
Total	54.6	48.4	11.3	10.0 - 12.7
Men	64.2	55.7	13.1	11.1 - 15.1
Women	46.8	42.4	9.3	7.5 - 11.1
White	55.3	51.1	7.6	6.1 - 9.1
Men	65.8	60.0	8.9	6.7 - 11.1
Women	46.9	44.0	6.2	4.2 - 8.2
Black	53.7	44.7	16.7	13.9 - 19.5
Men	62.1	50.2	19.1	14.9 - 23.3
Women	47.0	40.4	14.2	10.4 - 18.0
Phoenix central city				
Total	72.6	66.3	8.6	7.2 - 10.1
Men	80.2	72.3	9.9	7.8 - 11.9
Women	64.7	60.1	7.1	5.1 - 9.1

[Continued]

★ 701 ★

Employment Status of Civilians in Selected Central Cities:
Annual Averages, 1992
[Continued]

Area and population group	Civilian labor force participation rate	Employment population ratio	Unemployment	
			Rate	Error range of rate[1]
White	72.8	66.9	8.2	6.7 - 9.7
Men	80.3	72.8	9.4	7.4 - 11.5
Women	65.1	60.8	6.6	4.6 - 8.6
Hispanic origin	68.3	59.3	13.2	8.3 - 18.1
Men	84.2	72.8	13.5	7.2 - 19.9
Women	52.3	45.7	12.7	4.9 - 20.5
St. Louis central city				
Total	59.1	50.1	15.2	11.3 - 19.1
Men	68.5	59.0	13.9	8.8 - 19.0
Women	50.7	42.3	16.7	10.7 - 22.7
White	65.5	60.5	7.7	3.9 - 11.4
Men	76.4	71.3	6.6	2.0 - 11.3
Black	50.1	36.0	28.1	19.0 - 37.2
San Antonio central city				
Total	60.6	55.8	8.0	6.2 - 9.7
Men	69.7	64.1	8.1	5.7 - 10.4
Women	52.1	48.1	7.8	5.2 - 10.4
White	60.3	55.7	7.7	5.9 - 9.5
Men	69.6	64.1	8.0	5.5 - 10.4
Women	51.8	48.0	7.4	4.7 - 10.0
Hispanic origin	60.0	54.5	9.1	6.6 - 11.7
Men	69.6	62.8	9.7	6.2 - 13.2
Women	51.2	46.8	8.5	4.8 - 12.1
San Diego central city				
Total	62.5	57.5	8.0	6.3 - 9.8
Men	72.8	66.8	8.2	5.8 - 10.6
Women	53.8	49.6	7.8	5.3 - 10.3
White	62.4	57.6	7.7	5.8 - 9.6
Men	71.7	68.1	7.8	5.3 - 10.4
Women	54.1	50.1	7.5	4.7 - 10.2

[Continued]

★ 701 ★

Employment Status of Civilians in Selected Central Cities: Annual Averages, 1992

[Continued]

Area and population group	Civilian labor force participation rate	Employment population ratio	Unemployment	
			Rate	Error range of rate[1]
Hispanic origin	64.8	57.9	10.6	5.8 - 15.4
Men	84.5	74.0	12.4	5.8 - 19.0
San Francisco central city				
Total	69.7	63.4	9.2	7.2 - 11.2
Men	79.5	72.2	9.2	6.6 - 11.9
Women	60.0	54.6	9.1	6.1 - 12.1
White	73.3	65.9	10.1	7.4 - 12.8
Men	84.1	74.9	10.9	7.3 - 14.4
Women	61.5	56.0	9.0	5.0 - 12.9
Hispanic origin	75.9	61.6	18.9	12.1 - 25.6
Men	87.9	72.2	17.8	8.7 - 26.9

Source: "Selected Metropolitan Areas and Cities: Civilian Labor Force Participation Rates. Employment-Population Ratios, and Unemployment Rates by Sex, Race, and Hispanic Origin and Marital Status, 1992 Annual Averages," U.S. Department of Labor, *Geographical Profile of Employment and Unemployment, 1992*, pp. 107-111. *Notes:* Data for demographic groups are not shown when they do not meet BLS publication standards of reliability for the particular area based on the sample in that area. 1. Error ranges are calculated at the 90-percent confidence interval, which means that if repeated samples were drawn from the same population and an error range constructed around each sample estimate, in 9 out of 10 cases the true value based on a complete census of the population would be contained within these error ranges. 2. Less than 0.05 percent.

★ 702 ★

Labor Force

Employment Status of the Population, 1992-1994, and Monthly Averages, 1993 - I

Numbers in thousands.

Employment status	Annual average		1993					
	1992	1993	Jan.	Feb.	Mar.	Apr.	May	June
White								
Civilian noninstitutional population[1]	162,658	163,921	163,343	163,429	163,543	163,649	163,748	163,857
Civilian labor force	108,526	109,359	108,779	108,746	108,922	108,791	109,234	109,373
Participation rate	66.7	66.7	66.6	66.5	66.6	66.5	66.7	66.7
Employed	101,479	102,812	102,029	102,076	102,251	102,190	102,612	102,721
Employment-population ratio[2]	62.4	62.7	62.5	62.5	62.5	62.4	62.7	62.7
Unemployed	7,047	6,547	6,750	6,670	6,671	6,601	6,622	6,652
Unemployment rate	6.5	6.0	6.2	6.1	6.1	6.1	6.1	6.1
Black								
Civilian noninstitutional population[1]	21,958	22,329	22,157	22,184	22,217	22,249	22,280	22,313

[Continued]

★ 702 ★

Employment Status of the Population, 1992-1994, and Monthly Averages, 1993 - I
[Continued]

Employment status	Annual average		1993					
	1992	1993	Jan.	Feb.	Mar.	Apr.	May	June
Civilian labor force	13,891	13,943	13,817	14,014	13,862	13,868	13,944	13,922
Participation rate	63.3	62.4	62.4	63.2	62.4	62.3	62.6	62.4
Employed	11,933	12,146	11,864	12,157	11,991	11,965	12,140	12,076
Employment-population ratio[2]	54.3	54.4	53.5	54.8	54.0	53.8	54.5	54.1
Unemployed	1,958	1,796	1,953	1,857	1,871	1,903	1,804	1,846
Unemployment rate	14.1	12.9	14.1	13.3	13.5	13.7	12.9	13.3
Hispanic origin								
Civilian noninstitutional population[1]	15,244	15,753	15,500	15,540	15,585	15,635	15,681	15,729
Civilian labor force	10,131	10,377	10,225	10,273	10,311	10,232	10,247	10,285
Participation rate	66.5	65.9	66.0	66.1	66.2	65.4	65.3	65.4
Employed	8,971	9,272	9,064	9,113	9,152	9,154	9,226	9,221
Employment-population ratio[2]	58.9	58.9	58.5	58.6	58.7	58.5	58.8	58.6
Unemployed	1,160	1,104	1,161	1,160	1,159	1,078	1,021	1,064
Unemployment rate	11.4	10.6	11.4	11.3	11.2	10.5	10.0	10.3

Source: "Employment Status of the Population, by Sex, Age, Race, and Hispanic Origin, Monthly Data Seasonally Adjusted." *Monthly Labor Review* 117 (March 1994): 61-62. *Notes:* Detail for the above race and Hispanic-origin groups will not sum to totals because data for the "other races" groups are not presented and Hispanics are included in both the white and black population groups. 1. The population figures are not seasonally adjusted. 2. Civilian employment as a percent of the civilian noninstitutional population.

★ 703 ★
Labor Force

Employment Status of the Population, 1992-1994, and Monthly Averages, 1993 - II

Numbers in thousands.

Employment status	1993						1994
	July	Aug.	Sept.	Oct.	Nov.	Dec.	Jan.
White							
Civilian noninstitutional population[1]	163,971	164,074	164,190	164,309	164,421	164,516	165,014
Civilian labor force	109,393	109,646	109,492	110,009	109,804	110,016	110,802
Participation rate	66.7	66.8	66.7	67.0	66.8	66.9	67.1
Employed	102,835	103,179	103,094	103,273	103,662	103,807	104,355
Employment-population ratio[2]	62.7	62.9	62.8	62.9	63.0	63.1	63.2
Unemployed	6,558	6,467	6,398	6,736	6,142	6,209	6,447
Unemployment rate	6.0	5.9	5.8	6.1	5.6	5.6	5.8
Black							
Civilian noninstitutional population[1]	22,346	22,375	22,408	22,442	22,475	22,504	22,723
Civilian labor force	13,920	13,969	13,952	13,945	14,057	14,011	14,368
Participation rate	62.3	62.4	62.3	62.1	62.5	62.3	63.2
Employed	12,134	12,225	12,202	12,292	12,297	12,397	12,482
Employment-population ratio[2]	54.3	54.6	54.5	54.8	54.7	55.1	54.9
Unemployed	1,786	1,744	1,750	1,653	1,760	1,614	1,887

[Continued]

★ 703 ★

Employment Status of the Population, 1992-1994, and Monthly Averages, 1993 - II

[Continued]

Employment status	1993						1994
	July	Aug.	Sept.	Oct.	Nov.	Dec.	Jan.
Unemployment rate	12.8	12.5	12.5	11.9	12.5	11.5	13.1
Hispanic origin							
Civilian noninstitutional population[1]	15,777	15,824	15,871	15,917	15,967	16,014	17,849
Civilian labor force	10,375	10,331	10,433	10,586	10,575	10,625	11,746
Participation rate	65.8	65.3	65.7	66.5	66.2	66.3	65.8
Employed	9,250	9,311	9,394	9,384	9,476	9,513	10,495
Employment-population ratio[2]	58.6	58.8	59.2	59.0	59.3	59.4	58.8
Unemployed	1,125	1,020	1,039	1,202	1,099	1,112	1,251
Unemployment rate	10.8	9.9	10.0	11.4	10.4	10.5	10.6

Source: "Employment Status of the Population, by Sex, Age, Race, and Hispanic Origin, Monthly Data Seasonally Adjusted." *Monthly Labor Review* 117 (March 1994): 61-62. *Notes:* Data for 1994 are not directly comparable with data for 1993 and earlier years. Detail for the above race and Hispanic-origin groups will not sum to totals because data for the "other races" groups are not presented and Hispanics are included in both the white and black population groups. 1. The population figures are not seasonally adjusted. 2. Civilian employment as a percent of the civilian noninstitutional population.

★ 704 ★

Labor Force

Employment Status: Civilian Noninstitutional Population 16 to 19, 20 Years and Over, 1992 and 1993

Numbers in thousands.

Employment status, sex and age	Total		White		Black		Hispanic origin	
	1992	1993	1992	1993	1992	1993	1992	1993
Men, 16 years and over								
Civilian noninstitutional population	91,541	92,620	78,351	79,080	9,888	10,078	7,569	7,825
Civilian labor force	69,184	69,633	59,830	60,150	6,892	6,911	6,091	6,256
Percent of population	75.6	75.2	76.4	76.1	69.7	68.6	80.5	80.0
Employed	63,805	64,700	55,709	56,397	5,846	5,957	5,388	5,603
Agriculture	2,534	2,438	2,346	2,254	138	128	413	417
Nonagricultural industries	61,270	62,263	53,363	54,143	5,708	5,829	4,975	5,186
Unemployed	5,380	4,932	4,121	3,753	1,046	954	703	653
Unemployment rate	7.8	7.1	6.9	6.2	15.2	13.8	11.5	10.4
Not in labor force	22,356	22,987	18,521	18,929	2,997	3,167	1,478	1,569
Men, 20 years and over								
Civilian noninstitutional population	84,891	85,907	73,031	73,711	8,858	9,031	6,814	7,063
Civilian labor force	65,638	66,069	56,811	57,115	6,472	6,498	5,700	5,871
Percent of population	77.3	76.9	77.8	77.5	73.1	72.0	83.7	83.1
Employed	61,019	61,865	53,245	53,897	5,603	5,710	5,107	5,318
Agriculture	2,355	2,263	2,174	2,091	132	120	387	394
Nonagricultural industries	58,664	59,602	51,071	51,806	5,471	5,590	4,720	4,924
Unemployed	4,619	4,204	3,566	3,218	869	789	593	553

[Continued]

★ 704 ★

Employment Status: Civilian Noninstitutional Population 16 to 19, 20 Years and Over, 1992 and 1993

[Continued]

Employment status, sex and age	Total		White		Black		Hispanic origin	
	1992	1993	1992	1993	1992	1993	1992	1993
Unemployment rate	7.0	6.4	6.3	5.6	13.4	12.1	10.4	9.4
Not in labor force	19,253	19,838	16,220	16,596	2,386	2,532	1,114	1,192
Women, 16 years and over								
Civilian noninstitutional population	100,035	100,930	84,307	84,841	12,069	12,251	7,674	7,928
Civilian labor force	57,798	58,407	48,696	49,208	6,999	7,031	4,040	4,120
Percent of population	57.8	57.9	57.8	58.0	58.0	57.4	52.6	52.0
Employed	53,793	54,606	45,770	46,415	6,087	6,189	3,584	3,669
Agriculture	673	636	642	610	15	14	52	50
Nonagricultural industries	53,121	53,970	45,128	45,805	6,072	6,175	3,531	3,619
Unemployed	4,005	3,801	2,926	2,793	912	842	456	451
Unemployment rate	6.9	6.5	6.0	5.7	13.0	12.0	11.3	10.9
Not in labor force	42,237	42,522	35,610	35,633	5,070	5,220	3,635	3,808
Women, 20 years and over								
Civilian noninstitutional population	93,524	94,388	79,120	79,631	11,025	11,200	6,940	7,176
Civilian labor force	54,594	55,146	45,970	46,413	6,631	6,668	3,753	3,846
Percent of population	58.4	58.4	58.1	58.3	60.1	59.5	54.1	53.6
Employed	51,181	51,912	43,473	44,028	5,856	5,962	3,373	3,467
Agriculture	627	599	597	574	15	13	47	46
Nonagricultural industries	50,553	51,313	42,876	43,454	5,841	5,949	3,325	3,422
Unemployed	3,413	3,234	2,497	2,385	775	706	380	378
Unemployment rate	6.3	5.9	5.4	5.1	11.7	10.6	10.1	9.8
Not in labor force	38,930	39,242	33,150	33,218	4,394	4,532	3,187	3,330
Both sexes, 16 to 19 years								
Civilian noninstitutional population	13,161	13,255	10,506	10,579	2,074	2,099	1,490	1,515
Civilian labor force	6,751	6,826	5,744	5,831	787	776	678	660
Percent of population	51.3	51.5	54.7	55.1	37.9	37.0	45.5	43.6
Employed	5,398	5,530	4,761	4,887	474	474	492	487
Agriculture	225	212	216	199	7	9	31	28
Nonagricultural industries	5,174	5,317	4,545	4,689	467	466	461	459
Unemployed	1,352	1,296	983	943	313	302	186	173
Unemployment rate	20.0	19.0	17.1	16.2	39.8	38.9	27.5	26.2
Not in labor force	6,411	6,429	4,762	4,748	1,287	1,323	812	855

Source: "Employment Status of the Civilian Noninstitutional Population by Sex, Age, Race, and Hispanic Origin," *Employment and Earnings* 41 (January 1994), p. 228. *Notes:* Detail for the above race and Hispanic-origin groups will not sum to totals because data for the "other races" group are not presented and Hispanics are included in both the white and black populations.

★ 705 ★

Labor Force

Employment Status: Civilian Noninstitutional Population 16 to 19, 20 Years and Over: Quarterly Averages, 1992 and 1993

Numbers in thousands.

Employment status, sex and age	Total		White		Black		Hispanic origin	
	IV 1992	IV 1993	IV 1992	IV 1993	IV 1992	IV 1993	IV 1992	IV 1993
Men, 16 years and over								
Civilian noninstitutional population	91,952	93,030	78,630	79,358	9,959	10,152	7,660	7,930
Civilian labor force	68,970	69,472	59,610	60,026	6,865	6,819	6,158	6,323
Percent of population	75.0	74.7	75.8	75.6	68.9	67.2	80.4	79.7
Employed	64,023	65,128	55,861	56,703	5,885	6,005	5,498	5,695
Agriculture	2,472	2,377	2,285	2,196	143	123	445	410
Nonagricultural industries	61,551	62,751	53,576	54,507	5,742	5,882	5,053	5,285
Unemployed	4,948	4,344	3,749	3,323	980	813	660	628
Unemployment rate	7.2	6.3	6.3	5.5	14.3	11.9	10.7	9.9
Not in labor force	22,982	23,558	19,020	19,332	3,094	3,333	1,502	1,607
Men, 20 years and over								
Civilian noninstitutional population	85,262	86,258	73,303	73,959	8,925	9,092	6,898	7,172
Civilian labor force	65,585	66,143	56,758	57,152	6,466	6,482	5,776	5,960
Percent of population	76.9	76.7	77.4	77.3	72.5	71.3	83.7	83.1
Employed	61,316	62,437	53,479	54,316	5,656	5,801	5,211	5,417
Agriculture	2,312	2,258	2,134	2,082	135	121	409	394
Nonagricultural industries	59,005	60,179	51,345	52,234	5,521	5,680	4,802	5,024
Unemployed	4,269	3,706	3,279	2,835	810	682	565	543
Unemployment rate	6.5	5.6	5.8	5.0	12.5	10.5	9.8	9.1
Not in labor force	19,677	20,115	16,545	16,808	2,459	2,610	1,122	1,212
Women, 16 years and over								
Civilian noninstitutional population	100,367	101,285	84,505	85,057	12,137	12,322	7,762	8,036
Civilian labor force	57,975	59,018	48,811	49,682	7,031	7,152	3,998	4,210
Percent of population	57.8	58.3	57.8	58.4	57.9	58.0	51.5	52.4
Employed	54,163	55,443	46,067	47,050	6,127	6,375	3,513	3,737
Agriculture	618	613	589	590	12	14	41	51
Nonagricultural industries	53,546	54,831	45,479	46,460	6,115	6,361	3,472	3,686
Unemployed	3,811	3,574	2,744	2,632	904	778	485	473
Unemployment rate	6.6	6.1	5.6	5.3	12.9	10.9	12.1	11.2
Not in labor force	42,392	42,267	35,694	35,375	5,106	5,170	3,764	3,827
Women, 20 years and over								
Civilian noninstitutional population	93,860	94,710	79,324	79,821	11,092	11,265	7,020	7,276
Civilian labor force	54,947	55,852	46,213	46,936	6,700	6,808	3,702	3,909
Percent of population	58.5	59.0	58.3	58.8	60.4	60.4	52.7	53.7
Employed	51,662	52,778	43,849	44,667	5,922	6,151	3,305	3,518
Agriculture	584	583	555	561	12	14	38	48
Nonagricultural industries	51,078	52,195	43,293	44,107	5,910	6,138	3,267	3,471
Unemployed	3,285	3,074	2,365	2,269	778	657	398	391
Unemployment rate	6.0	5.5	5.1	4.8	11.6	9.7	10.7	10.0
Not in labor force	38,913	38,858	33,110	32,885	4,392	4,456	3,318	3,367

[Continued]

★ 705 ★

Employment Status: Civilian Noninstitutional Population 16 to 19, 20 Years and Over: Quarterly Averages, 1992 and 1993

[Continued]

Employment status, sex and age	Total		White		Black		Hispanic origin	
	IV 1992	IV 1993	IV 1992	IV 1993	IV 1992	IV 1993	IV 1992	IV 1993
Both sexes, 16 to 19 years								
Civilian noninstitutional population	13,196	13,347	10,508	10,635	2,079	2,117	1,503	1,518
Civilian labor force	6,413	6,495	5,449	5,620	730	680	678	664
Percent of population	48.6	48.7	51.9	52.8	35.1	32.1	45.1	43.7
Employed	5,208	5,356	4,600	4,769	434	428	496	496
Agriculture	193	148	184	142	8	3	39	20
Nonagricultural industries	5,014	5,208	4,416	4,627	425	425	456	476
Unemployed	1,205	1,139	849	850	296	252	182	168
Unemployment rate	18.8	17.5	15.6	15.1	40.6	37.1	26.9	25.2
Not in labor force	6,783	6,852	5,058	5,015	1,349	1,437	826	855

Source: "Employment Status of the Civilian Noninstitutional Population by Sex, Age, Race, and Hispanic Origin," *Employment and Earnings* 41 (January 1994), p. 67. *Notes:* Detail for the above race and Hispanic-origin groups will not sum to totals because data for the "other races" group are not presented and Hispanics are included in both the white and black populations.

★ 706 ★

Labor Force

Employment Status: Civilian Noninstitutional Population, January 1993 and January 1994

Numbers in thousands.

Employment status, and race	Total		Men, 20 years and over		Women, 20 years and over		Both sexes, 16 to 19 years	
	Jan. 1993	Jan. 1994	Jan. 1993	Jan. 1994	Jan. 1993	Jan. 1994	Jan. 1993	Jan. 1994
Total								
Civilian noninstitutional population	192,644	195,953	85,445	86,778	94,007	95,109	13,191	14,066
Civilian labor force	126,034	129,393	65,346	66,412	54,600	56,177	6,088	6,804
Percent of population	65.4	66.0	76.5	76.5	58.1	59.1	46.2	48.4
Employed	116,123	119,901	60,271	61,678	51,016	52,715	4,837	5,507
Agriculture	2,753	2,892	2,073	2,096	530	654	150	142
Nonagricultural industries	113,370	117,009	58,197	59,583	50,486	52,061	4,687	5,365
Unemployed	9,911	9,492	5,075	4,733	3,584	3,462	1,251	1,297
Unemployment rate	7.9	7.3	7.8	7.1	6.6	6.2	20.6	19.1
Not in labor force	66,610	66,561	20,099	20,366	39,408	38,933	7,103	7,262
White								
Civilian noninstitutional population	163,343	165,014	73,414	74,062	79,406	79,769	10,523	11,183
Civilian labor force	107,795	109,750	56,610	57,059	45,986	46,928	5,199	5,763
Percent of population	66.0	66.5	77.1	77.0	57.9	58.8	49.4	51.5
Employed	100,296	102,628	52,650	53,458	43,365	44,408	4,281	4,762
Agriculture	2,584	2,715	1,941	1,944	500	629	143	141
Nonagricultural industries	97,712	99,914	50,709	51,514	42,865	43,779	4,138	4,621
Unemployed	7,498	7,122	3,959	3,602	2,621	2,520	918	1,000
Unemployment rate	7.0	6.5	7.0	6.3	5.7	5.4	17.7	17.4
Not in labor force	55,548	55,264	16,804	17,003	33,420	32,841	5,323	5,420
Black								
Civilian noninstitutional population	22,157	22,723	8,953	9,105	11,121	11,424	2,083	2,194
Civilian labor force	13,648	14,197	6,417	6,501	6,527	6,896	704	800

[Continued]

★ 706 ★

Employment Status: Civilian Noninstitutional Population, January 1993 and January 1994
[Continued]

Employment status, and race	Total		Men, 20 years and over		Women, 20 years and over		Both sexes, 16 to 19 years	
	Jan. 1993	Jan. 1994	Jan. 1993	Jan. 1994	Jan. 1993	Jan. 1994	Jan. 1993	Jan. 1994
Percent of population	61.6	62.5	71.7	71.4	58.7	60.4	33.8	36.4
Employed	11,663	12,274	5,510	5,621	5,723	6,097	430	556
Agriculture	120	103	96	89	16	14	7	-
Nonagricultural industries	11,544	12,171	5,414	5,532	5,706	6,083	423	556
Unemployed	1,984	1,923	907	880	804	800	274	243
Unemployment rate	14.5	13.5	14.1	13.5	12.3	11.6	38.9	30.4
Not in labor force	8,509	8,526	2,536	2,604	4,594	4,528	1,379	1,395

Source: "Employment Status of the Civilian Noninstitutional Population by Race, Sex, and Age," *Employment and Earnings* 41 (January 1994), p. 54. *Notes:* Data for 1984 are not directly comparable with data for 1993 and earlier years.

★ 707 ★
Labor Force

Employment Status: Civilian Noninstitutional Population, January 1993-January 1994 - I

Numbers in thousands.

Employment status, race, sex, age, and Hispanic origin	1993						
	Jan.	Feb.	Mar.	Apr.	May	June	July
White							
Men, 20 years and over							
Civilian labor force	56,921	56,922	57,036	56,961	57,082	57,135	57,136
Percent of population	77.5	77.5	77.6	77.4	77.5	77.5	77.5
Employed	53,613	53,613	53,649	53,698	53,818	53,878	53,840
Employment population ratio[2]	73.0	73.0	73.0	73.0	73.1	73.1	73.0
Unemployed	3,308	3,309	3,387	3,263	3,264	3,257	3,296
Unemployment rate	5.8	5.8	5.9	5.7	5.7	5.7	5.8
Women, 20 years and over							
Civilian labor force	46,099	46,037	46,112	46,042	46,291	46,458	46,446
Percent of population	58.1	58.0	58.0	57.9	58.2	58.4	58.3
Employed	43,608	43,639	43,773	43,666	43,916	44,008	44,093
Employment population ratio[2]	54.9	54.9	55.1	54.9	55.2	55.3	55.4
Unemployed	2,491	2,398	2,339	2,376	2,375	2,450	2,353
Unemployment rate	5.4	5.2	5.1	5.2	5.1	5.3	5.1
Both sexes, 16 to 19 years							
Civilian labor force	5,759	5,787	5,774	5,788	5,861	5,780	5,811
Percent of population	54.7	55.0	54.8	54.9	55.5	54.7	54.9
Employed	4,080	4,824	4,829	4,826	4,878	4,835	4,902
Employment population ratio[2]	45.7	45.8	45.8	45.8	46.2	45.8	46.3
Unemployed	951	963	945	962	983	945	909
Unemployment rate	16.5	16.6	16.4	16.6	16.8	16.3	15.6
Men	17.9	17.8	17.1	18.5	17.2	18.4	17.7
Women	15.0	15.3	15.5	14.5	16.3	14.0	13.4

[Continued]

★ 707 ★

Employment Status: Civilian Noninstitutional Population, January 1993-January 1994 - I

[Continued]

Employment status, race, sex, age, and Hispanic origin	1993						
	Jan.	Feb.	Mar.	Apr.	May	June	July
Black							
Civilian noninstitutional population[1]	22,157	22,184	22,217	22,249	22,280	22,313	22,346
Civilian labor force	13,817	14,014	13,862	13,868	13,944	13,922	13,920
Percent of population	62.4	63.2	62.4	62.3	62.6	62.4	62.3
Employed	11,864	12,157	11,991	11,965	12,140	12,076	12,134
Employment population ratio[2]	53.5	54.8	54.0	53.8	54.5	54.1	54.3
Unemployed	1,953	1,857	1,871	1,903	1,804	1,846	1,786
Unemployment rate	14.1	13.3	13.5	13.7	12.9	13.3	12.8
Men, 20 years and over							
Civilian labor force	6,475	6,544	6,489	6,416	6,486	6,492	6,509
Percent of population	72.3	73.0	72.2	71.3	72.0	71.9	72.0
Employed	5,638	5,747	5,644	5,599	5,695	5,677	5,742
Employment population ratio[2]	63.0	64.1	62.8	62.2	63.2	62.9	63.5
Unemployed	837	797	845	817	791	815	767
Unemployment rate	12.9	12.2	13.0	12.7	12.2	12.6	11.8
Women, 20 years and over							
Civilian labor force	6,545	6,672	6,605	6,655	6,641	6,658	6,605
Percent of population	58.9	59.9	59.2	59.6	59.4	59.5	58.9
Employed	5,741	5,923	5,904	5,930	5,951	5,948	5,879
Employment population ratio[2]	51.6	53.2	53.0	53.1	53.2	53.1	52.5
Unemployed	804	749	701	725	690	710	726
Unemployment rate	12.3	11.2	10.6	10.9	10.4	10.7	11.0
Both sexes, 16 to 19 years							
Civilian labor force	797	798	768	797	817	772	806
Percent of population	38.3	38.3	36.8	38.2	39.1	36.9	38.5
Employed	485	487	443	436	494	451	513
Employment population ratio[2]	23.3	23.4	21.2	20.9	23.6	21.6	24.5
Unemployed	312	311	325	361	323	321	293
Unemployment rate	39.1	39.0	42.3	45.3	39.5	41.6	36.4
Men	39.7	39.5	44.1	46.8	40.2	38.8	37.9
Women	38.5	38.4	40.1	43.2	38.7	44.8	34.7
Hispanic origin							
Civilian noninstitutional population[1]	15,500	15,540	15,585	15,635	15,681	15,729	15,777
Civilian labor force	10,225	10,273	10,311	10,232	10,247	10,285	10,375
Percent of population	66.0	66.1	66.2	65.4	65.3	65.4	65.8
Employed	9,064	9,113	9,152	9,154	9,226	9,221	9,250
Employment population ratio[2]	58.5	58.6	58.7	58.5	58.8	58.6	58.6

[Continued]

★ 707 ★

Employment Status: Civilian Noninstitutional Population, January 1993-January 1994 - I

[Continued]

Employment status, race, sex, age, and Hispanic origin	1993						
	Jan.	Feb.	Mar.	Apr.	May	June	July
Unemployed	1,161	1,160	1,159	1,078	1,021	1,064	1,125
Unemployment rate	11.4	11.3	11.2	10.5	10.0	10.3	10.8

Source: "Employment Status of the Civilian Noninstitutional Population by Race, Sex, Age, and Hispanic Origin," *Employment and Earnings* 41 (February 1994), pp. 43-44. *Notes:* Detail for the above race and Hispanic-origin groups will not sum to totals because data for the "other races" group are not presented and Hispanics are included in both the white and black population groups. Data for 1994 are not directly comparable with data for 1993 and earlier years. 1. The population figures are not adjusted for seasonal variation. 2. Employment as a percent of the civilian noninstitutional population.

★ 708 ★

Labor Force

Employment Status: Civilian Noninstitutional Population, January 1993-January 1994 - II

Numbers in thousands.

Employment status, race, sex, age, and Hispanic origin	1993					1994
	Aug.	Sept.	Oct.	Nov.	Dec.	Jan.
White						
Men, 20 years and over						
Civilian labor force	57,196	57,097	57,390	57,123	57,280	57,457
Percent of population	77.5	77.3	77.7	77.2	77.4	77.6
Employed	53,986	53,948	54,144	54,279	54,283	54,438
Employment population ratio[2]	73.2	73.1	73.3	73.4	73.3	73.5
Unemployed	3,210	3,149	3,246	2,844	2,997	3,019
Unemployment rate	5.6	5.5	5.7	5.0	5.2	5.3
Women, 20 years and over						
Civilian labor force	46,586	46,544	46,710	46,768	46,872	47,025
Percent of population	58.5	58.4	58.5	58.6	58.7	59.0
Employed	44,263	44,207	44,223	44,392	44,554	44,631
Employment population ratio[2]	55.5	55.4	55.4	55.6	55.8	56.0
Unemployed	2,323	2,337	2,487	2,376	2,318	2,393
Unemployment rate	5.0	5.0	5.3	5.1	4.9	5.1
Both sexes, 16 to 19 years						
Civilian labor force	5,864	5,851	5,909	5,913	5,864	6,321
Percent of population	55.3	55.1	55.6	55.6	55.1	56.5
Employed	4,930	4,939	4,906	4,991	4,970	5,286
Employment population ratio[2]	46.5	46.5	46.2	46.9	46.7	47.3
Unemployed	934	912	1,003	922	894	1,034
Unemployment rate	15.9	15.6	17.0	15.6	15.2	16.4
Men	17.7	16.8	17.9	17.7	16.9	18.5
Women	14.0	14.3	16.0	13.3	13.4	14.0

[Continued]

★ 708 ★

Employment Status: Civilian Noninstitutional Population, January 1993-January 1994 - II

[Continued]

Employment status, race, sex, age, and Hispanic origin	1993					1994
	Aug.	Sept.	Oct.	Nov.	Dec.	Jan.
Black						
Civilian noninstitutional population[1]	22,375	22,408	22,422	22,475	22,504	22,723
Civilian labor force	13,969	13,952	13,945	14,057	14,011	14,368
Percent of population	62.4	62.3	62.1	62.5	62.3	63.2
Employed	12,225	12,202	12,292	12,297	12,397	12,482
Employment population ratio[2]	54.6	54.5	54.8	54.7	55.1	54.9
Unemployed	1,744	1,750	1,653	1,760	1,614	1,887
Unemployment rate	12.5	12.5	11.9	125	11.5	13.1
Men, 20 years and over						
Civilian labor force	6,552	6,507	6,482	6,529	6,469	6,563
Percent of population	72.4	71.8	71.5	71.8	70.9	72.1
Employed	5,764	5,717	5,770	5,725	5,787	5,753
Employment population ratio[2]	63.7	63.1	63.6	63.0	63.5	63.2
Unemployed	788	790	712	804	682	810
Unemployment rate	12.0	12.1	11.0	12.3	10.5	12.3
Women, 20 years and over						
Civilian labor force	6,644	6,686	6,731	6,766	6,801	6,917
Percent of population	59.2	59.5	59.8	60.1	60.3	60.5
Employed	5,947	6,001	6,059	6,111	6,143	6,121
Employment population ratio[2]	53.0	53.4	53.9	54.2	54.5	53.6
Unemployed	697	685	672	655	658	796
Unemployment rate	10.5	10.2	10.0	9.7	9.7	11.5
Both sexes, 16 to 19 years						
Civilian labor force	773	759	732	762	741	889
Percent of population	36.8	35.9	34.5	35.9	35.2	40.5
Employed	514	484	463	461	467	607
Employment population ratio[2]	24.5	22.9	21.8	21.7	22.2	27.7
Unemployed	259	275	269	301	274	281
Unemployment rate	33.5	36.2	36.7	39.5	37.0	31.7
Men	34.9	39.7	40.6	39.2	38.8	38.1
Women	32.0	32.3	32.8	39.7	35.2	25.5
Hispanic origin						
Civilian noninstitutional population[1]	15,824	15,871	15,917	15,967	16,014	17,849
Civilian labor force	10,331	10,433	10,586	10,575	10,625	11,746
Percent of population	65.3	65.7	66.5	66.2	66.3	65.8
Employed	9,311	9,394	9,384	9,476	9,513	10,495
Employment population ratio[2]	58.8	59.2	59.0	59.3	59.4	58.8

[Continued]

★ 708 ★

Employment Status: Civilian Noninstitutional Population, January 1993-January 1994 - II

[Continued]

Employment status, race, sex, age, and Hispanic origin	1993					1994
	Aug.	Sept.	Oct.	Nov.	Dec.	Jan.
Unemployed	1,020	1,039	1,202	1,099	1,112	1,251
Unemployment rate	9.9	10.0	11.4	10.4	10.5	10.6

Source: "Employment Status of the Civilian Noninstitutional Population by Race, Sex, Age, and Hispanic Origin," *Employment and Earnings* 41 (February 1994), pp. 43-44. *Notes:* Detail for the above race and Hispanic-origin groups will not sum to totals because data for the "other races" group are not presented and Hispanics are included in both the white and black population groups. Data for 1994 are not directly comparable with data for 1993 and earlier years. 1. The population figures are not adjusted for seasonal variation. 2. Employment as a percent of the civilian noninstitutional population.

★ 709 ★

Labor Force

Employment Status: Civilian Noninstitutional Population: Annual Averages 1993 and 1994

Numbers in thousands.

Employment status, and race	Total		Men, 20 years and over		Women, 20 years and over		Both sexes, 16 to 19 years	
	1992	1993	1992	1993	1992	1993	1992	1993
Total								
Civilian noninstitutional population	191,576	193,550	84,891	85,907	93,524	94,388	13,161	13,255
Civilian labor force	126,982	128,040	65,638	66,069	54,594	55,146	6,751	6,826
Percent of population	66.3	66.2	77.3	76.9	58.4	58.4	51.3	51.5
Employed	117,598	119,306	61,019	61,865	51,181	51,912	5,398	5,530
Agriculture	3,207	3,074	2,355	2,263	627	599	225	212
Nonagricultural industries	114,391	116,232	58,664	59,602	50,553	51,313	5,174	5,317
Unemployed	9,384	8,734	4,619	4,204	3,413	3,234	1,352	1,296
Unemployment rate	7.4	6.8	7.0	6.4	6.3	5.9	20.0	19.0
Not in labor force	64,593	65,509	19,253	19,838	38,930	39,242	6,411	6,429
White								
Civilian noninstitutional population	162,658	163,921	73,031	73,711	79,120	79,631	10,506	10,579
Civilian labor force	108,526	109,359	56,811	57,115	45,970	46,413	5,744	5,831
Percent of population	66.7	66.7	77.8	77.5	58.1	58.3	54.7	55.1
Employed	101,479	102,812	53,245	53,897	43,473	44,028	4,761	4,887
Agriculture	2,987	2,864	2,174	2,091	597	574	216	199
Nonagricultural industries	98,492	99,948	51,071	51,806	42,876	43,454	4,545	4,689
Unemployed	7,047	6,547	3,566	3,218	2,497	2,385	983	943
Unemployment rate	6.5	6.0	6.3	5.6	5.4	5.1	17.1	16.2
Not in labor force	54,132	54,562	16,220	16,596	33,150	33,218	4,762	4,748
Black								
Civilian noninstitutional population	21,958	22,329	8,858	9,031	11,025	11,200	2,074	2,099
Civilian labor force	13,891	13,943	6,472	6,498	6,631	6,668	787	776
Percent of population	63.3	62.4	73.1	72.0	60.1	59.5	37.9	37.0
Employed	11,933	12,146	5,603	5,710	5,856	5,962	474	474
Agriculture	153	142	132	120	15	13	7	9
Nonagricultural industries	11,780	12,004	5,471	5,590	5,841	5,949	467	466
Unemployed	1,958	1,796	869	789	775	706	313	302
Unemployment rate	14.1	12.9	13.4	12.1	11.7	10.6	39.8	38.9
Not in labor force	8,067	8,386	2,386	2,532	4,394	4,532	1,287	1,323

Source: "Employment Status of the Civilian Noninstitutional Population by Race, Sex, and Age," *Employment and Earnings* 41 (January 1994), p. 188.

★ 710 ★

Labor Force

Length of Work Week of Persons Employed: Annual Averages, 1992 - I

In thousands.

| Population group and area | Usually work full-time | | | | | | | |
	Total	Slack work or material shortages	Job started or terminated	Holiday	Bad weather	Own illness	On vacation	Other[1]
Total								
Northeast	2,278	302	30	1,039	98	241	320	249
New England	634	86	11	295	15	71	91	66
Middle Atlantic	1,644	216	20	744	83	170	229	183
Midwest	2,651	335	57	928	161	322	479	370
East North Central	1,830	231	36	675	125	214	303	246
West North Central	822	104	21	253	36	108	176	124
South	3,681	633	108	1,125	217	470	550	578
South Atlantic	1,937	363	53	579	107	254	299	281
East South Central	605	105	18	175	46	75	79	108
West South Central	1,139	165	36	371	64	141	172	189
West	2,419	497	64	764	82	292	368	352
Mountain	617	98	21	186	21	76	112	104
Pacific	1,802	399	43	578	61	215	256	248
White								
Northeast	2,026	265	28	915	91	212	295	221
New England	597	80	10	275	15	66	87	64
Middle Atlantic	1,429	185	18	640	76	145	207	157
Midwest	2,410	302	53	836	153	287	448	331
East North Central	1,633	206	33	597	118	185	280	213
West North Central	777	96	20	239	35	102	168	118
South	2,911	461	85	889	167	366	478	466
South Atlantic	1,472	248	38	442	80	188	259	217
East South Central	475	76	16	137	36	59	65	87
West South Central	963	136	31	310	51	120	153	161
West	2,108	448	57	642	72	252	329	308
Mountain	579	91	19	174	19	71	108	97
Pacific	1,529	357	38	468	53	181	222	211
Black								
Northeast	205	29	2	102	6	25	19	23
New England	28	4	2	17	2	4	2	2
Middle Atlantic	177	25	1	85	6	21	17	21
Midwest	198	27	3	75	7	31	24	31
East North Central	167	21	2	66	6	26	19	27
West North Central	3	3	3	3	3	3	3	3

[Continued]

★ 710 ★

Length of Work Week of Persons Employed: Annual Averages, 1992 - I

[Continued]

Population group and area	Usually work full-time							
	Total	Slack work or material shortages	Job started or terminated	Holiday	Bad weather	Own illness	On vacation	Other[1]
South	693	156	21	211	45	95	64	102
South Atlantic	428	105	14	126	25	62	36	60
East South Central	128	28	3	38	10	16	13	19
West South Central	138	23	4	47	9	17	15	23
West	97	14	3	37	3	14	11	16
Pacific	79	11	2	32	3	11	9	13

Source: "Census Regions and Divisions: Persons at Work 1 to 34 Hours by Sex, Age, Race, Reason for Working Less Than 35 Hours, and Usual Status, 1992 Annual Averages," U.S. Department of Labor, *Geographical Profile of Employment and Unemployment, 1992*, pp. 27-28. *Notes:* 1. Includes industrial disputes. 2. Less than 500 persons or less than 0.05 percent. 3. Data are not shown when the labor force base does not meet BLS publication standards of reliability for the particular area, based on the sample in that area.

★ 711 ★

Labor Force

Length of Work Week of Persons Employed: Annual Averages, 1992 - II

In thousands.

Population group and area	Usually work part-time				
	Total	Slack work or could only find part-time work	Does not want full-time work[1]	Full-time work less than 35 hours	Other
Total					
Northeast	3,963	848	2,552	345	218
New England	1,221	270	803	79	68
Middle Atlantic	2,742	578	1,748	266	150
Midwest	5,149	1,058	3,462	358	271
East North Central	3,501	768	2,271	280	182
West North Central	1,648	290	1,191	78	89
South	5,878	1,430	3,506	593	349
South Atlantic	3,032	712	1,816	325	180
East South Central	999	238	603	99	59
West South Central	1,847	480	1,087	169	110
West	4,128	1,023	2,535	353	217
Mountain	1,074	260	699	59	56
Pacific	3,054	763	1,836	294	161

[Continued]

★ 711 ★

Length of Work Week of Persons Employed: Annual Averages, 1992 - II
[Continued]

| Population group and area | Usually work part-time | | | | |
	Total	Slack work or could only find part-time work	Does not want full-time work[1]	Full-time work less than 35 hours	Other
White					
Northeast	3,626	736	2,389	306	195
New England	1,164	250	774	74	66
Middle Atlantic	2,462	486	1,615	232	129
Midwest	4,711	905	3,235	324	247
East North Central	3,154	644	2,100	248	161
West North Central	1,556	261	1,135	75	85
South	4,741	980	3,003	477	281
South Atlantic	2,372	467	1,510	254	141
East South Central	815	159	524	83	50
West South Central	1,553	354	969	140	91
West	3,636	887	2,252	310	187
Mountain	1,016	240	671	53	52
Pacific	2,620	647	1,580	257	135
Black					
Northeast	251	92	110	30	19
New England	40	17	19	3	2
Middle Atlantic	210	76	90	27	18
Midwest	339	132	159	29	19
East North Central	280	110	126	27	17
West North Central	59	21	34	2	2
South	1,016	412	433	109	62
South Atlantic	596	223	270	66	37
East South Central	178	78	75	16	9
West South Central	242	111	88	27	17
West	138	48	65	15	11
Pacific	117	40	55	13	9

Source: "Census Regions and Divisions: Persons at Work 1 to 34 Hours by Sex, Age, Race, Reason for Working Less Than 35 Hours, and Usual Status, 1992 Annual Averages," U.S. Department of Labor, *Geographical Profile of Employment and Unemployment, 1992*, pp. 27-28. *Notes:* 1. Does not want, or unavailable for, full-time work Less than 500 persons or less than 0.05 percent.

★ 712 ★

Labor Force

Median Ages of Labor Force by Select Characteristics, 1962-1990 and Projected, 1995, 2000, and 2005

Age in years.

Group	1962	1970	1975	1980	1985	1990	1995	2000	2005
Total	40.5	39.0	35.8	34.3	35.2	36.6	38.0	39.4	40.6
Men	40.5	39.4	36.5	35.1	35.6	36.7	38.0	39.4	40.5
Women	40.4	38.3	34.8	33.9	34.7	36.4	38.0	39.5	40.6
White	40.9	39.3	35.6	34.8	35.4	36.8	38.3	39.8	41.0
Black[1]	38.3	36.6	34.1	33.3	33.8	34.9	36.2	37.4	38.3
Asian and other[2]	[3]	[3]	[3]	33.8	34.9	36.5	37.2	38.0	38.6
Hispanic origin[4]	[5]	[5]	[5]	30.7	32.4	33.2	34.2	35.0	35.6

Source: "Median Ages of the Labor Force, by Sex, Race, and Hispanic Origin, Selected Historical Years and Projected Years 1995, 2000, and 2005." U.S. Department of Labor, Employment and Training Administration, *Training and Employment Report of the Secretary of Labor*, Covering the Period July 1988-September 1990, p. 102. Primary source: U.S. Department of Labor, Bureau of Labor Statistics. *Notes:* 1. For 1962 and 1970: Black and other. 2. The "Asian and other" group includes (1)Asians and PAcific Islanders and (2)American Indians and Alaskan natives. The historic data are derived by subtracting "Black" from the "Black and other" group; projections are made directly. 3. Because data for blacks were not tabulated separately before 1972, data for the "Asian and other" group were not available before that year. 4. Persons of Hispanic origin may be of any race. 5. Data on Hispanics were not available before 1980.

★ 713 ★

Labor Force

Nonagricultural Employees, January 1994

Numbers in thousands. Not seasonally adjusted.

Industry and class of worker	January 1994							
	Total at work	Worked 1 to 34 hours				Worked 35 hours or more	Average hours	
		Total	For economic reasons	For noneconomic reasons			Total at work	Persons who usually work full time
				Usually work full time	Usually work part time			
Race								
White								
16 years and over	95,836	26,041	3,961	6,955	15,125	69,795	38.8	43.2
Men	51,796	10,146	2,030	3,503	4,614	41,650	41.9	44.7
Women	44,040	15,894	1,931	3,452	10,511	28,145	35.2	41.1
Black								
16 years and over	11,661	3,212	770	975	1,468	8,449	37.3	40.9
Men	5,525	1,220	333	431	456	4,305	39.2	41.9
Women	6,136	1,993	437	544	1,012	4,143	35.6	39.9

Source: "Persons at Work in Nonagricultural Industries by Age, Sex, Race, Marital Status, and Usual Full- or Part-time Status." *Employment and Earnings* 41 (February 1994), p. 64. *Notes:* Data for 1994 are not directly comparable with data for 1993 and earlier years.

★ 714 ★

Labor Force

Nonagricultural Employees: Annual Averages, 1993

Numbers in thousands.

Sex, age, race, and marital status	1993							
	Total at work	On part time for economic reasons	On voluntary part time	On full-time schedules			Average hours, total at work	Average hours, workers on full-time schedules
				Total	40 hours or less	41 hours or more		
Race								
White								
16 years and over	94,835	4,951	13,070	76,814	46,714	30,100	39.5	44.0
Men	51,777	2,293	3,926	45,558	24,449	21,109	42.5	45.5
Women	43,058	2,658	9,144	31,256	22,266	8,991	35.9	41.7
Black								
16 years and over	11,408	914	1,066	9,428	7,290	2,138	38.3	41.8
Men	5,580	437	382	4,761	3,447	1,314	39.7	42.9
Women	5,828	476	684	4,667	3,843	824	36.9	40.6

Source: "Persons at Work in Nonagricultural Industries by Sex, Age, Race, Marital Status, and Usual Full- or Part-time Status." *Employment and Earnings* 41 (January 1994), p. 222.

★ 715 ★

Labor Force

Percent Employed by Race and Age, 1980 and 1990

Age cohort	1980		1990	
	African American[1]	White	African American	White
20-24 years	65.7	77.1	68.8 (+4.7%)	77.9 (+1.0%)
35-44 years	78.3	80.4	83.0 (+6.0%)	85.9 (+6.8%)
Total population	59.8	63.7	63.1 (+5.5%)	66.4 (+4.2%)

Source: "Percent in Labor Force by Race and Selected Age Cohorts: 1980 and 1990," *The State of Black America 1994*, p. 220. Primary source: Prepared by the National Urban League from U.S. Bureau of Labor Statistics, *Employment and Earnings*, Vol. 28, No. 4, April 1981, and Vol. 37, No. 4, (April 1990), Table A-4, Washington, DC, 1981 and 1990. *Note:* 1. Figures for "black and other."

★ 716 ★

Labor Force

Persons Employed But Not At Work: Characteristics and Annual Averages, 1992

In thousands.

Population group and area	Total	Reason not at work			
		Vacation	Illness	Bad weather	Other[1]
Total					
Northeast	1,286	750	266	15	254
New England	369	220	74	4	70
Middle Atlantic	917	530	191	11	184
Midwest	1,535	864	326	33	312
East North Central	1,109	629	246	24	210
West North Central	426	235	80	10	101
South	1,970	1,062	443	48	416
South Atlantic	1,017	554	229	16	218
East South Central	339	174	83	11	71
West South Central	614	334	131	21	127
West	1,292	737	223	30	302
Mountain	307	184	57	5	62
Pacific	984	554	165	25	240
White					
Northeast	1,131	670	230	15	216
New England	348	208	69	4	67
Middle Atlantic	782	461	161	11	149
Midwest	1,377	789	279	31	278
East North Central	978	569	205	22	182
West North Central	399	220	74	9	96
South	1,599	894	335	37	333
South Atlantic	785	447	161	12	165
East South Central	281	147	65	8	61
West South Central	532	300	109	16	107
West	1,136	654	190	28	263
Mountain	292	176	54	5	57
Pacific	843	478	136	24	205
Black					
Northeast	121	58	30	[2]	33
Middle Atlantic	106	50	26	[2]	30
Midwest	131	60	42	2	27
East North Central	112	50	37	1	23
South	341	153	104	11	73
South Atlantic	217	100	66	4	47

[Continued]

★ 716 ★

Persons Employed But Not At Work: Characteristics and Annual Averages, 1992
[Continued]

Population group and area	Total	Reason not at work			
		Vacation	Illness	Bad weather	Other[1]
East South Central	56	26	18	3	10
West South Central	68	27	20	4	16
West	55	26	16	[2]	13
Pacific	50	24	15	[2]	11
Hispanic origin					
Northeast	53	27	14	1	10
Middle Atlantic	45	24	11	1	9
South	124	63	25	3	32
South Atlantic	39	18	8	1	12
West South Central	84	45	17	3	19
West	178	85	37	12	44
Mountain	34	19	7	1	6
Pacific	144	66	30	10	38

Source: "Census Regions and Divisions: Employed Persons With a Job But Not At Work by Sex, Age, Race, Hispanic Origin, and Reason Not At Work, 1992 Annual Averages," U.S. Department of Labor, *Geographical Profile of Employment and Unemployment, 1992*, pp. 29-30. Notes: Data for demographic groups are not shown when they do not meet BLS publication standards of reliability for the particular area based on the sample in that area. Items may not add to totals or compute to displayed percentages because of rounding. Detail for Hispanic-origin groups will not add to totals because data for the "other races" group are not presented and Hispanics are included in both the white and black population groups. 1. Includes industrial disputes 2. Less than 500 persons.

★ 717 ★
Labor Force

Persons Employed Less Than Usual Status, by State: Annual Averages, 1992 - I

In thousands.

Population group and State	Usually work full time							
	Total	Slack work or material shortages	Job started or terminated	Holiday	Bad weather	Own illness	On vacation	Other[1]
White								
Alabama	121	19	5	37	8	13	19	20
Alaska	19	3	1	5	[3]	2	4	3
Arizona	144	29	5	47	3	18	22	20
Arkansas	74	11	2	17	7	10	13	15
California	1,162	292	27	368	42	127	151	155
Colorado	148	14	5	43	10	18	28	29
Connecticut	138	18	2	63	4	17	22	12

[Continued]

★ 717 ★

Persons Employed Less Than Usual Status, by State: Annual Averages, 1992 - I
[Continued]

Population group and State		Usually work full time						
	Total	Slack work or material shortages	Job started or terminated	Holiday	Bad weather	Own illness	On vacation	Other[1]
Delaware	25	4	[3]	8	1	2	6	3
District of Columbia	12	1	[3]	5	[3]	2	2	2
Florida	390	103	13	100	7	53	46	67
Georgia	169	20	3	56	8	23	29	30
Hawaii	19	4	1	7	1	2	1	2
Idaho	46	9	2	12	1	6	9	7
Illinois	487	68	9	202	20	50	73	64
Indiana	222	29	4	89	19	23	29	29
Iowa	124	15	2	35	6	18	26	22
Kansas	109	9	3	36	8	13	21	19
Kentucky	120	17	4	30	12	16	18	23
Louisiana	124	15	3	43	8	17	19	19
Maine	62	10	1	22	3	7	8	10
Maryland	183	18	4	72	11	20	38	20
Massachusetts	268	33	4	130	5	29	39	28
Michigan	340	47	8	107	38	40	56	44
Minnesota	236	25	7	73	5	30	59	37
Mississippi	69	12	1	20	6	8	11	11
Missouri	197	33	6	61	10	30	36	21
Montana	39	6	1	11	1	3	8	8
Nebraska	60	7	1	19	3	6	16	8
Nevada	48	12	2	13	1	7	7	7
New Hampshire	51	9	2	20	1	5	9	7
New Jersey	282	42	4	138	11	26	35	27
New Mexico	62	10	1	21	2	7	11	10
New York	597	77	7	280	21	59	86	68
North Carolina	258	43	7	74	18	32	47	37
North Dakota	25	4	1	7	1	2	6	5
Ohio	402	40	7	149	34	50	70	52
Oklahoma	108	21	3	36	6	11	17	14
Oregon	118	23	3	25	3	18	29	17
Pennsylvania	550	66	7	223	45	60	87	62
Rhode Island	53	6	1	32	1	5	5	4
South Carolina	98	16	3	25	5	14	22	14
South Dakota	27	3	1	8	1	3	6	5
Tennessee	165	28	6	49	10	22	18	34
Texas	657	89	23	214	30	82	104	113
Utah	74	8	2	22	[3]	9	19	12
Vermont	26	4	1	8	1	4	5	3

[Continued]

★ 717 ★

Persons Employed Less Than Usual Status, by State: Annual Averages, 1992 - I
[Continued]

Population group and State	Usually work full time							
	Total	Slack work or material shortages	Job started or terminated	Holiday	Bad weather	Own illness	On vacation	Other[1]
Virginia	275	32	7	83	23	34	61	36
Washington	212	35	6	63	6	32	37	33
West Virginia	62	12	1	18	7	7	9	7
Wisconsin	183	23	5	51	6	21	52	25
Wyoming	19	4	1	5	1	2	3	3
Black								
Alabama	4	4	4	4	4	4	4	4
Arkansas	4	4	4	4	4	4	4	4
California	71	10	2	30	3	10	8	11
Delaware	4	4	4	4	4	4	4	4
District of Columbia	17	3	3	8	3	2	1	3
Florida	79	28	3	18	2	11	3	13
Georgia	67	15	1	20	3	10	9	9
Illinois	69	7	1	35	1	9	5	12
Louisiana	4	4	4	4	4	4	4	4
Maryland	79	9	1	39	5	9	8	9
Massachusetts	4	4	4	4	4	4	4	4
Michigan	41	6	1	13	3	5	6	8
Mississippi	41	9	1	11	4	5	3	7
New Jersey	41	5	3	24	1	4	3	4
New York	94	13	1	49	2	11	8	10
North Carolina	70	19	2	19	5	11	5	9
Ohio	40	6	3	12	2	6	7	6
Pennsylvania	41	6	3	12	2	7	6	7
South Carolina	42	12	2	8	5	7	2	5
Tennessee	4	4	4	4	4	4	4	4
Texas	84	9	3	32	4	9	13	14
Virginia	67	17	4	13	5	11	7	11

Source: "States: Persons At Work 1 to 34 Hours by Sex, Race, Reason for Working Less Than 35 Hours, and Usual Status, 1992 Annual Averages," U.S. Department of Labor, *Geographical Profile of Employment and Unemployment, 1992*, pp. 81-82. *Notes:* Items may not add to totals because of rounding. 1. Includes industrial disputes. 2. Does not want, or unavailable for, full-time work. 3. Less than 500 persons. 4. Data are not shown when the labor force base does not meet BLS publication standards of reliability for the particular area, based on the sample in that area.

★ 718 ★

Labor Force

Persons Employed Less Than Usual Status, by State: Annual Averages, 1992 - II

In thousands.

Population group and State	Total	Usually work part time			
		Slack work or only find part-time	Does not want full-time work[1]	Full-time work less than 35 hours	Other
White					
Alabama	175	26	127	14	7
Alaska	28	8	15	2	2
Arizona	235	67	143	14	11
Arkansas	137	25	92	11	10
California	1,902	487	1,112	208	95
Colorado	231	55	148	11	16
Connecticut	266	54	182	17	14
Delaware	42	6	29	4	3
District of Columbia	12	4	7	1	2
Florida	790	179	462	94	55
Georgia	278	38	191	32	17
Hawaii	29	3	21	4	1
Idaho	85	22	56	3	4
Illinois	806	162	530	69	45
Indiana	406	86	262	40	18
Iowa	279	50	201	11	17
Kansas	197	29	151	8	8
Kentucky	257	55	160	29	12
Louisiana	225	62	128	18	17
Maine	118	31	72	8	7
Maryland	261	44	182	20	16
Massachusetts	529	115	347	35	32
Michigan	707	173	448	55	31
Minnesota	428	77	306	17	27
Mississippi	100	17	64	14	5
Missouri	381	62	276	25	18
Montana	74	17	48	3	5
Nebraska	148	22	113	6	8
Nevada	75	16	47	9	3
New Hampshire	104	25	69	6	4
New Jersey	474	79	323	49	24
New Mexico	111	31	66	9	6
New York	1,085	210	680	130	65
North Carolina	339	50	228	41	20
North Dakota	63	12	45	3	3
Ohio	804	154	532	68	50
Oklahoma	179	35	121	15	7
Oregon	255	64	163	14	15

[Continued]

★ 718 ★

Persons Employed Less Than Usual Status, by State: Annual Averages, 1992 - II

[Continued]

Population group and State		Usually work part time			
	Total	Slack work or only find part-time	Does not want full-time work[1]	Full-time work less than 35 hours	Other
Pennsylvania	903	198	612	53	40
Rhode Island	91	14	66	5	6
South Carolina	177	30	120	22	6
South Dakota	59	8	43	4	4
Tennessee	284	61	173	25	25
Texas	1,011	232	627	95	57
Utah	165	22	137	1	5
Vermont	55	10	38	4	3
Virginia	360	77	239	27	18
Washington	405	84	269	30	22
West Virginia	113	39	54	14	5
Wisconsin	432	69	329	16	18
Wyoming	41	11	26	3	2
Black					
Alabama	58	28	22	6	2
Arkansas	22	9	8	3	2
California	102	35	48	12	7
Delaware	8	3	4	1	1
District of Columbia	15	5	8	1	1
Florida	126	56	52	10	8
Georgia	116	42	52	15	6
Illinois	106	40	45	13	7
Louisiana	77	38	24	9	5
Maryland	81	29	35	9	8
Massachusetts	15	5	7	1	1
Michigan	67	26	33	4	5
Mississippi	50	23	18	6	3
New Jersey	47	17	21	7	2
New York	119	42	49	18	10
North Carolina	104	34	49	16	5
Ohio	64	23	30	7	4
Pennsylvania	44	17	20	2	5
South Carolina	67	24	32	8	2
Tennessee	50	21	24	3	2

[Continued]

★ 718 ★

Persons Employed Less Than Usual Status, by State: Annual Averages, 1992 - II

[Continued]

Population group and State		Usually work part time			
	Total	Slack work or only find part-time	Does not want full-time work[1]	Full-time work less than 35 hours	Other
Texas	122	52	47	14	9
Virginia	73	27	36	5	5

Source: "States: Persons At Work 1 to 34 Hours by Sex, Race, Reason for Working Less Than 35 Hours, and Usual Status, 1992 Annual Averages," U.S. Department of Labor, *Geographical Profile of Employment and Unemployment, 1992*, pp. 81-82. *Notes:* Items may not add to totals because of rounding. 1. Does not want, or unavailable for, full-time work. 2. Less than 500 persons.

★ 719 ★

Labor Force

Private Nonagricultural Wage and Salary Workers: Annual Averages, 1992 - I

Population group and industry	Northeast			Midwest		
	Total	New England	Middle Atlantic	Total	East North Central	West North Central
White						
Total (in thousands)	15,913	4,778	11,134	20,115	14,023	6,092
Percent	100.0	100.0	100.0	100.0	100.0	100.0
Mining	.2	.1	.3	.4	.4	.4
Construction	5.0	4.3	5.4	5.4	5.3	5.4
Manufacturing	22.0	23.8	21.2	26.7	28.9	21.6
Durable goods	12.7	15.7	11.4	17.1	19.1	12.3
Lumber and wood products	.4	.5	.3	.6	.6	.6
Furniture and fixtures	.4	.3	.4	.8	.9	.6
Stone, clay, and glass products	.6	.3	.7	.7	.9	.5
Machinery, except electrical	1.0	.7	1.1	1.5	1.9	.6
Fabricated metal products	1.3	1.6	1.2	2.0	2.3	1.5
Machinery, except electrical equipment	3.0	3.8	2.7	3.6	3.9	3.1
Electrical machinery, equipment, and supplies	2.2	3.0	1.8	2.3	2.5	1.9
Transportation equipment	1.6	2.7	1.1	4.1	5.0	2.2
Motor vehicles	.4	.3	.5	3.3	4.3	1.1
Professional and photographic equipment, and watches, etc.[1]	1.7	2.0	1.5	1.0	.9	1.1
Nondurable goods	9.3	8.1	9.8	9.6	9.8	9.3
Food and kindred products	1.3	1.0	1.5	2.5	2.1	3.4
Textile mill products	.5	.7	.4	.1	.1	.1
Apparel and other textile products	1.2	0.7	1.4	0.3	0.3	0.4
Paper and allied products	.8	1.0	.7	1.1	1.3	.7

[Continued]

★ 719 ★

Private Nonagricultural Wage and Salary Workers: Annual Averages, 1992 - I
[Continued]

Population group and industry	Northeast			Midwest		
	Total	New England	Middle Atlantic	Total	East North Central	West North Central
Rubber and miscellaneous plastics products	2.4	2.2	2.5	2.3	2.3	2.3
Chemicals and allied products	1.8	1.1	2.1	1.6	1.9	.9
Rubber and miscellaneous plastic products	.7	.8	.7	1.4	1.5	1.2
Transportation, communications, and public utilities	7.2	6.2	7.6	6.4	5.9	7.7
Transportation	4.0	3.3	4.3	3.9	3.6	4.5
Communications and other public utilities	3.2	2.9	3.3	2.6	2.3	3.1
Wholesale and retail trade	24.2	24.0	24.2	26.4	25.4	28.7
Wholesale trade	4.8	4.2	5.0	5.3	4.9	6.2
Retail trade	19.4	19.8	19.2	21.1	20.4	22.5
Finance, insurance, and real estate	9.7	9.1	10.0	7.3	7.2	7.5
Services excluding private households	31.7	32.5	31.4	27.5	26.9	28.7
Professional services	22.1	23.5	21.5	18.4	17.8	19.8
Educational services	3.4	3.8	3.2	2.2	2.1	2.4
Medical services, including hospitals	10.6	11.7	10.1	9.7	9.5	10.3
Black						
Total (in thousands)	1,464	204	1,260	1,545	1,279	267
Percent	100.0	100.0	100.0	100.0	100.0	100.00
Mining	[2]	[3]	[2]	.1	.1	[3]
Construction	3.0	2.6	3.0	1.7	1.8	1.0
Manufacturing	17.4	25.4	16.1	25.0	26.1	19.5
Durable goods	8.9	17.6	7.4	15.6	16.8	9.8
Lumber and wood products	[2]	[3]	[2]	.2	.2	[3]
Furniture and fixtures	.3	.4	.2	.2	.3	[3]
Stone, clay, and glass products	.6	.2	.7	.4	.5	.2
Machinery, except electrical	.4	.6	.4	2.0	2.2	1.1
Fabricated metal products	.8	2.0	.6	1.2	1.4	.7
Machinery, except electrical equipment	2.0	3.3	1.7	1.3	1.4	.6
Electrical machinery, equipment, and supplies	1.7	3.4	1.4	1.4	1.4	1.1
Transportation equipment	1.2	4.5	.7	7.6	8.3	4.4
Motor vehicles	.5	.5	.5	6.9	7.9	2.5
Professional and photographic equipment, and watches, etc.[1]	1.3	2.7	1.0	.9	.8	1.4
Nondurable goods	8.6	7.8	8.7	9.4	9.3	9.7
Food and kindred products	1.2	1.7	1.1	2.7	2.6	3.6
Textile mill products	.4	[3]	.5	.1	.1	.1
Apparel and other textile products	1.2	.6	1.3	.5	.5	.4
Paper and allied products	.7	.6	.7	.8	.7	1.6

[Continued]

★ 719 ★

Private Nonagricultural Wage and Salary Workers: Annual Averages, 1992 - I

[Continued]

Population group and industry	Northeast			Midwest		
	Total	New England	Middle Atlantic	Total	East North Central	West North Central
Rubber and miscellaneous plastics products	2.2	2.1	2.2	1.9	2.1	1.2
Chemicals and allied products	1.9	1.1	2.1	1.9	1.9	1.7
Rubber and miscellaneous plastic products	.5	1.2	.4	1.0	1.0	1.0
Transportation, communications, and public utilities	9.6	8.3	9.8	8.3	7.9	9.9
Transportation	5.4	5.3	5.5	4.5	4.3	5.4
Communications and other public utilities	4.2	3.0	4.4	3.8	3.7	4.4
Wholesale and retail trade	16.4	16.7	16.4	22.1	21.8	23.4
Wholesale trade	3.3	1.8	3.6	3.4	3.3	3.9
Retail trade	13.1	14.9	12.8	18.6	18.4	19.5
Finance, insurance, and real estate	11.3	9.2	11.6	7.7	7.7	7.4
Services excluding private households	42.3	37.8	43.0	35.3	34.6	38.7
Professional services	27.9	27.5	28.0	22.7	22.5	23.5
Educational services	2.3	2.4	2.3	2.2	2.1	3.1
Medical services, including hospitals	18.0	19.0	17.8	13.7	13.4	15.4
Hispanic origin						
Total (in thousands)	985	134	851	560	479	81
Percent	100.0	100.0	100.0	100.0	100.0	100.0
Mining	.1	.1	.1	.2	.2	[3]
Construction	3.5	2.8	3.6	3.0	2.8	4.1
Manufacturing	26.7	34.8	25.4	38.7	40.5	28.4
Durable goods	11.1	23.3	9.2	22.3	24.0	11.8
Lumber and wood products	.3	.6	.2	.6	.6	.7
Furniture and fixtures	.7	1.2	.7	.9	1.1	[3]
Stone, clay, and glass products	0.8	[2]	0.9	0.6	0.5	0.9
Machinery, except electrical	.5	2.1	.3	4.1	4.8	.2
Fabricated metal products	1.3	1.1	1.4	4.1	4.3	2.8
Machinery, except electrical equipment	1.6	2.6	1.5	2.2	2.4	.8
Electrical machinery, equipment, and supplies	2.0	5.7	1.4	3.6	4.0	.7
Transportation equipment	1.1	2.4	.9	4.4	4.4	4.5
Motor vehicles	.6	.8	.6	3.9	4.3	1.2
Professional and photographic equipment, and watches, etc.[1]	1.2	2.4	1.0	1.2	1.2	1.2
Nondurable goods	15.6	11.5	16.2	16.5	16.5	16.5
Food and kindred products	1.9	1.7	1.9	5.7	5.0	10.0
Textile mill products	1.2	1.9	1.1	[2]	[2]	[3]
Apparel and other textile products	5.0	1.4	5.6	.9	1.0	[3]
Paper and allied products	1.3	1.6	1.3	1.9	2.3	.1

[Continued]

★ 719 ★

Private Nonagricultural Wage and Salary Workers: Annual Averages, 1992 - I

[Continued]

Population group and industry	Northeast			Midwest		
	Total	New England	Middle Atlantic	Total	East North Central	West North Central
Rubber and miscellaneous plastics products	2.3	1.4	2.4	2.1	2.0	3.1
Chemicals and allied products	2.0	.6	2.2	2.3	2.2	2.3
Rubber and miscellaneous plastic products	1.1	1.1	1.1	2.6	2.9	1.0
Transportation, communications, and public utilities	6.3	3.5	6.8	4.5	4.2	6.3
Transportation	4.4	2.4	4.7	3.0	2.6	5.8
Communications and other public utilities	2.0	1.1	2.1	1.5	1.7	.4
Wholesale and retail trade	23.4	23.5	23.3	23.3	22.6	27.6
Wholesale trade	3.7	2.2	4.0	3.3	3.3	3.8
Retail trade	19.6	21.3	19.4	20.0	19.3	23.9
Finance, insurance, and real estate	8.9	5.1	9.6	5.8	4.9	11.3
Services excluding private households	31.0	30.1	31.2	24.5	24.9	22.4
Professional services	16.7	20.7	16.1	11.0	10.9	11.5
Educational services	1.6	1.9	1.5	1.7	1.6	2.0
Medical services, including hospitals	10.1	12.8	9.7	5.5	5.6	5.4

Source: "Census Regions and Divisions, Percent Distribution of Employed Private Nonagricultural Wage and Salary Workers by Industry, Sex, Race, and Hispanic Origin, 1992 Annual Averages," U.S. Department of Labor, *Geographical Profile of Employment and Unemployment, 1992*, pp. 22-23. *Notes:* Items may not add to totals or compute to displayed percentages because of rounding. Detail for race and Hispanic-origin groups will not add to totals because data for the "other races" group are not presented and Hispanics are included in both the white and black population groups. 1. Includes toys, amusement, and sporting goods. 2. Less than 500 persons employed or less than 0.05 percent of total employment. 3. Data are not shown when the labor force base does not meet BLS publication standards of reliability for the particular area, based on the sample in that area.

★ 720 ★

Labor Force

Private Nonagricultural Wage and Salary Workers: Annual Averages, 1992 - II

Population group and industry	South				West		
	Total	South Atlantic	East South Central	West South Central	Total	Mountain	Pacific
White							
Total (in thousands)	23,637	11,955	4,107	7,575	15,177	4,199	10,978
Percent	100.0	100.0	100.0	100.0	100.0	100.0	100.0
Mining	1.6	.6	1.3	3.3	.8	2.0	.4
Construction	7.2	7.4	6.5	7.3	6.6	7.3	6.4
Manufacturing	20.5	19.2	28.4	18.3	19.3	14.8	21.1
Durable goods	10.4	9.0	15.4	9.9	12.2	9.5	13.3
Lumber and wood products	.8	.6	1.4	.6	1.0	.8	1.1
Furniture and fixtures	.9	.9	1.8	.4	.5	.4	.6
Stone, clay, and glass products	.6	.5	.7	.7	.5	.5	.5

[Continued]

★ 720 ★

Private Nonagricultural Wage and Salary Workers: Annual Averages, 1992 - II
[Continued]

Population group and industry	South				West		
	Total	South Atlantic	East South Central	West South Central	Total	Mountain	Pacific
Machinery, except electrical	.7	.6	1.4	.6	.4	.4	.5
Fabricated metal products	1.1	.8	1.6	1.2	1.1	.8	1.1
Machinery, except electrical equipment	2.0	1.7	2.2	2.4	2.2	2.1	2.3
Electrical machinery, equipment, and supplies	1.7	1.4	2.5	1.6	2.2	1.8	2.3
Transportation equipment	1.7	1.4	3.1	1.4	2.9	1.5	3.5
Motor vehicles	.7	.4	1.9	.3	.4	.2	.4
Professional and photographic equipment, and watches, etc.[1]	.7	.7	.5	.7	1.1	.9	1.2
Nondurable goods	10.2	10.3	13.0	8.4	7.1	5.3	7.8
Food and kindred products	1.8	1.6	1.8	2.0	2.0	1.7	2.0
Textile mill products	1.4	2.4	.9	.1	.1	.1	.2
Apparel and other textile products	1.5	1.2	3.7	0.8	1.0	0.4	1.2
Paper and allied products	.9	.9	1.6	.6	.4	.2	.4
Rubber and miscellaneous plastics products	1.6	1.7	1.8	1.4	1.9	1.6	2.0
Chemicals and allied products	1.6	1.5	1.4	1.8	.9	.7	.9
Rubber and miscellaneous plastic products	.8	.8	1.3	.7	.6	.4	.6
Transportation, communications, and public utilities	7.9	7.8	7.5	8.3	7.6	8.1	7.4
Transportation	4.4	4.3	4.2	4.8	4.3	4.5	4.2
Communications and other public utilities	3.5	3.5	3.3	3.5	3.3	3.6	3.1
Wholesale and retail trade	26.9	26.8	25.4	27.7	27.4	27.8	27.3
Wholesale trade	5.7	5.5	5.3	6.2	5.3	4.9	5.4
Retail trade	21.2	21.4	20.1	21.5	22.2	22.9	21.9
Finance, insurance, and real estate	7.7	8.3	6.5	7.4	8.2	8.0	8.3
Services excluding private households	28.3	29.9	24.4	27.8	29.9	31.9	29.2
Professional services	17.6	18.6	16.2	16.8	17.2	17.8	17.0
Educational services	1.7	1.9	1.3	1.6	1.8	1.6	1.8
Medical services, including hospitals	8.8	8.7	9.3	8.6	7.9	8.3	7.8
Black							
Total (in thousands)	4,735	2,973	804	959	654	115	539
Percent	100.0	100.0	100.0	100.0	100.0	100.0	100.0
Mining	.4	.2	.4	.9	.5	2.2	.1
Construction	4.2	4.4	4.3	3.8	3.9	3.9	3.9
Manufacturing	27.6	28.2	33.3	21.2	17.0	9.6	18.5
Durable goods	11.4	9.7	17.3	11.5	9.9	4.7	11.0
Lumber and wood products	1.6	1.5	3.1	.8	.1	[2]	.1
Furniture and fixtures	.8	.7	1.7	.3	.2	[3]	.2
Stone, clay, and glass products	1.0	.9	1.5	1.0	.1	.4	[2]
Machinery, except electrical	.7	.5	1.4	.9	.1	.3	.1
Fabricated metal products	1.0	.7	1.9	1.1	.7	.6	.7
Machinery, except electrical equipment	1.6	1.4	1.6	2.5	1.6	.6	1.8
Electrical machinery, equipment, and supplies	2.0	1.9	2.1	2.2	1.7	1.1	1.8
Transportation equipment	1.8	1.6	2.3	1.7	4.9	1.0	5.7
Motor vehicles	.7	.6	1.4	.5	.3	[3]	.4

[Continued]

★ 720 ★

Private Nonagricultural Wage and Salary Workers: Annual Averages, 1992 - II

[Continued]

Population group and industry	South				West		
	Total	South Atlantic	East South Central	West South Central	Total	Mountain	Pacific
Professional and photographic equipment, and watches, etc.[1]	.6	.5	.6	.9	.5	.6	.5
Nondurable goods	16.3	18.5	16.0	9.6	7.1	5.0	7.5
Food and kindred products	3.8	3.9	4.5	2.8	1.8	.7	2.0
Textile mill products	3.7	5.5	1.0	.5	.1	[3]	.1
Apparel and other textile products	3.0	3.2	4.5	1.2	.6	[3]	.8
Paper and allied products	1.2	1.0	2.2	.9	.4	[3]	.5
Rubber and miscellaneous plastics products	1.2	1.3	1.2	.9	1.8	1.3	1.9
Chemicals and allied products	1.4	1.5	1.0	1.6	.8	1.4	.7
Rubber and miscellaneous plastic products	1.2	1.2	.9	1.1	.6	.5	.6
Transportation, communications, and public utilities	8.1	8.4	6.2	8.8	10.4	11.8	10.1
Transportation	4.8	4.9	3.9	5.2	5.5	6.3	5.4
Communications and other public utilities	3.3	3.4	2.3	3.5	4.8	5.5	4.7
Wholesale and retail trade	24.6	23.2	25.9	27.7	21.3	21.1	21.3
Wholesale trade	3.4	3.4	3.8	3.4	3.8	2.8	4.1
Retail trade	21.1	19.8	22.1	24.3	17.5	18.5	17.3
Finance, insurance, and real estate	5.4	5.8	4.1	5.3	9.4	7.9	9.7
Services excluding private households	29.7	29.8	25.8	32.4	37.6	43.5	36.4
Professional services	17.7	17.3	16.3	20.1	20.1	19.0	20.3
Educational services	1.5	1.7	1.3	2.1	3.4	1.8	
Medical services, including hospitals	11.0	10.6	13.3	9.7	8.7	10.0	
Hispanic origin							
Total (in thousands)	2,279	853	[3]	1,405	2,996	511	4,485
Percent	100.0	100.0	100.0	100.0	100.0	100.0	100.0
Mining	1.2	[2]	[3]	2.0	.5	1.5	.3
Construction	8.9	7.3	[3]	10.0	7.6	10.1	7.1
Manufacturing	18.9	17.3	[3]	19.5	27.0	17.0	29.0
Durable goods	8.4	6.8	[3]	9.2	13.9	9.0	14.9
Lumber and wood products	.4	.2	[3]	.5	.9	.8	.9
Furniture and fixtures	1.0	1.4	[3]	.7	1.2	.5	1.3
Stone, clay, and glass products	0.6	0.1	[3]	0.9	0.7	0.4	0.7
Machinery, except electrical	.5	.1	[3]	.8	.5	.3	:5
Fabricated metal products	.9	.5	[3]	1.2	1.7	.8	1.9
Machinery, except electrical equipment	1.5	1.2	[3]	1.5	2.0	1.6	2.1
Electrical machinery, equipment, and supplies	1.2	1.0	[3]	1.2	2.6	2.6	2.6
Transportation equipment	1.1	1.3	[3]	1.0	2.9	1.1	3.2
Motor vehicles	.3	.3	[3]	.3	.7	.1	.8
Professional and photographic equipment, and watches, etc.[1]	.7	.3	[3]	.9	1.0	.5	1.1
Nondurable goods	10.5	10.5	[3]	10.3	13.1	8.0	14.1
Food and kindred products	4.5	4.7	[3]	4.3	4.0	3.9	4.0
Textile mill products	.3	.5	[3]	.2	.4	.2	.4
Apparel and other textile products	2.2	2.7	[3]	1.8	3.6	1.2	4.1

[Continued]

★ 720 ★

Private Nonagricultural Wage and Salary Workers: Annual Averages, 1992 - II

[Continued]

Population group and industry	South				West		
	Total	South Atlantic	East South Central	West South Central	Total	Mountain	Pacific
Paper and allied products	.4	.4	3	.4	.5	.5	.4
Rubber and miscellaneous plastics products	1.0	1.1	3	.9	1.9	1.6	2.0
Chemicals and allied products	.8	.4	3	1.0	1.1	.3	1.3
Rubber and miscellaneous plastic products	.7	.4	3	.9	1.3	.2	1.6
Transportation, communications, and public utilities	6.5	7.7	3	6.0	6.4	7.0	6.3
Transportation	4.6	5.6	3	4.1	4.0	3.8	4.0
Communications and other public utilities	1.9	2.1	3	1.9	2.4	3.2	2.2
Wholesale and retail trade	31.1	29.1	3	32.4	28.6	28.9	28.6
Wholesale trade	5.9	6.2	3	5.7	5.0	4.1	5.2
Retail trade	25.2	22.8	3	26.8	23.7	24.8	23.4
Finance, insurance, and real estate	6.1	7.1	3	5.6	5.3	5.5	5.3
Services excluding private households	27.2	31.5	3	24.6	24.6	30.1	23.4
Professional services	12.9	14.5	3	11.8	10.6	13.0	10.1
Educational services	1.2	1.5	3	1.0	.9	1.0	.9
Medical services, including hospitals	7.3	7.8	3	6.8	6.1	7.6	5.8

Source: "Census Regions and Divisions, Percent Distribution of Employed Private Nonagricultural Wage and Salary Workers by Industry, Sex, Race, and Hispanic Origin, 1992 Annual Averages," U.S. Department of Labor, *Geographical Profile of Employment and Unemployment, 1992*, pp. 22-23. *Notes:* Items may not add to totals or compute to displayed percentages because of rounding. Detail for race and Hispanic-origin groups will not add to totals because data for the "other races" group are not presented and Hispanics are included in both the white and black population groups. 1. Includes toys, amusement, and sporting goods. 2. Less than 500 persons employed or less than 0.05 percent of total employment. 3. Data are not shown when the labor force base does not meet BLS publication standards of reliability for the particular area, based on the sample in that area.

★ 721 ★

Labor Force

Three Projections of Civilian Labor Force, 2005

Group	Participation rate (percent)			Level (thousands)		
	High	Moderate	Low	High	Moderate	Low
Total	71.5	69.0	66.1	156,169	150,732	141,774
16 to 24 years	72.6	69.5	65.8	25,138	24,048	22,153
25 to 54 years	89.4	87.3	85.1	107,105	104,562	99,553
55 years and over	37.4	34.6	31.5	23,926	22,122	20,068
Men	77.3	75.4	72.9	81,360	79,338	75,184
Women	66.1	63.0	59.8	74,809	71,394	66,590
White	72.3	69.7	66.7	130,453	125,785	118,370
Black	67.7	65.6	63.2	18,341	17,766	16,940

[Continued]

★ 721 ★

Three Projections of Civilian Labor Force, 2005

[Continued]

Group	Participation rate (percent)			Level (thousands)		
	High	Moderate	Low	High	Moderate	Low
Asian and other[1]	68.2	66.4	64.3	7,375	7,181	6,464
Hispanic[2]	74.6	69.9	67.3	17,906	16,790	16,163

Source: "Three Projections of the Civilian Labor Force by Sex, Age, Race, and Hispanic Origin, 2005." U.S. Department of Labor, Employment and Training Administration, *Training and Employment Report of the Secretary of Labor,* Covering the Period July 1988-September 1990, p. 103. Primary source: U.S. Department of Labor, Bureau of Labor Statistics. *Notes:* 1. The "Asian and other" group includes (1)Asians and Pacific Islanders and (2)American Indians and Alaskan natives. The historic data are derived by subtracting "Black" from the "Black and other" group; projections are made directly. 2. Persons of Hispanic origin may be of any race.

★ 722 ★

Labor Force

Unemployed Persons and Duration of Unemployment: Annual Averages, 1992

Population group and area	Total unemployed		Duration of unemployment						
	Number (in thousands)	Percent	Less than 5 weeks	5 to 14 weeks	15 weeks and over	15 to 26 weeks	27 weeks and over	27 to 51 weeks	52 weeks and over
White									
Northeast	1,651	100.0	25.9	27.9	46.1	18.1	28.0	13.4	14.5
New England	505	100.0	23.7	25.9	50.4	17.7	32.7	13.8	18.9
Middle Atlantic	1,146	100.0	27.0	28.8	44.2	18.3	25.9	13.3	12.6
Midwest	1,577	100.0	36.2	30.4	33.4	14.9	18.5	8.5	10.0
East North Central	1,198	100.0	35.5	29.7	34.8	15.1	19.7	9.0	10.7
West North Central	379	100.0	38.2	32.6	29.2	14.4	14.8	7.1	7.7
South	1,999	100.0	38.7	29.7	31.7	13.6	18.1	8.7	9.4
South Atlantic	979	100.0	35.1	29.2	35.6	14.8	20.9	10.7	10.2
East South Central	344	100.0	39.0	31.2	29.9	13.4	16.5	7.3	9.1
West South Central	677	100.0	43.6	29.5	26.9	11.9	14.9	6.6	8.3
West	1,820	100.0	38.3	29.0	32.7	14.6	18.1	8.6	9.4
Mountain	395	100.0	44.0	29.1	26.9	12.7	14.3	8.2	6.1
Pacific	1,425	100.0	36.7	29.0	34.3	15.2	19.1	8.8	10.3
Black									
Northeast	363	100.0	22.3	27.1	50.6	19.0	31.6	11.3	20.2
New England	45	100.0	21.4	24.9	53.7	19.5	34.2	14.9	19.3
Middle Atlantic	319	100.0	22.4	27.4	50.2	19.0	31.2	10.8	20.4
Midwest	418	100.0	39.5	30.8	29.7	12.6	17.0	6.0	11.1
East North Central	352	100.0	40.7	29.8	29.5	11.8	17.7	5.8	12.0
West North Central	66	100.0	33.1	36.2	30.7	17.3	13.3	6.9	6.4
South	1,016	100.0	37.1	29.0	33.8	14.9	19.0	7.5	11.5

[Continued]

★ 722 ★

Unemployed Persons and Duration of Unemployment: Annual Averages, 1992

[Continued]

Population group and area	Total unemployed		Duration of unemployment						
	Number (in thousands)	Percent	Less than 5 weeks	5 to 14 weeks	15 weeks and over	15 to 26 weeks	27 weeks and over	27 to 51 weeks	52 weeks and over
South Atlantic	577	100.0	34.3	28.7	37.0	16.7	20.3	7.7	12.5
East South Central	167	100.0	38.9	32.9	28.2	12.1	16.1	6.6	9.6
West South Central	272	100.0	42.0	27.3	30.6	12.6	18.0	7.5	10.5
West	160	100.0	29.5	36.4	34.0	12.7	21.3	8.9	12.4
Pacific	140	100.0	27.9	37.2	34.9	12.2	22.7	9.9	12.8
Hispanic origin									
Northeast	195	100.0	26.6	28.0	45.4	19.5	26.0	10.9	15.1
New England	28	100.0	20.5	27.7	51.8	17.9	33.9	16.0	17.9
Middle Atlantic	167	100.0	27.6	28.0	44.4	19.7	24.6	10.0	14.6
Midwest	79	100.0	36.9	32.4	30.7	13.9	16.8	8.0	8.8
East North Central	71	100.0	35.2	32.3	32.5	14.5	18.0	8.2	9.8
South	311	100.0	43.6	27.9	28.5	13.1	15.4	8.6	6.8
South Atlantic	114	100.0	38.2	26.5	35.3	16.9	18.4	11.2	7.2
West South Central	194	100.0	46.7	28.9	24.4	10.8	13.6	7.0	6.6
West	575	100.0	40.3	28.4	31.3	14.2	17.2	7.6	9.5
Mountain	75	100.0	46.0	28.5	25.5	10.1	15.4	8.8	6.6
Pacific	500	100.0	39.4	28.4	32.2	14.8	17.4	7.5	10.0

Source: "Census Regions and Divisions: Percent Distribution of Unemployed Persons by Sex, Age, Race, Hispanic Origin, and Duration of Unemployment, 1992 Annual Averages," U.S. Department of Labor, *Geographical Profile of Employment and Unemployment, 1992*, p. 34. *Notes:* Data for demographic groups are not shown when they do not meet BLS publication standards of reliability for the particular area based on the sample in that area. Items may not add to totals or compute to displayed percentages because of rounding. Detail for Hispanic-origin groups will not add to totals because data for the "other races" group are not presented and Hispanics are included in both the white and black population groups.

★ 723 ★

Labor Force

Unemployed Persons and Reasons for Unemployment: Annual Averages, 1992

Population group and area	Total unemployed		Reason for unemployment				
	Number (in thousands)	Percent	Job losers		Job leavers	Reentrants	New entrants
			Total	On layoff			
Northeast	2,071	100.0	65.9	16.2	6.8	19.3	8.0
New England	562	100.0	68.2	15.9	6.9	17.9	6.9
Middle Atlantic	1,509	100.0	65.0	16.4	6.8	19.9	8.4
Midwest	2,041	100.0	55.3	18.0	11.0	23.6	10.1
East North Central	1,583	100.0	56.4	18.9	9.9	23.2	10.6
West North Central	458	100.0	51.6	14.6	14.9	24.8	8.6

[Continued]

★ 723 ★

Unemployed Persons and Reasons for Unemployment: Annual Averages, 1992

[Continued]

Population group and area	Total unemployed		Reason for unemployment				
	Number (in thousands)	Percent	Job losers		Job leavers	Reentrants	New entrants
			Total	On layoff			
South	3,088	100.0	50.4	9.5	12.1	27.1	10.3
South Atlantic	1,592	100.0	51.8	9.8	11.4	27.2	9.7
East South Central	513	100.0	511.3	14.5	12.0	25.0	11.8
West South Central	983	100.0	47.6	6.5	13.5	28.2	10.7
West	2,184	100.0	56.9	11.4	10.7	23.3	9.1
Mountain	442	100.0	49.5	11.8	14.2	28.3	7.9
Pacific	1,742	100.0	58.8	11.3	9.8	22.0	9.4

Source: "Census Regions and Divisions: Percent Distribution of Unemployed Persons by Sex, Age, Race, Hispanic Origin, and Reason for Unemployment, 1992 Annual Averages," U.S. Department of Labor, *Geographical Profile of Employment and Unemployment, 1992*, p. 32. *Notes:* Data for demographic groups are not shown when they do not meet BLS publication standards of reliability for the particular area based on the sample in that area. Items may not add to totals or compute to displayed percentages because of rounding. Detail for Hispanic-origin groups will not add to totals because data for the "other races" group are not presented and Hispanics are included in both the white and black population groups.

★ 724 ★

Labor Force

Work Week of Persons Employed by State: Annual Averages, 1992

In thousands.

Population group and State	Total at work	Hours of work								Average hours	
		1 to 14 hours	15 to 29 hours	30 to 34 hours	35 hours and over					Total	Full-time schedules[1]
					Total	35 to 39 hours	40 hours	41 to 48 hours	49 hours and over		
White											
Alabama	1,328	55	133	109	1,031	71	530	142	289	40.3	48.0
Alaska	183	10	20	17	136	14	62	19	42	40.8	50.1
Arizona	1,447	65	170	143	1,069	84	521	153	311	39.7	48.5
Arkansas	879	39	101	71	667	46	340	98	182	39.8	47.9
California	11,020	500	1,404	1,161	7,956	582	4,365	997	2,011	38.7	47.7
Colorado	1,447	67	166	146	1,068	89	493	153	333	39.8	49.1
Connecticut	1,384	72	199	133	980	133	459	138	250	38.0	47.9
Delaware	273	11	32	24	205	21	102	27	56	39.6	48.0
District of Columbia	105	3	11	10	81	8	32	10	31	41.7	50.5
Florida	4,803	195	572	412	3,624	292	1,821	477	1,034	39.8	47.5
Georgia	1,991	82	209	156	1,544	104	740	224	476	40.6	48.5
Hawaii	172	8	23	17	124	10	62	16	35	39.3	48.9
Idaho	452	29	60	43	321	28	140	48	104	39.3	49.7
Illinois	4,563	229	569	494	3,269	287	1,636	480	867	38.7	48.5
Indiana	2,356	119	263	245	1,729	153	811	280	485	39.4	48.5
Iowa	1,384	87	191	125	981	83	405	167	326	39.7	50.1
Kansas	1,126	64	139	103	820	67	367	125	261	39.6	49.5
Kentucky	1,434	74	171	132	1,057	117	484	163	293	39.4	47.9
Louisiana	1,309	76	155	119	960	78	451	131	300	40.2	49.9
Maine	572	35	76	69	392	46	187	59	100	38.0	48.6

[Continued]

★ 724 ★

Work Week of Persons Employed by State: Annual Averages, 1992

[Continued]

Population group and State	Total at work	Hours of work								Average hours	
		1 to 14 hours	15 to 29 hours	30 to 34 hours	35 hours and over					Total	Full-time schedules[1]
					Total	35 to 39 hours	40 hours	41 to 48 hours	49 hours and over		
Maryland	1,654	74	201	170	1,209	107	592	171	340	39.3	49.1
Massachusetts	2,570	143	364	290	1,774	175	896	245	457	37.8	48.5
Michigan	3,510	203	483	361	2,463	210	1,173	389	692	38.6	48.9
Minnesota	2,124	143	299	222	1,460	136	615	256	453	38.7	50.2
Mississippi	712	26	77	66	543	42	269	74	159	40.5	48.6
Missouri	2,206	100	280	198	1,627	119	763	238	508	39.9	48.9
Montana	345	27	51	35	233	21	107	30	75	38.5	50.5
Nebraska	765	45	91	73	557	46	226	82	202	40.5	50.1
Nevada	541	18	55	50	418	29	252	44	93	39.4	46.3
New Hampshire	552	32	68	55	398	38	185	63	112	38.8	48.1
New Jersey	2,943	122	372	262	2,187	267	1,145	255	521	38.7	47.3
New Mexico	580	33	73	66	407	33	208	49	117	38.9	49.3
New York	6,092	288	824	569	4,410	605	2,243	520	1,042	38.2	47.4
North Carolina	2,424	104	266	228	1,827	150	891	302	483	39.6	48.2
North Dakota	275	20	42	26	187	15	78	26	68	39.5	51.1
Ohio	4,351	223	547	435	3,145	252	1,489	507	897	39.2	48.8
Oklahoma	1,207	54	122	112	920	73	471	120	257	40.3	48.4
Oregon	1,284	78	165	130	911	84	417	145	265	38.9	48.5
Pennsylvania	4,840	274	642	538	3,387	366	1,742	473	806	37.9	48.5
Rhode Island	429	24	68	53	284	33	144	41	66	37.0	48.9
South Carolina	1,150	45	130	99	876	75	435	144	221	39.7	47.4
South Dakota	318	21	37	28	232	18	87	37	90	41.1	51.1
Tennessee	1,866	79	208	161	1,417	123	730	202	363	39.6	47.5
Texas	6,700	284	730	654	5,032	381	2,349	785	1,517	40.2	48.9
Utah	723	52	114	73	484	43	236	65	140	37.6	49.2
Vermont	283	16	38	27	202	23	95	32	52	38.6	48.0
Virginia	2,361	105	280	251	1,725	151	826	236	512	39.6	49.5
Washington	2,099	127	265	226	1,482	125	746	223	388	38.4	48.4
West Virginia	616	33	81	61	441	47	236	61	97	38.1	47.3
Wisconsin	2,280	134	287	193	1,665	118	760	282	506	39.8	49.1
Wyoming	209	14	26	20	149	12	64	21	52	40.2	50.5
Black											
Alabama	380	12	46	45	277	24	194	22	37	37.6	45.7
Alaska	9	[2]	1	1	7	1	3	1	1	40.4	49.9
Arkansas	122	5	15	17	84	8	52	9	15	37.6	46.9
California	662	35	73	66	489	34	342	40	73	37.5	46.2
Colorado	69	3	8	7	51	4	33	5	9	38.4	47.4
Connecticut	132	7	14	18	94	19	58	6	11	36.6	46.4
Delaware	55	1	8	4	42	4	26	4	7	38.2	45.2
District of Columbia	129	3	14	15	97	7	74	5	12	37.8	46.3
Florida	799	30	100	75	594	55	409	45	85	37.6	45.0
Georgia	836	30	97	57	653	69	425	62	96	38.1	44.6
Illinois	612	22	86	66	437	50	276	42	69	37.3	46.8
Indiana	163	6	23	12	121	13	82	14	12	36.8	43.9
Kansas	64	2	8	7	47	3	33	4	7	37.6	46.2
Kentucky	99	6	14	10	68	10	41	9	9	35.8	45.7

[Continued]

★ 724 ★

Work Week of Persons Employed by State: Annual Averages, 1992
[Continued]

Population group and State	Total at work	1 to 14 hours	15 to 29 hours	30 to 34 hours	35 hours and over					Average hours	
					Total	35 to 39 hours	40 hours	41 to 48 hours	49 hours and over	Total	Full-time schedules[1]
Louisiana	345	23	51	34	237	30	145	23	39	36.4	45.6
Maryland	614	19	67	74	455	46	297	45	68	38.4	47.6
Massachusetts	91	3	12	9	66	7	45	5	9	37.4	46.0
Michigan	393	18	50	40	284	21	187	34	43	37.9	47.0
Mississippi	310	12	43	37	219	22	146	21	29	37.2	46.7
Missouri	177	7	29	17	124	11	87	11	15	36.5	45.0
Nebraska	18	2	2	1	14	1	12	2	1	37.4	42.0
Nevada	31	2	2	3	26	1	20	2	3	39.2	43.9
New Jersey	399	12	38	38	311	44	204	22	40	38.2	45.4
New York	965	24	110	80	751	166	446	54	85	37.5	44.4
North Carolina	658	30	82	63	483	49	303	58	73	37.7	46.0
Ohio	394	14	46	44	289	23	193	26	47	37.9	46.2
Oklahoma	77	5	15	8	49	4	34	5	6	34.8	44.9
Pennsylvania	337	12	37	36	252	29	167	21	36	38.0	46.7
Rhode Island	12	2	3	2	7	2	4	1	1	34.7	48.6
South Carolina	414	17	54	39	305	40	198	34	33	37.0	44.5
Tennessee	281	10	38	32	201	21	136	15	29	37.2	46.4
Texas	798	27	96	84	592	49	368	60	114	38.8	47.5
Virginia	548	18	61	60	408	43	234	52	79	38.9	47.1
Washington	65	3	7	5	49	3	28	9	9	38.4	45.5
West Virginia	22	1	3	3	15	1	10	2	2	36.1	44.1
Wisconsin	72	2	9	7	54	4	33	7	9	38.3	45.6
Hispanic origin											
Arizona	244	10	35	26	174	18	106	26	24	37.3	46.3
California	3,078	112	385	337	2,244	185	1,522	235	302	37.5	45.3
Colorado	151	4	19	18	111	8	70	11	21	38.5	47.1
Connecticut	69	3	10	9	48	6	31	5	5	36.4	47.3
District of Columbia	14	1	2	2	10	1	6	1	2	36.8	46.5
Florida	783	23	99	61	600	41	386	55	119	38.7	45.1
Hawaii	12	2	2	2	9	1	5	1	2	39.3	48.5
Idaho	24	1	2	3	18	3	6	3	6	42.6	50.1
Illinois	394	14	32	45	302	15	210	38	38	38.3	46.0
Maryland	68	3	6	6	53	4	30	5	14	40.9	48.9
Massachusetts	58	3	9	8	38	5	22	3	8	37.1	48.6
Michigan	63	4	9	9	40	3	24	4	9	37.5	49.8
Nevada	60	1	4	5	50	2	39	3	5	39.0	42.8
New Jersey	310	6	29	28	247	31	161	22	33	38.8	45.0
New Mexico	185	10	23	25	127	11	78	13	25	37.7	47.9
New York	657	14	76	56	511	93	311	45	62	38.0	44.7
North Carolina	41	2	4	4	31	4	17	5	5	38.6	46.7
Ohio	34	2	4	3	25	2	15	3	5	38.7	48.7
Oklahoma	40	2	4	4	30	3	16	5	6	39.0	46.0
Oregon	54	2	5	7	41	5	20	6	10	39.7	46.4
Pennsylvania	54	3	8	6	38	5	21	4	8	37.5	47.1

[Continued]

★ 724 ★

Work Week of Persons Employed by State: Annual Averages, 1992

[Continued]

Population group and State	Total at work	Hours of work								Average hours	
		1 to 14 hours	15 to 29 hours	30 to 34 hours	35 hours and over					Total	Full-time schedules[1]
					Total	35 to 39 hours	40 hours	41 to 48 hours	49 hours and over		
Rhode Island	15	[2]	2	2	11	[2]	9	1	1	37.4	46.9
Texas	1,713	71	208	200	1,234	123	704	184	223	38.0	47.0
Utah	26	1	4	2	19	3	8	2	6	40.1	49.1
Virginia	77	1	9	11	55	9	30	6	11	38.6	47.1
Wyoming	8	[2]	1	[2]	6	1	3	1	2	41.8	48.9

Source: "States: Persons At Work by Sex, Race, Hispanic Origin, and Hours of Work, 1992 Annual Averages," U.S. Department of Labor, *Geographical Profile of Employment and Unemployment, 1992*, pp. 77-78. *Notes:* Data for demographic groups are not shown when they do not meet BLS publication standards of reliability for the particular area based on the sample in that area. Items may not add to totals or compute to displayed percentages because of rounding. Detail for Hispanic-origin groups will not add to totals because data for the "other races" group are not presented and Hispanics are included in both the white and black population groups. 1. Refers to persons who worked 35 hours or more during the survey week. 2. Less than 500 persons.

Labor force

★ 725 ★

Employed Persons in Census Regions and Divisions: Characteristics and Annual Averages, 1992

Population group and area	Total at work	Hours of work								Average hours	
		1 to 14 hours	15 to 29 hours	30 to 34 hours	35 hours and over					Total	Full-time schedule[1]
					Total	35 to 39 hours	40 hours	41 to 48 hours	49 hours and over		
White											
Northeast	19,666	1,006	2,650	1,996	14,013	1,686	7,096	1,826	3,406	38.1	47.9
New England	5,790	322	812	627	4,029	447	1,966	579	1,036	38.0	48.3
Middle Atlantic	13,876	684	1,838	1,369	9,985	1,238	5,130	1,247	2,369	38.2	47.8
Midwest	25,257	1,388	3,228	2,504	18,137	1,503	8,411	2,867	5,356	39.2	49.1
East North Central	17,059	909	2,150	1,729	12,272	1,019	5,869	1,937	3,447	39.1	48.7
West North Central	8,198	480	1,079	775	5,865	483	2,542	930	1,909	39.6	49.8
South	30,813	1,339	3,478	2,835	23,161	1,887	11,298	3,368	6,608	39.9	48.4
South Atlantic	15,377	651	1,781	1,412	11,533	956	5,675	1,652	3,249	39.7	48.2
East South Central	5,340	235	589	467	4,049	353	2,012	582	1,103	39.8	47.9
West South Central	10,095	453	1,108	956	7,579	578	3,611	1,134	2,256	40.2	48.9
West	20,501	1,026	2,592	2,125	14,758	1,156	7,671	1,964	3,967	38.9	48.2
Mountain	5,743	304	716	575	4,149	340	2,019	564	1,225	39.2	48.9
Pacific	14,758	722	1,876	1,550	10,609	816	5,652	1,400	2,742	38.7	48.0
Black											
Northeast	1,940	58	214	183	1,484	267	926	110	182	37.6	45.2
New England	240	10	29	29	171	28	109	13	22	36.8	46.3
Middle Atlantic	1,700	48	185	154	1,313	239	817	97	160	37.7	45.1

[Continued]

★ 725 ★

Employed Persons in Census Regions and Divisions: Characteristics and Annual Averages, 1992

[Continued]

Population group and area	Total at work	Hours of work								Average hours	
		1 to 14 hours	15 to 29 hours	30 to 34 hours	35 hours and over					Total	Full-time schedule[1]
					Total	35 to 39 hours	40 hours	41 to 48 hours	49 hours and over		
Midwest	1,945	76	261	201	1,408	130	925	143	210	37.5	46.2
East North Central	1,633	63	216	169	1,185	112	771	122	180	37.6	46.3
West North Central	312	13	45	31	223	18	154	20	31	36.9	45.5
South	6,487	249	804	657	4,777	482	3,092	471	732	37.8	46.0
South Atlantic	4,074	148	486	390	3,051	314	1,976	307	454	38.0	45.7
East South Central	1,071	41	141	124	765	77	517	67	104	37.2	46.1
West South Central	1,342	60	177	143	961	91	599	97	174	37.8	46.8
West	913	46	99	90	678	48	459	65	106	37.8	46.4
Mountain	151	8	16	15	113	9	74	13	18	38.3	47.2
Pacific	762	39	83	75	565	40	385	52	88	37.8	46.3
Hispanic origin											
Northeast	1,170	29	134	109	898	140	556	80	121	38.1	45.3
New England	148	6	22	19	101	11	63	9	17	37.0	48.0
Middle Atlantic	1,021	23	112	90	797	129	493	71	104	38.2	44.9
Midwest	638	28	66	72	472	31	311	58	72	38.1	46.6
East North Central	539	23	51	62	402	23	270	51	58	38.2	46.6
West North Central	99	5	14	10	70	7	41	8	14	37.8	46.4
South	2,874	104	346	302	2,121	200	1,241	275	404	38.4	46.5
South Atlantic	1,053	29	128	91	805	67	499	80	159	38.9	45.6
West South Central	1,795	75	216	208	1,296	132	735	192	238	38.1	46.9
West	3,908	144	485	432	2,846	240	1,884	307	414	37.6	45.6
Mountain	702	27	88	79	508	46	312	59	91	38.1	46.8
Pacific	3,206	117	398	353	2,339	194	1,573	249	324	37.5	45.3

Source: "Census Regions and Divisions: Persons at Work by Sex, Race, Age, Hispanic Origin, and Hours of Work, 1992 Annual Averages," U.S. Department of Labor, *Geographical Profile of Employment and Unemployment, 1992,* p. 26. *Notes:* Data for demographic groups are not shown when they do not meet BLS publication standards of reliability for the particular area based on the sample in that area. Items may not add to totals or compute to displayed percentages because of rounding. Detail for Hispanic-origin groups will not add to totals because data for the "other races" group are not presented and Hispanics are included in both the white and black population groups. 1. Refers to persons who worked 35 hours or more during the survey week.

Men Unemployed

★ 726 ★

Unemployment Rate for Women, by Race and Age, 1983 and 1991

Age cohort	1983		1991	
	African American	White	African American	White
20-24 years	31.4	13.8	22.4 (-28.7%)	10.2 (-26.1%)
35-44 years	13.5	6.4	9.6 (-28.9%)	5.0 (-21.9%)
Total population	20.3	8.8	12.9 (-36.5%)	6.4 (-27.3%)

Source: "Male Unemployment Rate by Race and Selected Age Cohorts: 1983 and 1991." *The State of Black America 1994*, p. 225. Primary source: Prepared by the National Urban League from U.S. Bureau of the Census, *Statistical Abstract of the United States: 1985* (105th edition), Table 658, and *Statistical Abstract of the United States: 1992* (112th edition), Table 622, Washington, DC, 1983 and 1992.

Men in the Labor Force

★ 727 ★

Percent of Men Employed by Race and Age, 1980 and 1990

Age cohort	1980		1990	
	African American[1]	White	African American	White
20-24 years	77.0	86.0	75.5 (-1.9%)	84.5 (-1.7%)
35-44 years	90.0	96.0	88.0 (-2.2%)	95.4 (-0.6%)
Total population	69.4	77.7	69.2 (-0.3%)	76.5 (-1.5%)

Source: "Percent of Males in Labor Force by Race and Selected Age Cohorts: 1980 and 1990," *The State of Black America 1994*, p. 222. Primary source: Prepared by the National Urban League from U.S. Bureau of Labor Statistics, *Employment and Earnings*, Vol. 28, No. 4, April 1981, and Vol. 37, No. 4, April 1990, Table A-4, Washington, DC, 1985 and 1993. *Note:* 1. Figures for "black and other."

Occupations

★ 728 ★

Civilians Employed, by Detailed Industry, 1993. Annual Averages Part 1 – Agriculture

Numbers in thousands.

Industry	1993			
	Total employed	Percent of total		
		Women	Black	Hispanic origin
Agriculture	3,074	20.7	4.6	15.2
Agricultural production, crops	904	16.1	5.1	24.7
Agricultural production, livestock	1,163	21.3	2.0	3.9
Veterinary services	164	72.5	3.0	1.7
Landscape and horticultural services	678	8.7	9.2	21.8
Agricultural services, n.e.c.	166	39.6	3.7	29.0

Source: "Employed Civilians by Detailed Industry, Sex, Race, and Hispanic Origin." *Employment and Earnings* 41 (January 1994), p. 216. *Notes:* Generally, data for occupations with fewer than 50,000 employed are not published separately but are included in the totals for the appropriate categories shown.

★ 729 ★

Occupations

Civilians Employed, by Detailed Industry, 1993. Annual Averages Part 2 – Mining, and Construction

Numbers in thousands.

Industry	1993			
	Total employed	Percent of total		
		Women	Black	Hispanic origin
Mining	669	16.2	3.7	5.0
Coal mining	114	5.3	4.5	.4
Oil and gas extraction	371	11.4	6.4	5.1
Nonmetallic mining and quarrying, except fuel	135	11.4	6.4	5.1
Construction	7,220	8.6	6.4	8.7

Source: "Employed Civilians by Detailed Industry, Sex, Race, and Hispanic Origin." *Employment and Earnings* 41 (January 1994), p. 216. *Notes:* Generally, data for occupations with fewer than 50,000 employed are not published separately but are included in the totals for the appropriate categories shown.

★ 730 ★
Occupations

Civilians Employed, by Detailed Industry, 1993. Annual Averages Part 3 – Manufacturing

Numbers in thousands.

Industry	1993			
	Total employed	Percent of total		
		Women	Black	Hispanic origin
Manufacturing	19,557	32.3	10.1	9.0
Durable goods	11,325	26.6	8.7	8.0
Lumber and wood products, except furniture	705	15.4	11.7	5.6
Logging	139	4.7	15.2	.9
Sawmills, planing mills, and millwork	349	14.9	13.2	6.0
Wood buildings and mobile homes	75	19.9	5.0	4.6
Miscellaneous wood products	142	24.5	8.2	9.4
Furniture and fixtures	624	29.8	8.5	12.4
Stone, clay, glass, and concrete products	530	20.1	12.7	9.2
Glass and glass products	167	27.6	12.3	5.7
Cement, concrete, gypsum, and plaster products	174	9.5	14.0	10.2
Structural clay, pottery, and related products	88	26.5	15.4	11.9
Miscellaneous nonmetallic mineral and stone products	102	20.6	8.6	10.7
Metal industries	1,938	18.9	7.9	9.8
Primary metal industries	727	13.9	10.9	7.6
Blast furnaces, steelworks, rolling, and finishing mills	345	10.5	14.4	6.6
Iron and steel foundries	94	9.9	8.9	6.6
Primary aluminum industries	133	17.8	8.7	9.3
Other primary metal industries	155	20.3	6.2	8.7
Fabricated metal industries	1,212	21.9	6.1	11.2
Cutlery, hand tools, and general hardware	99	39.3	8.7	7.8
Fabricated structural metal products	480	16.5	5.2	10.2
Metal forging and stampings	131	22.1	5.4	7.4
Ordnance	73	27.9	7.7	6.0
Miscellaneous and not specified fabricated metal products	380	22.2	6.7	15.1
Machinery and computing equipment	2,224	22.0	5.5	5.5
Engines and turbines	73	16.1	5.8	1.6
Farm machinery and equipment	100	21.9	9.1	2.9
Construction and material handling machines	204	10.7	4.0	4.4
Metal working machinery	262	15.6	3.0	4.4
Computers and related equipment	518	34.5	7.1	6.3
Machinery, except electrical, n.e.c. and not specified	1,019	19.7	4.8	6.0
Electrical machinery, equipment and supplies	1,760	40.3	8.1	8.4
Household appliances	144	38.2	13.7	5.6
Radio, T.V., and communication equipment	4001	37.1	8.8	6.6
Electrical machinery, equipment, and supplies, n.e.c. and not specified	1,215	41.6	7.2	9.4
Transportation equipment	2,289	23.0	11.3	6.3
Motor vehicles and motor vehicle equipment	1,160	24.6	13.1	4.9
Aircraft and parts	503	20.3	9.7	8.8
Ship and boat building and repairing	221	15.2	18.0	5.4
Guided missiles, space vehicles, and parts	319	27.7	5.1	7.4
Professional and photographic equipment, and watches	686	41.2	8.9	8.1

[Continued]

★ 730 ★

Civilians Employed, by Detailed Industry, 1993. Annual Averages Part 3 – Manufacturing

[Continued]

Industry	1993			
	Total employed	Percent of total		
		Women	Black	Hispanic origin
Scientific and controlling instruments	211	35.9	6.5	4.8
Medical, dental, and optical instruments and supplies	361	46.4	6.8	10.9
Photographic equipment and supplies	104	33.9	12.6	4.5
Toys, amusements, and sporting goods	147	42.7	4.8	13.4
Miscellaneous and not specified manufacturing industries	422	41.0	8.6	14.1
Nondurable goods	8,232	40.1	12.0	10.4
Food and kindred products	1,763	33.0	14.2	14.9
Meat products	463	33.6	22.8	19.8
Dairy products	155	26.5	5.2	6.2
Canned, frozen, and preserved fruits and vegetables	226	42.3	9.7	24.1
Grain mill products	141	24.7	8.9	5.6
Bakery products	106	48.0	12.6	15.0
Beverage industries	219	25.0	10.5	8.4
Miscellaneous and not specified food and kindred products	224	33.3	13.5	14.9
Tobacco manufacturers	54	36.3	29.4	4.6
Textile mill products	620	46.8	23.7	6.3
Knitting mills	130	65.0	14.3	12.0
Carpets and rugs	52	49.8	35.6	2.4
Yarn, thread, and fabric mills	365	42.9	25.5	4.8
Apparel and other finished textile products	1,004	71.4	15.2	21.8
Apparel and accessories, except knit	850	73.0	14.7	22.9
Miscellaneous fabricated textile products	154	63.1	17.6	15.7
Paper and allied products	721	24.9	11.4	6.2
Pulp, paper, and paperboard mills	292	18.0	7.5	2.2
Miscellaneous paper and pulp products	207	34.2	14.0	7.8
Paperboard containers and boxes	222	25.2	14.0	10.0
Printing, publishing, and allied products	1,784	42.4	6.4	5.8
Newspaper publishing and printing	513	42.7	7.9	4.4
Printing, publishing, and allied industries, except newspapers	1,271	42.3	5.8	6.4
Chemicals and allied products	1,205	33.4	10.6	6.9
Plastics, synthetics, and resins	137	26.9	12.7	13.7
Drugs	310	45.7	12.2	6.6
Soaps and cosmetics	163	48.7	16.0	9.9
Industrial and miscellaneous chemicals	515	24.7	7.7	4.7
Petroleum and coal products	176	17.6	8.1	6.8
Petroleum refining	154	16.6	7.8	6.6
Rubber and miscellaneous plastics products	783	32.4	9.1	9.6
Tires and inner tubes	84	14.1	15.3	.9
Other rubber products, and plastics footwear and belting	140	35.4	7.6	7.0
Miscellaneous plastics products	559	34.4	8.6	11.5

[Continued]

★ 730 ★

Civilians Employed, by Detailed Industry, 1993. Annual Averages Part 3 – Manufacturing

[Continued]

Industry	1993			
	Total employed	Percent of total		
		Women	Black	Hispanic origin
Leather and leather products	122	53.7	7.8	14.2
Footwear, except rubber and plastic	65	55.1	5.9	10.0

Source: "Employed Civilians by Detailed Industry, Sex, Race, and Hispanic Origin." *Employment and Earnings* 41 (January 1994), pp. 216-217.
Notes: Generally, data for industries with fewer than 50,000 employed are not published separately but are included in the totals for the appropriate categories shown.

★ 731 ★

Occupations

Civilians Employed, by Detailed Industry, 1993. Annual Averages Part 4 – Transportation, Communications, and Other Public Utilities

Numbers in thousands.

Industry	1993			
	Total employed	Percent of total		
		Women	Black	Hispanic origin
Transportation, communications, and other public utilities	8,481	28.5	13.8	7.1
Transportation	5,317	25.5	14.3	8.1
Railroads	301	9.6	12.9	4.1
Bus service and urban transit	533	28.1	26.9	8.7
Taxicab service	121	8.9	27.1	15.0
Trucking service	2,014	13.3	11.7	7.0
Warehousing and storage	156	25.3	13.6	15.1
U.S. Postal Service	850	36.0	21.3	7.3
Water transportation	192	20.2	12.3	9.1
Air transportation	761	36.7	8.8	8.5
Services incidental to transportation	375	61.3	5.1	11.0
Communications	1,568	45.9	13.4	5.7
Radio and television broadcasting and cable	394	38.9	8.7	6.6
Telephone communications	1,141	48.5	15.3	5.5
Utilities and sanitary services	1,597	21.6	12.5	5.2
Electric light and power	653	23.8	8.8	3.9
Gas and steam supply systems	199	25.8	9.1	5.7
Electric and gas, and other combinations	171	22.4	14.0	4.5
Water supply and irrigation	207	17.7	13.7	5.5
Sanitary services	349	16.0	19.7	7.9

Source: "Employed Civilians by Detailed Industry, Sex, Race, and Hispanic Origin." *Employment and Earnings* 41 (January 1994), p. 217.
Notes: Generally, data for industries with fewer than 50,000 employed are not published separately but are included in the totals for the appropriate categories shown.

★ 732 ★

Occupations

Civilians Employed, by Detailed Industry, 1993. Annual Averages Part 5 – Wholesale and Retail Trade

Numbers in thousands.

Industry	1993			
	Total employed	Percent of total		
		Women	Black	Hispanic origin
Wholesale and retail trade	24,769	46.7	8.4	8.4
Wholesale trade	4,606	28.9	6.0	7.9
Durable goods	2,475	27.7	5.0	6.2
Motor vehicles and equipment	200	24.5	4.7	6.3
Furniture and furnishings	70	34.0	5.0	6.1
Lumber and construction materials	170	23.6	7.4	6.7
Professional and commercial equipment and supplies	442	32.9	6.2	5.3
Metals and minerals, except petroleum	94	24.3	5.4	4.6
Electrical goods	325	32.2	4.4	6.5
Hardware, plumbing and heating supplies	251	24.2	4.4	4.9
Machinery, equipment, and supplies	599	26.4	3.9	4.5
Scrap and waste materials	180	15.3	7.2	14.0
Miscellaneous wholesale trade, durable goods	143	36.7	3.6	7.8
Nondurable goods	2,131	30.3	7.1	9.9
Paper and paper products	145	38.0	5.6	4.2
Drugs, chemicals, and allied products	193	40.1	6.1	7.6
Apparel, fabrics, and notions	128	44.1	8.3	14.4
Groceries and related products	800	25.3	8.8	14.0
Farm products-raw materials	83	25.5	3.6	3.2
Petroleum products	150	27.5	4.3	5.0
Alcoholic beverages	143	17.8	10.0	5.6
Farm supplies	152	25.0	2.9	5.2
Miscellaneous nondurable goods and not specified wholesale trade	337	38.0	6.4	9.9
Retail trade	20,163	50.8	8.9	8.5
Lumber and building material retailing	509	24.9	5.2	5.6
Hardware stores	241	34.6	4.4	3.5
Retail nurseries and garden stores	100	37.2	4.1	4.2
Department stores	2,078	69.8	11.9	8.5
Variety stores	145	72.1	13.1	5.1
Miscellaneous general merchandise stores	131	60.4	13.5	4.9
Grocery stores	3,034	50.0	8.8	7.4
Retail bakeries	158	59.0	3.5	8.3
Food stores, n.e.c.	208	50.2	4.7	12.7
Motor vehicle dealers	1,100	17.9	7.5	6.8
Auto and home supply stores	436	18.5	6.1	7.3
Gasoline service stations	414	29.5	5.4	6.2
Miscellaneous vehicle dealers	92	22.4	1.7	2.4
Apparel and accessory stores, except shoe	869	75.8	9.3	9.5
Shoe stores	156	54.1	13.0	10.9
Furniture and home furnishings stores	558	37.9	6.5	9.9
Household appliance stores	107	27.4	4.3	7.1

[Continued]

★ 732 ★

Civilians Employed, by Detailed Industry, 1993. Annual Averages Part 5 – Wholesale and Retail Trade

[Continued]

Industry	1993			
	Total employed	Percent of total		
		Women	Black	Hispanic origin
Radio, TV, and computer stores	340	29.0	6.4	5.3
Music stores	133	36.2	7.4	8.3
Eating and drinking places	6,052	52.8	11.4	11.6
Drug stores	569	61.3	7.4	6.3
Liquor stores	121	39.8	8.9	7.3
Sporting goods, bicycles, and hobby stores	343	43.5	6.0	5.1
Book and stationery stores	226	53.6	6.6	2.7
Jewelry stores	160	60.0	3.9	10.0
Gift, novelty,a nd souvenir shops	204	78.7	3.2	3.1
Sewing, needlework, and piece goods stores	65	82.0	8.8	5.8
Catalog and mail order houses	134	69.5	8.7	5.9
Vending machine operators	78	33.2	5.8	6.2
Direct selling establishments	338	67.3	4.7	8.5
Fuel dealers	125	27.7	1.3	1.4
Retail florists	163	73.1	4.6	6.2
Miscellaneous retail stores and not specified retail trade	721	53.5	5.9	6.7

Source: "Employed Civilians by Detailed Industry, Sex, Race, and Hispanic Origin." *Employment and Earnings* 41 (January 1994), pp. 217-218. *Notes:* Generally, data for industries with fewer than 50,000 employed are not published separately but are included in the totals for the appropriate categories shown.

★ 733 ★

Occupations

Civilians Employed, by Detailed Industry, 1993. Annual Averages Part 6 – Retail Trade

Numbers in thousands.

Industry	1993			
	Total employed	Percent of total		
		Women	Black	Hispanic origin
Retail trade	20,163	50.8	8.9	8.5
Lumber and building material retailing	509	24.9	5.2	5.6
Hardware stores	241	34.6	4.4	3.5
Retail nurseries and garden stores	100	37.2	4.1	4.2
Department stores	2,078	69.8	11.9	8.5
Variety stores	145	72.1	13.1	5.1
Miscellaneous general merchandise stores	131	60.4	13.5	4.9
Grocery stores	3,034	50.0	8.8	7.4
Retail bakeries	158	59.0	3.5	8.3
Food stores, n.e.c.	208	50.2	4.7	12.7

[Continued]

★ 733 ★

Civilians Employed, by Detailed Industry, 1993. Annual Averages Part 6 – Retail Trade

[Continued]

Industry	1993			
	Total employed	Percent of total		
		Women	Black	Hispanic origin
Motor vehicle dealers	1,100	17.9	7.5	6.8
Auto and home supply stores	436	18.5	6.1	7.3
Gasoline service stations	414	29.5	5.4	6.2
Miscellaneous vehicle dealers	92	22.4	1.7	2.4
Apparel and accessory stores, except shoe	869	75.8	9.3	9.5
Shoe stores	156	54.1	13.0	10.9
Furniture and home furnishings stores	558	37.9	6.5	9.9
Household and appliance stores	107	27.4	4.3	7.1
Radio, TV, and computer stores	340	29.0	6.4	5.3
Music stores	133	36.2	7.4	8.3
Eating and drinking places	6,052	52.8	11.4	11.6
Drug stores	569	61.3	7.4	6.3
Liquor stores	121	39.8	8.9	7.3
Sporting goods, bicycles, and hobby stores	343	43.5	6.0	5.1
Book and stationery stores	226	53.6	6.6	2.7
Jewelry stores	160	60.0	3.9	10.0
Gift, novelty, and souvenir shops	204	78.7	3.2	3.1
Sewing, needlework, and piece goods stores	65	82.0	8.8	5.8
Catalog and mail order houses	134	69.5	8.7	5.9
Vending machine operators	78	33.2	5.8	6.2
Direct selling establishments	338	67.3	4.7	8.5
Fuel dealers	125	27.7	1.3	1.4
Retail florists	163	73.1	4.6	6.2
Miscellaneous retail stores and not specified retail trade	721	53.5	5.9	6.7

Source: "Employed Civilians by Detailed Industry, Sex, Race, and Hispanic Origin." *Employment and Earnings* 41 (January 1994), p. 218.
Notes: Generally, data for industries with fewer than 50,000 employed are not published separately but are included in the totals for the appropriate categories shown.

★ 734 ★

Occupations

Civilians Employed, by Detailed Industry, 1993. Annual Averages Part 7 – Finance, Insurance, and Real Estate

Numbers in thousands.

Industry	1993			
	Total employed	Percent of total		
		Women	Black	Hispanic origin
Finance, insurance, and real estate	7,962	58.6	8.4	6.0
Banking	1,927	70.6	10.1	6.8
Savings institutions, including credit unions	318	78.5	4.6	7.7
Credit agencies, n.e.c.	506	61.5	8.9	6.6
Security, commodity brokerage, and investment companies	697	38.8	6.5	3.2
Insurance	2,383	60.6	8.2	4.4
Real estate, including real estate-insurance offices	2,131	48.3	8.0	7.8

Source: "Employed Civilians by Detailed Industry, Sex, Race, and Hispanic Origin." *Employment and Earnings* 41 (January 1994), p. 218.
Notes: Generally, data for industries with fewer than 50,000 employed are not published separately but are included in the totals for the appropriate categories shown.

★ 735 ★

Occupations

Civilians Employed, by Detailed Industry, 1993. Annual Averages Part 8 – Services

Numbers in thousands.

Industry	1993			
	Total employed	Percent of total		
		Women	Black	Hispanic origin
Services	41,817	61.7	11.4	6.9
Private households	1,114	86.9	17.8	20.3
Other service industries	40,703	61.0	11.2	6.6
Business and repair services	6,838	35.1	10.4	9.0
Advertising	284	52.3	5.3	7.0
Services to dwellings and other buildings	749	43.5	18.1	17.4
Personnel supply services	698	61.0	18.7	7.4
Computer and data processing services	957	36.3	5.2	3.7
Detective and protective services	475	17.8	20.8	7.7
Business services, n.e.c.	1,526	51.0	8.2	6.9
Automotive rental and leasing, without drivers	177	32.8	10.3	10.6
Automotive parking and carwashes	163	18.6	16.4	19.2
Automotive repair and related services	1,137	9.7	6.8	11.7
Electrical repair shops	139	15.9	4.4	7.6
Miscellaneous repair services	534	13.6	5.3	7.8

[Continued]

★ 735 ★

Civilians Employed, by Detailed Industry, 1993. Annual Averages Part 8 – Services

[Continued]

Industry	1993			
	Total employed	Percent of total		
		Women	Black	Hispanic origin
Personal services, except private household	3,329	64.3	12.9	10.6
Hotels and motels	1,295	57.3	16.5	15.3
Lodging places, except hotels and motels	140	51.4	4.0	2.9
Laundry, cleaning, and garment services	479	55.0	15.7	14.0
Beauty shops	862	89.0	9.5	6.0
Barber shops	88	23.7	26.5	8.5
Funeral service and crematories	90	33.0	6.5	5.2
Dressmaking shops	51	93.6	7.5	3.3
Entertainment and recreational services	2,060	41.8	9.4	6.8
Theaters and motion pictures	517	38.2	10.1	6.0
Video tape rental	118	60.1	6.8	6.5
Bowling centers	61	51.3	4.8	1.7
Miscellaneous entertainment and recreation services	1,364	41.1	9.6	7.4
Professional and related services	28,293	68.6	11.5	5.5
Hospitals	5,032	76.2	15.6	5.6
Health services, except hospitals	5,521	78.2	12.7	6.3
Offices and clinics of physicians	1,450	73.9	4.4	6.9
Offices and clinics of dentists	567	74.5	1.9	5.9
Offices and clinics of chiropractors	116	58.8	.3	6.1
Offices and clinics of optometrists	67	69.2	3.1	5.8
Offices and clinics of health practitioners, n.e.c.	125	68.6	3.7	3.7
Nursing and personal care facilities	1,752	85.3	22.5	5.6
Health services, n.e.c.	1,443	78.1	15.7	7.2
Educational services	9,485	68.2	10.6	5.5
Elementary and secondary schools	6,372	74.5	11.6	5.9
Colleges and universities	2,633	52.6	9.1	5.1
Vocational schools	82	54.7	4.7	5.8
Libraries	187	80.7	8.9	2.0
Educational services, n.e.c.	212	69.6	5.0	2.9
Social services	2,770	81.6	16.6	6.8
Job training and vocational rehabilitation services	236	51.6	18.2	3.5
Child day care services	883	96.3	16.2	6.0
Family child care homes	306	98.9	9.3	10.3
Residential care facilities, without nursing	454	71.9	20.1	6.2
Social services, n.e.c.	890	73.9	17.3	7.5
Other professional services	5,486	46.0	5.3	3.7
Legal services	1,253	55.1	5.2	4.1
Museums, art galleries, and zoos	101	55.8	6.1	4.5
Labor unions	74	40.3	12.1	9.1
Religious organizations	848	45.5	7.1	4.9
Membership organizations, n.e.c.	380	64.2	9.1	4.0

[Continued]

★ 735 ★

Civilians Employed, by Detailed Industry, 1993. Annual Averages Part 8 – Services

[Continued]

Industry	1993			
	Total employed	Percent of total		
		Women	Black	Hispanic origin
Engineering, architectural, and surveying services	778	20.2	3.3	4.0
Accounting, auditing, and bookkeeping services	637	53.3	2.4	2.4
Research, development and testing services	602	40.3	6.7	3.2
Management and public relations services	527	41.4	4.8	2.1
Miscellaneous professional and related services	287	55.2	2.8	2.2
Forestry and fisheries	183	21.6	2.4	9.9
Forestry	100	32.1	2.1	14.7
Fishing, hunting, and trapping	83	8.8	2.8	4.1

Source: "Employed Civilians by Detailed Industry, Sex, Race, and Hispanic Origin." *Employment and Earnings* 41 (January 1994), pp. 218-219. *Notes:* N.e.c. is an abbreviation for "not elsewhere classified" and designates broad categories of industries which cannot be more specifically identified. Generally, data for industries with fewer than 50,000 employed are not published separately but are included in the totals for the appropriate categories shown.

★ 736 ★

Occupations

Civilians Employed, by Detailed Industry, 1993. Annual Averages Part 9 – Public Administration

Numbers in thousands.

Industry	1993			
	Total employed	Percent of total		
		Women	Black	Hispanic origin
Public administration	5,756	42.9	15.0	5.3
Executive and legislative offices	156	58.4	10.4	2.1
General government, n.e.c.	612	48.3	17.2	5.6
Justice, public order, and safety	2,169	31.5	14.8	5.3
Public finance, taxation, and monetary policy	407	57.7	13.2	5.7
Administration of human resources programs	760	68.5	19.3	6.7
Administration of environmental quality and housing programs	266	36.5	7.4	4.1
Administration of economic programs	597	43.9	13.5	4.0
National security and international affairs	789	36.1	14.8	5.9

Source: "Employed Civilians by Detailed Industry, Sex, Race, and Hispanic Origin." *Employment and Earnings* 41 (January 1994), p. 219. *Notes:* N.e.c. is an abbreviation for "not elsewhere classified" and designates broad categories of industries which cannot be more specifically identified. Generally, data for industries with fewer than 50,000 employed are not published separately but are included in the totals for the appropriate categories shown.

★ 737 ★

Occupations

Civilians Employed, by Detailed Occupation, 1993. Part 1 – Managerial and Professional Specialty

Numbers in thousands.

| Occupation | 1993 | | | |
| | Total employed | Percent of total | | |
		Women	Black	Hispanic origin
Managerial and professional specialty	32,280	47.8	6.6	4.0
Executive, administrative, and managerial occupations	15,376	42.0	6.2	4.5
Officials and administrators, public administration	581	45.2	11.3	4.5
Financial managers	529	46.2	4.4	4.2
Personnel and labor relations managers	96	60.7	7.9	4.6
Purchasing managers	109	34.9	8.0	5.3
Managers, marketing, advertising, and public relations	496	31.2	3.1	3.5
Administrators, education and related fields	635	59.9	13.0	3.8
Managers, medicine and health	450	70.5	6.5	4.2
Managers, food serving and lodging establishments	1,198	43.4	8.1	7.8
Managers, properties and real estate	481	45.7	6.6	6.3
Funeral directors	51	18.6	7.1	3.3
Management-related occupations	4,155	52.7	7.5	4.6
Accountants and auditors	1,387	49.2	7.0	4.2
Underwriters	113	69.9	4.9	4.8
Other financial officers	697	48.7	5.7	4.8
Management analysts	250	33.7	3.9	2.0
Personnel, training, and labor relations specialists	401	64.1	12.5	6.0
Buyers, wholesale and retail trade, except farm products	226	48.7	4.8	4.7
Construction inspectors	64	5.0	10.8	4.6
Inspectors and compliance officers, except construction	213	25.8	10.9	6.8

Source: "Employed Civilians by Detailed Occupation, Sex, Race, and Hispanic Origin." *Employment and Earnings* 41 (January 1994), p. 205.

★ 738 ★

Occupations

Civilians Employed, by Detailed Occupation, 1993. Part 2 – Professional Specialty

Numbers in thousands.

| Occupation | 1993 | | | |
| | Total employed | Percent of total | | |
		Women	Black	Hispanic origin
Professional specialty	16,904	53.2	7.0	3.6
Engineers, architects, and surveyors	1,859	9.3	3.7	3.6
Architects	123	18.6	3.1	2.3
Engineers	1,716	8.6	3.7	3.6
Aerospace engineers	83	7.5	2.1	3.9
Chemical engineers	58	10.0	2.5	4.9
Civil engineers	221	9.4	4.7	3.8
Electrical and electronic engineers	533	7.6	4.5	3.4
Industrial engineers	201	16.4	3.4	4.4
Mechanical engineers	296	5.2	4.4	3.3
Mathematical and computer scientists	1,051	32.4	6.0	2.5
Computer systems analysts and scientists	769	29.9	5.8	2.4
Operations and systems researchers and analysts	236	39.7	6.3	3.0
Natural scientists	531	30.1	3.6	1.9
Chemists, except biochemists	133	26.6	4.3	3.0
Geologists and geodesists	54	14.0	1.0	2.1
Biological and life scientists	114	40.4	3.9	1.4
Medical scientists	82	45.5	5.8	2.5
Health diagnosing occupations	909	20.5	3.0	3.9
Physicians	605	21.8	3.7	4.6
Dentists	152	10.5	1.9	3.0
Health assessment and treating occupations	2,602	86.4	8.3	3.5
Registered nurses	1,859	94.4	8.4	3.2
Pharmacists	187	38.1	6.1	2.7
Dietitians	94	92.8	17.5	6.0
Therapists	416	74.9	6.9	4.1
Respiratory therapists	92	58.4	10.0	6.9
Physical therapists	115	72.5	3.0	5.0
Speech therapists	83	91.8	6.7	1.2
Teachers, college and university	772	42.5	4.8	3.1
Teachers, except college and university	4,397	75.1	8.6	3.6
Prekindergarten and kindergarten	501	97.7	11.7	5.0
Elementary school	1,663	85.9	9.3	3.9
Secondary school	1,237	57.5	6.9	3.1
Special education	286	84.0	10.1	2.3
Counselors, educational and vocational	224	67.6	14.3	6.9
Librarians, archivists, and curators	223	83.5	6.2	3.8
Librarians	195	88.3	7.0	3.5
Social scientists and urban planners	399	57.0	5.9	3.0
Economists	117	47.6	4.8	3.5
Psychologists	241	64.1	7.1	3.1
Social, recreation, and religious workers	096	50.5	15.6	5.0
Social workers	586	68.9	21.4	6.0

[Continued]

★ 738 ★

Civilians Employed, by Detailed Occupation, 1993. Part 2 – Professional Specialty

[Continued]

Occupation	1993			
	Total employed	Percent of total		
		Women	Black	Hispanic origin
Recreation workers	89	75.1	14.8	4.7
Clergy	350	11.4	8.7	3.1
Lawyers and judges	815	22.8	2.8	2.1
Lawyers	777	22.9	2.7	2.1
Writers, artists, entertainers and athletes	2,026	46.6	5.3	4.7
Authors	139	57.2	2.4	1.9
Technical writers	63	52.8	2.7	2.3
Designers	541	52.6	3.7	4.4
Musicians and composers	174	32.8	8.8	5.8
Actors and directors	96	38.3	10.4	4.7
Painters, sculptors, craft artists, and artist printmakers	222	48.0	3.5	4.1
Photographers	135	26.2	6.5	7.1
Editors and reporters	266	48.5	5.0	3.4
Public relations specialists	155	59.6	7.0	3.5
Athletes	80	23.9	10.1	3.9

Source: "Employed Civilians by Detailed Occupation, Sex, Race, and Hispanic Origin." *Employment and Earnings* 41 (January 1994), pp. 205-206. *Notes:* Generally, data for occupations with fewer than 50,000 employed are not published separately but are included in the totals for the appropriate categories shown.

★ 739 ★

Occupations

Civilians Employed, by Detailed Occupation, 1993. Part 3 – Technical, Sales, and Administrative Support

Numbers in thousands.

Occupation	1993			
	Total employed	Percent of total		
		Women	Black	Hispanic origin
Technical, sales, and administrative support	36,814	63.8	9.3	6.3
Technicians and related support	4,014	50.5	9.6	5.0
Health technologists and technicians	1,522	81.0	12.4	5.8
Clinical laboratory technologists and technicians	315	76.1	12.1	6.1
Dental hygienists	76	99.3	.4	2.0
Health record technologists and technicians	63	88.8	20.4	5.7
Radiologic technicians	146	70.2	8.3	7.0
Licensed practical nurses	425	94.6	17.2	3.4
Engineering and related technologists and technicians	870	17.8	7.4	4.9
Electrical and electronic technicians	297	15.5	7.4	5.6

[Continued]

★ 739 ★

Civilians Employed, by Detailed Occupation, 1993. Part 3 – Technical, Sales, and Administrative Support

[Continued]

Occupation	1993			
	Total employed	Percent of total		
		Women	Black	Hispanic origin
Drafting occupations	244	18.1	6.9	5.9
Surveying and mapping technicians	73	5.0	4.8	2.8
Science technicians	261	37.5	7.2	5.2
Biological technicians	85	59.7	6.1	3.8
Chemical technicians	74	26.0	7.1	3.9
Technicians, except health, engineering, and science	1,361	39.9	8.4	4.0
Airplane pilots and navigators	101	3.9	5.5	2.4
Computer programmers	578	31.5	6.7	3.5
Legal assistants	254	79.6	8.6	4.9

Source: "Employed Civilians by Detailed Occupation, Sex, Race, and Hispanic Origin." *Employment and Earnings* 41 (January 1994), p. 206. *Notes:* Generally, data for occupations with fewer than 50,000 employed are not published separately but are included in the totals for the appropriate categories shown.

★ 740 ★

Occupations

Civilians Employed, by Detailed Occupation, 1993. Part 4 – Sales Occupations

Numbers in thousands.

Occupation	1993			
	Total employed	Percent of total		
		Women	Black	Hispanic origin
Sales occupations	14,245	48.1	6.7	5.9
Supervisors and proprietors	4,016	36.4	4.4	5.3
Sales representatives, finance and business services	2,317	40.5	4.7	3.7
Insurance sales	583	33.3	5.1	3.8
Real estate sales	710	51.4	2.5	4.1
Securities and financial services sales	355	28.1	4.1	2.2
Advertising and related sales	161	50.8	4.5	3.1
Sales occupations, other business services	508	39.0	7.6	4.0
Sales representatives, commodities, except retail	1,538	21.0	2.9	3.9
Sales representatives, mining, manufacturing, and wholesale	1,514	21.3	3.0	4.0
Sales workers, retail and personal services	6,281	64.9	9.7	7.5
Sales workers, motor vehicles and boats	280	7.7	7.2	5.3
Sales workers, apparel	474	79.6	11.1	9.4
Sales workers, shoes	101	59.2	14.1	11.9
Sales workers, furniture and home furnishings	171	44.8	3.9	6.6
Sales workers, radio, television, hi-fi, and appliances	201	27.4	8.5	6.7
Sales workers, hardware and building supplies	250	22.1	1.8	3.8
Sales workers, parts	155	7.0	5.5	8.9
Sales workers, other commodities	1,428	70.2	7.1	5.7

[Continued]

★ 740 ★

Civilians Employed, by Detailed Occupation, 1993. Part 4 – Sales Occupations
[Continued]

Occupation	1993			
	Total employed	Percent of total		
		Women	Black	Hispanic origin
Sales counter clerks	195	63.8	8.2	8.1
Cashiers	2,581	78.4	13.2	8.7
Street and door-to-door sales workers	319	69.9	7.3	7.6
News vendors	125	34.6	5.4	6.1
Sales related occupations	93	60.5	5.3	6.1
Demonstrators, promoters, and models	58	72.2	7.2	8.4

Source: "Employed Civilians by Detailed Occupation, Sex, Race, and Hispanic Origin." *Employment and Earnings* 41 (January 1994), p. 206. *Notes:* Generally, data for occupations with fewer than 50,000 employed are not published separately but are included in the totals for the appropriate categories shown.

★ 741 ★

Occupations

Civilians Employed, by Detailed Occupation, 1993. Part 5 – Administrative Support Occupations and Clerical

Numbers in thousands.

Occupation	1993			
	Total employed	Percent of total		
		Women	Black	Hispanic origin
Administrative support occupations, including clerical	18,555	78.8	11.2	6.8
Supervisors, administrative support	778	58.4	11.9	6.8
Supervisors, general office	484	66.1	13.5	5.7
Supervisors, financial records processing	96	72.2	5.0	4.2
Supervisors, distribution, scheduling, and adjusting clerks	165	30.6	12.1	12.0
Computer equipment operators	603	61.9	13.8	6.2
Computer operators	597	61.9	13.7	6.1
Secretaries, stenographers, and typists	4,174	96.2	8.9	5.9
Secretaries	3,586	98.9	7.7	5.8
Stenographers	94	93.6	2.9	5.1
Typists	494	94.3	18.8	7.4
Information clerks	1,678	88.8	9.3	7.9

Source: "Employed Civilians by Detailed Occupation, Sex, Race, and Hispanic Origin." *Employment and Earnings* 41 (January 1994), pp. 206-207. *Notes:* Generally, data for occupations with fewer than 50,000 employed are not published separately but are included in the totals for the appropriate categories shown.

★ 742 ★

Occupations

Civilians Employed, by Detailed Occupation, 1993. Part 6 – Service Occupations

Numbers in thousands.

Occupation	1993			
	Total employed	Percent of total		
		Women	Black	Hispanic origin
Service occupation	16,522	59.5	17.3	11.2
Private household	912	95.1	17.1	21.6
Child care workers	345	97.2	9.0	15.0
Cleaners and servants	534	94.0	21.6	25.7
Protective service	2,152	17.2	17.4	6.6
Supervisors	185	7.8	12.2	4.8
Police and detectives	96	10.3	6.6	6.1
Guards	53	8.6	23.1	5.0
Firefighting and fire prevention occupations	208	3.7	7.6	4.5
Firefighting occupations	188	3.3	7.5	5.0
Police and detectives	923	16.0	18.0	5.4
Police and detectives, public service	511	12.0	14.5	5.9
Sheriffs, bailiffs, and other law enforcement officers	117	19.5	13.4	4.6
Correctional institution officers	295	21.6	25.8	5.0
Guards	836	23.9	20.4	8.8
Guards and police, except public services	711	17.2	22.6	9.5

Source: "Employed Civilians by Detailed Occupation, Sex, Race, and Hispanic Origin." *Employment and Earnings* 41 (January 1994), p. 207. *Notes:* Generally, data for occupations with fewer than 50,000 employed are not published separately but are included in the totals for the appropriate categories shown.

★ 743 ★

Occupations

Civilians Employed, by Detailed Occupation, 1993. Part 7 – Service Occupations, Excluding Private Household and Protective Service

Numbers in thousands.

Occupation	1993			
	Total employed	Percent of total		
		Women	Black	Hispanic origin
Service occupations, except private household and protective service	13,547	63.9	17.3	11.2
Food preparation and service occupations	5,691	58.4	12.8	11.7
Supervisors, food preparation and service	337	68.0	10.4	6.9
Bartenders	321	53.3	3.8	3.1
Waiters and waitresses	1,414	80.0	4.6	7.5
Cooks	1,992	44.2	19.0	13.9

[Continued]

★ 743 ★

Civilians Employed, by Detailed Occupation, 1993. Part 7 – Service Occupations, Excluding Private Household and Protective Service

[Continued]

Occupation	1993			
	Total employed	Percent of total		
		Women	Black	Hispanic origin
Food counter, fountain and related occupations	367	69.2	12.6	8.1
Kitchen workers, food preparation	260	75.3	15.0	10.8
Waiters and waitresses' assistants	368	43.7	11.4	16.1
Miscellaneous food preparation	632	47.4	17.4	19.6
Health service occupations	2,213	87.4	27.3	7.7
Dental assistants	181	97.8	3.4	10.4
Health aides, except nursing	312	78.9	22.2	4.7
Nursing aides, orderlies, and attendants	1,719	87.9	30.7	7.9
Cleaning and building service occupations	2,959	42.2	22.4	16.2
Supervisors	149	40.7	18.4	11.0
Maids and housemen	661	81.7	27.3	18.6
Janitors and cleaners	2,086	30.7	21.5	16.1
Personal service occupations	2,594	80.7	12.9	7.6
Supervisors	115	69.0	9.4	4.2
Barbers	86	22.3	27.5	8.5
Hairdressers and cosmetologists	758	90.1	9.4	6.3
Attendants, amusement and recreation facilities	161	39.4	7.4	5.8
Public transportation attendants	104	80.4	8.8	7.4
Welfare service aides	73	82.1	21.2	16.6
Family child care providers	302	99.0	9.4	10.2
Early childhood teachers' assistants	418	96.6	15.7	6.4

Source: "Employed Civilians by Detailed Occupation, Sex, Race, and Hispanic Origin." *Employment and Earnings* 41 (January 1994), pp. 207-208. *Notes:* Generally, data for occupations with fewer than 50,000 employed are not published separately but are included in the totals for the appropriate categories shown.

★ 744 ★

Occupations

Civilians Employed, by Detailed Occupation, 1993. Part 8 – Precision Production, Craft, and Repair

Numbers in thousands.

Occupation	1993			
	Total employed	Percent of total		
		Women	Black	Hispanic origin
Precision production, craft, and repair	13,326	8.6	7.4	9.2
Mechanics and repairers	4,416	3.5	7.3	7.9
Supervisors	220	7.5	6.7	3.7
Mechanics and repairers, except supervisors	4,196	3.3	7.3	8.1

[Continued]

★ 744 ★

Civilians Employed, by Detailed Occupation, 1993. Part 8 – Precision Production, Craft, and Repair

[Continued]

Occupation	1993			
	Total employed	Percent of total		
		Women	Black	Hispanic origin
Vehicle and mobile equipment mechanics and repairers	1,800	1.0	6.1	9.9
Automobile mechanics	854	.6	6.4	10.8
Bus, truck, and stationary engine mechanics	344	.3	7.5	7.1
Aircraft engine mechanics	139	4.1	5.1	11.5
Small engine repairers	70	2.1	1.0	4.4
Automobile body and related repairers	192	.7	5.4	15.6
Heavy equipment mechanics	153	.6	6.4	4.7
Industrial machinery repairers	548	2.6	8.2	6.9
Electrical and electronic equipment repairers	655	9.5	9.0	5.8
Electronic repairers, communications and industrial equipment	165	7.8	9.6	8.8
Data processing equipment repairers	152	10.7	10.2	5.0
Telephone installers and repairers	188	12.5	9.9	3.4
Heating, air conditioning, and refrigeration mechanics	262	1.0	4.3	7.7
Miscellaneous mechanics and repairers	906	4.5	8.8	7.0
Office machine repairers	59	3.7	13.5	6.5

Source: "Employed Civilians by Detailed Occupation, Sex, Race, and Hispanic Origin." *Employment and Earnings* 41 (January 1994), p. 208.
Notes: Generally, data for occupations with fewer than 50,000 employed are not published separately but are included in the totals for the appropriate categories shown.

★ 745 ★

Occupations

Civilians Employed, by Detailed Occupation, 1993. Part 9 – Construction Trades

Numbers in thousands.

Occupation	1993			
	Total employed	Percent of total		
		Women	Black	Hispanic origin
Construction trades	5,004	1.9	6.5	9.5
Supervisors	736	2.0	3.9	4.9
Construction trades, except supervisors	4,269	1.9	7.0	10.2
Brickmasons and stonemasons	186	.6	15.2	10.7
Tile setters, hard and soft	56	2.6	3.2	21.9
Carpet installers	94	2.1	6.9	15.6
Carpenters	1,276	.9	4.5	7.7
Drywall installers	135	1.3	7.3	16.9
Electricians	666	1.1	6.1	5.1
Electrical power installers and repairers	110	.9	7.8	4.7

[Continued]

★ 745 ★

Civilians Employed, by Detailed Occupation, 1993. Part 9 – Construction Trades

[Continued]

Occupation	1993			
	Total employed	Percent of total		
		Women	Black	Hispanic origin
Painters, construction and maintenance	548	5.2	7.5	15.6
Plumbers, pipefitters, and steamfitters	435	.7	8.1	8.2
Concrete and terrazzo finishers	75	.0	14.5	25.3
Insulation workers	60	7.6	14.8	6.8
Roofers	222	1.2	9.0	18.9

Source: "Employed Civilians by Detailed Occupation, Sex, Race, and Hispanic Origin." *Employment and Earnings* 41 (January 1994), p. 208. *Notes:* Generally, data for occupations with fewer than 50,000 employed are not published separately but are included in the totals for the appropriate categories shown.

★ 746 ★

Occupations

Civilians Employed, by Detailed Occupation, 1993. Part 10 – Precision Production Occupations

Numbers in thousands.

Occupation	1993			
	Total employed	Percent of total		
		Women	Black	Hispanic origin
Precision occupations	3,758	23.6	8.8	10.5
Supervisors	1,227	16.9	8.0	8.6
Precision metalworking	634	7.5	6.1	8.1
Tool and die makers	133	1.7	3.1	3.1
Machinists	441	4.0	7.2	7.7
Sheet-metal workers	106	7.1	7.4	9.3
Precision woodworking occupations	127	11.7	8.5	10.3
Cabinet makers and bench carpenters	85	7.8	6.2	9.9
Precision textile, apparel, and furnishings machine workers	219	55.8	9.0	18.3
Dressmakers	97	94.1	8.0	9.0
Upholsterers	63	16.5	9.7	23.4
Precision workers, assorted materials	505	56.6	11.9	13.5
Optical goods workers	60	51.5	7.2	13.1
Dental laboratory and medical appliance technicians	51	33.1	3.9	14.7
Electrical and electronic equipment assemblers	315	67.5	14.6	14.9
Precision food production occupations	442	33.2	12.3	16.4
Butchers and meat cutters	269	21.3	14.1	21.2
Bakers	119	45.9	8.9	11.8
Precision inspectors, testers, and related workers	136	26.2	8.8	7.0
Inspectors, testers, and graders	127	25.6	9.1	7.2

[Continued]

★ 746 ★

Civilians Employed, by Detailed Occupation, 1993. Part 10 – Precision Production Occupations

[Continued]

Occupation	1993			
	Total employed	Percent of total		
		Women	Black	Hispanic origin
Plant and system operators	267	4.9	9.0	3.6
Water and sewage treatment plant operators	57	7.6	6.2	1.0
Stationary engineers	124	2.9	9.8	3.7

Source: "Employed Civilians by Detailed Occupation, Sex, Race, and Hispanic Origin." *Employment and Earnings* 41 (January 1994), pp. 208-209. *Notes:* Generally, data for occupations with fewer than 50,000 employed are not published separately but are included in the totals for the appropriate categories shown.

★ 747 ★

Occupations

Civilians Employed, by Detailed Occupation, 1993. Part 11 – Operators, Fabricators, and Laborers

Numbers in thousands.

Occupation	1993			
	Total employed	Percent of total		
		Women	Black	Hispanic origin
Operators, fabricators, and laborers	17,038	24.5	14.9	12.1
Machine operators, assemblers, and inspectors	7,415	38.7	14.7	13.8
Machine operators and tenders, except precision	4,757	38.8	15.5	14.5
Metalworking and plastic working machine operators	381	15.6	9.6	11.1
Punching and stamping press machine operators	99	27.3	12.9	10.4
Grinding, abrading, buffing, and polishing machine operators	124	12.1	7.9	16.1
Metal and plastic processing machine operators	149	21.6	11.2	12.9
Molding and casting machine operators	93	28.6	11.5	10.3
Woodworking machine operators	136	12.9	13.2	7.0
Sawing machine operators	85	10.6	16.4	6.6
Printing machine operators	403	25.8	6.0	7.5
Printing press operators	283	16.4	6.9	8.6
Typesetting and compositors	54	66.4	1.5	3.0
Textile, apparel, and furnishings machine operators	1,159	74.4	20.8	19.9
Winding and twisting machine operators	51	76.7	37.6	3.4
Textile sewing machine operators	616	85.8	18.5	24.1
Pressing machine operators	147	62.7	24.0	20.4
Laundering and dry cleaning machine operators	216	58.8	21.7	17.7
Machine operators, assorted materials	2,506	30.4	15.8	14.3
Packaging and filling machine operators	343	58.3	18.8	21.7
Mixing and blending machine operators	113	11.0	19.5	13.1
Separating, filtering, and clarifying machine operators	54	12.7	8.6	4.3
Painting and paint spraying machine operators	203	12.0	12.6	14.7

[Continued]

★ 747 ★

Civilians Employed, by Detailed Occupation, 1993. Part 11 – Operators, Fabricators, and Laborers

[Continued]

Occupation	1993			
	Total employed	Percent of total		
		Women	Black	Hispanic origin
Furnace, kiln and oven operators, except food	81	5.2	17.8	5.5
Slicing and cutting machine operators	195	28.1	12.7	19.7
Photographic process machine operators	80	51.7	8.7	8.9
Fabricators, assemblers, and hand working occupations	1,882	32.7	12.7	12.0
Welders and cutters	539	3.5	10.1	10.6
Assemblers	1,129	43.7	14.9	12.2
Production inspectors, testers, samplers, and weighers	777	52.4	15.0	13.7
Production inspectors, checkers, and examiners	560	54.4	14.8	11.1
Graders and sorters, except agricultural	156	53.3	15.7	23.0

Source: "Employed Civilians by Detailed Occupation, Sex, Race, and Hispanic Origin." *Employment and Earnings* 41 (January 1994), p. 209. *Notes:* Generally, data for occupations with fewer than 50,000 employed are not published separately but are included in the totals for the appropriate categories shown.

★ 748 ★

Occupations

Civilians Employed, by Detailed Occupation, 1993. Part 12 – Transportation and Material Moving Occupations

Numbers in thousands.

Occupation	1993			
	Total employed	Percent of total		
		Women	Black	Hispanic origin
Transportation and material moving occupations	5,004	9.3	14.0	8.6
Motor vehicle operators	3,825	10.8	14.2	8.9
Supervisors	84	18.3	8.4	7.4
Truck drivers	2,786	4.5	12.3	8.8
Drivers-sales workers	178	8.6	6.2	6.2
Bus drivers	506	45.8	23.1	8.0
Taxicab drivers and chauffeurs	225	9.4	23.6	12.3
Transpiration occupations,except motor vehicles	170	4.1	10.2	3.1
Rail transportation	108	4.0	11.7	1.5
Water transportation	62	4.2	7.6	5.7
Material moving equipment operators	1,009	4.5	13.9	8.5
Operating engineers	193	1.0	7.1	4.9
Crane and tower operators	85	1.1	10.6	5.6
Excavating and loading machine operators	111	.4	5.8	7.1

[Continued]

★ 748 ★

Civilians Employed, by Detailed Occupation, 1993. Part 12 – Transportation and Material Moving Occupations
[Continued]

| Occupation | 1993 | | | |
| | Total employed | Percent of total | | |
		Women	Black	Hispanic origin
Grader, dozer, and scraper operators	84	1.8	9.3	2.6
Industrial truck and tractor equipment operators	432	7.3	20.9	12.0

Source: "Employed Civilians by Detailed Occupation, Sex, Race, and Hispanic Origin." *Employment and Earnings* 41 (January 1994), p. 209. *Notes:* Generally, data for occupations with fewer than 50,000 employed are not published separately but are included in the totals for the appropriate categories shown.

★ 749 ★

Occupations

Civilians Employed, by Detailed Occupation, 1993. Part 13 – Handlers, Equipment Cleaners, Helpers, and Laborers

Numbers in thousands.

| Occupation | 1993 | | | |
| | Total employed | Percent of total | | |
		Women	Black	Hispanic origin
Handlers, equipment cleaners, helpers, and laborers	4,619	18.3	16.1	12.9
Helpers, construction and extractive occupations	116	2.0	9.6	18.6
Helpers, construction trades	106	1.9	9.9	19.9
Construction laborers	658	3.5	14.9	16.7
Freight, stock, and material handlers	1,850	20.7	16.7	9.9
Stock handlers and baggers	1,050	27.0	13.2	9.5
Machine feeders and offbearers	78	38.1	22.3	4.4
Garage and service station related occupations	196	6.5	13.4	7.7
Vehicle washers and equipment cleaners	256	13.0	19.1	20.8
Hand packers and packagers	321	58.3	15.4	13.9
Laborers, except construction	1,127	17.0	16.7	13.7

Source: "Employed Civilians by Detailed Occupation, Sex, Race, and Hispanic Origin." *Employment and Earnings* 41 (January 1994), p. 209. *Notes:* Generally, data for occupations with fewer than 50,000 employed are not published separately but are included in the totals for the appropriate categories shown.

★ 750 ★

Occupations

Civilians Employed, by Detailed Occupation, 1993. Part 14. Farming, Forestry, and Fishing

Numbers in thousands.

Occupation	1993			
	Total employed	Percent of total		
		Women	Black	Hispanic origin
Farming, forestry, and fishing	3,326	15.4	6.3	16.0
Farm operators and managers	1,170	14.3	0.9	2.4
Farmers	1,032	14.4	0.9	1.9
Farm managers	138	13.1	.7	6.7
Other agricultural and related occupations	1,963	17.0	9.5	24.8
Farm occupations, except managerial	871	20.6	6.8	29.4
Farm workers	801	20.0	7.0	28.5
Related agricultural occupations	1,092	14.2	11.6	21.1
Supervisors	71	2.5	8.5	17.7
Groundskeepers and gardeners, except farm	863	5.6	12.8	21.3
Animal caretakers, except farm	107	65.5	6.0	3.7
Forestry and logging occupations	132	5.9	10.4	11.8
Timber cutting and logging occupations	93	1.2	13.7	4.7
Fisheries, hunters, and trappers	61	4.4	1.7	4.8
Fishers	51	4.3	1.5	5.8

Source: "Employed Civilians by Detailed Occupation, Sex, Race, and Hispanic Origin,: *Employment and Earnings* 41 (January 1994), pp. 209-210. *Notes:* Generally, data for occupations with fewer than 50,000 employed are not published separately but are included in the totals for the appropriate categories shown.

★ 751 ★

Occupations

Civilians Employed, by Industry, Race, and Occupation: Annual Averages, 1993 - I

In thousands.

Industry and race	1993								
	Total employed	Managerial and professional specialty		Technical, sales and administrative support			Service		Precision production, craft, and repair
		Executive, adminis- trative and managerial	Professional specialty	Technicians and related support	Sales	Administrative support, including clerical	Private household	Other service[1]	
White									
Agriculture	2,864	96	85	36	13	111	-	13	41
Mining	630	99	69	22	4	70	-	7	218
Construction	6,599	893	123	44	73	368	-	25	3,927
Manufacturing	16,861	2,291	1,545	612	702	1,891	-	243	3,257
Durable goods	9,922	1,347	995	409	273	1,051	-	128	2,250
Nondurable goods	6,939	944	550	203	428	840	-	115	1,007
Transportation and public utilities	7,027	861	434	289	206	1,848	-	198	1,118

[Continued]

★ 751 ★

Civilians Employed, by Industry, Race, and Occupation: Annual Averages, 1993 - I
[Continued]

Industry and race	1993								
	Total employed	Managerial and professional specialty		Technical, sales and administrative support			Service		Precision production, craft, and repair
		Executive, adminis-trative and managerial	Professional specialty	Technicians and related support	Sales	Administrative support, including clerical	Private household	Other service[1]	
Wholesale and retail trade	21,696	1,937	388	157	9,165	2,073	-	3,956	1,260
Wholesale trade	4,186	495	75	44	1,723	690	-	31	269
Retail trade	17,510	1,442	313	112	7,443	1,383	-	3,925	991
Finance, insurance, and real estate	7,037	1,902	209	136	1,806	2,520	-	220	148
Services	35,414	4,721	11,393	1,918	819	5,834	721	6,388	1,793
Private households	874	3	6	2	1	9	721	53	10
Other service industries	34,540	4,718	11,388	1,916	818	5,824	-	6,335	1,783
Professional services	24,005	2,969	10,152	1,631	152	4,360	-	3,787	360
Public administration	4,685	1,088	724	223	21	1,121	-	1,199	192
Black									
Agriculture	142	2	1	1	1	2	-	1	2
Mining	25	2	2	-	-	2	-	-	11
Construction	461	24	5	1	1	16	-	8	241
Manufacturing	1,968	81	56	48	27	199	-	59	330
Durable goods	983	35	29	26	7	93	-	29	196
Nondurable goods	985	46	26	22	20	106	-	30	134
Transportation and public utilities	1,171	79	33	30	29	369	-	50	120
Wholesale and retail trade	2,076	127	27	9	716	176	-	536	77
Wholesale trade	276	15	5	1	48	51	-	6	18
Retail trade	1,799	112	21	8	668	126	-	530	59
Finance, insurance, and real estate	666	122	26	17	79	316	-	56	30
Services	4,776	363	939	246	93	729	156	1,751	152
Private households	198	-	1	-	-	1	156	25	1
Other service industries	4,578	363	937	246	92	728	-	1,727	151
Professional services	3,241	254	852	221	20	557	-	1,163	41
Public administration	861	159	94	34	3	271	-	242	22

Source: "Employed Civilians by Selected Social and Economic Categories, Sex, Race, and Hispanic Origin." *Employment and Earnings* 41 (January 1994), p. 214. *Note:* 1. Includes protective service, not shown separately.

★ 752 ★

Occupations

Civilians Employed, by Industry, Race, and Occupation: Annual Averages, 1993 - II

In thousands.

Industry and race	1993			
	Operators, fabricators, and laborers			Farming, forestry, and fishing
	Machine operators, assemblers, and inspectors	Transportation and material moving	Handlers, equipment cleaners, helpers, and laborers	
White				
Agriculture	7	45	18	2,399
Mining	24	95	22	-
Construction	68	456	603	20
Manufacturing	4,850	594	797	78
Durable goods	2,734	300	364	70
Nondurable goods	2,116	294	433	8
Transportation and public utilities	105	1,617	337	14
Wholesale and retail trade	253	887	1,562	58
Wholesale trade	97	425	300	37
Retail trade	156	462	1,262	21
Finance, insurance, and real estate	16	12	14	55
Services	640	444	353	390
Private households	-	5	15	49
Other service industries	640	439	337	342
Professional services	174	251	83	87
Public administration	31	36	26	23
Black				
Agriculture	1	5	1	126
Mining	-	6	2	-
Construction	9	49	103	4
Manufacturing	894	99	162	13
Durable goods	439	53	63	12
Nondurable goods	455	46	99	1
Transportation and public utilities	20	323	115	4
Wholesale and retail trade	26	114	262	5
Wholesale trade	12	56	58	4
Retail trade	14	58	204	1
Finance, insurance, and real estate	5	1	4	11
Services	132	88	85	41
Private households	-	-	5	9
Other service industries	132	89	80	33
Professional services	40	59	20	14
Public administration	6	14	10	6

Source: "Employed Civilians by Selected Social and Economic Categories, Sex, Race, and Hispanic Origin." *Employment and Earnings* 41 (January 1994), p. 214.

★ 753 ★

Occupations

Civilians Employed, by Occupation, 1992 and 1993

Percent distribution.

Occupation and race	Total		Men		Women	
	1992	1993	1992	1993	1992	1993
White						
Total, 16 years and over (thousands)	101,479	102,812	55,709	56,397	45,770	46,415
Percent	100.0	100.0	100.0	100.0	100.0	100.0
Managerial and professional specialty	27.5	28.1	26.8	27.1	28.5	29.3
Executive, administrative, and managerial	13.2	13.5	14.3	14.5	12.0	12.3
Professional specialty	14.3	14.6	12.5	12.6	16.5	17.0
Technical, sales, and administrative support	31.7	31.2	21.1	20.8	44.6	43.9
Technicians and related support	3.6	3.3	3.4	3.1	3.8	3.7
Sales occupations	12.4	12.5	12.0	12.0	12.9	13.0
Administrative support, including clerical	15.7	15.4	5.7	5.7	27.9	27.2
Service occupations	12.4	12.6	9.1	9.2	16.5	16.7
Private household	.7	.7	.1	.1	1.4	1.5
Protective service	1.6	1.7	2.5	2.6	.6	.6
Service, except private household and protective	10.1	10.2	6.5	6.6	14.5	14.7
Precession production, craft, and repair	11.6	11.6	19.5	19.6	2.0	2.0
Operators, fabricators, and laborers	13.6	13.5	18.9	18.9	7.2	7.0
Machine operators, assemblers, and inspectors	6.0	5.8	6.8	6.7	4.9	4.7
Transportation and material moving occupations	4.0	4.1	6.6	6.7	.8	.8
Handlers, equipment cleaners, helpers, and laborers	3.7	3.6	5.4	5.4	1.5	1.5
Farming, forestry, and fishing	3.1	3.0	4.7	4.5	1.1	1.0
Black						
Total, 16 years and over (thousands)	11,933	12,146	5,846	5,957	6,087	6,189
Percent	100.0	100.0	100.0	100.0	100.0	100.0
Managerial and professional specialty	16.8	17.6	14.1	14.7	19.5	20.5
Executive, administrative, and managerial	7.2	7.9	7.1	7.7	7.2	8.1
Professional specialty	9.7	9.7	7.0	6.9	12.3	12.4
Technical, sales, and administrative support	28.0	28.1	17.3	18.1	38.2	37.8
Technicians and related support	3.4	3.2	3.1	2.7	3.8	3.6
Sales occupations	7.3	7.8	5.7	6.5	8.7	9.1
Administrative support, including clerical	17.3	17.1	8.5	8.9	25.7	25.1
Service occupations	23.5	23.5	19.0	19.4	27.9	27.5
Private household	1.4	1.3	.1	.1	2.6	2.4
Protective service	3.2	3.1	4.9	4.9	1.5	1.3
Service, except private household and protective	19.0	19.2	14.0	14.4	23.9	23.8
Precession production, craft, and repair	8.4	8.1	14.8	14.0	2.1	2.5
Operators, fabricators, and laborers	21.4	20.9	31.3	30.6	12.0	11.5
Machine operators, assemblers, and inspectors	9.4	9.0	9.8	9.6	9.2	8.4
Transportation and material moving occupations	6.1	5.8	11.3	10.6	1.0	1.1
Handlers, equipment cleaners, helpers, and laborers	5.9	6.1	10.2	10.4	1.8	2.0
Farming, forestry, and fishing	1.9	1.7	3.6	3.2	.3	.3

Source: "Employed Civilians by Occupation, Race, and Sex." *Employment and Earnings* 41 (January 1994), p. 204.

★ 754 ★

Occupations

Civilians Employed, by Occupation, Race, and Gender: January 1993 and January 1994

Percent distribution.

Occupation and race	Total		Men		Women	
	Jan. 1993	Jan. 1994	Jan. 1993	Jan. 1994	Jan. 1993	Jan. 1994
White						
Total, 16 years and over (thousands)	100,296	102,628	54,815	55,878	45,481	46,750
Percent	100.0	100.0	100.0	100.0	100.0	100.0
Managerial and professional specialty	28.6	28.6	27.7	27.5	29.7	30.0
Executive, administrative, and managerial	13.7	13.5	14.8	14.2	12.4	12.6
Professional specialty	14.9	15.2	12.9	13.4	17.3	17.4
Technical, sales, and administrative support	31.4	31.3	21.0	21.1	44.0	43.5
Technicians and related support	3.5	3.3	3.3	3.0	3.8	3.7
Sales occupations	12.3	12.6	12.0	12.2	12.6	13.0
Administrative support, including clerical	15.6	15.4	5.7	5.9	27.6	26.8
Service occupations	12.7	12.5	9.7	9.4	16.3	16.3
Private household	.7	.6	-	[1]	1.4	1.2
Protective service	1.7	1.6	2.7	2.5	.5	.5
Service, except private household and protective	10.3	10.3	6.9	6.8	14.4	14.5
Precision production, craft, and repair	11.3	11.6	19.1	19.1	1.9	2.0
Operators, fabricators, and laborers	13.4	13.5	18.6	18.8	7.2	7.2
Machine operators, assemblers, and inspectors	6.0	5.9	6.8	6.8	4.9	4.8
Transportation and material moving occupations	3.9	4.1	6.4	6.7	.9	.9
Handlers, equipment cleaners, helpers, and laborers	3.6	3.6	5.3	5.3	1.4	1.5
Farming, forestry, and fishing	2.6	2.7	4.0	4.1	.9	1.1
Black						
Total, 16 years and over (thousands)	11,663	12,274	5,727	5,853	5,936	6,421
Percent	100.0	100.0	100.0	100.0	100.0	100.0
Managerial and professional specialty	17.6	18.5	15.0	16.3	20.1	20.5
Executive, administrative, and managerial	7.8	8.5	7.7	8.8	7.9	8.3
Professional specialty	9.8	10.0	7.3	7.6	12.3	12.1
Technical, sales, and administrative support	28.1	28.2	18.1	17.0	37.9	38.4
Technicians and related support	3.5	3.0	3.1	2.9	3.8	3.1
Sales occupations	7.9	8.2	6.3	6.0	9.5	10.1
Administrative support, including clerical	16.7	17.0	8.7	8.0	24.5	25.2
Service occupations	23.5	23.7	19.7	19.6	27.1	27.4
Private household	1.3	1.2	.1	-	2.5	2.2
Protective service	3.2	3.4	5.1	5.3	1.3	1.7
Service, except private household and protective	19.0	19.1	14.6	14.3	23.2	23.6
Precision production, craft, and repair	8.4	8.5	14.5	14.9	2.6	2.7
Operators, fabricators, and laborers	20.9	20.0	30.1	30.2	12.0	10.7
Machine operators, assemblers, and inspectors	8.6	8.7	8.7	9.9	8.4	7.5
Transportation and material moving occupations	6.2	5.5	11.4	10.4	1.2	1.0

[Continued]

★ 754 ★

Civilians Employed, by Occupation, Race, and Gender: January 1993 and January 1994

[Continued]

Occupation and race	Total		Men		Women	
	Jan. 1993	Jan. 1994	Jan. 1993	Jan. 1994	Jan. 1993	Jan. 1994
Handlers, equipment cleaners, helpers, and laborers	6.1	5.9	10.1	9.9	2.4	2.2
Farming, forestry, and fishing	1.5	1.0	2.6	1.9	.4	.2

Source: "Employed Persons by Occupation, Race and Sex Categories." *Employment and Earnings* 41 (February 1994), p. 59. *Notes:* Data for 1994 are not directly comparable with data for 1993 and earlier years. 1. Less than 0.05 percent.

★ 755 ★

Occupations

Civilians Employed, by Selected Social and Economic Categories, 1992 and 1993

Annual averages. In thousands.

Category	Total		White		Black		Hispanic origin	
	1992	1993	1992	1993	1992	1993	1992	1993
Characteristic								
Total (all civilian workers)	117,598	119,306	101,479	102,812	11,933	12,146	8,971	9,272
Men	63,805	64,700	55,709	56,397	5,846	5,957	5,388	5,603
Women	53,793	54,606	45,770	46,415	6,087	6,189	3,584	3,669
Occupation								
Managerial and professional specialty	31,153	32,280	27,948	28,859	2,009	2,140	1,205	1,306
Executive, administrative, and managerial	14,767	15,376	13,438	13,888	855	959	622	694
Professional specialty	16,386	16,904	14,510	14,971	1,154	1,181	583	613
Technical, sales, and administrative support	36,808	36,814	32,162	32,082	3,337	3,416	2,229	2,305
Technicians and related support	4,253	4,014	3,631	3,437	410	387	193	200
Sales occupations	13,919	14,245	12,564	12,809	865	948	783	836
Administrative support, including clerical	18,636	18,555	15,968	15,836	2,062	2,081	1,253	1,269
Service occupations	16,096	16,522	12,611	12,969	2,809	2,859	1,817	1,848
Private household	876	912	682	721	163	156	171	197
Protective service	2,096	2,152	1,664	1,728	377	374	150	142
Service, except private household and protective	13,124	13,457	10,265	10,521	2,269	2,329	1,496	1,508
Precision production, craft, and repair	13,128	13,326	11,767	11,955	996	985	1,204	1,226
Mechanics and repairers	4,441	4,416	4,001	3,977	337	321	348	347
Construction trades	4,790	5,004	4,369	4,576	332	327	437	473
Other precision production, craft, and repair	3,897	3,906	3,397	3,402	328	337	419	405
Operators, fabricators, and laborers	16,957	17,038	13,844	13,910	2,556	2,535	1,993	2,054
Machine operators, assemblers, and inspectors	7,524	7,415	6,070	5,992	1,128	1,092	1,015	1,024
Transportation and material moving occupations	4,878	5,004	4,059	4,186	724	699	412	431
Handlers, equipment cleaners, helpers, and laborers	4,556	4,619	3,715	3,732	705	743	567	598
Construction laborers	654	658	555	536	82	98	109	110
Other handlers, equipment cleaners, helpers, and laborers	3,901	3,962	3,160	3,195	623	646	458	489
Farming, forestry, and fishing	3,456	3,326	3,147	3,037	225	211	523	534

[Continued]

★ 755 ★

Civilians Employed, by Selected Social and Economic Categories, 1992 and 1993

[Continued]

Category	Total 1992	Total 1993	White 1992	White 1993	Black 1992	Black 1993	Hispanic origin 1992	Hispanic origin 1993
Major industry and class of worker								
Agriculture								
Wage and salary workers	1,696	1,637	1,539	1,484	114	103	406	407
Self-employed workers	1,398	1,332	1,337	1,275	39	39	59	61
Unpaid workers	113	105	111	104	-	-	1	-
Nonagricultural industries								
Wage and salary workers	105,5440	107,011	90,403	91,545	11,362	11,570	8,053	8,310
Government	18,086	18,504	14,713	14,996	2,734	2,816	1,028	1,119
Private industries	87,454	88,507	75,690	76,549	8,627	8,754	7,025	7,191
Private households	1,116	1,105	848	867	229	198	205	225
Other industries	86,338	87,402	74,842	75,682	8,398	8,557	6,820	6,966
Self-employed workers	8,619	9,003	7,878	8,211	413	429	439	482
Unpaid family workers	232	218	211	192	5	5	14	12
Full- and part-time status[1]								
Full-time schedules	95,000	96,443	81,793	82,888	9,803	10,006	7,271	7,491
Part time for economic reasons	6,385	6,348	5,206	5,157	938	942	832	850
Part time for noneconomic reasons	16,213	16,515	14,480	14,767	1,193	1,198	868	932

Source: "Employed Civilians by Selected Social and Economic Categories, Sex, Race, and Hispanic Origin." *Employment and Earnings* 41 (January 1994), pp. 230. *Notes:* Detail for the above race and Hispanic-origin groups will not sum to totals because data for the "other races" group are not presented and Hispanics are included in both the white and black population groups. 1. Employed persons "with a job but not at work" are distributed according to whether they usually work full or part time.

★ 756 ★

Occupations

Civilians Employed: Nonagricultural Industries, 1993

In thousands.

Sex, age, and race	Mining	Construction	Manufacturing Total	Manufacturing Durable goods	Manufacturing Nondurable goods	Transportation and public utilities	Wholesale and retail trade	Finance, insurance, and real estate	Services[1]	Public administration
Total										
Total, 16 years and over	669	7,220	19,557	11,325	8,232	8,481	24,769	7,962	40,703	5,756
16 to 19 years	6	179	362	170	191	92	3,018	126	1,326	39
20 years and over	664	7,042	19,196	11,155	8,041	8,389	21,751	7,836	39,377	5,717
20 to 24 years	34	669	1,599	852	747	544	3,867	766	3,936	297
25 years and over	629	6,374	17,597	10,303	7,294	7,846	17,884	7,070	35,441	5,420
25 to 54 years	572	5,600	15,316	8,991	6,325	6,958	15,189	5,986	30,131	4,710
55 years and over	57	774	2,281	1,312	969	888	2,695	1,084	5,310	710
Men, 16 years and over	561	6,603	13,249	8,314	4,934	6,062	13,202	3,298	15,855	3,288
16 to 19 years	4	166	251	125	126	66	1,504	41	577	17
20 years and over	557	6,437	12,997	8,190	4,808	5,995	11,698	3,256	15,278	3,271
20 to 24 years	29	623	1,083	621	461	357	1,982	252	1,607	159
25 years and over	528	5,815	11,915	7,568	4,346	5,638	9,717	3,004	13,671	3,112
25 to 54 years	479	5,123	10,333	6,572	3,760	4,947	8,278	2,438	11,416	2,691
55 years and over	49	692	1,582	996	586	691	566	2,255	421	
Women, 16 years and over	108	617	6,309	3,011	3,298	2,419	11,566	4,664	24,848	2,468
16 to 19 years	1	13	110	46	65	25	1,514	84	749	22
20 years and over	107	605	6,198	2,965	3,233	2,394	10,053	4,580	24,099	2,446
20 to 24 years	6	46	516	230	286	187	1,886	514	2,330	138
25 years and over	102	559	5,682	2,736	2,948	2,206	8,167	4,066	21,770	2,308

[Continued]

★ 756 ★

Civilians Employed: Nonagricultural Industries, 1993

[Continued]

Sex, age, and race	1993									
	Mining	Construction	Manufacturing			Transportation and public utilities	Wholesale and retail trade	Finance, insurance, and real estate	Services[1]	Public administration
			Total	Durable goods	Nondurable goods					
25 to 54 years	93	477	4,983	2,419	2,565	2,010	6,911	3,548	18,715	2,019
55 years and over	9	82	699	317	383	196	1,256	518	3,055	289
White										
Men, 16 years and over	527	6,025	11,635	7,365	4,270	5,058	11,533	2,926	13,544	2,776
16 to 19 years	4	156	227	114	113	56	1,317	35	497	14
20 years and over	523	5,868	11,409	7,251	4,157	5,002	10,216	2,891	13,047	2,762
20 to 24 years	28	571	944	548	396	289	1,698	213	1,326	134
25 years and over	495	5,297	10,464	6,703	3,761	4,714	8,519	2,678	11,721	2,629
25 to 54 years	449	4,666	9,045	5,806	3,239	4,130	7,209	2,155	9,724	2,267
55 years and over	46	631	1,419	897	522	584	1,310	523	1,997	362
Women, 16 years and over	103	574	5,225	2,557	2,669	1,969	10,163	4,111	20,996	1,908
16 to 19 years	1	12	98	42	57	20	1,339	77	662	14
20 years and over	102	562	5,127	2,515	2,612	1,949	8,824	4,034	20,335	1,894
20 to 24 years	4	43	431	197	234	151	1,614	444	1,987	101
25 years and over	98	519	4,696	2,318	2,378	1,798	7,210	3,591	18,348	1,794
25 to 54 years	89	440	4,091	2,040	2,051	1,625	6,039	3,101	15,752	1,546
55 years and over	9	79	605	278	327	173	1,171	490	2,596	248
Black										
Men, 16 years and over	22	430	1,181	668	512	808	1,118	256	1,594	399
16 to 19 years	-	5	19	6	12	7	140	4	61	2
20 years and over	22	425	1,162	662	500	801	978	252	1,533	397
20 to 24 years	-	34	97	50	48	57	200	29	207	20
25 years and over	21	391	1,064	612	452	744	778	223	1,326	378
25 to 54 years	19	342	939	535	404	656	692	193	1,143	334
55 years and over	2	49	125	77	48	88	86	30	183	44
Women, 16 years and over	3	30	787	314	473	363	957	410	2,985	462
16 to 19 years	-	-	10	3	7	4	126	5	67	7
20 years and over	3	30	777	311	466	359	832	405	2,918	455
20 to 24 years	1	2	61	23	39	27	206	52	272	31
25 years and over	3	28	715	288	427	331	626	354	2,646	424
25 to 54 years	3	27	655	260	395	313	578	334	2,273	393
55 years and over	-	1	60	28	32	18	48	20	373	31

Source: "Employed Civilians in Nonagricultural Industries by Sex, Age, and Race." *Employment and Earnings* 41 (January 1994), p. 215. *Note:* 1. Excludes private households.

★ 757 ★

Occupations

Employment by Industry, 1970 to 1992

In thousands, except percent.

Industry	1992			
	Total	Percent		
		Female	Black	Hispanic[1]
Total employed	117,598	45.7	10.1	7.6
Agriculture	3,210	21.0	4.8	14.5
Mining	664	16.0	3.6	7.0
Construction	7,013	8.9	6.3	8.3
Manufacturing	19,972	32.9	10.4	8.7
Transportation, communication, and other public utilities	8,245	28.3	13.4	6.7

[Continued]

★ 757 ★

Employment by Industry, 1970 to 1992
[Continued]

Industry	Total	1992		
		Percent		
		Female	Black	Hispanic[1]
Wholesale and retail trade	24,354	46.7	8.1	8.4
Wholesale trade	4,765	29.2	6.3	7.4
Retail trade	19,589	50.9	8.5	8.6
Finance, insurance, real estate	7,764	59.0	8.8	5.9
Banking and other finances	3,272	64.4	8.8	6.2
Insurance and real estate	4,495	55.0	8.7	5.7
Services[2]	40,758	61.7	11.4	6.8
Business and repair services[2]	6,553	35.3	10.7	9.4
Advertising	259	51.1	5.0	5.6
Services to dwellings and buildings	744	43.5	17.9	18.4
Personnel supply services	687	64.6	19.0	7.2
Computer and data processing	882	35.7	6.0	3.7
Detective/protective services	457	19.4	22.3	8.7
Automobile services	1,470	12.9	9.0	12.6
Personal services[2]	4,400	69.3	14.9	12.6
Private households	1,127	86.2	20.4	18.4
Hotels and lodging places	1,433	55.1	16.3	13.4
Entertainment and recreation	1,957	40.3	8.3	7.4
Professional and related services[2]	27,677	68.5	11.3	5.2
Hospitals	4,915	76.7	15.8	5.8
Health services, except hospitals	5,356	78.5	12.6	5.7
Elementary, secondary schools	6,178	73.7	11.7	5.2
Colleges and universities	2,587	53.0	9.0	4.4
Social services	2,727	81.8	15.0	6.8
Legal services	1,247	54.4	5.4	4.6
Public administration[3]	5,620	42.7	14.4	5.6

Source: "Employed by Industry, 1970 to 1992." U.S. Bureau of the Census, *Statistical Abstract of the United States*, 1993, p. 409. Primary source: U.S. Bureau of Labor Statistics, *Employment and Earnings*, monthly, January issues. *Notes:* 1. Persons of Hispanic origin may be of any race. 2. Includes industries not shown separately. 3. Includes workers involved in uniquely governmental activities, e.g., judicial and legislative.

★ 758 ★

Occupations

Hired Farmworkers and Earnings, 1990 and 1991

Represents average number of persons 15 years old and over in the civilian noninstitutional population who were employed at hired farmwork at any time during the year. Based on Current Population Survey.

Characteristic	Workers (1,000)		Median weekly earnings[1]	
	1990	1991	1990	1991
All workers	886	884	$200	$210
White[2]	540	533	201	222
Black and other races[2]	85	101	175	170
Hispanic	260	250	213	220

Source: "Hired Farmworkers—Workers and Weekly Earnings: 1990 and 1991." U.S. Bureau of the Census, *Statistical Abstract of the United States*, 1993, p. 665. Primary source: U.S. Dept. of Agriculture, Economic Research Service, unpublished data. *Notes:* 1. The weekly earnings the farmworker usually earns at his farmwork job before deductions and includes any overtime pay or commissions. 2. Excludes persons of Hispanic origin.

★ 759 ★

Occupations

Occupation of Longest Job of Year-Round Worker, 25 Years Old and Over, by Earnings and Education, 1991. Part I

Numbers in thousands. Persons as of March 1992.

Occupation of longest job	Total		Not a high school graduate		High school graduate		Some college or associate degree		Bachelor's degree or more	
	Male	Female	Male	Female	Male	Female	Male	Female	Male	Female
Black										
Executive, administrative, and managerial workers	330	333	18	17	42	63	101	110	169	144
Median earnings (dollars)	34,239	27,459	30,523	14,229	19,750	22,862	26,945	29,334	39,387	30,352
Standard error (dollars)	1,824	1,554	4,719	3,621	1,506	2,757	4,237	3,007	2,058	1,599
Professional specialty workers	299	521	9	11	28	46	61	118	201	346
Median earnings (dollars)	32,658	29,146	25,730	12,303	19,336	25,516	30,882	24,670	37,186	31,022
Standard error (dollars)	2,122	1,067	2,695	2,480	3,659	8,571	3,714	1,973	2,197	1,222
Technical and related support workers	120	134	8	8	33	39	43	62	36	26
Median earnings (dollars)	27,259	22,268	18,338	21,772	33,282	21,810	26,131	19,590	30,796	30,406
Standard error (dollars)	1,556	2,419	4,503	2,266	3,614	2,332	1,484	3,776	5,989	6,559
Sales workers	185	187	19	16	55	81	51	65	61	25
Median earnings (dollars)	22,269	19,546	30,845	15,252	15,984	15,971	20,436	21,579	24,906	24,142
Standard error (dollars)	1,178	1,861	4,623	7,756	3,312	948	1,365	1,524	3,316	5,463
Administrative support workers, including clerical	349	1,222	27	59	143	492	128	521	52	149
Median earnings (dollars)	27,162	20,036	23,570	18,203	21,695	19,407	31,132	20,066	26,132	22,444
Standard error (dollars)	1,562	438	4,315	1,834	1,918	700	1,116	675	3,440	1,350
Private household workers	-	22	-	18	-	2	-	1	-	-
Median earnings (dollars)	-	6,762	-	5,343	-	(S)	-	(S)	-	-
Standard error (dollars)	-	2,325	-	2,026	-	(S)	-	(S)	-	-
Protective service workers	164	64	21	7	73	25	49	27	21	5
Median earnings (dollars)	25,667	21,462	13,253	20,869	19,912	18,602	29,347	33,085	31,803	(S)
Standard error (dollars)	2,493	2,086	1,221	2,455	6,063	3,351	3,834	9,980	3,643	(S)
Service workers, except private household	414	764	113	251	190	410	89	100	21	4
Median earnings (dollars)	16,136	12,273	15,567	10,903	16,291	13,458	16,315	13,149	20,502	(S)
Standard error (dollars)	744	385	1,390	575	1,537	891	1,480	2,164	6,427	(S)
Farming, fishing, and forestry workers	64	7	33	4	26	3	3	-	2	-
Median earnings (dollars)	10,118	11,895	9,377	(S)	10,628	(S)	(S)	-	(S)	-
Standard error (dollars)	1,528	8,956	2,491	(S)	1,546	(S)	(S)	-	(S)	-

[Continued]

★ 759 ★

Occupation of Longest Job of Year-Round Worker, 25 Years Old and Over, by Earnings and Education, 1991. Part I

[Continued]

Occupation of longest job	Total		Not a high school graduate		High school graduate		Some college or associate degree		Bachelor's degree or more	
	Male	Female	Male	Female	Male	Female	Male	Female	Male	Female
Precision production, craft, and repair workers	619	60	115	16	295	29	173	13	36	2
Median earnings (dollars)	25,489	20,202	16,577	15,936	26,206	18,465	29,106	32,575	26,871	(S)
Standard error (dollars)	863	2,898	692	5,220	952	2,304	2,363	6,774	3,584	(S)
Machine operators, assemblers, and inspectors	408	313	113	97	218	160	65	48	11	7
Median earnings (dollars)	20,943	14,568	20,942	15,039	19,121	13,363	30,233	16,413	28,676	16,250
Standard error (dollars)	922	995	1,430	2,091	1,701	1,127	2,841	2,104	9,676	786
Transportation and material moving workers	438	29	126	5	246	15	54	10	12	-
Median earnings (dollars)	21,754	22,551	19,204	(S)	22,233	22,140	30,457	14,048	18,412	-
Standard error (dollars)	1,012	1,709	1,875	(S)	2,081	1,760	6,261	9,802	14,402	-
Handlers, equipment cleaners, helpers, and laborers	280	54	76	17	137	36	47	2	20	-
Median earnings (dollars)	17,404	14,371	16,088	9,734	17,149	16,863	18,093	(S)	26,954	-
Standard error (dollars)	1,053	2,144	2,079	2,068	920	5,561	2,358	(S)	6,270	-

Source: "Occupation of Longest Job in 1991 of Year-Round, Full-Time Workers 25 Years Old and Over, by Total Money Median Earnings." Claudette E. Bennett, *The Black Population in the United States: March 1992*, p. 63. *Notes:* Data where base is less than 75,000 may not meet statistical standards for reliability of derived figures.

★ 760 ★

Occupations

Occupation of Longest Job of Year-Round Worker, 25 Years Old and Over, by Earnings and Education, 1991. Part II

Numbers in thousands. Persons as of March 1992.

Occupation of longest job	Total		Not a high school graduate		High school graduate		Some college or associate degree		Bachelor's degree or more	
	Male	Female	Male	Female	Male	Female	Male	Female	Male	Female
White										
Executive, administrative, and managerial workers	6,865	4,228	213	105	1,212	1,302	1,705	1,285	3,734	1,536
Median earnings (dollars)	42,433	27,452	27,265	19,135	33,190	22,386	39,285	26,377	51,067	3,910
Standard error (dollars)	599	390	4,070	2,501	1,238	523	1,215	481	398	1,163
Professional specialty workers	5,342	4,285	34	38	277	286	759	819	4,273	3,142
Median earnings (dollars)	43,591	30,863	42,897	12,395	34,663	19,576	36,261	30,529	46,377	31,576
Standard error (dollars)	804	249	4,898	5,204	1,528	1,136	604	504	491	284
Technical and related support workers	1,370	1,171	24	35	340	315	616	529	390	291
Median earnings (dollars)	33,293	23,053	27,851	21,057	32,514	20,767	33,457	22,472	34,661	28,587
Standard error (dollars)	787	620	8,047	4,137	934	643	1,146	742	1,683	1,787
Sales workers	4,905	2,457	223	173	1,465	980	1,472	729	1,745	576
Median earnings (dollars)	32,000	18,972	22,204	12,363	27,112	15,680	31,699	20,257	41,491	30,677
Standard error (dollars)	319	663	1,533	1,066	486	453	491	822	812	1,159
Administrative support workers, including clerical	2,176	7,546	136	256	808	3,884	715	2,538	517	868
Median earnings (dollars)	29,397	20,196	23,294	17,040	27,248	19,433	30,049	21,028	31,630	21,598
Standard error (dollars)	713	142	1,783	996	1,033	239	914	230	799	479
Private household workers	12	97	-	52	2	37	-	5	10	3
Median earnings (dollars)	8,994	8,957	-	8,229	(S)	10,232	-	(S)	8,750	(S)
Standard error (dollars)	654	858	-	836	(S)	1,799	-	(S)	599	(S)
Protective service workers	1,097	152	66	13	406	47	450	78	174	14
Median earnings (dollars)	31,432	24,158	19,064	15,151	26,811	20,300	32,641	24,735	39,563	35,638
Standard error (dollars)	633	1,430	1,373	2,676	999	5,634	1,200	1,450	1,502	6,807
Service workers, except private household	1,693	2,130	429	496	799	1,068	348	460	118	106
Median earnings (dollars)	18,448	12,142	15,567	11,354	19,216	11,729	21,476	14,397	18,395	16,622
Standard error (dollars)	438	168	500	324	558	206	740	557	1,111	1,060
Farming, fishing, and forestry workers	1,359	157	352	38	597	63	268	35	142	22
Median earnings (dollars)	15,820	9,811	12,622	10,414	15,707	8,683	19,306	6,001	20,527	11,449
Standard error (dollars)	438	1,563	923	1,493	621	2,742	1,153	3,462	3,460	2,782
Precision production, craft, and repair workers	7,313	574	1,133	136	3,728	272	1,966	129	487	37
Median earnings (dollars)	29,634	20,018	21,846	14,279	29,281	19,700	32,955	26,648	35,865	24,910
Standard error (dollars)	415	1,079	379	1,214	519	1,369	783	2,507	933	4,310
Machine operators, assemblers, and inspectors	2,543	1,317	556	436	1,407	711	488	140	92	29

[Continued]

★ 760 ★

Occupation of Longest Job of Year-Round Worker, 25 Years Old and Over, by Earnings and Education, 1991. Part II

[Continued]

Occupation of longest job	Total		Not a high school graduate		High school graduate		Some college or associate degree		Bachelor's degree or more	
	Male	Female	Male	Female	Male	Female	Male	Female	Male	Female
Median earnings (dollars)	25,918	15,279	19,598	12,674	26,413	16,218	28,912	17,500	30,176	24,059
Standard error (dollars)	331	325	737	549	379	488	872	1,933	2,320	5,503
Transportation and material moving workers	2,387	117	537	28	1,252	65	487	19	111	5
Median earnings (dollars)	26,395	18,874	21,962	16,363	26,573	19,761	30,525	19,663	30,166	(S)
Standard error (dollars)	418	1,417	622	1,681	545	1,543	1,109	2,186	3,551	(S)
Handlers, equipment cleaners, helpers, and laborers	1,265	349	343	90	676	203	209	41	37	15
Median earnings (dollars)	20,084	16,296	15,906	13,125	21,512	16,681	23,437	21,541	25,747	11,503
Standard error (dollars)	589	689	712	1,916	773	824	1,955	857	5,362	1,639

Source: "Occupation of Longest Job in 1991 of Year-Round, Full-Time Workers 25 Years Old and Over, by Total Money Median Earnings." Claudette E. Bennett, *The Black Population in the United States: March 1992,* p. 64. *Notes:* Data where base is less than 75,000 may not meet statistical standards for reliability of derived figures.

★ 761 ★

Occupations

Occupation of Longest Job of Year-Round Worker, 25 Years Old and Over, by Earnings and Education, 1991. Part III

Numbers in thousands. Persons as of March 1992.

Occupation of longest job	Total		Not a high school graduate		High school graduate		Some college or associate degree		Bachelor's degree or more	
	Male	Female	Male	Female	Male	Female	Male	Female	Male	Female
Black-to-White ratio										
Executive, administrative, and managerial workers	(X)	(X)	(X)	(X)	(X)	(X)	(X)	(X)	(X)	(X)
Median earnings (dollars)	80.7	100.0	111.9	74.4	59.5	102.1	68.6	111.2	77.1	89.5
Standard error (dollars)	(X)	(X)	(X)	(X)	(X)	(X)	(X)	(X)	(X)	(X)
Professional specialty workers	(X)	(X)	(X)	(X)	(X)	(X)	(X)	(X)	(X)	(X)
Median earnings (dollars)	74.9	94.4	60.0	99.3	55.8	130.3	85.2	80.8	80.2	98.2
Standard error (dollars)	(X)	(X)	(X)	(X)	(X)	(X)	(X)	(X)	(X)	(X)
Technical and related support workers	(X)	(X)	(X)	(X)	(X)	(X)	(X)	(X)	(X)	(X)
Median earnings (dollars)	81.9	96.6	65.8	103.4	102.4	105.0	78.1	87.2	88.9	106.4
Standard error (dollars)	(X)	(X)	(X)	(X)	(X)	(X)	(X)	(X)	(X)	(X)
Sales workers	(X)	(X)	(X)	(X)	(X)	(X)	(X)	(X)	(X)	(X)
Median earnings (dollars)	69.6	103.0	138.9	123.4	59.0	101.9	64.5	106.5	60.0	78.7
Standard error (dollars)	(X)	(X)	(X)	(X)	(X)	(X)	(X)	(X)	(X)	(X)
Administrative support workers, including clerical	(X)	(X)	(X)	(X)	(X)	(X)	(X)	(X)	(X)	(X)
Median earnings (dollars)	92.4	99.2	101.2	106.8	79.6	99.9	103.6	95.4	82.6	103.9
Standard error (dollars)	(X)	(X)	(X)	(X)	(X)	(X)	(X)	(X)	(X)	(X)
Private household workers	(X)	(X)	(X)	(X)	(X)	(X)	(X)	(X)	(X)	(X)
Median earnings (dollars)	-	75.5	-	64.9	-	-	-	-	-	-
Standard error (dollars)	(X)	(X)	(X)	(X)	(X)	(X)	(X)	(X)	(X)	(X)
Protective service workers	(X)	(X)	(X)	(X)	(X)	(X)	(X)	(X)	(X)	(X)
Median earnings (dollars)	81.7	88.8	69.5	137.7	74.3	91.6	89.9	133.8	80.4	-
Standard error (dollars)	(X)	(X)	(X)	(X)	(X)	(X)	(X)	(X)	(X)	(X)
Service workers, except private household	(X)	(X)	(X)	(X)	(X)	(X)	(X)	(X)	(X)	(X)
Median earnings (dollars)	87.5	101.1	100.0	96.0	84.8	114.7	76.0	91.3	111.5	-
Standard error (dollars)	(X)	(X)	(X)	(X)	(X)	(X)	(X)	(X)	(X)	(X)
Farming, fishing, and forestry workers	(X)	(X)	(X)	(X)	(X)	(X)	(X)	(X)	(X)	(X)
Median earnings (dollars)	64.0	121.2	74.3	0.0	67.7	-	-	-	-	-
Standard error (dollars)	(X)	(X)	(X)	(X)	(X)	(X)	(X)	(X)	(X)	(X)
Precision production, craft, and repair workers	(X)	(X)	(X)	(X)	(X)	(X)	(X)	(X)	(X)	(X)
Median earnings (dollars)	86.0	100.9	75.9	111.6	89.5	93.7	88.3	122.2	74.9	-
Standard error (dollars)	(X)	(X)	(X)	(X)	(X)	(X)	(X)	(X)	(X)	(X)
Machine operators, assemblers, and inspectors	(X)	(X)	(X)	(X)	(X)	(X)	(X)	(X)	(X)	(X)
Median earnings (dollars)	80.8	95.3	106.9	118.7	72.4	82.4	104.6	93.8	95.0	67.5
Standard error (dollars)	(X)	(X)	(X)	(X)	(X)	(X)	(X)	(X)	(X)	(X)
Transportation and material moving workers	(X)	(X)	(X)	(X)	(X)	(X)	(X)	(X)	(X)	(X)
Median earnings (dollars)	82.4	119.5	87.4	-	83.7	112.0	99.8	71.4	61.0	-

[Continued]

★ 761 ★

Occupation of Longest Job of Year-Round Worker, 25 Years Old and Over, by Earnings and Education, 1991. Part III
[Continued]

Occupation of longest job	Total		Not a high school graduate		High school graduate		Some college or associate degree		Bachelor's degree or more	
	Male	Female	Male	Female	Male	Female	Male	Female	Male	Female
Standard error (dollars)	(X)	(X)	(X)	(X)	(X)	(X)	(X)	(X)	(X)	(X)
Handlers, equipment cleaners, helpers, and laborers	(X)	(X)	(X)	(X)	(X)	(X)	(X)	(X)	(X)	(X)
Median earnings (dollars)	86.7	88.2	101.1	74.2	79.7	101.1	77.2	-	104.7	-
Standard error (dollars)	(X)	(X)	(X)	(X)	(X)	(X)	(X)	(X)	(X)	(X)

Source: "Occupation of Longest Job in 1991 of Year-Round, Full-Time Workers 25 Years Old and Over, by Total Money Median Earnings." Claudette E. Bennett, *The Black Population in the United States: March 1992*, p. 65. *Notes:* Data where base is less than 75,000 may not meet statistical standards for reliability of derived figures.

★ 762 ★

Occupations

Occupational Distribution of Civilian Labor Force, 1992

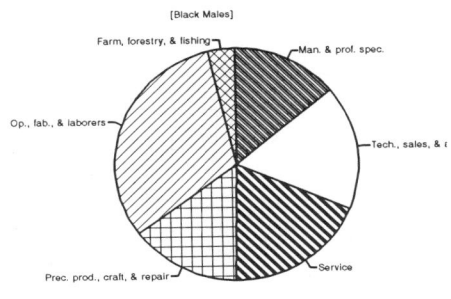

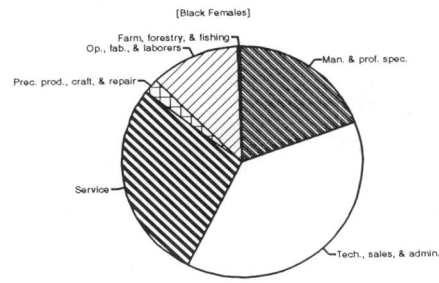

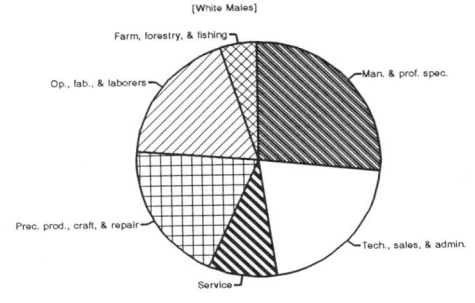

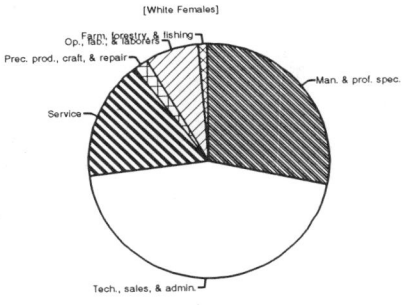

Annual averages.

	Black		White	
	Male	Female	Male	Female
Managerial and professional specialty	14.1	19.5	26.8	28.5
Technical, sales, and administrative support	17.3	38.2	21.1	44.6
Service	19.0	27.9	9.1	16.5
Precision production, craft, and repair	14.8	2.1	19.5	2.0

[Continued]

★ 762 ★

Occupational Distribution of Civilian Labor Force, 1992

[Continued]

	Black		White	
	Male	Female	Male	Female
Operators, fabricators, and laborers	31.3	12.0	18.9	7.2
Farming, forestry, and fishing	3.6	0.3	4.7	1.1

Source: "Occupational Distribution of the Employed Civilian Labor Force, by Sex and Race: 1992," Claudette E. Bennett, *The Black Population in the United States: March 1992*, p. 11.

★ 763 ★

Occupations

Occupations of Employed Civilians, 1983 and 1992. Part I.

For civilian noninstitutional population 16 years old and over. Annual average of monthly figures. Based on Current Population Survey. Persons of Hispanic origin may be of any race.

Occupation	1983				1992			
	Total employed (1,000)	Percent of total			Total employed (1,000)	Percent of total		
		Female	Black	Hispanic		Female	Black	Hispanic
Total	100,834	43.7	9.3	5.3	117,598	45.7	10.1	7.6
Managerial and professional specialty	23,592	40.9	5.6	2.6	31,153	47.3	6.5	3.9
Executive, administrative, and managerial[1]	10,772	32.4	4.7	2.8	14,767	41.5	5.8	4.2
Officials and administrators, public	417	38.5	8.3	3.8	619	43.8	11.5	3.7
Financial managers	357	38.6	3.5	3.1	517	46.3	3.0	3.9
Personnel and labor relations managers	106	43.9	4.9	2.6	104	58.9	5.9	3.4
Purchasing managers	82	23.6	5.1	1.4	113	33.2	5.8	2.2
Managers, marketing, advertising and public relations	396	21.8	2.7	1.7	516	33.6	2.9	3.4
Administrators, education and related fields	415	41.4	11.3	2.4	605	56.8	9.8	3.7
Managers, medicine and health	91	57.0	5.0	2.0	364	65.8	8.8	2.9
Managers, properties and real estate	305	42.8	5.5	5.2	436	45.5	7.2	5.6
Management-related occupations[1]	2,966	40.3	5.8	3.5	3,961	52.1	6.9	4.6
Accountants and auditors	1,105	38.7	5.5	3.3	1,365	51.2	5.8	4.4
Professional specialty[1]	12,820	48.1	6.4	2.5	16,386	52.6	7.0	3.5
Architects	103	12.7	1.6	1.5	138	15.3	1.3	4.3
Engineers[1]	1,572	5.8	2.7	2.2	1,751	8.5	3.9	2.9
Aerospace engineers	80	6.9	1.5	2.1	89	6.4	3.3	1.3
Chemical engineers	67	6.1	3.0	1.4	70	6.3	0.7	2.8
Civil engineers	211	4.0	1.9	3.2	217	7.9	3.4	3.9
Electrical and electronic	450	6.1	3.4	3.1	527	8.4	5.2	2.4
Industrial engineers	210	11.0	3.3	2.4	204	14.0	4.7	3.5
Mechanical	259	2.8	3.2	1.1	303	5.3	3.7	2.7
Mathematical and computer scientists[1]	463	29.6	5.4	2.6	935	33.5	6.8	3.1
Computer systems analysts, scientists	276	27.8	6.2	2.7	693	29.6	5.9	2.9
Operations and systems researchers and analysts	142	31.3	4.9	2.2	192	45.2	9.3	4.2
Natural scientists[1]	357	20.5	2.6	2.1	459	27.2	3.0	3.0
Chemists, except biochemists	98	23.3	4.3	1.2	120	30.1	2.8	3.4
Geologists and geodesists	65	18.0	1.1	2.6	52	11.8	0.3	1.0
Biological and life scientists	55	40.8	2.4	1.8	95	33.8	2.8	4.6
Health diagnosing occupations[1]	735	13.3	2.7	3.3	914	18.3	2.6	3.5
Physicians	519	15.8	3.2	4.5	614	2.4	3.3	4.3
Dentists	126	6.7	2.4	1.0	162	8.5	1.1	1.6
Health assessment and treating occupations	1,900	85.8	7.1	2.2	2,517	86.8	8.5	3.4

[Continued]

★ 763 ★

Occupations of Employed Civilians, 1983 and 1992. Part I.
[Continued]

Occupation	1983 Total employed (1,000)	1983 Percent of total Female	1983 Percent of total Black	1983 Percent of total Hispanic	1992 Total employed (1,000)	1992 Percent of total Female	1992 Percent of total Black	1992 Percent of total Hispanic
Registered nurses	1,372	95.8	6.7	1.8	1,805	94.3	8.5	2.9
Pharmacists	158	26.7	3.8	2.6	198	37.8	5.6	2.2
Dietitians	71	90.8	21.0	3.7	90	89.8	21.3	8.5
Therapists[1]	247	76.3	7.6	2.7	381	79.6	6.1	4.8
Inhalation therapists	69	69.4	6.5	3.7	77	67.9	8.3	7.3
Physical therapists	55	77.0	9.7	1.5	104	76.5	4.4	7.6
Speech therapists	51	90.5	1.5	-	79	92.1	1.7	0.3
Physicians' assistants	51	36.3	7.7	4.4	[2]	[2]	[2]	[2]
Teachers, college and university	606	36.3	4.4	1.8	737	40.9	4.7	2.7
Teachers, except college and university[1]	3,365	70.9	9.1	2.7	4,216	74.8	9.3	3.4
Prekindergarten and kindergarten	299	98.2	11.8	3.4	481	98.6	12.7	5.9
Elementary school	1,350	83.3	11.1	3.1	1,643	85.4	10.4	3.3
Secondary school	1,209	51.8	7.2	2.3	1,172	55.5	7.6	2.7
Special education	81	82.2	10.2	2.3	269	84.7	10.0	2.1
Counselors, educational and vocational	184	53.1	13.9	3.2	234	65.8	11.4	5.6
Librarians, archivists, and curators	213	84.4	7.8	1.6	207	83.6	5.6	3.0
Librarians	193	87.3	7.9	1.8	182	87.6	5.8	2.7
Social scientists and urban planners[1]	261	46.8	7.1	2.1	387	54.1	6.7	3.1
Economists	98	37.9	6.3	2.7	122	43.3	7.9	3.0
Psychologists	135	57.1	8.6	1.1	223	62.5	6.2	3.3
Social, recreation, and religious workers[1]	831	43.1	12.1	3.8	1,069	50.5	14.8	5.9
Social workers	407	64.3	18.2	6.3	590	68.9	20.2	7.4
Recreation workers	65	71.9	15.7	2.0	93	72.3	15.7	3.5
Clergy	293	5.6	4.9	1.4	317	8.4	6.4	4.0
Lawyers and judges	651	15.8	2.7	1.0	788	21.4	3.1	2.0
Lawyers	612	15.3	2.6	0.9	753	21.4	2.7	1.9
Writers, artists, entertainers, and athletes[1]	1,544	42.7	4.8	2.9	2,019	47.2	4.9	4.5
Authors	62	46.7	2.1	0.9	125	54.5	2.7	0.9
Technical writers	[2]	[2]	[2]	[2]	54	53.9	3.8	4.2
Designers	393	52.7	3.1	2.7	542	54.8	3.0	4.3
Musicians and composers	155	28.0	7.9	4.4	188	29.4	11.2	6.6
Actors and directors	60	30.8	6.6	3.4	96	40.3	6.6	3.8
Painters, sculptors, craft-artists, and artist printmakers	186	47.4	2.1	2.3	222	50.3	2.8	3.9
Photographers	113	20.7	4.0	3.4	129	26.4	6.4	4.6
Editors and reporters	204	48.4	2.9	2.1	264	49.7	4.4	2.8
Public relations specialists	157	50.1	6.2	1.9	161	58.5	5.5	2.4
Announcers	[2]	[2]	[2]	[2]	53	16.9	6.8	7.1
Athletes	58	17.6	9.4	1.7	71	25.8	9.7	5.3

Source: "Employed Civilians, by Occupation, Sex, Race, and Hispanic Origin: 1983 and 1992," U.S. Bureau of the Census, *Statistical Abstract of the United States,* 1993, p. 405. Primary source: U.S. Bureau of Labor Statistics, *Employment and Earnings,* monthly, January issues. *Notes:* - Represents or rounds to zero. NA Not available. 1. Includes other occupations, not shown separately. 2. Level of total employment below 50,000.

★ 764 ★

Occupations

Occupations of Employed Civilians, 1983 and 1992. Part II.

For civilian noninstitutional population 16 years old and over. Annual average of monthly figures. Based on Current Population Survey. Persons of Hispanic origin may be of any race.

Occupation	1983				1992			
	Total employed (1,000)	Percent of total			Total employed (1,000)	Percent of total		
		Female	Black	Hispanic		Female	Black	Hispanic
Technical, sales, and administrative support	31,265	64.6	7.6	4.3	36,808	63.9	9.1	6.1
Technicians and related support	3,053	48.2	8.2	3.1	4,253	49.0	9.7	4.5
Health technologists and technicians[1]	1,111	84.3	12.7	3.1	1,517	81.8	123	5.2
Clinical laboratory technologists and technicians	255	76.2	10.5	2.9	301	77.6	12.6	6.2
Dental hygienists	66	98.6	1.6	-	74	99.0	0.6	0.9
Health record technologists and technicians	[2]	[2]	[2]	[2]	57	88.9	12.9	8.2
Radiologic technicians	101	71.7	8.6	4.5	146	74.0	6.7	5.1
Licensed practical nurses	443	97.0	17.7	3.1	453	94.8	17.2	3.1
Engineering and related technologists and technicians[1]	822	18.4	6.1	3.5	920	17.5	7.0	4.8
Electrical and electronic technicians	260	12.5	8.2	4.6	330	12.2	9.2	5.3
Drafting occupations	273	17.5	5.5	2.3	255	19.0	4.4	3.9
Surveying and mapping technicians	[2]	[2]	[2]	[2]	72	7.8	0.3	6.6
Science technicians[1]	202	29.1	6.6	2.8	240	36.1	8.3	4.3
Biological technicians	52	37.7	2.9	2.0	79	51.7	5.2	5.9
Chemical technicians	82	26.9	9.5	3.5	77	26.7	10.0	3.6
Technicians, except health, engineering, and science[1]	917	35.3	5.0	2.7	1,576	37.9	8.9	3.8
Airplane pilots and navigators	69	2.1	-	1.6	97	2.3	2.2	2.5
Computer programmers	443	32.5	4.4	2.1	550	33.0	6.8	2.5
Legal assistants	128	74.0	4.3	3.6	241	76.7	10.0	5.2
Sales occupations	11,818	47.5	4.7	3.7	13,919	47.9	6.2	5.6
Supervisors and proprietors	2,958	28.4	3.6	3.4	3,886	34.8	4.0	4.8
Sales representatives, finance and business services[1]	1,853	37.2	2.7	2.2	2,247	40.0	4.7	3.6
Insurance sales	551	25.1	3.8	2.5	573	31.7	6.1	3.9
Real estate sales	570	48.9	1.3	1.5	719	50.4	2.6	3.7
Securities and financial services sales	212	23.6	3.1	1.1	322	29.6	3.7	2.5
Advertising and related sales	124	47.9	4.5	3.3	164	49.3	6.0	3.0
Sales representatives, commodities, except retail	1,442	15.1	2.1	2.2	1,570	21.8	2.4	3.3
Sales workers, retail and personal services	5,511	69.7	6.7	4.8	6,129	65.5	9.1	7.5
Cashiers	2,009	84.4	10.1	5.4	2,519	79.3	12.8	8.5
Sales-related occupations	54	58.7	2.8	1.3	93	65.7	6.0	4.0
Administrative support, including clerical	16,395	79.9	9.6	5.0	18,636	79.3	11.1	6.7
Supervisors	676	53.4	9.3	5.0	759	56.7	11.7	7.2
Computer equipment operators	605	63.9	12.5	6.0	664	63.6	13.0	6.7
Computer operators	597	63.7	12.1	6.0	661	63.6	13.1	6.6
Secretaries, stenographers, and typists[1]	4,861	98.2	7.3	4.5	4,315	98.4	9.3	5.7
Secretaries	3,891	99.0	5.8	4.0	3,700	99.0	8.2	5.4
Typists	906	95.6	13.8	6.4	547	95.1	17.2	7.7
Information clerks	1,174	88.9	8.5	5.5	1,653	88.7	8.3	7.9
Receptionists	602	96.8	7.5	6.6	892	97.3	7.5	8.0
Records processing occupations, except financial[1]	866	82.4	13.9	4.8	916	79.6	15.4	7.4
Order clerks	188	78.1	10.6	4.4	219	76.0	16.8	7.0
Personnel clerks, except payroll and time keeping	64	91.1	14.9	4.6	68	89.9	12.2	6.3
Library clerks	147	81.9	15.4	2.5	152	77.7	10.1	4.6
File clerks	287	83.5	16.7	6.1	292	81.3	17.9	8.6
Records clerks	157	82.8	11.6	5.6	172	79.0	15.9	9.4
Financial records processing[1]	2,457	89.4	4.6	3.7	2,335	90.6	5.3	5.4
Bookkeepers, accounting, and auditing clerks	1,970	91.0	4.3	3.3	1,841	90.7	4.3	5.2
Payroll and time keeping clerks	192	82.2	5.9	5.0	171	90.8	7.1	5.1
Billing clerks	146	88.4	6.2	3.9	180	92.4	10.3	5.4
Cost and rate clerks	96	75.6	5.9	5.3	67	81.1	7.4	9.6
Duplicating, mail and other office machine operators	68	62.6	16.0	6.1	72	53.6	19.5	6.9
Communications equipment operators	256	89.1	17.0	4.4	226	88.3	19.5	6.2

[Continued]

★ 764 ★

Occupations of Employed Civilians, 1983 and 1992. Part II.

[Continued]

Occupation	1983 Total employed (1,000)	1983 Percent of total Female	1983 Percent of total Black	1983 Percent of total Hispanic	1992 Total employed (1,000)	1992 Percent of total Female	1992 Percent of total Black	1992 Percent of total Hispanic
Telephone operators	244	90.4	17.0	4.3	215	89.9	18.9	6.3
Mail and message distributing occupations	799	31.6	18.1	4.5	902	37.3	19.0	7.6
Postal clerks, except mail carriers	248	36.7	26.2	5.2	252	46.1	26.6	8.4
Mail carrier, postal service	259	17.1	12.5	2.7	325	27.8	14.5	5.0
Mail clerks, except postal service	170	50.0	15.8	5.9	180	49.5	21.3	9.6
Messengers	122	26.2	16.7	5.2	144	28.2	13.1	9.8
Material recording, scheduling, and distributing[1,3]	1,562	37.5	10.9	6.6	1,846	43.5	12.7	8.2
Dispatchers	157	45.7	11.4	4.3	213	55.4	10.2	5.6
Production coordinators	182	44.0	6.1	2.2	194	49.2	9.2	3.3
Traffic, shipping, and receiving clerks	421	22.6	9.1	11.1	550	31.6	14.6	10.6
Stock and inventory clerks	532	38.7	13.3	5.5	544	40.3	12.8	8.7
Weighers, measurers, and checkers	79	47.2	16.9	5.8	65	47.2	14.7	14.0
Expediters	112	57.5	8.4	4.3	209	66.7	11.8	5.6
Adjusters and investigators	675	69.9	11.1	5.1	1,300	75.1	11.9	6.1
Insurance adjusters, examiners, and investigators	199	65.0	11.5	3.3	373	76.1	11.6	4.7
Investigators and adjusters, except insurance	301	70.1	11.3	4.8	686	75.6	12.1	6.7
Eligibility clerks, social welfare	69	88.7	12.9	9.4	78	86.9	13.9	8.6
Bill and account collectors	106	66.4	8.5	6.5	163	64.6	11.2	5.8
Miscellaneous administrative support[1]	2,397	85.2	12.5	5.9	3,642	82.9	12.8	7.3
General office clerks	648	80.6	12.7	5.2	717	83.5	12.6	8.1
Bank tellers	480	91.0	7.5	4.3	468	89.8	7.7	6.1
Data entry keyers	311	93.6	18.6	5.6	571	84.9	17.1	7.8
Statistical clerks	96	75.7	7.5	3.4	58	78.6	15.8	6.2
Teachers' aides	348	93.7	17.8	12.6	485	91.9	15.2	11.4

Source: "Employed Civilians, by Occupation, Sex, Race, and Hispanic Origin: 1983 and 1992," U.S. Bureau of the Census, *Statistical Abstract of the United States,* 1993, p. 406. Primary source: U.S. Bureau of Labor Statistics, *Employment and Earnings,* monthly, January issues. *Notes:* - Represents or rounds to zero. NA Not available. 1. Includes other occupations, not shown separately. 2. Level of total employment below 50,000. 3. Includes clerks.

★ 765 ★

Occupations

Occupations of Employed Civilians, 1983 and 1992. Part III.

For civilian noninstitutional population 16 years old and over. Annual average of monthly figures. Based on Current Population Survey. Persons of Hispanic origin may be of any race.

Occupation	1983 Total employed (1,000)	1983 Percent of total Female	1983 Percent of total Black	1983 Percent of total Hispanic	1992 Total employed (1,000)	1992 Percent of total Female	1992 Percent of total Black	1992 Percent of total Hispanic
Service occupations	13,857	60.1	16.6	6.8	16,096	59.7	17.5	11.3
Private household[1]	980	96.1	27.8	8.5	876	95.9	18.6	19.6
Child care workers	408	96.9	7.9	3.6	353	97.1	10.2	12.6
Cleaners and servants	512	95.8	42.4	11.8	484	94.8	23.6	24.6
Protective service	1,672	12.8	13.6	4.6	2,096	16.7	18.0	7.2
Supervisors, protective service	127	4.7	7.7	3.1	172	9.5	13.2	7.7
Supervisors, police and detectives	58	4.2	9.3	1.2	81	10.8	8.1	10.0
Firefighting and fire prevention	189	1.0	6.7	4.1	205	3.3	6.5	3.7
Firefighting occupations	170	1.0	7.3	3.8	190	2.4	6.5	3.5
Police and detectives	645	9.4	13.1	4.0	862	15.8	16.6	6.7
Police and detectives, public service	412	5.7	9.5	4.4	459	10.6	10.8	7.3
Sheriffs, bailiffs, and other law enforcement officers	87	13.2	11.5	4.0	108	20.8	16.3	7.4

[Continued]

★ 765 ★

Occupations of Employed Civilians, 1983 and 1992. Part III.

[Continued]

Occupation	1983 Total employed (1,000)	1983 Percent of total Female	Black	Hispanic	1992 Total employed (1,000)	1992 Percent of total Female	Black	Hispanic
Correctional institution officers	146	17.8	24.0	2.8	295	21.9	25.7	5.5
Guards	711	20.6	17.0	5.6	860	22.3	23.1	8.5
Guards and police, except public service	602	13.0	18.9	6.2	730	15.4	25.1	9.1
Service except private household and protective	11,205	64.0	16.0	6.9	13,124	64.1	17.3	11.4
Food preparation and service occupations[1]	4,860	63.3	10.5	6.8	5,459	59.0	13.0	12.2
Bartenders	338	48.4	2.7	4.4	291	55.0	2.9	4.9
Waiters and waitresses	1,357	87.8	4.1	3.6	1,369	79.6	4.7	7.6
Cooks	1,452	50.0	15.8	6.5	1,877	46.1	18.5	14.7
Food counter, fountain, and related occupations	326	76.0	9.1	6.7	316	72.6	12.3	8.5
Kitchen workers, food preparation	138	77.0	13.7	8.1	246	76.0	15.9	11.4
Waiters and waitresses' assistants	364	38.8	12.6	14.2	364	37.5	14.3	19.0
Health service occupations	1,739	89.2	23.5	4.8	2,105	88.8	27.7	7.1
Dental assistants	154	98.1	6.1	5.7	169	98.6	3.9	10.6
Health aides, except nursing	316	86.8	16.5	4.8	363	81.3	22.7	6.1
Nursing aides, orderlies, and attendants	1,269	88.7	27.3	4.7	1,574	89.4	31.4	7.0
Cleaning and building service occupations[1]	2,736	38.8	24.4	9.2	2,988	41.1	22.6	16.4
Maids and housemen	531	81.2	32.3	10.1	638	82.3	27.3	19.5
Janitors and cleaners	2,031	28.6	22.6	8.9	2,118	30.0	21.3	15.9
Personal service occupations[1]	1,870	79.2	11.1	6.0	2,573	81.7	11.9	7.5
Barbers	92	12.9	8.4	12.1	86	20.9	15.3	7.6
Hairdressers and cosmetologists	622	88.7	7.0	5.7	758	90.8	9.6	6.7
Attendants, amusement and recreation facilities	131	40.2	7.1	4.3	153	47.9	9.3	6.5
Public transportation attendants	63	74.3	11.3	5.9	91	82.5	10.8	8.6
Welfare service aides	77	92.5	24.2	10.5	92	87.7	18.3	16.5
Family child care providers	(NA)	(NA)	(NA)	(NA)	320	98.7	7.5	9.7
Early childhood teachers' assistants	(NA)	(NA)	(NA)	(NA)	385	95.9	14.4	6.3
Precision production, craft, and repair	12,328	8.1	6.8	6.2	13,128	8.6	7.6	9.1
Mechanics and repairers	4,158	3.0	6.8	5.3	4,441	3.3	7.6	7.8
Mechanics and repairers, except supervisors[1]	3,906	2.8	7.0	5.5	4,209	3.1	7.7	8.1
Vehicle and mobile equipment mechanics/repairers[1]	1,683	0.8	6.9	6.0	1,845	1.1	6.8	10.1
Automobile mechanics	800	0.5	7.8	6.0	894	0.8	7.4	10.7
Aircraft engine mechanics	95	2.5	4.0	7.6	143	4.6	8.0	15.3
Electrical and electronic equipment repairers[1]	674	7.4	7.3	4.5	654	7.4	8.7	6.1
Data processing equipment repairers	98	9.3	6.1	4.5	146	10.4	7.8	6.8
Telephone installers and repairers	247	9.9	7.8	3.7	58	3.4	12.9	6.4
Construction trades	4,289	1.8	6.6	6.0	4,790	1.9	6.9	9.1
Construction trades, except supervisors[1]	3,784	1.9	7.1	6.1	4,137	2.0	7.3	9.8
Carpenters	1,160	1.4	5.0	5.0	1,236	1.0	5.4	8.1
Extractive occupations	196	2.3	3.3	6.0	130	0.6	3.4	7.3
Precision production occupations	3,685	21.5	7.3	7.4	3,765	23.7	8.6	10.8
Operators, fabricators, and laborers	16,091	26.6	14.0	8.3	16,957	25.0	15.1	11.8
Machine operators, assemblers, and inspectors[1]	7,744	42.1	14.0	9.4	7,524	39.7	15.0	13.5
Textile, apparel, and furnishings machine operators[1]	1,414	82.1	18.7	12.5	1,200	76.4	23.7	18.0
Textile sewing machine operators	806	94.0	15.5	14.5	676	87.4	20.2	22.4
Pressing machine operators	141	66.4	27.1	14.2	123	67.1	27.1	23.6
Fabricators, assemblers, and hand working occupations	1,715	33.7	11.3	8.7	1,904	33.2	11.6	12.1
Production inspectors, testers, samplers, and weighers	794	53.8	13.0	7.7	773	53.6	14.6	12.3
Transportation and material moving occupations	4,201	7.8	13.0	5.9	4,878	8.8	14.9	8.4
Motor vehicle operators	2,978	9.2	13.5	6.0	3,706	10.2	15.7	8.6
Trucks, heavy and light	2,195	3.1	123	5.7	2,694	4.6	13.6	8.5
Transportation occupations, except motor vehicles	212	2.4	6.7	3.0	162	3.6	10.2	4.7
Material moving equipment operators	1,011	4.8	12.9	6.3	1,009	4.5	12.4	8.6
Industrial truck and tractor operators	369	5.6	19.6	8.2	421	6.6	17.4	12.5
Handlers, equipment cleaners, helpers, and laborers[1]	4,147	16.8	15.1	8.6	4,556	18.0	15.4	12.5
Freight, stock, and material handlers	1,488	15.4	15.3	7.1	1,737	18.6	16.6	9.8

[Continued]

★ 765 ★

Occupations of Employed Civilians, 1983 and 1992. Part III.
[Continued]

Occupation	1983 Total employed (1,000)	1983 Percent of total Female	1983 Percent of total Black	1983 Percent of total Hispanic	1992 Total employed (1,000)	1992 Percent of total Female	1992 Percent of total Black	1992 Percent of total Hispanic
Laborers, except construction	1,024	19.4	16.0	8.6	1,177	18.2	15.8	12.4
Farming, forestry, and fishing	3,700	16.0	7.5	8.2	3,456	15.9	6.5	15.1
Farm operators and managers	1,450	12.1	1.3	0.7	1,232	15.4	0.9	1.7
Other agricultural and related occupations	2,072	19.9	11.7	14.0	2,054	17.2	9.7	23.9
Farm workers	1,149	24.8	11.6	15.9	866	21.6	8.2	27.8
Forestry and logging occupations	126	1.4	12.8	2.1	109	5.4	13.0	7.4
Fishers, hunters, and trappers	53	4.5	1.8	2.5	60	3.9	0.9	5.5

Source: "Employed Civilians, by Occupation, Sex, Race, and Hispanic Origin: 1983 and 1992," U.S. Bureau of the Census, *Statistical Abstract of the United States,* 1993, p. 407. Primary source: U.S. Bureau of Labor Statistics, *Employment and Earnings,* monthly, January issues. *Notes:* NA Not available. 1. Includes other occupations, not shown separately.

Self-Employment

★ 766 ★

Incorporated and Unincorporated Self-Employed Women, 1975 and 1990

In percent, unless otherwise indicated.

Characteristic	1975 Incorporated	1975 Unincorporated	1990 Incorporated	1990 Unincorporated
Numbers of women (in thousands)	143	1,539	737	3,259
Race and Hispanic origin				
White	98.9	93.7	92.1	91.6
Black	[1]	4.1	3.1	4.0
Hispanic origin[2]	2.9	2.3	2.2	4.9

Source: "Distribution of Incorporated and Unincorporated Self-employed Women by Selected Characteristics, 1975 and 1990." *Monthly Labor Review* 117 (March 1994): 28. *Notes:* 1. Less than 0.1 percent. 2. Persons of Hispanic origin may be of any race.

★ 767 ★

Self-Employment

Self-Employed and Wage-and-Salary Workers: Percent Distribution, 1975 and 1990

Characteristic	Women		Men	
	1975	1990	1975	1990
Race and Hispanic origin				
Self-employed				
White	94.2	91.7	95.5	92.1
Black	3.8	3.9	2.8	4.2
Other	2.1	4.4	1.7	3.7
Hispanic origin[1]	2.4	4.4	2.6	4.7
Wage-and-salary				
White	87.0	84.5	89.3	86.2
Black	11.3	12.2	9.2	10.5
Other	1.7	3.3	1.6	3.4
Hispanic origin	3.9	6.8	4.3	8.3
Self-employment rate[2]				
Non-Hispanic white	4.5	7.4	10.8	13.7
Non-Hispanic black	1.4	2.3	3.3	5.4
Other non-Hispanic	5.0	9.0	10.9	13.9
Hispanic origin	2.5	4.5	6.2	7.4

Source: "Percent Distribution of Self-Employed and Wage-and-Salary Workers, by Gender and Other Selected Characteristics, 1975 and 1990." *Monthly Labor Review* 117 (March 1994): 23. *Notes:* 1. Persons of Hispanic origin may be of any race. 2. For each group, the self-employed as a percentage of nonagricultural employment.

Senior Citizens

★ 768 ★

Civilian Noninstitutional Population 25 and Over, Employed 1990

In thousands; annual averages.

Age, sex, and race	Civilian non-institutional population	Civilian labor force					Not in labor force
		Total	Percent of population	Employed	Unemployed		
					Number	Percent of labor force	
Total							
Men							
25 to 54 years	51,641	48,259	93.4	46,071	2,188	4.5	3,393
25 to 34 years	21,037	19,813	94.2	18,732	1,081	5.5	1,224

[Continued]

★ 768 ★

Civilian Noninstitutional Population 25 and Over, Employed 1990
[Continued]

Age, sex, and race	Civilian non-institutional population	Civilian labor force					Not in labor force
		Total	Percent of population	Employed	Unemployed		
					Number	Percent of labor force	
35 to 44 years	18,285	17,268	94.4	16,575	693	4.0	1,017
45 to 54 years	12,319	11,177	90.7	10,764	413	3.7	1,142
55 to 64 years	10,023	6,785	67.7	6,530	255	3.8	3,238
55 to 59 years	5,030	4,014	79.8	3,856	158	3.9	1,016
60 to 64 years	4,993	2,771	55.5	2,674	98	3.5	2,222
65 years and over	12,392	2,033	16.4	1,972	61	3.0	10,359
65 to 69 years	4,592	1,192	26.0	1,154	38	3.2	3,400
70 to 74 years	3,476	534	15.4	517	17	3.3	2,942
75 years and over	4,324	307	7.1	301	5	1.8	4,017
Women							
25 to 54 years	53,856	39,881	74.1	38,068	1,813	4.5	13,975
25 to 34 years	21,715	15,990	73.6	15,099	890	5.6	5,725
35 to 44 years	19,057	14,576	76.5	13,967	608	4.2	4,482
45 to 54 years	13,084	9,316	71.2	9,001	315	3.4	3,768
55 to 64 years	11,206	5,075	45.3	4,935	141	2.8	6,131
55 to 59 years	11,206	5,075	45.3	4,935	141	2.8	6,131
60 to 64 years	5,675	2,016	35.5	1,965	51	2.5	3,658
65 years and over	17,337	1,502	8.7	1,455	46	3.1	15,836
65 to 69 years	5,539	941	17.0	909	32	3.4	4,598
70 to 74 years	4,458	365	8.2	356	9	2.4	4,093
75 years and over	7,341	196	2.7	190	6	3.0	7,145
White							
Men							
25 to 54 years	44,326	41,848	94.4	40,197	1,652	3.9	2,478
25 to 34 years	17,837	16,991	95.3	16,199	792	4.7	845
35 to 44 years	15,800	15,059	95.3	14,531	529	3.5	741
45 to 54 years	10,690	9,798	91.7	9,467	331	3.4	891
55 to 64 years	8,826	6,059	68.7	5,842	217	3.6	2,767
55 to 59 years	4,402	3,574	81.2	3,442	132	3.7	828
60 to 64 years	4,424	2,485	56.2	2,400	84	3.4	1,939
65 years and over	11,129	1,865	16.8	1,812	53	2.8	9,264
65 to 69 years	4,106	1,094	26.6	1,062	32	2.9	3,012
70 to 74 years	3,117	486	15.6	470	16	3.2	2,632
75 years and over	3,905	285	7.3	280	5	1.7	3,620
Women							
25 to 54 years	45,127	33,570	74.4	32,256	1,314	3.9	11,557
25 to 34 years	17,958	13,327	74.2	12,722	605	4.5	4,631
35 to 44 years	16,056	12,312	76.7	11,858	454	3.7	3,744
45 to 54 years	11,113	7,930	71.4	7,676	254	3.2	3,183
55 to 64 years	9,715	4,427	45.6	4,309	118	2.7	5,288
55 to 59 years	4,755	2,649	55.7	2,575	74	2.8	2,106

[Continued]

★ 768 ★

Civilian Noninstitutional Population 25 and Over, Employed 1990

[Continued]

Age, sex, and race	Civilian non-institutional population	Civilian labor force					Not in labor force
		Total	Percent of population	Employed	Unemployed		
					Number	Percent of labor force	
60 to 64 years	4,959	1,778	35.8	1,734	44	2.5	3,182
65 years and over	15,514	1,325	8.5	1,288	37	2.8	14,190
65 to 69 years	4,876	828	17.0	805	24	2.9	4,048
70 to 74 years	4,017	324	8.1	316	8	2.5	3,693
75 years and over	6,621	173	2.6	167	5	3.1	6,448
Black							
Men							
25 to 54 years	5,424	4,739	87.4	4,293	446	9.4	685
25 to 34 years	2,422	2,151	88.8	1,903	248	11.5	271
35 to 44 years	1,824	1,606	88.1	1,470	136	8.5	218
45 to 54 years	1,178	982	83.3	920	62	6.3	196
55 to 64 years	957	554	57.9	524	30	5.5	403
55 to 59 years	499	336	67.4	317	20	5.9	163
60 to 64 years	458	218	47.5	207	10	4.8	240
65 years and over	1,012	131	13.0	125	6	4.6	881
65 to 69 years	386	73	19.0	69	4	5.9	313
70 to 74 years	287	41	14.2	39	1	3.4	246
75 years and over	339	17	5.0	17	-	(B)	322
Women							
25 to 54 years	6,654	4,904	73.7	4,480	424	8.7	1,750
25 to 34 years	2,922	2,113	72.3	1,860	252	11.9	810
35 to 44 years	2,257	1,752	77.6	1,625	127	7.2	505
45 to 54 years	1,475	1,040	70.5	995	45	4.4	435
55 to 64 years	1,172	506	43.2	488	19	3.7	666
55 to 59 years	606	313	51.6	300	12	3.9	294
60 to 64 years	566	194	34.3	188	6	3.2	372
65 years and over	1,493	148	9.9	139	9	5.8	1,346
65 to 69 years	525	93	17.7	86	7	7.8	433
70 to 74 years	358	35	9.7	34	1	2.6	323
75 years and over	610	20	3.2	19	1	(B)	590

Source: "Employment Status of the Civilian Noninstitutional Population 25 Years and Over, by Age, Sex, and Race: 1990," Cynthia Taeuber, *Sixty-Five Plus in America*, p. 8-9. Primary source: U.S. Bureau of Labor Statistics, *Employment and Earnings*, Vol. 38, No. 1, January 1991, Table 3. *Note:* - Represents zero or rounds to zero. B Base is less than 35,000.

★ 769 ★

Senior Citizens

Persons 50 Years and Over Employed, 1950 to 1980

Age and sex	1950	1960	1970	1980[1]
Total				
Male				
50 to 54 years	90.6	92.2	91.4	88.5
55 to 59 years	86.7	87.7	86.8	80.6
60 to 64 years	79.4	77.6	73.0	60.4
65 to 69 years	59.8	43.8	39.0	29.2
70 to 74 years	38.7	28.7	22.4	18.3
75 to 79 years	24.2	19.5	14.2	16.7
80 to 84 years	13.2	11.5	9.1	10.4
85 years and over	6.9	7.0	[4]	6.6
Female				
50 to 54 years	30.8	45.8	52.0	56.3
55 to 59 years	25.9	39.7	47.4	48.4
60 to 64 years	20.5	29.5	36.1	34.0
65 to 69 years	12.8	16.6	17.2	15.0
70 to 74 years	6.6	9.6	9.1	7.8
75 to 79 years	3.5	5.6	5.5	6.1
80 to 84 years	1.7	3.0	3.5	3.7
85 years and over	1.2	2.0	[4]	2.5
White				
Male				
50 to 54 years	91.0	92.8	92.2	89.6
55 to 59 years	87.0	88.5	87.6	81.8
60 to 64 years	79.7	78.4	73.7	61.0
65 to 69 years	60.0	44.1	39.3	29.5
70 to 74 years	38.6	28.8	22.7	18.5
75 to 79 years	23.9	19.6	14.3	17.0
80 to 84 years	12.9	11.5	9.0	10.5
85 years and over	6.6	6.9	[4]	6.6
Female				
50 to 54 years	29.8	45.1	51.5	56.1
55 to 59 years	25.2	39.1	47.1	48.2
60 to 64 years	20.0	29.1	35.9	33.8
65 to 69 years	12.5	16.3	17.0	14.8
70 to 74 years	6.5	9.4	8.9	7.7
75 to 79 years	3.4	5.5	5.3	6.0
80 to 84 years	1.6	3.0	3.4	3.6
85 years and over	1.2	1.9	[4]	2.5
Black[2]				
Male				
50 to 54 years	86.9	86.0	83.7	78.3
55 to 59 years	82.9	80.8	77.9	69.4
60 to 64 years	76.0	68.9	65.9	53.7
65 to 69 years	58.1	40.6	35.4	26.1

[Continued]

★ 769 ★

Persons 50 Years and Over Employed, 1950 to 1980
[Continued]

Age and sex	1950	1960	1970	1980[1]
70 to 74 years	40.2	27.3	19.6	16.3
75 to 79 years	27.6	19.2	13.0	13.7
80 to 84 years	16.7	12.1	9.7	8.8
85 years and over	9.8	8.0	[4]	6.6
Female				
50 to 54 years	40.9	52.5	56.5	58.4
55 to 59 years	34.9	44.7	50.2	50.2
60 to 64 years	27.6	34.1	38.8	36.9
65 to 69 years	16.4	19.5	19.4	16.9
70 to 74 years	8.4	11.5	11.6	9.3
75 to 79 years	5.1	7.0	7.5	6.9
80 to 84 years	2.4	4.0	5.7	4.2
85 years and over	2.1	3.1	[4]	3.2
Hispanic origin[3]				
Male				
50 to 54 years	(NA)	(NA)	88.6	86.5
55 to 59 years	(NA)	(NA)	84.1	78.8
60 to 64 years	(NA)	(NA)	70.3	62.6
65 to 69 years	(NA)	(NA)	36.8	31.7
70 to 74 years	(NA)	(NA)	19.7	18.7
75 to 79 years	(NA)	(NA)	13.6	13.9
80 to 84 years	(NA)	(NA)	8.5	9.6
85 years and over	(NA)	(NA)	[4]	6.8
Female				
50 to 54 years	(NA)	(NA)	42.0	50.5
55 to 59 years	(NA)	(NA)	34.7	42.4
60 to 64 years	(NA)	(NA)	24.3	30.3
65 to 69 years	(NA)	(NA)	11.2	12.3
70 to 74 years	(NA)	(NA)	6.3	6.9
75 to 79 years	(NA)	(NA)	5.0	4.2
80 to 84 years	(NA)	(NA)	3.6	3.0
85 years and over	(NA)	(NA)	[4]	2.7

Source: "Labor Force Participation Rates of Persons 50 Years and Over, by Age, Sex, Race, and Hispanic Origin: 1950 to 1980," Cynthia Taeuber, *Sixty-Five Plus in America*, p. 3-4. Primary source: U.S. Bureau of the Census, Decennial censuses, 1950 to 1980; for 1980, detailed age data for population 75 years and over from special tabulations prepared for the National Institute on Aging (Summary Tape File 5A, table 18.) *Notes:* NA Not available. 1. The figures for age groups 75 years and over are employment rates and do not include unemployed persons in the labor force. 2. Data for 1950 and 1960 are shown for Nonwhite. 3. Persons of Hispanic origin may be of any race. 4. Data for the population 85 and over in 1970 are not shown here because the count of persons 100 years and over was distorted by a problem with the design of the questionnaire.

Training Programs

★ 770 ★

Apprenticeship Program Data, 1989 and 1990

Apprenticeship data elements	FY 1990	FY 1989
Apprentices receiving training		
Total[1]	361,000	350,000
Percent minority	22.5	21.6
Percent female	7.1	7.2
Number of civilian apprenticeship programs	44,000	44,400
Military apprentices[2]	41,500	39,700
Percent minority	35.8	35.6
Percent female	6.5	6.1
Number of reviews conducted		
EEO compliance reviews	1,600	1,680
On-site quality reviews	1,857	1,988
Apprenticeship actions		
New registrations	98,200	96,900
Completions	39,400	43,400

Source: "Selected Apprenticeship Program Data." U.S. Department of Labor, Employment and Training Administration, *Training and Employment Report of the Secretary of Labor,* Covering the Period July 1988-September 1990, p. 24. Primary source: U.S. Department of Labor, Employment and Training Administration. *Notes:* 1. Includes new registrations, cancellations, and completions. Excludes military apprentices. 2. Data for apprentices on-board at end of year.

★ 771 ★

Training Programs

JPTA Programs, Titles II-A, II-B, and III, Distribution and Outcomes, 1987

Characteristic	Title II-A	Title II-B	Title III
Sex			
Male	48%	51%	62%
Female	52%	49%	38%
Age			
14-15 years	4%	38%	-
16-21 years	42%	62%	4%
22-54 years	53%	-	88%

[Continued]

★ 771 ★

JPTA Programs, Titles II-A, II-B, and III, Distribution and Outcomes, 1987

[Continued]

Characteristic	Title II-A	Title II-B	Title III
55 years and over	2%	-	8%
Education			
Dropout	27%	4%	15%
Student	19%	86%	1%
High school graduate	54%	10%	84%
Race/ethnic group			
White	50%	31%	72%
Black	33%	41%	17%
Hispanic	14%	24%	9%
Native American	1%	1%	1%
Asian	2%	3%	2%
Limited English ability	5%	11%	3%
Handicapped	1%	13%	3%
Single head of household	20%	2%	12%
Economically disadvantaged	94%	100%	31%
Outcome			
Entered-employment rate	61%	-	70%
Average hourly wage at placement	$4.85	-	$7.11

Source: "Participant Characteristics (Percent Distribution) and Outcomes, JPTA Titles III_a, II-B, and III, Program Year 1987." U.S. Department of Labor, Employment and Training Administration, *Training and Employment Report of the Secretary of Labor,* Covering the Period July 1987-September 1988, p. 7. Primary source: U.S. Department of Labor, Employment and Training Administration. *Notes:* Titles II-A and III data are based on characteristics of terminees, that is, persons who left JTPA programs during Program Year 1987. Title II-A data are for programs operated by SDAs. Items may not add to 100 due to rounding.

★ 772 ★

Training Programs

JPTA Programs: Selected Participant Characteristics and Program Activity, 1988 and 1989

Percent distribution.

Characteristic	PY 1989	PY 1988
Sex		
Male	57%	60%
Female	43	40
Age		
29 and under	25	27
30-54 years	67	65

[Continued]

★ 772 ★

JPTA Programs: Selected Participant Characteristics and Program Activity, 1988 and 1989
[Continued]

Characteristic	PY 1989	PY 1988
55 and over	7	8
Education		
Less than high school	17	16
High school graduate	83	83
Race/ethnicity		
White	69	71
Black	17	17
Hispanic	12	9
Native American	1	1
Asian	2	2
Unemployment		
Insurance claimant	49	47
Limited English	4	3
Handicapped	3	3
Single Head of Household	12	11
Program activity		
Classroom training	36	30
On-the-job training	21	20
Job search assistance	28	34
Other	15	16

Source: "JPTA Titles III: Selected Participant Characteristics and Program Activity." U.S. Department of Labor, Employment and Training Administration, *Training and Employment Report of the Secretary of Labor*, Covering the Period July 1988-September 1990, p. 14. Primary source: Statistics on characteristics are from JTPA Annual Status Report and the Worker Adjustment Annual Program Report; data on program activity are from the job Training Quarterly Survey. U.S. Department of Labor, Employment and Training Administration. *Notes:* All data reflects characteristics/activities of terminees. Figures may not add to 100 percent due to rounding.

★ 773 ★

Training Programs

JPTA Programs: Selected Participant Characteristics, 1988 and 1989

Percent distribution.

Characteristic	Title II-A: Adult		Title II-A: Youth		Title II-B	
	PY89	PY88	PY89	PY88	1990	1989
Sex						
Male	44	45	49	49	50	50
Female	56	55	51	51	50	50
Age						
14-15	-	-	15	10	42	41
16-21	-	-	85	90	58	59
22-54	97	97	-	-	-	-
55 and over	3	3	-	-	-	-
Education						
Dropout	27	28	26	27	4	4
Student	-	-	47	44	87	87
HS graduate	73	72	27	29	9	9
Race/ethnicity						
White	53	53	45	45	30	31
Black	30	29	35	35	40	40
Hispanic	14	14	16	16	26	24
Native American	2	2	2	2	1	1
Asian	2	2	2	2	3	3
Limited English	5	5	3	3	6	4
Handicapped	10	10	14	14	14	13
Single Head of Household	32	30	10	10	3	3

Source: "JPTA Titles II-A and II-B: Selected Participant Characteristics." U.S. Department of Labor, Employment and Training Administration, *Training and Employment Report of the Secretary of Labor*, Covering the Period July 1988-September 1990, p. 11. Primary source: JTPA Annual Status Report for Title II-A and JTPA Summer Youth Performance Report for Title II-B. U.S. Department of Labor, Employment and Training Administration. *Notes:* Title II-A data are based on characteristics of terminees—persons who left JTPA during Program Years 1988 and 1989 and are for programs operated by SDAs. Title II-B data, also for programs operated by SDAs, are for the summers of 1989 and 1990. Items may not add to 100 percent due to rounding.

★ 774 ★

Training Programs

Senior Community Service Employment Program Participants, 1987-1989

Characteristic	PY 1989	PY 1988	PY 1987
Ethnic group			
White	62.4	63.3	62.7
Black	23.9	23.3	23.8
Hispanic	9.0	8.8	8.7
Indian/Alaskan	1.7	1.6	1.5
Asian/Pacific Islander	3.0	3.1	3.2

Source: "Senior Community Service Employment Program, Selected Program Characteristics." U.S. Department of Labor, Employment and Training Administration, *Training and Employment Report of the Secretary of Labor*, Covering the Period July 1987-September 1988, p. 25. Primary source: U.S. Department of Labor, Employment and Training Administration. *Note:* Percentages may not add to 100 due to rounding.

Unemployment

★ 775 ★

Discouraged Persons Not in the Labor Force, 1993

In thousands.

Reason and sex	1993 Race and Hispanic origin		
	White	Black	Hispanic origin
Total			
Personal factors			
Employers think too young or old	134	15	13
Lacks education or training	96	32	28
Other personal handicap	68	14	7
Job market factors			
Could not find work	272	154	65
Thinks no job available	212	68	28
Men			
Personal factors			
Employers think too young or old	62	8	5
Lacks education or training	36	16	8
Other personal handicap	32	6	1

[Continued]

★ 775 ★

Discouraged Persons Not in the Labor Force, 1993
[Continued]

Reason and sex	1993 Race and Hispanic origin		
	White	Black	Hispanic origin
Job market factors			
Could not find work	138	72	30
Thinks no job available	101	23	18
Women			
Personal factors			
Employers think too young or old	71	8	8
Lacks education or training	61	16	20
Other personal handicap	35	8	6
Job market factors			
Could not find work	133	82	35
Thinks no job available	111	46	11

Source: "Persons Not in Labor Force Who Desire Work but Think They Cannot Get Jobs by Reason, Sex, Age, Race, and Hispanic Origin." *Employment and Earnings* 41 (January 1994): 226. *Notes:* Detail for the above race and Hispanic-origin groups will not sum to totals because data for the "other races" group are not presented and Hispanics are included in both the white and black population groups.

★ 776 ★

Unemployment

Displaced Workers

As of January. In percent, except total. For persons 20 years old and over with tenure of 3 years or more who lost or left a job between January 1987 and January 1992 because of plant closings or moves, slack work or the abolishment of their positions. Based on Current Population Survey and subject to sampling error.

Sex, age, race, and Hispanic origin	Total (1,000)	Employment status			Reason for job loss		
		Employed	Unemployed	Not in labor force	Plant or company closed down or moved	Slack work	Position or shift abolished
Total[1]	5,584	64.9	22.2	12.9	52.1	31.6	16.3
Males	3,447	66.6	24.5	8.9	49.4	34.7	15.9
20 to 24 years old	127	55.6	32.5	11.8	45.0	49.4	5.6
25 to 54 years old	2,728	71.6	24.5	3.9	49.5	34.0	16.5
55 to 64 years old	488	52.3	24.2	23.5	49.0	34.6	16.4
65 years old and over	103	15.8	15.3	68.9	53.3	35.7	11.0
Females	2,137	62.2	18.6	19.2	56.6	26.4	17.0
20 to 24 years old	76	72.7	7.3	20.1	43.8	46.1	10.0
25 to 54 years old	1,688	65.8	19.9	14.3	54.4	26.7	18.9
55 to 64 years old	262	51.4	17.4	31.2	70.5	19.4	10.1
65 years old and over	111	26.5	8.7	64.8	66.3	25.1	8.6

[Continued]

★ 776 ★

Displaced Workers
[Continued]

Sex, age, race, and Hispanic origin	Total (1,000)	Employment status			Reason for job loss		
		Employed	Unemployed	Not in labor force	Plant or company closed down or moved	Slack work	Position or shift abolished
White	4,828	65.7	21.2	13.0	52.5	30.6	16.9
Male	3,003	67.6	23.3	9.1	49.4	34.1	16.5
Female	1,825	62.7	17.8	19.5	57.6	24.8	17.6
Black	626	58.7	28.6	12.7	51.3	36.3	12.4
Male	356	58.9	33.4	7.7	52.3	36.6	11.1
Female	270	58.5	22.2	19.3	50.0	35.9	14.1
Hispanic origin[2]	511	60.4	27.4	12.3	57.0	34.5	8.5
Male	323	64.6	27.2	8.2	57.6	35.7	6.7
Female	188	53.0	27.7	19.3	56.0	32.5	11.6

Source: "Displaced Workers, by Selected Characteristics: 1992." U.S. Bureau of the Census, *Statistical Abstract of the United States*, 1993, p. 412. Primary source: U.S. Bureau of Labor Statistics, *News*, USDL 92-530. *Notes:* 1. Includes other races, not shown separately. 2. Persons of Hispanic origin may be of any race.

★ 777 ★

Unemployment

Duration of Unemployment, 1992 and 1993

Numbers in thousands.

Weeks of unemployment	Total		White		Black		Hispanic origin	
	1992	1993	1992	1993	1992	1993	1992	1993
Duration								
Total, 16 years and over	9,384	8,734	7,047	6,547	1,958	1,796	1,160	1,104
Less than 5 weeks	3,270	3,160	2,468	2,404	671	615	448	429
5 to 14 weeks	2,760	2,522	2,062	1,895	580	518	330	310
15 weeks and over	3,354	3,052	2,517	2,248	706	664	382	365
15 to 26 weeks	1,424	1,274	1,072	954	293	267	171	163
27 weeks and over	1,930	1,778	1,445	1,294	413	397	210	203
Average (mean) duration, in weeks	17.9	18.1	17.8	17.7	18.6	19.3	16.3	16.7
Median duration, in weeks	8.8	8.4	8.7	8.2	9.0	9.1	7.8	7.9
Percent distribution								
Total unemployed	100.0	100.0	100.0	100.0	100.0	100.0	100.0	100.0
Less than 5 weeks	34.9	36.2	35.0	36.7	34.3	34.2	38.6	38.8
5 to 14 weeks	29.4	28.9	29.3	28.9	29.7	28.8	28.5	28.1
15 weeks and over	35.7	34.9	35.7	34.3	36.1	36.9	32.9	33.1

[Continued]

★ 777 ★

Duration of Unemployment, 1992 and 1993

[Continued]

Weeks of unemployment	Total		White		Black		Hispanic origin	
	1992	1993	1992	1993	1992	1993	1992	1993
15 to 26 weeks	15.2	14.6	15.2	14.6	15.0	14.9	14.8	14.7
27 weeks and over	20.6	20.4	20.5	19.8	21.1	22.1	18.1	18.4

Source: "Unemployed Persons by Duration of Unemployment, Race, and Hispanic Origin." *Employment and Earnings* 41 (January 1994): 233. *Notes:* Detail for the above race and Hispanic-origin groups will not sum to totals because data for the "other races" group are not presented and Hispanics are included in both the white and black population groups.

★ 778 ★

Unemployment

Jobseekers Unemployed, and Jobsearch Methods, 1993

Sex, age, and race	1993								
	Thousands of persons		Methods used as a percent of total jobseekers						Average number of methods used
	Total unemployed	Total jobseekers	Public employment agency	Private employment agency	Employer directly	Placed or answered ads	Friends or relatives	Other	
Total, 16 years and over	8,734	7,514	22.2	9.0	72.8	42.0	23.4	5.5	1.75
White, 16 years and over	6,547	5,514	21.4	9.0	73.3	43.8	23.9	5.9	1.77
Men	3,753	3,061	23.3	9.5	74.1	42.7	26.3	6.5	1.82
Women	2,793	2,453	19.0	8.3	72.4	45.2	20.8	5.3	1.71
Black, 16 years and over	1,796	1,641	25.5	9.1	71.9	35.7	20.3	4.3	1.67
Men	954	862	26.8	8.8	72.9	35.1	22.9	4.4	1.71
Women	842	779	24.1	9.4	70.7	36.2	17.5	4.1	1.62

Source: "Unemployed Jobseekers by Sex, Age, Race, and Jobsearch Methods Used." *Employment and Earnings* 41 (January 1994): 200. *Notes:* The jobseekers total is less than the total unemployed because it does not include persons on layoff or waiting to begin a new job within 30 days, groups for whom jobseeking information is not collected. The percent using each method will always total more than 100 because many jobseekers use more than one method.

★ 779 ★

Unemployment

Persons Not in Labor Force, by Reason, Gender, and Race, 1990-1993 - I

In thousands.

Reason, sex, and race	1990 IV	1991 I	II	III	IV	1992 I	II	III	IV
Total									
Total not in labor force	63,722	64,089	64,099	64,759	64,870	64,587	64,236	64,452	65,089
White									
Total not in labor force	53,572	53,797	53,748	54,247	54,254	54,074	53,921	54,082	54,446

[Continued]

★ 779 ★

Persons Not in Labor Force, by Reason, Gender, and Race, 1990-1993 - I

[Continued]

Reason, sex, and race	1990 IV	1991 I	1991 II	1991 III	1991 IV	1992 I	1992 II	1992 III	1992 IV
Do not want a job now	49,565	49,658	49,986	50,085	49,936	49,548	49,475	49,584	49,922
Want a job now	3,917	4,097	3,973	4,213	4,328	4,387	4,518	4,521	4,564
Reason not looking									
School attendance	897	999	930	1,075	1,000	1,078	1,196	1,033	1,162
Ill health, disability	729	736	620	767	766	737	766	823	902
Home responsibilities	832	880	841	863	924	963	913	886	885
Think cannot get a job	619	657	628	705	758	718	771	786	707
Other reasons[1]	839	824	855	803	879	890	873	993	908
Black									
Total not in labor force	7,879	7,952	8,022	8,120	8,196	8,091	8,036	7,980	8,165
Do not want a job now	6,416	6,686	6,622	6,792	6,788	6,662	6,560	6,605	6,726
Want a job now	1,420	1,278	1,429	1,354	1,362	1,450	1,471	1,355	1,403
Reason not looking									
School attendance	442	338	359	371	349	358	446	353	376
Ill health, disability	195	211	258	220	226	253	230	244	248
Home responsibilities	283	268	274	260	326	310	294	256	275
Think cannot get a job	275	266	295	295	266	334	304	284	303
Other reasons[1]	225	196	242	208	195	196	197	218	201

Source: "Persons Not in the Labor Force by Reason, Sex, and Race, Seasonally Adjusted." *Employment and Earnings* 41 (January 1994): 62.
Notes: Data have been revised based on the experience through December 1993. 1. Includes small number of men not looking for work because of "home responsibilities."

★ 780 ★

Unemployment

Persons Not in Labor Force, by Reason, Gender, and Race, 1990-1993 - II

In thousands.

Reason, sex, and race	1993 I	1993 II	1993 III	1993 IV
Total				
Total not in labor force	65,441	65,398	65,618	65,602
White				
Total not in labor force	54,622	54,618	54,568	54,472
Do not want a job now	50,015	50,009	49,943	49,854
Want a job now	4,565	4,634	4,657	4,609

[Continued]

★ 780 ★

Persons Not in Labor Force, by Reason, Gender, and Race, 1990-1993 - II

[Continued]

Reason, sex, and race	1993			
	I	II	III	IV
Reason not looking				
School attendance	1,018	1,023	1,124	1,132
Ill health, disability	795	895	824	782
Home responsibilities	966	945	945	952
Think cannot get a job	774	820	794	749
Other reasons[1]	1,012	951	971	994
Black				
Total not in labor force	8,288	8,370	8,429	8,470
Do not want a job now	6,874	6,969	6,966	7,109
Want a job now	1,428	4,396	1,505	1,318
Reason not looking				
School attendance	374	392	454	331
Ill health, disability	250	219	286	258
Home responsibilities	313	314	332	276
Think cannot get a job	298	307	263	276
Other reasons[1]	193	166	170	177

Source: "Persons Not in the Labor Force by Reason, Sex, and Race, Seasonally Adjusted." *Employment and Earnings* 41 (January 1994): 62. *Notes:* Data have been revised based on the experience through December 1993. 1. Includes small number of men not looking for work because of "home responsibilities."

★ 781 ★

Unemployment

Persons Unemployed, Duration of Unemployment, and Percent, 1992 and 1993

Sex, age, race, and marital status	1993							Percent of unemployed in group			
	Thousands of persons					Weeks		Unemployed less than 5 weeks		Unemployed 15 weeks and over	
	Total	Less than 5 weeks	5 to 14 weeks	15 to 26 weeks	27 weeks and over	Average (mean) duration	Median duration	1992	1993	1992	1993
Total, 16 years and over	8,734	3,160	2,522	1,274	1,778	18.1	8.4	34.9	36.2	35.7	34.9
White, 16 years and over	6,547	2,404	1,895	954	1,294	17.7	8.2	35.0	36.7	35.7	34.3
Men	3,753	1,259	1,083	585	825	19.5	9.3	31.9	33.5	38.9	37.6
Women	2,793	1,145	812	369	468	15.3	6.8	39.4	41.0	31.3	30.0
Black, 16 years and over	1,796	615	518	267	397	19.3	9.1	34.3	34.2	36.1	36.9
Men	954	287	266	153	248	21.8	10.6	30.8	30.1	40.2	42.0
Women	842	328	252	114	149	16.5	7.4	38.3	38.9	31.4	31.2

Source: "Unemployed Persons by Sex, Age, Race, Marital Status, and Duration of Unemployment." *Employment and Earnings* 41 (January 1994): 233.

★ 782 ★

Unemployment

Reasons for Unemployment, 1992 and 1993

Numbers in thousands.

Reasons for unemployment	Total		White		Black		Hispanic origin	
	1992	1993	1992	1993	1992	1993	1992	1993
Number of unemployed								
Total, 16 years and over	9,384	8,734	7,047	6,547	1,958	1,796	1,160	1,104
Job losers	5,291	4,769	4,117	3,684	985	896	707	644
On layoff	1,246	1,104	1,054	932	160	146	125	107
Other job losers	4,045	3,664	3,064	2,751	825	751	583	537
Job leavers	975	946	759	740	176	166	94	114
Reentrants	2,228	2,145	1,596	1,541	545	501	227	215
New entrants	890	874	574	582	252	233	131	131
Percent distribution								
Total unemployed	100.0	100.0	100.0	100.0	100.0	100.0	100.0	100.0
Job losers	56.4	54.6	58.4	56.3	50.3	49.9	61.0	58.3
On layoff	13.3	12.6	15.0	14.2	8.2	8.1	10.8	9.7
Other job losers	43.1	42.0	43.5	42.0	42.2	41.8	50.2	48.6
Job leavers	10.4	10.8	10.8	11.3	9.0	9.2	8.1	10.4
Reentrants	23.7	24.6	22.6	23.5	27.8	27.9	19.5	19.5
New entrants	9.5	10.0	8.1	8.9	12.9	13.0	11.3	11.8
Unemployed as a percent of the civilian labor force								
Job losers	4.2	3.7	3.8	3.4	7.1	6.4	7.0	6.2
Job leavers	.8	.7	.7	.7	1.3	1.2	.9	1.1
Reentrants	1.8	1.7	1.5	1.4	3.9	3.6	2.2	2.1
New entrants	.7	.7	.5	.5	1.8	1.7	1.3	1.3

Source: "Unemployed Persons by Reason, Race, and Hispanic Origin." *Employment and Earnings* 41 (January 1994): 233. *Notes:* Detail for the above race and Hispanic-origin groups will not sum to totals because data for the "other races" group are not presented and Hispanics are included in both the white and black population groups.

★ 783 ★

Unemployment

Selected Unemployment Indicators, Annual 1992 and 1993, and Monthly, 1993 to January 1994 - I

Unemployment rates.

Selected categories	Annual averages		1993					
	1992	1993	Jan.	Feb.	Mar.	Apr.	May	June
Characteristic								
Total, all workers	7.4	6.8	7.1	7.0	7.0	7.0	6.9	6.9
Both sexes, 16 to 19 years	20.0	19.0	19.6	19.6	19.5	20.3	19.8	19.5
Men, 20 years and over	7.0	6.4	6.5	6.6	6.7	6.5	6.5	6.5

[Continued]

★ 783 ★

Selected Unemployment Indicators, Annual 1992 and 1993, and Monthly, 1993 to January 1994 - I

[Continued]

Selected categories	Annual averages		1993					
	1992	1993	Jan.	Feb.	Mar.	Apr.	May	June
Women, 20 years and over	6.3	5.9	6.3	6.0	5.7	6.0	5.9	5.9
White, total	6.5	6.0	6.2	6.1	6.1	6.1	6.1	6.1
Both sexes, 16 to 19 years	17.1	16.2	16.5	16.6	16.4	16.6	16.8	16.3
Men, 16 to 19 years	18.4	17.6	17.9	17.8	17.1	18.5	17.2	18.4
Women, 16 to 19 years	15.7	14.6	15.0	15.3	15.5	14.5	16.3	14.0
Men, 20 years and over	6.3	5.6	5.8	5.8	5.9	5.7	5.7	5.7
Women, 20 years and over	5.4	5.1	5.4	5.2	5.1	5.2	5.1	5.3
Black, total	14.1	12.9	14.1	13.3	13.5	13.7	12.9	13.3
Both sexes, 16 to 19 years	39.8	28.9	39.1	39.0	42.3	45.3	39.5	41.6
Men, 16 to 19 years	42.0	40.1	39.7	39.5	44.1	46.8	40.2	38.8
Women, 16 to 19 years	37.2	37.5	38.5	38.4	40.1	43.2	38.7	44.8
Men, 20 years and over	13.4	12.1	12.9	12.2	13.0	12.7	12.2	12.6
Women, 20 years and over	11.7	10.6	12.3	11.2	10.6	10.9	10.4	10.7
Hispanic origin, total	11.4	10.6	11.4	11.3	11.2	10.5	10.0	10.3
Married men, spouse present	5.0	4.4	4.5	4.6	4.7	4.5	4.5	4.4
Married women, spouse present	5.0	4.6	4.9	4.4	4.4	4.8	4.5	4.7
Women who maintain families	9.9	9.5	10.4	10.1	9.0	9.6	9.8	9.7
Full-time workers	7.4	6.8	7.1	7.0	6.9	6.9	6.9	6.9
Part-time workers	7.4	7.1	7.5	7.3	7.2	7.6	6.9	7.1

Source: "Selected Unemployment Indicators, Monthly Data Seasonally Adjusted." *Monthly Labor Review* 117 (March 1994): 63.

★ 784 ★

Unemployment

Selected Unemployment Indicators, Annual 1992 and 1993, and Monthly, 1993 to January 1994 - II

Unemployment rates.

Selected categories	1993						1994
	July	Aug.	Sept.	Oct.	Nov.	Dec.	Jan.
Characteristic							
Total, all workers	6.8	6.7	6.7	6.7	6.5	6.4	6.7
Both sexes, 16 to 19 years	18.4	18.4	17.9	18.9	18.3	17.8	18.4
Men, 20 years and over	6.5	6.4	6.3	6.2	5.9	5.8	5.9
Women, 20 years and over	5.8	5.7	5.8	5.8	5.7	5.7	6.0
White, total	6.0	5.9	5.8	6.1	5.6	5.6	5.8
Both sexes, 16 to 19 years	15.6	15.9	15.6	17.0	15.6	15.2	16.4
Men, 16 to 19 years	17.7	17.7	16.8	17.9	17.7	16.9	18.5
Women, 16 to 19 years	13.4	14.0	14.3	16.0	13.3	13.4	14.0

[Continued]

★ 784 ★

Selected Unemployment Indicators, Annual 1992 and 1993, and Monthly, 1993 to January 1994 - II

[Continued]

Selected categories	1993						1994
	July	Aug.	Sept.	Oct.	Nov.	Dec.	Jan.
Men, 20 years and over	5.8	5.6	5.5	5.7	5.0	5.2	5.3
Women, 20 years and over	5.1	5.0	5.0	5.3	5.1	4.9	5.1
Black, total	12.8	12.5	12.5	11.9	12.5	11.5	13.1
Both sexes, 16 to 19 years	36.4	33.5	36.2	36.7	39.5	37.0	31.7
Men, 16 to 19 years	37.9	34.9	39.7	40.6	39.2	38.8	38.1
Women, 16 to 19 years	34.7	32.0	32.3	32.8	39.7	35.2	25.5
Men, 20 years and over	11.8	12.0	12.1	11.0	12.3	10.5	12.3
Women, 20 years and over	11.0	10.5	10.2	10.0	9.7	9.7	11.5
Hispanic origin, total	10.8	9.9	10.0	11.4	10.4	10.5	10.6
Married men, spouse present	4.5	4.4	4.2	4.4	4.0	3.9	4.1
Married women, spouse present	4.7	4.5	4.6	4.8	4.4	4.3	4.4
Women who maintain families	9.6	9.0	9.0	9.3	9.0	10.2	9.4
Full-time workers	6.8	6.7	6.6	6.6	6.3	6.4	6.8
Part-time workers	6.7	6.8	6.9	7.2	6.9	6.6	6.2

Source: "Selected Unemployment Indicators, Monthly Data Seasonally Adjusted." *Monthly Labor Review* 117 (March 1994): 63.

★ 785 ★

Unemployment

Unemployed Persons and Duration Unemployment, by State; Annual Averages, 1992

Population group and state	Total unemployed		Duration of unemployment				
	Number (in thousands)	Percent	Less than 5 weeks	5 to 14 weeks	15 weeks and over	27 weeks and over	52 weeks and over
White							
Alabama	78	100.0	45.1	29.3	25.6	12.3	5.9
Alaska	17	100.0	40.3	29.8	29.9	10.2	4.0
Arizona	117	100.0	42.5	26.5	31.1	16.4	5.4
Arkansas	53	100.0	40.4	33.8	25.8	10.3	6.5
California	1,123	100.0	35.6	28.7	35.7	20.4	11.0
Colorado	91	100.0	48.3	28.0	23.7	13.7	6.6
Connecticut	115	100.0	22.9	23.2	54.0	36.2	19.4
Delaware	12	100.0	30.6	29.2	40.2	26.1	11.0
Florida	395	100.0	34.7	30.3	35.0	19.7	9.1
Georgia	104	100.0	41.3	27.2	31.5	18.7	10.4
Hawaii	11	100.0	41.1	38.9	20.1	5.6	3.8
Idaho	32	100.0	44.6	35.1	20.3	8.8	3.9
Illinois	307	100.0	33.1	28.6	38.3	22.4	13.7
Indiana	151	100.0	40.5	31.2	28.3	14.0	8.2

[Continued]

★ 785 ★

Unemployed Persons and Duration Unemployment, by State; Annual Averages, 1992

[Continued]

Population group and state	Total unemployed		Duration of unemployment				
	Number (in thousands)	Percent	Less than 5 weeks	5 to 14 weeks	15 weeks and over	27 weeks and over	52 weeks and over
Iowa	68	100.0	36.7	37.9	25.4	11.2	4.2
Kansas	43	100.0	42.4	34.5	23.2	11.7	6.0
Kentucky	103	100.0	29.7	29.9	40.4	25.9	15.6
Louisiana	85	100.0	37.6	27.9	34.5	21.8	11.0
Maine	46	100.0	29.9	30.4	39.7	23.8	14.2
Maryland	83	100.0	29.7	23.8	46.4	31.1	16.5
Massachusetts	237	100.0	21.2	25.7	53.2	35.0	21.4
Michigan	305	100.0	35.2	30.2	34.6	20.3	11.0
Minnesota	107	100.0	38.2	27.4	34.4	16.6	9.6
Mississippi	46	100.0	41.6	35.7	22.8	14.3	8.8
Missouri	117	100.0	33.0	35.1	31.9	18.2	9.4
Montana	23	100.0	46.0	29.0	25.0	12.7	7.3
Nebraska	23	100.0	52.4	26.9	20.7	9.7	5.7
Nevada	38	100.0	40.0	29.7	30.3	17.3	5.9
New Hampshire	46	100.0	28.2	27.9	43.9	24.6	13.2
New Jersey	246	100.0	23.0	29.1	47.9	29.0	14.1
New Mexico	43	100.0	38.8	32.4	28.7	15.6	7.8
New York	531	100.0	26.0	27.7	46.3	27.5	14.3
North Carolina	116	100.0	36.8	31.9	31.3	17.9	9.1
North Dakota	12	100.0	47.3	28.8	23.8	9.7	4.9
Ohio	321	100.0	32.9	29.6	37.5	22.0	11.1
Oklahoma	68	100.0	38.3	28.8	32.9	18.3	10.8
Oregon	107	100.0	42.6	30.0	27.4	13.7	7.1
Pennsylvania	369	100.0	30.9	30.3	38.8	21.5	9.3
Rhode Island	42	100.0	234.9	25.7	49.5	33.9	16.5
South Carolina	54	100.0	42.1	26.6	31..3	16.3	4.7
South Dakota	9	100.0	49.3	31.2	19.5	11.6	5.3
Tennessee	116	100.0	42.0	31.8	26.2	11.7	5.6
Texas	471	100.0	45.8	29.5	24.7	13.8	7.7
Utah	39	100.0	45.1	29.3	25.6	11.9	6.6
Vermont	21	100.0	30.6	30.5	38.9	22.2	15.3
Virginia	130	100.0	36.9	29.6	33.5	18.9	8.3
Washington	168	100.0	39.8	29.6	30.7	16.0	9.2
West Virginia	81	100.0	26.5	29.5	44.0	28.6	17.1
Wisconsin	113	100.0	43.7	29.7	26.7	11.7	4.3
Wyoming	13	100.0	47.6	32.0	20.4	10.0	5.6
Black							
Alabama	63	100.0	34.9	38.1	27.0	14.8	7.7
Arkansas	29	100.0	37.6	28.2	34.2	18.5	8.7
California	123	100.0	28.0	37.1	34.8	22.9	12.9
District of Columbia	19	100.0	22.7	32.3	45.0	24.8	15.8

[Continued]

★ 785 ★

Unemployed Persons and Duration Unemployment, by State; Annual Averages, 1992

[Continued]

Population group and state	Total unemployed		Duration of unemployment				
	Number (in thousands)	Percent	Less than 5 weeks	5 to 14 weeks	15 weeks and over	27 weeks and over	52 weeks and over
Florida	130	100.0	39.1	25.6	35.3	18.7	11.7
Georgia	115	100.0	33.5	25.3	41.2	26.0	15.7
Illinois	140	100.0	37.5	26.5	36.1	21.0	14.4
Louisiana	68	100.0	48.1	27.7	24.2	13.5	9.5
Maryland	82	100.0.	26.2	38.1	35.7	20.5	12.6
Massachusetts	21	100.0	28.2	24.3	47.5	34.0	26.7
Michigan	91	100.0	47.6	27.6	24.8	16.2	10.3
Mississippi	49	100.0	36.5	31.4	32.1	17.5	10.0
New Jersey	75	100.0	20.7	30.8	48.5	29.6	19.2
New York	169	100.0	19.2	25.4	55.4	34.0	23.0
North Carolina	85	100.0	41.3	28.3	30.4	14.4	7.4
Ohio	71	100.0	39.2	33.1	27.7	18.0	12.9
Pennsylvania	75	100.0	31.3	28.6	40.1	26.5	15.6
South Carolina	55	100.0	36.2	28.7	35.1	18.0	13.4
Texas	165	100.0	40.8	27.1	32.2	19.6	11.0
Virginia	78	100.0	30.4	29.6	40.0	18.5	11.0
Hispanic							
California	478	100.0	38.7	28.7	32.6	17.6	10.1
Florida	92	100.0	38.5	27.5	33.9	17.9	7.0
Illinois	49	100.0	33.5	30.9	35.5	20.1	10.4
New Jersey	42	100.0	27.4	31.8	40.8	22.8	12.1
New York	115	100.0	25.4	26.4	48.2	26.5	16.3
Texas	191	100.0	46.5	29.1	24.4	13.7	6.7

Source: "States: Percent Distribution of Unemployed Persons by Sex, Age, Race, Hispanic Origin, and Duration of Unemployment, 1992 Annual Averages," U.S. Department of Labor, *Geographical Profile of Employment and Unemployment, 1992*, p. 91. *Notes:* Data for demographic groups are not shown when they do not meet BLS publication standards of reliability for the particular area based on the sample in that area. Items may not add to totals or compute to displayed percentages because of rounding. Detail for Hispanic-origin groups will not add to totals because data for the "other races" group are not presented and Hispanics are included in both the white and black population groups.

★ 786 ★
Unemployment

Unemployed Persons and Reason for Unemployment, by State; Annual Averages, 1992

Population group and state	Total unemployed		Reason for unemployment				
	Number (in thousands)	Percent	Job losers		Job leavers	Reentrants	New entrants
			Total	On layoff			
White							
Alabama	78	100.0	55.7	16.5	16.2	18.8	9.3
Alaska	17	100.0	49.0	9.5	17.8	29.4	3.8
Arizona	117	100.0	51.1	10.3	12.7	26.7	9.5
Arkansas	53	100.0	47.4	14.6	16.2	27.3	9.1
California	1,123	100.0	62.2	11.9	8.4	19.8	9.6
Colorado	91	100.0	44.7	11.7	17.3	30.4	7.7
Connecticut	115	100.0	74.2	13.0	7.4	14.2	4.2
Delaware	12	100.0	59.7	27.3	13.9	22.2	4.1
Florida	395	100.0	54.4	8.7	11.3	26.6	7.8
Georgia	104	100.0	55.1	10.3	11.8	27.3	5.8
Hawaii	11	100.0	56.1	15.9	10.0	30.8	3.1
Idaho	32	100.0	55.0	23.7	14.3	24.6	6.0
Illinois	307	100.0	60.8	14.0	10.4	20.0	8.8
Indiana	151	100.0	53.6	17.0	12.3	22.9	11.3
Iowa	68	100.0	50.6	21.9	13.2	25.0	11.0
Kansas	43	100.0	44.3	11.6	20.6	28.3	6.8
Kentucky	103	100.0	56.7	15.7	8.4	24.2	10.6
Louisiana	85	100.0	55.0	12.3	12.0	24.1	8.9
Maine	46	100.0	67.7	18.9	3.2	22.5	6.6
Maryland	83	100.0	51.0	12.6	13.4	30.8	4.9
Massachusetts	237	100.0	68.0	16.9	7.5	18.1	6.5
Michigan	305	100.0	61.0	28.7	9.9	20.8	8.3
Minnesota	107	100.0	54.4	19.9	15.5	22.8	7.2
Mississippi	46	100.0	50.0	16.4	11.5	27.8	10.8
Missouri	117	100.0	59.4	12.1	13.5	20.2	6.9
Montana	23	100.0	43.9	14.8	13.3	35.6	7.2
Nebraska	23	100.0	38.7	8.7	22.8	30.4	8.1
Nevada	38	100.0	60.8	9.4	15.4	20.5	3.3
New Hampshire	46	100.0	63.4	14.3	7.5	19.8	9.3
New Jersey	246	100.0	73.1	17.1	5.8	14.4	6.8
New Mexico	43	100.0	53.6	7.5	12.9	28.5	5.1
New York	531	100.0	65.5	14.7	7.0	20.2	7.3
North Carolina	116	100.0	50.0	10.8	18.0	25.2	6.9
North Dakota	12	100.0	41.4	19.3	11.7	35.2	11.7
Ohio	321	100.0	59.4	22.6	9.0	22.5	9.1
Oklahoma	68	100.0	51.6	6.2	14.7	25.9	7.8
Oregon	107	100.0	50.9	13.2	16.6	26.9	5.6
Pennsylvania	369	100.0	63.2	25.3	7.8	20.9	8.1
Rhode Island	42	100.0	68.4	26.1	7.2	17.1	7.3
South Carolina	54	100.0	60.8	13.0	12.6	17.9	8.7
South Dakota	9	100.0	38.3	12.8	21.5	36.4	3.8

[Continued]

★ 786 ★

Unemployed Persons and Reason for Unemployment, by State; Annual Averages, 1992

[Continued]

Population group and state	Total unemployed		Reason for unemployment				
	Number (in thousands)	Percent	Job losers		Job leavers	Reentrants	New entrants
			Total	On layoff			
Tennessee	116	100.0	51.3	19.1	14.7	25.6	8.4
Texas	471	100.0	46.1	5.9	14.7	28.9	10.4
Utah	39	100.0	45.8	17.2	16.8	26.1	11.3
Vermont	21	100.0	65.0	19.5	7.6	20.4	7.0
Virginia	130	100.0	54.6	10.8	9.9	28.7	6.7
Washington	168	100.0	57.9	13.8	13.2	25.8	3.1
West Virginia	81	100.0	55.8	18.0	14.0	20.4	9.9
Wisconsin	113	100.0	56.9	23.0	13.7	21.5	7.9
Wyoming	13	100.0	43.8	11.3	8.9	42.0	5.3
Black							
Alabama	63	100.0	48.3	7.3	11.2	25.2	15.3
Arkansas	29	100.0	44.5	6.6	8.5	28.4	18.6
California	123	100.0	49.7	5.6	7.2	31.7	11.5
District of Columbia	19	100.0	68.3	7.6	4.1	22.2	5.4
Florida	130	100.0	48.4	7.7	7.1	30.5	14.0
Georgia	115	100.0	52.4	9.5	10.1	26.3	11.2
Illinois	140	100.0	47.8	7.5	7.0	29.2	16.0
Louisiana	68	100.0	41.7	5.3	7.9	35.8	14.6
Maryland	82	100.0	40.2	6.1	8.1	41.1	10.6
Massachusetts	21	100.0	54.3	7.4	5.7	20.4	19.6
Michigan	91	100.0	46.2	18.4	5.3	34.6	14.0
Mississippi	49	100.0	48.9	11.2	8.1	24.3	18.7
New Jersey	75	100.0	70.8	10.4	4.1	17.8	7.2
New York	169	100.0	61.3	7.0	6.1	20.2	12.3
North Carolina	85	100.0	43.8	9.3	14.9	28.9	12.4
Ohio	71	100.0	48.5	13.9	9.5	28.8	13.2
Pennsylvania	75	100.0	51.3	12.0	6.7	30.1	11.9
South Carolina	55	100.0	47.8	5.3	9.7	28.4	14.1
Texas	165	100.0	49.5	4.0	12.5	27.0	11.0
Virginia	78	100.0	46.4	7.7	9.1	23.9	20.7
Hispanic							
California	478	100.0	65.2	13.6	6.3	16.6	11.8
Florida	92	100.0	64.6	5.2	7.2	19.1	9.1
Illinois	49	100.0	61.9	10.0	8.7	19.4	10.0
New Jersey	42	100.0	71.6	19.8	6.5	12.2	9.7

[Continued]

★ 786 ★

Unemployed Persons and Reason for Unemployment, by State; Annual Averages, 1992

[Continued]

Population group and state	Total unemployed		Reason for unemployment				
	Number (in thousands)	Percent	Job losers		Job leavers	Reentrants	New entrants
			Total	On layoff			
New York	115	100.0	65.1	9.4	6.3	17.0	11.6
Texas	191	100.0	49.8	5.2	12.6	25.8	11.9

Source: "States: Percent Distribution of Unemployed Persons by Sex, Age, Race, Hispanic Origin, and Reason for Unemployment, 1992 Annual Averages," U.S. Department of Labor, *Geographical Profile of Employment and Unemployment, 1992*, pp. 86-87. *Notes:* Data for demographic groups are not shown when they do not meet BLS publication standards of reliability for the particular area based on the sample in that area. Items may not add to totals or compute to displayed percentages because of rounding. Detail for Hispanic-origin groups will not add to totals because data for the "other races" group are not presented and Hispanics are included in both the white and black population groups.

★ 787 ★

Unemployment

Unemployed Persons by Family Relationship and Presence of Unemployed Family Members, 1992 and 1993

Numbers in thousands.

Family relationship, race, and Hispanic origin	1992				1993			
		Percent of unemployed				Percent of unemployed		
	Total	With no employed person in family	With at least one employed person in family	With at least one person in family employed full time	Total	With no employed person in family	With at least one employed person in family	With at least one person in family employed full time
Total								
Total unemployed in families[1]	7,461	31.7	68.3	60.1	6,916	32.0	68.0	60.3
White								
Total unemployed in families[1]	5,565	29.1	70.9	62.7	5,141	29.0	71.0	62.8
Husbands	1,630	35.2	64.8	51.2	1,422	33.6	66.4	52.6
With children under 18 years of age	954	39.2	60.8	46.3	842	36.9	63.1	47.9
Wives	1,241	18.6	81.4	76.5	1,156	16.5	83.5	79.0
With children under 18 years of age	751	16.3	83.7	79.0	670	13.8	86.2	82.3
Relatives in married-couple families	1,386	11.6	88.4	83.4	1,273	11.9	88.1	83.4
Women who maintain families	384	79.6	20.4	12.6	389	80.2	19.8	11.7
With children under 18 years of age	313	89.2	10.8	5.3	307	89.2	10.8	4.3
Relatives in families maintained by women	556	31.9	68.1	58.4	554	34.2	65.8	55.7
Men who maintain families	169	70.1	29.9	24.0	162	71.7	28.3	21.4
With children under 18 years of age	88	90.8	9.2	7.4	94	91.9	8.1	4.8
Relatives in families maintained by men	199	27.6	72.4	66.0	185	27.9	72.1	64.7
Black								
Total unemployed in families[1]	1,577	41.4	58.6	50.5	1,457	42.8	57.2	50.9
Husbands	211	34.1	65.9	56.0	184	32.1	67.9	58.3
With children under 18 years of age	145	35.3	64.7	55.2	121	32.1	67.9	59.1
Wives	161	20.8	79.2	72.2	145	22.5	77.5	71.0
With children under 18 years of age	108	19.0	81.0	75.1	96	18.9	81.1	75.1
Relatives in married-couple families	289	13.7	86.3	78.0	279	15.4	84.6	79.3
Women who maintain families	314	90.2	9.8	6.2	299	89.8	10.2	6.8
With children under 18 years of age	269	93.9	6.1	3.6	260	94.1	5.9	3.3
Relatives in families maintained by women	494	35.2	64.8	54.4	439	37.0	63.0	55.2
Men who maintain families	46	69.4	30.6	27.2	49	77.6	22.4	20.0

[Continued]

★ 787 ★

Unemployed Persons by Family Relationship and Presence of Unemployed Family Members, 1992 and 1993

[Continued]

Family relationship, race, and Hispanic origin	1992				1993			
		Percent of unemployed				Percent of unemployed		
	Total	With no employed person in family	With at least one employed person in family	With at least one person in family employed full time	Total	With no employed person in family	With at least one employed person in family	With at least one person in family employed full time
With children under 18 years of age	24	[2]	[2]	[2]	29	[2]	[2]	[2]
Relatives in families maintained by men	63	31.0	69.0	56.4	61	31.6	68.4	60.9
Hispanic origin								
Total unemployed in families[1]	952	36.2	63.8	57.1	912	32.5	67.5	59.8
Husbands	254	47.1	52.9	43.6	234	41.7	58.3	46.1
With children under 18 years of age	192	50.3	49.7	39.6	183	44.6	55.4	43.1
Wives	178	23.3	76.7	70.9	175	17.6	82.4	76.8
With children under 18 years of age	126	24.5	75.5	70.4	122	18.2	81.8	77.2
Relatives in married-couple families	226	14.5	85.5	81.7	225	11.3	88.7	84.3
Women who maintain families	86	82.6	17.4	11.5	81	75.1	24.9	18.7
With children under 18 years of age	76	87.0	13.0	8.8	63	84.3	15.7	9.7
Relatives in families maintained by women	116	35.0	65.0	54.8	115	39.1	60.9	51.6
Men who maintain families	41	62.6	37.4	34.3	37	68.3	31.7	29.4
With children under 18 years of age	21	[2]	[2]	[2]	20	[2]	[2]	[2]
Relatives in families maintained by men	51	26.5	73.5	66.9	45	27.1	72.9	62.6

Source: "Unemployed Persons by Family Relationship, Race, Hispanic Origin, and Presence of Employed Family Member." *Employment and Earnings* 41 (January 1994): 237. *Notes:* Detail for the above race and Hispanic-origin groups will not sum to totals because data for the "other races" group are not presented and Hispanics are included in both the white and black population groups. 1. Excludes persons living alone or with nonrelatives, persons in married-couple families where the husband or wife is in the Armed Forces, and persons in unrelated subfamilies. Estimates for husbands, wives, and women who maintain families are somewhat different from marital status estimates shown in other tables in this publication because of differences in definitions and weighing patterns used in aggregating the data. 2. Data not shown where base is less than 35,000.

★ 788 ★

Unemployment

Unemployed Persons by Marital Status, Race, Age, and Gender, 1992 and 1993

Marital status, race, and age	Men				Women			
	Thousands of persons		Unemployment rates		Thousands of persons		Unemployment rates	
	1992	1993	1992	1993	1992	1993	1992	1993
Total, 16 years and over	5,380	4,932	7.8	7.1	4,005	3,801	6.9	6.5
Married, spouse present	2,124	1,878	5.0	4.4	1,584	1,465	5.0	4.6
Widowed, divorced, or separated	756	707	9.8	9.0	876	850	7.5	7.2
Single (never married)	2,499	2,347	13.1	12.3	1,545	1,487	10.7	10.2
White, 16 years and over	4,121	3,753	6.9	6.2	2,926	2,793	6.0	5.7
Married, spouse present	1,775	1,549	4.7	4.1	1,324	1,225	4.7	4.3
Widowed, divorced, or separated	580	542	9.1	8.3	655	619	7.1	6.6
Single (never married)	1,766	1,662	11.3	10.7	947	950	8.4	8.3
Black, 16 years and over	1,046	954	15.2	13.8	912	842	13.0	12.0
Married, spouse present	260	229	8.3	7.2	187	165	7.8	7.0
Widowed, divorced, or separated	150	135	13.7	12.8	193	195	9.8	9.8
Single (never married)	636	590	24.0	21.9	532	482	20.2	18.0

[Continued]

★ 788 ★

Unemployed Persons by Marital Status, Race, Age, and Gender, 1992 and 1993

[Continued]

Marital status, race, and age	Men				Women			
	Thousands of persons		Unemployment rates		Thousands of persons		Unemployment rates	
	1992	1993	1992	1993	1992	1993	1992	1993
Total, 25 years and over	3,734	3,396	6.4	5.8	2,751	2,621	5.7	5.4
Married, spouse present	2,003	1,769	4.9	4.3	1,406	1,302	4.7	4.3
Widowed, divorced, or separated	726	678	9.6	8.9	915	784	7.3	6.9
Single (never married)	1,006	949	10.2	9.5	531	534	7.6	7.5
White, 25 years and over	2,932	2,644	5.8	5.2	2,048	1,968	5.1	4.8
Married, spouse present	1,669	1,463	4.6	4.0	1,173	1,093	4.4	4.1
Widowed, divorced, or separated	554	519	8.9	8.2	6.7	527	6.8	6.3
Single (never married)	709	663	8.9	8.3	268	303	5.2	5.9
Black, 25 years and over	655	592	11.7	10.5	588	531	10.1	9.1
Married, spouse present	247	211	8.1	6.9	166	141	7.3	6.3
Widowed, divorced, or separated	147	130	13.6	12.5	182	178	9.5	9.2
Single (never married)	262	252	17.7	16.3	240	212	14.8	12.9

Source: "Unemployed Persons by Marital Status, Race, Age, and Sex." *Employment and Earnings* 41 (January 1994): 193.

★ 789 ★

Unemployment

Unemployed Persons: Characteristics, January 1993 and January 1994

Not seasonally adjusted.

Marital status, race, and age	Men				Women			
	Thousands of persons		Unemployment rates		Thousands of persons		Unemployment rates	
	Jan. 1993	Jan. 1994	Jan. 1993	Jan. 1994	Jan. 1993	Jan. 1994	1993	1994
Total, 16 years and over	5,790	5,526	8.5	7.9	4,121	3,966	7.2	6.7
Married, spouse present	2,302	2,149	5.4	5.0	1,670	1,570	5.3	4.8
Widowed, divorced, or separated	880	686	11.6	9.0	984	883	8.5	7.6
Single (never married)	2,608	2,690	14.1	14.0	1,467	1,512	10.4	10.2
White, 16 years and over	4,496	4,216	7.6	7.0	3,002	2,907	6.2	5.9
Married, spouse present	1,916	1,767	5.1	4.6	1,325	1,300	4.7	4.5
Widowed, divorced, or separated	666	501	10.5	7.9	761	653	8.2	7.1
Single (never married)	1,915	1,948	12.7	12.5	916	952	8.3	8.4
Black, 16 years and over	1,057	1,027	15.6	14.9	927	896	13.5	12.2
Married, spouse present	278	255	8.9	7.9	256	184	11.0	7.7
Widowed, divorced, or separated	191	143	18.8	14.3	184	202	9.6	9.8
Single (never married)	589	629	22.3	23.6	487	510	18.7	17.7
Total, 25 years and over	4,069	3,724	7.0	6.3	2,930	2,786	6.1	5.6
Married, spouse present	2,143	1,996	5.2	4.8	1,465	1,378	4.9	4.4
Widowed, divorced, or separated	846	645	11.5	8.6	926	817	8.2	7.3
Single (never married)	1,080	1,082	11.0	10.7	539	590	7.6	8.0
White, 25 years and over	3,210	2,870	6.3	5.6	2,186	2,050	5.4	4.9
Married, spouse present	1,791	1,641	4.9	4.5	1,166	1,130	4.4	4.1

[Continued]

★ 789 ★

Unemployed Persons: Characteristics, January 1993 and January 1994

[Continued]

Marital status, race, and age	Men				Women			
	Thousands of persons		Unemployment rates		Thousands of persons		Unemployment rates	
	Jan. 1993	Jan. 1994	Jan. 1993	Jan. 1994	Jan. 1993	Jan. 1994	1993	1994
Widowed, divorced, or separated	632	460	10.3	7.5	716	596	7.9	6.7
Single (never married)	787	769	10.0	9.5	304	324	5.9	6.1
Black, 25 years and over	700	634	12.5	11.2	612	610	10.7	10.1
Married, spouse present	252	231	8.3	7.4	219	167	10.0	7.3
Widowed, divorced, or separated	191	142	19.0	14.5	172	195	9.1	9.7
Single (never married)	257	261	16.7	16.8	220	248	13.6	14.3

Source: "Unemployed Persons by Marital Status, Race, Age,a nd Sex." *Employment and Earnings* 41 (February 1994): 235. *Notes:* Data for 1994 are not directly comparable with data for 1993 and earlier years.

★ 790 ★

Unemployment

Unemployed Workers: Trends, 1980 to 1992

In thousands, except as indicated. For civilian noninstitutional population 16 years old and over. Annual averages of monthly figures.

Item and characteristic	1980	1985	1986	1987	1988	1989	1990	1991	1992
Unemployed									
Total[1]	7,637	8,312	8,237	7,425	6,701	8,528	6,874	8,426	9,384
Labor force time lost (percent)[2]	7.9	8.1	7.9	7.1	6.3	5.9	6.2	7.6	8.3
White[3]	5,884	6,191	6,140	5,501	4,944	4,770	5,091	6,447	7,047
16 to 19 years old	1,291	1,074	1,070	995	910	863	856	977	983
20 to 24 years old	1,364	1,235	1,149	1,017	874	856	844	1,063	1,084
Black[3]	1,553	1,864	1,840	1,684	1,547	1,544	1,527	1,679	1,958
16 to 19 years old	343	357	347	312	288	300	258	270	313
20 to 24 years old	426	455	453	397	340	322	335	362	401
Hispanic[3,4]	620	811	857	751	732	750	769	963	1,160
16 to 19 years old	145	141	141	136	148	132	131	149	185
20 to 24 years old	138	171	183	152	145	158	135	172	193
Full-time workers	6,269	6,793	6,708	5,979	5,357	5,211	5,541	6,932	7,746
Part-time workers	1,369	1,519	1,529	1,446	1,343	1,317	1,332	1,494	1,638
Unemployment rate (percent)[5]									
Total[1]	7.1	7.2	7.0	6.2	5.5	5.3	5.5	6.7	7.4
White[3]	6.3	6.2	6.0	5.3	4.7	4.5	4.7	6.0	6.5
16 to 19 years old	15.5	15.7	15.6	14.4	13.1	12.7	13.4	16.4	17.1
20 to 24 years old	9.9	9.2	8.7	8.0	7.1	7.2	7.2	9.2	9.4
Black[3]	14.3	15.1	14.5	13.0	11.7	11.4	11.3	12.4	14.1
16 to 19 years old	38.5	40.2	39.3	34.7	32.4	32.4	31.1	36.3	39.8
20 to 24 years old	23.6	24.5	24.1	21.8	19.6	18.0	19.9	21.6	23.9
Hispanic[3,4]	10.1	10.5	10.6	8.8	8.2	8.0	8.0	8.9	11.4
16 to 19 years old	22.5	24.3	24.7	22.3	22.0	19.4	19.5	22.9	27.5

[Continued]

★ 790 ★

Unemployed Workers: Trends, 1980 to 1992

[Continued]

Item and characteristic	1980	1985	1986	1987	1988	1989	1990	1991	1992
20 to 24 years old	12.1	12.6	12.9	10.6	9.8	10.7	9.1	11.6	13.2
Experienced workers[6]	6.9	6.8	6.6	5.8	5.2	5.0	5.3	6.5	7.1
Women maintaining families[1]	9.2	10.5	9.9	9.3	8.2	8.1	8.2	9.1	9.9
White	7.3	8.1	7.8	6.8	6.0	6.1	6.3	7.2	7.8
Black	14.0	16.4	15.4	15.4	13.7	13.0	13.1	13.9	14.7
Married men, wife present[1]	4.2	4.3	4.4	3.9	3.3	3.0	3.4	4.4	5.0
White	3.9	4.0	4.0	3.6	3.0	2.8	3.1	4.2	4.7
Black	7.4	8.0	8.0	6.5	5.8	5.8	6.2	6.5	8.3

Source: "Unemployed Workers—Summary: 1980 to 1992." U.S. Bureau of the Census, *Statistical Abstract of the United States*, 1993, p. 413. Primary source: U.S. Bureau of Labor Statistics, *Employment and Earnings*, monthly, January issues; and unpublished data. *Notes:* 1. Includes other races, not shown separately. 2. Aggregate hours lost by the unemployed and persons on part time for economic reasons as a percent of potentially available labor force hours. 3. Includes other ages, not shown separately. 4. Persons of Hispanic origin may be of any race. 5. Unemployed as percent of civilian labor force in specified group. 6. Wage and salary workers.

★ 791 ★

Unemployment

Unemployed by Reason and Characteristics, Annual Averages, 1992 and 1993

In thousands.

Reason, race, and Hispanic origin	Total		Age						Sex			
			16 to 24 years		25 to 59 years		60 years and over		Men		Women	
	1992	1993	1992	1993	1992	1993	1992	1993	1992	1993	1992	1993
Total not in labor force	54,132	54,562	7,757	7,676	17,155	17,411	29,219	29,475	18,521	18,929	35,610	35,633
Do not want a job now	49,634	49,956	6,334	6,298	14,679	14,824	28,621	28,833	16,867	17,175	32,768	32,781
Current activity												
Going to school	5,077	5,118	4,409	658	694	10	15	2,469	2,008	2,095	1,957	2,081
Ill, disabled	3,965	4,176	124	123	2,187	2,308	1,654	1,745	2,008	2,095	1,957	2,081
Keeping house	18,871	18,081	995	1,003	9,066	8,871	8,810	8,207	272	313	18,599	17,767
Retired	18,065	18,827	-	-	443	457	17,621	18,370	10,346	10,575	7,719	8,252
Other activity	3,658	3,753	806	762	2,326	2,494	526	497	1,772	1,683	1,886	2,070
Want a job now	4,494	4,600	1,418	1,382	2,482	2,583	594	635	1,657	1,744	2,837	2,856
Reason for not looking												
School attendance	1,118	1,062	893	824	220	232	5	5	550	527	568	535
Ill health, disability	801	824	55	40	600	610	146	173	391	417	410	407
Home responsibility	911	952	172	186	703	718	36	48	-	-	911	952
Think cannot get a job	742	781	117	145	457	455	168	182	328	370	415	411
Other reasons[1]	922	982	181	187	501	568	240	227	389	430	534	551
Black												
Total not in labor force	8,067	8,386	2,097	2,119	3,032	3,242	2,938	3,025	2,997	3,167	5,070	5,220
Do not want a job now	6,640	6,980	1,565	1,583	2,236	2,481	2,840	2,915	2,473	2,658	4,168	4,322
Current activity												
Going to school	1,101	1,219	985	1,081	116	137	-	1	521	566	581	653
Ill, disabled	1,000	1,152	38	39	580	700	382	413	485	565	516	588
Keeping house	2,047	1,999	266	214	1,013	1,075	769	709	96	105	1,951	1,894
Retired	1,702	1,805	-	-	46	56	1,656	1,749	926	985	776	820
Other activity	790	805	275	249	481	514	33	43	444	438	345	368

[Continued]

★ 791 ★

Unemployed by Reason and Characteristics, Annual Averages, 1992 and 1993

[Continued]

Reason, race, and Hispanic origin	Total		Age						Sex			
			16 to 24 years		25 to 59 years		60 years and over		Men		Women	
	1992	1993	1992	1993	1992	1993	1992	1993	1992	1993	1992	1993
Want a job now	1,427	1,406	532	536	797	761	98	110	524	508	903	898
Reason for not looking												
School attendance	389	382	303	300	86	82	-	-	168	178	221	204
Ill health, disability	245	256	18	19	193	197	34	40	113	116	131	140
Home responsibility	282	309	77	94	202	210	4	5	-	-	282	309
Think cannot get a job	306	284	83	67	200	176	23	40	144	124	162	160
Other reasons[1]	205	176	51	56	116	96	37	24	98	90	107	86
Hispanic origin												
Total not in labor force	5,113	5,377	1,333	1,417	2,399	2,478	1,380	1,482	1,478	1,569	3,635	3,808
Do not want a job now	4,397	4,610	1,085	1,148	1,980	2,023	1,332	1,439	1,218	1,292	3,179	3,317
Current activity												
Going to school	694	737	610	663	84	72	-	2	319	331	375	406
Ill, disabled	441	473	20	16	294	303	127	154	237	245	204	228
Keeping house	2,154	2,177	324	340	1,349	1,393	480	444	35	45	2,119	2,132
Retired	723	832	-	-	14	13	709	819	417	490	307	343
Other activity	384	391	129	129	239	242	15	19	209	182	174	209
Want a job now	742	750	262	259	428	439	52	52	262	261	480	489
Reason for not looking												
School attendance	180	167	145	125	34	41	1	1	79	76	100	91
Ill health, disability	111	104	7	5	90	81	14	18	55	56	56	48
Home responsibility	174	200	39	48	131	149	4	4	-	-	174	200
Think cannot get a job	160	140	38	31	104	91	18	19	71	61	88	79
Other reasons[1]	117	138	33	51	69	77	15	10	56	67	61	71

Source: "Persons Not in Labor Force by Reason, Race, Hispanic Origin, Age, and Sex." *Employment and Earnings* 41 (January 1994): 225. *Notes:* Detail for the above race and Hispanic-origin groups will not sum to totals because data for the other races are not presented and Hispanics are included in both the white and black population groups. 1. Includes small number of men not looking for work because of "home responsibilities."

★ 792 ★

Unemployment

Unemployment Rate by Race and Age, 1983 and 1991

Age cohort	1983		1991	
	African American	White	African American	White
20-24 years	31.6	12.1	21.6 (-31.6%)	9.2 (-24.0%)
35-44 years	12.4	6.3	8.6 (-30.6%)	4.7 (-25.4%)
Total population	19.5	8.4	12.4 (-36.4%)	6.0 (-28.6%)

Source: "Unemployment Rate by Race and Selected Age Cohorts: 1983 and 1991." *The State of Black America 1994*, p. 223. Primary source: Prepared by the National Urban League from U.S. Bureau of the Census, *Statistical Abstract of the United States: 1985* (105th edition), Table 658, and *Statistical Abstract of the United States: 1992* (112th edition), Table 622, Washington, DC, 1983 and 1992.

★ 793 ★

Unemployment

Unemployment Rates: Trends, 1992 and 1993

Civilian workers.

Sex and age	Total		White		Black		Hispanic origin	
	1992	1993	1992	1993	1992	1993	1992	1993
Total, 16 years and over	7.4	6.8	6.5	6.0	14.1	12.9	11.4	10.6
16 to 19 years	20.0	19.0	17.1	16.2	39.8	38.9	27.5	26.2
16 to 17 years	23.0	21.3	20.1	19.0	44.8	39.7	35.7	35.3
18 to 19 years	18.1	17.5	15.1	14.3	37.1	38.4	23.4	21.9
20 to 24 years	111.3	10.5	9.4	8.7	23.9	22.0	13.2	13.1
25 years and over	6.1	5.6	5.5	5.0	10.9	9.8	9.8	9.0
25 to 54 years	6.3	5.8	5.6	5.2	11.6	10.2	9.9	9.1
55 years and over	4.8	4.3	4.6	4.0	6.0	6.7	8.6	7.7
Men, 16 years and over	7.8	7.1	6.9	6.2	15.2	13.8	11.5	10.4
16 to 19 years	21.5	20.4	18.4	17.6	42.0	40.1	28.2	26.1
16 to 17 years	24.4	22.8	21.3	20.1	47.5	42.7	36.6	34.6
18 to 19 years	19.5	18.8	16.4	15.9	39.1	38.6	24.1	22.0
20 to 24 years	12.2	11.3	10.4	9.5	24.5	23.0	13.7	12.6
25 years and over	6.4	5.8	5.8	5.2	11.7	10.5	9.8	8.8
25 to 54 years	6.6	5.9	5.9	5.3	12.3	10.8	9.8	8.9
55 years and over	5.2	4.7	4.9	4.2	7.4	8.4	9.9	8.7
Women, 16 years and over	6.9	6.5	6.0	5.7	13.0	12.0	11.3	10.9
16 to 19 years	18.5	17.4	15.7	14.6	37.2	37.5	26.4	26.4
16 to 17 years	21.4	19.6	18.9	17.7	41.7	36.1	34.4	36.2
18 to 19 years	16.5	16.0	13.6	12.5	34.8	38.2	22.4	21.7
20 to 24 years	10.2	9.6	8.3	7.8	23.1	20.9	12.4	14.0
25 years and over	5.7	5.4	5.1	4.8	10.1	9.1	9.7	9.1
25 to 54 years	6.0	5.6	5.2	5.0	10.8	9.7	10.1	9.4
55 years and over	4.2	3.8	4.1	3.7	4.4	4.9	6.6	6.4

Source: "Unemployment Rates by Sex, Age, Race, and Hispanic Origin." *Employment and Earnings* 41 (January 1994): 232.

★ 794 ★

Unemployment

Unemployment in Families by Type of Family and Presence of Employed Family Members, 1992 and 1993

Numbers in thousands.

Type of family, race, and Hispanic origin	1992					1993				
		With unemployment					With unemployment			
			Percent of families					Percent of families		
	Total families	Total	With no employed person in family	With at least one employed person in family	With at least one person in family employed full time	Total families	Total	With no employed person in family	With at least one employed person in family	With at least one person in family employed full time
Total										
Total families	66,785	6,643	30.1	69.9	61.6	67,378	6,178	30.4	69.6	61.8
White										
Total families	56,925	5,046	27.4	72.6	64.2	57,258	4,684	27.0	73.0	64.7
With children under 18 years of age	26,788	2,746	30.0	70.0	61.8	26,935	2,560	29.5	70.5	62.3
Married-couple families	46,627	3,850	20.5	79.5	71.1	46,820	3,495	18.9	81.1	73.0
With children under 18 years of age	21,174	2,154	20.0	80.0	71.4	21,163	1,945	18.1	81.9	73.4
Families maintained by women	7,773	865	51.3	48.7	39.8	7,860	865	52.5	47.5	38.2
With children under 18 years of age	4,507	473	66.0	34.0	26.6	4,572	477	65.0	35.0	27.0
Families maintained by men	2,526	330	45.5	54.5	48.2	2,578	324	46.9	53.1	46.0
With children under 18 years of age	1,107	119	68.9	31.1	28.6	1,200	138	68.1	31.9	27.5
Black										
Total families	7,624	1,325	41.2	58.8	50.7	7,725	1,216	43.7	56.3	50.2
With children under 18 years of age	4,389	760	48.7	51.3	44.9	4,436	696	51.6	48.4	43.7
Married-couple families	3,499	541	19.8	80.2	71.7	3,495	500	20.6	79.4	72.6
With children under 18 years of age	1,803	319	18.8	81.2	73.7	1,779	287	17.1	82.9	76.7
Families maintained by women	3,577	684	57.5	42.5	34.9	3,673	614	61.1	38.9	33.1
With children under 18 years of age	2,355	407	70.7	29.3	23.9	2,419	369	75.7	24.3	20.3
Families maintained by men	548	100	46.0	54.0	45.0	556	102	52.0	48.0	43.1
With children under 18 years of age	230	35	67.6	32.4	29.4	238	39	76.9	23.1	20.5
Hispanic origin										
Total families	5,166	822	34.3	65.7	58.8	5,373	796	307	69.3	61.7
With children under 18 years of age	3,334	545	38.4	61.6	54.8	3,416	525	34.1	65.9	58.3
Married-couple families	3,4884	557	26.6	73.4	66.6	3,636	545	21.8	78.2	70.1
With children under 18 years of age	2,294	393	27.5	72.5	65.4	2,341	386	23.3	76.7	68.9
Families maintained by women	1,244	186	54.8	45.2	36.6	1,302	178	52.8	47.2	40.4
With children under 18 years of age	864	123	67.5	32.5	26.0	880	109	63.3	36.7	29.4
Families maintained by men	438	79	39.2	60.8	55.7	436	73	42.5	57.5	52.1
With children under 18 years of age	176	29	1	1	1	195	30	1	1	1

Source: "Unemployment in Families by Type of Family, Race, Hispanic Origin, and Presence of Employed Family Member." *Employment and Earnings* 41 (January 1994): 236.
Notes: Detail for the above race and Hispanic-origin groups will not sum to totals because data for the "other races" group are not presented and Hispanics are included in both the white and black population groups. 1. Data not shown where base is less than 35,000.

★ 795 ★

Unemployment

Work-seeking Intentions and Work History of Persons Not in the Labor Force, 1992 and 1993

In thousands.

Work-seeking intentions, work history, and sex	Total		Race			
			White		Black	
	1992	1993	1992	1993	1992	1993
Total						
Do not intend to seek work	55,138	56,004	46,818	47,317	6,347	6,622
Intend to seek work in the next 12 months	9,455	9,506	7,310	7,238	1,720	1,764
Never worked	1,620	1,647	1,131	1,107	390	429
Last worked over 5 years ago	1,111	1,204	827	866	245	277
Last worked 1 to 5 years ago	2,394	2,496	1,810	1,879	482	489
Worked during the previous 12 months	4,330	4,159	3,542	3,387	604	568
Men						
Do not intend to seek work	18,747	19,297	15,711	16,084	2,367	2,533
Intend to seek work in the next 12 months	3,609	3,690	2,812	2,834	629	633
Never worked	730	723	520	500	170	169
Last worked over 5 years ago	253	286	180	193	62	75
Last worked 1 to 5 years ago	751	779	569	584	148	144
Worked during the previous 12 months	1,875	1,903	1,543	1,557	250	245
Women						
Do not intend to seek work	36,391	36,706	31,107	31,233	3,980	4,089
Intend to seek work in the next 12 months	5,846	5,816	4,498	4,404	1,091	1,131
Never worked	890	925	611	606	220	261
Last worked over 5 years ago	858	918	647	673	183	202
Last worked 1 to 5 years ago	1,643	1,717	2,241	1,295	334	345
Worked during the previous 12 months	2,456	2,256	1,999	1,830	354	323

Source: "Work-seeking Intentions of Persons Not in the Labor Force and Work History of Those Who Intend to Seek Work Within the Next 12 Months by Sex, Age, and Race." *Employment and Earnings* 41 (January 1994): 227.

Unions

★ 796 ★

Labor Unions by Selected Characteristics, 1983 and 1992. Part 1

Annual averages of monthly data. Covers employed wage and salary workers 16 years old and over. Excludes self-employed workers whose businesses are incorporated although they technically qualify as wage and salary workers. Based on Current Population Survey.

| Characteristic | Employed wage and salary workers | | | | | | | | | |
| | Total (1,000) | | Union members[1] (1,000) | | Represented by unions[2] (1,000) | | Percent union members | | Percent represented by union | |
	1983	1992	1983	1992	1983	1992	1983	1992	1983	1992
Total	88,290	103,688	17,717	16,390	20,532	18,540	20.1	15.8	23.3	17.9
White	77,046	88,624	14,844	13,416	17,182	15,148	19.3	15.1	22.3	17.1
Men	42,168	46,732	10,134	8,516	11,364	9,349	24.0	18.2	26.9	20.0
Women	34,877	41,892	4,710	4,900	5,818	5,799	13.5	11.7	16.7	13.8
Black	8,979	11,416	2,440	2,433	2,850	2,763	27.2	21.3	31.7	24.2
Men	4,477	5,480	1,420	1,309	1,615	1,448	31.7	23.9	36.1	26.4
Women	4,502	5,936	1,020	1,125	1,235	1,315	22.7	19.0	27.4	22.1
Hispanic[3]	(NA)	8,341	(NA)	1,244	(NA)	1,415	(NA)	14.9	(NA)	17.0
Men	(NA)	4,954	(NA)	834	(NA)	926	(NA)	16.8	(NA)	18.7
Women	(NA)	3,386	(NA)	410	(NA)	490	(NA)	12.1	(NA)	14.5

Source: "Union Members, by Selected Characteristics: 1983 and 1992." U.S. Bureau of the Census, *Statistical Abstract of the United States*, 1993, p. 436. Primary source: U.S. Bureau of Labor Statistics, *Employment and Earnings*, January issues. *Notes:* NA Not available. 1. Members of a labor union or an employee association similar to a labor union. 2. Members of a labor union or an employee association similar to a union as well as workers who report no union affiliation but whose jobs are covered by a union or an employee association contract. 3. Persons of Hispanic origin may be of any race.

★ 797 ★
Unions

Labor Unions by Selected Characteristics, 1983 and 1992. Part 2

Annual averages of monthly data. Covers employed wage and salary workers 16 years old and over. Excludes self-employed workers whose businesses are incorporated although they technically qualify as wage and salary workers. Based on Current Population Survey.

| Characteristic | Median usual weekly earnings (dols.)[3] | | | | | | | | | | | |
| | Total | | | Union members[1] | | | Represented by unions[2] | | | Not represented by unions | | |
	1983	1989	1992	1983	1989	1992	1983	1989	1992	1983	1989	1992
Total	313	399	445	388	497	547	383	494	541	288	372	413
White	319	409	462	396	506	568	391	503	562	295	384	426
Men	387	482	518	423	539	601	421	537	599	362	452	495
Women	254	334	388	314	427	496	313	423	492	240	317	370
Black	261	319	357	331	425	468	324	423	464	222	290	322
Men	293	348	380	366	478	498	360	470	494	244	305	338
Women	231	301	336	292	385	423	287	390	422	209	276	311
Hispanic[4]	(NA)	296	324	(NA)	420	481	(NA)	417	472	(NA)	276	303

[Continued]

★ 797 ★

Labor Unions by Selected Characteristics, 1983 and 1992. Part 2

[Continued]

| Characteristic | Median usual weekly earnings (dols.)[3] | | | | | | | | | | | |
| | Total | | | Union members[1] | | | Represented by unions[2] | | | Not represented by unions | | |
	1983	1989	1992	1983	1989	1992	1983	1989	1992	1983	1989	1992
Men	(NA)	315	345	(NA)	457	511	(NA)	451	504	(NA)	291	313
Women	(NA)	269	303	(NA)	369	397	(NA)	368	394	(NA)	255	289

Source: "Union Members, by Selected Characteristics: 1983 and 1992." U.S. Bureau of the Census, *Statistical Abstract of the United States*, 1993, p. 436. Primary source: U.S. Bureau of Labor Statistics, *Employment and Earnings*, January issues. *Notes:* NA Not available. 1. Members of a labor union or an employee association similar to a labor union. 2. Members of a labor union or an employee association similar to a union as well as workers who report no union affiliation but whose jobs are covered by a union or an employee association contract. 3. For full-time employed wage and salary workers; 1983 revised since originally published. 4. Persons of Hispanic origin may be of any race.

★ 798 ★

Unions

Median Weekly Earnings of Union and Non-union Employees, 1992 and 1993

| Age, sex, race, and Hispanic origin | 1992 | | | | 1993 | | | |
	Total	Members of unions[1]	Represented by unions[2]	Non-union	Total	Members of unions[1]	Represented by unions[2]	Non-union
Race, Hispanic origin, and sex								
White								
16 years and over	462	568	562	426	478	589	585	444
Men	518	601	599	495	531	619	618	505
Women	388	496	492	370	403	514	510	382
Black								
16 years and over	357	468	464	322	370	490	485	330
Men	380	498	494	338	392	514	510	345
Women	336	423	422	311	349	454	447	320
Hispanic origin								
16 years and over	324	481	472	303	335	481	478	311
Men	345	511	504	313	352	511	509	318
Women	303	397	394	289	314	413	415	297

Source: "Median Weekly Earnings of Full-time Wage and Salary Workers by Age, Sex, Race, Hispanic Origin, and Union Affiliation." *Employment and Earnings* 41 (January 1, 1994), p. 250. *Notes:* Data refer to the sole or principal job of full-time workers. Excluded are self-employed workers whose businesses are incorporated although they technically qualify as wage and salary workers. Detail for the above race and Hispanic-origin groups will not sum to totals because data for the "other races" group are not presented and Hispanics are included in both the white and black population groups. 1. Data refer to members of a labor union or an employee association similar to a union. 2. Data refer to members of a labor union or an employee association similar to a union as well as workers who report no union affiliation but whose jobs are covered by a union or an employee association contract.

Vietnam-era Veterans Employed

★ 799 ★

Employment Status of Male Vietnam-era Veterans and Nonveterans, Annual Averages, 1992 and 1993

Numbers in thousands.

Employment status and age	Veterans						Nonveterans					
	White		Black		Hispanic origin		White		Black		Hispanic origin	
	1992	1993	1992	1993	1992	1993	1992	1993	1992	1993	1992	1993
Total, 35 to 49 years												
Civilian noninstitutional population	5,597	5,312	557	515	252	215	16,564	17,380	1,939	2,050	1,793	1,965
Civilian labor force	5,280	4,987	491	456	233	199	15,616	16,356	1,648	1,726	1,633	1,788
Employed	5,002	4,736	444	418	217	183	14,792	15,591	1,460	1,549	1,474	1,627
Unemployed	279	251	47	38	16	16	824	765	188	177	159	161
Unemployment rate	5.3	5.0	9.6	8.3	7.0	8.3	5.3	4.7	11.4	10.2	9.8	9.0
35 to 39 years												
Civilian noninstitutional population	796	623	119	94	50	31	7,516	7,745	858	891	805	874
Civilian labor force	742	581	106	84	46	28	7,169	7,359	743	771	747	813
Employed	689	533	99	74	41	26	6,765	7,000	644	687	671	737
Unemployed	52	48	7	9	5	3	404	359	99	84	75	76
Unemployment rate	7.1	8.2	6.7	11.0	11.4	[1]	5.6	4.9	13.3	10.9	10.1	9.3
40 to 44 years												
Civilian noninstitutional population	2,403	2,036	242	197	111	91	5,300	5,756	634	703	592	626
Civilian labor force	2,266	1,914	212	175	101	83	5,000	5,428	536	588	544	570
Employed	2,146	1,814	187	157	94	76	4,753	5,188	488	532	494	523
Unemployed	120	100	25	19	7	7	247	240	48	56	49	48
Unemployment rate	5.3	5.2	11.6	10.6	7.2	8.8	4.9	4.4	8.9	9.5	9.1	8.3
45 to 49 years												
Civilian noninstitutional population	2,398	2,653	197	224	91	94	3,749	3,879	447	457	396	466
Civilian labor force	2,272	2,492	172	196	86	88	3,448	3,569	369	366	343	405
Employed	2,166	2,389	157	187	82	81	3,275	3,402	327	329	308	367
Unemployed	106	103	15	10	4	6	173	167	41	37	35	38
Unemployment rate	4.7	4.1	8.8	5.1	4.4	7.3	5.0	4.7	11.2	10.1	10.1	9.4

Source: "Employment Status of Male Vietnam-era Veterans and Nonveterans, by Age, Race, and Hispanic Origin." *Employment and Earnings* 41 (January 1994): 235. *Notes:* Male Vietnam-era veterans are men who served in the Armed Forces between August 5, 1964 and May 7, 1975. Nonveterans are men who have never served in the Armed Forces. Detail for the above race and Hispanic-origin groups will not to totals because data for the "other races" group are not presented and Hispanics are included in both the white and black population groups. 1. Data not shown where base is less than 35,000.

★ 800 ★

Vietnam-era Veterans Employed

Employment Status of Male Vietnam-era Veterans and Nonveterans: Quarterly Averages, 1992 and 1993

Numbers in thousands. Not seasonally adjusted.

Employment status and age	Veterans						Nonveterans					
	White		Black		Hispanic origin		White		Black		Hispanic origin	
	IV 1992	IV 1993	IV 1992	IV 1993	IV 1992	IV 1993	IV 1992	IV 1993	IV 1992	IV 1993	IV 1992	IV 1993
Total, 35 to 49 years												
Civilian noninstitutional population	5,501	5,173	533	483	252	234	16,900	17,677	2,017	2,114	1,878	1,952
Civilian labor force	5,165	4,845	457	426	223	220	15,963	16,628	1,707	1,756	1,747	1,771
Employed	4,899	4,616	410	392	207	192	15,184	15,956	1,532	1,576	1,604	1,632
Unemployed	266	228	47	33	15	28	779	672	174	179	143	139
Unemployment rate	5.2	4.7	10.3	7.8	6.9	12.7	4.9	4.0	10.2	10.2	8.2	7.9
35 to 39 years												
Civilian noninstitutional population	721	548	110	87	39	21	7,633	7,820	890	886	817	871
Civilian labor force	660	516	97	75	35	19	7,269	7,405	768	748	776	808
Employed	610	483	89	67	28	15	6,890	7,083	665	672	716	753
Unemployed	50	33	7	7	7	5	380	322	103	76	60	55
Unemployment rate	7.6	6.4	7.7	9.9	20.1	[1]	5.2	4.3	13.4	10.1	7.8	6.8
40 to 44 years												
Civilian noninstitutional population	2,255	1,916	221	164	113	99	5,466	5,917	666	753	649	628
Civilian labor force	2,107	1,789	186	144	97	97	5,186	5,566	558	629	589	568
Employed	1,986	1,682	162	128	93	83	4,954	5,350	517	571	540	507
Unemployed	121	107	25	16	4	14	232	216	41	58	49	61
Unemployment rate	5.7	6.0	13.2	11.1	4.2	14.2	4.5	3.9	7.3	9.2	8.3	10.7
45 to 49 years												
Civilian noninstitutional population	2,525	2,709	202	231	100	114	3,802	3,940	462	476	412	453
Civilian labor force	2,398	2,540	174	207	91	104	3,508	3,657	381	380	382	396
Employed	2,303	2,451	159	197	86	94	3,341	3,523	350	334	348	372
Unemployed	95	88	15	10	4	10	167	134	31	46	34	24
Unemployment rate	4.0	3.5	8.6	4.7	4.7	9.3	4.8	3.7	8.1	12.1	8.9	6.0

Source: "Employment Status of Male Vietnam-era Veterans and Nonveterans by Age, Race, and Hispanic Origin." *Employment and Earnings* 41 (January 1994), p. 73.
Notes: Male Vietnam-era veterans are men who served in the Armed Forces between August 5, 1964 and May 7, 1975. Nonveterans are men who have never served in the Armed Forces. Detail for the above race and Hispanic-origin groups will not sum to totals because data for the "other races" group are not presented and Hispanics are included in both the white and black population groups. 1. Data not shown where base is less than 60,000.

Women Unemployed

★ 801 ★

Unemployment Rate for Women, by Race and Age, 1983 and 1991

Age cohort	1983		1991	
	African American	White	African American	White
20-24 years	31.8	10.3	20.7 (-34.9%)	8.0 (-22.3%)
35-44 years	11.4	6.2	7.6 (-33.3%)	4.3 (-30.6%)
Total population	18.6	7.9	11.9 (-36.0%)	5.5 (-30.4%)

Source: "Female Unemployment Rate by Race and Selected Age Cohorts: 1983 and 1991." *The State of Black America 1994*, p. 224. Primary source: Prepared by the National Urban League from U.S. Bureau of the Census, *Statistical Abstract of the United States: 1985* (105th edition), Table 658, and *Statistical Abstract of the United States: 1992* (112th edition), Table 622, Washington, DC, 1983 and 1992.

Women in the Labor Force

★ 802 ★

Labor Force Participation Rates for Wives, Husband Present, and Youngest Child, 1975 to 1992

As of March, except as indicated. For civilian noninstitutional population, 16 years old and over. Based on Current Population Survey.

Presence and age of child	Total				White				Black			
	1975	1985	1990	1992	1975	1985	1990	1992	1975	1985	1990	1992
Wives, total	44.4	54.2	58.2	59.3	43.6	53.3	57.6	58.7	54.1	63.8	64.7	66.8
No children under 18	43.8	48.2	51.1	51.9	43.6	43.6	47.5	50.8	47.6	55.2	52.9	56.8
With children under 18	44.9	60.8	66.3	67.8	43.6	59.9	65.6	67.3	58.4	71.7	75.6	76.0
Under 6, total	36.7	53.4	58.9	59.9	34.7	52.1	57.8	59.1	54.9	69.6	73.1	73.5
Under 3	32.7	50.5	55.5	57.5	30.7	49.4	54.9	56.7	50.1	66.2	67.5	70.8
1 year or under	30.8	49.4	53.9	56.7	29.2	48.6	53.3	55.9	50.0	63.7	64.4	58.7
2 years	37.1	54.0	60.9	60.9	35.1	52.7	60.3	59.8	56.4	69.9	75.4	75.0
3 to 5 years	42.2	58.4	64.1	63.5	40.1	56.6	62.5	62.7	61.2	73.8	80.4	77.1
3 years	41.2	55.1	63.1	63.5	39.0	52.7	62.3	61.8	62.7	72.3	74.5	77.6
4 years	41.2	59.7	65.1	62.1	38.7	58.4	63.2	61.0	64.9	70.6	80.6	76.9
5 years	44.4	62.1	64.5	65.7	43.8	59.9	62.0	65.2	56.3	79.1	86.2	80.3

[Continued]

★ 802 ★

Labor Force Participation Rates for Wives, Husband Present, and Youngest Child, 1975 to 1992

[Continued]

Presence and age of child	Total				White				Black			
	1975	1985	1990	1992	1975	1985	1990	1992	1975	1985	1990	1992
6 to 13 years	51.8	68.2	73.0	74.9	50.7	67.7	72.6	74.5	65.7	73.3	77.6	79.7
14 to 17 years	53.5	67.0	75.1	76.6	53.4	66.6	74.9	76.9	52.3	74.4	78.8	74.0

Source: "Labor Force Participation Rates for Wives, Husband Present, by Age of Own Youngest Child: 1975 to 1992," U.S. Bureau of the Census, *Statistical Abstract of the United States*, 1993, p. 400. Primary source: U.S. Bureau of Labor Statistics, Bulletin 2340; and unpublished data.

★ 803 ★

Women in the Labor Force

Percent of Women Employed by Race and Age, 1980 and 1990

Age cohort	1980		1990	
	African American[1]	White	African American	White
20-24 years	56.5	68.5	63.2 (+11.9%)	71.7 (+4.7%)
35-44 years	69.1	65.6	79.0 (+14.3%)	76.6 (+16.8%)
Total population	52.0	51.0	58.2 (+11.9%)	57.1 (+12.0%)

Source: "Percent of Females in Labor Force by Race and Selected Age Cohorts: 1980 and 1990," *The State of Black America 1994*, p. 221. Primary source: Prepared by the National Urban League from U.S. Bureau of Labor Statistics, *Employment and Earnings*, Vol. 28, No. 4, April 1981, and Vol. 37, No. 4, April 1990, Table A-4, Washington, DC, 1985 and 1993. *Note:* 1. Figures for "black and other."

Work Schedules

★ 804 ★

Workers and Job-Related Work at Home, 1991

As of May. For persons 16 years old and over doing job-related work at home as part of their primary job in nonagriculture industries.

Characteristic	Total at work (1,000)	Persons doing job related work at home						
		Total			Class of worker--percent[4]			
		Number (1,000)[1]	Rate[2]	Mean hours worked (number)[3]	Wage and salary workers			Self-employed
					Total[1]	Paid for the work	Not paid for the work	
Total, 1991	109,126	19,967	18.3	9.1	71.6	9.4	60.9	27.8
Race and Hispanic origin								
White	94,387	18,520	19.6	9.0	71.2	9.6	60.3	28.3
Black	11,020	970	8.8	10.4	82.5	7.5	73.5	17.1
Hispanic[5]	7,977	667	8.4	9.8	69.6	8.5	58.8	29.7

Source: "Workers Doing Job-Related Work at Home, by Selected Characteristics: 1991," U.S. Bureau of the Census, *Statistical Abstract of the United States*, 1993, p. 404. Primary source: U.S. Bureau of Labor Statistics, *Monthly Labor Review*, forthcoming article. *Notes:* 1. Includes those that did not report pay status and unpaid family members. 2. Persons working at home as a percent of the total work at home. 3. For definition of mean, see Guide to Tabular Presentation. 4. Excludes unpaid family members. 5. Persons of Hispanic origin may be of any race.

★ 805 ★

Work Schedules

Workers on Flexible Schedules, 1985 and 1991 and by Characteristics, 1991

In thousands, except percent. As of May. For employed persons 16 years old and over who usually work full-time and who were at work during the survey reference week. A flexible schedule allows workers to vary the time they begin and end their work day. Based on Current Population Survey.

Characteristic	All workers			Workers with flexible schedules					
				Number			Percent		
	Total	Male	Female	Total	Male	Female	Total	Male	Female
Total, 1985	73,395	43,779	29,616	9,061	5,760	3,300	12.3	13.2	11.1
Total, 1991	80,452	46,308	34,145	12,118	7,168	4,950	15.1	15.5	14.5
Race									
White	68,795	40,267	28,528	10,630	6,416	4,214	15.5	15.9	14.8
Black	8,943	4,522	4,421	1,083	525	558	12.1	11.6	12.6
Hispanic[1]	6,598	4,172	2,425	702	427	275	10.6	10.2	11.3

Source: "Workers on Flexible Schedules, 1985 and 1991 and by Selected Characteristics: 1991," U.S. Bureau of the Census, *Statistical Abstract of the United States*, 1993, p. 403. Primary source: U.S. Bureau of Labor Statistics, *News*, USDL 92-491, August 4, 1992; and unpublished data. *Note:* 1. Persons of Hispanic origin may be of any race.

★ 806 ★

Work Schedules

Workers on Shift Schedules, 1985 and 1991

Characteristic	Total employed[1]	Work schedules--Percent distribution						
		Regular daytime schedules	Shift workers					
			Total	Evening	Night	Rotating	Irregular[2]	Other
Total, 1985[3]	73,395	84.1	15.9	6.3	2.7	4.3	(NA)	2.6
Total, 1991[3]	80,452	81.8	17.8	5.1	3.7	3.4	3.7	2.0
Race and Hispanic origin								
White	68,795	82.6	17.1	4.6	3.4	3.3	3.8	2.1
Black	8,943	76.0	23.3	8.4	5.6	4.7	2.9	1.8
Hispanic[4]	6,598	80.3	19.1	6.4	4.6	2.7	3.1	2.4

Source: "Workers on Shift Schedules, by Selected Characteristics: 1985 and 1991," U.S. Bureau of the Census, *Statistical Abstract of the United States*, 1993, p. 404. Primary source: U.S. Bureau of Labor Statistics, *News*, USDL 92-491, August 4, 1992; and unpublished data. *Notes:* 1. Includes a small number of workers who did not report data on shift work. 2. Employer arranged. 3. Data for 1985 are not strictly comparable to those for 1991 because of the addition of the "irregular" category in the May 1991 survey. Includes other races, not shown separately. 4. Persons of Hispanic origin may be of any race.

Work-Related Training

★ 807 ★

Persons Receiving Work-Related Training, 1990

In thousands, except as indicated. As of spring. For persons 18 to 64 years old. Based on Survey of Income and Program Participation.

Characteristic	Total	Race		Hispanic origin[1]
		White	Black	
All persons	152,815	129,575	17,891	12,463
Persons receiving work training	39,238	33,984	4,241	2,218
Uses training on current or most recent job	26,563	23,595	2,316	1,486
Location				
Apprenticeship	1,749	1,616	76	114
Business/vo-tech school	10,213	8,659	1,301	537
Community college	4,077	3,722	275	220
Four-year college	2,738	2,488	198	89
High school vo-tech program	2,158	1,875	250	183
Training program at work	13,330	11,774	1,164	598
Military	2,229	1,968	180	139
Previous job	1,821	1,636	134	134
Other	9,720	8,304	1,169	497

[Continued]

★ 807 ★

Persons Receiving Work-Related Training, 1990

[Continued]

Characteristic	Total	Race		Hispanic origin[1]
		White	Black	
Program paid for by--				
Self or family	11,540	10,348	914	588
Employer	17,834	16,059	1,379	880
Federal, State, or local government	10,429	8,079	2,010	748
Someone else	1,000	859	99	84
Length of training program (average number of weeks)	22	23	21	22

Source: "Persons Receiving Work-Related Training, by Selected Characteristics: 1990," U.S. Bureau of the Census, *Statistical Abstract of the United States*, 1993, p. 423. Primary source: U.S. Bureau of the Census, *Current Population Reports*, series P70- 32. *Note:* 1. Persons of Hispanic origin may be of any race.

Chapter 11
MILITARY AFFAIRS

★ 808 ★

Armed Forces Overseas, by Age, Sex, Race, and Hispanic Origin, 1992

Age	Race						Hispanic origin[1]		
	White			Black					
	Total	Male	Female	Total	Male	Female	Total	Male	Female
All ages	276,896	247,706	29,190	90,851	74,200	16,651	19,254	17,204	2,050
15 to 19 years	14,123	12,627	1,496	4,055	3,371	684	1,031	915	116
15 years	-	-	-	-	-	-	-	-	-
16 years	-	-	-	-	-	-	-	-	-
17 years	3	3	-	2	1	1	-	-	-
18 years	2,210	1,939	271	701	572	129	192	168	24
19 years	11,910	10,685	1,225	3,352	2,798	554	839	747	92
20 to 24 years	95,314	84,990	10,324	27,692	22,533	5,159	6,979	6,127	852
25 to 29 years	62,983	56,012	6,971	23,015	18,361	4,654	3,968	3,551	437
30 to 34 years	48,677	43,222	5,455	18,904	15,277	3,627	3,502	3,120	382
35 to 39 years	33,825	30,381	3,444	12,287	10,367	1,920	2,516	2,327	189
40 to 44 years	15,638	14,488	1,150	4,068	3,548	520	990	927	63
45 to 49 years	5,166	4,884	282	742	660	82	209	199	10
50 to 54 years	935	873	62	82	77	5	30	29	1
55 to 59 years	204	200	4	5	5	-	8	8	-
60 to 64 years	31	29	2	1	1	-	1	1	-

Source: "Estimates of Armed Forces Overseas, by Age, Sex, Race, and Hispanic Origin," *Population Projections of the United States, by Age, Sex, Race, and Hispanic Origin: 1993 to 2050,* Current Population Reports, P25-1104, pp. D2 and D-3. *Notes:* For data about Native Americans, Asians, and Pacific Islanders, see original. 1. Persons of Hispanic origin may be of any race.

★ 809 ★

Armed Forces

Black Senior Officers in the Department of Defense, 1992

[All services—numbers and percentages as of June 30, 1992].

Rank	Army	Air Force	Marine Corps	Navy	DoD Overall
Flag/General (O-7 to O-10)	25 6.5%	6 1.9%	1 1.5%	3 1.2%	35 3.4%
Captain/Colonel (O-6)	219 4.6%	98 1.9%	11 1.7%	39 1.1%	367 2.6%
Commander/Lt. Colonel (O-5)	662 6.3%	400 3.3%	46 2.9%	191 2.4%	1,299 4.1%
Lt. Commander? Major (O-4)	1,935 10.8%	1,199 6.6%	135 4.3%	448 3.3%	3,717 7.1%

Source: "Black Senior Officers in the Department of Defense." Stillwell, Paul, ed., *The Golden Thirteen: Recollections of the First Black Naval Officers,* Annapolis, Md.: Naval Institute Press, 1993. Primary source: The Special Assistant for Minority Affairs, Navy Recruiting Command.

★ 810 ★

Armed Forces

Resident Armed Forces, by Age, Sex, Race, and Hispanic Origin, 1992

Age	Race White Total	Race White Male	Race White Female	Race Black Total	Race Black Male	Race Black Female	Hispanic origin[1] Total	Hispanic origin[1] Male	Hispanic origin[1] Female
All ages	1,225,638	1,106,697	116,941	309,101	258,004	51,097	83,014	74,612	8,402
15 to 19 years	89,614	80,034	9,580	19,328	16,113	3,215	8,177	7,203	974
15 years	-	-	-	-	-	-	-	-	-
16 years	-	-	-	-	-	-	-	-	-
17 years	2,250	1,881	369	560	434	126	231	189	42
18 years	24,654	21,807	2,847	5,230	4,268	962	2,359	2,030	329
19 years	62,710	56,346	6,364	13,538	11,411	2,127	5,587	4,964	603
20 to 24 years	397,895	358,892	39,003	101,060	84,101	16,959	31,773	26,323	3,450
25 to 29 years	263,947	237,048	26,899	73,568	60,130	13,438	16,417	14,676	1,741
30 to 34 years	207,279	187,279	19,847	57,117	47,222	9,895	12,585	11,401	1,184
35 to 39 years	152,471	136,399	14,072	39,276	33,725	5,553	9,132	8,416	716
40 to 44 years	79,593	73,785	5,806	15,033	13,343	1,690	3,913	3,637	276
45 to 49 years	26,328	26,929	1,399	3,141	2,848	293	858	812	46
50 to 54 years	5,390	5,110	280	526	475	53	135	124	11

[Continued]

★ 810 ★

Resident Armed Forces, by Age, Sex, Race, and Hispanic Origin, 1992
[Continued]

Age	Race						Hispanic origin[1]		
	White			Black					
	Total	Male	Female	Total	Male	Female	Total	Male	Female
55 to 59 years	1,027	961	46	41	40	1	18	14	4
60 to 64 years	247	240	7	7	7	-	6	6	-

Source: "Estimates of Resident Armed Forces, by Age, Sex, Race, and Hispanic Origin," *Population Projections of the United States, by Age, Sex, Race, and Hispanic Origin: 1993 to 2050,* Current Population Reports, P25-1104, pp. D4 and D-5. *Notes:* For data about Native Americans, Asians, and Pacific Islanders, see original. 1. Persons of Hispanic origin may be of any race.

Navy

★ 811 ★

Black Enlisted Personnel in the U.S. Navy, 1992

[As of June 30, 1992].

Rate	Male	Female	Total	Percentage
Master Chief (E-9)	258	6	264	5.3%
Senior Chief (E-8)	672	37	709	6.7%
Chief (E-7)	2,948	229	3,177	8.8%
First Class (E-6)	10,982	1,558	12,540	13.8%
Second Class (E-5)	15,950	3,514	19,464	17.8%
Third Class (E-4)	17,733	3,661	21,394	20.4%
AN/FN/SN (E-3)	14,167	2,897	17,064	23.8%
AA/FA/SA (E-2)	7,159	824	7,983	19.0%
AR/FR/SR (E-1)	3,680	559	4,239	11.2%

Source: "Black Enlisted—U.S. Navy." Stillwell, Paul, ed., *The Golden Thirteen: Recollections of the First Black Naval Officers,* Annapolis, Md.: Naval Institute Press, 1993, p. 288. Primary source: The Special Assistant for Minority Affairs, Navy Recruiting Command.

★ 812 ★

Navy

Black Officers in the U.S. Navy, 1992

[As of June 30, 1992].

Rank	Male	Female	Total	Percentage
Flag (O-7 to O-9)	3	0	3	1.2%
Captain (O-6)	39	2	41	1.1%
Cammander (O-5)	169	40	209	2.5%
Lt. Commander (O-4)	382	99	481	3.3%
Lieutenant (O-3)	818	291	1,109	4.4%
Lieutenant (j.g.)(O-2)	447	123	570	5.7%
Ensign (O-1)	360	98	458	5.9%
CWO-4 (WO-4)	48	0	48	8.8%
CWO-3 (WO-3)	61	0	61	7.4%
CWO-2 (WO-2)	147	8	155	10.2%

Source: "Black Officers—U.S. Navy." Stillwell, Paul, ed., *The Golden Thirteen: Recollections of the First Black Naval Officers,* Annapolis, Md.: Naval Institute Press, 1993, p. 288. Primary source: The Special Assistant for Minority Affairs, Navy Recruiting Command.

★ 813 ★

Navy

Black Totals in the U.S. Navy, 1992

[As of June 30, 1992].

	Male	Female	Total	Percentage
Officer	2,474	661	3,135	4.3%
Enlisted	73,549	13,285	86,834	17.6%
Total Force	76,023	13,946	89,969	15.9%

Source: "Black Totals—U.S. Navy." Stillwell, Paul, ed., *The Golden Thirteen: Recollections of the First Black Naval Officers,* Annapolis, Md.: Naval Institute Press, 1993, p. 288. Primary source: The Special Assistant for Minority Affairs, Navy Recruiting Command.

Reserves

★ 814 ★

Ready Reserve Personnel Profile, 1989 and 1992

Item	Race					Percent distribution			
	Total	White	Black	Asian	American Indian	White	Black	Asian	American Indian
1989, total	1,676,15	1,282,456	372,765	13,237	7,647	76.5	22.2	0.8	0.5
1990, total	1,558,867	1,269,278	271,470	14,608	3,511	81.4	17.4	0.9	0.2
1991, total	1,154,515	906,748	190,214	12,711	5,652	78.5	16.5	1.1	0.5
Male	1,003,325	806,894	146,853	11,179	4,824	80.4	14.6	1.1	0.5
Officers	146,657	133,644	7,768	1,466	263	91.1	5.3	1.0	0.2
Enlisted	856,668	673,250	139,085	9,713	4,561	78.6	16.2	1.1	0.5
Female	151,190	99,854	43,360	1,532	828	66.0	28.7	1.0	0.5
Officers	25,772	20,541	3,838	243	47	79.7	14.9	0.9	0.2
Enlisted	125,418	79,313	39,522	1,289	781	63.2	31.5	1.0	0.6
1992, total	1,114,905	725,963	172,770	11,479	4,842	65.1	15.5	1.0	0.4
Male	966,503	660,227	141,741	11,501	4,360	68.3	14.7	1.2	0.5
Officers	141,770	12,874	7,684	1,451	261	9.1	5.4	1.0	0.2
Enlisted	824,733	647,352	134,057	10,050	4,099	78.5	16.3	1.2	0.5
Female	148,783	98,389	42,466	1,686	804	66.1	28.5	1.1	0.5
Officers	24,832	19,780	3,753	257	61	79.7	15.1	1.0	0.2
Enlisted	123,951	78,609	38,713	1,429	743	63.4	31.2	1.2	0.6

Source: "Ready Reserve Personnel Profile—Race, Age, and Sex: 1989 and 1992," *Statistical Record of America, 1993*, p. 3360. Primary source: U.S. Department of Defense, *Official Guard and Reserve Manpower Strengths and Statistics*, annual.

Chapter 12
POLITICS AND ELECTIONS

County Officials

★ 815 ★

Male County Officials by Race and Ethnicity, 1991 - I

Position	Total reporting (A)	Total males		Race					
				White		Black		American Indian	
		No. (B)	% of (A)	No.	% of (B)	No.	% of (B)	No.	% of (B)
Board chairman	1,585	1,408	88.8	1,308	92.9	39	2.8	13	0.9
Chief appointed administrative officer	743	619	83.3	572	92.4	13	2.1	4	0.6
Clerk to the governing board	1,481	452	30.5	417	92.3	9	2.0	2	0.4
Chief financial officer	1,233	678	55.0	625	92.2	6	0.9	5	0.7
County health officer	1,012	621	61.4	554	89.2	10	1.6	4	0.6
Planning director	790	681	86.2	643	94.4	5	0.7	4	0.6
County engineer	831	821	98.8	776	94.5	5	0.6	6	0.7
Director health/human services	892	516	57.8	455	88.2	14	2.7	4	0.8
Chief law enforcement official	1,591	1,586	99.7	1,471	92.7	17	1.1	16	1.0
Purchasing director	642	417	65.0	375	89.9	14	3.4	3	0.7
Personnel director	761	431	56.6	376	87.2	27	6.3	2	0.5

Source: "Male County Officials by Race and Ethnicity," *The Municipal Year Book, 1993*, p. 293. Published by permission.

★ 816 ★

County Officials

Male County Officials by Race and Ethnicity, 1991 - II

Position	Race							
	Asian		Other		Race not reported		Hispanic	
	No.	% of (B)	No.	% of (B)	No.	% of (B)	No.	% of (B)
Board chairman	2	0.1	2	0.1	44	3.1	13	0.9
Chief appointed administrative officer	4	0.6	4	0.6	22	3.6	6	1.0
Clerk to the governing board	2	0.4	2	0.4	20	4.4	6	1.3
Chief financial officer	6	0.9	3	0.4	33	4.9	7	1.0
County health officer	6	1.0	7	1.1	40	6.4	12	1.9
Planning director	4	0.6	1	0.1	24	3.5	5	0.7
County engineer	3	0.4	3	0.4	28	3.4	5	0.6
Director health/human services	3	0.6	3	0.6	37	7.2	17	3.3
Chief law enforcement official	4	0.3	6	0.4	72	4.5	26	1.6
Purchasing director	3	0.7	5	1.2	17	4.1	10	2.4
Personnel director	5	1.2	5	1.2	16	3.7	10	2.3

Source: "Male County Officials by Race and Ethnicity," *The Municipal Year Book, 1993*, p. 293. Published by permission.

★ 817 ★

County Officials

Women County Officials by Race and Ethnicity, 1991 - I

Position	Total reporting (A)	Total females		Race					
				White		Black		American Indian	
		No. (B)	% of (A)	No.	% of (B)	No.	% of (B)	No.	% of (B)
Board chairman	1,585	177	11.2	164	92.7	8	4.5	0	0.0
Chief appointed administrative officer	743	124	16.7	118	95.2	3	2.4	1	0.8
Clerk to the governing board	1,481	1,029	69.5	974	94.7	15	1.5	13	1.3
Chief financial officer	1,233	555	45.0	530	95.5	6	1.1	6	1.1
County health officer	1,012	391	38.6	367	93.9	10	2.6	1	0.3
Planning director	790	109	13.8	105	96.3	3	2.8	0	0.0
County engineer	831	10	1.2	8	80.0	0	0.0	0	0.0
Director health/human services	892	376	42.2	339	90.2	23	6.1	2	0.5
Chief law enforcement official	1,591	5	0.3	5	100.0	0	0.0	0	0.0
Purchasing director	642	225	35.0	203	90.2	13	5.8	3	1.3
Personnel director	761	330	43.4	291	88.2	24	7.3	4	1.2

Source: "Female County Officials by Race and Ethnicity," *The Municipal Year Book, 1993*, p. 293. Published by permission.

★ 818 ★

County Officials

Women County Officials by Race and Ethnicity, 1991 - II

Position	Race							
	Asian		Other		Race not reported		Hispanic	
	No.	% of (B)	No.	% of (B)	No.	% of (B)	No.	% of (B)
Board chairman	0	0.0	0	0.0	5	2.8	1	0.6
Chief appointed administrative officer	1	0.8	0	0.0	1	0.8	4	3.2
Clerk to the governing board	1	0.1	3	0.3	23	2.2	16	1.6
Chief financial officer	1	0.2	1	0.2	11	2.0	8	1.4
County health officer	1	0.3	3	0.8	9	2.3	1	0.3
Planning director	0	0.0	1	0.9	0	0.0	1	0.9
County engineer	0	0.0	0	0.0	2	20.0	0	0.0
Director health/human services	1	0.3	2	0.5	9	2.4	6	1.6
Chief law enforcement official	0	0.0	0	0.0	0	0.0	0	0.0
Purchasing director	1	0.4	2	0.9	3	1.3	5	2.2
Personnel director	1	0.3	1	0.3	9	2.7	11	3.3

Source: "Female County Officials by Race and Ethnicity," *The Municipal Year Book, 1993*, p. 293. Published by permission.

Elected and Appointed Officials

★ 819 ★

Black Elected Officials as Percentage of All Elected Officials, 1993

State	Blacks as a percentage of voting age population	Elected officials			
		Total	Black	% Black	Net change
Alabama	22.7	4,315	699	16.2	-3
Alaska	3.9	1,757	3	0.2	0
Arizona	2.7	3,183	15	0.5	-1
Arkansas	13.7	8,331	380	4.6	38
California	7.0	19,236	273	1.4	13
Colorado	3.7	8,035	20	0.2	2
Connecticut	7.4	8,489	62	0.7	-1
Delaware	15.3	1,227	23	1.9	-1
District of Columbia	62.4	325	198	60.9	-10
Florida	11.4	5,256	200	3.8	17
Georgia	24.6	6,556	545	18.3	32
Hawaii	2.3	160	0	0.0	0

[Continued]

★ 819 ★

Black Elected Officials as Percentage of All Elected Officials, 1993
[Continued]

State	Blacks as a percentage of voting age population	Elected officials			
		Total	Black	% Black	Net change
Idaho	-	4,678	0	0.0	0
Illinois	13.4	38,836	465	1.2	8
Indiana	7.1	11,355	72	0.6	2
Iowa	1.5	17,044	11	0.1	-2
Kansas	5.2	16,410	21	0.1	2
Kentucky	6.6	7,388	63	0.9	-2
Louisiana	27.9	4,966	636	12.8	57
Maine	-	6,978	1	0.0	0
Maryland	23.5	1,943	140	7.2	5
Massachusetts	4.4	13,631	30	0.2	2
Michigan	12.8	19,293	333	1.7	13
Minnesota	1.8	18,887	16	0.1	3
Mississippi	31.6	4,944	751	15.2	35
Missouri	9.7	17,115	185	1.1	8
Montana	-	5,646	0	0.0	0
Nebraska	3.2	15,064	6	0.0	2
Nevada	5.8	1,174	10	0.9	0
New Hampshire	0.6	6,721	2	0.0	0
New Jersey	12.4	9,345	211	2.3	10
New Mexico	1.9	2,096	3	0.1	-1
New York	14.7	25,999	299	1.2	12
North Carolina	20.1	5,531	468	8.5	10
North Dakota	0.5	15,141	0	0.0	0
Ohio	9.8	19,750	219	1.1	4
Oklahoma	6.6	9,290	123	1.3	-3
Oregon	1.4	8,367	10	0.1	-1
Pennsylvania	8.5	29,586	158	0.5	6
Rhode Island	3.3	1,120	12	1.1	1
South Carolina	26.9	3,692	450	12.2	37
South Dakota	-	9,249	3	0.0	0
Tennessee	14.4	6,841	168	2.5	1
Texas	11.2	26,932	472	1.8	172
Utah	0.7	2,588	0	0.0	-1
Vermont	-	8,021	2	0.0	0
Virginia	17.6	3,112	155	5.0	-1
Virgin Islands	61.4	41	31	75.6	-4
Washington	2.8	8,032	19	0.2	-1
West Virginia	3.0	2,838	21	0.7	-2
Wisconsin	4.1	18,242	30	0.2	6

[Continued]

★ 819 ★

Black Elected Officials as Percentage of All Elected Officials, 1993

[Continued]

State	Blacks as a percentage of voting age population	Elected officials			
		Total	Black	% Black	Net change
Wyoming	0.7	2,340	1	0.0	-1
Total	11.0[1]	497,196	8,015[2]	1.6	463

Source: "Black Elected Officials as a Percentage of All Elected Officials, January 1993," *Black Elected Officials, 1993,* p. xxiii. Published by permission. *Notes:* 1. al includes two statehood senators and one statehood representative from the District of Columbia. 2. Total voting age percentage does not include the Virgin Islands.

★ 820 ★

Elected and Appointed Officials

Black Elected Officials, by Office, 1970 to 1992, and Region and State, 1992

Region, division, and state	Total	U.S. and State legisla-tures[1]	City and county offices[2]	Law enforcement[3]	Education[4]
1970 (Feb.)	1,479	179	719	213	368
1975 (Apr.)	3,522	299	1,885	387	951
1978 (July)	4,544	316	2,616	458	1,154
1979 (July)	4,636	315	2,675	491	1,155
1980 (July)	4,963	326	2,871	534	1,232
1981 (July)	5,109	343	2,914	559	1,293
1982 (July)	5,241	342	3,017	573	1,309
1983 (July)	5,719	386	3,283	620	1,430
1984 (Jan.)	5,865	396	3,367	657	1,445
1985 (Jan.)	6,312	407	3,689	685	1,531
1986 (Jan.)	6,384	410	3,800	676	1,498
1987 (Jan.)	6,646	428	3,949	727	1,542
1988 (Jan.)	6,793	424	4,089	738	1,542
1989 (Jan.)	7,191	441	4,388	760	1,602
1990 (Jan.)	7,335	440	4,481	769	1,645
1991 (Jan.)	7,445	476	4,493	847	1,629
1992 (Jan.)	7,517	499	4,557	847	1,614
Northeast	747	89	288	116	254
N.E.	107	31	62	3	11
ME	1	-	1	-	-
NH	2	2	-	-	-
VT	2	2	-	-	-
MA	28	6	19	1	2
RI	11	8	3	-	-
CT	63	13	39	2	9

[Continued]

★ 820 ★

Black Elected Officials, by Office, 1970 to 1992, and Region and State, 1992

[Continued]

Region, division, and state	Total	U.S. and State legislatures[1]	City and county offices[2]	Law enforcement[3]	Education[4]
M.A.	640	58	226	113	243
NY	287	27	64	61	135
NJ	201	13	109	-	79
PA	152	18	53	52	29
Midwest	1,315	98	739	153	325
E.N.C.	1,086	73	592	129	292
OH	215	15	121	30	49
IN	70	9	49	4	8
IL	457	25	283	23	126
MI	320	18	128	68	106
WI	24	6	11	4	3
W.N.C.	229	25	147	24	33
MN	13	1	2	7	3
IA	13	1	7	1	4
MO	177	18	127	14	18
SD	3	-	3	-	-
NE	4	1	1	-	2
KS	19	4	7	2	6
South	5,110	280	3,432	474	924
S.A.	2,113	147	1,504	136	326
DE	24	3	14	1	6
MD	135	32	75	22	6
DC	208	4[5]	196	-	8
VA	156	13	128	15	-
WV	23	2	17	4	-
NC	458	19	331	31	77
SC	413	22	255	12	124
GA	513	38	362	27	86
FL	183	14	126	24	19
E.S.C.	1,650	66	1,166	180	238
KY	65	3	48	4	10
TN	167	14	103	25	25
AL	702	24	529	65	84
MS	716	25	486	86	119
W.S.C.	1,347	67	762	158	360
AR	342	12	188	39	103
LA	579	33	319	81	146
OK	126	6	97	2	21
TX	300	16	158	36	90
West	345	32	98	104	111
Mt	51	12	11	16	12
WY	2	1	-	-	1
CO	18	4	3	9	2
NM	4	-	1	2	1

[Continued]

★ 820 ★

Black Elected Officials, by Office, 1970 to 1992, and Region and State, 1992

[Continued]

Region, division, and state	Total	U.S. and State legislatures[1]	City and county offices[2]	Law enforcement[3]	Education[4]
AZ	16	4	3	3	6
UT	1	-	-	1	-
NV	10	3	4	1	2
Pac	294	20	87	88	99
WA	20	2	9	6	3
OR	11	4	3	4	-
CA	260	13	73	78	96
AK	3	1	2	-	-

Source: "Black Elected Officials, by Office, 1970 to 1992, and by Region and State, 1992," U.S. Bureau of the Census, *Statistical Abstract of the United States*, 1993, p. 280. Published by permission. Primary source: Joint Center for Political and Economic Studies, Washington, DC, *Black Elected Officials: A National Roster*, annual (copyright). Notes: - Represents zero. 1. Includes elected State administrators. 2. County commissioners and councilmen, mayors, vice mayors, aldermen, regional officials, and other. 3. Judges, magistrates, constables, marshalls, sheriffs, justices of the peace, and other. 4. Members of State education agencies, college boards, school boards, and other. 5. Includes two shadow senators and one shadow representative.

★ 821 ★

Elected and Appointed Officials

Change in Officials by Category of Office, 1970-1993, Part I

Year	Total BEOs		Federal		State		Substate region		Country	
	N	% change	N	% change	N	% change	N	% change	N	% change
1970	1,469	-	10	-	169	-	-	-	92	-
1971	1,860	26.6	14	40.0	202	19.5	-	-	120	30.4
1972	2,264	21.7	14	0.0	210	4.0	-	-	176	46.7
1973	2,621	15.8	16	14.3	240	14.3	-	-	211	19.9
1974	2,991	14.1	17	6.3	239	-0.4	-	-	242	14.7
1975	3,503	17.1	18	5.9	281	17.6	-	-	305	26.0
1976	3,979	13.6	18	0.0	281	0.0	30	-	355	16.4
1977	4,311	8.3	17	-5.6	299	6.4	33	10.0	381	7.3
1978	4,503	4.5	17	0.0	299	0.0	26	-21.2	410	7.6
1979	4,607	2.3	17	0.0	313	4.7	25	-3.8	398	-2.9
1980	4,912	6.6	17	0.0	323	3.2	25	0.0	451	13.3
1981	5,038	2.6	18	5.9	341	5.6	30	20.0	449	0.4
1982	5,160	2.4	18	0.0	336	-1.5	35	16.7	465	3.6
1983	5,606	8.6	21	16.7	379	12.8	29	-17.1	496	6.7
1984[1]	5,700	1.7	21	0.0	389	2.6	30	3.4	518	4.4
1985	6,056	6.2	20	-4.8	396	1.8	32	6.7	611	18.0
1986	6,424	6.1	20	0.0	400	1.0	31	-3.2	681	11.4
1987	6,681	4.0	23	15.0	417	4.3	23	-25.8	724	6.3
1988	6,829	2.2	23	0.0	413	-1.0	22	-4.3	742	2.5
1989	7,226	5.8	24	4.2	424	2.7	18	18.2	793	6.9

[Continued]

★ 821 ★

Change in Officials by Category of Office, 1970-1993, Part I

[Continued]

Year	Total BEOs		Federal		State		Substate region		Country	
	N	% change	N	% change	N	% change	N	% change	N	% change
1990	7,370	2.0	24	0.0	423	-0.2	18	0.0	810	2.1
1991	7,480	1.5	26	8.3	458	8.3	15	-16.7	810	0.0
1992[2]	7,552	1.0	26	0.0	484	5.7	15	0.0	857	5.8
1993[2]	8,015	6.1	39[3]	50.0	533	10.1	13	-13.3	913	6.5

Source: "Annual Change in Number of Black Elected Officials by Category of Office, 1970-1993," *Black Elected Officials*, 1993, p. xxii. Published by permission. *Notes:* 1. The 1984 figures reflect blacks who took office during the seven- month period between July 1, 1983 and January 30, 1984. 2. Includes two statehood senators and one statehood representative from the District of Columbia. 3. Congressmember Bennie Thompson (D-MS) was elected to fill Mike Espy's seat after January 31, 1993. Mike Espy gave up his seat before January 31, 1993 to serve as Secretary of Agriculture. When this document went to press there were 40 elected black federal officials.

★ 822 ★

Elected and Appointed Officials

Change in Officials by Category of Office, 1970-1993, Part II

Year	Municipal		Judicial/law enforcement		Education	
	N	% change	N	% change	N	% change
1970	623	-	213	-	362	-
1971	785	26.0	274	28.6	465	28.5
1972	932	18.7	263	-4.0	669	43.9
1973	1,053	13.0	334	27.0	767	14.6
1974	1,360	29.2	340	1.8	793	3.4
1975	1,573	15.7	387	13.8	9.9	18.4
1976	1,889	20.1	412	6.5	994	5.9
1977	2,083	10.3	447	8.5	1,051	5.7
1978	2,159	3.6	454	1.6	1,138	8.3
1979	2,224	3.0	486	7.0	1,144	0.5
1980	2,356	5.9	526	8.2	1,214	6.1
1981	2,384	1.2	549	4.4	1,267	4.4
1982	2,477	3.9	563	2.6	1,266	-0.1
1983	2,697	10.0	607	7.8	1,377	8.8
1984[1]	2,735	1.4	636	4.8	1,371	-0.4
1985	2,898	6.0	661	4.0	1,438	4.9
1986	3,112	7.4	676	2.3	1,504	4.6
1987	3,219	3.4	728	7.7	1,547	2.9
1988	3,341	3.8	738	1.4	1,550	0.2
1989	3,595	7.6	760	2.9	1,612	4.0
1990	3,671	2.1	769	1.2	1,655	2.7
1991	3,683	0.3	847	10.1	1,638	-1.0

[Continued]

★ 822 ★

Change in Officials by Category of Office, 1970-1993, Part II
[Continued]

Year	Municipal		Judicial/law enforcement		Education	
	N	% change	N	% change	N	% change
1992[2]	3,697	0.4	847	0.0	1,623	-0.9
1993[2]	3,903	5.6	922	8.9	1,689	4.1

Source: "Annual Change in Number of Black Elected Officials by Category of Office, 1970-1993," *Black Elected Officials*, 1993, p. xxii. Published by permission. *Notes:* 1. The 1984 figures reflect blacks who took office during the seven- month period between July 1, 1983 and January 30, 1984. 2. Includes two statehood senators and one statehood representative from the District of Columbia.

★ 823 ★

Elected and Appointed Officials

City Councils by Method of Election: Racial Composition, 1991

Election method	No. reporting	Composition of council (%)				
		White	Black	Hispanic	Asian	Native American
Nominated and elected at large	2,465	94.1	3.0	2.4	0.2	0.3
Nominated by district, elected at large	174	92.3	6.3	1.0	0.0	0.4
Nominated and elected by district	490	91.3	6.9	1.4	0.2	0.2
Other	152	90.6	4.5	3.0	0.0	1.9

Source: "Racial/Ethnic Composition, by Method of Election." Susan A. MacManus, "Women and Racial/Ethnic Minorities in Mayoral and Council Positions," *The Municipal Yearbook, 1993*, p. 82. Published by permission.

★ 824 ★

Elected and Appointed Officials

City Councils by Size: Racial Composition, 1991

Size of council	No. reporting	Composition of council (%)				
		White	Black	Hispanic	Asian	Native American
Less than 4	232	97.8	1.2	0.9	0.1	0.0
4-6	2,699	91.8	4.8	2.8	0.3	0.4
7-9	1,638	92.3	5.5	1.8	0.2	0.3
More than 9	180	90.5	7.9	1.2	0.3	0.1

Source: "Racial/Ethnic Composition, by Size of Council." Susan A. MacManus, "Women and Racial/Ethnic Minorities in Mayoral and Council Positions," *The Municipal Yearbook, 1993*, p. 81. Published by permission.

★ 825 ★

Elected and Appointed Officials

City Councils by Term: Racial Composition, 1991

Election method	No. reporting	Composition of council (%)				
		White	Black	Hispanic	Asian	Native American
Staggered	3,806	93.3	3.7	2.5	0.2	0.3
Simultaneous	876	88.9	9.5	1.2	0.1	0.2

Source: "Racial Composition of Council With Staggered and Simultaneous Terms." Susan A. MacManus, "Women and Racial/Ethnic Minorities in Mayoral and Council Positions," *The Municipal Yearbook, 1993,* p. 81. Published by permission.

★ 826 ★

Elected and Appointed Officials

City Councils in Types of Cities: Racial Composition, 1991

Election method	No. reporting	Composition of council (%)				
		White	Black	Hispanic	Asian	Native American
Partisan	1,179	94.0	4.8	0.9	0.1	0.1
Nonpartisan	3,471	92.0	4.7	2.7	0.2	0.4

Source: "Racial/Ethnic Composition of Council in Partisan and Nonpartisan Cities." Susan A. MacManus, "Women and Racial/Ethnic Minorities in Mayoral and Council Positions," *The Municipal Yearbook, 1993,* p. 81. Published by permission.

★ 827 ★

Elected and Appointed Officials

City Councils: Racial Composition, 1991

	No. reporting	White (%)	Black (%)	Hispanic (%)	Asian (%)	Native American (%)
Total, all cities	4,719	92.5	4.8	2.3	0.2	0.3
Population group						
Over 1,000,000	2	52.2	37.8	6.7	3.3	0.0
500,000-1,000,000	7	63.4	11.6	14.2	10.8	0.0
250,000-499,999	31	72.7	16.7	9.8	0.8	0.0
100,000-249,999	97	80.6	13.3	5.0	0.6	0.5
50,000-99,999	243	89.3	7.4	3.0	0.4	0.0

[Continued]

★ 827 ★

City Councils: Racial Composition, 1991
[Continued]

	No. reporting	White (%)	Black (%)	Hispanic (%)	Asian (%)	Native American (%)
25,000-49,999	488	90.9	5.9	2.8	0.3	0.2
10,000-24,999	1,071	92.1	5.3	2.3	0.2	0.1
5,000-9,999	1,174	92.8	4.2	2.5	0.2	0.3
2,500-4,999	1,234	94.6	3.5	1.3	0.1	0.5
Under 2,500	372	95.0	2.9	1.6	0.1	0.4
Geographic division						
New England	473	98.8	0.9	0.2	0.1	0.0
Mid-Atlantic	676	95.8	3.4	0.6	0.1	0.2
East North Central	903	96.5	2.8	0.4	0.1	0.2
West North Central	543	98.1	1.4	0.3	0.1	0.1
South Atlantic	614	85.7	13.4	0.8	0.1	0.2
East South Central	268	87.8	11.8	0.4	0.0	0.0
West South Central	481	82.6	8.2	8.2	0.1	1.0
Mountain	257	90.2	1.2	8.1	0.5	0.1
Pacific Coast	504	90.1	1.9	6.1	1.1	0.9
Metro status						
Central	361	82.9	12.6	3.8	0.6	0.1
Suburban	2,530	94.1	3.3	2.3	0.2	0.2
Independent	1,828	92.1	5.4	1.9	0.1	0.5
Form of government						
Mayor-council	2,099	92.8	5.2	1.5	0.2	0.3
Council-manager	2,267	91.4	4.9	3.2	0.3	0.3
Commission	88	91.9	4.6	2.4	0.0	1.1
Town meeting	229	99.7	0.2	0.1	0.0	0.0
Rep. town meeting	36	98.9	0.6	0.6	0.0	0.0

Source: "Racial Composition of Councils." Susan A. MacManus, "Women and Racial/Ethnic Minorities in Mayoral and Council Positions," *The Municipal Yearbook, 1993*, p. 80. Published by permission.

★ 828 ★

Elected and Appointed Officials

Congressional Black Caucus: Legislative Initiators, 1993

Proposing a bill is no guarantee that it will become law. During the first four months of 1993[1], the CBC sponsored 153 bills or resolutions and co-sponsored and other 2,396. Following the CBC's most active legislative initiators.

	Sponsor	Co-sponsor
Sen. Carol Moseley-Braun D-Ill.	3	52
Rep. Cardiss Collins D-Ill	36	56
Rep. Major R. Owens D-N.Y.	15	81
Rep. Charles B. Rangel D-N.Y.	10	111
Rep. John Conyers Jr. D-Mich	9	66

Source: "CBC: Down by Law." "Who is the Black Caucus?" *Black Enterprise* 24 (October 1993), p. 26. Published by permission. Primary source: *Black Congressional Monitor*, Washington, D.C., May 1993. *Note:* 1. January 5, 1993 to April 19, 1993.

★ 829 ★

Elected and Appointed Officials

Congressional Districts Represented by Blacks, January 1993

Member of Congress	District	Principal city	% Black Pop.	% White Pop.	% Hispanic Pop.
Hilliard (D-AL)	7	Montgomery	64	36	0
Dellums (D-CA)	9	Oakland	29	44	11
Dixon (D-CA)	32	Los Angeles	40	25	26
Tucker (D-CA)	37	Compton	34	13	41
Waters (D-CA)	35	Los Angeles	44	10	39
Franks (R-CT)	5	Waterbury	4	90	5
Holmes-Norton (D-DC)	At-Large	Washington, DC	62	31	5
Brown (D-FL)	3	Jacksonville	51	45	3
Meek (D-FL)	17	Miami	54	20	24
Hastings (D-FL)	23	Ft. Lauderdale	46	44	9
Bishop (D-GA)	2	Columbus	52	45	2
Lewis (D-GA)	5	Atlanta	58	39	2
McKinney (D-GA)	11	Atlanta	60	38	1
Collins (D-IL)	7	Chicago	60	32	4
Rush (D-IL)	1	Chicago	68	28	3
Reynolds (D-IL)	2	Chicago	66	27	6
Fields (D-LA)	4	Baton Rouge	63	36	1
Jefferson (D-LA)	2	New Orleans	56	38	4
Mfume (D-MD)	7	Baltimore	68	29	1
Wynn (D-MD)	4	Silver Spring	56	33	6
Conyers (D-MI)	14	Detroit	65	33	1
Rose Collins (D-MI)	15	Detroit	68	27	4

[Continued]

★ 829 ★

Congressional Districts Represented by Blacks, January 1993

[Continued]

Member of Congress	District	Principal city	% Black Pop.	% White Pop.	% Hispanic Pop.
Clay (D-MO)	1	St. Louis	48	50	1
Wheat (D-MO)	5	Kansas City	21	74	3
Payne (D-NJ)	10	Newark	57	28	12
Flake (D-NY)	6	Jamaica	55	22	16
Towns (D-NY)	10	New York	59	19	19
Owens (D-NY)	11	Brooklyn	72	14	11
Rangel (D-NY)	15	New York	47	6	44
Clayton (D-NC)	1	Fayetteville	53	45	1
Watt (D-NC)	12	Charlotte	53	45	1
Stokes (D-OH)	11	Cleveland	55	43	1
Blackwell (D-PA)	2	Philadelphia	58	37	2
Clyburn (D-SC)	6	Florence	58	40	1
Ford (D-TN)	9	Memphis	54	44	1
Johnson (D-TX)	30	Dallas	47	36	15
Washington (D-TX)	18	Houston	49	34	14
Scott (D-VA)	3	Richmond	61	36	1

Source: "Congressional Districts Represented by Blacks, January 1993," *Black Elected Officials, 1993*, p. xxvi. Published by permission. *Notes:* Numbers may not add to 100% since Asian Americans are not included and, according to the Census Bureau, persons of Spanish origin may be of any race.

★ 830 ★

Elected and Appointed Officials

Distribution of Elected Officials by Region and Office, January 1992

	Northeast	Midwest	South	West	Total
Federal	7	8	7	4	26
State	82	90	270	28	470
County	24	87	732	14	857
Mayors	24	65	237	12	338
Other municipal	239	587	2,463	70	3,359
Judicial/law enforcement	116	153	474	104	847
Education	254	325	924	111	1,614
Total	746	1,315	5,107	343	7,517[1]
Percent	9.9	17.5	67.9	4.6	100.0

Source: "Distribution of Black Elected Officials by Region and Category of Office, January 1992." *Black Elected Officials, 1993*, p. lxiii. Published by permission. *Notes:* The 35 BEOs from the Virgin Islands are not included in the census. 1. Includes two statehood senators, one statehood representative, and three regional officers.

★ 831 ★
Elected and Appointed Officials

Educational Level of Mayor, by Race/Ethnicity, 1991

Highest education level achieved	Race/ethnicity				
	White (%)	Asian (%)	Black (%)	Hispanic (%)	Native American (%)
No high school diploma	0.8	0.0	00	1.3	0.0
High school graduate	20.7	0.0	10.0	29.1	25.0
Some college	21.6[1]	0.0	20.0	35.4[1]	50.0
Bachelor's degree	26.9	44.4	25.0	7.6	12.5
Some master's work	4.1	11.1	6.3	3.8	0.0
Master's degree	12.6	22.2	17.5	13.9	6.3
Law degree	7.4	11.1	8.8	5.1	0.0
Ph.D.	1.4	0.0	6.3	1.3	0.0
Medical degree	1.5	11.1	2.5	1.3	0.0
Other	2.2	0.0	2.5	0.0	6.3

Source: "Educational Level of Mayor, by Race\Ethnicity." Susan A. MacManus, "Women and Racial/Ethnic Minorities in Mayoral and Council Positions," *The Municipal Yearbook, 1993*, p. 75. Published by permission. *Notes:* Numbers responding were: white, 4,042; Asian, 9; black, 80; Hispanic, 79; Native American, 16. 1. Includes post-high school technical.

★ 832 ★
Elected and Appointed Officials

Elective County Offices, January 1993

State	Members, county governing bodies	Members, other county bodies	Total members
Alabama	78	16	94
Alaska	-	-	-
Arizona	-	-	-
Arkansas	-	-	-
California	5	3	8
Colorado	1	-	1
Connecticut	-	-	-
Delaware	1	1	2
District of Columbia	-	-	-
Florida	26	2	28
Georgia	100	5	105
Hawaii	-	-	-
Idaho	-	-	-
Illinois	41	3	44
Indiana	13	-	13
Iowa	2	-	2

[Continued]

★ 832 ★

Elective County Offices, January 1993
[Continued]

State	Members, county governing bodies	Members, other county bodies	Total members
Kansas	3	-	3
Kentucky	1	-	1
Louisiana	139	-	139
Maine	-	-	-
Maryland	7	-	7
Massachusetts	-	-	-
Michigan	25	-	25
Minnesota	-	-	-
Mississippi	85	73	158
Missouri	2	-	2
Montana	-	-	-
Nebraska	1	-	1
Nevada	1	-	1
New Hampshire	-	-	-
New Jersey	6	1	7
New Mexico	-	-	-
New York	16	-	16
North Carolina	45	4	49
North Dakota	-	-	-
Ohio	1	-	1
Oklahoma	2	1	3
Oregon	1	-	1
Pennsylvania	2	-	2
Rhode Island	-	-	-
South Carolina	76	5	81
South Dakota	-	-	-
Tennessee	49	-	49
Texas	17	1	17
Utah	-	-	-
Vermont	-	-	-
Virginia	44	3	47
Washington	1	-	1
West Virginia	-	-	-
Wisconsin	4	-	4
Wyoming	-	-	-
Total	795	118	913

Source: "Blacks in Elective County Offices, January 1993." *Black Elected Officials, 1993*, p. xxxi. Published by permission.

★ 833 ★

Elected and Appointed Officials

Elective Judicial and Law Enforcement Offices, January 1993

State	Judges, state courts of last resort	Judges, other courts	Magistrates, Justices of The Peace, constables	Other judicial officials	Police chiefs sheriffs, and marshals	Total
Alabama	1	14	30	7	6	58
Alaska	-	-	-	-	-	-
Arizona	0	3	0	0	0	3
Arkansas	0	5	46	0	0	51
California	0	80	0	1	1	82
Colorado	1	8	0	1	0	10
Connecticut	0	0	2	0	0	2
Delaware	-	-	-	-	-	-
District of Columbia	-	-	-	-	-	-
Florida	1	26	0	0	1	28
Georgia	2	24	1	3	2	32
Hawaii	-	-	-	-	-	-
Idaho	-	-	-	-	-	-
Illinois	0	37	0	0	0	37
Indiana	0	4	0	0	0	4
Iowa	0	0	0	1	0	1
Kansas	0	3	0	0	0	3
Kentucky	0	2	2	1	0	5
Louisiana	0	42	42	3	17	104
Maine	-	-	-	-	-	-
Maryland	0	19	0	4	0	23
Massachusetts	0	0	1	1	0	2
Michigan	1	66	1	0	0	68
Minnesota	1	8	0	0	1	10
Mississippi	1	34	35	10	8	88
Missouri	0	11	0	2	1	14
Montana	-	-	-	-	-	-
Nebraska	-	-	-	-	-	-
Nevada	0	1	0	0	0	1
New Hampshire	-	-	-	-	-	-
New Jersey	-	-	-	-	-	-
New Mexico	0	2	0	0	0	2
New York	0	69	0	1	0	70
North Carolina	1	25	0	2	3	31
North Dakota	-	-	-	-	-	-
Ohio	0	29	0	1	0	30
Oklahoma	0	1	0	0	0	1
Oregon	0	4	0	0	0	4
Pennsylvania	1	30	21	0	0	52
Rhode Island	-	-	-	-	-	-
South Carolina	0	4	5	0	5	15
South Dakota	-	-	-	-	-	-
Tennessee	0	12	9	2	1	24
Texas	1	11	28	0	0	40
Utah	-	-	-	-	-	-

[Continued]

★ 833 ★

Elective Judicial and Law Enforcement Offices, January 1993
[Continued]

State	Judges, state courts of last resort	Judges, other courts	Magistrates, Justices of The Peace, constables	Other judicial officials	Police chiefs sheriffs, and marshals	Total
Vermont	-	-	-	-	-	-
Virginia	0	0	0	10	5	15
Washington	1	4	0	0	0	5
West Virginia	0	2	1	0	0	3
Wisconsin	0	3	0	0	1	4
Wyoming	-	-	-	-	-	-
Total	12	583	224	51	52	922

Source: "Blacks in Elective Judicial and Law Enforcement Offices, January 1993." *Black Elected Officials, 1993*, p. xxxvi. Published by permission.

★ 834 ★

Elected and Appointed Officials

Elective Municipal Offices, January 1993

State	Mayors	Municipal governing bodies	Members, Members, municipal boards	Members, advisory commissions	Other municipal officials	Total
Alabama	35	400	-	-	-	435
Alaska	-	2	-	-	-	2
Arizona	1	2	-	-	-	3
Arkansas	26	166	-	-	22	214
California	11	38	8	-	5	62
Colorado	-	3	-	-	-	3
Connecticut	3	27	5	-	3	38
Delaware	2	10	-	-	-	12
District of Columbia	1	11	-	173	-	185
Florida	9	96	-	-	-	105
Georgia	21	241	3	-	1	266
Hawaii	-	-	-	-	-	-
Idaho	-	-	-	-	-	-
Illinois	19	149	48	-	22	238
Indiana	1	27	-	7	2	37
Iowa	-	4	-	-	-	4
Kansas	-	4	-	-	-	4
Kentucky	1	45	-	-	-	46
Louisiana	24	180	-	-	2	206
Maine	1	-	-	-	-	1
Maryland	8	63	-	-	1	72
Massachusetts	1	10	7	-	-	18
Michigan	15	72	6	-	15	108

[Continued]

★ 834 ★

Elective Municipal Offices, January 1993
[Continued]

State	Mayors	Municipal governing bodies	Members, Members, municipal boards	Members, advisory commissions	Other municipal officials	Total
Minnesota	-	2	-	-	-	2
Mississippi	32	302	-	-	3	337
Missouri	17	110	-	-	5	132
Montana	-	-	-	-	-	-
Nebraska	-	1	-	-	-	1
Nevada	-	3	-	-	-	3
New Hampshire	-	-	-	-	-	-
New Jersey	12	94	-	-	-	106
New Mexico	-	-	-	-	-	-
New York	4	35	6	-	2	47
North Carolina	27	251	-	-	1	279
North Dakota	-	-	-	-	-	-
Ohio	11	100	3	-	9	123
Oklahoma	16	55	-	-	21	92
Oregon	-	1	-	-	-	1
Pennsylvania	4	48	-	-	1	53
Rhode Island	-	3	-	-	-	3
South Carolina	18	169	1	-	-	188
South Dakota	1	1	-	-	-	2
Tennessee	4	51	-	-	-	55
Texas	23	281	1	-	-	305
Utah	-	-	-	-	-	-
Vermont	-	-	-	-	-	-
Virginia	5	71	-	-	3	79
Washington	1	7	-	-	-	8
West Virginia	1	16	-	-	-	17
Wisconsin	1	10	-	-	-	11
Wyoming	-	-	-	-	-	-
Total	356	3,161	88	180	118	3,903

Source: "Blacks in Elective Municipal Offices, January 1993." *Black Elected Officials, 1993*, p. xxxiv. Published by permission.

★ 835 ★
Elected and Appointed Officials

Ethnicity of Mayor, 1991

Classification	No. reporting (A)	Ethnic background of mayor									
		Native American		Hispanic		Asian		White		Black	
		No.	% of (A)	No.	% of (A)	No.	% of (A)	No.	% of (A)	No.	% of (A)
Total, all cities	4,825	16	0.3	87	1.8	9	0.2	4,612	95.6	101	2.1
Population group											
Over 1,000,000	2	0	0.0	0	0.0	0	0.0		0.0	2	100.0
500,000-1,000,000	8	0	0.0	0	0.0	0	0.0	8	100.0	0	0.0
250,000-499,999	32	0	0.0	2	6.3	0	0.0	24	75.0	6	18.8
100,000-249,999	99	0	0.0	1	1.0	1	1.0	89	89.9	8	8.1
50,000-99,999	252	0	.0.	6	2.4	2	0.8	233	92.5	11	4.4
25,000-49,999	506	0	0.0	10	2.0	1	0.2	485	95.8	10	2.0
10,000-24,999	1,090	1	0.1	21	1.9	1	0.1	1,041	95.5	26	2.4
5,000-9,999	1,187	5	0.4	26	2.2	2	0.2	1,138	95.9	16	1.3
2,500-4,999	1,272	6	0.5	15	1.2	1	0.1	1,232	96.9	189	1.4
Under 2,500	377	4	1.1	6	1.6	1	0.3	362	96.0	4	1.1
Geographic division											
New England	446	1	0.2	0	0.0	0	0.0	441	98.9	4	0.9
Mid-Atlantic	694	0	0.0	5	0.7	0	0.0	676	97.4	13	1.9
East North Central	939	1	0.1	6	0.6	0	0.0	913	97.2	19	2.0
West North Central	556	0	0.0	1	0.2	1	0.2	548	98.6	6	1.1
South Atlantic	629	2	0.3	5	0.8	0	0.0	590	93.8	32	5.1
East South Central	279	0	0.0	1	0.4	0	0.0	271	97.1	7	2.5
West South Central	500	5	1.0	25	5.0	1	0.2	462	92.4	7	1.4
Mountain	262	1	0.4	18	6.9	2	0.8	239	91.2	2	0.8
Pacific Coast	520	6	1.2	26	5.0	5	1.0	472	90.8	11	2.1
Metro status											
Central	377	0	0.0	8	2.1	2	0.5	346	91.8	21	5.6
Suburban	2,577	6	0.2	51	2.0	5	0.2	2,464	95.6	51	2.0
Independent	1,871	10	0.5	28	1.5	2	0.1	1,802	96.3	29	1.5
Form of government											
Mayor-council	2,175	9	0.4	29	1.3	4	0.2	2,095	96.3	38	1.7
Council-manager	2,326	5	0.2	57	2.5	5	0.2	2,199	94.5	60	2.6
Commission	92	1	1.1	1	1.1	0	0.0	87	94.6	3	3.3
Town meeting	200	1	0.5	0	0.0	0	0.0	199	99.5	0	0.0
Rep. town meeting	32	0	0.0	0	0.0	0	0.0	32	100.0	0	0.0

Source: "Ethnicity of Mayor." Susan A. MacManus, "Women and Racial/Ethnic Minorities in Mayoral and Council Positions," *The Municipal Yearbook, 1993*, p. 75. Published by permission.

★ 836 ★

Elected and Appointed Officials

Male County Officials by Race and Ethnicity, 1991 - I

Position	Total reporting (A)	Total males		Race					
				White		Black		American Indian	
		No. (B)	% of (A)	No.	% of (B)	No.	% of (B)	No.	% of (B)
Board chairman	2,866	2,648	92.4	2,526	95.4	59	2.2	7	0.3
Chief appointed administrative officer	1,189	968	81.4	920	95.0	12	1.2	3	0.3
Clerk to the governing board	2,795	984	35.2	929	94.4	13	1.3	3	0.3
Chief financial officer	2,537	1,247	49.2	1,178	94.5	15	1.2	4	0.3
County health officer	1,970	1,392	70.7	1,302	93.5	21	1.5	1	0.1
Planning director	1,323	1,160	87.7	1,110	95.7	20	1.7	3	0.3
County engineer	1,586	1,580	99.6	1,521	96.3	3	0.2	2	0.1
Director welfare/human services	1,433	805	56.2	730	90.7	28	3.5	1	0.1
Chief law enforcement official	2,837	2,825	99.6	2,732	96.7	30	1.1	7	0.2
Purchasing director	1,209	774	64.0	728	94.1	21	2.7	3	0.4
Personnel director	1,073	666	62.1	609	91.4	32	4.8	3	0.5

Source: "Male County Officials by Race and Ethnicity," *Municipal Yearbook*, 1991, p. 287. Published by permission.

★ 837 ★

Elected and Appointed Officials

Male County Officials by Race and Ethnicity, 1991 - II

Position	Race				Race not reported		Hispanic	
	Asian		Other					
	No.	% of (B)	No.	% of (B)	No.	% of (B)	No.	% of (B)
Board chairman	3	0.1	5	0.2	48	1.8	32	1.2
Chief appointed administrative officer	4	0.4	3	0.3	26	2.7	16	1.7
Clerk to the governing board	3	0.3	2	0.2	34	3.5	16	1.6
Chief financial officer	5	0.4	5	0.4	40	3.2	17	1.4
County health officer	14	1.0	13	0.9	41	2.9	20	1.4
Planning director	5	0.4	4	0.3	18	1.6	13	1.1
County engineer	5	0.3	4	0.3	45	2.8	13	0.8
Director welfare/human services	4	0.5	5	0.6	37	4.6	24	3.0
Chief law enforcement official	3	0.1	5	0.2	48	1.7	42	1.5
Purchasing director	3	0.4	1	0.1	18	2.3	15	1.9
Personnel director	6	0.9	3	0.5	13	2.0	11	1.7

Source: "Male County Officials by Race and Ethnicity," *Municipal Yearbook*, 1991, p. 287. Published by permission.

★ 838 ★

Elected and Appointed Officials

Male County Officials by Race and Ethnicity, 1992 - I

Position	Total reporting (A)	Total males		White		Black		American Indian	
		No. (B)	% of (A)	No.	% of (B)	No.	% of (B)	No.	% of (B)
Board chairman	2,881	2,643	91.7	2,519	95.3	69	2.6	8	0.3
Chief appointed administrative officer	1,217	991	81.4	941	95.0	14	1.4	3	0.3
Clerk to the governing board	2,819	953	33.8	903	94.8	12	1.3	4	0.4
Chief financial officer	2,560	1,243	48.6	1,181	95.0	16	1.3	6	0.5
County health officer	1,978	1,379	69.7	1,296	94.0	22	1.6	1	0.1
Planning director	1,329	1,155	86.9	1,114	96.5	17	1.5	2	0.2
County engineer	1,617	1,607	99.4	1,549	96.4	4	0.2	3	0.2
Director health/human services	1,473	809	54.9	734	90.7	31	3.8	1	0.1
Chief law enforcement official	2,863	2,857	99.8	2,772	97.0	36	1.3	7	0.2
Purchasing director	1,225	783	63.9	732	93.5	22	2.8	4	0.5
Personnel director	1,093	671	61.4	611	91.1	36	5.4	4	0.6

Source: "Male County Officials by Race and Ethnicity," *Municipal Yearbook*, 1992, p. 285. Published by permission.

★ 839 ★

Elected and Appointed Officials

Male County Officials by Race and Ethnicity, 1992 - II

Position	Asian		Other		Race not reported		Hispanic	
	No.	% of (B)	No.	% of (B)	No.	% of (B)	No.	% of (B)
Board chairman	4	0.2	7	0.3	36	1.4	36	1.4
Chief appointed administrative officer	3	0.3	4	0.4	26	2.6	18	1.8
Clerk to the governing board	4	0.4	3	0.3	27	2.8	17	1.8
Chief financial officer	6	0.5	5	0.4	29	2.3	18	1.4
County health officer	13	0.9	12	0.9	35	2.5	21	1.5
Planning director	5	0.4	3	0.3	14	1.2	13	1.1
County engineer	4	0.2	3	0.2	44	2.7	14	0.9
Director health/human services	4	0.5	4	0.5	35	4.3	22	2.7
Chief law enforcement official	3	0.1	3	0.1	36	1.3	42	1.5
Purchasing director	4	0.5	1	0.1	20	2.6	16	2.0
Personnel director	5	0.7	4	0.6	11	1.6	16	2.4

Source: "Male County Officials by Race and Ethnicity," *Municipal Yearbook*, 1992, p. 285. Published by permission.

★ 840 ★

Elected and Appointed Officials

Male Municipal Officials by Race and Ethnicity, 1991 - I

Position	Total reporting (A)	Total males		Race					
				White		Black		American Indian	
		No. (B)	% of (A)	No.	% of (B)	No.	% of (B)	No.	% of (B)
Elected mayor/president	7,065	6,246	88.4	5,667	90.7	117	1.9	18	0.3
Chief appointed administrative officer/manager	5,056	4,426	87.5	4,127	93.2	48	1.1	7	0.2
Assistant manager/assistant CAO	1,524	1,026	67.3	894	87.1	45	4.4	7	0.7
Clerk	6,692	1,786	26.7	1,482	83.0	30	1.7	4	0.2
Chief financial officer	5,225	3,366	64.4	3,165	94.0	43	1.3	4	0.1
Director of economic development	848	698	82.3	303	43.4	13	1.9	1	0.1
Treasurer	4,791	2,425	50.6	2,169	89.4	32	1.3	8	0.3
Director of public works	5,854	5,779	98.7	5,422	93.8	112	1.9	25	0.4
Engineer	2,378	2,342	98.5	2,039	87.1	8	0.3	3	0.1
Police chief	6,580	6,550	99.5	6,184	94.4	137	2.1	31	0.5
Fire chief	6,183	6,178	99.9	5,920	95.8	57	0.9	10	0.2
Planning director	3,621	3,136	86.6	2,935	93.6	57	1.8	4	0.1
Personnel director	3,146	2,051	65.2	1,868	91.1	77	3.8	5	0.2
Risk manager	473	358	75.7	287	80.2	10	2.8	2	0.6
Director of parks and recreation	2,315	2,010	86.8	1,778	88.5	57	2.8	4	0.2
Superintendent of parks	2,006	1,930	96.2	1,767	91.6	39	2.0	5	0.3
Director of recreation	2,431	1,849	76.1	1,681	90.9	53	2.9	4	0.2
Librarian	2,759	564	20.4	498	88.3	7	1.2	1	0.2
Director of data processing	614	436	71.0	402	92.2	8	1.8	0	0.0
Purchasing director	2,778	1,968	70.8	1,821	92.5	45	2.3	1	0.1

Source: "Male Municipal Officials by Race and Ethnicity," *Municipal Yearbook*, 1991, p. 206. Published by permission.

★ 841 ★

Elected and Appointed Officials

Male Municipal Officials by Race and Ethnicity, 1991 - II

Position	Race				Race not reported		Hispanic	
	Asian		Other					
	No.	% of (B)	No.	% of (B)	No.	% of (B)	No.	% of (B)
Elected mayor/president	15	0.2	22	0.4	407	6.5	94	1.5
Chief appointed administrative officer/manager	6	0.1	12	0.3	226	5.1	81	1.8
Assistant manager/assistant CAO	5	0.5	5	0.5	70	6.8	29	2.8
Clerk	0	0.0	12	0.7	258	14.4	31	1.7
Chief financial officer	24	0.7	19	0.6	111	3.3	55	1.6
Director of economic development	3	0.4	5	0.7	373	53.4	7	1.0
Treasurer	10	0.4	11	0.5	195	8.0	27	1.1
Director of public works	20	0.3	32	0.6	168	2.9	120	2.1

[Continued]

★ 841 ★

Male Municipal Officials by Race and Ethnicity, 1991 - II

[Continued]

Position	Race				Race not reported		Hispanic	
	Asian		Other					
	No.	% of (B)	No.	% of (B)	No.	% of (B)	No.	% of (B)
Engineer	31	1.3	20	0.9	241	10.3	35	1.5
Police chief	5	0.1	26	0.4	167	2.5	113	1.7
Fire chief	2	0.0	17	0.3	172	2.8	68	1.1
Planning director	15	0.5	21	0.7	104	3.3	65	2.1
Personnel director	6	0.3	16	0.8	79	3.9	56	2.7
Risk manager	2	0.6	2	0.6	55	15.4	12	3.4
Director of parks and recreation	5	0.2	13	0.6	153	7.6	44	2.2
Superintendent of parks	7	0.4	19	1.0	93	4.8	70	3.6
Director of recreation	4	0.2	12	0.6	95	5.1	38	2.1
Librarian	3	0.5	3	0.5	52	9.2	6	1.1
Director of data processing	2	0.5	3	0.7	21	4.8	11	2.5
Purchasing director	8	0.4	16	0.8	77	3.9	60	3.0

Source: "Male Municipal Officials by Race and Ethnicity," *Municipal Yearbook*, 1991, p. 206. Published by permission.

★ 842 ★

Elected and Appointed Officials

Male Municipal Officials by Race and Ethnicity, 1992 - I

Position	Total reporting (A)	Total males		Race					
				White		Black		American Indian	
		No. (B)	% of (A)	No.	% of (B)	No.	% of (B)	No.	% of (B)
Elected mayor/president	7,081	6,256	88.3	5,795	92.6	121	1.9	18	0.3
Chief appointed administrative officer/manager	5,110	4,435	86.8	4,151	93.6	47	1.1	8	0.2
Assistant manager/assistant CAO	1,540	1,016	66.0	898	88.4	48	4.7	4	0.4
Clerk	6,691	1,683	25.2	1,435	85.3	29	1.7	4	0.2
Chief financial officer	5,222	3,323	63.6	3,146	94.7	40	1.2	2	0.1
Director of economic development	1,057	873	82.6	541	62.0	24	2.7	3	0.3
Treasurer	4,778	2,393	50.1	2,173	90.8	31	1.3	7	0.3
Director of public works	5,869	5,791	98.7	5,451	94.1	112	1.9	23	0.4
Engineer	2,530	2,493	98.5	2,192	87.9	14	0.6	3	0.1
Police chief	6,560	6,524	99.5	6,164	94.5	156	2.4	32	0.5
Fire chief	6,172	6,165	99.9	5,930	96.2	58	0.9	10	0.2
Planning director	3,612	3,110	86.1	2,916	93.8	58	1.9	3	0.1
Personnel director	3,154	2,023	64.1	1,850	91.4	80	4.0	3	0.1
Risk manager	733	529	72.2	441	83.4	15	2.8	3	0.6
Director of parks and recreation	2,451	2,104	85.8	1,871	88.9	64	3.0	3	0.1
Superintendent of parks	2,033	1,956	96.2	1,801	92.1	43	2.2	3	0.2
Director of recreation	2,416	1,787	74.0	1,625	90.9	53	3.0	4	0.2
Librarian	2,825	577	20.4	502	87.0	6	1.0	0	0.0

[Continued]

★ 842 ★

Male Municipal Officials by Race and Ethnicity, 1992 - I

[Continued]

Position	Total reporting (A)	Total males		Race					
				White		Black		American Indian	
		No. (B)	% of (A)	No.	% of (B)	No.	% of (B)	No.	% of (B)
Director of data processing	820	571	69.6	523	91.6	7	1.2	1	0.2
Purchasing director	2,786	1,961	70.4	1,813	92.5	47	2.4	1	0.1

Source: "Male Municipal Officials by Race and Ethnicity," *Municipal Yearbook*, 1992, p. 201. Published by permission.

★ 843 ★

Elected and Appointed Officials

Male Municipal Officials by Race and Ethnicity, 1992 - II

Position	Race				Race not reported		Hispanic	
	Asian		Other					
	No.	% of (B)	No.	% of (B)	No.	% of (B)	No.	% of (B)
Elected mayor/president	9	0.1	23	0.4	290	4.6	89	1.4
Chief appointed administrative officer/manager	7	0.2	14	0.3	208	4.7	85	1.9
Assistant manager/assistant CAO	4	0.4	5	0.5	57	5.6	31	3.1
Clerk	1	0.1	13	0.8	201	11.9	28	1.7
Chief financial officer	24	0.7	18	0.5	93	2.8	46	1.4
Director of economic development	4	0.5	4	0.5	297	34.0	12	1.4
Treasurer	9	0.4	16	0.7	157	6.6	28	1.2
Director of public works	17	0.3	31	0.5	157	2.7	121	2.1
Engineer	32	1.3	23	0.9	229	9.2	41	1.6
Police chief	7	0.1	26	0.4	139	2.1	105	1.6
Fire chief	1	0.0	19	0.3	147	2.4	76	1.2
Planning director	20	0.6	16	0.5	97	3.1	63	2.0
Personnel director	4	0.2	16	0.8	70	3.5	51	2.5
Risk manager	2	0.4	3	0.6	65	12.3	17	3.2
Director of parks and recreation	4	0.2	13	0.6	149	7.1	49	2.3
Superintendent of parks	6	0.3	20	1.0	83	4.2	72	3.7
Director of recreation	4	0.2	11	0.6	90	5.0	39	2.2
Librarian	4	07	3	0.5	62	10.7	9	1.6
Director of data processing	2	0.4	4	0.7	34	6.0	13	2.3
Purchasing director	8	0.4	17	0.9	75	3.8	60	3.1

Source: "Male Municipal Officials by Race and Ethnicity," *Municipal Yearbook*, 1992, p. 201. Published by permission.

★ 844 ★

Elected and Appointed Officials

Mayors of Cities with Populations Over 50,000, January 1993

Name	Term expiration	City	Population[1]	Percent Black
David Dinkins	12/93	New York, NY	7,322,564	28.7
Thomas Bradley	7/93	Los Angeles, CA	3,485,390	14.0
Coleman Young	12/93	Detroit, MI	1,027,974	75.1
Kurt Schmoke	12/95	Baltimore, MD	736,014	59.2
Willie Herenton	10/95	Memphis, TN	610,337	55.0
Sharon Pratt Kelly	12/94	Washington, DC	606,900	66.8
Norman B. Rice	12/93	Seattle, WA	516,259	10.0
Michael R. White	12/92	Cleveland, OH	505,616	46.6
Sidney Barthelemy	3/94	New Orleans, LA	496,938	61.9
Wellington Webb	6/95	Denver, CO	467,610	13.0
Emanuel Cleaver	4/95	Kansas City, MO	435,146	30.0
Maynard Jackson	12/93	Atlanta, GA	396,017	67.1
Elihu Harris	12/94	Oakland, CA	372,242	43.9
Dwight Tillery	12/92	Cincinnati, OH	364,040	38.0
Sharpe James	6/94	Newark, NJ	275,221	58.5
Richard Arrington	12/95	Birmingham, AL	265,968	63.3
Walter T. Kenney	6/92	Richmond, VA	203,056	55.2
Richard C. Dixon	12/93	Dayton, OH	182,044	40.4
Stanley Woodrow	12/95	Flint, MI	140,761	48.0
Carrie Perry	12/93	Hartford, CT	139,739	38.9
Charles E. Box	4/93	Rockford, IL	139,426	15.0
John C. Daniels, Jr.	12/93	New Haven, CT	130,474	36.1
Thomas Barnes	12/95	Gary, IN	116,646	80.6
Edward Vincent	11/94	Inglewood, CA	109,602	51.9
Noel Taylor	6/92	Roanoke, VA	96,397	24.3
Kenneth S. Reeves	12/93	Cambridge, MA	95,802	13.0
Coy Payne	3/92	Chandler, AZ	90,533	3.0
Bernice Wood[2]	6/93	Compton, CA	90,454	54.8
Douglas H. Palmer	6/94	Trenton, NJ	88,675	49.0
Aaron Thompson	6/93	Camden, NJ	87,492	56.4
George Livingston	11/93	Richmond, CA	87,425	43.8
Cardell Cooper	12/93	East Orange, NJ	73,552	89.9
Betty Payne	4/93	Evanston Twsp. IL	73,233	22.9
James H. Sills, Jr.	12/96	Wilmington, DE	71,529	52.4
Wallace E. Holland	12/93	Pontiac, MI	71,166	42.2
Henry Nickelberry	12/93	Saginaw, MI	69,512	40.3
Ronald A. Blackwood	12/95	Mt. Vernon, NY	67,153	55.3
Michael Steele	6/94	Irvington, NJ	59,774	69.9

Source: "Black Mayors of Cities with Populations Over 50,000, January 1993." *Black Elected Officials, 1993*, p. xxxiii. Published by permission. *Notes:* Mayors are listed by the population size of their respective cities, in decreasing order. 1. U.S. Bureau of the Census, 1990 population. 2. Bernice Wood was appointed to fill the mayor's post abdicated by Walter Tucker III, who was elected to California's 37th congressional district.

★ 845 ★

Elected and Appointed Officials

State Representatives, January 1993

State	Total House	Black members	Percent Black
Alabama	105	18	17.1
Alaska	40	1	2.5
Arizona	60	3	5.0
Arkansas	100	10	10.0
California	80	7	8.8
Colorado	65	3	4.6
Connecticut	151	9	6.0
Delaware	41	2	4.9
District of Columbia	-	-	-
Florida	120	14	11.7
Georgia	180	31	17.2
Hawaii	51	-	-
Idaho	70	-	-
Illinois	118	12	10.2
Indiana	100	7	7.0
Iowa	100	1	1.0
Kansas	125	4	3.2
Kentucky	100	3	3.0
Louisiana	105	23	21.9
Maine	151	-	-
Maryland	141	23	16.3
Massachusetts	160	7	4.4
Michigan	110	11	10.0
Minnesota	134	1	0.7
Mississippi	122	32	26.2
Missouri	163	13	8.0
Montana	100	-	-
Nebraska	-	-	-
Nevada	42	2	4.8
New Hampshire	400	2	0.5
New Jersey	80	10	12.5
New Mexico	70	-	-
New York	150	21	14.0
North Carolina	120	18	15.0
North Dakota	98	-	-
Ohio	99	12	12.1
Oklahoma	101	3	3.0
Oregon	60	2	3.3
Pennsylvania	203	14	6.9
Rhode Island	100	8	8.0
South Carolina	124	18	14.5
South Dakota	70	-	-
Tennessee	99	12	12.1
Texas	150	14	9.3
Utah	75	-	-
Vermont	150	2	1.3
Virginia	100	7	7.0

[Continued]

★ 845 ★

State Representatives, January 1993
[Continued]

State	Total House	Black members	Percent Black
Washington	98	1	1.0
West Virginia	100	1	1.0
Wisconsin	99	6	6.1
Wyoming	64	-	-
Total	5,440	388	7.1

Source: "Blacks in State Houses, January 1993." *Black Elected Officials, 1993*, p. xxviii. Published by permission.

★ 846 ★

Elected and Appointed Officials

State Senators, January 1993

State	Total Senate	Black members	Percent Black
Alabama	35	4	11.4
Alaska	20	-	-
Arizona	30	1	3.3
Arkansas	35	3	8.6
California	40	2	5.0
Colorado	35	1	2.9
Connecticut	36	3	8.3
Delaware	21	1	4.8
District of Columbia	-	-	-
Florida	40	5	12.5
Georgia	56	9	16.1
Hawaii	25	-	-
Idaho	35	-	-
Illinois	59	8	13.6
Indiana	50	4	8.0
Iowa	50	-	-
Kansas	40	2	5.0
Kentucky	38	1	2.6
Louisiana	39	8	20.5
Maine	35	-	-
Maryland	47	7	14.9
Massachusetts	40	1	2.5
Michigan	38	3	7.9
Minnesota	67	-	-
Mississippi	52	10	19.2
Missouri	34	3	8.8
Montana	50	-	-
Nebraska	49	1	2.0
Nevada	21	1	4.8

[Continued]

★ 846 ★

State Senators, January 1993

[Continued]

State	Total Senate	Black members	Percent Black
New Hampshire	24	-	-
New Jersey	40	2	5.0
New Mexico	42	-	-
New York	61	5	8.2
North Carolina	50	7	14.0
North Dakota	49	-	-
Ohio	33	3	9.1
Oklahoma	48	2	4.2
Oregon	30	1	3.3
Pennsylvania	50	3	6.0
Rhode Island	50	1	2.0
South Carolina	46	7	15.2
South Dakota	35	1	2.9
Tennessee	33	3	9.1
Texas	31	2	6.5
Utah	29	-	-
Vermont	30	-	-
Virginia	40	5	12.5
Washington	49	1	2.0
West Virginia	34	-	-
Wisconsin	33	2	6.1
Wyoming	30	-	-
Total	1,984	123	6.2

Source: "Blacks in State Senates, January 1993." *Black Elected Officials, 1993*, p. xxvii. Published by permission.

★ 847 ★

Elected and Appointed Officials

Statewide Elective Offices, January 1993

Name	Title	State	% White VAP[1]	%Black VAP[1]	% Other VAP[1]
Oscar W. Adams	Associate Justice State Supreme Court	AL	76.3	22.7	1.0
Gregory K. Scott[2]	Justice State Supreme Court	CO	89.4	3.7	6.9
Francisco Borges	State Treasurer	CT	88.7	7.4	3.9
Leander J. Shaw, Jr.	Justice State Supreme Court	FL	85.6	11.4	3.0

[Continued]

★ 847 ★

Statewide Elective Offices, January 1993
[Continued]

Name	Title	State	% White VAP[1]	%Black VAP[1]	% Other VAP[1]
Robert Benham	Associate Justice State Supreme Court	GA	73.5	24.6	1.9
Leah Sears-Collins	Justice State Supreme Court	GA	73.5	24.6	1.9
Carol Moseley Braun	U.S. Senator	IL	80.5	13.4	6.1
Roland W. Burris	State Attorney General	IL	80.5	13.4	6.1
Dwayne Marc Brown	Clerk, State Supreme & Appellate Courts	IN	91.4	7.1	1.5
Pamela Carter	Attorney General	IN	91.4	7.1	1.5
Richard H. Austin	Secretary of State	MI	84.9	12.8	2.3
Conrad Mallett, Jr.	Associate Justice State Supreme Court	MI	84.9	12.8	2.3
Alan Page	Associate Justice State Supreme Court	MN	95.6	1.8	2.6
Ralph Campbell	State Auditor	NC	77.7	20.1	2.4
Henry E. Frye	Associate Justice State Supreme Court	NC	77.7	20.1	2.4
J.C. Watts	State Corporation Commissioner	OK	84.3	6.6	9.0
Jim Hill	State Treasurer	OR	93.6	1.4	5.0
Robert C. Nix	Chief Justice State Supreme Court	PA	89.6	8.5	1.9
Morris Overstreet	Judge, Court of Criminal Appeals	TX	77.4	11.2	11.4
L. Douglas Wilder	Governor	VA	78.9	17.6	3.5
Charles Smith	Associate Judge State Supreme Court	WA	89.9	2.8	7.4

Source: "Blacks in Statewide Elective Offices, January 1993." *Black Elected Officials, 1993,* p. xxx. Published by permission. *Notes:* 1. Voting age population. 2. Appointed to an elective post.

★ 848 ★

Elected and Appointed Officials

Total Officials Elected: Gender and Ratio, 1970-1993

Year	Male BEOs	Female BEOs	Male/Female BEO Ratio
1970	1,469	160	9.2
1971	1,860	225	8.3
1972	2,264	153	14.8
1973	2,621	345	7.6
1974	2,991	416	7.2
1975	2,973	530	5.7
1976	3,295	684	4.8
1977	3,529	782	4.5
1978	3,660	843	4.3
1979	3,725	882	4.2
1980	3,936	976	4.0
1981	4,017	1,021	3.9
1982	4,079	1,081	3.8
1983	4,383	1,223	3.6
1984[1]	4,441	1,259	3.5
1985	4,697	1,359	3.4
1986	4,942	1,482	3.3
1987	5,117	1,564	3.3
1988	5,204	1,625	3.2
1989	5,412	1,814	3.0
1990	5,420	1,950	2.8
1991	5,427	2,053	2.6
1992	5,431	2,121	2.6
1993	5,683	2,332	2.4

Source: "Total Male and Female Black Elected Officials, by Year, 1970-1993." *Black Elected Officials, 1993,* p. xxv. Published by permission. *Notes:* 1. The 1984 figures reflect the number of blacks who took office during the seven-month period between July 1, 1983, and January 30, 1984.

★ 849 ★

Elected and Appointed Officials

Women County Officials by Race and Ethnicity, 1991 - I

Position	Total reporting (A)	Total females		Race					
				White		Black		American Indian	
		No. (B)	% of (A)	No.	% of (B)	No.	% of (B)	No.	% of (B)
Board chairman	2,866	218	7.6	211	96.8	4	1.8	1	0.5
Chief appointed administrative officer	1,189	221	18.6	211	95.5	5	2.3	0	0.0
Clerk to the governing board	2,795	1,811	64.8	1,754	96.9	26	1.4	9	0.5
Chief financial officer	2,537	1,290	50.8	1,254	97.2	11	0.9	2	0.2

[Continued]

★ 849 ★

Women County Officials by Race and Ethnicity, 1991 - I
[Continued]

Position	Total reporting (A)	Total females		White		Black		American Indian	
		No. (B)	% of (A)	No.	% of (B)	No.	% of (B)	No.	% of (B)
County health officer	1,970	578	29.3	552	95.5	11	1.9	1	0.2
Planning director	1,323	163	12.3	159	97.5	2	1.2	0	0.0
County engineer	1,586	6	0.4	6	100.0	0	0.0	0	0.0
Director welfare/human services	1,433	628	43.8	569	90.6	40	6.4	2	0.3
Chief law enforcement official	2,837	12	0.4	12	100.0	0	0.0	0	0.0
Purchasing director	1,209	435	36.0	401	92.2	23	5.3	3	0.7
Personnel director	1,073	407	37.9	369	90.7	29	7.1	1	0.2

Source: "Female County Officials by Race and Ethnicity," *Municipal Yearbook*, 1991, p. 287. Published by permission.

★ 850 ★

Elected and Appointed Officials

Women County Officials by Race and Ethnicity, 1991 - II

Position	Race				Race not reported		Hispanic	
	Asian		Other					
	No.	% of (B)	No.	% of (B)	No.	% of (B)	No.	% of (B)
Board chairman	0	0.0	2	0.9	0	0.0	5	2.3
Chief appointed administrative officer	1	0.5	1	0.5	3	1.4	6	2.7
Clerk to the governing board	1	0.1	8	0.4	13	0.7	27	1.5
Chief financial officer	1	0.1	4	0.3	18	1.4	17	1.3
County health officer	2	0.3	4	0.7	8	1.4	4	0.7
Planning director	0	0.0	0	0.0	2	1.2	4	2.5
County engineer	0	0.0	0	0.0	0	0.0	0	0.0
Director welfare/human services	2	0.3	3	0.5	12	1.9	10	1.6
Chief law enforcement official	0	0.0	0	0.0	0	0.0	0	0.0
Purchasing director	1	0.2	2	0.5	5	1.1	11	2.5
Personnel director	2	0.5	4	1.0	2	0.5	8	2.0

Source: "Female County Officials by Race and Ethnicity," *Municipal Yearbook*, 1991, p. 287. Published by permission.

★ 851 ★

Elected and Appointed Officials

Women County Officials by Race and Ethnicity, 1992 - I

Position	Total reporting (A)	Total females		Race					
				White		Black		American Indian	
		No. (B)	% of (A)	No.	% of (B)	No.	% of (B)	No.	% of (B)
Board chairman	2,881	238	8.3	229	96.2	7	2.9	2	0.8
Chief appointed administrative officer	1,217	226	18.6	217	96.0	4	1.8	0	0.0
Clerk to the governing board	2,819	1,866	66.2	1,808	96.9	29	1.6	8	0.4
Chief financial officer	2,560	1,317	51.4	1,284	97.5	11	0.8	0	0.0
County health officer	1,978	599	30.3	573	95.7	12	2.0	0	0.0
Planning director	1,329	174	13.1	170	97.7	2	1.1	0	0.0
County engineer	1,617	10	0.6	10	100.0	0	0.0	0	0.0
Director welfare/human services	1,473	664	45.1	600	90.4	45	6.8	2	0.3
Chief law enforcement official	2,863	6	0.2	6	100.0	0	0.0	0	0.0
Purchasing director	1,225	442	36.1	408	92.3	24	5.4	2	0.5
Personnel director	1,093	422	38.6	382	90.5	33	7.8	0	0.0

Source: "Female County Officials by Race and Ethnicity," *Municipal Yearbook*, 1992, p. 285. Published by permission.

★ 852 ★

Elected and Appointed Officials

Women County Officials by Race and Ethnicity, 1992 - II

Position	Race				Race not reported		Hispanic	
	Asian		Other					
	No.	% of (B)	No.	% of (B)	No.	% of (B)	No.	% of (B)
Board chairman	0	0.0	0	0.0	0	0.0	1	0.4
Chief appointed administrative officer	2	0.9	1	0.4	2	0.9	6	2.7
Clerk to the governing board	1	0.1	9	0.5	11	0.6	34	1.8
Chief financial officer	2	0.2	4	0.3	16	1.2	16	1.2
County health officer	4	0.7	3	0.5	7	1.2	4	0.7
Planning director	0	0.0	0	0.0	2	1.1	2	1.1
County engineer	0	0.0	0	0.0	0	0.0	0	0.0
Director welfare/human services	2	0.3	5	0.8	10	1.5	13	2.0
Chief law enforcement official	0	0.0	0	0.0	0	0.0	0	0.0
Purchasing director	1	0.2	2	0.5	5	1.1	10	2.3
Personnel director	2	0.5	4	0.9	1	0.2	7	1.7

Source: "Female County Officials by Race and Ethnicity," *Municipal Yearbook*, 1992, p. 285. Published by permission.

★ 853 ★

Elected and Appointed Officials

Women Elected County Officer, January 1993

State	Members, county governing bodies	Members, other county bodies	Total
Alabama	1	7	8
Alaska	-	-	-
Arizona	-	-	-
Arkansas	-	-	-
California	2	0	2
Colorado	-	-	-
Connecticut	-	-	-
Delaware	1	1	2
District of Columbia	-	-	-
Florida	8	1	9
Georgia	14	2	16
Hawaii	-	-	-
Idaho	-	-	-
Illinois	7	0	7
Indiana	3	0	3
Iowa	0	0	0
Kansas	1	0	1
Kentucky	-	-	-
Louisiana	9	0	9
Maine	-	-	-
Maryland	1	0	1
Massachusetts	-	-	-
Michigan	10	0	10
Minnesota	-	-	-
Mississippi	1	36	37
Missouri	1	0	1
Montana	-	-	-
Nebraska	1	0	1
Nevada	1	0	1
New Hampshire	-	-	-
New Jersey	1	1	2
New Mexico	-	-	-
New York	3	0	3
North Carolina	10	2	12
North Dakota	-	-	-
Ohio	-	-	-
Oklahoma	1	1	2
Oregon	1	0	1
Pennsylvania	1	0	1
Rhode Island	-	-	-
South Carolina	16	2	18
South Dakota	-	-	-
Tennessee	6	0	6
Texas	1	1	2
Utah	-	-	-

[Continued]

★ 853 ★

Women Elected County Officer, January 1993

[Continued]

State	Members, county governing bodies	Members, other county bodies	Total
Vermont	-	-	-
Virginia	4	2	6
Washington	-	-	-
West Virginia	-	-	-
Wisconsin	1	0	1
Wyoming	-	-	-
Total	106	56	162

Source: "Black Women in Elective County Offices, January 1993." *Black Elected Officials, 1993*, p. xxxii. Published by permission.

★ 854 ★

Elected and Appointed Officials

Women Elected Education Officers, January 1993

State	Members, state education agency	members, university and college boards	Members, local school boards	Other education boards	Total
Alabama	1	0	24	1	26
Alaska	-	-	-	-	-
Arizona	0	0	2	0	2
Arkansas	0	0	21	0	21
California	0	4	39	0	43
Colorado	0	0	1	0	1
Connecticut	0	0	7	0	7
Delaware	0	0	3	0	3
District of Columbia	0	0	4	0	4
Florida	0	0	4	0	4
Georgia	0	1	28	0	29
Hawaii	-	-	-	-	-
Idaho	-	-	-	-	-
Illinois	0	0	61	0	61
Indiana	0	0	2	0	2
Iowa	-	-	-	-	-
Kansas	0	0	4	0	4
Kentucky	0	0	2	0	2
Louisiana	0	0	36	0	36
Maine	-	-	-	-	-
Maryland	0	0	4	0	4
Massachusetts	0	0	1	0	1
Michigan	1	6	33	0	40

[Continued]

★ 854 ★

Women Elected Education Officers, January 1993

[Continued]

State	Members, state education agency	members, university and college boards	Members, local school boards	Other education boards	Total
Minnesota	0	0	2	0	2
Mississippi	0	0	35	0	35
Missouri	0	1	6	0	7
Montana	-	-	-	-	-
Nebraska	0	0	2	0	2
Nevada	0	1	1	0	2
New Hampshire	-	-	-	-	-
New Jersey	0	0	40	0	40
New Mexico	0	0	1	0	1
New York	0	0	77	0	77
North Carolina	0	0	25	0	25
North Dakota	-	-	-	-	-
Ohio	0	1	20	0	21
Oklahoma	0	0	9	0	9
Oregon	-	-	-	-	-
Pennsylvania	0	0	16	0	16
Rhode Island	-	-	-	-	-
South Carolina	0	0	41	0	41
South Dakota	-	-	-	-	-
Tennessee	0	0	14	0	14
Texas	0	1	24	0	25
Utah	-	-	-	-	-
Vermont	-	-	-	-	-
Virginia	-	-	-	-	-
Washington	0	0	1	0	1
West Virginia	-	-	-	-	-
Wisconsin	0	0	2	0	2
Wyoming	-	-	-	-	-
Total	2	15	592	1	610

Source: "Black Women in Elective Education Offices, January 1993." *Black Elected Officials, 1993*, p. lxii. Published by permission.

★ 855 ★

Elected and Appointed Officials

Women Elected Judicial and Law Enforcement Officers, January 1993

State	Judges, State courts of last resort	Judges, other courts	Magistrates, justices of the peace, constables	Other judicial officials	Police chiefs sheriffs, and marshals	Total
Alabama	0	2	3	2	0	7
Alaska	-	-	-	-	-	-
Arizona	-	-	-	-	-	-
Arkansas	0	2	9	0	0	11
California	0	22	0	0	0	22
Colorado	-	-	-	-	-	-
Connecticut	0	0	2	0	0	2
Delaware	-	-	-	-	-	-
District of Columbia	-	-	-	-	-	-
Florida	0	5	0	0	0	5
Georgia	1	10	0	1	1	13
Hawaii	-	-	-	-	-	-
Idaho	-	-	-	-	-	-
Illinois	0	14	0	0	0	14
Indiana	0	1	0	0	0	1
Iowa	0	0	0	1	0	1
Kansas	0	1	0	0	0	1
Kentucky	0	1	0	0	0	1
Louisiana	0	10	3	2	0	15
Maine	-	-	-	-	-	-
Maryland	0	3	0	3	0	6
Massachusetts	-	-	-	-	-	-
Michigan	0	28	0	0	0	28
Minnesota	0	2	0	0	0	2
Mississippi	0	8	0	4	0	12
Missouri	0	1	0	0	0	1
Montana	-	-	-	-	-	-
Nebraska	-	-	-	-	-	-
Nevada	-	-	-	-	-	-
New Hampshire	-	-	-	-	-	-
New Jersey	-	-	-	-	-	-
New Mexico	0	1	0	0	0	1
New York	0	19	0	0	0	19
North Carolina	0	7	0	0	0	7
North Dakota	-	-	-	-	-	-
Ohio	0	14	0	1	0	15
Oklahoma	-	-	-	-	-	-
Oregon	0	1	0	0	0	1
Pennsylvania	0	8	1	0	0	9
Rhode Island	-	-	-	-	-	-
South Carolina	0	1	1	1	0	3
South Dakota	-	-	-	-	-	-
Tennessee	0	1	1	0	0	2
Texas	0	4	4	0	0	8
Utah	-	-	-	-	-	-

[Continued]

★ 855 ★

Women Elected Judicial and Law Enforcement Officers, January 1993
[Continued]

State	Judges, State courts of last resort	Judges, other courts	Magistrates, justices of the peace, constables	Other judicial officials	Police chiefs sheriffs, and marshals	Total
Vermont	-	-	-	-	-	-
Virginia	0	0	0	3	0	3
Washington	0	1	0	0	0	1
West Virginia	0	0	1	0	0	1
Wisconsin	0	1	0	0	0	1
Wyoming	-	-	-	-	-	-
Total	1	168	25	18	1	213

Source: "Black Women in Elective Judicial and Law Enforcement Offices, January 1993." *Black Elected Officials, 1993*, p. xxxvii. Published by permission.

★ 856 ★

Elected and Appointed Officials

Women Elected Municipal Officers, January 1993

State	Mayors	Members, governing boards	Members, members, municipal boards	Members, advisory commissions	Other municipal officials	Total
Alabama	7	116	0	0	0	123
Alaska	0	1	0	0	0	1
Arizona	-	-	-	-	-	-
Arkansas	5	50	0	0	20	75
California	2	12	5	0	3	22
Colorado	0	1	0	0	0	1
Connecticut	1	13	1	0	2	17
Delaware	0	1	0	0	0	1
District of Columbia	1	3	0	102	0	106
Florida	1	23	0	0	0	24
Georgia	3	51	0	0	1	55
Hawaii	-	-	-	-	-	-
Idaho	-	-	-	-	-	-
Illinois	6	41	16	0	10	73
Indiana	0	6	0	4	1	11
Iowa	0	2	0	0	0	2
Kansas	-	-	-	-	-	-
Kentucky	1	13	0	0	0	14
Louisiana	4	31	0	0	0	35
Maine	-	-	-	-	-	-
Maryland	2	26	0	0	1	29
Massachusetts	0	2	1	0	0	3
Michigan	3	21	3	0	10	37

[Continued]

★ 856 ★

Women Elected Municipal Officers, January 1993

[Continued]

State	Mayors	Members, governing boards	Members, members, municipal boards	Members, advisory commissions	Other municipal officials	Total
Minnesota	0	1	0	0	0	1
Mississippi	9	71	0	0	3	83
Missouri	2	36	0	0	2	40
Montana	-	-	-	-	-	-
Nebraska	-	-	-	-	-	-
Nevada	-	-	-	-	-	-
New Hampshire	-	-	-	-	-	-
New Jersey	0	26	0	0	0	26
New Mexico	-	-	-	-	-	-
New York	1	12	1	0	2	16
North Carolina	5	54	0	0	1	60
North Dakota	-	-	-	-	-	-
Ohio	2	32	1	0	4	39
Oklahoma	6	22	0	0	17	45
Oregon	-	-	-	-	-	-
Pennsylvania	1	15	0	0	1	17
Rhode Island	-	-	-	-	-	-
South Carolina	3	49	0	0	0	52
South Dakota	-	-	-	-	-	-
Tennessee	0	9	0	0	0	9
Texas	1	76	0	0	0	77
Utah	-	-	-	-	-	-
Vermont	-	-	-	-	-	-
Virginia	2	13	0	0	2	17
Washington	0	2	0	0	0	2
West Virginia	0	2	0	0	0	2
Wisconsin	1	2	0	0	0	3
Wyoming	-	-	-	-	-	-
Total	69	835	28	106	80	1,118

Source: "Black Women in Elective Municipal Offices, January 1993." *Black Elected Officials, 1993*, p. xxxv. Published by permission.

★ 857 ★

Elected and Appointed Officials

Women Elected Officials as Percentage of All Blacks Elected, January 1993

State	Black elected officials	Female BEOs	Female Percentage
Alabama	699	172	24.6
Alaska	3	2	66.7
Arizona	15	4	26.7
Arkansas	380	117	30.8
California	273	102	37.4
Colorado	20	4	20.0
Connecticut	62	29	46.8
Delaware	23	7	30.4
District of Columbia	198	113	57.1
Florida	200	55	27.5
Georgia	545	140	25.7
Hawaii	0	0	0.0
Idaho	0	0	0.0
Illinois	465	163	35.1
Indiana	72	21	29.2
Iowa	11	3	27.3
Kansas	21	7	33.3
Kentucky	63	17	27.0
Louisiana	636	106	16.7
Maine	1	0	0.0
Maryland	140	50	35.7
Massachusetts	30	8	26.7
Michigan	333	127	38.1
Minnesota	16	5	31.3
Mississippi	751	176	23.4
Missouri	185	52	28.1
Montana	0	0	0
Nebraska	6	4	66.7
Nevada	10	2	20.0
New Hampshire	2	1	50.0
New Jersey	211	72	34.1
New Mexico	3	2	66.7
New York	299	123	41.1
North Carolina	468	114	24.4
North Dakota	0	0	0.0
Ohio	219	80	36.5
Oklahoma	123	56	45.5
Oregon	10	4	40.0
Pennsylvania	158	47	29.7
Rhode Island	12	4	33.3
South Carolina	450	131	29.1
South Dakota	3	0	0.0
Tennessee	168	35	20.8
Texas	472	119	25.2

[Continued]

★ 857 ★

Women Elected Officials as Percentage of All Blacks Elected, January 1993

[Continued]

State	Black elected officials	Female BEOs	Female Percentage
Utah	0	0	0.0
Vermont	2	1	50.0
Virginia	31	8	25.8
Washington	19	5	26.3
West Virginia	21	3	14.3
Wisconsin	30	10	33.3
Wyoming	1	0	0.0
Total	8,015[1]	2,332	29.1

Source: "Black Elected Female Officials as a Percentage of All Black Elected Officials, January 1993." *Black Elected Officials, 1993*, p. xxiv. Published by permission. *Notes:* 1. Includes two statehood senators and one statehood representative from the District of Columbia.

★ 858 ★

Elected and Appointed Officials

Women Elected Officials by Region and Office, January 1992

	Northeast	Midwest	South	West	Total
Federal	0	2	1	1	4
State	26	25	67	14	132
County	4	24	106	5	139
Mayors	3	17	54	1	75
Other municipal	76	183	690	22	971
Judicial/law enforcement	21	52	82	22	177
Education	141	141	278	50	610
Total	271	444	1,278	115	2,110[1]
Percent	12.8	21.0	60.6	5.4	100.0

Source: "Distribution of Black Female Elected Officials by Region and Category of Office, January 1992." *Black Elected Officials, 1993*, p. lxiii. Published by permission. *Notes:* The 11 women BEOs from the Virgin Islands are not included in this table, because the Virgin Islands are not included in the census. 1. Includes one statehood senator and one regional officer.

★ 859 ★
Elected and Appointed Officials

Women Municipal Officials by Race and Ethnicity, 1991 - I

Position	Total reporting (A)	Total females		Race					
				White		Black		American Indian	
		No. (B)	% of (A)	No.	% of (B)	No.	% of (B)	No.	% of (B)
Elected mayor/president	7,065	819	11.6	780	95.2	20	2.4	2	0.2
Chief appointed administrative officer/manager	5,056	630	12.5	603	95.7	16	2.5	1	0.2
Assistant manager/assistant CAO	1,524	498	32.7	470	94.4	18	3.6	2	0.4
Clerk	6,692	4,906	73.3	4,699	95.8	88	1.8	20	0.4
Chief financial officer	5,225	1,859	35.6	1,779	95.7	27	1.5	7	0.4
Director of economic development	848	150	17.7	119	79.3	5	3.3	1	0.7
Treasurer	4,791	2,366	49.4	2,281	96.4	26	1.1	8	0.3
Director of public works	5,854	75	1.3	67	89.3	5	6.7	0	0.0
Engineer	2,378	36	1.5	34	94.4	1	2.8	0	0.0
Police chief	6,580	30	0.5	29	96.7	0	0.0	0	0.0
Fire chief	6,183	5	0.1	4	80.0	0	0.0	0	0.0
Planning director	3,621	485	13.4	454	93.6	13	2.7	3	0.6
Personnel director	3,146	1,095	34.8	1,007	92.0	58	5.3	6	0.5
Risk manager	473	115	24.3	106	92.2	5	4.3	1	0.9
Director of parks and recreation	2,315	305	13.2	284	93.1	12	3.9	2	0.7
Superintendent of parks	2,006	76	3.8	72	94.7	3	3.9	0	0.0
Director of recreation	2,431	582	23.9	550	94.5	20	3.4	4	0.7
Librarian	2,759	2,195	79.6	2,115	96.4	15	0.7	3	0.1
Director of data processing	614	178	29.0	166	93.3	8	4.5	0	0.0
Purchasing director	2,778	810	29.2	755	93.2	41	5.1	3	0.4

Source: "Female Municipal Officials by Race and Ethnicity," *Municipal Yearbook*, 1991, p. 206. Published by permission.

★ 860 ★
Elected and Appointed Officials

Women Municipal Officials by Race and Ethnicity, 1991 - II

Position	Race				Race not reported		Hispanic	
	Asian		Other					
	No.	% of (B)	No.	% of (B)	No.	% of (B)	No.	% of (B)
Elected mayor/president	0	0.0	2	0.2	15	1.8	9	1.1
Chief appointed administrative officer/manager	2	0.3	0	0.0	8	1.3	5	0.8
Assistant manager/assistant CAO	1	0.2	1	0.2	6	1.2	8	1.6
Clerk	12	0.2	14	0.3	73	1.5	96	2.0
Chief financial officer	9	0.5	4	0.2	33	1.8	25	1.3
Director of economic development	0	0.0	1	0.7	24	16.0	2	1.3
Treasurer	5	0.2	9	0.4	37	1.6	32	1.4
Director of public works	0	0.0	0	0.0	3	4.0	0	0.0

[Continued]

★ 860 ★

Women Municipal Officials by Race and Ethnicity, 1991 - II
[Continued]

Position	Race				Race not reported		Hispanic	
	Asian		Other					
	No.	% of (B)	No.	% of (B)	No.	% of (B)	No.	% of (B)
Engineer	1	2.8	0	0.0	0	0.0	0	0.0
Police chief	0	0.0	0	0.0	1	3.3	0	0.0
Fire chief	0	0.0	0	0.0	1	20.0	0	0.0
Planning director	3	0.6	1	0.2	11	2.3	5	1.0
Personnel director	4	0.4	8	0.7	12	1.1	30	2.7
Risk manager	0	0.0	3	2.6	0	0.0	4	3.5
Director of parks and recreation	0	0.0	1	0.3	6	2.0	5	1.6
Superintendent of parks	1	1.3	0	0.0	0	0.0	0	0.0
Director of recreation	3	0.5	2	0.3	3	0.5	4	0.7
Librarian	9	0.4	7	0.3	46	2.1	19	0.9
Director of data processing	1	0.6	0	0.0	3	1.7	2	1.1
Purchasing director	1	0.1	5	0.6	5	0.6	19	2.3

Source: "Female Municipal Officials by Race and Ethnicity," *Municipal Yearbook,* 1991, p. 206. Published by permission.

★ 861 ★
Elected and Appointed Officials

Women Municipal Officials by Race and Ethnicity, 1992 - I

Position	Total reporting (A)	Total females		Race					
				White		Black		American Indian	
		No. (B)	% of (A)	No.	% of (B)	No.	% of (B)	No.	% of (B)
Elected mayor/president	7,081	825	11.7	784	95.0	23	2.8	1	0.1
Chief appointed administrative officer/manager	5,110	675	13.2	646	95.7	17	2.5	2	0.3
Assistant manager/assistant CAO	1,540	524	34.0	488	93.1	23	4.4	2	0.4
Clerk	6,691	5,008	74.8	4,800	95.8	92	1.8	19	0.4
Chief financial officer	5,222	1,899	36.4	1,809	95.3	36	1.9	7	0.4
Director of economic development	1,057	184	17.4	159	86.4	7	3.8	0	0.0
Treasurer	4,778	2,385	49.9	2,304	96.6	31	1.3	7	0.3
Director of public works	5,869	78	1.3	69	88.5	6	7.7	1	1.3
Engineer	2,530	37	1.5	36	97.3	1	2.7	0	0.0
Police chief	6,560	36	0.5	36	100.0	0	0.0	0	0.0
Fire chief	6,172	7	0.1	6	85.7	0	0.0	0	0.0
Planning director	3,612	502	13.9	472	94.0	16	3.2	2	0.4
Personnel director	3,154	1,131	35.9	1,034	91.4	67	5.9	7	0.6
Risk manager	733	204	27.8	186	91.2	9	4.4	2	1.0
Director of parks and recreation	2,451	347	14.2	324	93.4	13	3.7	4	1.2
Superintendent of parks	2,033	77	3.8	73	94.8	3	3.9	0	0.0
Director of recreation	2,416	629	26.0	600	95.4	19	3.0	4	0.6
Librarian	2,825	2,248	79.6	2,168	96.4	22	1.0	4	0.2

[Continued]

★ 861 ★

Women Municipal Officials by Race and Ethnicity, 1992 - I
[Continued]

Position	Total reporting (A)	Total females		White		Black		American Indian	
		No. (B)	% of (A)	No.	% of (B)	No.	% of (B)	No.	% of (B)
Director of data processing	820	249	30.4	238	95.6	9	3.6	0	0.0
Purchasing director	2,786	825	29.6	769	93.2	42	5.1	2	0.2

Source: "Women Municipal Officials by Race and Ethnicity," *Municipal Yearbook*, 1992, p. 201. Published by permission.

★ 862 ★
Elected and Appointed Officials

Women Municipal Officials by Race and Ethnicity, 1992 - II

Position	Race				Race not reported		Hispanic	
	Asian		Other					
	No.	% of (B)	No.	% of (B)	No.	% of (B)	No.	% of (B)
Elected mayor/president	2	0.2	3	0.4	12	1.5	14	1.7
Chief appointed administrative officer/manager	2	0.3	1	0.1	7	1.0	11	1.6
Assistant manager/assistant CAO	2	0.4	2	0.4	7	1.3	12	2.3
Clerk	11	0.2	17	0.3	69	1.4	107	2.1
Chief financial officer	10	0.5	5	0.3	32	1.7	38	2.0
Director of economic development	1	0.5	1	0.5	16	8.7	3	1.6
Treasurer	2	0.1	9	0.4	32	1.3	38	1.6
Director of public works	0	0.0	0	0.0	2	2.6	0	0.0
Engineer	0	0.0	0	0.0	0	0.0	0	0.0
Police chief	0	0.0	0	0.0	0	0.0	1	2.8
Fire chief	0	0.0	0	0.0	1	14.3	0	0.0
Planning director	4	0.8	1	0.2	7	1.4	6	1.2
Personnel director	4	0.4	10	0.9	9	0.8	34	3.0
Risk manager	1	0.5	4	2.0	2	1.0	8	3.9
Director of parks and recreation	0	0.0	1	0.3	5	1.4	4	1.2
Superintendent of parks	1	1.3	0	0.0	0	0.0	0	0.0
Director of recreation	2	0.3	2	0.3	2	0.3	4	0.6
Librarian	12	0.5	7	0.3	35	1.6	21	0.9
Director of data processing	1	0.4	0	0.0	1	0.4	1	0.4
Purchasing director	1	0.1	5	0.6	6	0.7	21	2.5

Source: "Women Municipal Officials by Race and Ethnicity," *Municipal Yearbook*, 1992, p. 201. Published by permission.

★ 863 ★

Elected and Appointed Officials

Women in State Legislatures, January 1993

State	Total Black members	Female Black House members	Female Black Senate members	Female Black total	Female percent of Black legislators
Alabama	22	3	0	3	13.6
Alaska	1	1	0	1	100.0
Arizona	4	1	1	2	50.0
Arkansas	13	6	0	6	46.2
California	9	4	2	6	66.7
Colorado	4	2	0	2	50.0
Connecticut	12	3	1	4	33.3
Delaware	3	0	0	0	0.0
District of Columbia	-	-	-	-	-
Florida	19	6	1	7	36.8
Georgia	40	15	1	16	40.0
Hawaii	0	0	0	0	0.0
Idaho	0	0	0	0	0.0
Illinois	20	5	3	8	40.0
Indiana	11	2	2	4	36.4
Iowa	1	0	0	0	0.0
Kansas	6	2	0	2	33.3
Kentucky	4	0	0	0	0.0
Louisiana	31	6	1	7	21.9
Maine	0	0	0	0	0.0
Maryland	30	8	2	10	33.3
Massachusetts	8	3	1	4	50.0
Michigan	14	2	0	2	14.3
Minnesota	1	0	0	0	0.0
Mississippi	42	6	2	8	19.0
Missouri	16	3	0	3	18.8
Montana	0	0	0	0	0.0
Nebraska	1	0	0	0	0.0
Nevada	3	0	0	0	0.0
New Hampshire	2	1	0	1	50.0
New Jersey	12	2	1	3	25.0
New Mexico	0	0	0	0	0.0
New York	26	6	2	8	30.8
North Carolina	25	3	0	3	12.0
North Dakota	0	0	0	0	0.0
Ohio	15	5	0	5	33.3
Oklahoma	5	1	2	3	60.0
Oregon	3	2	0	2	66.7
Pennsylvania	17	1	1	2	11.8
Rhode Island	9	4	0	4	44.4
South Carolina	25	5	1	6	24.0
South Dakota	1	0	0	0	0.0
Tennessee	15	4	1	5	33.3
Texas	16	5	0	5	31.3
Utah	0	0	0	0	0.0

[Continued]

★ 863 ★

Women in State Legislatures, January 1993
[Continued]

State	Total Black members	Female Black House members	Female Black Senate members	Female Black total	Female percent of Black legislators
Vermont	2	1	0	1	50.0
Virginia	12	3	2	5	41.7
Washington	2	0	1	1	50.0
West Virginia	1	0	0	0	0.0
Wisconsin	8	2	1	3	37.5
Wyoming	0	0	0	0	0.0
Total	511	123	29	152	29.8

Source: "Black Women in State Legislatures, January 1993." *Black Elected Officials, 1993*, p. xxxix. Published by permission.

Municipal Officials

★ 864 ★

Male Municipal Officials by Race and Ethnicity, 1991 - I

Position	Total reporting (A)	Total females		Race					
				White		Black		American Indian	
		No. (B)	% of (A)	No.	% of (B)	No.	% of (B)	No.	% of (B)
Elected mayor/president	4,397	3,842	87.4	3,386	88.1	75	2.0	27	0.7
Chief appointed administrative officer/manager	3,498	3,125	89.3	2,736	87.6	31	1.0	17	0.5
Assistant manager/assistant CAO	1,096	739	67.4	603	81.6	26	3.5	5	0.7
Clerk	4,001	1,099	27.5	666	60.6	13	1.2	3	0.3
Chief financial officer	2,996	1,973	65.9	1,589	80.5	18	0.9	9	0.5
Director of economic development	955	807	84.5	657	81.4	25	3.1	7	0.9
Treasurer	3,015	1,588	52.7	1,205	75.9	14	0.9	10	0.6
Director of public works	3,671	3,633	99.0	2,966	81.6	48	1.3	25	0.7
Engineer	1,915	1,891	98.7	1,568	82.9	15	0.8	5	0.3
Police chief	4,157	4,140	99.6	3,349	80.9	76	1.8	32	0.8
Fire chief	3,563	3,561	99.9	2,903	81.5	36	1.0	19	0.5
Planning director	1,998	1,695	84.8	1,356	80.0	34	2.0	5	0.3
Personnel director	2,054	1,221	59.4	967	79.2	48	3.9	8	0.7
Risk manager	943	673	71.4	549	81.6	16	2.4	5	0.7
Director of parks and recreation	2,090	1,793	85.8	1,433	79.9	36	2.0	7	0.4
Superintendent of parks	1,380	1,332	96.5	1,075	80.7	32	2.4	2	0.2
Director of recreation	1,357	990	73.0	748	75.6	31	3.1	7	0.7
Librarian	1,763	575	32.6	261	45.4	2	0.3	1	0.2

[Continued]

★ 864 ★

Male Municipal Officials by Race and Ethnicity, 1991 - I

[Continued]

Position	Total reporting (A)	Total females		White		Black		American Indian	
		No. (B)	% of (A)	No.	% of (B)	No.	% of (B)	No.	% of (B)
Director of data processing/info. serv.	883	621	70.3	495	79.7	9	1.4	4	0.6
Purchasing director	1,478	1,015	68.7	797	78.5	26	2.6	9	0.9

Source: "Male Municipal Officials by Race and Ethnicity," *The Municipal Year Book, 1993*, p. 216. Published by permission.

★ 865 ★

Municipal Officials

Male Municipal Officials by Race and Ethnicity, 1991 - II

Position	Race							
	Asian		Other		Race not reported		Hispanic	
	No.	% of (B)	No.	% of (B)	No.	% of (B)	No.	% of (B)
Elected mayor/president	8	0.2	18	0.5	328	8.5	71	1.8
Chief appointed administrative officer/manager	11	0.4	14	0.4	316	10.1	63	2.0
Assistant manager/assistant CAO	4	0.5	7	0.9	94	12.7	27	3.7
Clerk	1	0.1	5	0.5	411	37.4	15	1.4
Chief financial officer	20	1.0	13	0.7	324	16.4	41	2.1
Director of economic development	2	0.2	8	1.0	108	13.4	23	2.9
Treasurer	5	0.3	8	0.5	346	21.8	17	1.1
Director of public works	14	0.4	24	0.7	556	15.3	70	1.9
Engineer	22	1.2	25	1.3	256	13.5	31	1.6
Police chief	5	0.1	17	0.4	661	16.0	61	1.5
Fire chief	2	0.1	10	0.3	591	16.6	56	1.6
Planning director	14	0.8	13	0.8	273	16.1	32	1.9
Personnel director	5	0.4	12	1.0	181	14.8	36	2.9
Risk manager	7	1.0	3	0.4	93	13.8	20	3.0
Director of parks and recreation	5	0.3	10	0.6	302	16.8	42	2.3
Superintendent of parks	6	0.5	11	0.8	206	15.5	49	3.7
Director of recreation	4	0.4	7	0.7	193	19.5	26	2.6
Librarian	3	0.5	2	0.3	306	53.2	10	1.7
Director of data processing/info. serv.	5	0.8	4	0.6	104	16.7	22	3.5
Purchasing director	8	0.8	12	1.2	163	16.1	37	3.6

Source: "Male Municipal Officials by Race and Ethnicity," *The Municipal Year Book, 1993*, p. 216. Published by permission.

★ 866 ★
Municipal Officials

Women Municipal Officials by Race and Ethnicity, 1991 - I

Position	Total reporting (A)	Total females		White		Black		American Indian	
		No. (B)	% of (A)	No.	% of (B)	No.	% of (B)	No.	% of (B)
Elected mayor/president	4,397	555	12.6	527	95.0	8	1.4	4	0.7
Chief appointed administrative officer/manager	3,498	373	10.7	351	94.1	7	1.9	4	1.1
Assistant manager/assistant CAO	1,096	357	32.6	332	93.0	14	3.9	2	0.6
Clerk	4,001	2,902	72.5	2,705	93.2	60	2.1	25	0.9
Chief financial officer	2,996	1,023	34.1	952	93.1	23	2.2	11	1.1
Director of economic development	955	148	15.5	132	89.2	7	4.7	1	0.7
Treasurer	3,015	1,427	47.3	1,338	93.8	20	1.4	11	0.8
Director of public works	3,671	38	1.0	33	86.8	2	5.3	1	2.6
Engineer	1,915	24	1.3	24	100.0	0	0.0	0	0.0
Police chief	4,157	17	0.4	14	82.4	0	0.0	1	5.9
Fire chief	3,563	2	0.1	1	50.0	0	0.0	0	0.0
Planning director	1,998	303	15.2	282	93.1	9	3.0	2	0.7
Personnel director	2,054	833	40.6	744	89.3	54	6.5	11	1.3
Risk manager	943	270	28.6	249	92.2	11	4.1	0	0.0
Director of parks and recreation	2,090	297	14.2	272	91.6	14	4.7	4	1.3
Superintendent of parks	1,380	48	3.5	43	89.6	1	2.1	0	0.0
Director of recreation	1,357	367	27.0	339	92.4	12	3.3	2	0.5
Librarian	1,763	1,188	67.4	1,127	94.9	14	1.2	7	0.6
Director of data processing/info. serv.	883	262	29.7	246	93.9	7	2.7	1	0.4
Purchasing director	1,478	463	31.3	415	89.6	23	5.0	6	1.3

Source: "Female Municipal Officials by Race and Ethnicity," *The Municipal Year Book, 1993*, p. 216. Published by permission.

★ 867 ★
Municipal Officials

Women Municipal Officials by Race and Ethnicity, 1991 - II

Position	Asian		Other		Race not reported		Hispanic	
	No.	% of (B)	No.	% of (B)	No.	% of (B)	No.	% of (B)
Elected mayor/president	2	0.4	3	0.5	11	2.0	13	2.3
Chief appointed administrative officer/manager	1	0.3	0	0.0	10	2.7	6	1.6
Assistant manager/assistant CAO	3	0.8	1	0.3	5	1.4	6	1.7
Clerk	12	0.4	21	0.7	79	2.7	81	2.8
Chief financial officer	11	1.1	3	0.3	23	2.2	24	2.3
Director of economic development	2	1.4	0	0.0	6	4.1	5	3.4
Treasurer	7	0.5	6	0.4	45	3.2	26	1.8
Director of public works	0	0.0	0	0.0	2	5.3	0	0.0

[Continued]

★ 867 ★

Women Municipal Officials by Race and Ethnicity, 1991 - II

[Continued]

Position	Race							
	Asian		Other		Race not reported		Hispanic	
	No.	% of (B)	No.	% of (B)	No.	% of (B)	No.	% of (B)
Engineer	0	0.0	0	0.0	0	0.0	0	0.0
Police chief	0	0.0	0	0.0	2	11.8	1	5.9
Fire chief	0	0.0	0	0.0	1	50.0	0	0.0
Planning director	2	0.7	2	0.7	6	2.0	5	1.7
Personnel director	2	0.2	6	0.7	16	1.9	27	3.2
Risk manager	2	0.7	3	1.1	5	1.9	8	3.0
Director of parks and recreation	0	0.0	1	0.3	6	2.0	4	1.3
Superintendent of parks	0	0.0	2	4.2	2	4.2	3	6.3
Director of recreation	2	0.5	3	0.8	9	2.5	9	2.5
Librarian	6	0.5	5	0.4	29	2.4	15	1.3
Director of data processing/info. serv.	4	1.5	1	0.4	3	1.1	4	1.5
Purchasing director	3	0.6	3	0.6	13	2.	18	3.9

Source: "Female Municipal Officials by Race and Ethnicity," *The Municipal Year Book, 1993*, p. 216. Published by permission.

Political Parties

★ 868 ★

Political Party Identification, 1970-1990, and Characteristics, 1990

In percent. Covers citizens of voting-age living in private housing units in the contiguous United States. Data are from the National Election Studies and are based on a sample and subject to sampling variability; for details, see source.

Year and selected characteristic	Total	Strong Democrat	Weak Democrat	Independent Democrat	Independent	Independent Republican	Weak Republican	Strong Republican	Apolitical
Race									
White	100	17	19	11	11	13	16	11	1
Black	100	40	23	16	8	7	3	2	2

Source: "Political Party Identification of the Adult Population, by degree of Attachment, 1970 to 1990, and by Selected Characteristics, 1990," U.S. Bureau of the Census, *Statistical Abstract of the United States*, 1993, p. 282. Published by permission. Primary source: Center for Political Studies, University of Michigan, Ann Arbor, MI, unpublished data. Data prior to 1988 published in Warren E. Miller and Santa A. Traugott, *American National Election Studies Data Sourcebook, 1952-1986*, Harvard University Press, Cambridge, MA, 1989 (copyright).

Senior Citizens

★ 869 ★

Voters in the 1988 Elections

[Percent]

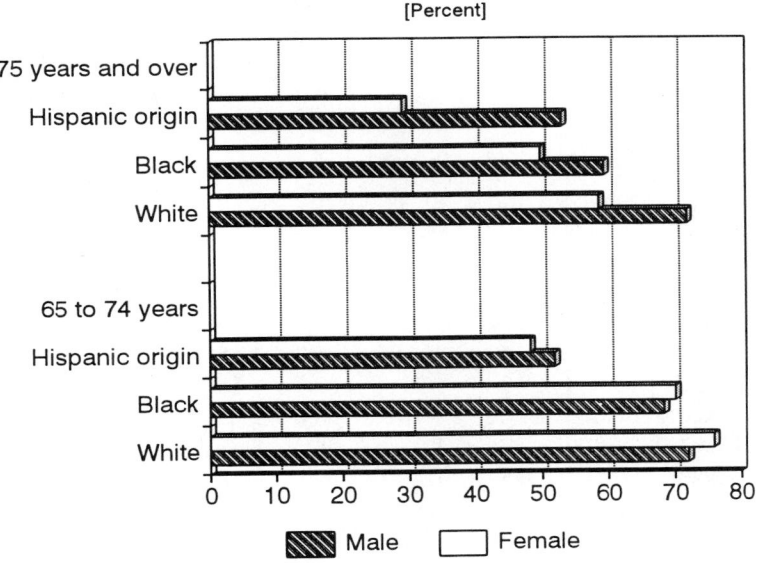

Male   Female

Age and race	Male	Female
65 to 74 years		
White	72.1	75.9
Black	68.5	70.2
Hispanic origin[1]	52.0	48.2
75 years and over		
White	71.9	58.7
Black	59.4	49.9
Hispanic origin[1]	53.1	29.1

Source: "Percentage of Persons 65 Years and Over Who Reported Voting in the 1988 Elections, by Age, Sex, Race, and Hispanic Origin," Cynthia Taeuber, *Sixty- Five Plus in America*, p. 6-21. Primary source: Jerry Jennings, U.S. Bureau of the Census, *Voting and Registration in the Election of November 1988*, Current Population Reports, Series P-20, No. 440. U.S. Government Printing Office, Washington, DC, October 1989, table 2. *Note:* 1. Hispanic origin may be of any race.

Voters and Voting

★ 870 ★

Mayoral Voting in Council, by Ethnicity, 1991

Classification	Black				White				Hispanic			
	No. reporting (A)	All[1] % of (A)	Tie[2] % of (A)	No[3] % of (A)	No. reporting (B)	All % of (B)	Tie % of (B)	No % of (B)	No. reporting (C)	All % of (C)	Tie % of (C)	No % of (C)
Total, all cities	101	53.5	28.7	14.9	4,568	50.0	39.1	8.8	87	62.1	29.9	5.7
Population group												
Over 1,000,000	2	0.0	0.0	100.0	0	0.0	0.0	0.0	0	0.0	0.0	0.0
500,000-1,000,000	8	0.0	0.0	0.0	8	25.0	12.5	62.5	0	0.0	0.0	0.0
250,000-499,999	6	50.0	0.0	50.0	24	66.7	0.0	29.2	2	50.0	0.0	50.0
100,000-249,999	8	50.0	0.0	37.5	89	65.2	13.5	19.1	1	0.0	0.0	0.0
50,000-99,888	11	63.6	0.0	27.3	231	63.6	12.6	21.2	6	100.0	0.0	0.0
25,000-49,999	10	60.0	30.0	10.0	478	60.0	22.2	14.0	10	80.0	10.0	10.0
10,000-24,999	26	65.4	26.9	7.7	1,032	54.8	31.1	10.9	21	61.9	28.6	9.5
5,000-9,999	16	56.3	37.5	6.3	1,125	49.2	43.2	6.1	26	65.4	30.8	3.8
2,500-4,999	18	38.9	55.6	0.0	1,220	39.2	55.3	4.3	15	40.0	53.3	0.0
Under 2,500	4	25.0	75.0	0.0	361	48.8	43.5	6.4	6	50.0	50.0	0.0
Geographic division												
New England	4	25.0	25.0	50.0	412	69.7	10.4	18.7	0	0.0	0.0	0.0
Mid-Atlantic	13	46.2	15.4	30.8	672	46.6	44.5	6.4	5	80.0	0.0	20.0
East North Central	19	52.6	26.3	15.8	907	39.5	42.8	13.2	6	50.0	33.3	16.7
West North Central	6	33.3	66.7	0.0	547	40.2	43.3	15.2	1	100.0	0.0	0.0
South Atlantic	32	56.3	34.4	6.3	588	49.7	44.7	3.9	5	60.0	0.0	40.0
East South Central	7	42.9	42.9	14.3	271	46.5	42.8	8.5	1	100.0	0.0	0.0
West South Central	7	42.9	42.9	14.3	461	42.7	52.3	3.7	25	48.0	48.0	0.0
Mountain	2	50.0	0.0	50.0	239	49.4	46.4	3.8	18	33.3	61.1	5.6
Pacific Coast	11	90.9	0.0	9.1	471	79.0	18.9	1.5	26	92.3	3.8	0.0
Metro status												
Central	21	42.9	4.8	47.6	344	50.3	19.5	26.7	8	62.5	0.0	25.0
Suburban	51	60.8	25.5	9.8	2,436	55.4	35.3	7.2	51	70.6	23.5	5.9
Independent	29	48.3	51.7	0.0	1,788	42.5	48.2	7.6	28	46.4	50.0	0.0
Form of government												
Mayor-council	38	13.2	50.0	31.6	433	20.8	59.9	16.3	29	20.7	62.1	17.2
Council-manager	60	78.3	15.0	5.0	1,594	72.6	23.8	2.2	57	84.2	12.3	0.0
Commission	3	66.7	33.3	0.0	80	92.0	8.0	0.0	1	0.0	100.0	0.0
Town meeting	0	0.0	0.0	0.0	153	86.4	5.1	7.3	0	0.0	0.0	0.0
Rep. town meeting	0	0.0	0.0	0.0	23	82.1	10.7	7.1	0	0.0	0.0	0.0

Source: "Mayoral Voting in Council, by Ethnicity." Susan A. MacManus, "Women and Racial/Ethnic Minorities in Mayoral and Council Positions," p. 76. Published by permission. *Notes:* Row percentages may not total 100 because the "other" category is not reported in this table. 1. Mayor can vote on all issues. 2. Mayor votes only to break a tie. 3. Mayor is never allowed to vote.

★ 871 ★

Voters and Voting

Registration, by Region: November 1968 to 1992

Numbers in thousands. Civilian noninstitutional population.

Region, race, Hispanic origin, sex, and age	Presidential elections of--						
	1992	1988	1984	1980	1976	1972	1968
United States							
Total, voting age	185,684	178,098	169,963	157,085	146,548	136,203	116,535
Percent registered	68.2	66.6	68.3	66.9	66.7	72.3	74.3
White	70.1	67.9	69.6	68.4	68.3	73.4	75.4
Black	63.9	64.5	66.3	60.0	58.5	65.5	66.2
Hispanic origin[1]	35.0	35.5	40.1	36.3	37.8	44.4	(NA)
Male	66.9	65.2	67.3	66.6	67.1	73.1	76.0
Female	69.3	67.8	69.3	67.1	66.4	71.6	72.8
18 to 24 years	52.5	48.2	51.3	49.2	51.3	58.9	56.0[2]
25 to 44 years	64.8	63.0	66.6	65.6	65.5	71.3	72.4
45 to 64 years	75.3	75.5	76.6	75.8	75.5	79.7	81.1
65 years and over	78.0	78.4	76.9	74.6	71.4	75.6	75.6
North and west							
Total, voting age	122,025	117,373	112,376	106,524	99,403	93,653	81,594
Percent registered	68.7	67.1	69.0	67.9	67.7	73.9	76.5
White	70.9	68.5	70.5	69.3	69.0	74.9	77.2
Black	63.0	65.9	67.2	60.6	60.9	67.0	71.8
South							
Total, voting age	63,659	60,725	57,587	50,561	47,145	42,550	34,941
Percent registered	67.2	65.6	66.9	64.8	64.6	68.7	69.2
White	68.5	66.6	67.8	66.2	66.7	69.8	70.8
Black	64.7	63.3	65.6	59.3	56.4	64.0	61.6

Source: "Reported Registration, by Region, Race, Hispanic Origin, Sex, Race, and Age: November 1968 to 1992," U.S. Bureau of the Census, Voting and Registration in the Election of November 1992, p, vi. Primary source: Current Population Reports, Series P-20, Nos. 192, 253, 293, 322, 344, 370, 383, 405, 414, 440, and table 2 of this report. *Notes:* NA Not available. 1. Persons of Hispanic origin may be of any race. 2. Prior to 1972, includes persons 18 to 20 years old in Georgia and Kentucky, 19 and 20 in Alaska, and 20 years old in Hawaii.

★ 872 ★

Voters and Voting

Vote Cast and Governor Elected, by State, 1986 to 1992

In thousands, except percent. D = Democrat, R = Republican, I = Independent.

Division and state	Candidate elected at most recent election	1986 Total vote[1]	1986 Percent leading party	1988 Total vote[1]	1988 Percent leading party	1990 Total vote[1]	1990 Percent leading party	1992 Total vote[1]	1992 Percent leading party
N.E.									
ME	John R. McKernan Jr.	427	R-39.9	(X)	(X)	522	R-46.7	(X)	(X)
NH	Steve Merrill	251	R-53.7	442	R-60.4	295	R-60.3	516	R-56.0
VT	Howard B. Dean	197	D-47.0	243	D-55.4	211	R-51.8	286	D-74.7
MA	William F. Weld	1,684	D-68.7	(X)	(X)	2,343	R-50.2	(X)	(X)
RI	Bruce G. Sundlun	323	R-64.7	401	R-50.8	357	D-74.1	425	D-61.5
CT	Lowell P. Weicker Jr.	994	D-57.9	(X)	(X)	1,141	I-40.4	(X)	(X)
M.A.									
NY	Mario M. Cuomo	4,294	D-64.6	(X)	(X)	4,057	D-53.2	(X)	(X)
NJ[2]	James J. Florio	1,973	R-69.6	(X)	(X)	2,254	D-61.2	(X)	(X)
PA	Robert P. Casey	3,388	D-50.7	(X)	(X)	3,053	D-67.7	(X)	(X)
E.N.C.									
OH	George V. Voinovich	3,067	D-60.6	(X)	(X)	3,478	R-55.7	(X)	(X)
IN	Evan Bayh	(X)	(X)	2,141	D-53.2	(X)	(X)	2,229	D-62.0
IL	Jim Edgar	3,144	R-52.7	(X)	(X)	3,257	R-50.7	(X)	(X)
MI	John Engler	2,397	D-68.1	(X)	(X)	2,565	R-49.8	(X)	(X)
WI	Tommy G. Thompson	1,527	R-52.7	(X)	(X)	1,380	R-58.2	(X)	(X)
W.N.C.									
MN	Arne Carlson	1,416	D-55.8	(X)	(X)	1,807	R-49.6	(X)	(X)
IA	Terry E. Branstad	911	R-51.9	(X)	(X)	976	R-60.6	(X)	(X)
MO	Mel Carnahan	(X)	(X)	2,086	R-64.2	(X)	(X)	2,344	D-58.7
ND	Edward T. Schafer	(X)	(X)	299	D-59.9	(X)	(X)	305	R-57.9
SD	George S. Mickelson	294	R-51.8	(X)	(X)	257	R-58.9	(X)	(X)
NE	Ben Nelson	564	R-52.9	(X)	(X)	587	D-49.9	(X)	(X)
KS	Joan Finney	841	R-51.9	(X)	(X)	783	D-48.6	(X)	(X)
S.A.									
DE	Thomas R. Carper	(X)	(X)	240	R-70.7	(X)	(X)	277	D-64.7
MD	William Donald Schaefer	1,101	D-82.4	(X)	(X)	1,111	D-59.8	(X)	(X)
VA[2]	L. Douglas Wilder	1,343	D-55.2	(X)	(X)	1,789	D-50.1	(X)	(X)
WV	Gaston Caperton	(X)	(X)	650	D-58.9	(X)	(X)	657	D-56.0
NC	James B. Hunt	(X)	(X)	2,180	R-56.1	(X)	(X)	2,595	D-52.7
SC	Carroll A. Campbell Jr.	754	R-51.0	(X)	(X)	761	R-69.5	(X)	(X)
GA	Zell Miller	1,175	D-70.5	(X)	(X)	1,450	D-52.9	(X)	(X)
FL	Lawton Chiles	3,386	R-54.6	(X)	(X)	3,531	D-56.5	(X)	(X)
E.S.C.									
KY[3]	Bereton C. Jones	781	D-64.6	(X)	(X)	835	D-64.7	(X)	(X)
TN	Ned McWherter	1,210	D-54.2	(X)	(X)	790	D-60.8	(X)	(X)
AL	Guy Hunt	1,236	R-56.3	(X)	(X)	1,216	R-52.1	(X)	(X)
MS[3]	Kirk Fordice	722	D-53.4	(X)	(X)	711	R-50.8	(X)	(X)
W.S.C.									
AR	Jim Guy Tucker[4]	689	D-63.9	(X)	(X0	696	D-57.5	(X)	(X0
LA	Edwin W. Edwards	1,559[5]	33.1[5]	(X)	(X)	1,728[6]	D-61.2[6]	(X)	(X)
OK	David Walters	910	R-47.5	(X)	(X)	911	D-57.4	(X)	(X)
TX	Ann W. Richards	3,441	R-52.7	(X)	(X)	3,893	D-49.5	(X)	(X)

[Continued]

★ 872 ★

Vote Cast and Governor Elected, by State, 1986 to 1992

[Continued]

Division and state	Candidate elected at most recent election	1986		1988		1990		1992	
		Total vote[1]	Percent leading party	Total vote[1]	Percent leading party	Total vote[1]	Percent leading party	Total vote[1]	Percent leading party
Mt									
MT	Marc Racicot	(X)	(X)	367	R-51.9	(X)	(X)	408	R-51.3
ID	Cecil D. Andrus	387	D-49.9	(X)	(X)	321	D-68.2	(X)	(X)
WY	Mike Sullivan	165	D-54.0	(X)	(X)	160	D-65.4	(X)	(X)
CO	Roy Romer	1,059	D-58.2	(X)	(X)	1,011	D-61.9	(X)	(X)
NM	Bruce King	395	R-53.0	(X)	(X)	411	D-54.6	(X)	(X)
AZ	Fife Symington	867	R-39.7	(X)	(X)	941[7]	R-52.4[7]	(X)	(X)
UT	Mike Leavitt	(X)	(X)	649	R-40.1	(X)	(X)	763	R-42.2
NV	Robert J. Miller	260	D-71.9	(X)	(X)	321	D-64.8	(X)	(X)
Pac.									
WA	Mike Lowry	(X)	(X)	1,875	D-62.2	(X)	(X)	2,271	D-52.2
OR	Barbara Roberts	1,060	D-51.9	(X)	(X)	1,113	D-45.7	(X)	(X)
CA	Pete Wilson	7,444	R-60.5	(X)	(X)	7,699	R-49.2	(X)	(X)
AK	Walter J. Hickel	180	D-47.3	(X)	(X)	195	I-38.9	(X)	(X)
HI	John Waihee III	334	D-52.0	(X)	(X)	340	D-59.8	(X)	(X)

Source: "Vote Cast for and Governor Elected, by State: 1986 to 1992," U.S. Bureau of the Census, *Statistical Abstract of the United States*, 1993, p. 278. Published by permission. Primary source: Elections Research Center, Chevy Chase, MD, *America Votes*, biennial and unpublished data (copyright). *Notes:* X Not applicable. 1. Includes minor party and scattered votes. 2. Voting years 1985 and 1989. 3. Voting years 1987 and 1991. 4. Jim Guy Tucker assumed the governorship upon the resignation of Bill Clinton following the 1992 presidential election. 5. Primary election in Oct. 1987, held on a nonparty basis. Winner was Democratic. Runner-up withdrew from runoff. 6. Result of runoff election in 1991. 7. Result of runoff election in February 1991.

★ 873 ★

Voters and Voting

Vote Cast for Leading Minority Party Presidential Candidates, 1936 to 1992

Prior to 1960, excludes Alaska and Hawaii; prior to 1964, excludes DC. Vote cast for major party candidates include the votes of minor parties cast for those candidates. See also *Historical Statistics, Colonial Times to 1970*, series Y 79-83 and Y 135.

Year	Candidate	Party	Popular vote (1,000)	Candidate	Party	Popular vote (1,000)
1936	William Lemke	Union	892	Norman Thomas	Socialist	188
1940	Norman Thomas	Socialist	116	Roger Babson	Prohibition	59
1944	Norman Thomas	Socialist	79	Claude Watson	Prohibition	75
1948	Strom Thurmond	States' Rights	1,176	Henry Wallace	Progressive	1,157
1952	Vincent Hallinan	Progressive	140	Stuart Hamblen	Prohibition	73
1956	T. Coleman Andrews	States' Rights	111	Eric Hass	Socialist Labor	44
1960	Eric Hass	Socialist Labor	48	Rutherford Decker	Prohibition	46
1964	Eric Hass	Socialist Labor	45	Clifton DeBerry	Socialist Workers	33
1968	George Wallace	American Independent	9,906	Henning Blomen	Socialist Labor	53
1972	John Schmitz	American	1,099	Benjamin Spock	People's	79
1976	Eugene McCarthy	Independent	757	Roger McBride	Libertarian	173
1980	John Anderson	Independent	5,720	Ed Clark	Libertarian	921
1984	David Bergland	Libertarian	228	Lyndon H. LaRouche	Independent	79

[Continued]

★ 873 ★

Vote Cast for Leading Minority Party Presidential Candidates, 1936 to 1992

[Continued]

Year	Candidate	Party	Popular vote (1,000)	Candidate	Party	Popular vote (1,000)
1988	Ron Paul	Libertarian	432	Lenora B. Fulani	New Alliance	217
1992	H. Ross Perot	Independent	19,742	Andre Marrou	Libertarian	292

Source: "Vote Cast for Leading Minority Party Candidates for President: 1936 to 1992," U.S. Bureau of the Census, *Statistical Abstract of the United States*, 1993, p. 263. Published by permission. Primary source: Source of table: Elections Research Center, Chevy Chase, MD, *America at the Polls 2*, 1965, and *America Votes*, biennial (copyright).

★ 874 ★

Voters and Voting

Votes Cast in U.S. Senatorial Elections: 1990 and 1992, and Incumbents, 1993

Incumbent senators and year term expires. D = Democrat; R = Republican.

Division and state	Name, party and year	Name, party and year	1990 Total (1,000)	1990 Percent for leading party	1992 Total (1,000)	1992 Percent for leading party
N.E.						
ME	George J. Mitchell (D) 1995	William S. Cohen (R) 1997	520	R-61.3	(X)	(X)
NH	Judd Gregg (R) 1999	Robert C. Smith (R) 1997	291	R-65.1	518	R-48.1
VT	Patrick J. Leahy (D) 1999	James M. Jeffords (R) 1995	(X)	(X)	286	D-54.2
MA	Edward M. Kennedy (D) 1995	John F. Kerry (D) 1997	2,316	D-57.1	(X)	(X)
RI	Claiborne Pell (D) 1997	John H. Chafee (R) 1995	364	D-61.8	(X)	(X)
CT	Christopher J. Dodd (D) 1999	Joseph I. Lieberman (D) 1995	(X)	(X)	1,501	D-58.8
M.A.						
NY	Daniel P. Moynihan (D) 1995	Alfonse M. D'Amato (R) 1999	(X)	(X)	6,459	R-49.0
NJ	Bill Bradley (D) 1997	Frank R. Latenberg (D) 1995	1,938	D-50.4	(X)	(X)
PA	Harris Wofford (D) 1995	Arien Specter (R) 1999	3,383[1]	D-55.0[1]	4,802	R-49.1
E.N.C.						
OH	John Glenn (D) 1999	Howard Metzenbaum (D) 1995	(X)	(X)	4,794	D-51.0
IN	Dan Coats (R) 1999	Richard G. Lugar (R) 1995	1,504[2]	R-53.6[2]	2,211	R-57.3
IL	Carol Moseley Braun (D) 1999	Paul Simon (D) 1997	3,251	D-65.1	4,940	D-53.3
MI	Carl Levin (D) 1997	Donald W. Riegle Jr. (D) 1995	2,560	D-57.5	(X)	(X)
WI	Herb Kohl (D) 1995	Russell Feingold (D) 1999	(X)	(X)	2,455	D-52.6
W.N.C.						
MN	Paul David Wellstone (D) 1997	Dave Durenberger (R) 1995	1,808	D-50.4	(X)	(X)
IA	Tom Harkin (D) 1997	Charles E. Grassley (R) 1999	984	D-54.5	1,292	R-69.6
MO	Christopher S. Bond (R) 1999	John C. Danforth (R) 1995	(X)	(X)	2,355	R-51.9
ND	Byron L. Dorgan (D) 1999	Kent Conrad (D) 1995[3]	(X)	(X)	304	D-59.0
SD	Thomas A. Daschle (D) 1999	Larry Pressler (R) 1997	259	R-52.4	334	D-64.9
NE	J. James Exon (D) 1997	J. Robert Kerrey (D) 1995	594	D-58.9	(X)	(X)
KS	Bob Dole (R) 1999	Nancy Kassebaum (R) 1997	786	R-73.6	1,126	R-62.7
S.A.						
DE	Joseph R. Biden, Jr. (D) 1997	William V. Roth, Jr. (R) 1995	180	D-62.7	(X)	(X)

[Continued]

★ 874 ★

Votes Cast in U.S. Senatorial Elections: 1990 and 1992, and Incumbents, 1993

[Continued]

Division and state	Name, party and year	Name, party and year	1990		1992	
			Total (1,000)	Percent for leading party	Total (1,000)	Percent for leading party
MD	Barbara A. Mikulski (D) 1999	Paul S. Sarbanes (D) 1995	(X)	(X)	1,842	D-71.0
VA	Charles S. Robb (D) 1995	John W. Warner (R) 1997	1,084	R-80.9	(X)	(X)
WV	Robert C. Byrd (D) 1995	John D. Rockefeller IV (D) 1997	404	D-68.3	(X)	(X)
NC	Lauch Faircloth (R) 1999	Jesse Helms (R) 1997	2,070	R-52.5	2,578	R-50.3
SC	Ernest F. Hollings (D) 1999	Strom Thurmond (R) 1997	751	R-64.2	1,180	D-50.1
GA	Paul Coverdall (R) 1999	Sam Nunn (D) 1997	1,034	D-100.0	1,254[4]	R-50.6[4]
FL	Bob Graham (D) 1999	Connie Mack (R) 1995	(X)	(X)	4,962	D-65.4
E.S.C.						
KY	Wendell H. Ford (D) 1999	Mitch McConnell (R) 1997	916	R-52.2	1,331	D-62.9
TN	Harlan Mathews (D) 1997[5]	Jim Sasser (D) 1995	784	D-67.7	(X)	(X)
AL	Howell Heflin (D) 1997	Richard C. Shelby (D) 1999	1,186	D-60.5	1,578	D-64.8
MS	Thad Cochran (R) 1997	Trent Lott (R) 1995	274	R-100.0	(X)	(X)
W.S.C.						
AR	Dale Bumpers (D) 1999	David Pryor (D) 1997	495	D-99.8	920	D-60.2
LA	John B. Breaux (D) 1999	J. Bennett Johnston (D) 1997	1,396[6]	D-56.5[6]	843[6]	D-73.1[6]
OK	David L. Boren (D) 1997	Don Nickles (R) 1999	884	D-83.2	1,294	R-58.5
TX	Kay Bailey Hutchison (R) 1995[7]	Phil Gramm (R) 1997	3,822	R-60.2	(X)	(X)
Mt						
MT	Max Baucus (D) 1997	Conrad Burns (R) 1995	319	D-68.1	(X)	(X)
ID	Larry E. Craig (R) 1997	Dirk Kempthorne (R) 1999	316	R-61.3	479	R-56.5
WY	Alan K. Simpson (R) 1997	Malcom Wallop (R) 1995	158	R-63.9	(X)	(X)
CO	Ben N. Campbell (D) 1999	Hank Brown (R) 1997	1,022	R-55.7	1,552	D-51.8
NM	Jeff Bingaman (D) 1995	Pete V. Domenici (R) 1997	407	R-72.9	(X)	(X)
AZ	Dennis DeConcini (D) 1995	John McCain (R) 1999	(X)	(X)	1,382	R-55.8
UT	Robert F. Bennett (R) 1999	Orrin G. Hatch (R) 1995	(X)	(X)	758	R-55.4
NV	Harry Reid (D) 1999	Richard H. Bryan (D) 1995	(X)	(X)	496	D-51.0
Pac						
WA	Patty Murray (D) 1999	Slade Gorton (R) 1995	(X)	(X)	2,219	D-54.0
OR	Mark O. Hatfield (R) 1997	Bob Packwood (R) 1999	1,099	R-53.7	1,376	R-52.1
CA	Barbara Boxer (D) 1999	Dianne Feinstein (D) 1995[8]	(X)	(X)	10,800	D-47.9
AK	Frank H. Murkowski (R) 1999	Ted Stevens (R) 1997	190	R-66.2	240	R-53.0
HI	Daniel K. Akaka (D) 1995	Daniel K. Inouye (D) 1999	350[9]	D-54.0[9]	364	D-57.3

Source: "Votes Cast for United States Senators, 1990 and 1992, and Incumbent Senators, 1993-States," U.S. Bureau of the Census, *Statistical Abstract of the United States,* 1993, p. 267. Published by permission. Primary source: Elections Research Center, Chevy Chase, MD, *America Votes,* biennial, (copyright). Notes: X Not applicable. 1. Special election in November 1991 to fill vacancy caused by death of Senator John Heinz. 2. Special election to fill the unexpected term of Vice-President Quayle. 3. In a special election in 1992 to fill an unexpired term, the Democratic candidate received 63.2 percent of 163,311 total votes cast. 4. Results of run-off election. In the first election no candidate received more than 50 percent of the vote. First election: total votes cast, 2,251,587; Democratic candidate, 1,108,416; Republican candidate, 1,073,282. 5. Appointed to fill the unexpired term of Vice-President Gore. 6. Louisiana holds an open-primary election with candidates from all parties running on the same ballot. Any candidate who receives a majority is elected. In 1990 and 1992 the indicated Senators received more than 50 percent of the vote in the October open-primary. 7. Winner of a special election in June 1993 to fill the unexpired term of Senator Bentsen. 8. In a special election in 1992 to fill an unexpired term, the Democratic Candidates received 54.3 percent of 10,782,743 votes cast. 9. Special election to fill the unexpired term of Senator Matsunaga.

★ 875 ★

Voters and Voting

Voting Time and Method in the Midwest: Election of November 1992

November 1992. Numbers in thousands.

Region, race, Hispanic origin, age	Total reported voted	Reported time of day	Before noon	Noon to 4 P.M.	4 P.M. to 6 P.M.	After 6 P.M.	Voted absentee	Did not report on time or method of voting
All races								
Total age 18 and over	29,830	28,930	12,545	5,742	5,339	3,485	1,819	900
18 to 24 years	3,013	2,721	789	602	606	425	300	292
25 to 44 years	12,457	12,159	4,825	2,317	2,706	1,976	336	297
45 to 64 years	8,724	8,520	3,890	1,632	1,637	983	378	203
65 years and over	5,636	5,529	3,041	1,191	391	101	805	107
White								
Total age 18 and over	27,136	26,389	11,460	5,216	4,872	3,148	1,692	747
18 to 24 years	2,657	2,404	669	522	540	386	287	253
25 to 44 years	11,315	11,064	4,399	2,081	2,471	1,794	319	251
45 to 64 years	7,927	7,775	3,547	1,491	1,510	874	354	152
65 years and over	5,236	5,146	2,845	1,123	352	94	732	91
Black								
Total age 18 and over	2,393	2,257	976	473	405	288	114	137
18 to 24 years	317	281	117	68	51	36	9	36
25 to 44 years	1,015	971	380	210	209	159	12	45
45 to 64 years	684	643	298	128	110	87	19	42
65 years and over	376	363	181	67	36	5	74	14
Hispanic origin[1]								
Total age 18 and over	354	341	106	80	81	70	3	13
18 to 24 years	33	32	7	9	7	8	-	1
25 to 44 years	183	173	50	33	49	39	2	10
45 to 64 years	99	97	25	29	24	18	-	2
65 years and over	39	39	23	8	-	5	2	-

Source: "Reported Time of Day and Method of Voting, by Race, Hispanic Origin, and Age," U.S. Bureau of the Census, *Voting and Registration in the Election of November 1992*, p. 68. *Note:* 1. Persons of Hispanic origin may be of any race.

★ 876 ★

Voters and Voting

Voting Time and Method in the Northeast: Election of November 1992

November 1992. Numbers in thousands.

Region, race, Hispanic origin, age	Total reported voted	Reported time of day	Before noon	Noon to 4 P.M.	4 P.M. to 6 P.M.	After 6 P.M.	Voted absentee	Did not report on time or method of voting
All races								
Total age 18 and over	23,448	22,574	8,925	4,278	3,656	4,928	787	875
18 to 24 years	2,044	1,834	389	332	319	536	257	210
25 to 44 years	9,763	9,439	3,379	1,577	1,675	2,640	168	324
45 to 64 years	7,089	6,848	2,709	1,211	1,283	1,483	161	241
65 years and over	4,553	4,453	2,448	1,158	378	270	200	99
White								
Total age 18 and over	21,305	20,639	8,205	3,931	3,349	4,408	746	667
18 to 24 years	1,817	1,643	347	289	287	474	246	174
25 to 44 years	8,758	8,522	3,061	1,443	1,508	2,356	153	236
45 to 64 years	6,441	6,258	2,484	1,099	1,195	1,326	154	183
65 years and over	4,289	4,216	2,313	1,100	359	250	193	73
Black								
Total age 18 and over	1,905	1,720	653	311	254	466	35	186
18 to 24 years	199	168	39	39	26	56	9	31
25 to 44 years	887	810	282	115	145	257	12	78
45 to 64 years	565	513	203	101	68	135	7	52
65 years and over	254	229	130	56	16	19	7	25
Hispanic origin[1]								
Total age 18 and over	675	634	229	133	104	160	8	41
18 to 24 years	92	84	19	18	20	23	3	8
25 to 44 years	335	314	104	67	52	87	4	21
45 to 64 years	202	190	81	29	32	47	1	12
65 years and over	46	46	25	18	-	3	-	-

Source: "Reported Time of Day and Method of Voting, by Race, Hispanic Origin, and Age," U.S. Bureau of the Census, *Voting and Registration in the Election of November 1992*, p. 67. *Note:* 1. Persons of Hispanic origin may be of any race.

★ 877 ★

Voters and Voting

Voting Time and Method in the South: Election of November 1992

November 1992. Numbers in thousands.

Region, race, Hispanic origin, age	Total reported voted	Reported time of day	Before noon	Noon to 4 P.M.	4 P.M. to 6 P.M.	After 6 P.M.	Voted absentee	Did not report on time or method of voting
All races								
Total age 18 and over	37,590	36,205	16,020	7,626	6,385	3,646	2,529	1,385
18 to 24 years	3,363	3,094	993	742	672	329	358	269
25 to 44 years	15,255	14,726	5,671	3,033	3,118	2,133	771	529
45 to 64 years	11,692	11,270	5,359	2,166	2,051	1,043	650	422
65 years and over	7,281	7,116	3,997	1,685	544	140	750	165
White								
Total age 18 and over	31,034	30,077	13,458	6,404	5,026	2,891	2,299	957
18 to 24 years	2,621	2,438	764	582	517	253	323	183
25 to 44 years	12,316	11,981	4,743	2,471	2,398	1,678	691	335
45 to 64 years	9,777	9,470	4,507	1,856	1,658	848	601	307
65 years and over	6,319	6,188	3,444	1,495	453	112	684	131
Black								
Total age 18 and over	6,151	5,744	2,423	1,147	1,263	693	197	408
18 to 24 years	689	608	211	149	149	71	28	81
25 to 44 years	2,786	2,598	884	534	691	425	64	188
45 to 64 years	1,750	1,645	788	290	355	170	41	105
65 years and over	927	893	539	174	88	28	64	33
Hispanic origin[1]								
Total age 18 and over	1,499	1,432	505	252	255	253	167	66
18 to 24 years	170	158	44	43	25	31	14	12
25 to 44 years	636	612	196	110	118	128	60	24
45 to 64 years	472	450	159	70	94	80	46	22
65 years and over	221	213	105	29	18	14	47	8

Source: "Reported Time of Day and Method of Voting, by Race, Hispanic Origin, and Age," U.S. Bureau of the Census, *Voting and Registration in the Election of November 1992*, p. 68. *Note:* 1. Persons of Hispanic origin may be of any race.

★ 878 ★

Voters and Voting

Voting Time and Method in the West: Election of November 1992

November 1992. Numbers in thousands.

Region, race, Hispanic origin, age	Total reported voted	Reported time of day	Before noon	Noon to 4 P.M.	4 P.M. to 6 P.M.	After 6 P.M.	Voted absentee	Did not report on time or method of voting
All races								
Total age 18 and over	22,997	22,365	8,940	3,632	3,585	3,090	3,118	632
18 to 24 years	2,021	1,837	510	403	402	348	174	184
25 to 44 years	9,914	9,646	3,442	1,544	1,910	1,823	929	266
45 to 64 years	6,896	6,771	2,913	1,028	1,081	797	952	124
65 years and over	4,166	4,109	2,075	657	192	122	1,063	57
White								
Total age 18 and over	20,930	20,392	8,208	3,322	3,156	2,792	2,914	538
18 to 24 years	1,837	1,673	464	366	355	329	158	164
25 to 44 years	8,989	8,762	3,165	1,417	1,684	1,646	850	227
45 to 64 years	6,224	6,124	2,654	926	944	704	896	100
65 years and over	3,880	3,833	1,925	612	173	113	1,010	47
Black								
Total age 18 and over	922	882	307	135	201	135	104	40
18 to 24 years	91	79	15	19	28	7	9	12
25 to 44 years	418	403	131	53	94	90	35	16
45 to 64 years	273	261	99	38	64	35	27	12
65 years and over	139	139	62	25	15	3	34	-
Hispanic origin[1]								
Total age 18 and over	1,710	1,639	635	272	310	306	116	71
18 to 24 years	197	166	46	39	39	41	1	31
25 to 44 years	825	796	259	122	177	178	59	29
45 to 64 years	524	520	241	81	88	76	33	4
65 years and over	164	157	90	30	5	9	23	7

Source: "Reported Time of Day and Method of Voting, by Race, Hispanic Origin, and Age," U.S. Bureau of the Census, *Voting and Registration in the Election of November 1992*, p. 69. *Note:* 1. Persons of Hispanic origin may be of any race.

★ 879 ★

Voters and Voting

Voting Time and Method: Election of November 1992

November 1992. Numbers in thousands.

Region, race, Hispanic origin, age	Total reported voted	Reported time of day	Before noon	Noon to 4 P.M.	4 P.M. to 6 P.M.	After 6 P.M.	Voted absentee	Did not report on time or method of voting
All races								
Total age 18 and over	113,866	110,074	46,431	21,277	18,964	15,149	8,253	3,792
18 to 24 years	10,442	9,486	2,681	2,078	1,999	1,639	1,089	955
25 to 44 years	47,388	45,972	17,317	8,470	9,409	8,572	2,204	1,417
45 to 64 years	34,400	33,409	14,871	6,038	6,052	4,307	2,141	991
65 years and over	21,636	21,207	11,561	4,691	1,504	632	2,819	429
White								
Total age 18 and over	100,405	97,496	41,332	18,873	16,402	13,238	7,651	2,909
18 to 24 years	8,932	8,157	2,244	1,758	1,699	1,442	1,014	775
25 to 44 years	41,379	40,329	15,369	7,412	8,060	7,475	2,013	1,050
45 to 64 years	30,370	29,627	13,192	5,371	5,307	3,752	2,005	743
65 years and over	19,724	19,383	10,527	4,331	1,337	569	2,619	342
Black								
Total age 18 and over	11,371	10,602	4,358	2,067	2,144	1,582	451	769
18 to 24 years	1,296	1,135	382	276	253	170	55	160
25 to 44 years	5,107	4,781	1,677	912	1,139	930	122	326
45 to 64 years	3,273	3,062	1,388	556	597	427	94	211
65 years and over	1,696	1,624	912	323	155	56	179	72
Hispanic origin[1]								
Total age 18 and over	4,238	4,046	1,475	737	750	789	295	192
18 to 24 years	492	440	117	109	92	104	18	53
25 to 44 years	1,979	1,895	609	333	396	432	124	84
45 to 64 years	1,297	1,257	506	210	239	222	81	39
65 years and over	470	455	243	85	23	32	72	15

Source: "Reported Time of Day and Method of Voting, by Race, Hispanic Origin, and Age," U.S. Bureau of the Census, *Voting and Registration in the Election of November 1992*, p. 67. *Note:* 1. Persons of Hispanic origin may be of any race.

★ 880 ★

Voters and Voting

Voting and Registration by State in the Election of November 1992: Part I – Northeastern States

November 1992. Numbers in thousands.

State, sex, race, and hispanic origin	All persons	Total registered	Percent	Standard error	Total voted	Percent	Standard error
Connecticut							
Total	2,428	1,863	76.8	1.63	1,740	71.7	1.74
Male	1,135	856	75.4	2.43	802	70.6	2.57
Female	1,293	1,007	77.9	2.19	938	72.6	2.36
White	2,170	1,735	80.0	1.63	1,628	75.0	1.76
Black	222	115	51.5	7.69	100	45.1	7.68
Hispanic origin[1]	127	63	49.7	13.24	54	42.8	13.10
Maine							
Total	938	801	85.4	1.21	695	74.1	1.50
Male	462	394	85.3	1.73	348	75.2	2.11
Female	476	407	85.4	1.70	348	73.1	2.14
White	922	792	85.9	1.20	690	74.9	1.50
Black	6	4	(B)	(B)	2	(B)	(B)
Hispanic origin[1]	2	1	(B)	(B)	1	(B)	(B)
Massachusetts							
Total	4,501	3,262	72.5	.81	2,960	65.8	.86
Male	2,087	1,497	71.7	1.20	1,386	66.4	1.25
Female	2,414	1,765	73.1	1.09	1,574	65.2	1.18
White	4,233	3,141	74.2	.82	2,854	67.4	.87
Black	184	95	51.7	5.41	83	45.3	5.39
Hispanic origin[1]	124	28	22.7	7.17	18	14.4	6.01
New Hampshire							
Total	853	596	69.9	1.73	551	64.6	1.81
Male	424	288	67.9	2.50	270	63.8	2.58
Female	429	309	71.9	2.40	281	65.4	2.54
White	840	592	70.5	1.74	548	65.2	1.82
Black	5	4	(B)	(B)	3	(B)	(B)
Hispanic origin[1]	1	1	(B)	(B)	1	(B)	(B)
New Jersey							
Total	5,838	3,972	68.0	.82	3,572	61.2	.86
Male	2,765	1,841	66.6	1.21	1,667	60.3	1.26
Female	3,074	2,131	69.3	1.12	1,905	62.0	1.18
White	4,837	3,436	71.0	.88	3,150	65.1	.93
Black	769	482	62.7	2.85	381	49.6	2.95
Hispanic origin[1]	564	207	36.8	4.31	173	30.6	4.12

[Continued]

★ 880 ★

Voting and Registration by State in the Election of November 1992: Part I – Northeastern States

[Continued]

State, sex, race, and hispanic origin	All persons	Total registered	Percent	Standard error	Total voted	Percent	Standard error
New York							
Total	13,401	8,314	62.0	.65	7,613	58.8	.66
Male	6,308	3,824	60.6	.96	3,500	55.5	.97
Female	7,093	4,490	63.3	.89	4,113	58.0	.91
White	10,790	7,136	66.1	.71	6,590	61.1	.73
Black	2,052	1,020	49.7	2.08	890	43.4	2.06
Hispanic origin[1]	1,172	449	38.3	3.47	382	32.6	3.34
Pennsylvania							
Total	9,170	5,967	65.1	.82	5,491	59.9	.84
Male	4,353	2,844	65.3	1.18	2,597	59.7	1.22
Female	4,817	3,123	64.8	1.13	2,895	60.1	1.16
White	8,257	5,426	65.7	.86	5,033	61.0	.88
Black	767	507	66.1	3.40	436	56.9	3.55
Hispanic origin[1]	107	52	48.5	12.48	39	36.5	12.02
Rhode Island							
Total	757	560	74.0	1.64	522	69.0	1.73
Male	355	269	75.7	2.34	251	70.5	2.49
Female	401	291	72.5	2.30	271	67.6	2.41
White	716	544	76.0	1.64	510	71.2	1.74
Black	28	13	(B)	(B)	9	(B)	(B)
Hispanic origin[1]	23	6	(B)	(B)	6	(B)	(B)
Vermont							
Total	443	337	76.1	1.59	304	68.6	1.73
Male	217	166	76.4	2.26	146	67.4	2.49
Female	226	171	75.9	2.23	158	69.8	2.39
White	440	335	76.2	1.59	302	68.7	1.73
Black	-	-	(B)	(B)	-	(B)	(B)
Hispanic origin[1]	3	2	(B)	(B)	2	(B)	(B)

Source: "Reported Voting and Registration, by Sex, Race, and Hispanic Origin, for States," U.S. Bureau of the Census, Voting and Registration in the Election of November 1992, pp. 23-30. *Note:* 1. Persons of Hispanic origin may be of any race.

★ 881 ★

Voters and Voting

Voting and Registration by State in the Election of November 1992: Part II – Pacific States

November 1992. Numbers in thousands.

State, sex, race, and hispanic origin	All persons	Total registered	Percent	Standard error	Total voted	Percent	Standard error
Alaska							
Total	347	260	75.1	1.46	238	68.6	1.56
Male	164	124	75.2	2.12	113	68.7	2.27
Female	183	137	75.0	2.01	125	68.5	2.16
White	274	216	78.8	1.55	200	73.2	1.68
Black	14	9	(B)	(B)	9	(B)	(B)
Hispanic origin[1]	8	5	(B)	(B)	4	(B)	(B)
California							
Total	22,340	12,864	57.6	.64	11,789	52.8	.65
Male	10,925	6,131	56.1	.92	5,640	51.6	.93
Female	11,415	6,733	59.0	.89	6,149	53.9	.91
White	18,759	11,349	60.5	.69	10,482	55.9	.70
Black	1,268	812	64.0	3.16	712	56.1	3.27
Hispanic origin[1]	5,443	1,384	25.4	1.80	1,135	20.9	1.68
Oregon							
Total	2,205	1,649	74.7	1.65	1,525	69.1	1.66
Male	1,076	768	71.4	2.33	895	64.6	2.46
Female	1,130	881	77.9	2.08	830	73.4	2.22
White	2,103	1,594	75.8	1.58	1,482	70.5	1.68
Black	36	22	(B)	(B)	14	(B)	(B)
Hispanic origin[1]	97	40	(B)	(B)	35	(B)	(B)
Hawaii							
Total	813	493	60.6	1.76	449	55.2	1.79
Male	381	222	58.2	2.60	205	53.7	2.63
Female	431	271	62.7	2.39	244	56.6	2.45
White	264	187	70.8	2.88	169	63.9	3.04
Black	16	9	(B)	(B)	8	(B)	(B)
Hispanic origin[1]	17	9	(B)	(B)	7	(B)	(B)
Washington							
Total	3,699	2,645	71.5	1.51	2,453	66.3	1.58
Male	1,817	1,262	69.4	2.20	1,159	63.8	2.29
Female	1,882	1,384	73.5	2.07	1,293	68.7	2.17
White	3,366	2,510	74.6	1.53	2,331	69.2	1.62

[Continued]

★ 881 ★

Voting and Registration by State in the Election of November 1992: Part II – Pacific States

[Continued]

State, sex, race, and hispanic origin	All persons	Total registered	Percent	Standard error	Total voted	Percent	Standard error
Black	132	51	38.4	10.41	41	31.2	9.91
Hispanic origin[1]	42	18	(B)	(B)	13	(B)	(B)

Source: "Reported Voting and Registration, by Sex, Race, and Hispanic Origin, for States," U.S. Bureau of the Census, Voting and Registration in the Election of November 1992, pp. 23-30. *Note:* 1. Persons of Hispanic origin may be of any race.

★ 882 ★

Voters and Voting

Voting and Registration by State in the Election of November 1992: Part III – Southern States

November 1992. Numbers in thousands.

State, sex, race, and hispanic origin	All persons	Total registered	Percent	Standard error	Total voted	Percent	Standard error
Alabama							
Total	3,007	2,317	77.1	1.42	1,913	63.6	1.63
Male	1,415	1,094	77.3	2.07	888	62.7	2.39
Female	1,592	1,223	76.8	1.96	1,026	64.4	2.23
White	2,210	1,753	79.3	1.60	1,456	65.9	1.87
Black	775	556	71.8	3.63	450	58.1	3.96
Hispanic origin[1]	13	11	(B)	(B)	9	(B)	(B)
Arkansas							
Total	1,745	1,161	66.5	1.60	1,012	58.0	1.67
Male	820	539	65.7	2.34	472	57.6	2.44
Female	924	622	67.3	2.18	540	58.4	2.29
White	1,434	972	67.8	1.74	870	60.7	1.82
Black	291	182	62.4	4.85	135	46.4	4.99
Hispanic origin[1]	15	8	(B)	(B)	8	(B)	(B)
Delaware							
Total	519	367	70.7	1.66	343	66.1	1.75
Male	244	172	70.6	2.45	159	65.1	2.56
Female	275	195	70.7	2.30	184	66.9	2.38
White	417	308	73.9	1.81	291	69.7	1.89
Black	85	49	(B)	(B)	44	(B)	(B)
Hispanic origin[1]	7	3	(B)	(B)	3	(B)	(B)

[Continued]

★ 882 ★

Voting and Registration by State in the Election of November 1992: Part III – Southern States

[Continued]

State, sex, race, and hispanic origin	All persons	Total registered	Percent	Standard error	Total voted	Percent	Standard error
District of Columbia							
Total	398	295	73.9	1.83	263	66.1	1.98
Male	189	134	70.9	2.76	118	62.2	2.94
Female	209	161	76.7	2.43	146	69.6	2.65
White	134	109	81.3	2.80	102	75.9	3.08
Black	256	183	71.4	2.85	159	62.1	3.06
Hispanic origin[1]	14	6	(B)	(B)	4	(B)	(B)
Florida							
Total	10,342	6,486	62.7	.78	5,772	55.8	.80
Male	4,858	2,925	60.2	1.15	2,608	53.7	1.17
Female	5,484	3,561	64.9	1.06	3,164	57.7	1.10
White	8,742	5,643	64.5	.84	5,062	57.9	.87
Black	1,410	772	54.7	2.63	652	46.3	2.64
Hispanic origin[1]	1,349	472	35.0	3.35	411	30.5	3.23
Georgia							
Total	4,723	2,928	62.0	1.64	2,556	54.1	1.68
Male	2,270	1,360	59.9	2.38	1,177	51.9	2.43
Female	2,453	1,568	63.9	2.25	1,379	56.2	2.32
White	3,136	2,109	67.3	1.94	1,840	58.7	2.04
Black	1,505	812	53.9	3.60	709	47.1	3.61
Hispanic origin[1]	74	4	(B)	(B)	4	(B)	(B)
Kentucky							
Total	2,724	1,769	64.9	1.63	1,569	57.6	1.69
Male	1,319	872	66.1	2.33	778	59.0	2.42
Female	1,405	897	63.6	2.29	791	56.3	2.36
White	2,496	1,635	65.5	1.70	1,454	58.2	1.76
Black	217	133	61.4	7.13	115	53.0	7.32
Hispanic origin[1]	17	3	(B)	(B)	3	(B)	(B)
Louisiana							
Total	2,972	2,290	77.0	1.52	2,040	68.6	1.68
Male	1,394	1,077	77.3	2.21	947	67.9	2.47
Female	1,579	1,213	76.8	2.09	1,093	69.3	2.29
White	2,196	1,674	76.2	1.79	1,502	68.3	1.96
Black	721	593	82.3	3.39	515	71.5	4.01
Hispanic origin[1]	41	22	(B)	(B)	22	(B)	(B)

[Continued]

★ 882 ★

Voting and Registration by State in the Election of November 1992: Part III – Southern States

[Continued]

State, sex, race, and hispanic origin	All persons	Total registered	Percent	Standard error	Total voted	Percent	Standard error
Maryland							
Total	3,605	2,584	71.7	1.61	2,392	66.4	1.88
Male	1,696	1,192	70.2	2.38	1,069	63.0	2.51
Female	1,907	1,392	73.0	2.18	1,323	69.4	2.26
White	2,644	1,975	74.7	1.81	1,840	69.6	1.91
Black	797	530	66.4	4.33	480	60.2	4.49
Hispanic origin[1]	93	36	(B)	(B)	36	(B)	(B)
Mississippi							
Total	1,839	1,458	79.3	1.34	1,226	66.7	1.55
Male	858	674	78.8	1.98	571	68.8	2.28
Female	983	784	79.7	1.81	655	66.6	2.13
White	1,230	986	80.2	1.61	854	69.4	1.86
Black	602	472	78.5	2.87	372	61.9	3.39
Hispanic origin[1]	3	1	(B)	(B)	1	(B)	(B)
North Carolina							
Total	4,976	3,418	68.7	.80	2,986	60.0	.84
Male	2,358	1,593	67.6	1.17	1,400	59.4	1.23
Female	2,619	1,825	69.7	1.09	1,587	60.6	1.16
White	3,797	2,688	70.8	.90	2,368	62.4	.96
Black	1,107	708	64.0	2.12	599	54.1	2.20
Hispanic origin[1]	71	15	(B)	(B)	14	(B)	(B)
Oklahoma							
Total	2,295	1,704	74.3	1.49	1,549	67.5	1.59
Male	1,073	775	72.2	2.23	707	65.8	2.36
Female	1,222	930	76.1	1.99	842	69.0	2.16
White	2,047	1,546	75.5	1.55	1,419	69.3	1.66
Black	143	94	65.5	7.84	75	52.4	8.24
Hispanic origin[1]	54	19	(B)	(B)	18	(B)	(B)
South Carolina							
Total	2,850	1,730	67.0	1.45	1,496	58.0	1.52
Male	1,188	789	66.4	2.14	673	56.7	2.25
Female	1,392	941	67.6	1.96	822	59.1	2.06
White	1,842	1,274	69.2	1.68	1,134	61.6	1.77
Black	693	430	62.0	3.49	338	48.8	3.59
Hispanic origin[1]	21	8	(B)	(B)	8	(B)	(B)

[Continued]

★ 882 ★

Voting and Registration by State in the Election of November 1992: Part III – Southern States

[Continued]

State, sex, race, and hispanic origin	All persons	Total registered	Percent	Standard error	Total voted	Percent	Standard error
Tennessee							
Total	3,714	2,413	65.0	1.54	2,065	55.6	1.60
Male	1,689	1,094	64.8	2.28	966	57.2	2.37
Female	2,024	1,316	65.1	2.08	1,099	54.3	2.18
White	3,132	1,986	63.4	1.69	1,718	54.8	1.75
Black	529	409	77.4	4.32	332	62.9	5.00
Hispanic origin[1]	22	2	(B)	(B)	2	(B)	(B)
Texas							
Total	12,267	7,956	64.9	.84	6,817	55.6	.87
Male	5,933	3,752	63.2	1.22	3,236	54.5	1.26
Female	6,334	4,204	66.4	1.16	3,582	56.5	1.21
White	10,597	7,001	66.1	.90	6,066	57.2	.94
Black	1,361	864	63.5	3.07	682	50.1	3.19
Hispanic origin[1]	2,806	1,203	42.9	2.86	927	33.1	2.71
Virginia							
Total	4,585	2,999	65.4	1.44	2,805	61.2	1.47
Male	2,143	1,410	65.8	2.10	1,330	62.0	2.15
Female	2,442	1,589	65.1	1.97	1,476	60.4	2.03
White	3,628	2,439	67.2	1.59	2,300	63.4	1.64
Black	799	516	64.5	4.19	472	59.0	4.31
Hispanic origin[1]	85	27	(B)	(B)	27	(B)	(B)
West Virginia							
Total	1,367	888	64.9	1.61	784	57.3	1.67
Male	652	418	64.1	2.34	372	57.0	2.42
Female	715	470	65.7		412	57.6	2.31
White	1,317	856	65.0	1.64	758	57.6	1.70
Black	43	28	(B)	(B)	21	(B)	(B)
Hispanic origin[1]	4	3	(B)	(B)	2	(B)	(B)

Source: "Reported Voting and Registration, by Sex, Race, and Hispanic Origin, for States," U.S. Bureau of the Census, Voting and Registration in the Election of November 1992, pp. 23-30. *Note:* 1. Persons of Hispanic origin may be of any race.

★ 883 ★

Voters and Voting

Voting and Registration by State in the Election of November 1992: Part IV – Western States

November 1992. Numbers in thousands.

State, sex, race, and hispanic origin	All persons	Total registered	Percent	Standard error	Total voted	Percent	Standard error
Arizona							
Total	2,671	1,882	70.5	1.57	1,728	64.7	1.65
Male	1,328	908	68.3	2.28	849	63.9	2.35
Female	1,343	974	72.6	2.17	878	65.4	2.31
White	2,549	1,831	71.8	1.59	1,684	66.1	1.67
Black	58	28	(B)	(B)	28	(B)	(B)
Hispanic origin[1]	442	197	44.7	6.62	156	35.3	6.37
Colorado							
Total	2,451	1,831	74.7	1.57	1,688	68.8	1.67
Male	1,138	814	71.5	2.39	759	66.8	2.49
Female	1,314	1,017	77.4	2.06	928	70.7	2.24
White	2,221	1,715	77.2	1.59	1,589	71.5	1.71
Black	119	75	63.0	9.53	64	53.4	9.85
Hispanic origin[1]	218	145	66.5	8.95	136	62.2	9.20
Idaho							
Total	726	521	71.8	1.48	486	66.9	1.55
Male	362	245	67.6	2.18	227	62.9	2.25
Female	365	277	75.9	1.98	258	70.8	2.11
White	711	512	72.0	1.49	477	67.0	1.56
Black	1	-	(B)	(B)	-	(B)	(B)
Hispanic origin[1]	30	10	(B)	(B)	8	(B)	(B)
Montana							
Total	570	440	77.3	1.42	407	71.4	1.53
Male	282	214	76.1	2.06	193	68.4	2.24
Female	289	226	78.4	1.96	214	74.3	2.08
White	544	426	78.3	1.43	395	72.6	1.55
Black	1	-	(B)	(B)	-	(B)	(B)
Hispanic origin[1]	6	3	(B)	(B)	3	(B)	(B)
Nevada							
Total	955	606	63.4	1.63	556	58.1	1.67
Male	484	298	61.6	2.31	271	55.9	2.36
Female	471	307	65.2	2.29	285	60.4	2.35
White	869	557	64.1	1.70	511	58.9	1.74
Black	51	33	(B)	(B)	29	(B)	(B)
Hispanic origin[1]	94	25	(B)	(B)	23	(B)	(B)

[Continued]

★ 883 ★

Voting and Registration by State in the Election of November 1992: Part IV – Western States

[Continued]

State, sex, race, and hispanic origin	All persons	Total registered	Percent	Standard error	Total voted	Percent	Standard error
New Mexico							
Total	1,079	724	67.1	1.59	675	62.6	1.64
Male	520	346	66.6	2.30	323	62.2	2.37
Female	559	378	67.6	2.20	352	62.9	2.27
White	970	660	68.0	1.67	618	63.8	1.72
Black	23	16	(B)	(B)	13	(B)	(B)
Hispanic origin[1]	334	190	56.8	4.73	172	51.3	4.78
Utah							
Total	1,097	865	78.8	1.45	793	72.3	1.59
Male	523	406	77.7	2.14	374	71.5	2.32
Female	575	459	79.8	1.97	419	72.9	2.18
White	1,068	847	79.3	1.46	779	73.0	1.60
Black	9	7	(B)	(B)	4	(B)	(B)
Hispanic origin[1]	31	16	(B)	(B)	14	(B)	(B)
Wyoming							
Total	331	225	68.1	1.86	213	64.2	1.92
Male	165	106	64.3	2.71	98	59.4	2.78
Female	166	119	71.9	2.54	115	68.9	2.61
White	325	224	68.9	1.87	212	65.1	1.92
Black	1	-	(B)	(B)	-	(B)	(B)
Hispanic origin[1]	12	6	(B)	(B)	6	(B)	(B)

Source: "Reported Voting and Registration, by Sex, Race, and Hispanic Origin, for States," U.S. Bureau of the Census, Voting and Registration in the Election of November 1992, pp. 23-30. *Note:* 1. Persons of Hispanic origin may be of any race.

★ 884 ★

Voters and Voting

Voting and Registration by State in the Election of November 1992: Part V – Midwestern States

November 1992. Numbers in thousands.

State, sex, race, and hispanic origin	All persons	Total registered	Percent	Standard error	Total voted	Percent	Standard error
Illinois							
Total	8,676	6,251	72.1	.61	5,650	65.1	.86
Male	4,118	2,913	70.7	1.19	2,631	63.9	1.25
Female	4,558	3,339	73.2	1.10	3,019	66.2	1.17

[Continued]

★ 884 ★

Voting and Registration by State in the Election of November 1992: Part V – Midwestern States

[Continued]

State, sex, race, and hispanic origin	All persons	Total registered	Percent	Standard error	Total voted	Percent	Standard error
White	7,197	5,237	72.8	.88	4,775	66.3	.93
Black	1,202	903	75.2	2.52	784	65.3	2.78
Hispanic origin[1]	646	223	34.6	4.91	171	26.4	4.55
Indiana							
Total	4,183	2,846	68.0	1.63	2,636	63.0	1.69
Male	1,916	1,265	66.0	2.45	1,152	60.1	2.53
Female	2,267	1,581	69.7	2.18	1,483	65.4	2.26
White	3,764	2,575	68.4	1.72	2,414	64.1	1.77
Black	378	244	64.4	6.74	194	51.4	7.04
Hispanic origin[1]	45	30	(B)	(B)	30	(B)	(B)
Iowa							
Total	2,100	1,635	77.9	1.37	1,486	70.8	1.50
Male	997	767	77.0	2.01	694	69.6	2.20
Female	1,103	868	78.7	1.86	792	71.8	2.05
White	2,044	1,610	78.8	1.36	1,466	71.7	1.50
Black	27	15	(B)	(B)	10	(B)	(B)
Hispanic origin[1]	15	8	(B)	(B)	5	(B)	(B)
Kansas							
Total	1,818	1,401	77.1	1.39	1,306	71.9	1.49
Male	906	689	76.1	2.00	638	70.5	2.14
Female	912	712	78.1	1.94	668	73.2	2.07
White	1,704	1,333	78.2	1.41	1,245	73.1	1.52
Black	88	57	(B)	(B)	52	(B)	(B)
Hispanic origin[1]	45	26	(B)	(B)	20	(B)	(B)
Michigan							
Total	6,811	5,078	74.6	.78	4,486	65.9	.85
Male	3,304	2,436	73.7	1.14	2,154	65.2	1.23
Female	3,507	2,642	75.3	1.08	2,332	66.5	1.18
White	5,812	4,374	75.3	.84	3,882	66.8	.92
Black	867	659	76.0	2.61	589	65.6	2.90
Hispanic origin[1]	111	58	52.5	11.08	48	43.3	11.00
Minnesota							
Total	3,237	2,789	86.1	1.20	2,403	74.2	1.53
Male	1,517	1,318	86.9	1.72	1,145	75.5	2.19
Female	1,721	1,471	85.5	1.69	1,258	73.1	2.12

[Continued]

★ 884 ★

Voting and Registration by State in the Election of November 1992: Part V – Midwestern States

[Continued]

State, sex, race, and hispanic origin	All persons	Total registered	Percent	Standard error	Total voted	Percent	Standard error
White	3,081	2,688	87.2	1.19	2,335	75.8	1.53
Black	88	77	(B)	(B)	51	(B)	(B)
Hispanic origin[1]	56	7	(B)	(B)	5	(B)	(B)
Missouri							
Total	3,803	2,821	74.2	1.54	2,518	66.2	1.67
Male	1,815	1,277	70.4	2.33	1,137	62.6	2.47
Female	1,988	1,544	77.7	2.03	1,381	69.5	2.25
White	3,403	2,590	76.1	1.59	2,320	68.2	1.74
Black	325	199	61.4	7.11	172	52.9	7.29
Hispanic origin[1]	66	31	(B)	(B)	28	(B)	(B)
Nebraska							
Total	1,167	853	73.1	1.44	771	66.1	1.54
Male	566	408	72.2	2.09	368	65.1	2.22
Female	601	445	74.0	1.98	403	67.1	2.13
White	1,119	833	74.4	1.45	756	67.5	1.55
Black	36	16	(B)	(B)	11	(B)	(B)
Hispanic origin[1]	18	7	(B)	(B)	7	(B)	(B)
North Dakota							
Total	444	403	90.8	.96	316	71.1	1.51
Male	209	188	90.1	1.46	149	71.4	2.20
Female	236	216	91.4	1.28	167	70.9	2.08
White	425	388	91.3	.96	307	72.1	1.53
Black	2	1	(B)	(B)	1	(B)	(B)
Hispanic origin[1]	1	-	(B)	(B)	-	(B)	(B)
Ohio							
Total	8,080	5,611	69.6	.81	5,187	64.3	.84
Male	3,839	2,633	68.6	1.18	2,434	63.4	1.23
Female	4,221	2,976	70.5	1.11	2,752	65.2	1.16
White	7,181	5,052	70.4	.85	4,682	65.2	.89
Black	804	527	85.6	3.20	473	58.9	3.31
Hispanic origin[1]	58	29	(B)	(B)	28	(B)	(B)
South Dakota							
Total	500	401	80.1	1.24	351	70.2	1.42
Male	244	192	78.7	1.82	171	69.9	2.04
Female	256	208	81.5	1.69	180	70.6	1.98

[Continued]

★ 884 ★

Voting and Registration by State in the Election of November 1992: Part V – Midwestern States

[Continued]

State, sex, race, and hispanic origin	All persons	Total registered	Percent	Standard error	Total voted	Percent	Standard error
White	474	385	81.2	1.25	338	71.4	1.44
Black	-	-	(B)	(B)	-	(B)	(B)
Hispanic origin[1]	1	1	(B)	(B)	1	(B)	(B)
Wisconsin							
Total	3,611	3,048	84.4	1.16	2,720	75.3	1.38
Male	1,758	1,475	83.9	1.69	1,297	73.8	2.02
Female	1,853	1,573	84.9	1.60	1,424	76.8	1.88
White	3,405	2,898	85.1	1.17	2,616	76.8	1.39
Black	126	110	87.5	6.87	74	59.2	10.20
Hispanic origin[1]	42	18	(B)	(B)	12	(B)	(B)

Source: "Reported Voting and Registration, by Sex, Race, and Hispanic Origin, for States," U.S. Bureau of the Census, Voting and Registration in the Election of November 1992, pp. 23-30. *Note:* 1. Persons of Hispanic origin may be of any race.

★ 885 ★

Voters and Voting

Voting and Registration in the Election of November 1992. Part I – United States

November 1992. Numbers in thousands.

Region, race, Hispanic origin, sex, and age	All persons	Reported registered		Reported voted		Reported that they did not vote[1]				
								Not registered		
		Number	Percent	Number	Percent	Total	Registered	Total[2]	Not a U.S. citizen	Do not know and not reported on registration[3]
All races										
Both sexes										
Total 18 years and over	185,684	126,578	68.2	113,866	61.3	71,818	12,712	59,106	11,900	10,638
18 to 20 years	9,727	4,696	48.3	3,749	38.5	5,979	947	5,032	857	821
21 to 24 years	14,644	8,091	55.3	6,693	45.7	7,951	1,398	6,553	1,288	1,075
25 to 34 years	41,603	25,223	60.6	22,120	53.2	19,483	3,103	16,380	3,675	2,223
35 to 44 years	39,716	27,503	69.2	25,269	63.6	14,447	2,234	12,213	2,800	2,144
45 to 54 years	28,058	20,785	74.1	19,292	68.8	8,766	1,493	7,273	1,435	1,466
55 to 64 years	21,089	16,231	77.0	15,107	71.6	5,982	1,124	4,858	955	1,107
65 to 74 years	18,445	14,685	79.6	13,607	73.8	4,839	1,078	3,760	538	952
75 years and over	12,401	9,364	75.5	8,030	64.8	4,371	1,334	3,037	352	850
Male										
Total 18 years and over	88,557	59,254	66.9	53,312	60.2	32,245	5,942	29,303	6,065	5,327
18 to 20 years	4,876	2,274	46.6	1,786	36.6	3,090	488	2,602	444	405
21 to 24 years	7,158	3,797	53.0	3,086	43.1	4,072	711	3,361	722	597
25 to 34 years	20,465	11,869	58.0	10,324	50.4	10,141	1,545	8,597	2,031	1,251
35 to 44 years	19,509	13,153	67.4	11,995	61.5	7,514	1,158	6,357	1,444	1,059
45 to 54 years	13,608	9,956	73.2	9,287	68.2	4,321	668	3,652	697	799
55 to 64 years	10,012	7,725	77.2	7,203	71.9	2,809	522	2,287	398	527
65 to 74 years	8,289	6,745	81.4	6,318	76.2	1,971	427	1,544	218	410
75 years and over	4,640	3,736	80.5	3,313	71.4	1,327	424	904	113	279

[Continued]

★ 885 ★

Voting and Registration in the Election of November 1992. Part I– United States

[Continued]

Region, race, Hispanic origin, sex, and age	All persons	Reported registered		Reported voted		Reported that they did not vote[1]				
								Not registered		
		Number	Percent	Number	Percent	Total	Registered	Total[2]	Not a U.S. citizen	Do not know and not reported on registration[3]
Female										
Total 18 years and over	97,126	67,324	69.3	60,554	62.3	36,573	6,770	29,803	5,834	5,311
18 to 20 years	4,851	2,421	49.9	1,962	40.5	2,889	459	2,430	413	416
21 to 24 years	7,486	4,294	57.4	3,607	48.2	3,879	687	3,192	566	478
25 to 34 years	21,138	13,354	63.2	11,795	55.8	9,342	1,558	7,784	1,645	972
35 to 44 years	20,207	14,350	71.0	13,274	65.7	6,933	1,076	5,856	1,356	1,085
45 to 54 years	14,450	10,829	74.9	10,005	69.2	4,445	825	3,620	738	667
55 to 64 years	11,077	8,506	76.8	7,904	71.4	3,173	602	2,571	558	580
65 to 74 years	10,157	7,940	78.2	7,289	71.8	2,868	652	2,216	320	541
75 years and over	7,761	5,628	72.5	4,717	60.8	3,044	911	2,133	240	571
White										
Both sexes										
Total 18 years and over	157,837	110,684	70.1	100,405	63.6	57,432	10,279	47,153	8,284	8,289
18 to 20 years	7,743	3,938	50.9	3,187	41.2	4,555	750	3,805	566	585
21 to 24 years	11,939	6,808	57.0	5,745	48.1	6,194	1,064	5,130	912	787
25 to 34 years	34,640	21,631	62.4	19,193	55.4	15,447	2,437	13,010	2,596	1,633
35 to 44 years	33,644	23,964	71.2	22,186	65.9	11,459	1,778	9,680	1,901	1,643
45 to 54 years	24,071	18,243	75.8	17,013	70.7	7,058	1,230	5,829	963	1,156
55 to 64 years	18,208	14,270	78.4	13,357	73.4	4,851	913	3,938	706	887
65 to 74 years	16,400	13,281	81.0	12,365	75.4	4,035	916	3,119	382	836
75 years and over	11,192	8,550	76.4	7,360	65.8	3,832	1,190	2,642	257	762
Male										
Total 18 years and over	75,915	52,383	69.0	47,552	62.6	28,364	4,831	23,533	4,298	4,116
18 to 20 years	3,876	1,897	48.9	1,510	39.0	2,366	387	1,979	302	279
21 to 24 years	5,874	3,238	55.1	2,710	46.1	3,164	528	2,637	527	435
25 to 34 years	17,286	10,360	59.9	9,130	52.8	8,156	1,230	6,927	1,460	933
35 to 44 years	16,752	11,617	69.3	10,677	63.7	6,075	940	5,135	1,006	809
45 to 54 years	11,826	8,886	75.1	8,315	70.3	3,511	571	2,940	470	633
55 to 64 years	8,706	6,859	78.8	6,428	73.8	2,278	432	1,846	292	406
65 to 74 years	7,385	6,115	82.8	5,754	77.9	1,631	360	1,271	156	364
75 years and over	4,210	3,411	81.0	3,028	71.9	1,181	383	798	85	258
Female										
Total 18 years and over	81,922	58,301	71.2	52,853	64.5	29,068	5,448	23,621	3,986	4,173
18 to 20 years	3,867	2,041	52.8	1,677	43.4	2,190	364	1,826	264	305
21 to 24 years	6,064	3,571	58.9	3,035	50.0	3,029	536	2,494	385	352
25 to 34 years	17,354	11,271	64.9	10,063	58.0	7,291	1,208	6,083	1,136	701
35 to 44 years	16,892	12,347	73.1	11,509	68.1	5,383	838	4,545	896	834
45 to 54 years	12,245	9,356	76.4	8,698	71.0	3,548	658	2,889	493	523
55 to 64 years	9,502	7,411	78.0	6,929	72.9	2,573	481	2,091	414	482
65 to 74 years	9,015	7,166	79.5	6,610	73.3	2,404	556	1,848	226	472
75 years and over	6,982	5,139	73.6	4,331	62.0	2,651	807	1,844	173	504
Black										
Both sexes										
Total 18 years and over	21,039	13,442	63.9	11,371	54.0	9,668	2,070	7,598	1,044	1,870
18 to 20 years	1,501	638	42.5	474	31.6	1,027	164	863	103	176
21 to 24 years	2,042	1,106	54.2	821	40.2	1,221	284	936	116	230
25 to 34 years	5,295	3,107	58.7	2,518	47.6	2,77	589	2,188	348	479
35 to 44 years	4,515	2,976	65.9	2,589	57.3	1,926	387	1,538	274	419
45 to 54 years	2,861	2,040	71.3	1,821	63.7	1,040	219	821	118	235
55 to 64 years	2,181	1,611	73.9	1,452	66.6	729	159	570	52	170
65 to 74 years	1,634	1,219	74.6	1,083	66.3	551	137	415	17	94

[Continued]

★ 885 ★

Voting and Registration in the Election of November 1992. Part I – United States

[Continued]

Region, race, Hispanic origin, sex, and age	All persons	Reported registered		Reported voted		Reported that they did not vote[1]				
								Not registered		
		Number	Percent	Number	Percent	Total	Registered	Total[2]	Not a U.S. citizen	Do not know and not reported on registration[3]
75 years and over	1,010	744	73.7	613	60.7	397	131	266	16	68
Male										
Total 18 years and over	9,425	5,727	60.8	4,786	50.8	4,639	941	3,698	548	965
18 to 20 years	745	307	41.2	218	29.2	527	89	438	47	96
21 to 24 years	940	471	50.1	319	33.9	621	152	469	60	134
25 to 34 years	2,388	1,283	53.7	1,009	42.2	1,379	274	1,105	216	260
35 to 44 years	2,027	1,295	63.9	1,113	54.9	914	181	732	138	207
45 to 54 years	1,267	838	66.2	757	59.7	510	82	428	59	125
55 to 64 years	984	710	72.1	640	65.0	344	70	274	18	94
65 to 74 years	724	533	73.7	475	65.6	249	58	190	8	37
75 years and over	350	290	82.7	256	73.0	95	34	61	2	12
Female										
Total 18 years and over	11,614	7,715	66.4	6,585	56.7	5,029	1,129	3,899	496	905
18 to 20 years	756	331	43.8	256	33.9	500	75	425	56	79
21 to 24 years	1,102	635	57.6	503	45.6	599	132	467	56	96
25 to 34 years	2,907	1,824	62.7	1,509	51.9	1,398	315	1,083	133	219
35 to 44 years	2,488	1,682	67.6	1,476	59.3	1,012	206	806	136	212
45 to 54 years	1,595	1,202	75.4	1,065	66.8	530	137	393	59	110
55 to 64 years	1,197	901	75.3	812	67.8	385	89	296	34	75
65 to 74 years	910	686	75.3	608	66.8	303	78	225	9	57
75 years and over	659	454	68.9	357	54.2	302	97	205	14	56
Hispanic origin[4]										
Both sexes										
Total 18 years and over	14,688	5,137	35.0	4,238	28.9	10,450	899	9,551	5,910	995
18 to 20 years	1,159	267	23.1	183	15.8	976	85	891	483	93
21 to 24 years	1,636	429	26.2	310	18.9	1,327	119	1,207	722	120
25 to 34 years	4,177	1,239	29.7	1,000	23.9	3,177	239	2,938	1,930	237
35 to 44 years	3,322	1,160	34.9	979	29.5	2,343	181	2,162	1,353	251
45 to 54 years	1,954	851	43.5	731	37.4	1,223	120	1,103	678	143
55 to 64 years	1,257	624	49.6	566	45.0	691	58	633	401	72
65 to 74 years	767	385	50.3	319	41.6	447	66	381	199	54
75 years and over	417	182	43.6	151	36.1	267	31	235	145	26
Male										
Total 18 years and over	7,273	2,337	32.1	1,946	26.8	5,327	391	4,936	3,132	519
18 to 20 years	563	110	19.5	72	12.8	491	38	454	259	46
21 to 24 years	855	178	20.8	124	14.5	731	54	677	413	82
25 to 34 years	2,182	562	25.8	464	21.3	1,717	98	1,619	1,100	121
35 to 44 years	1,669	540	32.4	447	26.8	1,222	93	1,129	722	127
45 to 54 years	983	416	42.4	363	36.9	620	53	566	354	78
55 to 64 years	535	276	51.6	252	47.1	283	24	259	164	37
65 to 74 years	340	186	54.8	159	46.7	181	27	154	71	18
75 years and over	147	69	46.8	66	44.9	81	3	78	49	10
Female										
Total 18 years and over	7,415	2,800	37.8	2,291	30.9	5,124	508	4,615	2,778	476
18 to 20 years	595	158	26.5	111	18.6	485	47	438	224	47
21 to 24 years	781	250	32.1	186	23.8	596	65	531	309	38
25 to 34 years	1,995	676	33.9	535	26.8	1,460	141	1,319	830	116
35 to 44 years	1,654	620	37.5	533	32.2	1,121	87	1,033	631	123
45 to 54 years	971	435	44.7	368	37.9	603	67	537	324	65
55 to 64 years	722	348	48.2	314	43.5	408	34	374	237	35

[Continued]

★ 885 ★

Voting and Registration in the Election of November 1992. Part I – United States
[Continued]

Region, race, Hispanic origin, sex, and age	All persons	Reported registered		Reported voted		Reported that they did not vote[1]				
								Not registered		
						Total	Registered	Total[2]	Not a U.S. citizen	Do not know and not reported on registration[3]
		Number	Percent	Number	Percent					
65 to 74 years	427	199	46.7	161	37.6	266	39	227	128	35
75 years and over	270	113	41.8	85	31.3	186	29	157	96	16

Source: "Reported Voting and Registration, by Race, Hispanic Origin, Sex, and Age, for the United States and Regions," U.S. Bureau of the Census, *Voting and Registration in the Election of November 1992*, pp. 4-5. *Notes:* 1. Includes persons reported as "did not vote," "do not know," and "not reported" on voting. 2. In addition to those reported as "not registered," total includes those "not a U.S. citizen," and "do not know" and "not reported" on registration. 3. Includes "do not know" and "not reported" on citizenship. 4. Persons of Hispanic origin may be of any race.

★ 886 ★

Voters and Voting

Voting and Registration in the Election of November 1992. Part II – New England

November 1992. Numbers in thousands.

Division, race, Hispanic origin, and age	All persons	Reported registered		Reported voted		Reported that they did not vote[1]				
								Not registered		
						Total	Registered	Total[2]	Not a U.S. citizen	Do not know and not reported on registration[3]
		Number	Percent	Number	Percent					
All races										
Total, 18 years and over	9,920	7,420	74.8	6,772	68.3	3,148	648	2,500	510	598
18 to 20 years	447	227	50.8	188	42.1	258	39	220	29	44
21 to 24 years	775	468	60.3	417	53.8	358	51	307	48	52
25 to 34 years	2,323	1,594	68.6	1,463	62.9	861	131	730	154	123
35 to 44 years	2,182	1,684	77.1	1,566	71.8	616	117	499	126	106
45 to 54 years	1,445	1,165	80.6	1,088	75.3	357	77	280	67	63
55 to 64 years	1,084	902	83.2	855	78.9	229	47	182	41	59
65 to 74 years	1,015	869	85.7	790	77.8	225	79	145	21	76
75 years and over	649	512	78.9	405	62.4	244	107	137	22	75
White										
Total, 18 years and over	9,322	7,139	76.6	6,533	70.1	2,789	607	2,182	369	548
18 to 20 years	407	220	54.0	181	44.5	226	39	187	16	38
21 to 24 years	699	434	62.2	387	55.5	311	47	264	35	40
25 to 34 years	2,171	1,542	71.0	1,420	65.4	751	122	629	102	111
35 to 44 years	2,027	1,598	78.8	1,485	73.3	542	113	429	89	98
45 to 54 years	1,384	1,135	82.0	1,067	77.1	317	68	249	57	61
55 to 64 years	1,023	864	84.4	823	80.4	201	42	159	37	74
65 to 74 years	985	847	85.9	775	78.7	210	71	139	17	71
75 years and over	626	500	79.9	394	63.0	231	105	126	16	71
Black										
Total, 18 years and over	445	231	51.8	198	44.4	247	33	214	77	34
18 to 20 years	32	7	(B)	7	(B)	25	-	25	9	4
21 to 24 years	53	25	(B)	23	(B)	29	2	27	4	7
25 to 34 years	116	44	37.6	35	30.4	81	8	72	31	10
35 to 44 years	116	71	61.4	67	57.5	49	5	45	22	7
45 to 54 years	44	20	(B)	13	(B)	31	7	24	5	1
55 to 64 years	47	32	(B)	29	(B)	18	3	14	2	3

[Continued]

★ 886 ★

Voting and Registration in the Election of November 1992. Part II – New England

[Continued]

| Division, race, Hispanic origin, and age | All persons | Reported registered | | Reported voted | | Reported that they did not vote[1] | | | | |
| | | | | | | Total | Registered | Not registered | | |
		Number	Percent	Number	Percent			Total[2]	Not a U.S. citizen	Do not know and not reported on registration[3]
65 to 74 years	24	20	(B)	13	(B)	11	7	4	1	1
75 years and over	13	11	(B)	10	(B)	3	1	2	1	1
Hispanic origin[4]										
Total, 18 years and over	280	100	35.9	81	29.1	199	19	180	76	14
18 to 20 years	21	3	(B)	3	(B)	19	-	19	3	2
21 to 24 years	35	8	(B)	8	(B)	27	-	27	9	-
25 to 34 years	91	33	36.1	24	25.8	68	9	58	34	3
35 to 44 years	74	32	(B)	25	(B)	48	6	42	19	3
45 to 54 years	38	19	(B)	18	(B)	20	1	19	6	5
55 to 64 years	10	5	(B)	3	(B)	7	2	5	-	-
65 to 74 years	9	2	(B)	2	(B)	7	-	7	4	1
75 years and over	2	-	(B)	-	(B)	2	-	2	1	-

Source: "Reported Voting and Registration, by Race, Hispanic Origin, Sex, and Age, for Divisions," U.S. Bureau of the Census, *Voting and Registration in the Election of November 1992*, p. 14. *Notes:* 1. Includes persons reported as "did not vote," "do not know," and "not reported" on voting. 2. In addition to those reported as "not registered," total includes those "not a U.S. citizen," and "do not know" and "not reported" on registration. 3. Includes "do not know" and "not reported" on citizenship. 4. Persons of Hispanic origin may be of any race.

★ 887 ★

Voters and Voting

Voting and Registration in the Election of November 1992. Part III – Middle Atlantic

November 1992. Numbers in thousands.

| Division, race, Hispanic origin, and age | All persons | Reported registered | | Reported voted | | Reported that they did not vote[1] | | | | |
| | | | | | | Total | Registered | Not registered | | |
		Number	Percent	Number	Percent			Total[2]	Not a U.S. citizen	Do not know and not reported on registration[3]
All races										
Total, 18 years and over	28,409	18,253	64.3	16,677	58.7	11,733	1,577	10,156	2,058	2,032
18 to 20 years	1,411	631	44.8	498	35.3	913	134	780	108	149
21 to 24 years	2,065	1,093	52.9	941	45.6	1,124	152	972	215	186
25 to 34 years	6,168	3,494	56.7	3,118	50.5	3,050	377	2,673	635	408
35 to 44 years	5,972	3,903	65.4	3,616	60.6	2,356	287	2,069	489	354
45 to 54 years	4,242	2,926	69.0	2,761	65.1	1,481	165	1,317	267	288
55 to 64 years	3,441	2,530	73.5	2,385	69.3	1,056	144	912	175	260
65 to 74 years	3,048	2,215	72.7	2,079	68.2	969	136	833	102	209
75 years and over	2,062	1,461	70.9	1,279	62.0	783	183	600	67	177
White										
Total, 18 years and over	23,884	15,999	67.0	14,773	61.9	9,111	1,226	7,885	1,098	1,581
18 to 20 years	1,074	529	49.2	432	40.2	642	97	546	32	104
21 to 24 years	1,639	923	56.3	816	49.8	823	108	715	100	133
25 to 34 years	5,095	2,992	58.7	2,692	52.8	2,403	300	2,103	359	310
35 to 44 years	4,930	3,379	68.6	3,162	64.1	1,768	218	1,550	249	258
45 to 54 years	3,569	2,554	71.6	2,431	68.1	1,138	123	1,015	127	226
55 to 64 years	2,963	2,229	75.2	2,120	71.6	843	109	734	115	216

[Continued]

★ 887 ★

Voting and Registration in the Election of November 1992. Part III – Middle Atlantic

[Continued]

Division, race, Hispanic origin, and age	All persons	Reported registered		Reported voted		Reported that they did not vote[1]				
						Total	Registered	Not registered		
		Number	Percent	Number	Percent			Total[2]	Not a U.S. citizen	Do not know and not reported on registration[3]
65 to 74 years	2,733	2,048	75.0	1,936	70.8	797	113	684	72	169
75 years and over	1,881	1,344	71.4	1,184	62.9	697	159	538	45	166
Black										
Total, 18 years and over	3,588	2,009	56.0	1,708	47.6	1,880	301	1,579	501	388
18 to 20 years	272	87	32.1	54	19.7	218	34	185	49	37
21 to 24 years	325	149	45.8	115	35.4	210	34	176	66	41
25 to 34 years	853	465	54.5	391	45.8	463	75	388	145	89
35 to 44 years	815	457	56.1	395	48.5	420	63	357	121	87
45 to 54 years	506	312	61.6	282	55.7	224	30	194	73	51
55 to 64 years	398	269	67.6	241	60.6	157	28	129	30	40
65 to 74 years	263	155	58.8	138	52.2	126	17	108	11	34
75 years and over	156	114	73.4	93	59.7	63	21	41	6	11
Hispanic origin[4]										
Total, 18 years and over	1,843	708	38.4	593	32.2	1,249	114	1,135	598	176
18 to 20 years	110	31	28.3	29	26.4	81	2	79	25	11
21 to 24 years	194	77	39.6	52	27.0	142	25	117	63	22
25 to 34 years	485	163	33.6	130	26.7	355	33	322	193	31
35 to 44 years	412	170	41.2	156	38.0	255	13	242	138	30
45 to 54 years	282	119	42.2	104	37.0	178	15	163	67	36
55 to 64 years	176	86	48.9	77	43.9	99	9	90	59	14
65 to 74 years	109	33	30.0	24	22.1	85	9	77	32	16
75 years and over	74	29	(B)	20	(B)	54	9	45	20	15

Source: "Reported Voting and Registration, by Race, Hispanic Origin, Sex, and Age, for Divisions," U.S. Bureau of the Census, *Voting and Registration in the Election of November 1992*, p. 15. *Notes:* 1. Includes persons reported as "did not vote," "do not know," and "not reported" on voting. 2. In addition to those reported as "not registered," total includes those "not a U.S. citizen," and "do not know" and "not reported" on registration. 3. Includes "do not know" and "not reported" on citizenship. 4. Persons of Hispanic origin may be of any race.

★ 888 ★

Voters and Voting

Voting and Registration in the Election of November 1992. Part IV – Northeast

November 1992. Numbers in thousands.

Region, race, Hispanic origin, sex, and age	All persons	Reported registered		Reported voted		Reported that they did not vote[1]				
						Total	Registered	Not registered		
		Number	Percent	Number	Percent			Total[2]	Not a U.S. citizen	Do not know and not reported on registration[3]
All races										
Both sexes										
Total 18 years and over	38,329	25,673	67.0	23,448	61.2	14,881	2,225	12,658	2,568	2,630
18 to 20 years	1,858	858	46.2	686	36.9	1,172	172	999	137	194
21 to 24 years	2,840	1,560	54.9	1,358	47.8	1,482	203	1,280	263	238
25 to 34 years	8,491	5,088	59.9	4,580	53.9	3,911	508	3,403	789	531
35 to 44 years	8,154	5,587	68.5	5,183	63.6	2,972	404	2,568	615	461
45 to 54 years	5,687	4,090	71.9	3,849	67.7	1,838	242	1,597	335	350
55 to 64 years	4,525	3,431	75.8	3,240	71.6	1,285	191	1,094	217	319

[Continued]

★ 888 ★

Voting and Registration in the Election of November 1992. Part IV – Northeast

[Continued]

Region, race, Hispanic origin, sex, and age	All persons	Reported registered		Reported voted		Reported that they did not vote[1]				
								Not registered		
		Number	Percent	Number	Percent	Total	Registered	Total[2]	Not a U.S. citizen	Do not know and not reported on registration[3]
65 to 74 years	4,063	3,085	75.9	2,869	70.6	1,194	216	978	123	285
75 years and over	2,711	1,974	72.8	1,684	62.1	1,027	290	737	89	252
Male										
Total 18 years and over	18,106	11,979	66.2	10,966	60.6	7,141	1,014	6,127	1,257	1,286
18 to 20 years	913	404	44.3	320	35.0	593	84	509	70	97
21 to 24 years	1,422	776	54.6	666	46.8	756	110	645	134	123
25 to 34 years	4,149	2,427	58.5	2,159	52.0	1,990	268	1,722	408	289
35 to 44 years	3,985	2,658	66.7	2,460	61.7	1,525	198	1,327	317	247
45 to 54 years	2,752	1,960	71.2	1,864	67.8	887	96	792	160	176
55 to 64 years	2,105	1,601	76.1	1,515	72.0	590	86	504	91	141
65 to 74 years	1,846	1,422	77.0	1,335	72.3	511	87	424	47	135
75 years and over	935	732	78.3	647	69.2	288	85	203	29	78
Female										
Total 18 years and over	20,223	13,694	67.7	12,483	61.7	7,740	1,211	6,529	1,311	1,344
18 to 20 years	945	454	48.1	367	38.8	578	88	490	67	96
21 to 24 years	1,419	784	55.3	692	48.8	727	93	634	129	115
25 to 34 years	4,342	2,661	61.3	2,421	55.8	1,921	240	1,681	380	242
35 to 44 years	4,169	2,929	70.2	2,723	65.3	1,446	206	1,240	298	214
45 to 54 years	2,935	2,131	72.6	1,985	67.6	951	146	805	174	174
55 to 64 years	2,420	1,830	75.6	1,725	71.3	695	105	590	125	178
65 to 74 years	2,217	1,663	75.0	1,534	69.2	683	129	554	77	150
75 years and over	1,775	1,241	69.9	1,037	58.4	739	205	534	60	174
White										
Both sexes										
Total 18 years and over	33,206	23,138	69.7	21,305	64.2	11,900	1,833	10,068	1,467	2,129
18 to 20 years	1,482	749	50.5	613	41.4	868	135	733	49	141
21 to 24 years	2,337	1,358	58.1	1,203	51.5	1,134	154	980	135	173
25 to 34 years	7,266	4,533	62.4	4,112	56.6	3,154	422	2,732	461	421
35 to 44 years	6,956	4,977	71.5	4,647	66.8	2,310	330	1,979	337	356
45 to 54 years	4,953	3,689	74.5	3,499	70.6	1,455	191	1,264	184	287
55 to 64 years	3,986	3,094	77.6	2,942	73.8	1,044	151	893	152	269
65 to 74 years	3,718	2,895	77.9	2,711	72.9	1,007	184	823	89	244
75 years and over	2,507	1,843	73.5	1,579	63.0	928	265	664	60	237
Male										
Total 18 years and over	15,762	10,894	69.1	10,047	63.7	5,715	847	4,868	723	1,015
18 to 20 years	729	343	47.1	279	38.2	450	65	385	31	67
21 to 24 years	1,176	671	57.1	588	50.0	588	83	505	79	86
25 to 34 years	3,582	2,197	61.3	1,971	55.0	1,611	226	1,385	242	232
35 to 44 years	3,431	2,415	70.4	2,243	65.4	1,187	171	1,016	170	184
45 to 54 years	2,396	1,770	73.9	1,691	70.6	705	79	626	86	143
55 to 64 years	1,878	1,466	78.1	1,397	74.4	481	69	412	61	117
65 to 74 years	1,696	1,344	79.2	1,269	74.8	427	74	352	33	114
75 years and over	875	689	78.7	609	69.6	266	80	186	21	74
Female										
Total 18 years and over	17,444	12,244	70.2	11,259	64.5	6,185	985	5,200	744	1,113
18 to 20 years	753	406	53.9	335	44.5	418	71	347	18	74
21 to 24 years	1,161	687	59.1	616	53.0	546	71	475	56	87
25 to 34 years	3,683	2,336	63.4	2,140	58.1	1,543	196	1,347	219	189
35 to 44 years	3,526	2,562	72.7	2,403	68.2	1,122	159	963	167	173
45 to 54 years	2,558	1,920	75.1	1,808	70.7	750	112	638	98	144

[Continued]

★ 888 ★

Voting and Registration in the Election of November 1992. Part IV – Northeast

[Continued]

Region, race, Hispanic origin, sex, and age	All persons	Reported registered		Reported voted		Reported that they did not vote[1]				
								Not registered		
		Number	Percent	Number	Percent	Total	Registered	Total[2]	Not a U.S. citizen	Do not know and not reported on registration[3]
55 to 64 years	2,108	1,627	77.2	1,546	73.3	563	82	481	92	152
65 to 74 years	2,022	1,551	76.7	1,441	71.3	580	110	471	56	130
75 years and over	1,632	1,155	70.7	970	59.4	663	185	478	39	163
Black										
Both sexes										
Total 18 years and over	4,033	2,239	55.5	1,905	47.2	2,128	334	1,794	578	423
18 to 20 years	304	94	31.0	60	19.9	243	34	210	58	41
21 to 24 years	378	174	46.1	138	36.7	239	36	203	71	48
25 to 34 years	969	509	52.5	426	43.9	544	83	461	176	99
35 to 44 years	931	529	56.8	462	49.6	469	67	402	144	94
45 to 54 years	550	332	60.4	295	53.6	255	37	218	78	52
55 to 64 years	445	302	67.7	271	60.8	174	31	144	32	42
65 to 74 years	287	175	60.8	151	52.4	137	24	113	13	35
75 years and over	169	125	74.2	103	60.9	66	22	44	7	11
Male										
Total 18 years and over	1,833	950	51.8	806	44.0	1,027	144	883	285	227
18 to 20 years	150	50	33.2	31	20.9	119	18	100	27	24
21 to 24 years	185	88	47.4	68	36.9	117	19	97	30	28
25 to 34 years	446	210	47.2	170	38.2	275	40	235	93	51
35 to 44 years	427	215	50.3	191	44.8	236	24	212	77	55
45 to 54 years	268	155	57.7	142	52.8	126	13	113	38	28
55 to 64 years	181	120	66.3	106	58.5	75	14	61	12	18
65 to 74 years	125	70	55.7	60	48.0	65	10	55	7	19
75 years and over	51	43	(B)	38	(B)	14	6	8	2	5
Female										
Total 18 years and over	2,200	1,290	58.6	1,100	50.0	1,101	190	911	292	195
18 to 20 years	154	44	28.8	29	18.9	125	15	109	31	17
21 to 24 years	192	86	44.9	70	36.4	122	16	106	40	20
25 to 34 years	524	299	57.0	256	48.8	268	43	225	83	47
35 to 44 years	504	314	62.3	271	53.7	233	43	190	66	39
45 to 54 years	282	177	62.9	153	54.3	129	24	105	40	24
55 to 64 years	264	182	68.8	165	62.4	99	17	83	20	25
65 to 74 years	162	105	64.8	91	55.8	72	15	57	6	17
75 years and over	118	82	70.0	66	55.7	52	17	35	5	7
Hispanic origin[4]										
Both sexes										
Total 18 years and over	2,123	808	38.1	675	31.8	1,448	133	1,314	674	190
18 to 20 years	132	34	25.6	32	24.0	100	2	98	28	13
21 to 24 years	229	85	37.1	60	26.4	169	25	144	73	22
25 to 34 years	576	196	34.0	153	26.6	423	43	380	227	34
35 to 44 years	485	201	41.4	182	37.4	304	20	284	157	33
45 to 54 years	320	138	43.1	122	38.1	198	16	182	73	40
55 to 64 years	186	91	49.1	80	43.1	106	11	94	59	14
65 to 74 years	118	34	29.1	26	21.7	92	9	84	36	17
75 years and over	76	29	38.1	20	26.4	56	9	47	22	15
Male										
Total 18 years and over	959	323	33.7	280	29.2	679	43	636	337	87
18 to 20 years	69	13	(B)	12	(B)	57	1	56	23	8
21 to 24 years	110	29	26.5	17	15.9	92	12	81	38	14
25 to 34 years	259	70	26.9	56	21.7	203	13	189	125	13

[Continued]

★ 888 ★

Voting and Registration in the Election of November 1992. Part IV – Northeast

[Continued]

Region, race, Hispanic origin, sex, and age	All persons	Reported registered		Reported voted		Reported that they did not vote[1]				
								Not registered		
		Number	Percent	Number	Percent	Total	Registered	Total[2]	Not a U.S. citizen	Do not know and not reported on registration[3]
35 to 44 years	231	92	39.6	84	36.2	148	8	140	73	15
45 to 54 years	152	60	39.3	55	36.1	97	5	92	38	23
55 to 64 years	64	34	(B)	33	(B)	31	2	30	19	5
65 to 74 years	53	20	(B)	18	(B)	36	3	33	8	6
75 years and over	21	6	(B)	6	(B)	16	-	16	11	4
Female										
Total 18 years and over	1,163	485	41.7	395	33.9	768	90	678	338	102
18 to 20 years	63	21	(B)	20	(B)	43	1	42	6	6
21 to 24 years	119	56	46.8	43	36.0	76	13	64	34	9
25 to 34 years	317	126	39.8	97	30.6	220	29	191	102	21
35 to 44 years	254	110	43.1	98	38.6	156	12	145	83	18
45 to 54 years	168	78	46.5	67	39.9	101	11	90	35	18
55 to 64 years	122	57	46.8	47	39.0	74	10	65	39	9
65 to 74 years	65	14	(B)	8	(B)	57	6	51	28	11
75 years and over	55	23	(B)	15	(B)	41	9	32	10	11

Source: "Reported Voting and Registration, by Race, Hispanic Origin, Sex, and Age, for the United States and Regions," U.S. Bureau of the Census, *Voting and Registration in the Election of November 1992*, pp. 6-7. *Notes:* 1. Includes persons reported as "did not vote," "do not know," and "not reported" on voting. 2. In addition to those reported as "not registered," total includes those "not a U.S. citizen," and "do not know" and "not reported" on registration. 3. Includes "do not know" and "not reported" on citizenship. 4. Persons of Hispanic origin may be of any race.

★ 889 ★

Voters and Voting

Voting and Registration in the Election of November 1992. Part V – East North Central

November 1992. Numbers in thousands.

Division, race, Hispanic origin, and age	All persons	Reported registered		Reported voted		Reported that they did not vote[1]				
								Not registered		
		Number	Percent	Number	Percent	Total	Registered	Total[2]	Not a U.S. citizen	Do not know and not reported on registration[3]
All races										
Total, 18 years and over	31,341	22,834	72.9	20,679	66.0	10,662	2,155	8,507	901	2,117
18 to 20 years	1,695	892	52.6	718	42.3	978	175	803	68	161
21 to 24 years	2,470	1,559	63.1	1,335	54.1	1,135	224	910	94	201
25 to 34 years	6,834	4,603	67.4	4,013	58.7	2,821	591	2,230	248	444
35 to 44 years	6,615	4,877	73.7	4,533	68.5	2,082	344	1,739	223	467
45 to 54 years	4,790	3,747	78.2	3,502	73.1	1,288	245	1,043	122	271
55 to 64 years	3,528	2,829	80.2	2,645	75.0	882	183	699	74	198
65 to 74 years	3,103	2,538	81.8	2,381	76.7	722	157	565	51	181
75 years and over	2,306	1,788	77.5	1,553	67.3	754	235	518	20	193
White										
Total, 18 years and over	27,358	20,137	73.6	18,369	67.1	8,989	1,768	7,220	638	1,749
18 to 20 years	1,394	753	54.0	609	43.7	785	143	642	50	125
21 to 24 years	2,055	1,319	64.2	1,134	55.2	921	185	736	64	152
25 to 34 years	5,893	4,021	68.2	3,547	60.2	2,346	474	1,872	149	351

[Continued]

★ 889 ★

Voting and Registration in the Election of November 1992. Part V – East North Central
[Continued]

Division, race, Hispanic origin, and age	All persons	Reported registered		Reported voted		Reported that they did not vote[1]				
						Total	Registered	Not registered		
								Total[2]	Not a U.S. citizen	Do not know and not reported on registration[3]
		Number	Percent	Number	Percent					
35 to 44 years	5,771	4,309	74.7	4,036	69.9	1,735	273	1,462	159	392
45 to 54 years	4,234	3,304	78.0	3,097	73.2	1,137	207	930	91	222
55 to 64 years	3,121	2,515	80.6	2,361	75.6	760	155	606	63	169
65 to 74 years	2,791	2,285	81.9	2,157	77.3	634	128	507	44	163
75 years and over	2,099	1,632	77.8	1,427	68.0	671	204	467	18	175
Black										
Total, 18 years and over	3,377	2,443	72.4	2,095	62.0	1,281	348	933	49	339
18 to 20 years	252	123	48.7	95	37.7	157	28	129	2	35
21 to 24 years	349	221	63.2	184	52.7	165	36	129	9	44
25 to 34 years	793	545	68.8	435	54.9	357	110	247	24	84
35 to 44 years	713	518	72.7	453	63.6	259	65	195	10	71
45 to 54 years	454	385	84.8	349	77.0	104	35	69	1	43
55 to 64 years	341	268	78.4	246	72.0	96	22	74	2	27
65 to 74 years	274	233	85.0	210	76.7	64	23	41	-	17
75 years and over	200	151	75.4	122	60.7	79	29	49	-	18
Hispanic origin[4]										
Total, 18 years and over	902	358	39.8	288	31.9	614	71	543	302	90
18 to 20 years	74	16	(B)	11	(B)	63	5	58	23	13
21 to 24 years	86	18	21.1	10	11.2	77	9	68	38	9
25 to 34 years	237	78	32.8	53	22.2	185	25	160	94	33
35 to 44 years	254	99	38.9	91	35.9	163	8	155	76	26
45 to 54 years	116	60	51.7	51	43.7	65	9	56	48	6
55 to 64 years	68	47	(B)	39	(B)	29	8	21	14	-
65 to 74 years	51	39	(B)	32	(B)	19	7	12	3	3
75 years and over	15	2	(B)	2	(B)	13	-	13	6	-

Source: "Reported Voting and Registration, by Race, Hispanic Origin, Sex, and Age, for Divisions," U.S. Bureau of the Census, *Voting and Registration in the Election of November 1992*, p. 16. *Notes:* 1. Includes persons reported as "did not vote," "do not know," and "not reported" on voting. 2. In addition to those reported as "not registered," total includes those "not a U.S. citizen," and "do not know" and "not reported" on registration. 3. Includes "do not know" and "not reported" on citizenship. 4. Persons of Hispanic origin may be of any race.

★ 890 ★

Voters and Voting

Voting and Registration in the Election of November 1992. Part VI – Midwest

November 1992. Numbers in thousands.

Region, race, Hispanic origin, sex, and age	All persons	Reported registered		Reported voted		Reported that they did not vote[1]				
						Total	Registered	Not registered		
								Total[2]	Not a U.S. citizen	Do not know and not reported on registration[3]
		Number	Percent	Number	Percent					
All races										
Both sexes										
Total 18 years and over	44,410	33,137	74.6	29,830	67.2	14,580	3,307	11,272	1,098	2,610
18 to 20 years	2,429	1,343	55.3	1,068	44.0	1,361	275	1,086	80	219
21 to 24 years	3,591	2,318	64.6	1,946	54.2	1,645	372	1,273	117	275
25 to 34 years	9,793	6,744	68.9	5,863	59.9	3,929	881	3,049	327	534

[Continued]

★ 890 ★

Voting and Registration in the Election of November 1992. Part VI – Midwest

[Continued]

Region, race, Hispanic origin, sex, and age	All persons	Reported registered		Reported voted		Reported that they did not vote[1]				
								Not registered		
						Total	Registered	Total[2]	Not a U.S. citizen	Do not know and not reported on registration[3]
		Number	Percent	Number	Percent					
35 to 44 years	9,339	7,120	76.2	6,593	70.6	2,746	526	2,219	260	537
45 to 54 years	6,670	5,338	80.0	4,990	74.8	1,680	348	1,332	146	334
55 to 64 years	4,876	3,994	81.9	3,734	76.6	1,143	260	882	81	249
65 to 74 years	4,397	3,646	82.9	3,403	77.4	994	243	751	62	226
75 years and over	3,315	2,635	79.5	2,233	67.4	1,082	402	680	25	236
Male										
Total 18 years and over	21,187	15,561	73.4	13,971	65.9	7,217	1,591	5,626	573	1,309
18 to 20 years	1,253	678	54.1	523	41.8	730	155	575	44	118
21 to 24 years	1,754	1,118	63.7	928	52.9	826	190	636	60	146
25 to 34 years	4,819	3,194	66.3	2,744	57.0	2,074	450	1,625	177	306
35 to 44 years	4,617	3,433	74.4	3,151	68.3	1,465	282	1,184	147	270
45 to 54 years	3,218	2,534	78.7	2,373	73.7	846	161	685	74	183
55 to 64 years	2,356	1,938	82.2	1,822	77.3	535	116	419	38	113
65 to 74 years	1,951	1,654	84.8	1,535	78.7	415	118	297	26	91
75 years and over	1,220	1,013	83.1	894	73.3	326	120	207	7	82
Female										
Total 18 years and over	23,222	17,576	75.7	15,859	68.3	7,363	1,717	5,646	525	1,301
18 to 20 years	1,176	665	56.5	544	46.3	631	121	511	36	101
21 to 24 years	1,837	1,200	65.3	1,018	55.4	819	182	637	57	129
25 to 34 years	4,974	3,550	71.4	3,119	62.7	1,855	431	1,424	150	228
35 to 44 years	4,722	3,687	78.1	3,442	72.9	1,280	245	1,036	113	267
45 to 54 years	3,452	2,804	81.2	2,617	75.8	835	187	648	72	151
55 to 64 years	2,520	2,056	81.6	1,912	75.9	608	144	464	43	137
65 to 74 years	2,447	1,993	81.4	1,868	76.4	578	124	454	36	135
75 years and over	2,095	1,621	77.4	1,339	63.9	756	282	473	19	154
White										
Both sexes										
Total 18 years and over	39,607	29,964	75.7	27,136	68.5	12,471	2,828	9,643	770	2,164
18 to 20 years	2,080	1,190	57.2	955	45.9	1,124	235	890	58	171
21 to 24 years	3,062	2,023	66.1	1,702	55.6	1,360	321	1,039	77	205
25 to 34 years	8,590	6,037	70.3	5,310	61.8	3,280	726	2,554	201	422
35 to 44 years	8,346	6,446	77.2	6,005	72.0	2,341	441	1,899	187	451
45 to 54 years	6,038	4,853	80.4	4,545	75.3	1,494	309	1,185	107	270
55 to 64 years	4,384	3,606	82.3	3,382	77.2	1,002	224	778	67	220
65 to 74 years	4,037	3,358	83.2	3,149	78.0	888	209	679	51	208
75 years and over	3,070	2,451	79.9	2,087	68.0	983	364	619	22	218
Male										
Total 18 years and over	19,036	14,221	74.7	12,846	67.5	6,190	1,375	4,815	394	1,097
18 to 20 years	1,079	606	56.1	472	43.7	607	134	474	30	95
21 to 24 years	1,505	988	65.6	830	55.1	676	158	518	39	106
25 to 34 years	4,278	2,908	68.0	2,523	59.0	1,755	385	1,371	108	255
35 to 44 years	4,168	3,162	75.9	2,923	70.1	1,245	239	1,007	98	227
45 to 54 years	2,959	2,342	79.1	2,197	74.3	761	144	617	55	155
55 to 64 years	2,142	1,766	82.4	1,657	77.4	484	108	376	35	93
65 to 74 years	1,780	1,515	85.1	1,415	79.5	366	100	265	25	85
75 years and over	1,124	936	83.3	829	73.8	295	107	188	3	82
Female										
Total 18 years and over	20,571	15,743	76.5	14,289	69.5	6,281	1,453	4,828	376	1,067
18 to 20 years	1,000	584	58.4	483	48.3	517	101	416	27	76
21 to 24 years	1,556	1,035	66.5	872	56.0	684	163	522	38	99

[Continued]

★ 890 ★

Voting and Registration in the Election of November 1992. Part VI – Midwest

[Continued]

Region, race, Hispanic origin, sex, and age	All persons	Reported registered		Reported voted		Reported that they did not vote[1]				
						Total	Registered	Not registered		
								Total[2]	Not a U.S. citizen	Do not know and not reported on registration[3]
		Number	Percent	Number	Percent					
25 to 34 years	4,312	3,129	72.6	2,787	64.6	1,525	342	1,183	93	167
35 to 44 years	4,177	3,285	78.6	3,082	73.8	1,095	203	893	89	224
45 to 54 years	3,080	2,512	81.6	2,347	76.2	732	164	568	52	115
55 to 64 years	2,242	1,840	82.1	1,725	76.9	517	115	402	33	127
65 to 74 years	2,257	1,843	81.6	1,734	76.8	523	109	414	26	123
75 years and over	1,945	1,515	77.9	1,258	64.7	688	257	430	19	136
Black										
Both sexes										
Total 18 years and over	3,943	2,809	71.2	2,393	60.7	1,550	415	1,135	50	389
18 to 20 years	278	128	46.0	95	34.2	183	33	150	2	41
21 to 24 years	441	267	60.5	222	50.4	219	45	174	11	63
25 to 34 years	977	642	65.8	503	51.4	474	140	334	24	97
35 to 44 years	799	588	73.6	513	64.1	287	75	211	10	78
45 to 54 years	495	413	83.5	378	76.3	117	35	82	1	49
55 to 64 years	411	335	81.5	307	74.6	105	28	76	2	27
65 to 74 years	309	259	83.7	236	76.4	73	23	50	-	17
75 years and over	232	176	75.7	140	60.3	92	36	56	-	18
Male										
Total 18 years and over	1,723	1,168	67.8	974	56.6	748	193	555	34	183
18 to 20 years	131	58	44.0	40	30.7	91	18	74	2	20
21 to 24 years	196	112	57.3	84	42.9	112	28	84	5	35
25 to 34 years	431	257	59.6	199	46.1	232	58	174	19	45
35 to 44 years	359	240	67.0	198	55.2	161	42	119	9	42
45 to 54 years	184	152	82.5	139	75.2	46	13	32	-	18
55 to 64 years	185	153	82.8	145	78.5	40	8	32	-	19
65 to 74 years	148	121	81.7	107	72.1	41	14	27	-	5
75 years and over	87	74	84.4	62	71.5	25	11	14	-	-
Female										
Total 18 years and over	2,221	1,641	73.9	1,419	63.9	802	222	580	16	206
18 to 20 years	147	70	47.8	55	37.3	92	15	77	-	22
21 to 24 years	245	154	63.0	138	56.4	107	16	90	6	28
25 to 34 years	546	385	70.6	304	55.7	242	82	160	5	52
35 to 44 years	440	348	79.0	315	71.4	126	33	93	2	36
45 to 54 years	311	261	84.1	239	77.0	72	22	49	1	31
55 to 64 years	226	182	80.4	161	71.4	65	20	44	2	8
65 to 74 years	161	138	85.5	130	80.3	32	8	23	-	12
75 years and over	145	102	70.5	78	53.6	67	25	43	-	18
Hispanic origin[4]										
Both sexes										
Total 18 years and over	1,104	439	39.8	354	32.1	750	85	665	367	105
18 to 20 years	86	20	23.5	14	16.8	72	6	66	27	16
21 to 24 years	110	29	26.6	19	17.3	91	10	81	39	15
25 to 34 years	312	103	33.0	74	23.9	237	26	209	120	35
35 to 44 years	312	120	38.5	109	34.8	204	11	192	99	31
45 to 54 years	132	68	51.8	59	44.8	73	9	64	55	6
55 to 64 years	74	50	(B)	40	(B)	34	10	24	14	-
65 to 74 years	60	47	(B)	37	(B)	23	10	12	3	3
75 years and over	19	2	(B)	2	(B)	17	-	17	10	-
Male										
Total 18 years and over	550	204	37.1	170	30.8	380	34	346	204	57

[Continued]

★ 890 ★

Voting and Registration in the Election of November 1992. Part VI – Midwest

[Continued]

Region, race, Hispanic origin, sex, and age	All persons	Reported registered		Reported voted		Reported that they did not vote[1]				
								Not registered		
		Number	Percent	Number	Percent	Total	Registered	Total[2]	Not a U.S. citizen	Do not know and not reported on registration[3]
18 to 20 years	38	7	(B)	3	(B)	35	4	31	13	7
21 to 24 years	54	12	(B)	7	(B)	47	5	42	24	11
25 to 34 years	149	43	28.8	32	21.2	117	11	106	62	16
35 to 44 years	151	44	29.3	43	28.2	109	2	107	60	19
45 to 54 years	75	36	48.0	34	45.2	41	2	39	36	3
55 to 64 years	44	28	(B)	23	(B)	21	6	15	9	-
65 to 74 years	36	32	(B)	27	(B)	8	5	3	-	2
75 years and over	4	2	(B)	2	(B)	2	-	2	-	-
Female										
Total 18 years and over	554	235	42.4	185	33.3	370	51	319	163	48
18 to 20 years	48	14	(B)	12	(B)	37	2	35	14	9
21 to 24 years	55	17	(B)	12	(B)	43	5	38	15	4
25 to 34 years	163	60	36.9	43	26.3	120	17	103	58	19
35 to 44 years	161	76	47.1	66	41.0	95	10	85	39	12
45 to 54 years	57	32	(B)	25	(B)	32	7	25	19	3
55 to 64 years	30	21	(B)	17	(B)	13	4	9	5	-
65 to 74 years	24	15	(B)	10	(B)	15	5	9	3	2
75 years and over	15	-	(B)	-	(B)	15	-	15	10	-

Source: "Reported Voting and Registration, by Race, Hispanic Origin, Sex, and Age, for the United States and Regions," U.S. Bureau of the Census, *Voting and Registration in the Election of November 1992*, pp. 8-9. *Notes:* 1. Includes persons reported as "did not vote," "do not know," and "not reported" on voting. 2. In addition to those reported as "not registered," total includes those "not a U.S. citizen," and "do not know" and "not reported" on registration. 3. Includes "do not know" and "not reported" on citizenship. 4. Persons of Hispanic origin may be of any race.

★ 891 ★

Voters and Voting

Voting and Registration in the Election of November 1992. Part VII – South

November 1992. Numbers in thousands.

Region, race, Hispanic origin, sex, and age	All persons	Reported registered		Reported voted		Reported that they did not vote[1]				
								Not registered		
		Number	Percent	Number	Percent	Total	Registered	Total[2]	Not a U.S. citizen	Do not know and not reported on registration[3]
All races										
Both sexes										
Total 18 years and over	63,659	42,762	67.2	37,590	59.0	26,068	5,172	20,896	2,838	3,229
18 to 20 years	3,321	1,586	47.8	1,223	36.8	2,099	363	1,735	192	221
21 to 24 years	4,993	2,682	53.7	2,140	42.9	2,852	541	2,311	273	355
25 to 34 years	14,025	8,235	58.7	7,043	50.2	6,982	1,193	5,790	894	686
35 to 44 years	13,421	9,132	68.0	8,212	61.2	5,209	920	4,289	661	684
45 to 54 years	9,775	7,143	73.1	6,490	66.4	3,285	653	2,631	340	494
55 to 64 years	7,528	5,718	76.0	5,201	69.1	2,327	517	1,810	227	342
65 to 74 years	6,520	5,181	79.5	4,689	71.9	1,831	492	1,339	137	256
75 years and over	4,076	3,084	75.7	2,592	63.6	1,484	492	992	115	191
Male										
Total 18 years and over	30,100	19,870	66.0	17,470	58.0	12,630	2,401	10,229	1,459	1,658

[Continued]

★ 891 ★

Voting and Registration in the Election of November 1992. Part VII – South

[Continued]

Region, race, Hispanic origin, sex, and age	All persons	Reported registered		Reported voted		Reported that they did not vote[1]				
								Not registered		
		Number	Percent	Number	Percent	Total	Registered	Total[2]	Not a U.S. citizen	Do not know and not reported on registration[3]
18 to 20 years	1,670	760	45.5	566	33.9	1,104	194	910	104	101
21 to 24 years	2,377	1,179	49.6	892	37.5	1,485	288	1,197	141	223
25 to 34 years	6,763	3,769	55.7	3,202	47.3	3,561	567	2,994	517	402
35 to 44 years	6,500	4,340	66.8	3,874	59.6	2,627	467	2,160	341	325
45 to 54 years	4,748	3,426	72.2	3,138	66.1	1,611	289	1,322	175	285
55 to 64 years	3,556	2,724	76.6	2,479	69.7	1,077	245	832	86	168
65 to 74 years	2,907	2,373	81.7	2,185	75.2	722	189	533	58	101
75 years and over	1,578	1,298	82.3	1,134	71.9	443	164	280	38	50
Female										
Total 18 years and over	33,559	22,892	68.2	20,121	60.0	13,438	2,771	10,667	1,379	1,573
18 to 20 years	1,651	826	50.0	657	39.8	994	169	825	88	121
21 to 24 years	2,616	1,503	57.4	1,249	47.7	1,367	254	1,113	132	132
25 to 34 years	7,262	4,467	61.5	3,841	52.9	3,421	626	2,795	377	284
35 to 44 years	6,921	4,792	69.2	4,338	62.7	2,583	454	2,129	320	358
45 to 54 years	5,026	3,717	74.0	3,352	66.7	1,674	364	1,309	165	210
55 to 64 years	3,972	2,994	75.4	2,722	68.5	1,250	272	978	141	174
65 to 74 years	3,613	2,807	77.7	2,504	69.3	1,109	303	806	79	155
75 years and over	2,498	1,786	71.5	1,458	58.4	1,040	328	713	77	141
White										
Both sexes										
Total 18 years and over	51,001	34,954	68.5	31,034	60.8	19,967	3,920	16,047	2,044	2,299
18 to 20 years	2,426	1,192	49.1	926	38.2	1,500	266	1,234	134	151
21 to 24 years	3,801	2,047	53.9	1,695	44.6	2,106	352	1,754	204	261
25 to 34 years	10,819	6,452	59.6	5,605	51.8	5,215	847	4,367	610	445
35 to 44 years	10,700	7,411	69.3	6,712	62.7	3,989	699	3,290	437	464
45 to 54 years	7,984	5,918	74.1	5,403	67.7	2,581	516	2,066	267	360
55 to 64 years	6,225	4,780	76.8	4,375	70.3	1,851	406	1,445	183	252
65 to 74 years	5,522	4,463	80.8	4,054	73.4	1,467	408	1,059	108	211
75 years and over	3,523	2,691	76.4	2,265	64.3	1,259	426	833	101	155
Male										
Total 18 years and over	24,394	16,474	67.5	14,669	60.1	9,725	1,806	7,920	1,053	1,174
18 to 20 years	1,221	567	46.5	425	34.8	795	142	654	79	62
21 to 24 years	1,834	919	50.1	740	40.3	1,095	179	916	101	173
25 to 34 years	5,313	3,023	56.9	2,631	49.5	2,682	392	2,290	346	255
35 to 44 years	5,301	3,562	57.2	3,202	60.4	2,099	360	1,739	243	227
45 to 54 years	3,932	2,911	74.0	2,676	68.1	1,256	235	1,021	137	205
55 to 64 years	2,942	2,296	78.1	2,104	71.5	838	192	646	65	122
65 to 74 years	2,464	2,050	83.2	1,892	76.8	572	158	414	46	86
75 years and over	1,386	1,146	82.7	999	72.1	387	148	239	35	43
Female										
Total 18 years and over	26,607	18,480	69.5	16,365	61.5	10,242	2,114	8,128	991	1,125
18 to 20 years	1,205	625	51.9	501	41.5	705	125	580	55	89
21 to 24 years	1,966	1,128	57.4	955	48.6	1,011	173	838	103	88
25 to 34 years	5,506	3,429	62.3	2,974	54.0	2,533	456	2,077	265	190
35 to 44 years	5,399	3,848	71.3	3,510	65.0	1,889	339	1,551	193	237
45 to 54 years	4,052	3,007	74.2	2,726	67.3	1,325	281	1,045	130	154
55 to 64 years	3,283	2,484	75.7	2,271	69.2	1,013	213	799	118	130
65 to 74 years	3,058	2,413	78.9	2,163	70.7	895	250	644	61	126
75 years and over	2,138	1,544	72.2	1,266	59.2	871	278	593	66	111

[Continued]

★ 891 ★

Voting and Registration in the Election of November 1992. Part VII – South

[Continued]

Region, race, Hispanic origin, sex, and age	All persons	Reported registered		Reported voted		Reported that they did not vote[1]				
								Not registered		
		Number	Percent	Number	Percent	Total	Registered	Total[2]	Not a U.S. citizen	Do not know and not reported on registration[3]
Black										
Both sexes										
Total 18 years and over	11,334	7,331	64.7	6,151	54.3	5,183	1,180	4,004	338	836
18 to 20 years	807	365	45.2	275	34.1	532	90	442	34	62
21 to 24 years	1,071	590	55.1	414	38.6	657	176	481	31	85
25 to 34 years	2,878	1,708	59.3	1,373	47.7	1,504	334	1,170	124	222
35 to 44 years	2,387	1,621	67.9	1,413	59.2	974	208	766	95	194
45 to 54 years	1,574	1,112	70.6	982	62.4	591	129	462	35	120
55 to 64 years	1,162	862	74.2	767	66.0	395	95	300	6	78
65 to 74 years	924	692	74.9	610	66.1	313	81	232	4	41
75 years and over	532	382	71.8	316	59.4	216	66	150	9	35
Male										
Total 18 years and over	5,099	3,180	62.4	2,622	51.4	2,477	558	1,919	183	440
18 to 20 years	405	177	43.8	127	31.4	277	50	227	10	36
21 to 24 years	485	238	49.0	141	29.1	344	96	247	23	46
25 to 34 years	1,298	716	55.1	546	42.1	752	170	582	92	137
35 to 44 years	1,051	737	70.2	637	60.6	414	101	313	38	87
45 to 54 years	719	462	64.3	411	57.2	308	51	256	19	72
55 to 64 years	550	396	72.0	352	64.1	197	44	154	-	44
65 to 74 years	409	308	75.3	277	67.8	132	31	101	1	13
75 years and over	184	146	79.6	130	70.9	53	16	38	-	7
Female										
Total 18 years and over	6,236	4,151	66.6	3,529	56.6	2,706	621	2,085	155	396
18 to 20 years	403	188	46.7	148	36.8	255	40	215	25	27
21 to 24 years	586	352	60.1	272	46.5	313	80	234	8	39
25 to 34 years	1,580	992	62.8	827	52.4	753	165	588	31	85
35 to 44 years	1,336	884	66.1	776	58.1	560	108	452	57	107
45 to 54 years	855	649	75.9	571	66.8	284	78	206	16	49
55 to 64 years	612	466	76.1	415	67.8	197	51	146	6	34
65 to 74 years	515	384	74.5	333	64.7	182	50	131	3	28
75 years and over	349	236	67.6	186	53.3	163	50	113	9	28
Hispanic origin[4]										
Both sexes										
Total 18 years and over	4,687	1,842	39.3	1,499	32.0	3,189	344	2,845	1,591	314
18 to 20 years	338	102	30.1	61	18.2	276	40	236	118	26
21 to 24 years	467	142	30.4	109	23.3	358	33	325	154	47
25 to 34 years	1,248	396	31.7	304	24.4	944	91	852	491	65
35 to 44 years	992	400	40.3	332	33.4	661	68	593	314	79
45 to 54 years	708	327	46.2	274	38.7	434	53	381	227	50
55 to 64 years	431	215	49.9	198	45.9	233	17	216	125	27
65 to 74 years	311	169	54.3	140	45.1	171	29	142	80	15
75 years and over	192	92	48.1	81	42.0	112	12	100	81	5
Male										
Total 18 years and over	2,330	855	36.7	704	30.2	1,627	152	1,475	846	176
18 to 20 years	187	50	26.8	27	14.7	159	23	137	69	15
21 to 24 years	223	53	23.7	38	16.9	185	15	170	78	35
25 to 34 years	668	185	27.8	153	23.0	514	32	482	292	39
35 to 44 years	511	191	37.4	160	31.2	351	31	320	177	41
45 to 54 years	367	162	44.2	137	37.3	230	25	205	125	32
55 to 64 years	162	96	59.1	90	55.5	72	6	66	42	10
65 to 74 years	136	81	59.7	62	45.5	74	19	55	33	-

[Continued]

★ 891 ★

Voting and Registration in the Election of November 1992. Part VII – South

[Continued]

Region, race, Hispanic origin, sex, and age	All persons	Reported registered		Reported voted		Reported that they did not vote[1]				
						Total	Registered	Not registered		
		Number	Percent	Number	Percent			Total[2]	Not a U.S. citizen	Do not know and not reported on registration[3]
75 years and over	76	37	48.4	37	48.4	39	-	39	29	5
Female										
Total 18 years and over	2,357	987	41.9	795	33.7	1,562	192	1,370	745	139
18 to 20 years	151	51	34.2	34	22.5	117	18	99	48	12
21 to 24 years	243	89	36.6	71	29.1	173	18	154	76	12
25 to 34 years	581	210	36.2	151	26.0	430	59	370	199	26
35 to 44 years	481	209	43.4	172	35.7	309	37	273	137	38
45 to 54 years	341	165	48.4	137	40.1	204	28	176	102	18
55 to 64 years	269	119	44.3	108	40.1	161	11	150	83	18
65 to 74 years	175	87	50.1	78	44.8	96	9	87	47	15
75 years and over	116	56	47.8	44	37.8	72	12	61	52	-

Source: "Reported Voting and Registration, by Race, Hispanic Origin, Sex, and Age, for the United States and Regions," U.S. Bureau of the Census, *Voting and Registration in the Election of November 1992*, pp. 10-11. *Notes:* 1. Includes persons reported as "did not vote," "do not know," and "not reported" on voting. 2. In addition to those reported as "not registered," total includes those "not a U.S. citizen," and "do not know" and "not reported" on registration. 3. Includes "do not know" and "not reported" on citizenship. 4. Persons of Hispanic origin may be of any race.

★ 892 ★

Voters and Voting

Voting and Registration in the Election of November 1992. Part VII – West

November 1992. Numbers in thousands.

Region, race, Hispanic origin, sex, and age	All persons	Reported registered		Reported voted		Reported that they did not vote[1]				
						Total	Registered	Not registered		
		Number	Percent	Number	Percent			Total[2]	Not a U.S. citizen	Do not know and not reported on registration[3]
All races										
Both sexes										
Total 18 years and over	39,286	25,005	63.6	22,997	58.5	16,289	2,008	14,281	5,396	2,169
18 to 20 years	2,119	908	42.9	772	36.4	1,348	136	1,211	448	187
21 to 24 years	3,220	1,531	47.6	1,249	38.8	1,971	282	1,689	635	207
25 to 34 years	9,294	5,155	55.5	4,633	49.9	4,661	522	4,139	1,666	472
35 to 44 years	8,801	5,664	64.4	5,281	60.0	3,520	383	3,137	1,263	463
45 to 54 years	5,926	4,214	71.1	3,963	66.9	1,963	250	1,712	614	287
55 to 64 years	4,160	3,088	74.2	2,932	70.5	1,227	155	1,072	431	197
65 to 74 years	3,466	2,773	80.0	2,645	76.3	820	128	692	216	185
75 years and over	2,299	1,672	72.7	1,521	66.1	778	151	627	123	171
Male										
Total 18 years and over	19,164	11,843	61.8	10,907	56.9	8,257	936	7,321	2,777	1,076
18 to 20 years	1,040	432	41.6	377	36.3	663	55	608	227	89
21 to 24 years	1,606	724	45.1	600	37.4	1,006	124	882	387	105
25 to 34 years	4,735	2,479	52.4	2,219	46.9	2,516	260	2,256	928	254
35 to 44 years	4,407	2,722	61.8	2,510	57.0	1,897	212	1,685	639	217
45 to 54 years	2,890	2,036	70.4	1,913	66.2	977	123	854	288	156
55 to 64 years	1,994	1,462	73.3	1,387	69.5	608	75	532	183	105
65 to 74 years	1,586	1,296	81.7	1,263	79.6	323	33	290	87	82

[Continued]

★ 892 ★

Voting and Registration in the Election of November 1992. Part VII – West

[Continued]

Region, race, Hispanic origin, sex, and age	All persons	Reported registered		Reported voted		Reported that they did not vote[1]				
						Total	Registered	Not registered		
		Number	Percent	Number	Percent			Total[2]	Not a U.S. citizen	Do not know and not reported on registration[3]
75 years and over	906	692	76.4	637	70.3	269	55	214	39	68
Female										
Total 18 years and over	20,122	13,163	65.4	12,091	60.1	8,031	1,072	6,960	2,619	1,093
18 to 20 years	1,079	476	44.1	395	36.6	685	81	603	221	98
21 to 24 years	1,614	807	50.0	649	40.2	966	159	807	248	103
25 to 34 years	4,559	2,676	58.7	2,414	53.0	2,145	262	1,883	738	218
35 to 44 years	4,395	2,943	67.0	2,771	63.1	1,624	172	1,452	624	246
45 to 54 years	3,036	2,178	71.7	2,050	67.5	986	127	858	327	131
55 to 64 years	2,165	1,626	75.1	1,546	71.4	620	80	540	248	91
65 to 74 years	1,880	1,477	78.6	1,382	73.5	498	95	403	129	102
75 years and over	1,393	980	70.4	884	63.4	509	96	413	84	103
White										
Both sexes										
Total 18 years and over	34,023	22,628	66.5	20,930	61.5	13,093	1,698	11,395	4,003	1,697
18 to 20 years	1,755	806	45.9	693	39.5	1,062	114	949	326	121
21 to 24 years	2,738	1,381	50.4	1,144	41.8	1,594	237	1,357	496	148
25 to 34 years	7,965	4,609	57.9	4,167	52.3	3,798	442	3,356	1,324	345
35 to 44 years	7,642	5,130	67.1	4,822	63.1	2,820	308	2,512	940	372
45 to 54 years	5,095	3,781	74.2	3,567	70.0	1,529	215	1,314	404	239
55 to 64 years	3,613	2,790	77.2	2,658	73.6	955	133	822	303	146
65 to 74 years	3,123	2,565	82.1	2,451	78.5	673	115	558	135	173
75 years and over	2,092	1,565	74.8	1,429	68.3	662	136	527	74	152
Male										
Total 18 years and over	16,724	10,793	64.5	9,990	59.7	6,734	803	5,930	2,128	830
18 to 20 years	847	381	45.0	334	39.5	513	47	466	163	55
21 to 24 years	1,358	660	48.6	552	40.7	806	108	698	308	71
25 to 34 years	4,113	2,232	54.3	2,005	48.7	2,108	228	1,881	764	191
35 to 44 years	3,852	2,479	64.3	2,308	59.9	1,544	170	1,373	494	172
45 to 54 years	2,540	1,864	73.4	1,750	68.9	789	113	676	191	130
55 to 64 years	1,744	1,331	76.3	1,269	72.8	475	62	413	132	74
65 to 74 years	1,445	1,206	83.5	1,179	81.6	266	28	239	52	80
75 years and over	825	640	77.6	592	71.7	233	48	185	24	58
Female										
Total 18 years and over	17,300	11,835	68.4	10,940	63.2	6,360	895	5,465	1,875	867
18 to 20 years	908	425	46.9	358	39.5	549	67	482	164	67
21 to 24 years	1,380	721	52.2	592	42.9	789	129	659	188	78
25 to 34 years	3,852	2,376	61.7	2,162	56.1	1,690	214	1,476	560	154
35 to 44 years	3,790	2,651	70.0	2,514	66.3	1,276	137	1,139	447	200
45 to 54 years	2,556	1,918	75.0	1,816	71.1	739	101	638	213	109
55 to 64 years	1,868	1,459	78.1	1,388	74.3	480	71	409	171	72
65 to 74 years	1,678	1,359	81.0	1,272	75.8	406	87	319	83	93
75 years and over	1,267	925	73.0	837	66.1	429	87	342	50	94
Black										
Both sexes										
Total 18 years and over	1,729	1,063	61.5	922	53.3	807	141	666	78	222
18 to 20 years	111	51	45.4	43	38.9	68	7	61	9	31
21 to 24 years	153	76	49.3	48	31.0	106	28	78	4	34
25 to 34 years	472	248	52.7	216	45.9	255	32	223	24	62
35 to 44 years	398	239	59.9	202	50.7	196	36	160	25	53
45 to 54 years	243	183	75.5	167	68.6	76	17	60	4	13

[Continued]

★ 892 ★

Voting and Registration in the Election of November 1992. Part VII – West

[Continued]

Region, race, Hispanic origin, sex, and age	All persons	Reported registered		Reported voted		Reported that they did not vote[1]				
								Not registered		
		Number	Percent	Number	Percent	Total	Registered	Total[2]	Not a U.S. citizen	Do not know and not reported on registration[3]
55 to 64 years	163	112	69.2	107	65.7	56	6	50	12	23
65 to 74 years	114	94	82.7	85	75.2	28	8	20	1	1
75 years and over	76	61	80.1	54	71.1	22	7	15	-	4
Male										
Total 18 years and over	771	429	55.7	384	49.8	387	45	342	45	115
18 to 20 years	59	22	(B)	19	(B)	40	3	37	9	18
21 to 24 years	74	33	(B)	25	(B)	49	8	40	3	24
25 to 34 years	214	100	46.8	94	44.0	120	6	114	11	28
35 to 44 years	191	102	53.5	87	45.8	103	15	89	14	23
45 to 54 years	96	69	72.3	65	68.2	30	4	27	2	7
55 to 64 years	69	41	(B)	37	(B)	32	5	28	6	14
65 to 74 years	42	35	(B)	31	(B)	11	4	7	-	1
75 years and over	28	27	(B)	26	(B)	2	1	1	-	-
Female										
Total 18 years and over	958	634	66.2	538	56.2	420	96	33	33	108
18 to 20 years	53	29	(B)	24	(B)	28	4	24	-	14
21 to 24 years	80	42	53.2	22	28.2	57	20	37	1	10
25 to 34 years	258	148	57.5	122	47.5	135	26	110	13	35
35 to 44 years	207	136	65.8	115	55.3	93	22	71	11	30
45 to 54 years	147	114	77.6	101	68.9	46	13	33	2	7
55 to 64 years	94	71	76.0	70	74.9	24	1	22	6	9
65 to 74 years	72	59	(B)	54	(B)	17	5	13	-	-
75 years and over	48	34	(B)	28	(B)	20	6	14	-	4
Hispanic origin[4]										
Both sexes										
Total 18 years and over	6,774	2,047	30.2	1,710	25.2	5,064	337	4,727	3,278	385
18 to 20 years	603	112	18.5	75	12.5	528	37	491	310	38
21 to 24 years	831	173	20.8	122	14.6	709	51	658	456	36
25 to 34 years	2,041	544	26.7	468	22.9	1,573	77	1,496	1,092	103
35 to 44 years	1,532	439	28.7	357	23.3	1,175	82	1,093	783	107
45 to 54 years	793	317	40.0	276	34.8	517	41	476	322	47
55 to 64 years	566	268	47.4	248	43.8	318	20	298	203	30
65 to 74 years	278	135	48.6	117	41.9	162	19	143	79	19
75 years and over	129	58	45.2	48	37.0	82	11	71	32	6
Male										
Total 18 years and over	3,433	955	27.8	793	23.1	2,640	162	2,479	1,745	199
18 to 20 years	269	40	14.8	30	11.0	240	10	229	154	16
21 to 24 years	468	84	18.0	62	13.2	406	22	384	273	23
25 to 34 years	1,106	265	23.9	223	20.2	883	41	842	620	54
35 to 44 years	775	213	27.5	161	20.7	615	53	562	411	52
45 to 54 years	388	158	40.8	137	35.3	251	21	230	155	20
55 to 64 years	266	118	44.3	107	40.2	159	11	148	94	22
65 to 74 years	115	52	45.6	52	45.1	63	1	62	30	11
75 years and over	46	24	(B)	22	(B)	24	3	21	9	1
Female										
Total 18 years and over	3,341	1,092	32.7	917	27.5	2,423	175	2,248	1,533	186
18 to 20 years	334	72	21.5	46	13.6	288	26	262	156	21
21 to 24 years	363	88	24.4	60	16.5	303	29	274	183	13
25 to 34 years	934	280	29.9	244	26.2	690	35	655	472	50
35 to 44 years	757	226	29.9	197	26.0	560	29	531	372	55

[Continued]

★ 892 ★

Voting and Registration in the Election of November 1992. Part VII – West

[Continued]

| Region, race, Hispanic origin, sex, and age | All persons | Reported registered | | Reported voted | | Reported that they did not vote[1] | | | | |
| | | | | | | Total | Registered | Not registered | | |
		Number	Percent	Number	Percent			Total[2]	Not a U.S. citizen	Do not know and not reported on registration[3]
45 to 54 years	405	159	39.3	139	34.3	266	20	246	167	27
55 to 64 years	301	151	50.1	141	47.0	159	9	150	109	8
65 to 74 years	163	83	50.7	65	39.7	99	18	80	49	8
75 years and over	84	34	40.7	26	31.2	58	8	50	23	5

Source: "Reported Voting and Registration, by Race, Hispanic Origin, Sex, and Age, for the United States and Regions," U.S. Bureau of the Census, *Voting and Registration in the Election of November 1992*, pp. 12-13. *Notes:* 1. Includes persons reported as "did not vote," "do not know," and "not reported" on voting. 2. In addition to those reported as "not registered," total includes those "not a U.S. citizen," and "do not know" and "not reported" on registration. 3. Includes "do not know" and "not reported" on citizenship. 4. Persons of Hispanic origin may be of any race.

★ 893 ★

Voters and Voting

Voting and Registration in the Election of November 1992. Part IX – West North Central

November 1992. Numbers in thousands.

| Division, race, Hispanic origin, and age | All persons | Reported registered | | Reported voted | | Reported that they did not vote[1] | | | | |
| | | | | | | Total | Registered | Not registered | | |
		Number	Percent	Number	Percent			Total[2]	Not a U.S. citizen	Do not know and not reported on registration[3]
All races										
Total, 18 years and over	13,069	10,304	78.8	9,151	70.0	3,917	1,152	2,765	197	493
18 to 20 years	733	451	61.5	350	47.7	383	101	283	11	58
21 to 24 years	1,121	759	67.7	611	54.5	510	148	363	23	74
25 to 34 years	2,959	2,141	72.3	1,851	62.5	1,109	290	819	78	90
35 to 44 years	2,742	2,243	82.3	2,060	75.6	663	183	481	38	70
45 to 54 years	1,881	1,591	84.6	1,488	79.1	392	103	290	24	63
55 to 64 years	1,349	1,165	86.4	1,088	80.7	260	77	183	7	51
65 to 74 years	1,294	1,108	85.6	1,023	79.0	271	85	186	10	45
75 years and over	1,008	847	84.0	680	67.5	328	166	162	5	42
White										
Total, 18 years and over	12,249	9,827	80.2	8,767	71.6	3,482	1,060	2,423	133	416
18 to 20 years	685	437	63.8	346	50.5	339	91	248	8	46
21 to 24 years	1,007	704	69.9	568	56.4	439	136	304	13	53
25 to 34 years	2,697	2,015	74.7	1,763	65.4	934	253	682	52	71
35 to 44 years	2,575	2,138	83.0	1,969	76.5	606	168	437	29	58
45 to 54 years	1,805	1,550	85.9	1,448	80.2	357	102	255	16	48
55 to 64 years	1,263	1,090	86.4	1,022	80.9	241	69	172	5	51
65 to 74 years	1,246	1,073	86.1	992	79.6	254	81	173	7	45
75 years and over	971	820	84.4	660	67.9	311	160	152	4	42
Black										
Total, 18 years and over	567	365	64.5	298	52.6	269	67	201	1	50
18 to 20 years	26	5	(B)	-	(B)	26	5	21	-	6
21 to 24 years	91	46	50.3	38	41.2	54	8	45	1	19
25 to 34 years	184	97	52.8	67	36.4	117	30	87	-	13
35 to 44 years	87	70	81.2	59	68.6	27	11	16	-	7

[Continued]

★ 893 ★

Voting and Registration in the Election of November 1992. Part IX − West North Central

[Continued]

Division, race, Hispanic origin, and age	All persons	Reported registered		Reported voted		Reported that they did not vote[1]				
						Total	Registered	Not registered		
								Total[2]	Not a U.S. citizen	Do not know and not reported on registration[3]
		Number	Percent	Number	Percent					
45 to 54 years	41	28	(B)	28	(B)	13	-	13	-	6
55 to 64 years	70	67	(B)	61	(B)	9	6	3	-	-
65 to 74 years	35	26	(B)	26	(B)	9	-	9	-	-
75 years and over	32	25	(B)	18	(B)	14	7	7	-	-
Hispanic origin[4]										
Total, 18 years and over	203	81	39.8	66	32.7	137	15	122	65	15
18 to 20 years	13	5	(B)	4	(B)	9	1	8	4	3
21 to 24 years	23	11	(B)	9	(B)	14	2	12	2	6
25 to 34 years	74	25	(B)	22	(B)	52	3	49	26	2
35 to 44 years	58	21	(B)	17	(B)	41	4	37	22	6
45 to 54 years	16	8	(B)	8	(B)	7	-	7	7	-
55 to 64 years	6	2	(B)	1	(B)	5	1	4	-	-
65 to 74 years	8	8	(B)	5	(B)	3	3	-	-	-
75 years and over	4	-	(B)	-	(B)	4	-	4	4	-

Source: "Reported Voting and Registration, by Race, Hispanic Origin, Sex, and Age, for Divisions," U.S. Bureau of the Census, *Voting and Registration in the Election of November 1992*, p. 18. *Notes:* 1. Includes persons reported as "did not vote," "do not know," and "not reported" on voting. 2. In addition to those reported as "not registered," total includes those "not a U.S. citizen," and "do not know" and "not reported" on registration. 3. Includes "do not know" and "not reported" on citizenship. 4. Persons of Hispanic origin may be of any race.

★ 894 ★

Voters and Voting

Voting and Registration in the Election of November 1992. Part X − Pacific

November 1992. Numbers in thousands.

Division, race, Hispanic origin, and age	All persons	Reported registered		Reported voted		Reported that they did not vote[1]				
						Total	Registered	Not registered		
								Total[2]	Not a U.S. citizen	Do not know and not reported on registration[3]
		Number	Percent	Number	Percent					
All races										
Total, 18 years and over	29,405	17,911	60.9	16,453	56.0	12,952	1,458	11,494	4,987	1,737
18 to 20 years	1,611	632	39.2	542	33.6	1,070	90	979	423	153
21 to 24 years	2,501	1,116	44.6	912	36.5	1,589	204	1,385	587	167
25 to 34 years	7,035	3,730	53.0	3,325	47.3	3,710	406	3,304	1,548	374
35 to 44 years	6,612	4,058	61.4	3,782	57.2	2,830	276	2,554	1,162	381
45 to 54 years	4,402	3,008	68.3	2,824	64.2	1,578	184	1,394	568	238
55 to 64 years	3,006	2,149	71.5	2,036	67.7	970	113	857	393	148
65 to 74 years	2,531	1,991	78.7	1,907	75.3	624	84	539	192	144
75 years and over	1,708	1,227	71.8	1,127	66.0	581	100	481	114	132
White										
Total, 18 years and over	24,766	15,855	64.0	14,664	59.2	10,101	1,191	8,910	3,665	1,321
18 to 20 years	1,286	540	42.0	470	36.6	815	70	745	307	99
21 to 24 years	2,083	985	47.3	819	39.3	1,264	166	1,098	459	117
25 to 34 years	5,898	3,270	55.4	2,938	49.8	2,960	332	2,628	1,222	261
35 to 44 years	5,568	3,590	64.5	3,379	60.7	2,189	210	1,979	859	295

[Continued]

★ 894 ★

Voting and Registration in the Election of November 1992. Part X – Pacific

[Continued]

Division, race, Hispanic origin, and age	All persons	Reported registered		Reported voted		Reported that they did not vote[1]				
								Not registered		
		Number	Percent	Number	Percent	Total	Registered	Total[2]	Not a U.S. citizen	Do not know and not reported on registration[3]
45 to 54 years	3,659	2,630	71.9	2,475	67.6	1,184	155	1,029	367	194
55 to 64 years	2,541	1,910	75.2	1,816	71.5	724	94	630	268	108
65 to 74 years	2,214	1,801	81.4	1,725	77.9	489	77	412	116	133
75 years and over	1,518	1,129	74.4	1,043	68.7	475	86	389	68	113
Black										
Total, 18 years and over	1,466	903	61.6	783	53.4	683	120	563	77	192
18 to 20 years	94	47	49.4	39	41.8	55	7	48	9	24
21 to 24 years	133	63	47.6	41	30.8	92	22	70	3	33
25 to 34 years	379	203	53.6	174	45.8	205	29	176	23	55
35 to 44 years	359	209	58.3	177	49.4	182	32	150	25	50
45 to 54 years	216	164	75.6	149	68.8	68	15	53	4	12
55 to 64 years	117	78	66.7	74	62.8	44	5	39	12	15
65 to 74 years	102	84	82.2	80	78.4	22	4	18	1	–
75 years and over	65	55	(B)	16	6	10	–	4		
Hispanic origin[4]										
Total, 18 years and over	5,607	1,455	25.9	1,194	21.3	4,413	261	4,152	3,041	328
18 to 20 years	519	78	15.1	50	9.7	469	28	440	292	33
21 to 24 years	717	131	18.3	94	13.2	622	37	586	420	29
25 to 34 years	1,703	397	23.3	332	19.5	1,371	65	1,306	1,015	90
35 to 44 years	1,285	315	24.5	249	19.4	1,036	66	970	727	90
45 to 54 years	642	226	35.2	196	30.5	446	30	416	300	36
55 to 64 years	424	170	40.0	156	36.8	268	14	255	186	30
65 to 74 years	212	92	43.3	75	35.6	136	16	120	71	16
75 years and over	105	46	43.3	41	38.8	64	5	60	30	4

Source: "Reported Voting and Registration, by Race, Hispanic Origin, Sex, and Age, for Divisions," U.S. Bureau of the Census, *Voting and Registration in the Election of November 1992,* p. 22. *Notes:* 1. Includes persons reported as "did not vote," "do not know," and "not reported" on voting. 2. In addition to those reported as "not registered," total includes those "not a U.S. citizen," and "do not know" and "not reported" on registration. 3. Includes "do not know" and "not reported" on citizenship. 4. Persons of Hispanic origin may be of any race.

★ 895 ★

Voters and Voting

Voting and Registration in the Election of November 1992. Part XI – South Atlantic

November 1992. Numbers in thousands.

Division, race, Hispanic origin, and age	All persons	Reported registered		Reported voted		Reported that they did not vote[1]				
								Not registered		
		Number	Percent	Number	Percent	Total	Registered	Total[2]	Not a U.S. citizen	Do not know and not reported on registration[3]
All races										
Total, 18 years and over	33,095	21,693	65.5	19,398	58.6	13,698	2,296	11,402	1,637	1,663
18 to 20 years	1,526	741	48.5	581	38.1	945	159	785	79	105
21 to 24 years	2,536	1,333	52.5	1,106	43.6	1,430	227	1,203	137	190
25 to 34 years	7,169	4,080	56.9	3,532	49.3	3,637	548	3,089	529	327
35 to 44 years	6,866	4,473	65.1	4,052	59.0	2,815	422	2,393	395	329

[Continued]

★ 895 ★

Voting and Registration in the Election of November 1992. Part XI – South Atlantic

[Continued]

Division, race, Hispanic origin, and age	All persons	Reported registered		Reported voted		Reported that they did not vote[1]				
								Not registered		
						Total	Registered	Total[2]	Not a U.S. citizen	Do not know and not reported on registration[3]
		Number	Percent	Number	Percent					
45 to 54 years	5,151	3,673	71.3	3,419	66.4	1,732	255	1,478	178	247
55 to 64 years	4,036	2,949	73.1	2,704	67.0	1,332	244	1,088	147	190
65 to 74 years	3,585	2,777	77.5	2,556	71.3	1,029	221	808	99	161
75 years and over	2,226	1,668	74.9	1,448	65.0	778	220	558	74	114
White										
Total, 18 years and over	25,657	17,401	67.8	15,695	61.2	9,962	1,706	8,256	1,078	1,108
18 to 20 years	1,027	526	51.2	411	40.0	616	115	501	49	73
21 to 24 years	1,856	989	53.3	856	46.2	999	133	866	93	129
25 to 34 years	5,232	3,060	58.5	2,707	51.7	2,524	352	2,172	309	187
35 to 44 years	5,285	3,562	67.4	3,223	61.0	2,063	339	1,723	227	205
45 to 54 years	4,038	2,946	73.0	2,749	68.1	1,288	196	1,092	127	165
55 to 64 years	3,244	2,425	74.8	2,229	68.7	1,015	196	819	125	130
65 to 74 years	3,034	2,410	79.4	2,227	73.4	807	183	624	83	129
75 years and over	1,942	1,484	76.4	1,292	66.5	650	192	458	65	90
Black										
Total, 18 years and over	6,697	4,027	60.1	3,474	51.9	3,222	553	2,669	284	517
18 to 20 years	450	195	43.5	157	34.9	293	39	254	20	27
21 to 24 years	606	313	51.7	230	38.0	376	83	293	22	55
25 to 34 years	1,756	979	55.7	788	44.9	968	191	778	111	135
35 to 44 years	1,383	859	62.1	783	56.6	600	75	525	83	112
45 to 54 years	994	665	66.9	611	61.5	383	54	329	32	79
55 to 64 years	718	482	67.1	437	60.9	281	45	236	6	54
65 to 74 years	511	355	69.5	317	62.0	194	38	156	1	30
75 years and over	279	179	64.4	151	54.1	128	29	99	9	24
Hispanic origin[4]										
Total, 18 years and over	1,718	574	33.4	509	29.6	1,209	65	1,144	790	66
18 to 20 years	88	18	20.9	11	12.0	77	8	70	41	8
21 to 24 years	139	37	26.4	31	22.0	108	6	102	68	11
25 to 34 years	488	125	25.6	106	21.7	382	19	363	229	16
35 to 44 years	329	104	31.6	91	27.7	238	13	225	157	19
45 to 54 years	270	118	43.6	107	39.8	162	10	152	106	10
55 to 64 years	184	72	39.5	70	38.2	113	2	111	84	2
65 to 74 years	144	70	48.9	68	47.3	76	2	73	58	2
75 years and over	77	30	38.4	25	32.6	52	4	48	47	-

Source: "Reported Voting and Registration, by Race, Hispanic Origin, Sex, and Age, for Divisions," U.S. Bureau of the Census, *Voting and Registration in the Election of November 1992*, p. 18. *Notes:* 1. Includes persons reported as "did not vote," "do not know," and "not reported" on voting. 2. In addition to those reported as "not registered," total includes those "not a U.S. citizen," and "do not know" and "not reported" on registration. 3. Includes "do not know" and "not reported" on citizenship. 4. Persons of Hispanic origin may be of any race.

★ 896 ★

Voters and Voting

Voting and Registration in the Election of November 1992. Part XII – East South Central

November 1992. Numbers in thousands.

Division, race, Hispanic origin, and age	All persons	Reported registered		Reported voted		Reported that they did not vote[1]				
								Not registered		
		Number	Percent	Number	Percent	Total	Registered	Total[2]	Not a U.S. citizen	Do not know and not reported on registration[3]
All races										
Total, 18 years and over	11,284	7,957	70.5	6,774	60.0	4,510	1,184	3,327	69	584
18 to 20 years	629	312	49.6	248	39.3	382	65	317	5	47
21 to 24 years	934	527	56.4	406	43.5	527	121	407	15	52
25 to 34 years	2,413	1,498	62.1	1,221	50.6	1,192	278	914	24	127
35 to 44 years	2,404	1,733	72.1	1,521	63.3	882	211	671	7	151
45 to 54 years	1,670	1,296	77.6	1,149	68.8	520	147	373	8	92
55 to 64 years	1,366	1,100	80.5	971	71.1	395	129	266	-	54
65 to 74 years	1,107	901	81.4	774	70.0	332	127	205	8	25
75 years and over	762	589	77.3	483	63.4	279	106	173	2	36
White										
Total, 18 years and over	9,068	6,360	70.1	5,482	60.4	3,587	878	2,708	27	433
18 to 20 years	437	215	49.1	178	40.7	259	37	222	2	24
21 to 24 years	709	386	54.4	303	42.7	406	83	324	6	39
25 to 34 years	1,952	1,187	60.8	975	49.9	977	213	765	13	105
35 to 44 years	1,890	1,330	70.4	1,191	63.0	700	139	560	3	109
45 to 54 years	1,404	1,104	78.7	985	70.2	419	119	299	3	61
55 to 64 years	1,149	908	79.0	811	70.6	338	97	241	-	44
65 to 74 years	902	744	82.4	640	71.0	262	103	158	-	22
75 years and over	625	487	77.9	399	63.9	226	88	138	-	29
Black										
Total, 18 years and over	2,122	1,571	74.0	1,270	59.8	852	301	551	16	130
18 to 20 years	190	98	51.3	70	36.7	120	28	92	2	23
21 to 24 years	213	138	64.9	104	48.6	110	35	75	7	11
25 to 34 years	433	308	71.0	243	56.0	191	65	126	3	17
35 to 44 years	495	390	78.8	320	64.5	176	71	105	2	40
45 to 54 years	248	188	76.0	160	64.7	87	28	59	-	21
55 to 64 years	210	189	89.7	156	74.2	54	33	22	-	9
65 to 74 years	200	158	79.1	134	67.2	65	24	42	3	3
75 years and over	133	102	76.9	84	63.3	49	18	31	-	6
Hispanic origin[4]										
Total, 18 years and over	54	17	(B)	15	(B)	39	2	37	11	7
18 to 20 years	4	2	(B)	2	(B)	2	-	2	-	-
21 to 24 years	-	-	(B)	-	(B)	-	-	-	-	-
25 to 34 years	21	8	(B)	6	(B)	15	2	13	8	-
35 to 44 years	7	-	(B)	-	(B)	7	-	7	-	3
45 to 54 years	17	3	(B)	3	(B)	14	-	14	3	4
55 to 64 years	2	2	(B)	2	(B)	-	-	-	-	-
65 to 74 years	-	-	(B)	-	(B)	-	-	-	-	-
75 years and over	3	3	(B)	3	(B)	-	-	-	-	-

Source: "Reported Voting and Registration, by Race, Hispanic Origin, Sex, and Age, for Divisions," U.S. Bureau of the Census, *Voting and Registration in the Election of November 1992*, p. 20. *Notes:* 1. Includes persons reported as "did not vote," "do not know," and "not reported" on voting. 2. In addition to those reported as "not registered," total includes those "not a U.S. citizen," and "do not know" and "not reported" on registration. 3. Includes "do not know" and "not reported" on citizenship. 4. Persons of Hispanic origin may be of any race.

★ 897 ★
Voters and Voting

Voting and Registration in the Election of November 1992. Part XIII – West South Central

November 1992. Numbers in thousands.

Division, race, Hispanic origin, and age	All persons	Reported registered		Reported voted		Reported that they did not vote[1]				
								Not registered		
		Number	Percent	Number	Percent	Total	Registered	Total[2]	Not a U.S. citizen	Do not know and not reported on registration[3]
All races										
Total, 18 years and over	19,279	13,112	68.0	11,419	59.2	7,860	1,693	6,168	1,132	982
18 to 20 years	1,166	533	45.7	394	33.8	772	140	633	108	70
21 to 24 years	1,523	822	54.0	628	41.2	895	194	700	121	113
25 to 34 years	4,443	2,657	59.8	2,290	51.5	2,153	367	1,786	341	232
35 to 44 years	4,151	2,926	70.5	2,639	63.6	1,512	287	1,225	259	204
45 to 54 years	2,954	2,174	73.6	1,922	65.1	1,032	252	780	155	155
55 to 64 years	2,126	1,669	78.5	1,526	71.8	600	143	457	80	98
65 to 74 years	1,827	1,502	82.2	1,358	74.3	469	144	325	30	69
75 years and over	1,089	827	76.0	661	60.7	427	166	262	39	41
White										
Total, 18 years and over	16,276	11,193	68.8	9,858	60.6	6,419	1,335	5,083	939	757
18 to 20 years	963	452	47.0	337	35.0	626	115	511	83	54
21 to 24 years	1,236	672	54.4	536	43.4	700	136	564	105	93
25 to 34 years	3,636	2,205	60.6	1,923	52.9	1,713	282	1,431	288	154
35 to 44 years	3,525	2,519	71.5	2,298	65.2	1,227	220	1,006	206	150
45 to 54 years	2,542	1,868	73.5	1,668	65.6	875	200	674	137	133
55 to 64 years	1,832	1,448	79.0	1,335	72.8	498	113	385	59	78
65 to 74 years	1,585	1,309	82.6	1,187	74.9	399	122	276	24	60
75 years and over	957	720	75.3	574	60.0	383	147	236	36	36
Black										
Total, 18 years and over	2,516	1,733	68.9	1,407	55.9	1,109	326	783	38	190
18 to 20 years	168	72	43.1	49	29.2	119	23	96	12	12
21 to 24 years	251	138	54.9	80	31.7	171	58	113	3	18
25 to 34 years	688	421	61.2	343	49.8	346	79	267	11	70
35 to 44 years	508	372	73.2	310	61.0	198	62	136	10	41
45 to 54 years	333	259	77.8	211	63.5	121	48	74	3	20
55 to 64 years	234	191	81.8	174	74.4	60	17	43	-	16
65 to 74 years	213	179	83.7	159	74.8	54	19	35	-	8
75 years and over	120	100	83.2	81	67.3	39	19	20	-	5
Hispanic origin[4]										
Total, 18 years and over	2,916	1,252	42.9	975	33.4	1,941	277	1,664	790	239
18 to 20 years	246	81	33.1	49	20.0	197	32	164	77	19
21 to 24 years	328	105	32.1	78	23.8	250	27	223	86	36
25 to 34 years	739	263	35.6	193	26.1	546	70	476	253	49
35 to 44 years	656	296	45.1	240	36.7	415	55	360	157	57
45 to 54 years	422	207	49.0	164	38.8	258	43	215	118	36
55 to 64 years	246	141	57.3	126	51.3	120	15	105	42	25
65 to 74 years	167	99	59.0	72	43.2	95	26	69	22	12
75 years and over	112	60	53.5	53	47.0	60	7	52	34	5

Source: "Reported Voting and Registration, by Race, Hispanic Origin, Sex, and Age, for Divisions," U.S. Bureau of the Census, *Voting and Registration in the Election of November 1992*, p. 20. *Notes:* 1. Includes persons reported as "did not vote," "do not know," and "not reported" on voting. 2. In addition to those reported as "not registered," total includes those "not a U.S. citizen," and "do not know" and "not reported" on registration. 3. Includes "do not know" and "not reported" on citizenship. 4. Persons of Hispanic origin may be of any race.

★ 898 ★
Voters and Voting

Voting and Registration in the Election of November 1992. Part XIV – Mountain

November 1992. Numbers in thousands.

Division, race, Hispanic origin, and age	All persons	Reported registered		Reported voted		Reported that they did not vote[1]				
						Total	Registered	Not registered		
		Number	Percent	Number	Percent			Total[2]	Not a U.S. citizen	Do not know and not reported on registration[3]
All races										
Total, 18 years and over	9,881	7,094	71.8	6,544	66.2	3,337	550	2,787	409	433
18 to 20 years	508	276	54.4	230	45.3	278	46	232	25	34
21 to 24 years	719	416	57.8	337	46.9	382	78	304	49	40
25 to 34 years	2,260	1,425	63.1	1,309	57.9	951	116	835	118	98
35 to 44 years	2,190	1,606	73.4	1,499	68.5	691	107	583	101	82
45 to 54 years	1,524	1,205	79.1	1,139	74.8	385	66	319	46	49
55 to 64 years	1,154	939	81.3	897	77.7	257	42	215	38	49
65 to 74 years	935	782	83.6	739	79.0	196	43	153	25	41
75 years and over	592	445	75.2	394	66.7	197	51	146	9	39
White										
Total, 18 years and over	9,258	6,773	73.2	6,266	67.7	2,992	507	2,485	338	376
18 to 20 years	469	266	56.7	223	47.4	247	44	203	19	22
21 to 24 years	655	396	60.4	325	49.6	330	71	259	37	31
25 to 34 years	2,067	1,339	64.8	1,229	59.4	838	110	729	102	84
35 to 44 years	2,074	1,540	74.3	1,443	69.6	631	98	533	82	77
45 to 54 years	1,436	1,151	80.1	1,092	76.0	344	59	285	38	45
55 to 64 years	1,072	880	82.1	842	78.5	231	39	192	35	38
65 to 74 years	909	764	87.0	726	79.8	183	38	146	19	40
75 years and over	574	436	75.9	387	67.3	187	49	138	6	39
Black										
Total, 18 years and over	263	160	60.9	139	52.8	124	21	103	1	30
18 to 20 years	17	4	(B)	4	(B)	13	-	13	-	7
21 to 24 years	21	12	(B)	7	(B)	14	6	8	1	1
25 to 34 years	93	45	48.7	43	46.1	50	2	48	1	8
35 to 44 years	39	29	(B)	25	(B)	14	4	10	-	3
45 to 54 years	26	20	(B)	18	(B)	9	2	7	-	2
55 to 64 years	45	34	(B)	33	(B)	12	1	11	-	8
65 to 74 years	12	10	(B)	6	(B)	6	5	2	-	1
75 years and over	10	6	(B)	4	(B)	6	1	5	-	-
Hispanic origin[4]										
Total, 18 years and over	1,167	593	50.8	516	44.2	651	76	575	237	58
18 to 20 years	85	33	39.5	25	29.6	59	8	51	19	4
21 to 24 years	114	42	36.5	27	23.8	87	15	72	36	7
25 to 34 years	337	147	43.6	135	40.1	202	12	190	77	13
35 to 44 years	247	124	50.2	108	43.7	139	16	123	56	17
45 to 54 years	152	91	60.3	80	53.0	71	11	60	22	11
55 to 64 years	142	99	69.3	92	64.8	50	6	44	17	-
65 to 74 years	67	44	(B)	41	(B)	25	2	23	8	2
75 years and over	24	13	(B)	7	(B)	17	6	11	2	2

Source: "Reported Voting and Registration, by Race, Hispanic Origin, Sex, and Age, for Divisions," U.S. Bureau of the Census, *Voting and Registration in the Election of November 1992*, p. 21. *Notes:* 1. Includes persons reported as "did not vote," "do not know," and "not reported" on voting. 2. In addition to those reported as "not registered," total includes those "not a U.S. citizen," and "do not know" and "not reported" on registration. 3. Includes "do not know" and "not reported" on citizenship. 4. Persons of Hispanic origin may be of any race.

★ 899 ★

Voters and Voting

Voting and Registration in the Election of November 1992: Employment Status

November 1992. Numbers in thousands.

Race, sex, Hispanic origin, employment status, and class or worker	All persons	Reported registered		Reported voted		Reported that they did not vote[1]				
						Total	Registered	Not registered		
		Number	Percent	Number	Percent			Total[2]	Not a U.S. citizen	Do not know and not reported on registration[3]
White										
Both sexes										
Total	157,837	110,684	70.1	100,405	63.6	57,432	10,279	47,153	8,284	8,289
Civilian labor force	106,314	74,977	70.5	68,621	64.5	37,692	6,356	31,336	5,474	5,086
Employed	100,200	71,617	71.5	65,678	65.5	34,522	5,940	28,582	4,833	4,735
Agricultural industries	2,840	1,791	63.1	1,626	57.3	1,212	163	1,049	419	131
Self-employed workers[4]	1,424	1,151	80.8	1,058	74.3	366	93	273	55	68
Wage and salary workers	1,416	640	45.2	570	40.3	846	70	776	364	63
Nonagricultural industries	97,360	69,827	71.7	64,050	65.8	33,310	5,777	27,533	4,413	4,604
Private wage and salary workers	74,164	50,843	68.6	46,122	62.2	28,042	4,721	23,320	3,848	3,675
Government workers	15,008	12,758	85.0	12,087	80.5	2,920	671	2,250	239	515
Self-employed workers[4]	8,188	6,225	76.0	5,840	71.3	2,348	385	1,963	327	414
Unemployed	6,114	3,360	55.0	2,944	48.1	3,170	416	2,754	641	351
Not in labor force	51,523	35,707	69.3	31,784	61.7	19,739	3,923	15,817	2,810	3,203
Male										
Total	75,915	52,383	69.0	47,552	62.6	28,364	4,831	23,533	4,298	4,116
Civilian labor force	58,477	39,863	68.2	36,364	62.2	22,113	3,499	18,614	3,591	2,930
Employed	54,894	37,956	69.1	34,694	63.2	20,201	3,263	16,938	3,191	2,710
Agricultural industries	2,227	1,350	60.6	1,226	55.0	1,001	124	877	380	95
Self-employed workers[4]	1,113	899	80.7	829	74.4	284	70	215	49	47
Wage and salary workers	1,114	451	40.5	397	35.6	717	54	663	330	48
Nonagricultural industries	52,667	36,606	69.5	33,468	63.5	19,199	3,139	16,060	2,811	2,614
Private wage and salary workers	40,554	27,072	66.8	24,518	60.5	16,037	2,554	13,482	2,452	2,058
Government workers	6,859	5,699	83.1	5,374	78.3	1,485	326	1,159	121	253
Self-employed workers[4]	5,254	3,835	73.0	3,576	68.1	1,678	259	1,419	238	303
Unemployed	3,582	1,907	53.2	1,670	46.6	1,912	237	1,676	401	221
Not in labor force	17,439	12,520	71.8	11,188	64.2	1,332	4,919	706	1,186	
Female										
Total	81,922	58,301	71.2	52,853	64.5	29,068	5,448	23,621	3,986	4,173
Civilian labor force	47,837	35,114	73.4	32,257	67.4	15,579	2,857	12,723	1,882	2,156
Employed	45,306	33,661	74.3	30,984	68.4	14,322	2,677	11,644	1,642	2,025
Agricultural industries	613	441	72.0	402	65.6	211	39	172	39	36
Self-employed workers[4]	311	252	81.1	229	73.7	82	23	59	6	21
Wage and salary workers	302	189	62.5	173	57.3	129	16	113	34	15
Nonagricultural industries	44,693	33,220	74.3	30,582	68.4	14,111	2,638	11,473	1,602	1,990
Private wage and salary workers	33,609	23,771	70.7	21,604	64.3	12,005	2,167	9,838	1,396	1,617
Government workers	8,149	7,059	86.6	6,714	82.4	1,435	345	1,090	118	263
Self-employed workers[4]	2,934	2,390	81.5	2,264	77.1	670	127	544	88	110
Unemployed	2,531	1,453	57.4	1,273	50.3	1,258	180	1,078	241	131
Not in labor force	34,085	23,187	68.0	20,596	60.4	13,489	2,591	10,898	2,104	2,017
Black										
Both sexes										
Total	21,039	13,442	63.9	11,371	54.0	9,668	2,070	7,598	1,044	1,870
Civilian labor force	13,629	8,993	66.0	7,771	57.0	5,857	1,221	4,636	745	1,152
Employed	11,851	8,042	67.9	7,001	59.1	4,851	1,042	3,809	636	966
Agricultural industries	144	70	48.6	65	44.9	79	5	74	9	12
Self-employed workers[4]	31	19	(B)	19	(B)	13	-	13	-	4
Wage and salary workers	113	51	45.5	46	40.9	67	5	61	9	9
Nonagricultural industries	11,708	7,972	68.1	6,936	59.2	4,772	1,036	3,735	627	954
Private wage and salary workers	8,588	5,514	64.2	4,660	54.3	3,928	853	3,074	537	743
Government workers	2,667	2,167	81.2	2,005	75.2	662	162	501	60	172
Self-employed workers[4]	452	292	64.6	271	59.9	181	21	160	30	38
Unemployed	1,777	950	53.5	771	43.4	1,006	180	827	110	186
Not in labor force	7,411	4,449	60.0	3,600	48.6	3,811	849	2,962	299	718
Male										
Total	9,425	5,727	60.8	4,786	50.8	4,639	941	3,698	548	965
Civilian labor force	6,737	4,155	61.7	3,490	51.8	3,247	665	2,582	438	662
Employed	5,836	3,713	63.6	3,135	53.7	2,702	578	2,123	378	544
Agricultural industries	129	61	47.8	56	43.7	72	5	67	7	12

[Continued]

★ 899 ★

Voting and Registration in the Election of November 1992: Employment Status

[Continued]

Race, sex, Hispanic origin, employment status, and class or worker	All persons	Reported registered		Reported voted		Reported that they did not vote[1]				
								Not registered		
		Number	Percent	Number	Percent	Total	Registered	Total[2]	Not a U.S. citizen	Do not know and not reported on registration[3]
Self-employed workers[4]	29	16	(B)	16	(B)	13	-	13	-	4
Wage and salary workers	100	45	45.5	40	40.2	60	5	54	7	9
Nonagricultural industries	5,708	3,651	64.0	3,079	53.9	2,629	573	2,056	372	532
Private wage and salary workers	4,321	2,628	60.8	2,144	49.6	2,177	483	1,693	308	420
Government workers	1,098	850	77.5	773	70.4	325	77	247	40	77
Self-employed workers[4]	289	174	60.1	161	55.8	128	12	115	23	35
Unemployed	901	442	49.1	355	39.4	545	87	458	60	117
Not in labor force	2,688	1,572	58.5	1,296	48.2	1,392	276	1,117	110	304
Female										
Total	11,614	7,715	66.4	6,585	56.7	5,029	1,129	3,899	496	905
Civilian labor force	6,892	4,838	70.2	4,281	62.1	2,610	556	2,054	307	491
Employed	6,015	4,329	72.0	3,866	64.3	2,149	464	1,686	257	422
Agricultural industries	15	8	(B)	8	(B)	7	-	7	2	-
Self-employed workers[4]	2	2	(B)	2	(B)	-	-	-	-	-
Wage and salary workers	13	6	(B)	6	(B)	7	-	7	2	-
Nonagricultural industries	6,000	4,321	72.0	3,857	64.3	2,142	464	1,679	255	422
Private wage and salary workers	4,267	2,886	67.6	2,516	59.0	1,751	370	1,381	229	323
Government workers	1,570	1,317	83.9	1,232	78.5	338	85	253	20	96
Self-employed workers[4]	163	118	72.6	109	67.1	54	9	45	6	3
Unemployed	877	508	58.0	416	47.4	461	93	368	50	69
Not in labor force	4,722	2,877	60.9	2,304	48.8	2,418	573	1,845	189	414
Hispanic origin[5]										
Both sexes										
Total	14,688	5,137	35.0	4,238	26.9	10,450	899	9,551	5,910	995
Civilian labor force	9,915	3,513	35.4	2,921	29.5	6,993	592	6,401	4,019	838
Employed	8,800	3,226	36.7	2,682	30.5	6,117	543	5,574	3,489	570
Agricultural industries	506	51	10.0	44	8.6	463	7	456	397	16
Self-employed workers[4]	70	16	(B)	10	(B)	60	6	54	43	8
Wage and salary workers	436	35	7.9	34	7.8	402	1	402	354	8
Nonagricultural industries	8,293	3,175	38.3	2,639	31.8	5,655	536	5,118	3,092	555
Private wage and salary workers	6,835	2,290	33.5	1,843	27.0	4,992	447	4,545	2,827	457
Government workers	1,013	699	69.0	641	63.3	372	58	314	90	61
Self-employed workers[4]	445	187	41.9	155	34.8	290	32	258	175	36
Unemployed	1,115	287	25.8	239	21.4	876	48	828	530	68
Not in labor force	4,773	1,624	34.0	1,316	27.6	3,457	308	3,150	1,891	357
Male										
Total	7,273	2,337	32.1	1,946	26.8	5,327	391	4,936	3,132	519
Civilian labor force	6,003	1,838	30.6	1,532	25.5	4,471	306	4,165	2,736	391
Employed	5,347	1,692	31.6	1,405	26.3	3,943	287	3,658	2,400	354
Agricultural industries	456	42	9.3	35	7.7	421	7	414	361	16
Self-employed workers[4]	66	1f	(B)	10	(B)	56	6	50	39	8
Wage and salary workers	390	26	6.7	25	6.5	365	1	364	322	8
Nonagricultural industries	4,891	1,649	33.7	1,370	28.0	3,522	280	3,242	2,040	338
Private wage and salary workers	4,097	1,207	29.5	975	23.8	3,122	232	2,890	1,865	285
Government workers	490	329	67.1	303	61.8	187	26	161	41	25
Self-employed workers[4]	304	113	37.3	92	30.2	212	22	191	133	28
Unemployed	655	146	22.4	127	19.3	528	20	509	335	37
Not in labor force	1,271	499	39.3	415	32.7	856	84	771	396	128
Female										
Total	7,415	2,800	37.8	2,291	30.9	5,124	508	4,615	2,778	476
Civilian labor force	3,912	1,675	42.8	1,390	35.5	2,522	285	2,237	1,283	247
Employed	3,452	1,534	44.4	1,278	37.0	2,175	257	1,918	1,089	216
Agricultural industries	50	8	(B)	8	(B)	42	-	42	36	-
Self-employed workers[4]	4	-	(B)	-	(B)	4	-	4	4	-
Wage and salary workers	46	8	(B)	8	(B)	37	-	37	32	-
Nonagricultural industries	3,402	1,526	44.8	1,269	37.3	2,133	257	1,876	1,053	216
Private wage and salary workers	2,738	1,083	39.5	868	31.7	1,870	215	1,655	962	172
Government workers	523	370	70.7	338	64.7	185	32	153	49	36
Self-employed workers[4]	141	73	51.9	63	44.6	78	10	68	43	9

[Continued]

★ 899 ★

Voting and Registration in the Election of November 1992: Employment Status

[Continued]

Race, sex, Hispanic origin, employment status, and class or worker	All persons	Reported registered		Reported voted		Reported that they did not vote[1]				
						Total	Registered	Not registered		
								Total[2]	Not a U.S. citizen	Do not know and not reported on registration[3]
		Number	Percent	Number	Percent					
Unemployed	460	141	30.6	112	24.4	347	29	319	195	31
Not in labor force	3,503	1,124	32.1	901	25.7	2,602	223	2,379	1,495	229

Source: "Reported Voting and Registration by Race, Hispanic Origin, Sex, Employment Status, and Class of Worker," U.S. Bureau of the Census, *Voting and Registration in the Election of November 1992*, pp. 48-50. *Notes:* 1. Includes persons reported as "did not vote," "do not know," and "not reported" on voting. 2. In addition to those reported as "not registered," total includes those "not a U.S. citizen," and "not reported" on registration. 3. Includes "do not know" and "not reported" on citizenship. 4. Includes unpaid family workers. 5. Persons of Hispanic origin may be of any race.

★ 900 ★

Voters and Voting

Voting and Registration in the Election of November 1992: Metropolitan – Nonmetropolitan Residence

November 1992. Numbers in thousands.

Race, Hispanic origin, and residence	All persons	Reported registered		Reported voted		Reported that they did not vote[1]				
						Total	Registered	Not registered		
								Total[2]	Not a U.S. citizen	Do not know and not reported on registration[3]
		Number	Percent	Number	Percent					
All races										
Total	185,684	126,578	68.2	113,866	61.3	71,818	12,712	59,106	11,900	10,638
Metropolitan	144,593	97,460	67.4	88,222	61.0	56,371	9,238	47,133	11,332	8,880
In central cities	55,855	36,118	64.7	32,301	57.8	23,554	3,817	19,736	6,024	3,696
Outside central cities	88,738	61,342	69.1	55,921	63.0	32,817	5,421	27,397	5,308	5,183
Nonmetropolitan	41,091	29,118	70.9	25,644	62.4	15,447	3,474	11,973	568	1,758
White										
Total	157,837	110,684	70.1	100,405	63.6	57,432	10,279	47,153	8,284	8,289
Metropolitan	120,848	84,094	69.6	76,807	63.6	44,041	7,287	36,755	7,811	6,800
In central cities	41,176	27,598	67.0	25,011	60.7	16,165	2,587	13,578	3,961	2,371
Outside central cities	79,672	56,495	70.9	51,796	65.0	27,876	4,699	23,177	3,850	4,429
Nonmetropolitan	36,988	26,590	71.9	23,598	63.8	13,391	2,992	10,399	473	1,489
Black										
Total	21,039	13,442	63.9	11,371	54.0	9,668	2,070	7,598	1,044	1,870
Metropolitan	17,677	11,341	64.2	9,703	54.9	7,974	1,637	6,337	1,018	1,638
In central cities	11,689	7,596	65.0	6,505	55.6	5,185	1,091	4,094	708	1,127
Outside central cities	5,988	3,745	62.5	3,199	53.4	2,789	546	2,243	310	510
Nonmetropolitan	3,362	2,101	62.5	1,668	49.6	1,694	433	1,261	26	233
Hispanic origin[4]										
Total	14,688	5,137	35.0	4,238	28.9	10,450	899	9,551	5,910	995
Metropolitan	13,569	4,655	34.3	3,855	28.4	9,714	800	8,914	5,655	944
In central cities	7,303	2,529	34.6	2,053	28.1	5,250	476	4,774	3,020	486

[Continued]

★ 900 ★

Voting and Registration in the Election of November 1992: Metropolitan – Nonmetropolitan Residence

[Continued]

| Race, Hispanic origin, and residence | All persons | Reported registered | | Reported voted | | Reported that they did not vote[1] | | | | |
| | | | | | | | | Not registered | | |
		Number	Percent	Number	Percent	Total	Registered	Total[2]	Not a U.S. citizen	Do not know and not reported on registration[3]
Outside central cities	6,266	2,126	33.9	1,802	28.8	4,464	325	4,140	2,635	459
Nonmetropolitan	1,119	481	43.0	383	34.2	736	99	637	256	51

Source: "Reported Voting and Registration, by Race, Hispanic Origin, and Metropolitan-Nonmetropolitan Residence," U.S. Bureau of the Census, *Voting and Registration in the Election of November 1992*, p. 31. *Notes:* 1. Includes persons reported as "did not vote," "do not know," and "not reported" on voting. 2. In addition to those reported as "not registered," total includes those "not a U.S. citizen," and "do not know" and "not reported" on registration. 3. Includes "do not know" and "not reported" on citizenship. 4. Persons of Hispanic origin may be of any race.

★ 901 ★

Voters and Voting

Voting and Registration in the Election of November 1992: Years of School Completed

November 1992. Numbers in thousands.

| Race, Hispanic origin, sex and years of school completed | All persons | Reported registered | | Reported voted | | Reported that they did not vote[1] | | | | |
| | | | | | | | | Not registered | | |
		Number	Percent	Number	Percent	Total	Registered	Total[2]	Not a U.S. citizen	Do not know and not reported on registration[3]
All races										
Both sexes										
Total	185,684	126,578	68.2	113,866	61.3	71,818	12,712	59,106	11,900	10,638
Less than 5th grade	3,466	1,005	29.0	746	21.5	2,720	258	2,462	1,280	230
5th to 8th grade	11,925	5,745	48.2	4,660	39.1	7,265	1,085	6,179	2,160	812
9th to 12th grade, no diploma	20,970	10,575	50.4	8,638	41.2	12,332	1,937	10,395	1,729	1,533
High school graduate	65,281	42,355	64.9	37,517	57.5	27,765	4,839	22,926	2,896	4,180
Some college or associate degree	46,691	35,226	75.4	32,069	68.7	14,622	3,157	11,465	1,946	2,474
Bachelor's degree	25,055	21,055	84.0	20,009	79.9	5,046	1,047	3,999	1,169	963
Advanced degree	12,296	10,617	86.3	10,227	83.2	2,069	389	1,680	719	446
Male										
Total	88,557	59,254	66.9	53,312	60.2	35,245	5,942	29,303	6,065	5,327
Less than 5th grade	1,718	550	32.0	423	24.6	1,294	127	1,168	575	101
5th to 8th grade	5,710	2,737	47.9	2,284	40.0	3,427	453	2,974	1,118	402
9th to 12th grade, no diploma	9,867	4,754	48.2	3,910	39.6	5,956	844	5,113	923	827
High school graduate	29,336	18,244	62.2	16,047	54.7	13,289	2,197	11,092	1,350	2,035
Some college or associate degree	21,963	16,241	73.9	14,680	66.8	7,283	1,561	5,723	964	1,208
Bachelor's degree	12,701	10,526	82.9	9,991	78.7	2,709	534	2,175	654	486
Advanced degree	7,263	6,204	85.4	5,977	82.3	1,286	226	1,060	481	268
Female										
Total	97,126	67,324	69.3	60,554	62.3	36,573	6,770	29,803	5,834	5,311
Less than 5th grade	1,749	455	26.0	323	18.5	1,426	131	1,294	705	129
5th to 8th grade	6,214	3,009	48.4	2,376	38.2	3,838	632	3,206	1,042	410
9th to 12th grade, no diploma	11,103	5,821	52.4	4,728	42.6	6,376	1,094	5,282	806	706
High school graduate	35,945	24,112	67.1	21,470	59.7	14,475	2,642	11,834	1,547	2,145
Some college or associate degree	24,728	18,985	76.8	17,389	70.3	7,338	1,596	5,743	982	1,266

[Continued]

★ 901 ★

Voting and Registration in the Election of November 1992: Years of School Completed

[Continued]

Race, Hispanic origin, sex and years of school completed	All persons	Reported registered		Reported voted		Reported that they did not vote[1]				
								Not registered		
		Number	Percent	Number	Percent	Total	Registered	Total[2]	Not a U.S. citizen	Do not know and not reported on registration[3]
Bachelor's degree	12,354	10,530	85.2	10,017	81.1	2,337	513	1,824	515	477
Advanced degree	5,033	4,413	87.7	4,250	84.4	783	163	620	238	178
White										
Both sexes										
Total	157,837	110,684	70.1	100,405	63.6	57,432	10,279	47,153	8,284	8,289
Less than 5th grade	2,545	646	25.4	478	18.8	2,067	168	1,899	1,065	173
5th to 8th grade	9,894	4,681	47.3	3,798	38.4	6,096	883	5,213	1,814	670
9th to 12th grade, no diploma	16,398	8,324	50.8	6,861	41.8	9,537	1,463	8,074	1,316	1,083
High school graduate	55,728	36,921	66.3	32,998	59.2	22,730	3,923	18,807	1,950	3,280
Some college or associate degree	40,104	31,042	77.4	28,467	71.0	11,637	2,575	9,062	1,183	1,963
Bachelor's degree	22,304	19,346	86.7	18,421	82.6	3,882	925	2,958	601	766
Advanced degree	10,865	9,724	89.5	9,382	86.4	1,483	342	1,140	354	354
Male										
Total	75,915	52,383	69.0	47,552	62.6	28,364	4,831	23,533	4,298	4,116
Less than 5th grade	1,296	346	26.7	268	20.7	1,028	78	950	506	80
5th to 8th grade	4,769	2,258	47.3	1,878	39.4	2,891	380	2,511	950	337
9th to 12th grade, no diploma	7,908	3,907	49.4	3,242	41.0	4,667	665	4,001	733	566
High school graduate	25,024	15,873	63.4	14,092	56.3	10,932	1,781	9,151	941	1,574
Some college or associate degree	18,991	14,427	76.0	13,180	69.4	5,812	1,248	4,564	581	959
Bachelor's degree	11,460	9,819	85.7	9,341	81.5	2,119	478	1,641	343	393
Advanced degree	6,466	5,753	89.0	5,552	85.9	915	201	713	245	207
Female										
Total	81,922	58,301	71.2	52,853	64.5	29,068	5,448	23,621	3,986	4,173
Less than 5th grade	1,249	300	24.0	210	16.8	1,039	90	949	559	92
5th to 8th grade	5,125	2,423	47.3	1,920	37.5	3,205	503	2,702	865	333
9th to 12th grade, no diploma	8,489	4,417	52.0	3,619	42.6	4,870	797	4,073	583	517
High school graduate	30,704	21,048	68.6	18,906	61.6	11,798	2,142	9,656	1,009	1,706
Some college or associate degree	21,113	16,615	78.7	15,288	72.4	5,825	1,327	4,498	602	1,004
Bachelor's degree	10,844	9,527	87.9	9,080	83.7	1,763	447	1,316	258	373
Advanced degree	4,398	3,972	90.3	3,830	87.1	568	141	427	110	147
Black										
Both sexes										
Total	21,039	13,442	63.9	11,371	54.0	9,668	2,070	7,598	1,044	1,870
Less than 5th grade	619	303	48.9	228	36.9	391	74	316	30	40
5th to 8th grade	1,562	951	60.9	775	49.6	788	177	611	126	101
9th to 12th grade, no diploma	3,911	2,105	53.8	1,669	42.7	2,242	436	1,805	153	394
High school graduate	7,719	4,746	61.5	3,944	51.1	3,775	802	2,973	360	783
Some college or associate degree	5,019	3,548	70.7	3,062	61.0	1,957	486	1,471	266	393
Bachelor's degree	1,481	1,182	79.8	1,116	75.3	365	67	299	71	104
Advanced degree	729	606	83.1	579	79.4	151	28	123	39	55
Male										
Total	9,425	5,727	60.8	4,786	50.8	4,639	941	3,698	548	965
Less than 5th grade	330	186	56.4	140	42.4	190	46	144	14	15
5th to 8th grade	743	436	58.7	373	50.1	370	64	307	68	43
9th to 12th grade, no diploma	1,657	781	47.2	618	37.3	1,038	163	875	78	227
High school graduate	3,488	2,055	58.9	1,694	48.6	1,795	362	1,433	171	409
Some college or associate degree	2,213	1,516	68.5	1,259	56.9	953	257	696	138	189
Bachelor's degree	629	473	75.1	438	69.5	192	35	157	45	47
Advanced degree	366	279	76.3	265	72.4	101	14	87	34	35

[Continued]

★ 901 ★

Voting and Registration in the Election of November 1992: Years of School Completed
[Continued]

Race, Hispanic origin, sex and years of school completed	All persons	Reported registered		Reported voted		Reported that they did not vote[1]				
								Not registered		
		Number	Percent	Number	Percent	Total	Registered	Total[2]	Not a U.S. citizen	Do not know and not reported on registration[3]
Female										
Total	11,614	7,715	66.4	6,585	56.7	5,029	1,129	3,899	496	905
Less than 5th grade	290	117	40.4	89	30.6	201	28	173	15	24
5th to 8th grade	819	515	62.9	402	49.1	417	113	304	58	57
9th to 12th grade, no diploma	2,254	1,324	58.7	1,051	46.6	1,203	273	930	75	167
High school graduate	4,231	2,691	63.6	2,250	53.2	1,981	441	1,540	189	374
Some college or associate degree	2,806	2,032	72.4	1,802	64.2	1,004	229	774	128	205
Bachelor's degree	851	709	83.3	678	79.6	173	31	142	26	57
Advanced degree	364	327	90.0	314	86.3	50	13	36	5	20
Hispanic origin[4]										
Both sexes										
Total	14,688	5,137	35.0	4,238	28.9	10,450	899	9,551	5,910	995
Less than 5th grade	1,488	263	17.7	186	12.5	1,302	77	1,225	970	90
5th to 8th grade	2,785	519	18.6	419	15.0	2,367	100	2,267	1,688	180
9th to 12th grade, no diploma	2,681	655	24.4	492	18.3	2,189	163	2,026	1,145	220
High school graduate	4,010	1,585	39.5	1,282	32.0	2,728	303	2,425	1,230	264
Some college or associate degree	2,584	1,427	55.2	1,218	47.1	1,366	209	1,157	576	185
Bachelor's degree	796	468	58.8	439	55.1	357	29	328	212	40
Advanced degree	343	220	64.0	202	59.0	141	17	124	91	15
Male										
Total	7,273	2,337	32.1	1,946	26.8	5,327	391	4,936	3,132	519
Less than 5th grade	715	100	14.0	84	11.7	632	16	615	478	39
5th to 8th grade	1,389	211	15.2	172	12.4	1,217	39	1,178	900	99
9th to 12th grade, no diploma	1,374	290	21.1	222	16.2	1,151	68	1,083	642	118
High school graduate	1,946	727	37.4	589	30.3	1,357	139	1,218	635	138
Some college or associate degree	1,240	663	53.5	563	45.4	677	100	577	296	87
Bachelor's degree	397	225	56.7	210	52.8	188	16	172	114	23
Advanced degree	212	121	56.8	107	50.6	105	13	92	68	15
Female										
Total	7,415	2,800	37.8	2,291	30.9	5,124	508	4,615	2,778	476
Less than 5th grade	773	163	21.1	103	13.3	671	61	610	492	51
5th to 8th grade	1,396	308	22.1	247	17.7	1,150	61	1,088	788	81
9th to 12th grade, no diploma	1,307	365	27.9	269	20.6	1,038	95	943	503	102
High school graduate	2,064	857	41.5	693	33.6	1,371	164	1,206	595	125
Some college or associate degree	1,344	764	56.9	655	48.8	689	109	580	280	98
Bachelor's degree	399	243	60.8	229	57.4	170	14	156	98	18
Advanced degree	131	99	75.6	95	72.4	36	4	32	23	-

Source: "Reported Voting and Registration by Race, Hispanic Origin, Sex, and Years of School Completed" U.S. Bureau of the Census, *Voting and Registration in the Election of November 1992*, pp. 39-40. *Notes:* 1. Includes persons reported as "did not vote," "do not know," and "not reported" on voting. 2. In addition to those reported as "not registered," total includes those "not a U.S. citizen," and "do not know" and "not reported" on registration. 3. Includes "do not know" and "not reported" on citizenship. 4. Persons of Hispanic origin may be of any race.

★ 902 ★

Voters and Voting

Voting and Registration of Family Members: Election of November 1992.

November 1992. Numbers in thousands.

Race, Hispanic origin, and family income	All persons	Reported registered		Reported voted		Reported that they did not vote[1]				
						Total	Registered	Not registered		
		Number	Percent	Number	Percent			Total[2]	Not a U.S. citizen	Do not know and not reported on registration[3]
All races										
Total	148,286	102,071	68.8	92,492	62.4	55,794	9,580	46,214	9,973	8,007
Under $5,000	4,489	1,953	43.5	1,453	32.4	3,036	499	2,536	588	381
$5,000 to $9,999	9,420	4,640	49.3	3,722	39.5	5,699	919	4,780	1,148	616
$10,000 to $14,999	13,529	7,486	55.3	6,333	46.8	7,196	1,154	6,042	1,695	784
$15,000 to $19,999	22,974	14,500	63.1	12,790	55.7	10,184	1,710	8,474	1,979	1,037
$20,000 to $24,999	22,598	15,547	66.8	14,128	62.5	8,470	1,419	7,051	1,468	972
$25,000 to 34,999	26,259	19,713	75.1	18,245	69.5	8,013	1,468	6,546	1,107	1,085
$35,000 to $49,999	23,452	18,945	80.8	17,762	75.7	5,690	1,182	4,508	781	779
$50,000 and over	15,112	12,722	84.2	12,067	79.9	3,044	655	2,390	543	670
Income not reported	10,453	6,565	62.8	5,991	57.3	4,462	574	3,887	664	1,682
White										
Total	126,099	89,527	71.0	81,819	64.9	44,280	7,708	36,572	6,979	6,139
Under $5,000	2,683	1,129	42.1	867	32.3	1,815	261	1,554	426	190
$5,000 to $9,999	6,580	3,236	49.2	2,610	39.7	3,970	626	3,344	817	407
$10,000 to $14,999	10,662	5,890	55.2	5,074	47.6	5,587	816	4,771	1,312	549
$15,000 to $19,999	19,273	12,360	64.1	10,958	56.9	8,315	1,402	6,913	1,442	791
$20,000 to $24,999	19,688	13,837	70.3	12,629	64.1	7,080	1,208	5,852	1,047	761
$25,000 to 34,999	23,349	17,898	76.7	16,615	71.2	6,734	1,282	5,451	711	893
$35,000 to $49,999	21,322	17,528	82.2	16,482	77.3	4,840	1,046	3,794	503	648
$50,000 and over	13,945	12,016	86.2	11,413	81.8	2,531	602	1,929	331	564
Income not reported	8,598	5,634	65.5	5,170	60.1	3,428	464	2,964	391	1,336
Black										
Total	16,471	10,504	63.8	8,917	54.1	7,554	1,586	5,967	810	1,470
Under $5,000	1,568	758	48.3	541	34.5	1,026	217	810	47	169
$5,000 to $9,999	2,414	1,309	54.2	1,042	43.2	1,372	266	1,105	118	191
$10,000 to $14,999	2,348	1,475	62.8	1,172	49.9	1,176	303	873	118	202
$15,000 to $19,999	2,839	1,882	66.3	1,616	56.9	1,223	266	957	172	204
$20,000 to $24,999	2,165	1,475	68.1	1,296	59.9	869	179	690	122	149
$25,000 to 34,999	1,993	1,448	72.6	1,304	65.4	689	144	545	95	128
$35,000 to $49,999	1,225	995	81.2	903	73.7	322	92	230	33	63
$50,000 and over	472	368	78.0	339	71.9	133	29	104	10	51
Income not reported	1,447	795	54.9	703	48.6	743	91	652	95	311
Hispanic origin[4]										
Total	12,360	4,274	34.6	3,525	28.5	8,836	749	8,086	5,079	834
Under $5,000	794	200	25.1	108	13.7	686	91	595	349	79
$5,000 to $9,999	1,680	515	30.6	402	23.9	1,277	113	1,165	714	129
$10,000 to $14,999	2,138	433	20.2	322	15.1	1,816	111	1,706	1,178	138
$15,000 to $19,999	2,573	782	30.4	624	24.3	1,949	158	1,791	1,191	123
$20,000 to $24,999	1,768	676	38.3	605	34.2	1,163	72	1,091	712	97
$25,000 to 34,999	1,415	688	48.6	596	42.1	819	92	727	409	68
$35,000 to $49,999	1,073	570	53.2	509	47.5	564	61	503	237	86

[Continued]

★ 902 ★

Voting and Registration of Family Members: Election of November 1992.

[Continued]

Race, Hispanic origin, and family income	All persons	Reported registered		Reported voted		Reported that they did not vote[1]				
						Total	Registered	Not registered		
		Number	Percent	Number	Percent			Total[2]	Not a U.S. citizen	Do not know and not reported on registration[3]
$50,000 and over	429	263	61.3	235	54.7	194	28	166	85	39
Income not reported	490	147	30.1	123	25.1	367	24	343	204	75

Source: "Reported Voting and Registration of Family Members, by Race, Hispanic Origin, and Family Income," U.S. Bureau of the Census, *Voting and Registration in the Election of November 1992*, p. 54. *Notes:* 1. Includes persons reported as "did not vote," "do not know," and "not reported" on voting. 2. In addition to those reported as "not registered," total includes those "not a U.S. citizen," and "do not know" and "not reported" on registration. 3. Includes "do not know" and "not reported" on citizenship. 4. Persons of Hispanic origin may be of any race.

★ 903 ★

Voters and Voting

Voting and Registration of Persons 18-24 Years in the Election of November 1992: Enrollment Status

November 1992. Numbers in thousands.

Race, Hispanic origin, sex and enrollment status	All persons	Reported registered		Reported voted		Reported that they did not vote[1]				
						Total	Registered	Not registered		
		Number	Percent	Number	Percent			Total[2]	Not a U.S. citizen	Do not know and not reported on registration[3]
All races										
Both sexes										
Total, 18 to 24 years old	24,371	12,787	52.5	10,442	42.8	13,930	2,346	11,584	2,145	1,896
Enrolled in school	9,450	5,996	63.4	5,050	53.4	4,400	946	3,454	755	778
18 to 20 years old	5,420	3,135	57.8	2,608	48.1	2,812	526	2,285	431	478
21 to 24 years old	4,030	2,861	71.0	2,442	60.6	1,588	419	1,169	324	300
In high school	1,399	527	37.7	436	31.2	963	91	872	184	121
In college	8,051	5,469	67.9	4,614	57.3	3,437	855	2,582	571	658
Full-time	6,884	4,674	67.9	3,894	56.6	2,990	780	2,210	485	586
Part-time	1,167	795	68.1	720	61.7	447	75	372	86	71
Not enrolled in school	14,922	6,791	45.5	5,391	36.1	9,530	1,400	8,130	1,391	1,118
18 to 20 years old	4,307	1,561	36.2	1,140	26.5	3,167	421	2,746	426	343
21 to 24 years old	10,614	5,230	49.3	4,251	40.1	6,363	979	5,384	965	775
Male										
Total, 18 to 24 years old	12,034	6,072	50.5	4,872	40.5	7,162	1,199	5,963	1,166	1,002
Enrolled in school	4,762	2,885	60.6	2,367	49.7	2,395	519	1,877	407	430
18 to 20 years old	2,731	1,485	54.4	1,213	44.4	1,518	273	1,246	220	250
21 to 24 years old	2,031	1,400	68.9	1,154	56.8	877	246	631	187	180
In high school	875	311	35.6	258	29.5	617	54	564	108	77
In college	3,887	2,574	66.2	2,109	54.3	1,778	465	1,313	300	352
Full-time	3,369	2,223	66.0	1,791	53.2	1,577	432	1,146	273	317
Part-time	518	351	67.7	317	61.2	201	33	168	27	35
Not enrolled in school	7,272	3,186	43.8	2,506	34.5	4,766	681	4,086	759	572
18 to 20 years old	2,145	789	36.8	573	26.7	1,572	216	1,356	224	156
21 to 24 years old	5,127	2,398	46.8	1,933	37.7	3,195	465	2,730	534	417

[Continued]

★ 903 ★

Voting and Registration of Persons 18-24 Years in the Election of November 1992: Enrollment Status

[Continued]

Race, Hispanic origin, sex and enrollment status	All persons	Reported registered		Reported voted		Reported that they did not vote[1]				
								Not registered		
		Number	Percent	Number	Percent	Total	Registered	Total[2]	Not a U.S. citizen	Do not know and not reported on registration[3]
Female										
Total, 18 to 24 years old	12,337	6,716	54.4	5,569	45.1	6,768	1,146	5,622	979	894
Enrolled in school	4,688	3,111	66.4	2,684	57.2	2,004	427	1,577	347	349
18 to 20 years old	2,689	1,649	61.3	1,395	51.9	1,293	254	1,039	211	228
21 to 24 years old	1,999	1,461	73.1	1,288	64.4	711	173	538	136	120
In high school	524	215	41.1	178	34.0	345	37	308	76	44
In college	4,164	2,895	69.5	2,505	60.2	1,659	390	1,269	271	305
Full-time	3,515	2,451	69.7	2,102	59.8	1,413	349	1,064	212	269
Part-time	649	444	68.5	403	62.1	246	41	205	60	36
Not enrolled in school	7,649	3,605	47.1	2,886	37.7	4,764	719	4,044	632	545
18 to 20 years old	2,163	772	35.7	567	26.2	1,596	205	1,390	201	187
21 to 24 years old	5,487	2,833	51.6	2,319	42.3	3,168	514	2,654	430	358
White										
Both sexes										
Total, 18 to 24 years old	19,681	10,746	54.6	8,932	45.4	10,749	1,814	8,935	1,478	1,372
Enrolled in school	7,607	5,130	67.4	4,396	57.8	3,211	733	2,477	398	560
18 to 20 years old	4,340	2,679	61.7	2,281	52.6	2,058	398	1,660	235	339
21 to 24 years old	3,268	2,450	75.0	2,115	64.7	1,153	335	817	163	221
In high school	989	398	40.3	354	35.7	636	45	591	121	74
In college	6,618	4,731	71.5	4,043	61.1	2,575	689	1,887	276	485
Full-time	5,638	4,042	71.7	3,413	60.5	2,226	629	1,597	232	430
Part-time	979	690	70.4	630	64.3	350	60	290	44	55
Not enrolled in school	12,074	5,616	46.5	4,536	37.6	7,538	1,080	6,458	1,080	812
18 to 20 years old	3,403	1,258	37.0	906	26.6	2,497	352	2,145	331	246
21 to 24 years old	8,671	4,358	50.3	3,630	41.9	5,041	728	4,313	749	566
Male										
Total, 18 to 24 years old	9,750	5,134	52.7	4,220	43.3	5,530	915	4,616	830	714
Enrolled in school	3,841	2,501	65.1	2,107	54.9	1,733	394	1,339	225	293
18 to 20 years old	2,159	1,261	58.4	1,054	48.8	1,105	207	898	129	160
21 to 24 years old	1,681	1,240	73.8	1,053	62.7	628	187	441	95	132
In high school	611	235	38.5	204	33.4	407	31	376	71	41
In college	3,229	2,266	70.2	1,903	58.9	1,326	363	964	154	251
Full-time	2,785	1,958	70.3	1,621	58.2	1,163	337	826	139	227
Part-time	445	308	69.2	282	63.4	163	26	137	15	25
Not enrolled in school	5,909	2,633	44.6	2,112	35.7	3,797	521	3,276	605	421
18 to 20 years old	1,717	636	37.0	456	26.6	1,260	179	1,081	173	119
21 to 24 years old	4,193	1,997	47.6	1,656	39.5	2,537	341	2,195	432	302
Female										
Total, 18 to 24 years old	9,931	5,611	56.5	4,712	47.4	5,219	899	4,320	649	658
Enrolled in school	3,767	2,628	69.8	2,289	60.8	1,478	339	1,138	173	267
18 to 20 years old	2,180	1,418	65.0	1,227	56.3	953	191	762	106	179
21 to 24 years old	1,586	1,210	76.3	1,061	66.9	525	149	376	68	88
In high school	378	163	43.1	149	39.5	229	14	215	51	33
In college	3,389	2,466	72.8	2,140	63.1	1,249	326	923	123	234
Full-time	2,854	2,083	73.0	1,792	62.8	1,062	292	771	93	203
Part-time	535	382	71.5	348	65.1	187	34	153	30	30
Not enrolled in school	6,165	2,983	48.4	2,423	39.3	3,742	560	3,182	475	391
18 to 20 years old	1,687	622	36.9	450	26.7	1,237	173	1,064	158	127
21 to 24 years old	4,478	2,361	52.7	1,973	44.1	2,505	387	2,118	317	264

[Continued]

★ 903 ★

Voting and Registration of Persons 18-24 Years in the Election of November 1992: Enrollment Status

[Continued]

Race, Hispanic origin, sex and enrollment status	All persons	Reported registered		Reported voted		Reported that they did not vote[1]				
								Not registered		
		Number	Percent	Number	Percent	Total	Registered	Total[2]	Not a U.S. citizen	Do not know and not reported on registration[3]
Black										
Both sexes										
Total, 18 to 24 years old	3,543	1,744	49.2	1,296	36.6	2,247	448	1,799	219	405
Enrolled in school	1,224	689	56.3	528	43.1	697	161	535	106	142
18 to 20 years old	758	370	48.9	265	35.0	492	105	387	65	93
21 to 24 years old	467	319	68.3	262	56.2	204	56	148	41	49
In high school	338	122	36.2	78	23.1	260	44	216	27	38
In college	886	567	63.9	450	50.8	436	117	320	79	104
Full-time	764	482	63.0	373	48.8	391	109	282	69	90
Part-time	122	85	69.6	77	63.0	45	8	37	10	14
Not enrolled in school	2,319	1,055	45.5	468	33.1	1,551	287	1,264	113	263
18 to 20 years old	743	268	36.0	209	28.1	534	59	476	38	83
21 to 24 years old	1,575	787	50.0	559	35.5	1,016	228	788	75	180
Male										
Total, 18 to 24 years old	1,685	778	46.2	537	31.8	1,148	241	907	107	230
Enrolled in school	590	283	47.9	191	32.4	399	92	307	53	98
18 to 20 years old	400	172	43.0	116	29.0	284	56	228	29	65
21 to 24 years old	190	111	58.2	76	39.7	115	35	80	24	34
In high school	223	72	32.2	51	22.8	172	21	151	20	28
In college	368	211	57.4	141	38.3	227	70	157	33	70
Full-time	325	178	54.7	109	33.6	216	69	147	33	63
Part-time	43	34	(B)	32	(B)	11	2	9	-	8
Not enrolled in school	1,095	495	45.2	345	31.5	749	150	600	54	132
18 to 20 years old	345	135	39.0	102	29.6	243	33	210	18	32
21 to 24 years old	749	360	48.0	243	32.4	506	117	389	36	100
Female										
Total, 18 to 24 years old	1,858	966	52.0	759	40.9	1,099	207	892	112	175
Enrolled in school	634	406	64.1	336	53.0	298	70	228	52	44
18 to 20 years old	358	198	55.4	149	41.8	208	49	160	36	28
21 to 24 years old	276	208	75.3	187	67.6	89	21	68	17	16
In high school	115	51	43.9	27	23.6	88	23	65	7	10
In college	518	355	68.6	309	59.6	209	46	163	45	34
Full-time	439	304	69.2	264	60.0	176	40	135	35	28
Part-time	79	51	64.9	45	57.1	34	6	28	10	6
Not enrolled in school	1,224	560	45.7	423	34.5	801	137	664	59	131
18 to 20 years old	398	133	33.4	107	26.9	291	26	265	20	51
21 to 24 years old	826	427	51.7	316	38.3	510	111	399	39	80
Hispanic origin[4]										
Both sexes										
Total, 18 to 24 years old	2,795	696	24.9	492	17.6	2,303	204	2,099	1,206	213
Enrolled in school	742	287	38.6	223	30.1	519	63	455	237	71
18 to 20 years old	472	154	32.6	120	25.5	351	33	318	169	42
21 to 24 years old	270	133	49.2	103	38.1	167	30	137	68	28
In high school	207	29	14.1	25	11.8	183	5	178	99	22
In college	535	258	48.2	199	37.2	336	59	277	138	49
Full-time	395	186	47.1	148	37.6	246	38	209	98	42
Part-time	140	72	51.2	50	36.0	90	21	68	40	7
Not enrolled in school	2,053	409	19.9	269	13.1	1,784	140	1,644	969	142
18 to 20 years old	687	114	16.5	62	9.1	625	51	573	315	51
21 to 24 years old	1,366	296	21.6	207	15.1	1,159	89	1,070	654	91

[Continued]

★ 903 ★

Voting and Registration of Persons 18-24 Years in the Election of November 1992: Enrollment Status

[Continued]

Race, Hispanic origin, sex and enrollment status	All persons	Reported registered		Reported voted		Reported that they did not vote[1]				
								Not registered		
		Number	Percent	Number	Percent	Total	Registered	Total[2]	Not a U.S. citizen	Do not know and not reported on registration[3]
Male										
Total, 18 to 24 years old	1,418	288	20.3	196	13.8	1,222	92	1,130	673	128
Enrolled in school	354	110	31.2	82	23.3	271	28	243	114	40
18 to 20 years old	220	56	25.5	44	20.2	176	12	164	85	19
21 to 24 years old	133	54	40.7	38	28.4	95	16	79	29	20
In high school	116	12	10.7	10	8.4	106	3	104	52	15
In college	237	98	41.2	73	30.6	165	25	140	62	25
Full-time	179	70	39.0	55	30.6	124	15	109	50	23
Part-time	58	28	(B)	18	(B)	41	10	30	12	2
Not enrolled in school	1,065	178	16.7	114	10.7	951	64	887	558	88
18 to 20 years old	343	54	15.6	28	8.0	315	26	289	174	26
21 to 24 years old	722	124	17.2	86	11.9	636	38	598	384	62
Female										
Total, 18 to 24 years old	1,377	408	29.6	296	21.5	1,080	112	969	533	85
Enrolled in school	388	176	45.4	141	36.3	247	36	212	122	31
18 to 20 years old	251	98	38.8	76	30.2	175	22	154	83	23
21 to 24 years old	137	79	57.5	65	47.4	72	14	58	39	8
In high school	91	17	18.4	15	16.2	76	2	74	47	7
In college	297	160	53.7	126	42.4	171	34	138	76	24
Full-time	216	116	53.8	94	43.4	122	23	100	48	19
Part-time	82	44	53.4	33	39.9	49	11	38	28	5
Not enrolled in school	988	232	23.4	155	15.7	833	76	757	411	54
18 to 20 years old	344	60	17.4	35	10.1	310	25	284	141	24
21 to 24 years old	644	172	26.6	121	18.7	523	51	472	270	30

Source: "Reported Voting and Registration of Persons 18 to 24 Years Old, by Race, Hispanic Origin, Sex, and Enrollment Status," U.S. Bureau of the Census, *Voting and Registration in the Election of November 1992*, pp. 32- 33. *Notes:* 1. Includes persons reported as "did not vote," "do not know," and "not reported" on voting. 2. In addition to those reported as "not registered," total includes those "not a U.S. citizen," and "do not know" and "not reported" on registration. 3. Includes "do not know" and "not reported" on citizenship. 4. Persons of Hispanic origin may be of any race.

★ 904 ★

Voters and Voting

Voting and Registration of the Employed: Election of November 1992. Part I

November 1992. Numbers in thousands.

Race, Hispanic origin, sex, and major occupation group	All persons	Reported registered		Reported voted		Reported that they did not vote[1]				
								Not registered		
		Number	Percent	Number	Percent	Total	Registered	Total[2]	Not a U.S. citizen	Do not know and not reported on registration[3]
White										
Both sexes										
Total employed	100,200	71,617	71.5	65,678	65.5	34,522	5,940	28,582	4,833	4,735
Managerial and professional	28,432	24,166	85.0	22,937	80.7	5,495	1,229	4,266	700	969
Executive, administrative, and managerial	13,584	11,231	82.7	10,610	78.1	2,974	621	2,353	328	512
Professional specialty	14,848	12,935	87.1	12,327	83.0	2,520	608	1,913	371	457
Technical, sales, and administrative support	31,522	23,789	75.5	22,014	69.8	9,508	1,776	7,732	799	1,423
Technicians and related support	3,624	2,802	77.3	2,617	72.2	1,007	185	822	92	121

[Continued]

★ 904 ★

Voting and Registration of the Employed: Election of November 1992. Part I

[Continued]

Race, Hispanic origin, sex, and major occupation group	All persons	Reported registered		Reported voted		Reported that they did not vote[1]				
								Not registered		
		Number	Percent	Number	Percent	Total	Registered	Total[2]	Not a U.S. citizen	Do not know and not reported on registration[3]
Sales	12,092	8,756	72.4	8,061	66.7	4,031	695	3,336	375	650
Administrative support, include clerical	15,805	12,231	77.4	11,335	71.7	4,470	896	3,574	332	652
Service occupations	11,782	7,164	60.8	6,276	53.3	5,505	888	4,617	983	727
Private household	626	289	46.1	248	39.5	379	42	337	127	44
Service, except household	11,155	6,875	61.6	6,029	54.0	5,126	846	4,280	856	683
Farming. forestry, and fishing	2,922	1,796	61.5	1,613	55.2	1,308	183	1,126	445	137
Precision product, craft, and repair	11,738	7,150	60.9	6,348	54.1	5,390	802	4,588	693	622
Operators, fabricators, and laborers	13,805	7,552	54.7	6,490	47.0	7,315	1,062	6,253	1,213	857
Machine operators, assemblers, and inspectors	6,036	3,168	52.5	2,752	45.6	3,284	416	2,868	705	335
Transportation and material moving	4,167	2,514	60.3	2,168	52.0	1,999	346	1,653	216	244
Handlers, equipment cleaners, helpers, and laborers	3,602	1,869	51.9	1,570	43.6	2,032	300	1,733	292	278
Male										
Total employed	54,894	37,956	69.1	34,694	63.2	20,201	3,263	16,938	3,191	2,710
Managerial and professional	15,161	12,729	84.0	12,100	79.8	3,060	629	2,431	461	510
Executive, administrative, and managerial	8,026	6,602	82.3	6,258	78.0	1,768	343	1,424	222	283
Professional specialty	7,135	6,128	85.9	5,842	81.9	1,293	286	1,007	239	227
Technical, sales, and administrative support	11,485	8,657	75.4	8,077	70.3	3,409	580	2,828	355	559
Technicians and related support	1,897	1,492	78.6	1,400	73.8	497	92	406	52	67
Sales	6,451	4,840	75.0	4,518	70.0	1,934	323	1,611	198	355
Administrative support, include clerical	3,137	2,325	74.1	2,159	68.8	978	166	812	105	137
Service occupations	4,668	2,867	61.4	2,521	54.0	2,147	346	1,801	476	304
Private household	16	6	(B)	6	(B)	10	-	10	5	-
Service, except household	4,652	2,861	61.5	2,515	54.1	2,137	346	1,791	471	304
Farming. forestry, and fishing	2,418	1,445	59.8	1,295	53.6	1,123	150	973	398	111
Precision product, craft, and repair	10,787	6,564	60.8	5,822	54.0	4,965	741	4,223	643	571
Operators, fabricators, and laborers	10,376	5,695	54.9	4,879	47.0	5,497	816	4,681	857	654
Machine operators, assemblers, and inspectors	3,674	1,987	54.1	1,738	47.3	1,937	249	1,688	392	201
Transportation and material moving	3,779	2,237	59.2	1,912	50.6	1,867	325	1,542	213	207
Handlers, equipment cleaners, helpers, and laborers	2,922	1,470	50.3	1,229	42.1	1,693	241	1,452	253	246
Female										
Total employed	45,306	33,661	74.3	30,984	68.4	14,322	2,677	11,644	1,642	2,025
Managerial and professional	13,271	11,437	86.2	10,837	81.7	2,434	600	1,835	239	459
Executive, administrative, and managerial	5,558	4,630	83.3	4,352	78.3	1,207	278	929	106	228
Professional specialty	7,713	6,807	88.3	6,486	84.1	1,227	322	906	133	231
Technical, sales, and administrative support	20,036	15,133	75.5	13,937	69.6	6,099	1,196	4,904	444	864
Technicians and related support	1,726	1,310	75.9	1,217	70.5	509	93	416	40	54
Sales	5,641	3,916	69.4	3,544	62.8	2,097	372	1,725	177	295
Administrative support, include clerical	12,669	9,906	78.2	9,176	72.4	3,493	730	2,762	227	515
Service occupations	7,114	4,297	60.4	3,755	52.8	3,359	542	2,816	507	422
Private household	610	283	46.4	242	39.6	369	42	327	122	44
Service, except household	6,503	4,014	61.7	3,514	54.0	2,990	501	2,489	385	379
Farming. forestry, and fishing	504	351	69.7	318	63.2	185	33	153	46	26
Precision product, craft, and repair	951	586	61.6	525	55.2	426	61	365	50	51
Operators, fabricators, and laborers	3,429	1,857	54.2	1,611	47.0	1,818	246	1,572	356	203
Machine operators, assemblers, and inspectors	2,361	1,181	50.0	1,014	42.9	1,347	167	1,180	312	134

[Continued]

★ 904 ★

Voting and Registration of the Employed: Election of November 1992. Part I
[Continued]

Race, Hispanic origin, sex, and major occupation group	All persons	Reported registered		Reported voted		Reported that they did not vote[1]				
								Not registered		
		Number	Percent	Number	Percent	Total	Registered	Total[2]	Not a U.S. citizen	Do not know and not reported on registration[3]
Transportation and material moving	388	277	71.3	256	66.0	132	21	111	4	37
Handlers, equipment cleaners, helpers, and laborers	680	399	58.7	341	50.1	339	58	281	39	32

Source: "Reported Voting and Registration of Employed Persons, by Race, Hispanic Origin, Sex, and Major Occupation Group," U.S. Bureau of the Census, *Voting and Registration in the Election of November 1992*, p. 52. *Notes:* 1. Includes persons reported as "did not vote," "do not know," and "not reported" on voting. 2. In addition to those reported as "not registered," total includes those "not a U.S. citizen," and "do not know" and "not reported" on registration.

★ 905 ★

Voters and Voting

Voting and Registration of the Employed: Election of November 1992. Part II

November 1992. Numbers in thousands.

Race, Hispanic origin, sex, and major occupation group	All persons	Reported registered		Reported voted		Reported that they did not vote[1]				
								Not registered		
		Number	Percent	Number	Percent	Total	Registered	Total[2]	Not a U.S. citizen	Do not know and not reported on registration[3]
Black										
Both sexes										
Total employed	11,851	8,042	67.9	7,001	59.1	4,851	1,042	3,809	636	966
Managerial and professional	2,126	1,713	80.6	1,603	75.4	523	110	413	77	142
Executive, administrative, and managerial	896	724	80.8	683	76.2	213	41	172	32	55
Professional specialty	1,230	989	80.4	920	74.8	310	69	241	45	88
Technical, sales, and administrative support	3,169	2,257	71.2	2,019	63.7	1,150	238	912	137	212
Technicians and related support	383	288	75.1	248	64.8	135	40	95	11	20
Sales	855	535	62.5	462	54.1	393	72	320	58	75
Administrative support, include clerical	1,931	1,435	74.3	1,309	67.8	622	126	496	68	117
Service occupations	2,713	1,749	64.5	1,447	53.3	1,266	302	964	223	219
Private household	152	91	59.9	84	55.4	68	7	61	26	11
Service, except household	2,562	1,658	64.7	1,363	53.2	1,199	295	903	197	208
Farming, forestry, and fishing	222	115	51.9	103	46.5	119	12	107	12	25
Precision product, craft, and repair	1,067	684	64.1	594	55.6	474	90	384	69	104
Operators, fabricators, and laborers	2,555	1,525	59.7	1,235	48.3	1,320	290	1,030	117	264
Machine operators, assemblers, and inspectors	1,124	739	65.8	594	52.9	530	145	385	42	78
Transportation and material moving	709	415	58.5	349	49.2	361	66	295	29	75
Handlers, equipment cleaners, helpers, and laborers	722	371	51.4	292	40.5	430	79	351	45	111
Male										
Total employed	5,836	3,713	63.6	3,135	53.7	2,702	578	2,123	378	544
Managerial and professional	846	850	76.8	604	71.4	242	46	196	45	66
Executive, administrative, and managerial	430	344	80.1	330	76.7	100	15	85	14	31
Professional specialty	416	305	73.4	275	66.0	142	31	111	31	35
Technical, sales, and administrative support	964	642	66.5	559	58.0	405	82	323	68	44
Technicians and related support	175	132	75.5	108	61.7	67	24	43	5	3
Sales	302	185	61.1	160	53.1	142	24	118	35	19
Administrative support, include clerical	487	325	66.7	291	59.7	196	34	162	28	21
Service occupations	1,101	686	62.3	537	48.8	564	149	415	95	112
Private household	7	7	(B)	7	(B)	-	-	-	-	-
Service, except household	1,095	679	62.1	531	48.5	564	149	415	95	112

[Continued]

★ 905 ★

Voting and Registration of the Employed: Election of November 1992. Part II

[Continued]

Race, Hispanic origin, sex, and major occupation group	All persons	Reported registered		Reported voted		Reported that they did not vote[1]				
								Not registered		
		Number	Percent	Number	Percent	Total	Registered	Total[2]	Not a U.S. citizen	Do not know and not reported on registration[3]
Farming. forestry, and fishing	197	102	51.7	90	45.6	107	12	95	10	25
Precision product, craft, and repair	891	564	63.3	489	54.9	402	75	327	66	89
Operators, fabricators, and laborers	1,837	1,070	58.2	855	46.6	982	215	767	94	209
Machine operators, assemblers, and inspectors	556	374	67.2	293	52.6	263	81	182	22	32
Transportation and material moving	653	380	58.2	314	48.1	339	66	273	29	72
Handlers, equipment cleaners, helpers, and laborers	627	316	50.3	248	39.6	379	68	312	42	105
Female										
Total employed	6,015	4,329	72.0	3,866	64.3	2,149	464	1,686	257	422
Managerial and professional	1,279	1,063	83.1	998	78.0	281	64	217	32	76
Executive, administrative, and managerial	466	379	81.5	353	75.8	113	26	86	18	24
Professional specialty	814	683	84.0	645	79.3	168	38	130	14	52
Technical, sales, and administrative support	2,205	1,616	73.3	1,460	66.2	745	155	589	69	168
Technicians and related support	208	156	74.8	140	67.4	68	15	53	6	17
Sales	553	350	63.3	302	54.6	251	48	203	23	56
Administrative support, include clerical	1,444	1,110	76.9	1,018	70.5	426	92	334	40	95
Service occupations	1,612	1,063	66.0	910	56.4	702	153	549	128	107
Private household	145	84	58.1	77	53.4	68	7	61	26	11
Service, except household	1,467	979	66.7	832	56.7	635	147	488	102	96
Farming. forestry, and fishing	24	13	(B)	13	(B)	11	-	11	2	-
Precision product, craft, and repair	177	120	67.9	105	59.4	72	15	57	3	15
Operators, fabricators, and laborers	718	455	63.4	380	52.9	338	75	263	23	55
Machine operators, assemblers, and inspectors	568	365	64.3	301	53.1	266	64	202	20	46
Transportation and material moving	56	35	(B)	35	(B)	21	-	21	-	3
Handlers, equipment cleaners, helpers, and laborers	94	55	58.7	44	46.5	50	11	39	3	6

Source: "Reported Voting and Registration of Employed Persons, by Race, Hispanic Origin, Sex, and Major Occupation Group," U.S. Bureau of the Census, *Voting and Registration in the Election of November 1992*, p. 53. *Notes:* 1. Includes persons reported as "did not vote," "do not know," and "not reported" on voting. 2. In addition to those reported as "not registered," total includes those "not a U.S. citizen," and "do not know" and "not reported" on registration.

★ 906 ★

Voters and Voting

Voting and Registration of the Employed: Election of November 1992. Part III

November 1992. Numbers in thousands.

Race, Hispanic origin, sex, and major occupation group	All persons	Reported registered		Reported voted		Reported that they did not vote[1]				
								Not registered		
		Number	Percent	Number	Percent	Total	Registered	Total[2]	Not a U.S. citizen	Do not know and not reported on registration[3]
Hispanic origin[4]										
Both sexes										
Total employed	8,800	3,226	36.7	2,682	30.5	6,117	543	5,574	3,489	570
Managerial and professional	1,188	775	65.2	700	58.9	489	76	413	205	65
Executive, administrative, and managerial	636	375	58.9	333	52.4	303	42	261	120	48
Professional specialty	553	401	72.5	367	66.3	186	34	152	85	17
Technical, sales, and administrative support	2,137	1,116	52.2	925	43.3	1,212	191	1,021	434	147
Technicians and related support	203	109	53.9	102	50.4	101	7	94	47	4

[Continued]

★ 906 ★

Voting and Registration of the Employed: Election of November 1992. Part III

[Continued]

Race, Hispanic origin, sex, and major occupation group	All persons	Reported registered		Reported voted		Reported that they did not vote[1]					
								Not registered			
		Number	Percent	Number	Percent	Total	Registered	Total[2]	Not a U.S. citizen	Do not know and not reported on registration[3]	
Sales	755	336	44.5	265	35.1	490	71	419	199	68	
Administrative support, include clerical	1,179	671	56.9	558	47.3	622	113	509	187	75	
Service occupations	1,654	461	27.9	366	22.2	1,287	94	1,193	814	121	
Private household	135	13	9.8	4	3.3	131	9	122	108	8	
Service, except household	1,518	448	29.5	362	23.8	1,156	86	1,071	706	113	
Farming. forestry, and fishing	558	57	10.3	42	7.5	516	16	500	421	24	
Precision product, craft, and repair	1,209	338	28.0	266	22.0	943	72	871	547	94	
Operators, fabricators, and laborers	2,054	478	23.3	383	18.7	1,671	95	1,576	1,068	119	
Machine operators, assemblers, and inspectors	1,101	209	19.0	175	15.9	926	34	892	640	62	
Transportation and material moving	421	131	31.2	111	26.3	310	20	290	176	21	
Handlers, equipment cleaners, helpers, and laborers	532	138	26.0	98	18.4	434	40	394	253	35	
Male											
Total employed	5,347	1,692	31.6	1,405	26.3	3,943	287	3,656	2,400	354	
Managerial and professional	646	405	62.7	349	54.1	297	56	241	138	34	
Executive, administrative, and managerial	379	225	59.3	194	51.1	186	31	154	89	22	
Professional specialty	266	180	67.7	155	58.3	111	25	86	50	12	
Technical, sales, and administrative support	776	361	48.5	314	40.5	462	47	415	217	49	
Technicians and related support	100	56	56.7	54	53.8	46	3	43	29	4	
Sales	359	147	40.9	128	35.7	231	19	212	106	32	
Administrative support, include clerical	317	158	49.7	133	41.8	185	25	159	81	14	
Service occupations	811	202	24.9	172	21.2	639	29	610	418	75	
Private household	5	-	(B)	-	(B)	5	-	5	5	-	
Service, except household	806	202	25.0	172	21.4	634	29	604	413	75	
Farming. forestry, and fishing	499	55	10.9	39	7.8	460	16	444	375	24	
Precision product, craft, and repair	1,094	299	27.3	231	21.1	863	68	795	504	81	
Operators, fabricators, and laborers	1,522	370	24.3	299	19.7	1,222	71	1,152	748	90	
Machine operators, assemblers, and inspectors	650	129	19.8	113	17.3	537	16	521	356	39	
Transportation and material moving	401	124	30.9	105	26.1	296	19	277	172	17	
Handlers, equipment cleaners, helpers, and laborers	471	118	25.0	82	17.4	389	35	354	220	33	
Female											
Total employed	3,452	1,534	44.4	1,278	37.0	2,175	257	1,918	1,089	216	
Managerial and professional	543	370	68.2	351	64.6	192	20	172	67	31	
Executive, administrative, and managerial	256	150	58.5	139	54.3	117	11	106	31	26	
Professional specialty	286	221	77.0	211	73.8	75	9	66	35	5	
Technical, sales, and administrative support	1,361	755	55.5	611	44.9	750	144	606	217	98	
Technicians and related support	103	53	51.1	49	47.1	55	4	50	18	-	
Sales	396	189	47.9	137	34.7	258	52	206	93	36	
Administrative support, include clerical	862	513	59.5	425	49.3	437	88	349	106	62	
Service occupations	842	259	30.8	194	23.0	648	65	583	396	46	
Private household	130	13	10.2	4	3.4	126	9	117	103	8	
Service, except household	712	246	34.5	190	26.6	522	56	466	293	38	
Farming. forestry, and fishing	59	3	(B)	3	(B)	56	-	56	46	-	
Precision product, craft, and repair	114	39	33.9	35	30.4	80	4	76	43	13	
Operators, fabricators, and laborers	533	108	20.3	84	15.8	449	24	425	320	29	
Machine operators, assemblers, and inspectors	451	80	17.7	62	13.8	389	18	371	284	23	

[Continued]

★ 906 ★

Voting and Registration of the Employed: Election of November 1992. Part III

[Continued]

Race, Hispanic origin, sex, and major occupation group	All persons	Reported registered		Reported voted		Reported that they did not vote[1]				
								Not registered		
		Number	Percent	Number	Percent	Total	Registered	Total[2]	Not a U.S. citizen	Do not know and not reported on registration[3]
Transportation and material moving	20	8	(B)	6	(B)	14	1	13	4	4
Handlers, equipment cleaners, helpers, and laborers	61	20	(B)	16	(B)	45	5	41	33	1

Source: "Reported Voting and Registration of Employed Persons, by Race, Hispanic Origin, Sex, and Major Occupation Group," U.S. Bureau of the Census, *Voting and Registration in the Election of November 1992*, p. 54. *Notes:* 1. Includes persons reported as "did not vote," "do not know," and "not reported" on voting. 2. In addition to those reported as "not registered," total includes those "not a U.S. citizen," and "do not know" and "not reported" on registration. 3. Includes "do not know" and "not reported" on citizenship. 4. Persons of Hispanic origin may be of any race.

★ 907 ★

Voters and Voting

Voting and Registration: Characteristics of Respondents: November 1992 - I

November 1992. Numbers in thousands.

Race, Hispanic origin, sex, and type of respondent	All persons	Reported registered	Reported voted	Total	Reported that they did not vote Registered			
					Yes	No	Do not know	Not reported
All races								
All respondents[1]								
Both sexes	185,684	126,578	113,866	49,986	11,172	36,435	2,027	353
Male	88,557	59,254	53,312	24,152	5,111	17,835	1,036	170
Female	97,126	67,324	60,554	25,835	6,061	18,600	991	182
Reported by self								
Both sexes	99,135	72,569	65,582	27,310	6,556	19,914	659	181
Male	37,273	26,704	24,302	10,213	2,269	7,654	223	67
Female	61,862	45,866	41,280	17,097	4,287	12,260	436	114
Reported by other								
Both sexes	80,226	53,403	47,733	22,373	4,573	16,287	1,363	150
Male	48,261	32,251	28,736	13,772	2,822	10,049	811	90
Female	31,965	21,152	18,996	8,601	1,751	6,238	553	60
White								
All respondents[1]								
Both sexes	157,837	110,684	100,405	41,416	9,084	30,462	1,586	284
Male	75,915	52,383	47,552	20,211	4,223	15,052	800	136
Female	81,922	58,301	52,853	21,206	4,861	15,411	786	148
Reported by self								
Both sexes	84,653	63,158	57,484	22,542	5,275	16,595	521	151
Male	31,952	23,490	21,533	8,432	1,839	6,357	180	56
Female	52,700	39,669	35,951	14,110	3,436	10,238	341	95

[Continued]

★ 907 ★

Voting and Registration: Characteristics of Respondents: November 1992 - I
[Continued]

Race, Hispanic origin, sex, and type of respondent	All persons	Reported registered	Reported voted	Total	Reported that they did not vote Registered			
					Yes	No	Do not know	Not reported
Reported by other								
Both sexes	68,022	47,005	42,441	18,638	3,779	13,680	1,060	119
Male	41,465	28,625	25,772	11,640	2,370	8,581	618	72
Female	26,557	18,380	16,670	6,998	1,409	5,099	442	47
Black								
All respondents[1]								
Both sexes	21,039	13,442	11,371	6,844	1,767	4,669	354	54
Male	9,425	5,727	4,786	3,135	742	2,177	194	22
Female	11,614	7,715	6,585	3,709	1,025	2,492	160	32
Reported by self								
Both sexes	11,478	8,130	6,981	3,936	1,125	2,670	117	24
Male	3,994	2,696	2,318	1,422	367	1,014	35	6
Female	7,484	5,434	4,663	2,514	758	1,657	82	18
Reported by other								
Both sexes	8,700	5,236	4,326	2,851	631	1,960	237	22
Male	5,038	3,003	2,445	1,687	370	1,147	159	11
Female	3,662	2,233	1,881	1,163	261	813	78	11
Hispanic[2]								
All respondents[2]								
Both sexes	14,688	5,137	4,238	3,622	811	2,619	154	38
Male	7,273	2,337	1,946	1,720	345	1,274	79	22
Female	7,415	2,800	2,291	1,903	466	1,345	75	16
Reported by self								
Both sexes	6,964	2,891	2,430	1,922	446	1,405	50	20
Male	2,767	1,026	889	708	126	562	16	4
Female	4,196	1,866	1,541	1,213	319	844	34	16
Reported by other								
Both sexes	7,065	2,215	1,784	1,665	358	1,185	104	18
Male	4,181	1,300	1,047	998	216	701	63	18
Female	2,884	915	736	667	142	484	41	-

Source: "Reported Voting and Registration, by Race, Hispanic Origin, Sex, and Type of Respondent," U.S. Bureau of the Census, *Voting and Registration in the Election of November 1992*, p. 70. *Notes:* - represents zero. 1. Includes not reported on type of respondent, not shown separately. 2. Persons of Hispanic origin may be of any race.

★ 908 ★

Voters and Voting

Voting and Registration: Characteristics of Respondents: November 1992 - II

November 1992. Numbers in thousands.

Race, Hispanic origin, sex, and type of respondent	Do not know (voting)					Not reported (voting)		
	Total	Registered				Total[1]	Not reported on registration	Not a U.S. citizen
		Yes	No	Do not know	Not reported			
All races								
All respondents[2]								
Both sexes	3,089	819	44	2,174	53	6,843	6,121	11,900
Male	1,854	531	29	1,260	34	3,174	2,874	6,065
Female	1,235	288	14	914	19	3,669	3,247	5,834
Reported by self								
Both sexes	159	12	1	143	4	860	441	5,224
Male	61	7	-	50	3	306	181	2,390
Female	98	5	1	92	-	554	259	2,834
Reported by other								
Both sexes	2,910	806	43	2,013	48	698	406	6,513
Male	1,778	523	29	1,196	30	383	213	3,592
Female	1,131	283	13	817	18	316	193	2,921
White								
All respondents[2]								
Both sexes	2,049	532	32	1,457	28	5,683	5,021	8,284
Male	1,230	340	24	848	19	2,625	2,357	4,298
Female	819	193	8	609	9	3,058	2,663	3,986
Reported by self								
Both sexes	126	4	1	118	3	781	386	3,720
Male	48	-	-	44	3	274	157	1,665
Female	77	3	1	73	-	507	229	2,055
Reported by other								
Both sexes	1,906	527	31	1,325	23	581	324	4,455
Male	1,167	338	24	790	15	316	171	2,570
Female	739	189	7	535	8	265	153	1,885
Black								
All respondents[2]								
Both sexes	891	255	11	608	17	888	840	1,044
Male	532	171	5	348	9	424	397	548
Female	359	84	6	261	8	464	444	496
Reported by self								
Both sexes	28	7	-	22	-	67	50	465
Male	10	7	-	3	-	28	23	217
Female	19	-	-	19	-	40	26	248

[Continued]

★ 908 ★

Voting and Registration: Characteristics of Respondents: November 1992 - II
[Continued]

Race, Hispanic origin, sex, and type of respondent	Do not know (voting)					Not reported (voting)		
	Total	Registered				Total[1]	Not reported on registration	Not a U.S. citizen
		Yes	No	Do not know	Not reported			
Reported by other								
Both sexes	863	248	11	587	17	106	76	554
Male	522	164	5	345	9	63	40	320
Female	340	84	6	242	8	43	36	235
Hispanic[3]								
All respondents[2]								
Both sexes	298	54	5	236	3	620	563	5,910
Male	181	28	5	146	3	294	269	3,132
Female	117	27	-	90	-	326	294	2,778
Reported by self								
Both sexes	9	-	-	9	-	53	21	2,549
Male	4	-	-	4	-	21	9	1,145
Female	5	-	-	5	-	33	12	1,404
Reported by other								
Both sexes	287	54	5	225	3	52	22	3,278
Male	175	28	5	140	3	24	5	1,936
Female	112	27	-	85	-	28	17	1,341

Source: "Reported Voting and Registration, by Race, Hispanic Origin, Sex, and Type of Respondent," U.S. Bureau of the Census, *Voting and Registration in the Election of November 1992*, p. 70. *Notes:* - represents zero. 1. Includes 721,000 persons who reported that they were registered and 89,000 who reported that they were not registered. 2. Includes not reported on type of respondent, not shown separately. 3. Persons of Hispanic origin may be of any race.

★ 909 ★

Voters and Voting

Voting and Registration: Residence and Tenure: Election of November 1992. Part I

November 1992. Numbers in thousands.

Race, Hispanic origin, tenure, and duration of residence	All persons	Reported registered		Reported voted		Reported that they did not vote[1]				
						Total	Registered	Not registered		
		Number	Percent	Number	Percent			Total[2]	Not a U.S. citizen	Do not know and not reported on registration[3]
White										
18 years and over										
In all units	157,837	110,684	70.1	100,405	63.6	57,432	10,279	47,153	8,284	8,289
Duration of residence										
Less than 1 month	2,567	1,260	49.9	1,019	39.7	1,547	261	1,287	276	103
1 to 6 months	15,493	8,634	55.7	7,232	46.7	8,260	1,401	6,859	1,617	523
7 to 11 months	6,540	3,780	57.8	3,292	50.3	3,248	488	2,760	624	161
1 to 2 years	23,139	14,635	63.2	13,191	57.0	9,948	1,444	8,504	1,941	510
3 to 4 years	18,408	13,078	71.0	11,946	64.9	6,462	1,131	5,331	1,213	421

[Continued]

★ 909 ★

Voting and Registration: Residence and Tenure: Election of November 1992. Part I

[Continued]

Race, Hispanic origin, tenure, and duration of residence	All persons	Reported registered		Reported voted		Reported that they did not vote[1]				
								Not registered		
		Number	Percent	Number	Percent	Total	Registered	Total[2]	Not a U.S. citizen	Do not know and not reported on registration[3]
5 years or longer	85,909	68,563	79.8	63,068	73.4	22,841	5,495	17,346	2,509	1,829
Not reported	5,781	714	12.4	656	11.3	5,125	58	5,067	104	4,742
Owner occupied units	114,077	87,382	76.6	80,490	70.6	33,587	6,892	26,695	3,191	5,677
Duration of residence										
Less than 1 month	819	526	64.2	455	55.6	363	70	293	18	35
1 to 6 months	5,098	3,401	66.7	3,012	59.1	2,086	389	1,697	235	157
7 to 11 months	2,597	1,831	70.5	1,653	63.6	944	178	766	124	57
1 to 2 years	12,115	8,632	71.2	7,980	65.9	4,136	652	3,484	559	213
3 to 4 years	12,719	9,714	76.4	8,967	70.5	3,752	747	3,004	534	277
5 years or longer	76,701	62,718	81.8	57,906	75.5	18,795	4,812	13,982	1,693	1,597
Not reported	4,029	560	13.9	517	12.8	3,511	43	3,468	3,342	
Renter-occupied units	41,333	21,924	53.0	18,723	45.3	22,610	3,200	19,409	4,871	2,485
Duration of residence										
Less than 1 month	1,682	723	43.0	541	32.2	1,141	182	958	242	68
1 to 6 months	10,080	5,089	50.5	4,103	40.7	5,977	986	4,991	1,331	347
7 to 11 months	3,801	1,861	49.0	1,563	41.1	2,238	298	1,940	500	96
1 to 2 years	10,516	5,742	54.6	4,972	47.3	5,543	770	4,774	1,319	272
3 to 4 years	5,404	3,180	58.9	2,814	52.1	2,590	366	2,223	650	135
5 years or longer	8,187	5,183	63.3	4,601	56.2	3,586	582	3,004	767	211
Not reported	1,663	144	8.6	129	7.7	1,535	15	1,520	64	1,356
18 to 24 years										
In all units	19,681	10,746	54.6	8,932	45.4	10,749	1,814	8,935	1,478	1,372
Duration of residence										
Less than 1 month	736	315	42.8	233	31.6	503	83	421	101	38
1 to 6 months	4,260	2,136	50.1	1,666	39.1	2,594	469	2,124	408	187
7 to 11 months	1,354	688	50.8	554	40.9	800	134	666	128	47
1 to 2 years	3,384	1,712	50.6	1,418	41.9	1,966	294	1,672	424	104
3 to 4 years	1,606	864	53.8	736	45.9	869	127	742	142	71
5 years or longer	7,680	4,972	64.7	4,268	55.6	3,412	704	2,708	247	393
Not reported	662	59	8.9	57	8.6	605	2	603	27	533
Owner-occupied units	11,131	6,777	60.9	5,783	52.0	5,348	994	4,354	395	859
Duration of residence										
Less than 1 month	186	90	48.6	73	39.4	113	17	96	4	14
1 to 6 months	843	414	49.2	333	39.5	510	82	428	40	52
7 to 11 months	407	258	63.3	209	51.2	199	49	149	20	12
1 to 2 years	1,287	712	55.3	607	47.2	680	105	575	99	43
3 to 4 years	1,060	610	57.5	523	49.3	537	87	450	63	41
5 years or longer	6,960	4,650	66.8	3,998	57.4	2,963	652	2,310	165	366
Not reported	389	43	11.1	41	10.5	348	2	346	4	331
Renter-occupied units	8,173	3,801	46.5	3,020	36.9	5,153	781	4,372	1,050	481
Duration of residence										
Less than 1 month	522	206	39.4	148	28.4	373	57	316	92	24
1 to 6 months	3,328	1,689	50.8	1,307	39.3	2,021	382	1,639	365	124
7 to 11 months	915	411	44.9	330	36.0	586	81	504	108	35
1 to 2 years	2,004	954	47.6	773	38.6	1,231	181	1,050	317	55
3 to 4 years	519	246	47.5	208	40.1	311	38	273	70	28
5 years or longer	643	280	43.5	238	37.1	404	41	363	78	26
Not reported	242	15	6.4	15	6.4	226	-	226	19	189

[Continued]

★ 909 ★

Voting and Registration: Residence and Tenure: Election of November 1992. Part I

[Continued]

Race, Hispanic origin, tenure, and duration of residence	All persons	Reported registered		Reported voted		Reported that they did not vote[1]				
								Not registered		
		Number	Percent	Number	Percent	Total	Registered	Total[2]	Not a U.S. citizen	Do not know and not reported on registration[3]
25 to 44 years										
In all units	68,285	45,595	66.8	41,379	60.6	26,906	4,216	22,690	4,498	3,276
Duration of residence										
Less than 1 month	1,401	740	52.8	594	42.4	807	146	661	138	50
1 to 6 months	8,758	4,944	56.5	4,243	48.4	4,515	701	3,814	978	278
7 to 11 months	3,878	2,222	57.3	1,976	51.0	1,902	246	1,656	400	87
1 to 2 years	14,361	9,088	63.2	8,231	57.2	6,150	857	5,292	1,194	301
3 to 4 years	11,172	7,921	70.9	7,219	64.6	3,953	702	3,251	750	236
5 years or longer	26,466	20,372	77.0	18,831	71.2	7,635	1,541	6,094	988	551
Not reported	2,229	307	13.8	285	12.8	1,944	21	1,922	51	1,774
Owner-occupied units	44,685	33,262	74.4	30,720	68.7	13,965	2,542	11,423	1,496	1,981
Duration of residence										
Less than 1 month	458	307	67.1	273	59.7	185	34	151	14	17
1 to 6 months	3,126	2,153	68.9	1,934	61.9	1,192	219	973	152	72
7 to 11 months	1,507	1,073	71.2	990	65.7	517	83	434	73	33
1 to 2 years	7,599	5,405	71.1	5,004	65.9	2,595	401	2,193	329	129
3 to 4 years	7,666	5,858	76.4	5,390	70.3	2,276	467	1,808	319	161
5 years or longer	22,933	18,231	79.5	16,912	73.7	6,021	1,318	4,703	591	473
Not reported	1,396	235	16.8	216	15.5	1,180	19	1,161	17	1,096
Renter-occupied units	22,423	11,667	52.0	10,058	44.9	12,365	1,609	10,756	2,849	1,247
Duration of residence										
Less than 1 month	916	424	46.3	312	34.1	604	112	492	112	32
1 to 6 months	5,448	2,701	49.6	2,236	41.0	3,213	465	2,747	778	199
7 to 11 months	2,297	1,106	48.1	947	41.2	1,350	158	1,192	327	49
1 to 2 years	6,500	3,551	54.6	3,100	47.7	3,400	451	2,949	817	166
3 to 4 years	3,323	1,927	58.0	1,705	51.3	1,618	222	1,396	420	72
5 years or longer	3,133	1,888	60.3	1,689	53.9	1,443	199	1,245	370	68
Not reported	805	70	8.7	68	8.4	737	2	735	25	661
45 to 64 years										
In all units	42,279	32,513	76.9	30,370	71.8	11,909	2,143	9,766	1,669	2,043
Duration of residence										
Less than 1 month	327	182	55.5	152	46.5	175	29	146	27	9
1 to 6 months	1,877	1,151	61.3	995	53.0	882	156	726	188	48
7 to 11 months	987	647	65.6	574	58.1	414	74	340	75	21
1 to 2 years	3,824	2,702	70.6	2,504	65.5	1,321	198	1,123	258	81
3 to 4 years	3,813	2,893	75.9	2,708	71.0	1,105	184	921	242	86
5 years or longer	29,832	24,731	82.9	23,244	77.9	6,587	1,487	5,101	858	460
Not reported	1,619	208	12.8	193	11.9	1,426	15	1,411	21	1,339
Owner-occupied units	35,450	28,669	80.9	26,970	76.1	8,480	1,699	6,781	957	1,570
Duration of residence										
Less than 1 month	146	115	78.6	96	65.6	50	19	31	-	2
1 to 6 months	868	632	72.8	570	65.7	298	62	236	37	23
7 to 11 months	503	371	73.9	339	67.5	163	32	131	23	7
1 to 2 years	2,327	1,824	78.4	1,722	74.0	604	102	503	102	30
3 to 4 years	2,845	2,304	81.0	2,175	76.4	671	130	541	112	61
5 years or longer	27,498	23,254	84.6	21,910	79.7	5,588	1,344	4,244	676	388
Not reported	1,264	169	13.4	158	12.5	1,106	11	1,095	7	1,060
Renter-occupied units	6,368	3,566	56.0	3,149	49.5	3,219	417	2,803	693	448
Duration of residence										
Less than 1 month	176	64	36.5	54	30.7	122	10	112	27	7

[Continued]

★ 909 ★

Voting and Registration: Residence and Tenure: Election of November 1992. Part I
[Continued]

Race, Hispanic origin, tenure, and duration of residence	All persons	Reported registered		Reported voted		Reported that they did not vote[1]				
						Total	Registered	Not registered		
		Number	Percent	Number	Percent			Total[2]	Not a U.S. citizen	Do not know and not reported on registration[3]
1 to 6 months	977	502	51.3	410	42.0	567	91	476	151	23
7 to 11 months	451	255	56.4	213	47.2	239	42	197	52	11
1 to 2 years	1,398	815	58.3	725	51.9	673	90	583	150	41
3 to 4 years	930	569	61.1	516	55.4	415	53	362	124	23
5 years or longer	2,101	1,332	63.4	1,205	57.4	896	126	770	175	72
Not reported	334	30	9.0	26	7.8	307	4	303	14	271
65 years and over										
In all units	27,592	21,831	79.1	19,724	71.5	7,867	2,106	5,761	639	1,598
Duration of residence										
Less than 1 month	102	43	42.2	40	39.5	62	3	59	10	7
1 to 6 months	596	404	67.5	329	55.0	269	75	195	42	11
7 to 11 months	321	222	69.2	189	58.8	133	34	99	21	7
1 to 2 years	1,550	1,133	73.1	1,038	67.0	512	95	418	66	25
3 to 4 years	1,818	1,401	77.0	1,282	70.5	535	118	417	80	28
5 years or longer	21,931	18,487	84.3	16,724	76.3	5,206	1,763	3,443	416	425
Not reported	1,271	141	11.1	121	9.6	1,150	19	1,130	5	1,096
Owner-occupied units	22,811	18,673	81.9	17,017	74.6	5,794	1,656	4,137	344	1,267
Duration of residence										
Less than 1 month	29	14	(B)	14	(B)	16	-	16	-	2
1 to 6 months	261	201	77.0	175	66.9	86	26	60	6	9
7 to 11 months	181	129	71.6	115	63.7	66	14	51	8	5
1 to 2 years	903	690	76.4	646	71.6	257	44	213	28	11
3 to 4 years	1,147	943	82.1	879	76.6	269	64	205	40	15
5 years or longer	19,309	16,584	85.9	15,086	78.1	4,223	1,496	2,725	260	370
Not reported	980	113	11.5	102	10.5	878	11	867	-	855
Renter-occupied units	4,369	2,890	66.1	2,496	57.1	1,873	394	1,479	280	309
Duration of residence										
Less than 1 month	68	29	(B)	27	(B)	41	3	38	10	5
1 to 6 months	326	197	60.4	150	45.9	177	47	129	36	1
7 to 11 months	137	90	65.6	73	53.0	64	17	47	13	1
1 to 2 years	613	423	69.0	375	61.1	238	48	190	35	9
3 to 4 years	632	438	69.4	385	61.0	246	53	193	37	12
5 years or longer	2,310	1,684	72.9	1,468	63.6	842	216	626	145	46
Not reported	283	28	9.9	19	6.7	264	9	255	5	235

Source: "Reported Voting and Registration, by Race, Hispanic Origin, Age, Duration of Residence, and Tenure," U.S. Bureau of the Census, *Voting and Registration in the Election of November 1992*, pp. 59-61. *Notes:* 1. Includes persons reported as "did not vote," "do not know," and "not reported" on voting. 2. In addition to those reported as "not registered," total includes those "not a U.S. citizen," and "do not know" and "not reported" on registration. 3. Includes "do not know" and "not reported" on citizenship.

★ 910 ★

Voters and Voting

Voting and Registration: Residence and Tenure: Election of November 1992. Part II

November 1992. Numbers in thousands.

Race, Hispanic origin, tenure, and duration of residence	All persons	Reported registered		Reported voted		Reported that they did not vote[1]				
								Not registered		
		Number	Percent	Number	Percent	Total	Registered	Total[2]	Not a U.S. citizen	Do not know and not reported on registration[3]
Black										
18 years and over										
In all units	21,039	13,442	63.9	11,371	54.0	9,668	2,070	7,598	1,044	1,870
Duration of residence										
Less than 1 month	517	279	54.0	205	39.7	312	74	238	16	42
1 to 6 months	2,264	1,220	53.9	889	39.3	1,374	331	1,044	159	131
7 to 11 months	1,073	584	54.4	452	42.1	621	132	489	56	66
1 to 2 years	3,267	1,898	58.1	1,585	48.5	1,681	313	1,368	252	174
3 to 4 years	2,437	1,545	63.4	1,290	52.9	1,147	255	891	242	105
5 years or longer	10,419	7,800	74.9	6,854	65.8	3,566	946	2,619	295	488
Not reported	1,063	115	10.8	95	9.0	968	19	948	23	863
Owner occupied units	10,643	7,730	72.6	6,798	63.9	3,845	932	2,913	318	821
Duration of residence										
Less than 1 month	101	65	64.4	56	55.8	44	9	36	3	1
1 to 6 months	391	243	62.1	190	48.6	201	53	148	15	32
7 to 11 months	227	139	61.0	111	49.1	115	27	88	22	15
1 to 2 years	881	552	62.7	502	57.1	378	49	329	65	54
3 to 4 years	873	609	69.7	510	58.5	363	98	264	54	28
5 years or longer	7,673	6,041	78.7	5,364	69.9	2,310	677	1,632	149	321
Not reported	498	83	16.6	64	12.9	433	18	415	10	370
Renter-occupied units	10,087	5,543	55.0	4,445	44.1	5,642	1,098	4,544	723	1,009
Duration of residence										
Less than 1 month	409	209	51.0	149	36.4	260	60	200	13	41
1 to 6 months	1,831	963	52.6	687	37.5	1,144	276	868	144	99
7 to 11 months	822	430	52.2	326	39.7	496	103	393	35	51
1 to 2 years	2,362	1,326	56.1	1,070	45.3	1,292	256	1,036	187	121
3 to 4 years	1,522	914	60.0	761	50.0	761	153	608	188	74
5 years or longer	2,599	1,670	64.2	1,422	54.7	1,177	248	930	144	149
Not reported	541	32	5.9	31	5.7	510	1	509	12	474
18 to 24 years										
In all units	3,543	1,744	49.2	1,296	36.6	2,247	448	1,799	219	405
Duration of residence										
Less than 1 month	111	27	24.7	18	16.6	92	9	83	2	3
1 to 6 months	583	272	46.7	177	30.4	406	95	310	39	47
7 to 11 months	223	102	45.6	63	28.2	160	39	121	14	7
1 to 2 years	612	282	46.1	215	35.1	397	67	330	62	34
3 to 4 years	359	173	48.2	108	30.1	251	65	186	55	31
5 years or longer	1,465	867	59.2	696	47.5	768	171	598	46	130
Not reported	192	21	11.0	19	9.8	173	2	171	2	152
Owner-occupied units	1,561	897	57.5	716	45.8	846	181	664	59	193
Duration of residence										
Less than 1 month	15	1	(B)	1	(B)	14	-	14	-	1
1 to 6 months	108	51	47.2	41	38.2	67	10	57	3	19
7 to 11 months	34	11	(B)	6	(B)	28	5	23	8	5
1 to 2 years	135	84	62.1	74	54.5	61	10	51	2	15
3 to 4 years	111	60	54.5	34	31.1	76	26	50	16	8
5 years or longer	1,087	680	62.5	550	50.6	537	130	407	30	94
Not reported	71	10	(B)	9	(B)	63	1	61	-	51
Renter-occupied units	1,931	822	42.6	564	29.2	1,367	256	1,109	160	205

[Continued]

★ 910 ★

Voting and Registration: Residence and Tenure: Election of November 1992. Part II

[Continued]

Race, Hispanic origin, tenure, and duration of residence	All persons	Reported registered		Reported voted		Reported that they did not vote[1]				
								Not registered		
		Number	Percent	Number	Percent	Total	Registered	Total[2]	Not a U.S. citizen	Do not know and not reported on registration[3]
Duration of residence										
Less than 1 month	94	26	27.9	17	18.3	77	9	68	2	2
1 to 6 months	458	218	47.5	134	29.2	325	84	241	36	29
7 to 11 months	181	83	45.8	50	27.9	130	32	98	6	2
1 to 2 years	477	198	41.5	141	29.6	336	57	279	60	19
3 to 4 years	248	112	45.4	73	29.6	174	39	135	39	24
5 years or longer	359	174	48.5	138	38.6	220	36	185	16	31
Not reported	115	11	9.6	10	8.6	105	1	104	2	98
25 to 44 years										
In all units	9,810	6,083	62.0	5,107	52.1	4,703	977	3,727	622	898
Duration of residence										
Less than 1 month	321	196	61.1	140	43.4	182	57	125	13	36
1 to 6 months	1,367	785	56.6	585	42.2	802	199	603	108	69
7 to 11 months	690	383	55.5	314	45.4	377	70	307	36	49
1 to 2 years	1,980	1,186	59.9	988	49.9	992	198	794	148	100
3 to 4 years	1,469	958	65.2	833	56.7	636	125	511	150	50
5 years or longer	3,471	2,518	72.6	2,201	63.4	1,270	318	953	156	196
Not reported	491	57	11.7	47	9.6	444	11	434	12	398
Owner-occupied units	4,164	2,939	70.6	2,561	61.5	1,603	378	1,225	180	344
Duration of residence										
Less than 1 month	65	48	(B)	41	(B)	23	6	17	2	-
1 to 6 months	227	164	72.4	123	54.4	103	41	63	13	7
7 to 11 months	148	93	62.6	78	52.6	70	15	55	10	9
1 to 2 years	555	338	60.8	306	55.0	250	32	218	49	25
3 to 4 years	492	361	73.4	316	64.7	174	43	131	29	8
5 years or longer	2,462	1,892	76.9	1,662	67.5	800	230	570	72	135
Not reported	215	43	20.2	33	15.3	183	11	172	5	160
Renter-occupied units	5,501	3,062	55.7	2,484	45.2	3,017	578	2,439	441	537
Duration of residence										
Less than 1 month	252	144	57.0	98	39.1	153	45	108	11	36
1 to 6 months	1,138	611	53.7	453	39.8	685	158	527	95	62
7 to 11 months	526	282	53.6	227	43.2	299	55	244	26	40
1 to 2 years	1,411	838	59.4	675	47.8	736	163	573	99	75
3 to 4 years	944	580	61.4	502	53.2	442	78	365	120	38
5 years or longer	961	594	61.8	514	53.5	447	80	367	82	54
Not reported	269	14	5.2	14	5.2	255	-	255	7	231
45 to 64 years										
In all units	5,042	3,651	72.4	3,273	64.9	1,769	378	1,391	170	405
Duration of residence										
Less than 1 month	79	53	67.1	45	57.0	34	8	26	1	3
1 to 6 months	239	135	56.6	111	46.3	128	25	104	11	10
7 to 11 months	128	71	55.7	56	43.8	72	15	57	5	10
1 to 2 years	485	295	60.9	274	56.6	210	21	190	38	34
3 to 4 years	466	329	70.7	276	59.3	190	53	137	29	17
5 years or longer	3,370	2,740	81.3	2,488	73.8	882	252	630	76	109
Not reported	276	27	9.9	23	8.5	253	4	249	8	223
Owner-occupied units	3,185	2,502	78.6	2,289	71.9	896	213	683	65	202
Duration of residence										
Less than 1 month	20	15	(B)	13	(B)	7	2	5	1	-
1 to 6 months	47	22	(B)	20	(B)	27	2	25	-	5

[Continued]

★ 910 ★

Voting and Registration: Residence and Tenure: Election of November 1992. Part II

[Continued]

Race, Hispanic origin, tenure, and duration of residence	All persons	Reported registered		Reported voted		Reported that they did not vote[1]				
								Not registered		
		Number	Percent	Number	Percent	Total	Registered	Total[2]	Not a U.S. citizen	Do not know and not reported on registration[3]
7 to 11 months	35	25	(B)	19	(B)	17	7	10	4	1
1 to 2 years	139	94	67.5	91	65.5	48	3	45	14	11
3 to 4 years	215	149	69.4	122	56.8	93	27	66	6	12
5 years or longer	2,574	2,174	84.5	2,006	77.9	568	168	399	35	61
Not reported	155	22	14.3	18	11.7	137	4	133	5	114
Renter-occupied units	1,789	1,108	61.9	951	53.1	838	157	682	104	192
Duration of residence										
Less than 1 month	59	38	(B)	32	(B)	27	6	21	-	3
1 to 6 months	188	111	59.2	89	47.1	99	23	77	11	5
7 to 11 months	93	46	49.6	37	40.2	55	9	47	1	9
1 to 2 years	338	193	57.2	177	52.5	160	16	144	24	23
3 to 4 years	242	174	71.9	148	61.1	94	26	68	24	5
5 years or longer	756	541	71.5	463	61.2	293	78	215	41	43
Not reported	115	5	4.5	5	4.5	109	-	109	3	104
65 years and over										
In all units	2,644	1,963	74.3	1,696	64.1	948	268	680	33	162
Duration of residence										
Less than 1 month	6	2	(B)	2	(B)	4	-	4	-	-
1 to 6 months	55	28	(B)	16	(B)	39	12	27	2	5
7 to 11 months	32	28	(B)	20	(B)	12	8	4	1	-
1 to 2 years	190	136	71.5	108	57.0	82	28	54	5	6
3 to 4 years	143	85	59.7	73	51.3	70	12	58	8	7
5 years or longer	2,114	1,875	79.2	1,469	69.5	644	206	439	16	54
Not reported	104	9	8.4	6	6.0	98	2	95	-	90
Owner-occupied units	1,733	1,392	80.3	1,232	71.1	500	160	341	14	81
Duration of residence										
Less than 1 month	1	1	(B)	1	(B)	-	-	-	-	-
1 to 6 months	9	5	(B)	5	(B)	4	-	4	-	2
7 to 11 months	10	10	(B)	9	(B)	1	1	-	-	-
1 to 2 years	51	36	(B)	32	(B)	19	4	15	-	3
3 to 4 years	55	38	(B)	36	(B)	20	2	17	3	-
5 years or longer	1,551	1,295	83.5	1,145	73.9	406	149	256	12	30
Not reported	56	7	(B)	5	(B)	52	2	49	-	46
Renter-occupied units	865	551	63.6	446	51.6	419	104	314	18	76
Duration of residence										
Less than 1 month	5	2	(B)	2	(B)	4	-	4	-	-
1 to 6 months	46	23	(B)	11	(B)	35	12	24	2	3
7 to 11 months	22	18	(B)	11	(B)	11	7	4	1	-
1 to 2 years	137	97	71.2	76	55.7	61	21	39	5	3
3 to 4 years	88	48	54.1	38	43.0	50	10	40	6	7
5 years or longer	524	361	68.9	307	58.5	217	55	163	5	21
Not reported	43	2	(B)	2	(B)	41	-	41	-	41
Hispanic origin[4]										
18 years and over										
In all units	14,688	5,137	35.0	4,238	28.9	10,450	899	9,551	5,910	995
Duration of residence										
Less than 1 month	441	108	24.4	68	15.4	373	40	333	232	9
1 to 6 months	2,370	575	24.2	415	17.5	1,958	160	1,796	1,223	51
7 to 11 months	871	242	27.8	190	21.9	680	52	628	435	29
1 to 2 years	2,979	789	26.5	648	21.7	2,331	141	2,190	1,470	86

[Continued]

★ 910 ★

Voting and Registration: Residence and Tenure: Election of November 1992. Part II

[Continued]

Race, Hispanic origin, tenure, and duration of residence	All persons	Reported registered		Reported voted		Reported that they did not vote[1]				
								Not registered		
		Number	Percent	Number	Percent	Total	Registered	Total[2]	Not a U.S. citizen	Do not know and not reported on registration[3]
3 to 4 years	1,930	662	34.3	588	30.4	1,344	76	1,267	864	71
5 years or longer	5,370	2,723	50.7	2,294	42.7	3,076	428	2,647	1,586	192
Not reported	728	39	5.3	37	5.0	691	2	689	101	557
Owner-occupied units	6,523	3,044	46.7	2,588	39.7	3,935	455	3,479	1,898	491
Duration of residence										
Less than 1 month	44	28	(B)	24	(B)	21	4	16	11	-
1 to 6 months	394	155	39.3	112	28.5	282	43	239	141	9
7 to 11 months	203	81	39.8	77	38.2	125	3	122	74	8
1 to 2 years	864	258	29.9	218	25.2	646	40	606	376	31
3 to 4 years	862	361	41.8	317	36.7	545	44	502	325	40
5 years or longer	3,823	2,139	55.9	1,819	47.6	2,004	319	1,685	936	144
Not reported	332	23	7.0	21	6.4	311	2	309	36	262
Renter-occupied units	7,914	2,048	25.9	1,813	20.4	6,301	435	5,865	3,855	501
Duration of residence										
Less than 1 month	382	78	20.4	44	11.6	338	34	304	211	9
1 to 6 months	1,917	413	21.5	298	15.5	1,620	116	1,504	1,037	41
7 to 11 months	662	159	24.0	113	17.0	549	46	503	360	23
1 to 2 years	2,051	517	25.2	416	20.3	1,635	101	1,533	1,052	55
3 to 4 years	1,051	297	28.3	266	25.3	785	31	754	529	31
5 years or longer	1,475	569	38.6	462	31.3	1,013	107	906	615	48
Not reported	375	14	3.8	14	3.8	361	-	361	52	294

Source: "Reported Voting and Registration, by Race, Hispanic Origin, Age, Duration of Residence, and Tenure," U.S. Bureau of the Census, *Voting and Registration in the Election of November 1992*, pp. 62-64. *Notes:* 1. Includes persons reported as "did not vote," "do not know," and "not reported" on voting. 2. In addition to those reported as "not registered," total includes those "not a U.S. citizen," and "do not know" and "not reported" on registration. 3. Includes "do not know" and "not reported" on citizenship. 4. Persons of Hispanic origin may be of any race.

★ 911 ★

Voters and Voting

Voting and Registration: Residence and Tenure: Election of November 1992. Part III

November 1992. Numbers in thousands.

Race, Hispanic origin, tenure, and duration of residence	All persons	Reported registered		Reported voted		Reported that they did not vote[1]				
								Not registered		
		Number	Percent	Number	Percent	Total	Registered	Total[2]	Not a U.S. citizen	Do not know and not reported on registration[3]
Hispanic origin[4]										
18 years and over										
In all units	14,688	5,137	35.0	4,238	28.9	10,450	899	9,551	5,910	995
Duration of residence										
Less than 1 month	441	108	24.4	68	15.4	373	40	333	232	9
1 to 6 months	2,370	575	24.2	415	17.5	1,956	160	1,796	1,223	51
7 to 11 months	871	242	27.8	190	21.9	680	52	628	435	29
1 to 2 years	2,979	789	26.5	648	21.7	2,331	141	2,190	1,470	86
3 to 4 years	1,930	662	34.3	586	30.4	1,344	76	1,267	864	71
5 years or longer	5,370	2,723	50.7	2,294	42.7	3,076	428	2,647	1,586	192

[Continued]

★ 911 ★

Voting and Registration: Residence and Tenure: Election of November 1992. Part III

[Continued]

Race, Hispanic origin, tenure, and duration of residence	All persons	Reported registered		Reported voted		Reported that they did not vote[1]				
								Not registered		
		Number	Percent	Number	Percent	Total	Registered	Total[2]	Not a U.S. citizen	Do not know and not reported on registration[3]
Not reported	728	39	5.3	37	5.0	691	2	689	101	557
Owner-occupied units	6,523	3,044	46.7	2,588	39.7	3,935	455	3,479	1,898	491
Duration of residence										
Less than 1 month	44	28	(B)	24	(B)	21	4	16	11	-
1 to 6 months	394	155	39.3	112	28.5	282	43	239	141	9
7 to 11 months	203	81	39.8	77	38.2	125	3	122	74	6
1 to 2 years	864	258	29.9	218	25.2	646	40	606	376	31
3 to 4 years	862	361	41.8	317	36.7	545	44	502	325	40
5 years or longer	3,823	2,139	55.9	1,819	47.6	2,004	319	1,685	936	144
Not reported	332	23	7.0	21	6.4	311	2	309	36	262
Renter-occupied units	7,914	2,048	25.9	1,613	20.4	6,301	435	5,865	3,855	501
Duration of residence										
Less than 1 month	382	78	20.4	44	11.6	338	34	304	211	9
1 to 6 months	1,917	413	21.5	298	15.5	1,620	116	1,504	1,037	41
7 to 11 months	662	159	24.0	113	17.0	549	46	503	360	23
1 to 2 years	2,051	517	25.2	416	20.3	1,635	101	1,533	1,052	55
3 to 4 years	1,051	297	28.3	266	25.3	785	31	754	529	31
5 years or longer	1,475	569	38.6	462	31.3	1,013	107	906	615	48
Not reported	375	14	3.8	14	3.8	361	-	361	52	294
18 to 24 years										
In all units	2,795	696	24.9	492	17.6	2,303	204	2,099	1,206	213
Duration of residence										
Less than 1 month	164	22	13.7	10	6.3	153	12	141	92	1
1 to 6 months	669	130	19.4	99	14.8	570	31	539	327	10
7 to 11 months	197	53	26.8	30	15.1	167	23	144	102	9
1 to 2 years	626	101	16.2	61	9.7	566	41	525	369	15
3 to 4 years	269	67	25.0	53	19.8	216	14	202	122	20
5 years or longer	728	315	43.3	232	31.9	496	83	413	162	55
Not reported	142	7	5.1	7	5.1	135	-	135	31	101
Owner-occupied units	948	336	35.4	250	26.4	698	86	612	289	103
Duration of residence										
Less than 1 month	6	4	(B)	4	(B)	2	-	2	2	-
1 to 6 months	82	20	24.4	17	21.2	64	3	62	33	1
7 to 11 months	41	17	(B)	15	(B)	27	2	24	14	2
1 to 2 years	147	25	16.9	13	8.5	135	12	122	83	10
3 to 4 years	124	35	28.7	27	21.5	97	9	88	54	11
5 years or longer	497	231	46.6	172	34.6	325	60	265	94	41
Not reported	52	3	(B)	3	(B)	48	-	48	10	39
Renter-occupied units	1,802	351	19.5	241	13.4	1,560	109	1,451	892	109
Duration of residence										
Less than 1 month	152	17	11.0	6	4.1	146	10	136	86	1
1 to 6 months	580	108	18.6	82	14.1	498	26	472	291	9
7 to 11 months	152	33	22.0	15	9.8	137	18	118	88	8
1 to 2 years	471	76	16.1	48	10.1	424	28	395	279	5
3 to 4 years	140	30	21.8	27	19.1	113	4	109	65	9
5 years or longer	223	83	37.0	60	27.0	163	22	141	66	14
Not reported	84	4	4.7	4	4.7	80	-	80	17	62
25 to 44 years										
In all units	7,499	2,399	32.0	1,979	26.4	5,520	420	5,100	3,283	488

[Continued]

★ 911 ★

Voting and Registration: Residence and Tenure: Election of November 1992. Part III
[Continued]

Race, Hispanic origin, tenure, and duration of residence	All persons	Reported registered		Reported voted		Reported that they did not vote[1]				
								Not registered		
		Number	Percent	Number	Percent	Total	Registered	Total[2]	Not a U.S. citizen	Do not know and not reported on registration[3]
Duration of residence										
Less than 1 month	218	67	30.8	43	19.5	175	25	151	112	8
1 to 6 months	1,364	343	25.2	237	17.4	1,126	106	1,020	720	30
7 to 11 months	535	147	27.4	125	23.4	410	21	388	267	12
1 to 2 years	1,839	487	26.5	422	23.0	1,417	65	1,352	881	58
3 to 4 years	1,197	411	34.3	364	30.4	833	46	786	542	40
5 years or longer	1,990	922	46.3	765	38.4	1,225	157	1,068	711	55
Not reported	357	23	6.3	23	6.3	335	-	335	49	285
Owner-occupied units	2,943	1,271	43.2	1,079	36.6	1,865	192	1,672	940	213
Duration of residence										
Less than 1 month	30	15	(B)	11	(B)	19	4	15	9	-
1 to 6 months	242	104	43.0	72	29.9	170	32	138	85	2
7 to 11 months	112	44	39.8	43	38.7	68	1	67	37	3
1 to 2 years	551	167	30.3	153	27.7	398	14	384	219	21
3 to 4 years	551	234	42.5	208	37.7	343	27	317	201	28
5 years or longer	1,318	696	52.8	581	44.1	737	115	622	376	47
Not reported	140	11	7.9	11	7.9	129	-	129	15	113
Renter-occupied units	4,396	1,108	25.2	881	20.0	3,516	228	3,288	2,227	273
Duration of residence										
Less than 1 month	180	52	29.1	32	17.8	148	20	128	96	8
1 to 6 months	1,070	235	21.9	160	15.0	910	74	835	594	27
7 to 11 months	422	102	24.2	82	19.5	340	20	320	229	10
1 to 2 years	1,242	315	25.3	264	21.3	978	51	927	631	37
3 to 4 years	639	175	27.5	156	24.4	483	20	463	335	12
5 years or longer	637	219	34.3	177	27.7	460	42	418	316	8
Not reported	208	10	5.0	10	5.0	197	-	197	26	172
45 to 64 years										
In all units	3,211	1,475	45.9	1,297	40.4	1,914	178	1,736	1,078	215
Duration of residence										
Less than 1 month	54	18	(B)	15	(B)	39	3	36	25	-
1 to 6 months	281	88	31.5	70	24.7	212	19	193	143	11
7 to 11 months	110	37	34.0	32	29.1	78	5	73	48	7
1 to 2 years	390	143	36.7	122	31.2	268	22	247	178	8
3 to 4 years	363	139	38.4	123	34.0	239	16	223	157	10
5 years or longer	1,844	1,040	56.4	928	50.3	916	112	805	511	59
Not reported	168	9	5.2	7	4.1	161	2	159	17	120
Owner-occupied units	1,925	1,030	53.5	917	47.6	1,009	113	896	535	121
Duration of residence										
Less than 1 month	9	9	(B)	9	(B)	-	-	-	-	-
1 to 6 months	65	27	(B)	21	(B)	45	6	39	23	6
7 to 11 months	40	17	(B)	17	(B)	22	-	22	16	2
1 to 2 years	120	48	40.3	37	31.0	83	11	72	56	-
3 to 4 years	140	67	47.7	58	41.8	81	8	73	54	1
5 years or longer	1,448	853	58.9	767	53.0	681	86	596	374	38
Not reported	103	9	8.5	7	6.6	96	2	94	12	73
Renter-occupied units	1,254	431	34.3	366	29.2	888	65	823	533	93
Duration of residence										
Less than 1 month	45	9	(B)	6	(B)	39	3	36	25	-
1 to 6 months	216	62	28.6	49	22.7	167	13	154	120	5
7 to 11 months	70	20	(B)	15	(B)	56	5	50	31	5

[Continued]

★ 911 ★

Voting and Registration: Residence and Tenure: Election of November 1992. Part III
[Continued]

Race, Hispanic origin, tenure, and duration of residence	All persons	Reported registered		Reported voted		Reported that they did not vote[1]				
								Not registered		
		Number	Percent	Number	Percent	Total	Registered	Total[2]	Not a U.S. citizen	Do not know and not reported on registration[3]
1 to 2 years	259	87	33.6	77	29.6	182	10	172	119	8
3 to 4 years	222	73	32.8	65	29.3	157	8	149	101	8
5 years or longer	382	180	47.3	155	40.5	227	26	201	131	20
Not reported	60	-	(B)	-	(B)	60	-	60	5	47
65 years and over										
In all units	1,184	567	47.9	470	39.7	714	97	617	344	80
Duration of residence										
Less than 1 month	5	-	(B)	-	(B)	5	-	5	3	-
1 to 6 months	57	13	(B)	9	(B)	48	4	44	34	-
7 to 11 months	28	5	(B)	3	(B)	25	2	23	18	-
1 to 2 years	124	58	46.5	43	34.8	81	15	66	41	5
3 to 4 years	101	45	44.5	45	44.5	56	-	56	43	1
5 years or longer	808	446	55.2	369	45.7	438	76	362	201	23
Not reported	61	-	(B)	-	(B)	61	-	61	4	51
Owner-occupied units	706	407	57.6	343	48.6	363	64	299	134	54
Duration of residence										
Less than 1 month	-	-	(B)	-	(B)	-	-	-	-	-
1 to 6 months	5	4	(B)	2	(B)	3	2	1	1	-
7 to 11 months	10	2	(B)	2	(B)	8	-	8	7	-
1 to 2 years	45	18	(B)	15	(B)	30	2	28	18	-
3 to 4 years	48	24	(B)	24	(B)	24	-	24	16	-
5 years or longer	560	359	64.0	299	53.4	261	59	202	93	17
Not reported	37	-	(B)	-	(B)	37	-	37	-	37
Renter-occupied units	461	159	34.4	125	27.1	336	34	303	203	25
Duration of residence										
Less than 1 month	5	-	(B)	-	(B)	5	-	5	3	-
1 to 6 months	52	9	(B)	7	(B)	45	2	43	33	-
7 to 11 months	19	3	(B)	1	(B)	17	2	15	11	-
1 to 2 years	79	40	51.0	28	35.4	51	12	38	23	5
3 to 4 years	51	19	(B)	19	(B)	33	-	33	27	1
5 years or longer	233	87	37.5	70	30.1	163	17	146	101	6
Not reported	23	-	(B)	-	(B)	23	-	23	4	14

Source: "Reported Voting and Registration, by Race, Hispanic Origin, Age, Duration of Residence, and Tenure," U.S. Bureau of the Census, *Voting and Registration in the Election of November 1992,* pp. 64-66. *Notes:* 1. Includes persons reported as "did not vote," "do not know," and "not reported" on voting. 2. In addition to those reported as "not registered," total includes those "not a U.S. citizen," and "do not know" and "not reported" on registration. 3. Includes "do not know" and "not reported" on citizenship. 4. Persons of Hispanic origin may be of any race.

★ 912 ★

Voters and Voting

Voting in Presidential Election Years: November 1964 to 1992

Numbers in thousands. Civilian noninstitutional population.

Region, race, Hispanic origin, sex, and age	Presidential elections of--							
	1964	1968	1972	1976	1980	1984	1988	1992
United States								
Total, voting age	110,604	116,535	136,203	146,548	157,085	169,963	178,098	185,684
Percent voted	69.3	67.8	63.0	59.2	59.2	59.9	57.4	61.3
White	70.7	69.1	64.5	60.9	60.9	61.4	59.1	63.6
Black	58.5[2]	57.6	52.1	48.7	50.5	55.8	51.5	54.0
Hispanic origin[1]	(NA)	(NA)	37.5	31.8	29.9	32.6	28.8	28.9
Male	71.9	69.8	64.1	59.6	59.1	59.0	56.4	60.2
Female	67.0	66.0	62.0	58.8	59.4	60.8	58.3	62.3
18 to 24 years	50.9[3]	50.4[3]	49.6	42.2	39.9	40.8	36.2	42.8
25 to 44 years	69.0	66.6	62.7	58.7	58.7	58.4	54.0	58.3
45 to 64 years	75.9	74.9	70.8	68.7	69.3	69.8	67.9	70.0
65 years and over	66.3	65.8	63.5	62.2	65.1	67.7	68.8	70.1
North and West								
Total, voting age	78,174	81,594	93,653	99,403	106,524	112,376	117,373	122,025
Percent voted	74.6	71.0	66.4	61.2	61.0	61.6	58.9	62.5
White	74.7	71.8	67.5	62.6	62.4	63.0	60.4	64.9
Black	72.0[2]	64.8	56.7	52.2	52.8	58.9	55.6	53.8
South								
Total, voting age	32,429	34,941	42,550	47,145	50,561	57,587	60,725	63,659
Percent voted	56.7	60.1	55.4	54.9	55.6	56.8	54.5	59.0
White	59.5	61.9	57.0	57.1	57.4	58.1	56.4	60.8
Black	44.0[2]	51.6	47.8	45.7	48.2	53.2	48.0	54.3

Source: "Reported Voting and Registration in Presidential Election Years, by Region, Race, Hispanic Origin, Sex, and Age: November 1964 to 1992," U.S. Bureau of the Census, *Voting and Registration in the Election of November 1992,* p. v. Primary source: Current Population Reports, Series P-20, Nos. 174, 228, 293, 344, 383, 414, 440, and table 2 of this report. *Notes:* NA Not available. 1. Persons of Hispanic origin may be of any race. 2. Black and other races in 1964. 3. Prior to 1972, includes persons 18 to 20 years old in Georgia and Kentucky, 19 and 20 in Alaska, and 20 years old in Hawaii.

★ 913 ★

Voters and Voting

Voting-Age Population, Registered and Voting: 1978 to 1992 - I

As of November. Covers civilian noninstitutional population 18 years old and over. Includes aliens. Figures are based on Current Population Survey (see text, section 1, and Appendix III) and differ from those in table 455 based on population estimates and official vote counts.

Characteristic	Voting-age population (mil.)								Percent reporting they registered Presidential election years			
	1978	1980	1982	1984	1986	1988	1990	1992	1980	1984	1988	1992
White	133.4	137.7	143.6	146.8	149.9	152.9	155.6	157.8	68.4	69.6	67.9	70.1
Black	15.6	16.4	17.6	18.4	19.0	19.7	20.4	21.0	60.0	66.3	64.5	63.9
Hispanic[2]	6.8	8.2	8.8	9.5	11.8	12.9	13.8	14.7	36.3	40.1	35.5	35.0

Source: "Voting-Age Population, Percent Reported Registered, and Voted: 1978 to 1992," U.S. Bureau of the Census, *Statistical Abstract of the United States*, 1993, p. 283. Primary source: U.S. Bureau of the Census, *Current Population Reports*, P20-466, and earlier reports. *Notes:* 1. Includes other races not shown separately. 2. Hispanic persons may be of any race. 3. For composition of regions, see table 31. 4. Represents those who completed ninth to twelfth grade, but have no high school diploma. 5. High school graduate. 6. Some college or associate degree. 7. Bachelor's or advance degree.

★ 914 ★

Voters and Voting

Voting-Age Population, Registered and Voting: 1978 to 1992 - II

As of November. Covers civilian noninstitutional population 18 years old and over. Includes aliens. Figures are based on Current Population Survey (see text, section 1, and Appendix III) and differ from those in table 455 based on population estimates and official vote counts.

Characteristic	Percent reporting they registered Congressional election years				Percent reporting they voted							
					Presidential election years				Congressional election years			
	1978	1982	1986	1990	1980	1984	1988	1992	1978	1982	1986	1990
White	63.8	65.6	65.3	63.8	60.9	61.4	59.1	63.6	47.3	49.9	47.0	46.7
Black	57.1	59.1	64.0	58.8	50.5	55.8	51.5	54.0	37.2	43.0	43.2	39.2
Hispanic[2]	32.9	35.3	35.9	32.3	29.9	32.6	28.8	28.9	23.5	25.3	24.2	21.0

Source: "Voting-Age Population, Percent Reported Registered, and Voted: 1978 to 1992," U.S. Bureau of the Census, *Statistical Abstract of the United States*, 1993, p. 283. Primary source: U.S. Bureau of the Census, *Current Population Reports*, P20-466, and earlier reports. *Notes:* 1. Includes other races not shown separately. 2. Hispanic persons may be of any race. 3. For composition of regions, see table 31. 4. Represents those who completed ninth to twelfth grade, but have no high school diploma. 5. High school graduate. 6. Some college or associate degree. 7. Bachelor's or advance degree.

Chapter 13
POPULATION

Ancestry Groups

★ 915 ★

Leading Ancestry Groups, Foreign Born, 1990

Region, Division, and State	Foreign born population			Ancestry groups						Person 5 years old and over speaking foreign language	
	Number (1,000)	Percent of total population	Year of entry 1980-90 (1,000)	Leading ancestry group		Second leading group		Third leading group		Number (1,000)	Percent of total population
				Group	Number (1,000)	Group	Number (1,000)	Group	Number (1,000)		
U.S.	19,767	7.9	8,661	German	57,947	Irish	38,736	English	32,652	31,845	13.8
Northeast	5,231	10.3	2,059	German	9,929	Irish	9,420	Italian	7,504	7,824	16.5
N.E.	1,043	7.9	368	Irish	2,949	English	2,330	French	1,591	1,703	13.9
ME	36	3.0	7	English	372	French	224	Irish	217	105	9.2
NH	41	3.7	10	English	266	Irish	232	French	205	89	8.7
VT	18	3.1	3	English	147	French	133	Irish	101	30	5.8
MA	574	9.5	223	Irish	1,571	English	921	Italian	844	852	15.2
RI	95	9.5	35	Irish	214	Italian	199	English	161	159	17.0
CT	279	8.5	90	Italian	628	Irish	614	English	463	466	15.2
M.A.	4,188	11.1	1,691	German	8,622	Irish	6,471	Italian	5,668	6,122	17.5
NY	2,852	15.9	1,190	German	2,899	Italian	2,838	Irish	2,800	3,909	23.3
NJ	967	12.5	385	Italian	1,457	Irish	1,415	German	1,408	1,406	19.5
PA	369	3.1	116	German	4,315	Irish	2,256	Italian	1,373	807	7.3
Midwest	2,131	3.6	754	German	22,477	Irish	9,643	English	7,294	3,921	7.1
E.N.C.	1,783	4.2	607	German	14,776	Irish	6,655	English	5,083	3,125	8.0
OH	260	2.4	71	German	4,068	Irish	1,896	English	1,449	546	5.4
IN	94	1.7	31	German	2,085	Irish	965	English	767	246	4.8
IL	952	8.3	371	German	3,326	Irish	1,861	Afro American	1,426	1,499	14.2
MI	355	3.8	94	German	2,666	Irish	1,320	English	1,315	570	6.6
WI	122	2.5	41	German	2,631	Irish	612	Polish	506	264	5.8
W.N.C.	348	2.0	147	German	7,702	Irish	2,988	English	2,211	796	4.9
MN	113	2.6	51	German	2,021	Norwegian	757	Irish	574	227	5.6
IA	43	1.6	19	German	1,395	Irish	527	English	389	100	3.9
MO	84	1.6	30	German	1,843	Irish	1,038	English	743	178	3.8
ND	9	1.5	3	German	325	Norwegian	189	Irish	54	47	7.9
SD	8	1.1	2	German	355	Norwegian	106	Irish	88	42	6.5
NE	28	1.8	10	German	795	Irish	272	English	209	70	4.8
KS	63	2.5	31	German	968	Irish	436	English	406	132	5.7
South	4,582	5.4	2,027	German	14,630	Irish	12,951	Afro American	12,936	8,670	10.9
S.A.	2,723	6.3	1,169	German	7,882	Irish	6,377	English	6,307	3,708	9.1
DE	22	3.3	4	Irish	139	German	18	English	123	42	6.9
MD	313	6.6	148	German	1,218	Afro American	966	Irish	769	395	8.9
DC	59	9.7	34	Afro American	315	German	39	Irish	34	71	12.5
VA	312	5.0	159	German	1,186	English	1,051	Afro American	970	419	7.3
WV	16	0.9	4	German	469	Irish	348	English	270	44	2.6
NC	115	1.7	52	Afro American	1,228	German	1,111	English	987	241	3.9
SC	50	1.4	18	Afro American	870	German	500	Irish	486	113	3.5
GA	173	2.7	90	Afro American	1,421	Irish	971	English	890	285	4.8
FL	1,663	12.9	660	German	2,410	Irish	1,899	English	1,846	2,098	17.3
E.S.C.	157	1.0	65	Irish	2,581	German	2,177	English	1,978	392	2.8
KY	34	0.9	14	German	798	Irish	696	American	586	86	2.5
TN	59	1.2	26	Irish	875	German	724	English	692	132	2.9
AL	44	1.1	18	Afro American	839	American	687	Irish	617	108	2.9
MS	20	0.8	8	Afro American	775	Irish	393	American	317	67	2.8
W.S.C.	1,702	6.4	793	German	4,572	Irish	3,993	English	3,091	4,569	18.6
AR	25	1.1	10	Irish	464	German	400	Afro American	307	61	2.8
LA	87	2.1	35	Afro American	1,097	French	550	Irish	518	392	10.1
OK	65	2.1	30	German	714	Irish	642	American Indian	469	146	5.0

[Continued]

★ 915 ★

Leading Ancestry Groups, Foreign Born, 1990

[Continued]

Region, Division, and State	Foreign born population			Ancestry groups						Person 5 years old and over speaking foreign language	
	Number (1,000)	Percent of total population	Year of entry 1980-90 (1,000)	Leading ancestry group		Second leading group		Third leading group		Number (1,000)	Percent of total population
				Group	Number (1,000)	Group	Number (1,000)	Group	Number (1,000)		
TX	1,524	9.0	718	Mexican	3,403	German	2,950	Irish	2,369	3,970	25.4
West	7,823	14.8	3,821	German	10,911	English	8,110	Irish	6,721	11,430	23.5
Mt	715	5.2	298	German	3,477	English	2,843	Irish	1,921	1,901	15.1
MT	14	1.7	3	German	285	Irish	139	English	137	37	5.0
ID	29	2.9	13	English	291	German	279	Irish	142	59	6.4
WY	8	1.7	2	German	158	English	101	Irish	73	24	5.7
CO	142	4.3	57	German	1,064	English	582	Irish	538	321	10.5
NM	81	5.3	31	German	234	Mexican	216	Spanish	191	494	35.5
AZ	278	7.6	117	German	878	English	586	Irish	530	700	20.8
UT	59	3.4	26	English	750	German	299	Danish	163	120	7.8
NV	105	8.7	48	German	280	English	207	Irish	200	146	13.2
Pac	7,108	18.2	3,523	German	7,433	English	5,267	Irish	4,800	9,529	26.4
WA	322	6.6	129	German	1,390	English	897	Irish	768	403	9.0
OR	139	4.9	61	German	879	English	575	Irish	467	192	7.3
CA	6,459	21.7	3,256	Mexican	5,322	German	4,935	English	3,646	8,619	31.5
AK	25	4.5	11	German	127	English	77	Irish	74	60	12.1
HI	163	14.7	67	Japanese	262	Filipino	176	Hawaiian	157	255	24.8

Source: "Foreign-Born Population, Leading ANcestry Groups, and Persons Speaking a Language Other Than English At Home—States: 1990," U.S. Bureau of the Census, *Statistical Abstract of the United States*, 1993, p. 52. Primary source: U.S. Bureau of the Census, *1990 Census of Population and Housing, Summary Social, Economic, and Housing Characteristics, United States* (1990 CPH-5-1) and *1990 Census of Population, Supplementary Reports, Detailed Ancestry Groups for States* (1990 CP-S-1-2).

★ 916 ★

Ancestry Groups

Selected Ancestry Groups, by Region, 1990

As of April 1. Covers persons who reported single and multiple ancestry groups. Persons who reported a multiple ancestry group may be included in more than one category. Major classifications of ancestry groups do not represent strict geographic or cultural definitions. Based on a sample and subject to sampling variability.

Ancestry group	Total (1,000)	Percent distribution by region			
		Northeast	Midwest	South	West
North America					
Acadian	668	1	2	91	5
Afro-American	23,777	15	21	54	10
American Indian	8,708	9	22	47	23
American	12,396	10	18	61	11
Canadian	550	34	18	21	28
French Canadian	2,167	45	20	20	15
United States	644	16	18	53	13
White	1,800	7	13	53	28

Source: "Population, by Selected Ancestry Group and Region: 1990," U.S. Bureau of the Census, *Statistical Abstract of the United States*, 1993, p. 51. Primary source: U.S. bureau of the Census, *1990 Census of Population, Supplementary Reports, detailed Ancestry Groups for States* (1990 CP-S-1-2).

Characteristics

★ 917 ★

Social Characteristics of Population by Gender, Region, and Race, March 1992. Part I.

Numbers in thousands.

Characteristic	All races			Black			White		
	Both sexes	Male	Female	Both sexes	Male	Female	Both sexes	Male	Female
United States									
Age									
Total	251,447	122,528	128,919	31,439	14,781	16,658	210,257	102,965	107,293
Percent	100.0	100.0	100.0	100.0	100.0	100.0	100.0	100.0	100.0
Under 5 years	7.8	8.2	7.4	10.1	11.0	9.3	7.4	7.7	7.0
5 to 9 years	7.4	7.8	7.0	9.1	9.8	8.4	7.1	7.4	6.7
10 to 14 years	7.2	7.5	6.8	9.1	9.9	8.5	6.8	7.1	6.5
15 to 19 years	6.6	6.8	6.4	8.3	8.8	7.9	6.3	6.5	6.1
20 to 24 years	7.1	7.2	7.0	7.9	7.8	8.0	7.0	7.1	6.9
25 to 29 years	8.0	8.2	7.8	8.5	8.4	8.6	7.9	8.1	7.7
30 to 34 years	8.9	9.1	8.7	8.8	8.6	8.9	8.9	9.1	8.7
35 to 44 years	15.7	15.9	15.6	14.2	13.7	14.6	15.9	16.3	15.6
45 to 54 years	10.7	10.7	10.8	8.9	8.4	9.3	11.1	11.1	11.0
55 to 64 years	8.4	8.2	8.6	6.9	6.6	7.1	8.7	8.5	8.9
65 to 74 years	7.3	6.7	7.9	5.3	5.0	5.6	7.8	7.1	8.4
75 years and over	4.8	3.7	5.9	3.0	2.2	3.7	5.2	4.0	6.4
16 years and over	76.3	75.1	77.4	70.0	67.6	72.2	77.5	76.4	78.5
18 years and over	73.7	72.4	74.9	66.7	64.0	69.1	75.0	73.8	76.1
21 years and over	69.7	68.4	71.0	61.8	59.0	64.3	71.2	70.0	72.3
55 years and over	20.6	18.6	22.4	15.2	13.8	16.4	21.7	19.6	23.7
65 years and over	12.2	10.4	13.8	8.3	7.2	9.3	13.0	11.1	14.8
Median age (years)	33.3	32.4	34.3	28.2	26.7	29.6	34.3	33.3	35.2
Marital status									
Total, 15 years and over	195,256	93,760	101,496	22,542	10,252	12,290	165,571	80,049	85,522
Percent	100.0	100.0	100.0	100.0	100.0	100.0	100.0	100.0	100.0
Never married	26.5	30.2	23.1	41.8	45.0	39.1	24.2	28.1	20.6
Married, spouse present	54.8	57.1	52.7	32.4	36.1	29.4	57.9	59.9	56.0
Married, spouse absent	3.3	2.8	3.6	7.6	6.4	8.6	2.6	2.3	2.9
Widowed	7.1	2.7	11.2	8.2	4.2	11.5	7.0	2.5	11.3
Divorced	8.4	7.2	9.4	10.1	8.4	11.5	8.2	7.2	9.2
Educational attainment									
Total, 25 to 34 years old	42,496	21,124	21,371	5,423	2,505	2,918	35,320	17,736	17,584
Percent completed									
Less than 9th grade	4.0	4.3	3.6	2.6	2.7	2.5	4.1	4.5	3.8
High school graduate or more	86.5	85.9	87.0	81.8	82.2	81.4	87.1	86.3	87.8
Some college or associate degree	25.6	24.2	26.9	26.1	23.6	28.3	25.6	24.3	26.8

[Continued]

★ 917 ★

Social Characteristics of Population by Gender, Region, and Race, March 1992. Part I.
[Continued]

Characteristic	All races			Black			White		
	Both sexes	Male	Female	Both sexes	Male	Female	Both sexes	Male	Female
Bachelor's degree or more	23.2	23.3	23.1	12.0	12.2	11.8	24.2	24.1	24.3
Type of family									
All families	67,175	(X)	(X)	7,716	(X)	(X)	57,225	(X)	(X)
Percent	100.0	(X)	(X)	100.0	(X)	(X)	100.0	(X)	(X)
Married couple	78.1	(X)	(X)	47.1	(X)	(X)	82.3	(X)	(X)
Female householder, no spouse present	17.4	(X)	(X)	46.4	(X)	(X)	13.5	(X)	(X)
Male householder, no spouse present	4.5	(X)	(X)	6.5	(X)	(X)	4.1	(X)	(X)
Families with householder 55 years and over									
All families	20,432	(X)	(X)	1,912	(X)	(X)	18,015	(X)	(X)
Percent	100.0	(X)	(X)	100.0	(X)	(X)	100.0	(X)	(X)
Married couple	82.9	(X)	(X)	55.5	(X)	(X)	86.0	(X)	(X)
Female householder, no spouse present	13.5	(X)	(X)	36.2	(X)	(X)	11.0	(X)	(X)
Male householder, no spouse present	3.6	(X)	(X)	8.4	(X)	(X)	3.0	(X)	(X)

Source: "Selected Social Characteristics of the Population, by Sex, Region, and Race: March 1992." Claudette E. Bennett, *The Black Population in the United States: March 1992*, p. 22.

★ 918 ★
Characteristics

Social Characteristics of Population by Gender, Region, and Race, March 1992. Part II.

Numbers in thousands.

Characteristic	All races			Black			White		
	Both sexes	Male	Female	Both sexes	Male	Female	Both sexes	Male	Female
South									
Age									
Total	86,004	41,667	44,337	17,090	8,036	9,054	67,256	32,800	34,456
Percent	100.0	100.0	100.0	100.0	100.0	100.0	100.0	100.0	100.0
Under 5 years	7.4	7.8	7.1	9.7	10.6	8.9	6.8	7.0	6.6
5 to 9 years	7.3	7.7	6.9	9.2	9.6	8.8	6.8	7.2	6.4
10 to 14 years	7.3	7.8	6.9	9.4	10.6	8.3	6.8	7.1	6.6
15 to 19 years	6.6	7.0	6.3	8.5	9.1	8.0	6.1	6.4	5.8
20 to 24 years	7.3	7.4	7.2	7.9	8.1	7.8	7.1	7.2	7.1
25 to 29 years	8.0	8.1	7.8	8.4	7.9	8.7	7.8	8.1	7.6
30 to 34 years	8.8	9.0	8.6	8.3	8.4	8.3	8.9	9.1	8.7
35 to 44 years	15.5	15.6	15.3	14.6	14.1	15.0	15.7	16.1	15.3
45 to 54 years	10.7	10.7	10.8	8.4	7.5	9.2	11.3	11.5	11.2
55 to 64 years	8.6	8.3	8.8	6.9	6.5	7.2	9.1	8.9	9.3
65 to 74 years	7.5	6.8	8.2	5.4	5.1	5.7	8.1	7.4	8.9

[Continued]

★ 918 ★

Social Characteristics of Population by Gender, Region, and Race, March 1992. Part II.
[Continued]

Characteristic	All races			Black			White		
	Both sexes	Male	Female	Both sexes	Male	Female	Both sexes	Male	Female
75 years and over	4.9	3.8	5.9	3.3	2.4	4.0	5.4	4.2	6.5
16 years and over	76.6	75.4	77.8	70.0	67.4	72.4	78.3	77.4	79.3
18 years and over	74.0	72.5	75.4	66.6	63.5	69.3	76.0	74.8	77.1
21 years and over	69.9	68.4	71.2	61.5	58.3	64.4	72.1	71.1	73.0
55 years and over	21.0	19.0	22.9	15.6	14.0	16.9	22.6	20.4	24.7
65 years and over	12.4	10.7	14.1	8.7	7.5	9.8	13.5	11.6	15.4
Median age (years)	33.4	32.4	34.5	28.1	26.2	29.7	34.8	33.9	35.8
Marital status									
Total, 15 years and over	67,023	31,969	35,054	12,251	5,550	6,701	53,507	25,802	27,705
Percent	100.0	100.0	100.0	100.0	100.0	100.0	100.0	100.0	100.0
Never married	24.5	28.0	21.2	40.0	44.0	36.8	20.8	24.5	17.5
Married, spouse present	55.7	58.4	53.3	34.3	38.0	31.2	60.6	62.9	58.5
Married, spouse absent	3.6	3.2	4.0	7.4	6.4	8.2	2.8	2.5	3.0
Widowed	7.5	2.7	11.9	8.5	4.2	12.0	7.3	2.4	11.9
Divorced	8.7	7.6	9.6	9.8	7.4	11.8	8.4	7.7	9.1
Educational attainment									
Total, 25 to 34 years old	14,426	7,118	7,308	2,852	1,310	1,542	11,233	5,620	5,613
Percent completed									
Less than 9th grade	4.1	4.7	3.5	2.5	2.7	2.3	4.5	5.1	3.8
High school graduate or more	85.0	84.1	85.9	82.6	83.5	81.8	85.5	84.2	86.8
Some college or associate degree	24.9	24.3	25.4	24.3	22.4	25.9	25.3	25.1	25.5
Bachelor's degree or more	21.0	19.8	22.1	11.1	10.0	12.0	23.0	21.4	24.6
Type of family									
All families	23,679	(X)	(X)	4,253	(X)	(X)	19,023	(X)	(X)
Percent	100.0	(X)	(X)	100.0	(X)	(X)	100.0	(X)	(X)
Married couple	77.5	(X)	(X)	49.2	(X)	(X)	83.8	(X)	(X)
Female householder, no spouse present	18.3	(X)	(X)	44.3	(X)	(X)	12.6	(X)	(X)
Male householder, no spouse present	4.2	(X)	(X)	6.4	(X)	()X	3.6	(X)	(X)
Families with householder 55 years and over									
All families	7,352	(X)	(X)	1,126	(X)	(X)	6,150	(X)	(X)
Percent	100.0	(X)	(X)	100.0	(X)	(X)	100.0	(X)	(X)
Married couple	81.1	(X)	(X)	55.4	(X)	(X)	85.8	(X)	(X)
Female householder, no spouse present	15.2	(X)	(X)	35.8	(X)	(X)	11.4	(X)	(X)
Male householder, no spouse present	3.7	(X)	(X)	8.7	(X)	(X)	2.7	(X)	(X)

Source: "Selected Social Characteristics of the Population, by Sex, Region, and Race: March 1992." Claudette E. Bennett, *The Black Population in the United States: March 1992*, p. 23.

★ 919 ★

Characteristics

Social Characteristics of Population by Gender, Region, and Race, March 1992. Part III.

Numbers in thousands.

Characteristic	All races			Black			White		
	Both sexes	Male	Female	Both sexes	Male	Female	Both sexes	Male	Female
North and West									
Age									
Total	165,444	80,861	84,582	14,349	6,745	7,604	143,002	70,165	72,837
Percent	100.0	100.0	100.0	100.0	100.0	100.0	100.0	100.0	100.0
Under 5 years	8.0	8.4	7.6	10.5	11.3	9.8	7.6	8.0	7.2
5 to 9 years	7.5	7.8	7.1	8.9	10.0	8.0	7.2	7.5	6.9
10 to 14 years	7.1	7.4	6.8	8.8	9.0	8.7	6.8	7.1	6.5
15 to 19 years	6.6	6.8	6.4	8.0	8.4	7.7	6.4	6.6	6.2
20 to 24 years	7.0	7.1	6.9	7.9	7.5	8.3	6.9	7.0	6.8
25 to 29 years	8.0	8.2	7.8	8.6	8.9	8.4	8.0	8.2	7.8
30 to 34 years	8.9	9.1	8.8	9.3	8.8	9.7	8.9	9.1	8.6
35 to 44 years	15.9	16.1	15.7	13.7	13.3	14.1	16.0	16.3	15.8
45 to 54 years	10.8	10.7	10.8	9.4	9.4	9.4	10.9	10.9	11.0
55 to 64 years	8.3	8.1	8.5	6.9	6.7	7.1	8.5	8.3	8.7
65 to 74 years	7.2	6.7	7.7	5.1	4.9	5.3	7.6	7.0	8.1
75 years and over	4.8	3.6	5.9	2.7	1.9	3.4	5.1	3.9	6.4
16 years and over	76.1	75.0	77.3	69.9	67.7	71.9	77.0	75.9	78.1
18 years and over	73.5	72.3	74.7	66.8	64.5	68.8	74.5	73.3	75.6
21 years and over	69.6	68.3	70.9	62.2	59.8	64.2	70.7	69.5	71.9
55 years and over	20.3	18.4	22.2	14.7	13.5	15.8	21.2	19.2	23.2
65 years and over	12.0	10.3	13.6	7.8	6.8	8.7	12.7	10.9	14.5
Median age (years)	33.3	32.4	34.2	28.3	27.1	29.5	34.0	33.1	35.0
Marital status									
Total, 15 years and over	128,233	61,791	66,442	10,291	4,702	5,589	112,064	54,248	57,817
Percent	100.0	100.0	100.0	100.0	100.0	100.0	100.0	100.0	100.0
Never married	27.5	31.3	24.0	43.9	46.2	41.9	25.8	29.8	22.1
Married, spouse present	54.3	56.4	52.4	30.2	33.8	27.2	56.6	58.5	54.8
Married, spouse absent	3.1	2.6	3.4	7.8	6.4	9.0	2.5	2.1	2.9
Widowed	6.9	2.7	10.8	7.7	4.0	10.8	6.9	2.6	10.9
Divorced	8.2	7.0	9.3	10.4	9.6	11.1	8.1	6.9	9.3
Educational attainment									
Total, 25 to 34 years old	28,069	14,006	14,063	2,571	1,195	1,376	24,088	12,117	11,971
Percent completed									
Less than 9th grade	3.9	4.1	3.7	2.6	2.6	2.7	4.0	4.2	3.7
High school graduate or more	87.2	86.9	87.5	80.9	80.8	81.0	87.8	87.3	88.3
Some college or associate degree	25.9	24.2	27.7	28.2	25.0	31.1	25.7	24.0	27.5
Bachelor's degree or more	24.3	25.1	23.6	13.0	14.7	11.6	24.7	25.3	24.2
Type of family									
All families	43,496	(X)	(X)	3,463	(X)	(X)	38,203	(X)	(X)

[Continued]

★ 919 ★

Social Characteristics of Population by Gender, Region, and Race, March 1992. Part III.

[Continued]

Characteristic	All races			Black			White		
	Both sexes	Male	Female	Both sexes	Male	Female	Both sexes	Male	Female
Percent	100.0	(X)	(X)	100.0	(X)	(X)	100.0	(X)	(X)
Married couple	78.4	(X)	(X)	44.4	(X)	(X)	81.6	(X)	(X)
Female householder, no spouse present	16.9	(X)	(X)	49.0	(X)	(X)	14.0	(X)	(X)
Male householder, no spouse present	4.7	(X)	(X)	6.7	(X)	(X)	4.4	(X)	(X)
Families with householder 55 years and over									
All families	13,080	(X)	(X)	785	(X)	(X)	11,865	(X)	(X)
Percent	100.0	(X)	(X)	100.0	(X)	(X)	100.0	(X)	(X)
Married couple	84.0	(X)	(X)	55.5	(X)	(X)	86.1	(X)	(X)
Female householder, no spouse present	12.5	(X)	(X)	36.7	(X)	(X)	10.7	(X)	(X)
Male householder, no spouse present	3.6	(X)	(X)	7.8	(X)	(X)	3.2	(X)	(X)

Source: "Selected Social Characteristics of the Population, by Sex, Region, and Race: March 1992." Claudette E. Bennett, *The Black Population in the United States: March 1992*, p. 24.

★ 920 ★

Characteristics

Social Characteristics of Population: Summary, March 1992 and 1980

Numbers in thousands.

Characteristic	1992						1980					
	Black			White			Black			White		
	Both sexes	Male	Female	Both sexes	Male	Female	Both sexes	Male	Female	Both sexes	Male	Female
Age												
Total persons	31,439	14,781	16,658	210,257	102,965	107,293	26,033	12,133	13,900	191,905	93,468	98,437
Percent												
16 years and over	70.0	67.6	72.2	77.5	76.4	78.5	68.5	66.0	70.6	76.3	75.1	77.4
18 years and over	66.7	64.0	69.1	75.0	73.8	76.1	63.9	61.1	66.3	72.7	71.3	74.0
21 years and over	61.8	59.0	64.3	71.2	70.0	72.3	57.4	54.6	59.9	67.1	65.6	68.5
65 years and over	8.3	7.2	9.3	13.0	11.1	14.8	7.8	7.0	8.6	11.4	9.6	13.1
Median age (years)	28.2	26.7	29.6	34.3	33.3	35.2	24.8	23.5	26.1	30.8	29.7	31.9
Education												
Total, 25 years and over	17,445	7,803	9,641	137,657	66,063	71,594	12,927	5,717	7,209	114,763	54,389	60,374
Percent completed												
High school graduate or more	67.7	67.0	68.2	80.9	91.1	80.7	51.2	51.1	51.3	70.5	71.0	70.1
Bachelor's degree or more	11.9	11.9	12.0	22.1	25.2	19.1	7.9	7.7	8.1	17.8	22.1	14.0
Total, 25 to 34 years old	5,423	2,505	2,918	35,320	17,736	17,584	4,097	1,856	2,241	31,435	15,667	15,768
Percent completed												
High school graduate or more	81.8	82.2	81.4	87.1	86.3	87.8	75.4	75.3	75.5	86.8	87.2	86.2
Bachelor's degree or more	12.0	12.2	11.8	24.2	24.1	24.3	12.4	12.3	12.5	25.4	28.9	21.8
Marital status												
Total, 15 years and over	22,542	10,252	12,290	165,571	80,049	85,522	18,400	8,292	10,108	149,769	71,887	77,882
Percent	100.0	100.0	100.0	100.0	100.0	100.0	100.0	100.0	100.0	100.0	100.0	100.0

[Continued]

★ 920 ★

Social Characteristics of Population: Summary, March 1992 and 1980

[Continued]

Characteristic	1992						1980					
	Black			White			Black			White		
	Both sexes	Male	Female	Both sexes	Male	Female	Both sexes	Male	Female	Both sexes	Male	Female
Never married	41.8	45.0	39.1	24.2	28.1	20.6	37.0	41.1	33.7	24.4	28.1	21.0
Married	40.0	42.5	37.9	60.5	62.2	58.9	46.5	48.9	44.6	62.8	65.0	60.7
Widowed	8.2	4.2	11.5	7.0	2.5	11.3	8.8	3.7	13.0	7.3	2.3	11.9
Divorced	10.1	8.4	11.5	8.2	7.2	9.2	7.6	6.3	8.7	5.6	4.7	6.4
Type of family												
All families	7,716	(X)	(X)	57,225	(X)	(X)	6,184	(X)	(X)	52,243	(X)	(X)
Percent	100.0	(X)	(X)	100.0	(X)	(X)	100.0	(X)	(X)	100.0	(X)	(X)
Married couple	47.1	(X)	(X)	82.3	(X)	(X)	55.5	(X)	(X)	85.7	(X)	(X)
Female householder, no spouse present	46.4	(X)	(X)	13.5	(X)	(X)	40.3	(X)	(X)	11.6	(X)	(X)
Male householder, no spouse present	6.5	(X)	(X)	4.1	(X)	(X)	4.1	(X)	(X)	2.8	(X)	(X)

Source: "Selected Summary Social Characteristics of the Population, by Sex and Race: March 1992 and 1980." Claudette E. Bennett, *The Black Population in the United States: March 1992*, p. 5.

Distribution

★ 921 ★

Cities with 100,000 or More Inhabitants in 1990, and Population 1970 to 1990

Population: as of April 1. Data refer to boundaries in effect on January 1, 1990. Minus sign (-) indicates decrease.

City	Population[1]								
	1970, total (1,000)	1980, total (1,000)	Total (1,000)	Rank	Percent change, 1980-90	Black	American Indian, Eskimo, Aleut	Asian, Pacific Islander	Hispanic[2]
Abilene, TX	90	96	107	180	8.5	7.0	0.4	1.3	15.5
Akron, OH	275	237	223	71	-6.0	24.5	0.3	1.2	0.7
Albany, NY	116	102	100	196	-1.7	20.6	0.3	2.3	3.1
Alburquerque, NM	245	332	385	38	15.5	3.0	3.0	1.7	34.5
Alexandria, VA	111	103	111	164	7.7	21.9	0.3	4.2	9.7
Allentown, PA	110	104	105	183	1.5	5.0	0.2	1.3	11.7
Amarillo, TX	127	149	158	110	5.6	6.0	0.8	1.9	14.7
Anaheim, CA	166	219	266	59	21.4	2.5	0.5	9.4	31.4
Anchorage, AK	48	174	226	69	29.8	6.4	6.4	4.8	4.1
Ann Arbor, MI	100	108	110	170	1.5	9.0	0.4	7.7	2.6
Arlington, TX	90	160	262	61	63.5	8.4	0.5	3.9	8.9
Atlanta, GA	495	425	394	36	-7.3	67.1	0.1	0.9	1.9
Aurora, CO	75	159	222	72	40.1	11.4	0.6	3.8	6.6
Austin, TX	254	346	466	27	34.6	12.4	0.4	3.0	23.0
Bakersfield, CA	70	106	175	97	65.5	9.4	1.1	3.6	20.5
Baltimore, MD	905	787	736	12	-6.4	59.2	0.3	1.1	1.0

[Continued]

★ 921 ★

Cities with 100,000 or More Inhabitants in 1990, and Population 1970 to 1990

[Continued]

City	Population[1]								
	1970, total (1,000)	1980, total (1,000)	Total (1,000)	Rank	Percent change, 1980-90	Black	American Indian, Eskimo, Aleut	Asian, Pacific Islander	Hispanic[2]
Baton Rouge, LA	166	220	220	73	-0.4	43.9	0.1	1.7	1.6
Beaumont, TX	118	118	114	155	-3.2	41.3	0.2	1.7	4.3
Berkeley, CA	114	103	103	190	-0.6	18.8	0.6	14.8	8.4
Birmingham, AL	301	284	266	60	-6.5	63.3	0.1	0.6	(Z)
Boise City, ID	75	102	126	145	23.0	0.6	0.6	1.6	2.7
Boston, MA	641	563	574	20	2.0	25.6	0.3	5.3	10.8
Bridgeport, CT	157	143	142	123	-0.6	26.6	0.3	2.3	26.5
Buffalo, NY	463	358	328	50	-8.3	30.7	0.8	1.0	4.9
Cedar Rapids, IA	111	110	109	173	-1.3	2.9	0.2	1.0	1.1
Charlotte, NC	241	315	396	35	25.5	31.8	0.4	1.8	1.4
Chattanooga, TN	120	170	152	113	-10.0	33.7	0.2	1.0	0.6
Chesapeake, VA	90	114	152	114	32.8	27.4	0.3	1.2	1.3
Chicago, IL	3,369	3,005	2,784	3	-7.4	39.1	0.3	3.7	19.6
Chula Vista, CA	68	84	135	131	61.0	4.6	0.6	8.9	37.3
Cincinnati, OH	454	385	364	45	-5.5	37.9	0.2	1.1	0.7
Cleveland, OH	751	574	506	23	-11.9	46.6	0.3	1.0	4.6
Colorado Springs, CO	136	215	281	54	30.7	7.0	0.8	2.4	9.1
Columbia, SC	114	101	103	188	2.2	43.7	0.3	1.4	2.0
Columbus, GA[3]	155	169	179	93	5.5	38.1	0.3	1.4	3.0
Columbus, OH	540	565	633	16	12.0	22.6	0.2	2.4	1.1
Concord, CA	85	104	111	163	7.3	2.4	0.7	8.7	11.5
Corpus Christi, TX	205	232	257	64	10.9	4.8	0.4	0.9	50.4
Dallas, TX	844	905	1,008	8	11.4	29.5	0.5	2.2	20.9
Dayton, OH	243	194	182	89	-5.9	40.4	0.2	0.6	0.7
Denver, CO	515	493	468	26	-5.1	12.8	1.2	2.4	23.0
Des Moines, IA	201	191	193	80	1.1	7.1	0.4	2.4	2.4
Detroit, MI	1,514	1,203	1,028	7	-14.6	75.7	0.4	0.8	2.8
Durham, NC	95	101	137	130	35.1	45.7	0.2	2.0	1.2
Elizabeth, NJ	113	106	110	168	3.6	19.8	0.3	2.7	39.1
El Monte, CA	70	79	106	181	33.5	1.0	0.6	11.8	72.5
El Paso, TX	322	425	515	22	21.2	3.4	0.4	1.2	69.0
Erie, PA	129	119	109	175	-8.7	12.0	0.2	0.5	2.4
Escondido, CA	37	64	109	176	68.8	1.5	0.8	3.7	23.4
Eugene, OR	79	106	113	159	6.6	1.3	0.9	3.5	2.7
Evansville, IN	139	130	126	144	-3.2	9.5	0.2	0.6	0.6
Flint, MI	193	160	141	125	-11.8	47.9	0.7	0.5	2.9
Fort Lauderdale, FL	140	153	149	116	-2.6	28.1	0.2	0.9	7.2
Fort Wayne, IN	178	172	173	99	0.4	16.7	0.3	1.0	2.7
Fort Worth, TX	393	385	448	28	16.2	22.0	0.4	2.0	19.5
Fremont, CA	101	132	173	98	31.4	3.8	0.7	19.4	13.3
Fresno, CA	166	217	354	47	62.9	8.3	1.1	12.5	29.9
Fullerton, CA	86	102	114	156	11.6	2.2	0.5	12.2	21.3
Garden Grove, CA	121	123	143	120	16.0	1.5	0.6	20.5	23.5

[Continued]

★ 921 ★

Cities with 100,000 or More Inhabitants in 1990, and Population 1970 to 1990

[Continued]

City	Population[1]								
	1970, total (1,000)	1980, total (1,000)	Total (1,000)	Rank	Percent change, 1980-90	Black	American Indian, Eskimo, Aleut	Asian, Pacific Islander	Hispanic[2]
Garland, TX	81	139	181	91	30.1	8.9	0.2	4.5	11.6
Gary, IN	175	152	117	154	-23.2	80.6	0.2	0.2	5.7
Glendale, AZ	36	97	148	117	52.4	3.0	0.9	2.1	15.5
Glendale, CA	133	139	180	92	29.5	1.3	0.3	14.1	21.0
Grand Rapids, MI	198	182	189	83	4.0	18.5	0.8	1.1	5.0
Greensboro, NC	144	156	184	88	18.2	33.9	0.5	1.4	1.0
Hampton, VA	121	123	134	133	9.1	38.9	0.3	1.7	2.0
Hartford, CT	158	136	140	128	2.5	38.9	0.3	1.4	31.6
Hayward, CA	93	94	111	162	19.0	9.8	0.0	15.5	23.9
Hialeah, FL	102	145	188	85	29.4	1.9	0.1	0.5	87.6
Hollywood, FL	107	121	122	148	0.3	8.5	0.2	1.3	11.9
Honolulu, HI[4]	325	365	377	39	3.3	1.3	0.3	70.5	4.6
Houston, TX	1,234	1,595	1,631	4	2.2	28.1	0.3	4.1	27.6
Huntington Beach, CA	116	171	182	90	6.5	0.9	0.6	8.3	11.2
Huntsville, AL	139	143	160	109	12.2	24.4	0.5	2.1	1.2
Independence, MO	112	112	112	160	0.5	1.4	0.6	1.0	2.0
Indianapolis, IN[3]	737	701	731	13	4.3	22.6	0.2	0.9	1.1
Inglewood, CA	90	94	110	169	16.4	51.9	0.4	2.5	38.5
Irvine, CA	[5]	62	110	167	77.6	1.8	0.2	18.1	6.3
Irving, TX	97	110	155	112	41.0	7.5	0.6	4.6	16.3
Jackson, MS	154	203	197	78	-3.1	55.7	0.1	0.5	0.4
Jacksonville, FL[3]	504	541	635	15	17.9	25.2	0.3	1.9	2.6
Jersey City, NJ	260	224	229	67	2.2	29.7	0.3	11.4	24.2
Kansas City, KS	168	161	150	115	-7.1	29.3	0.7	1.2	7.1
Kansas City, MO	507	448	435	31	-2.9	29.6	0.5	1.2	3.9
Knoxville, TN	175	175	165	102	-5.7	15.8	0.2	1.0	0.7
Lakewood, CO	93	114	126	143	11.1	1.0	0.7	1.9	9.1
Lansing, MI	131	130	127	142	-2.4	18.6	1.0	1.8	7.9
Laredo, TX	69	91	123	147	34.4	0.1	0.2	0.4	93.9
Las Vegas, NV	126	165	258	63	56.8	11.4	0.9	3.6	12.5
Lexington-Fayette, KY	108	204	225	70	10.4	13.4	.2	1.6	1.1
Lincoln, NE	150	172	192	81	11.7	2.4	0.6	1.7	2.0
Little Rock, AR	132	159	176	96	10.5	34.0	0.3	0.9	0.8
Livonia, MI	110	105	101	193	-3.8	0.3	0.2	1.3	1.3
Long Beach, CA	359	361	429	32	18.8	13.7	0.6	13.6	23.6
Los Angeles, CA	2,812	2,969	3,485	2	17.4	14.0	0.5	9.8	39.9
Louisville, KY	362	299	270	58	-9.8	29.7	0.2	0.7	0.7
Lowell, MA	94	92	103	189	11.9	2.4	0.2	11.1	10.1
Lubbock, TX	149	174	186	87	6.8	8.6	0.3	1.4	22.5
Macon, GA	122	117	107	179	-8.2	52.2	0.1	0.4	0.6
Madison, WI	172	171	191	82	11.8	4.2	0.4	3.9	2.0
Memphis, TN	624	646	610	18	-5.5	54.8	0.2	0.8	0.7
Mesa, AZ	63	152	288	53	89.0	1.9	1.0	1.5	10.9

[Continued]

★ 921 ★

Cities with 100,000 or More Inhabitants in 1990, and Population 1970 to 1990

[Continued]

City	Population[1]								
	1970, total (1,000)	1980, total (1,000)	Total (1,000)	Rank	Percent change, 1980-90	Black	American Indian, Eskimo, Aleut	Asian, Pacific Islander	Hispanic[2]
Mesquite, TX	55	67	101	191	51.3	5.8	0.5	2.6	8.8
Miami, FL	335	347	359	46	3.5	27.4	0.2	0.6	62.5
Milwaukee, WI	717	636	628	17	-1.3	30.5	0.9	1.9	6.3
Minneapolis, MN	434	371	368	43	-0.7	13.0	3.3	4.3	2.1
Mobile, AL	190	200	196	79	-2.1	38.9	0.2	1.0	1.0
Modesto, CA	62	107	165	103	54.0	2.7	1.0	7.8	16.3
Montgomery, AL	133	178	188	86	5.4	42.3	0.2	0.7	0.8
Moreno Valley, CA	5	5	119	151	(X)	13.8	0.7	6.6	22.9
Nashville-Davidson, TN[3]	426	456	488	25	6.9	24.3	0.2	1.4	0.9
Newark, NJ	382	329	275	56	-16.4	58.5	0.2	1.2	26.1
New Haven, CT	138	126	130	138	3.5	36.1	0.3	2.4	13.2
New Orleans, LA	593	558	497	24	-10.9	61.9	0.2	1.9	3.5
Newport News, VA	138	145	171	100	18.3	33.6	0.3	2.3	2.8
New York, NY	7,896	7,072	7,323	1	3.5	28.7	0.4	7.0	24.4
Bronx Borough	1,472	1,169	1,204	(X)	3.0	37.3	0.5	3.0	43.5
Brooklyn Borough	2,602	2,231	2,301	(X)	3.1	37.9	0.3	4.8	20.1
Manhattan Borough	1,539	1,428	1,488	(X)	4.1	22.0	0.4	7.4	26.0
Queens Borough	1,987	1,891	1,952	(X)	3.2	21.7	0.4	12.2	19.5
Staten Island Borough	295	352	379	(X)	7.7	8.1	0.2	4.5	8.0
Norfolk, VA	308	267	261	62	-2.2	39.1	0.4	2.6	2.9
Oakland, CA	362	339	372	40	9.7	43.9	0.6	14.8	13.9
Oceanside, CA	40	77	128	140	67.1	7.9	0.7	6.1	22.6
Oklahoma City, OK	368	404	445	29	10.1	16.0	4.2	2.4	5.0
Omaha, NE	347	314	336	48	7.0	13.1	0.7	1.0	3.1
Ontario, CA	64	89	133	134	49.9	7.3	0.7	3.9	41.7
Orange, CA	77	91	111	166	21.0	1.4	0.5	7.9	22.8
Orlando, FL	99	128	165	104	28.4	26.9	0.3	1.6	8.7
Overland Park, KS	78	82	112	161	36.7	1.8	0.3	1.9	2.0
Oxnard, CA	71	108	143	121	31.8	5.2	0.8	8.6	54.4
Passadena, TX	113	118	132	137	11.4	19.0	0.4	8.1	27.3
Paterson, NJ	145	138	141	124	2.1	36.0	0.3	1.4	41.0
Peoria, IL	127	124	114	157	-8.6	20.9	0.2	1.7	1.6
Philadelphia, PA	1,949	1,688	1,586	5	-6.1	39.9	0.2	2.7	5.6
Phoenix, AZ	584	790	983	9	24.5	5.2	1.9	1.7	20.0
Pittsburgh, PA	520	424	370	41	-12.8	25.8	0.2	1.6	0.9
Plano, TX	18	72	128	141	76.8	4.1	0.3	4.0	6.2
Pomona, CA	87	93	132	136	42.0	14.4	0.6	6.7	51.3
Portland, OR	380	368	437	30	18.8	7.7	1.2	5.3	3.2
Portsmouth, VA	111	105	104	186	-0.6	47.3	0.3	0.8	1.3
Providence, RI	179	157	161	107	2.5	14.8	0.9	5.9	15.5
Raleigh, NC	123	150	211	74	40.4	27.6	0.3	2.5	1.4
Ranch Cucamonga, CA	5	55	101	192	83.5	5.9	0.6	5.4	20.0
Reno, NV	73	101	134	132	32.8	2.9	1.4	4.9	11.1

[Continued]

★ 921 ★

Cities with 100,000 or More Inhabitants in 1990, and Population 1970 to 1990

[Continued]

City	Population[1]								
	1970, total (1,000)	1980, total (1,000)	Total (1,000)	Rank	Percent change, 1980-90	Black	American Indian, Eskimo, Aleut	Asian, Pacific Islander	Hispanic[2]
Richmond, VA	249	219	203	76	-7.5	55.2	0.2	0.9	0.9
Riverside, CA	140	171	227	68	32.8	7.4	0.8	5.2	26.0
Rochester, NY	295	242	230	66	-4.7	31.5	0.5	1.8	8.7
Rockford, IL	147	140	140	127	0.2	15.0	0.3	1.5	4.2
Sacramento, CA	257	276	369	42	34.0	15.3	1.2	15.0	16.2
St. Louis, MO	622	453	397	34	-12.4	47.5	0.2	0.9	1.3
St. Paul, MN	310	270	272	57	0.7	7.4	1.4	7.1	4.2
St. Petersburg, FL	216	239	240	65	0.7	19.6	0.2	1.7	2.6
Salem, OR	69	89	108	078	21.0	1.5	1.6	2.4	6.1
Salinas, CA	59	80	109	174	35.2	3.0	0.9	8.1	50.6
Salt Lake City, UT	176	163	160	108	-1.9	1.7	1.6	4.7	9.7
San Antonio, TX	654	786	936	10	19.1	7.0	0.4	1.1	55.6
San Bernardino, CA	107	119	164	105	38.2	16.0	1.0	4.0	34.6
San Diego, CA	697	876	1,111	6	26.8	9.4	0.6	11.8	20.7
San Francisco, CA	716	679	724	14	6.6	10.9	0.5	29.1	13.9
San Jose, CA	460	629	782	11	24.3	4.7	0.7	19.5	26.6
Santa Ana, CA	156	204	294	52	44.0	2.6	0.5	9.7	65.2
Santa Clarita, CA	5	5	111	165	(X)	1.5	0.6	4.2	13.4
Santa Rosa, CA	50	83	113	158	37.1	1.8	1.2	3.4	9.5
Savannah, GA	118	142	138	129	-2.6	51.3	0.2	1.1	1.4
Scottsdale, AZ	68	89	130	139	46.8	0.8	0.6	1.2	4.8
Seattle, WA	531	494	516	21	4.5	10.1	1.4	11.8	3.6
Shreveport, LA	182	206	199	77	-4.1	44.8	0.2	0.5	1.1
Simi Valley, CA	60	78	100	195	29.3	1.5	0.6	5.5	12.7
Sioux Falls, SD	72	81	101	194	24.0	0.7	1.6	0.7	0.6
South Bend, IN	126	110	106	182	-3.8	20.9	0.4	0.9	3.4
Spokane, WA	171	171	177	94	3.4	1.9	2.0	2.1	2.1
Springfield, IL	92	100	105	184	5.2	13.0	0.2	1.0	0.8
Springfield, MA	164	152	157	111	3.1	19.2	0.2	1.0	16.9
Springfield, MO	120	133	140	126	5.5	2.5	0.7	0.9	1.0
Stamford, CT	109	102	108	177	5.5	17.8	0.1	2.6	9.8
Sterling Heights, MI	61	109	118	152	8.1	0.4	0.2	2.9	1.1
Stockton, CA	110	150	211	75	42.3	9.6	1.0	22.8	25.0
Sunnyvale, CA	96	107	117	153	10.0	3.4	0.5	19.3	13.2
Syracuse, NY	197	170	164	106	-3.7	20.3	1.3	2.2	2.9
Tacoma, WA	154	159	177	95	11.5	11.4	2.0	6.9	3.8
Tallahassee, FL	73	82	125	146	53.0	29.1	0.2	1.8	3.0
Tampa, FL	278	272	280	55	3.1	25.0	0.3	1.4	15.0
Tempe, AZ	64	107	142	122	32.7	3.2	1.3	4.1	10.9
Thousand Oaks, CA	36	77	104	185	35.4	1.2	0.4	4.8	9.6
Toledo, OH	383	355	333	49	-6.1	19.7	0.3	1.0	4.0
Topeka, KS	125	119	120	149	1.0	10.6	1.3	0.8	5.8
Torrance, CA	135	130	133	135	2.5	1.5	0.4	21.9	10.1

[Continued]

★ 921 ★

Cities with 100,000 or More Inhabitants in 1990, and Population 1970 to 1990

[Continued]

City	Population[1]								
	1970, total (1,000)	1980, total (1,000)	Total (1,000)	Rank	Percent change, 1980-90	Black	American Indian, Eskimo, Aleut	Asian, Pacific Islander	Hispanic[2]
Tucson, AZ	263	331	405	33	22.6	4.3	1.6	2.2	29.3
Tulsa, OK	330	361	367	44	1.8	13.6	4.7	1.4	2.6
Vallejo, CA	72	80	109	171	36.0	21.2	0.7	23.0	10.8
Virginia Beach, VA	172	262	393	37	49.9	13.9	0.4	4.3	3.1
Waco, TX	95	101	104	187	2.3	23.1	0.3	0.9	16.3
Warren, MI	179	161	145	118	-10.1	0.7	0.5	1.3	1.1
Washington, DC	757	638	607	19	-4.9	65.8	0.2	1.8	5.4
Waterbury, CT	108	103	109	172	5.5	13.0	0.3	0.7	13.4
Wichita, KS	277	280	304	51	8.6	11.3	1.2	2.6	5.0
Winston-Salem, NC	134	132	143	119	8.8	39.3	0.2	0.8	0.9
Worcester, MA	177	162	170	101	4.9	4.5	0.3	2.8	9.6
Yonkers, NY	204	195	188	84	-3.7	14.1	0.2	3.0	16.7

Source: "Cities With 100,000 or More Inhabitants in 1990—Population, 1970 to 1990, and Land Area, 1990." U.S. Bureau of the Census, *Statistical Abstract of the United States*, 1993, p. 42. Primary source: U.S. Bureau of the Census, *Census of Population: 1970*, vol. I, chapters A and B; *1980 Census of Population*, vol. 1, chapters A and B; *1990 Census of Population and Housing, Summary Population and Housing Characteristics*, (CPH-1). *Notes:* X Not applicable. Z Less than .05 percent. 1. Population totals include corrections through October 1992; the data by race and Hispanic origin are not corrected. 2. Hispanic persons may be of any race. 3. Represents the portion of a consolidated city that is not within one or more separately incorporated places. 4. Data represent a census designated place as delineated by state and local officials. 5. Not incorporated.

★ 922 ★

Distribution

Distribution of Population by Residence, Region, and Race, 1980, 1986, 1990, 1992 - I

Region and residence	1992				1990			
	All races	Black	White	Black as a percent of total	All races	Black	White	Black as a percent of total
Residence								
United States								
Total	251,447	31,439	210,257	12.5	246,191	30,392	206,983	12.3
Percent	100.0	100.0	100.0	(X)	100.0	100.0	100.0	(X)
All metropolitan areas	78.0	84.7	76.6	13.6	77.7	83.8	76.4	13.3
Inside central cities	30.2	55.7	26.0	23.0	30.5	56.8	26.2	22.9
Outside central cities	47.8	29.0	50.7	7.6	47.2	27.0	50.2	7.1
Nonmetropolitan areas	22.0	15.3	23.4	8.7	22.3	16.2	23.6	9.0
South								
Total	86,004	17,090	67,256	19.9	84,107	16,512	66,051	19.6
Percent	100.0	100.0	100.0	(X)	100.0	100.0	100.0	(X)
All metropolitan areas	72.1	73.7	71.6	20.3	71.4	71.7	71.1	19.7

[Continued]

★ 922 ★

Distribution of Population by Residence, Region, and Race, 1980, 1986, 1990, 1992 - I

[Continued]

Region and residence	1992				1990			
	All races	Black	White	Black as a percent of total	All races	Black	White	Black as a percent of total
Inside central cities	27.9	43.0	24.1	30.6	28.1	42.4	24.4	29.7
Outside central cities	44.2	30.7	47.5	13.8	43.3	29.3	46.7	13.3
Nonmetropolitan areas	27.9	26.3	28.4	18.8	28.6	28.3	28.9	19.4
North and West								
Total	165,444	14,349	143,002	8.7	162,085	13,881	140,931	8.6
Percent	100.0	100.0	100.0	(X)	100.0	100.0	100.0	(X)
All metropolitan areas	81.1	97.9	79.0	10.5	81.0	98.1	78.9	10.4
Inside central cities	31.4	70.8	26.8	19.5	31.8	73.8	27.0	19.9
Outside central cities	49.6	27.1	52.2	4.7	49.2	24.2	51.9	4.2
Nonmetropolitan areas	18.9	2.1	21.0	1.0	19.0	1.9	21.1	0.9

Source: "Distribution of the Population, by Type of Residence, Region, and Race: March 1992, 1990, 1986, and 1980." Claude E. Bennett, *The Black Population in the United States, March 1992*, p. 4.

★ 923 ★

Distribution

Distribution of Population by Residence, Region, and Race, 1980, 1986, 1990, 1992 - II

Region and residence	1986				1980			
	All races	Black	White	Black as a percent of total	All races	Black	White	Black as a percent of total
Residence								
United States								
Total	236,749	28,538	201,019	12.1	223,160	26,033	191,905	11.7
Percent	100.0	100.0	100.0	(X)	100.0	100.0	100.0	(X)
All metropolitan areas	77.4	83.5	76.3	13.0	67.8	76.7	66.3	13.2
Inside central cities	31.5	59.4	27.1	22.8	27.8	55.5	23.8	23.3
Outside central cities	45.9	24.0	49.2	6.3	40.0	21.2	42.5	6.2
Nonmetropolitan areas	22.6	16.5	23.7	8.8	32.2	23.3	33.7	8.4
South								
Total	80,652	15,455	64,016	19.2	74,046	13,599	59,597	18.4
Percent	100.0	100.0	100.0	(X)	100.0	100.0	100.0	(X)
All metropolitan areas	71.5	71.5	71.4	19.2	55.7	60.0	54.6	19.8
Inside central cities	29.9	46.2	25.8	29.6	25.0	40.3	21.6	29.6
Outside central cities	41.6	25.3	45.6	11.6	30.7	19.7	33.0	11.8
Nonmetropolitan areas	28.5	25.8	25.6	19.2	44.3	40.0	45.4	16.6

[Continued]

★ 923 ★

Distribution of Population by Residence, Region, and Race, 1980, 1986, 1990, 1992 - II

[Continued]

Region and residence	1986				1980			
	All races	Black	White	Black as a percent of total	All races	Black	White	Black as a percent of total
North and West								
Total	156,097	13,083	137,003	8.4	149,114	12,435	132,307	8.3
Percent	100.0	100.0	100.0	(X)	100.0	100.0	100.0	(X)
All metropolitan areas	80.4	97.6	78.5	10.2	73.8	95.1	71.5	10.7
Inside central cities	32.3	75.1	27.7	19.5	29.2	72.1	24.7	20.6
Outside central cities	48.1	22.5	50.8	3.9	44.6	22.9	46.8	4.3
Nonmetropolitan areas	19.6	2.4	21.5	1.0	26.2	4.9	28.5	1.6

Source: "Distribution of the Population, by Type of Residence, Region, and Race: March 1992, 1990, 1986, and 1980." Claude E. Bennett, The Black Population in the United States, March 1992, p. 4.

★ 924 ★

Distribution

Distribution of Population: Racial and Hispanic Metropolitan Areas, 1990

As of April 1. For Black, Hispanic origin, and Asian and Pacific Islander populations, areas selected had 100,000 or more of specified group; for American Indian, Eskimo, and Aleut population, areas selected are ten areas with largest number of that group. CMSA = consolidated metropolitan statistical area. MSA = metropolitan statistical area. Areas as defined by U.S. Office of Management and Budget, December 31, 1992.

Metropolitan area	Number of specified group (1,000)	Percent of total metro area	Metropolitan area	Number of specified group (1000)	Percent of total metro area
BLACK			HISPANIC ORIGIN[1]		
New York-Northern, New Jersey-Long Island, NY-NJ-CT-PA CMSA	3,439	17.8	Los Angeles-Riverside-Orange County, CA CMSA	4,779	32.9
Washington-Baltimore, DC-MD-VA-WV CMSA	1,696	25.2	New York-Northern New Jersey-Long island, NY-NJ-CT-PA CMSA	2,843	14.7
Chicago-Gary-Kenosha, IL-IN-WI CMSA	1,564	19.0	Miami-Fort Lauderdale, FL CMSA	1,062	33.3
Los Angeles-Riverside-Orange County, CA CMSA	1,230	8.5	San Francisco-Oakland-San Jose, CA CMSA	970	15.5
Philadelphia-Wilmington-Atlantic City, PA-NJ-DE-MD CMSA	1,083	18.4	Chicago-Gary-Kenosha, IL-IN-Wi CMSA	898	10.9
Detroit-Ann Arbor-Flint, MI CMSA	1,061	20.5	Houston-Galveston-Brazoria, TX CMSA	773	20.7
			San Antonio, TX MSA	628	47.4
Atlanta, GA MSA	747	25.2	Dallas-Fort Worth, TX CMSA	526	13.0
Houston-Galveston- Brazoria, TX CMSA	668	17.9	San Diego, CA MSA	511	20.4
Miami-Fort Lauderdale, FL CMSA	591	18.5	El Paso, TX MSA	412	69.6
Dallas-Fort Worth, TX CMSA	566	14.0	Phoenix-Mesa, AZ MSA	380	17.0
San Francisco-Oakland-San Jose, CA CMSA	538	8.6	McAllen-Edinburg-Mission, TX MSA	327	85.2
New Orleans, LA MSA	447	34.8	Fresno, CA MSA	267	35.3
Cleveland-Akron, OH CMSA	445	15.6	Washington-Baltimore, DC-MD-VA-WV CMSA	259	3.9
St. Louis, MO-IL MSA	424	17.0	Denver-Boulder-Greeley, CO CMSA	254	12.8
			Boston-Brockton-Nashua, MA-NH-ME-CT CMSA	239	4.4
Memphis, TN-AR-MS MSA	410	40.7			
Norfolk-Virginia Beach-Newport News, VA-NC MSA	409	28.3	Philadelphia-Wilmington-Atlantic City, PA-NJ-DE-MD CMSA	224	3.8
Boston-Brockton-Nashua, MA-NH-ME-CT CMSA	261	4.8	Albuquerque, NM MSA	218	37.1
Richmond-Petersburg, VA MSA	252	29.2	Brownsville-Harlingen-San Benito, TX MSA	213	81.9
Birmingham, AL MSA	241	28.7	Corpus Christi, TX MSA	182	52.0
Charlotte-Gastonia-Rock Hill, NC-SC MSA	232	19.9	Austin-San Marcos, TX MSA	177	20.9
Milwaukee-Racine, WI CMSA	214	13.3	Sacramento-Yolo, CA CMSA	172	11.6
			Tucson, AZ MSA	163	24.5
Raleigh-Durham-Chapel Hill, NC MSA	207	24.2	Bakersfield, CA MSA	152	28.0
Cincinnati-Hamilton, OH-KY-IN MSA	204	11.2	Tampa-St. Petersburg-Clearwater, FL MSA	139	6.7
Greensboro--Winston-Salem-High Point, NC MSA	203	19.3	Laredo, TX MSA	125	93.9
Kansas City, MO-KS MSA	201	12.7	Visalia-Tulare-Porterville, CA MSA	121	38.8
Tampa-St. Petersburg-Clearwater, FL MSA	186	9.0	Salinas, CA MSA	120	33.6
Indianapolis, IN MSA	182	13.2	Stockton-Lodi, CA MSA	113	23.4
			Detroit-Ann Arbor-Flint, MI CMSA	105	2.0
Jacksonville, FL MSA	181	20.0	Orlando, FL MSA	101	8.2
Pittsburgh, PA MSA	180	7.5			

[Continued]

★ 924 ★

Distribution of Population: Racial and Hispanic Metropolitan Areas, 1990

[Continued]

Metropolitan area	Number of specified group (1,000)	Percent of total metro area	Metropolitan area	Number of specified group (1000)	Percent of total metro area
Jackson, MS MSA	168	42.5	ASIAN AND PACIFIC ISLANDER		
Columbus, OH MSA	163	12.1	Los Angeles-Riverside-Orange County, CA CMSA	1,339	9.2
San Diego, CA MSA	159	6.4	San Francisco-Oakland-San Jose, CA CMSA	927	14.8
			New York-Northern New Jersey-Long Island, NY-NJ-CT-PA CMSA	898	4.6
Charleston-North Charleston, SC MSA	153	30.2	Honolulu, HI MSA	526	63.0
Nashville, TN MSA	152	15.5	Chicago-Gary-Kenosha, IL-IN-Wi CMSA	258	3.1
Orlando, FL MSA	147	12.0	Washington-Baltimore, DC-MD-VA-WV CMSA	248	3.7
Greenville-Spartanburg-Anderson, SC MSA	145	17.4	San Diego, CA MSA	198	7.9
			Seattle-Tacoma-Bremerton, WA CMSA	181	6.1
Baton Rouge, LA MSA	143	30.5	Boston-Brockton-Nashua, MA-NH-ME-CT CMSA	137	2.5
Columbia, SC MSA	138	30.4	Houston-Galveston-Brazoria, TX CMSA	132	3.5
Seattle-Tacoma-Bremerton, WA CMSA	133	4.5	Philadelphia-Wilmington-Atlantic City, PA-NJ-DE-MD CMSA	119	2.0
Mobile, AL MSA	131	27.4	Sacramento-Yolo, CA CMSA	115	7.7
Shreveport-Bossier City, LA MSA	130	34.6	AMERICAN INDIAN, ESKIMO, ALEUT		
Dayton-Springfield, OH MSA	126	13.3	Los Angeles-Riverside-Orange County, CA CMSA	87	0.6
Augusta-Aiken, GA-SC MSA	125	31.6	Phoenix-Mesa, AZ MSA	49	2.2
Louisville, KY-IN MSA	122	12.9	New York-Northern New Jersey-Long Island, NY-NJ-CT-PA CMSA	48	0.2
Buffalo-Niagara Falls, NY MSA	122	10.3	Tulsa, OK MSA	48	6.8
			Oklahoma City, OK MSA	46	4.8
West Palm Beach-Boca Raton, FL MSA	108	12.5	San Francisco-Oakland-San Jose, CA CMSA	41	0.7
Montgomery, AL MSA	105	36.0	Seattle-Tacoma-Bremerton, WA CMSA	38	1.3
Sacramento-Yolo CA CMSA	102	6.9	Albuquerque, NM MSA	30	5.1
Little Rock-North Little Rock, AR MSA	102	19.9	Minneapolis-St. Paul, MN-Wi MSA	24	1.0
Macon, GA MSA	102	35.0	Detroit-Ann Arbor-Flint, MI CMSA	21	0.4
Oklahoma City, OK MSA	101	10.5			

Source: "Metropolitan Areas With Large Numbers of Selected Racial Groups and of Hispanic Origin Population." U.S. Bureau of the Census, *Statistical Abstract of the United States*, 1993, p. 40. Primary source: U.S. Bureau of the Census, unpublished data. *Note:* 1. Persons of Hispanic origin may be of any race.

★ 925 ★

Distribution

Distribution of Population: Residence, Region, Age, Gender, and Race, March 1992. Part I

Numbers in thousands.

Region, age, sex, and residence	Number			Percent distribution		
	All races	Black	White	All races	Black	White
Region and age						
Total, all persons	251,447	31,439	210,257	100.0	100.0	100.0
South	86.004	17,090	67,256	34.2	54.4	32.0
North and West	165,444	14,349	143,002	65.8	45.6	68.0
Northeast	50,841	5,510	43,802	20.2	17.5	20.8
Midwest	60,423	6,330	52,775	24.0	20.1	25.1
West	54,179	2,510	46,425	21.5	8.0	22.1
Male	122,528	14,781	102,965	100.0	100.0	100.0
South	41,667	8,036	32,800	34.0	54.4	31.9
North and West	80,861	6,745	70,165	66.0	45.6	68.1
Northeast	24,540	2,555	21,198	20.0	17.3	20.6
Midwest	29,296	2,939	25,698	23.9	19.9	25.0
West	27,025	1,251	23,269	22.1	8.5	22.6
Female	128,919	16,658	107,293	100.0	100.0	100.0
South	44,337	9,054	34,456	34.4	54.4	32.1

[Continued]

★ 925 ★

Distribution of Population: Residence, Region, Age, Gender, and Race, March 1992. Part I

[Continued]

Region, age, sex, and residence	Number			Percent distribution		
	All races	Black	White	All races	Black	White
North and West	84,582	7,604	72,837	65.6	45.6	67.9
Northeast	26,301	2,955	22,605	20.4	17.7	21.1
Midwest	31,128	3,391	27,077	24.1	20.4	25.2
West	27,154	1,259	23,156	21.1	7.6	21.6
Total, persons 55 years and over	51,740	4,772	45,577	100.0	100.0	100.0
South	18,086	2,660	15,207	35.0	55.7	33.4
North and West	33,654	2,113	30,370	65.0	44.3	66.6
Northeast	11,183	855	10,150	21.6	17.9	22.3
Midwest	12,459	926	11,395	24.1	19.4	25.0
West	10,012	331	8,825	19.4	6.9	19.4
Male	22,835	2,035	20,162	100.0	100.0	100.0
South	7,917	1,126	6,701	34.7	55.3	33.2
North and West	14,918	910	13,460	65.3	44.7	66.8
Northeast	4,894	373	4,436	21.4	18.3	22.0
Midwest	5,449	379	5,012	23.9	18.6	24.9
West	4,575	158	4,012	20.0	7.7	19.9
Female	28,904	2,737	25,415	100.0	100.0	100.0
South	10,169	1,534	8,505	35.2	56.0	33.5
North and West	18,736	1,203	16,910	64.8	44.0	66.5
Northeast	6,289	483	5,713	21.8	17.6	22.5
Midwest	7,010	547	6,383	24.3	20.0	25.1
West	5,437	174	4,813	18.8	6.4	18.9
Residence and age						
United States						
Total, all persons	251,447	31,439	210,257	100.0	100.0	100.0
All metropolitan areas	196,138	26,636	161,135	78.0	84.7	76.6
Inside central cities	76,004	17,505	54,570	30.2	55.7	26.0
Outside central cities	120,134	9,131	106,565	47.8	29.0	50.7
Nonmetropolitan areas	55,310	4,803	49,122	22.0	15.3	23.4
Male	122,528	14,781	102,965	100.0	100.0	100.0
All metropolitan areas	95,466	12,544	78,828	77.9	84.9	76.6
Inside central cities	36,396	8,132	26,336	29.7	55.0	25.6
Outside central cities	59,070	4,412	52,491	48.2	29.8	51.0
Nonmetropolitan areas	27,062	2,237	24,137	22.1	15.1	23.4
Female	128,919	16,658	107,293	100.0	100.0	100.0
All metropolitan areas	100,672	14,092	82,308	78.1	84.6	76.7
Inside central cities	39,607	9,372	28,234	30.7	56.3	26.3
Outside central cities	61,064	4,720	54,074	47.4	28.3	50.4
Nonmetropolitan areas	28,248	2,566	24,985	21.9	15.4	23.3

[Continued]

★ 925 ★

Distribution of Population: Residence, Region, Age, Gender, and Race, March 1992. Part I

[Continued]

Region, age, sex, and residence	Number			Percent distribution		
	All races	Black	White	All races	Black	White
Total, persons 55 years and over	51,740	4,772	45,577	100.0	100.0	100.0
All metropolitan areas	38,540	3,837	33,513	74.5	80.4	73.5
Inside central cities	15,059	2,814	11,629	29.1	59.0	25.5
Outside central cities	23,481	1,023	21,883	45.4	21.4	48.0
Nonmetropolitan areas	13,199	935	12,064	25.5	19.6	26.5
Male	22,835	2,035	20,162	100.0	100.0	100.0
All metropolitan areas	16,921	1,646	14,718	74.1	80.9	73.0
Inside central cities	6,315	1,168	4,864	27.7	57.4	24.1
Outside central cities	10,606	478	9,854	46.4	23.5	48.9
nonmetropolitan areas	5,914	390	5,444	25.9	19.1	27.0
Female	28,904	2,737	25,415	100.0	100.0	100.0
All metropolitan areas	21,619	2,191	18,795	74.8	80.1	74.0
Inside central cities	8,744	1,647	6,766	30.3	60.2	26.6
Outside central cities	12,875	544	12,029	44.5	19.9	47.3
Nonmetropolitan areas	7,285	546	6,621	25.2	19.9	26.0

Source: "Distribution of the Population, by Type of Residence, Region, Age, Sex, and Race: March 1992." Claude E. Bennett, *The Black Population in the United States, March 1992*, pp. 27-28.

★ 926 ★

Distribution

Distribution of Population: Residence, Region, Age, Gender, and Race, March 1992. Part II

Numbers in thousands.

Region, age, sex, and residence	Number			Percent distribution		
	All races	Black	White	All races	Black	White
Residence and age						
South						
Total	86,004	17,090	67,256	100.0	100.0	100.0
All metropolitan areas	62,013	12,589	48,153	72.1	73.7	71.6
Inside central cities	24,002	7,347	16,199	27.9	43.0	24.1
Outside central cities	38,011	5,242	31,954	44.2	30.7	47.5
Nonmetropolitan areas	23,991	4,501	19,102	27.9	26.3	28.4
Male	41,667	8,036	32,800	100.0	100.0	100.0
All metropolitan areas	30,116	5,949	23,537	72.3	74.0	71.8
Inside central cities	11,491	3,440	7,812	27.6	42.8	23.8
Outside central cities	18,625	2,509	15,725	44.7	31.2	47.9
Nonmetropolitan areas	11,551	2,087	9,263	27.7	26.0	28.2

[Continued]

★ 926 ★

Distribution of Population: Residence, Region, Age, Gender, and Race, March 1992. Part II

[Continued]

Region, age, sex, and residence	Number			Percent distribution		
	All races	Black	White	All races	Black	White
Female	44,337	9,054	34,456	100.0	100.0	100.0
All metropolitan areas	31,897	6,641	24,616	71.9	73.3	71.4
Inside central cities	12,511	3,907	8,387	28.2	43.2	24.3
Outside central cities	19,386	2,733	16,229	43.7	30.2	47.1
Nonmetropolitan areas	12,440	2,414	9,840	28.1	26.7	28.6
Total, persons 55 years and over	18,086	2,660	15,207	100.0	100.0	100.0
All metropolitan areas	12,172	1,766	10,256	67.3	66.4	67.4
Inside central cities	4,735	1,151	3,541	26.2	43.3	23.3
Outside central cities	7,437	615	6,715	41.1	23.1	44.2
Nonmetropolitan areas	5,913	894	4,951	32.7	33.6	32.6
Male	7,917	1,126	6,701	100.0	100.0	100.0
All metropolitan areas	5,337	754	4,517	67.4	67.0	67.4
Inside central cities	1,980	478	1,484	25.0	42.5	22.1
Outside central cities	3,358	276	3,033	42.4	24.5	45.3
Nonmetropolitan areas	2,580	371	2,185	32.6	33.0	32.6
Female	10,169	1,534	8,505	100.0	100.0	100.0
All metropolitan areas	6,835	1,011	5,739	67.2	65.9	67.5
Inside central cities	2,755	673	2,057	27.1	43.9	24.2
Outside central cities	4,080	339	3,682	40.1	22.1	43.3
Nonmetropolitan areas	3,334	522	2,767	32.8	34.1	32.5
North and West						
Total	165,444	14,349	143,002	100.0	100.0	100.0
All metropolitan areas	134,125	14,047	112,982	81.1	97.9	79.0
Inside central cities	52,002	10,158	38,371	31.4	70.8	26.8
Outside central cities	82,123	3,889	74,611	49.6	27.1	52.2
Nonmetropolitan areas	31,319	302	30,020	18.9	2.1	21.0
Male	80,861	6,745	70,165	100.0	100.0	100.0
All metropolitan areas	65,350	6,595	55,290	80.8	97.8	78.8
Inside central cities	24,905	4,693	18,524	30.8	69.6	26.4
Outside central cities	40,445	1,903	36,766	50.0	28.2	52.4
Nonmetropolitan areas	15,511	149	14,875	19.2	2.2	21.2
Female	84,582	7,604	72,837	100.0	100.0	100.0
All metropolitan areas	68,775	7,451	57,692	81.3	98.0	79.2
Inside central cities	27,096	5,465	19,847	32.0	71.9	27.2
Outside central cities	41,678	1,986	37,845	49.3	26.1	52.0
Nonmetropolitan areas	15,808	153	15,145	18.7	2.0	20.8
Total, persons 55 years and over	33,654	2,113	30,370	100.0	100.0	100.0

[Continued]

★ 926 ★

Distribution of Population: Residence, Region, Age, Gender, and Race, March 1992. Part II

[Continued]

Region, age, sex, and residence	Number			Percent distribution		
	All races	Black	White	All races	Black	White
All metropolitan areas	26,368	2,071	23,257	78.4	98.0	76.6
Inside central cities	10,324	1,663	8,088	30.7	78.7	26.6
Outside central cities	16,044	408	15,169	47.7	19.3	49.9
Nonmetropolitan areas	7,286	42	7,113	21.6	2.0	23.4
Male	14,918	910	13,460	100.0	100.0	100.0
All metropolitan areas	11,584	891	10,201	77.6	98.0	75.8
Inside central cities	4,335	689	3,380	29.1	75.8	25.1
Outside central cities	7,249	202	6,821	48.6	22.2	50.7
Nonmetropolitan areas	3,335	18	3,259	22.4	2.0	24.2
Female	18,736	1,203	16,910	100.0	100.0	100.0
All metropolitan areas	14,784	1,180	13,056	78.9	98.1	77.2
Inside central cities	5,989	974	4,708	32.0	80.9	27.8
Outside central cities	8,795	206	8,348	46.9	17.1	49.4
Nonmetropolitan areas	3,952	23	3,854	21.1	2.0	22.8

Source: "Distribution of the Population, by Type of Residence, Region, Age, Sex, and Race: March 1992." Claude E. Bennett, *The Black Population in the United States, March 1992*, pp. 27-28.

★ 927 ★

Distribution

Diverse Edge Cities, 1992

Ten of the most diverse edge cities percentage of black population (with select downtowns for comparisons).

	Percentage
(Atlanta DTN)	86.4
1. Southfield-Northland Mall area, MI	72.7
2. Lanham/Landover Area, MD	71.5
(Detroit DTN)	67.8
3. Memphis Airport Area, TN	67.4
4. The Research Triangle Park Area, NC	59.1
5. Security Boulevard, MD	49.8
(Oakland DTN)	42.1
6. Eisenhower Valley, VA	38.8
7. Crown Center, KS	36.1
(Washington, DC, DTN)	36.0
8. Greenspring/I-45, TX	33.5

[Continued]

★ 927 ★

Diverse Edge Cities, 1992
[Continued]

	Percentage
9. Silver Spring, MD	29.6
10. Texas Medical Center-Rice University, TX	28.6

Source: "Ten of the Most Diverse Edge Cities, Percentage of Black Population," *The State of Black America 1994*, p. 16. Primary source: *The Edge City News*, 1992. *Notes:* Attained by dividing the black population in the area by the total population in the area.

★ 928 ★

Distribution

Largest Metropolitan Areas: Racial and Hispanic Population, 1990

As of April 1. Areas as defined by U.S. Office of Management and Budget, December 31, 1992.

Metropolitan area[1]	Total population (1000)	Percent of total metropolitan population			
		Black	American Indian Eskimo, Aleut	Asian and Pacific Islander	Hispanic origin[2]
New York-Northern New Jersey-Long Island, NY-NJ-CT-PA CMSA	19,342	17.8	0.2	4.6	14.7
Los Angeles-Riverside-Orange County, CA CMSA	14,532	8.5	0.6	7.2	32.9
Chicago-Gary-Kenosha, IL-IN-WI CMSA	8,240	19.0	0.2	3.1	10.9
Washington-Baltimore, DC-MD-VA-WV CMSA	6,727	25.2	0.3	3.7	3.9
San Francisco-Oakland-San Jose, CA CMSA	6,253	8.6	0.7	14.8	15.5
Philadelphia-Wilmington-Atlantic City, PA-NJ-DE-MD CMSA	5,893	18.4	0.2	2.0	3.8
Boston-Brockton-Nashua, MA-NH-ME-CT CMSA	5,455	4.8	0.2	2.5	4.4
Detroit-Ann Arbor-Flint, MI CMSA	5,187	20.5	0.4	1.4	2.0
Dallas-Fort Worth, TX CMSA	4,037	14.0	0.5	2.4	13.0
Houston-Galveston-Brazoria, TX CMSA	3,731	17.9	0.3	3.5	20.7
Miami-Fort Lauderdale, FL CMSA	3,193	18.5	0.2	1.4	33.3
Seattle-Tacoma-Bremerton, WA CMSA	2,970	4.5	1.3	6.1	3.0
Atlanta, GA MSA	2,960	25.2	0.2	1.8	2.0
Cleveland-Akron, OH CMSA	2,860	15.6	0.2	1.0	1.9
Minneapolis-St. Paul, MN-WI MSA	2,539	3.5	1.0	2.6	1.5
San Diego, CA MSA	2,496	6.4	0.8	7.9	20.4
St. Louis, MO-IL MSA	2,493	17.0	0.2	1.0	1.1
Pittsburgh, PA MSA	2,395	7.5	0.1	0.7	0.6
Phoenix-Mesa, AZ MSA	2,238	3.5	2.2	1.6	17.0
Tampa-St. Petersburg-Clearwater, FL MSA	2,068	9.0	0.3	1.1	6.7
Denver-Boulder-Greeley, CO CMSA	1,980	5.0	0.7	2.2	12.8
Cincinnati-Hamilton, OH-KY-IN CMSA	1,818	11.2	0.1	0.8	0.5
Portland-Salem, OR-WA CMSA	1,793	2.5	1.0	3.2	4.0
Milwaukee-Racine, WI CMSA	1,607	13.3	0.5	1.2	3.8
Kansas City, MO-KS MSA	1,583	12.7	0.5	1.1	2.9
Sacramento-Yolo, CA CMSA	1,481	6.9	1.1	7.7	11.6
Norfolk-Virginia Beach-Newport News, VA-NC MSA	1,443	28.3	0.3	2.4	2.3
Indianapolis, IN MSA	1,380	13.2	0.2	0.8	0.9
Columbus, OH MSA	1,345	12.1	0.2	1.6	0.8
San Antonio, TX MSA	1,325	6.7	0.4	1.2	47.4
New Orleans, LA MSA	1,285	34.8	0.3	1.7	4.2
Orlando, FL MSA	1,225	12.0	0.3	1.7	8.2
Buffalo-Niagara Falls, NY MSA	1,189	10.3	0.6	0.9	2.0

[Continued]

★ 928 ★

Largest Metropolitan Areas: Racial and Hispanic Population, 1990

[Continued]

Metropolitan area[1]	Total population (1000)	Percent of total metropolitan population			
		Black	American Indian Eskimo, Aleut	Asian and Pacific Islander	Hispanic origin[2]
Charlotte-Gastonia-Rock Hill, NC-SC MSA	1,162	19.9	0.4	1.0	0.9
Hartford, CT MSA	1,158	8.3	0.2	1.5	6.9
Providence-Fall River-Warwick, RI-MA-MSA	1,134	3.3	0.3	1.8	4.2
Salt Lake City-Ogden, UT MSA	1,072	1.0	0.8	2.4	5.8
Rochester, NY MSA	1,062	8.9	0.3	1.3	3.0
Greensboro-Winston-Salem-High Point, NC MSA	1,050	19.3	0.3	0.7	0.7
Memphis, TN-AR-MS MSA	1,007	40.7	0.2	0.8	0.8
Nashville, TN MSA	985	15.5	0.2	1.0	0.8
Oklahoma City, OK MSA	959	10.5	4.8	1.9	3.6
Dayton-Springfield, OH MSA	951	13.3	0.2	1.0	0.8
Louisville, KY-IN MSA	949	12.9	0.2	0.6	0.6
Grand Rapids-Muskegon-Holland, MI MSA	938	6.9	0.6	0.9	3.1
Jacksonville, FL MSA	907	20.0	0.3	1.7	2.5
Richmond-Petersburg, VA MSA	866	29.2	0.3	1.4	1.1
West Palm Beach-Boca Raton, FL MSA	864	12.5	0.1	1.0	7.7
Albany-Schenectady-Troy, NY MSA	861	4.6	0.2	1.2	1.7
Raleigh-Durham-Chapel Hill, NC MSA	856	24.2	0.3	1.6	1.3
Las Vegas, NV-AZ MSA	853	8.4	1.1	3.1	10.4
Austin-San Marcos, TX MSA	846	9.4	0.4	2.2	20.9
Birmingham, Al MSA	840	28.7	0.2	0.5	0.4
Honolulu, HI MSA	836	3.1	0.4	63.0	6.8
Greenville-Spartanburg-Anderson, SC MSA	831	17.4	0.1	0.6	0.7
Fresno, CA MSA	756	4.8	1.1	7.7	35.3
Syracuse, NY MSA	742	5.7	0.6	1.1	1.4
Tulsa, OK MSA	709	8.2	6.8	0.9	2.1
Tucson, AZ MSA	667	3.1	3.0	1.8	24.5
Omaha, NE-IA MSA	640	8.0	0.5	1.0	2.6
Scranton-Wilkes-Barre-Hazleton, PA MSA	638	0.9	0.1	0.5	0.6
Toledo, OH MSA	614	11.4	0.2	1.0	3.3
Youngstown-Warren, OH MSA	601	9.4	0.2	0.4	1.3
Allentown-Bethlehem-Easton, PA MSA	595	2.0	0.1	1.1	4.6
El Paso, TX MSA	592	3.7	0.4	1.1	69.6
Alburquerque, NM MSA	589	2.5	5.1	1.4	37.1
Harrisburg-Lebanon-Carlisle, PA MSA	588	6.7	0.1	1.1	1.7
Springfield, MA MSA	588	6.3	0.2	1.4	8.5
Knoxville, TN MSA	586	6.1	0.2	0.8	0.5
Bakersfield, CA MSA	543	5.5	1.3	3.0	28.0
Little Rock-North Little Rock, AR MSA	513	19.9	0.4	0.7	0.8
Charleston-North Charleston, SC MSA	507	30.2	0.3	1.2	1.5
Sarasota-Bradenton, FL MSA	489	5.8	0.2	0.5	3.1
Wichita, KS MSA	485	7.6	1.1	1.9	4.1
Stockton-Lodi, CA MSA	481	5.6	1.1	12.4	23.4

Source: "Largest Metropolitan Areas—Racial and Hispanic Origin Populations: 1990," U.S. Bureau of the Census, *Statistical Abstract of the United States*, 1993, p. 41. Primary source: U.S. Bureau of the Census, unpublished data. *Notes:* 1. Metropolitan areas are shown in rank order of total population of consolidated metropolitan statistical areas (CMSA0 and metropolitan areas (MSA). 2. Persons of Hispanic origin may be of any race.

★ 929 ★

Distribution

Resident Population, by Race and Hispanic Origin: States, 1990. Part I

In thousands. As of April 1.

Region, division and state	Total[1]	White	Black	American Indian, Eskimo, Aleut				Asian, Pacific Islander			
				Total	American Indian	Eskimo	Aleut	Total[2]	Chinese	Filipino	Japanese
U.S.	248,710	199,686	29,986	1,959	1,878	57	24	7,274	1,645	1,407	848
Northeast	50,809	42,069	5,613	125	122	2	2	1,335	445	143	74
New England	13,207	12,033	628	33	32	(Z)	(Z)	232	72	15	15
Maine	1,228	1,208	5	6	6	(Z)	(Z)	7	1	1	1
New Hampshire	1,109	1,087	7	2	2	(Z)	(Z)	9	2	1	1
Vermont	563	555	2	2	2	(Z)	(Z)	3	1	(Z)	(Z)
Massachusetts	6,016	5,405	300	12	12	(Z)	(Z)	143	54	6	9
Rhode Island	1,003	917	39	4	4	(Z)	(Z)	18	3	2	1
Connecticut	3,187	2,859	274	7	6	(Z)	(Z)	51	11	5	4
Middle Atlantic	37,602	30,036	4,986	92	90	1	2	1,104	373	128	59
New York	17,990	13,385	2,859	63	31	1	1	694	284	62	35
New Jersey	7,730	6,130	1,037	15	15	(Z)	(Z)	273	59	53	17
Pennsylvania	11,882	10,520	1,090	15	14	(Z)	(Z)	137	30	12	7
Midwest	59,669	52,018	5,716	338	334	2	2	768	133	113	63
East North Central	42,009	35,764	4,817	150	147	1	1	573	103	97	50
Ohio	10,847	9,522	1,155	20	20	(Z)	(Z)	91	19	10	10
Indiana	5,544	5,021	432	13	12	(Z)	(Z)	38	7	5	5
Illinois	11,431	8,953	1,694	22	21	(Z)	(Z)	285	50	64	22
Michigan	9,295	7,756	1,292	56	55	(Z)	(Z)	105	19	14	11
Wisconsin	4,892	4,513	245	39	39	(Z)	(Z)	54	7	4	3
West North Central	17,660	16,254	899	188	187	1	1	195	30	17	13
Minnesota	4,375	4,130	95	50	49	(Z)	(Z)	78	9	4	4
Iowa	2,777	2,683	48	7	7	(Z)	(Z)	25	4	2	2
Missouri	5,117	4,486	548	20	20	(Z)	(Z)	41	9	6	3
North Dakota	639	604	4	26	26	(Z)	(Z)	3	1	1	(Z)
South Dakota	696	638	3	51	51	(Z)	(Z)	3	(Z)	1	(Z)
Nebraska	1,578	1,481	57	12	12	(Z)	(Z)	12	2	1	2
Kansas	2,478	2,232	143	22	22	(Z)	(Z)	32	5	3	2
South	85,446	65,582	15,829	563	557	3	3	1,122	204	159	67
South Atlantic	43,567	33,391	8,924	172	170	1	1	631	114	108	39
Delaware	666	535	112	2	2	(Z)	(Z)	9	2	1	1
Maryland	4,781	3,394	1,190	13	13	(Z)	(Z)	140	31	19	7
District of Columbia	607	180	400	1	1	(Z)	(Z)	11	3	2	1
Virginia	6,187	4,792	1,163	15	15	(Z)	(Z)	159	21	35	8
West Virginia	1,793	1,726	56	2	2	(Z)	(Z)	7	1	2	1
North Carolina	6,629	5,008	1,456	80	80	(Z)	(Z)	52	9	5	5
South Carolina	3,487	2,407	1,040	8	8	(Z)	(Z)	22	3	6	2
Georgia	6,478	4,600	1,747	13	13	(Z)	(Z)	76	13	6	6
Florida	12,938	10,749	1,760	36	35	(Z)	(Z)	154	31	32	9
East South Central	15,176	12,049	2,977	41	40	(Z)	(Z)	84	15	9	9
Kentucky	3,685	3,392	263	6	6	(Z)	(Z)	18	3	2	3
Tennessee	4,877	4,048	778	10	10	(Z)	(Z)	32	6	3	3
Alabama	4,041	2,976	1,021	17	16	(Z)	(Z)	22	4	2	2
Mississippi	2,573	1,633	915	9	8	(Z)	(Z)	13	3	2	1
West South Central	26,703	20,142	3,929	350	347	1	1	407	76	43	20
Arkansas	2,351	1,945	374	13	13	(Z)	(Z)	13	2	2	1
Louisiana	4,220	2,839	1,299	19	18	(Z)	(Z)	41	5	4	2

[Continued]

★ 929 ★

Resident Population, by Race and Hispanic Origin: States, 1990. Part I

[Continued]

Region, division and state	Total[1]	White	Black	American Indian, Eskimo, Aleut				Asian, Pacific Islander			
				Total	American Indian	Eskimo	Aleut	Total[2]	Chinese	Filipino	Japanese
Oklahoma	3,146	2,584	234	252	252	(Z)	(Z)	34	5	3	2
Texas	16,987	12,775	2,022	66	64	1	1	319	63	34	15
West	52,786	40,017	2,828	933	866	51	17	4,048	863	991	643
Mountain	13,659	11,762	374	481	478	1	1	217	40	32	34
Montana	799	741	2	48	48	(Z)	(Z)	4	1	1	1
Idaho	1,007	950	3	14	14	(Z)	(Z)	9	1	1	3
Wyoming	454	427	4	9	9	(Z)	(Z)	3	1	(Z)	1
Colorado	3,294	2,905	133	28	27	(Z)	(Z)	60	9	5	11
New Mexico	1,515	1,146	30	134	134	(Z)	(Z)	14	3	2	2
Arizona	3,665	2,963	111	204	203	(Z)	(Z)	55	14	8	6
Utah	1,723	1,616	12	24	24	(Z)	(Z)	33	5	2	7
Nevada	1,202	1,013	79	20	19	(Z)	(Z)	38	7	12	4
Pacific	39,127	28,255	2,454	453	387	49	16	3,831	823	960	609
Washington	4,867	4,309	150	81	78	2	2	211	34	44	34
Oregon	2,842	2,637	46	38	37	1	1	69	14	7	12
California	29,760	20,524	2,209	242	236	3	4	2,846	705	732	313
Alaska	550	415	22	86	31	44	10	20	1	8	2
Hawaii	1,108	370	27	5	5	(Z)	(Z)	685	69	169	247

Source: "Resident Population, by Race and Hispanic Origin—States: 1990." U.S. Bureau of the Census, *Statistical Abstract of the United States,* 1993, p. 30. Primary source: U.S. Bureau of the Census, *1990 Census of Population, General Population Characteristics, United States* (CP-1-1). *Notes:* Z Less than 500. 1. Includes other races, not shown separately. 2. Includes other Asian and Pacific Islander races not shown separately.

★ 930 ★

Distribution

Resident Population, by Race and Hispanic Origin: States, 1990. Part II

In thousands. As of April 1.

Region, division and state	Asian, Pacific Islander						Hispanic origin[1]				
	Asian Indian	Korean	Vietnamese	Hawaiian	Samoan	Guamanian	Total	Mexican	Puerto Rican	Cuban	Other Hispanic
U.S.	815	799	615	211	63	49	22,354	13,496	2,728	1,044	5,086
Northeast	285	182	61	4	2	4	3,754	175	1,872	184	1,524
New England	36	21	22	1	(Z)	1	568	29	316	16	207
Maine	1	1	1	(Z)	(Z)	(Z)	7	2	1	(Z)	3
New Hampshire	2	2	1	(Z)	(Z)	(Z)	11	2	3	1	5
Vermont	1	1	(Z)	(Z)	(Z)	(Z)	4	1	1	(Z)	2
Massachusetts	20	12	15	1	(Z)	(Z)	288	13	151	8	116
Rhode Island	2	1	1	(Z)	(Z)	(Z)	46	2	13	1	29
Connecticut	12	5	4	(Z)	(Z)	(Z)	213	8	147	6	51
Middle Atlantic	249	161	39	3	1	3	3,186	146	1,556	167	1,317
New York	141	96	16	1	1	2	2,214	93	1,087	74	960
New Jersey	79	39	7	1	(Z)	1	740	29	320	85	306
Pennsylvania	28	27	16	1	(Z)	(Z)	232	24	149	7	52
Midwest	146	109	52	6	2	3	1,727	1,153	258	37	279
East North Central	123	80	26	3	1	2	1,438	944	244	30	220
Ohio	21	11	5	1	(Z)	(Z)	140	58	46	4	32
Indiana	7	5	2	1	(Z)	(Z)	99	67	14	2	16
Illinois	64	42	10	1	(Z)	1	904	624	146	18	116
Michigan	24	16	6	1	(Z)	(Z)	202	138	19	5	40

[Continued]

★ 930 ★

Resident Population, by Race and Hispanic Origin: States, 1990. Part II

[Continued]

Region, division and state	Asian, Pacific Islander						Hispanic origin[1]				
	Asian Indian	Korean	Vietnamese	Hawaiian	Samoan	Guamanian	Total	Mexican	Puerto Rican	Cuban	Other Hispanic
Wisconsin	7	6	2	(Z)	(Z)	(Z)	93	58	19	2	15
West North Central	23	29	26	2	1	1	289	209	14	6	60
Minnesota	8	12	9	(Z)	(Z)	(Z)	54	35	3	2	14
Iowa	3	5	3	(Z)	(Z)	(Z)	33	24	1	(Z)	7
Missouri	6	6	4	1	1	(Z)	62	38	4	2	17
North Dakota	(Z)	1	(Z)	(Z)	(Z)	(Z)	5	3	(Z)	(Z)	1
South Dakota	(Z)	1	(Z)	(Z)	(Z)	(Z)	5	3	(Z)	(Z)	1
Nebraska	1	2	2	(Z)	(Z)	(Z)	37	30	1	(Z)	6
Kansas	4	4	7	(Z)	(Z)	(Z)	94	76	4	1	13
South	196	153	169	12	4	8	6.767	4,344	406	735	1,282
South Atlantic	114	101	62	7	2	5	2,133	315	338	702	778
Delaware	2	1	(Z)	(Z)	(Z)	(Z)	16	3	8	1	4
Maryland	28	30	9	1	(Z)	1	125	18	18	6	83
District of Columbia	2	1	1	(Z)	(Z)	(Z)	33	3	2	1	26
Virginia	20	30	21	1	(Z)	1	160	33	24	6	97
West Virginia	2	1	(Z)	(Z)	(Z)	(Z)	8	3	1	(Z)	5
North Carolina	10	7	5	1	(Z)	1	77	33	15	4	26
South Carolina	4	3	2	(Z)	(Z)	(Z)	31	11	6	2	11
Georgia	14	15	8	1	(Z)	1	109	49	17	8	34
Florida	31	12	16	2	1	1	1,574	161	247	674	492
East South Central	15	12	10	1	(Z)	1	95	39	13	5	39
Kentucky	3	3	2	(Z)	(Z)	(Z)	22	9	4	1	9
Tennessee	6	5	2	1	(Z)	(Z)	33	14	4	2	13
Alabama	4	3	2	(Z)	(Z)	(Z)	25	10	4	1	10
Mississippi	2	1	4	(Z)	(Z)	(Z)	16	7	1	(Z)	7
West South Central	67	40	97	4	1	3	4,539	3,990	55	28	466
Arkansas	1	1	2	(Z)	(Z)	(Z)	20	12	1	(Z)	6
Louisiana	5	3	18	(Z)	(Z)	(Z)	93	23	6	9	55
Oklahoma	5	5	7	1	(Z)	(Z)	86	63	5	1	17
Texas	56	32	70	3	1	2	4,340	3,891	43	18	388
West	189	355	334	189	55	34	10,106	7,824	192	88	2,002
Mountain	15	28	20	7	3	2	1,992	1,440	26	12	514
Montana	(Z)	1	(Z)	(Z)	(Z)	(Z)	12	8	(Z)	(Z)	3
Idaho	(Z)	1	1	(Z)	(Z)	(Z0	53	13	1	(Z)	9
Wyoming	(Z)	(Z)	(Z)	(Z)	(Z)	(Z)	26	19	(Z)	(Z)	7
Colorado	4	11	7	1	(Z)	1	424	282	7	2	133
New Mexico	2	1	1	(Z)	(Z)	(Z)	579	329	3	1	247
Arizona	6	6	5	2	(Z)	1	688	616	8	2	62
Utah	2	3	3	1	2	(Z)	85	57	2	(Z)	25
Nevada	2	4	2	2	(Z)	(Z)	124	85	4	6	29
Pacific	173	327	314	182	52	32	8,114	6,384	166	76	1,488
Washington	8	30	19	5	4	4	215	156	9	2	47
Oregon	4	9	9	2	1	1	113	86	3	1	23
California	160	260	280	34	32	25	7,688	6,119	126	72	1,371
Alaska	(Z)	4	1	1	1	(Z)	18	9	2	(Z)	6
Hawaii	1	24	5	139	15	2	81	14	26	1	41

Source: "Resident Population, by Race and Hispanic Origin—States: 1990." U.S. Bureau of the Census, *Statistical Abstract of the United States*, 1993, p. 31. Primary source: U.S. Bureau of the Census, *1990 Census of Population, General Population Characteristics, United States* (CP-1-1). *Notes:* Z Less than 500. 1. Persons of Hispanic origin may be of any race.

★ 931 ★

Distribution

Resident Population, by Region, Race, and Hispanic Origin, 1990

Race and Hispanic origin	Population (1,000)					Percent distribution				
	United States	Northeast	Midwest	South	West	United States	Northeast	Midwest	South	West
Total	248,710	50,809	59,669	85,446	52,786	100.0	20.4	24.0	34.4	21.2
White	199,686	42,069	52,018	65,582	40,017	100.0	21.1	26.0	32.8	20.0
Black	29,986	5,613	5,716	15,829	2,828	100.0	18.7	19.1	52.8	9.4
American Indian, Eskimo, Aleut	1,959	125	338	563	933	100.0	6.4	17.2	28.7	47.6
American Indian	1,878	122	334	557	866	100.0	6.5	17.8	29.7	46.1
Eskimo	57	2	2	3	51	100.0	2.9	3.5	4.9	88.8
Aleut	24	2	2	3	17	100.0	8.1	8.1	11.5	72.3
Asian or Pacific Islander	7,274	1,335	768	1,122	4,048	100.0	18.4	10.6	15.4	55.7
Chinese	1,645	445	133	204	863	100.0	27.0	8.1	12.4	52.4
Filipino	1,407	143	113	159	991	100.0	10.2	8.1	11.3	70.5
Japanese	848	74	63	67	643	100.0	8.8	7.5	7.9	75.9
Asian Indian	815	285	146	196	189	100.0	35.0	17.9	24.0	23.1
Korean	799	182	109	153	355	100.0	22.8	13.7	19.2	44.4
Vietnamese	615	61	52	169	334	100.0	9.8	8.5	27.4	54.3
Laotian	149	16	28	29	76	100.0	10.7	18.6	19.6	51.0
Cambodian	147	30	13	19	85	100.0	20.5	8.8	13.1	57.7
Thai	91	12	13	24	43	100.0	12.9	14.2	26.0	46.8
Hmong	90	2	37	2	50	100.0	1.9	41.3	1.8	55.0
Pakistani	81	26	15	22	17	100.0	34.3	18.9	26.5	20.4
Hawaiian	211	4	6	12	189	100.0	2.0	2.6	5.8	89.6
Samoan	63	2	2	4	55	100.0	2.4	3.6	6.4	87.6
Guamanian	49	4	3	8	34	100.0	7.3	6.4	16.8	69.5
Other Asian or Pacific Islander	263	49	34	54	126	100.0	18.5	12.9	20.6	48.0
Other races	9,805	1,667	829	2,350	4,960	100.0	17.0	8.5	24.0	50.6
Hispanic origin[1]	22,354	3,754	1,727	6,767	10,106	100.0	16.8	7.7	30.3	45.2
Mexican	13,496	175	1,153	4,344	7,824	100.0	1.3	8.5	32.2	58.0
Puerto Rican	2,728	1,872	258	406	192	100.0	68.6	9.4	14.9	7.0
Cuban	1,044	184	37	735	88	100.0	17.6	3.5	70.5	8.5
Other Hispanic	226,356	47,055	57,942	78,679	42,680	100.0	20.8	25.6	34.8	18.9
Not of Hispanic origin	226,356	47,055	57,942	78,679	42,680	100.0	20.8	25.6	34.8	18.9

Source: "Resident Population, by Region, Race, and Hispanic Origin: 1990." U.S. Bureau of the Census, *Statistical Abstract of the United States*, 1993, p. 32. Primary source: U.S. Bureau of the Census, *1990 Census of Population, General Population Characteristics, United States* (CP-1-1). *Note:* 1. Persons of Hispanic origin may be of any race.

Economic Characteristics

★ 932 ★

Economic Characteristics of Persons and Families: Gender and Race, 1991

Numbers in thousands.

Characteristic	All races	Black	White
Poverty Status of families with householder 55 years old and over			
All families	20,432	1,912	18,015
Number below poverty level	1,478	447	959
Percent below poverty level	7.2	23.4	5.3
Married couple	16,946	1,060	15,491
Number below poverty level	919	155	703
Percent below poverty level	5.4	14.7	4.5
Female householder, no spouse present	2,750	691	1,975
Number below poverty level	474	257	208
Percent below poverty level	17.2	37.2	10.5
Male householder, no spouse present	735	160	548
Number below poverty level	84	34	48
Percent below poverty level	11.5	21.3	8.7

Source: "Selected Economic Characteristics of Persons and Families, by Sex and Race: March 1992." Claudette E. Bennett, *The Black Population in the United States: March 1992*, p. 26.

★ 933 ★

Economic Characteristics

Economic Characteristics of Persons and Families: Gender and Race, 1992

In 1991 dollars.

Income and earnings	1991			1989			1982			1979		
	Black	White	Ratio: Black to White	Black	White	Ratio: Black to White	Black	White	Ratio: Black to White	Black	White	Ratio: Black to White
Median income												
Households (dollars)	18,807	31,569	0.60	19,862	33,398	0.59	17,051	30,085	0.57	18,650	31,766	0.59
Standard error (dollars)	395	153	(X)	404	163	(X)	274	137	(X)	341	151	(X)
Families (dollars)	21,548	37,782	0.57	22,197	39,514	0.56	19,373	35,052	0.55	21,302	37,619	0.57
Standard error (dollars)	445	210	(X)	489	199	(X)	472	174	(X)	388	164	(X)
Persons												
Male (dollars)	12,962	21,395	0.61	13,850	22,916	0.60	12,591	21,011	0.60	14,019	22,648	0.62
Standard error (dollars)	335	112	(X)	335	122	(X)	331	138	(X)	283	118	(X)
Female (dollars)	8,814	10,722	0.82	8,650	10,777	0.80	7,498	8,501	0.88	7,358	8,085	0.91
Standard error (dollars)	195	75	(X)	263	81	(X)	180	60	(X)	162	70	(X)
Median income by type of family												
Married couple (dollars)	33,307	41,506	0.80	33,666	43,066	0.78	29,329	37,673	0.78	30,923	40,039	0.77

[Continued]

★ 933 ★

Economic Characteristics of Persons and Families: Gender and Race, 1992

[Continued]

Income and earnings	1991			1989			1982			1979		
	Black	White	Ratio: Black to White	Black	White	Ratio: Black to White	Black	White	Ratio: Black to White	Black	White	Ratio: Black to White
Standard error (dollars)	758	213	(X)	733	247	(X)	510	170	(X)	593	169	(X)
Female householder, no spouse present (dollars)	11,414	19,552	0.58	12,774	20,810	0.61	10,625	19,228	0.55	12,635	21,041	0.60
Standard error (dollars)	414	386	(X)	395	417	(X)	322	353	(X)	342	311	(X)
Male householder, no spouse present (dollars)	24,508	28,924	0.85	20,205	33,487	0.60	20,887	30,511	0.68	22,850	32,506	0.70
Standard error (dollars)	2,118	785	(X)	880	813	(X)	1,396	855	(X)	1,717	1,025	(X)
Median earnings of persons												
Both sexes (dollars)	13,771	17,687	0.78	14,436	18,373	0.79	12,778	16,082	0.79	13,403	17,229	0.78
Standard error (dollars)	281	118	(X)	295	89	(X)	288	76	(X)	191	109	(X)
Male (dollars)	15,494	22,732	0.68	16,827	24,338	0.69	14,971	22,722	0.66	16,436	24,974	0.66
Standard error (dollars)	309	208	(X)	336	126	(X)	278	132	(X)	449	169	(X)
Female (dollars)	12,210	12,994	0.94	12,658	12,877	0.98	11,114	10,883	1.02	10,865	11,001	0.99
Standard error (dollars)	221	136	(X)	253	96	(X)	229	87	(X)	283	101	(X)
Median earnings of year-round, full-time workers												
Both sexes (dollars)	20,453	25,721	0.80	20,964	26,538	0.79	19,343	25,359	0.76	20,601	26,489	0.78
Standard error (dollars)	229	92	(X)	270	143	(X)	309	134	(X)	248	129	(X)
Male (dollars)	22,075	30,266	0.73	22,436	31,349	0.72	22,087	30,776	0.72	23,261	32,033	0.73
Standard error (dollars)	326	125	(X)	301	249	(X)	373	141	(X)	462	129	(X)
Female (dollars)	18,720	20,792	0.90	19,100	20,784	0.92	17,284	18,749	0.92	17,345	18,829	0.92
Standard error (dollars)	349	105	(X)	305	153	(X)	215	134	(X)	328	101	(X)
Per capita money income												
Per capita income (dollars)	9,170	15,510	0.59	9,608	16,362	0.59	7,636	13,573	0.56	8,179	13,940	0.59

Source: "Selected Economic Characteristics of Persons and Families, by Sex and Race: March 1992," Claudette E. Bennett, *The Black Population in the United States: March 1992*, p. 25.

Education Level

★ 934 ★

Education of Persons 25 Years Old and Older, by Gender, Region, and Race, March 1991. Part I

Numbers in thousands.

Characteristic	Black			White		
	Both sexes	Male	Female	Both sexes	Male	Female
United States						
Total, 25 years old and over	17,445	7,803	9,641	137,657	66,063	71,594
Percent	100.0	100.0	100.0	100.0	100.0	100.0
Elementary						
Total	13.3	14.4	12.3	9.0	9.2	8.9

[Continued]

★ 934 ★

Education of Persons 25 Years Old and Older, by Gender, Region, and Race, March 1991. Part I

[Continued]

Characteristic	Black			White		
	Both sexes	Male	Female	Both sexes	Male	Female
None to 4th grade	3.9	4.7	3.2	1.8	1.9	1.7
5th to 8th grade	9.4	9.7	9.1	7.2	7.2	7.2
High school						
Total	54.7	55.0	54.5	46.5	43.5	49.2
9th to 12th grade (no diploma)	19.1	18.5	19.5	10.1	9.8	10.4
High school graduate	35.7	36.4	35.0	36.4	33.7	38.8
College						
Total	32.0	30.6	33.1	44.5	47.4	41.9
Some college or associate degree	20.1	18.7	21.2	22.5	22.2	22.7
Bachelor's degree or more	11.9	11.9	12.0	22.1	25.2	19.1
Percent high school graduate or more	67.7	67.0	68.2	80.9	91.1	80.7
Total, 25 to 34 years old	5,423	2,505	2,918	35,320	17,736	17,584
Percent	100.0	100.0	100.0	100.0	100.0	100.0
Elementary						
Total	2.6	2.7	2.5	4.1	4.5	3.8
None to 4th grade	0.9	1.0	0.9	1.0	1.1	0.9
5th to 8th grade	1.6	1.7	1.6	3.1	3.4	2.8
High school						
Total	59.3	61.5	57.4	46.1	47.1	45.1
9th to 12th grade (no diploma)	15.6	15.1	16.1	8.8	9.2	8.4
High school graduate	43.6	46.3	41.3	37.3	37.9	36.7
College						
Total	38.1	35.8	40.1	49.8	48.4	51.2
Some college or associate degree	26.1	23.6	28.3	25.6	24.3	26.8
Bachelor's degree or more	12.0	12.2	11.8	24.2	24.1	24.3
Percent high school graduate or more	81.8	82.2	81.4	87.1	86.3	87.8
Total, 35 to 44 years old	4,462	2,027	2,435	33,501	16,738	16,763
Percent	100.0	100.0	100.0	100.0	100.0	100.0
Elementary						
Total	3.6	4.8	2.7	4.1	4.4	3.9
None to 4th grade	0.7	1.1	0.5	1.2	1.3	1.2
5th to 8th grade	2.9	3.7	2.2	2.9	3.1	2.8
High school						
Total	54.6	55.3	53.9	40.6	38.9	42.3
9th to 12th grade (no diploma)	15.5	15.1	15.8	6.7	6.7	6.6
High school graduate	39.1	40.2	38.1	33.9	32.2	35.6

[Continued]

★ 934 ★

Education of Persons 25 Years Old and Older, by Gender, Region, and Race, March 1991. Part I

[Continued]

Characteristic	Black			White		
	Both sexes	Male	Female	Both sexes	Male	Female
College						
Total	41.8	39.9	43.4	55.3	56.8	53.8
Some college or associate degree	25.1	23.5	26.4	27.7	27.5	28.0
Bachelor's degree or more	16.7	16.4	17.0	27.6	29.3	25.8
Percent high school graduate or more	80.9	80.1	81.5	89.2	89.0	89.4

Source: "Educational Attainment of Persons 25 Years Old and Over, by Sex, Region, and Race: March 1992," Claudette E. Bennett, *The Black Population in the United States: March 1992*, p. 37.

★ 935 ★

Education Level

Education of Persons 25 Years Old and Older, by Gender, Region, and Race, March 1991. Part II

Numbers in thousands.

Characteristic	Black			White		
	Both sexes	Male	Female	Both sexes	Male	Female
South						
Total, 25 years old and over	9,443	4,173	5,270	44,599	21,348	23,251
Percent	100.0	100.0	100.0	100.0	100.0	100.0
Elementary						
Total	16.1	17.7	14.8	10.4	11.0	9.9
None to 4th grade	5.4	6.6	4.5	2.3	2.6	1.9
5th to 8th grade	10.6	11.1	10.2	8.1	8.4	7.9
High school						
Total	54.6	55.0	54.2	46.8	43.4	49.9
9th to 12th grade (no diploma)	19.6	18.9	20.2	11.4	11.0	11.8
High school graduate	35.0	36.1	34.1	35.4	32.4	38.1
College						
Total	29.4	27.4	31.0	42.8	45.6	40.2
Some college or associate degree	17.9	16.3	19.2	22.3	22.2	22.3
Bachelor's degree or more	11.5	11.1	11.8	20.5	23.4	17.9
Percent high school graduate or more	64.3	63.4	65.1	78.2	78.1	78.3
Total, 25 to 34 years old	2,852	1,310	1,542	11,233	5,620	5,613
Percent	100.0	100.0	100.0	100.0	100.0	100.0
Elementary						
Total	2.5	2.7	2.3	4.5	5.1	3.8
None to 4th grade	0.9	0.7	1.1	1.0	1.3	0.6
5th to 8th grade	1.6	2.0	1.2	3.5	3.8	3.2

[Continued]

★ 935 ★

Education of Persons 25 Years Old and Older, by Gender, Region, and Race, March 1991. Part II

[Continued]

Characteristic	Black			White		
	Both sexes	Male	Female	Both sexes	Male	Female
High school						
Total	62.2	64.9	59.8	47.3	48.5	46.1
9th to 12th grade (no diploma)	14.9	13.8	15.9	10.0	10.7	9.3
High school graduate	47.3	51.1	44.0	37.3	37.7	36.8
College						
Total	35.3	32.4	37.8	48.2	46.4	50.1
Some college or associate degree	24.3	22.4	25.9	25.3	25.1	25.5
Bachelor's degree or more	11.1	10.0	12.0	23.0	21.4	24.6
Percent high school graduate or more	82.6	83.5	81.8	85.5	84.2	86.8
Total, 35 to 44 years old	2,493	1,133	1,359	10,553	5,267	5,286
Percent	100.0	100.0	100.0	100.0	100.0	100.0
Elementary						
Total	4.2	5.5	3.0	4.5	4.8	4.1
None to 4th grade	1.1	1.5	0.7	1.1	1.3	0.9
5th to 8th grade	3.1	4.0	2.4	3.4	3.5	3.2
High school						
Total	56.0	56.2	55.9	42.4	40.2	44.6
9th to 12th grade (no diploma)	17.0	15.4	18.3	8.3	8.7	8.0
High school graduate	39.1	40.8	37.6	34.0	31.5	36.5
College						
Total	39.8	38.2	41.1	53.2	55.0	51.3
Some college or associate degree	23.6	22.1	24.8	27.5	27.4	27.5
Bachelor's degree or more	16.2	16.1	16.3	25.7	27.6	23.8
Percent high school graduate or more	78.9	79.1	78.7	87.2	86.6	87.9

Source: "Educational Attainment of Persons 25 Years Old and Over, by Sex, Region, and Race: March 1992," Claudette E. Bennett, *The Black Population in the United States: March 1992*, pp. 37-38.

★ 936 ★

Education Level

Education of Persons 25 Years Old and Older, by Gender, Region, and Race, March 1991. Part III

Numbers in thousands.

Characteristic	Black			White		
	Both sexes	Male	Female	Both sexes	Male	Female
North and West						
Total, 25 years old and over	8,001	3,630	4,371	93,058	44,715	48,344
Percent	100.0	100.0	100.0	100.0	100.0	100.0
Elementary						
Total	10.0	10.7	9.4	8.4	8.3	8.4
None to 4th grade	2.1	2.6	1.7	1.6	1.6	1.6
5th to 8th grade	7.9	8.1	7.7	6.8	6.7	6.8
High school						
Total	54.9	55.0	54.9	46.3	43.5	48.9
9th to 12th grade (no diploma)	18.4	18.1	18.7	9.5	9.2	9.8
High school graduate	36.5	36.8	36.2	36.8	34.3	39.1
College						
Total	35.1	34.3	35.7	45.3	48.2	42.6
Some college or associate degree	22.6	21.5	23.5	22.5	22.2	22.9
Bachelor's degree or more	12.5	12.8	12.2	22.8	26.0	19.7
Percent high school graduate or more	71.6	71.1	71.9	82.1	82.5	81.6
Total, 25 to 34 years old	2,571	1,195	1,376	24,088	12,117	11,971
Percent	100.0	100.0	100.0	100.0	100.0	100.
Elementary						
Total	2.6	2.6	2.7	4.0	4.2	3.7
None to 4th grade	0.9	1.3	0.7	1.0	1.0	1.1
5th to 8th grade	1.7	1.4	2.0	2.9	3.2	2.6
High school						
Total	56.1	57.7	54.7	45.5	46.5	44.6
9th to 12th grade (no diploma)	18.5	16.6	16.3	8.2	8.5	8.0
High school graduate	39.6	41.1	38.4	37.3	38.0	36.6
College						
Total	41.3	39.6	42.7	50.5	49.3	51.7
Some college or associate degree	28.2	25.0	31.1	25.7	24.0	27.5
Bachelor's degree or more	13.0	14.7	11.6	24.7	25.3	24.2
Total, 35 to 44 years old	1,970	894	1,076	22,948	11,471	11,477
Percent	100.0	100.0	100.0	100.0	100.0	100.0
Elementary						
Total	2.9	3.8	2.2	4.0	4.2	3.8
None to 4th grade	0.3	0.5	0.2	1.3	1.3	1.3
5th to 8th grade	2.6	3.3	2.0	2.7	2.8	2.6

[Continued]

★ 936 ★

Education of Persons 25 Years Old and Older, by Gender, Region, and Race, March 1991. Part III

[Continued]

Characteristic	Black			White		
	Both sexes	Male	Female	Both sexes	Male	Female
High school						
Total	52.7	54.2	51.5	39.7	38.2	41.2
9th to 12th grade (no diploma)	13.6	14.8	12.6	5.9	5.8	6.0
High school graduate	39.1	39.5	38.8	33.8	32.5	35.2
College						
Total	44.3	42.0	46.3	56.3	57.6	54.9
Some college or associate degree	27.0	25.2	28.5	27.8	27.5	28.2
Bachelor's degree or more	17.3	16.8	17.8	28.4	30.1	26.7
Percent high school graduate or more	83.5	81.4	85.1	90.1	90.1	90.2

Source: "Educational Attainment of Persons 25 Years Old and Over, by Sex, Region, and Race: March 1992," Claudette E. Bennett, *The Black Population in the United States: March 1992*, pp. 38-39.

★ 937 ★

Education Level

Percent of Persons 25 Years Old and Over, by Education, Gender, and Race, 1991

[Percent]

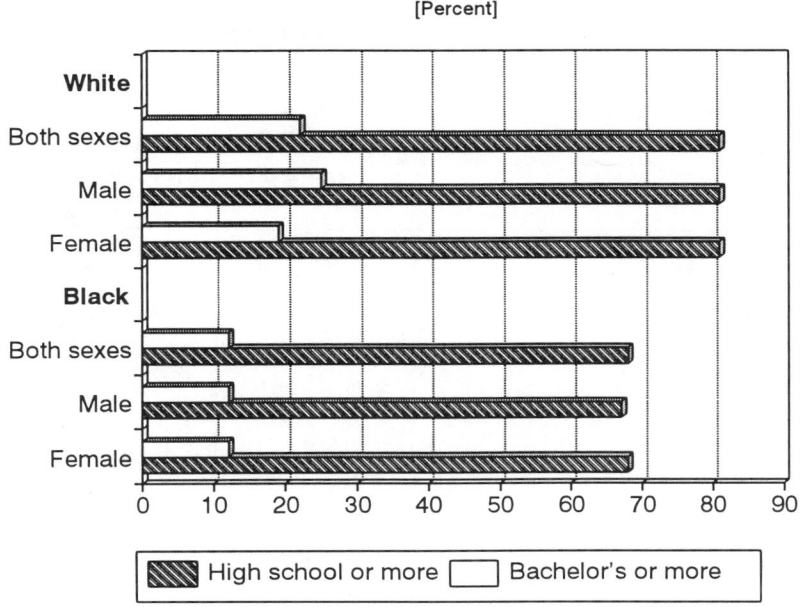

Education	Black			White		
	Both sexes	Male	Female	Both sexes	Male	Female
High school graduate or more	68	67	68	81	81	81
Bachelor's degree or more	12	12	12	22	25	19

Source: "Percent of Persons 25 Years Old and Over, by Education, Sex and Race: March 1992." Claudette E. Bennett, *The Black Population in the United States: March 1992*, p. 8.

Household Characteristics

★ 938 ★

Characteristics of Households: Type, Region, and Race, March 1992. Part I.

Numbers in thousands.

Characteristic	Black						White					
		Family households			Nonfamily households			Family households			Nonfamily households	
	Total	Married couple	Female householder, no spouse present	Male householder, no spouse present	Female householder	Male householder	Total	Married couple	Female householder, no spouse present	Male householder, no spouse present	Female householder	Male householder
United States												
Total, all households	11,083	3,631	3,582	504	1,773	1,594	81,682	47,124	7,727	2,374	13,981	10,476
Size of household												
Percent	100.0	100.0	100.0	100.0	100.0	100.0	100.0	100.0	100.0	100.0	100.0	100.0
One person	26.3	-	-	-	90.0	82.8	25.1	-	-	-	89.3	76.6
Two persons	26.1	27.7	35.7	47.8	8.6	13.1	33.2	40.9	45.7	49.2	9.1	18.1
Three persons	18.7	24.6	27.8	24.9	1.0	2.3	16.9	21.7	32.0	26.7	0.9	3.5
Four persons	15.0	24.4	18.9	14.6	0.3	1.2	15.2	23.0	13.7	14.6	0.5	1.4
Five persons	7.9	13.6	9.4	9.0	0.1	0.2	6.4	9.8	5.6	5.4	0.1	0.3
Six persons	3.3	5.7	4.0	2.1	-	0.3	2.0	3.0	1.7	2.8	-	0.1
Seven or more persons	2.7	3.9	4.2	1.6	-	0.1	1.1	1.6	1.3	1.5	-	-
Age of householder												
Percent	100.0	100.0	100.0	100.0	100.0	100.0	100.0	100.0	100.0	100.0	100.0	100.0
15 to 34 years	30.4	24.6	40.0	29.1	21.6	31.8	25.3	23.1	30.3	32.4	19.1	38.3
35 to 44 years	23.8	28.5	26.8	25.2	12.1	19.0	22.4	25.4	27.7	26.5	9.6	21.6
45 to 54 years	15.5	17.7	13.9	13.9	15.1	14.9	16.3	18.7	16.5	18.0	10.1	13.2
55 years and over	30.4	29.2	19.3	31.7	51.2	34.2	36.0	32.9	25.6	23.1	61.2	26.9
Related children under 18 years												
Percent	100.0	100.0	100.0	100.0	100.0	100.0	100.0	100.0	100.0	100.0	100.0	100.0
No related children	53.6	41.4	22.6	51.6	100.0	100.0	65.3	52.9	35.7	50.0	100.0	100.0
With related children	46.4	58.6	77.4	48.4	-	-	34.7	47.1	64.3	50.0	-	-
One child	19.0	23.6	31.1	27.1	-	-	14.5	18.4	32.1	29.5	-	-
Two children	14.7	20.3	23.0	14.0	-	-	13.3	18.8	21.2	15.3	-	-
Three children	7.8	9.7	13.7	4.4	-	-	5.0	7.2	7.7	3.9	-	-
Four or more children	4.9	5.0	9.6	2.8	-	-	1.9	2.7	3.3	1.3	-	-
Own children under 18 years												
Percent	100.0	100.0	100.0	100.0	100.0	100.0	100.0	100.0	100.0	100.0	100.0	100.0
No own children	59.9	46.9	34.8	63.4	100.0	100.0	66.9	54.3	41.9	56.3	100.0	100.0
With own children	40.1	53.0	65.2	36.6	-	-	33.1	45.7	58.1	43.7	-	-
One child	16.9	21.1	27.6	22.6	-	-	13.7	17.7	28.8	26.1	-	-
Two children	12.9	18.9	19.3	10.3	-	-	12.8	18.4	19.2	13.7	-	-
Three children	6.7	9.1	11.0	2.5	-	-	4.8	7.0	7.2	3.2	-	-
Four or more children	3.7	3.9	7.2	1.2	-	-	1.8	2.5	2.9	0.7	-	-

Source: "Selected Characteristics of Households, by Type, Region, and Race of Householder: March 1992." Claudette E. Bennett, *The Black Population in the United States: March 1992*, p. 31.

★ 939 ★
Household Characteristics

Characteristics of Households: Type, Region, and Race, March 1992. Part II.

Numbers in thousands.

Characteristic	Black						White					
		Family households			Nonfamily households			Family households			Nonfamily households	
	Total	Married couple	Female householder, no spouse present	Male householder, no spouse present	Female householder	Male householder	Total	Married couple	Female householder, no spouse present	Male householder, no spouse present	Female householder	Male householder
South												
Total, all households	5,972	2,094	1,886	273	949	770	26,582	15,936	2,396	690	4,401	3,158
Size of household												
Percent	100.0	100.0	100.0	100.0	100.0	100.0	100.0	100.0	100.0	100.0	100.0	100.0
One person	24.9	-	-	-	89.0	83.9	24.3	-	-	-	90.5	78.8
Two persons	25.9	27.4	35.3	44.5	9.8	12.1	34.6	42.3	50.7	54.5	8.3	16.2
Three persons	19.4	24.8	28.7	24.0	1.2	2.5	18.1	23.7	30.6	24.4	0.7	3.5
Four persons	15.5	24.2	19.5	15.1	-	1.5	15.0	22.3	11.9	14.5	0.4	1.3
Five persons	8.1	14.2	8.3	11.4	-	-	5.5	8.3	4.8	3.7	0.1	0.2
Six persons	3.7	5.9	4.6	2.7	-	-	1.6	2.4	1.0	2.2	-	-
Seven or more persons	2.4	3.5	3.5	2.3	-	-	0.8	1.1	1.0	0.7	-	0.1
Age of householder												
Percent	100.0	100.0	100.0	100.0	100.0	100.0	100.0	100.0	100.0	100.0	100.0	100.0
15 to 34 years	29.6	25.5	37.6	26.2	21.2	32.8	25.7	24.2	27.2	32.4	19.5	39.1
35 to 44 years	24.3	28.5	27.8	27.0	13.0	17.3	21.4	24.0	25.6	27.2	8.8	21.3
45 to 54 years	14.5	16.2	13.2	10.7	14.9	13.8	16.3	18.7	17.8	15.9	9.1	13.1
55 years and over	31.6	29.8	21.4	36.0	50.9	36.1	36.7	33.1	29.4	24.4	62.7	26.5
Related children under 18 years												
Percent	100.0	100.0	100.0	100.0	100.0	100.0	100.0	100.0	100.0	100.0	100.0	100.0
No related children	52.5	40.4	23.6	46.2	100.0	100.0	65.6	53.9	39.4	50.0	100.0	100.0
With related children	47.5	59.6	76.4	53.8	-	-	34.4	46.1	60.6	50.0	-	-
One child	19.1	23.9	30.3	25.4	-	-	15.7	19.9	33.6	28.0	-	-
Two children	15.5	24.2	23.7	17.1	-	-	13.2	18.4	18.8	17.2	-	-
Three children	7.9	10.1	12.9	6.6	-	-	4.2	5.9	6.0	4.6	-	-
Four or more children	4.9	4.9	9.4	4.6	-	-	1.4	2.0	2.1	0.2	-	-
Own children under 18 years												
Percent	100.0	100.0	100.0	100.0	100.0	100.0	100.0	100.0	100.0	100.0	100.0	100.0
No own children	59.6	46.1	37.6	62.0	100.0	100.0	67.6	55.8	46.3	57.6	100.0	100.0
With own children	40.4	53.9	62.4	38.0	-	-	32.4	44.2	53.7	42.4	-	-
One child	17.0	21.8	26.5	20.8	-	-	14.5	19.6	29.6	23.7	-	-
Two children	13.1	18.8	18.9	11.3	-	-	12.6	17.9	17.0	14.4	-	-
Three children	6.9	9.6	10.5	3.7	-	-	4.0	5.7	5.5	4.3	-	-
Four or more children	3.4	3.7	6.4	2.2	-	-	1.2	1.9	1.5	-	-	-

Source: "Selected Characteristics of Households, by Type, Region, and Race of Householder: March 1992." Claudette E. Bennett, *The Black Population in the United States: March 1992*, pp. 31-32.

★ 940 ★
Household Characteristics

Characteristics of Households: Type, Region, and Race, March 1992. Part III.

Numbers in thousands.

Characteristic	Black						White					
		Family households			Nonfamily households			Family households			Nonfamily households	
	Total	Married couple	Female householder, no spouse present	Male householder, no spouse present	Female householder	Male householder	Total	Married couple	Female householder, no spouse present	Male householder, no spouse present	Female householder	Male householder
North and West												
Total, all households	5,111	1,536	1,696	231	824	824	55,100	31,188	5,331	1,684	9,580	7,318
Size of household												
Percent	100.0	100.0	100.0	100.0	100.0	100.0	100.0	100.0	100.0	100.0	100.0	100.0
One person	27.9	-	-	-	91.3	81.7	25.5	-	-	-	88.8	75.7

[Continued]

★ 940 ★

Characteristics of Households: Type, Region, and Race, March 1992. Part III.

[Continued]

Characteristic	Black						White					
		Family households			Nonfamily households			Family households			Nonfamily households	
	Total	Married couple	Female householder, no spouse present	Male householder, no spouse present	Female householder	Male householder	Total	Married couple	Female householder, no spouse present	Male householder, no spouse present	Female householder	Male householder
Two persons	26.2	28.2	36.2	51.7	7.2	13.9	32.5	40.1	43.4	46.9	9.4	18.9
Three persons	17.8	24.2	26.8	25.9	0.8	2.2	16.3	20.6	32.6	27.6	1.0	3.5
Four persons	14.3	24.7	18.1	14.0	0.6	1.0	15.3	23.3	14.6	14.6	0.6	1.4
Five persons	7.7	12.8	10.5	6.0	0.1	0.4	6.9	10.6	6.0	6.1	0.1	0.3
Six persons	2.9	5.5	3.3	1.5	-	0.7	2.2	3.4	2.0	3.1	-	0.1
Seven or more persons	3.1	4.6	5.0	0.8	-	0.2	1.3	1.9	1.4	1.8	-	-
Age of householder												
Percent	100.0	100.0	100.0	100.0	100.0	100.0	100.0	100.0	100.0	100.0	100.0	100.0
15 to 34 years	31.2	23.5	42.7	32.6	22.1	30.9	25.1	22.5	31.6	32.4	19.0	37.9
35 to 44 years	23.1	28.4	25.6	23.0	11.0	20.6	23.0	26.1	28.7	26.2	10.0	21.7
45 to 54 years	16.7	19.8	14.7	17.7	15.4	16.0	16.3	18.7	15.9	18.8	10.5	13.2
55 years and over	28.9	28.4	17.0	26.7	51.5	32.5	35.7	32.8	23.8	22.5	60.5	27.1
Related children under 18 years												
Percent	100.0	100.0	100.0	100.0	100.0	100.0	100.0	100.0	100.0	100.0	100.0	100.0
No related children	54.8	42.6	21.5	58.1	100.0	100.0	65.1	52.3	34.0	50.0	100.0	100.0
With related children	45.2	57.4	78.5	41.9	-	-	34.9	47.7	66.0	50.0	-	-
One child	18.8	23.1	31.9	29.1	-	-	14.0	17.7	31.3	30.2	-	-
Two children	13.9	20.0	22.3	10.3	-	-	13.3	19.0	22.2	14.5	-	-
Three children	7.7	9.2	14.5	1.9	-	-	5.4	7.9	8.5	3.5	-	-
Four or more children	4.8	5.1	9.7	0.6	-	-	2.2	3.1	3.9	1.7	-	-
Own children under 18 years												
Percent	100.0	100.0	100.0	100.0	100.0	100.0	100.0	100.0	100.0	100.0	100.0	100.0
No own children	60.2	48.2	31.7	65.2	100.0	100.0	66.6	53.6	39.9	55.8	100.0	100.0
With own children	39.8	51.8	68.3	34.8	-	-	33.4	46.4	60.1	44.2	-	-
One child	16.8	20.2	28.9	24.6	-	-	13.3	17.2	28.4	27.1	-	-
Two children	12.7	18.9	19.8	9.1	-	-	12.9	18.6	20.2	13.4	-	-
Three children	6.4	8.5	11.6	1.1	-	-	5.2	7.7	7.9	2.7	-	-
Four or more children	3.9	4.3	8.0	-	-	-	2.0	2.9	3.5	1.0	-	-

Source: "Selected Characteristics of Households, by Type, Region, and Race of Householder: March 1992." Claudette E. Bennett, *The Black Population in the United States: March 1992*, pp. 32-33.

★ 941 ★

Household Characteristics

Households With and Without Children: Geographical Characteristics, 1991

Numbers in thousands, excluding percents.

Characteristic	All households					Households with children				
	Total	White	Black	Hispanic origin	Non-Hispanic origin	Total	White	Black	Hispanic origin	Non-Hispanic origin
United States	93,147	79,140	10,832	6,239	86,907	34,588	27,864	5,124	3,437	31,150
Metropolitan/Nonmetropolitan areas										
Percent	100	100	100	100	100	100	100	100	100	100
Inside metropolitan statistical areas	78	77	86	90	77	78	76	86	90	77
Inside central cities	32	27	59	52	30	30	24	57	51	28
Suburbs	46	49	27	38	47	48	51	29	38	49
Outside metropolitan statistical areas	22	23	14	10	23	22	24	14	10	23
Regions										
Percent	100	100	100	100	100	100	100	100	100	100
Northeast	20	21	18	18	21	19	19	17	17	19
Midwest	24	25	20	7	26	24	25	21	6	26
South	35	33	53	32	35	35	33	54	30	36
West	21	21	9	43	19	22	23	8	47	19

[Continued]

★ 941 ★

Households With and Without Children: Geographical Characteristics, 1991
[Continued]

Characteristic	All households					Households with children				
	Total	White	Black	Hispanic origin	Non-Hispanic origin	Total	White	Black	Hispanic origin	Non-Hispanic origin
Place size										
Total (in places)	65,441	53,723	9,117	5,409	60,033	23,284	17,721	4,275	2,953	20,330
Percent with										
Less than 10,000 persons	20	23	9	11	21	21	24	9	11	22
10,000 to 49,999 persons	31	33	21	24	32	32	34	23	24	33
50,000 to 99,999 persons	12	13	9	14	12	12	12	9	13	12
100,000 to 249,999 persons	11	11	14	13	11	12	11	15	13	12
250,000 to 499,999 persons	8	7	12	10	8	7	6	12	9	7
500,000 to 999,999 persons	7	6	13	8	7	7	5	12	8	7
1,000,000 persons or more	10	7	21	21	9	10	6	20	22	8
Inside poverty areas										
Total	8,661	4,345	3,872	1,409	7,252	3,651	1,584	1,809	806	2,844
Central cities	5,215	1,973	2,876	1,029	4,186	2,240	718	1,313	608	1,633
Suburbs	955	612	314	228	727	408	243	148	127	281
Nonmetropolitan areas	2,491	1,760	682	152	2,339	1,002	624	349	71	931
Percent	100	100	100	100	100	100	100	100	100	100
Central cities	60	45	74	73	58	61	45	73	75	57
Suburbs	11	14	8	16	10	11	15	8	16	10
Nonmetropolitan areas	29	41	18	11	32	27	39	19	9	33

Source: "Selected Geographical Characteristics, All Households and Households With Children, by Race and Hispanic Origin of Householder: 1991." Jeanne Woodward, *Housing America's Children*, p. 2.

★ 942 ★

Household Characteristics

Distribution of Households with Children, 1991

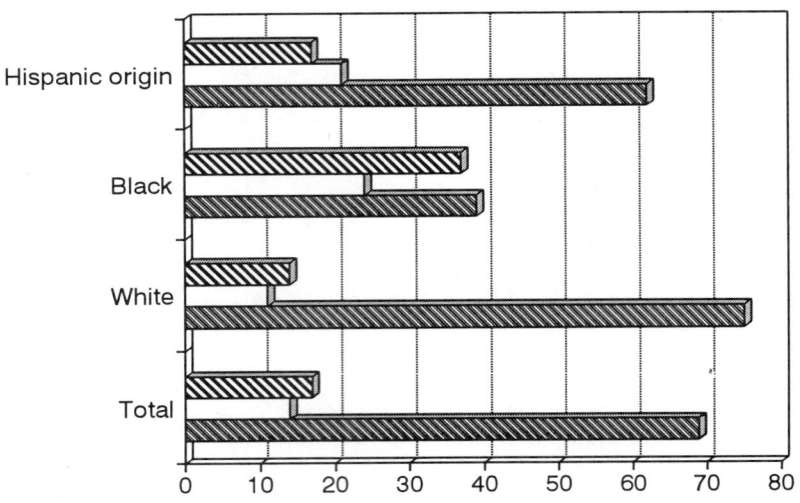

In percent.

Characteristic	Total	White	Black	Hispanic origin
Married couples	69	75	39	62
Other, 2 (+) adults	14	11	24	21
1 adult	17	14	37	17

Source: "Distribution of Households with Children by Type, Race, and Hispanic Origin of Householder: 1991." Jeanne Woodward, *Housing America's Children*, p. 8. *Note:* Hispanic-origin may be of any race.

★ 943 ★

Household Characteristics

Households: Race, Hispanic and Type, 1992

As of March. Based on Current Population Survey.

Characteristic	Number of households (1,000)			Percent distribution			Persons per household		
	Total	Black	Hispanic[1]	Total	Black	Hispanic[1]	Total	Black	Hispanic[1]
Total	95,669	11,083	6,379	100	100	100	2.62	2.81	3.45
Age of householder									
15 to 24 years old	4,859	702	555	5	6	9	2.35	2.58	3.08
25 to 29 years old	8,810	1,203	785	9	11	12	2.68	2.98	3.56

[Continued]

★ 943 ★

Households: Race, Hispanic and Type, 1992
[Continued]

Characteristic	Number of households (1,000)			Percent distribution			Persons per household		
	Total	Black	Hispanic[1]	Total	Black	Hispanic[1]	Total	Black	Hispanic[1]
30 to 34 years old	11,197	1,458	1,035	12	13	16	3.06	2.90	3.80
35 to 44 years old	21,774	2,636	1,640	23	24	26	3.26	3.34	3.94
45 to 54 years old	15,547	1,718	979	16	16	15	2.89	2.99	3.56
55 to 64 years old	12,559	1,418	728	13	13	11	2.34	2.62	2.98
65 to 74 years old	12,043	1,245	417	13	11	7	1.86	2.03	2.44
75 years old and over	8,878	703	238	9	6	4	1.57	1.84	2.05
Region									
Northeast	19,314	1,907	1,135	20	17	18	2.63	2.84	3.03
Midwest	23,327	2,238	456	24	20	7	2.59	2.78	3.32
South	33,073	5,972	1,980	35	54	31	2.60	2.85	3.38
West	19,955	966	2,807	21	9	44	2.71	2.50	3.70
Size of household									
One person	23,974	2,915	887	25	26	14	1.00	1.00	1.00
Two persons	30,734	2,887	1,364	32	26	21	2.00	2.00	2.00
Three persons	16,398	2,069	1,247	17	19	20	3.00	3.00	3.00
Four persons	14,710	1,661	1,272	15	15	20	4.00	4.00	4.00
Five persons	6,389	878	836	7	8	13	5.00	5.00	5.00
Six persons	2,126	368	406	2	3	6	6.00	6.00	6.00
Seven or more	1,338	304	367	1	3	6	(NA)	(NA)	(NA)
Marital status of householder									
Never married (single)	14,461	2,931	1,118	15	26	18	(NA)	(NA)	(NA)
Married, spouse present	52,457	3,631	3,532	55	33	55	(NA)	(NA)	(NA)
Married, spouse absent	4,571	1,273	575	5	11	9	(NA)	(NA)	(NA)
Widowed	11,895	1,527	444	12	14	9	(NA)	(NA)	(NA)
Divorced	12,286	1,721	711	13	16	11	(NA)	(NA)	(NA)

Source: "Household Characteristics, by Race, Hispanic Origin, and Type: 1992," U.S. Bureau of the Census, *Statistical Abstract of the United States, 1993*, p. 57. Primary source: U.S. Bureau of the Census, *Current Population Reports*, P20-467 and P60-180. *Notes:* NA Not available. 1. HIspanic persons may be of any race.

Household Size

★ 944 ★

Change in Household Size, 1960 to 1992

Year and race	Number of households (thousands)	Average annual percent change	Average size of household
Black			
1960	4,779	(X)	3.82
1970	6,180	2.57	3.54
1980	8,586	3.29	3.02
1990	10,486	2.00	2.88
1991	10,671	1.75	2.87
1992	11,083	3.79	2.83
White			
1960	47,868	(X)	3.23
1970	56,529	1.66	3.06
1980	70,766	2.25	2.71
1990	80,163	1.25	2.58
1991	80,968	1.00	2.58
1992	81,682	0.88	2.57

Source: "Number of Households and Average Size, by Race: 1960 to 1992." Claudette E. Bennett, *The Black Population in the United States: March 1992*, p. 6. *Note:* 1960 and 1970 data are from the decennial censuses.

Households

★ 945 ★

Household Tenure, by Residence, region, Age, and Race, March 1992

Numbers in thousands.

Characteristic	Black			White		
	United States	South	North and West	United States	South	North and West
Tenure by residence for householders 15 years and over						
Total	11,083	5,972	5,111	81,682	26,582	55,100
Percent	100.0	100.0	100.0	100.0	100.0	100.0
Own or buying home	42.3	47.7	35.98	67.5	70.2	66.2

[Continued]

★ 945 ★

Household Tenure, by Residence, region, Age, and Race, March 1992
[Continued]

Characteristic	Black			White		
	United States	South	North and West	United States	South	North and West
Renting	55.8	49.6	63.1	30.8	27.8	32.3
Occupier paid no cash rent	2.0	2.8	1.0	1.7	2.0	1.6
All metropolitan areas	9,402	4,392	5,010	62,641	19,078	43,563
Percent	100.0	100.0	100.0	100.0	100.0	100.0
Own or buying home	39.9	44.2	36.1	65.3	67.6	64.2
Renting	58.7	54.0	62.9	33.5	31.0	34.5
Occupier paid no cash rent	1.4	1.8	1.0	1.3	1.4	1.2
Central cities	6,343	2,613	3,730	22,691	6,749	15,942
Percent	100.0	100.0	100.0	100.0	100.0	100.0
Own or buying home	36.5	39.8	34.1	52.9	56.4	51.4
Renting	62.5	58.8	65.0	46.2	42.7	47.7
Occupier paid no cash rent	1.1	1.4	0.9	0.9	1.0	0.9
Metropolitan, not in central cities	3,059	1,779	1,280	39,950	12,329	27,621
Percent	100.0	100.0	100.0	100.0	100.0	100.0
Own or buying home	47.0	50.8	41.9	72.3	73.7	71.6
Renting	51.0	46.8	56.8	26.3	24.7	27.0
Occupier paid no cash rent	1.9	2.4	1.3	1.5	1.6	1.4
Outside metropolitan areas	1,680	1,580	100	19,040	7,504	11,537
Percent	100.0	100.0	100.0	100.0	100.0	100.0
Own or buying home	55.4	57.3	26.4	74.8	76.7	73.5
Renting	39.3	37.3	70.2	22.1	19.6	23.7
Occupier paid no cash rent	5.3	5.4	3.4	3.2	3.7	2.8
Tenure by residence for householders 55 years and over						
Total	3,365	1,888	1,477	29,389	9,745	19,644
Percent	100.0	100.0	100.0	100.0	100.0	100.0
Own or buying home	60.9	65.8	54.7	80.6	85.1	78.4
Renting	36.7	30.8	44.2	17.8	13.3	19.9
Occupier paid no cash rent	2.4	3.5	1.1	1.6	1.5	1.6
All metropolitan areas	2,700	1,252	1,448	21,685	6,556	15,128
Percent	100.0	100.0	100.0	100.0	100.0	100.0
Own or buying home	58.5	62.5	55.0	79.0	84.5	76.6
Renting	40.0	35.6	43.9	19.7	14.4	22.0
Occupier paid no cash rent	1.5	1.9	1.1	1.3	1.0	1.4
Central cities	1,990	800	1,190	7,801	2,301	5,500
Percent	100.0	100.0	100.0	100.0	100.0	100.0
Own or buying home	55.8	58.8	53.8	69.7	78.6	65.9
Renting	43.0	39.6	45.3	29.3	20.5	32.9

[Continued]

★ 945 ★

Household Tenure, by Residence, region, Age, and Race, March 1992

[Continued]

Characteristic	Black			White		
	United States	South	North and West	United States	South	North and West
Occupier paid no cash rent	1.2	1.6	0.9	1.1	0.9	1.2
Metropolitan, not in central cities	710	451	258	13,883	4,255	9,628
Percent	100.0	100.0	100.0	100.0	100.0	100.0
Own or buying home	65.9	69.1	60.2	84.3	87.7	82.7
Renting	31.8	28.5	37.5	14.4	11.2	15.8
Occupier paid no cash rent	2.4	2.4	2.3	1.4	1.1	1.5
Outside metropolitan areas	665	636	29	7,705	3,189	4,516
Percent	100.0	100.0	100.0	100.0	100.0	100.0
Own or buying home	70.7	72.1	40.2	85.2	86.3	84.5
Renting	23.0	21.3	59.8	12.2	11.1	13.0
Occupier paid no cash rent	6.3	6.6	-	2.5	2.6	2.5

Source: "Tenure of Households, by residence, region, and Age and Race of Householder: March 1992." Claudette E. Bennett, *The Black Population in the United States: March 1992,* p. 19.

★ 946 ★

Households

Households: Characteristics and Size, 1970 to 1992

As of March. Based on Current Population Survey.

Characteristic of householder and size of household	1970	1975	1980	1985	1990	1992
	Number (mil.)					
Total[1]	63.4	71.1	80.8	86.8	93.3	95.7
White	56.6	62.9	70.8	75.3	80.2	81.7
Black	6.2	7.3	8.6	9.5	10.5	11.1
Hispanic[2]	(NA)	(NA)	3.7	4.9	5.9	6.4
	Percent distribution					
Total[1]	100	100	100	100	100	100
White	90	89	88	87	86	85
Black	10	10	11	11	11	12
Hispanic[2]	(NA)	(NA)	5	6	6	7

Source: "Households, by Characteristics of Householder and Size of Household: 1970 to 1992." U.S. Bureau of the Census, *Statistical Abstract of the United States,* 1993, p. 56. Primary source: U.S. Bureau of the Census, *Current Population Reports,* P20-467, and earlier reports; and unpublished data. *Notes:* NA Not available. 1. Includes other races, not shown separately. 2. Hispanic persons may be of any race.

Immigration

★ 947 ★

Natural Increase and Civilian Immigration, 1981-1991

[Rate per 1,000]

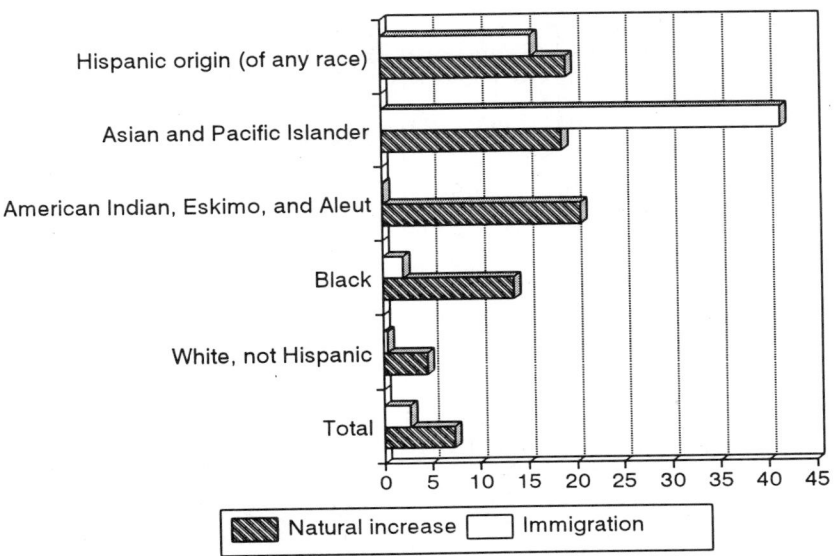

Resident population. Rate per 1,000 population. Consistent with 1990 census, as enumerated. Race data for 1990 modified to assign a specified race to each person.

	Natural increase	Net civilian immigration
Total	7.3	2.8
White, not Hispanic	4.6	0.5
Black	13.6	2.2
American Indian, Eskimo, and Aleut	20.6	0.2
Asian and Pacific Islander	18.7	41.5
Hispanic origin (of any race)	19.2	15.6

Source: "Average Annual Rates of Increase and Net Civilian Immigration, by Race and Hispanic Origin: July 1, 1981 to July 1, 1991." Frederick W. Hollomann, "National Population Trends," in U.S. Bureau of the Census, Current Population Reports, *Population Profile of the United States: 1992*, p. 3.

Marital Status

★ 948 ★

Marital Status of Persons 15 Years Old and Over: Characteristics, March 1992. Part I.

Numbers in thousands.

Race, region, and marital status	Total, 15 years and over		15 to 24 years		25 to 34 years		35 to 44 years	
	Male	Female	Male	Female	Male	Female	Male	Female
Black								
United States								
Total	10,252	12,290	2,449	2,648	2,505	2,918	2,027	2,435
Percent	100.0	100.0	100.0	100.0	100.0	100.0	100.0	100.0
Never married	45.0	39.1	94.3	89.1	56.4	50.5	26.1	25.1
Married, spouse present	36.1	29.4	4.4	7.6	31.5	31.2	51.8	40.9
Married, spouse absent	6.4	8.6	0.6	2.3	5.9	9.1	8.9	12.2
Widowed	4.2	11.5	-	0.1	0.3	0.3	1.2	2.8
Divorced	8.4	11.5	0.7	0.9	5.9	8.9	11.9	19.0
South								
Total	5,550	6,701	1,377	1,430	1,310	1,542	1,133	1,359
Percent	100.0	100.0	100.0	100.0	100.0	100.0	100.0	100.0
Never married	44.0	36.8	92.6	86.3	53.2	46.4	26.6	22.6
Married, spouse present	38.0	31.2	5.6	9.9	35.0	33.5	53.4	42.5
Married, spouse absent	6.4	8.2	0.7	2.3	5.2	10.3	7.9	11.4
Widowed	4.2	12.0	-	-	0.5	0.2	1.0	3.1
Divorced	7.4	11.8	1.1	1.5	6.1	9.6	11.2	20.5
North and West								
Total	4,702	5,589	1,072	1,218	1,195	1,376	894	1,076
Percent	100.0	100.0	100.0	100.0	100.0	100.0	100.0	100.0
Never married	46.2	41.9	96.4	92.5	59.8	55.0	25.4	28.2
Married, spouse present	33.8	27.2	2.9	4.9	27.7	28.6	49.8	38.9
Married, spouse absent	6.4	9.0	0.4	2.3	6.6	7.7	10.3	13.3
Widowed	4.0	10.8	-	0.2	0.2	0.5	1.6	2.4
Divorced	9.6	11.1	0.2	0.2	5.7	8.2	12.9	17.2
White								
United States								
Total	80,049	85,522	13,986	13,927	17,736	17,584	16,738	11,427
Percent	100.0	100.0	100.0	100.0	100.0	100.0	100.0	100.0
Never married	28.1	20.6	88.3	77.5	35.8	21.4	12.7	8.6
Married, spouse present	59.9	56.0	10.2	18.9	55.0	64.0	72.1	72.9
Married, spouse absent	2.3	2.9	0.8	2.0	2.8	4.5	3.2	3.8
Widowed	2.5	11.3	-	0.1	0.1	0.5	0.5	1.3
Divorced	7.2	9.2	0.7	1.4	6.4	9.5	11.5	13.4
South								
Total	25,802	27,705	4,453	4,454	5,620	5,613	5,267	5,286
Percent	100.0	100.0	100.0	100.0	100.0	100.0	100.0	100.0
Never married	24.5	17.5	84.8	71.8	28.5	16.7	9.8	6.6

[Continued]

★ 948 ★

Marital Status of Persons 15 Years Old and Over: Characteristics, March 1992. Part I.

[Continued]

Race, region, and marital status	Total, 15 years and over		15 to 24 years		25 to 34 years		35 to 44 years	
	Male	Female	Male	Female	Male	Female	Male	Female
Married, spouse present	62.9	58.5	12.8	23.5	60.2	68.1	73.5	75.6
Married, spouse absent	2.5	3.0	1.1	2.7	3.7	4.6	3.6	3.6
Widowed	2.4	11.9	-	0.1	0.1	0.4	0.3	1.2
Divorced	7.7	9.1	1.2	1.9	7.5	10.2	12.8	13.0
North and West								
Total	54,248	57,817	9,533	9,473	12,117	11,971	11,471	11,477
Percent	100.0	100.0	100.0	100.0	100.0	100.0	100.0	100.0
Never married	29.8	22.1	89.9	80.3	39.1	23.6	14.0	9.5
Married, spouse present	58.5	54.8	8.9	16.8	52.5	62.1	71.5	71.6
Married, spouse absent	2.1	2.9	0.7	1.7	2.3	4.5	3.0	4.0
Widowed	2.6	10.9	-	0.1	0.1	0.6	0.5	1.4
Divorced	6.9	9.3	0.5	1.2	5.9	9.2	10.9	13.5

Source: "Marital Status of Persons 15 Years Old and Over, by Age, Sex, Region, and Race: March 1992." Claudette E. Bennett, *The Black Population in the United States: March 1992*, p. 30.

★ 949 ★

Marital Status

Marital Status of Persons 15 Years Old and Over: Characteristics, March 1992. Part II.

Numbers in thousands.

Race, region, and marital status	45 to 54 years		55 to 64 years		65 years and over	
	Male	Female	Male	Female	Male	Female
Black						
United States						
Total	1,235	1,552	978	1,188	1,058	1,549
Percent	100.0	100.0	100.0	100.0	100.0	100.0
Never married	14.4	11.5	12.0	8.2	6.5	5.7
Married, spouse present	53.8	41.3	52.9	40.1	54.1	24.8
Married, spouse absent	11.4	15.2	11.0	9.3	5.9	5.4
Widowed	2.4	10.1	8.9	25.6	26.2	56.2
Divorced	18.1	21.9	15.3	16.8	7.4	7.9
South						
Total	604	835	525	649	601	885
Percent	100.0	100.0	100.0	100.0	100.0	100.0
Never married	14.0	11.7	9.1	8.7	5.5	5.8
Married, spouse present	57.2	42.3	51.4	41.6	59.3	26.3
Married, spouse absent	13.8	12.7	14.4	8.2	4.5	4.8
Widowed	1.3	12.2	10.6	25.7	25.7	55.7
Divorced	13.7	21.2	14.4	15.7	5.0	7.4

[Continued]

★ 949 ★

Marital Status of Persons 15 Years Old and Over: Characteristics, March 1992. Part II.

[Continued]

Race, region, and marital status	45 to 54 years		55 to 64 years		65 years and over	
	Male	Female	Male	Female	Male	Female
North and West						
Total	632	717	453	540	457	664
Percent	100.0	100.0	100.0	100.0	100.0	100.0
Never married	14.8	11.1	15.3	7.5	7.9	5.4
Married, spouse present	50.4	40.3	54.5	38.3	47.2	22.9
Married, spouse absent	9.0	18.1	7.0	10.6	7.7	6.2
Widowed	3.4	7.8	6.8	25.4	26.7	57.0
Divorced	22.3	22.7	16.4	18.2	10.5	8.4
White						
United States						
Total	11,427	11,832	8,731	9,549	11,431	15,866
Percent	100.0	100.0	100.0	100.0	100.0	100.0
Never married	6.7	4.5	4.9	3.5	4.0	4.8
Married, spouse present	77.9	72.4	81.9	69.7	75.8	41.3
Married, spouse absent	2.6	3.2	2.2	2.5	1.5	1.0
Widowed	0.8	3.9	2.9	13.5	14.0	47.6
Divorced	11.9	15.9	8.1	10.9	4.7	5.3
South						
Total	3,761	3,846	2,905	3,193	3,796	5,312
Percent	100.0	100.0	100.0	100.0	100.0	100.0
Never married	5.8	3.2	3.2	1.5	2.7	3.5
Married, spouse present	80.0	73.9	84.2	70.5	77.5	42.1
Married, spouse absent	2.0	3.5	2.1	2.4	1.6	1.1
Widowed	0.7	3.8	2.6	15.5	13.3	48.5
Divorced	11.5	15.5	7.9	10.1	5.0	4.8
North and West						
Total	7,666	7,986	5,826	6,356	7,635	10,554
Percent	100.0	100.0	100.0	100.0	100.0	100.0
Never married	7.2	5.1	5.7	4.5	4.7	5.5
Married, spouse present	76.9	71.7	80.8	69.2	75.0	40.9
Married, spouse absent	2.9	3.1	2.2	2.6	1.5	1.0
Widowed	0.9	4.0	3.0	12.4	14.3	47.1
Divorced	12.1	16.1	8.3	11.3	4.5	5.5

Source: "Marital Status of Persons 15 Years Old and Over, by Age, Sex, Region, and Race: March 1992." Claudette E. Bennett, *The Black Population in the United States: March 1992*, p. 30.

★ 950 ★

Marital Status

Marital Status: Gender, Race, and Hispanic Origin, 1970-1992

In millions, except percent. As of March, except as noted. Persons 18 years old and over. Excludes members of Armed Forces except those living off post or with their families on post. Except as noted, based on Current Population Survey.

Marital status, race and Hispanic origin	Total				Male				Female			
	1970	1980	1990	1992	1970	1980	1990	1992	1970	1980	1990	1992
Total[1]	132.5	159.5	181.8	185.3	62.5	75.7	86.9	88.7	70.0	83.8	95.0	96.6
Never married	21.4	32.3	40.4	41.8	11.8	18.0	22.4	23.2	9.6	14.3	17.9	18.6
Married	95.0	104.6	112.6	113.3	47.1	51.8	55.8	56.2	47.9	52.8	56.7	57.1
Widowed	11.8	12.7	13.8	13.9	2.1	2.0	2.3	2.5	9.7	10.8	11.5	11.3
Divorced	4.3	9.9	15.1	16.3	1.6	3.9	6.3	6.8	2.7	6.0	8.8	9.6
Percent of total	100.0	100.0	100.0	100.0	100.0	100.0	100.0	100.0	100.0	100.0	100.0	100.0
Never married	16.2	20.3	22.2	22.6	18.9	23.8	25.8	26.2	13.7	17.1	18.9	19.2
Married	71.7	65.5	61.9	61.1	75.3	68.4	64.3	63.3	68.5	63.0	59.7	59.1
Widowed	8.9	8.0	7.6	7.5	3.3	2.6	2.7	2.9	13.9	12.8	12.1	11.7
Divorced	3.2	6.2	8.3	8.8	2.5	5.2	7.2	7.6	3.9	7.1	9.3	9.9
Percent standardized for age[2]												
Never married	14.1	16.5	20.6	21.7	16.5	18.7	23.3	24.4	12.1	14.5	18.2	19.2
Married	74.2	69.3	63.7	62.5	77.6	72.9	66.5	64.9	70.8	65.9	61.2	60.2
Widowed	8.3	7.6	6.9	6.7	3.3	2.7	2.7	2.8	13.0	12.1	10.8	10.3
Divorced	3.4	6.6	8.7	9.1	2.6	5.6	7.6	7.9	4.1	7.6	9.8	10.3
White												
Total	118.2	139.5	155.5	157.6	55.9	66.7	74.8	76.0	62.2	72.8	80.6	81.6
Never married	18.4	26.4	31.6	32.2	10.2	15.0	18.0	18.4	8.2	11.4	13.6	13.8
Married	85.8	93.8	99.5	100.1	42.7	46.7	49.5	50.0	43.1	47.1	49.9	50.3
Widowed	10.3	10.9	11.7	11.7	1.7	1.6	1.9	2.0	8.6	9.3	9.8	9.6
Divorced	3.7	8.3	12.6	13.6	1.3	3.4	5.4	5.8	2.3	5.0	7.3	7.9
Percent of total	100.0	100.0	100.0	100.0	100.0	100.0	100.0	100.0	100.0	100.0	100.0	100.0
Never married	15.6	18.9	20.3	20.5	18.2	22.5	24.1	24.2	13.2	15.7	16.9	16.9
Married	72.6	67.2	64.0	63.5	76.3	70.0	66.2	65.5	69.3	64.7	61.9	61.6
Widowed	8.7	7.8	7.5	7.4	3.1	2.5	2.6	2.7	13.8	12.8	12.2	11.8
Divorced	3.1	6.0	8.1	8.6	2.4	5.0	7.2	7.6	3.8	6.8	9.0	9.6
Black												
Total	13.0	16.6	20.3	21.0	5.9	7.4	9.1	9.5	7.1	9.2	11.2	11.5
Never married	2.7	5.1	7.1	7.8	1.4	2.5	3.5	3.8	1.2	2.5	3.6	4.0
Married	8.3	8.5	9.3	9.0	3.9	4.1	4.5	4.4	4.4	4.5	4.8	4.7
Widowed	1.4	1.6	1.7	1.8	0.3	0.3	0.3	0.4	1.1	1.3	1.4	1.4
Divorced	0.6	1.4	2.1	2.3	0.2	0.5	0.8	0.9	0.4	0.9	1.3	1.4
Percent of total	100.0	100.0	100.0	100.0	100.0	100.0	100.0	100.0	100.0	100.	100.0	100.0
Never married	20.6	30.5	35.1	37.4	24.3	34.3	38.4	40.4	17.4	27.4	32.5	35.0
Married	64.1	51.4	45.8	43.0	66.9	54.6	49.2	46.0	61.7	48.7	43.0	40.6
Widowed	11.0	9.8	8.5	8.8	5.2	4.2	3.7	4.5	15.8	14.3	12.4	12.3
Divorced	4.4	8.4	10.6	10.8	3.6	7.0	8.8	9.1	5.0	9.5	12.0	12.2

[Continued]

★ 950 ★

Marital Status: Gender, Race, and Hispanic Origin, 1970-1992
[Continued]

Marital status, race and Hispanic origin	Total				Male				Female			
	1970	1980	1990	1992	1970	1980	1990	1992	1970	1980	1990	1992
Hispanic[3]												
Total	5.1	7.9	13.6	14.4	2.4	3.8	6.7	7.2	2.6	4.1	6.8	7.2
Never married	0.9	1.9	3.7	4.0	0.5	1.0	2.2	2.3	0.4	0.9	1.5	1.7
Married	3.6	5.2	8.4	8.7	1.8	2.5	4.1	4.3	1.8	2.6	4.3	4.4
Widowed	0.3	0.4	0.5	0.6	0.1	0.1	0.1	0.1	0.2	0.3	0.4	0.5
Divorced	0.2	0.5	1.0	1.1	0.1	0.2	0.4	0.4	0.1	0.3	0.6	0.6
Percent of total	100.0	100.0	100.0	100.0	100.0	100.0	100.0	100.0	100.0	100.0	100.0	100.0
Never married	18.6	24.1	27.2	27.9	21.2	27.3	32.1	32.4	16.2	21.1	22.5	23.5
Married	71.8	65.6	61.7	60.3	73.8	67.1	60.9	60.0	70.0	64.3	62.4	60.7
Widowed	5.6	4.4	4.0	4.4	2.3	1.6	1.5	1.7	8.7	7.1	6.5	7.2
Divorced	3.9	5.8	7.0	7.3	2.7	4.0	5.5	5.9	5.1	7.6	8.5	8.7

Source: "Marital Status of the Population, by Sex, Race, and Hispanic Origin: 1970 to 1992." U.S. Bureau of the Census, *Statistical Abstract of the United States,* 1993, p. 53. Primary source: U.S. Bureau of the Census, *1970 Census of Population*, volume 1, part 1, and *Current Population Reports*, P20-468; and earlier reports. *Notes:* 1. Includes persons of other races, not shown separately. 2. 1960 age distribution used as standard; standardization improves comparability over time by removing effects of changes in age distribution of population. 3. Hispanic persons may be of any race. 1970 data as of April and based on census.

★ 951 ★

Marital Status

Married Couples of Same or Mixed Races and Origins, 1970 to 1992

In thousands. As of March, except as noted. Persons 15 years old and over. Persons of Hispanic origin may be of any race. Except as noted, based on Current Population Survey.

Race and origin of spouses	1970[1]	1980	1992
Married couples, total	44,598	49,714	53,512
Race			
Same race couples	43,922	48,264	50,873
White/white	40,578	44,910	47,358
Black/black	3,344	3,354	3,515
Interracial couples	310	651	1,161
Black/white	65	167	246
Black husband/White wife	41	122	163
White husband/Black wife	24	45	83
White/other race[2]	233	450	883
Black/other race[2]	12	34	32
All other couples	366	799	1,478
Hispanic origin			
Hispanic/Hispanic	1,368	1,906	3,297

[Continued]

★ 951 ★

Married Couples of Same or Mixed Races and Origins,
1970 to 1992
[Continued]

Race and origin of spouses	1970[1]	1980	1992
Hispanic/other origin (not Hispanic)	584	891	1,155
All other couples (not Hispanic origin)	42,645	46,917	49,060

Source: "Married Couples of Same or Mixed Races and Origins: 1970 to 1992." U.S. Bureau of the Census, *Statistical Abstract of the United States*, 1993, p. 54. Primary source: U.S. Bureau of the Census, *Current Population Reports*, P20-468; and earlier reports. *Notes:* 1. As of April and based on Census of population. 2. Excluding white and black.

Population Characteristics

★ 952 ★

Resident Population: Characteristics 1790 to 1991, and Projections, 1995 to 2050

In thousands. See also *Historical Statistics, Colonial Times to 1970*, series A 73-81 and A 143-149.

Date	Sex			Race					Hispanic origin[1]	Residence[2]	
			White	Black	Other						
	Male	Female			Total	American Indians and Alaska Natives	Asian and Pacific Islanders			Urban	Rural
1790 (Aug. 2)[3]	(NA)	(NA)	3,172	757	(NA)	(NA)	(NA)	(NA)	202	3,728
1800 (Aug. 4)[3]	(NA)	(NA)	4,306	1,002	(NA)	(NA)	(NA)	(NA)	322	4,986
1850 (June 1)[3]	11,838	11,354	19,553	3,639	(NA)	(NA)	(NA)	(NA)	3,544	19,648
1860 (June 1)[3]	16,085	15,358	26,923	4,442	79	(NA)	(NA)	(NA)	6,217	25,227
1870 (June 1)[3]	19,494	19,065	33,589	4,880	89	(NA)	(NA)	(NA)	9,902	28,656
1880 (June 1)[3]	25,519	24,637	43,403	6,581	172	(NA)	(NA)	(NA)	14,130	36,026
1890 (June 1)[3]	32,237	30,711	55,101	7,489	358	(NA)	(NA)	(NA)	22,106	40,841
1900 (June 1)[3]	38,816	37,178	66,809	8,834	351	(NA)	(NA)	(NA)	30,160	45,835
1910 (Apr. 15)[3]	47,332	44,640	81,732	9,828	413	(NA)	(NA)	(NA)	41,999	49,973
1920 (Jan. 1)[3]	53,900	51,810	94,821	10,463	427	(NA)	(NA)	(NA)	54,158	51,553
1930 (Apr. 1)[3]	62,137	60,638	110,287	11,891	597	(NA)	(NA)	(NA)	68,955	53,820
1940 (Apr. 1)[3]	66,062	65,608	118,215	12,866	589	(NA)	(NA)	(NA)	74,424	57,246
1950 (Apr. 1)[3]	74,833	75,864	134,942	15,042	713	(NA)	(NA)	(NA)	96,468	54,230
1950 (Apr. 1)	75,187	76,139	135,150	15,045	1,131	(NA)	(NA)	(NA)	96,847	54,479
1960 (Apr. 1)	88,331	90,992	158,832	18,872	1,620	(NA)	(NA)	(NA)	125,269	54,054
1970 (Apr. 1)[4]	98,926	104,309	178,098	22,581	2,557	(NA)	(NA)	(NA)	149,647	53,565
1980 (Apr. 1)[5,6]	110,053	116,493	194,713	26,683	5,150	1,420	3,729	14,609	167,051	59,495
1990 (Apr. 1)[5]	121,239	127,471	208,704	30,483	9,523	2,065	7,458	22,354	187,053	61,656
1991 (July 1)[7]	122,979	129,198	210,899	31,164	10,113	2,117	7,996	23,350	(NA)	(NA)

[Continued]

★ 952 ★

Resident Population: Characteristics 1790 to 1991, and Projections, 1995 to 2050

[Continued]

Date	Sex		Race					Hispanic origin[1]	Residence[2]	
			White	Black	Other					
					Total	American Indians and Alaska Natives	Asian and Pacific Islanders		Urban	Rural
	Male	Female								
1995(July 1)[8]	128,292	134,461	217,511	33,147	12,096	2,247	9,849	26,522	(NA)	(NA)
2000 (July 1)[8]	134,338	140,477	224,594	35,525	14,696	2,409	12,287	30,602	(NA)	(NA)
2005 (July 1)[8]	140,097	146,227	230,993	37,907	17,423	2,583	14,840	34,842	(NA)	(NA)
2010 (July 1)[8]	146,012	152,097	237,412	40,429	20,268	2,772	17,496	39,312	(NA)	(NA)
2015 (July 1)[8]	152,178	158,192	244,073	43,074	23,223	2,971	20,252	44,030	(NA)	(NA)
2020 (July 1)[8]	158,308	164,294	250,587	45,743	26,271	3,175	23,096	48,952	(NA)	(NA)
2025 (July 1)[8]	164,054	170,162	256,425	48,388	29,403	3,386	26,017	54,018	(NA)	(NA)
2050 (July 1)[8]	187,271	195,403	274,761	62,181	45,732	4,641	41,091	80,675	(NA)	(NA)

Source: "Resident Population—Selected Characteristics, and Median Age, 1790 to 1991, and Projections, 1995 to 2050," U.S. Bureau of the Census, *Statistical Abstract of the United States*, 1993, p. 14. Primary source: U.S. Bureau of the Census, *U.S. Census of Population: 1940*. Vol. II, Part 1, and Vol. IV, Part1; 1950, Vol. II. Part I; 1970, Vol. 1, Part B; and *Current Population Reports*, P25-1092 and P25-1095. *Notes:* NA Not available. 1. Persons of Hispanic origin may be of any race. 2. Beginning 1950, current definition. For explanation of change, see text, section 1. 3. Excludes Alaska and Hawaii. 4. The revised 1970 resident population count is 203,302,031; which incorporates changes due to errors found after tabulations were completed. The race and sex data shown here reflect the official 1970 census count while the residence data come from the tabulated count; see text, section 1. 5. The race data shown have been modified; see text, section 1 for explanation. 6. Total population count has been revised since the 1980 census publications. Numbers at age, race, Hispanic origin, and sex have not been corrected. 7. Estimated. 8. Middle series projection; see table 3.

Population Trends and Projections

★ 953 ★

Age, Race, and Hispanic Origin of Population, 1995-2050. Part I.

In thousands. As of July 1. Resident population. Middle series.

Year	Total	Under 5 years	5 to 13 years	14 to 17 years	18 to 24 years	25 to 34 years	35 to 44 years	45 to 64 years	65 years and over	85 years and over	100 years and over	Median age
White Projections												
1995	218,334	15,841	27,167	11,544	20,339	34,027	35,081	44,207	30,129	3,271	43	35.1
2000	226,267	14,945	28,534	12,409	20,477	30,534	37,139	50,874	31,357	3,917	63	36.7
2005	233,343	14,587	28,244	13,078	22,149	28,900	34,879	59,001	32,505	4,581	97	38.1
2010	240,297	14,893	27,184	13,203	23,396	29,715	31,394	65,543	34,967	5,357	135	39.0
2020	254,791	16,039	28,247	12,472	22,715	33,203	30,616	65,748	45,748	6,106	235	39.6
2030	267,457	16,073	29,667	13,424	22,995	31,995	34,119	59,892	59,295	7,646	365	40.6
2040	277,232	16,572	29,791	13,708	24,494	32,964	32,960	62,885	63,858	11,907	486	41.5
2050	285,591	17,216	30,982	13,887	24,568	34,671	33,980	65,304	64,982	16,061	908	41.2
Black Projections												
1995	33,117	3,243	5,285	2,285	3,764	5,475	5,088	5,246	2,732	268	7	28.9
2000	35,469	3,214	5,836	2,431	3,900	5,235	5,610	6,324	2,919	326	10	29.6

[Continued]

★ 953 ★

Age, Race, and Hispanic Origin of Population, 1995-2050. Part I.

[Continued]

Year	Total	Under 5 years	5 to 13 years	14 to 17 years	18 to 24 years	25 to 34 years	35 to 44 years	45 to 64 years	65 years and over	85 years and over	100 years and over	Median age
2005	37,793	3,310	6,025	2,741	4,222	5,176	5,506	7,691	3,121	367	18	30.1
2010	40,224	3,518	6,114	2,842	4,687	5,409	5,268	8,951	3,435	413	25	30.2
2020	45,409	3,953	6,852	2,974	4,969	6,422	5,455	9,920	4,863	490	40	30.9
2030	50,596	4,277	7,536	3,382	5,474	6,766	6,462	9,872	6,827	615	58	31.8
2040	55,917	4,731	8,227	3,663	6,094	7,533	6,837	11,092	7,739	972	80	31.8
2050	61,586	5,171	9,066	4,026	6,613	8,303	7,610	12,431	8,368	1,405	135	32.1

Source: "Percent Distribution of Births, Deaths, and Net Immigration, by Race and Hispanic Origin: 1995 to 2050." Jennifer Cheeseman Day, *Population Projections of the United States, by Age, Sex, Race, and Hispanic Origin: 1993 to 2050*, P25-1104, p. xxv. Primary source: U.S. Bureau of the Census, Current Population Reports, P25-1092.

★ 954 ★

Population Trends and Projections

Age, Race, and Hispanic Origin of Population, 1995-2050. Part II.

In thousands. As of July 1. Resident population. Middle series.

Year	Total	Under 5 years	5 to 13 years	14 to 17 years	18 to 24 years	25 to 34 years	35 to 44 years	45 to 64 years	65 years and over	85 years and over	100 years and over	Median age
American Indian, Eskimo, and Aleut												
Projections												
1995	2,226	222	410	167	256	364	327	341	138	14	0	26.6
2000	2,380	215	428	192	271	356	352	405	161	21	1	27.3
2005	2,543	231	420	210	312	363	345	474	187	28	2	27.5
2010	2,719	252	435	198	342	399	340	535	218	36	3	27.9
2020	3,090	277	509	223	337	474	381	578	310	51	7	29.2
2030	3,473	303	552	256	398	486	454	610	415	69	10	29.5
2040	3,894	341	614	275	434	572	468	712	480	104	15	29.8
2050	4,346	373	684	310	480	622	549	782	547	142	23	30.1
Asian and Pacific Islander												
Projections												
1995	9,756	876	1,400	594	1,105	1,805	1,655	1,671	650	44	1	30.2
2000	12,125	1,056	1,751	779	1,262	2,112	2,023	2,258	884	69	2	30.8
2005	14,608	1,205	2,153	918	1,553	2,353	2,344	2,922	1,158	106	4	31.5
2010	17,191	1,354	2,479	1,143	1,795	2,656	2,657	3,623	1,484	163	6	31.8
2020	22,653	1,688	3,092	1,449	2,434	3,454	3,209	4,899	2,427	312	18	32.7
2030	28,467	2,036	3,774	1,760	2,937	4,325	4,005	5,992	3,638	513	43	33.6
2040	34,461	2,334	4,436	2,102	3,490	5,058	4,870	7,237	4,936	856	79	34.6
2050	40,508	2,622	5,013	2,406	4,050	5,865	5,600	8,741	6,212	1,286	143	35.5

Source: "Percent Distribution of Births, Deaths, and Net Immigration, by Race and Hispanic Origin: 1995 to 2050." Jennifer Cheeseman Day, *Population Projections of the United States, by Age, Sex, Race, and Hispanic Origin: 1993 to 2050*, P25-1104, p. xxv. Primary source: U.S. Bureau of the Census, Current Population Reports, P25-1092.

★ 955 ★

Population Trends and Projections

Age, Race, and Hispanic Origin of Population, 1995-2050. Part III

In thousands. As of July 1. Resident population. Middle series.

Year	Total	Under 5 years	5 to 13 years	14 to 17 years	18 to 24 years	25 to 34 years	35 to 44 years	45 to 64 years	65 years and over	85 years and over	100 years and over	Median age
Hispanic origin[1]												
Projections												
1995	26,798	3,090	4,560	1,817	3,204	5,021	3,894	3,681	1,532	141	2	26.5
2000	31,166	3,293	5,542	2,102	3,547	5,145	4,830	4,780	1,925	203	4	27.3
2005	35,702	3,579	6,196	2,582	4,070	5,301	5,396	6,211	2,368	267	8	27.6
2010	40,525	3,983	6,651	2,909	4,863	5,834	5,519	7,848	2,918	367	13	28.1
2020	51,217	4,947	8,150	3,376	5,852	7,678	6,226	10,241	4,747	595	27	29.1
2030	62,810	5,716	9,773	4,165	6,929	8,951	8,075	11,586	7,616	906	53	30.3
2040	75,130	6,584	11,201	4,838	8,297	10,544	9,366	14,081	10,220	1,583	86	31.1
2050	88,071	7,429	12,805	5,520	9,429	12,346	10,982	17,109	12,452	2,602	161	32.1
White, not Hispanic												
Projections												
1995	193,900	13,020	23,032	9,892	17,413	29,455	31,542	40,841	28,705	3,138	41	36.2
2000	197,872	11,936	23,506	10,507	17,245	25,852	32,741	46,511	29,575	3,726	59	38.1
2005	200,842	11,326	22,605	10,751	18,458	24,077	29,965	53,340	30,321	4,331	89	39.9
2010	203,441	11,273	21,141	10,564	19,001	24,421	26,370	58,387	32,285	5,013	123	41.0
2020	208,280	11,543	20,874	9,425	17,402	26,261	24,964	56,406	41,405	5,552	210	42.3
2030	210,480	10,883	20,821	9,671	16,732	23,888	26,813	49,337	52,333	6,808	315	43.7
2040	209,148	10,604	19,669	9,348	16,995	23,449	24,475	50,100	54,509	10,448	407	45.0
2050	205,849	10,485	19,421	8,923	16,058	23,528	24,063	49,777	53,597	13,661	760	45.2

Source: "Percent Distribution of Births, Deaths, and Net Immigration, by Race and Hispanic Origin: 1995 to 2050." Jennifer Cheeseman Day, *Population Projections of the United States, by Age, Sex, Race, and Hispanic Origin: 1993 to 2050,* P25-1104, p. xxvi. Primary source: U.S. Bureau of the Census, Current Population Reports, P25-1092. *Note:* 1. Persons of Hispanic origin may be of any race.

★ 956 ★

Population Trends and Projections

Births, Deaths, and Net Immigration, 1995 to 2050

Year	Total	Race				Hispanic origin[3]	Not of Hispanic origin, by race			
		White	Black	American Indian[1]	Asian[2]		White	Black	American Indian[1]	Asian[2]
Births										
Middle series										
1995	100.0	77.7	17.0	1.0	4.2	15.9	63.2	16.1	0.8	3.9
2000	100.0	76.1	17.7	1.1	5.2	17.5	60.1	16.6	0.9	4.8
2005	100.0	74.8	18.1	1.2	6.0	19.0	57.5	17.0	1.0	5.5
2010	100.0	73.8	18.5	1.2	6.5	20.3	55.4	17.3	1.0	6.0
2020	100.0	72.3	18.9	1.2	7.6	22.9	51.6	17.5	1.0	7.0
2030	100.0	70.0	19.9	1.3	8.9	25.5	46.8	18.3	1.1	8.2
2040	100.0	68.5	20.5	1.4	9.6	27.6	43.5	18.8	1.2	8.8
2050	100.0	67.1	21.2	1.4	10.3	29.4	40.5	19.4	1.3	9.4

[Continued]

★ 956 ★

Births, Deaths, and Net Immigration, 1995 to 2050

[Continued]

Year	Total	Race				Hispanic origin[3]	Not of Hispanic origin, by race			
		White	Black	American Indian[1]	Asian[2]		White	Black	American Indian[1]	Asian[2]
Lowest series										
2050	100.0	71.5	17.7	1.3	9.4	27.7	46.2	16.2	1.1	8.8
Highest series										
2050	100.0	69.1	18.0	1.2	11.8	33.6	38.8	15.9	1.0	10.8
Deaths										
Middle series										
1995	100.0	85.6	12.8	0.5	1.2	4.5	81.4	12.6	0.4	1.1
2000	100.0	84.7	13.3	0.5	1.5	5.3	79.8	13.0	0.4	1.4
2005	100.0	84.1	13.5	0.5	1.9	6.0	78.6	13.2	0.4	1.8
2010	100.0	83.4	13.7	0.5	2.4	6.7	77.2	13.3	0.5	2.2
2020	100.0	81.6	14.4	0.6	3.3	8.3	74.0	14.0	0.6	3.2
2030	100.0	80.3	14.8	0.7	4.2	9.8	71.4	14.2	0.6	4.0
2040	100.0	79.5	14.7	0.7	5.1	11.4	69.1	14.1	0.6	4.8
2050	100.0	78.3	14.9	0.7	6.1	13.5	66.0	14.1	0.6	5.8
Lowest series										
2050	100.0	79.2	14.8	0.8	5.2	13.3	67.0	14.0	0.7	5.0
Highest series										
2050	100.0	78.8	13.6	0.7	6.9	14.3	65.8	12.8	0.6	6.5
Net immigration										
Middle series										
1993-2050	100.0	54.8	9.2	0.0	36.0	36.6	21.9	6.9	0.0	34.5
Lowest series										
1995-2050	100.0	40.1	7.7	0.0	52.1	40.3	2.9	6.0	0.0	50.9
Highest series										
1995-2050	100.0	57.9	9.5	0.0	32.7	36.2	25.6	7.1	0.0	31.2

Source: "Percent Distribution of Births, Deaths, and Net Immigration, by Race and Hispanic Origin: 1995 to 2050." Jennifer Cheeseman Day, *Population Projections of the United States, by Age, Sex, Race, and Hispanic Origin: 1993 to 2050*, P25-1104, p. xxxi. Primary source: U.S. Bureau of the Census, Current Population Reports, P25-1092. *Notes:* 1. American Indian represents American Indian, Eskimo, and Aleut. 2. Asian represents Asian and Pacific Islander. 3. Persons of Hispanic origin may be of any race.

★ 957 ★

Population Trends and Projections

Change in Population, by Race and Hispanic Origin, 1990 to 2050

As of July1. Resident population.

Year	Total	Race				Hispanic origin[3]	Not of Hispanic origin, by race			
		White	Black	American Indian[1]	Asian[2]		White	Black	American Indian	Asian[2]
Total percent change										
1990 to 2050	57.2	36.5	101.1	109.4	435.1	290.5	9.2	91.7	104.9	436.4
Average annual percent change										
1990 to 1995	1.09	0.86	1.57	1.40	5.07	3.45	0.56	1.47	1.30	5.11
1995 to 2000	0.95	0.71	1.37	1.34	4.35	3.02	0.41	1.28	1.29	4.39
2000 to 2005	0.85	0.62	1.27	1.32	3.73	2.72	0.30	1.18	1.27	3.75
2005 to 2010	0.83	0.59	1.25	1.34	3.26	2.53	0.26	1.16	1.29	3.27
2010 to 2020	0.82	0.59	1.21	1.28	2.76	2.34	0.24	1.13	1.23	2.76
2020 to 2030	0.71	0.49	1.08	1.17	2.28	2.04	0.11	1.00	1.14	2.28
2030 to 2040	0.60	0.36	1.00	1.14	1.91	1.79	-0.06	0.93	1.13	1.90
2040 to 2050	0.54	0.30	0.97	1.10	1.62	1.59	-0.16	0.90	1.10	1.60

Source: "Population Change, by Race and Hispanic Origin: 1990 to 2050." Jennifer Cheeseman Day, *Population Projections of the United States, by Age, Sex, Race, and Hispanic Origin: 1993 to 2050*, P25-1104, p. xxiii. Primary source: U.S. Bureau of the Census, Current Population Reports, P25-1092. *Notes:* 1. American Indian represents American Indian, Eskimo, and Aleut. 2. Asian represents Asian and Pacific Islander. 3. Persons of Hispanic origin may be of any race.

★ 958 ★

Population Trends and Projections

Changes in Population, 1990-1992

April 1, 1990 to January 1, 1992 population increase.

	Number	%
Total	4,958,000	2.0
White, not Hispanic	1,870,000	1.0
Black	990,000	3.2
American Indian, Eskimo, and Aleut	76,000	3.7
Asian and Pacific Islander	748,000	10.0
Hispanic origin (of any race)	1,368,000	6.1

Source: Untitled table. Frederick W. Hollomann, "National Population Trends," in U.S. Bureau of the Census, Current Population Reports, *Population Profile of the United States: 1992*, p. 2. *Notes:* Total population on January 1, 1992: 253,668,000. Resident population. Consistent with the 1990 census, as enumerated. Race data for 1990 modified to assign a specified race to each person.

★ 959 ★

Population Trends and Projections

Components of Population Change for the United States, 1993 to 2050. Part I White

Numbers in thousands. Resident population.

Calendar year	Rate per 1,000 mid-year population						Population change during calendar year					
	July 1 population	Net change	Natural increase	Births	Deaths	Net immigration	July 1 population	Net change	Natural increase	Births	Deaths	Net immigration
1993	214,778	8.6	6.3	14.9	8.6	2.2	213,845	1,837	1,355	3,203	1,848	482
1994	216,586	8.2	6.0	14.6	8.6	2.2	215,682	1,778	1,296	3,166	1,870	482
1995	218,334	7.9	5.7	14.3	8.7	2.2	217,460	1,718	1,236	3,128	1,892	482
1996	220,023	7.5	5.4	14.1	8.7	2.2	219,179	1,661	1,179	3,092	1,914	482
1997	221,656	7.3	5.1	13.8	8.7	2.2	220,840	1,607	1,125	3,060	1,935	482
1998	223,238	7.0	4.8	13.6	8.8	2.2	222,447	1,558	1,076	3,033	1,956	482
1999	224,773	6.7	4.6	13.4	8.8	2.1	224,005	1,515	1,033	3,010	1,977	482
2000	226,267	6.5	4.4	13.2	8.8	2.1	225,520	1,478	995	2,992	1,997	482
2001	227,726	6.4	4.2	13.1	8.8	2.1	226,998	1,447	964	2,980	2,015	482
2002	229,161	6.2	4.1	13.0	8.9	2.1	228,444	1,421	9.9	2,972	2,033	482
2003	230,570	6.1	4.0	12.9	8.9	2.1	229,865	1,401	918	2,970	2,051	482
2004	231,962	6.0	3.9	12.8	8.9	2.1	231,266	1,387	904	2,973	2,069	482
2005	233,343	5.9	3.8	12.8	8.9	2.1	232,652	1,379	897	2,983	2,086	482
2006	234,720	5.9	3.8	12.8	9.0	2.1	234,032	1,378	896	2,998	2,103	482
2007	236,099	5.9	3.8	12.8	9.0	2.0	235,410	1,383	901	3,020	2,119	482
2008	237,486	5.9	3.8	12.8	9.0	2.0	236,793	1,393	910	3,046	2,136	482
2009	238,885	5.9	3.9	12.9	9.0	2.0	238,185	1,406	923	3,075	2,152	482
2010	240,297	5.9	3.9	12.9	9.0	2.0	239,591	1,420	938	3,106	2,168	482
2011	241,724	5.9	3.9	13.0	9.0	2.0	241,011	1,434	952	3,137	2,185	482
2012	243,165	5.9	4.0	13.0	9.1	2.0	242,445	1,446	964	3,165	2,201	482
2013	244,617	6.0	4.0	13.0	9.1	2.0	243,891	1,456	974	3,192	2,218	482
2014	246,077	5.9	4.0	13.1	9.1	2.0	245,347	1,462	980	3,215	2,235	482
2015	247,542	5.9	4.0	13.1	9.1	1.9	246,810	1,465	983	3,235	2,252	482
2016	249,008	5.9	3.9	13.1	9.1	1.9	248,275	1,464	982	3,252	2,271	482
2017	250,470	5.8	3.9	13.0	9.1	1.9	249,739	1,458	976	3,266	2,290	482
2018	251,924	5.7	3.8	13.0	9.2	1.9	251,197	1,448	966	3,276	2,310	482
2019	253,366	5.7	3.8	13.0	9.2	1.9	252,645	1,433	951	3,282	2,331	482
2020	254,791	5.6	3.7	12.9	9.2	1.9	254,078	1,414	932	3,286	2,354	482
2021	256,195	5.4	3.5	12.8	9.3	1.9	255,493	1,391	909	3,287	2,378	482
2022	257,574	5.3	3.4	12.8	9.3	1.9	256,884	1,365	883	3,286	2,404	482
2023	258,924	5.2	3.3	12.7	9.4	1.9	258,249	1,335	853	3,284	2,431	482
2024	260,244	5.0	3.2	12.6	9.5	1.9	259,584	1,303	821	3,281	2,460	482
2025	261,531	4.9	3.0	12.5	9.5	1.8	260,887	1,270	788	3,278	2,490	482
2026	262,784	4.7	2.9	12.5	9.6	1.8	262,157	1,236	754	3,275	2,521	482
2027	264,003	4.6	2.7	12.4	9.7	1.8	263,393	1,202	720	3,274	2,554	482
2028	265,187	4.4	2.6	12.3	9.8	1.8	264,595	1,168	686	3,274	2,589	482
2029	266,338	4.3	2.4	12.3	9.9	1.8	265,763	1,135	652	3,276	2,624	482
2030	267,457	4.1	2.3	12.3	9.9	1.8	266,898	1,103	620	3,281	2,660	482
2031	268,544	4.0	2.2	12.2	10.0	1.8	268,000	1,072	590	3,288	2,697	482
2032	269,601	3.9	2.1	12.2	10.1	1.8	269,073	1,044	562	3,297	2,735	482
2033	270,632	3.8	2.0	12.2	10.2	1.8	270,117	1,018	536	3,308	2,772	482
2034	271,637	3.7	1.9	12.2	10.3	1.8	271,135	994	511	3,320	2,809	482
2035	272,619	3.6	1.8	12.2	10.4	1.8	272,128	971	489	3,335	2,846	482
2036	273,579	3.5	1.7	12.2	10.5	1.8	273,099	950	468	3,351	2,883	482
2037	274,519	3.4	1.6	12.3	10.6	1.8	274,049	931	448	3,367	2,919	482
2038	275,440	3.3	1.6	12.3	10.7	1.8	274,980	913	430	3,384	2,953	482
2039	276,344	3.2	1.5	12.3	10.8	1.7	275,892	896	414	3,400	2,987	482
2040	277,232	3.2	1.4	12.3	10.9	1.7	276,788	881	399	3,417	3,018	482
2041	278,106	3.1	1.4	12.3	11.0	1.7	277,669	867	385	3,432	3,047	482
2042	278,967	3.1	1.3	12.4	11.0	1.7	278,536	855	373	3,447	3,074	482
2043	279,816	3.0	1.3	12.4	11.1	1.7	279,392	844	362	3,460	3,098	482
2044	280,656	3.0	1.3	12.4	11.1	1.7	280,236	835	353	3,473	3,120	482

[Continued]

★ 959 ★

Components of Population Change for the United States, 1993 to 2050. Part I White

[Continued]

Calendar year	Rate per 1,000 mid-year population						Population change during calendar year					
	July 1 population	Net change	Natural increase	Births	Deaths	Net immigration	July 1 population	Net change	Natural increase	Births	Deaths	Net immigration
2045	281,486	2.9	1.2	12.4	11.1	1.7	281,071	828	346	3,483	3,138	482
2046	282,311	2.9	1.2	12.4	11.2	1.7	281,899	823	341	3,493	3,152	482
2047	283,132	2.9	1.2	12.4	11.2	1.7	282,722	820	338	3,501	3,163	482
2048	283,951	2.9	1.2	12.4	11.2	1.7	283,542	819	337	3,508	3,172	482
2049	284,770	2.9	1.2	12.3	11.2	1.7	284,361	820	337	3,515	3,177	482
2050	285,591	2.9	1.2	12.3	11.1	1.7	285,180	822	340	3,521	3,181	482

Source: "Annual Projections and Components of Change, for the United States: 1993 to 2050 (Middle Series). Part B. White Population." Jennifer Cheeseman Day, *Population Projections of the United States, by Age, Sex, Race, and Hispanic Origin: 1993 to 2050,* P25-1104, p. 2.

★ 960 ★

Population Trends and Projections

Components of Population Change for the United States, 1993 to 2050. Part II Black

Numbers in thousands. Resident population.

Calendar year	Rate per 1,000 mid-year population						Population change during calendar year					
	July 1 population	Net change	Natural increase	Births	Deaths	Net immigration	July 1 population	Net change	Natural increase	Births	Deaths	Net immigration
1993	32,137	15.5	13.0	21.4	8.4	2.5	31,886	498	417	688	271	81
1994	32,631	15.0	12.5	21.0	8.5	2.5	32,384	490	409	686	277	81
1995	33,117	14.6	12.1	20.7	8.5	2.4	32,874	483	402	685	283	81
1996	33,597	14.2	11.8	20.4	8.6	2.4	33,357	477	396	685	289	81
1997	34,071	13.8	11.5	20.1	8.7	2.4	33,834	472	391	686	295	81
1998	34,541	13.5	11.2	19.9	8.7	2.3	34,306	468	387	688	301	81
1999	35,006	13.3	10.9	19.7	8.8	2.3	34,774	464	383	691	308	81
2000	35,469	13.0	10.8	19.6	8.8	2.3	35,238	463	382	695	313	81
2001	35,932	12.9	10.6	19.4	8.8	2.3	35,700	463	382	699	317	81
2002	36,395	12.7	10.5	19.3	8.8	2.2	36,163	463	382	703	321	81
2003	36,858	12.6	10.4	19.2	8.8	2.2	36,626	465	384	709	325	81
2004	37,324	12.5	10.3	19.2	8.8	2.2	37,091	467	386	716	329	81
2005	37,793	12.5	10.3	19.1	8.8	2.1	37,558	471	390	724	334	81
2006	38,266	12.4	10.3	19.2	8.8	2.1	38,029	476	395	733	338	81
2007	38,745	12.5	10.4	19.2	8.8	2.1	38,505	483	402	744	342	81
2008	39,231	12.5	10.4	19.3	8.8	2.1	38,968	490	409	756	347	81
2009	39,724	12.5	10.5	19.3	8.9	2.0	39,478	497	416	767	352	81
2010	40,224	12.5	10.5	19.4	8.9	2.0	39,974	503	422	779	357	81
2011	40,731	12.5	10.5	19.4	8.9	2.0	40,478	509	428	790	362	81
2012	41,242	12.4	10.5	19.4	8.9	2.0	40,986	513	433	799	367	81
2013	41,758	12.4	10.4	19.4	8.9	1.9	41,500	517	436	808	372	81
2014	42,276	12.3	10.4	19.3	8.9	1.9	42,017	519	439	816	378	81
2015	42,797	12.2	10.3	19.3	9.0	1.9	42,536	521	440	824	384	81
2016	43,318	12.1	10.2	19.2	9.0	1.9	43,058	522	441	831	390	81
2017	43,841	11.9	10.1	19.1	9.0	1.8	43,580	523	442	838	396	81
2018	44,364	11.8	10.0	19.0	9.1	1.8	44,103	523	442	844	402	81
2019	44,887	11.6	9.8	18.9	9.1	1.8	44,625	522	441	851	409	81
2020	45,409	11.5	9.7	18.9	9.2	1.8	45,148	522	441	857	416	81
2021	45,930	11.3	9.6	18.8	9.2	1.8	45,669	521	440	863	423	81
2022	46,450	11.2	9.4	18.7	9.3	1.7	46,190	520	439	869	430	81
2023	46,969	11.0	9.3	18.6	9.3	1.7	46,709	519	438	875	437	81
2024	47,487	10.9	9.2	18.6	9.4	1.7	47,228	518	437	882	445	81

[Continued]

★ 960 ★

Components of Population Change for the United States, 1993 to 2050. Part II Black

[Continued]

Calendar year	Rate per 1,000 mid-year population						Population change during calendar year					
	July 1 population	Net change	Natural increase	Births	Deaths	Net immigration	July 1 population	Net change	Natural increase	Births	Deaths	Net immigration
2025	48,005	10.8	9.1	18.5	9.4	1.7	47,746	517	436	889	452	81
2026	48,522	10.7	9.0	18.5	9.5	1.7	48,263	517	436	896	460	81
2027	49,039	10.6	8.9	18.4	9.5	1.6	48,781	518	437	904	468	81
2028	49,557	10.5	8.8	18.4	9.6	1.6	49,298	518	437	913	475	81
2029	50,076	10.4	8.8	18.4	9.6	1.6	49,817	519	438	922	483	81
2030	50,596	10.3	8.7	18.4	9.7	1.6	50,336	521	440	931	491	81
2031	51,117	10.2	8.6	18.4	9.8	1.6	50,857	522	441	940	499	81
2032	51,640	10.2	8.6	18.4	9.8	1.6	51,379	524	443	949	506	81
2033	52,166	10.1	8.5	18.4	9.8	1.6	51,903	526	446	959	513	81
2034	52,693	10.0	8.5	18.4	9.9	1.5	52,430	529	448	969	520	81
2035	53,224	10.0	8.5	18.4	9.9	1.5	52,959	532	451	978	527	81
2036	53,757	9.9	8.4	18.4	9.9	1.5	53,490	534	453	988	534	81
2037	54,292	9.9	8.4	18.4	10.0	1.5	54,025	537	456	997	541	81
2038	54,831	9.8	8.4	18.4	10.0	1.5	54,562	540	459	1,006	547	81
2039	55,373	9.8	8.3	18.3	10.0	1.5	55,102	543	462	1,016	554	81
2040	55,917	9.8	8.3	18.3	10.0	1.4	55,645	546	465	1,025	559	81
2041	56,465	9.7	8.3	18.3	10.0	1.4	56,191	550	469	1,034	565	81
2042	57,017	9.7	8.3	18.3	10.0	1.4	56,741	554	473	1,043	570	81
2043	57,572	9.7	8.3	18.3	10.0	1.4	57,294	558	477	1,052	575	81
2044	58,132	9.7	8.3	18.2	10.0	1.4	57,852	562	481	1,060	580	81
2045	58,695	9.6	8.3	18.2	10.0	1.4	58,414	566	485	1,069	584	81
2046	59,264	9.6	8.3	18.2	9.9	1.4	58,980	570	490	1,078	589	81
2047	59,836	9.6	8.3	18.2	9.9	1.4	59,550	575	494	1,087	593	81
2048	60,414	9.6	8.3	18.1	9.9	1.3	60,125	580	500	1,096	597	81
2049	60,997	9.6	8.3	18.1	9.8	1.3	60,706	586	505	1,105	600	81
2050	61,586	9.6	8.3	18.1	9.8	1.3	61,292	592	511	1,115	604	81

Source: "Annual Projections and Components of Change, for the United States: 1993 to 2050 (Middle Series). Part C. Black Population." Jennifer Cheeseman Day, *Population Projections of the United States, by Age, Sex, Race, and Hispanic Origin: 1993 to 2050,* P25-1104, p. 3.

★ 961 ★

Population Trends and Projections

Components of Population Change for the United States, 1993 to 2050. Part III American Indian, Eskimo, and Aleut

Numbers in thousands. Resident population.

Calendar year	Rate per 1,000 mid-year population						Population change during calendar year					
	July 1 population	Net change	Natural increase	Births	Deaths	Net immigration	July 1 population	Net change	Natural increase	Births	Deaths	Net immigration
1993	2,165	14.2	14.1	18.4	4.3	0.1	2,150	31	30	40	9	0
1994	2,196	13.9	13.8	18.1	4.3	0.1	2,181	31	30	40	10	0
1995	2,226	13.7	13.5	17.9	4.4	0.1	2,211	30	30	40	10	0
1996	2,257	13.5	13.3	17.8	4.4	0.1	2,242	30	30	40	10	0
1997	2,287	13.3	13.2	17.7	4.5	0.1	2,272	30	30	40	10	0
1998	2,318	13.2	13.1	17.6	4.5	0.1	2,302	31	30	41	10	0
1999	2,349	13.2	13.1	17.6	4.6	0.1	2,333	31	31	41	11	0
2000	2,380	13.2	13.1	17.7	4.6	0.1	2,364	31	31	42	11	0
2001	2,411	13.2	13.1	17.8	4.7	0.1	2,396	32	32	43	11	0
2002	2,444	13.2	13.1	17.9	4.7	0.1	2,427	32	32	44	12	0
2003	2,476	13.2	13.1	17.9	4.8	0.1	2,460	33	33	44	12	0

[Continued]

★ 961 ★

Components of Population Change for the United States, 1993 to 2050. Part III American Indian, Eskimo, and Aleut

[Continued]

Calendar year	Rate per 1,000 mid-year population						Population change during calendar year					
	July 1 population	Net change	Natural increase	Births	Deaths	Net immigration	July 1 population	Net change	Natural increase	Births	Deaths	Net immigration
2004	2,509	13.3	13.2	18.0	4.9	0.1	2,493	33	33	45	12	0
2005	2,543	13.3	13.2	18.1	4.9	0.1	2,526	34	34	46	13	0
2006	2,577	13.4	13.3	18.3	5.0	0.1	2,560	34	34	47	13	0
2007	2,612	13.4	13.3	18.4	5.1	0.1	2,594	35	35	48	13	0
2008	2,647	13.5	13.3	18.5	5.1	0.1	2,629	36	35	49	14	0
2009	2,683	13.5	13.4	18.5	5.2	0.1	2,665	36	36	50	14	0
2010	2,719	13.5	13.3	18.6	5.2	0.1	2,701	37	36	51	14	0
2011	2,756	13.4	13.3	18.6	5.3	0.1	2,738	37	37	51	15	0
2012	2,793	13.3	13.2	18.5	5.3	0.1	2,774	37	37	52	15	0
2013	2,830	13.1	13.0	18.4	5.4	0.1	2,812	37	37	52	15	0
2014	2,867	13.0	12.9	18.3	5.4	0.1	2,849	37	37	53	16	0
2015	2,904	12.8	12.7	18.2	5.5	0.1	2,886	37	37	53	16	0
2016	2,942	12.6	12.5	18.1	5.6	0.1	2,923	37	37	53	16	0
2017	2,979	12.5	12.4	18.0	5.6	0.1	2,960	37	37	54	17	0
2018	3,016	12.3	12.2	17.9	5.7	0.1	2,997	37	37	54	17	0
2019	3,053	12.2	12.1	17.8	5.7	0.1	3,035	37	37	54	17	0
2020	3,090	12.0	12.0	17.7	5.8	0.1	3,072	37	37	55	18	0
2021	3,128	11.9	11.8	17.6	5.8	0.1	3,109	37	37	55	18	0
2022	3,165	11.8	11.7	17.6	5.9	0.1	3,146	37	37	56	19	0
2023	3,202	11.7	11.7	17.6	5.9	0.1	3,184	38	37	56	19	0
2024	3,240	11.7	11.6	17.5	6.0	0.1	3,221	38	38	57	19	0
2025	3,278	11.6	11.5	17.5	6.0	0.1	3,259	38	38	58	20	0
2026	3,316	11.6	11.5	17.6	6.1	0.1	3,297	38	38	58	20	0
2027	3,355	11.6	11.5	17.6	6.1	0.1	3,336	39	39	59	20	0
2028	3,394	11.6	11.5	17.6	6.1	0.1	3,375	39	39	60	21	0
2029	3,434	11.6	11.5	17.7	6.2	0.1	3,414	40	39	61	21	0
2030	3,473	11.6	11.5	17.7	6.2	0.1	3,453	40	40	61	22	0
2031	3,514	11.5	11.5	17.7	6.3	0.1	3,494	41	40	62	22	0
2032	3,555	11.5	11.5	17.8	6.3	0.1	3,534	41	41	63	22	0
2033	3,596	11.5	11.4	17.8	6.3	0.1	3,575	41	41	64	23	0
2034	3,637	11.5	11.4	17.8	6.4	0.1	3,617	42	42	65	23	0
2035	3,679	11.5	11.4	17.8	6.4	0.1	3,658	42	42	66	24	0
2036	3,722	11.4	11.4	17.8	6.5	0.1	3,701	43	42	66	24	0
2037	3,764	11.4	11.3	17.8	6.5	0.1	3,743	43	43	67	24	0
2038	3,807	11.3	11.3	17.8	6.5	0.1	3,786	43	43	68	25	0
2039	3,851	11.3	11.2	17.8	6.6	0.1	3,829	43	43	68	25	0
2040	3,894	11.2	11.1	17.7	6.6	0.1	3,873	44	43	69	26	0
2041	3,938	11.2	11.1	17.7	6.6	0.1	3,916	44	44	70	26	0
2042	3,982	11.1	11.0	17.7	6.7	0.1	3,960	44	44	70	27	0
2043	4,027	11.0	11.0	17.7	6.7	0.1	4,004	44	44	71	27	0
2044	4,071	11.0	10.9	17.6	6.7	0.1	4,049	45	44	72	27	0
2045	4,116	10.9	10.9	17.6	6.7	0.1	4,094	45	45	73	28	0
2046	4,161	10.9	10.8	17.6	6.8	0.1	4,139	45	45	73	28	0
2047	4,207	10.9	10.8	17.6	6.8	0.1	4,184	46	45	74	29	0
2048	4,253	10.9	10.8	17.6	6.8	0.1	4,230	46	46	75	29	0
2049	4,299	10.8	10.8	17.6	6.8	0.1	4,276	47	46	76	29	0
2050	4,346	10.8	10.8	17.6	6.8	0.1	4,323	47	47	76	30	0

Source: "Annual Projections and Components of Change, for the United States: 1993 to 2050 (Middle Series). Part D. American Indian, Eskimo, and Aleut Population." Jennifer Cheeseman Day, *Population Projections of the United States, by Age, Sex, Race, and Hispanic Origin: 1993 to 2050*, P25-1104, p. 4.

★ 962 ★

Population Trends and Projections

Components of Population Change for the United States, 1993 to 2050. Part IV Asian and Pacific Islander

Numbers in thousands. Resident population.

Calendar year	Rate per 1,000 mid-year population						Population change during calendar year					
	July 1 population	Net change	Natural increase	Births	Deaths	Net immigration	July 1 population	Net change	Natural increase	Births	Deaths	Net immigration
1993	8,846	50.7	14.9	17.5	2.6	35.8	8,624	449	132	155	23	317
1994	9,298	48.9	14.9	17.5	2.7	34.1	9,072	455	138	163	25	317
1995	9,756	47.2	14.8	17.5	2.7	32.5	9,527	461	144	170	26	317
1996	10,219	45.6	14.6	17.4	2.8	31.0	9,988	466	149	178	28	317
1997	10,688	44.1	14.5	17.3	2.8	29.6	10,454	471	155	185	30	317
1998	11,162	42.7	14.3	17.2	2.8	28.4	10,925	476	160	192	32	317
1999	11,641	41.4	14.2	17.0	2.9	27.2	11,401	481	165	196	34	317
2000	12,125	40.1	14.0	16.9	2.9	26.1	11,883	486	169	205	36	317
2001	12,613	38.9	13.6	16.8	3.0	25.1	12,369	490	174	212	38	317
2002	13,106	37.7	13.6	16.7	3.1	24.2	12,859	495	178	218	40	317
2003	13,602	36.7	13.4	16.5	3.1	23.3	13,354	499	182	225	43	317
2004	14,103	35.6	13.2	16.4	3.2	22.4	13,853	503	186	231	45	317
2005	14,608	34.7	13.0	16.3	3.3	21.7	14,355	507	190	238	48	317
2006	15,116	33.8	12.8	16.2	3.3	20.9	14,862	511	194	244	50	317
2007	15,629	32.9	12.7	16.0	3.4	20.3	15,373	515	198	251	53	317
2008	16,145	32.1	12.5	16.0	3.5	19.6	15,887	519	202	258	56	317
2009	16,666	31.4	12.4	15.9	3.5	19.0	16,406	523	206	265	59	317
2010	17,191	30.6	12.2	15.8	3.6	18.4	16,928	527	210	272	62	317
2011	17,719	29.9	12.1	15.7	3.6	17.9	17,455	531	214	279	65	317
2012	18,252	29.3	11.9	15.7	3.7	17.3	17,985	535	218	286	68	317
2013	18,788	28.7	11.8	15.6	3.8	16.9	18,520	538	222	293	71	317
2014	19,329	28.1	11.7	15.5	3.8	16.4	19,058	542	226	300	74	317
2015	19,873	27.5	11.6	15.5	3.9	15.9	19,601	546	230	307	78	317
2016	20,421	26.9	11.4	15.4	4.0	15.5	20,147	550	234	315	81	317
2017	20,973	26.4	11.3	15.4	4.0	15.1	20,697	554	237	322	85	317
2018	21,529	25.9	11.2	15.3	4.1	14.7	21,251	558	241	330	89	317
2019	22,089	25.4	11.1	15.3	4.2	14.3	21,809	562	245	337	92	317
2020	22,653	25.0	11.0	15.2	4.2	14.0	22,371	566	249	345	96	317
2021	23,220	24.5	10.9	15.2	4.3	13.6	22,936	569	253	353	100	317
2022	23,791	24.1	10.8	15.1	4.4	13.3	23,506	573	256	360	104	317
2023	24,366	23.6	10.7	15.1	4.4	13.0	24,078	576	260	368	106	317
2024	24,943	23.2	10.5	15.0	4.5	12.7	24,654	579	263	375	113	317
2025	25,524	22.8	10.4	15.0	4.6	12.4	25,234	582	266	383	117	317
2026	26,108	22.4	10.3	14.9	4.6	12.1	25,816	585	268	390	121	317
2027	26,694	22.0	10.2	14.9	4.7	11.9	26,401	588	271	397	126	317
2028	27,283	21.6	10.0	14.8	4.8	11.6	26,989	590	273	404	130	317
2029	27,874	21.2	9.9	14.7	4.8	11.4	27,579	592	275	410	135	317
2030	28,467	20.9	9.7	14.6	4.9	11.1	28,171	594	277	417	140	317
2031	29,062	20.5	9.6	14.6	5.0	10.9	28,764	595	279	423	145	317
2032	29,658	20.1	9.4	14.5	5.0	10.7	29,360	597	280	430	149	317
2033	30,255	19.8	9.3	14.4	5.1	10.5	29,957	598	281	436	154	317
2034	30,854	19.4	9.2	14.3	5.2	10.3	30,555	599	282	442	160	317
2035	31,453	19.1	9.0	14.2	5.2	10.1	31,154	600	283	448	165	317
2036	32,054	18.7	8.9	14.2	5.3	9.9	31,753	601	284	454	170	317
2037	32,655	18.4	8.7	14.1	5.4	9.7	32,354	601	285	460	175	317
2038	33,256	18.1	8.6	14.0	5.4	9.5	32,955	602	285	466	181	317
2039	33,859	17.8	8.4	13.9	5.5	9.4	33,557	602	286	472	186	317
2040	34,461	17.5	8.3	13.9	5.6	9.2	34,160	603	286	478	192	317
2041	35,064	17.2	8.2	13.8	5.6	9.0	34,763	603	287	484	197	317
2042	35,668	16.9	8.1	13.7	5.7	8.9	35,366	604	287	490	203	317

[Continued]

★ 962 ★

Components of Population Change for the United States, 1993 to 2050. Part IV Asian and Pacific Islander

[Continued]

Calendar year	Rate per 1,000 mid-year population						Population change during calendar year					
	July 1 population	Net change	Natural increase	Births	Deaths	Net immigration	July 1 population	Net change	Natural increase	Births	Deaths	Net immigration
2043	36,272	16.7	7.9	13.7	5.8	8.7	35,970	604	288	496	209	317
2044	36,876	16.4	7.8	13.6	5.8	8.6	36,574	604	288	502	215	317
2045	37,481	16.1	7.7	13.6	5.9	8.4	37,179	605	288	509	220	317
2046	38,086	15.9	7.6	13.5	5.9	8.3	37,783	605	289	515	226	317
2047	38,691	15.6	7.5	13.5	6.0	8.2	38,389	605	289	521	232	317
2048	39,297	15.4	7.4	13.4	6.0	8.1	38,994	606	289	527	238	317
2049	39,903	15.2	7.2	13.3	6.1	7.9	39,600	606	289	533	243	317
2050	40,508	15.0	7.1	13.3	6.1	7.8	40,205	606	289	539	249	317

Source: "Annual Projections and Components of Change, for the United States: 1993 to 2050 (Middle Series). Part E. Asian and Pacific Islander Population." Jennifer Cheeseman Day, *Population Projections of the United States, by Age, Sex, Race, and Hispanic Origin: 1993 to 2050,* P25-1104, p. 5.

★ 963 ★

Population Trends and Projections

Components of Population Change for the United States, 1993 to 2050. Part V.

Numbers in thousands. Resident population.

Calendar year	Rate per 1,000 mid-year population						Population change during calendar year					
	July 1 population	Net change	Natural increase	Births	Deaths	Net immigration	July 1 population	Net change	Natural increase	Births	Deaths	Net immigration
1993	25,085	33.9	21.0	24.7	3.6	12.8	24,662	850	528	619	91	322
1994	25,939	33.0	20.6	24.3	3.7	12.4	25,512	856	534	629	95	322
1995	26,798	32.1	20.1	23.8	3.7	12.0	26,368	862	539	639	100	322
1996	27,662	31.3	19.7	23.4	3.8	11.6	27,230	866	544	648	105	322
1997	28,530	30.5	19.2	23.1	3.8	11.3	28,096	871	548	658	109	322
1998	29,403	29.8	18.8	22.7	3.9	11.0	28,966	876	554	668	114	322
1999	30,281	29.1	18.5	22.4	4.0	10.6	29,842	881	559	679	120	322
2000	31,166	28.5	18.1	22.1	4.0	10.3	30,723	888	565	690	128	322
2001	32,056	27.9	17.9	21.9	4.0	10.1	31,611	895	572	702	129	322
2002	32,955	27.4	17.6	21.7	4.1	9.8	32,506	902	580	714	134	322
2003	33,861	26.9	17.4	21.5	4.1	9.5	33,408	911	589	727	139	322
2004	34,777	26.5	17.2	21.3	4.1	9.3	34,319	920	598	741	143	322
2005	35,702	26.1	17.0	21.2	4.2	9.0	35,239	931	609	757	148	322
2006	36,638	25.7	16.9	21.1	4.2	8.8	36,170	943	621	774	153	322
2007	37,588	25.4	16.9	21.1	4.2	8.6	37,113	957	634	793	159	322
2008	38,551	25.2	16.8	21.1	4.3	8.4	38,070	971	649	813	164	322
2009	39,530	25.0	16.8	21.1	4.3	8.2	39,041	987	664	834	170	322
2010	40,525	24.7	16.8	21.1	4.3	8.0	40,028	1,002	680	855	175	322
2011	41,535	24.5	16.7	21.1	4.4	7.8	41,030	1,018	695	876	181	322
2012	42,560	24.3	16.7	21.1	4.4	7.6	42,047	1,032	710	897	187	322
2013	43,599	24.0	16.6	21.0	4.4	7.4	43,080	1,047	724	917	193	322
2014	44,653	23.7	16.5	21.0	4.5	7.2	44,126	1,060	738	937	199	322
2015	45,719	23.5	16.4	20.9	4.5	7.0	45,186	1,073	750	956	206	322
2016	46,798	23.2	16.3	20.8	4.5	6.9	46,259	1,084	762	974	212	322
2017	47,888	22.9	16.1	20.7	4.6	6.7	47,343	1,095	773	992	219	322
2018	48,988	22.6	16.0	20.6	4.6	6.6	48,438	1,105	783	1,009	226	322
2019	50,098	22.2	15.8	20.5	4.6	6.4	49,543	1,115	792	1,025	233	322
2020	51,217	21.9	15.6	20.3	4.7	6.3	50,658	1,123	801	1,041	240	322
2021	52,345	21.6	15.5	20.2	4.7	6.2	51,781	1,131	809	1,057	248	322

[Continued]

★ 963 ★

Components of Population Change for the United States, 1993 to 2050. Part V.

[Continued]

Calendar year	Rate per 1,000 mid-year population						Population change during calendar year					
	July 1 population	Net change	Natural increase	Births	Deaths	Net immigration	July 1 population	Net change	Natural increase	Births	Deaths	Net immigration
2022	53,479	21.3	15.3	20.0	4.8	6.0	52,912	1,138	816	1,072	256	322
2023	54,622	21.0	15.1	19.9	4.8	5.9	54,050	1,146	823	1,087	263	322
2024	55,771	20.7	14.9	19.8	4.9	5.8	55,196	1,153	830	1,102	272	322
2025	56,927	20.4	14.7	19.6	4.9	5.7	56,349	1,159	837	1,117	280	322
2026	58,089	20.1	14.5	19.5	5.0	5.5	57,508	1,166	844	1,133	289	322
2027	59,259	19.8	14.4	19.4	5.0	5.4	58,674	1,173	851	1,148	297	322
2028	60,435	19.5	14.2	19.3	5.1	5.3	59,847	1,180	858	1,164	307	322
2029	61,619	19.3	14.0	19.2	5.1	5.2	61,027	1,187	865	1,181	316	322
2030	62,810	19.0	13.9	19.1	5.2	5.1	62,214	1,194	872	1,198	326	322
2031	64,008	18.8	13.7	19.0	5.2	5.0	63,409	1,202	880	1,215	335	322
2032	65,213	18.5	13.6	18.9	5.3	4.9	64,610	1,210	887	1,233	345	322
2033	66,247	18.3	13.5	18.8	5.4	4.9	65,820	1,217	985	1,251	356	322
2034	67,648	18.1	13.3	18.8	5.4	4.8	67,037	1,225	903	1,269	366	322
2035	68,877	17.9	13.2	18.7	5.5	4.7	68,262	1,233	910	1,287	377	322
2036	70,113	17.7	13.1	18.6	5.5	4.6	69,495	1,240	918	1,306	388	322
2037	71,357	17.5	13.0	18.6	5.6	4.5	70,735	1,247	925	1,324	399	322
2038	72,608	17.3	12.8	18.5	5.6	4.4	71,982	1,254	932	1,342	410	322
2039	73,866	17.1	12.7	18.4	5.7	4.4	73,237	1,261	939	1,361	422	322
2040	75,130	16.9	12.6	18.3	5.8	4.3	74,498	1,268	945	1,378	433	322
2041	76,401	16.7	12.5	18.3	5.8	4.2	75,766	1,274	951	1,396	445	322
2042	77,678	16.5	12.3	18.2	5.9	4.1	77,039	1,279	957	1,414	456	322
2043	78,960	16.3	12.2	18.1	5.9	4.1	78,319	1,285	963	1,431	468	322
2044	80,247	16.1	12.1	18.0	6.0	4.0	79,603	1,290	698	1,448	480	322
2045	81,539	15.9	11.9	18.0	6.0	4.0	80,893	1,295	973	1,464	492	322
2046	82,837	15.7	11.8	17.9	6.1	3.9	82,188	1,299	977	1,480	503	322
2047	84,138	15.5	11.7	17.8	6.1	3.8	83,487	1,304	982	1,496	515	322
2048	85,445	15.3	11.5	17.7	6.2	3.8	84,792	1,309	986	1,512	526	322
2049	86,756	15.1	11.4	17.6	6.2	3.7	86,100	1,313	991	1,528	537	322
2050	88,071	15.0	11.3	17.5	6.2	3.7	87,413	1,318	996	1,544	548	322

Source: "Annual Projections and Components of Change, for the United States: 1993 to 2050 (Middle Series). Part F. Hispanic Origin Population." Jennifer Cheeseman Day, *Population Projections of the United States, by Age, Sex, Race, and Hispanic Origin: 1993 to 2050*, P25-1104, p. 6.

★ 964 ★

Population Trends and Projections

Components of Population Change for the United States, 1993 to 2050. Part VI White, Not Hispanic

Numbers in thousands. Resident population.

Calendar year	Rate per 1,000 mid-year population						Population change during calendar year					
	July 1 population	Net change	Natural increase	Births	Deaths	Net immigration	July 1 population	Net change	Natural increase	Births	Deaths	Net immigration
1993	191,899	5.6	4.5	13.7	9.2	1.0	191,350	1,065	872	2,637	1,764	193
1994	192,932	5.2	4.2	13.4	9.2	1.0	192,415	1,000	808	2,590	1,782	193
1995	193,900	4.8	3.8	13.1	9.3	1.0	193,416	936	744	2,544	1,800	193
1996	194,805	4.5	3.5	12.8	9.3	1.0	194,352	875	682	2,499	1,817	193
1997	195,650	4.2	3.2	12.6	9.4	1.0	195,227	817	625	2,459	1,834	193
1998	196,440	3.9	2.9	12.3	9.4	1.0	196,045	764	572	2,422	1,851	193
1999	197,178	3.6	2.7	12.1	9.5	1.0	196,809	716	523	2,390	1,867	193
2000	197,872	3.4	2.4	11.9	9.5	1.0	197,525	674	481	2,363	1,882	193

[Continued]

★ 964 ★

Components of Population Change for the United States, 1993 to 2050. Part VI White, Not Hispanic

[Continued]

Calendar year	Rate per 1,000 mid-year population						Population change during calendar year					
	July 1 population	Net change	Natural increase	Births	Deaths	Net immigration	July 1 population	Net change	Natural increase	Births	Deaths	Net immigration
2001	198,526	3.2	2.2	11.8	9.6	1.0	198,199	636	444	2,340	1,896	193
2002	199,144	3.0	2.1	11.7	9.6	1.0	198,835	6.4	411	2,321	1,910	193
2003	199,733	2.9	1.9	11.6	9.6	1.0	199,439	577	384	2,307	1,924	193
2004	200,298	2.8	1.8	11.5	9.7	1.0	200,015	554	362	2,298	1,937	193
2005	200,842	2.7	1.7	11.4	9.7	1.0	200,570	538	345	2,294	1,949	193
2006	201,373	2.6	1.7	11.4	9.7	1.0	201,108	526	333	2,295	1,961	193
2007	201,895	2.6	1.6	11.4	9.8	1.0	201,634	519	326	2,300	1,973	193
2008	202,411	2.5	1.6	11.4	9.8	1.0	202,153	516	323	2,308	1,985	193
2009	202,926	2.5	1.6	11.4	9.8	0.9	202,669	515	322	2,319	1,996	193
2010	203,441	2.5	1.6	11.5	9.9	0.9	203,184	515	323	2,330	2,008	193
2011	203,957	2.5	1.6	11.5	9.9	0.9	203,699	516	323	2,342	2,019	193
2012	204,472	2.5	1.6	11.5	9.9	0.9	204,215	515	322	2,352	2,030	193
2013	204,986	2.5	1.6	11.5	10.0	0.9	204,729	511	319	2,360	2,041	193
2014	205,495	2.5	1.5	11.5	10.0	0.9	205,241	506	313	2,365	2,052	193
2015	205,997	2.4	1.5	11.5	10.0	0.9	205,746	497	304	2,368	2,064	193
2016	206,489	2.3	1.4	11.5	10.1	0.9	206,243	485	292	2,368	2,076	193
2017	206,966	2.3	1.3	11.4	10.1	0.9	206,727	469	276	2,366	2,089	193
2018	207,427	2.2	1.2	11.4	10.1	0.9	207,196	450	257	2,360	2,103	193
2019	207,866	2.1	1.1	11.3	10.2	0.9	207,646	427	234	2,352	2,118	193
2020	208,280	1.9	1.0	11.2	10.2	0.9	208,073	400	207	2,342	2,134	193
2021	208,666	1.8	0.8	11.2	10.3	0.9	208,473	370	177	2,329	2,152	193
2022	209,020	1.6	0.7	11.1	10.4	0.9	208,843	336	144	2,314	2,170	193
2023	209,339	1.4	0.5	11.0	10.5	0.9	209,179	300	108	2,298	2,190	193
2024	209,620	1.3	0.3	10.9	10.6	0.9	209,479	262	70	2,281	2,212	193
2025	209,863	1.1	0.1	10.8	10.6	0.9	209,742	223	30	2,265	2,234	193
2026	210,067	0.9	0.0	10.7	10.7	0.9	209,965	183	-10	2,248	2,258	193
2027	210,230	0.7	-0.2	10.6	10.9	0.9	210,148	143	-50	2,233	2,283	193
2028	210,353	0.5	-0.4	10.5	11.0	0.9	210,291	103	-90	2,219	2,309	193
2029	210,436	0.3	-0.6	10.5	11.1	0.9	210,394	63	-129	2,207	2,336	193
2030	210,480	0.1	-0.8	10.4	11.2	0.9	210,458	25	-168	2,196	2,364	193
2031	210,486	-0.1	-1.0	10.4	11.4	0.9	210,483	-12	-205	2,187	2,392	193
2032	210,456	-0.2	-1.1	10.4	11.5	0.9	210,471	-47	-240	2,180	2,420	193
2033	210,392	-0.4	-1.3	10.3	11.6	0.9	210,424	-80	-273	2,175	2,448	193
2034	210,297	-0.5	-1.4	10.3	11.8	0.9	210,344	-111	-304	2,172	2,476	193
2035	210,171	-0.7	-1.6	10.3	11.9	0.9	210,234	-140	-333	2,170	2,503	193
2036	210,016	-0.8	-1.7	10.3	12.0	0.9	210,093	-168	-361	2,169	2,530	193
2037	209,835	-0.9	-1.8	10.3	12.2	0.9	209,926	-194	-386	2,169	2,555	193
2038	209,629	-1.0	-2.0	10.3	12.3	0.9	209,732	-218	-411	2,169	2,580	193
2039	209,399	-1.1	-2.1	10.4	12.4	0.9	209,514	-240	-433	2,170	2,603	193
2040	209,148	-1.2	-2.2	10.4	12.5	0.9	209,273	-261	-454	2,170	2,624	193
2041	208,876	-1.3	-2.3	10.4	12.7	0.9	209,012	-280	-473	2,169	2,642	193
2042	208,587	-1.4	-2.4	10.4	12.7	0.9	208,732	-298	-490	2,168	2,659	193
2043	208,281	-1.5	-2.4	10.4	12.8	0.9	208,434	-313	-506	2,166	2,672	193
2044	207,961	-1.6	-2.5	10.4	12.9	0.9	208,121	-327	-520	2,163	2,683	193
2045	207,627	-1.6	-2.6	10.4	13.0	0.9	207,794	-338	-531	2,159	2,690	193
2046	207,284	-1.7	-2.6	10.4	13.0	0.9	207,456	-348	-540	2,154	2,694	193
2047	206,932	-1.7	-2.6	10.4	13.0	0.9	207,108	-354	-547	2,148	2,695	193
2048	206,575	-1.7	-2.7	10.4	13.0	0.9	206,754	-359	-552	2,141	2,693	193

[Continued]

★ 964 ★

Components of Population Change for the United States, 1993 to 2050. Part VI White, Not Hispanic

[Continued]

Calendar year	Rate per 1,000 mid-year population						Population change during calendar year					
	July 1 population	Net change	Natural increase	Births	Deaths	Net immigration	July 1 population	Net change	Natural increase	Births	Deaths	Net immigration
2049	206,213	-1.8	-2.7	10.3	13.0	0.9	206,394	-363	-556	2,133	2,689	193
2050	205,849	-1.8	-2.7	10.3	13.0	0.9	206,031	-365	-557	2,125	2,682	193

Source: "Annual Projections and Components of Change, for the United States: 1993 to 2050 (Middle Series). Part G. White, Not Hispanic Population." Jennifer Cheeseman Day, *Population Projections of the United States, by Age, Sex, Race, and Hispanic Origin: 1993 to 2050*, P25-1104, p. 7.

★ 965 ★

Population Trends and Projections

Components of Population Change for the United States, 1993 to 2050. Part VII Black, Not Hispanic

Numbers in thousands. Resident population.

Calendar year	Rate per 1,000 mid-year population						Population change during calendar year					
	July 1 population	Net change	Natural increase	Births	Deaths	Net immigration	July 1 population	Net change	Natural increase	Births	Deaths	Net immigration
1993	30,768	14.6	12.6	21.2	8.7	2.0	30,542	448	387	653	266	61
1994	31,212	14.1	12.1	20.8	8.7	2.0	30,990	440	379	650	272	61
1995	31,648	13.7	11.7	20.5	8.8	1.9	31,430	432	371	649	279	61
1996	32,076	13.3	11.4	20.2	8.8	1.9	31,862	426	365	648	283	61
1997	32,499	12.9	11.1	20.0	8.9	1.9	32,288	420	359	649	289	61
1998	32,917	12.6	10.8	19.7	9.0	1.9	32,708	416	355	650	295	61
1999	33,331	12.4	10.5	19.6	9.0	1.8	33,124	412	351	652	302	61
2000	33,741	12.1	10.3	19.4	9.1	1.8	33,536	410	349	655	307	61
2001	34,151	12.0	10.2	19.3	9.1	1.8	33,946	409	348	659	310	61
2002	34,560	11.8	10.1	19.2	9.1	1.8	34,355	409	348	662	314	61
2003	34,969	11.7	10.0	19.1	9.1	1.7	34,765	410	349	667	318	61
2004	35,380	11.6	9.9	19.0	9.1	1.7	35,175	412	350	672	322	61
2005	35,793	11.6	9.9	19.0	9.1	1.7	35,586	415	353	679	326	61
2006	36,209	11.6	9.9	19.0	9.1	1.7	36,001	419	358	688	330	61
2007	36,631	11.6	9.9	19.0	9.1	1.7	36,420	424	363	697	334	61
2008	37,058	11.6	10.0	19.1	9.1	1.7	36,844	430	369	707	338	61
2009	37,491	11.6	10.0	19.1	9.1	1.6	37,274	436	375	718	343	61
2010	37,930	11.6	10.0	19.2	9.2	1.6	37,711	442	381	728	347	61
2011	38,375	11.6	10.0	19.2	9.2	1.6	38,153	447	385	737	352	61
2012	38,824	11.6	10.0	19.2	9.2	1.6	38,599	450	389	746	357	61
2013	39,275	11.5	10.0	19.2	9.2	1.6	39,049	453	392	753	362	61
2014	39,729	11.4	9.9	19.1	9.2	1.5	39,502	455	393	760	367	61
2015	40,184	11.3	9.8	19.1	9.3	1.5	39,957	455	394	767	373	61
2016	40,640	11.2	9.7	19.0	9.3	1.5	40,412	456	395	773	378	61
2017	41,096	11.1	9.6	18.9	9.3	1.5	40,868	456	394	779	384	61
2018	41,551	11.0	9.5	18.9	9.4	1.5	41,324	455	394	784	390	61
2019	42,006	10.8	9.4	18.8	9.4	1.5	41,779	454	393	789	396	61
2020	42,459	10.7	9.2	18.7	9.5	1.4	42,233	453	392	794	403	61
2021	42,912	10.5	9.1	18.6	9.5	1.4	42,685	451	390	799	409	61
2022	43,362	10.4	9.0	18.6	9.6	1.4	43,137	450	389	804	416	61
2023	43,811	10.2	8.8	18.5	9.6	1.4	43,587	448	387	810	423	61
2024	44,259	10.1	8.7	18.4	9.7	1.4	44,035	447	386	816	430	61
2025	44,705	10.0	8.6	18.4	9.8	1.4	44,482	446	385	822	437	61

[Continued]

★ 965 ★

Components of Population Change for the United States, 1993 to 2050. Part VII Black, Not Hispanic

[Continued]

Calendar year	Rate per 1,000 mid-year population						Population change during calendar year					
	July 1 population	Net change	Natural increase	Births	Deaths	Net immigration	July 1 population	Net change	Natural increase	Births	Deaths	Net immigration
2026	45,151	9.9	8.5	18.3	9.8	1.4	44,928	446	384	828	444	61
2027	45,597	9.8	8.4	18.3	9.9	1.3	45,374	445	384	835	451	61
2028	46,042	9.7	8.3	18.3	9.9	1.3	45,819	446	384	842	458	61
2029	46,488	9.6	8.3	18.3	10.0	1.3	46,265	446	385	850	465	61
2030	46,934	9.5	8.2	18.3	10.1	1.3	46,711	447	386	858	472	61
2031	47,381	9.5	8.2	18.3	10.1	1.3	47,156	448	387	866	479	61
2032	47,830	9.4	8.1	18.3	10.2	1.3	47,606	449	388	874	486	61
2033	48,280	9.3	8.1	18.3	10.2	1.3	48,055	451	390	883	493	61
2034	48,732	9.3	8.0	18.3	10.2	1.3	48,506	453	392	891	499	61
2035	49,186	9.2	8.0	18.3	10.3	1.2	48,959	455	394	900	506	61
2036	49,642	9.2	8.0	18.3	10.3	1.2	49,414	457	396	908	512	61
2037	50,100	9.2	7.9	18.3	10.3	1.2	49,871	459	398	916	518	61
2038	50,561	9.1	7.9	18.3	10.4	1.2	50,330	462	401	924	524	61
2039	51,024	9.1	7.9	18.3	10.4	1.2	50,792	464	403	932	529	61
2040	51,489	9.1	7.9	18.3	10.4	1.2	51,256	467	406	940	534	61
2041	51,958	9.0	7.9	18.2	10.4	1.2	51,723	470	409	948	539	61
2042	52,430	9.0	7.9	18.2	10.4	1.2	52,194	473	412	956	544	61
2043	52,905	9.0	7.9	18.2	10.4	1.2	52,667	477	416	964	548	61
2044	53,384	9.0	7.9	18.2	10.3	1.1	53,144	481	420	971	552	61
2045	53,866	9.0	7.9	18.2	10.36	1.1	53,625	485	424	979	556	61
2046	54,353	9.0	7.9	18.2	10.3	1.1	54,110	489	428	987	559	61
2047	54,844	9.0	7.9	18.1	10.3	1.1	54,598	493	432	995	563	61
2048	55,340	9.0	7.9	18.1	10.2	1.1	55,092	498	437	1,003	566	61
2049	55,840	9.0	7.9	18.1	10.2	1.1	55,590	503	442	1,011	569	61
2050	56,346	9.0	7.9	18.1	10.2	1.1	56,093	508	447	1,020	573	61

Source: "Annual Projections and Components of Change, for the United States: 1993 to 2050 (Middle Series). Part H. Black, Not Hispanic Population." Jennifer Cheeseman Day, *Population Projections of the United States, by Age, Sex, Race, and Hispanic Origin: 1993 to 2050*, P25-1104, p. 8.

★ 966 ★

Population Trends and Projections

Components of Population Change for the United States, 1993 to 2050. Part VIII American Indian, Eskimo, and Aleut, Not Hispanic

Numbers in thousands. Resident population.

Calendar year	Rate per 1,000 mid-year population						Population change during calendar year					
	July 1 population	Net change	Natural increase	Births	Deaths	Net immigration	July 1 population	Net change	Natural increase	Births	Deaths	Net immigration
1993	1,876	13.7	13.6	18.2	4.6	0.1	1,863	26	26	34	9	0
1994	1,902	13.4	13.4	18.0	4.6	0.1	1,889	26	25	34	9	0
1995	1,927	13.2	13.1	17.8	4.6	0.1	1,915	25	25	34	9	0
1996	1,953	13.0	12.9	17.6	4.7	0.1	1,940	25	25	34	9	0
1997	1,978	12.9	12.8	17.5	4.7	0.1	1,965	25	25	35	9	0
1998	2,004	12.8	12.7	17.5	4.8	0.1	1,991	26	25	35	10	0
1999	2,029	12.7	12.7	17.5	4.8	0.1	2,016	26	26	35	10	0
2000	2,055	12.7	12.7	17.5	4.8	0.1	2,042	26	26	36	10	0
2001	2,082	12.7	12.7	17.6	4.9	0.1	2,068	27	26	37	10	0
2002	2,108	12.7	12.7	17.7	5.0	0.1	2,095	27	27	37	11	0

[Continued]

★ 966 ★

Components of Population Change for the United States, 1993 to 2050. Part VIII American Indian, Eskimo, and Aleut, Not Hispanic

[Continued]

Calendar year	Rate per 1,000 mid-year population						Population change during calendar year					
	July 1 population	Net change	Natural increase	Births	Deaths	Net immigration	July 1 population	Net change	Natural increase	Births	Deaths	Net immigration
2003	2,135	12.7	12.7	17.7	5.1	0.1	2,122	27	27	38	11	0
2004	2,163	12.7	12.7	17.8	5.1	0.1	2,149	28	27	39	11	0
2005	2,190	12.8	12.7	17.9	5.2	0.1	2,177	28	28	39	11	0
2006	2,219	12.8	12.7	18.0	5.3	0.1	2,205	28	28	40	12	0
2007	2,247	12.8	12.7	18.1	5.3	0.1	2,233	29	29	41	12	0
2008	2,276	12.8	12.8	18.2	5.4	0.1	2,262	29	29	41	12	0
2009	2,306	12.9	12.8	18.3	5.5	0.1	2,291	30	29	42	13	0
2010	2,336	12.8	12.8	18.3	5.5	0.1	2,321	30	30	43	13	0
2011	2,366	12.8	12.7	18.3	5.6	0.1	2,351	30	30	43	13	0
2012	2,396	12.7	12.6	18.2	5.6	0.1	2,381	30	30	44	13	0
2013	2,426	12.6	12.5	18.2	5.7	0.1	2,411	30	30	44	14	0
2014	2,457	12.4	12.4	18.1	5.7	0.1	2,442	31	30	44	14	0
2015	2,487	12.3	12.2	18.0	5.8	0.1	2,472	31	30	45	14	0
2016	2,518	12.1	12.1	17.9	5.8	0.1	2,503	31	30	45	15	0
2017	2,549	12.0	12.0	17.8	5.9	0.1	2,533	31	30	45	15	0
2018	2,579	11.9	11.8	17.8	5.9	0.1	2,564	31	31	46	15	0
2019	2,610	11.8	11.7	17.7	6.0	0.1	2,595	31	31	46	16	0
2020	2,641	11.7	11.6	17.6	6.0	0.1	2,625	31	31	47	16	0
2021	2,671	11.6	11.5	17.6	6.1	0.1	2,656	31	31	47	16	0
2022	2,702	11.5	11.4	17.6	6.1	0.1	2,687	31	31	47	17	0
2023	2,734	11.4	11.4	17.6	6.2	0.1	2,718	31	31	48	17	0
2024	2,765	11.4	11.3	17.6	6.2	0.1	2,749	31	31	49	17	0
2025	2,796	11.4	11.3	17.6	6.3	0.1	2,781	32	32	49	18	0
2026	2,828	11.3	11.3	17.6	6.3	0.1	2,812	32	32	50	18	0
2027	2,861	11.3	11.3	17.6	6.4	0.0	2,844	32	32	50	18	0
2028	2,893	11.3	11.3	17.7	6.4	0.0	2,877	33	333	51	18	0
2029	2,926	11.3	11.3	17.7	6.4	0.0	2,910	33	33	52	19	0
2030	2,960	11.3	11.3	17.8	6.5	0.0	2,943	34	33	53	19	0
2031	2,993	11.4	11.3	17.8	6.5	0.0	2,976	34	34	53	19	0
2032	3,028	11.4	11.3	17.8	6.5	0.0	3,010	34	34	54	20	0
2033	3,062	11.4	11.3	17.9	6.6	0.0	3,045	35	35	55	20	0
2034	3,097	11.3	11.3	17.9	6.6	0.0	3,080	35	35	55	20	0
2035	3,132	11.3	11.3	17.9	6.6	0.0	3,115	36	35	56	21	0
2036	3,168	11.3	11.3	17.9	6.6	0.0	3,150	36	36	57	21	0
2037	3,204	11.3	11.2	17.9	6.7	0.0	3,186	36	36	57	21	0
2038	3,240	11.3	11.2	17.9	6.7	0.0	3,222	36	36	58	22	0
2039	3,277	11.2	11.2	17.9	6.7	0.0	3,259	37	37	59	22	0
2040	3,314	11.2	11.1	17.9	6.7	0.0	3,295	37	37	59	22	0
2041	3,351	11.1	11.1	17.9	6.8	0.0	3,333	37	37	60	23	0
2042	3,389	11.1	11.1	17.9	6.8	0.0	3,370	38	37	61	23	0
2043	3,426	11.1	11.0	17.8	6.8	0.0	3,408	38	38	61	23	0
2044	3,465	11.0	11.0	17.8	6.8	0.0	3,445	38	38	62	24	0
2045	3,503	11.0	11.0	17.8	6.9	0.0	3,484	39	38	62	24	0
2046	3,542	11.0	11.0	17.8	6.9	0.0	3,522	39	39	63	24	0
2047	3,581	11.0	11.0	17.8	6.9	0.0	3,561	39	39	64	25	0
2048	3,621	11.0	11.0	17.8	6.9	0.0	3,601	40	40	65	25	0
2049	3,661	11.0	11.0	17.9	6.9	0.0	3,641	40	40	65	25	0
2050	3,701	11.0	11.0	17.9	6.9	0.0	3,681	41	41	66	26	0

Source: "Annual Projections and Components of Change, for the United States: 1993 to 2050 (Middle Series). Part I. American Indian, Eskimo, and Aleut, Not Hispanic Population." Jennifer Cheeseman Day, *Population Projections of the United States, by Age, Sex, Race, and Hispanic Origin: 1993 to 2050*, P25-1104, p. 9.

★ 967 ★

Population Trends and Projections

Components of Population Change for the United States, 1993 to 2050. Part IX Asian and Pacific Islander, Not Hispanic

Numbers in thousands. Resident population.

Calendar year	Rate per 1,000 mid-year population						Population change during calendar year					
	July 1 population	Net change	Natural increase	Births	Deaths	Net immigration	July 1 population	Net change	Natural increase	Births	Deaths	Net immigration
1993	8,296	51.3	14.7	17.3	2.6	36.6	8,087	426	122	143	22	304
1994	8,727	49.4	14.6	17.3	2.6	34.8	8,513	431	128	151	23	304
1995	9,161	47.7	14.5	17.2	2.7	33.2	8,944	437	133	158	25	304
1996	9,601	46.0	14.4	17.1	2.7	31.6	9,381	442	138	165	26	304
1997	10,045	44.5	14.3	17.0	2.8	30.2	9,823	447	143	171	28	304
1998	10,495	43.0	14.1	16.9	2.8	28.9	10,270	452	148	178	30	304
1999	10,949	41.7	13.9	16.8	2.9	27.7	10,722	456	153	184	31	304
2000	11,407	40.4	13.8	16.7	2.9	26.6	11,178	461	157	190	33	304
2001	11,870	39.2	13.6	16.5	3.0	25.6	11,639	465	161	196	35	304
2002	12,337	38.0	13.4	16.4	3.0	24.6	12,104	469	165	202	38	304
2003	12,807	36.9	13.2	16.3	3.1	23.7	12,572	472	169	208	40	304
2004	13,281	35.8	13.0	16.1	3.2	22.9	13,044	476	172	214	42	304
2005	13,759	34.8	12.8	16.0	3.2	22.1	13,520	479	176	220	45	304
2006	14,240	33.9	12.6	15.9	3.3	21.3	14,000	483	179	226	47	304
2007	14,725	33.0	12.4	15.8	3.4	20.6	14,482	486	183	232	50	304
2008	15,213	32.2	12.2	15.7	3.4	20.0	14,969	490	186	238	52	304
2009	15,704	31.4	12.1	15.6	3.5	19.3	15,459	493	190	245	55	304
2010	16,199	30.7	11.9	15.5	3.6	18.7	15,952	497	193	251	58	304
2011	16,698	30.0	11.8	15.4	3.6	18.2	16,449	500	197	257	61	304
2012	17,200	29.3	11.6	15.4	3.7	17.7	16,949	504	200	264	64	304
2013	17,706	28.7	11.5	15.3	3.8	17.2	17,453	507	204	271	67	304
2014	18,215	28.1	11.4	15.2	3.8	16.7	17,961	511	207	277	70	304
2015	18,728	27.5	11.3	15.2	3.9	16.2	18,472	515	211	284	73	304
2016	19,244	26.9	11.1	15.1	4.0	15.8	18,986	518	214	291	77	304
2017	19,764	26.4	11.0	15.1	4.0	15.4	19,504	522	218	298	80	304
2018	20,287	25.9	10.9	15.0	4.1	15.0	20,026	525	221	305	83	304
2019	20,814	25.4	10.8	15.0	4.2	14.6	20,551	529	225	312	87	304
2020	21,345	24.9	10.7	15.0	4.2	14.2	21,080	532	228	319	91	304
2021	21,879	24.5	10.6	14.9	4.3	13.9	21,612	536	232	326	94	304
2022	22,416	24.0	10.5	14.9	4.4	13.5	22,147	539	235	333	98	304
2023	22,956	23.6	10.4	14.8	4.5	13.2	22,686	542	238	340	102	304
2024	23,500	23.2	10.3	14.8	4.5	12.9	23,228	545	241	347	106	304
2025	24,046	22.8	10.1	14.7	4.6	12.6	23,773	547	244	354	110	304
2026	24,595	22.4	10.0	14.7	4.7	12.3	24,320	550	246	361	114	304
2027	25,146	22.0	9.9	14.6	4.7	12.1	24,870	552	248	367	119	304
2028	25,699	21.6	9.7	14.5	4.8	11.8	25,422	554	250	373	123	304
2029	26,254	21.2	9.6	14.5	4.9	11.6	25,976	556	252	380	128	304
2030	26,810	20.8	9.5	14.4	4.9	11.3	26,532	557	254	386	132	304
2031	27,368	20.4	9.3	14.3	5.0	11.1	27,089	559	255	391	137	304
2032	27,928	20.0	9.2	14.2	5.1	10.9	27,648	560	256	397	141	304
2033	28,488	19.7	9.0	14.1	5.1	10.7	28,208	561	257	403	146	304
2034	29,049	19.3	8.9	14.1	5.2	10.5	28,768	561	258	408	151	304
2035	29,610	19.0	8.7	14.0	5.3	10.3	29,329	562	258	414	156	304
2036	30,172	18.6	8.6	13.9	5.3	10.1	29,891	562	259	419	161	304
2037	30,735	18.3	8.4	13.8	5.4	9.9	30,453	563	259	425	166	304
2038	31,298	18.0	8.3	13.7	5.5	9.7	31,016	563	259	430	171	304
2039	31,861	17.7	8.1	13.7	5.5	9.5	31,579	563	260	436	176	304
2040	32,424	17.4	8.0	13.6	5.6	9.4	32,142	563	260	441	181	304
2041	32,987	17.1	7.9	13.5	5.7	9.2	32,706	564	260	447	187	304
2042	33,551	16.8	7.8	13.5	5.7	9.1	33,269	564	260	452	192	304

[Continued]

★ 967 ★

Components of Population Change for the United States, 1993 to 2050. Part IX Asian and Pacific Islander, Not Hispanic

[Continued]

Calendar year	Rate per 1,000 mid-year population						Population change during calendar year					
	July 1 population	Net change	Natural increase	Births	Deaths	Net immigration	July 1 population	Net change	Natural increase	Births	Deaths	Net immigration
2043	34,115	16.5	7.6	13.4	5.8	8.9	33,833	564	260	458	197	304
2044	34,679	16.3	7.5	13.4	5.8	8.8	34,397	564	260	463	203	304
2045	35,243	16.0	7.4	13.3	5.9	8.6	34,961	564	260	469	208	304
2046	35,807	15.8	7.3	13.2	6.0	8.5	35,525	564	260	474	214	304
2047	36,371	15.5	7.2	13.2	6.0	8.3	36,089	564	261	480	219	304
2048	36,935	15.3	7.1	13.1	6.1	8.2	36,653	564	261	485	225	304
2049	37,500	15.0	6.9	13.1	6.1	8.1	37,218	564	261	491	230	304
2050	38,064	14.8	6.8	13.0	6.2	8.0	37,782	564	261	496	235	304

Source: "Annual Projections and Components of Change, for the United States: 1993 to 2050 (Middle Series). Part J. Asian and Pacific Islander, Not Hispanic Population." Jennifer Cheeseman Day, *Population Projections of the United States, by Age, Sex, Race, and Hispanic Origin: 1993 to 2050*, P25-1104, p. 10.

★ 968 ★

Population Trends and Projections

Distribution of Population; Percent by Race and Hispanic Origin, 1995 to 2050

As of July1. Resident population.

Year	Total	Race				Hispanic origin[3]	Not of Hispanic origin, by race			
		White	Black	American Indian[1]	Asian[2]		White	Black	American Indian	Asian[2]
Projections										
Lowest series										
2050	100.0	75.0	14.7	1.2	9.1	20.2	56.5	13.6	1.0	8.7
Middle series										
1995	100.0	82.9	12.6	0.8	3.7	10.2	73.6	12.0	0.7	3.5
2000	100.0	81.9	12.8	0.9	4.4	11.3	71.6	12.2	0.7	4.1
2005	100.0	80.9	13.1	0.9	5.1	12.4	69.7	12.4	0.8	4.8
2010	100.0	80.0	13.4	0.9	5.7	13.5	67.7	12.6	0.8	5.4
2020	100.0	78.2	13.9	0.9	7.0	15.7	63.9	13.0	0.8	6.5
2030	100.0	76.4	14.5	1.0	8.1	17.9	60.1	13.4	0.8	7.7
2040	100.0	74.6	15.1	1.0	9.3	20.2	56.3	13.9	0.9	8.7
2050	100.0	72.8	15.7	1.1	10.3	22.5	52.5	14.4	0.9	9.7
Highest series										
2050	100.0	72.4	15.3	1.0	11.3	24.6	50.3	13.7	0.8	10.6

Source: "Percent Distribution of the Population, by Race and Hispanic Origin." Jennifer Cheeseman Day, *Population Projections of the United States, by Age, Sex, Race, and Hispanic Origin: 1993 to 2050*, P25-1104, p. xxii. *Notes:* 1. American Indian represents American Indian, Eskimo, and Aleut. 2. Asian represents Asian and Pacific Islander. 3. Persons of Hispanic origin may be of any race.

★ 969 ★

Population Trends and Projections

Growth of Total Population: Percent by Race and Hispanic Origin, 1990 to 2050

As of July1. Resident population.

Year	Total	Race				Hispanic origin[3]	Not of Hispanic origin, by race			
		White	Black	American Indian[1]	Asian[2]		White	Black	American Indian	Asian[2]
Projections										
1990 to 1995	100.0	65.5	17.8	1.1	15.6	30.3	38.1	16.0	0.9	14.7
1995 to 2000	100.0	61.9	18.4	1.2	18.5	34.1	31.0	16.3	1.0	17.5
2000 to 2005	100.0	58.7	19.3	1.4	20.6	37.7	24.7	17.0	1.1	19.5
2005 to 2010	100.0	57.3	20.0	1.4	21.3	39.7	21.4	17.6	1.2	20.1
2010 to 2020	100.0	56.8	20.3	1.5	21.4	41.9	19.0	17.8	1.2	20.2
2020 to 2030	100.0	52.7	21.6	1.6	24.2	48.2	9.1	18.6	1.3	22.7
2030 to 2040	100.0	45.4	24.7	2.0	27.9	57.3[4]	-	21.2[4]	1.6[4]	26.1[4]
2040 to 2050	100.0	40.7	27.6	2.2	29.5	63.0[4]	-	23.7[4]	1.9[4]	27.5[4]

Source: "Percent of Total Population Growth, by Race and Hispanic Origin: 1990 to 2050." Jennifer Cheeseman Day, *Population Projections of the United States, by Age, Sex, Race, and Hispanic Origin: 1993 to 2050*, P25-1104, p. xxiii. Primary source: Current Population Reports, P25-1095; tables 1 and 4. *Notes:* 1. American Indian represents American Indian, Eskimo, and Aleut. 2. Asian represents Asian and Pacific Islander. 3. Persons of Hispanic origin may be of any race. 4. These percentages do not add to 100 percent due to the declining size of the White, not Hispanic population.

★ 970 ★

Population Trends and Projections

Population: Race and Hispanic Origin: by Series, 1990 to 2050

As of July1. Resident population.

Year	Total	Race				Hispanic origin[3]	Not of Hispanic origin, by race			
		White	Black	American Indian[1]	Asian[2]		White	Black	American Indian	Asian[2]
Projections										
Lowest series										
2050	285,502	214,054	42,026	3,323	26,099	57,643	161,382	38,933	2,807	24,738
Middle series										
1995	263,434	218,334	33,117	2,226	9,756	26,798	193,900	31,648	1,927	9,161
2000	276,241	226,267	35,469	2,380	12,125	31,166	197,872	33,741	2,055	11,407
2005	288,286	233,343	37,793	2,543	14,608	35,702	200,842	35,793	2,190	13,759
2010	300,431	240,297	40,224	2,719	17,191	40,525	203,441	37,930	2,336	16,199
2020	325,942	254,791	45,409	3,090	22,653	51,217	208,280	42,459	2,641	21,345
2030	349,993	267,457	50,596	3,473	28,467	62,810	210,480	46,934	2,960	26,810
2040	371,505	277,232	55,917	3,894	34,461	75,130	209,148	51,489	3,314	32,424
2050	392,031	285,591	61,586	4,346	40,508	88,071	2054,849	56,346	3,701	38,064

[Continued]

★ 970 ★

Population: Race and Hispanic Origin: by Series, 1990 to 2050

[Continued]

Year	Total	Race				Hispanic origin[3]	Not of Hispanic origin, by race			
		White	Black	American Indian[1]	Asian[2]		White	Black	American Indian	Asian[2]
Higher series 2050	522,098	378,408	79,722	5,039	58,930	128,255	262,855	71,675	4,221	55,093

Source: "Population, by Race and Hispanic Origin: 1990 to 2050." Jennifer Cheeseman Day, *Population Projections of the United States, by Age, Sex, Race, and Hispanic Origin: 1993 to 2050*, P25-1104, p. xxii. Primary source: Current Population Reports, P25-1095, and tables 2 and 3. *Notes:* 1. American Indian represents American Indian, Eskimo, and Aleut. 2. Asian represents Asian and Pacific Islander. 3. Persons of Hispanic origin may be of any race.

★ 971 ★

Population Trends and Projections

Projections of Resident Population, 1995-2025. Part I.

As of July 1. Data are for middle series.

Age, sex, and race	Population (1,000)					Percent distribution		
	1995	2000	2005	2010	2025	2000	2010	2025
White, total	217,511	224,594	230,993	237,412	256,425	100.0	100.0	100.0
Under 5 years old	15,321	14,496	14,254	14,617	15,461	6.5	6.2	6.0
5 to 13 years old	27,201	28,043	27,303	26,455	28,361	12.5	11.1	11.1
14 to 17 years old	11,662	12,322	13,068	12,764	12,601	5.5	5.4	4.9
18 to 24 years old	19,862	20,615	21,997	23,107	21,828	9.2	9.7	8.5
25 to 34 years old	33,296	29,825	28,603	29,813	32,125	13.3	12.6	12.5
35 to 44 years old	35,336	36,660	34,110	30,634	32,436	16.3	12.9	12.6
45 to 54 years old	26,548	31,203	34,536	35,860	28,928	13.9	15.1	11.3
55 to 64 years old	18,242	20,554	25,265	29,772	32,180	9.2	12.5	12.5
65 to 74 years old	16,652	15,906	15,899	18,136	29,764	7.1	7.6	11.6
75 to 84 years old	10,093	11,111	11,538	11,191	16,798	4.9	4.7	6.6
85 years old and over	3,299	3,862	4,419	5,064	5,944	1.7	2.1	2.3
Male	106,684	110,359	113,689	117,043	126,877	49.1	49.3	49.5
Female	110,827	114,235	117,304	120,369	129,548	50.9	50.7	50.5
Black, total	33,147	35,525	37,907	40,429	48,388	100.0	100.0	100.0
Under 5 years old	3,177	3,183	3,299	3,529	4,088	9.0	8.7	8.4
5 to 13 years old	5,337	5,829	5,939	6,081	7,221	16.4	15.0	14.9
14 to 17 years old	2,296	2,418	2,786	2,803	3,189	6.8	6.9	6.6
18 to 24 years old	3,703	3,923	4,197	4,701	5,095	11.0	11.6	10.5
25 to 34 years old	5,372	5,131	5,163	5,426	6,572	14.4	13.4	13.6
35 to 44 years old	5,154	5,586	5,422	5,175	5,934	15.7	12.8	12.3
45 to 54 years old	3,237	4,110	4,910	5,323	4,998	11.6	13.2	10.3
55 to 64 years old	2,130	2,413	3,015	3,838	4,904	6.8	9.5	10.1
65 to 74 years old	1,625	1,690	1,804	2,070	4,041	4.8	5.1	8.4
75 to 84 years old	837	909	990	1,044	1,712	2.6	2.6	3.5
85 years old and over	278	333	382	439	635	0.9	1.1	1.3

[Continued]

★ 971 ★

Projections of Resident Population, 1995-2025. Part I.

[Continued]

Age, sex, and race	Population (1,000)					Percent distribution		
	1995	2000	2005	2010	2025	2000	2010	2025
Male	15,711	16,846	17,972	19,168	22,951	47.4	47.4	47.4
Female	17,436	18,680	19,936	21,261	25,437	52.6	52.6	52.6

Source: "Projections of Resident Population, by Age, Sex, and Race: 1995 to 2025." U.S. Bureau of the Census, *Statistical Abstract of the United States*, 1993, p. 24. Primary source: U.S. Bureau of the Census, Current Population Reports, P25-1092.

★ 972 ★

Population Trends and Projections

Projections of Resident Population, 1995-2025. Part II.

As of July 1. Data are for middle series.

Age, sex, and race	Population (1,000)					Percent distribution		
	1995	2000	2005	2010	2025	2000	2010	2025
American Indian, Eskimo, Aleut, total	2,247	2,409	2,583	2,772	3,386	100.0	100.0	100.0
Under 5 years old	220	224	242	264	308	9.3	9.5	9.1
5 to 13 years old	412	425	429	455	561	17.6	16.4	16.6
14 to 17 years old	174	193	208	204	254	8.0	7.4	7.5
18 to 24 years old	225	277	314	337	382	11.5	12.2	11.3
25 to 34 years old	362	356	367	407	4818	14.8	14.7	14.2
35 to 44 years old	332	352	345	341	436	14.6	12.3	12.9
45 to 54 years old	219	262	298	318	320	10.9	11.5	9.5
55 to 64 years old	132	153	186	225	271	6.4	8.1	8.0
65 to 74 years old	85	93	104	123	210	3.9	4.4	6.2
75 to 84 years old	42	51	58	64	108	2.1	2.3	3.2
85 years old and over	15	22	28	35	56	0.9	1.3	1.7
Male	1,113	1,193	1,279	1,372	1,677	49.5	49.5	49.5
Female	1,133	1,216	1,304	1,400	1,709	50.5	50.5	50.5
Asian, Pacific Islander, total	9,849	12,287	14,840	17,496	26,017	100.0	100.0	100.0
Under 5 years old	836	1,006	1,163	1,320	1,824	8.2	7.5	7.0
5 to 13 years old	1,422	1,755	2,110	2,435	3,405	14.3	13.9	13.1
14 to 17 years old	623	801	956	1,136	1,608	6.5	6.5	6.2
18 to 24 years old	1,084	1,303	1,603	1,863	2,700	10.6	10.6	10.4
25 to 34 years old	1,813	2,104	2,361	2,722	3,952	17.1	15.6	15.2
35 to 44 years old	1,678	2,064	2,406	2,704	3,737	16.8	15.5	14.4
45 to 54 years old	1,079	1,479	1,866	2,236	3,101	12.0	12.8	11.9
55 to 64 years old	649	867	1,181	1,542	2,536	7.1	8.8	9.7
65 to 74 years old	439	568	717	907	1,806	4.6	5.2	6.9
75 to 84 years old	184	269	368	468	947	2.2	2.7	3.6
85 years old and over	46	72	108	165	402	0.6	0.9	1.5

[Continued]

★ 972 ★

Projections of Resident Population, 1995-2025. Part II.

[Continued]

Age, sex, and race	Population (1,000)					Percent distribution		
	1995	2000	2005	2010	2025	2000	2010	2025
Male	4,784	5,940	7,157	8,428	12,549	48.3	48.2	48.2
Female	5,066	6,347	7,684	9,068	13,468	51.7	51.8	51.8

Source: "Projections of Resident Population, by Age, Sex, and Race: 1995 to 2025." U.S. Bureau of the Census, *Statistical Abstract of the United States*, 1993, p. 24. Primary source: U.S. Bureau of the Census, Current Population Reports, P25-1092.

★ 973 ★

Population Trends and Projections

Projections of Resident Population, 1995-2025. Part III.

As of July 1. Resident population. Data are for middle series.

Age, sex, and race	Population (1,000)					Percent distribution		
	1995	2000	2005	2010	2025	2000	2010	2025
Hispanic origin, total[1]	26,522	30,602	34,842	39,312	54,018	100.0	100.0	100.0
Under 5 years old	2,804	3,055	3,338	3,683	4,746	10.0	9.4	8.8
5 to 13 years old	4,535	5,226	5,720	6,212	8,133	17.1	15.8	15.1
14 to 17 years old	1,816	2,097	2,486	2,698	3,464	6.9	6.9	6.4
18 to 24 years old	3,224	3,544	4,059	4,676	5,919	11.6	11.9	11.0
25 to 34 years old	5,046	5,200	5,344	5,851	7,940	17.0	14.9	14.7
35 to 44 years old	3,887	4,821	5,410	5,560	6,991	15.8	14.1	12.9
45 to 54 years old	2,270	3,029	3,902	4,800	5,680	9.9	12.2	10.5
55 to 64 years old	1,408	1,729	2,266	2,993	5,261	5.6	7.6	9.7
65 to 74 years old	955	1,135	1,327	1,625	3,569	3.7	4.1	6.6
75 to 84 years old	437	578	750	889	1,689	1.9	2.3	3.1
85 years old and over	139	188	240	324	625	0.6	0.8	1.2
Male	13,470	15,492	17,585	19,792	27,059	50.6	50.3	50.1
Female	13,052	15,110	17,257	19,519	26,959	49.4	49.7	49.9
Non-Hispanic White, total	193,307	196,701	199,274	201,668	207,439	100.0	100.0	100.0
Under 5 years old	12,753	11,706	11,214	11,270	11,155	6.0	5.6	5.4
5 to 13 years old	23,083	23,290	22,097	20,815	21,007	11.8	10.3	10.1
14 to 17 years old	10,008	10,423	10,821	10,318	9,480	5.3	5.1	4.6
18 to 24 years old	16,915	17,385	18,317	18,871	16,475	8.8	9.4	7.9
25 to 34 years old	28,703	25,095	23,745	24,509	24,933	12.8	12.2	12.0
35 to 44 years old	31,800	32,270	29,187	25,580	26,118	16.4	12.7	12.6
45 to 54 years old	24,475	28,442	30,981	31,483	23,758	14.5	15.6	11.5
55 to 64 years old	16,949	18,970	23,195	27,041	27,379	9.6	13.4	13.2
65 to 74 years old	15,767	14,862	14,683	16,652	26,513	7.6	8.3	12.8
75 to 84 years old	9,685	10,574	10,842	10,370	15,252	5.4	5.1	7.4
85 years old and over	3,169	3,685	4,194	4,761	5,365	1.9	2.4	2.6
Male	94,375	96,213	97,645	99,002	102,265	48.9	49.1	49.3
Female	98,931	100,488	101,629	102,667	105,173	51.1	50.9	50.7

[Continued]

★ 973 ★

Projections of Resident Population, 1995-2025. Part III.
[Continued]

Age, sex, and race	Population (1,000)					Percent distribution		
	1995	2000	2005	2010	2025	2000	2010	2025
Non-Hispanic Black, total	31,702	33,834	35,957	38,201	45,237	100.0	100.0	100.0
Under 5 years old	3,024	3,014	3,111	3,316	3,807	8.9	8.7	8.4
5 to 13 years old	5,083	5,531	5,615	5,723	6,731	16.3	15.0	14.9
14 to 17 years old	2,198	2,298	2,638	2,647	2,977	6.8	6.9	6.6
18 to 24 years old	3,533	3,729	3,966	4,424	4,743	11.0	11.6	10.5
25 to 34 years old	5,093	4,846	4,870	5,098	6,110	14.3	13.3	13.5
35 to 44 years old	4,933	5,316	5,122	4,870	5,530	15.7	12.7	12.2
45 to 54 years old	3,113	3,940	4,689	5,055	4,687	11.6	13.2	10.4
55 to 64 years old	2,056	2,320	2,889	3,667	4,607	6.9	9.6	10.2
65 to 74 years old	1,579	1,631	1,730	1,977	3,824	4.8	5.2	8.5
75 to 84 years old	817	884	954	998	1,615	2.6	2.6	3.6
85 years old and over	273	326	372	425	605	1.0	1.1	1.3
Male	14,985	16,002	17,004	18,066	21,405	47.3	47.3	47.3
Female	16,717	17,833	18,954	20,135	23,832	52.7	52.7	52.7

Source: "Projections of Resident Population, by Age, Sex, and Race: 1995 to 2025." U.S. Bureau of the Census, *Statistical Abstract of the United States*, 1993, p. 25. Primary source: U.S. Bureau of the Census, Current Population Reports, P25-1092. *Note:* 1. Persons of Hispanic origin may be of any race.

★ 974 ★

Population Trends and Projections

Projections of Resident Population, 1995-2025. Part IV.

As of July 1. Resident population. Data are for middle series.

Age, sex, and race	Population (1,000)					Percent distribution		
	1995	2000	2005	2010	2025	2000	2010	2025
Non-Hispanic American Indian, Eskimo, Aleut, total	1,956	2,096	2,245	2,407	2,942	100.0	100.0	100.0
Under 5 years old	193	196	212	230	273	9.4	9.6	9.3
5 to 13 years old	351	369	375	398	492	17.6	16.5	16.7
14 to 17 years old	150	165	179	178	221	7.9	7.4	7.5
18 to 24 years old	220	238	268	288	332	11.4	12.0	11.3
25 to 34 years old	307	303	316	348	420	14.5	14.5	14.3
35 to 44 years old	289	302	292	289	369	14.4	12.0	12.5
45 to 54 years old	196	231	259	273	274	11.0	11.3	9.3
55 to 64 years old	120	139	167	198	230	6.6	8.2	7.8
65 to 74 years old	78	84	95	111	182	4.0	4.6	6.2
75 to 84 years old	40	47	53	59	97	2.2	2.5	3.3
85 years old and over	14	20	26	33	53	1.0	1.4	1.8
Male	963	1,032	1,106	1,186	1,453	49.2	49.3	49.4
Female	993	1,064	1,139	1,221	1,490	50.8	50.7	50.6
Non-Hispanic Asian, Pacific Islander, total	9,266	11,582	14,005	16,522	24,580	100.0	100.0	100.0

[Continued]

★ 974 ★

Projections of Resident Population, 1995-2025. Part IV.

[Continued]

Age, sex, and race	Population (1,000)					Percent distribution		
	1995	2000	2005	2010	2025	2000	2010	2025
Under 5 years old	778	938	1,084	1,230	1,701	8.1	7.4	6.9
5 to 13 years old	1,319	1,634	1,974	2,276	3,183	14.1	13.8	12.9
14 to 17 years old	582	750	895	1,066	1,506	6.5	6.5	6.1
18 to 24 years old	1,013	1,220	1,504	1,747	2,536	10.5	10.6	10.3
25 to 34 years old	1,694	1,971	2,219	2,561	3,725	17.0	15.5	15.2
35 to 44 years old	1,591	1,953	2,271	2,554	3,534	16.9	15.5	14.4
45 to 54 years old	1,029	1,412	1,779	2,127	2,947	12.2	12.9	12.0
55 to 64 years old	620	830	1,132	1,478	2,414	7.2	8.9	9.8
65 to 74 years old	421	545	689	872	1,731	4.7	5.3	7.0
75 to 84 years old	176	258	355	451	912	2.2	2.7	3.7
85 years old and over	44	69	105	159	390	0.6	1.0	1.6
Male	4,498	5,600	6,757	7,966	11,871	48.4	48.2	48.3
Female	4,768	5,983	7,248	8,556	12,709	51.7	51.8	51.7

Source: "Projections of Resident Population, by Age, Sex, and Race: 1995 to 2025." U.S. Bureau of the Census, *Statistical Abstract of the United States,* 1993, p. 25. Primary source: U.S. Bureau of the Census, Current Population Reports, P25-1092.

★ 975 ★

Population Trends and Projections

Projections: Age Range, Gender, and Race, July 1, 1995. Part I. A

Numbers in thousands. Resident population.

Date and age	Race											
	White			Black			American Indian, Eskimo, and Aleut			Asian and Pacific Islander		
	Total	Male	Female	Total	Male	Female	Total	Male	Female	Total	Male	Female
July 1, 1995												
All ages	218,334	107,140	111,195	33,117	15,697	17,420	2,226	1,103	1,123	9,756	4,745	5,011
Under 5 years	15,841	8,137	7,704	3,243	1,645	1,598	222	112	110	876	449	426
5 to 9 years	15,154	7,781	7,373	2,981	1,513	1,468	225	115	110	757	387	370
10 to 14 years	15,009	7,712	7,297	2,900	1,471	1,429	231	117	114	800	406	394
15 to 19 years	14,116	7,267	6,848	2,762	1,399	1,363	194	98	96	719	367	352
20 to 24 years	14,772	7,553	7,220	2,692	1,324	1,368	185	95	90	823	413	410
25 to 29 years	15,618	7,904	7,713	2,625	1,248	1,377	177	91	86	874	428	446
30 to 34 years	18,409	9,287	9,122	2,850	1,333	1,517	187	93	94	931	451	480
35 to 39 years	18,407	9,262	9,145	2,754	1,287	1,467	175	86	89	880	423	457
40 to 44 years	16,674	8,337	8,337	2,334	1,081	1,253	152	74	78	775	363	412
45 to 49 years	14,387	7,139	7,248	1,762	803	959	121	59	62	603	282	321
50 to 54 years	11,465	5,632	5,833	1,360	611	748	91	44	47	435	206	229
55 to 59 years	9,509	4,614	4,895	1,127	495	632	72	34	37	343	161	182
60 to 64 years	8,846	4,213	4,633	997	430	567	57	27	30	290	127	163
65 to 69 years	8,890	4,054	4,836	919	391	528	47	21	25	244	103	141
70 to 74 years	7,932	3,467	4,465	710	285	425	37	16	20	185	82	103
75 to 79 years	6,019	2,445	3,574	511	192	318	25	10	15	114	49	64
80 to 84 years	4,016	1,432	2,584	324	110	214	16	6	10	64	29	35
85 to 89 years	2,118	635	1,483	163	51	113	8	3	6	28	12	15
90 to 94 years	882	216	665	78	22	57	4	1	3	12	5	7
95 to 99 years	229	45	184	20	5	15	1	0	1	3	1	2
100 years and over	43	7	36	7	2	6	0	0	0	1	0	1

[Continued]

★ 975 ★

Projections: Age Range, Gender, and Race, July 1, 1995. Part I. A

[Continued]

Date and age	Race											
	White			Black			American Indian, Eskimo, and Aleut			Asian and Pacific Islander		
	Total	Male	Female	Total	Male	Female	Total	Male	Female	Total	Male	Female
16 years and over	169,450	82,027	87,423	23,414	10,774	12,640	1,506	737	768	7,173	3,426	3,747
18 years and over	163,783	79,099	84,684	22,303	10,207	12,096	1,427	697	729	6,887	3,279	3,608
15 to 44 years	97,995	49,610	48,386	16,015	7,671	8,345	1,070	537	533	5,002	2,446	2,557
65 years and over	30,129	12,302	17,827	2,732	1,057	1,674	138	58	80	650	281	369
85 years and over	3,271	904	2,368	268	79	190	14	5	10	44	18	25
Median age	35.1	33.9	36.2	28.9	27.1	30.4	26.6	25.8	27.5	30.2	29.2	31.1
Mean age	36.6	35.2	38.0	31.0	29.5	32.4	29.3	28.5	30.0	31.4	30.5	32.3

Source: "Projections of the Population, by Age, Sex, Race, and Hispanic Origin, for the United States, 1993 to 2020, (Middle Series)." Jennifer Cheeseman Day, *Population Projections of the United States, by Age, Sex, Race, and Hispanic Origin: 1993 to 2050*, P25-1104, p. 16.

★ 976 ★

Population Trends and Projections

Projections: Age Range, Gender, and Race, July 1, 1995. Part I. B

Numbers in thousands. Resident population.

Date and age	White			Black			American Indian, Eskimo, and Aleut			Asian and Pacific Islander		
	Total	Male	Female	Total	Male	Female	Total	Male	Female	Total	Male	Female
July 1, 1995												
All ages	226,267	111,245	115,022	35,469	16,802	18,667	2,380	1,177	1,203	12,125	5,877	6,248
Under 5 years	14,945	7,675	7,270	3,214	1,633	1,581	215	109	107	1,056	541	515
5 to 9 years	16,012	8,226	7,787	3,302	1,679	1,623	229	116	113	989	509	480
10 to 14 years	15,627	8,025	7,602	3,141	1,597	1,544	248	126	122	955	488	467
15 to 19 years	15,581	8,018	7,563	3,057	1,551	1,505	229	115	114	952	483	469
20 to 24 years	14,199	7,262	6,937	2,668	1,311	1,356	185	93	93	895	443	452
25 to 29 years	14,551	7,347	7,204	2,597	1,238	1,359	183	94	89	1,039	500	539
30 to 34 years	15,983	8,047	7,936	2,638	1,228	1,410	173	88	85	1,073	516	557
35 to 39 years	18,586	9,364	9,222	2,873	1,340	1,533	181	91	90	1,055	512	543
40 to 44 years	18,553	9,300	9,253	2,737	1,272	1,465	171	84	87	968	463	505
45 to 49 years	16,339	8,124	8,215	2,248	1,024	1,224	143	69	73	800	375	425
50 to 54 years	14,196	6,993	7,202	1,706	763	943	113	55	58	625	288	337
55 to 59 years	11,200	5,454	5,745	1,305	571	733	84	40	44	459	212	247
60 to 64 years	9,139	4,386	4,753	1,065	455	610	65	31	34	374	171	202
65 to 69 years	8,297	3,864	4,433	938	401	536	50	23	27	309	132	177
70 to 74 years	7,923	3,537	4,386	748	306	442	40	18	22	246	103	143
75 to 79 years	6,735	2,832	3,904	571	220	351	31	13	18	170	73	97
80 to 84 years	4,484	1,681	2,804	336	115	221	19	8	11	91	39	53
85 to 89 years	2,452	766	1,686	186	57	128	12	4	8	43	17	25
90 to 94 years	1,075	269	807	94	26	68	6	2	4	18	7	10
95 to 99 years	327	65	262	37	9	28	3	1	2	7	2	4
100 years and over	63	9	53	10	2	7	1	0	1	2	0	1
16 years and over	176,654	85,761	90,893	25,212	11,588	13,624	1,639	802	838	8,929	4,239	4,690
18 years and over	170,380	82,522	87,858	23,988	10,961	13,027	1,545	754	791	8,540	4,040	4,500
15 to 44 years	97,452	49,338	48,114	16,569	7,941	8,627	1,122	564	558	5,982	2,917	3,065
65 years and over	31,357	13,023	18,334	2,919	1,138	1,781	161	68	93	884	374	510
85 years and over	3,917	1,109	2,808	326	95	231	21	7	15	69	28	41

[Continued]

★ 976 ★

Projections: Age Range, Gender, and Race, July 1, 1995. Part I. B

[Continued]

Date and age	White			Black			American Indian, Eskimo, and Aleut			Asian and Pacific Islander		
	Total	Male	Female	Total	Male	Female	Total	Male	Female	Total	Male	Female
Median age	36.7	35.6	37.9	29.6	27.6	31.3	27.3	26.7	28.0	30.8	29.8	31.8
Mean age	37.4	36.0	38.7	31.7	30.1	33.1	30.2	29.4	31.0	32.1	31.1	33.0

Source: "Projections of the Population, by Age, Sex, Race, and Hispanic Origin, for the United States, 1993 to 2020, (Middle Series)." Jennifer Cheeseman Day, *Population Projections of the United States, by Age, Sex, Race, and Hispanic Origin: 1993 to 2050*, P25-1104, p. 26.

★ 977 ★

Population Trends and Projections

Projections: Age Range, Gender, and Race, July 1, 1995. Part II. A

Numbers in thousands. Resident population.

Date and age	Hispanic origin[1]			Not of Hispanic origin, by race					
				White			Black		
	Total	Male	Female	Total	Male	Female	Total	Male	Female
July 1, 1995									
All ages	26,798	13,610	13,188	193,900	94,716	99,184	31,648	14,958	16,689
Under 5 years	3,090	1,580	1,510	13,020	6,694	6,326	3,076	1,560	1,516
5 to 9 years	2,644	1,352	1,291	12,759	6,556	6,203	2,827	1,434	1,393
10 to 14 years	2,387	1,219	1,168	12,841	6,605	6,237	2,770	1,405	1,365
15 to 19 years	2,233	1,147	1,086	12,080	6,220	5,859	2,640	1,337	1,304
20 to 24 years	2,316	1,224	1,092	12,657	6,432	6,225	2,569	1,261	1,308
25 to 29 years	2,494	1,345	1,149	13,349	6,678	6,672	2,486	1,175	1,311
30 to 34 years	2,527	1,335	1,192	16,106	8,067	8,039	2,710	1,260	1,449
35 to 39 years	2,165	1,120	1,044	16,439	8,242	8,197	2,629	1,222	1,407
40 to 44 years	1,729	874	855	15,103	7,542	7,561	2,234	1,030	1,204
45 to 49 years	1,310	647	663	13,193	6,549	6,644	1,689	768	922
50 to 54 years	964	465	499	10,584	5,207	5,377	1,307	586	721
55 to 59 years	767	363	404	8,806	4,281	4,525	1,085	476	610
60 to 64 years	640	296	344	8,258	3,940	4,318	964	415	548
65 to 69 years	546	244	302	8,386	3,828	4,558	891	379	512
70 to 74 years	409	179	230	7,552	3,300	4,252	692	278	414
75 to 79 years	261	106	156	5,777	2,347	3,430	498	187	311
80 to 84 years	175	65	111	3,852	1,371	2,480	316	107	209
85 to 89 years	90	32	58	2,033	605	1,428	160	49	110
90 to 94 years	39	13	26	844	204	640	77	21	56
95 to 99 years	9	3	6	220	42	178	19	5	15
100 years and over	2	1	2	41	7	34	7	2	5
16 years and over	18,216	9,222	8,995	152,817	73,593	79,224	22,421	10,278	12,142
18 years and over	17,331	8,766	8,566	147,955	71,080	76,875	21,358	9,736	11,621
15 to 44 years	13,464	7,045	6,419	85,734	43,181	42,553	15,268	7,285	7,983
65 years and over	1,532	642	890	28,705	11,704	17,001	2,661	1,029	1,632
85 years and over	141	48	93	3,138	858	2,280	263	77	186

[Continued]

★ 977 ★

Projections: Age Range, Gender, and Race, July 1, 1995. Part II. A

[Continued]

| Date and age | Hispanic origin[1] | | | Not of Hispanic origin, by race | | | | | |
| | | | | White | | | Black | | |
	Total	Male	Female	Total	Male	Female	Total	Male	Female
Median age	26.5	26.1	27.0	36.2	35.1	37.4	29.0	27.1	30.5
Mean age	28.5	27.8	29.3	37.6	36.1	39.1	31.2	29.6	32.6

Source: "Projections of the Population, by Age, Sex, Race, and Hispanic Origin, for the United States, 1993 to 2020, (Middle Series)." Jennifer Cheeseman Day, *Population Projections of the United States, by Age, Sex, Race, and Hispanic Origin: 1993 to 2050*, P25-1104, p. 17. *Note:* 1. Persons of Hispanic origin may be of any race.

★ 978 ★

Population Trends and Projections

Projections: Age Range, Gender, and Race, July 1, 1995. Part II. B

Numbers in thousands. Resident population.

| Date and age | Not of Hispanic origin, by race | | | | | |
| | American Indian, Eskimo, Aleut | | | Asian and Pacific Islander | | |
	Total	Male	Female	Total	Male	Female
July 1, 1995						
All ages	1,927	948	979	9,161	4,453	4,708
Under 5 years	188	95	93	807	414	392
5 to 9 years	189	96	93	697	356	341
10 to 14 years	197	100	97	744	378	366
15 to 19 years	167	84	82	670	343	328
20 to 24 years	160	82	78	770	387	384
25 to 29 years	150	76	74	814	398	416
30 to 34 years	160	79	81	874	423	451
35 to 39 years	151	73	78	832	400	432
40 to 44 years	133	64	69	735	344	391
45 to 49 years	107	52	55	573	268	306
50 to 54 years	82	40	43	415	197	218
55 to 59 years	65	31	34	327	154	173
60 to 64 years	52	24	28	277	121	156
65 to 69 years	42	19	23	234	98	135
70 to 74 years	34	15	19	178	79	99
75 to 79 years	23	9	14	109	47	62
80 to 84 years	15	6	9	61	28	33
85 to 89 years	8	3	5	26	12	15
90 to 94 years	4	1	3	11	5	7
95 to 99 years	1	0	1	3	1	2
100 years and over	0	0	0	1	0	1
16 years and over	1,316	639	678	6,773	3,233	3,540
18 years and over	1,249	604	644	6,506	3,096	3,411
15 to 44 years	920	457	462	4,697	2,294	2,402
65 years and over	127	54	74	624	270	354
85 years and over	13	4	9	42	17	24

[Continued]

★ 978 ★

Projections: Age Range, Gender, and Race, July 1, 1995. Part II. B

[Continued]

| Date and age | Not of Hispanic origin, by race | | | | | |
| | American Indian, Eskimo, Aleut | | | Asian and Pacific Islander | | |
	Total	Male	Female	Total	Male	Female
Median age	27.1	26.1	28.1	30.4	29.4	31.4
Mean age	29.7	28.9	30.6	31.6	30.7	32.5

Source: "Projections of the Population, by Age, Sex, Race, and Hispanic Origin, for the United States, 1993 to 2020, (Middle Series)." Jennifer Cheeseman Day, *Population Projections of the United States, by Age, Sex, Race, and Hispanic Origin: 1993 to 2050*, P25-1104, p. 17.

★ 979 ★

Population Trends and Projections

Projections: Age Range, Gender, and Race, July 1, 2000. Part I

Numbers in thousands. Resident population.

| Date and age | Hispanic origin[1] | | | Not of Hispanic origin, by race | | | | | |
| | | | | White | | | Black | | |
	Total	Male	Female	Total	Male	Female	Total	Male	Female
July 1, 2000									
All ages	31,166	15,777	15,388	197,872	96,846	101,025	33,741	15,939	17,802
Under 5 years	3,293	1,683	1,610	11,936	6,137	5,799	3,033	1,540	1,493
5 to 9 years	3,232	1,655	1,578	13,068	6,719	6,350	3,123	1,587	1,536
10 to 14 years	2,846	1,456	1,390	13,060	6,711	6,348	2,972	1,511	1,462
15 to 19 years	2,624	1,346	1,278	13,200	6,795	6,405	2,910	1,477	1,433
20 to 24 years	2,489	1,282	1,208	11,930	6,090	5,840	2,531	1,243	1,288
25 to 29 years	2,488	1,297	1,191	12,286	6,160	6,125	2,462	1,171	1,291
30 to 34 years	2,657	1,414	1,244	13,566	6,756	6,810	2,491	1,154	1,338
35 to 39 years	2,609	1,367	1,242	16,208	8,115	8,093	2,728	1,265	1,463
40 to 44 years	2,221	1,139	1,082	16,533	8,263	8,270	2,610	1,207	1,403
45 to 49 years	1,725	863	862	14,770	7,338	7,432	2,149	975	1,174
50 to 54 years	1,321	644	677	12,990	6,405	6,585	1,632	728	904
55 to 59 years	969	461	509	10,313	5,032	5,281	1,251	547	705
60 to 64 years	765	356	409	8,438	4,059	4,379	1,023	436	587
65 to 69 years	627	283	344	7,722	3,604	4,118	903	386	517
70 to 74 years	510	224	285	7,452	3,329	4,123	723	296	427
75 to 79 years	371	158	213	6,390	2,684	3,707	554	213	341
80 to 84 years	215	82	132	4,284	1,604	2,681	326	112	215
85 to 89 years	123	42	81	2,336	727	1,610	181	56	125
90 to 94 years	56	18	38	1,022	251	771	92	25	67
95 to 99 years	20	6	14	308	59	249	36	9	27
100 years and over	4	1	3	59	8	50	9	2	7
16 years and over	21,271	10,716	10,555	157,252	75,963	81,289	24,043	11,010	13,032
18 years and over	20,228	10,180	10,048	151,923	73,211	78,712	22,877	10,413	12,463
15 to 44 years	15,088	7,844	7,244	83,722	42,178	41,543	15,732	7,516	8,215
65 years and over	1,925	816	1,110	29,575	12,266	17,308	2,825	1,099	1,726
85 years and over	203	67	135	3,726	1,046	2,680	319	92	226

[Continued]

★ 979 ★

Projections: Age Range, Gender, and Race, July 1, 2000. Part I

[Continued]

| Date and age | Hispanic origin[1] | | | Not of Hispanic origin, by race | | | | | |
| | | | | White | | | Black | | |
	Total	Male	Female	Total	Male	Female	Total	Male	Female
Median age	27.3	26.9	27.7	38.1	36.9	39.3	29.7	27.7	31.5
Mean age	29.3	28.6	30.1	38.5	37.1	39.9	31.8	30.2	33.2

Source: "Projections of the Population, by Age, Sex, Race, and Hispanic Origin, for the United States, 1993 to 2020, (Middle Series)." Jennifer Cheeseman Day, *Population Projections of the United States, by Age, Sex, Race, and Hispanic Origin: 1993 to 2050*, P25-1104, p. 27. *Note:* 1. Persons of Hispanic origin may be of any race.

★ 980 ★

Population Trends and Projections

Projections: Age Range, Gender, and Race, July 1, 2000. Part II

Numbers in thousands. Resident population.

| Date and age | Not of Hispanic origin, by race | | | | | |
| | American Indian, Eskimo, Aleut | | | Asian and Pacific Islander | | |
	Total	Male	Female	Total	Male	Female
July 1, 2000						
All ages	2,055	1,010	1,045	11,407	5,529	5,879
Under 5 years	185	93	92	984	504	480
5 to 9 years	194	98	96	913	471	443
10 to 14 years	209	106	102	885	452	433
15 to 19 years	195	98	97	890	452	438
20 to 24 years	159	80	80	837	415	422
25 to 29 years	158	81	78	976	470	507
30 to 34 years	146	73	73	1,007	484	523
35 to 39 years	155	76	78	996	483	512
40 to 44 years	147	71	76	918	439	479
45 to 49 years	125	60	65	761	357	404
50 to 54 years	100	48	52	596	275	321
55 to 59 years	75	36	40	438	203	235
60 to 64 years	59	28	31	357	164	193
65 to 69 years	46	21	25	296	126	169
70 to 74 years	36	16	20	236	99	137
75 to 79 years	28	12	16	163	70	93
80 to 84 years	18	7	11	88	37	50
85 to 89 years	11	4	7	41	17	24
90 to 94 years	6	2	4	17	7	10
95 to 99 years	3	1	2	6	2	4
100 years and over	1	0	1	2	0	1
16 years and over	1,426	692	735	8,443	4,009	4,434
18 years and over	1,346	651	695	8,079	3,823	4,256
15 to 44 years	961	479	482	5,623	2,743	2,881
65 years and over	148	62	86	849	360	489
85 years and over	20	6	14	66	26	40

[Continued]

★ 980 ★

Projections: Age Range, Gender, and Race, July 1, 2000. Part II

[Continued]

Date and age	Not of Hispanic origin, by race					
	American Indian, Eskimo, Aleut			Asian and Pacific Islander		
	Total	Male	Female	Total	Male	Female
Median age	27.7	26.9	28.6	31.1	30.0	32.1
Mean age	30.6	29.8	31.5	32.3	31.3	33.2

Source: "Projections of the Population, by Age, Sex, Race, and Hispanic Origin, for the United States, 1993 to 2020, (Middle Series)." Jennifer Cheeseman Day, *Population Projections of the United States, by Age, Sex, Race, and Hispanic Origin: 1993 to 2050*, P25-1104, p. 27. *Note:* 1. Persons of Hispanic origin may be of any race.

★ 981 ★

Population Trends and Projections

Projections: Age Range, Gender, and Race, July 1, 2005. Part I.

Date and age	Race											
	White			Black			American Indian, Eskimo, and Aleut			Asian and Pacific Islander		
	Total	Male	Female	Total	Male	Female	Total	Male	Female	Total	Male	Female
July 1, 2005												
All ages	233,343	114,911	118,433	37,793	17,886	19,906	2,543	1,256	1,287	14,608	7,068	7,540
Under 5 years	14,587	7,492	7,095	3,310	1,684	1,626	231	116	114	1,205	616	589
5 to 9 years	15,124	7,768	7,356	3,272	1,666	1,606	222	113	110	1,180	606	574
10 to 14 years	16,503	8,479	8,023	3,472	1,769	1,703	253	128	125	1,214	624	590
15 to 19 years	16,211	8,339	7,873	3,304	1,682	1,622	247	125	123	1,109	565	544
20 to 24 years	15,633	7,995	7,639	2,942	1,449	1,492	220	109	111	1,121	555	566
25 to 29 years	13,982	7,063	6,919	2,567	1,223	1,345	184	92	92	1,114	532	582
30 to 34 years	14,918	7,494	7,424	2,609	1,218	1,391	179	91	88	1,239	589	651
35 to 39 years	16,154	8,119	8,035	2,657	1,233	1,424	168	86	82	1,199	579	621
40 to 44 years	18,725	9,394	9,331	2,849	1,320	1,529	177	88	88	1,145	553	592
45 to 49 years	18,175	9,057	9,118	2,633	1,203	1,430	160	78	82	987	472	515
50 to 54 years	16,138	7,966	8,171	2,178	973	1,204	134	64	69	815	378	437
55 to 59 years	13,906	6,799	7,108	1,645	718	927	104	50	54	637	289	348
60 to 64 years	10,782	5,204	5,578	1,235	528	708	76	36	41	483	219	264
65 to 69 years	8,616	4,057	4,559	1,006	428	578	58	26	31	386	172	214
70 to 74 years	7,442	3,409	4,033	766	317	449	44	20	24	306	130	175
75 to 79 years	6,792	2,938	3,854	605	238	367	34	15	19	224	91	133
80 to 84 years	5,074	1,982	3,092	376	132	244	24	10	14	136	57	79
85 to 89 years	2,781	920	1,861	193	60	133	14	5	9	63	24	39
90 to 94 years	1,286	338	948	110	31	79	9	3	6	28	11	18
95 to 99 years	417	84	333	46	11	35	4	1	3	10	4	6
100 years and over	97	14	82	18	4	14	2	0	1	4	1	3
16 years and over	183,871	89,496	94,375	27,038	12,409	14,629	1,782	871	911	10,778	5,103	5,675
18 years and over	177,435	86,176	91,259	25,716	11,731	13,985	1,682	820	861	10,331	4,874	5,457
15 to 44 years	95,623	48,403	47,220	16,927	8,124	8,803	1,176	591	584	6,928	3,372	3,556
65 years and over	32,505	13,742	18,762	3,121	1,221	1,899	187	80	107	1,158	490	667
85 years and over	4,581	1,357	3,224	367	106	260	28	9	19	106	40	66
Median age	38.1	36.8	39.4	30.1	27.7	32.1	27.5	26.9	28.2	31.5	30.3	32.6
Mean age	38.2	37.0	39.5	32.3	30.7	33.8	31.0	30.2	31.8	32.8	31.7	33.8

Source: "Projections of the Population, by Age, Sex, Race, and Hispanic Origin, for the United States, 1993 to 2020, (Middle Series)." Jennifer Cheeseman Day, *Population Projections of the United States, by Age, Sex, Race, and Hispanic Origin: 1993 to 2050*, P25-1104, p. 36.

★ 982 ★

Population Trends and Projections

Projections: Age Range, Gender, and Race, July 1, 2005. Part II. - A

Date and age	Hispanic origin[1]			Not of Hispanic origin					
				White			Black		
	Total	Male	Female	Total	Male	Female	Total	Male	Female
July 1, 2010									
All ages	40,525	20,410	20,115	203,441	99,903	103,538	37,930	17,890	20,040
Under 5 years	3,983	2,038	1,945	11,273	5,798	5,475	3,289	1,675	1,613
5 to 9 years	3,721	1,904	1,817	11,392	5,859	5,533	3,157	1,609	1,547
10 to 14 years	3,654	1,869	1,784	12,289	6,322	5,968	3,233	1,649	1,584
15 to 19 years	3,709	1,906	1,804	13,763	7,086	6,677	3,435	1,751	1,684
20 to 24 years	3,339	1,716	1,623	13,261	6,769	6,492	2,986	1,475	1,511
25 to 29 years	3,019	1,533	1,486	12,635	6,368	6,267	2,659	1,268	1,391
30 to 34 years	2,815	1,418	1,398	11,786	5,914	5,873	2,425	1,130	1,295
35 to 39 years	2,731	1,396	1,335	12,607	6,293	6,314	2,482	1,153	1,329
40 to 44 years	2,788	1,457	1,332	13,763	6,828	6,935	2,485	1,140	1,345
45 to 49 years	2,631	1,351	1,280	15,961	7,921	8,040	2,597	1,176	1,421
50 to 54 years	2,197	1,102	1,096	15,967	7,886	8,081	2,426	1,081	1,345
55 to 59 years	1,717	838	879	14,245	6,986	7,259	1,997	866	1,130
60 to 64 years	1,303	618	684	12,214	5,939	6,276	1,483	630	854
65 to 69 years	939	431	508	9,361	4,457	4,904	1,116	474	642
70 to 74 years	698	314	384	7,142	3,329	3,813	787	232	464
75 to 79 years	533	233	300	5,950	2,660	3,289	595	236	359
80 to 84 years	381	157	224	4,818	1,946	2,872	383	137	246
85 to 89 years	216	82	134	3,001	1,035	1,966	208	66	142
90 to 94 years	98	32	65	1,407	391	1,016	111	31	80
95 to 99 years	40	12	29	483	100	383	54	13	41
100 years and over	13	3	10	123	16	106	25	6	19
16 years and over	28,442	14,227	14,216	165,888	80,585	85,302	27,583	12,615	14,968
18 years and over	26,982	13,475	13,507	160,463	77,783	82,681	26,232	11,921	14,311
15 to 44 years	18,401	9,424	8,977	77,816	39,259	38,557	16,471	7,917	8,554
65 years and over	2,918	1,264	1,654	32,285	13,934	18,350	3,278	1,286	1,992
85 years and over	367	129	238	5,013	1,542	3,471	397	116	282
Median age	28.1	27.5	28.7	41.0	39.7	42.4	30.4	28.1	32.6
Mean age	31.0	30.4	31.7	40.5	39.2	41.7	33.1	31.3	34.6

Source: "Projections of the Population, by Age, Sex, Race, and Hispanic Origin, for the United States, 1993 to 2020, (Middle Series)." Jennifer Cheeseman Day, *Population Projections of the United States, by Age, Sex, Race, and Hispanic Origin: 1993 to 2050*, P25-1104, p. 39. *Note:* 1. Persons of Hispanic origin may be of any race.

★ 983 ★

Population Trends and Projections

Projections: Age Range, Gender, and Race, July 1, 2005. Part II. - B

Date and age	Hispanic origin[1]			Not of Hispanic origin					
				White			Black		
	Total	Male	Female	Total	Male	Female	Total	Male	Female
July 1, 2005									
All ages	35,702	18,022	17,679	200,842	98,472	102,370	35,793	16,891	16,901
Under 5 years	3,579	1,830	1,749	11,326	5,824	5,502	3,109	1,581	1,528
5 to 9 years	3,435	1,757	1,678	11,992	6,166	5,826	3,079	1,567	1,512
10 to 14 years	3,448	1,765	1,682	13,376	6,787	6,498	3,278	1,669	1,609
15 to 19 years	3,093	1,588	1,504	13,422	6,905	6,517	3,117	1,586	1,531
20 to 24 years	2,873	1,475	1,397	13,025	6,650	6,375	2,781	1,369	1,412
25 to 29 years	2,651	1,349	1,302	11,573	5,830	5,742	2,420	1,151	1,269
30 to 34 years	2,650	1,365	1,285	12,504	6,243	6,261	2,466	1,149	1,317
35 to 39 years	2,735	1,442	1,293	13,666	6,801	6,865	2,506	1,156	1,349
40 to 44 years	2,661	1,381	1,280	16,299	8,132	8,167	2,701	1,244	1,457
45 to 49 years	2,202	1,117	1,085	16,171	8,039	8,132	2,506	1,139	1,368
50 to 54 years	1,728	854	874	14,564	7,188	7,376	2,078	924	1,154
55 to 59 years	1,319	634	685	12,702	6,218	6,484	1,571	683	888
60 to 64 years	962	450	512	9,903	4,791	5,112	1,181	503	678
65 to 69 years	748	341	408	7,933	3,746	4,187	963	406	555
70 to 74 years	585	261	325	6,094	3,169	3,735	735	303	432
75 to 79 years	463	199	264	6,363	2,753	3,611	582	228	354
80 to 84 years	304	124	181	4,790	1,866	2,924	363	127	236
85 to 89 years	152	54	98	2,639	869	1,770	187	58	129
90 to 94 years	78	25	54	1,212	315	897	107	30	77
95 to 99 years	29	8	20	390	76	314	45	11	34
100 years and over	8	2	6	89	12	77	17	4	13
16 years and over	24,580	12,332	12,248	161,485	78,733	83,252	25,665	11,737	13,929
18 years and over	23,345	11,696	11,649	156,160	75,486	80,675	24,419	11,098	13,321
15 to 44 years	16,662	8,601	8,061	80,489	40,562	39,927	15,991	7,655	8,335
65 years and over	2,368	1,014	1,354	30,321	12,806	17,515	2,999	1,170	1,829
85 years and over	267	90	177	4,331	1,272	3,058	356	103	253
Median age	27.6	27.1	28.2	39.9	38.6	41.0	30.2	27.8	32.3
Mean age	30.2	29.5	30.9	39.5	36.2	40.8	32.5	30.8	33.9

Source: "Projections of the Population, by Age, Sex, Race, and Hispanic Origin, for the United States, 1993 to 2020, (Middle Series)." Jennifer Cheeseman Day, *Population Projections of the United States, by Age, Sex, Race, and Hispanic Origin: 1993 to 2050*, P25-1104, p. 37.
Note: 1. Persons of Hispanic origin may be of any race.

★ 984 ★

Population Trends and Projections

Projections: Age Range, Gender, and Race, July 1, 2005. Part II. - C

Date and age	Not of Hispanic origin					
	American Indian, Eskimo, and Aleut			Asian and Pacific Islander		
	Total	Male	Female	Total	Male	Female
July 1, 2005						
All ages	2,190	1,075	1,115	13,759	6,660	7,099
Under 5 years	198	100	98	1,122	574	548
5 to 9 years	191	97	94	1,101	566	535
10 to 14 years	215	109	106	1,126	579	546
15 to 19 years	206	105	103	1,033	527	506
20 to 24 years	187	93	95	1,050	520	530
25 to 29 years	158	79	79	1,046	500	546
30 to 34 years	155	78	77	1,169	556	613
35 to 39 years	142	71	70	1,130	546	585
40 to 44 years	151	74	77	1,084	524	560
45 to 49 years	138	67	71	938	449	489
50 to 54 years	117	56	61	777	360	417
55 to 59 years	92	44	48	609	277	332
60 to 64 years	69	32	37	462	210	252
65 to 69 years	52	24	28	370	165	205
70 to 74 years	39	18	22	293	125	168
75 to 79 years	31	13	17	215	88	127
80 to 84 years	22	9	13	131	56	76
85 to 89 years	13	4	8	61	23	38
90 to 94 years	8	2	6	27	10	17
95 to 99 years	4	1	3	10	4	6
100 years and over	2	0	1	4	1	3
16 years and over	1,542	748	794	10,197	4,831	5,367
18 years and over	1,458	705	753	9,781	4,617	5,164
15 to 44 years	1,001	500	501	6,512	3,173	3,340
65 years and over	171	72	99	1,112	472	640
85 years and over	27	8	18	102	38	63
Median age	28.0	27.1	28.9	31.8	30.6	32.8
Mean age	31.4	30.5	32.3	33.0	31.9	34.0

Source: "Projections of the Population, by Age, Sex, Race, and Hispanic Origin, for the United States, 1993 to 2020, (Middle Series)."
Jennifer Cheeseman Day, *Population Projections of the United States, by Age, Sex, Race, and Hispanic Origin: 1993 to 2050*, P25-1104,
p. 37.

★ 985 ★

Population Trends and Projections

Projections: Age Range, Gender, and Race, July 1, 2010. Part I

Date and age	Race											
	White			Black			American Indian, Eskimo, and Aleut			Asian and Pacific Islander		
	Total	Male	Female	Total	Male	Female	Total	Male	Female	Total	Male	Female
July 1, 2010												
All ages	240,297	118,505	121,792	40,224	19,027	21,197	2,719	1,342	1,377	17,191	8,312	8,878
Under 5 years	14,893	7,650	7,243	3,518	1,793	1,725	252	127	125	1,354	691	663
5 to 9 years	14,775	7,590	7,185	3,369	1,718	1,650	239	121	118	1,339	687	652
10 to 14 years	15,606	8,019	7,587	3,442	1,756	1,686	247	125	122	1,430	734	696
15 to 19 years	17,126	8,815	8,311	3,649	1,861	1,788	252	126	126	1,372	702	670
20 to 24 years	16,276	8,323	7,953	3,185	1,576	1,610	238	119	119	1,276	635	641
25 to 29 years	3,202	1,632	1,570	589	286	304	47	23	23	263	129	134
30 to 34 years	14,346	7,211	7,135	2,580	1,204	1,376	180	89	91	1,316	621	695
35 to 39 years	15,093	7,572	7,522	2,629	1,224	1,406	175	89	85	1,366	653	713
40 to 44 years	16,301	8,160	8,141	2,639	1,217	1,422	165	84	80	1,291	622	670
45 to 49 years	18,361	9,157	9,204	2,744	1,250	1,493	167	83	83	1,159	560	599
50 to 54 years	17,969	8,891	9,078	2,553	1,144	1,408	150	73	77	996	471	525
55 to 59 years	15,810	7,751	8,058	2,096	915	1,181	123	59	64	817	374	443
60 to 64 years	13,403	6,505	6,898	1,558	664	894	95	45	50	651	291	359
65 to 69 years	10,217	4,851	5,366	1,171	499	672	68	31	37	487	215	272
70 to 74 years	7,781	3,617	4,165	826	341	485	50	23	27	377	166	210
75 to 79 years	6,440	2,875	3,565	624	249	375	37	16	21	278	115	163
80 to 84 years	5,172	2,092	3,080	401	144	257	26	11	15	180	72	108
85 to 89 years	3,203	1,112	2,091	217	70	147	17	6	11	96	36	60
90 to 94 years	1,499	421	1,077	115	32	83	10	3	7	43	15	28
95 to 99 years	521	111	410	55	14	42	6	2	4	17	6	11
100 years and over	135	19	115	25	6	20	3	1	3	6	2	4
16 years and over	191,766	93,571	98,196	29,184	13,395	15,789	1,931	944	987	12,778	6,052	6,726
18 years and over	185,015	90,085	94,930	27,750	12,658	15,092	1,833	894	939	12,215	5,762	6,453
15 to 44 years	94,512	47,844	46,868	17,511	8,432	9,079	1,228	616	612	7,960	3,876	4,084
65 years and over	34,967	15,098	19,869	3,435	1,354	2,081	218	93	125	1,484	628	856
85 years and over	5,357	1,663	3,694	413	121	292	36	12	25	163	59	104
Median age	39.0	37.7	40.2	30.2	28.0	32.3	27.9	27.4	28.5	31.8	30.5	33.0
Mean age	39.0	37.8	40.2	32.9	31.3	34.4	31.6	30.9	32.4	33.5	32.4	34.5

Source: "Projections of the Population, by Age, Sex, Race, and Hispanic Origin, for the United States, 1993 to 2020, (Middle Series)." Jennifer Cheeseman Day, *Population Projections of the United States, by Age, Sex, Race, and Hispanic Origin: 1993 to 2050*, P25-1104, p. 38.

★ 986 ★

Population Trends and Projections

Projections: Age Range, Gender, and Race, July 1, 2010. Part II

Date and age	Not of Hispanic origin					
	American Indian, Eskimo, and Aleut			Asian and Pacific Islander		
	Total	Male	Female	Total	Male	Female
July 1, 2010						
All ages	2,336	1,146	1,190	16,199	7,838	8,361
Under 5 years	214	108	107	1,258	642	615
5 to 9 years	204	103	101	1,249	641	607
10 to 14 years	212	107	105	1,337	686	650

[Continued]

★ 986 ★

Projections: Age Range, Gender, and Race, July 1, 2010. Part II

[Continued]

Date and age	Not of Hispanic origin					
	American Indian, Eskimo, and Aleut			Asian and Pacific Islander		
	Total	Male	Female	Total	Male	Female
15 to 19 years	214	107	107	1,277	654	623
20 to 24 years	200	100	100	1,191	594	597
25 to 29 years	186	92	94	1,259	605	653
30 to 34 years	155	77	78	1,240	587	654
35 to 39 years	151	77	74	1,292	619	674
40 to 44 years	139	70	69	1,221	588	632
45 to 49 years	142	70	72	1,100	532	568
50 to 54 years	130	62	67	948	449	499
55 to 59 years	107	51	57	780	357	423
60 to 64 years	84	39	45	623	279	344
65 to 69 years	61	28	33	467	207	260
70 to 74 years	45	20	25	362	160	202
75 to 79 years	34	15	19	267	111	156
80 to 84 years	24	10	14	174	70	104
85 to 89 years	16	6	10	93	35	58
90 to 94 years	10	3	7	42	14	27
95 to 99 years	6	1	4	16	6	11
100 years and over	3	1	2	6	2	4
16 years and over	1,662	806	856	12,085	5,729	6,356
18 years and over	1,578	764	814	11,558	5,457	6,100
15 to 44 years	1,044	521	522	7,480	3,647	3,833
65 years and over	198	84	114	1,426	604	822
85 years and over	34	11	23	157	57	100
Median age	28.3	27.5	29.0	32.1	30.8	33.3
Mean age	32.0	31.1	32.8	33.7	32.6	34.7

Source: "Projections of the Population, by Age, Sex, Race, and Hispanic Origin, for the United States, 1993 to 2020, (Middle Series)." Jennifer Cheeseman Day, *Population Projections of the United States, by Age, Sex, Race, and Hispanic Origin: 1993 to 2050*, P25-1104, p. 39.

★ 987 ★

Population Trends and Projections

Projections: Age Range, Gender, and Race, July 1, 2015. Part I.

Date and age	[126]Race White			Black			American Indian, Eskimo, and Aleut			Asian and Pacific Islander		
	Total	Male	Female	Total	Male	Female	Total	Male	Female	Total	Male	Female
July 1, 2015												
All ages	247,542	122,233	125,309	42,797	20,242	22,555	2,904	1,432	1,472	19,873	9,610	10,263
Under 5 years	15,586	8,008	7,578	3,772	1,927	1,846	269	135	133	1,516	773	743
5 to 9 years	15,080	7,749	7,331	3,579	1,829	1,750	261	132	129	1,497	768	730
10 to 14 years	15,251	7,839	7,413	3,543	1,811	1,732	265	134	131	1,610	825	785
15 to 19 years	16,208	8,344	7,864	3,618	1,849	1,769	246	123	123	1,589	811	778
20 to 24 years	17,160	8,783	8,377	3,507	1,739	1,768	342	121	122	1,532	768	764
25 to 29 years	15,982	8,074	7,908	3,054	1,464	1,589	237	118	119	1,497	726	772

[Continued]

★ 987 ★

Projections: Age Range, Gender, and Race, July 1, 2015. Part I.
[Continued]

Date and age	[126]Race White Total	Male	Female	Black Total	Male	Female	American Indian, Eskimo, and Aleut Total	Male	Female	Asian and Pacific Islander Total	Male	Female
30 to 34 years	15,744	7,914	7,830	2,839	1,329	1,511	215	106	109	1,541	733	809
35 to 39 years	14,530	7,295	7,235	2,604	1,212	1,392	175	87	88	1,445	687	757
40 to 44 years	15,246	7,618	7,627	2,615	1,211	1,404	171	87	84	1,459	697	762
45 to 49 years	16,002	7,962	8,040	2,545	1,155	1,390	155	79	76	1,303	628	675
50 to 54 years	18,161	8,993	9,169	2,664	1,192	1,472	156	77	79	1,163	556	607
55 to 59 years	17,628	8,669	8,960	2,462	1,079	1,383	138	67	72	988	462	526
60 to 64 years	15,299	7,457	7,842	1,997	854	1,143	113	53	60	821	371	450
65 to 69 years	12,748	6,101	6,647	1,486	634	852	85	39	45	643	281	362
70 to 74 years	9,278	4,361	4,916	965	401	564	59	27	32	470	206	264
75 to 79 years	6,792	3,093	3,698	677	270	406	43	19	24	341	146	195
80 to 84 years	4,948	2,077	2,871	414	151	263	29	12	17	224	91	133
85 to 89 years	3,316	1,200	2,116	232	76	155	19	7	12	129	46	83
90 to 94 years	1,774	529	1,245	131	37	93	13	4	9	67	23	44
95 to 99 years	628	143	486	60	15	45	7	2	5	26	8	18
100 years and over	182	26	155	33	7	26	5	1	4	11	3	8
16 years and over	198,510	97,034	101,476	31,181	14,305	16,876	2,057	1,004	1,053	14,920	7,076	7,844
18 years and over	192,093	93,720	98,373	29,743	13,564	16,178	1,957	954	1,003	14,276	6,745	7,531
15 to 44 years	94,870	48,029	46,842	18,238	8,804	9,434	1,287	642	645	9,063	4,422	4,642
65 years and over	39,665	17,529	22,135	3,997	1,592	2,405	259	112	147	1,911	806	1,105
85 years and over	5,900	1,897	4,002	455	136	319	44	15	29	233	81	152
Median age	39.4	38.0	40.8	30.6	28.2	32.7	28.5	27.9	29.1	32.2	30.9	33.5
Mean age	39.7	38.6	40.8	33.5	31.8	35.0	32.1	31.3	32.9	34.2	33.0	35.2

Source: "Projections of the Population, by Age, Sex, Race, and Hispanic Origin, for the United States, 1993 to 2020, (Middle Series)." Jennifer Cheeseman Day, *Population Projections of the United States, by Age, Sex, Race, and Hispanic Origin: 1993 to 2050*, P25-1104, p. 40.

★ 988 ★

Population Trends and Projections

Projections: Age Range, Gender, and Race, July 1, 2015. Part II - A

Date and age	Hispanic origin[1] Total	Male	Female	Not of Hispanic origin White Total	Male	Female	Black Total	Male	Female
July 1, 2015									
All ages	45,719	22,982	22,737	205,997	101,300	104,697	40,184	18,951	21,234
Under 5 years	4,489	2,298	2,191	11,508	5,921	5,588	3,511	1,792	1,719
5 to 9 years	4,128	2,114	2,014	11,337	5,833	5,505	3,338	1,705	1,633
10 to 14 years	3,948	2,022	1,927	11,677	6,008	5,668	3,313	1,693	1,621
15 to 19 years	3,924	2,015	1,909	12,648	6,515	6,134	3,389	1,731	1,658
20 to 24 years	3,944	2,026	1,918	13,581	6,940	6,641	3,282	1,625	1,657
25 to 29 years	3,464	1,760	1,705	12,859	6,481	6,378	2,847	1,363	1,484
30 to 34 years	3,187	1,602	1,585	12,857	6,454	6,403	2,661	1,244	1,417
35 to 39 years	2,898	1,450	1,448	11,896	5,969	5,927	2,444	1,137	1,308
40 to 44 years	2,786	1,412	1,374	12,708	6,324	6,383	2,465	1,139	1,325
45 to 49 years	2,757	1,426	1,331	13,491	6,657	6,833	2,393	1,080	1,313
50 to 54 years	2,621	1,331	1,290	15,767	7,773	7,994	2,518	1,119	1,399
55 to 59 years	2,179	1,081	1,098	15,641	7,681	7,960	2,336	1,017	1,319
60 to 64 years	1,692	816	875	13,758	6,711	7,046	1,897	806	1,091
65 to 69 years	1,268	592	676	11,593	5,561	6,033	1,410	598	811
70 to 74 years	876	398	477	8,477	3,997	4,481	915	378	537

[Continued]

★ 988 ★

Projections: Age Range, Gender, and Race, July 1, 2015. Part II - A

[Continued]

| Date and age | Hispanic origin[1] | | | Not of Hispanic origin | | | | | |
| | | | | White | | | Black | | |
	Total	Male	Female	Total	Male	Female	Total	Male	Female
75 to 79 years	637	282	355	6,208	2,835	3,374	641	254	386
80 to 84 years	439	184	255	4,542	1,906	2,636	392	142	250
85 to 89 years	272	105	167	3,062	1,102	1,960	219	72	148
90 to 94 years	140	49	91	1,643	482	1,160	125	35	89
95 to 99 years	52	16	36	580	128	452	58	14	43
100 years and over	20	5	15	163	22	141	32	7	25
16 years and over	32,378	16,151	16,227	169,062	82,295	86,767	29,346	13,415	15,931
18 years and over	30,814	15,346	15,468	164,062	79,711	84,352	27,998	12,721	15,277
15 to 44 years	20,203	10,263	9,939	76,549	38,683	37,866	17,087	8,239	8,848
65 years and over	3,703	1,631	2,072	36,269	16,032	20,237	3,791	1,501	2,290
85 years and over	483	175	309	5,448	1,734	3,714	434	128	306
Median age	28.4	27.8	29.1	42.0	40.5	43.4	30.8	28.4	33.0
Mean age	31.7	31.1	32.4	41.3	40.1	42.5	33.6	31.8	35.2

Source: "Projections of the Population, by Age, Sex, Race, and Hispanic Origin, for the United States, 1993 to 2020, (Middle Series)." Jennifer Cheeseman Day, *Population Projections of the United States, by Age, Sex, Race, and Hispanic Origin: 1993 to 2050*, P25-1104, p. 41. *Note:* 1. Persons of Hispanic origin may be of any race.

★ 989 ★

Population Trends and Projections

Projections: Age Range, Gender, and Race, July 1, 2015. Part II. - B

| Date and age | Not of Hispanic origin | | | | | |
| | American Indian, Eskimo, and Aleut | | | Asian and Pacific Islander | | |
	Total	Male	Female	Total	Male	Female
July 1, 2015						
All ages	2,487	1,220	1,268	18,728	9,064	9,664
Under 5 years	228	115	113	1,406	717	688
5 to 9 years	222	112	110	1,393	714	678
10 to 14 years	227	114	112	1,504	772	733
15 to 19 years	211	106	105	1,489	761	728
20 to 24 years	206	102	104	1,429	718	711
25 to 29 years	198	99	100	1,402	680	722
30 to 34 years	182	90	93	1,453	692	761
35 to 39 years	151	75	76	1,366	652	715
40 to 44 years	148	75	73	1,384	662	722
45 to 49 years	131	66	65	1,234	596	638
50 to 54 years	133	65	68	1,105	529	576
55 to 59 years	119	56	63	942	441	501
60 to 64 years	98	46	53	785	355	430
65 to 69 years	75	34	40	616	269	347
70 to 74 years	53	24	29	451	198	253

[Continued]

★ 989 ★

Projections: Age Range, Gender, and Race, July 1, 2015. Part II. - B
[Continued]

Date and age	Not of Hispanic origin					
	American Indian, Eskimo, and Aleut			Asian and Pacific Islander		
	Total	Male	Female	Total	Male	Female
75 to 79 years	39	17	22	328	141	187
80 to 84 years	26	11	15	216	88	128
85 to 89 years	17	7	11	125	45	80
90 to 94 years	12	4	8	65	22	42
95 to 99 years	7	2	5	25	8	17
100 years and over	5	1	4	11	3	7
16 years and over	1,766	856	910	14,117	6,703	7,414
18 years and over	1,680	813	868	13,513	6,393	7,121
15 to 44 years	1,096	546	550	8,524	4,165	4,358
65 years and over	234	100	134	1,835	775	1,060
85 years and over	41	13	27	225	78	147
Median age	28.8	28.0	29.5	32.5	31.2	33.8
Mean age	32.4	31.5	33.3	34.4	33.2	35.5

Source: "Projections of the Population, by Age, Sex, Race, and Hispanic Origin, for the United States, 1993 to 2020, (Middle Series)."
Jennifer Cheeseman Day, *Population Projections of the United States, by Age, Sex, Race, and Hispanic Origin: 1993 to 2050*, P25-1104, p. 41.

★ 990 ★

Population Trends and Projections

Projections: Age Range, Gender, and Race, July 1, 2020. Part I.

Date and age	[126]Race White			Black			American Indian, Eskimo, and Aleut			Asian and Pacific Islander		
	Total	Male	Female	Total	Male	Female	Total	Male	Female	Total	Male	Female
July 1, 2020												
All ages	166,045	254,791	125,933	126,858	21,480	23,929	3,090	1,523	1,568	22,653	10,961	11,691
Under 5 years	16,039	8,242	7,797	3,953	2,022	1,931	277	139	138	1,688	860	828
5 to 9 years	10,488	15,770	8,105	7,665	3,834	1,962	1,872	278	137	1,668	854	814
10 to 14 years	15,562	8,000	7,562	3,760	1,925	1,835	290	146	143	1,789	915	873
15 to 19 years	15,842	8,157	7,685	3,721	1,905	1,817	265	133	132	1,772	903	869
20 to 24 years	16,262	8,325	7,937	3,480	1,729	1,751	237	118	119	1,746	876	870
25 to 29 years	16,839	8,514	8,324	3,361	1,615	1,746	241	120	122	1,755	860	895
30 to 34 years	16,364	8,226	8,137	3,061	1,438	1,623	233	115	117	1,699	815	884
35 to 39 years	15,932	7,999	7,932	2,864	1,337	1,527	209	103	106	1,669	799	870
40 to 44 years	14,684	7,344	7,341	2,591	1,200	1,391	172	86	87	1,540	733	808
45 to 49 years	14,972	7,435	7,537	2,521	1,148	1,373	1641	82	79	1,466	702	764
50 to 54 years	15,841	7,823	8,018	2,475	1,102	1,373	146	74	72	1,302	622	681
55 to 59 years	17,842	8,784	9,058	2,572	1,125	1,447	144	71	73	1,147	543	604
60 to 64 years	17,093	8,366	8,727	2,352	1,011	1,341	127	61	67	984	454	530
65 to 69 years	14,561	7,007	7,554	1,900	814	1,086	101	46	54	800	353	447
70 to 74 years	11,633	5,527	6,105	1,227	512	715	74	34	40	613	267	347
75 to 79 years	8,163	3,778	4,385	794	320	474	51	22	28	425	182	243
80 to 84 years	5,286	2,276	3,010	452	165	287	34	15	19	276	117	160
85 to 89 years	3,216	1,214	2,003	240	80	160	21	8	13	162	59	103
90 to 94 years	1,884	590	1,293	141	41	99	14	5	9	91	30	61
95 to 99 years	771	186	585	69	17	52	9	3	6	41	13	28
100 years and over	235	35	200	40	9	31	7	2	5	18	5	13
16 years and over	204,326	99,993	104,333	33,113	15,186	17,926	2,189	1,068	1,121	17,143	8,145	8,998
18 years and over	198,030	96,741	101,289	31,630	14,421	17,208	2,081	1,013	1,067	16,425	7,777	8,648

[Continued]

★ 990 ★

Projections: Age Range, Gender, and Race, July 1, 2020. Part I.

[Continued]

Date and age	[126]Race White			Black			American Indian, Eskimo, and Aleut			Asian and Pacific Islander		
	Total	Male	Female	Total	Male	Female	Total	Male	Female	Total	Male	Female
15 to 44 years	95,922	48,566	47,357	19,078	9,224	9,854	1,357	674	683	10,182	4,986	5,196
65 years and over	45,748	20,612	25,136	4,863	1,960	2,904	310	135	175	2,427	1,025	1,402
85 years and over	6,106	2,025	4,081	490	148	342	51	17	34	312	107	205
Median age	39.6	38.4	40.9	30.9	28.8	33.1	29.2	28.6	29.7	32.7	31.3	34.0
Mean age	40.4	39.3	41.5	34.0	32.2	35.5	32.6	31.8	33.4	34.8	33.6	35.9

Source: "Projections of the Population, by Age, Sex, Race, and Hispanic Origin, for the United States, 1993 to 2020, (Middle Series)." Jennifer Cheeseman Day, *Population Projections of the United States, by Age, Sex, Race, and Hispanic Origin: 1993 to 2050*, P25-1104, p. 42.

★ 991 ★

Population Trends and Projections

Projections: Age Range, Gender, and Race, July 1, 2020. Part II - A

Date and age	Hispanic origin[1]			Not of Hispanic origin					
				White			Black		
	Total	Male	Female	Total	Male	Female	Total	Male	Female
July 1, 2020									
All ages	51,217	25,705	25,512	208,280	102,533	105,747	42,459	20,025	22,434
Under 5 years	4,947	2,534	2,413	11,543	5,940	5,604	3,665	1,873	1,792
5 to 9 years	4,633	2,374	2,259	11,572	5,954	5,617	3,561	1,821	1,740
10 to 14 years	4,364	2,236	2,129	11,622	5,961	5,640	3,500	1,791	1,709
15 to 19 years	4,225	2,170	2,055	12,019	6,0192	5,827	3,471	1,776	1,696
20 to 24 years	4,156	2,133	2,022	12,488	6,382	6,106	3,240	1,608	1,632
25 to 29 years	4,045	2,053	1,992	13,175	6,646	6,527	3,130	1,502	1,627
30 to 34 years	3,633	1,827	1,806	13,086	6,569	6,517	2,847	1,336	1,511
35 to 39 years	3,270	1,634	1,636	12,968	6,509	6,459	2,681	1,251	1,430
40 to 44 years	2,956	1,467	1,489	11,996	6,001	5,995	2,428	1,123	1,305
45 to 49 years	2,757	1,384	1,373	12,459	6,166	6,293	2,372	1,079	1,294
50 to 54 years	2,746	1,405	1,341	13,338	6,537	6,801	2,323	1,028	1,295
55 to 59 years	2,596	1,305	1,291	15,469	7,587	7,882	2,426	1,053	1,372
60 to 64 years	2,142	1,052	1,090	15,140	7,405	7,736	2,226	950	1,276
65 to 69 years	1,644	782	862	13,067	6,295	6,772	1,798	765	1,032
70 to 74 years	1,182	548	634	10,554	5,027	5,528	1,158	480	678
75 to 79 years	800	359	440	7,431	3,448	3,983	748	299	449
80 to 84 years	526	224	303	4,801	2,070	2,732	424	153	271
85 to 89 years	315	124	192	2,924	1,099	1,825	225	74	151
90 to 94 years	178	64	115	1,717	531	1,187	133	38	94
95 to 99 years	75	24	51	701	163	538	66	16	50
100 years and over	27	7	20	210	29	181	39	8	31
16 years and over	36,430	18,130	18,300	171,210	83,454	87,755	31,035	14,182	16,853
18 years and over	34,744	17,261	17,483	166,438	80,988	85,451	29,651	13,469	16,183
15 to 44 years	22,285	11,285	11,000	75,732	38,301	37,431	17,796	8,595	9,201
65 years and over	4,747	2,132	2,615	41,405	18,661	22,745	4,590	1,835	2,755
85 years and over	595	219	377	5,552	1,821	3,731	463	137	325

[Continued]

★ 991 ★

Projections: Age Range, Gender, and Race, July 1, 2020. Part II - A

[Continued]

Date and age	Hispanic origin[1]			Not of Hispanic origin					
				White			Black		
	Total	Male	Female	Total	Male	Female	Total	Male	Female
Median age	29.1	28.4	29.7	42.3	40.9	43.8	31.1	28.9	33.3
Mean age	32.4	31.7	33.1	42.2	41.0	43.3	34.1	32.3	35.7

Source: "Projections of the Population, by Age, Sex, Race, and Hispanic Origin, for the United States, 1993 to 2020, (Middle Series)." Jennifer Cheeseman Day, *Population Projections of the United States, by Age, Sex, Race, and Hispanic Origin: 1993 to 2050*, P25-1104, p. 43. *Note:* 1. Persons of Hispanic origin may be of any race.

★ 992 ★

Population Trends and Projections

Projections: Age Range, Gender, and Race, July 1, 2020. Part II - B

Date and age	Not of Hispanic origin					
	American Indian, Eskimo, and Aleut			Asian and Pacific Islander		
	Total	Male	Female	Total	Male	Female
July 1, 2020						
All ages	2,641	1,294	1,346	21,345	10,339	11,006
Under 5 years	236	119	117	1,565	796	767
5 to 9 years	236	119	117	1,549	794	755
10 to 14 years	246	124	122	1,667	854	813
15 to 19 years	226	113	113	1,659	847	812
20 to 24 years	203	101	102	1,639	823	816
25 to 29 years	205	101	103	1,642	805	837
30 to 34 years	194	96	98	1,596	767	830
35 to 39 years	177	88	90	1,578	758	820
40 to 44 years	147	73	74	1,460	697	763
45 to 49 years	139	70	69	1,393	669	724
50 to 54 years	123	61	62	1,235	591	645
55 to 59 years	123	59	64	1,092	517	574
60 to 64 years	109	51	58	939	434	505
65 to 69 years	87	40	48	766	338	428
70 to 74 years	65	30	36	589	256	332
75 to 79 years	45	20	25	408	175	233
80 to 84 years	30	13	17	266	112	153
85 to 89 years	19	7	12	156	57	99
90 to 94 years	13	5	9	88	29	59
95 to 99 years	8	2	6	40	13	27
100 years and over	6	1	5	17	5	12
16 years and over	1,874	908	966	16,222	7,718	8,503
18 years and over	1,782	862	920	15,550	7,373	8,177
15 to 44 years	1,153	572	581	9,574	4,697	4,878
65 years and over	276	118	157	2,330	985	1,344
85 years and over	47	16	31	301	103	198

[Continued]

★ 992 ★

Projections: Age Range, Gender, and Race, July 1, 2020. Part II - B

[Continued]

| Date and age | Not of Hispanic origin | | | | | |
| | American Indian, Eskimo, and Aleut | | | Asian and Pacific Islander | | |
	Total	Male	Female	Total	Male	Female
Median age	29.3	28.6	29.9	33.0	31.6	34.3
Mean age	32.8	31.9	33.7	35.0	33.9	36.1

Source: "Projections of the Population, by Age, Sex, Race, and Hispanic Origin, for the United States, 1993 to 2020, (Middle Series)."
Jennifer Cheeseman Day, *Population Projections of the United States, by Age, Sex, Race, and Hispanic Origin: 1993 to 2050*, P25-1104, p. 43.

★ 993 ★

Population Trends and Projections

Projections: Age Range, Gender, and Race, July 1, 2025. Part I.

| Date and age | [126]Race White | | | Black | | | American Indian, Eskimo, and Aleut | | | Asian and Pacific Islander | | |
	Total	Male	Female	Total	Male	Female	Total	Male	Female	Total	Male	Female
July 1, 2025												
All ages	261,531	129,322	132,209	48,005	22,713	25,291	3,278	1,614	1,664	25,524	12,363	13,161
Under 5 years	16,117	8,282	7,835	4,102	2,100	2,002	287	144	143	1,866	950	916
5 to 9 years	16,216	8,333	7,882	4,015	2,.057	1,958	287	145	142	1,849	946	903
10 to 14 years	16,257	8,357	7,900	4,022	2,061	1,961	309	156	153	1,980	1,013	968
15 to 19 years	16,150	8,316	7,734	3,944	2,021	1,923	289	145	144	1,953	994	959
20 to 24 years	15,904	8,142	7,762	3,579	1,781	1,798	255	127	129	1,927	967	960
25 to 29 years	15,977	8,080	7,897	3,333	1,604	1,729	236	117	119	1,971	968	1,003
30 to 34 years	17,225	8,665	8,559	3,365	1,584	1,781	237	117	120	1,956	948	1,008
35 to 39 years	16,547	8,308	8,239	3,086	1,446	1,640	227	113	114	1,826	882	944
40 to 44 years	16,087	8,044	8,043	2,847	1,322	1,525	205	101	104	1,765	846	919
45 to 49 years	14,425	7,168	7,257	2,501	1,139	1,362	162	81	82	1,546	738	806
50 to 54 years	14,834	7,310	7,524	2,455	1,097	1,358	151	76	75	1,459	693	767
55 to 59 years	15,578	7,650	7,929	2,391	1,041	1,350	135	68	67	1,281	606	675
60 to 64 years	17,335	8,500	8,835	2,467	1,060	1,407	133	64	68	1,136	531	605
65 to 69 years	16,302	7,886	8,416	2,241	968	1,273	114	53	61	952	429	523
70 to 74 years	13,332	6,383	6,949	1,574	662	912	88	40	48	758	333	426
75 to 79 years	10,297	4,833	5,464	1,014	413	602	64	29	35	552	234	318
80 to 84 years	6,429	2,826	3,603	533	197	336	40	17	23	345	145	200
85 to 89 years	3,497	1,361	2,136	263	88	175	24	9	15	201	76	125
90 to 94 years	1,870	616	1,254	147	44	103	16	6	10	116	38	77
95 to 99 years	846	215	631	77	20	57	10	3	7	57	17	40
100 years and over	306	48	258	49	10	39	9	2	6	29	8	21
16 years and over	209,734	102,699	107,035	35,061	16,083	18,978	2,334	1,138	1,196	19,425	9,248	10,177
18 years and over	203,283	99,367	103,916	33,482	15,268	15,268	18,215	1,078	1,137	18,634	8,842	9,791
15 to 44 years	97,890	49,555	48,335	20,153	9,758	10,396	1,450	720	730	11,398	5,606	5,792
65 years and over	52,879	24,167	28,771	5,897	2,401	3,496	365	160	205	3,010	1,281	1,728
85 years and over	6,518	2,239	4,279	535	162	374	59	20	39	402	140	262
Median age	40.1	38.9	41.4	31.5	29.2	33.6	29.5	28.9	30.1	33.1	31.8	34.3
Mean age	41.0	40.0	42.1	34.4	32.6	36.0	33.0	32.3	33.8	35.4	34.2	36.5

Source: "Projections of the Population, by Age, Sex, Race, and Hispanic Origin, for the United States, 1993 to 2020, (Middle Series)." Jennifer Cheeseman Day, *Population Projections of the United States, by Age, Sex, Race, and Hispanic Origin: 1993 to 2050*, P25-1104, p. 44.

★ 994 ★

Population Trends and Projections

Projections: Age Range, Gender, and Race, July 1, 2025. Part II - A

Date and age	Hispanic origin[1]			Not of Hispanic origin					
				White			Black		
	Total	Male	Female	Total	Male	Female	Total	Male	Female
July 1, 2025									
All ages	56,927	28,531	28,396	209,863	103,362	106,501	44,705	21,089	23,616
Under 5 years	5,337	2,735	2,603	11,267	5,797	5,470	3,790	1,938	1,852
5 to 9 years	5,089	2,608	2,481	11,602	5,969	5,632	3,715	1,901	1,813
10 to 14 years	4,879	2,500	2,379	11,855	5,101	5,754	3,730	1,910	1,820
15 to 19 years	4,648	2,388	2,260	11,955	6,159	5,796	3,663	1,876	1,788
20 to 24 years	4,452	2,286	2,166	11,870	6,066	5,804	3,319	1,649	1,670
25 to 29 years	4,247	2,154	2,093	12,128	6,120	6,007	3,088	1,484	1,604
30 to 34 years	4,217	2,120	2,097	13,403	6,734	6,669	3,126	1,470	1,656
35 to 39 years	3,716	1,859	1,857	13,193	6,621	6,572	2,866	1,341	1,524
40 to 44 years	3,327	1,650	1,677	13,069	6,539	6,531	2,660	1,235	1,426
45 to 49 years	2,922	1,436	1,486	11,766	5,853	5,913	2,340	1,065	1,275
50 to 54 years	2,746	1,362	1,383	12,328	6,059	6,270	2,307	1,028	1,279
55 to 59 years	2,718	1,376	1,341	13,098	6,388	6,710	2,240	968	1,271
60 to 64 years	2,548	1,270	1,279	15,006	7,335	7,670	2,321	989	1,332
65 to 69 years	2,078	1,006	1,071	14,411	6,968	7,443	2,112	905	1,207
70 to 74 years	1,531	725	807	11,937	5,722	6,215	1,481	618	863
75 to 79 years	1,081	496	585	9,310	4,380	4,930	950	383	568
80 to 84 years	663	287	377	5,819	2,562	3,257	497	181	316
85 to 89 years	380	152	229	3,145	1,220	1,924	244	80	164
90 to 94 years	209	76	133	1,676	545	1,131	137	40	97
95 to 99 years	97	32	65	756	185	571	72	18	54
100 years and over	39	11	28	269	38	232	47	10	37
16 years and over	40,682	20,206	20,476	172,777	84,277	88,500	32,723	14,956	17,767
18 years and over	38,820	19,247	19,574	168,006	81,811	86,195	31,257	14,200	17,058
15 to 44 years	24,609	12,458	12,151	75,618	38,239	37,379	18,722	9,054	9,668
65 years and over	6,079	2,786	3,293	47,323	21,621	25,703	5,541	2,234	3,306
85 years and over	725	271	455	5,846	1,988	3,858	500	148	352
Median age	29.8	29.1	30.5	42.9	41.6	44.3	31.7	29.3	33.9
Mean age	33.0	32.4	33.7	43.0	41.9	44.1	34.6	32.7	36.2

Source: "Projections of the Population, by Age, Sex, Race, and Hispanic Origin, for the United States, 1993 to 2020, (Middle Series)." Jennifer Cheeseman Day, *Population Projections of the United States, by Age, Sex, Race, and Hispanic Origin: 1993 to 2050*, P25-1104, p. 45. *Note:* 1. Persons of Hispanic origin may be of any race.

★ 995 ★

Population Trends and Projections

Projections: Age Range, Gender, and Race, July 1, 2025. Part II - B

| Date and age | Not of Hispanic origin | | | | | |
| | American Indian, Eskimo, and Aleut | | | Asian and Pacific Islander | | |
	Total	Male	Female	Total	Male	Female
July 1, 2025						
All ages	2,796	1,370	1,426	24,046	11,660	12,386
Under 5 years	246	123	122	1,732	882	850
5 to 9 years	244	213	121	1,717	879	838
10 to 14 years	262	132	130	1,843	943	899
15 to 19 years	245	123	123	1,825	930	894
20 to 24 years	218	108	110	1,806	907	899
25 to 29 years	202	100	102	1,852	911	941
30 to 34 years	201	99	102	1,835	891	944
35 to 39 years	189	94	95	1,721	833	888
40 to 44 years	174	86	88	1,673	804	869
45 to 49 years	139	69	70	1,468	703	765
50 to 54 years	130	65	65	1,389	661	728
55 to 59 years	113	55	57	1,216	576	640
60 to 64 years	113	54	59	1,082	507	575
65 to 69 years	97	45	53	910	411	499
70 to 74 years	77	35	42	727	319	408
75 to 79 years	56	25	31	530	225	305
80 to 84 years	36	16	20	331	140	192
85 to 89 years	22	9	14	194	74	120
90 to 94 years	15	5	10	112	37	74
95 to 99 years	9	3	6	55	17	38
100 years and over	8	2	6	28	8	20
16 years and over	1,992	965	1,027	18,378	8,763	9,615
18 years and over	1,892	914	977	17,639	8,384	9,256
15 to 44 years	1,229	609	620	10,713	5,278	5,435
65 years and over	320	138	182	2,887	1,231	1,656
85 years and over	55	19	36	389	136	253
Median age	29.6	28.8	30.3	33.4	32.1	34.6
Mean age	33.2	32.3	34.1	35.6	34.4	36.7

Source: "Projections of the Population, by Age, Sex, Race, and Hispanic Origin, for the United States, 1993 to 2020, (Middle Series)." Jennifer Cheeseman Day, *Population Projections of the United States, by Age, Sex, Race, and Hispanic Origin: 1993 to 2050*, P25-1104, p. 45.

★ 996 ★

Population Trends and Projections

Projections: Age Range, Gender, and Race, July 1, 2030. Part I.

Date and age	[126]Race White Total	Male	Female	Black Total	Male	Female	American Indian, Eskimo, and Aleut Total	Male	Female	Asian and Pacific Islander Total	Male	Female
July 1, 2030												
All ages	267,457	132,230	135,226	50,596	23,947	26,649	3,473	1,709	1,765	28,467	13,804	14,663
Under 5 years	16,073	8,257	7,816	4,277	2,190	2,087	303	152	151	2,036	1,036	1,001
5 to 9 years	16,294	8,371	7,923	4,167	2,135	2,032	298	150	147	2,036	1,041	995
10 to 14 years	16,710	8,587	8,123	4,211	2,159	2,053	319	161	158	2,183	1,115	1,068
15 to 19 years	16,862	8,681	8,181	4,218	2,163	2,056	309	154	154	2,147	1,092	1,055
20 to 24 years	16,219	8,303	7,916	3,795	1,891	1,905	280	139	141	2,105	1,056	1,049
25 to 29 years	15,636	7,908	7,728	3,427	1,652	1,776	254	126	128	5,153	1,060	1,093
30 to 34 years	16,359	8,231	8,128	3,339	1,574	1,765	232	115	118	2,172	1,057	1,115
35 to 39 years	17,410	8,746	8,664	3,392	1,593	1,799	231	115	117	2,082	1,016	1,066
40 to 44 years	16,709	8,353	8,356	3,070	1,431	1,639	223	111	112	1,923	929	994
45 to 49 years	15,800	7,848	7,952	2,750	1,256	1,494	193	95	98	1,761	847	915
50 to 54 years	14,303	7,051	7,253	2,438	1,089	1,349	153	75	78	1,538	728	810
55 to 59 years	14,610	7,159	7,450	2,378	1,039	1,339	140	70	70	1,430	673	757
60 to 64 years	15,179	7,428	7,751	2,306	987	1,319	124	62	62	1,263	591	672
65 to 69 years	16,567	8,036	8,531	2,352	1,016	1,336	119	57	62	1,093	500	593
70 to 74 years	14,991	7,230	7,761	1,862	792	1,070	100	46	54	899	403	496
75 to 79 years	11,896	5,645	6,252	1,310	537	773	76	34	42	683	293	390
80 to 84 years	8,195	3,668	4,527	687	256	431	50	22	28	450	188	262
85 to 89 years	4,326	1,726	2,599	311	106	206	29	11	18	253	96	157
90 to 94 years	2,089	714	1,375	163	48	115	19	7	12	145	50	94
95 to 99 years	866	231	635	82	21	61	11	3	8	73	22	50
100 years and over	365	59	12	46	10	3	8	43	12	32		
16 years and over	215,039	105,295	109,743	37,087	17,024	20,063	2,489	1,213	1,277	21,767	10,385	11,382
18 years and over	208,294	101,812	106,482	35,401	16,153	19,248	2,364	1,149	1,214	20,897	9,940	10,957
15 to 44 years	99,194	50,221	48,973	21,243	10,303	10,940	1,529	760	769	12,581	6,210	6,372
65 years and over	59,295	27,309	31,986	6,827	2,789	4,037	415	184	232	3,638	1,564	2,074
85 years and over	7,646	2,730	4,916	615	187	427	69	24	45	513	180	333
Median age	40.6	39.5	41.8	31.8	27.4	34.0	29.5	28.9	30.1	33.6	32.4	34.8
Mean age	41.6	40.6	42.7	34.8	33.0	36.4	33.4	32.6	34.2	35.9	34.7	37.1

Source: "Projections of the Population, by Age, Sex, Race, and Hispanic Origin, for the United States, 1993 to 2020, (Middle Series)." Jennifer Cheeseman Day, *Population Projections of the United States, by Age, Sex, Race, and Hispanic Origin: 1993 to 2050*, P25-1104, p. 46.

★ 997 ★

Population Trends and Projections

Projections: Age Range, Gender, and Race, July 1, 2030. Part II - A

Date and age	Hispanic origin[1] Total	Male	Female	Not of Hispanic origin White Total	Male	Female	Black Total	Male	Female
July 1, 2030									
All ages	62,810	31,437	31,372	210,480	103,640	106,840	46,934	22,147	24,787
Under 5 years	5,716	2,929	2,787	10,883	5,598	5,285	3,941	2,016	1,925
5 to 9 years	5,478	2,808	2,670	11,326	5,826	5,501	3,843	1,967	1,875
10 to 14 years	5,344	2,739	2,605	11,886	6,115	5,771	3,891	1,993	1,898
15 to 19 years	5,175	2,659	2,516	12,195	6,281	5,914	3,903	2,000	1,904
20 to 24 years	4,871	2,500	2,370	11,816	6,038	5,778	3,506	1,744	1,762
25 to 29 years	4,531	2,298	2,233	11,538	5,822	5,716	3,163	1,522	1,641
30 to 34 years	4,420	2,220	2,199	12,350	6,207	6,144	3,087	1,453	1,634

[Continued]

★ 997 ★

Projections: Age Range, Gender, and Race, July 1, 2030. Part II - A
[Continued]

Date and age	Hispanic origin[1]			Not of Hispanic origin					
				White			Black		
	Total	Male	Female	Total	Male	Female	Total	Male	Female
35 to 39 years	4,300	2,151	2,149	13,513	6,787	6,726	3,147	1,476	1,671
40 to 44 years	3,775	1,875	1,900	13,300	6,651	6,649	2,847	1,325	1,522
45 to 49 years	3,285	1,614	1,671	12,819	6,375	6,444	2,566	1,171	1,395
50 to 54 years	2,911	1,415	1,496	11,651	5,753	5,898	2,277	1,016	1,261
55 to 59 years	2,720	1,336	1,384	12,125	5,931	6,194	2,230	972	1,258
60 to 64 years	2,670	1,341	1,330	12,742	6,198	6,545	2,154	915	1,239
65 to 69 years	2,471	1,217	1,255	14,313	6,924	7,390	2,203	944	1,259
70 to 74 years	1,937	935	1,002	13,224	6,376	6,848	1,745	735	1,010
75 to 79 years	1,402	659	744	10,618	5,044	5,574	1,224	496	729
80 to 84 years	899	398	501	7,370	3,303	4,067	637	233	404
85 to 89 years	482	196	286	3,880	1,545	2,335	287	95	192
90 to 94 years	255	95	161	1,853	627	1,227	150	43	107
95 to 99 years	115	39	77	759	195	564	76	19	57
100 years and over	53	15	38	315	44	271	56	12	44
16 years and over	45,230	22,427	22,803	173,979	84,862	89,117	34,471	15,766	18,705
18 years and over	43,156	21,358	21,797	169,105	82,343	86,762	32,910	14,960	17,950
15 to 44 years	27,070	13,703	13,367	74,712	37,786	36,926	19,654	9,520	10,134
65 years and over	7,616	3,553	4,062	52,333	24,058	28,275	6,379	2,577	3,802
85 years and over	906	344	562	6,808	2,412	4,396	569	169	400
Median age	30.3	29.5	31.1	43.7	42.4	45.0	32.0	29.5	34.3
Mean age	33.7	33.0	34.4	43.8	42.7	44.9	34.9	33.0	36.6

Source: "Projections of the Population, by Age, Sex, Race, and Hispanic Origin, for the United States, 1993 to 2020, (Middle Series)." Jennifer Cheeseman Day, *Population Projections of the United States, by Age, Sex, Race, and Hispanic Origin: 1993 to 2050,* P25-1104, p. 47. *Note:* 1. Persons of Hispanic origin may be of any race.

★ 998 ★

Population Trends and Projections

Projections: Age Range, Gender, and Race, July 1, 2030. Part II - B

Date and age	Not of Hispanic origin					
	American Indian, Eskimo, and Aleut			Asian and Pacific Islander		
	Total	Male	Female	Total	Male	Female
July 1, 2030						
All ages	2,960	1,449	1,510	26,810	13,016	13,794
Under 5 years	260	131	129	1,890	962	928
5 to 9 years	254	128	126	1,891	968	924
10 to 14 years	271	137	134	2,031	1,038	992
15 to 19 years	261	131	131	2,002	1,020	983
20 to 24 years	237	117	119	1,969	989	980
25 to 29 years	217	107	109	2,022	996	1,025
30 to 34 years	198	98	101	2,046	997	1,049

[Continued]

★ 998 ★

Projections: Age Range, Gender, and Race, July 1, 2030. Part II - B
[Continued]

Date and age	Not of Hispanic origin					
	American Indian, Eskimo, and Aleut			Asian and Pacific Islander		
	Total	Male	Female	Total	Male	Female
35 to 39 years	196	97	99	1,959	959	1,001
40 to 44 years	186	92	94	1,816	880	936
45 to 49 years	164	81	83	1,671	806	865
50 to 54 years	130	64	66	1,462	694	768
55 to 59 years	120	60	61	1,362	643	719
60 to 64 years	104	51	53	1,201	563	638
65 to 69 years	101	47	54	1,043	478	565
70 to 74 years	86	39	47	860	386	474
75 to 79 years	66	29	37	655	281	374
80 to 84 years	44	19	25	432	181	251
85 to 89 years	26	10	16	244	93	151
90 to 94 years	17	6	11	140	49	91
95 to 99 years	10	3	7	71	22	49
100 years and over	10	2	8	42	11	31
16 years and over	2,119	1,026	1,093	20,584	9,836	10,748
18 years and over	2,012	972	1,040	19,773	9,420	10,352
15 to 44 years	1,295	642	653	11,815	5,841	5,973
65 years and over	361	157	204	3,487	1,501	1,986
85 years and over	64	22	42	496	175	322
Median age	29.5	28.8	30.3	33.9	32.7	35.1
Mean age	33.5	32.6	34.4	36.2	35.0	37.3

Source: "Projections of the Population, by Age, Sex, Race, and Hispanic Origin, for the United States, 1993 to 2020, (Middle Series)." Jennifer Cheeseman Day, *Population Projections of the United States, by Age, Sex, Race, and Hispanic Origin: 1993 to 2050*, P25-1104, p. 47.

★ 999 ★

Population Trends and Projections

Projections: Age Range, Gender, and Race, July 1, 2035. Part I.

Date and age	[126]Race White			Black			American Indian, Eskimo, and Aleut			Asian and Pacific Islander		
	Total	Male	Female	Total	Male	Female	Total	Male	Female	Total	Male	Female
July 1, 2035												
All ages	272,619	134,701	137,917	53,224	25,204	28,020	3,679	1,809	1,871	31,453	15,270	16,183
Under 5 years	16,213	8,326	7,887	4,497	2,304	2,194	323	162	161	2,190	1,113	1,077
5 to 9 years	16,255	8,348	7,907	4,348	2,229	2,119	314	159	156	2,214	1,131	1,083
10 to 14 years	16,794	8,627	8,167	4,373	2,242	2,131	330	167	164	2,392	1,221	1,171
15 to 19 years	17,334	8,920	8,413	4,419	2,267	2,153	319	159	159	2,352	1,195	1,157
20 to 24 years	16,929	8,666	8,263	4,061	2,025	2,036	299	148	150	2,295	1,151	1,144
25 to 29 years	15,939	8,061	7,878	3,634	1,753	1,880	279	138	140	2,332	1,150	1,182
30 to 34 years	16,022	8,062	7,960	3,437	1,622	1,815	250	124	127	2,353	1,148	1,206
35 to 39 years	16,554	8,317	8,237	3,373	1,586	1,786	227	112	114	2,296	1,125	1,171
40 to 44 years	17,583	8,794	8,789	3,3277	1,577	1,799	227	113	115	2,178	1,064	1,114
45 to 49 years	16,417	8,152	8,266	2,969	1,361	1,607	210	104	106	1,914	928	986
50 to 54 years	15,675	7,723	7,952	2,686	1,203	1,483	182	89	93	1,746	833	913
55 to 59 years	14,112	6,917	7,195	2,375	1,038	1,337	141	69	73	1,506	707	799

[Continued]

★ 999 ★

Projections: Age Range, Gender, and Race, July 1, 2035. Part I.

[Continued]

Date and age	[126]Race White Total	Male	Female	Black Total	Male	Female	American Indian, Eskimo, and Aleut Total	Male	Female	Asian and Pacific Islander Total	Male	Female
60 to 64 years	14,273	6,976	7,298	2,301	990	1,312	129	64	65	1,406	655	751
65 to 69 years	14,551	7,050	7,500	2,202	950	1,252	111	54	57	1,213	555	658
70 to 74 years	15,300	7,413	7,887	1,964	839	1,125	105	50	55	1,030	468	562
75 to 79 years	13,472	6,458	7,014	1,560	648	912	87	40	47	809	355	454
80 to 84 years	9,536	4,326	5,210	887	334	553	61	27	34	557	236	322
85 to 89 years	5,599	2,285	3,315	404	138	265	37	15	23	333	125	207
90 to 94 years	2,652	935	1,717	195	58	137	23	8	15	184	64	120
95 to 99 years	1,002	279	724	93	24	69	13	4	9	92	30	62
100 years and over	406	67	339	67	14	53	12	3	9	59	16	43
16 years and over	219,955	107,650	112,305	39,117	17,972	21,145	2,645	1,287	1,358	24,169	11,556	12,613
18 years and over	213,039	104,080	108,959	37,354	17,061	20,293	2,515	1,222	1,293	23,215	11,068	12,147
15 to 44 years	100,361	50,820	49,541	22,301	10,831	11,470	1,600	795	806	13,807	6,833	6,974
65 years and over	62,518	28,813	33,705	7,373	3,005	4,367	449	201	248	2,278	1,849	2,428
85 years and over	9,659	3,565	6,094	759	235	525	85	30	55	668	235	433
Median age	41.2	40.0	42.5	31.8	29.4	34.1	29.6	28.9	30.2	34.2	32.9	35.3
Mean age	42.1	41.0	43.2	35.0	33.2	36.7	33.7	32.9	34.5	36.5	35.3	37.6

Source: "Projections of the Population, by Age, Sex, Race, and Hispanic Origin, for the United States, 1993 to 2020, (Middle Series)." Jennifer Cheeseman Day, *Population Projections of the United States, by Age, Sex, Race, and Hispanic Origin: 1993 to 2050*, P25-1104, p. 48.

★ 1000 ★

Population Trends and Projections

Projections: Age Range, Gender, and Race, July 1, 2035. Part II - A

Date and age	Hispanic origin[1] Total	Male	Female	Not of Hispanic origin White Total	Male	Female	Black Total	Male	Female
July 1, 2035									
All ages	68,877	34,430	34,447	210,171	103,406	106,765	49,186	23,222	25,963
Under 5 years	6,134	3,144	2,991	10,649	5,475	5,174	4,133	2,115	2,018
5 to 9 years	5,858	3,002	2,855	10,949	5,629	5,320	3,999	2,048	1,951
10 to 14 years	5,742	2,943	2,799	11,611	5,971	5,640	4,028	2,064	1,964
15 to 19 years	5,652	2,905	2,748	12,233	6,298	5,935	4,074	2,088	1,986
20 to 24 years	5,391	2,767	2,624	12,059	6,161	5,898	3,739	1,861	1,877
25 to 29 years	4,932	2,500	2,432	11,489	5,797	5,692	3,342	1,610	1,732
30 to 34 years	4,706	2,365	2,341	11,763	5,911	5,851	3,166	1,492	1,674
35 to 39 years	4,506	2,254	2,252	12,468	6,264	6,204	3,114	1,463	1,651
40 to 44 years	4,361	2,167	2,194	13,629	6,821	6,808	3,129	1,460	1,669
45 to 49 years	3,723	1,832	1,891	13,053	6,488	6,565	2,749	1,259	1,490
50 to 54 years	3,270	1,589	1,681	12,704	6,271	6,433	2,502	1,120	1,383
55 to 59 years	2,884	1,388	1,496	11,482	5,643	5,839	2,214	966	1,248
60 to 64 years	2,676	1,304	1,372	11,829	5,776	6,053	2,152	923	1,230
65 to 69 years	2,591	1,286	1,305	12,192	5,874	6,318	2,047	876	1,171
70 to 74 years	2,307	1,133	1,174	13,191	6,375	6,816	1,829	773	1,056
75 to 79 years	1,779	854	925	11,848	5,678	6,170	1,451	594	857
80 to 84 years	1,169	531	639	8,464	3,840	4,625	820	302	518
85 to 89 years	657	274	383	4,993	2,032	2,961	369	123	246
90 to 94 years	327	124	203	2,350	821	1,529	178	52	127

[Continued]

★ 1000 ★

Projections: Age Range, Gender, and Race, July 1, 2035. Part II - A

[Continued]

Date and age	Hispanic origin[1]			Not of Hispanic origin					
				White			Black		
	Total	Male	Female	Total	Male	Female	Total	Male	Female
95 to 99 years	143	49	94	870	233	637	86	21	65
100 years and over	67	20	48	343	49	294	64	13	51
16 years and over	50,013	24,760	25,253	174,577	85,103	89,474	36,207	16,575	19,632
18 years and over	47,752	23,595	24,157	169,700	82,584	87,117	34,582	15,736	18,846
15 to 44 years	29,548	14,957	14,591	73,641	37,251	36,389	20,564	9,974	10,590
65 years and over	9,041	4,270	4,771	54,252	24,902	29,350	6,844	2,754	4,090
85 years and over	1,195	466	729	8,556	3,135	5,422	697	209	488
Median age	30.8	29.9	31.6	44.4	43.1	45.6	32.0	29.5	34.4
Mean age	34.2	33.5	35.0	44.4	43.3	45.6	35.1	33.2	36.9

Source: "Projections of the Population, by Age, Sex, Race, and Hispanic Origin, for the United States, 1993 to 2020, (Middle Series)."
Jennifer Cheeseman Day, *Population Projections of the United States, by Age, Sex, Race, and Hispanic Origin: 1993 to 2050*, P25-1104, p. 49.
Note: 1. Persons of Hispanic origin may be of any race.

★ 1001 ★

Population Trends and Projections

Projections: Age Range, Gender, and Race, July 1, 2035. Part II - B

Date and age	Not of Hispanic origin					
	American Indian, Eskimo, and Aleut			Asian and Pacific Islander		
	Total	Male	Female	Total	Male	Female
July 1, 2035						
All ages	3,132	1,533	1,599	29,610	14,393	15,217
Under 5 years	277	139	138	2,030	1,032	998
5 to 9 years	269	136	134	2,057	1,052	1,005
10 to 14 years	282	142	140	2,226	1,138	1,089
15 to 19 years	271	135	136	2,192	1,115	1,077
20 to 24 years	253	125	127	2,143	1,077	1,067
25 to 29 years	236	117	119	2,186	1,079	1,107
30 to 34 years	213	105	108	2,215	1,082	1,133
35 to 39 years	193	96	98	2,168	1,065	1,103
40 to 44 years	192	95	97	2,054	1,006	1,048
45 to 49 years	175	87	88	1,810	880	930
50 to 54 years	154	75	79	1,659	793	865
55 to 59 years	120	58	62	1,434	675	758
60 to 64 years	111	54	57	1,341	626	715
65 to 69 years	93	44	49	1,155	530	625
70 to 74 years	89	41	47	984	448	536
75 to 79 years	74	33	41	775	341	434
80 to 84 years	53	23	30	535	227	308
85 to 89 years	33	13	20	320	121	200
90 to 94 years	21	7	13	178	62	116

[Continued]

★ 1001 ★

Projections: Age Range, Gender, and Race, July 1, 2035. Part II - B

[Continued]

Date and age	Not of Hispanic origin					
	American Indian, Eskimo, and Aleut			Asian and Pacific Islander		
	Total	Male	Female	Total	Male	Female
95 to 99 years	12	4	8	89	29	60
100 years and over	12	3	9	58	16	42
16 years and over	2,247	1,087	1,159	22,842	10,939	11,903
18 years and over	2,137	1,032	1,105	21,953	10,484	11,469
15 to 44 years	1,358	673	685	12,959	6,424	6,535
65 years and over	386	169	217	4,095	1,773	2,322
85 years and over	78	27	51	645	227	418
Median age	29.5	28.8	30.3	34.5	33.3	35.6
Mean age	33.7	32.8	34.6	36.7	35.5	37.8

Source: "Projections of the Population, by Age, Sex, Race, and Hispanic Origin, for the United States, 1993 to 2020, (Middle Series)." Jennifer Cheeseman Day, *Population Projections of the United States, by Age, Sex, Race, and Hispanic Origin: 1993 to 2050*, P25-1104, p. 49.

★ 1002 ★

Population Trends and Projections

Projections: Age Range, Gender, and Race, July 1, 2040. Part I.

Date and age	[126]Race White			Black			American Indian, Eskimo, and Aleut			Asian and Pacific Islander		
	Total	Male	Female	Total	Male	Female	Total	Male	Female	Total	Male	Female
July 1, 2040												
All ages	277,232	136,884	140,348	55,917	26,502	29,416	3,894	1,913	1,982	34,461	16,750	17,711
Under 5 years	16,572	8,508	8,064	4,731	2,424	2,307	341	171	170	2,334	1,185	1,149
5 to 9 years	16,402	8,422	7,980	4,573	2,345	2,228	335	169	166	2,375	1,213	1,163
10 to 14 years	16,761	8,608	8,153	4,565	2,342	2,223	349	176	173	2,592	1,322	1,270
15 to 19 years	17,429	8,968	8,462	4,590	2,356	2,235	331	165	165	2,564	1,302	1,262
20 to 24 years	17,400	8,906	8,494	4,256	2,124	2,132	308	153	155	2,497	1,252	1,244
25 to 29 years	16,628	8,409	8,219	3,888	1,879	2,009	298	148	150	2,525	1,246	1,278
30 to 34 years	16,336	8,220	8,116	3,645	1,723	1,922	274	135	139	2,533	1,238	1,295
35 to 39 years	16,226	8,155	8,071	3,475	1,638	1,837	245	121	123	2,477	1,217	1,260
40 to 44 years	16,734	8,372	8,362	3,362	1,574	1,788	223	110	112	2,393	1,173	1,219
45 to 49 years	17,283	8,586	8,697	3,268	1,503	1,766	214	106	108	2,159	1,057	1,102
50 to 54 years	16,302	8,030	8,273	2,905	1,308	1,598	198	97	101	1,894	910	983
55 to 59 years	15,488	7,591	7,896	2,619	1,149	1,471	169	81	87	1,704	807	897
60 to 64 years	13,812	6,756	7,056	2,300	990	1,309	131	63	68	1,480	688	792
65 to 69 years	13,716	6,643	7,073	2,205	957	1,248	116	56	60	1,347	615	732
70 to 74 years	13,497	6,544	6,953	1,847	790	1,057	99	48	51	1,142	520	622
75 to 79 years	13,846	6,687	7,159	1,656	692	964	91	43	49	929	414	515
80 to 84 years	10,893	5,008	5,885	1,059	404	655	69	31	38	662	287	375
85 to 89 years	6,595	2,738	3,858	522	181	341	45	18	27	414	158	256
90 to 94 years	3,512	1,273	2,238	256	77	178	29	10	18	245	85	160
95 to 99 years	1,314	378	936	114	29	85	16	5	11	119	38	81
100 years and over	486	83	403	80	16	64	15	4	11	79	22	57
16 years and over	224,099	109,599	114,500	41,126	18,916	22,210	2,800	1,361	1,439	26,629	12,759	13,869
18 years and over	217,161	106,018	111,143	39,296	17,970	21,326	2,665	1,293	1,372	25,589	12,228	13,361
15 to 44 years	100,753	51,030	49,723	23,216	11,293	11,923	1,679	833	845	14,987	7,428	7,559
65 years and over	63,858	29,354	34,505	7,739	3,148	4,591	480	215	264	4,936	2,139	2,797
85 years and over	11,907	4,471	7,435	972	304	668	104	37	67	856	303	553

[Continued]

★ 1002 ★

Projections: Age Range, Gender, and Race, July 1, 2040. Part I.

[Continued]

Date and age	[126]Race White			Black			American Indian, Eskimo, and Aleut			Asian and Pacific Islander		
	Total	Male	Female	Total	Male	Female	Total	Male	Female	Total	Male	Female
Median age	41.5	40.1	42.8	31.8	29.4	34.1	29.8	29.1	30.4	34.6	33.5	35.7
Mean age	42.4	41.3	43.5	35.1	33.3	36.8	33.9	33.0	34.7	37.0	35.8	38.1

Source: "Projections of the Population, by Age, Sex, Race, and Hispanic Origin, for the United States, 1993 to 2020, (Middle Series)." Jennifer Cheeseman Day, *Population Projections of the United States, by Age, Sex, Race, and Hispanic Origin: 1993 to 2050*, P25-1104, p. 50.

★ 1003 ★

Population Trends and Projections

Projections: Age Range, Gender, and Race, July 1, 2040. Part II - A

Date and age	Hispanic origin[1]			Not of Hispanic origin					
				White			Black		
	Total	Male	Female	Total	Male	Female	Total	Male	Female
July 1, 2040									
All ages	75,130	37,509	37,621	209,148	102,807	106,341	51,489	24,332	27,157
Under 5 years	6,584	3,374	3,210	10,604	5,450	5,154	4,337	2,220	2,117
5 to 9 years	6,278	3,218	3,060	10,721	5,510	5,211	4,196	2,150	2,046
10 to 14 years	6,132	3,144	2,989	11,231	5,774	5,458	4,193	2,149	2,044
15 to 19 years	6,063	3,116	2,947	11,960	6,155	5,804	4,219	2,163	2,056
20 to 24 years	5,863	3,009	2,854	12,101	6,181	5,920	3,904	1,946	1,959
25 to 29 years	5,432	2,753	2,679	11,729	5,918	5,811	3,564	1,719	1,845
30 to 34 years	5,112	2,566	2,543	11,720	5,891	5,830	3,345	1,579	1,766
35 to 39 years	4,795	2,400	2,395	11,887	5,973	5,913	3,197	1,504	1,692
40 to 44 years	4,571	2,271	2,299	12,588	6,303	6,285	3,101	1,450	1,651
45 to 49 years	4,297	2,116	2,181	13,384	6,657	6,727	3,024	1,389	1,636
50 to 54 years	3,705	1,803	1,901	12,951	6,390	6,561	2,686	1,207	1,479
55 to 59 years	3,239	1,559	1,680	12,542	6,164	6,377	2,436	1,067	1,369
60 to 64 years	2,840	1,357	1,483	11,223	5,510	5,713	2,138	919	1,219
65 to 69 years	2,601	1,254	1,347	11,345	5,493	5,852	2,052	888	1,164
70 to 74 years	2,423	1,201	1,222	11,285	5,443	5,842	1,706	722	983
75 to 79 years	2,125	1,039	1,085	11,903	5,735	6,168	1,529	629	900
80 to 84 years	1,488	692	796	9,528	4,374	5,154	974	363	610
85 to 89 years	860	368	492	5,804	2,399	3,405	476	160	316
90 to 94 years	451	175	275	3,096	1,112	1,984	232	68	164
95 to 99 years	187	65	122	1,141	317	824	105	26	79
100 years and over	86	25	60	407	59	348	76	15	61
16 years and over	54,929	27,153	27,776	174,280	84,884	89,396	37,916	17,377	20,538
18 years and over	52,507	25,905	26,602	169,526	82,428	87,098	36,233	16,508	19,725
15 to 44 years	31,835	16,117	15,718	71,984	36,421	35,563	21,330	10,361	10,969
65 years and over	10,220	4,821	5,399	54,509	24,932	29,577	7,149	2,871	4,278
85 years and over	1,583	634	949	10,448	3,887	6,561	888	268	619

[Continued]

★ 1003 ★

Projections: Age Range, Gender, and Race, July 1, 2040. Part II - A

[Continued]

| Date and age | Hispanic origin[1] | | | Not of Hispanic origin | | | | | |
| | | | | White | | | Black | | |
	Total	Male	Female	Total	Male	Female	Total	Male	Female
Median age	31.1	30.3	32.1	45.0	43.7	46.4	31.9	29.5	34.3
Mean age	34.7	34.0	35.5	44.9	43.7	46.1	35.2	33.3	37.0

Source: "Projections of the Population, by Age, Sex, Race, and Hispanic Origin, for the United States, 1993 to 2020, (Middle Series)."
Jennifer Cheeseman Day, *Population Projections of the United States, by Age, Sex, Race, and Hispanic Origin: 1993 to 2050*, P25-1104, p. 51.
Note: 1. Persons of Hispanic origin may be of any race.

★ 1004 ★

Population Trends and Projections

Projections: Age Range, Gender, and Race, July 1, 2040. Part II - B

| Date and age | Not of Hispanic origin | | | | | |
| | American Indian, Eskimo, and Aleut | | | Asian and Pacific Islander | | |
	Total	Male	Female	Total	Male	Female
July 1, 2040						
All ages	3,314	1,621	1,693	32,424	15,780	16,644
Under 5 years	293	147	146	2,160	1,098	1,062
5 to 9 years	287	145	142	2,204	1,126	1,078
10 to 14 years	299	151	148	2,412	1,231	1,180
15 to 19 years	282	141	141	2,390	1,215	1,175
20 to 24 years	262	130	132	2,330	1,170	1,160
25 to 29 years	252	125	127	2,362	1,167	1,194
30 to 34 years	232	114	118	2,380	1,165	1,215
35 to 39 years	208	103	105	2,336	1,151	1,186
40 to 44 years	190	94	96	2,263	1,13	1,150
45 to 49 years	181	89	92	2,039	1,001	1,037
50 to 54 years	164	80	84	1,793	865	928
55 to 59 years	142	68	74	1,620	769	851
60 to 64 years	111	53	58	1,410	658	752
65 to 69 years	99	48	51	1,286	588	697
70 to 74 years	82	39	43	1,089	497	592
75 to 79 years	77	35	42	888	397	491
80 to 84 years	59	26	33	635	276	359
85 to 89 years	39	15	24	398	152	246
90 to 94 years	26	9	17	236	82	154
95 to 99 years	15	5	10	115	37	78
100 years and over	14	4	10	77	21	56
16 years and over	2,376	1,149	1,227	25,153	12,072	13,080
18 years and over	2,261	1,091	1,170	24,183	11,576	12,607
15 to 44 years	1,425	706	719	14,061	6,981	7,081
65 years and over	412	181	230	4,724	2,051	2,674
85 years and over	94	33	61	827	293	534

[Continued]

★ 1004 ★

Projections: Age Range, Gender, and Race, July 1, 2040. Part II - B
[Continued]

| Date and age | Not of Hispanic origin | | | | | |
| | American Indian, Eskimo, and Aleut | | | Asian and Pacific Islander | | |
	Total	Male	Female	Total	Male	Female
Median age	29.7	28.9	30.4	34.9	33.8	36.1
Mean age	33.9	32.9	34.8	37.2	36.0	38.3

Source: "Projections of the Population, by Age, Sex, Race, and Hispanic Origin, for the United States, 1993 to 2020, (Middle Series)." Jennifer Cheeseman Day, *Population Projections of the United States, by Age, Sex, Race, and Hispanic Origin: 1993 to 2050*, P25-1104, p. 51.

★ 1005 ★

Population Trends and Projections

Projections: Age Range, Gender, and Race, July 1, 2045. Part I.

| Date and age | [126]Race White | | | Black | | | American Indian, Eskimo, and Aleut | | | Asian and Pacific Islander | | |
	Total	Male	Female	Total	Male	Female	Total	Male	Female	Total	Male	Female
July 1, 2045												
All ages	281,486	138,916	142,571	58,695	27,855	30,840	4,116	2,020	2,096	37,481	18,241	19,240
Under 5 years	16,959	8,706	8,254	4,953	2,540	2,413	356	179	177	2,477	1,258	1,220
5 to 9 years	16,763	8,605	8,157	4,808	2,467	2,341	353	178	175	2,526	1,289	1,237
10 to 14 years	16,913	8,685	8,228	4,797	2,462	2,335	372	187	184	2,773	1,414	1,359
15 to 19 years	17,398	8,950	8,448	4,788	2,459	2,329	350	175	175	2,765	1,403	1,362
20 to 24 years	17,495	8,954	8,541	4,419	2,208	2,211	320	159	161	2,705	1,357	1,348
25 to 29 years	17,085	8,640	8,445	4,072	1,970	2,101	308	153	155	2,728	1,348	1,379
30 to 34 years	17,034	8,572	8,462	3,897	1,845	2,051	293	145	148	2,725	1,334	1,391
35 to 39 years	16,547	8,319	8,228	3,684	1,740	1,944	268	133	135	2,656	1,308	1,348
40 to 44 years	16,412	8,215	8,196	3,465	1,626	1,839	240	119	121	2,574	1,266	1,308
45 to 49 years	16,460	8,180	8,279	3,256	1,501	1,754	210	104	106	2,366	1,164	1,202
50 to 54 years	17,169	8,462	8,707	3,200	1,445	1,756	202	99	103	2,130	1,035	1,094
55 to 59 years	16,118	7,900	8,217	2,840	1,252	1,587	183	89	94	1,844	881	964
60 to 64 years	15,180	7,431	7,748	2,544	1,102	1,443	156	75	82	1,668	782	886
65 to 69 years	13,332	6,473	6,859	2,217	966	1,252	118	56	62	1,417	646	771
70 to 74 years	12,770	6,200	6,570	1,853	800	1,053	103	50	53	1,267	576	691
75 to 79 years	12,294	5,956	6,338	1,566	656	909	86	41	45	1,032	462	570
80 to 84 years	11,278	5,237	6,041	1,126	433	693	73	34	39	762	336	426
85 to 89 years	7,630	3,222	4,408	625	220	405	51	21	31	494	194	300
90 to 94 years	4,214	1,563	2,651	333	102	231	35	13	22	307	108	199
95 to 99 years	1,787	530	1,258	153	40	113	20	7	14	160	552	109
100 years and over	648	113	535	100	20	80	18	5	13	105	29	75
16 years and over	227,452	111,172	116,280	43,172	19,889	23,283	2,960	1,438	1,523	29,133	13,990	15,143
18 years and over	220,528	107,598	112,929	41,261	18,900	22,361	2,818	1,366	1,452	28,012	13,418	14,595
15 to 44 years	101,971	51,651	50,320	24,324	11,849	12,475	1,779	883	895	16,152	8,016	8,136
65 years and over	63,954	29,295	34,659	7,973	3,237	4,737	504	225	279	5,545	2,403	3,142
85 years and over	14,280	5,428	8,852	1,211	382	829	124	45	80	1,066	383	683
Median age	41.4	40.0	42.8	31.9	29.6	34.1	30.0	29.3	30.7	35.1	34.0	36.2
Mean age	42.6	41.4	43.7	35.2	33.3	36.9	34.0	33.2	34.9	37.4	36.3	38.6

Source: "Projections of the Population, by Age, Sex, Race, and Hispanic Origin, for the United States, 1993 to 2020, (Middle Series)." Jennifer Cheeseman Day, *Population Projections of the United States, by Age, Sex, Race, and Hispanic Origin: 1993 to 2050*, P25-1104, p. 52.

★ 1006 ★

Population Trends and Projections

Projections: Age Range, Gender, and Race, July 1, 2045. Part II - A

Date and age	Hispanic origin[1]			Not of Hispanic origin					
				White			Black		
	Total	Male	Female	Total	Male	Female	Total	Male	Female
July 1, 2045									
All ages	81,539	40,662	40,877	207,627	101,992	105,636	53,866	25,491	28,375
Under 5 years	7,022	3,599	3,423	10,595	5,444	5,151	4,532	2,321	2,211
5 to 9 years	6,728	3,449	3,278	10,677	5,487	5,191	4,401	2,256	2,145
10 to 14 years	6,562	3,365	3,196	11,001	5,654	5,347	4,396	2,255	2,141
15 to 19 years	6,462	3,321	3,141	11,573	5,955	5,618	4,390	2,252	2,137
20 to 24 years	6,268	3,216	3,051	11,831	6,042	5,790	4,042	2,017	2,026
25 to 29 years	5,886	2,983	2,903	11,775	5,941	5,834	3,720	1,797	1,924
30 to 34 years	5,615	2,822	2,794	11,966	6,015	5,951	3,565	1,686	1,880
35 to 39 years	5,203	2,605	2,598	11,848	5,956	5,892	3,377	1,593	1,785
40 to 44 years	4,863	2,419	2,444	12,009	6,016	5,993	3,183	1,492	1,691
45 to 49 years	4,503	2,218	2,285	12,371	6,158	6,213	2,998	1,380	1,617
50 to 54 years	4,273	2,082	2,191	13,288	6,562	6,726	2,957	1,332	1,624
55 to 59 years	3,668	1,769	1,899	12,796	6,289	6,507	2,621	1,153	1,468
60 to 64 years	3,190	1,525	1,665	12,279	6,034	6,244	2,360	1,020	1,340
65 to 69 years	2,766	1,309	1,457	10,816	5,275	5,541	2,052	891	1,160
70 to 74 years	2,437	1,174	1,263	10,543	5,120	5,423	1,714	737	977
75 to 79 years	2,240	1,108	1,132	10,248	4,941	5,307	1,433	591	842
80 to 84 years	1,782	846	936	9,640	4,459	5,181	1,026	385	641
85 to 89 years	1,100	484	616	6,617	2,777	3,840	565	192	373
90 to 94 years	595	239	357	3,666	1,344	2,322	301	88	213
95 to 99 years	261	94	167	1,546	443	1,104	139	34	104
100 years and over	114	34	79	543	81	462	94	19	76
16 years and over	59,939	29,588	30,351	173,113	84,254	88,859	39,652	18,204	21,448
18 years and over	57,357	28,257	29,100	168,515	81,880	86,635	37,900	17,299	20,602
15 to 44 years	34,297	17,366	16,930	71,002	35,924	35,077	22,278	10,836	11,442
65 years and over	11,296	5,288	6,008	53,619	24,440	29,179	7,324	2,938	4,387
85 years and over	2,070	851	1,219	12,372	4,645	7,728	1,100	333	766
Median age	31.6	30.7	32.6	45.2	43.8	46.7	32.0	29.6	34.3
Mean age	35.2	34.4	36.0	45.1	43.9	46.4	35.3	33.3	37.0

Source: "Projections of the Population, by Age, Sex, Race, and Hispanic Origin, for the United States, 1993 to 2020, (Middle Series)."
Jennifer Cheeseman Day, *Population Projections of the United States, by Age, Sex, Race, and Hispanic Origin: 1993 to 2050,* P25-1104, p. 53.
Note: 1. Persons of Hispanic origin may be of any race.

★ 1007 ★

Population Trends and Projections

Projections: Age Range, Gender, and Race, July 1, 2045. Part II - B

| Date and age | Not of Hispanic origin | | | | | |
| | American Indian, Eskimo, and Aleut | | | Asian and Pacific Islander | | |
	Total	Male	Female	Total	Male	Female
July 1, 2045						
All ages	3,503	1,712	1,791	35,243	17,174	18,069
Under 5 years	307	154	153	2,291	1,164	1,127
5 to 9 years	303	153	150	2,341	1,195	1,146
10 to 14 years	318	160	158	2,577	1,315	1,262
15 to 19 years	299	149	150	2,577	1,309	1,268
20 to 24 years	273	135	138	2,525	1,268	1,257
25 to 29 years	261	129	132	2,550	1,262	1,288
30 to 34 years	247	122	126	2,555	1,253	1,302
35 to 39 years	226	112	114	2,500	1,234	1,266
40 to 44 years	204	101	103	2,432	1,199	1,233
45 to 49 years	179	88	91	2,240	1,105	1,135
50 to 54 years	170	83	87	2,013	982	1,032
55 to 59 years	152	74	78	1,749	837	911
60 to 64 years	132	63	69	1,588	747	841
65 to 69 years	100	47	53	1,352	918	733
70 to 74 years	88	42	46	1,211	552	659
75 to 79 years	72	34	38	984	441	543
80 to 84 years	62	28	34	730	322	407
85 to 89 years	44	18	27	475	187	288
90 to 94 years	31	11	20	296	104	192
95 to 99 years	19	6	13	155	50	105
100 years and over	17	4	13	102	29	74
16 years and over	2,510	1,213	1,297	27,503	13,229	14,273
18 years and over	2,388	1,152	1,236	26,458	12,695	13,763
15 to 44 years	1,510	748	762	15,139	7,525	7,614
65 years and over	432	190	242	5,305	2,304	3,001
85 years and over	111	39	72	1,028	370	659
Median age	29.8	29.1	30.6	35.4	34.3	36.5
Mean age	34.0	33.0	34.9	37.7	36.5	38.8

Source: "Projections of the Population, by Age, Sex, Race, and Hispanic Origin, for the United States, 1993 to 2020, (Middle Series)."
Jennifer Cheeseman Day, *Population Projections of the United States, by Age, Sex, Race, and Hispanic Origin: 1993 to 2050*, P25-1104, p. 53.

★ 1008 ★

Population Trends and Projections

Projections: Age Range, Gender, and Race, July 1, 2050. Part I.

Date and age	[126]Race White Total	Male	Female	Black Total	Male	Female	American Indian, Eskimo, and Aleut Total	Male	Female	Asian and Pacific Islander Total	Male	Female
July 1, 2050												
All ages	285,591	140,947	144,644	61,586	29,279	32,307	4,346	2,131	2,215	40,508	19,741	20,768
Under 5 years	17,216	8,837	8,379	5,171	2,653	2,518	373	187	186	2,622	1,330	1,291
5 to 9 years	17,146	8,802	8,344	5,031	2,583	2,447	370	187	183	2,677	1,365	1,311
10 to 14 years	17,277	8,871	8,406	5,038	2,588	2,450	392	198	195	2,942	1,499	1,443
15 to 19 years	17,550	9,028	8,522	5,028	2,585	2,443	372	186	186	2,948	1,495	1,453
20 to 24 years	17,464	8,939	8,525	4,608	2,305	2,302	339	168	170	2,903	1,456	1,447
25 to 29 years	17,177	8,689	8,489	4,225	2,047	2,178	319	158	161	2,937	1,454	1,484
30 to 34 years	17,494	8,806	8,688	4,078	1,935	2,143	303	149	153	2,928	1,436	1,492
35 to 39 years	17,246	8,673	8,573	3,937	1,864	2,073	286	142	144	2,847	1,405	1,442
40 to 44 years	16,734	8,381	8,353	3,673	1,728	1,944	263	131	133	2,753	1,358	1,395
45 to 49 years	16,144	8,029	8,115	3,355	1,552	1,803	227	112	114	2,540	1,254	1,287
50 to 54 years	16,355	8,065	8,290	3,190	1,445	1,745	198	97	101	2,329	1,137	1,191
55 to 59 years	16,985	8,334	8,650	3,126	1,384	1,742	187	90	97	2,069	999	1,070
60 to 64 years	15,820	7,750	8,070	2,760	1,203	1,557	170	82	88	1,803	853	950
65 to 69 years	14,672	7,137	7,535	2,452	1,076	1,376	141	66	75	1,594	734	860
70 to 74 years	12,461	6,072	6,389	1,868	811	1,057	105	49	55	1,335	607	728
75 to 79 years	11,696	5,687	6,009	1,576	667	909	90	43	47	1,148	514	634
80 to 84 years	10,093	4,714	5,378	1,066	412	655	69	33	36	850	376	473
85 to 89 years	7,991	3,421	4,571	666	236	430	54	22	32	571	228	343
90 to 94 years	4,968	1,886	3,082	402	125	277	40	15	25	369	134	235
95 to 99 years	2,193	665	1,528	202	53	149	25	8	17	203	67	137
100 years and over	908	161	747	135	28	107	23	6	16	143	40	103
16 years and over	230,502	112,662	117,839	45,331	20,932	24,399	3,132	1,520	1,612	31,661	15,237	16,424
18 years and over	223,507	109,052	114,454	43,324	19,892	23,432	2,980	1,444	1,536	30,467	14,628	15,839
15 to 44 years	103,666	52,516	51,150	25,547	12,464	13,084	1,882	935	947	17,315	8,603	8,712
65 years and over	64,982	29,742	35,240	8,368	3,407	4,960	547	244	303	6,212	2,699	3,513
85 years and over	16,061	6,132	9,928	1,405	441	964	142	52	90	1,286	469	817
Median age	41.2	39.9	42.6	32.1	29.7	34.3	30.1	29.4	30.8	35.5	34.4	36.6
Mean age	42.6	41.4	43.8	35.3	33.4	37.0	34.2	33.3	35.0	37.9	36.7	39.0

Source: "Projections of the Population, by Age, Sex, Race, and Hispanic Origin, for the United States, 1993 to 2020, (Middle Series)." Jennifer Cheeseman Day, *Population Projections of the United States, by Age, Sex, Race, and Hispanic Origin: 1993 to 2050,* P25-1104, p. 54.

★ 1009 ★

Population Trends and Projections

Projections: Age Range, Gender, and Race, July 1, 2050. Part II - A

Date and age	Hispanic origin[1] Total	Male	Female	Not of Hispanic origin White Total	Male	Female	Black Total	Male	Female
July 1, 2050									
All ages	88,071	43,878	44,193	205,849	101,122	104,727	56,346	26,716	29,630
Under 5 years	7,429	3,808	3,620	10,485	5,387	5,098	4,723	2,421	2,302
5 to 9 years	7,166	3,675	3,491	10,665	5,480	5,185	4,596	2,357	2,238
10 to 14 years	7,022	3,601	3,421	10,954	5,629	5,324	4,607	2,364	2,243
15 to 19 years	6,903	3,548	3,354	11,333	5,831	5,501	4,599	2,362	2,237
20 to 24 years	6,663	3,419	3,243	11,448	5,845	5,603	4,205	2,100	2,105
25 to 29 years	6,274	3,180	3,094	11,517	5,811	5,706	3,849	1,862	1,988
30 to 34 years	6,072	3,052	3,020	12,011	6,039	5,972	3,720	1,761	1,958

[Continued]

★ 1009 ★

Projections: Age Range, Gender, and Race, July 1, 2050. Part II - A

[Continued]

Date and age	Hispanic origin[1]			Not of Hispanic origin					
				White			Black		
	Total	Male	Female	Total	Male	Female	Total	Male	Female
35 to 39 years	5,709	2,860	2,849	12,093	6,081	6,011	3,598	1,701	1,898
40 to 44 years	5,273	2,625	2,648	11,970	6,000	5,970	3,362	1,580	1,783
45 to 49 years	4,789	2,362	2,428	11,804	5,880	5,925	3,078	1,421	1,657
50 to 54 years	4,478	2,183	2,295	12,286	6,073	6,213	2,933	1,326	1,607
55 to 59 years	4,229	2,043	2,186	13,139	6,467	6,672	2,883	1,274	1,609
60 to 64 years	3,613	1,732	1,880	12,548	6,171	6,377	2,541	1,105	1,436
65 to 69 years	3,108	1,473	1,635	11,851	5,792	6,060	1,263	991	1,272
70 to 74 years	2,597	1,230	1,367	10,092	4,943	5,148	1,717	743	973
75 to 79 years	2,260	1,088	1,172	9,630	4,687	4,943	1,445	606	839
80 to 84 years	1,884	906	979	8,363	3,881	4,482	961	362	599
85 to 89 years	1,324	595	728	6,769	2,871	3,899	596	203	392
90 to 94 years	767	317	451	4,261	1,594	2,667	360	106	254
95 to 99 years	350	130	220	1,871	545	1,325	184	46	138
100 years and over	161	50	110	760	115	645	126	25	102
16 years and over	65,076	32,086	32,990	171,533	83,487	88,045	41,491	19,095	22,396
18 years and over	62,317	30,664	31,653	167,022	81,158	85,864	39,654	18,145	21,510
15 to 44 years	36,894	18,685	18,209	70,371	35,608	34,763	23,333	11,365	11,968
65 years and over	12,452	5,790	6,662	53,597	24,427	29,169	7,652	3,082	4,570
85 years and over	2,602	1,092	1,510	13,661	5,125	8,536	1,266	380	886
Median age	32.1	31.1	33.1	45.2	43.7	46.7	32.1	29.7	34.4
Mean age	35.6	34.7	36.4	45.3	44.0	46.6	35.3	33.4	37.1

Source: "Projections of the Population, by Age, Sex, Race, and Hispanic Origin, for the United States, 1993 to 2020, (Middle Series)." Jennifer Cheeseman Day, *Population Projections of the United States, by Age, Sex, Race, and Hispanic Origin: 1993 to 2050*, P25-1104, p. 55. *Note:* 1. Persons of Hispanic origin may be of any race.

★ 1010 ★

Population Trends and Projections

Projections: Age Range, Gender, and Race, July 1, 2050. Part II - B

Date and age	Not of Hispanic origin					
	American Indian, Eskimo, and Aleut			Asian and Pacific Islander		
	Total	Male	Female	Total	Male	Female
July 1, 2050						
All ages	3,701	1,808	1,893	38,064	18,573	19,491
Under 5 years	323	162	161	2,422	1,230	1,192
5 to 9 years	318	160	158	2,477	1,264	1,213
10 to 14 years	337	169	167	2,730	1,392	1,338
15 to 19 years	318	159	159	2,744	1,393	1,351
20 to 24 years	289	143	146	2,709	1,360	1,349
25 to 29 years	272	135	137	2,746	1,361	1,386
30 to 34 years	256	126	130	2,744	1,348	1,396

[Continued]

★ 1010 ★

Projections: Age Range, Gender, and Race, July 1, 2050. Part II - B
[Continued]

| Date and age | Not of Hispanic origin | | | | | |
| | American Indian, Eskimo, and Aleut | | | Asian and Pacific Islander | | |
	Total	Male	Female	Total	Male	Female
35 to 39 years	241	119	122	2,675	1,323	1,352
40 to 44 years	222	110	112	2,595	1,283	1,312
45 to 49 years	192	95	97	2,403	1,189	1,214
50 to 54 years	168	82	86	2,208	1,081	1,126
55 to 59 years	157	76				
60 to 64 years	141	67	73	1,711	812	899
65 to 69 years	118	55	63	1,518	701	817
70 to 74 years	89	42	47	1,274	581	693
75 to 79 years	77	37	40	1,097	493	605
80 to 84 years	58	27	31	812	360	451
85 to 89 years	46	19	27	548	220	328
90 to 94 years	35	13	22	356	129	226
95 to 99 years	23	7	15	197	65	132
100 years and over	22	6	16	140	39	101
16 years and over	2,656	1,283	1,373	29,870	14,399	15,471
18 years and over	2,525	1,218	1,308	28,758	13,831	14,927
15 to 44 years	1,599	792	806	16,213	8,068	8,146
65 years and over	467	205	262	5,941	2,588	3,353
85 years and over	125	45	81	1,239	452	787
Median age	29.9	29.1	30.7	35.8	34.8	36.9
Mean age	34.1	33.1	35.0	38.2	37.0	39.3

Source: "Projections of the Population, by Age, Sex, Race, and Hispanic Origin, for the United States, 1993 to 2020, (Middle Series)." Jennifer Cheeseman Day, *Population Projections of the United States, by Age, Sex, Race, and Hispanic Origin: 1993 to 2050*, P25-1104, p. 55.

Poverty

★ 1011 ★

Poverty Status of Persons, Families, and Children under 18, Numbers, 1960 to 1991

Year and race/ethnicity	All persons	Number below the poverty level, in thousands				
		In all families			In families with female householder, no husband present	
		Total	Householder	Related children under 18	Total	Related children under 18
White[1]						
1960	28,309	24,262	6,115	11,229	4,296	2,357
1970	17,484	13,323	3,708	6,138	3,761	2,247
1980	19,699	14,587	4,195	6,817	4,940	2,813
1988	20,715	15,001	4,471	7,095	5,950	3,385
1989	20,785	15,179	4,409	7,164	5,723	3,320
1990	22,326	15,916	4,622	7,696	6,210	3,597
1991	23,747	17,268	5,022	8,316	6,806	3,941
Black[1]						
1959	9,927	9,112	1,860	5,022	2,416	1,475
1970	7,548	6,683	1,481	3,922	3,656	2,383
1980	8,579	7,190	1,826	3,906	4,984	2,944
1988	9,356	7,650	2,090	4,148	5,601	3,130
1989	9,302	7,704	2,077	4,257	5,530	3,256
1990	9,837	8,160	2,193	4,412	6,005	3,543
1991	10,242	8,504	2,343	4,637	6,557	3,853
Hispanic origin[2]						
1980	3,491	3,143	751	1,718	1,319	809
1987	5,422	4,761	1,168	2,606	2,045	1,241
1988	5,357	4,700	1,141	2,576	2,052	1,208
1989	5,430	4,659	1,133	2,496	1,902	1,163
1990	6,006	5,091	1,244	2,750	2,115	1,314
1991	6,339	5,541	1,372	2,977	2,282	1,398

Source: "Poverty Status of Persons, Families, and Children under 18, by Race/Ethnicity: 1960 to 1991: 1988." *Digest of Educational Statistics,* 1993, p. 29. Primary source: U.S. Department of Commerce, Bureau of the Census, *Current Population Reports,* Series P-60, No. 181. (This table was prepared January 1993). *Notes:* 1. Includes persons of Hispanic origin. 2. Persons of Hispanic origin may be of any race.

★ 1012 ★

Poverty

Poverty Status of Persons, Families, and Children under 18, Percentage, 1960 to 1991

Year and race/ethnicity	All persons	In all families			In families with female householder, no husband present	
		Total	Householder	Related children under 18	Total	Related children under 18
White[1]						
1960	17.8	16.2	14.9	20.0	39.0	59.9
1970	9.9	8.1	8.0	10.5	28.4	43.1
1980	10.2	8.6	8.0	13.4	28.0	41.6
1988	10.1	8.6	7.9	14.0	29.2	43.0
1989	10.0	8.6	7.8	14.1	28.1	42.8
1990	10.7	9.0	8.1	15.1	29.8	45.9
1991	11.3	9.7	8.8	16.1	31.5	47.1
Black[1]						
1959	55.1	54.9	48.1	65.5	70.6	81.6
1970	33.5	32.2	29.5	41.5	58.7	67.7
1980	32.5	31.1	28.9	42.1	53.4	64.8
1988	31.3	30.0	28.2	42.8	51.9	61.8
1989	30.7	29.7	27.8	43.2	49.4	62.9
1990	31.9	31.0	29.3	44.2	50.6	64.7
1991	32.7	32.0	30.4	45.6	54.8	68.2
Hispanic origin[2]						
1975	26.9	26.3	25.1	33.1	57.2	68.4
1985	29.0	28.3	25.5	39.6	55.7	72.4
1988	26.7	26.0	23.7	37.3	55.0	65.5
1989	26.2	25.2	23.4	35.5	50.6	65.0
1990	28.1	26.9	25.0	37.7	53.0	68.4
1991	28.7	28.2	26.5	39.8	52.7	68.6

Source: "Poverty Status of Persons, Families, and Children under 18, by Race/Ethnicity: 1960 to 1991." *Digest of Educational Statistics*, 1993, p. 29. Primary source: U.S. Department of Commerce, Bureau of the Census, *Current Population Reports*, Series P-60, No. 181. (This table was prepared January 1993). *Notes:* 1. Includes persons of Hispanic origin. 2. Persons of Hispanic origin may be of any race.

Resident Population

★ 1013 ★

Components of Population Change: 1980 to 1991 and Projections, 1995 and 2000

Year	Total (Jan. 1-Dec. 31)						Rate per 1,000 midyear population				
	Population at start of period (1,000)	Net increase[1]		Natural increase		Net migration[3] (1,000)	Net growth rate[1]	Natural increase			Net migration rate[3]
		Total (1,000)	Percent[2]	Births (1,000)	Deaths (1,000)			Total	Birth rate	Death rate	
White											
1980[4]	194,713	1,270	0.7	2,203	1,277	429	8.6	6.3	15.0	8.7	2.9
1985	201,419	1,378	0.7	2,991	1,819	354	6.8	5.8	14.8	9.0	1.8
1989	206,874	1,502	0.7	3,132	1,854	410	7.2	6.2	15.1	8.9	2.0
1990	208,376	1,699	0.8	3,224	1,876	400	8.1	6.4	15.4	9.0	1.9
1991	210,075	1,773	0.8	3,144	1,869	498	8.4	6.0	14.9	8.9	2.4
Projections[5]											
1995	216,730	1,535	0.7	3,046	1,975	463	7.1	4.9	14.0	9.1	2.1
2000	223,926	1,323	0.6	2,931	2,071	463	5.9	3.8	13.0	9.2	2.1
Black											
1980[4]	26,683	289	1.1	445	173	62	14.4	13.5	22.1	8.6	3.1
1985	28,406	364	1.3	608	244	61	12.7	12.7	21.3	8.5	2.1
1989	29,939	438	1.5	709	268	58	14.5	14.7	23.5	8.9	1.9
1990	30,377	527	1.7	739	260	63	17.2	15.6	24.1	8.5	2.1
1991	30,904	569	1.8	735	259	93	18.3	15.3	23.6	8.3	3.0
Projections[5]											
1995	32,904	483	1.5	681	278	80	14.6	12.2	20.6	8.4	2.4
2000	35,290	472	1.3	695	304	80	13.3	11.0	19.6	8.6	2.3
American Indian, Eskimo, Aleut											
1980[4]	1,420	38	2.7	28	9	1	35.3	21.4	26.2	4.8	0.5
1985	1,687	64	3.8	43	7	-	37.3	20.7	24.8	4.2	0.2
1989	1,962	82	4.2	49	7	-	41.1	20.6	24.6	4.1	0.1
1990	2,044	53	2.6	50	7	-	25.3	20.2	24.2	4.0	-
1991	2,097	44	2.1	51	5	1	20.6	20.2	24.2	4.0	0.4
Projections[5]											
1995	2,231	32	1.4	41	9	-	14.2	14.2	18.4	4.2	0.1
2000	2,392	33	1.4	44	11	-	13.9	13.8	18.4	4.5	-
Asian, Pacific Islander											
1980[4]	3,729	302	8.1	67	8	232	103.4	20.1	22.9	2.9	79.3
1985	5,426	365	6.7	118	16	233	65.1	18.3	21.1	2.9	41.6
1989	6,929	417	6.0	150	21	244	58.4	18.1	21.1	2.9	34.2
1990	7,345	443	6.0	166	18	283	58.5	19.6	21.9	2.3	37.4
1991	7,789	418	5.4	164	20	274	52.2	18.0	20.5	2.5	34.2
Projections[5]											
1995	9,613	475	4.9	166	28	337	48.2	14.0	16.9	2.9	34.2
2000	12,038	500	4.2	202	39	337	40.7	13.3	16.4	3.1	27.4
Hispanic origin[6]											
1980[4]	14,609	613	4.2	261	45	339	54.5	19.2	23.2	4.0	30.2
1985	17,997	763	4.2	415	73	302	41.6	18.6	22.6	4.0	16.4
1989	21,207	915	4.3	537	82	291	42.3	21.0	24.8	3.8	13.4
1990	22,122	828	3.7	556	75	303	36.7	21.3	24.7	3.3	13.4
1991	22,950	772	3.4	575	110	306	33.0	19.9	24.6	4.7	13.1
Projections[5]											
1995	26,122	803	3.1	587	108	324	30.3	18.1	22.1	4.1	12.2
2000	30,189	830	2.7	642	135	324	27.1	16.6	21.0	4.4	10.6
White, non-Hispanic											
1980[4]	180,906	716	0.4	1,958	1,234	113	5.3	5.3	14.4	9.1	0.8
1985	184,649	729	0.4	2,607	1,750	82	3.9	4.6	14.1	9.5	0.4
1989	187,410	750	0.4	2,642	1,778	147	4.0	4.6	14.1	9.5	0.8

[Continued]

★ 1013 ★

Components of Population Change: 1980 to 1991 and Projections, 1995 and 2000

[Continued]

Year	Population at start of period (1,000)	Total (Jan. 1-Dec. 31)					Rate per 1,000 midyear population				
		Net increase[1]		Natural increase		Net migration[3] (1,000)	Net growth rate[1]	Natural increase			Net migration rate[3]
		Total (1,000)	Percent[2]	Births (1,000)	Deaths (1,000)			Total	Birth rate	Death rate	
1990	188,160	956	0.5	2,701	128	128	5.1	4.7	14.3	9.6	0.7
1991	189,116	1,054	0.6	2,598	1,768	224	5.6	4.4	13.7	9.3	1.2
Projections[5]											
1995	192,888	808	0.4	2,510	1,875	174	4.2	3.3	13.0	9.7	0.9
2000	196,406	573	0.3	2,346	1,947	174	2.9	2.0	11.9	9.9	0.9
Black, non-Hispanic											
1980[4]	26,142	253	1.0	434	171	47	12.8	13.3	22.0	8.7	2.4
1985	27,607	297	1.1	588	241	42	10.7	12.5	21.2	8.7	1.5
1989	28,844	346	1.2	680	264	40	11.9	14.4	23.4	9.1	1.4
1990	29,191	477	1.6	718	256	43	16.2	15.7	24.4	8.7	1.5
1991	29,667	536	1.8	716	254	74	17.9	15.5	24.0	8.5	2.5
Projections[5]											
1995	31,484	435	1.4	648	273	60	13.7	11.8	20.4	8.6	1.9
2000	33,624	421	1.3	658	297	60	12.4	10.7	19.5	8.8	1.8

Source: "Components of Population Change, by Race and Hispanic Origin: 1980 to 1991, and Projections, 1995-2000," U.S. Bureau of the Census, *Statistical Abstract of the United States*, 1993, p. 20. Primary source: U.S. Bureau of the Census, *Current Population Reports*, P25-1095. *Notes:* - Represents or rounds to zero. 1. Prior to April 1, 1990, includes "error of closure" (the amount necessary to make the components of change add to the net change between censuses), for which figures are not shown separately. 2. Percent of population at beginning of period. 3. Covers net international migration and movement of Armed forces, Federally affiliated civilian citizens, and their dependents. 4. Represents data for period April 1, 1980, to December 31, 1980. 5. Based on middle series of assumptions. See footnotes 1, table 3. 6. Persons of Hispanic origin may be of any race.

★ 1014 ★

Resident Population

Resident Population Projections, by Race, 1992 and 2050

Year and series	Total	Race				Hispanic origin[1]	Not of Hispanic origin, by race			
		White	Black	American Indian, Eskimo, Aleut	Asian, Pacific Islander		White	Black	American Indian, Eskimo, Aleut	Asian, Pacific Islander
Lowest series										
1992	254,678	212,485	31,649	2,150	8,394	24,062	190,506	30,353	1,873	7,883
1993	256,930	213,886	32,098	2,182	8,765	24,730	191,298	30,764	1,900	8,237
1994	258,932	215,101	32,512	2,212	9,107	25,350	191,948	31,144	1,926	8,565
1995	260,715	216,1551	32,900	2,241	9,422	25,926	192,470	31,501	1,952	8,866
2000	268,108	220,092	34,642	2,383	10,991	28,693	193,877	33,094	2,074	10,370
2005	273,605	222,430	36,110	2,522	12,543	31,301	193,830	34,424	2,193	11,857
2010	278,078	223,922	37,419	2,658	14,079	33,828	193,016	35,596	2,309	13,329
2020	285,200	225,437	39,656	2,916	17,190	38,866	189,931	37,560	2,532	16,310
2030	286,710	222,311	41,008	3,412	20,250	43,465	182,604	38,667	2,731	19,244
2040	282,286	214,276	41,596	3,342	23,072	47,391	170,987	39,047	2,912	21,948
2050	275,647	204,724	41,712	3,519	25,691	50,790	158,340	38,981	3,077	24,458
Middle series										
1992	254,922	212,648	31,673	2,150	8,451	24,136	190,604	30,372	1,873	7,937
1993	257,592	214,328	32,171	2,183	8,911	24,927	191,568	30,822	1,901	8,374

[Continued]

★ 1014 ★

Resident Population Projections, by Race, 1992 and 2050
[Continued]

Year and series	Total	Race				Hispanic origin[1]	Not of Hispanic origin, by race			
		White	Black	American Indian, Eskimo, Aleut	Asian, Pacific Islander		White	Black	American Indian, Eskimo, Aleut	Asian, Pacific Islander
1994	260,202	215,948	32,662	2,215	9,377	25,722	192,469	31,265	1,929	8,818
1995	262,754	217,511	33,147	2,247	9,849	26,522	193,307	31,702	1,956	9,266
2000	274,815	224,594	35,525	2,409	12,287	30,602	196,701	33,834	2,096	11,582
2005	286,324	230,993	37,907	2,583	14,840	34,842	199,274	35,957	2,245	14,005
2010	298,109	237,412	40,429	2,772	17,496	39,312	201,668	38,201	2,407	16,522
2020	322,602	250,587	45,743	3,175	23,096	48,952	206,162	42,911	2,756	21,821
2030	344,951	261,318	51,031	3,610	28,993	59,197	207,674	47,552	3,141	27,388
2040	364,349	268,778	56,445	4,099	32,027	69,827	205,587	52,285	3,581	33,070
2050	382,674	274,761	62,181	4,641	41,091	80,675	201,841	57,316	4,078	38,765
Highest series										
1992	255,147	212,792	31,699	2,150	8,506	24,208	190,685	30,392	1,873	7,989
1993	258,211	214,735	32,241	2,183	9,052	25,113	191,812	30,878	1,901	8,507
1994	261,399	216,749	32,798	2,216	9,636	26,068	192,965	31,376	1,930	9,061
1995	264,685	218,811	33,368	2,250	10,257	27,073	194,123	31,882	1,958	9,649
2000	281,306	229,063	36,307	2,422	13,514	32,343	199,639	34,485	2,106	12,733
2005	298,773	239,680	39,460	2,615	17,017	38,066	205,123	37,269	2,270	16,045
2010	317,895	251,352	42,947	2,833	20,763	44,328	211,185	40,346	2,455	19,580
2020	360,123	277,183	50,719	3,317	28,904	58,445	224,371	47,187	2,871	27,248
2030	405,130	304,030	59,296	3,879	37,925	74,717	236,654	54,682	3,360	35,716
2040	453,687	332,346	68,889	4,557	47,894	93,466	248,211	63,014	3,956	45,040
2050	506,740	363,006	79,609	5,359	58,766	114,904	259,721	72,277	4,667	55,171
Percent distribution										
Lowest series										
1992	100.0	83.4	12.4	0.8	3.3	9.4	74.8	11.9	0.7	3.1
1995	100.0	82.9	12.6	0.9	3.6	9.9	73.8	12.1	0.7	3.4
2000	100.0	82.1	12.9	0.9	4.1	10.7	72.3	12.3	0.8	3.9
2005	100.0	81.3	13.2	0.9	4.6	11.4	70.8	12.6	0.8	4.3
2010	100.0	80.5	13.5	1.0	5.1	12.2	69.4	12.8	0.8	4.8
2020	100.0	79.0	13.9	1.0	6.0	13.6	66.6	13.2	0.9	5.7
2030	100.0	77.5	14.3	1.1	7.1	15.2	63.7	13.5	1.0	6.7
2040	100.0	75.9	14.7	1.2	8.2	16.8	60.6	13.8	1.0	7.8
2050	100.0	74.3	15.1	1.3	9.3	18.4	57.4	14.1	1.1	8.9
Middle series										
1992	100.0	83.4	12.4	0.8	3.3	9.5	74.8	11.9	0.7	3.1
1995	100.0	82.8	12.6	0.9	3.7	10.1	73.6	12.1	0.7	3.5
2000	100.0	81.7	12.9	0.9	4.5	11.1	71.6	12.3	0.8	4.2
2005	100.0	80.7	13.2	0.9	5.2	12.2	69.6	12.6	0.8	4.9
2010	100.0	79.6	13.6	0.9	5.9	13.2	67.6	12.8	0.8	5.5
2020	100.0	77.7	14.2	1.0	7.2	15.2	63.9	13.3	0.9	6.8
2030	100.0	75.8	14.8	1.0	8.4	17.2	60.2	13.8	0.9	7.9

[Continued]

★ 1014 ★

Resident Population Projections, by Race, 1992 and 2050

[Continued]

Year and series	Total	Race				Hispanic origin[1]	Not of Hispanic origin, by race			
		White	Black	American Indian, Eskimo, Aleut	Asian, Pacific Islander		White	Black	American Indian, Eskimo, Aleut	Asian, Pacific Islander
2040	100.0	73.8	15.5	1.1	9.6	19.2	56.4	14.4	1.0	9.1
2050	100.0	71.8	16.2	1.2	10.7	21.1	52.7	15.0	1.1	10.1
Highest series										
1992	100.0	83.4	12.4	0.8	3.3	9.5	74.7	11.9	0.7	3.1
1995	100.0	82.7	12.6	0.9	3.9	10.2	73.3	12.0	0.7	3.6
2000	100.0	81.4	12.9	0.9	4.8	11.5	71.0	12.3	0.7	4.5
2005	100.0	80.2	13.2	0.9	5.7	12.7	68.7	12.5	0.8	5.4
2010	100.0	79.1	13.5	0.9	6.5	13.9	66.4	12.7	0.8	6.2
2020	100.0	77.0	14.1	0.9	8.0	16.2	62.3	13.1	0.8	7.6
2030	100.0	75.0	14.6	1.0	9.4	18.4	58.4	13.5	0.8	8.8
2040	100.0	73.3	15.2	1.0	10.6	20.6	54.7	13.9	0.9	9.9
2050	100.0	71.6	15.7	1.1	11.6	22.7	51.3	14.3	0.9	10.9
Percent change (middle series)										
1992-2000	7.8	5.6	12.2	12.0	45.4	26.8	3.2	11.4	11.9	45.9
2000-2010	8.5	5.7	13.8	15.1	42.4	28.5	2.5	12.9	14.8	42.7
2010-2020	8.2	5.5	13.1	14.5	32.0	24.5	2.2	12.3	14.5	32.1
2020-2030	6.9	4.3	11.6	13.7	25.5	20.9	0.7	10.8	14.0	25.5
2030-2040	5.6	2.9	10.6	13.5	20.8	18.0	-1.0	10.0	14.0	20.7
2040-2050	5.0	2.2	10.2	13.2	17.3	15.5	-1.8	9.6	13.9	17.2

Source: "Resident Population Projections, by Race and Hispanic Origin: 1992 to 2050," U.S. Bureau of the Census, *Statistical Abstract of the United States,* 1993, p. 19. Primary source: U.S. Bureau of the Census, *Current Population Reports,* P25-1092. *Note:* 1. Persons of Hispanic origin may be of any race.

★ 1015 ★

Resident Population

Resident Population, by Age and Hispanic Origin, 1980 to 1991. Part I

In thousands, except percent. As of April, except 1991 as of July.

Year and sex	Total, all years	Under 5 years	5-9 years	10-14 years	15-19 years	20-24 years	25-29 years	30-34 years	35-39 years	40-44 years	45-49 years
Hispanic origin											
1980	14,609	1,663	1,537	1,475	1,606	1,586	1,376	1,129	854	712	622
1990	22,354	2,467	2,178	1,989	2,084	2,320	2,337	2,045	1,642	1,276	936
1991	23,350	2,600	2,256	2,088	2,057	2,357	2,400	2,175	1,761	1,394	997
Male	11,890	1,330	1,151	1,068	1,076	1,288	1,285	1,136	899	697	488
Female	11,460	1,271	1,104	1,021	981	1,070	1,115	1,039	862	698	509
Non-Hispanic											
White											
1980	180,906	11,842	12,262	13,703	16,166	16,574	15,358	14,091	11,315	9,437	9,104

[Continued]

★ 1015 ★

Resident Population, by Age and Hispanic Origin, 1980 to 1991. Part I

[Continued]

Year and sex	Total, all years	Under 5 years	5-9 years	10-14 years	15-19 years	20-24 years	25-29 years	30-34 years	35-39 years	40-44 years	45-49 years
1990	188,300	12,721	12,516	11,854	12,447	13,522	15,508	16,332	15,162	13,839	10,971
1991	189,568	12,781	12,581	12,218	11,868	13,483	14,851	16,445	15,572	14,659	11,186
Male	92,399	6,559	6,463	6,281	6,094	6,831	7,449	8,236	7,791	7,313	5,538
Female	97,169	6,221	6,118	5,937	5,775	6,652	7,402	8,209	7,781	7,346	5,649
Black											
1980	26,142	2,399	2,455	2,635	2,944	2,689	2,292	1,865	1,438	1,233	1,127
1990	29,273	2,799	2,597	2,525	2,605	2,528	2,649	2,600	2,265	1,811	1,362
1991	29,910	2,967	2,625	2,612	2,539	2,541	2,618	2,640	2,361	1,977	1,396
Male	14,118	1,501	1,329	1,320	1,284	1,240	1,239	1,227	1,092	910	635
Female	15,792	1,466	1,296	1,292	1,256	1,302	1,379	1,413	1,269	1,067	762
American Indian, Eskimo, Aleut											
1980	1,326	136	135	145	158	138	116	100	79	66	55
1990	1,796	185	179	170	165	151	160	156	138	117	90
1991	1,845	203	181	178	161	155	157	158	141	123	93
Male	908	103	92	91	82	79	78	77	68	59	45
Female	936	100	89	88	79	75	78	81	74	63	48
Asian, Pacific Islander											
1980	3,563	308	311	285	294	332	378	376	279	221	182
1990	6,988	586	566	522	581	611	673	700	640	545	384
1991	7,504	671	596	575	580	658	692	741	683	602	421
Male	3,664	343	302	293	299	337	342	359	324	279	201
Female	3,840	327	293	282	281	320	350	383	359	323	220
1991, Percent											
Hispanic origin	100.0	11.1	9.7	8.9	8.8	10.1	10.3	9.3	7.5	6.0	4.3
Non-Hispanic											
White	100.0	6.7	6.6	6.4	6.3	7.1	7.8	8.7	8.2	7.7	5.9
Black	100.0	9.9	8.8	8.7	8.5	8.5	8.8	8.8	7.9	6.6	4.7
American Indian, Eskimo, Aleut	100.0	11.0	9.8	9.6	8.7	8.4	8.5	8.6	7.6	6.7	5.0
Asian, Pacific Islander	100.0	8.9	7.9	7.7	7.7	8.8	9.2	9.9	9.1	8.0	5.6

Source: "Resident Population, by Age and Hispanic Origin: 1980 to 1991," U.S. Bureau of the Census, *Statistical Abstract of the United States*, 1993, p. 22. Primary source: U.S. Bureau of the Census, *Current Population Reports*, P25-1095. *Note:* Hispanic persons may be of any race.

★ 1016 ★

Resident Population

Resident Population, by Age and Hispanic Origin, 1980 to 1991. Part II

In thousands, except percent. As of April, except 1991 as of July.

Year and sex	50-54 years	55-59 years	60-64 years	65-74 years	75-84 years	85 years and over	5-13 years	14-17 years	18-24 years
Hispanic origin									
1980	564	454	321	457	203	49	2,715	1,251	2,240
1990	750	633	550	715	339	91	3,782	1,575	3,215
1991	788	656	580	770	364	106	3,947	1,593	3,221
Male	380	310	268	337	141	37	2,016	820	1,747
Female	408	346	312	433	224	69	1,930	772	1,473
Non-Hispanic									
White									
1980	9,824	9,963	8,775	13,614	6,863	2,014	23,126	12,313	23,267
1990	9,057	8,548	8,871	15,511	8,767	2,675	22,106	9,225	19,008
1991	9,291	8,433	8,776	15,623	9,017	2,786	22,443	9,265	18,444
Male	4,560	4,082	4,148	6,898	3,398	760	11,531	4,773	9,364
Female	4,731	4,352	4,628	8,726	5,619	2,025	10,911	4,491	9,080
Black									
1980	1,114	1,024	861	1,327	582	157	4,530	2,331	3,862
1990	1,137	1,008	945	1,465	757	219	4,639	1,976	3,640
1991	1,168	1,017	951	1,501	770	228	4,727	1,987	3,603
Male	524	449	409	617	276	67	2,391	1,009	1,772
Female	644	568	541	883	493	160	2,337	978	1,832
American Indian, Eskimo, Aleut									
1980	49	43	32	46	20	6	249	127	199
1990	72	58	48	68	31	9	317	131	217
1991	74	59	49	70	33	10	326	131	218
Male	35	28	23	31	14	3	165	66	112
Female	38	31	26	39	21	6	160	64	106
Asian, Pacific Islander									
1980	158	131	96	136	60	14	540	226	455
1990	297	240	210	287	116	27	984	436	862
1991	324	258	226	316	130	31	1,062	448	899
Male	157	118	97	140	60	12	538	231	460
Female	167	140	129	177	71	18	521	218	438
1991, Percent									
Hispanic origin	3.4	2.8	2.5	3.3	1.6	0.5	16.9	6.8	13.8
Non-Hispanic									
White	4.9	4.4	4.6	8.2	4.8	1.5	11.8	4.9	9.7
Black	3.9	3.4	3.2	5.0	2.6	0.8	15.8	6.6	12.0

[Continued]

★ 1016 ★

Resident Population, by Age and Hispanic Origin, 1980 to 1991. Part II
[Continued]

Year and sex	50-54 years	55-59 years	60-64 years	65-74 years	75-84 years	85 years and over	5-13 years	14-17 years	18-24 years
American Indian, Eskimo, Aleut	4.0	3.2	2.7	3.8	1.8	0.5	17.7	7.1	11.8
Asian, Pacific Islander	4.3	3.4	3.0	4.2	1.7	0.4	14.2	6.0	12.0

Source: "Resident Population, by Age and Hispanic Origin: 1980 to 1991," U.S. Bureau of the Census, *Statistical Abstract of the United States*, 1993, p. 22. Primary source: U.S. Bureau of the Census, *Current Population Reports*, P25-1095. *Note:* Hispanic persons may be of any race.

★ 1017 ★

Resident Population

Resident Population, by Age and Race, 1980 to 1991. Part I

In thousands, except percent. As of April, except 1991 as of July.

Year, sex and race	Total all years	Under 5 years	5-9 years	10-14 years	15-19 years	20-24 years	25-29 years	30-34 years	35-39 years	40-44 years	45-49 years
All races											
1980	226,546	16,348	16,700	18,242	21,168	21,319	19,521	17,561	13,965	11,669	11,090
1990	248,710	18,758	18,035	17,060	17,882	19,132	21,328	21,833	19,846	17,589	13,744
1991	252,177	19,222	18,237	17,671	17,205	19,194	20,718	22,159	20,518	18,754	14,094
Male	122,979	9,836	9,337	9,051	8,834	9,775	10,393	11,034	10,174	9,258	6,907
Female	129,198	9,386	8,900	8,620	8,371	9,419	10,325	11,125	10,344	9,496	7,188
White											
1980	194,713	13,414	13,717	15,095	17,681	18,072	16,658	15,157	12,122	10,110	9,693
1990	208,704	14,960	14,502	13,670	14,351	15,637	17,638	18,190	16,652	15,001	11,826
1991	210,899	15,168	14,634	14,122	13,749	15,630	17,036	18,424	17,170	15,927	12,097
Male	103,268	7,780	7,512	7,254	7,078	8,006	8,620	9,272	8,608	7,948	5,984
Female	107,631	7,388	7,123	6,868	6,671	7,625	8,416	9,152	8,562	7,980	6,113
Black											
1980	26,683	2,459	2,509	2,691	3,007	2,749	2,342	1,904	1,469	1,260	1,150
1990	30,483	2,939	2,711	2,629	2,714	2,655	2,780	2,718	2,359	1,882	1,413
1991	31,164	3,099	2,747	2,722	2,647	2,671	2,753	2,763	2,463	2,055	1,451
Male	14,753	1,568	1,391	1,376	1,339	1,309	1,311	1,291	1,144	949	661
Female	16,412	1,531	1,356	1,346	1,307	1,361	1,442	1,472	1,318	1,106	790
American Indian, Eskimo, Aleut											
1980	1,420	149	147	156	170	149	125	107	84	69	58
1990	2,065	220	209	197	191	179	188	181	157	132	99
1991	2,117	234	212	207	186	183	184	183	162	138	103
Male	1,050	119	108	105	95	95	93	90	79	67	50
Female	1,068	115	105	102	91	88	91	93	83	71	53
Asian, Pacific Islander											
1980	3,729	326	328	300	310	349	396	393	291	230	188
1990	7,458	638	612	564	626	661	722	745	678	574	405
1991	7,996	720	644	621	623	710	744	788	723	633	444
Male	3,909	369	327	316	321	365	369	381	343	294	212
Female	4,087	352	317	305	345	375	407	380	339	232	

[Continued]

★ 1017 ★

Resident Population, by Age and Race, 1980 to 1991. Part I

[Continued]

Year, sex and race	Total all years	Under 5 years	5-9 years	10-14 years	15-19 years	20-24 years	25-29 years	30-34 years	35-39 years	40-44 years	45-49 years
Percent											
Total, 1991	100.0	7.6	7.2	7.0	6.8	7.6	8.2	8.8	8.1	7.4	5.6
White	100.0	7.2	6.9	6.7	6.5	7.4	8.1	8.7	8.1	7.6	5.7
Black	100.0	9.9	8.8	8.7	8.5	8.6	8.8	8.9	7.9	6.6	4.7
American Indian, Eskimo, Aleut	100.0	11.1	10.0	9.8	8.8	8.6	8.7	8.6	7.7	6.5	4.9
Asian, Pacific Islander	100.0	9.0	8.1	7.8	7.8	8.9	9.3	9.9	9.0	7.9	5.6

Source: "Resident Population, by Age and Race: 1980 to 1991," U.S. Bureau of the Census, *Statistical Abstract of the United States,* 1993, p. 21. Primary source: U.S. Bureau of the Census, *Current Population Reports,* P25-1095.

★ 1018 ★

Resident Population

Resident Population, by Age and Race, 1980 to 1991. Part II

In thousands, except percent. As of April, except 1991 as of July.

Year, sex and race	50-54 years	55-59 years	60-64 years	65-74 years	75-84 years	85 years and over	5-13 years	14-17 years	18-24 years
All races									
1980	11,710	11,615	10,088	15,581	7,729	2,240	31,159	16,247	30,022
1990	11,313	10,487	10,625	18,045	10,012	3,021	31,826	13,340	26,942
1991	11,645	10,423	10,582	18,280	10,314	3,160	32,500	13,423	26,385
Male	5,656	4,987	4,945	8,022	3,888	881	16,641	6,902	13,456
Female	5,989	5,436	5,637	10,258	6,426	2,279	15,859	6,521	12,929
White									
1980	10,360	10,394	9,078	14,045	7,057	2,060	25,692	13,491	25,381
1990	9,744	9,131	9,381	16,175	9,084	2,761	25,557	10,664	21,939
1991	10,013	9,037	9,312	16,338	9,356	2,886	26,036	10,718	21,380
Male	4,908	4,367	4,396	7,212	3,529	795	13,368	5,523	10,960
Female	5,105	4,669	4,916	9,125	5,827	2,090	12,669	5,196	10,422
Black									
1980	1,135	1,041	874	1,344	588	159	4,628	2,380	3,948
1990	1,178	1,041	972	1,498	772	223	4,837	2,056	3,816
1991	1,211	1,050	979	1,536	786	232	4,938	2,071	3,778
Male	544	464	422	631	282	69	2,498	1,053	1,865
Female	666	586	557	905	504	163	2,440	1,018	1,912
American Indian, Eskimo, Aleut									
1980	52	45	34	48	21	6	270	136	216
1990	79	64	53	73	34	9	368	152	256
1991	81	65	54	76	36	11	380	151	256
Male	39	31	26	34	14	4	193	77	133
Female	42	34	29	42	21	7	188	74	124

[Continued]

★ 1018 ★

Resident Population, by Age and Race, 1980 to 1991. Part II
[Continued]

Year, sex and race	50-54 years	55-59 years	60-64 years	65-74 years	75-84 years	85 years and over	5-13 years	14-17 years	18-24 years
Asian, Pacific Islander									
1980	163	135	101	143	63	15	570	239	478
1990	312	252	220	300	122	29	1,063	469	931
1991	340	271	237	329	135	32	1,146	484	969
Male	165	124	102	145	63	13	582	250	497
Female	175	147	135	184	74	19	564	234	471
Percent									
Total, 1991	4.6	4.1	4.2	7.2	4.1	1.3	12.9	5.3	10.5
White	4.7	4.3	4.4	7.7	4.4	1.4	12.3	5.1	10.1
Black	3.9	3.4	3.1	4.9	2.5	0.7	15.8	6.6	12.1
American Indian, Eskimo, Aleut	3.8	3.1	2.6	3.6	1.7	0.5	17.9	7.1	12.1
Asian, Pacific Islander	4.3	3.4	3.0	4.1	1.7	0.4	14.3	6.1	12.1

Source: "Resident Population, by Age and Race: 1980 to 1991," U.S. Bureau of the Census, *Statistical Abstract of the United States*, 1993, p. 21. Primary source: U.S. Bureau of the Census, *Current Population Reports*, P25-1095.

★ 1019 ★

Resident Population

Resident Population, by Race and Hispanic Origin, 1980 and 1990

As of April 1.

Race and Hispanic origin	Number (1,000)		Percent distribution		Change, 1980-90	
	1980[1]	1990	1980[1]	1990	Number (1,000)	Percent
All persons	226,546	248,710	100.0	100.0	22,164	9.8
Race						
White	188,372	199,686	83.1	80.3	11,314	6.0
Black	26,495	29,986	11.7	12.1	3,491	13.2
American Indian, Eskimo, or Aleut	1,420	1,959	0.6	0.8	539	37.9
American Indian	1,364	1,878	0.6	0.8	514	37.7
Eskimo	42	57	(Z)	(Z)	15	35.6
Aleut	14	24	(Z)	(Z)	10	67.5
Asian or Pacific Islander	3,500[2]	7,274	1.5	2.9	3,773	107.8
Chinese	806	1,645	0.4	0.7	839	104.1
Filipino	775	1,407	0.3	0.6	632	81.6
Japanese	701	848	0.3	0.3	147	20.9
Asian Indian	362	815	0.2	0.3	454	125.6
Korean	355	799	0.2	0.3	444	125.3
Vietnamese	262	615	0.1	0.2	353	134.8
Hawaiian	167	211	0.1	0.1	44	26.5
Samoan	42	63	(Z)	(Z)	21	50.1
Guamanian	32	49	(Z)	(Z)	17	53.4
Other Asian or Pacific Islander	(NA)	822	(NA)	0.3	(NA)	(NA)

[Continued]

★ 1019 ★

Resident Population, by Race and Hispanic Origin, 1980 and 1990
[Continued]

| Race and Hispanic origin | Number (1,000) | | Percent distribution | | Change, 1980-90 | |
	1980[1]	1990	1980[1]	1990	Number (1,000)	Percent
Other race	6,758	9,805	3.0	3.9	3,047	45.1
Hispanic origin						
Hispanic origin[3]	14,609	22,354	6.4	9.0	7,745	53.0
Mexican	8,740	13,496	3.9	5.4	4,755	54.4
Puerto Rican	2,014	2,728	0.9	1.1	714	35.4
Cuban	803	1,044	0.4	0.4	241	30.0
Other Hispanic	3,051	5,086	1.3	2.0	2,035	66.7
Not of Hispanic origin	211,937	226,356	93.6	91.0	14,419	6.8

Source: "Resident Population, by Race and Hispanic Origin: 1980 and 990," U.S. Bureau of the Census, *Statistical Abstract of the United States,* 1993, p. 18. Primary source: U.S. Bureau of the Census, press release CB91-216. *Notes:* NA Not available. Z Less than 0.05 percent. 1. See footnote 4, table 1. 2. Not entirely comparable with 1990 counts. The 1980 count shown here which is based on 100-percent tabulations includes only the nine specific Asian or Pacific Islander groups listed separately in the 1980 race item. The 1980 total Asian or Pacific Islander population of 3,726,440 from sample tabulations is comparable to the 1990 count; these figures include groups not listed separately in the race item on the 1980 census form. 3. Persons of Hispanic origin may be of any race.

★ 1020 ★

Resident Population

Resident Population, by Race and Hispanic Origin, 1980 to 1991

In thousands. As of July, except as indicated. These data are consistent with the 1980 and 1990 decennial enumerations and have been modified from the official census counts; see text, section 1 for explanation.

| Year | Total | Race | | | | Hispanic origin[1] | Not of Hispanic origin, by race | | | |
		White	Black	American Indian, Eskimo, Aleut	Asian, Pacific Islander		White	Black	American Indian, Eskimo, Aleut	Asian, Pacific Islander
1980 (April)	226,546	194,713	26,683	1,420	3,729	14,609	180,906	26,142	1,326	3,563
1980	227,225	195,185	26,771	1,433	3,837	14,869	181,140	26,215	1,336	3,665
1981	229,466	196,635	27,133	1,483	4,214	15,560	181,974	26,532	1,377	4,022
1982	231,664	198,037	27,508	1,537	4,581	16,240	182,782	26,856	1,420	4,367
1983	233,792	199,420	27,867	1,596	4,909	16,935	183,561	27,159	1,466	4,671
1984	235,825	200,708	28,212	1,656	5,249	17,640	184,243	27,444	1,512	4,986
1985	237,924	202,031	28,569	1,718	5,606	18,368	184,945	27,738	1,558	5,315
1986	240,133	203,430	28,942	1,783	5,978	19,154	185,678	28,040	1,606	5,655
1987	242,289	204,770	29,325	1,851	6,343	19,946	186,353	28,351	1,654	5,985
1988	244,499	206,129	29,723	1,923	6,724	20,786	187,012	28,669	1,703	6,329
1989	246,819	207,540	30,143	2,001	7,134	21,648	187,713	29,005	1,755	6,689
1990 (April)	248,710	208,704	30,483	2,065	7,458	22,354	188,300	29,273	1,796	6,988

[Continued]

★ 1020 ★

Resident Population, by Race and Hispanic Origin, 1980 to 1991

[Continued]

Year	Total	Race				Hispanic origin[1]	Not of Hispanic origin, by race			
		White	Black	American Indian, Eskimo, Aleut	Asian, Pacific Islander		White	Black	American Indian, Eskimo, Aleut	Asian, Pacific Islander
1990	249,415	209,150	30,620	2,075	7,570	22,554	188,559	29,400	1,806	7,096
1991	252,177	210,899	31,164	2,117	7,996	23,350	189,568	29,910	1,845	7,504

Source: "Resident Population, by Race and Hispanic Origin: 1980 to 1991," U.S. Bureau of the Census, *Statistical Abstract of the United States*, 1993, p. 18. Primary source: U.S. Bureau of the Census, *Current Population Reports*, P25-1095. *Note:* 1. Persons of Hispanic origin may be of any race.

★ 1021 ★

Resident Population

Resident population by Race and Hispanic Origin. by Single Years of Age, 1991

In thousands, except as indicated. As of July 1. Resident population.

Age	White	Black	American Indian Eskimo, Aleut	Asian, Pacific Islander	Hispanic origin[1]
Total	210,899	31,164	2,117	7,996	23,350
Under 5 years old	15,168	3,099	234	720	2,600
Under 1 yr. old	3,102	677	54	178	553
1 yr. old	3,129	645	48	146	553
2 yrs. old	3,013	613	45	134	515
3 yrs. old	2,957	583	44	134	499
4 yrs. old	2,966	580	43	128	481
5-9 yrs. old	14,634	2,747	212	644	2,256
5 yrs. old	2,967	562	43	130	473
6 yrs. old	2,963	549	42	128	457
7 yrs. old	2,870	536	42	127	443
8 yrs. old	2,820	523	42	127	431
9 yrs. old	3,015	577	43	132	451
10 - 14 yrs. old	14,122	2,722	207	621	2,088
10 yrs. old	2,966	560	44	134	451
11 yrs. old	2,929	565	42	126	431
12 yrs. old	2,774	546	41	123	406
13 yrs. old	2,733	521	40	120	403
14 yrs. old	2,720	531	39	119	399
15-19 yrs. old	13,749	2,647	186	623	2,057
15 yrs. old	2,624	511	38	120	395
16 yrs. old	2,692	513	37	120	398
17 yrs. old	2,682	516	37	125	401
18 yrs. old	2,700	526	35	121	415

[Continued]

★ 1021 ★

Resident population by Race and Hispanic Origin. by Single Years of Age, 1991

[Continued]

Age	White	Black	American Indian Eskimo, Aleut	Asian, Pacific Islander	Hispanic origin[1]
19 yrs. old	3,050	581	38	138	449
20-24 yrs. old	15,630	2,671	183	710	2,357
20 yrs. old	3,295	597	39	148	481
21 yrs. old	3,226	555	38	149	473
22 yrs. old	2,057	499	35	140	464
23 yrs. old	2,957	506	34	135	461
24 yrs. old	3,094	514	36	137	479
25-29 yrs. old	17,036	2,753	184	744	2,400
25 yrs. old	3,127	532	36	142	484
26 yrs. old	3,313	547	36	145	486
27 yrs. old	3,474	555	38	151	487
28 yrs. old	3,363	521	35	145	461
29 yrs. old	3,760	598	40	161	481
30-34 yrs. old	18,424	2,763	183	788	2,175
30 yrs. old	3,722	562	38	160	480
31 yrs. old	3,661	557	37	159	447
32 yrs. old	3,632	545	36	157	424
33 yrs. old	3,650	533	35	153	415
34 yrs. old	3,759	566	37	159	409
35-39 yrs. old	17,170	2,463	162	723	1,761
35 yrs. old	3,586	533	35	157	397
36 yrs. old	3,501	508	33	147	371
37 yrs. old	3,447	496	33	144	348
38 yrs. old	3,180	436	30	133	317
39 yrs. old	3,456	490	32	141	328
40-44 yrs. old	15,927	2,055	138	633	1,394
40 yrs. old	3,208	452	30	139	313
41 yrs. old	3,121	431	29	135	290
42 yrs. old	3,073	404	27	125	271
43 yrs. old	3,048	384	28	125	257
44 yrs. old	3,477	385	25	109	264
45-49 yrs. old	12,097	1,451	103	444	997
45 yrs. old	2,414	297	22	99	223
46 yrs. old	2,416	292	21	92	207
47 yrs. old	2,438	299	21	90	197
48 yrs. old	2,494	265	19	78	187
49 yrs. old	2,336	298	20	83	183
50-54 yrs. old	10,013	1,211	81	340	788
50 yrs. old	2,171	261	18	78	175
51 yrs. old	2,006	247	17	70	162

[Continued]

★ 1021 ★

Resident population by Race and Hispanic Origin. by Single Years of Age, 1991

[Continued]

Age	White	Black	American Indian Eskimo, Aleut	Asian, Pacific Islander	Hispanic origin[1]
52 yrs. old	1,980	236	16	66	155
53 yrs. old	1,964	237	15	63	151
54 yrs. old	1,892	230	15	63	145
55-59 yrs. old	9,037	1,050	65	271	656
55 yrs. old	1,825	229	14	60	140
56 yrs. old	1,909	217	14	55	143
57 yrs. old	1,789	212	13	55	130
58 yrs. old	1,697	188	12	49	120
59 yrs. old	1,816	204	12	52	123
60-64 yrs. old	9,312	979	54	237	580
60 yrs. old	1,858	203	12	51	126
61 yrs. old	1,847	194	11	48	120
62 yrs. old	1,822	196	11	47	115
63 yrs. old	1,900	189	10	45	113
64 yrs. old	1,884	198	10	45	107
65-69 yrs. old	8,926	873	44	194	458
65 yrs. old	1,820	198	10	44	102
66 yrs. old	1,838	182	9	40	97
67 yrs. old	1,806	173	9	38	92
68 yrs. old	1,703	161	8	37	85
69 yrs. old	1,758	159	8	35	81
70-74 yrs. old	7,412	663	32	135	312
70 yrs. old	1,688	157	8	34	77
71 yrs. old	1,551	144	7	30	68
72 yrs. old	1,484	130	7	28	61
73 yrs. old	1,372	118	6	23	54
74 yrs. old	1,317	114	5	22	51
75-79 yrs. old	5,679	489	23	88	223
75 yrs. old	1,271	108	5	21	49
76 yrs. old	1,211	99	5	19	47
77 yrs. old	1,125	97	5	18	44
78 yrs. old	1,086	97	4	16	44
79 yrs. old	985	87	4	15	39
80-84 yrs. old	3,677	297	13	47	141
80 yrs. old	881	73	3	12	35
81 yrs. old	802	67	3	10	31
82 yrs. old	732	56	3	9	27
83 yrs. old	676	51	2	8	25
84 yrs. old	585	49	2	8	22
85-89 yrs. old	1,915	147	7	21	72

[Continued]

★ 1021 ★

Resident population by Race and Hispanic Origin. by Single Years of Age, 1991

[Continued]

Age	White	Black	American Indian Eskimo, Aleut	Asian, Pacific Islander	Hispanic origin[1]
90-94 yrs. old	741	60	3	8	26
95-99 yrs. old	194	17	1	2	7
100 yrs. old and over	36	7	-	1	2
Median age (yr.)	34.1	28.1	26.0	29.6	25.7

Source: "White, Black, Asian Pacific Islander, American Indian, and Hispanic Origin Population, by Single Years of Age: 1991," U.S. Bureau of the Census, *Statistical Abstract of the United States,* 1993, p. 23. Primary source: U.S. Bureau of the Census, *Current Population Reports,* P25-1095. *Notes:* - Represents or rounds to zero. 1. Persons of Hispanic origin may be of any race.

★ 1022 ★

Resident Population

Resident population: Characteristics 1850 to 1991, and Projections, 1995 to 2050

In percent, except as indicated. For definition of median, see Guide to Tabular Presentation.

Date	Race		
	White	Black	Other
1850 (June 1)[1]	84.3	15.7	(NA)
1860 (June 1)[1]	85.6	14.1	0.3
1870 (June 1)[1]	87.1	12.7	0.2
1880 (June 1)[1]	86.5	13.1	0.3
1890 (June 1)[1]	87.5	11.9	0.6
1900 (June 1)[1]	87.9	11.6	0.5
1910 (Apr. 15)[1]	88.9	10.7	0.4
1920 (Jan. 1)[1]	89.7	9.9	0.4
1930 (Apr. 1)[1]	89.8	9.7	0.5
1940 (Apr. 1)[1]	89.8	9.8	0.4
1950 (Apr. 1)[1]	89.5	10.0	0.5
1950 (Apr. 1)	89.3	9.9	0.7
1960 (Apr. 1)	88.6	10.5	0.9
1970 (Apr. 1)	87.6	11.1	1.3
1980 (Apr. 1)[2,3]	85.9	11.8	2.3
1990 (Apr. 1)[2]	83.9	12.3	3.8
1991 (July 1)[4]	83.6	12.4	4.0
1995 (July 1)[5]	82.8	12.6	4.6
2000 (July 1)[5]	81.7	12.9	5.3

[Continued]

★ 1022 ★

Resident population: Characteristics 1850 to 1991, and Projections, 1995 to 2050

[Continued]

Date	Race		
	White	Black	Other
2025 (July 1)[5]	76.7	14.5	8.8
2550 (July 1)[5]	71.8	16.2	12.0

Source: "Resident Population Characteristics-Percent Distribution and abd Median Age, 1850 to 1991, and Projections, 1995 to 2050," U.S. Bureau of the Census, *Statistical Abstract of the United States*, 1993, p. 14. Primary source: U.S. Bureau of the Census, *U.S. Census of Population: 1940*, vol. II, part 1, and vol. IV, part 1; *1950*, vol. II, part 1; *1960*, vol. I, part I; *1970*, vol. I, part B; and *Current Population Reports*, P25-1092 and P25- 1095. *Notes:* NA Not available. 1. Excludes Alaska and Hawaii. 2. The race data shown have been modified; see text, section 1 for explanation. 3. See footnote 4, table 1. 4. Estimated. 5. Middle series projection; see table 3.

School-Age Population

★ 1023 ★

School-Age Resident Population, 1965-1992

In thousands.

Year	White[2]			Black[2]		
	Total[1]	Male	Female	Total[1]	Male	Female
1965	42,891	21,872	21,019	6,440	3,220	3,221
1968	44,422	22,677	21,744	6,903	3,453	3,450
1971	44,644	22,809	21,834	7,182	3,600	3,583
1974	43,454	22,210	21,244	7,213	3,618	3,596
1977	41,737	21,350	20,386	7,167	3,600	3,568
1980	39,001	19,981	19,021	6,996	3,524	3,473
1983	36,859	18,899	17,960	6,841	3,457	3,385
1986	36,531	18,746	17,786	6,957	3,527	3,430
1989	36,325	18,645	17,680	7,104	3,612	3,494
1990	36,323	18,668	17,655	6,917	3,502	3,415
1991	36,756	18,891	17,865	7,009	3,551	3,458
1992	37,245	19,140	18,105	7,145	3,622	3,523

Source: "Estimates of School-Age Resident Population by Race and Sex: July 1, 1960 to July 1, 1992." *Digest of Educational Statistics*, 1993, p. 24. Primary source: U.S. Department of Commerce, Bureau of the Census, *Current Population Reports*, Series P-25, Nos. 519, 917, 1000, 1022, 1057 1092, and unpublished data. (This table was prepared May 1993). *Notes:* Same data have been revised from previously published figures. Because of rounding, details may not add to totals. 1. Includes persons 5 to 17 years. 2. Includes persons of Hispanic origin.

Senior Citizens

★ 1024 ★

Black Population 85 Years and Over, 1980 to 2050

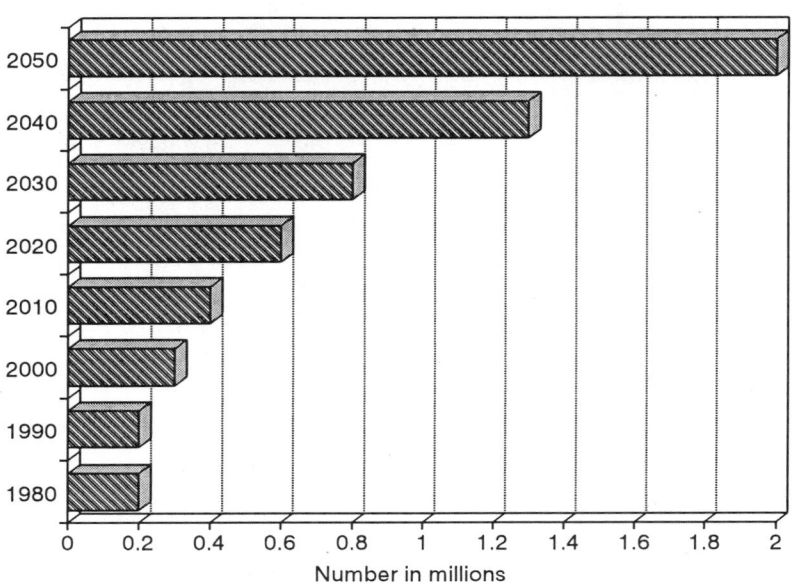

Number in millions

Year	Black population 85 years and over 1980 to 2050 (in millions)
1980	0.2
1990	0.2
2000	0.3
2010	0.4
2020	0.6
2030	0.8
2040	1.3
2050	2.0

Source: "Black Population 85 Years and Over: 1980 to 2050," Cynthia Taeuber, *Sixty-Five Plus in America*, p. 2-13. Primary source: U.S. Bureau of the Census, 1980 from 1980 Census of Population; 1990 from 1990 Census of Population and Housing, Series CPH-L-74, *Modified and Actual Age, Sex, Race, and Hispanic Origin Data*; 2000 to 2050 from Jennifer C. Day, *Population Projections of the United States, by Age, Sex, Race, and Hispanic Origin: 1992 to 2050*, Current Population Reports, P25-1092. U.S. Government Printing Office, Washington, DC, 1992 (middle series projections).

★ 1025 ★

Senior Citizens

Characteristics, by Gender, of Senior Citizens, 1980 to 1992

As of March, except as noted. Covers civilian noninstitutional population.

Characteristic	Total				Male				Female			
	1980	1985	1990	1992	1980	1985	1990	1992	1980	1985	1990	1992
Total[1] (million)	24.2	26.8	29.6	30.6	9.9	11.0	12.3	12.8	14.2	15.8	17.2	17.8
White (million)	21.9	24.2	26.5	27.3	9.0	9.9	11.0	11.4	12.9	14.3	15.4	15.9
Black (million)	2.0	2.2	2.5	2.6	0.8	0.9	1.0	1.1	1.2	1.3	1.5	1.5

Source: "Persons 65 Years Old and Over—Characteristics by Sex: 1980 to 1992," U.S. Bureau of the Census, *Statistical Abstract of the United States*, 1993, p. 45. Primary source: Except as noted, U.S. Bureau of the Census, Current Population Reports, P20-468 and earlier reports; P60-181 and earlier reports; and unpublished data. *Note:* 1. Includes other races not shown separately.

★ 1026 ★

Senior Citizens

High School and College Graduates 25 and Over, March 1989

Age	Four years of high school or more				Four years of college or more			
	Total	White	Black	Hispanic origin[1]	Total	White	Black	Hispanic origin[1]
25 years and over	76.9	78.4	64.6	50.9	21.1	21.8	11.8	9.9
25 to 29 years	85.5	86.0	82.2	61.0	23.4	24.4	12.7	10.1
30 to 34 years	87.6	88.4	83.0	58.6	25.0	25.8	14.0	11.8
35 to 39 years	87.5	88.8	80.3	58.9	28.0	29.0	17.4	10.7
40 to 44 years	85.6	87.0	75.6	52.6	27.7	28.6	15.8	11.3
45 to 49 years	80.9	82.8	65.5	46.8	23.8	24.2	14.8	9.9
50 to 54 years	75.4	78.2	54.2	41.8	19.8	20.5	8.5	10.1
55 to 59 years	72.0	74.4	50.5	39.5	18.2	19.0	8.7	8.2
60 to 64 years	66.2	69.4	39.4	33.7	14.2	15.3	5.0	6.0
65 years and over	54.9	57.9	24.6	27.7	11.1	11.7	4.6	5.9
65 to 69 years	63.0	66.6	30.8	33.0	12.6	13.3	4.8	8.8
70 to 74 years	57.0	60.4	21.3	25.3	10.6	11.3	2.9	3.0
75 years and over	46.2	48.6	20.6	23.0	10.1	10.6	5.4	4.5

Source: "Percentage of High School and College Graduates in Population 25 Years and Over, by Age, Race, and Hispanic Origin: March 1989," Cynthia Taeuber, *Sixty-Five Plus in America*, p. 6-17. Primary source: Robert Kominiski, U.S. Bureau of the Census, *Educational Attainment in the United States: March 1989 and 1988*, Current Population Reports, Series P-20, No. 451. U.S. Government Printing Office, Washington, DC, 1991, tables 1 and 2. *Note:* 1. Hispanic origin may be of any race.

★ 1027 ★

Senior Citizens

Income Distribution in 1989 of Householders 65 to 74 Years, 1990

Percentage of noninstitutional population.

Income	White	Black	Asian and Pacific Islander	Hispanic origin[1]
Under $10,000	20.9	44.2	20.7	35.5
$10,000 to $24,999	37.6	32.2	26.3	34.2
$25,000 to $49,999	27.7	17.1	28.3	21.3
$50,000 and over	13.9	6.5	24.6	9.1

Source: "Income Distribution in 1989 of Householders 65 to 74 Years: 1990." *Profiles of America's Elderly*, No. 3 (November 1993), p. 7. Primary source: U.S. Bureau of the Census, 1990 Census of Population, Summary Tape File 3C. *Notes:* The numbers of elderly American Indian, Eskimo, and Aleut are not shown as the population is too small for the data to be reliable. 1. May be of any race.

★ 1028 ★

Senior Citizens

Income of Householders 75 Years and Over, 1990

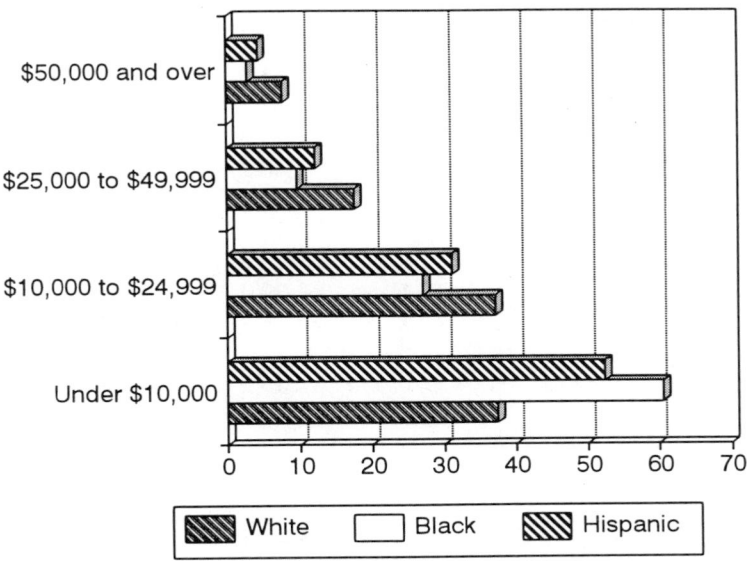

Percentage of noninstitutional population.

Income	White	Black	Hispanic origin[1]
Under $10,000	37.4	60.5	52.3
$10,000 to $24,999	37.2	27.1	31.1
$25,000 to $49,999	17.6	9.6	12.2
$50,000 and over	7.8	2.9	4.4

Source: "Income Distribution in 1989 of Householders 75 Years and Over: 1990." *Profiles of America's Elderly,* No. 3 (November 1993), p. 8. Primary source: U.S. Bureau of the Census, 1990 Census of Population, Summary Tape File 3C. *Notes:* The numbers of elderly American Indians, Eskimo, and Aleut, and of Asian and Pacific Islanders are not shown as the population group is too small to be reliable. 1. May be of any race.

★ 1029 ★

Senior Citizens

Marital Status of Persons 65 and Over, 1990

Age, race, and Hispanic origin[1]	Married, spouse present		Widowed	
	Male	Female	Male	Female
65 years and over	74.2	39.7	14.2	48.6
White	76.3	41.1	13.3	48.1
Black	54.2	25.3	23.4	53.7
Hispanic origin[1]	73.3	40.0	12.6	42.2

[Continued]

★ 1029 ★

Marital Status of Persons 65 and Over, 1990

[Continued]

Age, race, and Hispanic origin[1]	Married, spouse present		Widowed	
	Male	Female	Male	Female
65 to 74 years	78.2	51.1	9.2	36.1
White	80.3	53.2	8.2	35.1
Black	55.5	29.6	19.8	45.1
Hispanic origin[1]	77.4	48.2	7.9	31.5
75 to 84 years	71.2	27.7	19.5	62.0
White	72.8	28.3	19.0	62.1
Black	56.1	22.0	23.6	61.0
Hispanic origin[1]	67.2	26.3	18.8	60.9
85 years and over	46.9	10.1	43.4	79.8
White	48.7	10.3	42.0	79.1
Black	(B)	8.6	(B)	85.6
Hispanic origin[1]	(B)	(B)	(B)	(B)

Source: "Percentage of Persons 65 and Over, by Marital Status, Age, Race, and Hispanic Origin: 1990," Cynthia Taeuber, *Sixty-Five Plus in America*, p. 4- 6. Primary source: Arlene F. Saluter, U.S. Bureau of the Census, *Marital Status and Living Arrangements: March 1990*, Current Population Reports, Series P-20, No. 450. Washington, DC, U.S. Government Printing Office, May 1991, table 1. *Notes:* B Base is less than 75,000. 1. Hispanic origin may be of any race.

★ 1030 ★

Senior Citizens

Marital Status of Senior Population: United States, March 1990. Part I

In thousands.

Subject	White				Black				Hispanic origin[1]			
	65 years and over	65 to 74 years	75 to 84 years	85 years and over	65 years and over	65 to 74 years	75 to 84 years	85 years and over	65 years and over	65 to 74 years	75 to 84 years	85 years and over
United States, total												
Male, total	11,035	7,182	3,195	658	1,003	626	302	75	467	340	101	26
Never married	445	321	105	18	57	39	11	7	12	8	4	1
Married, spouse present	8,416	5,768	2,327	321	544	348	170	26	343	263	68	12
Married, spouse absent	183	102	53	27	84	47	35	2	34	27	6	1
Separated	109	75	25	8	65	36	29	-	24	19	6	-
Other	74	27	28	19	19	11	6	2	9	8	-	1
Widowed	1,473	590	607	277	235	124	71	40	59	27	19	13
Divorced	519	400	103	15	83	68	15	-	20	15	5	-
Unmarried	2,437	1,311	815	310	376	231	98	47	91	·50	27	14
Female, total	15,444	8,867	5,245	1,332	1,484	882	472	130	557	373	154	30
Never married	757	394	275	87	79	51	23	5	30	19	8	3
Married, spouse present	6,342	4,721	1,483	138	376	261	104	11	223	180	40	2
Married, spouse absent	199	140	54	5	93	65	26	2	21	15	7	-
Separated	91	77	11	3	88	60	26	2	16	11	6	-
Other	109	63	43	2	5	5	-	-	5	4	1	-
Widowed	7,423	3,115	3,255	1,053	797	398	288	111	235	118	94	24

[Continued]

★ 1030 ★

Marital Status of Senior Population: United States, March 1990. Part I
[Continued]

Subject	White				Black				Hispanic origin[1]			
	65 years and over	65 to 74 years	75 to 84 years	85 years and over	65 years and over	65 to 74 years	75 to 84 years	85 years and over	65 years and over	65 to 74 years	75 to 84 years	85 years and over
Divorced	722	497	177	48	138	106	31	1	48	42	5	1
Unmarried	8,902	4,006	3,707	1,189	1,015	556	342	117	313	179	106	27

Source: "Marital Status of Persons 15 Years and Over, by Age, Sex, Race, Hispanic Origin, Region, Metropolitan, and Nonmetropolitan: March 1990," Cynthia Taeuber, *Sixty-Five Plus in America*, p. 8-22-8-23. Adapted by the editors. Primary source: U.S. Bureau of the Census, unpublished tables consistent with Marital Status and Living Arrangements: March 1990, Current Population Reports, Series P-20, No. 450. U.S. Government Printing Office, Washington, DC, May 1991, table 1. *Notes:* - Represents zero or rounds to zero. 1. Hispanic origin may be of any race.

★ 1031 ★

Senior Citizens

Marital Status of Senior Population: Northeast, March 1990. Part II

In thousands.

Subject	White				Black				Hispanic origin[1]			
	65 years and over	65 to 74 years	75 to 84 years	85 years and over	65 years and over	65 to 74 years	75 to 84 years	85 years and over	65 years and over	65 to 74 years	75 to 84 years	85 years and over
Northeast												
Male, total	2,517	1,647	710	161	200	132	56	12	52	40	11	1
Never married	147	119	26	2	18	14	4	-	1	1	-	-
Married, spouse present	1,829	1,270	488	71	92	62	25	5	44	34	10	1
Married, spouse absent	48	32	9	6	25	15	10	-	2	1	1	-
Separated	30	27	1	3	20	14	7	-	2	1	1	-
Other	18	5	9	4	5	2	3	-	-	-	-	-
Widowed	415	163	172	81	42	18	17	7	4	2	1	1
Divorced	78	63	15	1	23	23	-	-	1	1	-	-
Unmarried	640	345	213	83	83	55	21	7	6	5	1	1
Female, total	3,634	2,059	1,221	354	251	148	82	21	101	60	32	9
Never married	337	167	128	42	27	17	9	1	15	9	5	1
Married, spouse present	1,402	1,068	299	35	64	45	19	-	29	25	2	1
Married, spouse absent	52	38	13	1	7	4	1	1	6	3	3	-
Separated	29	23	6	-	7	4	1	1	6	3	3	-
Other	23	15	7	1	-	-	-	-	-	-	-	-
Widowed	1,719	703	753	263	132	66	48	18	46	19	20	7
Divorced	124	83	29	12	21	16	5	-	5	3	2	-
Unmarried	2,181	953	910	318	180	99	61	20	66	32	27	8

Source: "Marital Status of Persons 15 Years and Over, by Age, Sex, Race, Hispanic Origin, Region, Metropolitan, and Nonmetropolitan: March 1990," Cynthia Taeuber, *Sixty-Five Plus in America*, p. 8-22-8-23. Adapted by the editors. Primary source: U.S. Bureau of the Census, unpublished tables consistent with Marital Status and Living Arrangements: March 1990, Current Population Reports, Series P-20, No. 450. U.S. Government Printing Office, Washington, DC, May 1991, table 1. *Notes:* - Represents zero or rounds to zero. 1. Hispanic origin may be of any race.

★ 1032 ★

Senior Citizens

Marital Status of Senior Population: Midwest, March 1990. Part III

In thousands.

Subject	White				Black				Hispanic origin[1]			
	65 years and over	65 to 74 years	75 to 84 years	85 years and over	65 years and over	65 to 74 years	75 to 84 years	85 years and over	65 years and over	65 to 74 years	75 to 84 years	85 years and over
Midwest												
Male, total	2,786	1,822	803	162	173	118	48	7	24	17	6	-
Never married	133	97	35	1	7	7	-	-	-	-	-	-
Married, spouse present	2,161	1,481	596	84	99	66	27	6	20	15	5	-
Married, spouse absent	41	17	20	4	20	9	9	2	1	1	-	-
Separated	26	14	11	-	14	5	9	-	-	-	-	-
Other	15	3	9	4	6	4	-	2	1	1	-	-
Widowed	335	137	129	69	39	27	12	-	2	1	1	-
Divorced	116	90	23	3	8	8	-	-	1	-	1	-
Unmarried	584	324	187	74	54	43	12	-	3	1	2	-
Female, total	4,042	2,212	1,464	367	265	161	78	25	20	15	2	3
Never married	211	108	79	24	2	2	-	-	2	-	-	2
Married, spouse present	1,649	1,191	428	30	77	50	25	2	6	6	-	-
Married, spouse absent	46	27	16	3	13	11	2	-	1	1	-	-
Separated	14	11	-	3	11	9	2	-	-	-	-	-
Other	33	16	16	-	2	2	-	-	1	1	-	-
Widowed	1,966	775	894	297	142	74	45	23	8	5	2	1
Divorced	170	111	46	13	32	25	7	-	3	3	-	-
Unmarried	2,348	994	1,020	334	176	101	52	23	13	8	2	3

Source: "Marital Status of Persons 15 Years and Over, by Age, Sex, Race, Hispanic Origin, Region, Metropolitan, and Nonmetropolitan: March 1990," Cynthia Taeuber, *Sixty-Five Plus in America*, p. 8-24-8-25. Adapted by the editors. Primary source: U.S. Bureau of the Census, unpublished tables consistent with Marital Status and Living Arrangements: March 1990, Current Population Reports, Series P-20, No. 450. U.S. Government Printing Office, Washington, DC, May 1991, table 1. *Notes:* - Represents zero or rounds to zero. 1. Hispanic origin may be of any race.

★ 1033 ★

Senior Citizens

Marital Status of Senior Population: South, March 1990. Part IV

In thousands.

Subject	White				Black				Hispanic origin[1]			
	65 years and over	65 to 74 years	75 to 84 years	85 years and over	65 years and over	65 to 74 years	75 to 84 years	85 years and over	65 years and over	65 to 74 years	75 to 84 years	85 years and over
South												
Male, total	3,582	2,284	1,081	216	544	320	180	43	200	129	56	14
Never married	75	44	27	4	25	11	7	7	4	2	2	-
Married, spouse present	2,819	1,906	803	111	311	184	114	14	133	93	34	6
Married, spouse absent	53	23	19	11	34	23	11	-	16	10	5	1
Separated	31	17	9	4	26	18	8	-	14	9	5	-
Other	23	6	10	7	8	5	2	-	2	2	-	1
Widowed	469	187	196	86	131	68	40	23	35	16	12	7
Divorced	164	124	36	4	42	34	8	-	11	8	3	-
Unmarried	709	355	259	94	199	113	56	30	50	26	17	7
Female, total	5,019	2,965	1,689	365	869	515	277	77	215	143	64	8
Never married	141	82	48	12	47	29	14	3	6	3	3	-
Married, spouse present	2,083	1,559	482	42	217	153	55	9	95	74	20	1
Married, spouse absent	61	41	20	-	56	46	10	-	7	5	3	-

[Continued]

★ 1033 ★

Marital Status of Senior Population: South, March 1990. Part IV

[Continued]

Subject	White				Black				Hispanic origin[1]			
	65 years and over	65 to 74 years	75 to 84 years	85 years and over	65 years and over	65 to 74 years	75 to 84 years	85 years and over	65 years and over	65 to 74 years	75 to 84 years	85 years and over
Separated	27	25	2	-	53	43	10	-	5	4	2	-
Other	34	16	18	-	3	3	-	-	2	1	1	-
Widowed	2,493	1,107	1,088	298	471	230	178	63	87	46	36	6
Divorced	240	175	52	13	79	58	19	1	19	16	3	1
Unmarried	2,875	1,364	1,188	323	596	317	212	68	113	65	41	7

Source: "Marital Status of Persons 15 Years and Over, by Age, Sex, Race, Hispanic Origin, Region, Metropolitan, and Nonmetropolitan: March 1990," Cynthia Taeuber, *Sixty-Five Plus in America*, p. 8-24-8-25. Adapted by the editors. Primary source: U.S. Bureau of the Census, unpublished tables consistent with Marital Status and Living Arrangements: March 1990, Current Population Reports, Series P-20, No. 450. U.S. Government Printing Office, Washington, DC, May 1991, table 1. *Notes:* - Represents zero or rounds to zero. 1. Hispanic origin may be of any race.

★ 1034 ★

Senior Citizens

Marital Status of Senior Population: West, March 1990. Part V.

In thousands.

Subject	White				Black				Hispanic origin[1]			
	65 years and over	65 to 74 years	75 to 84 years	85 years and over	65 years and over	65 to 74 years	75 to 84 years	85 years and over	65 years and over	65 to 74 years	75 to 84 years	85 years and over
West												
Male, total	2,150	1,429	601	120	86	56	18	12	192	154	27	11
Never married	90	61	18	11	7	7	-	-	7	5	1	1
Married, spouse present	1,606	1,111	440	55	42	35	4	2	146	121	19	5
Married, spouse absent	41	30	4	6	5	-	5	-	14	14	-	-
Separated	22	17	4	1	5	-	5	-	9	9	-	-
Other	19	13	-	5	-	-	-	-	6	6	-	-
Widowed	253	103	110	40	22	11	2	10	19	7	6	5
Divorced	160	123	29	8	11	3	7	-	6	6	-	-
Unmarried	503	287	157	59	39	21	9	10	32	18	8	6
Female, total	2,748	1,632	871	245	99	58	35	7	221	155	55	11
Never married	67	37	21	9	4	4	-	-	7	7	-	-
Married, spouse present	1,209	903	275	31	19	14	5	-	93	75	18	1
Married, spouse absent	40	34	5	1	17	4	13	-	7	6	1	-
Separated	21	18	4	-	17	4	13	-	5	4	1	-
Other	18	16	2	1	-	-	-	-	2	2	-	-
Widowed	1,244	530	519	195	52	28	17	7	93	48	36	10
Divorced	188	128	51	9	8	8	-	-	20	20	-	-
Unmarried	1,499	695	590	214	63	39	17	7	121	74	36	11

Source: "Marital Status of Persons 15 Years and Over, by Age, Sex, Race, Hispanic Origin, Region, Metropolitan, and Nonmetropolitan: March 1990," Cynthia Taeuber, *Sixty-Five Plus in America*, p. 8-26-8-27. Adapted by the editors. Primary source: U.S. Bureau of the Census, unpublished tables consistent with Marital Status and Living Arrangements: March 1990, Current Population Reports, Series P-20, No. 450. U.S. Government Printing Office, Washington, DC, May 1991, table 1. *Notes:* - Represents zero or rounds to zero. 1. Hispanic origin may be of any race.

★ 1035 ★

Senior Citizens

Marital Status of Senior Population: Metropolitan, March 1990. Part VI.

In thousands.

Subject	White				Black				Hispanic origin[1]			
	65 years and over	65 to 74 years	75 to 84 years	85 years and over	65 years and over	65 to 74 years	75 to 84 years	85 years and over	65 years and over	65 to 74 years	75 to 84 years	85 years and over
Metropolitan												
Male, total	7,976	5,242	2,253	481	794	501	234	59	414	301	91	23
Never married	341	255	71	15	53	37	10	7	10	7	3	-
Married, spouse present	5,981	4,155	1,611	214	412	266	128	18	304	233	60	10
Married, spouse absent	130	77	36	18	65	36	26	2	30	26	4	-
Separated	83	58	17	8	50	27	23	-	22	18	4	-
Other	47	19	18	10	15	10	3	2	8	8	-	-
Widowed	1,131	448	466	218	195	108	55	32	52	21	19	12
Divorced	393	308	70	15	69	54	15	-	19	14	4	-
Unmarried	1,865	1,011	606	248	318	199	80	39	81	42	26	12
Female, total	11,228	6,494	3,807	928	1,169	695	378	97	512	340	143	30
Never married	605	311	231	63	54	30	20	4	30	19	8	3
Married, spouse present	4,519	3,394	1,019	105	274	196	76	2	197	159	36	2
Married, spouse absent	146	100	43	2	86	62	22	2	20	13	7	-
Separated	72	61	11	-	81	57	22	2	15	10	6	-
Other	73	39	31	2	5	5	-	-	5	4	1	-
Widowed	5,388	2,298	2,372	718	638	321	229	88	219	108	87	23
Divorced	572	389	143	40	117	86	31	-	46	41	5	1
Unmarried	6,564	2,999	2,746	820	810	437	280	92	295	168	100	27

Source: "Marital Status of Persons 15 Years and Over, by Age, Sex, Race, Hispanic Origin, Region, Metropolitan, and Nonmetropolitan: March 1990," Cynthia Taeuber, *Sixty-Five Plus in America*, p. 8-26-8-27. Adapted by the editors. Primary source: U.S. Bureau of the Census, unpublished tables consistent with Marital Status and Living Arrangements: March 1990, Current Population Reports, Series P-20, No. 450. U.S. Government Printing Office, Washington, DC, May 1991, table 1. *Notes:* - Represents zero or rounds to zero. 1. Hispanic origin may be of any race.

★ 1036 ★

Senior Citizens

Marital Status of Senior Population: Nonmetropolitan, March 1990. Part VII.

In thousands.

Subject	White				Black				Hispanic origin[1]			
	65 years and over	65 to 74 years	75 to 84 years	85 years and over	65 years and over	65 to 74 years	75 to 84 years	85 years and over	65 years and over	65 to 74 years	75 to 84 years	85 years and over
Nonmetropolitan												
Male, total	3,059	1,940	942	177	209	125	68	15	53	39	10	4
Never married	104	67	34	3	4	3	1	-	2	1	1	1
Married, spouse present	2,435	1,613	715	106	132	82	42	8	39	31	7	1
Married, spouse absent	53	26	18	9	19	11	8	-	4	2	2	1
Separated	26	18	8	-	16	9	6	-	3	1	2	-
Other	27	8	10	9	4	1	2	-	1	1	-	1
Widowed	342	142	141	58	40	16	17	7	7	6	-	1
Divorced	126	92	34	-	14	14	-	-	1	1	-	-
Unmarried	572	301	209	62	58	33	18	7	10	7	1	2
Female, total	4,215	2,373	1,438	404	314	187	94	33	44	33	11	-
Never married	152	83	45	25	25	22	3	-	-	-	-	-
Married, spouse present	1,824	1,327	465	32	102	66	28	9	25	21	4	-
Married, spouse absent	54	40	12	3	7	3	4	-	1	1	-	-

[Continued]

★ 1036 ★

Marital Status of Senior Population: Nonmetropolitan, March 1990. Part VII.

[Continued]

Subject	White				Black				Hispanic origin[1]			
	65 years and over	65 to 74 years	75 to 84 years	85 years and over	65 years and over	65 to 74 years	75 to 84 years	85 years and over	65 years and over	65 to 74 years	75 to 84 years	85 years and over
Separated	18	16	-	3	7	3	4	-	1	1	-	-
Other	36	24	12	-	-	-	-	-	-	-	-	-
Widowed	2,035	817	883	336	159	77	58	23	16	10	6	-
Divorced	151	107	34	9	21	20	-	1	2	2	-	-
Unmarried	2,338	1,007	962	369	205	119	62	25	18	11	6	-

Source: "Marital Status of Persons 15 Years and Over, by Age, Sex, Race, Hispanic Origin, Region, Metropolitan, and Nonmetropolitan: March 1990," Cynthia Taeuber, *Sixty-Five Plus in America*, p. 8-28-8-29. Adapted by the editors. Primary source: U.S. Bureau of the Census, unpublished tables consistent with Marital Status and Living Arrangements: March 1990, Current Population Reports, Series P-20, No. 450. U.S. Government Printing Office, Washington, DC, May 1991, table 1. *Notes:* - Represents zero or rounds to zero. 1. Hispanic origin may be of any race.

★ 1037 ★

Senior Citizens

Metropolitan and Nonmetropolitan Area Residents 65 and Over, 1990

Area, race, Hispanic origin, and sex	65 years and over	65 to 69 years	70 to 74 years	75 to 79 years	80 years and over	80 to 84 years	85 years and over
Metropolitan							
Total	23,004,177	7,558,466	5,892,987	4,463,816	5,088,908	2,855,256	2,233,652
Male	9,160,888	3,362,778	2,490,975	1,726,348	1,580,787	971,978	608,809
Female	13,843,289	4,195,688	3,402,012	2,737,468	3,508,121	1,883,278	1,624,843
White	20,270,085	6,552,823	5,188,426	3,960,211	4,568,625	2,559,144	2,009,481
Male	8,074,947	2,932,036	2,205,054	1,533,836	1,404,021	865,502	538,519
Female	12,195,138	3,620,787	2,983,372	2,426,375	3,164,604	1,693,642	1,470,962
Black	1,983,740	704,538	509,664	372,955	396,583	223,160	173,423
Male	760,891	297,024	201,922	136,754	125,191	74,877	50,314
Female	1,222,849	407,514	307,742	236,201	271,392	148,283	123,109
American Indian, Eskimo, and Aleut	53,909	21,030	13,900	9,567	9,412	5,495	3,917
Male	21,943	9,349	5,846	3,628	3,120	1,858	1,262
Female	31,966	11,681	8,054	5,939	6,292	3,637	2,655
Asian and Pacific Islander	421,806	166,601	113,880	74,272	67,053	40,233	26,820
Male	188,569	74,014	50,385	33,731	30,439	19,363	11,076
Female	233,237	92,587	63,495	40,541	36,614	20,870	15,744
Other	274,637	113,474	67,117	46,811	47,235	27,224	20,011
Male	114,538	50,355	27,768	18,399	18,016	10,378	7,638
Female	160,099	63,119	39,349	28,412	29,219	16,846	12,373
Hispanic origin[1]	1,029,247	388,641	254,220	188,280	198,106	114,726	83,380
Male	421,722	172,993	104,961	72,498	71,270	42,328	28,942
Female	607,525	215,648	149,259	115,782	126,836	72,398	54,438
Nonmetropolitan							
Total	8,237,654	2,553,269	2,101,836	1,657,553	1,924,996	1,078,483	846,513
Male	3,404,285	1,169,529	918,331	673,420	643,005	394,116	248,889
Female	4,833,369	1,383,740	1,183,505	984,133	1,281,991	684,367	597,624
White	7,581,888	2,346,814	1,938,138	1,524,814	1,772,122	993,551	778,571
Male	3,139,962	1,081,193	850,675	620,563	587,531	361,600	225,931

[Continued]

★ 1037 ★

Metropolitan and Nonmetropolitan Area Residents 65 and Over, 1990

[Continued]

Area, race, Hispanic origin, and sex	65 years and over	65 to 69 years	70 to 74 years	75 to 79 years	80 years and over	80 to 84 years	85 years and over
Female	4,441,926	1,265,621	1,087,463	904,251	1,184,591	631,951	552,640
Black	524,811	158,507	130,751	108,315	127,238	70,478	56,760
Male	204,541	65,918	52,777	41,786	44,060	25,782	18,278
Female	320,270	92,589	77,974	66,529	83,178	44,696	38,482
American Indian, Eskimo, and Aleut	60,544	21,680	15,370	11,585	11,909	6,621	5,288
Male	26,146	9,949	6,654	4,832	4,711	2,699	2,012
Female	34,398	11,731	8,716	6,753	7,198	3,922	3,276
Asian and Pacific Islander	32,652	11,896	8,354	5,867	6,535	3,617	2,918
Male	15,878	5,504	3,915	3,070	3,389	2,066	1,323
Female	16,774	6,392	4,439	2,797	3,146	1,551	1,595
Other	37,759	14,372	9,223	6,972	7,192	4,216	2,976
Male	17,758	6,965	4,310	3,169	3,314	1,969	1,345
Female	20,001	7,407	4,913	3,803	3,878	2,247	1,631
Hispanic origin[1]	132,036	47,616	32,552	24,985	26,883	15,699	11,184
Male	59,687	22,585	14,844	10,857	11,401	6,846	4,555
Female	72,349	25,031	17,708	14,128	15,482	8,853	6,629

Source: "Population of the United States for Metropolitan/Nonmetropolitan Areas, by Age, Sex, Race, and Hispanic Origin: 1990," Cynthia Taeuber, *Sixty-Five Plus in America*, p. 5-13. Primary source: U.S. Bureau of the Census, 1990 Census of Population and Housing, Summary Tape File 1A. *Note:* 1. Hispanic origin may be of any race.

★ 1038 ★

Senior Citizens

Parent Support Ratio, 1990 and 2050

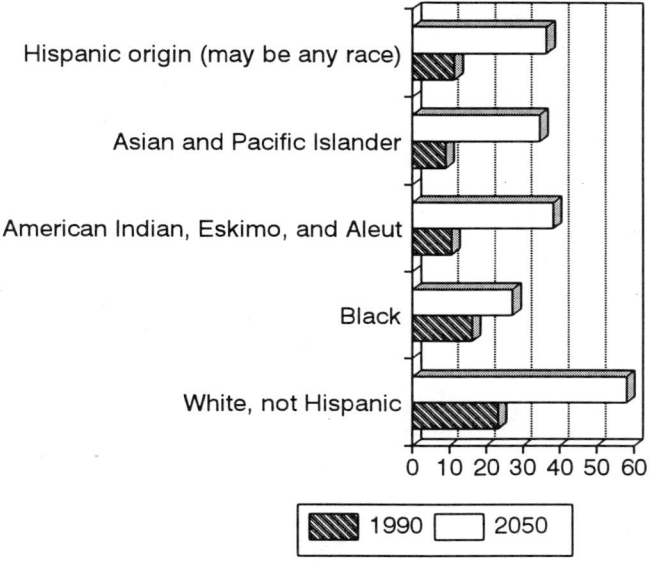

Number of persons 80 years and over per 100 persons aged 50 to 64 years in a specified race group or for Hispanic origin.

Race	1990	2050
White, not Hispanic	23.1	58.0
Black	16.0	27.2
American Indian, Eskimo, and Aleut	10.7	38.0
Asian and Pacific Islander	9.1	34.4
Hispanic origin (may be any race)	11.3	36.4

Source: "Parent Support Ratio, by Race and Hispanic Origin: 1990 and 2050." *Profiles of America's Elderly*, No. 3 (November 1993), p. 4. Primary source: U.S. Bureau of the Census, Fred Hollmann, *U.S. Population Estimates, by Age, Sex, Race, and Hispanic Origin: 1980 to 1991*, Current Population Reports, P25-1095, U.S. Government Printing Office, Washington, DC, November 1992; Jennifer C. Day, *Population Projections of the United States, by Age, Sex, Race, and Hispanic Origin: 1993 to 2050*, Current Population Reports, P25-1104, U.S. Government Printing Office, Washington, DC, 1993 (middle series projections).

★ 1039 ★

Senior Citizens

Parent and Sandwich Generation Support Ratios, 1990 and 2050

Ratio/race	1950	1990	2010	2030	2050
Parent support ratio[1]					
Total	3	9	10	15	27
White	3	10	11	16	30
Black	3	7	7	11	21
Other races	2	4	7	12	20
Hispanic origin[2]	(NA)	5	6	9	19
Sandwich generation[3]					
Total	144	228	167	296	269
White	148	235	173	314	287
Black	97[4]	195	136	255	231
Hispanic origin[2]	(NA)	159	114	211	210

Source: "Parent and Generation Support Ratios: 1950 to 2050," Cynthia Taeuber, *Sixty-Five Plus in America*, p. 2-15. Primary source: U.S. Bureau of the Census, 1950 from 1950 Census of Population, Volume 2, Part 1, Chapter C, table 112; 1990 from 1990 Census of Population and Housing, Series CPH-L-74, *Modified and Actual Age, Sex, Race, and Hispanic Origin Data;* 2010 to 2050 from *Population Projections of the United States, by Age, Sex, Race, and Hispanic Origin: 1992 to 2050*, Current Population Reports, P25-1092. U.S. Government Printing Office, Washington, DC, 1992 (middle series projections). *Notes:* NA Not available. 1. Ratio of persons 85 years old and over to persons 50 to 64 years old. 2. Hispanic origin may be of any race. 3. Ratio of persons aged 18 to 22 enrolled in college plus persons aged 65 to 79 persons aged 45 to 49 years. College enrollment for 1990-2050 is based on 1989 rates for 18 to 22 year olds (total, 37.3 percent; White, 38.7 percent; Blacks, 27.9 percent; Hispanics, 24.7 percent). 4. 1950 data are for "Blacks and other races" combined. Over 90 percent of "Black and other races" were Black in 1950.

★ 1040 ★

Senior Citizens

Percent White and Percent Black of the Total Population 65 Years and Over, 1980 to 2050

	Percent White of the total population 65 years and over 1980 to 2050	Black population 65 years and over 1980 to 2050 (in millions)
1980	89.8	2.1
1990	90.2	2.5
2000	88.5	2.9
2010	86.6	3.6
2020	84.9	5.2
2030	83.4	7.4

[Continued]

★ 1040 ★

Percent White and Percent Black of the Total Population 65 Years and Over, 1980 to 2050
[Continued]

	Percent White of the total population 65 years and over 1980 to 2050	Black population 65 years and over 1980 to 2050 (in millions)
2040	81.3	8.5
2050	79.1	9.4

Source: "Percent White of the Total Population 65 Years and Over: 1980 to 2050," and "Percent Black of the Total Population 65 Years and Over: 1980 to 2050," Cynthia Taeuber, *Sixty-Five Plus in America*, p. 2-12. Primary source: U.S. Bureau of the Census, 1980 from 1980 Census of Population; 1990 from 1990 Census of Population and Housing, Series CPH-L-74, *Modified and Actual Age, Sex, Race, and Hispanic Origin Data*; 2000 to 2050 from Jennifer C. Day, *Population Projections of the United States, by Age, Sex, Race, and Hispanic Origin; 1992 to 2050*, Current Population Reports, P25-1092. U.S. Government Printing Office, Washington, DC, 1992 (middle series projections).

★ 1041 ★

Senior Citizens

Percent of Seniors Educated, 1991

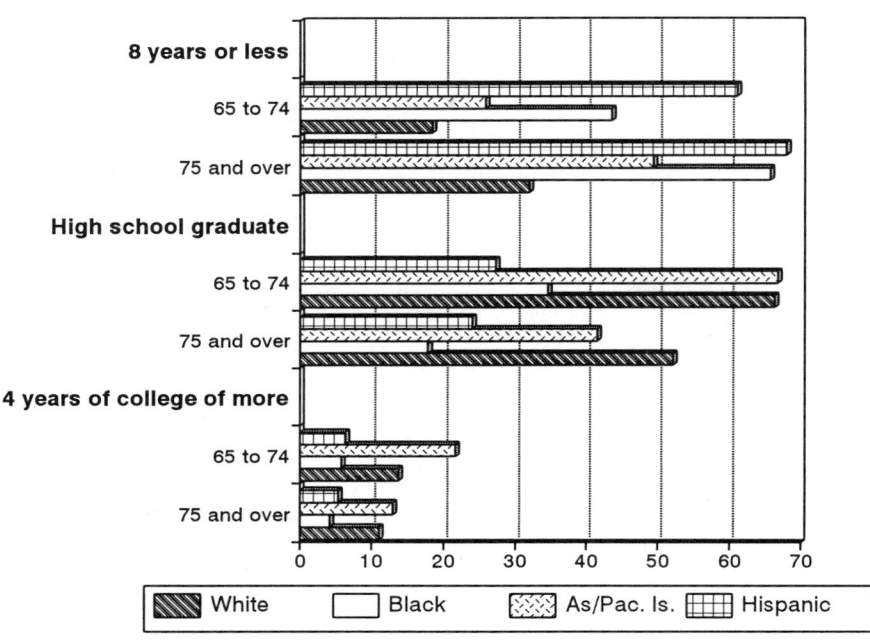

Percent of noninstitutional population.

Age	White	Black	Asian and Pacific Islander	Hispanic origin[1]
8 years or less				
65 to 74	18.4	43.5	25.9	61.1
75 and over	32.0	65.7	49.3	68.1
High school graduate				
65 to 74	66.4	34.5	66.8	27.3
75 and over	52.1	17.8	41.4	24.0
4 years of college or more				
65 to 74	13.8	5.8	21.7	6.4
75 and over	11.0	4.1	12.9	5.4

Source: "Years of School Completed, by Age, Race, and Hispanic Origin: 1991." *Profiles of America's Elderly,* No. 3 (November 1993), p. 7. Primary source: Robert Kominski and Andrea Adams, U.S. Bureau of the Census, *Educational Attainment in the United States: March 1991 and 1990,* Current Population Reports, Series P-20, No. 462, table 1, for White, Black and Hispanic-origin data. Data for Asian and Pacific Islanders are from unpublished special tabulations consistent with P-20, No. 462, from the March 1991 Current Population Survey, available from Claudette Bennett, Racial Statistics Branch, Population Division. *Notes:* The numbers of elderly American Indian, Eskimo, and Aleut are not shown as the population is too small for the data to be reliable. 1. May be of any race.

★ 1042 ★

Senior Citizens

Percentage Married with Spouse Present

	Male	Female
65 to 74		
White	80.3	53.2
Hispanic origin[1]	77.4	48.2
Black	55.5	29.6
75 to 84		
White	72.8	28.3
Hispanic origin[1]	67.2	26.3
Black	56.1	22.0
85 and over		
White	48.7	10.3
Black	(B)	8.6
Hispanic origin[1]	(B)	(B)

Source: "Percentage of Persons 65 Years and Over Who are Married with Spouse Present, by Age, Sex, Race, and hispanic Origin: March 1990," Cynthia Taeuber, *Sixty-Five Plus in America*, p. 3-6. Primary source: Arlene F. Saluter, U.S. Bureau of the Census, *Marital Status and Living Arrangements: March 1990*, Current Population Reports, Series P-20, No. 450. U.S. Government Printing Office, Washington, DC, May 1991, table 1. *Notes:* B Base is less than 75,000. 1. Hispanic origin may be of any race.

★ 1043 ★

Senior Citizens

Percentage of Persons 65 and Over Living Alone, March 1990

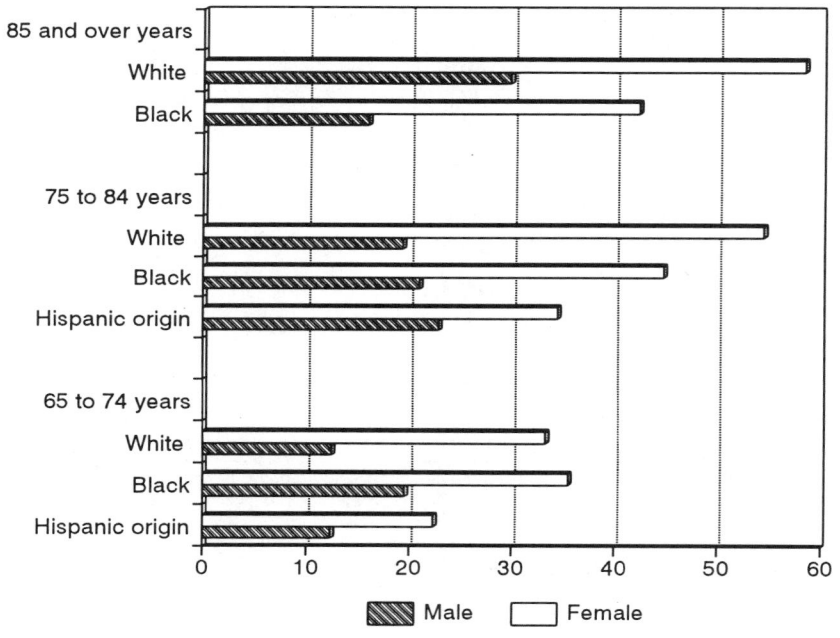

Male　　　□ Female

Noninstitutional population.

Age, and race	Male	Female
65 to 74 years		
White	12.5	33.2
Black	19.6	35.4
Hispanic origin[1]	12.4	22.3
75 to 84 years		
White	19.3	54.5
Black	20.9	44.7
Hispanic origin[1]	22.8	34.4
85 and over years		
White	29.8	58.4
Black	16.0	42.3
Hispanic origin[1]	(B)	(B)

Source: "Percentage of Persons 65 Years and Over Living Alone, by Age, Race, and Hispanic Origin: March 1990," Cynthia Taeuber, *Sixty-Five Plus in America*, p. 6-6. Primary source: Data for 1990: Arlene F. Saluter, U.S. Bureau of the Census, *Marital Status and Living Arrangements: March 1990*, Current Population Reports, Series P-20, No. 450. U.S. Government Printing Office, Washington, DC, May 1991, table 7. *Notes:* B Base is less than 75,000. 1. Hispanic origin may be of any race.

★ 1044 ★

Senior Citizens

Percentage of Persons 65 and Over Living Alone, March 1992

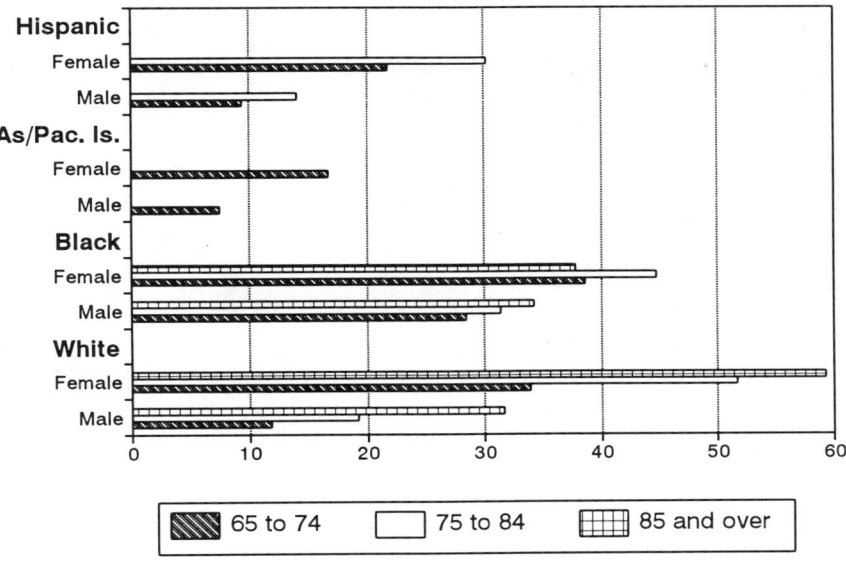

Noninstitutional population.

Age	White		Black		Asian and Pacific Islander		Hispanic origin[1]	
	Male	Female	Male	Female	Male	Female	Male	Female
65 to 74	11.8	33.9	28.4	38.6	7.5	16.7	9.4	21.8
75 to 84	19.2	51.7	31.4	44.8	(B)	(B)	14.1	30.2
85 and over	31.7	59.3	34.2	37.8	(B)	(B)	(B)	(B)

Source: "Percentage of Persons 65 and Over Living Alone: March 1992." *Profiles of America's Elderly*, No. 3 (November 1993), p. 6. Primary source: U.S. Bureau of the Census, Arlene F. Saluter, *Marital Status and Living Arrangements: March 1992*, Current Population Reports, Series P20- 468, U.S. Government Printing Office, Washington, DC, December 1992, table 7. Data for Asian and Pacific Islanders are from unpublished special tabulations consistent with P-20-468 from the March 1992 Current Population Survey, available from Claudette Bennett, Racial Statistics Branch, Population Division. *Notes:* (B) Base is less than 75,000. The number of elderly American Indian, Eskimo, and Aleut are not shown as the population is too small for the data to be reliable. 1. May be of any race.

★ 1045 ★

Senior Citizens

Percentage of Persons 65 and Over Who are Widowed, March 1992

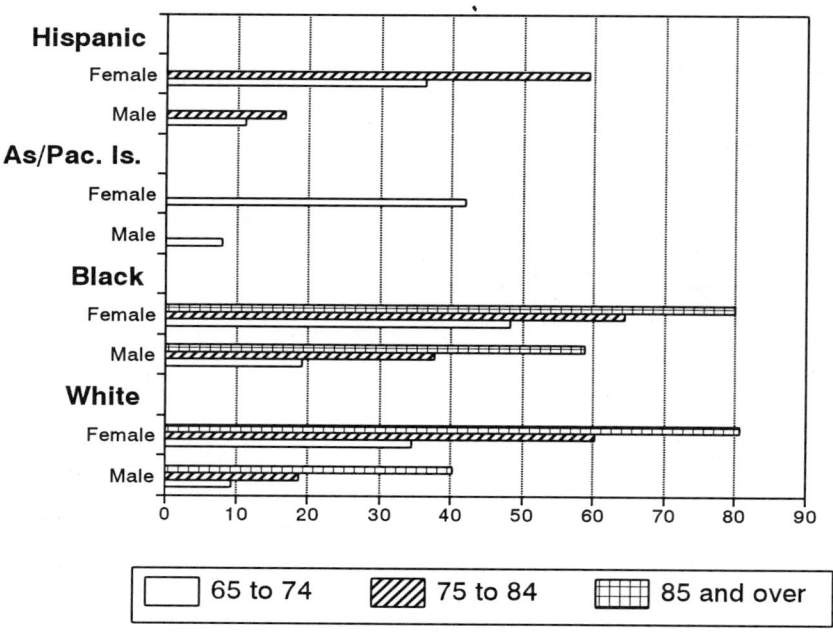

Noninstitutional population.

Age	White		Black		Asian and Pacific Islander		Hispanic origin[1]	
	Male	Female	Male	Female	Male	Female	Male	Female
65 to 74	9.3	34.4	19.0	48.3	7.9	41.9	11.0	36.3
75 to 84	18.6	60.2	37.6	64.4	(B)	(B)	16.6	59.2
85 and over	40.2	80.8	58.8	80.0	(B)	(B)	(B)	(B)

Source: "Percentage of Persons 65 and Over Who Are Widowed: March 1992." *Profiles of America's Elderly*, No. 3 (November 1993), p. 5. Primary source: U.S. Bureau of the Census, Arlene F. Saluter, *Marital Status and Living Arrangements: March 1992*, Current Population Reports, Series P20- 468, U.S. Government Printing Office, Washington, DC, December 1992, table 7. Data for Asian and Pacific Islanders are from unpublished special tabulations consistent with P-20-468 from the March 1992 Current Population Survey, available from Claudette Bennett, Racial Statistics Branch, Population Division. *Notes:* (B) Base is less than 75,000. The number of elderly American Indian, Eskimo, and Aleut are not shown as the population is too small for the data to be reliable. 1. May be of any race.

★ 1046 ★

Senior Citizens

Percentage of Persons 65 and Over Widowed, March 1990

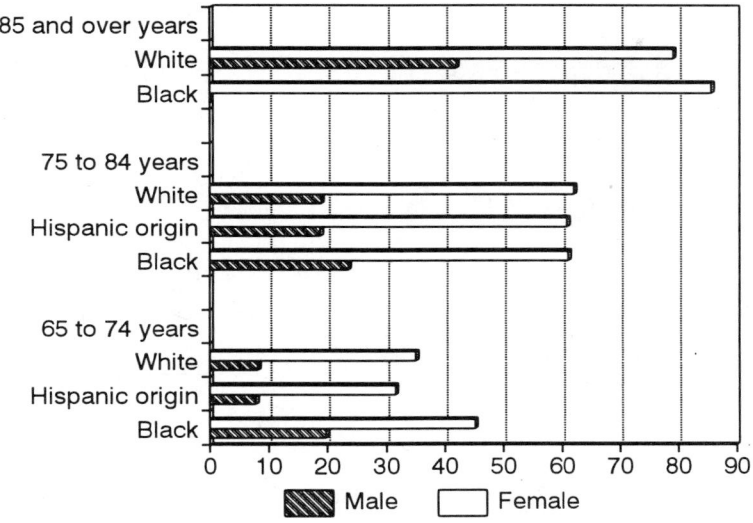

Noninstitutional population.

Age, and race	Male	Female
65 to 74 years		
White	8.2	35.1
Black	19.8	45.1
Hispanic origin[1]	7.9	31.5
75 to 84 years		
White	19.0	62.1
Black	23.6	61.0
Hispanic origin[1]	18.8	60.9
85 and over years		
White	42.0	79.1
Black	(B)	85.6
Hispanic origin[1]	(B)	(B)

Source: "Percentage of Persons 65 and Over Who are Widowed, by Age, Race, and Hispanic Origin: March 1990," Cynthia Taeuber, *Sixty-Five Plus in America*, p. 4-6. Primary source: Data for 1990: Arlene F. Saluter, U.S. Bureau of the Census, *Marital Status and Living Arrangements: March 1990*, Current Population Reports, Series P-20, No. 450. U.S. Government Printing Office, Washington, DC, May 1991, table 1. *Notes:* B Base is less than 75,000. 1. Hispanic origin may be of any race.

★ 1047 ★

Senior Citizens

Percentage of Population 65 Years and Over, Who are 80 or Older, 1990 and 2050

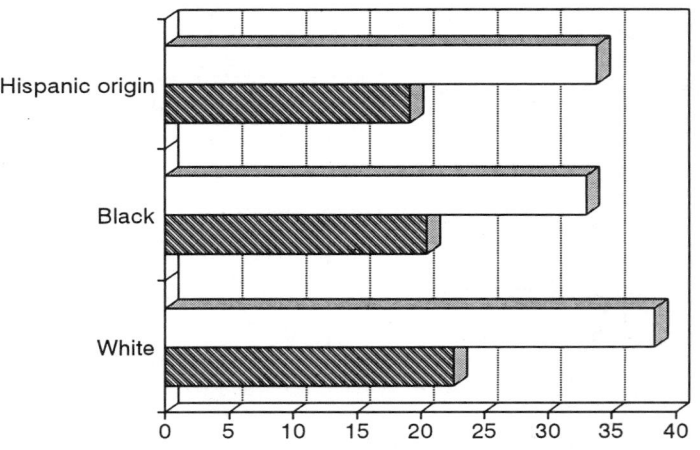

1990   2050

	1990	2050
White	22.6	38.3
Black	20.5	33.0
Hispanic origin[1]	19.2	33.8

Source: "Percentage of Population 65 Years and Over, Who are 80 and Older: 1980 to 2050," Cynthia Taeuber, *Sixty-Five Plus in America*, p. 2-15. Primary source: U.S. Bureau of the Census, 1980 from 1980 Census of Population; 1990 from 1990 Census of Population and Housing, Series CPH-L-74, *Modified and Actual Age, Sex, Race, and Hispanic Origin Data*; 2050 from Jennifer C. Day, *Population Projections of the United States, by Age, Sex, Race, and Hispanic Origin: 1992 to 2050*, Current Population Reports, P25-1092. U.S. Government Printing Office, Washington, DC, 1992 (middle series projections). *Note:* 1. Hispanic origin may be of any race.

★ 1048 ★

Senior Citizens

Percentage of Population 65 Years and Over, by Race and Hispanic Origin, 1990 and 2050

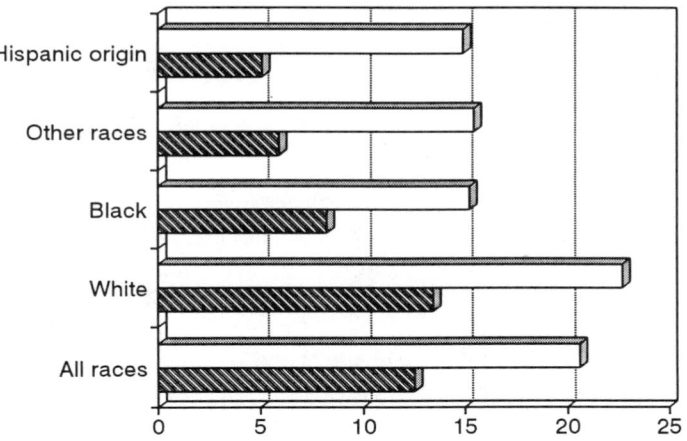

	1990	2050
All races	12.5	20.6
White	13.4	22.7
Black	8.2	15.2
Other races	5.9	15.4
Hispanic origin[1]	5.1	14.9

Source: "Percentage of Population 65 Years and Over, by Race and Hispanic Origin: 1980 to 2050," Cynthia Taeuber, Sixty-Five Plus in America, p. 2-14. Primary source: U.S. Bureau of the Census, 1980 from 1980 Census of Population; 1990 from 1990 Census of Population and Housing, Series CPH-L-74, *Modified and Actual Age, Sex, Race, and Hispanic Origin Data*; 2050 from Jennifer C. Day, *Population Projections of the United States, by Age, Sex, Race, and Hispanic Origin: 1992 to 2050*, Current Population Reports, P25-1092. U.S. Government Printing Office, Washington, DC, 1992 (middle series projections). *Note:* 1. Hispanic origin may be of any race.

★ 1049 ★

Senior Citizens

Persons 65 Years and Over by Race and Hispanic Origin, 1990 to 2050. Part I.

Numbers in thousands.

Year	Total, 65+ population Number	White			Black		
		Total	Not Hispanic		Total	Not Hispanic	
			Number	Percent of total 65+ population		Number	Percent of total 65+ population
1990	31,079	28,021	26,952	86.7	2,492	2,442	7.9
2000	35,322	31,357	29,575	83.7	2,919	2,825	8.0
2010	40,104	34,967	32,285	80.5	3,435	3,278	8.2
2020	53,348	45,748	41,405	77.6	4,863	4,590	8.6
2030	70,175	59,295	52,333	74.6	6,827	6,379	9.1
2040	77,014	63,858	54,509	70.8	7,739	7,149	9.3
2050	80,109	64,982	53,597	66.9	8,368	7,652	9.6

Source: "Persons 65 Years and Over by Race and Hispanic Origin: 1990 to 2050." Profiles of America's Elderly, No. 3 (November 1993), p. 3. Primary source: U.S. Bureau of the Census, 1990 from Current Population Reports, P25-1095, *U.S. Population Estimates, by Age, Sex, Race, and Hispanic Origin: 1980 to 1991,* November 1991. Data for 2000 shown in Current Population Reports, P25-1104, *Projections of the United States, by Age, Sex, Race, and Hispanic Origin: 1993 to 2050,* (middle series) U.S. Government Printing Office, Washington, DC, 1993.

★ 1050 ★

Senior Citizens

Persons 65 Years and Over by Race and Hispanic Origin, 1990 to 2050. Part II.

Numbers in thousands.

Year	American Indian, Eskimo, and Aleut			Asian and Pacific Islander			Hispanic origin (may be of any race)	
	Total	Not Hispanic		Total	Not Hispanic			
		Number	Percent of total 65+ population		Number	Percent of total 65+ population	Number	Percent of total 65+ population
1990	116	108	0.3	450	431	1.4	1,146	3.7
2000	161	148	0.4	884	849	2.4	1,925	5.4
2010	218	198	0.5	1,484	1,426	3.6	2,918	7.3
2020	310	276	0.5	2,427	2,330	4.4	4,747	8.9
2030	415	361	0.5	3,638	3,487	5.0	7,616	10.9
2040	480	412	0.5	4,936	4,724	6.1	10,220	13.3
2050	547	467	0.6	6,212	5,941	7.4	12,452	15.5

Source: "Persons 65 Years and Over by Race and Hispanic Origin: 1990 to 2050." Profiles of America's Elderly, No. 3 (November 1993), p. 3. Primary source: U.S. Bureau of the Census, 1990 from Current Population Reports, P25-1095, *U.S. Population Estimates, by Age, Sex, Race, and Hispanic Origin: 1980 to 1991,* November 1991. Data for 2000 shown in Current Population Reports, P25-1104, *Projections of the United States, by Age, Sex, Race, and Hispanic Origin: 1993 to 2050,* (middle series) U.S. Government Printing Office, Washington, DC, 1993.

★ 1051 ★

Senior Citizens

Persons 65 Years and Over, 1990-2050

In millions.

	1990	2050
All races		
65 years and over	31.1	78.9
65 to 79 years	24.1	49.5
80 years and over	6.9	29.4
White		
65 years and over	28.0	62.4
65 to 79 years	21.7	38.5
80 years and over	6.3	23.9
Black		
65 years and over	2.5	9.4
65 to 79 years	2.0	6.3
80 years and over	0.5	3.1
Other races[1]		
65 years and over	0.6	7.1
65 to 79 years	0.5	4.7
80 years and over	0.1	2.4
Hispanic origin[2]		
65 years and over	1.1	12.0
65 to 79 years	0.9	8.0
80 years and over	0.2	4.1

Source: "Persons 65 Years and Over, by Age, Race, and Hispanic Origin: 1990 and 2050," Cynthia Taeuber, *Sixty-Five Plus in America*, p. 2-11. Primary source: U.S. Bureau of the Census, 1990 from 1990 Census of Population and Housing, Series CPH-L-74, *Modified and Actual Age, Sex, Race, and Hispanic Origin Data*; 2050 from Jennifer C. Day, *Population Projections of the United States, by Age, Sex, Race, and Hispanic Origin: 1992 to 2050*, Current Population Reports, P25-1092. U.S. Government Printing Office, Washington, DC, 1992 (middle series projections). *Notes:* 1. Includes Asians and Pacific Islanders. 2. Hispanic origin may be of any race.

★ 1052 ★

Senior Citizens

Persons 65 Years and Over, by Specific Race and Hispanic Origin, 1980-1990

In millions.

	1980	1990
All races	25.6	31.1
White	22.9	28.0
Black	2.1	2.5
American Indian, Eskimo, or Aleut	0.1	0.1

[Continued]

★ 1052 ★

Persons 65 Years and Over, by Specific Race and Hispanic Origin, 1980-1990

[Continued]

	1980	1990
Asian or Pacific Islander	0.2	0.5
Hispanic origin[1]	0.7	1.1

Source: "Persons 65 Years and Over, by Specific Race and Hispanic Origin: 1980 and 1990," Cynthia Taeuber, *Sixty-Five Plus in America*, p. 2-10. Primary source: U.S. Bureau of the Census, 1980 Census of Population, General Social and Economic Characteristics, PC80-1-C1, U.S. Summary, U.S. Government Printing Office, Washington, DC, 1983, tables 120 and 130; 1990 Census of Population and Housing, Series CPH-L-74, *Modified and Actual Age, Sex, Race, and Hispanic Origin Data. Notes:* The data for 1980 does not distribute persons of unspecified races among the specified races as has been done in the 1990 data. Thus, those elderly who marked "other race" on the 1980 Census questionnaire are not included here. In data for 1990 from the CPH-L-74 series used here, persons who marked "other race" were assigned the race reported by a nearby person with an identical response to the Hispanic origin question. 1. Hispanic origin may be of any race.

★ 1053 ★
Senior Citizens

Persons 65 and Over Living Alone, March 1990

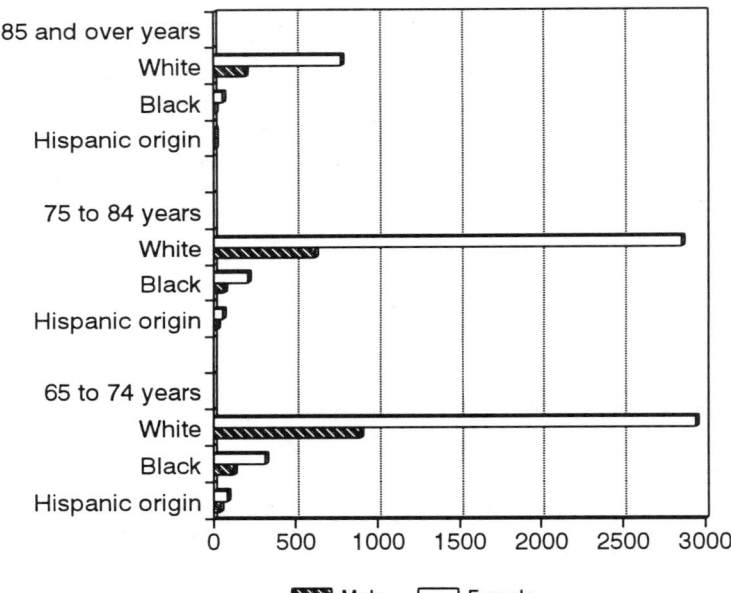

In thousands, noninstitutional population.

Age, and race	Male	Female
65 to 74 years		
White	896	2,945
Black	123	312
Hispanic origin[1]	42	83

[Continued]

★ 1053 ★

Persons 65 and Over Living Alone, March 1990
[Continued]

Age, and race	Male	Female
75 to 84 years		
White	617	2,857
Black	63	211
Hispanic origin[1]	23	53
85 and over years		
White	196	778
Black	12	55
Hispanic origin[1]	9	12

Source: "Number of Persons 65 Years and Over Living Alone, by Age, Race, and Hispanic Origin: March 1990," Cynthia Taeuber, *Sixty-Five Plus in America*, p. 5-6. Primary source: Data for 1990: Arlene F. Saluter, U.S. Bureau of the Census, *Marital Status and Living Arrangements: March 1990*, Current Population Reports, Series P-20, No. 450. U.S. Government Printing Office, Washington, DC, May 1991, table 1. *Note:* 1. Hispanic origin may be of any race.

★ 1054 ★

Senior Citizens

Persons 65 and Over in Places of 100,000 or More: 1990, Part I. - A

Rank based on total, all races, all ages.

Place	Rank	State	Total, all races			Black		American Indian, Eskimo and Aleut	
			All ages	65 years and over	85 years and over	65 years and over	85 years and over	65 years and over	85 years and over
New York city	1	NY	7,322,564	953,317	102,554	174,798	13,828	1,771	148
Los Angeles city	2	CA	3,485,398	347,713	35,419	51,893	4,535	958	84
Chicago city	3	IL	2,783,726	330,182	30,522	98,181	8,174	414	36
Houston city	4	TX	1,630,553	135,077	12,129	32,586	2,840	236	15
Philadelphia city	5	PA	1,585,577	240,714	22,801	70,253	5,822	361	29
San Diego city	6	CA	1,110,549	113,495	10,203	5,368	384	351	27
Detroit city	7	MI	1,027,974	124,933	12,506	77,444	7,005	295	24
Dallas city	8	TX	1,006,877	97,891	10,131	18,565	1,643	263	21
Phoenix city	9	AZ	983,403	95,226	7,832	3,379	343	469	31
San Antonio city	10	TX	935,933	98,365	9,421	6,895	667	235	17
San Jose city	11	CA	782,248	56,358	5,022	1,105	83	180	11
Baltimore city	12	MD	736,014	100,916	9,695	41,981	3,267	160	20
Indianapolis city	13	IN	731,327	83,628	8,505	15,330	1,551	125	12
San Francisco city	14	CA	723,959	105,380	12,148	9,932	763	196	16
Jacksonville city (remainder)	15	FL	635,230	67,343	5,792	14,182	1,199	94	5
Columbus city	16	OH	632,910	57,939	5,961	10,761	911	92	11
Milwaukee city	17	WI	628,088	78,145	8,781	8,578	611	197	12
Memphis city	18	TN	610,337	74,597	7,341	27,969	2,843	72	4
Washington city	19	DC	606,900	77,847	7,847	52,263	4,253	184	14
Boston city	20	MA	574,283	65,950	8,219	9,864	806	136	13
Seattle city	21	WA	516,259	78,400	9,271	4,577	311	391	20
El Paso city	22	TX	515,342	45,016	3,770	768	52	130	8

[Continued]

★ 1054 ★

Persons 65 and Over in Places of 100,000 or More: 1990, Part I. - A

[Continued]

Place	Rank	State	Total, all races			Black		American Indian, Eskimo and Aleut	
			All ages	65 years and over	85 years and over	65 years and over	85 years and over	65 years and over	85 years and over
Cleveland city	23	OH	505,616	70,753	6,612	27,266	2,272	119	9
New Orleans city	24	LA	496,938	64,658	6,656	27,279	2,346	64	7
Nashville-Davidson (remainder)	25	TN	488,374	55,826	5,747	10,087	1,008	64	6
Denver city	26	CO	467,610	64,805	7,648	5,207	486	267	25
Austin city	27	TX	465,622	34,577	3,709	3,858	420	66	6
Forth Worth city	28	TX	447,619	50,225	5,133	8,195	791	120	7
Oklahoma city	29	OK	444,719	52,779	5,310	5,653	570	1,277	86
Portland city	30	OR	437,319	63,657	7,551	2,730	164	270	16
Kansas City city	31	KS	435,146	56,166	6,393	11,910	1,119	139	10
Long Beach city	32	CA	429,433	46,463	5,426	2,087	162	165	12
Tucson city	33	AZ	405,390	51,198	5,228	1,360	120	315	38
St. Louis city	34	MO	396,685	66,001	8,389	23,236	2,515	98	11
Charlotte city	35	NC	395,934	38,802	3,603	8,188	642	56	3
Atlanta city	36	GA	394,017	44,432	5,071	24,039	2,157	43	4
Virginia Beach city	37	VA	393,069	23,214	1,846	2,062	164	31	1
Albuquerque city	38	NM	384,736	42,767	3,761	668	58	445	28
Oakland city	39	CA	372,242	44,855	5,303	16,038	1,243	154	13
Pittsburgh city	40	PA	369,879	66,336	6,624	12,115	1,249	67	7
Sacramento city	41	CA	369,365	44,619	4,464	3,748	272	260	14
Minneapolis city	42	MN	368,383	47,718	7,475	2,006	171	285	11
Tulsa city	43	OK	367,302	46,684	4,828	4,196	479	1,344	111
Honolulu CDP	44	HI	365,272	58,279	5,312	169	17	49	2
Cincinnati city	45	OH	364,040	50,726	6,230	14,220	1,269	51	7
Miami city	46	FL	358,548	59,347	6,284	7,084	527	42	5
Fresno city	47	CA	354,202	35,804	3,989	2,453	212	193	12
Omaha city	48	NE	335,795	43,297	4,980	3,284	328	114	4
Toledo city	49	OH	332,943	45,201	4,421	5,507	472	51	2
Buffalo city	50	NY	328,123	48,703	5,083	9,191	709	144	9

Source: "Population of Persons 65 Years and Over in Places of 100,000 or More, by Age, Race, and Hispanic Origin: 1990," Cynthia Taeuber, *Sixty-Five Plus in America*, p. 8-18. Primary source: U.S. Bureau of the Census, 1990 Census of Population and Housing Summary Tape File 1-A.

★ 1055 ★
Senior Citizens

Persons 65 and Over in Places of 100,000 or More: 1990, Part I. - B

Rank based on total, all races, all ages.

Place	Rank	State	Asian and Pacific Islander		Hispanic origin	
			65 years and over	85 years and over	65 years and over	85 years and over
New York city	1	NY	33,214	2,112	108,702	8,480
Los Angeles city	2	CA	31,891	2,374	50,058	4,365
Chicago city	3	IL	8,361	499	19,283	1,262
Houston city	4	TX	2,848	118	15,303	1,091
Philadelphia city	5	PA	2,225	128	4,076	273
San Diego city	6	CA	7,246	281	10,388	826
Detroit city	7	MI	498	31	1,897	168
Dallas city	8	TX	636	31	6,470	513
Phoenix city	9	AZ	864	64	7,856	576
San Antonio city	10	TX	513	31	38,524	3,272
San Jose city	11	CA	8,810	451	7,894	639
Baltimore city	12	MD	439	38	698	51
Indianapolis city	13	IN	248	16	366	28
San Francisco city	14	CA	27,168	2,163	9,094	1,013
Jacksonville city (remainder)	15	FL	517	21	996	78
Columbus city	16	OH	369	14	294	17
Milwaukee city	17	WI	410	15	1,375	103
Memphis city	18	TN	193	8	243	21
Washington city	19	DC	785	69	1,585	149
Boston city	20	MA	2,191	201	2,380	184
Seattle city	21	WA	6,062	505	988	83
El Paso city	22	TX	283	11	23,938	2,055
Cleveland city	23	OH	297	23	1,029	65
New Orleans city	24	LA	478	41	2,244	230
Nashville-Davidson (remainder)	25	TN	175	8	205	15
Denver city	26	CO	912	63	6,547	537
Austin city	27	TX	251	14	3,966	309
Forth Worth city	28	TX	201	13	2,979	234
Oklahoma city	29	OK	317	20	768	65
Portland city	30	OR	1,640	121	674	69
Kansas City city	31	KS	247	17	1,143	109
Long Beach city	32	CA	3,130	142	2,741	201
Tucson city	33	AZ	325	16	7,350	635
St. Louis city	34	MO	125	18	525	75
Charlotte city	35	NC	175	6	272	17
Atlanta city	36	GA	143	9	442	40
Virginia Beach city	37	VA	724	13	261	20
Albuquerque city	38	NM	250	10	9,066	746
Oakland city	39	CA	5,555	395	2,692	276
Pittsburgh city	40	PA	143	12	330	36

[Continued]

★ 1055 ★

Persons 65 and Over in Places of 100,000 or More: 1990, Part I. - B

[Continued]

Place	Rank	State	Asian and Pacific Islander		Hispanic origin	
			65 years and over	85 years and over	65 years and over	85 years and over
Sacramento city	41	CA	5,144	400	3,453	284
Minneapolis city	42	MN	508	32	319	38
Tulsa city	43	OK	156	4	431	30
Honolulu CDP	44	HI	43,075	3,922	1,318	107
Cincinnati city	45	OH	186	13	132	19
Miami city	46	FL	171	27	43,710	4,254
Fresno city	47	CA	2,003	130	4,664	389
Omaha city	48	NE	107	10	580	45
Toledo city	49	OH	140	12	613	52
Buffalo city	50	NY	80	4	723	65

Source: "Population of Persons 65 Years and Over in Places of 100,000 or More, by Age, Race, and Hispanic Origin: 1990," Cynthia Taeuber, *Sixty-Five Plus in America*, p. 8-18. Primary source: U.S. Bureau of the Census, 1990 Census of Population and Housing Summary Tape File 1-A. *Note:* 1. Hispanic may be of any race.

★ 1056 ★

Senior Citizens

Persons 65 and Over in Places of 100,000 or More: 1990, Part II - A

Rank based on total, all races, all ages.

Place	Rank	State	Total, all races			Black		American Indian, Eskimo and Aleut	
			All ages	65 years and over	85 years and over	65 years and over	85 years and over	65 years and over	85 years and over
Wichita city	51	KS	304,011	37,655	3,894	2,393	193	205	12
Santa Ana city	52	CA	293,742	16,522	1,776	379	24	85	7
Mesa city	53	AZ	288,091	35,713	2,933	241	31	88	5
Colorado Springs city	54	CO	281,140	25,781	2,545	605	39	109	12
Tampa city	55	FL	280,015	40,934	4,104	6,130	600	80	7
Newark city	56	NJ	275,221	25,547	2,139	12,882	943	54	6
St. Paul city	57	MN	272,235	37,412	5,309	1,112	111	120	12
Louisville city	58	KY	269,063	44,641	5,064	9,750	970	45	5
Anaheim city	59	CA	266,406	22,292	2,266	104	5	69	7
Birmingham city	60	AL	265,968	39,480	4,452	19,254	1,988	29	3
Arlington city	61	TX	261,721	13,012	1,038	171	7	57	7
Norfolk city	62	VA	261,229	27,458	2,313	8,970	660	43	1
Las Vegas city	63	NV	258,295	26,532	1,566	1,761	119	114	5
Corpus Christi city	64	TX	257,453	25,933	2,186	1,439	116	71	5
St. Petersburg city	65	FL	238,629	52,945	7,975	4,034	340	49	8
Rochester city	66	NY	231,636	28,135	4,036	3,542	265	73	6
Jersey City city	67	NJ	228,537	25,287	2,384	4,145	292	46	1
Riverside city	68	CA	226,505	20,266	2,116	975	85	101	3

[Continued]

★ 1056 ★

Persons 65 and Over in Places of 100,000 or More: 1990, Part II - A

[Continued]

Place	Rank	State	Total, all races			Black		American Indian, Eskimo and Aleut	
			All ages	65 years and over	85 years and over	65 years and over	85 years and over	65 years and over	85 years and over
Anchorage city	69	AK	226,338	8,258	374	328	14	420	24
Lexington-Fayette city	70	KY	225,366	22,312	2,308	2,778	285	17	2
Akron city	71	OH	223,019	33,171	3,470	5,179	401	42	6
Aurora city	72	CO	222,103	15,044	1,135	445	26	45	2
Baton Rouge city	73	LA	219,531	25,161	2,356	7,685	662	26	4
Stockton city	74	CA	210,943	22,107	2,213	1,741	124	131	5
Raleigh city	75	NC	207,951	18,332	1,814	3,814	358	20	2
Richmond city	76	VA	203,056	31,181	3,435	12,345	934	44	2
Shreveport city	77	LA	198,525	27,206	3,105	8,203	947	34	9
Jackson city	78	MS	196,637	22,851	2,247	8,080	815	12	2
Mobile city	79	AL	196,278	26,900	2,618	7,718	672	31	3
Des Moines city	80	IA	193,187	25,884	3,053	1,325	123	40	5
Lincoln city	81	NE	191,972	21,005	2,674	198	13	29	2
Madison city	82	WI	191,262	17,831	2,211	210	15	14	0
Grand Rapids city	83	MI	189,126	24,711	3,508	2,077	159	48	2
Yonkers city	84	NY	188,082	30,935	2,979	1,684	158	13	2
Hialeah city	85	FL	188,004	26,338	2,362	292	20	14	0
Montgomery city	86	AL	187,106	21,884	2,185	6,574	680	21	0
Lubbock city	87	TX	186,206	18,299	1,842	1,143	98	32	3
Greensboro city	88	NC	183,521	21,591	2,239	4,736	442	43	4
Dayton city	89	OH	182,044	23,929	2,060	7,931	568	28	0
Huntington Beach city	90	CA	181,519	15,088	1,211	24	2	33	4
Garland city	91	TX	180,650	9,970	715	381	33	30	2
Glendale city	92	CA	180,038	23,977	3,208	78	10	32	7
Columbus city (remainder)	93	GA	178,681	19,254	1,570	5,138	442	15	0
Spokane city	94	WA	177,196	28,788	3,539	296	17	160	8
Tacoma city	95	WA	176,664	24,258	3,235	1,118	72	178	13
Little Rock city	96	AR	175,795	22,071	2,497	4,461	545	32	2
Bakersfield city	97	CA	174,820	15,998	1,696	1,323	144	115	14
Fremont city	98	CA	173,339	11,541	951	123	4	42	3
Fort Wayne city	99	IN	173,072	23,091	2,807	1,637	117	19	0
Arlington CDP	100	VA	170,936	19,409	1,738	1,357	106	27	3

Source: "Population of Persons 65 Years and Over in Places of 100,000 or More, by Age, Race, and Hispanic Origin: 1990," Cynthia Taeuber, *Sixty-Five Plus in America*, p. 8-19 Primary source: U.S. Bureau of the Census, 1990 Census of Population and Housing Summary Tape File 1-A.

★ 1057 ★

Senior Citizens

Persons 65 and Over in Places of 100,000 or More: 1990, Part II. - B

Rank based on total, all races, all ages.

Place	Rank	State	Asian and Pacific Islander		Hispanic origin	
			65 years and over	85 years and over	65 years and over	85 years and over
Wichita city	51	KS	199	16	666	45
Santa Ana city	52	CA	1,424	65	4,216	320
Mesa city	53	AZ	112	4	992	78
Colorado Springs city	54	CO	292	8	1,115	85
Tampa city	55	FL	130	4	7,916	818
Newark city	56	NJ	138	10	3,421	259
St. Paul city	57	MN	556	22	540	55
Louisville city	58	KY	75	5	172	18
Anaheim city	59	CA	1,257	77	2,123	165
Birmingham city	60	AL	31	3	95	8
Arlington city	61	TX	188	11	330	18
Norfolk city	62	VA	321	11	245	18
Las Vegas city	63	NV	525	20	1,406	71
Corpus Christi city	64	TX	119	2	9,035	680
St. Petersburg city	65	FL	143	12	828	89
Rochester city	66	NY	112	4	770	49
Jersey City city	67	NJ	1,307	53	2,384	154
Riverside city	68	CA	411	14	2,143	173
Anchorage city	69	AK	538	17	141	10
Lexington-Fayette city	70	KY	66	3	81	6
Akron city	71	OH	62	2	124	17
Aurora city	72	CO	424	19	414	31
Baton Rouge city	73	LA	92	5	281	19
Stockton city	74	CA	3,346	295	2,873	204
Raleigh city	75	NC	102	8	110	7
Richmond city	76	VA	91	4	141	16
Shreveport city	77	LA	31	3	151	15
Jackson city	78	MS	31	3	86	7
Mobile city	79	AL	59	3	199	18
Des Moines city	80	IA	167	11	251	22
Lincoln city	81	NE	92	2	148	8
Madison city	82	WI	155	5	104	5
Grand Rapids city	83	MI	65	7	281	27
Yonkers city	84	NY	256	9	1,410	118
Hialeah city	85	FL	52	8	22,747	1,912
Montgomery city	86	AL	33	1	87	10
Lubbock city	87	TX	40	1	1,580	91
Greensboro city	88	NC	71	2	96	7
Dayton city	89	OH	56	6	71	5
Huntington Beach city	90	CA	825	46	660	74

[Continued]

★ 1057 ★

Persons 65 and Over in Places of 100,000 or More: 1990, Part II. - B

[Continued]

Place	Rank	State	Asian and Pacific Islander		Hispanic origin	
			65 years and over	85 years and over	65 years and over	85 years and over
Garland city	91	TX	263	4	416	24
Glendale city	92	CA	1,306	59	2,075	221
Columbus city (remainder)	93	GA	67	2	191	8
Spokane city	94	WA	344	24	195	16
Tacoma city	95	WA	553	24	278	18
Little Rock city	96	AR	62	3	58	6
Bakersfield city	97	CA	368	25	1,160	94
Fremont city	98	CA	1,705	67	1,023	71
Fort Wayne city	99	IN	47	5	200	11
Arlington CDP	100	VA	542	29	700	61

Source: "Population of Persons 65 Years and Over in Places of 100,000 or More, by Age, Race, and Hispanic Origin: 1990," Cynthia Taeuber, *Sixty-Five Plus in America*, p. 8-19. Primary source: U.S. Bureau of the Census, 1990 Census of Population and Housing Summary Tape File 1-A. *Note:* 1. Hispanic may be of any race.

★ 1058 ★

Senior Citizens

Persons 65 and Over in Places of 100,000 or More: 1990, Part III - A

Rank based on total, all races, all ages.

Place	Rank	State	Total, all races			Black		American Indian, Eskimo and Aleut	
			All ages	65 years and over	85 years and over	65 years and over	85 years and over	65 years and over	85 years and over
Newport News city	101	VA	170,045	15,804	1,192	4,400	254	17	1
Worcester city	102	MA	169,759	27,287	3,466	301	27	37	3
Knoxville city	103	TN	165,121	25,441	2,708	2,979	271	40	5
Modesto city	104	CA	164,730	17,268	1,769	260	22	99	4
Orlando city	105	FL	164,693	18,755	2,180	3,101	262	15	0
San Bernardino city	106	CA	164,164	16,396	1,616	1,658	153	99	9
Syracuse city	107	NY	163,860	24,394	3,327	1,784	189	86	7
Providence city	108	RI	160,728	21,802	2,659	1,379	111	90	17
Salt Lake City city	109	UT	159,936	23,192	2,832	228	13	43	3
Huntsville city	110	AL	159,789	15,982	1,351	2,137	240	30	1
Amarillo city	111	TX	157,615	18,974	1,876	771	73	77	5
Springfield city	112	MA	156,983	21,568	2,384	2,024	166	22	3
Irving city	113	TX	155,037	8,413	645	122	10	22	1
Chattanooga city	114	TN	152,466	23,269	2,584	5,713	577	29	6
Chesapeake city	115	VA	151,976	12,844	991	3,223	230	24	0
Kansas City city	116	KS	149,767	19,489	2,022	4,163	503	67	8
Metairie CDP	117	LA	149,428	21,013	1,474	537	35	11	0

[Continued]

★ 1058 ★

Persons 65 and Over in Places of 100,000 or More: 1990, Part III - A

[Continued]

Place	Rank	State	Total, all races			Black		American Indian, Eskimo and Aleut	
			All ages	65 years and over	85 years and over	65 years and over	85 years and over	65 years and over	85 years and over
Fort Lauderdale city	118	FL	149,377	26,562	3,351	2,156	145	21	2
Glendale city	119	AZ	148,134	11,675	1,154	143	7	42	1
Warren city	120	MI	144,864	21,555	1,677	35	2	30	1
Winston-Salem city	121	NC	143,485	20,331	2,355	5,497	531	21	0
Garden Grove city	122	CA	143,050	12,512	1,083	60	5	53	1
Oxnard city	123	CA	142,216	11,003	876	445	21	61	2
Tempe city	124	AZ	141,865	9,305	850	90	4	25	2
Bridgeport city	125	CT	141,686	19,245	2,064	2,190	133	16	0
Paterson city	126	NJ	140,891	13,551	1,197	2,706	170	24	0
Flint city	127	MI	140,761	15,100	1,619	4,294	300	55	7
Springfield city	128	MO	140,494	21,329	2,696	301	30	77	4
Hartford city	129	CT	139,739	13,809	1,536	3,453	240	32	4
Rockford city	130	IL	139,426	20,535	2,644	1,167	80	14	1
Savannah city	131	GA	137,560	18,957	1,576	7,758	635	16	2
Durham city	132	NC	136,611	15,443	1,677	5,405	522	17	2
Chula Vista city	133	CA	135,163	15,767	1,417	199	13	50	3
Reno city	134	NV	133,850	15,802	1,264	251	16	80	3
Hampton city	135	VA	133,793	12,801	898	3,985	290	32	2
Ontario city	136	CA	133,179	8,489	706	267	13	47	0
Torrance city	137	CA	133,107	15,900	1,546	56	7	28	2
Pomona city	138	CA	131,723	9,191	1,036	883	68	43	1
Pasadena city	139	CA	131,591	17,338	2,500	2,372	289	43	7
New Haven city	140	CT	130,474	16,067	1,965	2,925	225	40	7
Scottsdale city	141	AZ	130,069	21,199	1,901	42	1	54	6
Plano city	142	TX	128,713	4,577	437	152	15	11	1
Oceanside city	143	CA	128,398	18,010	1,193	186	8	43	3
Lansing city	144	MI	127,321	12,171	1,205	1,131	90	45	1
Lakewood city	145	CO	126,481	13,343	1,448	66	5	24	3
East Los Angeles CDP	146	CA	126,379	9,617	1,017	25	3	38	2
Evansville city	147	IN	126,272	21,661	2,464	1,369	142	33	2
Boise City city	148	ID	125,738	14,970	1,627	25	5	30	4
Tallahassee city	149	FL	124,773	10,946	1,035	2,546	245	9	1
Paradise CDP	150	NV	124,682	15,864	692	373	13	47	1

Source: "Population of Persons 65 Years and Over in Places of 100,000 or More, by Age, Race, and Hispanic Origin: 1990," Cynthia Taeuber, *Sixty-Five Plus in America*, p. 8-20. Primary source: U.S. Bureau of the Census, 1990 Census of Population and Housing Summary Tape File 1-A.

★ 1059 ★

Senior Citizens

Persons 65 and Over in Places of 100,000 or More: 1990, Part III. - B

Rank based on total, all races, all ages.

Place	Rank	State	Asian and Pacific Islander		Hispanic origin	
			65 years and over	85 years and over	65 years and over	85 years and over
Newport News city	101	VA	123	5	140	7
Worcester city	102	MA	96	4	508	35
Knoxville city	103	TN	29	3	58	4
Modesto city	104	CA	441	24	1,001	70
Orlando city	105	FL	98	10	1,111	88
San Bernardino city	106	CA	266	11	2,754	228
Syracuse city	107	NY	81	2	175	16
Providence city	108	RI	263	19	839	55
Salt Lake City city	109	UT	468	45	878	63
Huntsville city	110	AL	85	0	79	4
Amarillo city	111	TX	77	4	745	57
Springfield city	112	MA	64	1	744	52
Irving city	113	TX	172	11	412	24
Chattanooga city	114	TN	44	2	67	4
Chesapeake city	115	VA	73	5	95	7
Kansas City city	116	KS	73	1	614	52
Metairie CDP	117	LA	129	6	903	79
Fort Lauderdale city	118	FL	60	2	991	89
Glendale city	119	AZ	194	16	882	77
Warren city	120	MI	98	6	172	8
Winston-Salem city	121	NC	37	0	55	4
Garden Grove city	122	CA	1,400	71	917	72
Oxnard city	123	CA	925	45	2,924	200
Tempe city	124	AZ	136	13	506	41
Bridgeport city	125	CT	90	3	1,592	117
Paterson city	126	NJ	57	2	2,245	129
Flint city	127	MI	28	3	199	27
Springfield city	128	MO	45	4	60	9
Hartford city	129	CT	72	8	1,475	108
Rockford city	130	IL	68	5	213	18
Savannah city	131	GA	79	6	98	12
Durham city	132	NC	60	0	52	3
Chula Vista city	133	CA	833	55	2,903	197
Reno city	134	NV	334	13	556	39
Hampton city	135	VA	75	6	106	5
Ontario city	136	CA	188	7	1,459	111
Torrance city	137	CA	1,620	97	831	69
Pomona city	138	CA	409	12	1,777	125
Pasadena city	139	CA	974	58	1,522	128
New Haven city	140	CT	53	6	546	50

[Continued]

★ 1059 ★

Persons 65 and Over in Places of 100,000 or More: 1990, Part III. - B

[Continued]

Place	Rank	State	Asian and Pacific Islander		Hispanic origin	
			65 years and over	85 years and over	65 years and over	85 years and over
Scottsdale city	141	AZ	82	1	302	23
Plano city	142	TX	142	1	123	8
Oceanside city	143	CA	377	15	1,072	78
Lansing city	144	MI	68	2	351	18
Lakewood city	145	CO	155	9	493	68
East Los Angeles CDP	146	CA	417	32	8,255	879
Evansville city	147	IN	18	2	67	7
Boise City city	148	ID	63	5	188	23
Tallahassee city	149	FL	59	3	119	16
Paradise CDP	150	NV	279	9	716	35

Source: "Population of Persons 65 Years and Over in Places of 100,000 or More, by Age, Race, and Hispanic Origin: 1990," Cynthia Taeuber, *Sixty-Five Plus in America*, p. 8-20. Primary source: U.S. Bureau of the Census, 1990 Census of Population and Housing Summary Tape File 1-A. *Note:* 1. Hispanic may be of any race.

★ 1060 ★

Senior Citizens

Persons 65 and Over in Places of 100,000 or More: 1990, Part IV - A

Rank based on total, all races, all ages.

Place	Rank	State	Total, all races			Black		American Indian, Eskimo and Aleut	
			All ages	65 years and over	85 years and over	65 years and over	85 years and over	65 years and over	85 years and over
Laredo city	151	TX	122,899	10,020	1,019	4	0	9	2
Hollywood city	152	FL	121,697	28,101	3,817	557	55	15	0
Topeka city	153	KS	119,883	17,681	2,100	1,236	153	90	3
Pasadena city	154	TX	119,363	9,142	677	25	3	27	2
Moreno Valley city	155	CA	118,779	4,734	211	369	18	27	4
Sterling Heights city	156	MI	117,810	10,872	856	16	0	10	2
Sunnyvale city	157	CA	117,229	12,191	1,015	81	4	17	0
Gary city	158	IN	116,646	13,261	992	9,059	663	17	2
Beaumont city	159	TX	114,323	15,737	1,668	4,405	430	21	1
Fullerton city	160	CA	114,144	11,668	1,251	68	5	36	2
Peoria city	161	IL	113,504	16,381	1,756	1,359	93	12	2
Santa Rosa city	162	CA	113,313	18,472	1,960	64	5	72	1
Eugene city	163	OR	112,669	14,276	1,710	43	2	31	3
Independence city	164	MO	112,301	16,148	1,547	92	17	45	2
Overland Park city	165	KS	111,790	11,068	950	43	3	23	1
Hayward city	166	CA	111,498	11,910	1,076	296	29	57	4
Concord city	167	CA	111,348	10,543	1,009	63	3	41	5

[Continued]

★ 1060 ★

Persons 65 and Over in Places of 100,000 or More: 1990, Part IV - A
[Continued]

| Place | Rank | State | Total, all races | | | Black | | American Indian, Eskimo and Aleut | |
			All ages	65 years and over	85 years and over	65 years and over	85 years and over	65 years and over	85 years and over
Alexandria city	168	VA	111,183	11,406	1,380	1,615	133	7	2
Orange city	169	CA	110,658	9,631	1,070	33	2	33	2
Santa Clarita city	170	CA	110,642	6,916	563	42	6	48	3
Irvine city	171	CA	110,330	6,357	527	30	1	8	1
Elizabeth city	172	NJ	110,002	13,270	1,289	1,278	111	17	2
Inglewood city	173	CA	109,602	7,494	778	3,291	265	27	1
Ann Arbor city	174	MI	109,592	7,881	990	620	58	8	1
Vallejo city	175	CA	109,199	11,851	978	1,725	97	48	6
Waterbury city	176	CT	108,961	17,925	1,864	1,037	87	23	0
Salinas city	177	CA	108,777	9,048	853	129	10	68	4
Cedar Rapids city	178	IA	108,751	14,324	1,732	157	14	11	0
Erie city	179	PA	108,718	17,488	1,638	789	56	14	1
Escondido city	180	CA	108,635	14,074	1,778	23	3	58	7
Stamford city	181	CT	108,056	14,333	1,521	1,244	96	12	1
Salem city	182	OR	107,786	15,679	1,862	26	3	80	10
Citrus Heights CDP	183	CA	107,439	10,326	708	75	4	54	3
Abilene city	184	TX	106,654	12,568	1,533	527	59	19	1
Macon city	185	GA	106,612	15,521	1,401	5,416	491	17	4
El Monte city	186	CA	106,209	6,824	620	26	4	30	3
South Bend city	187	IN	105,511	17,740	1,932	2,067	161	20	1
Springfield city	188	IL	105,227	15,632	1,878	986	94	23	5
Allentown city	189	PA	105,090	17,767	1,950	186	6	9	0
Thousand Oaks city	190	CA	104,352	9,427	958	34	2	24	0
Portsmouth city	191	VA	103,907	14,399	1,083	5,687	415	21	1
Waco city	192	TX	103,590	15,450	1,854	2,424	273	20	0
Lowell city	193	MA	103,439	12,510	1,449	80	8	7	0
Berkeley city	194	CA	102,724	11,252	1,459	3,313	325	30	1
Mesquite city	195	TX	101,484	5,365	568	39	3	15	2
Rancho Cucamonga city	196	CA	101,409	5,125	332	138	9	20	3
Albany city	197	NY	101,082	15,495	2,338	1,267	104	19	2
Livonia city	198	MI	100,850	13,180	1,395	17	4	17	1
Sioux Falls city	199	SD	100,814	11,715	1,520	32	3	36	0
Simi Valley city	200	CA	100,217	5,273	389	38	0	18	0

Source: "Population of Persons 65 Years and Over in Places of 100,000 or More, by Age, Race, and Hispanic Origin: 1990," Cynthia Taeuber, *Sixty-Five Plus in America,* p. 8-21. Primary source: U.S. Bureau of the Census, 1990 Census of Population and Housing Summary Tape File 1-A.

★ 1061 ★
Senior Citizens

Persons 65 and Over in Places of 100,000 or More: 1990, Part IV. - B

Rank based on total, all races, all ages.

Place	Rank	State	Asian and Pacific Islander		Hispanic origin	
			65 years and over	85 years and over	65 years and over	85 years and over
Laredo city	151	TX	14	1	9,253	943
Hollywood city	152	FL	88	5	1,287	122
Topeka city	153	KS	59	3	515	37
Pasadena city	154	TX	46	1	952	56
Moreno Valley city	155	CA	338	10	616	27
Sterling Heights city	156	MI	163	8	76	3
Sunnyvale city	157	CA	1,374	59	1,047	85
Gary city	158	IN	13	0	571	38
Beaumont city	159	TX	80	5	349	26
Fullerton city	160	CA	604	28	878	85
Peoria city	161	IL	53	4	116	11
Santa Rosa city	162	CA	184	12	453	43
Eugene city	163	OR	61	5	116	6
Independence city	164	MO	42	2	75	7
Overland Park city	165	KS	82	3	77	8
Hayward city	166	CA	1,091	74	1,572	128
Concord city	167	CA	524	27	563	53
Alexandria city	168	VA	262	6	236	12
Orange city	169	CA	436	27	759	76
Santa Clarita city	170	CA	202	4	390	18
Irvine city	171	CA	671	27	210	10
Elizabeth city	172	NJ	136	9	2,524	164
Inglewood city	173	CA	194	14	1,050	72
Ann Arbor city	174	MI	139	11	74	13
Vallejo city	175	CA	1,802	95	684	57
Waterbury city	176	CT	22	0	565	49
Salinas city	177	CA	894	69	1,667	111
Cedar Rapids city	178	IA	25	0	72	6
Erie city	179	PA	24	1	86	4
Escondido city	180	CA	189	15	774	70
Stamford city	181	CT	141	9	476	39
Salem city	182	OR	100	5	167	16
Citrus Heights CDP	183	CA	140	3	287	11
Abilene city	184	TX	34	3	647	43
Macon city	185	GA	12	2	51	1
El Monte city	186	CA	828	34	2,404	186
South Bend city	187	IN	40	1	127	9
Springfield city	188	IL	38	4	81	6
Allentown city	189	PA	53	3	323	14
Thousand Oaks city	190	CA	247	14	327	36

[Continued]

★ 1061 ★

Persons 65 and Over in Places of 100,000 or More: 1990, Part IV. - B

[Continued]

Place	Rank	State	Asian and Pacific Islander		Hispanic origin	
			65 years and over	85 years and over	65 years and over	85 years and over
Portsmouth city	191	VA	45	4	74	8
Waco city	192	TX	19	1	965	79
Lowell city	193	MA	218	13	290	21
Berkeley city	194	CA	1,239	119	344	38
Mesquite city	195	TX	102	2	202	10
Rancho Cucamonga city	196	CA	224	9	644	41
Albany city	197	NY	64	3	139	17
Livonia city	198	MI	81	7	96	9
Sioux Falls city	199	SD	12	0	32	3
Simi Valley city	200	CA	226	11	380	28

Source: "Population of Persons 65 Years and Over in Places of 100,000 or More, by Age, Race, and Hispanic Origin: 1990," Cynthia Taeuber, *Sixty-Five Plus in America*, p. 8-21. Primary source: U.S. Bureau of the Census, 1990 Census of Population and Housing Summary Tape File 1-A. *Note:* 1. Hispanic may be of any race.

★ 1062 ★

Senior Citizens

Persons 80 Years and Over by Race and Hispanic Origin, 1990 to 2050. Part I.

Numbers in thousands.

Year	Total, 80+ population Number	White			Black		
		Total	Not Hispanic		Total	Not Hispanic	
			Number	Percent of total 80+ population		Number	Percent of total 80+ population
1990	6,930	6,327	6,121	88.3	511	502	7.2
2000	9,264	8,401	8,010	86.5	662	645	7.0
2010	11,748	10,529	9,831	83.7	814	780	6.6
2020	13,007	11,392	10,353	79.6	942	887	6.8
2030	18,225	15,841	14,178	77.8	1,302	1,206	6.6
2040	26,524	22,800	19,976	75.3	2,031	1,862	7.0
2050	30,971	26,154	22,024	71.1	2,471	2,227	7.2

Source: "Persons 80 Years and Over by Race and Hispanic Origin: 1990 to 2050." *Profiles of America's Elderly*, No. 3 (November 1993), p. 3. Primary source: U.S. Bureau of the Census, 1990 from Current Population Reports, P25-1095, *U.S. Population Estimates, by Age, Sex, Race, and Hispanic Origin: 1980 to 1991*, November 1991. Data for 2000 shown in Current Population Reports, P25-1104, *Projections of the United States, by Age, Sex, Race, and Hispanic Origin: 1993 to 2050*, (middle series) U.S. Government Printing Office, Washington, DC, 1993.

★ 1063 ★

Senior Citizens

Persons 80 Years and Over by Race and Hispanic Origin, 1990 to 2050. Part II.

Numbers in thousands.

Year	American Indian, Eskimo, and Aleut			Asian and Pacific Islander			Hispanic origin (may be of any race)	
		Not Hispanic			Not Hispanic			
	Total	Number	Percent of total 80+ population	Total	Number	Percent of total 80+ population	Number	Percent of total 80+ population
1990	21	20	0.3	71	67	1.0	219	3.2
2000	40	38	0.4	160	154	1.7	418	4.5
2010	62	58	0.5	343	331	2.8	748	6.4
2020	85	77	0.6	588	567	4.4	1,121	8.6
2030	119	108	0.6	963	928	5.1	1,805	9.9
2040	173	153	0.6	1,518	1,462	5.5	3,071	11.6
2050	211	183	0.6	2,136	2,051	6.6	4,486	14.5

Source: "Persons 80 Years and Over by Race and Hispanic Origin: 1990 to 2050." *Profiles of America's Elderly*, No. 3 (November 1993), p. 3. Primary source: U.S. Bureau of the Census, 1990 from Current Population Reports, P25-1095, *U.S. Population Estimates, by Age, Sex, Race, and Hispanic Origin: 1980 to 1991*, November 1991. Data for 2000 shown in Current Population Reports, P25-1104, *Projections of the United States, by Age, Sex, Race, and Hispanic Origin: 1993 to 2050*, (middle series) U.S. Government Printing Office, Washington, DC, 1993.

★ 1064 ★

Senior Citizens

Persons Needing Physical Assistance

Percent

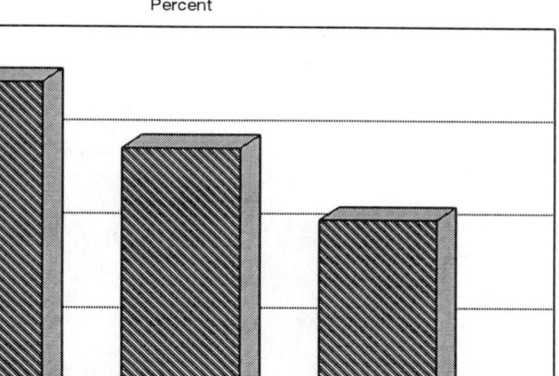

Civilian noninstitutional population.

	Percent
Black	22.7
Hispanic origin[1]	19.2
White	15.4

Source: "Percentage of Persons Needing Assistance With Everyday Activities, by Age and Sex, 1986," Cynthia Taeuber, *Sixty-Five Plus in America*, p. 3-13. Primary source: U.S. Bureau of the Census, *The Need for Personal Assistance With Everyday Activities: Recipients and Caregivers*, Current Population Reports, Series P-70, No. 19. U.S. Government Printing Office, Washington, DC, 1990, table B. *Note:* 1. Hispanic origin may be of any race.

★ 1065 ★

Senior Citizens

Population 65 Years and Over, by Age, Gender and Hispanic Origin, 1990.
Part I

Race and sex	Total, 65 years and over	65 to 69 years	70 to 74 years	Total, 75 years and over	75 to 79 years
All races					
Total	31,078,895	10,065,835	7,979,660	13,033,400	6,102,929
Male	12,492,766	4,507,539	3,399,275	4,585,952	2,388,895
Female	18,586,129	5,558,296	4,580,385	8,447,448	3,714,034
Males per 100 females	67.2	81.1	74.2	54.3	64.3

[Continued]

★ 1065 ★

Population 65 Years and Over, by Age, Gender and Hispanic Origin, 1990.
Part I
[Continued]

Race and sex	Total, 65 years and over	65 to 69 years	70 to 74 years	Total, 75 years and over	75 to 79 years
White					
Total	28,020,562	8,983,978	7,191,013	11,845,571	5,518,341
Male	11,284,407	4,047,535	3,079,801	4,157,071	2,165,061
Female	16,736,155	4,936,443	4,111,212	7,688,500	3,353,280
Males per 100 females	67.4	82.0	74.9	54.1	64.6
Black					
Total	2,492,221	859,694	638,077	994,450	483,535
Male	956,936	360,653	252,967	343,316	178,695
Female	1,535,285	499,041	385,110	651,134	304,840
Males per 100 females	62.3	72.3	65.7	52.7	58.6
American Indian, Eskimo, and Aleut					
Total	116,153	43,374	29,831	42,948	21,522
Male	48,874	19,658	12,759	16,457	8,552
Female	67,279	23,716	17,072	26,491	12,970
Males per 100 females	72.6	82.9	74.7	62.1	65.9
Asian and Pacific Islander					
Total	449,959	178,789	120,739	150,431	79,531
Male	202,549	79,693	53,748	69,108	36,587
Female	247,410	99,096	66,991	81,323	42,944
Males per 100 females	81.9	80.4	80.2	85.0	85.2
Hispanic origin					
Total	1,146,223	431,000	284,085	431,138	211,432
Male	474,830	192,949	118,696	163,185	82,364
Female	671,393	238,051	165,389	267,953	129,068
Males per 100 females	70.7	81.1	71.8	60.9	63.8

Source: "Persons 65 Years and Over, by Age, Sex, Race, and Hispanic Origin: 1990," Cynthia Taeuber, *Sixty-Five Plus in America*, p. 2-6. Primary source: U.S. Bureau of the Census. 1990 Census of Housing, Series CPH-L-74, *Modified and Actual Age, Sex, Race, and Hispanic Origin Data.*

★ 1066 ★

Senior Citizens

Population 65 Years and Over, by Age, Gender and Hispanic Origin, 1990. Part II

Race and sex	Total, 80 years and over	80 to 84 years	Total, 85 years and over	85 to 89 years	90 to 94 years	95 to 99 years	Total, 100 years and over
All races							
Total	6,930,471	3,909,046	3,021,425	2,034,661	747,979	202,977	35,808
Male	2,197,057	1,355,830	841,227	605,936	184,048	43,544	7,699
Female	4,733,414	2,553,216	2,180,198	1,428,725	563,931	159,433	28,109
Males per 100 females	46.4	53.1	38.6	42.4	32.6	27.3	27.4
White							
Total	6,327,230	3,566,268	2,760,962	1,858,176	689,928	183,505	29,353
Male	1,992,010	1,232,184	759,826	547,832	167,568	38,559	5,867
Female	4,335,220	2,334,084	2,001,136	1,310,344	522,360	144,946	23,486
Males per 100 females	45.9	52.8	38.0	41.8	32.1	26.6	25.0
Black							
Total	510,915	288,283	222,632	150,294	49,599	17,049	5,690
Male	164,621	98,351	66,270	46,949	13,485	4,277	1,559
Female	346,294	189,932	156,362	103,345	36,114	12,772	4,131
Males per 100 females	47.5	51.8	42.4	45.4	37.3	33.5	37.7
American Indian, Eskimo, and Aleut							
Total	21,426	12,236	9,190	6,287	1,982	659	262
Male	7,905	4,641	3,264	2,265	680	222	97
Female	13,521	7,595	5,926	4,022	1,302	437	165
Males per 100 females	58.5	61.1	55.1	56.3	52.2	50.8	58.8
Asian and Pacific Islander							
Total	70,900	42,259	28,641	19,904	6,470	1,764	503
Male	32,521	20,654	11,867	8,890	2,315	486	176
Female	38,379	21,605	16,774	11,014	4,155	1,278	327
Males per 100 females	84.7	95.6	70.7	80.7	55.7	38.0	53.8
Hispanic origin[1]							
Total	219,706	128,302	91,404	64,945	19,257	5,616	1,586
Male	80,821	48,430	32,391	23,695	6,405	1,726	565
Female	138,885	79,872	59,013	41,250	12,852	3,890	1,021
Males per 100 females	58.2	60.6	54.9	57.4	49.8	44.4	55.3

Source: "Persons 65 Years and Over, by Age, Sex, Race, and Hispanic Origin: 1990," Cynthia Taeuber, *Sixty-Five Plus in America*, p. 2-7. Primary source: U.S. Bureau of the Census. 1990 Census of Housing, Series CPH-L-74, *Modified and Actual Age, Sex, Race, and Hispanic Origin Data. Note:* 1. Hispanic origin may be of any race.

★ 1067 ★

Senior Citizens

Population 65 and Over, by State, 1990

State	Population
United States	2,508,551
Alabama	100,000 or more
Alaska	Under 10,000
Arizona	Under 10,000
Arkansas	10,000 to 99,999
California	100,000 or more
Colorado	Under 10,000
Connecticut	10,000 to 99,999
Delaware	Under 10,000
District of Columbia	10,000 to 99,9999
Florida	100,000 or more
Georgia	100,000 or more
Hawaii	Under 10,000
Idaho	Under 10,000
Illinois	100000 or more
Indiana	10,000 to 99,999
Iowa	Under 10,000
Kansas	10,000 to 99,999
Kentucky	10,000 to 99,999
Louisiana	100,000 or more
Maine	Under 10,000
Maryland	10,000 to 99,999
Massachusetts	10,000 to 99,999
Michigan	100,000 or more
Minnesota	Under 10,000
Mississippi	10,000 to 99,999
Missouri	10,000 to 99,999
Montana	Under 10,000
Nebraska	Under 10,000
Nevada	Under 10,000
New Hampshire	Under 10,000
New Jersey	10,000 to 99,999
New Mexico	Under 10,000
New York	100,000 or more
North Carolina	100,000 or more
North Dakota	Under 10,000
Ohio	100,000 or more
Oklahoma	10,000 to 99,999
Oregon	Under 10,000
Pennsylvania	100,000 or more
Rhode Island	10,000 to 99,999
South Carolina	10,000 to 99,999
South Dakota	Under 10,000
Tennessee	10,000 to 99,999
Texas	100,000 or more
Utah	Under 10,000
Vermont	Under 10,000
Virginia	100,000 or more

[Continued]

★ 1067 ★

Population 65 and Over, by State, 1990
[Continued]

State	Population
Washington	Under 10,000
West Virginia	Under 10,000
Wisconsin	10,000 to 99,999
Wyoming	Under 10,000

Source: "Black Population 65 Years and Over, by State: 1990," Cynthia Taeuber, *Sixty-Five Plus in America*, p. 5-15. Primary source: U.S. Bureau of the Census, 1990 Census of Population and Housing, Summary Tape File 1A.

★ 1068 ★

Senior Citizens

Population 85 and Over, by State, 1990

State	Population
United States	230,183
Alabama	10,000 or more
Alaska	Under 1,000
Arizona	Under 1,000
Arkansas	1,000 to 9,999
California	10,000 or more
Colorado	Under 1,000
Connecticut	1,000 to 9,999
Delaware	1,000 to 9,999
District of Columbia	1,000 to 9,999
Florida	10,000 or more
Georgia	10,000 or more
Hawaii	Under 1,000
Idaho	Under 1,000
Illinois	10,000 or more
Indiana	1,000 to 9,999
Iowa	Under 1,000
Kansas	1,000 to 9,999
Kentucky	1,000 to 9,999
Louisiana	10,000 or more
Maine	Under 1,000
Maryland	1,000 to 9,999
Massachusetts	1,000 to 9,999
Michigan	1,000 to 9,999
Minnesota	Under 1,000
Mississippi	10,000 or more
Missouri	1,000 to 9,999
Montana	Under 1,000
Nebraska	Under 1,000
Nevada	Under 1,000
New Hampshire	Under 1,000
New Jersey	1,000 to 9,999

[Continued]

★ 1068 ★

Population 85 and Over, by State, 1990
[Continued]

State	Population
New Mexico	Under 1,000
New York	10,000 or more
North Carolina	10,000 or more
North Dakota	Under 1,000
Ohio	1,000 to 9,999
Oklahoma	1,000 to 9,999
Oregon	Under 1,000
Pennsylvania	1,000 to 9,999
Rhode Island	1,000 to 9,999
South Carolina	1,000 to 9,999
South Dakota	Under 1,000
Tennessee	1,000 to 9,999
Texas	10,000 or more
Utah	Under 1,000
Vermont	Under 1,000
Virginia	1,000 to 9,999
Washington	Under 1,000
West Virginia	1,000 to 9,999
Wisconsin	Under 1,000
Wyoming	Under 1,000

Source: "Black Population 85 Years and Over, by State: 1990," Cynthia Taeuber, *Sixty-Five Plus in America*, p. 5-15. Primary source: U.S. Bureau of the Census, 1990 Census of Population and Housing, Summary Tape File 1A.

★ 1069 ★

Senior Citizens

Ratios of Youth and Elderly to Other Adults, 1990 and 2050

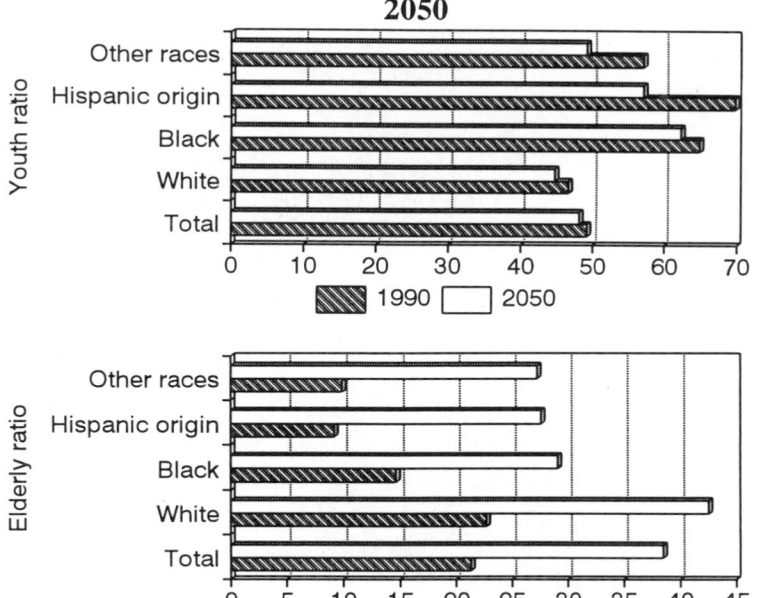

Number of persons of given age per 100 persons aged 20 to 64.

	Youth ratio		Elderly ratio	
	1990	2050	1990	2050
Total	49.2	48.1	21.3	38.5
White	46.7	44.8	22.7	42.5
Black	64.8	62.2	14.7	29.0
Hispanic origin[1]	69.8	57.1	9.2	27.6
Other races	57.1	49.2	9.9	27.2

Source: "Ratios of Youth and Elderly to Other Adults, by Race, and Hispanic Origin: 1950 to 2050," Cynthia Taeuber, *Sixty-Five Plus in America*, p. 2- 15. Primary source: U.S. Bureau of the Census, 1990 from 1990 Census of Population and Housing, Series CPH-L-74, *Modified and Actual Age, Sex, Race, and Hispanic Origin Data*; 2050 from Jennifer C. Day, *Population Projections of the United States, by Age, Sex, Race, and Hispanic origin: 1992 to 2050*, Current Population Reports, P25-1092. U.S. Government Printing Office, Washington, DC, 1992 (middle series projections). *Notes:* Youth Ratio is the number of persons under age 20 divided by the number of persons aged 20 to 64 times 100. Elderly Ratio is the number of persons age 65 years and over divided by the number of persons aged 20 to 64 times 100. 1. Hispanic origin may be of any race.

★ 1070 ★
Senior Citizens

Residents 65 and Over, by Region, 1990

Age, race, and Hispanic origin	United States	Northeast	Midwest	South	West
All persons					
65 years and over	31,241,831	6,995,156	7,749,130	10,724,182	5,773,363
65 to 84 years	28,161,666	6,285,347	6,909,267	9,732,160	5,234,892
85 years and over	3,080,165	709,809	839,863	992,022	538,471
White					
65 years and over	27,851,973	6,409,025	7,205,491	9,172,048	5,065,409
65 to 84 years	25,063,921	5,744,880	6,412,420	8,326,660	4,579,961
85 years and over	2,788,052	664,145	793,071	845,388	485,448
Black					
65 years and over	2,508,551	454,809	474,957	1,386,175	192,610
65 to 84 years	2,278,368	417,408	432,595	1,251,232	177,133
85 years and over	230,183	37,401	42,362	134,943	15,477
American Indian, Eskimo, and Aleut					
65 years and over	114,453	8,575	17,461	40,379	48,038
65 to 84 years	105,248	7,872	16,258	37,167	43,951
85 years and over	9,205	703	1,203	3,212	4,087
Asian and Pacific Islander					
65 years and over	454,458	64,960	30,427	39,525	319,546
65 to 84 years	424,720	61,132	28,681	37,491	297,416
85 years and over	29,738	3,828	1,746	2,034	22,130
Other races					
65 years and over	312,396	57,787	20,794	86,055	147,760
65 to 84 years	289,409	54,055	19,313	79,610	136,431
85 years and over	22,987	3,732	1,481	6,445	11,329
Hispanic origin[1]					
65 years and over	1,161,283	199,502	71,169	446,984	443,628
65 to 84 years	1,066,719	183,808	65,525	409,681	407,705
85 years and over	94,564	15,694	5,644	37,303	35,923

Source: "Persons 65 and Over for Regions, by Age, Sex, Race, and Hispanic Origin: 1990," Cynthia Taeuber, *Sixty-Five Plus in American*, p. 5-14. Primary source: U.S. Bureau of the Census, 1990 Census of Population and Housing, Summary Tape File 1A. *Note:* 1. Hispanic origin may be of any race.

★ 1071 ★

Senior Citizens

Size of Households for Householders 65 and Over, March 1990

In thousands; noninstitutional population.

Size of household	Number					Percent				85 years and over
	All ages	65 years and over	65 to 74 years	75 to 84 and over	85 years and over	65 years years	65 to 74 years	75 to 84 years		
All races										
All households	93,347	20,156	11,733	6,856	1,567	100.0	100.0	100.0	100.0	
One-person household	22,999	9,175	4,350	3,774	1,051	45.5	37.1	55.0	67.1	
Two-person households	30,114	8,927	5,847	2,641	439	44.3	49.8	38.5	28.0	
Three-person households	16,128	1,328	971	310	47	6.6	8.3	4.5	3.0	
Four-or-more-person households	24,107	728	566	131	31	3.6	4.8	1.9	2.0	
Persons per household	2.63	1.75	1.89	1.57	1.44	(X)	(X)	(X)	(X)	
White										
All households	80,163	18,144	10,477	6,244	1,423	100.0	100.0	100.0	100.0	
One-person household	19,879	8,290	3,841	3,475	974	45.7	36.7	55.7	68.4	
Two-person households	26,714	8,235	5,411	2,431	393	45.4	51.6	38.9	27.6	
Three-person households	13,585	1,108	809	261	38	6.1	7.7	4.2	2.7	
Four-or-more-person households	19,985	511	417	77	17	2.8	4.0	1.2	1.2	
Persons per household	2.58	1.71	1.85	1.54	1.4	(X)	(X)	(X)	(X)	
Black										
All households	10,486	1,696	1,028	543	125	100.0	100.0	100.0	100.0	
One-person household	2,610	776	435	274	67	45.8	42.3	50.5	53.6	
Two-person households	2,721	559	335	183	41	33.0	32.6	33.7	32.8	
Three-person households	2,043	179	130	41	8	10.6	12.6	7.6	6.4	
Four-or-more-person households	3,113	182	129	45	8	10.7	12.5	8.3	6.4	
Persons per household	2.88	2.08	2.24	1.86	1.78	(X)	(X)	(X)	(X)	
Hispanic origin[1]										
All households	5,933	671	463	168	40	100.0	100.0	100.0	(B)	
One-person household	856	222	125	76	21	33.1	27.0	45.2	(B)	
Two-person households	1,292	278	204	64	10	41.4	44.1	38.1	(B)	
Three-person households	1,139	84	67	13	4	12.5	14.5	7.7	(B)	
Four-or-more-person households	2,646	88	68	14	6	13.1	14.7	8.3	(B)	
Persons per household	3.47	2.21	2.32	1.95	(B)	(X)	(X)	(X)	(X)	

Source: "Size of Households for Householders 65 Years and Over Living Alone, by Age, Race, and Hispanic Origin: March 1990," Cynthia Taeuber, *Sixty-Five Plus in America*, p. 6-10. Primary source: Steve Rawlings, U.S. Bureau of the Census, *Household and Family Characteristics: March 1990 and 1989*, Current Population Reports, Series P- 20, No. 447. U.S. Government Printing Office, Washington, DC, 1990, table 17. *Notes:* B Base is less than 75,000. X Not applicable. 1. Hispanic may be of any race.

★ 1072 ★

Senior Citizens

Support Ratios of Elderly Persons: 1990 and 2050

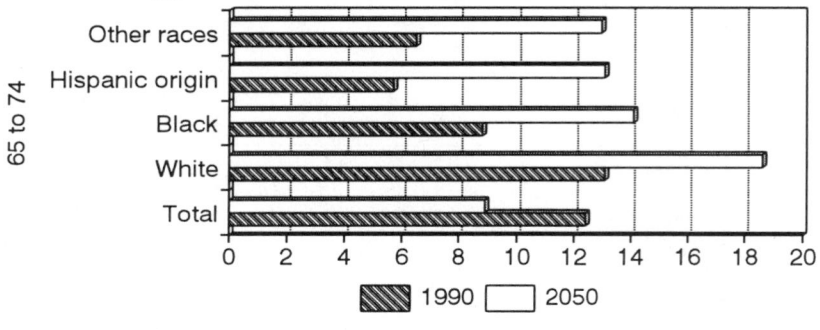

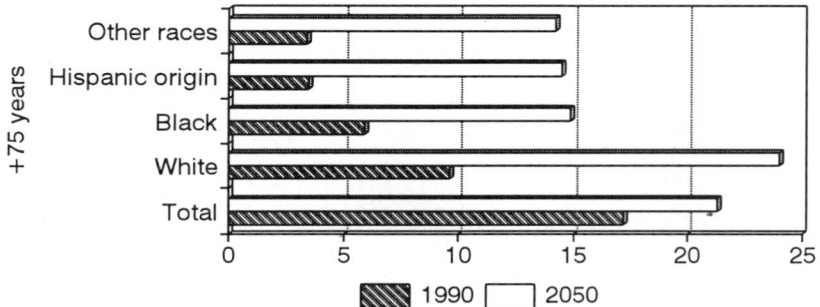

Number of persons of given age per 100 persons aged 20 to 64.

	65 to 74 years		75 years and older	
	1990	2050	1990	2050
Total	12.4	8.9	17.2	21.3
White	13.1	18.6	9.6	24.0
Black	8.8	14.1	5.9	14.9
Hispanic origin[1]	5.7	13.1	3.5	14.5
Other races	6.5	13.0	3.4	14.2

Source: "Support Ratios of Elderly Persons by Age, Race, and Hispanic Origin: 1950 to 2050," Cynthia Taeuber, *Sixty-Five Plus in America*, p. 2-19. Primary source: U.S. Bureau of the Census, 1990 from 1990 Census of Population and Housing, Series CPH-L-74, *Modified and Actual Age, Sex, Race, and Hispanic Origin Data*; 2050 from Jennifer C. Day, *Population Projections of the United States, by Age, Sex, Race, and Hispanic origin: 1992 to 2050*, Current Population Reports, P25-1092. U.S. Government Printing Office, Washington, DC, 1992 (middle series projections). *Note:* 1. Hispanic origin may be of any race.

★ 1073 ★

Senior Citizens

Two-Elderly-Generation Support Ratios, 1950 to 2050

Ratio of persons aged 85 years and over to persons aged 65 to 69 years.

Race	1950	1990	2010	2030	2050
Total	12	30	47	42	93
White	12	31	49	44	97
Black	11	26	36	31	77
Other races	14	17	34	46	78
Hispanic origin[1]	(NA)	21	35	32	72

Source: "Two-Elderly-Generation Support Rations: 1950 to 2050," Cynthia Taeuber, *Sixty-Five Plus in America*, p. 2-5. Primary source: U.S. Bureau of the Census, 1950 from 1950 Census of Population, Volume 2, Part 1, Chapter C, table 112; 1990 from 1990 Census of Population and housing, Series CPH-L-74, *Modified and Actual Age, Sex, Race, and Hispanic Origin Data*; 2010 to 2050 from *Population Projections of the United States, by Age, Sex, Race, and Hispanic Origin: 1992 to 2050*, Current Population Reports, P25-1092. U.S. Government Printing Office, Washington, DC, 1992 (middle series projections). *Notes:* NA Not available. 1. Hispanic origin may be of any race.

★ 1074 ★

Senior Citizens

Years of School Completed 25 and Over, March 1989

In thousands; noninstitutional population.

Age, race, and Hispanic origin	Total	0 to 8 years	High school, 1 to 3 years	High school 4 years	College, 1 to 3 years	College, 4 years or more	Percent high school graduates
All races							
Number							
25 years and over	154,155	17,922	17,719	59,336	26,614	32,565	76.9
25 to 64 years	125,133	9,451	13,093	49,701	23,546	29,342	82.0
65 years and over	29,022	8,471	4,626	9,635	3,068	3,223	54.9
65 to 69 years	10,018	2,161	1,548	3,837	1,210	1,262	63.0
70 to 74 years	7,729	1,937	1,386	2,747	838	821	57.0
75 years and over	11,276	4,374	1,691	3,051	1,019	1,140	46.2
Percent							
25 years and over	100.0	11.6	11.5	38.5	17.3	21.1	(X)
25 to 64 years	100.0	7.6	10.5	39.7	18.8	23.4	(X)
65 years and over	100.0	29.2	15.9	33.2	10.6	11.1	(X)
65 to 69 years	100.0	21.6	15.5	38.3	12.1	12.6	(X)
70 to 74 years	100.0	25.1	17.9	35.5	10.8	10.6	(X)
75 years and over	100.0	38.8	15.0	27.1	9.0	10.1	(X)
Black							
Number							
25 years and over	16,395	2,839	2,959	5,988	2,679	1,929	64.6
25 to 64 years	13,959	1,443	2,518	5,602	2,578	1,816	71.6

[Continued]

★ 1074 ★

Years of School Completed 25 and Over, March 1989
[Continued]

Age, race, and Hispanic origin	Total	0 to 8 years	High school, 1 to 3 years	High school 4 years	College, 1 to 3 years	College, 4 years or more	Percent high school graduates
65 years and over	2,436	1,396	441	386	101	113	24.6
65 to 69 years	908	414	214	192	43	44	30.8
70 to 74 years	624	359	131	84	31	18	21.3
75 years and over	904	623	95	111	25	49	20.6
Percent							
25 years and over	100.0	17.3	18.0	36.5	16.3	11.8	(X)
25 to 64 years	100.0	10.3	18.0	40.1	18.5	13.0	(X)
65 years and over	100.0	57.3	18.1	15.8	4.1	4.6	(X)
65 to 69 years	100.0	45.6	23.6	21.1	4.7	4.8	(X)
70 to 74 years	100.0	57.5	21.0	13.5	5.0	2.9	(X)
75 years and over	100.0	68.9	10.5	12.3	2.8	5.4	(X)
Hispanic origin[1]							
Number							
25 years and over	10,438	3,589	1,539	2,907	1,373	1,030	50.9
25 to 64 years	9,433	2,947	1,455	2,732	1,329	971	53.3
65 years and over	1,005	642	84	175	44	59	27.7
65 to 69 years	419	247	34	69	33	37	33.0
70 to 74 years	232	149	24	46	5	7	25.3
75 years and over	354	246	28	59	7	16	23.0
Percent							
25 years and over	100.0	34.4	14.7	27.9	13.2	9.9	(X)
25 to 64 years	100.0	31.2	15.4	29.0	14.1	10.3	(X)
65 years and over	100.0	63.9	8.4	17.4	4.4	5.9	(X)
65 to 69 years	100.0	58.9	8.1	16.5	7.9	8.8	(X)
70 to 74 years	100.0	64.2	10.3	19.8	2.2	3.0	(X)
75 years and over	100.0	69.5	7.9	16.7	2.0	4.5	(X)

Source: "Years of School Completed, by Age, Race, and Hispanic Origin: March 1989," Cynthia Taeuber, *Sixty-Five Plus in America*, p. 6-15. Primary source: Robert Kominiski, U.S. Bureau of the Census, *Educational Attainment in the United States: March 1989 and 1988*, Current Population Reports, Series P-20, No. 451. U.S. Government Printing Office, Washington, DC, 1991, table 1. *Notes:* X Not applicable. 1. Hispanic origin may be of any race.

Social and Economic Characteristics

★ 1075 ★

Social and Economic Characteristics of the White and Black Populations, 1980-1992

As of March, except labor force status, annual average. Excludes members of Armed Forces except those living off post or with their families on post. Based on Current population Survey.

| Characteristic | Number (1000) | | | | | | | | | Percent distribution, 1992 | |
| | All races[1] | | | White | | | Black | | | White | Black |
	1980	1990	1992	1980	1990	1992	1980	1990	1992		
Total persons	223,160	246,191	251,447	191,905	206,983	210,257	26,033	30,392	31,439	100.0	100.0
Under 5 years old	16,319	18,932	19,566	13,307	15,161	15,487	2,444	2,932	3,172	7.4	10.1
5 to 14 years old	34,979	35,467	36,625	28,828	28,405	29,199	5,190	5,546	5,725	13.9	18.2
15 to 44 years old	103,493	115,690	116,491	88,570	96,656	96,735	12,247	14,660	14,983	46.0	47.7
45 to 64 years old	44,174	46,536	48,175	39,302	40,282	41,538	4,112	4,766	4,953	19.8	15.8
65 years old and over	24,194	29,566	30,590	21,898	26,479	27,297	2,040	2,487	2,606	13.0	8.3
Years of school completed											
Persons 25 years old and over	130,409	156,538	160,838	114,763	134,687	137,657	12,927	16,751	17,445	100.0	100.0
Elementary: 0 to 8 years	22,817	17,591	15,439	18,739	14,131	12,426	3,559	2,701	2,317	9.0	13.3
High school: 1 to 3 years	18,086	17,461	17,672	15,064	14,080	13,911	2,748	2,969	3,324	10.1	19.1
4 years	47,934	60,119	57,864	43,149	52,449	50,049	3,980	6,239	6,220	36.4	35.7
College: 1 to 3 years	19,379	28,075	35,523	17,350	24,350	30,915	1,618	2,952	3,502	22.5	20.1
4 years or more	22,193	33,291	34,340	20,460	29,677	30,356	1,024	1,890	2,080	22.1	11.9
Labor force status[2]											
Civilians 16 years old and over	167,745	188,049	191,576	146,122	160,415	162,658	17,824	21,300	21,958	100.0	100.0
Civilian labor force	106,940	124,787	126,982	93,600	107,177	108,526	10,865	13,493	13,891	66.7	63.3
Employed	99,303	117,914	117,598	87,715	102,087	101,479	9,313	11,966	11,933	62.4	54.3
Unemployed	7,637	6,874	9,384	5,884	5,091	7,047	1,553	1,527	1,958	4.3	8.9
Unemployment rate[3]	7.1	5.5	7.4	6.3	4.7	6.5	14.3	11.3	14.1	(X)	(X)
Not in labor force	60,806	63,262	64,593	52,523	53,237	54,132	6,959	7,808	8,067	33.3	36.7
Family type											
Total families	59,550	66,090	67,173	52,243	56,590	57,224	6,184	7,470	7,716	100.0	100.0
With own children[4]	31,022	32,289	32,738	26,474	26,718	27,038	3,820	4,378	4,444	47.2	57.6
Married couple	49,112	52,317	52,457	44,751	46,981	47,124	3,433	3,750	3,631	82.3	47.1
With own children[4]	24,961	24,537	24,416	22,415	21,579	21,513	1,927	1,972	1,926	37.6	25.0
Female householder, no spouse present	8,705	10,890	11,693	6,052	7,306	7,726	2,495	3,275	3,582	13.5	46.4
With own children[4]	5,445	6,599	7,042	3,558	4,199	4,488	1,793	2,232	2,334	7.8	30.2
Male householder, no spouse present	1,733	2,884	3,025	1,441	2,303	2,374	256	446	504	4.1	6.5
With own children[4]	616	1,153	1,280	500	939	1,037	99	173	184	1.8	2.4
Family income in previous year in constant (1991) dollars											
Total families	59,550	66,090	67,173	52,243	56,590	57,224	6,184	7,470	7,716	100.0	100.0
Less than $5,000	1,489	2,115	2,442	993	1,302	1,453	439	725	877	2.5	11.4
$5,000 to $9,999	3,335	3,701	4,079	2,299	2,547	2,759	934	1,023	1,156	4.8	15.0
$10,000 to $14,999	4,407	4,758	4,844	3,500	3,678	3,808	853	941	859	6.7	11.1
$15,000 to $24,999	10,004	10,046	10,745	8,463	8,375	9,006	1,330	1,412	1,416	15.7	18.4
$25,000 to $34,999	9,707	9,980	10,502	8,725	8,715	9,073	866	1,023	1,115	15.9	14.5
$35,000 to $49,999	13,518	13,152	13,116	12,329	11,714	11,624	952	1,091	1,143	20.3	14.8
$50,000 or more	17,091	22,338	21,445	15,934	20,259	·19,502	810	1,255	1,148	34.1	14.9
Median income (dol.)[5]	36,051	37,579	35,939	37,619	39,514	37,783	21,302	22,197	21,548	(X)	(X)
Poverty[6]											
Families below poverty level	5,461	6,784	7,712	3,581	4,409	5,022	1,722	2,077	2,343	8.8	30.4
Persons below poverty level	26,072	31,528	35,708	17,214	20,785	23,747	8,050	9,302	10,242	11.3	32.7

[Continued]

★ 1075 ★

Social and Economic Characteristics of the White and Black Populations, 1980-1992
[Continued]

| Characteristic | Number (1000) | | | | | | | | | Percent distribution, 1992 | |
| | All races[1] | | | White | | | Black | | | White | Black |
	1980	1990	1992	1980	1990	1992	1980	1990	1992		
Housing tenure											
Total occupied units	80,776	93,347	95,669	70,766	80,163	81,675	8,586	10,486	11,083	100.0	100.0
Owner-occupied	54,891	59,846	61,310	49,913	54,094	55,117	4,173	4,445	4,683	67.5	42.3
Renter-occupied	24421	31,895	32,705	19,581	24,685	25,167	4,257	5,862	6,183	30.8	55.8
No cash rent	1,464	1,606	1,654	1,272	1,384	1,391	156	178	217	1.7	2.0

Source: "Social and Economic Characteristics of the White and Black Population: 1980 to 1992." U.S. Bureau of the Census, *Statistical Abstract of the United States,* 1993, p. 46. Primary source: Except as noted, U.S. Bureau of the Census, *Current Population Reports,* P20-448 and P20-464; P60-180 and P60-181; and unpublished data. *Notes:* X Not applicable. 1. Includes other races not shown separately. 2. Source: U.S. Bureau of Labor Statistics, *Employment and Earnings,* January issues. 3. Total unemployment as percent of civilian labor force. 4. Children under 18 years old. 5. For definition of median, see Guide to Tabular Presentation. 6. For explanation of poverty level, see text, section 14.

Women Heads of Household

★ 1076 ★

Female Family Householders Without Spouse: Characteristics, 1980-1992

As of March. Covers persons 15 years old and over. Based on Current Population Survey.

| Characteristic | Unit | White | | | Black | | | Hispanic origin[1] | | |
		1980	1990	1992	1980	1990	1992	1980	1990	1992
Female family householder	1,000	6,052	7,306	7,726	2,495	3,275	3,582	610	1,116	1,261
Marital status										
Never married (single)	Percent	11	15	17	27	39	42	23	27	29
Married, spouse absent	Percent	17	16	17	29	21	19	32	29	26
Widowed	Percent	33	26	23	22	17	16	15	16	16
Divorced	Percent	40	43	43	22	23	23	30	29	30
Presence of children under 18										
No own children	Percent	41	43	42	28	32	35	25	33	32
With own children	Percent	59	58	58	72	68	65	75	67	68

Source: "Female Family Householders With No Spouse Present—Characteristics, by Race and Hispanic Origin: 1980 to 1992." U.S. Bureau of the Census, *Statistical Abstract of the United States,* 1993, p. 62. Primary source: U.S. Bureau of the Census, *Current Population Reports,* P20-467, and earlier reports. *Note:* 1. Persons of Hispanic origin may be of any race.

★ 1077 ★

Women Heads of Household

Percent Female-Headed Families, by Race and Age, 1982 and 1991

Age cohort (Age of family head)	1982		1991	
	African American	White	African American	White
20-29 years	49.8	13.3	60.4 (+21.3%)	18.1 (+36.1%)
35-44 years	43.4	14.3	45.3 (+4.4%)	14.5 (+1.4%)
Total population	40.6	12.4	46.4 (+14.3%)	13.5 (+8.9%)

Source: "Percent of Female-Headed Families by Race and Selected Age Cohorts: 1982 and 1992." *The State of Black America 1994*, p. 232. Primary source: Calculated by the National Urban League from U.S. Bureau of the Census, Current Population Reports, Series P20-381, *Household and Family Characteristics: March 1982*, Table 3, and Steve W. Rawlings, *Household and Family Characteristics: March 1992* (Series P20-467), Table 3, Washington, DC, Government Printing Office, 1983 and 1993.

Chapter 14
THE PROFESSIONS

Engineering

★ 1078 ★

Engineering Degrees Conferred by Institutions of Higher Learning, by Racial/Ethnic Group and Sex, 1992

	Total	Men	Women
Bachelor's Degrees in Engineering			
All	63653	53,042	9,972
Black	2,374	1,525	750
Hispanic	2,708	2,076	508
Asian	6,479	4,845	1,242
Native American	163	126	31
Foreign National	4,389	3,721	490
Master's Degrees in Engineering			
All	28,540	23,928	4,414
Black	516	338	171
Hispanic	527	421	93
Asian	2,432	1,750	450
Native American	40	29	5
Foreign National	9,054	7,590	988
Doctorate Degrees in Engineering			
All	5,958	5,345	592
Black	39	32	6
Hispanic	62	52	8
Asian	394	307	46
Native American	10	7	3
Foreign National	3,088	2,807	210

Source: "Engineering Degrees Conferred by Institutions of Higher Education by Race/Ethnic Group and Sex of Student, 1992," *Black Issues in Higher Education*, Vol. 10, No. 3, 8 April 1993, p. 21. Primary source: *Engineering and Technology Degrees 1992*, Engineering Workforce Commission. *Notes:* Numbers of men and women may not add up to total because of "unknown" category. Hispanic includes Puerto Ricans; Native American includes Native Alaskans. Published by permission.

Law Schools

★ 1079 ★

Black Enrollment in Prestigious Law Schools, 1992-93

Law school	Black enrollees	% Black
George Mason University	58	12.8%
Harvard University	182	10.1
Vanderbilt University	53	9.9
University of Virginia	114	9.9
Stanford University	54	9.8
University of North Carolina	64	9.2
University of Pennsylvania	70	9.0
Columbia University	93	8.1
Yale University	46	8.1
Georgetown University	137	7.7
George Washington University	91	7.5
Duke University	46	7.2
University of Texas	103	7.0

Source: "Black Students Are Making Solid Headway into the Nation's Most Prestigious Law Schools," *Journal of Blacks in Higher Education*, Vol. 1, No. 1, Autumn 1993, p. 23. Primary source: American Bar Association, U.S. Department of Labor, Bureau of Labor Statistics, and telephone surveys of admission officers of law schools. Published by permission.

★ 1080 ★

Law Schools

Black Faculty at Selective Law Schools

Law school	Black full-time tenured and untenured faculty	% Black
Georgetown University	9	10.8%
Duke University	6	9.7
Columbia University	6	7.8
Harvard University	5	7.3
Stanford University	4	7.3
University of Pennsylvania	6	7.2
Yale University	5	6.3
George Washington University	8	6.2
University of North Carolina	4	6.2
University of Virginia	3	5.0

[Continued]

★ 1080 ★

Black Faculty at Selective Law Schools
[Continued]

Law school	Black full-time tenured and untenured faculty	% Black
Vanderbilt University	4	5.0
George Mason University	2	4.2
University of Texas	1	1.6

Source: "At Most Academically Selective Law Schools Faculty Ranks Are Increasingly Black," *Journal of Blacks in Higher Education*, Vol. 1, No. 1, Autumn 1993, p. 23. Primary source: American Bar Association, U.S. Department of Labor, Bureau of Labor Statistics, and telephone surveys of admission officers of law schools. Published by permission.

★ 1081 ★

Law Schools

Distribution of Full-Time Law School Faculty, by Title and Race, 1993-94

Dean	Professor emeritus (%)	Professor (%)	Associate professor (%)	Assistant professor (%)	Lecturer (%)	Dean (%)	Associate dean (%)	Assistant dean (%)	Other (%)
Black	2.72	30.75	25.94	20.92	3.97	2.09	4.18	6.07	3.33
White	8.69	53.0	12.79	7.41	4.34	2.56	4.33	2.91	3.99

Source: "Distribution of Black Full-Time Faculty by Title 1993-94, Distribution of White Full-Time Faculty by Title 1993-94," *Black Issues in Higher Education*, Vol. 10, No. 24, 27 January 1994, p. 44. Primary source: Association of American Law Schools. Printed by permission. Compiled by editors.

★ 1082 ★

Law Schools

Law School Degrees Conferred by Historically Black Colleges and Universities, 1992

	White		Black		Asian American		Mexican American		Total
	No.	%	No.	%	No.	%	No.	%	
Howard U.	82	(19.2)	322	(75.6)	13	(3.1)	0	(0.0)	426
Southern U.	155	(46.8)	183	(56.2)	0	(0.0)	0	(0.0)	331
N.C. Central	156	(51.0)	149	(48.7)	0	(0.0)	0	(0.0)	306
Texas Southern	128	(22.1)	309	(53.4)	14	(2.4)	117	(20.2)	579

Source: "J.D./ LLB's Conferred by HBCUs Academic Year 1992," *Black Issues in Higher Education*, Vol. 10, No. 7, 3 June 1993, p. 14. Primary source: American Bar Association, Table of Degrees Conferred, 1992. *Notes:* Data omitted for groups with less than 2% in any institution. Published by permission.

★ 1083 ★

Law Schools

Law School Degrees Conferred on Blacks by the Twenty Most Productive Institutions, 1989-90

Institution	State	Men	Women	Total	%
Howard Univ	DC	41	45	86	76.8
Georgetown Univ	DC	34	44	78	12.4
Southern Univ-Baton Rouge	LA	37	28	65	59.1
Harvard Univ	MA	26	36	62	11.3
Univ of Maryland-Baltimore	MD	14	19	43	15.2
Tulane Univ of Louisiana	LA	11	24	35	12.8
Texas Southern Univ	TX	22	13	35	35.7
North Carolina Central Univ	NC	11	20	31	50.8
Univ of Virginia-Main Campus	VA	20	10	30	7.6
Rutgers Univ-Newark Campus	NJ	11	18	29	12.7
Temple Univ	PA	13	14	27	8.4
Univ of Michigan-Ann Arbor	MI	12	15	27	6.8
Univ of California-Los Angeles	CA	9	17	26	8.1
New York Univ	NY	13	13	26	6.5
Univ of Florida	FL	10	14	24	6.3
SUNY-Buffalo	NY	10	12	22	9.5
CUNY-Queens	NY	9	12	21	15.7
Univ of California-Berkeley	CA	8	10	18	6.4
Univ of Iowa	IA	7	10	17	8.0
Univ of Pittsburgh-Main Campus	PA	14	3	17	7.3
Yale Univ	CT	7	9	16	8.0

Source: "Professional Degrees Conferred 1989-90, Law Degrees, African American," *Black Issues in Higher Education*, Vol. 10, No. 6, 20 May 1993, p. 100. Primary source: Department of Education.

★ 1084 ★

Law Schools

Law School Enrollment, by Sex and Race, Fall 1992

	Full time				Part time				Combined	
	Women	%	Men	%	Women	%	Men	%	Total	%
White	36,560	(79.3)	52,543	(85.5)	7,164	(83.8)	10,679	(88.3)	106,946	(83.4)
Black	4,255	(9.2)	3,049	(5.0)	808	(9.5)	586	(4.9)	8,698	(6.8)
Am. Indian	320	(0.7)	360	(0.6)	47	(0.5)	49	(0.4)	776	(0.6)
Asian	2,623	(5.7)	2,585	(4.2)	240	(2.8)	375	(3.1)	5,823	(4.5)
Mexican	919	(2.0)	1,184	(1.9)	52	(0.6)	103	(0.9)	2,258	(1.8)
Puerto Rican	241	(0.5)	253	(0.4)	52	(0.6)	41	(0.3)	587	(0.5)

[Continued]

★ 1084 ★

Law School Enrollment, by Sex and Race, Fall 1992

[Continued]

	Full time				Part time				Combined	
	Women	%	Men	%	Women	%	Men	%	Total	%
Other Hispanic	1,179	(2.6)	1,504	(2.4)	184	(2.2)	257	(2.1)	3,124	(2.4)
Total	46,097	(100.0)	61,478	(100.0)	8,547	(100.0)	12,090	(100.0)	128,212	(100.0)

Source: "Total Law School Enrollment Fall 1992," *Black Issues in Higher Education*, Vol. 10, No. 7, 3 June 1993, p. 14. Primary source: American Bar Association, Student Body Table C10-923, February 1993.

★ 1085 ★

Law Schools

Law School Professors, by Race, 1992-93

	Number
Total Full Law Professors	3,861
Ethnic response[1]	3,454
White	3,211
African American	135
Other minorities	108
Total Associate Law Professors	1,105
Ethnic response	992
White	793
African American	122
Other minorities	77
Total Assistant Law Professors	727
Ethnic response	628
White	463
African American	102
Other minorities	63

Source: "Law School Professors by Race 1992-93," *Black Issues in Higher Education*, Vol. 10, No. 8, 1 July, 1993, p. 14. Primary source: American Association of Law Schools. *Note:* 1. Total response to questions of ethnicity/race.

Medicine

★ 1086 ★

Enrollment in U.S. Medical Schools, 1985 to 1991

	1985	1986	1987	1988	1989	1990	1991	% change '90 to '91
U.S. citizen (and paramount resident)								
White	54,335	53,136	51,728	50,366	48,961	47,893	47,094	-1.7
Underrepresented minorities								
Black	3,849	3,892	3,968	3,995	4,145	4,241	4,334	2.2
American Indian/Alaskan Native	235	242	233	237	258	277	301	8.7
Mexican American/Chicano	1,143	1,153	1,144	1,128	1,087	1,109	1,205	8.7
Puerto Rican (Mainland)	428	435	467	438	452	457	485	6.1
Subtotal	5,655	5,722	5,812	5,798	5,942	6,084	6,325	4.0
Other U.S. students								
Asian or Pacific Islander	4,289	4,883	5,738	6,595	7,489	8,436	9,438	11.9
Puerto Rican (Commonwealth)	896	918	899	909	817	796	847	2.
Other Hispanic	991	1,015	1,038	1,091	1,181	1,176	1,138	-3.2
Unidentified	19	6	17	21	75	144	132	na
Subtotal	6,195	6,822	7,692	8,616	9,562	10,552	11,525	9.2
Foreign	400	445	503	520	551	634	658	3.8
Total enrollment	66,585	66,125	65,735	65,300	65,016	65,163	65,602	0.7

Source: "U.S. Medical School Enrollments," *Black Issues in Higher Education*, Vol. 9, No. 3, 9 April 1992, p. 38. Primary source: American Association of American Medical Colleges, Facts: Applicants, Matriculants and Graduates 1985-1991. Published by permission.

★ 1087 ★

Medicine

Enrollment in U.S. Medical Schools, by Sex and Racial/Ethnic Group, 1985 to 1991

	1985	1986	1987	1988	1989	1990	1991	% change '90 to '91
					Men			
U.S. citizen (and paramount resident)								
White	37,322	36,097	34,804	33,502	32,129	30,922	30,054	-2.8
Underrepresented minorities								
Black	2,053	1,978	1,912	1,845	1,887	1,874	1,902	1.5
American Indian/Alaskan Native	140	141	140	135	150	159	154	-3.1
Mexican American/Chicano	777	759	765	720	691	682	742	8.8
Puerto Rican (Mainland)	266	273	292	266	263	260	270	3.8
Subtotal	3,236	3,151	3,109	2,966	2,991	2,975	3,068	3.1

[Continued]

★ 1087 ★

Enrollment in U.S. Medical Schools, by Sex and Racial/Ethnic Group, 1985 to 1991
[Continued]

	1985	1986	1987	1988	1989	1990	1991	% change '90 to '91
Other U.S. men								
Subtotal	4,091	4,475	4,949	5,504	6,008	6,570	7,088	7.9
Foreign	286	302	329	343	375	410	430	4.9
Total men	44,935	44,025	43,191	42,315	41,503	40,877	40,640	-0.6
Women								
U.S. citizen (and paramount resident)								
White	17,013	17,039	16,924	16,864	16,832	16,971	17,040	0.4
Underrepresented minorities								
Black	1,796	1,914	2,056	2,150	2,258	2,367	2,432	2.7
American Indian/Alaskan Native	95	101	93	102	108	118	147	24.6
Mexican American/Chicano	366	394	379	408	396	427	463	8.4
Puerto Rican (Mainland)	162	162	175	172	189	197	215	9.1
Subtotal	2,419	2,571	2,703	2,832	2,951	3,109	3,257	4.8
Other U.S. women								
Subtotal	2,104	2,347	2,743	3,112	3,554	3,982	4,437	11.4
Foreign								
Total women	21,650	22,100	22,544	22,985	23,513	24,286	24,962	2.8

Source: "U.S. Medical School Enrollments," *Black Issues in Higher Education,* Vol. 9, No. 3, 9 April 1992, p. 38. Primary source: American Association of American Medical Colleges, Facts: Applicants, Matriculants and Graduates 1985-1991. Published by permission.

★ 1088 ★

Medicine

Graduates from U.S. Medical Schools, 1985 to 1991

	1985	1986	1987	1988	1989	1990	1991	% change '90 to '91
Racial/ethnic (self-description)								
White	13,788	13,413	13,144	13,002	12,606	12,075	11,792	-2.3
Underrepresented minorities								
Black	828	824	820	850	821	874	918	5.0
American Indian/Alaskan Native	65	49	63	58	57	52	46	-1.15
Mexican American/Chicano	242	233	226	241	245	249	261	4.8
Puerto Rican (Mainland)	89	89	81	112	124	93	101	8.6
Subtotal	1,224	1,195	1,190	1,261	1,247	1,288	1,326	4.6
Other graduates								
Asian or Pacific Islander	750	909	947	1,119	1,241	1,433	1,687	17.7
Puerto Rican (Commonwealth)	229	200	208	224	232	211	190	-10.0
Other Hispanic	247	269	250	248	255	266	280	5.3

[Continued]

★ 1088 ★

Graduates from U.S. Medical Schools, 1985 to 1991

[Continued]

	1985	1986	1987	1988	1989	1990	1991	% change '90 to '91
Unidentified	80	131	91	65	49	145	152	na
Subtotal	1,306	1,509	1,496	1,656	1,777	2,055	2,309	12.4
Total graduates	16,318	16,117	15,830	15,919	15,630	15,398	15,427	0.2

Source: "U.S. Medical School Graduates," *Black Issues in Higher Education*, Vol. 9, No. 3, 9 April 1992, p. 39. Primary source: American Association of American Medical Colleges, Facts: Applicants, Matriculants and Graduates 1985-1991. Published by permission.

★ 1089 ★

Medicine

Graduates from U.S. Medical Schools, by Sex and Racial/Ethnic Group, 1985 to 1991

	1985	1986	1987	1988	1989	1990	1991	% change '90 to '91
Men								
Racial/ethnic (self-description)								
White	9,769	9,401	9,029	8,898	8,577	8,158	7,750	-5.0
Underrepresented minorities								
Black	456	474	458	441	396	421	402	-4.5
American Indian/Alaskan Native	47	34	33	38	32	33	31	-6.1
Mexican American/Chicano	165	161	147	174	162	166	162	-2.4
Puerto Rican (Mainland)	51	52	53	69	78	56	68	21.4
Subtotal	719	721	691	722	688	676	663	-1.9
Other men								
Subtotal	926	1,038	1,003	1,084	1,164	1,333	1,461	9.6
Total men	11,414	11,160	10,723	10,704	10,409	10,164	9,854	-2.9
Women								
Racial/ethnic (self-description)								
White	4,019	4,012	4,115	4,104	4,029	3,917	4,042	3.2
Underrepresented minorities								
Black	372	350	362	409	425	453	516	13.9
American Indian/Alaskan Native	18	15	30	20	25	19	15	-21.1
Mexican American/Chicano	77	72	79	67	83	83	99	19.3
Puerto Rican (Mainland)	38	37	28	43	46	37	33	-10.8
Subtotal	505	474	499	539	579	592	663	12.0
Other women								
Subtotal	380	471	493	572	613	722	848	17.5
Total women	4,904	4,957	5,107	5,215	5,221	5,231	5,553	6.2

Source: "U.S. Medical School Enrollments," *Black Issues in Higher Education*, Vol. 9, No. 3, 9 April 1992, p. 39. Primary source: American Association of American Medical Colleges, Facts: Applicants, Matriculants and Graduates 1985-1991. Published by permission.

★ 1090 ★

Medicine

Medical Degrees Conferred on Blacks by the Twenty Most Productive Institutions, 1991-92

Institution	Men	Women	Total
Howard Univ	22	34	56
Univ of Michigan	16	12	28
Meharry	15	12	27
New York Med	11	16	27
Univ of Illinois	11	15	26
Wayne State	12	13	25
Temple Univ	8	16	24
SUNY-Brooklyn	10	12	22
Univ of Med and Dent of NJ-Johnson	6	11	17
Hahneman	7	10	17
Georgetown	7	9	16
Morehouse	8	7	15
California-San Francisco	6	7	13
Case Western	5	8	13
Cornell	6	6	12
Harvard	5	7	12
Medical Coll of Georgia	8	3	11
Johns Hopkins	5	6	11
Wash Univ-St. Louis	6	5	11
Univ of Med and Dent of NJ-NJ	3	7	10
Albert Einstein	6	4	10
SUNY-Syracuse	5	5	10
Texas-Galveston	5	5	10
Drew-UCLA	1	8	9

Source: "Professional Degrees Conferred 1991-92, Medical Degrees, African American," *Black Issues in Higher Education*, Vol. 10, No. 6, 20 May 1993, p. 102. Primary source: American Association of Medical Colleges, 1992. Published by permission.

Chapter 15
SOCIAL AND HUMAN SERVICES

Child Support and Alimony

★ 1091 ★

Child Support and Alimony: Selected Characteristics of Women, 1989

Alimony data are for ever-divorced and currently separated women. Women with own children under 21 years of age present from absent fathers. For 1989, women 15 years old and over as of April 1990; for previous years, women 18 years old and over as of April of the following year. Covers civilian noninstitutional population. Based on Current Population Survey.

Recipiency status of women	Unit	Total[1]	Age 18 to 29 years	Age 30 to 39 years	Age 40 years and over	Race White	Race Black	Hispanic[2]	Current marital status Divorced	Current marital status Married[3]	Current marital status Never married	Separated
Child support												
All women, total	1,000	9,955	3,086	4,175	2,566	6,905	2,770	1,112	3,056	2,531	2,950	1,352
Payments awarded	1,000	5,748	1,408	2,685	1,632	4,661	955	452	2,347	1,999	704	648
Percent of total	Percent	58	46	64	64	68	35	41	77	79	24	48
Supposed to receive child support in 1989	1,000	4,953	1,208	2,413	1,309	4,048	791	364	2,123	1,685	583	527
Percent received payment	Percent	75	76	74	76	77	70	70	77	72	73	80
Mean child support	Dollars	2,995	1,981	3,032	3,903	3,132	2,263	2,965	3,322	2,931	1,888	3,060
Percent of total income	Percent	19	20	18	19	19	16	20	17	20	20	21
Women with incomes below the poverty level in 1989	1,000	3,206	1,531	1,189	434	1,763	1,314	536	820	176	1,590	612
Payments awarded	1,000	1,387	608	568	195	962	384	177	577	127	389	288
Percent of total	Percent	43	40	48	45	55	29	33	70	72	25	47
Supposed to receive child support in 1989	1,000	1,190	507	500	168	827	325	148	525	106	334	221
Percent received payment	Percent	68	68	67	72	68	70	64	66	67	69	74
Mean child support	Dollars	1,889	1,515	2,167	2,316	1,972	1,674	1,824	2,112	2,275	1,553	1,717
Percent of total income	Percent	37	33	36	56	39	32	37	38	52	34	35
Alimony												
All women, total	1,000	20,610	2,464	6,093	12,051	17,245	2,863	1,499	8,888	7,738	(X)	2,790
Number awarded payments	1,000	3,189	184	610	2,394	2,801	305	171	1,472	1,170	(X)	316
Percent of total	Percent	16	8	10	20	16	11	11	17	15	(X)	11
Supposed to receive payments	1,000	922	85	267	569	787	98	63	567	170	(X)	164
Women with incomes below the poverty level in 1989	1,000	3,692	726	1,206	1,758	2,640	931	477	1,860	420	(X)	1,147
Number awarded payments	1,000	429	60	96	273	340	76	31	223	55	(X)	110
Percent of total	Percent	12	8	8	16	13	8	6	12	13	(X)	10
Supposed to receive payments	1,000	178	43	56	79	149	26	21	112	11	(X)	54

Source: "Child Support and Alimony—Selected Characteristics of Women: 1989," U.S. Bureau of the Census, *Statistical Abstract of the United States*, 1993, p.385. Primary source: U.S. Bureau of the Census, *Current Population Reports*, P60-173. *Notes:* X Not applicable. 1. Includes other items, not shown separately. 2. Hispanic women may be of any race. 3. Remarried women whose previous marriage ended in divorce.

Pension Coverage of Workers

★ 1092 ★

Pension Plan Coverage: Workers by Selected Characteristics, 1991

Covers workers as of March of following year who had earnings in year shown. Based on Current Population Survey.

Sex and age	Number with coverage (1,000)				Percent of total workers			
	Total[1]	White	Black	Hispanic[2]	Total[1]	White	Black	Hispanic[2]
Total	53,913	46,704	5,663	2,732	40	41	40	27
Male	30,751	27,075	2,803	1,640	43	43	40	27
Under 65 years old	30,163	26,536	2,762	1,622	44	44	41	27
15 to 24 years old	1,529	1,299	191	136	13	13	15	11
25 to 44 years old	17,764	15,530	1,682	1,028	47	48	44	29
45 to 64 years old	10,870	9,708	888	458	56	57	54	39
65 years old and over	589	539	42	18	21	22	20	18
Female	23,162	19,629	2,859	1,093	37	37	40	26
Under 65 years old	22,724	19,240	2,823	1,084	38	38	41	26
15 to 24 years old	1,313	1,117	174	86	12	12	14	10
25 to 44 years old	13,763	11,582	1,756	680	42	42	44	29
45 to 64 years old	7,649	6,541	894	318	47	47	53	37
65 years old and over	438	389	36	9	22	23	20	17

Source: "Pension Plan Coverage of Workers, by Selected Characteristics: 1991," U.S. Bureau of the Census, *Statistical Abstract of the United States,* 1993, p. 378. Primary source: U.S. Bureau of the Census, unpublished data. *Notes:* 1. Includes other races, not shown separately. 2. Hispanic persons may be of any race.

Retirement and Functional Limitations

★ 1093 ★

Functional Limitations of Retired Persons, 1984

Percents in 90-percent confidence intervals.

Functional limitations status	Did not work	Worked
Black		
No limitations	19.1 to 34.7	3.6 to 13.4
One or more limitations	51.8 to 69.0	0.7 to 7.7
One or more severe limitations	24.0 to 40.6	-
White[1]		
No limitations	41.5 to 46.7	7.0 to 10.0

[Continued]

★ 1093 ★

Functional Limitations of Retired Persons, 1984
[Continued]

Functional limitations status	Did not work	Worked
One or more limitations	37.4 to 42.6	3.7 to 5.9
One or more severe limitations	15.3 to 19.3	0.6 to 1.6

Source: "Functional Limitations Status of Persons 65 to 69 Years, With Retirement Income, by Employment Status, Sex and Race: 1984," Cynthia Taeuber, *Sixty-Five Plus in America*, p. 3-16. Primary source: U.S. Bureau of the Census, 1984 Survey of Income and Program Participation, Health-Wealth File, wave 3 (tabulations produced by Arnold Goldstein, Population Division). *Notes:* Percentage of age/race group based on 603,000 Blacks and 7,404,000 Whites. - Indicates zero sample cases. 1. Data are for all races other than Black.

★ 1094 ★

Retirement and Functional Limitations

Work Disability Status of Persons 65 to 72 Years Old, 1984

In thousands. Percents in 90-percent confidence intervals.

Race and sex	Total number	Percent	
		With a work disability	Prevented from working
Black males	448	47.7 to 67.9	34.7 to 55.1
White males[1]	5,415	36.7 to 42.5	24.8 to 30.0
Black females	653	49.2 to 66.8	43.2 to 61.0
White females[1]	6,707	33.1 to 38.5	25.3 to 30.3

Source: "Work Disability Status of Persons 65 to 72 Years, by Sex and Race: 1984," Cynthia Taeuber, *Sixty-Five Plus in America*, p. 3-16. Primary source: U.S. Bureau of the Census, 1984 Survey of Income and Program Participation, Health-Wealth File, wave 3 (tabulations produced by Arnold Goldstein, Population Division). *Note:* 1. Data are for all races other than Black.

Social Service Programs

★ 1095 ★

Participation Rates in Major Assistance Programs, 1988

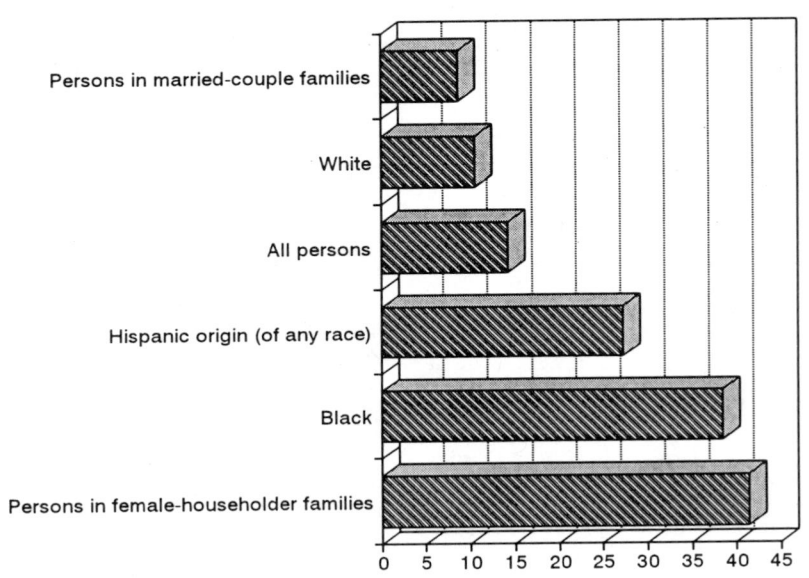

Characteristic	Percent
Persons in female-householder families	41.5
Black	38.5
Hispanic origin (of any race)	27.3
All persons	14.2
White	10.5
Persons in married-couple families	8.7

Source: Martin Shea, "Program Participation." *Population Profile of the United States, 1993*, p. 31.

Volunteer Workers

★ 1096 ★

Adult Volunteer Workers, 1989

For year ending in May. Covers civilian noninstitutional population, 16 years old and over. A volunteer is a person who performed unpaid work for an organization such as a church, the Boy or Girl Scouts, a school, Little League, etc. during the year. Persons who did not work on their own such as helping out neighbors or relatives are excluded. Based on Current Population Survey.

| Characteristic | Volunteer workers | | Percent distribution of volunteers, by type of organization[1] | | | | | | | |
	Number (1,000)	Percent of population	Total	Churches, other religious organizations	Schools, other educational institutions	Civic or political organizations	Hospitals, other health organizations	Social or welfare organizations	Sport or recreational organizations	Other organizations
Total[2]	38,042	20.4	100.0	37.4	15.1	13.2	10.4	9.9	7.8	6.3
White	34,823	21.9	100.0	36.6	15.1	13.5	10.7	9.8	8.0	6.3
Black	2,505	11.9	100.0	50.4	12.4	9.6	7.0	10.4	4.6	5.6
Hispanic origin[3]	1,289	9.4	100.0	42.2	18.3	9.6	8.5	8.9	6.9	5.6

Source: "Percent of Adult Population Doing Volunteer Work: 1989," U.S. Bureau of the Census, *Statistical Abstract of the United States*, 1993, p. 386. Primary source: U.S. Bureau of Labor Statistics, *News*, USDL 90-154, March 29, 1990. *Notes:* 1. Organization for which most of the work was done. 2. Includes other races, not shown separately. 3. Persons of Hispanic origin may be of any race.

Chapter 16
SPORTS AND LEISURE

Basketball

★ 1097 ★

Scoring Defense, Central Intercollegiate Athletic Association, 1992-93

Team	GM	FGM	FTM	PTS	AVG
Fayetteville State	27	680	340	1,841	68.2
Virginia Union	28	743	309	1,983	70.8
Norfolk State	29	731	483	2,090	72.1
N.C. Central	27	702	484	1,981	73.4
Hampton	27	681	521	2,010	74.4
Johnson C. Smith	28	846	418	2,270	81.1
Virginia State	27	771	568	2,208	81.8
Livingstone	24	688	489	1,973	82.2
St. Augustine's	25	726	464	2,060	82.4
Shaw	27	838	450	2,236	82.8
St. Paul's	28	793	614	2,340	83.6
Bowie State	27	819	502	2,276	84.3
Winston-Salem State	23	722	415	1,941	84.4
Elizabeth City State	28	897	499	2,433	86.9

Source: "CIAA Statistics," *Sportsview,* 4 (Winter 1993), p. 130.

★ 1098 ★

Basketball

Scoring Offense and Scoring Defense, Mid-Eastern Athletic Conference, 1992-93

Team	GM	FGM	FTM	PTS	AVG
Scoring Offense					
Morgan State	25	727	362	1,939	77.6
Coppin State	30	736	532	2,214	73.8
Delaware State	29	786	427	2,138	73.7
Florida A&M	28	792	379	2,062	73.6
S.C. State	29	756	489	2,122	73.1
N.C. A&T	26	683	398	1,891	72.7
Howard	28	658	494	1,931	70.0
Md.-Eastern Shore	27	632	419	1,794	66.4
Bethune Cookman	27	678	279	1,731	64.1
Scoring Defense					
Coppin State	30	729	409	1,994	66.5
S.C. State	29	775	398	2,115	72.9
Md.-Eastern Shore	27	726	433	1,995	73.9
N.C. A&T	26	688	449	1,949	74.9
Howard	28	730	534	2,125	75.9
Delaware State	29	796	519	2,273	78.4
Bethune Cookman	27	763	427	2,121	78.6
Florida A&M	28	804	524	2,309	82.5
Morgan State	25	712	597	2,141	85.6

Source: "MEAC Statistics," *Sportsview*, 4 (Winter 1993), p. 134.

★ 1099 ★

Basketball

Scoring Offense and Scoring Defense, Southern Intercollegiate Athletic Conference, 1992-93

Team	GM	FGM	FTM	PTS	AVG
Scoring Offense					
Alabama A&M	31	1,105	660	2,990	96.5
LeMoyne-Owen	28	917	540	2,519	90.0
Morehouse	26	841	481	2,248	86.5
Ft. Valley St.	28	806	532	2,264	80.9
Miles	24	699	319	1,899	79.1
Tuskegee	27	760	407	2,090	77.4
Albany St.	25	756	343	1,931	77.2
Paine	26	724	427	1,971	75.8
Savannah St.	27	749	399	2,030	75.2
Clark Atlanta	26	698	372	1,862	71.6

[Continued]

★ 1099 ★

Scoring Offense and Scoring Defense, Southern Intercollegiate Athletic Conference, 1992-93

[Continued]

Team	GM	FGM	FTM	PTS	AVG
Morris Brown	26	690	375	1,845	71.0
Scoring Defense					
Paine	26	661	462	1,865	71.7
Tuskegee	27	721	535	2,111	78.2
Albany St.	25	669	531	1,968	78.7
Savannah St.	27	801	434	2,147	79.5
Alabama A&M	31	864	668	2,522	81.4
LeMoyne-Owen	28	823	487	2,282	81.5
Ft. Valley St.	28	844	477	2,298	82.1
Morehouse	26	818	444	2,197	84.5
Clark Atlanta	26	810	471	2,216	85.2
Morris Brown	26	830	450	2,231	85.8
Miles	24	846	384	2,212	92.2

Source: "SLAC Statistics," *Sportsview,* 4 (Winter 1993), p. 138.

★ 1100 ★

Basketball

Scoring Offense and Scoring Defense, Southwestern Athletic Conference, 1992-93

Team	GM	FGM	FTM	PTS	AVG
Scoring Offense					
Southern	31	1,075	603	3,011	97.1
Alabama State	27	852	541	2,378	88.1
Jackson State	34	1,054	601	2,884	84.8
Grambling State	27	872	429	2,254	83.5
Miss. Valley	28	854	383	2,289	81.8
Prairie View	27	770	424	2,139	79.2
Alcorn State	27	842	345	2,132	79.0
Texas Southern	27	717	511	2,093	77.5
Scoring Defense					
Jackson State	34	948	573	2,651	78.0
Texas Southern	27	812	374	2,162	80.1
Miss. Valley	28	849	482	2,328	83.1
Grambling State	27	793	547	2,268	84.0
Alabama State	27	856	419	2,288	84.7
Southern	31	1,016	471	2,650	85.5
Alcorn State	27	838	483	2,313	85.7
Prairie View	27	982	620	2,763	102.3

Source: "SWAC Statistics," *Sportsview,* 4 (Winter 1993), p. 142.

★ 1101 ★

Basketball

Scoring Offense, Central Intercollegiate Athletic Association, 1992-93

Team	GM	FGM	FTM	PTS	AVG
Johnson C. Smith	28	848	656	2,497	89.2
Elizabeth City State	28	877	494	2,476	88.4
N.C. Central	27	822	526	2,302	85.3
Virginia Union	28	843	579	2,359	84.3
Virginia State	27	812	506	2,237	82.9
Shaw	27	780	486	2,224	82.4
Bowie State	27	791	505	2,196	81.3
Norfolk State	29	827	562	2,346	80.9
St. Augustine's	25	715	423	1,972	78.9
St. Paul's	28	838	345	2,182	77.9
Hampton	27	797	430	2,103	77.9
Fayettville State	27	693	478	2,004	74.2
Livingstone	24	659	369	1,771	73.8
Winston-Salem State	23	619	333	1,691	73.5

Source: "CIAA Statistics," *Sportsview*, 4 (Winter 1993), p. 130.

Collegiate Sports

★ 1102 ★

Graduation Rates in Selected Conferences, 1993, Part 1

	Graduation rates							
	Total		White		Black		Hisp	
	M	F	M	F	M	F	M	F
Atlantic Coast Conference								
Gene F. Corrigan, Commissioner								
Clemson University	44	66	53	67	29	75	-	-
Duke University	89	95	91	95	83	100	-	-
Florida State University	42	49	47	50	36	48	-	-
Georgia Tech	59	65	60	71	56	44	100	100
Univ. of Maryland Coll. Park	51	71	57	76	35	50	-	-
UNC-Chapel Hill	68	80	71	85	62	53	100	-
North Carolina State Univ.	46	72	57	71	22	75	50	-
University of Virginia	79	90	85	91	63	81	-	-
Wake Frost University	61	83	64	86	52	50	-	-

[Continued]

★ 1102 ★

Graduation Rates in Selected Conferences, 1993, Part 1
[Continued]

	Graduation rates							
	Total		White		Black		Hisp	
	M	F	M	F	M	F	M	F
Big Eight Conference								
Carl C. James, Commissioner								
Univ. of Colorado	61	42	67	40	51	20	40	100
Univ. of Kansas	47	54	53	56	32	44	100	100
Kansas State University	41	71	53	73	17	56	-	100
University of Missouri	57	81	66	84	36	68	100	100
Iowa State University	44	75	49	78	29	58	-	-
University of Nebraska	49	57	59	60	22	43	43	0
University of Oklahoma	36	50	43	53	30	47	0	0
Oklahoma State University	25	38	28	43	21	25	0	-

Source: "Grades of Glory: Statistics," *Black Issues in Higher Education*, 10 (December 2, 1993), p. 36.

★ 1103 ★

Collegiate Sports

Graduation Rates in Selected Conferences, 1993, Part 2

	Graduation rates							
	Total		White		Black		Hisp	
	M	F	M	F	M	F	M	F
Big Ten Conference								
James E. Delany, Commissioner								
Univ. of Illinois-Champaign	63	80	63	78	65	78	67	100
Indiana University	58	69	60	68	37	80	100	0
University of Iowa	61	65	67	67	42	63	-	50
University of Michigan	62	84	66	87	52	71	40	100
Michigan State University	57	81	66	84	36	68	100	100
University of Minnesota	40	68	45	71	23	50	-	-
Northwestern University	80	83	80	80	75	75	100	100
Ohio State University	59	70	65	77	71	67	100	100
Pennsylvania State Univ.	67	74	68	77	71	67	100	100
Purdue University	54	75	62	80	37	57	50	0
University of Wisconsin	39	56	43	63	0	14	-	0
Mid-Eastern Athletic Conference								
Kenneth A. Free, Commissioner								
Bethune-Cookman College	34	40	33	-	34	40	-	-
Coppin State College	-	-	-	-	-	-	-	-
Delaware State College	40	39	50	-	38	41	-	-
Florida A&M University	31	35	40	0	30	36	0	0
Howard University	41	48	-	-	38	49	-	-
Univ. of Maryland-E. Shore	20	28	0	0	19	30	0	0

[Continued]

★ 1103 ★

Graduation Rates in Selected Conferences, 1993, Part 2
[Continued]

	Graduation rates							
	Total		White		Black		Hisp	
	M	F	M	F	M	F	M	F
Morgan State University	44	45	67	100	43	45	-	-
No. Carolina A&T State	34	74	-	-	34	74	-	-

Source: "Grades of Glory: Statistics," *Black Issues in Higher Education*, 10 (December 2, 1993), pp. 36-37.

★ 1104 ★

Collegiate Sports

Graduation Rates in Selected Conferences, 1993, Part 3

	Graduation rates							
	Total		White		Black		Hisp	
	M	F	M	F	M	F	M	F
Ohio Valley Conference								
R. Daniel Beebe, Commissioner								
Austin Peay State Univ.	30	33	36	38	22	27	-	-
Eastern Kentucky Univ.	44	49	60	57	24	26	100	-
Middle Tenn. State Univ.	28	50	30	59	27	36	-	-
Morehead State University	44	67	56	68	21	-	0	-
Murray State University	45	48	57	51	25	33	-	-
SE Missouri State Univ	32	45	35	53	24	24	-	-
Univ. of Tennessee, Martin	-	-	-	-	-	-	-	-
Tennessee State University	24	34	0	-	24	34	-	-
Tennessee Tech Univ.	57	57	53	57	66	50	-	-
Pacific-10 Conference								
Elwood B. Hahn, Commissioner								
Univ. of Arizona	40	62	43	59	35	55	17	67
Arizona State Univ.	32	62	39	69	25	43	25	25
Univ. of California, Berkley	62	72	65	76	57	67	86	75
UCLA	57	64	66	63	41	73	29	75
Univ. of Oregon	45	62	47	60	34	83	67	100
Oregon State University	53	50	52	55	59	38	33	0
Univ. of Southern California	48	69	54	77	42	47	20	50
Stanford University	82	90	83	93	63	100	67	0
Univ. of Washington	47	73	55	74	31	71	-	-
Washington State Univ.	46	53	52	59	30	17	0	50

Source: "Grades of Glory: Statistics," *Black Issues in Higher Education*, 10 (December 2, 1993), p. 37.

★ 1105 ★

Collegiate Sports

Graduation Rates in Selected Conferences, 1993, Part 4

	Graduation rates							
	Total		White		Black		Hisp	
	M	F	M	F	M	F	M	F
Southeastern Conference								
Roy F. Kramer, Commissioner								
University of Alabama	35	58	40	64	25	40	33	-
University of Arkansas	40	48	43	45	36	57	33	-
Auburn University	45	72	58	77	27	55	0	-
Univ. of Florida	41	58	48	55	27	60	-	-
Univ. of Georgia	39	60	51	61	11	59	50	50
Univ. of Kentucky	53	63	55	63	46	55	-	-
LSU	29	38	37	33	15	37	-	0
Univ. of Mississippi	51	46	62	53	36	33	-	-
Mississippi State Univ.	51	57	64	57	32	56	0	-
Univ. of South Carolina	51	64	58	64	37	50	0	-
Univ. of Tenn.-Knoxville	44	61	53	59	29	67	33	-
Vanderbilt University	72	79	75	77	57	100	100	-
Southwestern Athletic Conference								
James Frank, Commissioner								
Alabama State University	25	45	-	-	25	43	-	-
Alcorn State University	21	35	-	-	21	35	-	-
Grambling State University	40	62	20	0	36	61	-	-
Jackson State University	35	53	0	-	35	53	-	-
Mississippi Valley State	42	43	-	-	42	43	-	-
Prairie View A&M	-	-	-	-	-	-	-	-
Southern (Baton Rouge)	26	28	0	-	26	28	-	-
Texas Southern University	12	24	0	-	10	25	13	-

Source: "Grades of Glory: Statistics," *Black Issues in Higher Education*, 10 (December 2, 1993), p. 38.

Professional Sports

★ 1106 ★

Racial Composition of Major Professional Sports Coaches and Managers, 1989 to 1993

Racial compositions (number in parentheses) of the NBA, NFL and MLB head coaches and managers (years mark the beginning of each season).

	Year				
	1989	1990	1991	1992	1993
National Basketball Association					
Black	22 (6)	22 (6)	7 (2)	26[1] (7)	
White	78 (21)	78 (21)	93 (25)	74 (20)	
Latino	0 (0)	0 (0)	0 (0)	0 (0)	
National Football League					
Black	4 (1)	4 (1)	7 (2)	7 (2)	
White	96 (27)	86 (27)	93 (26)	89 (25)	
Latino	0 (0)	0 (0)	0 (0)	4 (1)	
Major League Baseball					
Black	4 (1)	4 (1)	7 (2)	7 (2)	14 (4)
White	96 (27)	96 (27)	93 (26)	89 (25)	79 (22)
Latino	0 (0)	0 (0)	0 (0)	4 (1)	7 (2)

Source: "Coaches, managers," *The Tennessean*, Nashville, 9 July 1993. Primary source: Center for the Study of Sport in Society, Northeastern University. Published by permission. *Notes:* 1. Figures for the NBA reflect that Quinn Buckner took over his new position in Dallas at the conclusion of the season and Don Chaney was appointed head coach in Detroit.

★ 1107 ★

Professional Sports

Racial Composition of Major Professional Sports Teams, 1989 to 1993

Racial compositions, in percentages, of the NBA, NFL and MLB players (years mark the beginning of each season).

	Year				
	1989	1990	1991	1992	1993
National Basketball Association					
Black	75	72	75	77	
White	25	28	25	23	
Latino	0	0	0	0	
Other[1]	0	0	0	0	

[Continued]

★ 1107 ★

Racial Composition of Major Professional Sports Teams, 1989 to 1993

[Continued]

	Year				
	1989	1990	1991	1992	1993
National Football League					
Black	60	61	62	68	
White	40	39	36	30	
Latino	0	0	2	[1]	
Other	0	0	0	1	
Major League Baseball					
Black	NA	17	18	17	16
White	NA	70	68	68	67
Latino	NA	13	14	14	16
Other	NA	0	0	0	[1]

Source: "Players," *The Tennessean,* Nashville, 9 July 1993. Primary source: Center for the Study of Sport in Society, Northeastern University. Published by permission. *Notes:* 1. Indicates less than one percent. 2. There were 22 Pacific Islanders playing in the NFL during the 1992 season.

★ 1108 ★

Professional Sports

Six Best Paid Black Athletes, 1993

[In millions of dollars]

Athlete	Sport	Sport pay	Other	Total
Michael Jordan	NBA (retired)	$4	$32	$36
Riddick Bowe	Boxing	$23	$2	$25
George Foreman	Boxing	$12.5	$3.3	$15.8
Shaquille O'Neal	NBA	$3.3	$11.9	$15.2
Lennox Lewis	Boxing	$14	$1	$15
Cecil Fielder	Baseball	$12.4	$0.3	$12.7

Source: "Six Best-Paid Athletes," *Jet,* 85 (December 20, 1993), p. 51. Primary source: *Forbes* magazine. Published by permission.

Chapter 17
VITAL STATISTICS

Abortions

★ 1109 ★

Abortions Estimated by Rate and Ratio, 1972 to 1988

Refers to women 15 to 44 years old at time of abortion.

Year	Women 15-44 years old (1,000)	Number (1,000)	Ratio per 1,000 women	Ratio per 1,000 live births
All races				
1972	44,588	586.8	13.2	184
1975	47,606	1,034.2	21.7	331
1976	48,721	1,179.3	24.2	361
1977	49,814	1,316.7	26.4	400
1978	50,920	1,409.6	27.7	413
1979	52,016	1,497.7	28.8	420
1980	53,048	1,553.9	29.3	428
1981	53,901	1,577.3	29.3	430
1982	54,679	1,573.9	28.8	428
1983	55,340	1,575.0	28.5	436
1984	56,061	1,577.2	28.1	423
1985	56,754	1,588.6	28.0	422
1986[1]	57,483	1,574.0	27.4	416
1987	57,964	1,559.1	27.1	405
1988	58,192	1,590.8	27.3	401
White				
1972	38,532	455.3	11.8	175
1975	40,857	701.2	17.2	276
1976	41,721	784.9	18.8	296
1977	42,567	888.8	20.9	333
1978	43,427	969.4	22.3	356
1979	44,266	1,062.4	24.0	373
1980	44,942	1,093.6	24.3	376
1981	45,494	1,107.8	24.3	377
1982	46,049	1,095.2	23.8	373
1983	46,506	1,084.4	23.3	376

[Continued]

★ 1109 ★

Abortions Estimated by Rate and Ratio, 1972 to 1988

[Continued]

Year	Women 15-44 years old (1,000)	Number (1,000)	Ratio per 1,000 women	Ratio per 1,000 live births
1984	47,023	1,086.6	23.1	366
1985	47,512	1,075.6	22.6	360
1986[1]	48,010	1,044.7	21.8	350
1987	48,288	1,017.3	21.1	338
1988	48,325	1,025.7	21.2	333
Black and other				
1972	6,056	131.5	21.7	223
1975	6,749	333.0	49.3	565
1976	7,000	394.4	56.3	638
1977	7,247	427.9	59.0	679
1978	7,493	440.2	58.7	665
1979	7,750	435.3	56.2	625
1980	8,106	460.3	56.5	642
1981	8,407	469.6	55.9	645
1982	8,630	478.7	55.5	646
1983	8,834	490.6	55.5	670
1984	9,038	490.6	54.3	646
1985	9,242	512.9	55.5	659
1986[1]	9,473	529.3	55.9	661
1987	9,676	541.8	56.0	648
1988	9,867	565.1	57.3	638

Source: "Abortions—Estimated Number, Rate, and Ratio, by Race: 1972 to 1988." U.S. Bureau of the Census, *Statistical Abstract of the United States*, 1993, p. 83. Published by permission. Primary source: 1972, U.S. Centers for Disease Control, Atlanta, GA, *Abortion Surveillance, Annual Summary*, 1972, 1974, and The Alan Guttmacher Institute; 1975-1988, S.K. Henshaw and J. Van Vort, eds. *Abortion Factbook, 1992 Edition: Readings, Trends and State and Local Data to 1988*, The Alan Guttmacher Institute, New York, NY, 1992 (copyright). *Notes:* 1. Total abortions in 1986 have been estimated by interpolation between 1985 and 1987.

★ 1110 ★
Abortions

Abortions by Selected Characteristics, 1973 to 1988

Number of abortions from surveys conducted by source; characteristics from the U.S. Centers for Disease Control's (CDC) annual abortion surveillance summaries, with adjustments for changes in States reporting data to the CDC each year.

Characteristic	Number (1,000)				Percent distribution				Abortion ratio[1]			
	1973	1980	1985	1988	1973	1980	1985	1988	1973	1980	1985	1988
Race of women												
White	549	1,094	1,076	1,026	74	70	68	65	178	274	265	250
Black and other	196	460	513	565	26	30	32	36	252	392	397	389

Source: "Abortions, by Selected Characteristics: 197 to 1988." U.S. Bureau of the Census, *Statistical Abstract of the United States*, 1993, p. 83. Published by permission. Primary source: S.K. Henshaw and J. Van Vort, eds. *Abortion Factbook, 1992 Edition: Readings, Trends and State and Local Data to 1988*, The Alan Guttmacher Institute, New York, NY, 1992 (copyright). *Notes:* 1. Number of abortions per 1,000 abortions and live births. Live births are those which occurred from July 1 of year shown through June 30 of the following year (to match time of conception with abortions).

Birth Control

★ 1111 ★

Contraceptive Use by Women, 15 to 44 Years Old, 1982 and 1988

Based on the 1982 and 1988 National Survey of Family Growth.

Contraceptive status and method	All women, 1982	1988								
		All women[1]	Age			Race		Marital status		
			15-24 years	25-34 years	35-44 years	Black	White	Never married	Currently married	Formerly married
All women (1,000)	54,099	57,900	18,592	21,726	17,582	47,077	7,679	21,058	29,147	7,695
Percent distribution										
Sterile	27.2	29.7	3.1	27.0	61.3	30.5	29.6	5.2	44.0	42.6
Surgically sterile	25.7	28.3	2.4	26.0	58.7	29.2	27.8	4.3	42.4	40.9
Noncontraceptively sterile[2]	6.6	4.7	0.2	2.7	12.0	4.7	5.7	0.9	6.2	9.7
Contraceptively sterile[3]	19.0	23.6	2.2	23.3	46.7	24.5	22.1	3.4	36.2	31.3
Nonsurgically sterile[4]	1.5	1.4	0.7	0.9	2.7	1.3	1.8	1.0	1.6	1.7
Pregnant postpartum	5.0	4.8	5.0	7.6	1.1	4.8	5.0	2.4	7.1	2.5
Seeking pregnancy	4.2	3.8	2.7	5.8	2.4	3.7	3.9	1.3	6.0	2.0
Other nonusers	26.9	25.0	45.7	16.7	13.5	23.8	26.9	52.5	4.8	26.6
Not sexually active[5]	19.5	19.0	37.9	10.3	8.5	18.1	16.7	43.5	0.3	19.5
Sexually active	7.4	6.5	7.8	6.4	5.0	5.7	10.2	9.0	4.5	7.1
Nonsurgical contraceptors	36.7	36.7	43.5	43.0	21.6	37.2	34.6	38.5	38.1	26.3
Pill	15.6	18.5	29.7	21.6	3.0	18.4	21.6	24.7	15.1	14.5
IUD	4.0	1.2	0.1	1.4	2.1	1.1	1.7	0.6	1.5	2.1
Diaphragm	4.5	3.5	1.3	4.8	4.1	3.8	1.1	2.1	4.6	3.0
Condom	6.7	8.8	9.5	9.1	7.7	9.2	5.8	8.2	10.6	3.4
Foam	1.3	0.6	0.3	0.8	0.8	0.6	0.6	0.2	1.0	0.5

[Continued]

★ 1111 ★

Contraceptive Use by Women, 15 to 44 Years Old, 1982 and 1988

[Continued]

Contraceptive status and method	All women, 1982	1988								
		All women[1]	Age			Race		Marital status		
			15-24 years	25-34 years	35-44 years	Black	White	Never married	Currently married	Formerly married
Rhythm[6]	2.2	1.4	0.6	1.7	1.8	1.4	1.2	0.6	2.1	1.1
Other methods[7]	2.5	2.6	2.0	3.6	2.2	2.5	2.6	2.1	3.2	1.7

Source: "Contraceptive Use by Women, 15 to 44 Years Old, by Age, Race, Marital Status, and Method of Contraception: 1982 and 1988." U.S. Bureau of the Census, *Statistical Abstract of the United States*, 1993, p. 82. Primary source: U.S. National Center for Health Statistics, *Advance Data from Vital and Health Statistics*, No. 182.

conditions. 5. Those having intercourse in the last 3 months before the survey. 6. Periodic abstinence and natural family planning. 7. Withdrawal, douche, suppository, and less frequently used methods.

Birth Projections

★ 1112 ★

Birth Projection, by Women 18 to 34 Years Old, 1992

As of June. Covers women in the civilian noninstitutional population. Based on Current Population Survey.

Characteristic	Women reporting on birth expectations (1,000)	Rate per 1,000 women			Percentage expecting	
		Births to date	Future births expected	Lifetime births expected	No lifetime births	No future births
Total[1]	24,223	1,135	963	2,098	9.3	48.3
White	20,141	1,077	1,014	2,091	9.3	46.1
Black	3,217	1,508	628	2,136	9.3	63.4
Hispanic[2]	2,357	1,493	838	2,331	5.7	51.4

Source: "Lifetime Births Expected by Women, 18 to 34 Years Old, by Selected Characteristics: 1992," U.S. Bureau of the Census, *Statistical Abstract of the United States*, 1993, p. 81. Primary source: U.S. Bureau of the Census, *Current Population Reports*, series P20, unpublished data. *Notes:* 1. Includes other races not shown separately. 2. Persons of Hispanic origin may be of any race.

★ 1113 ★

Birth Projections

Birth Projections: Lifetime Births, 1971-1992

As of June.

Year	Lifetime births to all wives[1] Aged--			Lifetime births to white wives aged--			Lifetime births to Black wives aged--			Lifetime births to Hispanic[2] wives aged--		
	18 to 24 yrs. old	25 to 29 yrs. old	30 to 34 yrs. old	18 to 24 yrs. old	25 to 29 yrs. old	30 to 34 yrs. old	18 to 24 yrs. old	25 to 29 yrs. old	30 to 34 yrs. old	18 to 24 yrs. old	25 to 29 yrs. old	30 to 34 yrs. old
1971	2,375	2,619	2,989	2,353	2,577	2,936	2,623	3,112	3,714	(NA)	(NA)	(NA)
1975	2,173	2,260	2,610	2,147	2,233	2,564	2,489	2,587	3,212	2,223	2,607	3,238
1980	2,134	2,166	2,248	2,130	2,146	2,223	2,155	2,426	2,522	2,428	2,495	2,909
1985	2,183	2,236	2,167	2,177	2,227	2,139	2,242	2,259	2,521	2,367	2,628	2,712
1990	2,244	2,285	2,277	2,218	2,272	2,257	2,509	2,443	2,579	2,404	2,482	2,824
1992	2,279	2,271	2,218	2,274	2,259	2,208	2,353	2,389	2,362	2,511	2,437	2,600

Source: "Lifetime Births Expected per 1,000 Wives: 1971 to 1992." U.S. Bureau of the Census, *Statistical Abstract of the United States*, 1993, p. 81. *Notes:* NA Not available. 1. Includes other races not shown separately. 2. Persons of Hispanic origin may be of any race.

Births

★ 1114 ★

Birth Rates: Live Order and Race, 1970-1990

Births per 1,000 women 15 to 44 years old in specified racial group. Live-birth order refers to number of children born alive. Figures for births of order not distributed.

Live-birth order	All races[1]					Black				
	1970	1980	1985	1989	1990	1970	1980	1985	1989	1990
Total	87.9	68.4	66.3	69.2	70.9	115.4	88.1	82.4	90.8	91.9
First birth	34.2	29.5	27.6	28.4	29.0	43.3	35.2	32.5	34.9	34.6
Second birth	24.2	21.8	22.0	22.4	22.8	27.1	25.7	24.5	26.8	27.1
Third birth	13.6	10.3	10.4	11.3	11.7	16.1	14.5	14.0	16.0	16.4
Fourth birth	7.2	3.9	3.8	4.3	4.5	10.0	6.7	6.3	7.4	7.7
Fifth birth	3.8	1.5	1.4	1.6	1.7	6.4	3.0	2.7	3.1	3.3
Sixth and seventh	3.2	1.0	0.8	0.9	1.0	7.0	2.1	1.8	2.0	2.1
Eighth and over	1.8	0.4	0.3	0.3	0.3	5.6	0.9	0.6	0.6	0.6

Source: "Birth Rates, by Live-Order and Race: 1970 to 1990." U.S. Bureau of the Census, *Statistical Abstract of the United States*, 1993, p. 76. Primary source: U.S. National Center for Health Statistics, *Vital Statistics of the United States*, annual; and *Monthly Vital Statistics Reports*. *Note:* 1. Includes other races not shown separately.

★ 1115 ★

Births

Births Born Out-of-Wedlock, 1970, 1980, and 1988

[Percent]

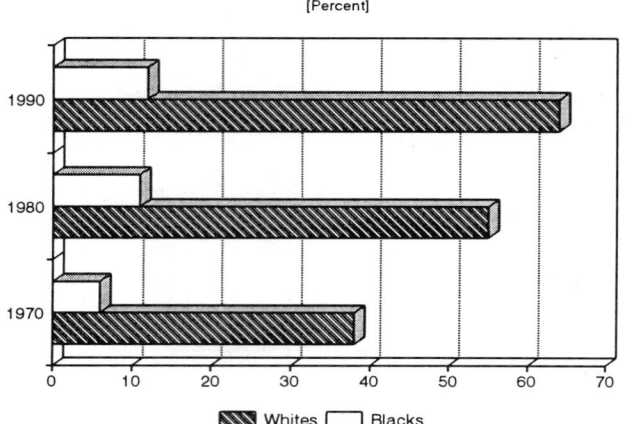

Whites ▨   Blacks ☐

	Percent		
	1970	1980	1988
Blacks	38.0	55.0	64.0
Whites	6.0	11.0	18.0

Source: "Babies Born Out-of-Wedlock, by Race, 1970, 1980, and 1988." *The State of Black America 1994*, p. 18. Primary source: National Center for Health Statistics, *Monthly Vital Statistics, Report 39*, No. 4., Supplement (1990), Table 18, and *Vital Statistics of the United States 1987* (Washington, DC: GPO, 1989), Table 1-31.

★ 1116 ★

Births

Births and Birth Rates, 1970 to 1990

Births in thousands and by race of child, except as indicated. Excludes births to nonresidents of the United States.

Item	1970	1980	1982	1983	1984	1985	1986	1987	1988	1989	1990
Live births[1]	3,731	3,612	3,681	3,639	3,669	3,761	3,757	3,809	3,910	4,041	4,158
White	3,091	2,899	2,942	2,904	2,924	2,991	2,970	2,992	3,046	3,132	3,225
Black	572	590	593	586	593	608	621	642	672	709	725
Birth rate per 1,000 population	18.4	15.9	15.9	15.6	15.6	15.8	15.6	15.7	16.0	16.4	16.7
White	17.4	14.9	14.9	14.6	14.6	14.8	14.6	14.6	14.8	15.1	15.5
Black	25.3	22.1	21.5	21.0	21.0	21.3	21.5	21.9	22.6	23.5	23.8
Male	19.4	16.8	16.7	16.4	16.4	16.7	16.5	16.5	16.8	17.2	17.6
Female	17.4	15.1	15.1	14.8	14.8	15.0	14.9	14.9	15.2	15.6	15.9
Plural birth ratio[2]	18.1[3]	19.3	19.9	20.3	20.3	21.0	21.6	22.0	22.4	23.0	23.3
White	17.3[3]	18.5	19.2	19.6	19.8	20.4	21.2	21.6	22.0	22.5[4]	22.9[4]

[Continued]

★ 1116 ★

Births and Birth Rates, 1970 to 1990
[Continued]

Item	1970	1980	1982	1983	1984	1985	1986	1987	1988	1989	1990
Black	22.8[3]	24.1	24.1	24.5	24.2	25.3	24.9	25.4	25.8	26.9[4]	27.0[4]
Birth rate per 1,000 women[5]	87.9	68.4	67.3	65.8	65.4	66.2	65.4	65.7	67.2	69.2	70.9
White[5]	84.1	64.7	63.9	62.4	62.2	63.0	61.9	62.0	63.0	64.7	66.9
Black[5]	115.4	88.1	84.1	81.7	81.4	82.2	82.4	83.8	86.6	90.4	91.9
Birth rate per 1,000 men[6]	71.5	57.0	56.4	55.1	55.0	55.6	54.8	55.0	55.8	57.2	58.4
White[6]	67.1	52.9	52.5	51.4	51.2	51.9	51.0	50.9	51.5	52.6	53.8
Black[6]	107.4	83.8	80.4	78.1	77.6	78.3	78.3	79.5	82.0	85.4	86.3

Source: "Births and Birth Rates: 1970 to 1992." U.S. Bureau of the Census, *Statistical Abstract of the United States*, 1993, p. 74. Primary source: U.S. National Center for Health Statistics, *Vital Statistics of the United States*, annual; *Monthly Vital Statistics Report*, and unpublished data. *Notes:* 1. Includes other races not shown separately. 2. Number of multiple births per 1,000 live births. 3. 1971. 4. Race of mother, data not directly comparable with prior years. 5. Per 1,000 women, 15 to 44 years old in specified group. 6. Rate computed by relating total births, regardless of age of father, to 1,000 men, 15 to 44 years old in specified group.

★ 1117 ★
Births

Live Births – by Number and Rate, by State, 1990 and 1991

Registered births. Excludes births to nonresidents of the United States, except as noted. By race of child, except as indicated.

Division and state	Number (1,000) 1990					1991 prel.[3]	Rate per 1,000 population[4] 1990					1991 prel.[3]
	All races[1]	White	Black	Hispanic[2] Total	Hispanic[2] Mexican		All races[1]	White	Black	Hispanic[2] Total	Hispanic[2] Mexican	
U.S.	4,158.2	3,225.3	724.6	595.1	385.6	4,111.0	16.7	15.5	22.4	26.7	28.7	16.2
New England	201.2	175.3	19.0	15.5	0.6	189.5	15.2	14.6	30.3	27.2	20.1	14.3
Maine	17.4	16.9	0.2	0.1	(Z)	16.6	14.1	14.0	31.7	18.6	13.0	13.2
New Hampshire	17.6	17.1	0.2	(NA)	(NA)	16.1	15.8	15.7	25.6	(NA)	(NA)	13.9
Vermont	8.3	8.2	(Z)	(Z)	(Z)	7.7	14.7	14.7	19.5	8.2	9.7	13.2
Massachusetts	92.7	78.3	10.2	8.4	0.3	86.3	15.4	14.5	34.1	29.3	22.5	14.5
Rhode Island	15.2	13.1	1.3	1.6	0.1	14.6	15.1	14.3	34.5	34.1	35.7	14.5
Connecticut	50.1	41.8	7.0	5.3	0.2	48.3	15.2	14.6	25.6	25.0	20.1	14.8
Middle Atlantic	591.8	450.1	118.7	75.7	5.1	578.8	15.7	15.0	23.8	23.8	34.9	15.2
New York	297.6	216.8	66.3	53.1	3.8	292.4	16.5	16.2	23.2	24.0	41.0	16.2
New Jersey	122.3	92.6	24.5	17.0	0.9	117.8	15.8	15.1	23.6	22.9	30.4	15.1
Pennsylvania	172.0	140.7	27.9	5.6	0.4	168.6	14.5	13.4	25.6	24.3	16.5	13.9
East North Central	675.5	536.7	123.5	34.7	23.8	662.4	16.1	15.0	25.6	24.1	25.2	15.5
Ohio	166.9	137.1	27.7	2.4	1.0	158.6	15.4	14.4	24.0	17.4	17.5	14.4
Indiana	86.2	74.6	10.3	1.9	1.4	84.7	15.6	14.9	23.8	18.8	20.3	14.9
Illinois	195.8	144.8	45.2	24.2	17.7	194.0	17.1	16.2	26.7	26.8	28.3	16.5
Michigan	153.7	117.5	32.9	4.3	2.6	153.4	16.5	15.2	25.5	21.5	18.5	16.4
Wisconsin	72.9	62.7	7.5	1.9	1.2	71.7	14.9	13.9	30.7	20.0	21.3	14.5

[Continued]

★ 1117 ★

Live Births – by Number and Rate, by State, 1990 and 1991
[Continued]

Division and state	Number (1,000)						Rate per 1,000 population[4]					
	1990					1991 prel.[3]	1990					1991 prel.[3]
	All races[1]	White	Black	Hispanic[2]			All races[1]	White	Black	Hispanic[2]		
				Total	Mexican					Total	Mexican	
West North Central	270.3	235.4	24.1	5.6	4.1	262.4	15.3	14.5	26.9	19.5	19.8	14.5
Minnesota	68.0	61.0	3.1	1.1	0.8	67.0	15.5	14.8	32.7	19.5	22.5	15.1
Iowa	39.4	37.4	1.3	0.6	0.4	36.0	14.2	13.9	26.6	19.1	15.5	12.6
Missouri	79.3	63.8	14.1	1.0	0.7	78.0	15.5	14.2	25.7	15.7	19.0	15.0
North Dakota	9.3	8.2	0.1	0.1	0.1	9.1	14.5	13.6	33.8	26.4	26.1	14.0
South Dakota	11.0	9.0	0.1	0.1	0.1	11.0	15.8	14.1	41.4	21.5	20.9	15.4
Nebraska	24.4	22.1	1.6	0.8	0.5	23.9	15.4	14.9	27.0	21.7	17.4	14.7
Kansas	39.0	34.0	3.8	2.0	1.6	37.3	15.7	15.3	26.8	20.9	21.0	14.6
South Atlantic	700.3	476.1	207.2	39.4	8.9	689.1	16.1	14.3	23.2	18.5	28.3	15.5
Delaware	11.1	8.2	2.7	0.3	0.1	11.2	16.7	15.4	23.8	20.4	27.2	16.0
Maryland	80.2	51.0	26.0	2.5	0.5	84.5	16.8	15.0	21.9	20.2	29.7	17.5
District of Columbia	11.9	1.7	9.2	0.9	(Z)	10.0	19.5	9.7	22.9	27.2	14.8	17.0
Virginia	99.4	71.1	24.7	3.5	0.6	96.6	16.1	14.8	21.3	21.6	19.1	15.4
West Virginia	22.6	21.5	1.0	0.1	(Z)	22.2	12.6	12.4	17.1	8.6	8.5	12.2
North Carolina	104.5	69.8	31.7	1.8	1.0	102.4	15.8	13.9	21.7	22.9	30.2	15.2
South Carolina	58.6	35.2	22.7	0.6	0.3	57.7	16.8	14.6	21.8	18.7	24.1	16.0
Georgia	112.7	69.7	41.3	2.3	1.3	110.0	17.4	15.1	23.6	20.8	26.7	16.6
Florida	199.3	147.9	48.1	27.6	5.0	194.5	15.4	13.8	27.3	17.5	31.0	14.6
East South Central	236.4	167.5	66.3	1.2	0.6	232.1	15.6	13.9	22.3	12.4	15.9	14.9
Kentucky	54.4	48.5	5.4	0.3	0.2	54.9	14.8	14.3	20.4	12.5	20.5	14.7
Tennessee	75.0	56.0	18.0	0.4	0.2	73.1	15.4	13.8	23.2	13.6	16.6	14.5
Alabama	63.5	40.8	2.0	0.3	0.2	605	15.7	13.7	21.6	14.0	16.8	14.6
Mississippi	43.6	22.2	20.9	0.1	(Z)	43.5	16.9	13.6	22.8	7.3	7.0	16.6
West South Central	472.7	366.1	90.2	117.2	103.5	482.0	17.7	18.2	22.9	25.8	25.9	17.7
Arkansas	36.5	27.2	8.7	0.4	0.2	34.6	15.5	14.0	23.2	20.8	19.7	14.2
Louisiana	72.2	40.7	30.0	0.9	0.2	74.6	17.1	14.3	23.1	10.1	8.6	17.2
Oklahoma	47.6	35.6	5.5	(NA)	(NA)	47.3	15.1	13.8	23.7	(NA)	(NA)	14.7
Texas	316.4	262.6	45.9	115.8	103.0	325.6	18.6	20.6	22.7	26.7	26.5	18.8
Mountain	242.8	212.4	10.2	48.7	31.8	243.4	17.8	18.1	27.2	24.4	22.1	17.5
Montana	11.6	9.7	0.1	0.3	0.1	11.5	14.5	13.1	31.5	24.0	17.3	14.3
Idaho	16.4	15.8	0.1	1.4	1.2	17.2	16.3	16.6	27.3	27.1	27.7	16.6
Wyoming	7.0	6.5	0.1	0.5	0.3	6.8	15.4	15.2	29.7	19.8	17.2	14.6
Colorado	53.5	48.1	3.4	9.3	4.7	54.0	16.2	16.6	25.5	21.9	16.5	16.1
New Mexico	27.4	22.2	0.8	12.2	2.8	28.2	18.1	19.3	25.8	21.1	8.6	18.0
Arizona	69.0	58.1	3.2	19.7	18.9	67.7	18.8	19.6	28.7	28.6	30.7	18.3
Utah	36.3	33.9	0.4	2.0	1.2	35.1	21.1	21.0	33.3	24.1	21.8	20.1
Nevada	21.6	18.1	2.2	3.3	2.5	23.0	18.0	17.9	27.4	26.2	29.1	18.8
Pacific	767.2	605.7	65.3	257.0	207.2	755.5	19.6	21.4	26.6	31.7	32.5	18.8
Washington	79.3	68.2	4.1	5.7	3.9	75.7	16.3	15.8	27.7	26.6	24.8	15.2

[Continued]

★ 1117 ★

Live Births – by Number and Rate, by State, 1990 and 1991

[Continued]

Division and state	Number (1,000)						Rate per 1,000 population[4]					
	1990					1991 prel.[3]	1990					1991 prel.[3]
	All races[1]	White	Black	Hispanic[2]			All races[1]	White	Black	Hispanic[2]		
				Total	Mexican					Total	Mexican	
Oregon	42.9	38.9	1.3	3.0	2.6	42.8	15.1	14.7	27.3	26.4	30.9	14.6
California	621.6	486.3	58.2	245.6	200.1	605.7	20.6	23.7	26.4	31.9	32.7	19.8
Alaska	11.9	7.6	0.7	0.3	0.2	11.2	21.6	18.3	31.9	18.4	21.5	21.1
Hawaii	20.5	4.7	1.0	2.4	0.3	20.0	18.5	12.7	35.3	29.9	24.2	17.5

Source: "Live Births—by Number and Rate, by State: 1900 and 1991," U.S. Bureau of the Census, *Statistical Abstract of the United States*, 1993, p. 75. Primary source: U.S. national Center for Health Statistics, *Vital Statistics of the United States*, annual, and *Monthly Vital Statistics Report*. *Notes:* NA Not available. Z Less than 50. 1. Includes other races not shown separately. 2. Persons of Hispanic origin may be of any race. Births by Hispanic origin of mother. 3. Includes births to nonresident. Provisional. 4. Based on resident population enumerated as of April 1 for 1990 and estimated as of July 1 for 1991.

★ 1118 ★

Births

Live Births: Place and Delivery, 1970-1990

Represents registered births. Excludes births to nonresidents of the United States.

Year	Births attended (1,000)			Median birth weight[3]			Percent of births with low birth weight[5]			Percent of births by period in which prenatal care began	
	In hospital[1]	Physician	Midwife and other[2]	Total[4]	White	Black	Total[3]	White	Black	1st trimester	3d trimester or no prenatal care
1970	3,708	5	18	7 lb.-4 oz.	7 lb.-5 oz.	6 lb.-14 oz.	7.9	6.8	13.9	68.0	7.9
1975	3,105	11	28	7 lb.-5 oz.	7 lb.-7 oz.	6 lb.-15 oz.	7.4	6.3	13.1	72.4	6.0
1980	3,576	12	24	7 lb.-7 oz.	7 lb.-8 oz.	7 lb.-0 oz.	6.8	5.7	12.5	76.3	5.1
1981	3,592	11	27	7 lb.-7 oz.	7 lb.-8 oz.	7 lb.-0 oz.	6.8	5.7	12.5	76.3	5.2
1982	3,642	10	28	7 lb.-7 oz.	7 lb.-8 oz.	7 lb.-0 oz.	6.8	5.6	12.4	76.1	5.5
1983	3,600	10	29	7 lb.-7 oz.	7 lb.-8 oz.	7 lb.-0 oz.	6.8	5.7	12.6	76.2	5.6
1984	3,631	10	28	7 lb.-7 oz.	7 lb.-9 oz.	7 lb.-0 oz.	6.7	5.6	12.4	76.5	5.6
1985	3,722	10	29	7 lb.-7 oz.	7 lb.-9 oz.	7 lb.-0 oz.	6.8	5.6	12.4	76.2	5.7
1986	3,720	9	27	7 lb.-7 oz.	7 lb.-9 oz.	7 lb.-0 oz.	6.8	5.6	12.5	75.9	6.0
1987	3,774	8	27	7 lb.-7 oz.	7 lb.-9 oz.	7 lb.-0 oz.	6.9	5.7	12.7	76.0	6.1
1988	3,872	9	28	7 lb.-7 oz.	7 lb.-9 oz.	7 lb.-0 oz.	6.9	5.6	13.0	75.9	6.1
1989	3,991	13	22	7 lb.-7 oz.	7 lb.-8 oz.	6 lb.-15 oz.	7.0	5.7	13.2	73.9	6.3
1990	4,110	14	21	7 lb.-7 oz.	7 lb.-8 oz.	7 lb.-0 oz.	7.0	5.7	13.3	74.2	6.0

Source: "Live Births, by Place of Delivery, Median and Low Birth Weight, and Prenatal Care: 1970 to 1990." U.S. Bureau of the Census, *Statistical Abstract of the United States*, 1993, p. 77. Primary source: U.S. National Center for Health Statistics, *Vital Statistics of the United States*, annual; *Monthly Vital Statistics Report*; and unpublished data. *Notes:* 1. Includes all births in hospitals or institutions and in clinics. 2. Includes births with attendant not specified. 3. Beginning 1989, median birth weight based on race of mother; prior to 1989, based on race of child. 4. Includes other races not shown separately. 5. Through 1975, births of 2,500 grams (5 lb.- 8 oz.) or less at birth; thereafter, less than 2,500 grams.

★ 1119 ★
Births

Live Births: Race and Hispanic Origin, 1985 and 1990

Represents registered births. Excludes births to nonresidents of the United States. Data are based on Hispanic-origin of mother and beginning 1990, race of mother. Prior to 1990, data are for race of child and are not comparable. Hispanic-origin data are available from only 23 States and the District of Columbia in 1985 and 48 States and DC in 1990. However, in 1985 approximately 90 percent of all births to Hispanic mothers occur to residents of the 23 States, in 1990 this percent is approximately 99.6.

Race and Hispanic origin	Number of births (1,000)		Births to teenage mothers, percent of total		Births to unmarried mothers, percent of total		Percent of mothers beginning prenatal care during--				Percent of births with low birth weight[1]	
							First trimester		Third trimester or no care			
	1985	1990	1985	1990	1985	1990	1985	1990	1985	1990	1985	1990
Total	3,761	4,158	12.7	12.8	22.0	26.6	76.2	74.2	5.7	6.0	6.8	7.0
White	2,991	3,290	10.8	10.9	14.5	16.9	79.4	77.7	4.7	4.9	5.6	5.7
Black	608	684	23.0	23.1	60.1	66.7	61.8	60.7	10.0	10.9	12.4	13.3
American Indian, Eskimo, Aleut	43	39	19.1	19.5	40.7	53.6	60.3	57.9	11.5	12.9	5.9	6.1
Asian and Pacific Islander[2]	116	142	5.5	5.7	10.1	(NA)	75.0	(NA)	6.1	(NA)	6.1	(NA)
Filipino	21	26	5.8	6.1	12.1	15.9	77.2	77.1	4.6	4.5	6.9	7.3
Chinese	18	23	1.1	1.2	3.7	5.0	82.4	81.3	4.2	3.4	5.0	4.7
Japanese	10	9	2.9	2.9	7.9	9.6	85.8	87.0	2.6	2.9	5.9	6.2
Hawaiian	7	6	15.9	18.4	(NA)	45.0	(NA)	65.8	(NA)	8.7	6.4	7.2
Hispanic origin[3]	373	595	16.5	16.8	29.5	36.7	61.2	60.2	12.5	12.0	6.2	6.1
Mexican	243	386	17.5	17.7	25.7	33.3	59.9	57.8	12.9	13.2	5.8	5.5
Puerto Rican	35	59	20.9	21.7	51.1	55.9	58.3	633.5	15.5	10.6	8.7	9.0
Cuban	10	11	7.1	7.7	16.1	18.2	82.5	84.8	3.7	2.8	6.0	5.7
Central and South American	41	83	8.2	9.0	34.9	41.2	60.6	61.5	12.5	10.9	5.7	5.8

Source: "Birth Rates, by Race and Type of Hispanic Origin—Selected Characteristics: 1985 and 1990." U.S. Bureau of the Census, *Statistical Abstract of the United States,* 1993, p. 77. *Notes:* NA Not available. 1. Births less than 2,500 grams (5 lb.- 8 oz.). 2. Includes other races not shown separately. 3. Hispanic persons may be of any race. Includes other types, not shown separately.

★ 1120 ★
Births

Women Giving Birth: Social and Economic Characteristics, 1992

As of June. Covers civilian noninstitutional population. Since the number of women who had a birth during the 12-month period was tabulated and not the actual numbers of births, some small underestimation of fertility for this period may exist due to the omission of: (1)Multiple births, (2)Two or more live births spaced within the 12-month period (the woman is counted only once), (3)Women who had births in the period and who did not survive to the survey date, (4)Women who were in institutions and therefore not in the survey universe. These losses may be somewhat offset by the inclusion in the CPS of births to immigrants who did not have their children born in the United States and births to nonresident women. These births would not have been recorded in the vital registration system. Based on Current Population Survey (CPS).

Characteristic	Total, 15 to 44 years old			15 to 29 years old			30 to 44 years old		
	Number of women (1,000)	Women who have had a child in the last year		Number of women (1,000)	Women who have had a child in the last year		Number of women (1,000)	Women who have had a child in the last year	
		Total births per 1,000 women	First births per 1,000 women		Total births per 1,000 women	First births per 1,000 women		Total births per 1,000 women	First births per 1,000 women
Total[1]	58,614	63	25	27,312	86	41	31,302	43	11
White	48,157	62	25	22,102	82	41	26,056	44	11
Black	8,017	69	22	4,070	106	39	3,947	31	5
Asian or Pacific Islander	1,827	64	26	832	69	30	996	59	16
Hispanic[2]	5,555	95	35	2,915	60	2,640	62	8	

Source: "Social and Economic Characteristics of Women Who Have Had a Child in the Last Year: 1992." U.S. Bureau of the Census, *Statistical Abstract of the United States,* 1993, p. 79. Primary source: U.S. Bureau of the Census, *Current Population Reports,* series P20- 454; and unpublished data. *Notes:* 1. Includes women of other races and women with family income not reported, not shown separately. 2. Persons of Hispanic origin may be of any race.

★ 1121 ★

Births

Women Giving Birth: Unmarried Women by Race and Age, 1970-1990

Excludes births to nonresidents of United States. Data for 1970 include estimates for States in which marital status data were not reported. Beginning in 1980, marital status is inferred from a comparison of the child's and parents' surnames on the birth certificate for those States that do not report on marital status. No estimates included for misstatements on birth records or failures to register births.

Race of child and age of mother	1970	1980	1985	1989	1990
Number (1,000)					
Total live births[1]	398.7	665.7	828.2	1,094.2	1,165.4
White	175.1	320.1	433.0	593.9	647.4
Black	215.1	325.7	365.5	457.5	472.7
Under 15 years old	9.5	9.0	9.4	10.6	10.7
15 to 19 years old	190.4	262.8	270.9	337.3	350.0
20 to 24 years old	126.7	237.3	300.4	378.1	403.9
25 to 29 years old	40.6	99.6	152.0	215.5	230.0
30 to 34 years old	19.1	41.0	67.3	106.3	118.2
35 years old and over	12.4	16.1	28.2	46.3	52.7
Percent distribution					
Total[1]	100	100	100	100	100
White	44	48	52	54	56
Black	54	49	44	42	41
Under 15 years old	2	1	1	1	1
15 to 19 years old	48	40	33	31	30
20 to 24 years old	32	36	36	35	35
25 to 29 years old	10	15	18	20	20
30 to 34 years old	5	6	8	10	10
35 years old and over	3	2	3	4	5
As percent of all births in racial groups					
Total[1]	11	18	22	27	26
White	6	11	15	19	20
Black	38	55	60	64	65
Birth rate[2]					
Total[1,3]	26.4	29.4	32.8	41.6	43.8
White[3]	13.9	17.6	21.8	29.2	31.8
Black[3]	95.5	82.9	79.0	93.8	93.9
15 to 19 years	22.4	27.6	31.4	40.1	42.5
20 to 24 years	38.4	40.9	46.5	61.2	65.1

[Continued]

★ 1121 ★

Women Giving Birth: Unmarried Women by Race and Age, 1970-1990

[Continued]

Race of child and age of mother	1970	1980	1985	1989	1990
25 to 29 years	37.0	34.0	39.9	52.8	56.0
30 to 34 years	27.1	21.1	25.2	34.9	37.6

Source: "Births to Unmarried Women, by Race of Child and Age of Mother: 1970 to 1992." U.S. Bureau of the Census, *Statistical Abstract of the United States*, 1993, p. 78. Primary source: U.S. National Center for Health Statistics, *Vital Statistics of the United States*, annual; *Monthly Vital Statistics Report*; and unpublished data. *Notes:* 1. Includes other races not shown separately. 2. Rate per 1,000 unmarried women (never-married, widowed, and divorced) estimated as of July 1. 3. Covers women aged 15 to 44 years.

★ 1122 ★

Births

Women Without Children: Childless Women and Children Ever Born, 1992

Characteristic	Total number of women (1,00)	Women by number of children ever born (percent)				Children ever born	
		Total	None	One	Two or more	Total number (1,000)	Per 1,000 women
All races[1]							
Women ever married	37,260	100	18	23	59	65,874	1,768
15 to 19 years old	339	100	44	46	10	228	673
20 to 24 years old	3,064	100	38	34	28	3,046	994
25 to 29 years old	6,780	100	28	30	43	9,237	1,362
30 to 34 years old	9,050	100	17	23	60	15,988	1,767
35 to 39 years old	9,337	100	12	19	69	18,866	2,020
40 to 44 years old	8,690	100	11	18	71	18,509	2,130
Women never married	21,354	100	81	10	9	7,440	348
15 to 19 years old	7,847	100	95	4	1	511	65
20 to 24 years old	6,023	100	80	13	7	1,827	303
25 to 29 years old	3,259	100	70	13	17	1,926	591
30 to 34 years old	2,199	100	64	15	22	1,728	786
35 to 39 years old	1,300	100	65	13	22	1,016	782
40 to 44 years old	726	100	75	11	15	433	596
White							
Women ever married	32,165	100	19	23	58	55,482	1,725
15 to 19 years old	313	100	45	47	8	200	638
20 to 24 years old	2,708	100	40	35	26	2,546	940
25 to 29 years old	5,929	100	28	30	41	7,827	1,320
30 to 34 years old	7,799	100	17	23	60	13,650	1,750
35 to 39 years old	7,993	100	13	19	68	15,753	1,971
40 to 44 years old	7,422	100	12	17	72	15,507	2,089
Women never married	15,993	100	88	7	5	3,006	188
15 to 19 years old	6,191	100	96	3	1	272	44

[Continued]

★ 1122 ★

Women Without Children: Childless Women and Children Ever Born, 1992
[Continued]

Characteristic	Total number of women (1,00)	Women by number of children ever born (percent)				Children ever born	
		Total	None	One	Two or more	Total number (1,000)	Per 1,000 women
20 to 24 years old	4,674	100	87	9	4	814	174
25 to 29 years old	2,287	100	80	10	10	832	364
30 to 34 years old	1,476	100	77	13	10	641	434
35 to 39 years old	837	100	82	8	10	307	367
40 to 44 years old	529	100	88	6	6	140	265
Black							
Women ever married	3,585	100	11	23	66	7,686	2,144
15 to 19 years old	16	100	(B)	(B)	(B)	16	(B)
20 to 24 years old	238	100	22	28	50	384	1,609
25 to 29 years old	608	100	19	27	54	1,072	1,763
30 to 34 years old	858	100	12	22	66	1,707	1,988
35 to 39 years old	934	100	7	22	71	2,255	2,416
40 to 44 years old	930	100	7	22	72	2,251	2,421
Women never married	4,432	100	53	20	27	4,215	951
15 to 19 years old	1,297	100	88	9	4	226	174
20 to 24 years old	1,097	100	47	29	23	966	881
25 to 29 years old	813	100	41	23	37	1,028	1,264
30 to 34 years old	634	100	31	21	49	1,039	1,639
35 to 39 years old	412	100	30	24	47	677	1,645
40 to 44 years old	179	100	34	26	41	278	1,555

Source: "Childless Women and Children Ever Born, by Race and Age, and Marital Status: 1992," U.S. Bureau of the Census, *Statistical Abstract of the United States*, 1993, p. 80. Primary source: U.S. Bureau of the Census, *Current Population Reports*, series P20, unpublished data. *Notes:* B Base figure too small to meet statistical standards for reliability. 1. Includes other races, not shown separately.

Deaths

★ 1123 ★

AIDS Deaths: Trends, 1982-1992

Data are shown by year of death and are subject to substantial retrospective changes. Based on reporting by State health department.

Characteristic	Number									Percent distribution	
	Total, 1982-1992[1]	1985	1986	1987	1988	1989	1990	1991	1992	Total	1992
Total[2]	166,467	6,682	11,537	15,451	19,657	26,157	28,060	30,593	22,675	100	100
Age											
Under 5 years old	1,796	96	127	223	246	277	290	248	192	1	1
5 to 12 years old	373	10	24	50	46	58	68	74	36	(Z)	(Z)

[Continued]

★ 1123 ★

AIDS Deaths: Trends, 1982-1992

[Continued]

Characteristic	Number									Percent distribution	
	Total, 1982-1992[1]	1985	1986	1987	1988	1989	1990	1991	1992	Total	1992
13 to 29 years old	30,518	1,329	2,286	3,012	3,794	4,801	5,021	5,292	3,809	18	17
30 to 39 years old	75,023	3,013	5,244	6,924	8,761	11,880	12,654	13,731	10,265	45	45
40 to 49 years old	39,177	1,396	2,469	3,285	4,326	6,063	6,825	7,743	5,855	24	26
50 to 59 years old	13,493	590	923	1,266	1,642	2,156	2,238	2,442	1,760	8	8
60 years old and over	6,087	248	464	691	842	922	964	1,063	758	4	3
Sex											
Male	148,863	6,177	10,557	13,921	17,551	23,394	24,936	27,048	20,110	89	89
Female	17,604	505	980	1,530	2,106	2,763	3,124	3,545	2,565	11	11
Race/ethnicity											
White, non-Hispanic	92,007	3,981	6,871	8,717	10,694	14,219	15,384	16,609	12,323	56	55
Black, non-Hispanic	50,121	1,777	3,044	4,525	6,015	7,969	8,584	9,468	7,092	30	32
Hispanic	22,727	884	1,535	2,087	2,764	3,703	3,823	4,157	2,971	14	13

Source: "Acquired Immunodeficiency Syndrone (AIDS) by Selected Characteristics: 1982 to 1992." U.S. Bureau of the Census, *Statistical Abstract of the United States*, 1993, p. 96. Primary source: U.S. Center for Disease Control, Atlanta GA, *Surveillance Report*, annual. *Notes:* Z Less than 0.5 percent. 1. Includes deaths prior to 1982. 2. Includes other race/ethnicity groups not shown separately.

★ 1124 ★

Deaths

Causes and Selected Characteristics of Deaths, 1990

In thousands. Excludes deaths of nonresidents of the United States. Deaths classified according to ninth revision of *International Classification of Diseases*.

Age, sex, and race	Total[1]	Heart disease	Cancer	Accidents and adverse effects	Cerebro-vascular diseases	Chronic obstructive pulmonary diseases[2]	Pneumonia, flu	Suicide	Chronic liver disease, cirrhosis	Diabetes mellitus	Homicide and legal intervention
All races[3]											
Both sexes, total[4]	2,148.5	720.1	505.3	92.0	144.1	86.7	79.5	30.9	25.8	47.7	24.9
Under 1 year old	38.4	0.8	0.1	0.9	0.1	0.1	0.6	-	(Z)	(Z)	0.3
1 to 4 years old	6.9	0.3	0.5	2.6	(Z)	0.1	0.2	-	(Z)	(Z)	0.4
5 to 14 years old	8.4	0.3	1.1	3.7	0.1	0.1	0.1	0.3	(Z)	(Z)	0.5
15 to 24 years old	36.7	0.9	1.8	16.2	0.2	0.2	0.2	4.9	(Z)	0.1	7.4
25 to 34 years old	60.1	3.3	5.4	16.0	0.9	0.3	0.8	6.6	0.9	0.7	7.6
35 to 44 years old	83.6	11.8	16.2	11.7	2.4	0.6	1.4	5.7	3.6	1.5	4.4
45 to 54 years old	118.6	30.2	39.8	7.4	4.7	2.3	1.8	3.7	4.5	2.8	1.9
55 to 64 years old	252.7	77.5	94.9	7.2	10.1	10.3	3.9	3.4	6.3	7.0	1.1
65 to 74 years old	477.9	161.4	157.4	8.4	26.1	27.5	10.7	3.2	6.3	13.3	0.7
75 to 84 years old	601.4	229.8	135.0	10.0	50.0	32.1	25.4	2.5	3.4	14.5	0.4
85 years old and over	463.1	203.6	53.0	7.8	49.4	13.1	34.4	0.7	0.7	7.7	0.1
White											
Both sexes, total[4]	1,853.3	637.4	441.6	76.9	124.5	80.2	70.8	28.1	21.5	38.7	12.2
Under 1 year old	24.9	0.5	0.1	0.6	0.1	(Z)	0.4	-	(Z)	(Z)	0.2
1 to 4 years old	4.9	0.2	0.4	1.9	(Z)	(Z)	0.1	-	(Z)	(Z)	0.2
5 to 14 years old	6.3	0.2	0.9	2.7	0.1	0.1	0.1	0.2	(Z)	(Z)	0.3
15 to 24 years old	26.9	0.6	1.5	13.8	0.2	0.1	0.2	4.2	(Z)	0.1	3.0
25 to 34 years old	42.8	2.2	4.3	13.0	0.6	0.2	0.4	5.7	0.6	0.5	3.5
35 to 44 years old	61.0	8.6	13.0	9.4	1.5	0.4	0.9	5.2	2.6	1.1	2.3
45 to 54 years old	92.2	23.6	32.4	6.0	3.1	1.8	1.2	3.4	3.4	2.0	1.2

[Continued]

★ 1124 ★

Causes and Selected Characteristics of Deaths, 1990
[Continued]

Age, sex, and race	Total[1]	Heart disease	Cancer	Accidents and adverse effects	Cerebro-vascular diseases	Chronic obstructive pulmonary diseases[2]	Pneumonia, flu	Suicide	Chronic liver disease, cirrhosis	Diabetes mellitus	Homicide and legal intervention
55 to 64 years old	209.1	64.2	80.5	6.0	7.5	9.2	3.1	3.2	5.3	5.2	0.7
65 to 74 years old	416.2	140.5	138.8	7.2	21.4	25.6	9.2	3.1	5.7	10.7	0.5
75 to 84 years old	541.7	208.0	121.5	9.0	44.4	30.3	23.0	2.4	3.2	12.4	0.3
85 years old and over	427.0	188.8	48.1	7.2	45.6	12.4	32.2	0.6	0.7	6.8	0.1
Black											
Both sexes, total[4]	265.5	75.1	57.1	12.4	17.4	5.7	7.6	2.1	3.8	8.1	12.1
Under 1 year old	12.3	0.3	(Z)	0.3	0.1	(Z)	0.2	-	(Z)	(Z)	0.1
1 to 4 years old	1.8	0.1	0.1	0.6	(Z)	(Z)	(Z)	-	(Z)	(Z)	0.2
5 to 14 years old	1.8	0.1	0.2	0.7	(Z)	0.1	(Z)	(Z)	(Z)	(Z)	0.2
15 to 24 years old	8.6	0.3	0.3	1.8	0.1	0.1	0.1	0.5	(Z)	(Z)	4.2
25 to 34 years old	15.8	1.0	0.9	2.4	0.3	0.1	0.3	0.7	0.2	0.2	4.0
35 to 44 years old	20.5	3.0	2.8	2.0	0.9	0.2	0.5	0.4	0.9	0.3	2.0
45 to 54 years old	23.9	6.1	6.5	1.2	1.4	0.4	0.5	0.2	0.9	0.8	0.7
55 to 64 years old	39.5	12.3	13.0	1.1	2.3	1.0	0.8	0.1	0.9	1.6	0.4
65 to 74 years old	55.7	19.0	16.7	1.0	4.1	1.7	1.4	0.1	0.5	2.4	0.2
75 to 84 years old	53.4	19.7	12.1	0.9	5.0	1.6	2.0	(Z)	0.2	2.0	0.1
85 years old and over	32.0	13.3	4.4	0.5	3.3	0.5	1.8	(Z)	(Z)	0.8	(Z)

Source: "Deaths by Selected Causes and Selected Characteristics: 1990." U.S. Bureau of the Census, *Statistical Abstract of the United States*, 1993, p. 92. Primary source: U.S. National Center for Health Statistics, *Vital Statistics of the United States*, annual. *Notes:* - Represents zero. Z Fewer than 50. 1. Includes other causes not shown separately. 2. Includes allied conditions. 3. Includes other races not shown separately. 4. Includes those deaths with age not stated.

★ 1125 ★

Deaths

Causes of Death: Characteristics, 1970-1990

Deaths per 100,000 population in specified group. Except as noted, excludes deaths of nonresidents of the United States.

Year, race, and age	Total[1]	Heart disease	Malignant neoplasms	Accidents and adverse effects	Cerebro-vascular diseases	Chronic obstructive pulmonary diseases[2]	Pneumonia, flu	Suicide	Chronic liver disease, cirrhosis	Diabetes mellitus	Homicide and legal intervention
All races[3]											
Both sexes											
1970, age-adjusted	714.3	253.6	129.9	53.7	66.3	[4]	22.1	11.8	14.7	14.1	9.1
1980, age-adjusted	585.8	202.0	132.8	42.3	40.8	15.9	12.9	11.4	12.2	10.1	10.8
1985, age-adjusted	548.9	181.4	134.4	34.8	32.5	18.8	13.5	11.5	9.7	9.7	8.3
1990, age-adjusted[5]	520.2	152.0	135.0	32.5	27.7	19.7	14.0	11.5	8.6	11.7	10.2
15 to 24 years old	99.2	2.5	4.9	43.9	0.6	0.5	0.6	13.2	0.1	0.3	19.9
25 to 34 years old	139.2	7.6	12.6	37.0	2.2	0.7	1.8	15.2	2.1	1.6	17.7
35 to 44 years old	223.2	31.4	43.3	31.3	6.5	1.6	3.8	15.3	9.7	4.0	11.8
45 to 54 years old	473.4	120.5	158.9	29.4	18.7	9.1	7.0	14.8	18.0	11.3	7.6
55 to 64 years old	1,196.9	367.3	449.6	34.3	48.0	48.9	18.6	16.0	29.9	33.0	5.0
65 to 74 years old	2,648.6	894.3	872.3	46.6	144.4	152.5	59.1	17.9	34.9	73.6	3.8
75 to 84 years old	6,007.2	2,295.7	1,348.5	100.3	499.3	321.1	253.5	24.9	34.1	145.2	4.3
85 years old and over	15,327.4	6,739.9	1,752.9	257.1	1,633.9	433.3	1,140.0	22.2	23.4	255.0	4.6
White											
Both sexes											
1970, age-adjusted	679.6	249.1	127.8	51.0	61.8	[4]	19.8	12.4	13.4	12.9	4.7
1980, age-adjusted	559.4	197.6	129.6	41.5	38.0	16.3	12.2	12.1	11.0	9.1	6.9
1985, age-adjusted	524.9	176.6	131.2	34.2	30.1	19.2	12.9	12.3	8.9	8.6	5.4
1990, age-adjusted	492.8	146.9	131.5	31.8	25.5	20.1	13.4	12.2	8.0	10.4	5.9
Male											
1970, age-adjusted	893.4	347.6	154.3	76.2	68.3	[4]	26.0	18.2	18.8	12.7	7.3
1980, age-adjusted	745.3	277.5	160.5	62.3	41.9	26.7	16.2	18.9	15.7	9.5	10.9
1985, age-adjusted	693.3	246.2	160.4	50.5	33.0	28.7	17.5	19.9	12.7	9.2	8.1
1990, age-adjusted	644.3	202.0	160.3	46.4	27.7	27.4	17.5	20.1	11.5	11.3	8.9

[Continued]

★ 1125 ★

Causes of Death: Characteristics, 1970-1990

[Continued]

Year, race, and age	Total[1]	Heart disease	Malignant neoplasms	Accidents and adverse effects	Cerebro-vascular diseases	Chronic obstructive pulmonary diseases[2]	Pneumonia, flu	Suicide	Chronic liver disease, cirrhosis	Diabetes mellitus	Homicide and legal intervention
Female											
1970, age-adjusted	501.7	167.8	107.6	27.2	56.2	[4]	15.0	7.2	8.7	12.8	2.2
1980, age-adjusted	411.1	134.6	107.7	21.4	35.2	9.2	9.4	5.7	7.0	8.7	3.2
1985, age-adjusted	391.0	121.7	110.5	18.4	27.9	12.9	9.9	5.3	5.6	8.1	2.9
1990, age-adjusted	369.9	103.1	111.2	17.6	23.8	15.2	10.6	4.8	4.8	9.5	2.8
Black											
Both sexes											
1970, age-adjusted	1,044.0	307.6	111.2	74.4	114.5	[4]	40.4	6.1	24.8	26.5	46.1
1980, age-adjusted	842.5	255.7	172.1	51.2	68.5	12.5	19.2	6.4	21.6	20.3	40.6
1985, age-adjusted	793.6	240.6	176.6	42.3	55.8	15.3	18.8	6.4	16.3	20.1	29.2
1990, age-adjusted	789.2	213.5	182.0	39.7	48.4	16.9	19.8	7.0	13.7	24.8	39.5
Male											
1970, age-adjusted	1,318.6	375.9	198.0	119.5	122.5	[4]	53.8	9.9	33.1	21.2	82.1
1980, age-adjusted	1,112.8	327.3	229.9	82.0	77.5	20.9	28.0	11.1	30.6	17.7	71.9
1985, age-adjusted	1,053.4	310.8	239.9	67.6	62.7	24.8	27.5	11.5	23.8	18.2	50.2
1990, age-adjusted	1,061.3	275.9	248.1	62.4	56.1	26.5	28.7	12.4	20.0	23.6	68.7
Female											
1970, age-adjusted	814.4	251.7	123.5	35.3	107.9	[4]	29.2	2.9	17.8	30.9	15.0
1980, age-adjusted	631.1	201.1	129.7	25.1	61.7	6.3	12.7	2.4	14.4	22.1	13.7
1985, age-adjusted	594.8	188.3	131.8	20.9	50.6	8.8	12.5	2.1	10.2	21.3	10.9
1990, age-adjusted	581.6	168.1	137.2	20.4	42.7	10.7	13.7	2.4	8.7	25.4	13.0

Source: "Death Rates, by Selected Causes and Age: 1970 to 1990." U.S. Bureau of the Census, *Statistical Abstract of the United States*, 1993, p. 94. Primary source: U.S. National Center for Health Statistics, *Vital Statistics of the United States*, annual; and *Monthly Vital Statistics Report*. *Notes:* 1. Includes other causes not shown separately. 2. Includes allied conditions. 3. Includes other races not shown separately. 4. Data not available on a comparable basis with later years. 5. Includes persons under 15 years old not shown separately.

★ 1126 ★

Deaths

Causes of Death: Trends and Characteristics of Death by Suicide, 1970-1990 - Part 1

Numbers in percent.

Age	Total[1]			Male					
				White			Black		
	1970	1980	1990	1970	1980	1990	1970	1980	1990
All ages[2]	11.6	11.9	12.4	18.0	19.9	22.0	8.0	10.3	12.0
10 to 14 years old	0.6	0.8	1.5	1.1	1.4	2.3	0.3	0.5	1.6
15 to 19 years old	5.9	8.5	11.1	9.4	15.0	19.3	4.7	5.6	11.5
20 to 24 years old	12.2	16.1	15.1	19.3	27.8	26.8	18.7	20.0	19.0
25 to 34 years old	14.1	16.0	15.2	19.9	25.6	25.6	19.2	21.8	21.9
35 to 44 years old	16.9	15.4	15.3	23.3	23.5	25.3	12.6	15.6	16.9
45 to 54 years old	20.0	15.9	14.8	29.5	24.2	24.8	13.8	12.0	14.8
55 to 64 years old	21.4	15.9	16.0	35.0	25.8	27.5	10.6	11.7	10.8
65 to 74 years over	20.8	16.9	17.9	38.7	32.5	34.2	8.7	11.1	14.7

[Continued]

★ 1126 ★

Causes of Death: Trends and Characteristics of Death by Suicide, 1970-1990 - Part 1

[Continued]

Age	Total[1]			Male					
				White			Black		
	1970	1980	1990	1970	1980	1990	1970	1980	1990
75 to 84 years over	21.2	19.1	24.9	45.5	45.5	60.2	8.9	10.5	14.4
85 years and over	19.0	19.2	22.2	45.8	52.8	70.3	8.7	18.9	(B)

Source: "Suicide Rates, by Sex, Race, and Age Group: 1970 to 1990." U.S. Bureau of the Census, *Statistical Abstract of the United States*, 1993, p. 99. Primary source: U.S. National Center for Health Statistics, *Monthly Vital Statistics Report*; and unpublished data. *Notes:* B Base figure too small to meet statistical standards for reliability of a derived figure. 1. Includes other races not shown separately. 2. Includes other age groups not shown separately.

★ 1127 ★

Deaths

Causes of Death: Trends and Characteristics of Death by Suicide, 1970-1990 - Part II

Numbers in percent.

Age	Female					
	White			Black		
	1970	1980	1990	1970	1980	1990
All ages[1]	7.1	5.9	5.3	2.6	2.2	2.3
10 to 14 years old	0.3	0.3	0.9	0.4	0.1	(B)
15 to 19 years old	2.9	3.3	4.0	2.9	1.6	1.9
20 to 24 years old	5.7	5.9	4.4	4.9	3.1	2.6
25 to 34 years old	9.0	7.5	6.0	5.7	4.1	3.7
35 to 44 years old	13.0	9.1	7.4	3.7	4.6	4.0
45 to 54 years old	13.5	10.2	7.5	3.7	2.8	3.2
55 to 64 years old	12.3	9.1	8.0	2.0	2.3	2.6
65 to 74 years over	9.6	7.0	7.2	2.9	1.7	2.6
75 to 84 years over	7.2	5.7	6.7	1.7	1.4	(B)
85 years and over	5.8	5.8	5.4	2.8	-	(B)

Source: "Suicide Rates, by Sex, Race, and Age Group: 1970 to 1990." U.S. Bureau of the Census, *Statistical Abstract of the United States*, 1993, p. 99. Primary source: U.S. National Center for Health Statistics, *Monthly Vital Statistics Report*; and unpublished data. *Notes:* - Represents zero. B Base figure too small to meet statistical standards for reliability of a derived figure. 1. Includes other age groups not shown separately.

★ 1128 ★

Deaths

Causes of Deaths: Accidents and Violence, 1970-1990

Rates are per 100,000 population. Excludes deaths of nonresidents of the United States. Beginning 1980, deaths classified according to the ninth revision of the *International Classification of Diseases*. For earlier years, classified according to the revision in use at the time.

Cause of death and age	White						Black					
	Male			Female			Male			Female		
	1970	1980	1990	1970	1980	1990	1970	1980	1990	1970	1980	1990
Total[1]	101.9	97.1	81.2	42.4	36.3	32.1	183.2	154.0	142.0	51.7	42.6	38.6
Motor vehicle accidents	39.1	35.9	26.7	14.8	12.8	11.6	44.3	31.1	28.1	13.4	8.3	9.4
All other accidents	38.2	30.4	23.6	18.3	14.4	12.4	63.3	46.0	32.7	22.5	18.6	13.4
Suicide	18.0	19.9	22.0	7.1	5.9	5.3	8.0	10.3	12.0	2.6	2.2	2.3
Homicide	6.8	10.9	9.0	2.1	3.2	2.8	67.6	66.6	69.2	13.3	13.5	13.5
15 to 24 years old	130.7	138.6	107.3	34.9	37.3	30.5	234.3	162.0	208.0	45.5	35.0	34.9
25 to 34 years old	96.6	118.4	97.4	23.8	29.0	26.0	384.4	256.9	218.1	76.0	49.4	48.1
35 to 44 years old	85.7	94.1	82.3	25.8	29.2	24.4	345.2	218.1	176.6	77.2	43.2	38.5
45 to 54 years old	87.5	90.8	73.5	30.4	31.8	25.3	303.3	207.3	138.5	65.5	40.2	30.7
55 to 64 years old	101.5	92.3	79.5	36.3	33.8	29.4	242.4	188.5	129.9	56.0	47.3	36.1
65 years old and over	216.9	163.9	150.7	122.4	87.2	80.1	220.0	215.8	175.5	107.9	102.9	81.6
65 to 74 years old	128.0	116.7	99.7	57.7	46.4	40.5	217.4	182.2	141.8	81.5	68.7	50.4
75 to 84 years old	229.3	209.2	195.7	149.0	101.5	89.4	236.0	261.4	206.1	140.1	137.5	95.8
85 years old and over	466.7	438.5	428.3	391.4	268.1	232.4	271.8	379.2	359.1	214.3	235.7	213.0

Source: "Death Rates from Accidents and Violence: 1970 to 1990." U.S. Bureau of the Census, *Statistical Abstract of the United States*, 1993, p. 98. *Note:* 1. Includes persons under 15 years old, not shown separately.

★ 1129 ★

Deaths

Deaths and Death Rates: Trends by Gender and Race, 1970-1990

Rates are per 1,000 population for specified groups. Excludes deaths of nonresidents of the United States and fetal deaths. The standard population for this table is the total population of the United States enumerated in 1940.

Sex and race	1970	1980	1981	1982	1983	1984	1985	1986	1987	1988	1989	1990
Deaths[1] (1,000)	1,921	1,990	1,978	1,975	2,019	2,039	2,086	2,105	2,123	2,168	2,150	2,148
Male[1] (1,000)	1,078	1,075	1,064	1,056	1,072	1,077	1,098	1,104	1,108	1,126	1,114	1,113
Female[1] (1,000)	843	915	914	918	947	963	989	1,001	1,015	1,042	1,036	1,035
White (1,000)	1,682	1,739	1,731	1,729	1,766	1,782	1,819	1,831	1,843	1,877	1,854	1,853
Male (1,000)	942	934	925	919	932	935	950	953	953	965	951	951
Female (1,000)	740	805	806	810	834	847	869	879	890	911	903	902
Black (1,000)	226	233	229	227	233	236	244	250	255	264	268	266
Male (1,000)	128	130	127	126	128	129	134	137	140	144	146	145
Female (1,000)	98	103	101	101	105	107	111	113	115	120	121	120
Death rates[1]	9.5	8.8	8.6	8.5	8.6	8.6	8.8	8.8	8.8	8.8	8.7	8.6

[Continued]

★ 1129 ★

Deaths and Death Rates: Trends by Gender and Race, 1970-1990
[Continued]

Sex and race	1970	1980	1981	1982	1983	1984	1985	1986	1987	1988	1989	1990
Male[1]	10.9	9.8	9.5	9.4	9.4	9.4	9.5	9.4	9.4	9.5	9.3	9.2
Female[1]	8.1	7.9	7.8	7.7	7.9	7.9	8.1	8.1	8.2	8.3	8.2	8.1
White	9.5	8.9	8.8	8.7	8.9	8.9	9.0	9.0	9.0	9.1	8.9	8.9
Male	10.9	9.8	9.7	9.5	9.6	9.5	9.6	9.6	9.5	9.6	9.4	9.3
Female	8.1	8.1	8.0	8.0	8.2	8.2	8.4	8.4	8.5	8.7	8.5	8.5
Black	10.0	8.8	8.4	8.2	8.4	8.4	8.5	8.6	8.7	8.9	8.9	8.8
Male	11.9	10.3	9.9	9.7	9.7	9.7	9.9	10.0	10.1	10.3	10.3	10.1
Female	8.3	7.3	7.1	7.0	7.2	7.2	7.3	7.4	7.5	7.6	7.6	7.5
Age-adjusted death rates[1]	7.1	5.9	5.7	5.5	5.5	5.5	5.5	5.4	5.4	5.4	5.3	5.2
Male[1]	9.3	7.8	7.5	7.3	7.3	7.2	7.2	7.2	7.1	7.1	6.9	6.8
Female[1]	5.3	4.3	4.2	4.1	4.1	4.1	4.1	4.1	4.0	4.1	4.0	3.9
White	6.8	5.6	5.4	5.3	5.3	5.3	5.2	5.2	5.1	5.1	5.0	4.9
Male	8.9	7.5	7.2	7.1	7.0	6.9	6.9	6.8	6.7	6.7	6.5	6.4
Female	5.0	4.1	4.0	3.9	3.9	3.9	3.9	3.9	3.8	3.9	3.8	3.7
Black	10.4	8.4	8.0	7.8	7.9	7.8	7.9	8.0	8.0	8.1	8.1	7.9
Male	13.2	11.1	10.7	10.4	10.4	10.4	10.5	10.6	10.6	10.8	10.8	10.6
Female	8.1	6.3	6.0	5.9	6.0	5.9	5.9	5.9	5.9	6.0	5.9	5.8

Source: "Deaths and Death Rates, by Sex and Race: 1970 to 1990." U.S. Bureau of the Census, *Statistical Abstract of the United States*, 1993, p. 87. Primary source: U.S. national Center for Health Statistics, *Vital Statistics of the United States*, annual; and *Monthly Vital Statistics Report*. *Note:* 1. Includes other races, not shown separately.

★ 1130 ★

Deaths

Deaths and Death Rates: Trends by Gender and Race, 1970-1991

Number of deaths per 100,000 population in specified group.

Sex, year, and race	All ages[1]	Under 1 yr. old	1-4 yr. old	5-14 yr. old	15-24 yr. old	25-34 yr. old	35-44 yr. old	45-54 yr. old	55-64 yr. old	65-74 yr. old	75-84 yr. old	85 yr. old and over
Male[2]												
1970	1,090	2,410	93	51	189	215	403	959	2,283	4,874	10,010	17,822
1980	977	1,429	73	37	172	196	299	767	1,815	4,105	8,817	18,801
1985	949	1,220	59	32	139	180	279	672	1,711	3,856	8,502	18,614
1990	918	1,083	52	29	147	204	310	610	1,553	3,492	7,889	18,057
1991, prel.[3]	909	1,007	49	29	161	201	311	598	1,504	3,307	7,663	17,151
White												
1970	1,087	2,113	84	48	171	177	344	883	2,203	4,810	10,099	18,552
1980	983	1,230	66	35	167	171	257	699	1,729	4,036	8,830	19,097
1985	964	1,057	53	30	134	159	243	612	1,626	3,771	8,486	18,980
1990	931	896	46	26	131	176	268	549	1,467	3,398	7,845	18,268
1991, prel.[3]	924	846	43	26	143	172	269	541	1,431	3,246	7,689	17,620
Black												
1970	1,187	4,299	151	67	321	560	957	1,778	3,257	5,803	9,455	12,222

[Continued]

★ 1130 ★

Deaths and Death Rates: Trends by Gender and Race, 1970-1991
[Continued]

Sex, year, and race	All ages[1]	Under 1 yr. old	1-4 yr. old	5-14 yr. old	15-24 yr. old	25-34 yr. old	35-44 yr. old	45-54 yr. old	55-64 yr. old	65-74 yr. old	75-84 yr. old	85 yr. old and over
1980	1,034	2,587	111	47	209	407	690	1,480	2,873	5,131	9,232	16,099
1985	989	2,220	90	42	174	352	630	1,293	2,780	5,172	9,262	15,774
1990	1,008	2,112	86	41	252	431	700	1,261	2,618	4,946	9,130	16,955
1991, prel.[3]	962	1,899	82	44	277	414	707	1,203	2,323	4,286	8,359	14,325
Female[2]												
1970	808	1,864	75	32	68	102	231	517	1,099	2,580	6,678	15,518
1980	785	1,142	55	24	58	76	159	413	934	2,145	5,440	14,747
1985	809	951	45	21	50	69	139	375	926	2,097	5,162	14,554
1990	812	856	41	19	49	74	138	343	879	1,991	4,883	14,274
1991, prel.[3]	802	791	45	19	52	74	136	326	855	1,972	4,862	13,328
White												
1970	813	1,615	66	30	62	84	193	463	1,015	2,471	6,699	15,980
1980	806	963	49	23	56	65	138	373	876	2,067	5,402	14,980
1985	840	799	40	20	48	59	122	342	869	2,027	5,112	14,745
1990	847	690	36	18	46	62	117	309	823	1,924	4,839	14,401
1991, prel.[3]	838	645	37	17	48	63	115	296	807	1,915	4,842	13,553
Black												
1970	829	3,369	129	44	112	231	533	1,044	1,986	3,861	6,692	10,707
1980	733	2,124	84	31	71	150	324	768	1,561	3,057	6,212	12,367
1985	734	1,821	71	29	60	138	277	668	1,533	2,968	6,078	12,703
1990	748	1,736	68	28	69	160	299	639	1,453	2,866	5,688	13,310
1991, prel.[3]	721	1,546	80	28	77	147	298	587	1,328	2,750	5,762	11,856

Source: "Death Rates, by Age, Sex, and Race: 1970 to 1991." U.S. Bureau of the Census, *Statistical Abstract of the United States*, 1993, p. 87. Primary source: U.S. National Center for Health Statistics, *Vital Statistics of the United States*, annual; *Monthly Vital Statistics Report*; and unpublished data. *Notes:* 1. Includes unknown age. 2. Includes other races not shown separately. 3. Includes deaths of nonresidents. Based on a 10-percent sample of deaths.

★ 1131 ★

Deaths

Deaths, by Age and Leading Cause, 1990

Excludes deaths of nonresidents of the United States. Deaths classified according to ninth revision of *International Classification of Diseases*.

Age and leading cause of death	Number of deaths			Death rate per 100,000 population		
	Total	Male	Female	Total	Male	Female
All ages[1]						
All races[2]	2,148,463	1,113,417	1,035,046	863.8	918.4	812.0
White	1,853,254	950,812	902,442	888.0	930.9	846.9
Black	265,498	145,359	120,139	871.0	1,008.0	747.9
Leading cause of death						
Heart disease	720,058	360,788	359,270	289.5	297.6	281.8
Malignant neoplasms (cancer)	505,322	268,283	237,039	203.2	221.3	186.0
Cerebrovascular disease (stroke)	144,088	56,697	87,391	57.9	46.8	68.6

[Continued]

★ 1131 ★

Deaths, by Age and Leading Cause, 1990

[Continued]

Age and leading cause of death	Number of deaths			Death rate per 100,000 population		
	Total	Male	Female	Total	Male	Female
Accidents	91,983	61,938	30,045	37.0	51.1	23.6
Chronic obstructive pulmonary disease	86,679	49,416	37,263	34.9	40.8	29.2
Pneumonia	79,513	36,898	42,615	32.0	30.4	33.4
Diabetes	47,664	20,266	27,398	19.2	16.7	21.5
Suicide	30,906	24,724	6,182	12.4	20.4	4.8
Chronic liver disease, cirrhosis	25,815	16,627	9,188	10.4	13.7	7.2
HIV infection[3]	25,188	22,386	2,802	10.1	18.5	2.2

Source: "Deaths, by Age and Leading Cause: 1990." U.S. Bureau of the Census, *Statistical Abstract of the United States*, 1993, p. 93. Primary source: U.S. National Center for Health Statistics, *Vital Statistics of the United States*, annual; and unpublished data. *Notes:* 1. Includes those deaths with age not stated. 2. Includes other races not shown separately. 3. Human immunodeficiency virus.

★ 1132 ★

Deaths

Firearm Mortality Among Selected Age Groups, 1990

Death rate per 100,000 population. Deaths classified according to the ninth revision of the *International Classification of Diseases*.

Item	Under 5 years old	5 to 9 years old	10 to 14 years old	15 to 19 years old	20 to 24 years old	25 to 34 years old
Male						
Total						
White	0.6	0.6	4.2	26.5	32.5	27.8
Black	1.2	1.5	10.2	119.9	157.6	108.5
Homicide						
White	0.4	0.2	1.3	9.7	12.9	10.8
Black	0.8	1.0	6.9	105.3	140.7	94.4
Suicide						
White	(X)	(X)	1.2	13.5	17.5	15.6
Black	(X)	(X)	1.1	8.8	13.2	12.2
Accidents						
White	0.3	0.5	2.9	1.5	1.6	1.1
Black	0.4	0.5	4.9	1.9	2.7	1.4
Female						
Total						
White	0.3	0.4	1.0	4.6	4.9	5.5
Black	1.1	1.2	3.7	12.2	14.4	14.6
Homicide						
White	0.2	0.3	0.4	2.0	2.3	2.4
Black	0.9	0.9	3.1	10.4	12.4	12.7
Suicide						
White	(X)	(X)	0.5	2.3	2.4	2.9
Black	(X)	(X)	0.4	1.3	1.3	1.4

[Continued]

★ 1132 ★

Firearm Mortality Among Selected Age Groups, 1990
[Continued]

Item	Under 5 years old	5 to 9 years old	10 to 14 years old	15 to 19 years old	20 to 24 years old	25 to 34 years old
Accidents						
White	0.1	0.1	0.1	0.2	0.2	0.2
Black	0.3	0.3	0.2	0.4	0.6	0.3

Source: "Firearm Mortality Among Children, Youth, and Young Adults, 1 to 34 Years Old: 1990." U.S. Bureau of the Census, *Statistical Abstract of the United States*, 1993, p. 99. Primary source: U.S. National Center for Health Statistics, *Advance Data from Vital and Health Statistics*, No. 231. *Note:* X Not applicable.

Fertility

★ 1133 ★

Fertility Rates: Trends and Projections, 1992-2010

Birth rates represent live births per 1,000 women in age group indicated. Projections are based on middle fertility assumptions.

Age group	All races[1]		White		Black		Asian and Pacific Islanders		American Indian Eskimo, Aleut		Hispanic[2]	
	1992	2010	1992	2010	1992	2010	1992	2010	1992	2010	1992	2010
Total fertility rate	2,054	2,092	1,953	1,981	2,468	2,459	2,335	2,271	2,874	2,855	2,655	2,588
Birth rates												
10 to 14 years old	1.4	1.5	0.6	0.7	5.3	5.2	0.6	0.6	1.9	1.9	2.3	2.2
15 to 19 years old	57.3	58.7	46.7	48.5	113.7	112.7	28.3	27.2	105.9	105.5	86.3	84.1
20 to 24 years old	118.0	120.0	109.8	112.1	161.6	160.8	95.8	93.1	210.5	209.0	163.2	159.1
25 to 29 years old	120.0	121.0	119.8	120.0	113.8	113.6	152.2	149.1	142.1	141.1	139.1	135.6
30 to 34 years old	79.2	80.5	79.7	79.9	66.6	66.6	123.8	120.7	76.2	75.8	88.6	86.4
35 to 39 years old	29.6	31.0	29.1	29.9	27.2	27.3	53.6	52.2	31.2	31.1	41.3	40.2
40 to 44 years old	4.9	5.4	4.7	5.0	5.2	5.3	11.3	11.0	6.7	6.8	9.7	9.5
45 to 49 years old	0.2	0.3	0.2	0.2	0.3	0.3	1.3	1.2	0.4	0.4	0.6	0.6

Source: "Projected Fertility Rates, by Race and Age Group: 1992 and 2010." U.S. Bureau of the Census, *Statistical Abstract of the United States*, 1993, p. 76. Primary source: U.S. Bureau of the Census, *Current Population Reports*, series P25- 1092; and unpublished data. *Notes:* 1. Includes other races not shown separately. 2. Persons of Hispanic origin may be of any race.

★ 1134 ★

Fertility

Fertility Rates: Trends, 1960-1989

Based on race of child and registered births only. Beginning 1970, excludes births to nonresidents of the United States. The total fertility rate is the number of births that 1,000 women would have in their lifetime if, at each year of age, they experienced the birth rates occurring in the specified year. A total fertility rate of 2,110 represents "replacement level" fertility for the total population under current mortality conditions (assuming no net immigration). The intrinsic rate of natural increase is the rate that would eventually prevail if a population were to experience, at each year of age, the birth rates and death rates occurring in the specified year and if those rates remained unchanged over a long period of time. Minus sign (-) indicates decrease.

Annual average and year	Total fertility rate			Intrinsic rate of natural increase		
	Total	White	Black and other	Total	White	Black and other
1960-64	3,449	3,326	4,326	18.6	17.1	27.7
1965-69	2,622	2,512	3,362	8.2	6.4	18.6
1970-74	2,094	1,997	2,680	-0.7	-2.5	9.1
1975-79	1,774	1,685	2,270	-6.6	-8.5	3.0
1980-84	1,819	1,731	2,262	-5.4	-7.3	3.0
1985-88	1,870	1,769	2,339	-4.2	-6.3	4.3
1970	2,480	2,385	3,067	6.0	4.5	14.4
1971	2,267	2,161	2,920	2.6	0.8	12.6
1972	2,010	1,907	2,628	-2.0	-3.9	8.6
1973	1,879	1,783	2,443	-4.5	-6.5	5.7
1974	1,835	1,749	2,339	-5.4	-7.2	4.0
1975	1,774	1,686	2,276	-6.7	-8.6	3.0
1976	1,738	1,652	2,223	-7.4	-9.3	2.1
1977	1,790	1,703	2,279	-6.2	-8.1	3.2
1978	1,760	1,668	2,265	-6.8	-8.8	2.9
1979	1,808	1,716	2,310	-5.7	-7.7	3.8
1980	1,840	1,749	2,323	-5.1	-7.0	4.0
1981	1,815	1,726	2,275	-5.5	-7.4	3.3
1982	1,829	1,742	2,265	-5.2	-7.0	3.0
1983	1,803	1,718	2,225	-5.7	-7.5	2.5
1984	1,806	1,719	2,224	-5.6	-7.4	2.4
1985	1,843	1,754	2,263	-4.8	-6.6	3.1
1986	1,836	1,742	2,282	-4.9	-6.8	3.3
1987	1,871	1,767	2,349	-4.2	-6.3	4.5
1988	1,932	1,814	2,463	-3.0	-5.3	6.3
1989	2,014	1,885	2,583	-1.4	-3.8	8.2

Source: "Total Fertility Rate and Intrinsic Rate of Natural Increase: 1960 to 1989." U.S. Bureau of the Census, *Statistical Abstract of the United States*, 1993, p. 76. Primary source: U.S. National Center for Health Statistics, *Vital Statistics of the United States*, annual; and unpublished data.

★ 1135 ★

Fertility

Fertility and Mortality: Projections 1992, 1993, and 2050

Middle series.

Subject	P25-1092		New	
	1992	2050	1993	2050
Total fertility rate[1]				
Total	2,052	2,119	2,074	2,150
White	1,951	2,009	1,973	2,054
Black	2,459	2,444	2,470	2,452
American Indian[2]	2,865	2,840	2,778	2,718
Asian[3]	2,327	2,078	2,514	2,134
Hispanic origin[4]	2,650	2,358	2,900	2,473
White, not Hispanic	1,850	1,850	1,840	1,840
Black, not Hispanic	2,450	2,450	2,450	2,450
American Indian, not Hispanic[2]	2,900	2,900	2,760	2,760
Asian, not Hispanic[3]	2,300	2,057	2,480	2,100
Life expectancy at birth[5]				
Total	75.8	82.1	76.0	82.6
White	76.4	83.0	76.8	83.8
Black	71.0	76.6	70.4	75.5
American Indian[2]	78.0	84.1	76.9	82.9
Asian[3]	83.2	88.0	82.9	88.0
Hispanic origin[4]	78.2	83.1	79.0	84.2
White, not Hispanic	76.4	83.1	76.6	83.8
Black, not Hispanic	70.8	76.0	70.2	74.8
American Indian, not Hispanic[2]	78.0	84.3	76.3	82.6
Asian, not Hispanic[3]	83.4	88.3	82.9	88.3
Yearly net immigration (thousands)				
Total	880	880	880	880
White	463	463	482	482
Black	80	80	81	81
American Indian[2]	0	0	0	0
Asian[3]	337	337	317	317
Hispanic origin[4]	324	324	322	322
White, not Hispanic	174	174	193	193
Black, not Hispanic	60	60	61	61
American Indian, not Hispanic[2]	0	0	0	0
Asian, not Hispanic[3]	323	323	304	304

Source: "Comparison of Principal Fertility, Mortality, and Net Immigration Assumptions Between Old and New Projections." Jennifer Cheeseman Day, *Population Projections of the United States, by Age, Sex, Race, and Hispanic Origin: 1993 to 2050*, p. ix. *Notes:* 1. Adjusted for under-registration of births and undercount in the 1990 census as measured by Demographic Analysis. 2. American Indian represents American Indian, Eskimo, and Aleut. 3. Asian represents Asian and Pacific Islander. 4. Persons of Hispanic origin may be of any race. 5. Adjusted for net census coverage error in the 1990 census as measured by Demographic Analysis.

Infant Mortality

★ 1136 ★

Infant Deaths: Trends in Infant Mortality Rates by State, 1980-1990

Deaths per 1,000 live births, by place of residence. Represents deaths of infants **under** 1 year old, exclusive of fetal deaths. Excludes deaths of nonresidents of the United States.

Division and state	Total[1]			White		Black	
	1980	1985	1990	1980	1990	1980	1990
U.S.	12.6	10.6	9.2	11.0	7.7	21.4	17.0
New England	10.5	9.2	7.2	10.1	6.8	17.7	12.3
Maine	9.2	9.1	6.2	9.4	6.2	-	6.1[2]
New Hampshire	9.9	9.3	7.1	9.9	7.2	22.5[2]	5.4[2]
Vermont	10.7	8.5	6.4	10.7	6.5	-	-
Massachusetts	10.5	9.1	7.0	10.1	6.7	16.8	10.4
Rhode Island	11.0	8.2	8.1	10.9	8.3	17.4[2]	9.7[2]
Connecticut	11.2	10.0	7.9	10.2	6.6	19.1	16.0
Middle Atlantic	12.8	10.8	9.5	11.1	7.5	21.1	17.7
New York	12.5	10.8	9.6	10.8	7.7	20.0	17.3
New Jersey	12.5	10.6	9.0	10.3	6.8	21.9	17.3
Pennsylvania	13.2	11.0	9.6	11.9	7.8	23.1	18.8
East North Central	13.0	10.9	10.1	10.9	8.0	24.4	20.0
Ohio	12.8	10.3	9.8	11.2	8.2	23.0	18.3
Indiana	11.9	10.9	9.6	10.5	8.9	23.4	16.0
Illinois	14.8	11.7	10.7	11.7	7.7	26.3	21.5
Michigan	12.8	11.4	10.7	10.6	7.9	24.2	21.0
Wisconsin	10.3	9.1	8.2	9.7	7.2	18.5	18.1
West North Central	11.3	9.5	8.4	10.5	7.5	21.3	17.3
Minnesota	10.0	8.8	7.3	9.6	6.7	20.0	19.7
Iowa	11.8	9.5	8.1	11.5	7.8	27.2	18.0
Missouri	12.4	10.2	9.4	11.1	7.8	20.7	17.5
North Dakota	12.1	8.5	8.0	11.7	7.9	27.5[2]	-
South Dakota	10.9	9.9	10.1	9.0	8.6	-	7.4[2]
Nebraska	11.5	9.6	8.3	10.7	7.2	25.2	16.8
Kansas	10.4	9.3	8.4	9.5	7.7	20.6	15.4
South Atlantic	14.5	12.1	10.7	11.6	8.0	21.6	17.4
Delaware	13.9	14.8	10.1	9.8	7.3	27.9	19.4
Maryland	14.0	11.9	9.5	11.6	6.5	20.4	16.3
District of Columbia	25.0	20.8	20.7	17.8	12.1	26.7	24.4
Virginia	13.6	11.5	10.2	11.9	7.5	19.8	18.8
West Virginia	11.8	10.7	9.9	11.4	9.6	21.5	16.6[2]
North Carolina	14.5	11.8	10.6	12.1	8.3	20.0	16.0
South Carolina	15.8	14.2	11.7	10.8	8.3	22.9	17.1
Georgia	14.5	12.7	12.4	10.8	9.1	21.0	18.0
Florida	14.6	11.3	9.6	11.8	7.6	22.8	16.2
East South Central	14.5	12.1	10.4	11.8	8.1	21.8	16.2
Kentucky	12.9	11.2	8.5	12.0	8.0	22.0	13.6

[Continued]

★ 1136 ★

Infant Deaths: Trends in Infant Mortality Rates by State, 1980-1990

[Continued]

Division and state	Total[1]			White		Black	
	1980	1985	1990	1980	1990	1980	1990
Tennessee	13.5	11.4	10.3	11.9	8.0	19.3	17.5
Alabama	15.1	12.6	10.8	11.6	8.3	21.6	15.9
Mississippi	17.0	13.7	12.1	11.1	8.5	23.7	16.1
West South Central	12.7	10.4	8.7	11.1	7.4	19.8	14.7
Arkansas	12.7	11.6	9.2	10.3	8.0	20.0	13.6
Louisiana	14.3	11.9	11.1	10.5	7.3	20.6	16.5
Oklahoma	12.7	10.9	9.2	12.1	9.4	21.8	13.2
Texas	12.2	9.8	8.1	11.2	7.1	18.8	13.9
Mountain	11.0	9.8	8.6	10.7	8.3	19.5	15.0
Montana	12.4	10.3	9.0	11.8	8.6	-	13.3[2]
Idaho	10.7	10.4	8.7	10.7	8.7	-	10.9[2]
Wyoming	9.8	12.2	8.6	9.3	8.8	25.9[2]	-
Colorado	10.1	9.4	8.8	9.8	8.4	19.1	16.5
New Mexico	11.5	10.6	9.0	11.3	9.3	23.1[2]	12.8[2]
Arizona	12.4	9.7	8.8	11.8	8.2	18.4	16.7
Utah	10.4	9.6	7.5	10.5	7.4	27.3[2]	13.0[2]
Nevada	10.7	8.5	8.4	10.0	8.1	20.6	12.5
Pacific	11.2	9.7	7.9	10.9	7.6	17.6	14.1
Washington	11.8	10.7	7.8	11.5	7.6	16.4	14.5
Oregon	12.2	9.9	8.3	12.2	8.1	15.9[2]	15.1[2]
California	11.1	9.5	7.9	10.6	7.6	18.0	14.2
Alaska	12.3	10.8	10.5	9.4	8.5	19.5[2]	11.2[2]
Hawaii	10.3	8.8	6.7	11.6	5.1	11.8[2]	11.5[2]

Source: "Infant Mortality Rates, by Race-States: 1980 to 1990." U.S. Bureau of the Census, *Statistical Abstract of the United States,* 1993, p. 90. *Notes:* - Represents zero. 1. Includes other races, not shown separately. 2. Based on a frequency of less than 20 infant deaths.

★ 1137 ★

Infant Mortality

Infant Mortality Rate for Selected Cities, 1988

City	All	White	Black
San Jose	7.8	7.8	[1]
San Diego	8.1	7.6	13.9
San Francisco	8.4	9.6	[1]
Dallas	9.3	8.2	12.3
San Antonio	9.5	8.9	[1]
Los Angeles	10.1	8.3	20.2
Phoenix	10.9	10.1	23.5
Jacksonville	11.3	9.6	14.7
Houston	11.3	10.4	17.0
Milwaukee	12.1	7.8	17.5
Indianapolis	12.6	9.7	20.1

[Continued]

★ 1137 ★

Infant Mortality Rate for Selected Cities, 1988
[Continued]

City	All	White	Black
New Orleans	12.7	[1]	15.5
New York	13.2	11.0	18.3
Boston	13.9	8.4	22.1
Columbus	14.2	12.5	18.0
Chicago	15.2	9.9	21.0
Cleveland	17.0	12.2	20.9
Philadelphia	17.5	12.0	22.5
Memphis	17.6	9.0	21.6
Baltimore	18.0	12.9	20.4
Detroit	21.0	13.6	23.1
Washington, D.C.	23.2	19.9	26.0

Source: "Infant Mortality, 1988 Rate per 1,000 Live Births for Cities Above 500,000." *Black Issues in Higher Education,* 9 (April 1992), p. 33. Primary source: National Center for Health Sciences. *Notes:* The United States ranked 22nd among developed countries in 1989, with a rate of 9.8. 1. Too few infant deaths in these populations to calculate an infant mortality rate.

★ 1138 ★

Infant Mortality

Infant, Maternal, and Neonatal Rates and Fetal Mortality Rates, 1970-1990 - I

Deaths per 1,000 live births, except as noted. Excludes deaths of nonresidents of the United States.

Item	1970	1975	1980	1981	1982	1983	1984
Infant deaths[1]	20.0	16.1	12.6	11.9	11.5	11.2	10.8
White	17.8	14.2	11.0	10.5	10.1	9.7	9.4
Black and other	30.9	24.2	19.1	17.8	17.3	16.8	16.1
Black	32.6	26.2	21.4	20.0	19.6	19.2	18.4
Maternal deaths[2]	21.5	12.8	9.2	8.5	7.9	8.0	7.8
White	14.4	9.1	6.7	6.3	5.8	5.9	5.4
Black and other	55.9	29.0	19.8	17.3	16.4	16.3	16.9
Black	59.8	31.3	21.5	20.4	18.2	18.3	19.7
Fetal deaths[3]	14.2	10.7	9.2	9.0	8.9	8.5	8.2
White	12.4	9.5	8.2	8.0	7.9	7.5	7.4
Black and other	22.6	16.0	13.4	12.8	12.7	12.4	11.5
Neonatal deaths[4]	15.1	11.6	8.5	8.0	7.7	7.3	7.0
White	13.8	10.4	7.5	7.1	6.8	6.4	6.2
Black and other	21.4	16.8	12.5	11.8	11.3	10.8	10.2
Black	22.8	18.3	14.1	13.4	13.1	12.4	11.8

Source: "Infant, Maternal, and Neonatal Rates and Fetal Mortality Ratios, by Race: 1970 to 1990." U.S. Bureau of the Census, *Statistical Abstract of the United States,* 1993, p. 89. *Notes:* 1. Represents deaths of infants under 1 year old, exclusive of fetal deaths. 2. Per 100,000 live births from deliveries and complications of pregnancy, childbirth, and the puerperium. Beginning 1979, deaths are classified according to the ninth revision of the *International Classification of Diseases;* earlier years classified according to the revision in use at the time. 3. Includes only those deaths with stated or presumed period of gestation of 20 weeks or more. 4. Represents deaths of infants under 28 days old, exclusive of fetal deaths.

★ 1139 ★

Infant Mortality

Infant, Maternal, and Neonatal Rates and Fetal Mortality Rates, 1970-1990 - II

Deaths per 1,000 live births, except as noted. Excludes deaths of nonresidents of the United States.

Item	1985	1986	1987	1988	1989	1990
Infant deaths[1]	10.6	10.4	10.1	10.0	9.8	9.2
White	9.3	8.9	8.6	8.5	8.2	7.7
Black and other	15.8	15.7	15.4	15.0	15.2	14.4
Black	18.2	18.0	17.9	17.6	17.7	17.0
Maternal deaths[2]	7.8	7.2	6.6	8.4	7.9	8.2
White	5.2	4.9	5.1	5.9	5.6	5.4
Black and other	18.1	16.0	12.0	17.4	16.5	19.1
Black	20.4	18.8	14.2	19.5	18.4	22.4
Fetal deaths[3]	7.9	7.7	7.7	7.5	7.5	7.5
White	7.0	6.8	6.7	6.4	6.4	6.4
Black and other	11.3	11.2	11.5	11.4	11.4	(NA)
Neonatal deaths[4]	7.0	6.7	6.5	6.3	6.2	5.8
White	6.1	5.8	5.5	5.4	5.2	4.9
Black and other	10.3	10.1	10.0	9.7	9.6	9.2
Black	12.1	11.7	11.7	11.5	11.3	10.9

Source: "Infant, Maternal, and Neonatal Rates and Fetal Mortality Ratios, by Race: 1970 to 1990." U.S. Bureau of the Census, *Statistical Abstract of the United States*, 1993, p. 89. *Notes:* NA Not available. 1. Represents deaths of infants under 1 year old, exclusive of fetal deaths. 2. Per 100,000 live births from deliveries and complications of pregnancy, childbirth, and the puerperium. Beginning 1979, deaths are classified according to the ninth revision of the *International Classification of Diseases*; earlier years classified according to the revision in use at the time. 3. Includes only those deaths with stated or presumed period of gestation of 20 weeks or more. 4. Represents deaths of infants under 28 days old, exclusive of fetal deaths.

Life Expectancy

★ 1140 ★

Life Expectancy at Age 65, 1993 to 2050. Part I.

The 1993 to 2050 values are consistent with population totals adjusted for undercount by Demographic Analysis.

Projections for calendar year	Life expectancy at 65									
	Total		White		Black		American Indian, Eskimo, and Aleut		Asian and Pacific Islander	
	Male	Female	Male	Female	Male	Female	Male	Female	Male	Female
Low assumption										
1993	15.5	19.4	15.6	19.5	14.0	18.1	18.9	23.5	20.0	24.4
1994	15.5	19.4	15.6	19.5	14.0	18.1	18.9	23.5	20.0	24.4
1995	15.5	19.4	15.6	19.5	14.0	18.1	16.9	23.6	20.0	24.4

[Continued]

★ 1140 ★

Life Expectancy at Age 65, 1993 to 2050. Part I.

[Continued]

	Life expectancy at 65									
Projections for calendar year	Total		White		Black		Race American Indian, Eskimo, and Aleut		Asian and Pacific Islander	
	Male	Female	Male	Female	Male	Female	Male	Female	Male	Female
1996	15.5	19.4	15.6	19.5	14.0	18.1	18.9	23.6	20.0	24.4
1997	15.5	19.4	15.6	19.5	14.0	18.1	18.9	23.6	20.0	24.4
1998	15.5	19.4	15.6	19.5	14.0	18.1	18.9	23.7	20.0	24.4
1999	15.5	19.4	15.6	19.5	14.0	18.1	18.9	23.7	20.0	24.4
2000	15.5	19.4	15.6	19.5	14.0	18.1	18.9	23.8	20.0	24.4
2001	15.5	19.5	15.6	19.5	14.0	18.1	18.9	23.8	20.0	24.5
2002	15.5	19.5	15.6	19.5	14.0	18.1	18.9	23.9	20.0	24.5
2003	15.6	19.5	15.6	19.5	14.0	18.1	18.9	23.9	20.0	24.5
2004	15.6	19.5	15.6	19.5	14.0	18.1	18.9	24.0	20.0	24.5
2005 and beyond	15.6	19.5	15.6	19.5	14.0	18.1	18.9	24.0	20.0	24.5
Middle assumption										
1993	15.7	19.5	15.7	19.6	14.1	18.1	19.1	23.7	20.2	24.6
1994	15.8	19.6	15.8	19.6	14.1	18.2	19.2	23.8	20.3	24.7
1995	15.9	19.6	16.0	19.7	14.2	18.2	19.4	24.0	20.4	24.7
1996	16.0	19.7	16.1	19.8	14.2	18.3	19.5	24.2	20.6	24.8
1997	16.1	19.8	16.2	19.9	14.3	18.3	19.7	24.4	20.7	24.9
1998	16.2	19.9	16.3	19.9	14.3	18.4	19.8	24.6	20.8	25.0
1999	16.3	19.9	16.4	20.0	14.4	18.4	19.9	24.7	20.9	25.1
2000	16.4	20.0	16.5	20.1	14.4	18.5	20.1	24.9	21.1	25.2
2001	16.5	20.1	16.6	20.1	14.5	18.5	20.2	25.0	21.1	25.3
2002	16.6	20.2	16.7	20.2	14.5	18.6	20.2	25.0	21.2	25.4
2003	16.7	20.2	16.8	20.3	14.6	18.6	20.3	25.1	21.2	25.4
2004	16.7	20.3	16.9	20.4	14.6	18.7	20.3	25.2	21.3	25.5
2005	16.8	20.4	16.9	20.5	14.6	18.7	20.4	25.2	21.4	25.5
2010	17.2	20.8	17.4	20.8	14.9	18.9	20.7	25.5	21.7	25.8
2015	17.6	21.1	17.8	21.2	15.1	19.1	21.0	25.8	22.0	26.1
2020	18.0	21.5	18.2	21.6	15.3	19.3	21.4	26.1	22.3	26.4
2025	18.5	21.9	18.7	22.0	15.5	19.5	21.7	26.4	22.6	26.7
2030	18.9	22.3	19.1	22.5	15.7	19.8	22.0	26.8	23.0	27.0
2035	19.4	22.7	19.6	22.9	16.0	20.0	22.4	27.1	23.3	27.3
2040	19.8	23.2	20.1	23.3	16.2	20.2	22.7	27.4	23.6	27.7
2045	20.3	23.6	20.6	23.8	16.4	20.4	23.1	27.7	24.0	28.0
2050	20.6	24.0	21.1	24.2	16.7	20.7	23.4	28.1	24.3	28.3
High assumption										
1993	15.7	19.6	15.7	19.7	14.1	18.2	19.0	23.6	20.1	24.6
1994	15.8	19.7	15.8	19.8	14.2	18.3	19.1	23.7	20.2	24.7
1995	15.9	19.8	15.9	19.9	14.2	18.5	19.2	23.8	20.3	24.8
1996	16.0	20.0	16.0	20.1	14.3	18.6	19.3	24.0	20.4	24.9
1997	16.1	20.1	16.1	20.2	14.4	18.7	19.4	24.1	20.5	25.0
1998	16.2	20.2	16.3	20.3	14.5	18.8	19.5	24.2	20.6	25.1
1999	16.3	20.4	16.4	20.5	14.6	18.9	19.6	24.4	20.7	25.2
2000	16.4	20.5	16.5	20.6	14.6	19.0	19.7	24.5	20.8	25.3
2001	16.5	20.6	16.6	20.7	14.7	19.2	19.8	24.6	20.9	25.5
2002	16.6	20.8	16.7	20.9	14.8	19.3	19.9	24.7	20.9	25.6
2003	16.7	20.9	16.8	21.0	14.9	19.4	20.0	24.9	21.0	25.7
2004	16.8	21.1	16.9	21.2	15.0	19.5	20.1	25.0	21.1	25.8
2005	16.9	21.2	17.0	21.3	15.1	19.6	20.2	25.1	21.2	25.9
2010	17.5	21.9	17.6	22.0	15.5	20.3	20.7	25.8	21.7	26.5
2015	18.1	22.7	18.2	22.8	15.9	20.9	21.2	26.5	22.2	27.1
2020	18.7	23.4	18.8	23.6	16.4	21.6	21.7	27.2	22.7	27.7
2025	19.3	24.2	19.4	24.4	16.8	22.3	22.3	28.0	23.2	28.3
2030	19.9	25.0	20.1	25.2	17.3	23.0	22.8	28.7	23.7	28.9

[Continued]

★ 1140 ★

Life Expectancy at Age 65, 1993 to 2050. Part I.

[Continued]

Projections for calendar year	Life expectancy at 65									
	Total		Race							
			White		Black		American Indian, Eskimo, and Aleut		Asian and Pacific Islander	
	Male	Female	Male	Female	Male	Female	Male	Female	Male	Female
2035	20.6	25.9	20.8	26.0	17.8	23.7	23.4	29.5	24.2	29.6
2040	21.3	26.7	21.5	26.9	18.3	24.5	24.0	30.3	24.7	30.2
2045	22.0	27.6	22.2	27.8	18.8	25.2	24.6	31.1	25.3	30.9
2050	22.7	28.6	22.9	28.8	19.4	26.0	25.2	32.0	25.8	31.6

Source: "Life Expectancy at Age 65, by Race, Hispanic Origin, and Sex: 1993 to 2050." Jennifer Cheeseman Day, *Population Projections of the United States, by Age, Sex, Race, and Hispanic Origin: 1993 to 2050,* P25-1104, p. B-2.

★ 1141 ★

Life Expectancy

Life Expectancy at Age 65, 1993 to 2050. Part II.

The 1993 to 2050 values are consistent with population totals adjusted for undercount by Demographic Analysis.

Projections for calendar year	Life expectancy at 65									
	Hispanic origin[1]		Not of Hispanic origin, by race							
			White		Black		American Indian, Eskimo, and Aleut		Asian and Pacific Islander	
	Male	Female	Male	Female	Male	Female	Male	Female	Male	Female
Low assumption										
1993	18.4	22.0	15.5	19.4	13.9	18.0	18.9	24.3	20.1	24.6
1994	18.4	22.0	15.5	19.4	13.9	18.0	18.9	24.3	20.1	24.6
1995	18.4	22.0	15.5	19.4	13.9	18.0	18.9	24.3	20.1	24.6
1996	18.4	22.0	15.5	19.4	13.9	18.0	18.9	24.3	20.1	24.6
1997	18.4	22.0	15.5	19.4	13.9	18.0	18.9	24.3	20.1	24.6
1998	18.4	22.0	15.5	19.4	13.8	18.0	18.9	24.3	20.1	24.6
1999	18.4	22.0	15.5	19.4	13.8	18.0	18.9	24.3	20.1	24.6
2000	18.3	22.0	15.5	19.4	13.8	18.0	18.9	24.3	20.1	24.6
2001	18.3	22.0	15.5	19.4	13.8	18.0	18.9	24.3	20.1	24.6
2002	18.3	22.0	15.5	19.4	13.8	18.0	18.9	24.3	20.1	24.6
2003	18.3	22.0	15.5	19.4	13.8	18.0	18.9	24.3	20.1	24.6
2004	18.3	22.0	15.5	19.4	13.8	18.0	18.9	24.3	20.1	24.6
2005 and beyond	18.3	22.0	15.5	19.4	13.8	18.0	18.9	24.3	20.1	24.6
Middle assumption										
1993	18.6	22.1	15.6	19.5	14.0	18.0	19.1	24.5	20.3	24.8
1994	18.7	22.2	15.7	19.6	14.0	18.1	19.3	24.5	20.4	24.8
1995	18.8	22.3	15.9	19.6	14.1	16.1	19.4	24.6	20.5	24.9
1996	18.9	22.4	16.0	19.7	14.1	18.2	19.6	24.7	20.6	25.0
1997	19.0	22.4	16.1	19.8	14.2	18.2	19.7	24.8	20.8	25.1
1998	19.1	22.5	16.2	19.8	14.2	18.3	19.9	24.9	20.9	25.2
1999	19.2	22.6	16.3	19.9	14.3	18.3	20.0	25.0	21.0	25.3
2000	19.3	22.7	16.4	20.0	14.3	18.4	20.1	25.1	21.1	25.4
2001	19.4	22.7	16.5	20.0	14.3	18.4	20.2	25.1	21.2	25.4
2002	19.4	22.8	16.6	20.1	14.4	18.5	20.3	25.2	21.3	25.5
2003	19.5	22.8	16.6	20.2	14.4	18.5	20.3	25.3	21.3	25.5
2004	19.6	22.9	16.7	20.2	14.4	18.5	20.4	25.3	21.4	25.6
2005	19.6	22.9	16.8	20.3	14.5	18.6	20.5	25.4	21.4	25.7
2010	19.9	2..2	17.2	20.7	14.7	18.7	20.8	25.7	21.8	26.0
2015	20.2	23.4	17.6	21.1	14.9	18.9	21.1	26.0	22.1	26.3
2020	20.5	23.7	18.0	21.5	15.0	19.1	21.5	26.4	22.4	26.6

[Continued]

★ 1141 ★

Life Expectancy at Age 65, 1993 to 2050. Part II.
[Continued]

Projections for calendar year	Life expectancy at 65									
	Hispanic origin[1]		Not of Hispanic origin, by race							
			White		Black		American Indian, Eskimo, and Aleut		Asian and Pacific Islander	
	Male	Female	Male	Female	Male	Female	Male	Female	Male	Female
2025	20.8	24.0	18.5	21.9	15.2	19.3	21.6	26.7	22.7	26.9
2030	21.1	24.2	18.9	22.3	15.4	19.5	22.1	27.0	23.0	27.2
2035	21.5	24.5	19.4	22.7	15.6	19.7	22.5	27.4	23.4	27.5
2040	21.8	24.6	19.8	23.1	15.8	19.9	22.6	27.7	23.7	27.6
2045	22.1	25.1	20.3	23.5	16.0	20.1	23.2	28.0	24.1	28.1
2050	22.5	25.3	20.8	24.0	16.2	20.3	23.6	28.4	24.4	28.4
High assumption										
1993	18.6	22.2	15.6	19.6	14.0	18.1	19.1	24.5	20.2	24.8
1994	18.7	22.3	15.7	19.7	14.1	18.3	19.2	24.6	20.3	24.9
1995	18.7	22.4	15.8	19.9	14.2	18.4	19.3	24.6	20.4	25.0
1996	18.8	22.5	15.9	20.0	14.2	18.5	19.4	24.9	20.5	25.1
1997	18.9	22.6	16.1	20.1	14.3	18.6	19.5	25.0	20.6	25.2
1998	19.0	22.7	16.2	20.3	14.4	18.7	19.6	25.2	20.7	25.3
1999	19.1	22.8	16.3	20.4	14.5	18.8	19.7	25.3	20.8	25.4
2000	19.2	22.9	16.4	20.6	14.5	18.9	19.8	25.4	20.9	25.5
2001	19.2	23.0	16.5	20.7	14.6	19.0	19.9	25.6	20.9	25.7
2002	19.3	23.1	16.6	20.8	14.7	19.1	20.0	25.7	21.0	25.8
2003	19.4	23.2	16.7	21.0	14.6	19.3	20.1	25.8	21.1	25.9
2004	19.5	23.3	16.8	21.1	14.8	19.4	20.2	26.0	21.2	26.0
2005	19.6	23.4	16.9	21.3	14.9	19.5	20.3	26.1	21.3	26.1
2010	20.0	23.9	17.5	22.0	15.3	20.1	20.8	26.8	21.8	26.7
2015	20.5	24.5	18.1	22.8	15.7	20.7	21.3	27.5	22.3	27.3
2020	20.9	25.0	18.7	23.6	16.2	21.3	21.9	28.2	22.8	27.9
2025	21.4	25.6	19.3	24.4	16.6	22.0	22.5	29.0	23.3	28.5
2030	21.9	26.2	20.0	25.3	17.0	22.6	23.0	29.7	23.8	29.1
2035	22.4	26.8	20.7	26.1	17.5	23.3	23.6	30.5	24.3	29.7
2040	22.9	27.4	21.3	27.0	18.0	24.0	24.2	31.3	24.8	30.4
2045	23.4	28.0	22.1	28.0	18.4	24.8	24.9	32.2	25.4	31.1
2050	23.9	28.7	22.8	29.0	18.9	25.5	25.5	33.0	25.9	31.7

Source: "Life Expectancy at Age 65, by Race, Hispanic Origin, and Sex: 1993 to 2050." Jennifer Cheeseman Day, *Population Projections of the United States, by Age, Sex, Race, and Hispanic Origin: 1993 to 2050,* P25-1104, p. B-2. *Note:* 1. Persons of Hispanic origin may be of any race.

★ 1142 ★

Life Expectancy

Life Expectancy, 1970-1991 and Projections, 1995-2010

In years. Excludes deaths of nonresidents of the United States.

Year	Total			White			Black and other			Black		
	Total	Male	Female	Total	Male	Female	Total	Male	Female	Total	Male	Female
1970	70.8	67.1	74.7	71.7	68.0	75.6	65.3	61.3	69.4	64.1	60.0	68.3
1975	72.6	68.8	76.6	73.4	69.5	77.3	68.0	63.7	72.4	66.8	62.4	71.3
1976	72.9	69.1	76.8	73.6	69.9	77.5	68.4	64.2	72.7	67.2	62.9	71.6
1977	73.3	69.5	77.2	74.0	70.2	77.9	68.9	64.7	73.2	67.7	63.4	72.0
1978	73.5	69.6	77.3	74.1	70.4	78.0	69.3	65.0	73.5	68.1	63.7	72.4
1979	73.9	70.0	77.8	74.6	70.8	78.4	69.8	65.4	74.1	68.5	64.0	72.9
1980	73.7	70.0	77.4	74.4	70.7	78.1	69.5	65.3	73.6	68.1	63.8	72.5

[Continued]

★ 1142 ★

Life Expectancy, 1970-1991 and Projections, 1995-2010

[Continued]

Year	Total			White			Black and other			Black		
	Total	Male	Female	Total	Male	Female	Total	Male	Female	Total	Male	Female
1981	74.1	70.4	77.8	74.8	71.1	78.4	70.3	66.2	74.4	68.9	64.5	73.2
1982	74.5	70.8	78.1	75.1	71.5	78.7	70.9	66.8	74.9	69.4	65.1	73.6
1983	74.6	71.0	78.1	75.2	71.6	78.7	70.9	67.0	74.7	69.4	65.2	73.5
1984	74.7	71.1	78.2	75.3	71.8	78.7	71.1	67.2	74.9	69.5	65.3	73.6
1985	74.7	71.1	78.2	75.3	71.8	78.7	71.0	67.0	74.8	69.3	65.0	73.4
1986	74.7	71.2	78.2	75.4	71.9	78.8	70.9	66.8	74.9	69.1	64.8	73.4
1987	74.9	71.4	78.3	75.6	72.1	78.9	71.0	66.9	75.0	69.1	64.7	73.4
1988	74.9	71.4	78.3	75.6	72.2	78.9	70.8	66.7	74.8	68.9	64.4	73.2
1989	75.1	71.7	78.5	75.9	72.5	79.2	70.9	66.7	74.9	68.8	64.3	73.3
1990	75.4	71.8	78.8	76.1	72.7	79.4	71.2	67.0	75.2	69.1	64.5	73.6
1991, prel.	75.7	72.2	79.1	76.4	73.0	79.7	72.2	68.1	76.2	70.0	65.6	74.3
Projections[1]												
1995	76.1	72.6	79.5	76.7	73.4	80.0	(NA)	(NA)	(NA)	70.9	66.6	75.2
2000	76.6	73.0	80.0	77.2	73.9	80.5	(NA)	(NA)	(NA)	71.1	66.5	75.8
2005	77.1	73.6	80.5	77.7	74.5	81.0	(NA)	(NA)	(NA)	71.6	66.9	76.4
2010	77.6	74.2	81.1	78.3	75.1	81.5	(NA)	(NA)	(NA)	72.2	67.3	77.0

Source: "Expectation of Life at Birth, 1970 to 1991, and Projections, 1995 to 2010." U.,S. Bureau of the Census, *Statistical Abstract of the United States,* 1993, p. 85. Primary source: Except as noted, U.S. National Center for Health Statistics, *Vital Statistics of the United States,* annual, and *Monthly Vital Statistics Reports. Notes:* NA Not available. 1. Based on middle mortality assumptions; for details, see source. Source: U.S. Bureau of the Census, *Current Population Reports,* series P25- 1092.

★ 1143 ★

Life Expectancy

Life Expectancy: Table Values, 1969-1990

Beginning 1970, excludes deaths of nonresidents of the United States.

Age and sex	White				Black		
	1969-71	1979-1981	1985	1990	1979-1981	1985	1990
Average expectation of life in years							
At birth							
Male	67.9	70.8	71.8	72.7	64.1	65.0	64.5
Female	75.5	78.2	78.7	79.4	72.9	73.4	73.6
Age 20							
Male	50.2	52.5	53.2	54.0	46.4	47.1	46.7
Female	57.2	59.4	59.8	60.3	54.9	55.2	55.3
Age 40							
Male	31.9	34.0	34.7	35.6	29.5	29.8	30.1
Female	38.1	40.2	40.4	41.0	36.3	36.4	36.8
Age 50							
Male	23.3	25.3	25.8	26.7	22.0	22.1	22.5
Female	29.1	31.0	31.1	31.6	27.8	27.8	28.2
Age 65							
Male	13.0	14.3	14.5	15.2	13.3	13.0	13.2

[Continued]

★ 1143 ★

Life Expectancy: Table Values, 1969-1990
[Continued]

Age and sex	White				Black		
	1969-71	1979-1981	1985	1990	1979-1981	1985	1990
Female	16.9	18.6	18.7	19.1	17.1	16.9	17.2
Expected deaths per 1,000 Alive at specified age							
At birth							
Male	20.1	12.3	10.6	8.6	23.0	19.9	19.7
Female	15.3	9.7	8.0	6.6	19.3	16.5	16.3
Age 20							
Male	1.9	1.8	1.4	1.4	2.2	1.9	2.7
Female	0.6	0.6	0.5	0.5	0.7	0.6	0.7
Age 40							
Male	3.4	2.6	2.5	2.7	6.9	6.5	7.1
Female	1.9	1.4	1.3	1.2	3.2	2.9	3.1
Age 50							
Male	8.9	7.1	6.2	5.6	14.9	13.3	12.8
Female	4.7	3.8	3.5	3.2	7.7	6.8	6.6
Age 65							
Male	33.9	27.4	25.2	23.0	38.5	28.5	36.8
Female	15.6	13.6	13.5	12.8	21.6	21.4	21.4
Number surviving to specified age per 1,000 born alive							
Age 20							
Male	965	975	979	981	961	966	963
Female	976	984	986	988	972	976	976
Age 40							
Male	926	940	946	946	885	897	880
Female	958	969	973	975	941	948	944
Age 50							
Male	877	901	911	912	801	820	801
Female	929	947	953	957	896	908	904
Age 65							
Male	663	724	744	760	551	571	571
Female	816	848	855	864	733	746	751

Source: "Selected Life Table Values: 1969 to 1990." U.S. Bureau of the Census, *Statistical Abstract of the United States*, 1993, p. 85. Primary source: U.S. National Center for Health Statistics, *U.S. Life Tables and Actuarial Tables, 1959-61, 1969-71, and 1979-81*; *Vital Statistics of the United States*, annual; and unpublished data.

Life Expectancy

Life and Death Expectation: Characteristics, 1990

Age in 1990 (years)	Expectation of life in years					Expected deaths per 1,000 alive at specified age[1]				
	Total	White		Black		Total	White		Black	
		Male	Female	Male	Female		Male	Female	Male	Female
At birth	75.4	72.7	79.4	64.5	73.6	9.27	8.55	6.59	19.68	16.30
1	75.1	72.3	78.9	64.8	73.8	0.66	0.67	0.50	1.20	0.87
2	74.1	71.4	78.0	63.9	72.9	0.49	0.48	0.38	0.88	0.70
3	73.1	70.4	77.0	62.9	71.9	0.38	0.36	0.30	0.67	0.57
4	72.2	69.4	76.0	62.0	71.0	0.31	0.30	0.25	0.55	0.45
5	71.2	68.5	75.0	61.0	70.0	0.27	0.27	0.21	0.47	0.37
6	70.2	67.5	74.1	60.1	69.0	0.25	0.26	0.19	0.43	0.30
7	69.2	66.5	73.1	59.1	68.1	0.22	0.25	0.17	0.39	0.26
8	68.2	65.5	72.1	58.1	67.1	0.20	0.22	0.15	0.33	0.23
9	67.3	64.5	71.1	57.1	66.1	0.17	0.18	0.14	0.26	0.22
10	66.3	63.5	70.1	56.1	65.1	0.15	0.15	0.13	0.20	0.22
11	65.3	62.6	69.1	55.1	64.1	0.15	0.15	0.14	0.21	0.24
12	64.3	61.6	68.1	54.2	63.1	0.20	0.21	0.16	0.32	0.27
13	63.3	60.6	67.1	53.2	62.1	0.31	0.36	0.21	0.58	0.31
14	62.3	59.6	66.2	52.2	61.2	0.46	0.58	0.28	0.93	0.36
15	61.3	58.6	65.2	51.3	60.2	0.64	0.82	0.36	1.34	0.42
16	60.4	57.7	64.2	50.3	59.2	0.80	1.04	0.43	1.73	0.49
17	59.4	56.7	63.2	49.4	58.2	0.92	1.21	0.48	2.07	0.55
18	58.5	55.8	62.3	48.5	57.3	0.99	1.31	0.50	2.33	0.60
19	57.5	54.9	61.3	47.6	56.3	1.02	1.36	0.49	2.54	0.65
20	56.6	54.0	60.3	46.7	55.3	1.04	1.39	0.47	2.74	0.70
21	55.7	53.0	59.3	45.9	54.4	1.07	1.43	0.46	2.95	0.75
22	54.7	52.1	58.4	45.0	53.4	1.10	1.47	0.45	3.13	0.82
23	53.8	51.2	57.4	44.1	52.5	1.13	1.49	0.46	3.27	0.90
24	52.8	50.3	56.4	43.3	51.5	1.15	1.52	0.48	3.37	0.99
25	51.9	49.3	55.4	42.4	50.6	1.18	1.54	0.50	3.46	1.09
26	51.0	48.4	54.5	41.6	49.6	1.20	1.55	0.52	3.56	1.19
27	50.0	47.5	53.5	40.7	48.7	1.23	1.58	0.54	3.70	1.30
28	49.1	46.6	52.5	39.9	47.7	1.28	1.63	0.56	3.90	1.41
29	48.1	45.6	51.6	39.0	46.8	1.33	1.69	0.59	4.14	1.53
30	47.2	44.7	50.6	38.2	45.9	1.39	1.76	0.61	4.39	1.65
31	46.3	43.8	49.6	37.4	45.0	1.46	1.83	0.65	4.64	1.78
32	45.3	42.9	48.7	36.5	44.0	1.53	1.91	0.68	4.90	1.91
33	44.4	41.9	47.7	35.7	43.1	1.61	2.00	0.73	5.17	2.04
34	43.5	41.0	46.7	34.9	42.2	1.69	2.09	0.77	5.45	2.16
35	42.6	40.1	45.8	34.1	41.3	1.79	2.21	0.83	5.76	2.30
36	41.6	39.2	44.8	33.3	40.4	1.90	2.32	0.89	6.07	2.45
37	40.7	38.3	43.8	32.5	39.5	2.00	2.43	0.95	6.37	2.60
38	39.8	37.4	42.9	31.7	38.6	2.08	2.52	1.02	6.63	2.74
39	38.9	36.5	41.9	30.9	37.7	2.16	2.60	1.09	6.87	2.89
40	38.0	35.6	41.0	30.1	36.8	2.24	2.69	1.18	7.11	3.05
41	37.0	34.7	40.0	29.3	35.9	2.36	2.81	1.28	7.40	3.23

[Continued]

★ 1144 ★

Life and Death Expectation: Characteristics, 1990
[Continued]

Age in 1990 (years)	Expectation of life in years					Expected deaths per 1,000 alive at specified age[1]				
	Total	White		Black		Total	White		Black	
		Male	Female	Male	Female		Male	Female	Male	Female
42	36.1	33.8	39.1	28.5	35.0	2.50	2.97	1.40	7.77	3.46
43	35.2	32.9	38.1	27.7	34.1	2.68	3.16	1.55	8.23	3.72
44	34.3	32.0	37.2	27.0	33.3	2.90	3.39	1.72	8.78	4.03
45	33.4	31.1	36.2	26.2	32.4	3.15	3.66	1.92	9.39	4.38
46	32.5	30.2	35.3	25.4	31.5	3.42	3.96	2.13	10.03	4.75
47	31.6	29.3	34.4	24.7	30.7	3.73	4.30	2.37	10.70	5.15
48	30.7	28.4	33.5	24.0	29.8	4.07	4.69	2.62	11.37	5.59
49	29.9	27.6	32.5	23.2	29.0	4.44	5.12	2.89	12.07	6.07
50	29.0	26.7	31.6	22.5	28.2	4.86	5.60	3.20	12.81	6.58
51	28.1	25.8	30.7	21.8	27.4	5.31	6.14	3.53	13.63	7.14
52	27.3	25.0	29.8	21.1	26.6	5.82	6.67	3.89	14.60	7.75
53	26.4	24.2	29.0	20.4	25.8	6.39	7.46	4.28	15.78	8.41
54	25.6	23.3	28.1	19.7	25.0	7.01	8.25	4.70	17.12	9.12
55	24.8	22.5	27.2	19.0	24.2	7.68	9.10	5.15	18.57	9.87
56	24.0	21.7	26.4	18.4	23.4	8.41	10.03	5.65	20.08	10.67
57	23.2	21.0	25.5	17.8	22.7	9.22	11.07	6.21	21.65	11.57
58	22.4	20.2	24.7	17.1	22.0	10.12	12.24	6.85	23.27	12.58
59	21.6	19.4	23.8	16.5	21.2	11.11	13.54	7.56	24.94	13.69
60	20.8	18.7	23.0	15.9	20.5	12.19	14.96	8.35	26.71	14.87
61	20.1	18.0	22.2	15.4	19.8	13.33	16.46	9.18	28.59	16.09
62	19.4	17.3	21.4	14.8	19.1	14.50	18.01	10.04	30.54	17.36
63	18.6	16.6	20.6	14.3	18.5	15.70	19.59	10.91	32.56	18.68
64	17.9	15.9	19.8	13.7	17.8	16.93	21.24	11.80	34.63	20.00
65	17.2	15.2	19.1	13.2	17.2	18.23	22.96	12.76	36.79	21.41
70	13.9	12.1	15.4	10.7	14.1	27.33	35.15	19.78	51.00	29.49
75	10.9	9.4	12.0	8.6	11.2	41.08	53.88	31.08	68.81	40.48
80	8.3	7.1	9.0	6.7	8.6	62.24	82.01	49.92	93.91	58.55
85 and over	6.1	5.2	6.4	5.0	6.3	1,000	1,000	1,000	1,000	1,000

Source: "Expectation of Life and Expected Deaths, by Race, Sex, and Age: 1990." U.S. Bureau of the Census, *Statistical Abstract of the United States*, 1993, p. 86. Primary source: U.S. National Center for Health Statistics, *Vital Statistics of the United States*, annual; and unpublished data. *Notes:* 1. Based on the proportion of the cohort who are alive at the beginning of an indicated age interval who will die before reaching the end of that interval. For example, out of every 1,000 people alive and exactly 50 years old at the beginning of the period, between 4 and 5 (4.86) will die before reaching their 51st birthdays.

Marital Status

★ 1145 ★

First Marriage Dissolution, and Remarriage: Women, 1988

Item	Number (1,000)	Years until remarriage (cummulative percent)					
		All	1	2	3	4	5
All races[1]							
Year of dissolution of first marriage							
All years	11,577	56.8	20.6	32.8	40.7	46.2	49.7
1980-84	3,504	47.5	16.3	28.1	36.4	41.1[2]	45.4[2]
1975-79	3,235	65.3	21.9	36.0	44.7	52.7	55.4
1970-74	1,887	83.2	24.9	38.6	47.9	56.4	61.2
1965-69	1,013	89.9	32.6	48.7	60.2	65.0	72.8
White							
Year of dissolution of first marriage							
All years	10,103	59.9	21.9	35.2	43.5	49.4	53.0
1980-84	3,030	51.4	18.2	31.1	40.3	45.2[2]	49.8[2]
1975-79	2,839	69.5	23.2	38.5	46.9	55.6	58.4
1970-74	1,622	87.5	24.9	39.8	49.8	59.3	64.3
1965-69	893	91.0	34.7	52.3	64.9	69.3	76.9
Black							
Year of dissolution of first marriage							
All years	1,166	34.0	10.9	16.5	19.6	22.7	25.0
1980-84	380	19.7	4.7[3]	10.6[3]	12.9[3]	14.8[2]	14.8[2]
1975-79	301	32.3	11.4[3]	15.6[3]	18.5	22.2	24.9
1970-74	227	59.0	22.3	29.4	35.3	38.7	42.3
1965-69	98	81.2	20.9[3]	27.3[3]	31.3[3]	40.8	52.1
Hispanic[4], all years	942	44.7	12.5	16.6	22.7	27.8	29.9

Source: "First Marriage Dissolution and Years Until Remarriage for Women, by Race, and Hispanic Origin: 1988," U.S. Bureau of the Census, *Statistical Abstract of the United States,* 1993, p. 102. Primary source: National Center for Health Statistics, *Advance Data from Vital and Health Statistics,* No. 194. *Notes:* 1. Includes other races. 2. The percent having remarried is biased downward because the women had not completed the indicated number of years since dissolution of first marriage at the time of the survey. 3. Figure does not meet standard of reliability or precision. 4. Hispanic persons may be of any race.

★ 1146 ★
Marital Status

Marriage Experience for Women: Characteristics, 1980 and 1990

Marital status and age	All races		White		Black		Hispanic[1]	
	1980	1990	1980	1990	1980	1990	1980	1990
Ever married								
20 to 24 years old	49.5	38.5	52.2	41.3	33.3	23.5	55.4	45.8
25 to 29 years old	78.6	69.0	81.0	73.2	62.3	45.0	80.2	69.6
30 to 34 years old	89.9	82.2	91.6	85.6	77.9	61.1	88.3	83.0
35 to 39 years old	94.3	89.4	95.3	91.4	87.4	74.9	91.2	88.9
40 to 44 years old	95.1	92.0	95.8	93.4	89.7	82.1	94.2	92.8
45 to 49 years old	95.9	94.4	96.4	95.1	92.5	89.7	94.4	91.7
50 to 54 years old	95.3	95.5	95.8	96.1	92.1	91.9	95.0	91.8
Divorced after								
first marriage								
20 to 24 years old	14.2	12.5	14.7	12.8	10.5	9.6	9.4	6.8
25 to 29 years old	20.7	19.2	21.0	19.8	20.2	17.8	13.9	13.5
30 to 34 years old	26.2	28.1	25.8	28.6	31.4	26.6	21.1	19.9
35 to 39 years old	27.2	34.1	26.7	34.6	32.9	35.8	21.9	29.7
40 to 44 years old	26.1	35.8	25.5	35.2	33.7	45.1	19.7	26.6
45 to 49 years old	23.1	35.2	22.7	35.5	29.0	39.8	23.9	24.6
50 to 54 years old	21.8	29.5	21.0	28.5	29.0	39.2	22.5	22.9
Remarried after								
divorce								
20 to 24 years old	45.5	38.1	47.0	39.3	(B)	(B)	(B)	(B)
25 to 29 years old	53.4	51.8	56.4	52.8	27.9	44.4	(B)	49.5
30 to 34 years old	60.9	59.6	63.3	61.4	42.0	42.0	58.3	45.9
35 to 39 years old	64.9	65.0	66.9	66.5	50.6	54.0	45.2	51.2
40 to 44 years old	67.4	67.1	68.6	69.5	58.4	50.3	(B)	53.9
45 to 49 years old	69.2	65.9	70.4	67.2	62.7	55.0	(B)	51.0
50 to 54 years old	72.0	63.0	72.6	65.4	72.7	50.2	(B)	62.2
Redivorced								
after remarriage								
20 to 24 years old	8.5	13.1	(NA)	(NA)	(NA)	(NA)	(NA)	(NA)
25 to 29 years old	15.6	17.8	(NA)	(NA)	(NA)	(NA)	(NA)	(NA)
30 to 34 years old	19.1	22.7	(NA)	(NA)	(NA)	(NA)	(NA)	(NA)
35 to 39 years old	24.7	28.5	(NA)	(NA)	(NA)	(NA)	(NA)	(NA)
40 to 44 years old	28.4	30.6	(NA)	(NA)	(NA)	(NA)	(NA)	(NA)
45 to 49 years old	25.1	36.4	(NA)	(NA)	(NA)	(NA)	(NA)	(NA)
50 to 54 years old	29.0	34.5	(NA)	(NA)	(NA)	(NA)	(NA)	(NA)

Source: "Marriage Experience for Women, by Age, Race, and Hispanic Origin: 1980 and 1990," U.S. Bureau of the Census, *Statistical Abstract of the United States*, 1993, p. 102. Primary source: U.S. Bureau of the Census, *Current Population Reports*, P23-180. *Notes:* B Base is less than 75,000. NA Not available. 1. Persons of Hispanic origin may be of any race.

Pregnancies

★ 1147 ★

Pregnancies, by Selected Characteristics, 1988

Item	Total	Under 15 years old	15 to 19 years old	20 to 24 years old	25 to 29 years old	30 to 34 years old	35 to 39 years old	40 years and over
Number (1,000)								
Total								
All pregnancies	6,341	27	988	1,774	1,821	1,195	456	79
Live births	3,910	11	478	1,067	1,239	804	270	41
Induced abortions	1,591	14	393	520	347	197	96	24
Fetal losses	840	3	117	187	234	194	91	14
White								
All pregnancies	4,696	11	673	1,270	1,406	927	350	59
Live births	3,046	4	315	805	1,011	661	218	32
Induced abortions	1,026	6	264	332	219	125	63	17
Fetal losses	626	2	94	133	177	141	69	10
Other races								
All pregnancies	1,643	16	315	504	415	268	106	20
Live births	863	7	163	263	229	142	52	9
Induced abortions	565	8	129	187	129	72	33	7
Fetal losses	214	1	23	54	58	53	21	4
Rate per 1,000 women								
Total								
All pregnancies	109.0	3.3	110.8	185.3	166.7	109.7	47.2	9.6
Live births	67.2	1.3	53.6	111.5	113.4	73.7	27.9	5.0
Induced abortions	27.3	1.7	44.0	54.2	31.8	18.1	9.9	3.0
Fetal losses	14.4	0.3	13.2	19.6	21.5	17.8	9.4	1.7
White								
All pregnancies	97.2	1.8	93.4	161.7	155.3	102.2	43.2	8.4
Live births	63.0	0.6	43.7	102.5	111.6	72.9	26.9	4.6
Induced abortions	21.2	0.9	36.6	42.3	24.1	13.8	7.8	2.4
Fetal losses	13.0	0.2	13.1	16.9	19.5	15.5	8.5	1.5
Other races								
All pregnancies	166.5	9.7	184.3	292.3	221.9	146.5	68.1	16.6
Live births	87.5	4.0	95.3	152.3	122.3	77.8	33.4	7.3
Induced abortions	57.3	4.9	75.5	108.5	68.8	39.6	21.0	6.2
Fetal losses	21.7	0.7	13.6	31.5	30.8	29.1	13.8	3.0
Percent distribution								
Total								
All pregnancies	100.0	100.0	100.0	100.0	100.0	100.0	100.0	100.0
Live births	61.7	39.2	48.4	60.2	68.1	67.2	59.1	51.6
Induced abortions	25.1	50.5	39.7	29.3	19.1	16.5	21.0	30.9
Fetal losses	13.3	10.3	11.9	10.6	12.9	16.2	19.9	17.4

[Continued]

★ 1147 ★

Pregnancies, by Selected Characteristics, 1988

[Continued]

Item	Total	Under 15 years old	15 to 19 years old	20 to 24 years old	25 to 29 years old	30 to 34 years old	35 to 39 years old	40 years and over
White								
All pregnancies	100.0	100.0	100.0	100.0	100.0	100.0	100.0	100.0
Live births	64.8	35.6	46.9	63.4	71.9	71.3	62.2	54.1
Induced abortions	21.8	50.2	39.2	26.2	15.6	13.5	18.1	28.7
Fetal losses	13.3	14.2	14.0	10.5	12.6	15.2	19.8	17.2
Other races								
All pregnancies	100.0	100.0	100.0	100.0	100.0	100.0	100.0	100.0
Live births	52.6	41.8	51.7	52.1	55.1	53.1	49.0	44.2
Induced abortions	34.4	50.7	41.0	37.1	31.0	27.0	30.8	37.6
Fetal losses	13.1	7.5	7.4	10.8	13.9	19.9	20.2	18.2

Source: "Pregnancies, by Outcome, Age of Women, and Race: 1988." U.S. Bureau of the Census, *Statistical Abstract of the United States*, 1993, p. 82. Primary source: U.S. National Center for Health Statistics, *Monthly Vital Statistics Report*, vol. 41, No. 6.

Senior Citizens

★ 1148 ★

Abridged Life Table, 1988

Period of life between two exact ages	Male		Female	
	White	Black	White	Black
0 to 1 year	72.3	64.9	78.9	73.4
65 to 70 years	14.9	13.4	18.7	16.9
70 to 75 years	11.8	10.9	15.0	13.8
75 to 80 years	9.1	8.6	11.7	10.9
80 to 85 years	6.8	6.8	8.7	8.4
85 years and over	5.1	5.5	6.3	6.6

Source: "Average Number of Years of Life Remaining at Beginning of Age Interval: Abridged Life Table for 1988," Cynthia Taeuber, *Sixty-Five Plus in America*, p. 3-3. Primary source: National Center for Health Statistics, Vital Statistics of the United States 1988, Vol. II, Part A, Life Tables, Table 6-1.

★ 1149 ★

Senior Citizens

Death Rates for Cerebrovascular Diseases, 1988

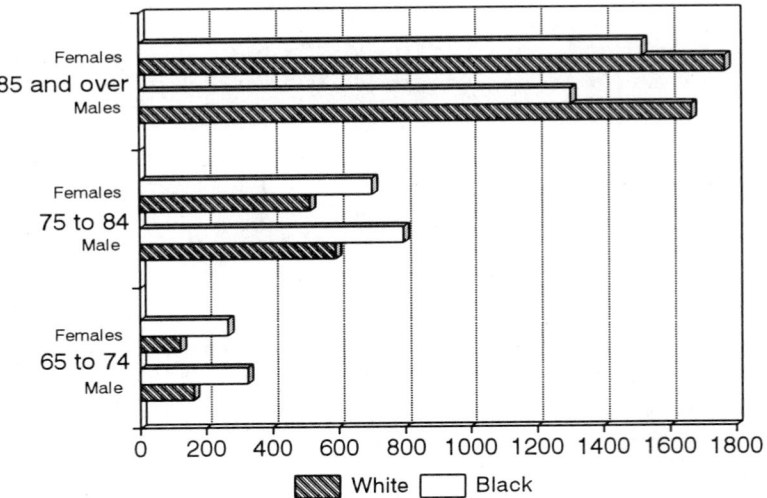

Deaths per 100,000 resident population.

Age	Male		Female	
	White	Black	White	Black
65 to 74	164	326	125	265
75 to 84	591	796	513	701
85 and over	1,667	1,303	1,767	1,518

Source: "Death Rates for Cerebrovascular Diseases for Persons 65 Years and Over, Age, Sex, and Race: 1988," Cynthia Taeuber, *Sixty-Five Plus in America*, p. 3-9. Primary source: National Center for Health Statistics, *Health, United States, 1990*, Hyattsville, MD: Public Health Service, 1991, Table 28.

★ 1150 ★

Senior Citizens

Death Rates for Heart Disease, 1960 and 1988

Deaths per 100,000 resident population.

Age, race, and sex	Deaths		Percent change, 1960 to 1988
	1960[1]	1988	
65 to 74 years			
White males	2,297.9	1,348.0	-41.34
Black males	2,281.4	1,616.7	-29.14
White females	1,229.8	656.2	-46.64
Black females	1,680.5	1,060.0	-36.92
75 to 84 years			
White males	4,839.9	3,257.6	-32.69
Black males	3,533.6	3,435.7	-2.77
White females	3,629.7	2,101.5	-42.10

[Continued]

★ 1150 ★

Death Rates for Heart Disease, 1960 and 1988
[Continued]

Age, race, and sex	Deaths		Percent change, 1960 to 1988
	1960[1]	1988	
Black females	2,926.9	2,625.6	-10.29
85 years and over			
White males	10,135.8	8,072.5	-20.36
Black males	6,037.9	6,165.7	2.12
White females	9,280.8	6,957.3	-25.04
Black females	5,650.0	5,648.1	-0.03

Source: "Death Rates for Diseases of the Heart, by Age, Sex, and Race: 1960 and 1988," Cynthia Taeuber, *Sixty-Five Plus in America*, p. 3-10. Primary source: National Center for Health Statistics, *Health, United States, 1990*, Hyattsville, MD: Public Health Service, 1991, Table 27. *Note:* 1. Includes deaths of nonresidents of the United States.

★ 1151 ★
Senior Citizens

Death Rates for Heart Disease, 1988

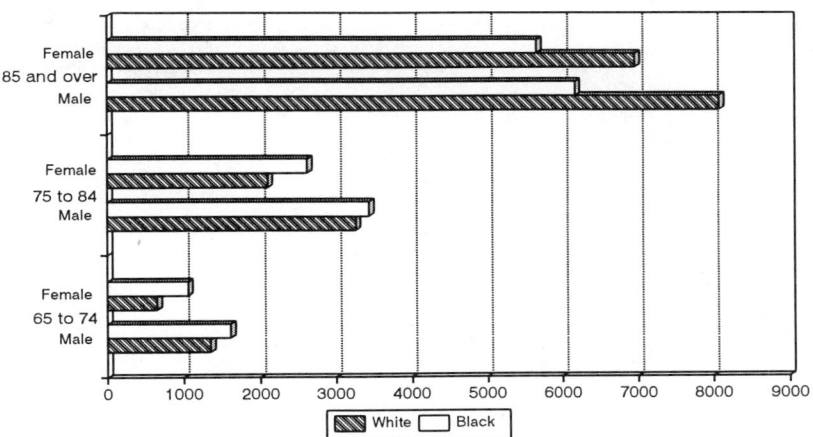

Deaths per 100,000 resident population.

Age	Male		Female	
	White	Black	White	Black
65 to 74	1,348	1,617	656	1,060
75 to 84	3,258	3,436	2,102	2,626
85 and over	8,073	6,166	6,957	5,648

Source: "Death Rates for Diseases of the Heart for Persons 65 Years and Over, by Age, Sex, and Race: 1988," Cynthia Taeuber, *Sixty-Five Plus in America*, p. 3-8. Primary source: National Center for Health Statistics, *Health, United States, 1990*, Hyattsville, MD: Public Health Service, 1991, Table 27.

★ 1152 ★

Senior Citizens

Death Rates for Malignant Neoplasms, 1960 and 1988

Deaths per 100,000 resident population.

Age, race, and sex	Deaths		Percent change, 1960 to 1988
	1960[1]	1988	
65 to 74 years			
White males	887.3	1,050.4	18.4
Black males	938.5	1,434.5	52.9
White females	562.1	660.0	17.4
Black females	541.6	728.3	34.5
75 to 84 years			
White males	1,413.7	1,839.7	30.1
Black males	1,053.3	2,344.5	122.6
White females	939.3	984.4	4.8
Black females	696.3	1,062.6	52.6
85 years and over			
White males	1,791.4	2,533.0	41.4
Black males	1,155.2	2,720.0	135.5
White females	1,304.9	1,300.1	-0.4
Black females	728.9	1,288.0	76.7

Source: "Death Rates for Malignant Neoplasms, by Age, Sex, and Race: 1960 and 1988," Cynthia Taeuber, *Sixty-Five Plus in America*, p. 3-11. Primary source: National Center for Health Statistics, *Health, United States, 1990*, Hyattsville, MD: Public Health Service, 1991, Table 29. *Note:* 1. Includes deaths of nonresidents of the United States.

★ 1153 ★

Senior Citizens

Death Rates for Malignant Neoplasms, 1988

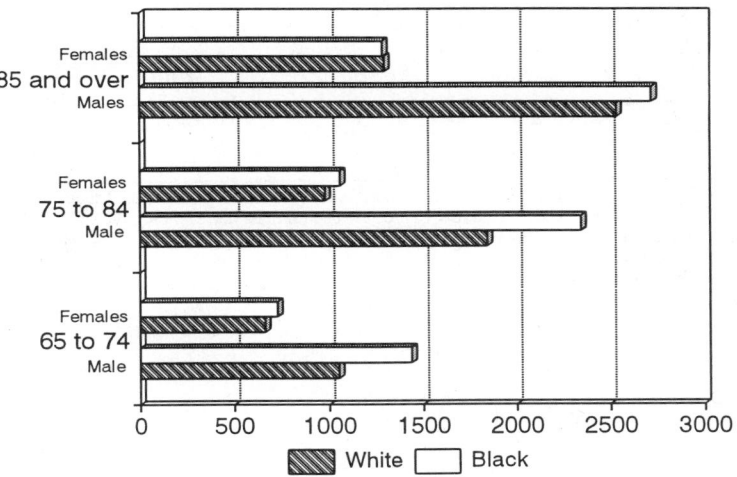

Deaths per 100,000 resident population.

Age	Male		Female	
	White	Black	White	Black
65 to 74	1,050	1,435	660	728
75 to 84	1,840	2,345	984	1,063
85 and over	2,533	2,720	1,300	1,288

Source: "Death Rates for Malignant Neoplasms for Persons 65 Years and Over, by Age, Sex, and Race: 1988," Cynthia Taeuber, *Sixty-Five Plus in America*, p. 3-8. Primary source: National Center for Health Statistics, *Health, United States, 1990*, Hyattsville, MD: Public Health Service, 1991, Table 27.

★ 1154 ★

Senior Citizens

Life Expectancy at Age 85: 1900 to 1988

Average number of years of life remaining.

Year	Male		Female	
	White	Black	White	Black
1900 to 1902	3.8	4.0	4.1	5.1
1909 to 1911	3.9	4.5	4.1	5.1
1919 to 1921	4.1	4.5	4.2	5.2
1929 to 1931	4.0	4.3	4.2	5.5
1939 to 1941	4.0	5.1	4.3	6.4
1949 to 1951	4.4	5.4	4.8	6.2
1959 to 1961	4.3	5.1	4.7	5.4
1969 to 1971[1]	4.6	6.0	5.5	7.1

[Continued]

★ 1154 ★

Life Expectancy at Age 85: 1900 to 1988
[Continued]

Year	Male		Female	
	White	Black	White	Black
1979 to 1981[1]	5.1	5.7	6.3	7.2
1988[1]	5.1	5.5	6.3	6.6

Source: "Life Expectancy at Age 85 Years, by Sex and Race: 1900 and 1988," Cynthia Taeuber, *Sixty-Five Plus in America*, p. 3. Primary source: National Center for Health Statistics. Data for 1900-1971 from Vital Statistics of the United States 1978, Volume II-Section 5, Life Tables. Data for 1979-1981 from U.S. Decennial Life Tables for 1979-1981, Volume 1, No. 1, U.S. Life Tables. Data for 1988 from Vital Statistics of the United States 1988, Volume II, Life Tables, Table 6-1. *Notes:* 1. Deaths of nonresidents of the United States were excluded beginning in 1970.

★ 1155 ★

Senior Citizens

Life Expectancy at Birth and 65 Years of Age, Selected Years 1900 to 1989

Specified age and year	All races			White		Black	
	Both sexes	Male	Female	Male	Female	Male	Female
At birth							
1900[1,2]	47.3	46.3	48.3	46.6	48.7	32.5[3]	33.5[3]
1950[2]	68.2	65.6	71.1	66.5	72.2	58.9	62.7
1960[2]	69.7	66.6	73.1	67.4	74.1	60.7	65.9
1970	70.9	67.1	74.8	68.0	75.6	60.0	68.3
1980	73.7	70.0	77.4	70.7	78.1	63.8	72.5
1989	75.3	71.8	78.6	72.7	79.2	64.8	73.5
At 65 years							
1900-1902[1,2]	11.9	11.5	12.2	11.5	12.2	10.4	11.4
1950[2]	13.9	12.8	15.0	12.8	15.1	12.9	14.9
1960[2]	14.3	12.8	15.8	12.9	15.9	12.7	15.1
1970	15.2	13.1	17.0	13.1	17.1	12.5	15.7
1980	16.4	14.1	18.3	14.2	18.4	13.0	16.8
1989	17.2	15.2	18.8	15.2	19.0	13.6	17.0

Source: "Life Expectancy at Birth and at 65 Years of Age, by Race and Sex: Selected Years 1900 to 1989," Cynthia Taeuber, *Sixty-Five Plus in America*, p. 1-3. Primary source: National Center for Health Statistics, *Health United States, 1990*, Hyattsville, MD: Public Health Service, 1991, Table 15. 1989 "At birth" data from, Monthly Vital Statistics Report, Vol. 40, No. 8(S)2, January 7, 1992. 1989 "At 65 years" data unpublished final data from Mortality Statistics Branch. *Notes:* 1. Death registration area only. The death registration area increased from 10 States and the District of Columbia in 19+00 to the coterminous United States in 1933. 2. Includes deaths of nonresidents of the United States. 3. Figure is for the all other population.

Reference Sources

Bennett, Claudette E. *The Black Population in the United States: March 1992*. U.S. Bureau of the Census, Current Population Reports, P20-471. Washington, D.C.: U.S. Government Printing Office, 1993.

Black Elected Officials: A National Roster. 21st ed. Washington, D.C.: Joint Center for Political and Economic Studies Press, 1993.

Black Enterprise. Monthly. Earl Graves Publishing Co., 130 Fifth Avenue, New York, NY 10011. Most issues contain statistical data on blacks and business.

Black Issues in Higher Education. Biweekly. Cox, Matthews and Associates, 10520 Warwich Avenue, Suite B-8, Fairfax, Virginia 22030.

Bowles, Elinor. *Cultural Centers of Color*. Washington, D.C.: National Endowment for the Arts, 1994.

Day, Jennifer Cheeseman. *Population Projections of the United States, by Age, Sex, Race, and Hispanic Origin: 1993 to 2050*. U.S. Bureau of the Census, Current Population Reports, P25-1104. Washington, D.C.: U.S. Government Printing Office, 1993.

Fordyce, Hugh R., and Alan H. Kirschner. *1992 Statistical Report*. United Negro College Fund. New York: UNCF, 1993.

Grail, Timothy S. *Our Nation's Housing in 1991*. U.S. Department of Housing and Urban Development and U.S. Bureau of the Census, Current Housing Reports, H121/93-2. Washington, D.C.: U.S. Government Printing Office, April 1993.

Hoffman, Charlene, Thomas D. Snyder, and Bill Sonnenberg. *Historically Black Colleges and Universities, 1976-90*. U.S. Department of Education, National Center for Education Statistics. Washington, D.C.: U.S. Government Printing Office, 1992.

Interagency Board for Nutrition Monitoring and Related Research. *Nutrition Monitoring in the United States: Chartbook I: Selected Findings From the National Nutrition Monitoring and Related Research Program*. Hyattsville, Maryland: Public Health Service, 1993.

Journal of Blacks in Higher Education, Inaugural Issue, Autumn 1993. Quarterly. Box 491, Stroudsburg, PA 18360-9952.

Maguire, Kathleen, Ann L. Pastore, and Timothy J. Flanagan, eds. *Sourcebook of Criminal Justice Statistics 1992*. U.S. Department of Justice, Bureau of Justice Statistics. Washington, D.C.: U.S. Government Printing Office, 1993.

Monthly Labor Review 117, January and March 1994. Bureau of Labor Statistics, Washington, D.C. 20212.

The Municipal Yearbook, 1993. Washington, D.C.: International City/County Management Association, 1993.

Naifeh, Mary L. *Housing of the Elderly: 1991*. U.S. Bureau of the Census, Current Housing Reports, Series H123/93-1. Washington, D.C.: U.S. Government Printing Office, 1993.

Profiles of America's Elderly, No. 3, November 1993. U.S. Department of Commerce, Bureau of the Census, and U.S. Department of Health and Human Services, National Institute on Aging.

Rasinski, Kenneth, Steven J. Ingels, Donald A. Rock, Judith M. Pollack, and Shi-Chang Wu. *America's High School Sophomores: A Ten Year Comparison*. U.S. Department of Education, National Center for Education Statistics. Washington, D.C.: U.S. Government Printing Office, 1993.

Sportsview. Premiere Issue. Sports View Publications, 1 Midtown Plaza, Suite 700, 1360 Peachtree Street, Atlanta, Georgia 30309.

The State of Black America 1994. New York: National Urban League, January 1994.

Taeuber, Cynthia. *Sixty-Five Plus in America*. U.S. Bureau of the Census, Current Population Reports, Special Studies, P23-178RV. Revised based on projection from P25-1092 issued November 1992. Washington: U.S. Government Printing Office, 1992, rev. 1993.

The Tennessean (Nashville), July 9, 1993. Daily. 1100 Broadway, Nashville, TN 37203.

U.S. Bureau of the Census. Americans with Disabilities: 1991-92. Current Population Reports P70-33. Washington, D.C.: U.S. Government Printing Office, 1993.

U.S. Bureau of the Census. *Money Income of Households, Families, and Persons in the United States: 1992.* Current Population Reports, Series P60-184. Washington, D.C.: U.S. Government Printing Office, September 1993.

U.S. Bureau of the Census. *Population Profile of the United States: 1993.* Current Population Reports, Series P23-185. Washington, D.C.: U.S. Government Printing Office, 1993.

U.S. Bureau of the Census. *Statistical Abstract of the United States.* 113 ed. Washington, D.C.: U.S. Government Printing Office, 1993.

U.S. Bureau of the Census. *Voting and Registration in the Election of November 1992.* Current Population Reports, Population Statistics P20-466. Washington, D.C.: U.S. Government Printing Office, 1993.

U.S. Department of Education, National Center for Education Statistics. *Digest of Educational Statistics 1993.* NCES 93-292. Washington, D.C.: U.S. Government Printing Office, 1993.

U.S. Department of Education, National Center for Education Statistics. *The Condition of Education, 1993.* Washington, D.C.: U.S. Government Printing Office, 1993.

U.S. Department of Heath and Human Welfare. *HIV/AIDS Surveillance Report* 5, no. 2 (July 1993). Public Health Services, Centers for Disease Control and Prevention.

U.S. Department of Justice, Bureau of Justice Statistics. *Drugs, Crime, and the Justice System.* NCJ-133652. Washington, D.C.: U.S. Government Printing Office, 1992.

U.S. Department of Labor, Bureau of Labor Statistics. *Consumer Expenditure Survey, 1990-91.* Bulletin 2425. Washington, D.C.: U.S. Government Printing Office, 1993.

U.S. Department of Labor, Bureau of Labor Statistics. *Employment and Earnings* 41, no. 1 (January 1994).

U.S. Department of Labor. *Employment and Earnings*. Washington, D.C.: U.S. Government Printing Office, February 1994.

U.S. Department of Labor. *Geographic Profile of Employment and Unemployment, 1992*. Bulletin 2428. Washington, D.C.: U.S. Government Printing Office, July 1993.

U.S. Department of Labor. *Training and Employment Report of the Secretary of Labor*. Covering the Period July 1987, 1988.

———September 1988. Washington, D.C.: U.S. Government Printing Office, 1992.

———September 1990. Washington, D.C.: U.S. Government Printing Office, 1992

The Wall Street Journal. December 21, 1993. 200 Burnett Road, Chicopee, MA 01020.

Woodward, Jeanne. *Housing America's Children in 1991*. U.S. Bureau of the Census, Current Housing Reports, Series H121/93-6. Washington, D.C.: U.S. Government Printing Office, 1993.

INDEX

Page numbers immediately follow the index terms. Values in brackets are table numbers.

Numbers following p. or pp. are page references. Numbers in [ ] are table references.

Numbers following p. or pp. are page references. Numbers in [ ] are table references.

1137

Index

Numbers following p. or pp. are page references. Numbers in [ ] are table references.

Numbers following p. or pp. are page references. Numbers in [ ] are table references.

Numbers following p. or pp. are page references. Numbers in [ ] are table references.

Numbers following p. or pp. are page references. Numbers in [ ] are table references.

Numbers following p. or pp. are page references. Numbers in [ ] are table references.

Numbers following p. or pp. are page references. Numbers in [ ] are table references.

Numbers following p. or pp. are page references. Numbers in [ ] are table references.

Numbers following p. or pp. are page references. Numbers in [ ] are table references.

1145